ENERGY

1 J ≡ 1 N·m	1 B = 778.169 ft·lbf
1 B = 1.055 056 kJ	1 J = 9.478×10^{-4} B
1 ft·lbf = 1.3558 J	1 cal ≡ 4.1840 J
1 IT cal ≡ 4.1868 J	

SPECIFIC ENERGY

1 B/lbm = 2.326 kJ/kg	1 kJ/kg = 0.4299 B/lbm
1 B/lbmol = 2.326 kJ/kmol	1 kJ/kmol = 0.4299 B/lbmol

SPECIFIC ENTROPY, SPECIFIC HEAT, GAS CONSTANT

1 B/lbm·R = 4.1868 kJ/kg·K	1 kJ/kg = 0.4299 B/lbm
1 B/lbmol·R = 4.1868 kJ/kmol·K	1 kJ/kmol·K = 0.2388 B/lbm·R

DENSITY

1 lbm/ft^3 = 16.018 kg/m^3	1 kg/m^3 = 0.062 428 lbm/ft^3

SPECIFIC VOLUME

1 ft^3/lbm = 0.062 428 m^3/kg	1 kg/m^3 = 16.018 ft^3/lbm

POWER

1 B/s = 1.055 056 kJ/s	1 hp = 2545 B/h
1 hp = 550 ft·lbf/s	1 kW = 1.3410 hp
1 W ≡ 1 J/s	

VELOCITY

1 mph = 1.467 ft/s	1 mph = 0.4470 m/s
1 ft/s = 0.3048 m/s	

TEMPERATURE

$T(°C) = \frac{5}{9}(T(F) - 32)$	$T(F) = \frac{5}{9} T(°C) + 32$
$T(°C) = T(K) - 273.15$	$T(F) = T(R) - 459.67$
1 K = 1.8 R or 1.8 T(K) = T(R)	

Constants

$\overline{R} = 8.31441$ kJ/kmol·K = 1.986 B/lbmol·R

$\mathfrak{F} = 96.487 \times 10^6$ coulomb/kmol of electrons

Avagadro's number = 6.022169×10^{26} molecules/kmol

$g_c = 32.1740$ ft·lbf/lbm·s^2

Engineering Thermodynamics

Engineering Thermodynamics

J.B. Jones

Mechanical Engineering Department
Virginia Polytechnic Institute and State University
Blacksburg, Virginia

R.E. Dugan

Science and Technology Division
Institute for Defense Analyses
Alexandria, Virginia

Prentice Hall, Englewood Cliffs, New Jersey, 07632

Library of Congress Cataloging-in-Publication Data

Jones, J. B. (James Beverly)
Engineering thermodynamics / J.B. Jones, R.E. Dugan.
p. cm.
Includes bibliographical references and index.
ISBN 0-02-361332-7
1. Thermodynamics. I. Dugan, R. E. (Regina E.) II. Title.
TJ265.J64 1996
621.402'1—dc20 95-9503
CIP

Editor-in-Chief: Marsha Horton
Acquisitions Editor: Bill Stenquist
Copyeditor: Virginia Dunn
Editorial-Production Service: Electronic Publishing Services Inc.
Buyer: Donna Sullivan
Editorial Assistant: Meg Weist
Cover Designer: Griffin III
Cover photographs courtesy of General Electric Corporation

Printed in the United States of America

10 9 8 7 6 5 4 3 2 1

ISBN 0-02-361332-7

Prentice Hall International (UK) Limited, *London*
Prentice Hall of Austria Pty. Limited, *Sydney*
Prentice Hall Canada Inc., *Toronto*
Prentice Hall Hispanoamericana, S.A., *Mexico*
Prentice Hall of India Private limited, *New Delhi*
Prentice Hall of Japan, Inc., *Tokyo*
Simon & Schuster Asia Pte. Ltd., *Singapore*
Editora Prentice-Hall do Brasil Ltda., *Rio de Janeiro*

Contents

Chapters 14 through 17 apply fundamentals from earlier chapters to power generation, refrigeration, and direct energy conversion. Material from all earlier chapters is used in these chapters.

Chapter C provides a brief overview of what has been discussed and what is yet to come in the book.

Chapter D ties together preceding and following chapters.

Chapter 8 discusses the concepts of maximum work, availability, and available fraction of energy in order to reach the highly useful concept of irreversibility.

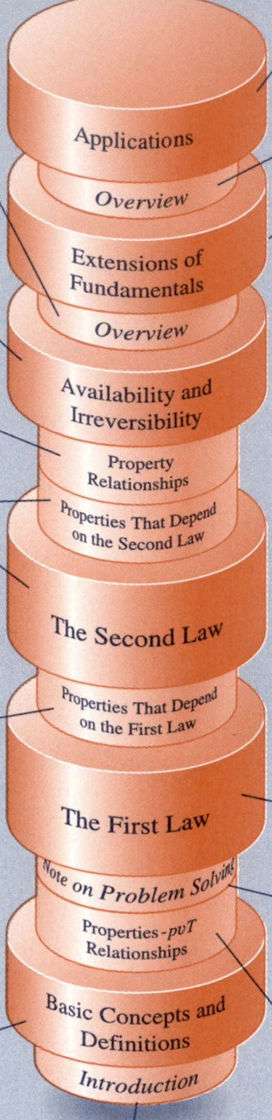

Chapter 9 on gas- and gas-vapor mixtures is one that students usually enjoy, because it can explain many phenomena of everyday life.

Chapter 10 explains two important characteristics of binary mixtures that are relevant to Chapters 15 and 16.

Chapters 11 and 12 on combustion and thermochemical equilibrium illustrate the use of the second law for determining possible end states. This material precedes the other application chapters because it shows the power of thermodynamics even more forcefully than the later chapters, so students should not miss experience with it.

Chapter 13 on thermodynamic aspects of fluid flow breaks away from heavy dependence on constant-specific-heat calculations for gases.

Chapter 7 presents a selection of general property relationships dependent on both the first and second laws and relevant to later chapters and other engineering applications.

Chapters 5 and 6 open discussions of the second law with the Clausius and Kelvin-Planck statements because students are familiar with engines and refrigerators. The parallelism between the operational definitions of entropy and stored energy is highlighted.

Chapter 4 treats relationships among internal energy, enthalpy, and other properties to open a tremendous number of applications to students. Many problems are at the end of Chapter 4 because students at this point can handle many applications.

Chapter 3 introduces the first law and gives an operational definition of stored energy.

Chapter B provides a methodology and many tips to help students develop their problem-solving abilities.

Chapter 1 emphasizes the steady flow case expression $-\int v\,dp$ along with the closed system expression $\int V\,dp$ to show that both come from mechanics.

Chapter 2 covers properties defined independently of the first and second laws so students can solve a wider range of problems immediately after the first law is introduced.

Chapter A introduces the breadth of thermodynamics applications and illustrates the operation of some engineering systems.

Applications
Overview
Extensions of Fundamentals
Overview
Availability and Irreversibility
Property Relationships
Properties That Depend on the Second Law
The Second Law
Properties That Depend on the First Law
The First Law
Note on Problem Solving
Properties-pvT Relationships
Basic Concepts and Definitions
Introduction

Preface

Why another book on thermodynamics? There are many thermodynamics books available. However, we believe that this book provides a unique combination of features that will make it inviting and effective for both instructors and students.

First, the presentation is rigorous. Terms are defined operationally, constraints and assumptions are plainly stated, and developments avoid shortcuts taken in the name of expediency. The topic sequence, as shown by the diagram on the facing page and the table of contents, has been developed from extensive experience to preserve both logical development and ease of learning.

To complement the rigorous presentation, we have taken great care to make the text readable and accessible by students. To this end, we have provided many pedagogical aids such as expanded diagrams that show how the topics covered in the book are related to one another, annotated references, and margin notes. The margin notes highlight important results, caution students about common pitfalls, test their understanding, and help them in finding topics. The example problems are extensive and have been designed to help students with concepts they are likely to find difficult.

The engineering emphasis is strong. We use many real systems that are likely to be encountered in engineering practice. We emphasize sound engineering principles by using a methodical rather than a rigid problem-solving approach. Both the example problems and the end-of-chapter problems reflect this. We encourage students to estimate their results and check to see if they are reasonable.

Because we recognize the essential role of thermodynamics and other engineering sciences in supporting engineering design, we provide problems characteristic of engineering design, and we integrate computer use throughout the book. We address the importance of this feature in detail below because the use of computers in engineering science courses is often misunderstood.

Why use computers in thermodynamics? We hear many discussions of the use of computers in fundamentals courses. Some capable teachers even deemphasize computer use because they believe it distracts students from the course material. Some say they have taught the fundamentals of thermodynamics for years without computers. True! However, engineering design courses have improved markedly in recent years so that they are now more extensive, more focused on customer needs, and more concerned with the entire product realization process. Consequently, engineering science courses now need to give students more preparation for design, including experience with open-ended problems, formulation of questions, parametric studies, what-if questions, incomplete information, and other characteristics of engineering design. This goal calls for computer use in fundamentals courses.

In addition, our experience and that of others suggests that appropriate computer use enhances learning. Computers make it possible to examine problems that are more extensive and more complex than those that can be treated otherwise. Further, computers free students from many tedious tasks such as looking up values in tables, interpolation, and plotting. Thus, while we agree that the fundamentals of thermodynamics can be studied without the use of computers, we suggest that they can also be studied without the use of a calculator, yet few would suggest doing so.

We have written this book on the premise that students will be using computers. We are aware that often students do not use computers extensively. A course description may suggest that computer use is encouraged or required, but a close look at the course material often shows that computers are not implemented as an integral part of the course. Realistically, to integrate computer use, the computer must be used frequently and at the discretion of the students, not just for a few "computer assignments" during the course. By demonstrating the effective use of computers throughout the text and by including flexible software, we are trying to provide incentives and tools that will encourage both students and instructors to integrate computers into their study of thermodynamics.

We seek in this fundamentals course to give students some early experience with engineering design, the central function of engineering. However, by the time students are studying thermodynamics, they generally have not developed the necessary background to handle many engineering design issues such as customer needs, cost, quality, reliability, disposal, environmental effects, etc. Thus, it is too early for broad engineering design projects. Nevertheless, in thermodynamics we can provide experience with open-ended, parametric, and what-if problems. In this book, we include many such problems. Often, we refer to them as design study problems because they represent only parts of entire design problems. Even so, these problems can overwhelm students, so in several cases we present a series of problems where the early ones are pieces of a final one. Students can often tackle the broader, more complex problem with greater facility and confidence after having solved some of the preliminary problems.

Further, as mentioned above, the use of computers allows students to solve many parametric problems that are simple expansions of basic problems but provide the flavor of engineering optimization. We do not specifically identify problems as "computer" or "open-ended" or "design" problems. This allows the student to develop engineering judgment which requires an ability to identify the character of problems posed and the methods most suitable for solution. To save the instructor time in selecting problems to assign, though, the Solutions Manual makes clear the nature of the solution required by each problem.

The software accompanying this book provides basic equation models and data models. This approach allows users to create models that suit their specific needs and focuses attention on the basic equations of thermodynamics. We avoid ready-made models for solving specific problems such as a Brayton cycle. Although such computer models can be quite handy, it is difficult for the user to adapt them to slightly different problems. For example, a program for analyzing regenerative steam power cycles can be helpful in determining the optimum pressures for feedwater heating. However, programs that do only this are usually of little value if a process steam extraction is added to the cycle. On the other hand, if the user has built the model from basic equations, he or she can readily add the basic equations that accommodate the process steam flow.

The data models provide on line data. By on line data we mean data that can be directly called, with no copying or reentering from the screen. This is in direct contrast to lookup data models that are a limited use of computers. On line data are orders of magnitude more effective and less prone to errors. The basic equation and data models we present are not only general purpose, they are open. That is, the user can examine all parts of the models in detail, and to help in this, explanations of model structure are often included.

To build our software, we have of necessity chosen a specific equation solver. It is TK Solver, produced by Universal Technical Systems, Inc., Rockford, Illinois. All of our models use TK Solver as a foundation. We recognize that there are always tradeoffs in choosing an equation solver, and some users may have opted for a different package with other strengths and weaknesses. Thus, in a few cases we have shown example problem solutions produced with other software to demonstrate that it is possible to build similar models to the extent the required capabilities are available in other equation solvers.

A manual with instructions on using the software is included with this book. Detailed illustrations in the manual are taken from example problems. For those who have never used TK Solver, learning to use it requires a little time and patience as does learning any new software package; but our experience is that this initial effort pays substantial dividends.

Well over 1500 problems of a wide variety are included, ranging in scope from simple exercises involving a single concept to complex problems involving several principles, multiple data sources, and exercise of judgment in modeling. Some problems may be solved in a few minutes. These problems might be assigned as "Solve any three problems within ten minutes." Other problems may require hours for a solution. Both kinds of problems and the ones between these extremes develop student understanding and skills. We believe such a variety helps the instructor find problems suitable for the many different needs that arise in a course.

We deliberately use a variety of abbreviations for units such as cc, cm^3, psia, and lbf/in^2 in problem statements. These various forms are used in practice and are likely to be used for many years to come, so students should be familiar with them. We appreciate the immediate comfort of working in one set of units with only a single set of abbreviations. However, the long-term benefit of having facility with other sets of units used in practice outweighs this immediate comfort.

The data tables in the book are easy to find and use. The edge tabs, identifying headers, units at column heads, and property diagrams showing table coverage are all designed for this purpose. The tables are based on modern data sources and are fully referenced. The values agree exactly with the values from the associated software.

Some other comments on content and style. Occasionally we introduce concepts that are helpful or necessary for our developments, even though we do not discuss them in detail. Examples are the mentions of the third law of thermodynamics and of reaction rates. Both of these are subjects of extensive study in themselves, and it is well for students to be apprised of them. Although a detailed study is beyond the scope of this book, we believe that opening a door encourages further investigation. Moreover, once students have been introduced to a concept and have seen how it is used, they are more interested in it when they next encounter it.

In our developments, we sometimes do not derive the most general case of an equation. We choose this approach since students can generalize expressions as well as they can narrow them for specific cases and some developments are unnecessarily complex for the general case.

Topics introduced in early chapters are used repeatedly in later chapters to build the student's capabilities continuously. Reminding students over and again of earlier principles and giving them many chances to practice reinforces the topics in a way that no other method does.

We hope that you will give us the benefit of your comments and suggestions.

Acknowledgments. Many people have helped to produce this book. Bill Stenquist, Executive Editor for Prentice Hall, has on many occasions awed us with his ability to solve problems, clear obstacles, and smooth the way. His knowledge, vision, and enthusiasm were essential to the completion of the project, and we both sincerely enjoyed working with him. Todd Piefer of UTS, Inc., put his stamp on us and on the software that accompanies this book. Todd is an expert in TK Solver, and he is tenacious. He made the software better and easier to use. Consistent with his demeanor, Todd demonstrated genuine commitment to the project.

Several other highly competent people at Prentice Hall, York Graphic Services, FineLine Illustrations, and UTS participated. We appreciate the patience and cooperation of Elaine Wetterau with whom we worked early in the project, Virginia E. Dunn who skillfully edited the manuscript, and Lilian Brady who did most of the subsequent editing. Lilian continually amazed us with both her editorial skills and her ability keep the many parts of the project synchronized. Her effort and commitment made this a significantly better book. Jerry Roselli of FineLine was especially creative and patient. Jerry and others at FineLine turned our rough sketches into exceptional figures.

Several reviewers, most of whom were unknown to us until the manuscript was completed, provided insightful comments and sage advice, and we are grateful to them. It was only after great debate that we ultimately decided not to follow some of their suggestions. We would like to thank Penrose Albright of the Institute for Defense Analyses, Donald E. Beasley of Clemson University, E. F. Brown of Virginia Polytechnic Institute and State University, Larry Dubois of the Advanced Research Projects Agency, John M. Kincaid of the State University of New York at Stony Brook, Dwight W. Senser of the University of Wyoming, J. Edward Sunderland of the University of Massachusetts, and Michael K. Wells of Montana State University for their perceptive and thorough reviews.

Eloise McCoy typed the entire manuscript. Some chapters we provided to Eloise were nearly ''encrypted'' as she had to assemble them from inputs on paper, computer diskettes, and dictation tapes. Her cheerful dedication complements her productivity, judgment, and work quality which are of the highest order.

Several people, mostly students, helped us in the preparation of the manuscript. They edited the manuscript, visited the library in search of data and references, performed calculations, assisted in the software development, and kept us honest in our commitment to make the book accessible to students. Ann K. Carrithers, David Michael DeHart, Donald E. Deliz, Bryan E. Gibbs, and Scott E. Miles earned our respect and gratitude continually. It is a pleasure to work with students with such remarkable commitment. John Broussard, David C. Copenhaver, John J. Datovech, Aaron Golub, Joell R. Hibshman II, Brian Krum, Michael Mulqueen, and Rajesh Vasisht also did excellent work primarily on the solutions manual. These students often worked under considerable pressure, and we are thankful for their perseverance. Ellen Jones-Walker assisted in checking material and otherwise, and James A. Hardell produced many of the charts. The services of both were of the highest quality.

J. B. Jones
Blacksburg
jbjones@vt.edu

Regina E. Dugan
Arlington
rdugan@ida.org

April 1995

Once again my wife, Jane Hardcastle Jones, has stood by me through the preparation of a book manuscript with the resulting missed vacations and family weekends and, in the present case, countless interruptions of the peace of our home by a growing number of communication means. My gratitude to her is boundless.

John R. Dixon of the University of Massachusetts, who writes much better than I do and undertakes more extensive projects, has inspired me for years. During the time that this book has absorbed me, he has been writing another. We confer on many things, including the rigors of authorship. In our parallel experiences he has supported me far more than he knows, and I am deeply grateful to him.

—JBJ

Many friends and family members helped me endure this demanding project: Brad Harmon, who demonstrated his tireless commitment to me and to this project with encouraging words, kind hugs, seemingly endless patience, and deliveries to the express mail box. I am grateful to have this giving and selfless man in my life. Terry DeDominicis, Mary Kennedy, Susanne Vielmetter, Kelly Goodwin, and Vivian George, who endured many conversations about ''the book'' and lifted my spirits on countless occasions. These women have, each in her own way, enriched my life.

—RED

Preface for Students

So, you're taking a course in thermodynamics? Thermodynamics is a fascinating field. It has the elegance of a few simple laws and the challenges of a breadth of applications. This book will provide you with a sound understanding of thermodynamics. We have taken care to present the fundamentals rigorously and have complemented this approach with many learning aids to help you in your study.

We encourage the use of computers throughout this book. Of course, a sound understanding of the fundamentals of thermodynamics can be acquired without using a computer. Thermodynamics does not *require* a computer; it does not *require* a calculator for that matter. After all, both authors of this book studied thermodynamics before computers were commonplace, and one of us studied thermodynamics using a slide rule. (Imagine, all this in addition to walking to class uphill — both ways — in the snow and bitter cold!)

Nevertheless, computers allow us to expand our problem-solving ability and apply the principles of thermodynamics to solve many more interesting problems. Thus, computer use in an engineering science course better prepares you for engineering design where computer use is essential. Computers offer the additional benefit of reducing many tedious tasks such as interpolation and plotting.

We challenge you to explore the software included with this text to enhance your study of thermodynamics. We encourage you to begin this process early, with simple problems. At first, these simple problems may be quicker to solve by hand, but later, as the problems become more complex, the time savings will be substantial.

The software provided with this book takes two forms: basic equation models and online data models. By online data we mean data that, when defined by other parameters in the problem, can be called directly into the problem without copying from another location. Using the basic equation models and the data models in combination provides you with a "toolbox" from which you build your own models. This approach is different from that of a "ready-made" software approach that provides complete models for specific problems and consequently affords less flexibility. To be sure, you must understand the thermodynamics involved; the models are not a substitute for knowledge of the principles. Thus, wherever possible, we have referenced these models back to the textbook so that if you need help, you will know where to find it. In addition, the models are completely "open" so that you can see every basic equation involved.

Our experience has been that once exposed to this software, students plunge right in and soon become quite proficient. We have built our software using TK Solver, a powerful simultaneous equation solver. You will find that TK Solver is useful not only in thermodynamics applications but in your other courses as well. Further, TK Solver is widely used in engineering practice.

In addition to the software, many features in this

book are designed to facilitate your study of thermodynamics. A "panorama" is introduced in Chapter A and appears at the opening of each chapter to help you understand how the various thermodynamic principles fit together. Margin notes highlight important equations and results, caution you of pitfalls, test your understanding, and help you find topics. A listing of the sections in each chapter on the chapter opening page helps you prepare for what you are about to study. A summary at the end of each chapter provides an overview of the important points in the chapter. Annotated references at the ends of many chapters allow you to choose the most appropriate reference from among those provided. The appendix tables are clearly marked with thumb tabs, match the data in the software, and have headings that include units on each page. The charts are drawn to large scales with relatively few lines, and the use of color makes them easier to read.

We recommend that you look over the panorama, section listing, introductory paragraphs, margin notes, and summary as a preview before you begin each chapter and as a review after you have completed the chapter.

In addition to the main chapters, we have included four brief chapters, indicated by letters, to help you. *Chapter A* shows how some actual engineering systems operate because we have found that many students are unfamiliar with such systems. *Chapter B* is a note on problem solving. Review this chapter frequently throughout your study of thermodynamics. It provides many tips that will help you to save time in solving problems, identify areas requiring further study, and avoid "spinning your wheels." *Chapters C* and *D* provide brief overviews of what has been discussed and what is yet to come in the book.

Good luck.

J. B. Jones
Regina E. Dugan

Symbols and Abbreviations

Symbol	*Meaning*	*First Appears in Section*
A	area; Helmholtz function, $U - TS$	1·4; 6·5
a	linear acceleration; specific Helmholtz function, $u - Ts$; speed of a pressure wave	1·4; 6·5; 13·3
b	Darrieus function, $h - T_0 s$	8·2
C	number of components (in the phase rule)	10·1
c	speed of sound; $(\partial p/\partial \rho)_s$	13·3
$\mathscr{C}$	charge	17·3·2
c_p	specific heat at constant pressure, $(\partial h/\partial T)_p$	4·1·2
c_T	constant-temperature coefficient, $(\partial h/\partial p)_T$	7·5·2
c_v	specific heat at constant volume; $(\partial u/\partial T)_v$	4·1·2
E	stored energy	3·2
e	specific stored energy, E/m	3·2
E_a	Arrhenius activation energy	12·5
F	force; maximum number of independent intensive properties (in the phase rule)	1·4; 10·1
F	force dimension	1·4
f	number of degrees of freedom of a molecule	4·2·4
$\mathfrak{F}$	Faraday constant, 96.487×10^6 coulomb/kmol of electrons	17·2·2
G	Gibbs function, $H - TS$	6·5
g	gravitational acceleration or the acceleration of a freely falling body; specific Gibbs function, $h - Ts$	1·5; 6·5
g_c	dimensional constant	1·4
H	enthalpy, $U + pV$	3·5·1
h	specific enthalpy, $u + pv$; height of a fluid column	3·5·1; 1·5·2
ΔH_f	enthalpy (change) of formation	11·3·4
Δh_f	specific enthalpy (change) of formation	11·3·4
ΔH_R	enthalpy (change) of reaction	11·3·2

Symbol	*Meaning*	*First Appears in Section*
Δh_R	specific enthalpy (change) of reaction	11·3·2
I	irreversibility; electric current	8·4; 17·3·2
i	specific irreversibility, I/m	8·4
j	valence	17·2·2
k	ratio of specific heats, c_p/c_v	4·2·1
k_i	Henry's Law constant	10·4
K_f	equilibrium constant of formation	12·2·2
K_p	equilibrium constant	12·2
KE	kinetic energy	3·3
ke	kinetic energy per unit mass	6·1
L	length	1·4
L	length dimension	1·4
M	molar mass; Mach number, V/c	2·3·1; 13·3
M	mass dimension	1·4
m	mass	1·2
$\dot{m}$	mass rate of flow	1·4
N	number of moles	2·3·1
n	polytropic exponent; number of molecules	4·2·2; 6·7
P	number of phases (in the phase rule); probability	10·1; 6·7
p	pressure	1·2
p_R	reduced pressure	7·6·2
p_r	relative pressure	6·2·1
PE	potential energy	3·3
pe	potential energy per unit mass	3·5·3
Q	heat	1·10
q	heat transfer per unit mass	3·5·2
$\dot{Q}$	time rate of heat transfer	1·10
R	gas constant	2·3·1
$\overline{R}$	universal gas constant	2·3·1
r	compression ratio	15·2·1
r_c	cutoff ratio	15·2·2
S	entropy	6·1
s	specific entropy; distance	6·1; 1·9
$\mathcal{S}$	solubility	10·4
T	temperature; torque	1·2; 1·9
T	temperature dimension	App I·1
T_R	reduced temperature	7·6·2
t	time	1·4
t	time dimension	1·4
U	internal energy	3·3
u	specific internal energy; velocity; velocity of a point on a rotor	3·3; 13·3; 14·4·1
ΔU_R	internal energy (change) of reaction	11·3·1

Symbol	*Meaning*	*First Appears in Section*
V	volume; velocity	1·2; 1·4
v	specific volume	1·2
$\mathcal{V}$	voltage (emf)	17·2·2
v_r	relative specific volume	6·2·1
Vol	volume	1·7
$\dot{Vol}$	volume rate of flow	3·5·3
W	work	1·9
w	weight; work per unit mass	1·5; 1·9
$\dot{W}$	power (work per unit time)	1·9
x	quality; mass fraction; a property in general	2·1; 9·1·1; 1·2
Y	stream availability	8·2
y	specific stream availability; mole fraction; a property in general	8·2; 9·1·1; 1·6
Z	compressibility factor, $Z \equiv pV/mRT$	2·3·2
z	elevation	1·5·2
Overbar	indicates that a property is on a molar basis (as in $\bar{h}$, kJ/mol)	

Greek letters

β	coefficient of performance; coefficient of volume expansion, $(\partial v/\partial T)_p/V$	3·6; 7·4
γ	specific weight	1·5
ε	fuel cell efficiency	17·2·2
η	efficiency	3·6
Θ	Debye's constant or characteristic temperature	7·4
θ	temperature on any nonthermodynamic scale; angular displacement	7·5·3; 1·9
κ_s	isentropic compressibility, $-(\partial v/\partial p)_s/V$	7·4
κ_T	isothermal compressibility, $-(\partial v/\partial p)_T/V$	7·4
μ	Joule-Thomson coefficient, $-(\partial T/\partial p)_h$	7·5
ν	stoichiometric coefficient (number of moles in chemical equation)	12·2
ρ	density	1·4
$\bar{\rho}$	molar density	12·5
Φ	availability of a closed system	8·2
ϕ	specific availability of a closed system, $\dfrac{\Phi}{m}$; $\displaystyle\int \frac{c_p dT}{T}$; relative humidity	8·2; 6·2·1; 9·4·2
ω	humidity ratio; shaft speed	9·4·2; 1·9

Subscripts

a	air
c	critical state; (see also g_c in list of symbols)
da	dry air
db	dry bulb
dg	dry gas
dp	dew point
f	final state; saturated liquid; fuel; of formation (as in Δh_f)
fg	difference between property of saturated liquid and property of saturated vapor at the same pressure and temperature
g	saturated vapor
H	high temperature (as in T_H and Q_H)
HP	heat pump
i	initial state; ice point; ideal or isentropic; intermediate (as intermediate pressure in multistage compression)
if	difference between property of saturated solid and saturated liquid at the same pressure and temperature
ig	difference between property of saturated solid and saturated vapor at the same pressure and temperature
int rev	internally reversible
L	low temperature (as in T_L and Q_L); liquid
m	mixture
p	of products (as in h_p)
R	reduced coordinate; energy reservoir; of reaction (as in ΔH_R)
r	relative; of reactants (as in h_r)
ref	reference temperature
s	steam point
sat	saturation
t	total or stagnation
th	throat
v	vapor
wb	wet bulb
σ	referring to an open-system boundary
1,2,3, . . .	referring to different states of a system or different locations in space

Superscripts

°	standard state pressure
*	state used in relating real-gas and ideal-gas properties; state at which $M(=V/c) = 1$

Abbreviation	*Meaning*
cfm	cubic feet per minute
cfs	cubic feet per second
db	dry bulb
dp	dew point
fps	feet per second
gph	gallon per hour
gpm	gallon per minute
hhv	higher heating value
kWh	kilowatt-hour
lhv	lower heating value
mph	miles per hour
pdl	poundal
PRV	pressure-reducing valve
psi	pound force per square inch
psia	pound force per square inch absolute
psig	pound force per square inch gage
wb	wet bulb

Chapter A

Introduction to Thermodynamics

Thermodynamics is the science of energy. Notably, it is characterized by a few basic principles that can be applied to a great diversity of problems in many fields. *Engineering thermodynamics* is that part of the science that applies to the design and analysis of devices and systems for energy conversion.

A sound understanding of thermodynamics will allow you to determine how energy is controlled and converted in devices ranging from ink jet printers to paper-making machines, from lawn mower engines to thousand-megawatt power plants, from single-can beverage coolers to spacecraft climate-control systems, and from welding torches to complex chemical plants. Thermodynamics is central to engineering.

As an example, thermodynamics is at the heart of every aspect of the design of an aircraft jet engine. To provide a force to move the aircraft, the jet engine must produce a high-velocity jet of gas. It does this by taking in atmospheric air, compressing it, burning fuel in it, expanding it through the turbine to drive the compressor, and then expanding it further through a nozzle. But how much work must the turbine provide? What flow rate of gas is required? What shaft speed will give the proper blade velocities to match the gas velocities? Answers to these questions tell us how large the blades and flow passages must be and how much stress the blade fastenings must withstand. Thousands of decisions must be made in the design of a system as complex as a jet engine, and virtually every one of them depends on the results of thermodynamic analyses.

As another example, the design of a respirator to assist people who have difficulty breathing also relies on thermodynamics. Respirators for infants, for adults, and for patients with special problems may be quite different in size and features; but all must supply oxygen or oxygen-enriched air at the proper temperature and humidity, regardless of the temperature and humidity of the gas entering the device. If the entering gas is too dry, moisture must be added. Furthermore, if the moisture added is in liquid form, some heat will be required to compensate for the drop in temperature that occurs when the moisture evaporates into the gas. But how much moisture? How much heat? If the capacity for providing the needed temperature and humidity is not sufficient, the respirator will not perform its intended function. If the capacity is excessive, the device will be unnecessarily large, heavy, or costly. Any one of these results is undesirable. Thermodynamics helps us answer these and many more respirator design questions.

The jet engine and the respirator are only two examples from a myriad of applications of thermodynamics. On pages 4–8, we describe in more detail some systems and devices that require thermodynamic analyses for their design and operation.

An Overview of Thermodynamics

The figure to the right is a panoramic view of thermodynamics that is repeated throughout the book. Its purpose is to show the relationships among the various topics in thermodynamics. For example, the panorama illustrates the relationship between thermodynamic principles and physical properties. Thermodynamics depends integrally on physical properties; with them we describe systems and solve problems. By the same token, many physical properties depend on thermodynamics for their very definition. We have arranged the chapters in this book to highlight this relationship. The solid arrows in the panorama show the primary sequence in the development of thermodynamic principles. The dashed arrows indicate how chapters on physical properties are inserted to enhance our study of thermodynamics. We discuss properties for the first time in Chapter 2 so that when you finish Chapter 3 on the first law, you will be prepared to solve interesting problems. Chapter 4 introduces properties that depend on the first law, and so on. By Chapter 8 you will have all the basic thermodynamic principles and all the physical property information necessary to solve a wide range of problems. The remaining chapters focus on just such problems.

The unnumbered chapters are brief discussions. Chapter B treats problem solving, and the others reflect on topics already discussed and preview those that follow.

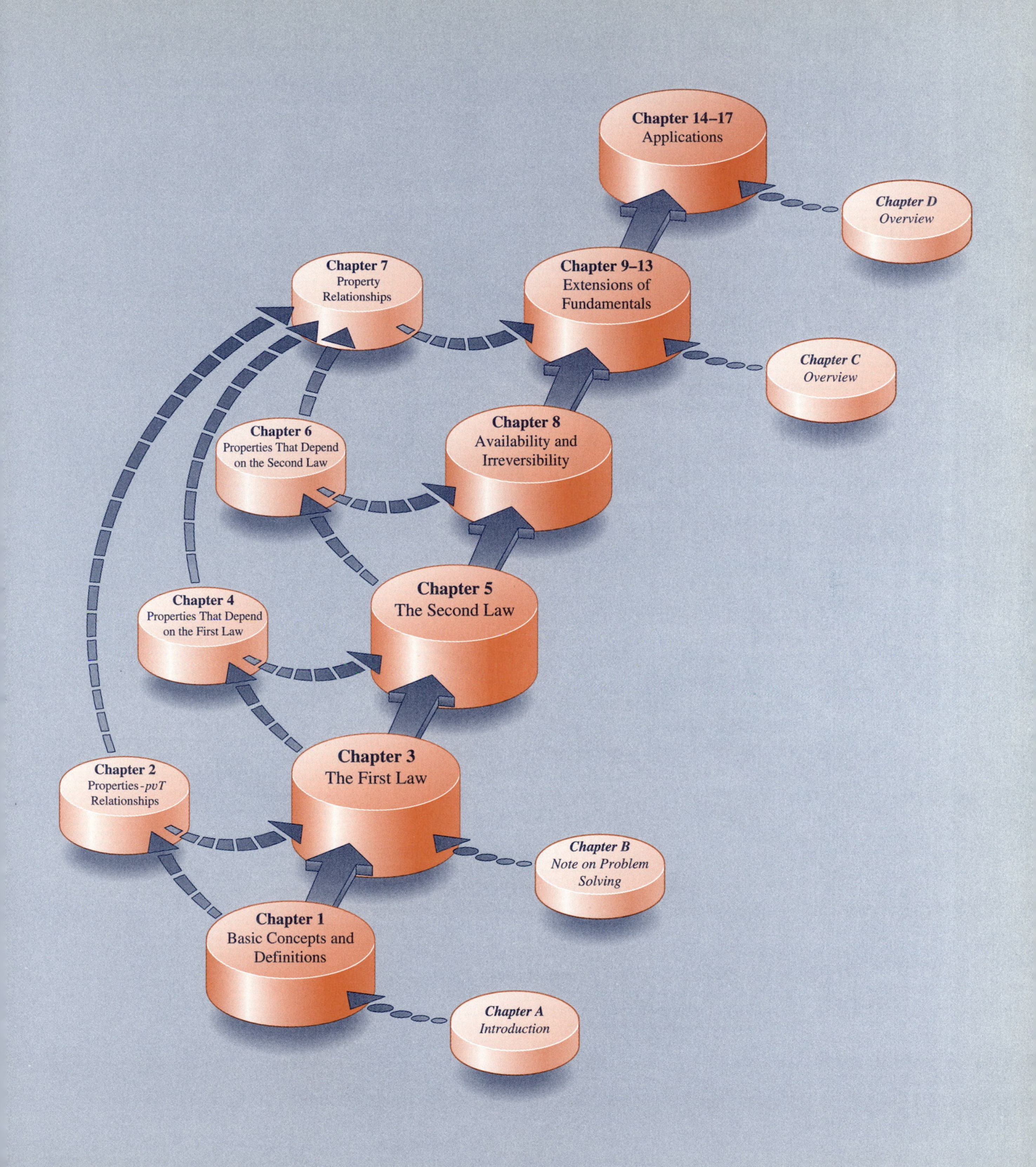

Chapter 14–17
Applications
Chapter D
Overview
Chapter 7
Property
Relationships
Chapter 9–13
Extensions of
Fundamentals
Chapter C
Overview
Chapter 8
Availability and
Irreversibility
Chapter 6
Properties That Depend
on the Second Law
Chapter 5
The Second Law
Chapter 4
Properties That Depend
on the First Law
Chapter 3
The First Law
Chapter 2
Properties -*pvT*
Relationships
Chapter B
Note on Problem
Solving
Chapter 1
Basic Concepts and
Definitions
Chapter A
Introduction

Vapor-Compression Refrigeration System

A refrigerator removes heat from a low-temperature region and rejects heat to cooling water or the atmosphere. The basic idea is to have the refrigerant sometimes at a low temperature, so that it will absorb heat from the space to be cooled, and at other times at a high temperature, so that heat can be rejected.

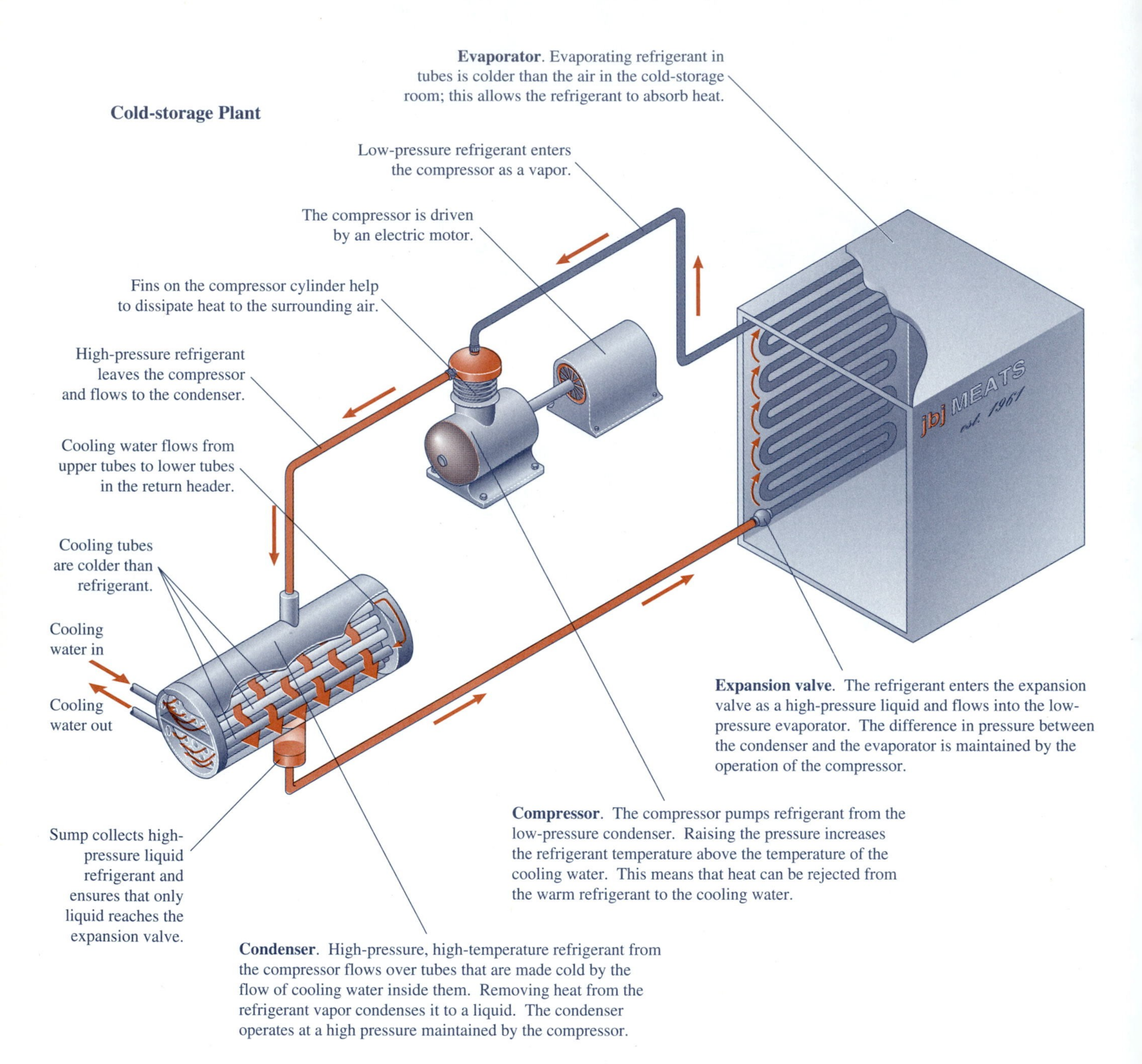

Evaporator. Evaporating refrigerant in tubes is colder than the air in the cold-storage room; this allows the refrigerant to absorb heat.
Cold-storage Plant
Low-pressure refrigerant enters the compressor as a vapor.
The compressor is driven by an electric motor.
Fins on the compressor cylinder help to dissipate heat to the surrounding air.
High-pressure refrigerant leaves the compressor and flows to the condenser.
Cooling water flows from upper tubes to lower tubes in the return header.
jbj MEATS est. 1961
Cooling tubes are colder than refrigerant.
Cooling water in
Cooling water out
Expansion valve. The refrigerant enters the expansion valve as a high-pressure liquid and flows into the low-pressure evaporator. The difference in pressure between the condenser and the evaporator is maintained by the operation of the compressor.
Compressor. The compressor pumps refrigerant from the low-pressure condenser. Raising the pressure increases the refrigerant temperature above the temperature of the cooling water. This means that heat can be rejected from the warm refrigerant to the cooling water.
Sump collects high-pressure liquid refrigerant and ensures that only liquid reaches the expansion valve.
Condenser. High-pressure, high-temperature refrigerant from the compressor flows over tubes that are made cold by the flow of cooling water inside them. Removing heat from the refrigerant vapor condenses it to a liquid. The condenser operates at a high pressure maintained by the compressor.

Schematic Diagram

We used simplified schematic diagrams in making thermodynamic analyses of refrigeration systems. Standard symbols represent the various components.

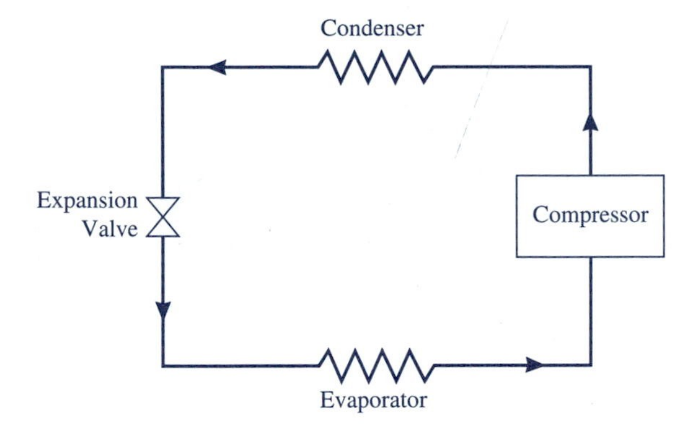

Household Refrigerator

The basic operation of a household refrigerator is the same as that of the cold storage plant. However, some components are slightly different.

In a household refrigerator, room air flow is used instead of cooling water to remove heat from the condensing refrigerant. That is why the coils on the back of your refrigerator feel hot.

Low-pressure vapor returns to the compressor suction.

High-pressure liquid refrigerant flows into a capillary tube that serves as an expansion valve. Refrigerant leaving the capillary tube flows into the evaporator as a low-pressure, low-temperature mixture of liquid and vapor. In the evaporator, the remaining liquid evaporates, absorbing heat from the food storage compartment.

The compressor and driving motor are in a sealed unit.

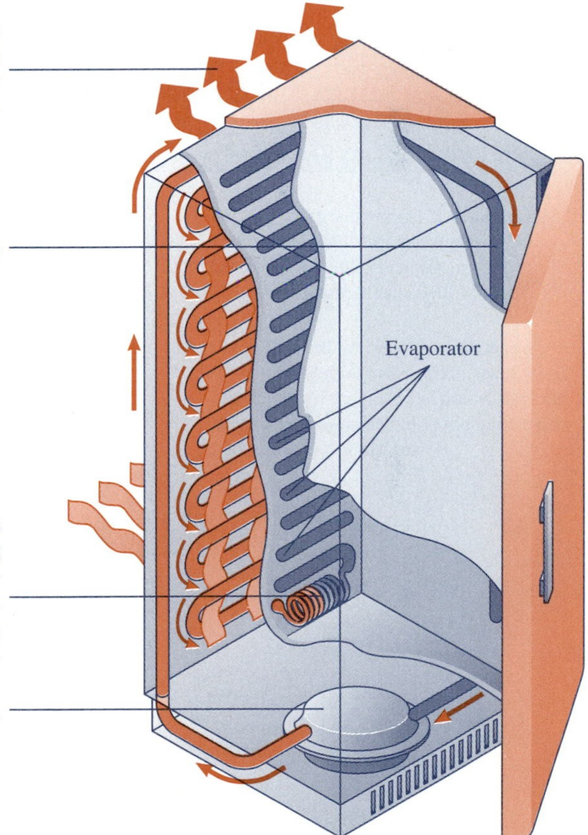

Steam Power Plant

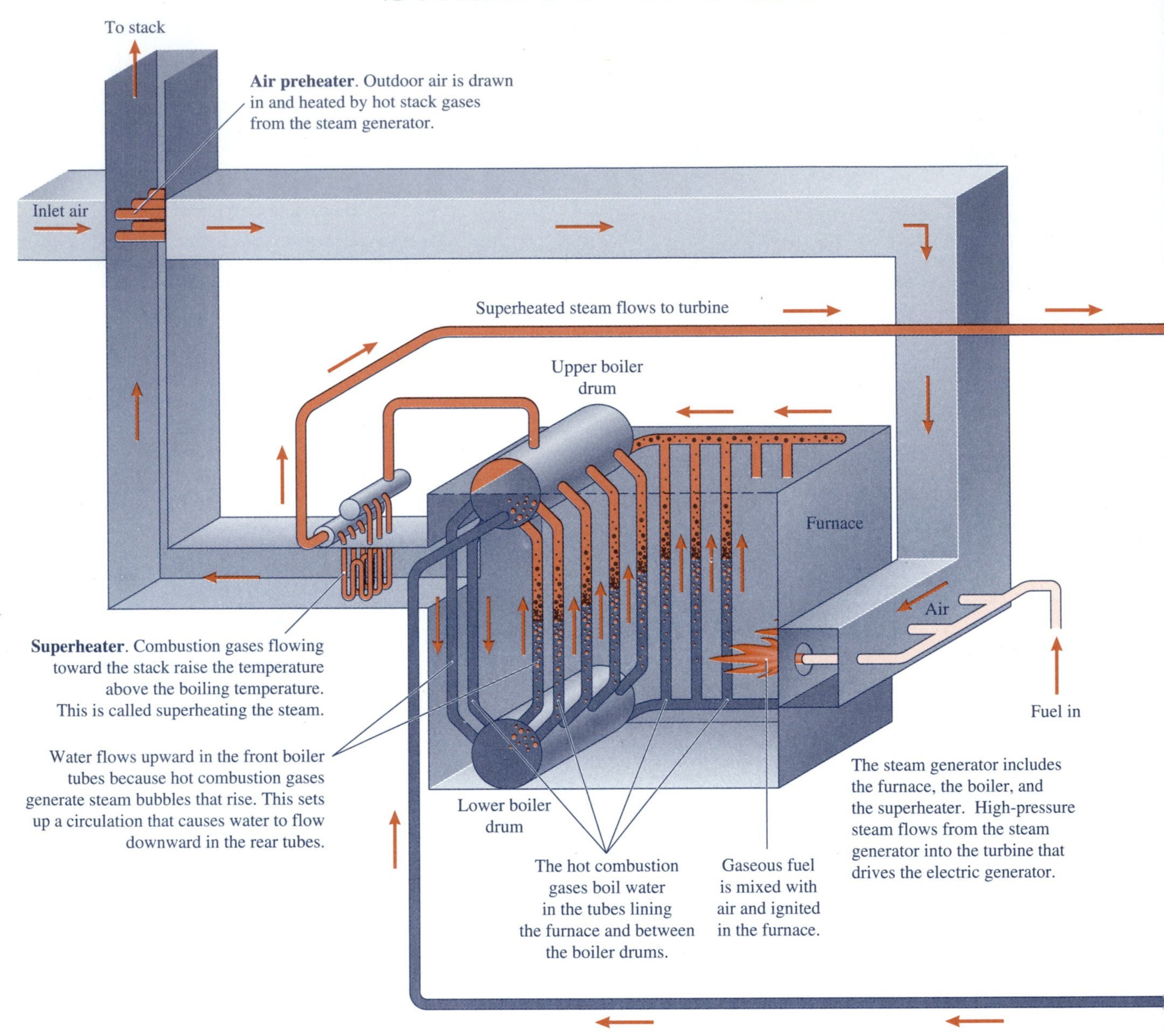

To stack
Air preheater. Outdoor air is drawn in and heated by hot stack gases from the steam generator.
Inlet air
Superheated steam flows to turbine
Upper boiler drum
Furnace
Air
Fuel in
Superheater. Combustion gases flowing toward the stack raise the temperature above the boiling temperature. This is called superheating the steam.
Water flows upward in the front boiler tubes because hot combustion gases generate steam bubbles that rise. This sets up a circulation that causes water to flow downward in the rear tubes.
Lower boiler drum
The hot combustion gases boil water in the tubes lining the furnace and between the boiler drums.
Gaseous fuel is mixed with air and ignited in the furnace.
The steam generator includes the furnace, the boiler, and the superheater. High-pressure steam flows from the steam generator into the turbine that drives the electric generator.

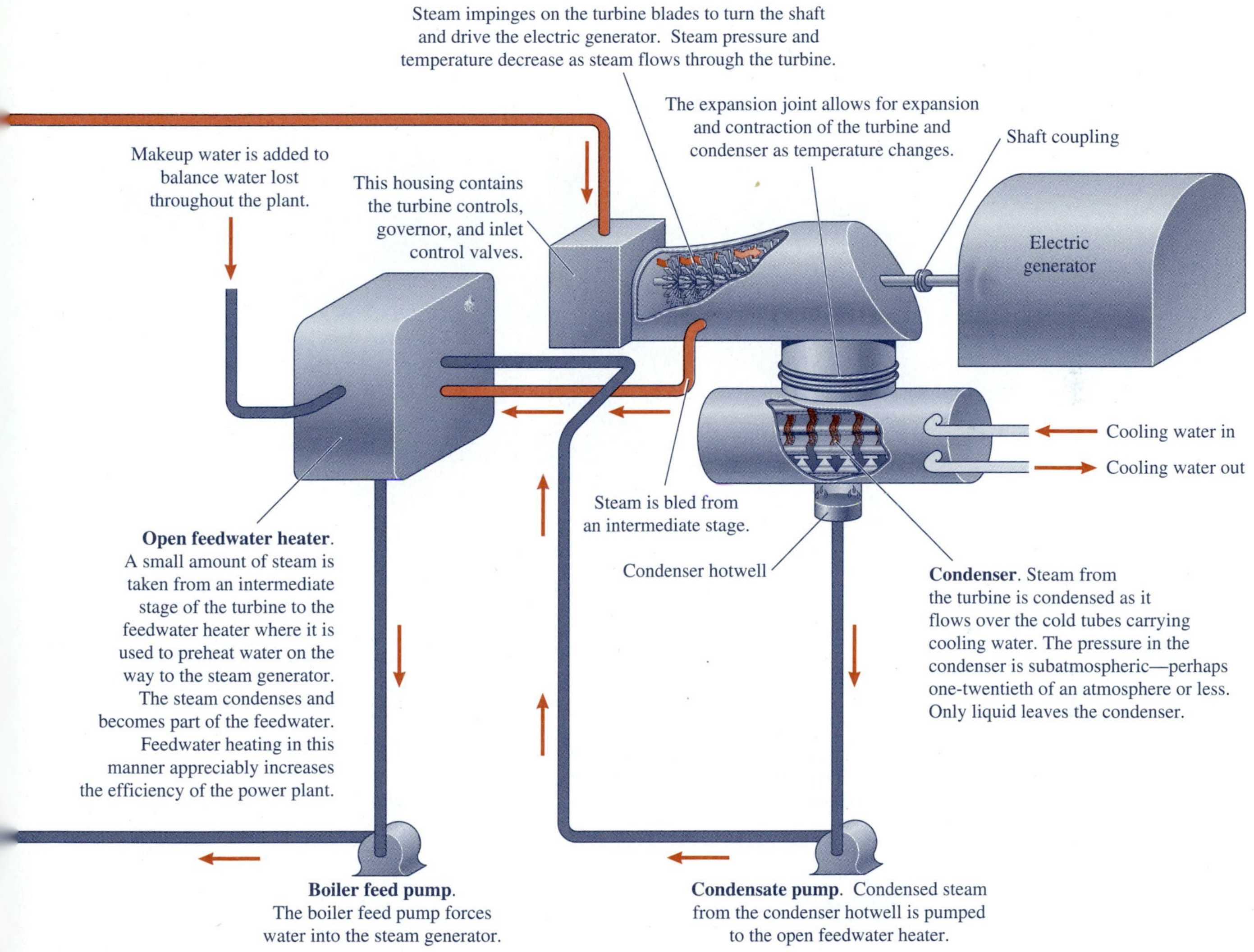

Steam impinges on the turbine blades to turn the shaft and drive the electric generator. Steam pressure and temperature decrease as steam flows through the turbine.

The expansion joint allows for expansion and contraction of the turbine and condenser as temperature changes.

Shaft coupling

Makeup water is added to balance water lost throughout the plant.

This housing contains the turbine controls, governor, and inlet control valves.

Electric generator

Cooling water in

Cooling water out

Steam is bled from an intermediate stage.

Condenser hotwell

Open feedwater heater. A small amount of steam is taken from an intermediate stage of the turbine to the feedwater heater where it is used to preheat water on the way to the steam generator. The steam condenses and becomes part of the feedwater. Feedwater heating in this manner appreciably increases the efficiency of the power plant.

Condenser. Steam from the turbine is condensed as it flows over the cold tubes carrying cooling water. The pressure in the condenser is subatmospheric—perhaps one-twentieth of an atmosphere or less. Only liquid leaves the condenser.

Boiler feed pump. The boiler feed pump forces water into the steam generator.

Condensate pump. Condensed steam from the condenser hotwell is pumped to the open feedwater heater.

Jet Engine

Air that enters the engine first flows through the compressor where its pressure and temperature increase. Fuel is injected into the air stream and burned in the combustion chamber. The high-temperature combustion gases flow through the turbine to produce work to drive the compressor that is mounted on the same shaft as the turbine. The gas leaving the turbine flows through the exhaust nozzle to form the high-velocity exhaust jet.

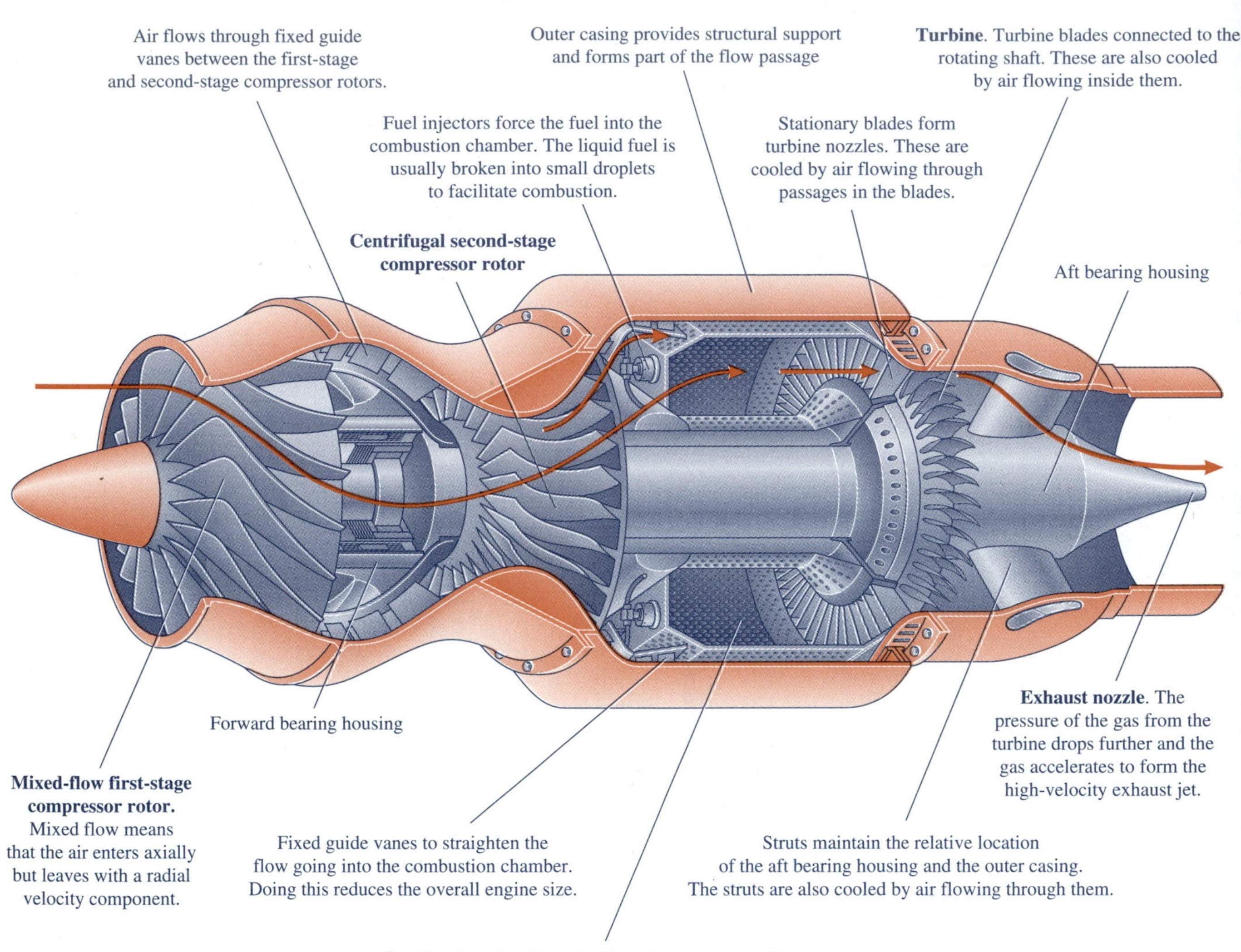

Air flows through fixed guide vanes between the first-stage and second-stage compressor rotors.
Outer casing provides structural support and forms part of the flow passage
Turbine. Turbine blades connected to the rotating shaft. These are also cooled by air flowing inside them.
Fuel injectors force the fuel into the combustion chamber. The liquid fuel is usually broken into small droplets to facilitate combustion.
Stationary blades form turbine nozzles. These are cooled by air flowing through passages in the blades.
Centrifugal second-stage compressor rotor
Aft bearing housing
Forward bearing housing
Exhaust nozzle. The pressure of the gas from the turbine drops further and the gas accelerates to form the high-velocity exhaust jet.
Mixed-flow first-stage compressor rotor. Mixed flow means that the air enters axially but leaves with a radial velocity component.
Fixed guide vanes to straighten the flow going into the combustion chamber. Doing this reduces the overall engine size.
Struts maintain the relative location of the aft bearing housing and the outer casing. The struts are also cooled by air flowing through them.
Combustion chamber. Air from the compressor flows through many holes in the combustion chamber walls to mix with fuel and provide a high-temperature gas stream to the turbine nozzles.

Some essential parts that are not shown are the fuel and lubricating oil pumps driven from the main shaft, the oil cooler, the control system, and the instrumentation system.

(Published with permission of AlliedSignal, Inc.)

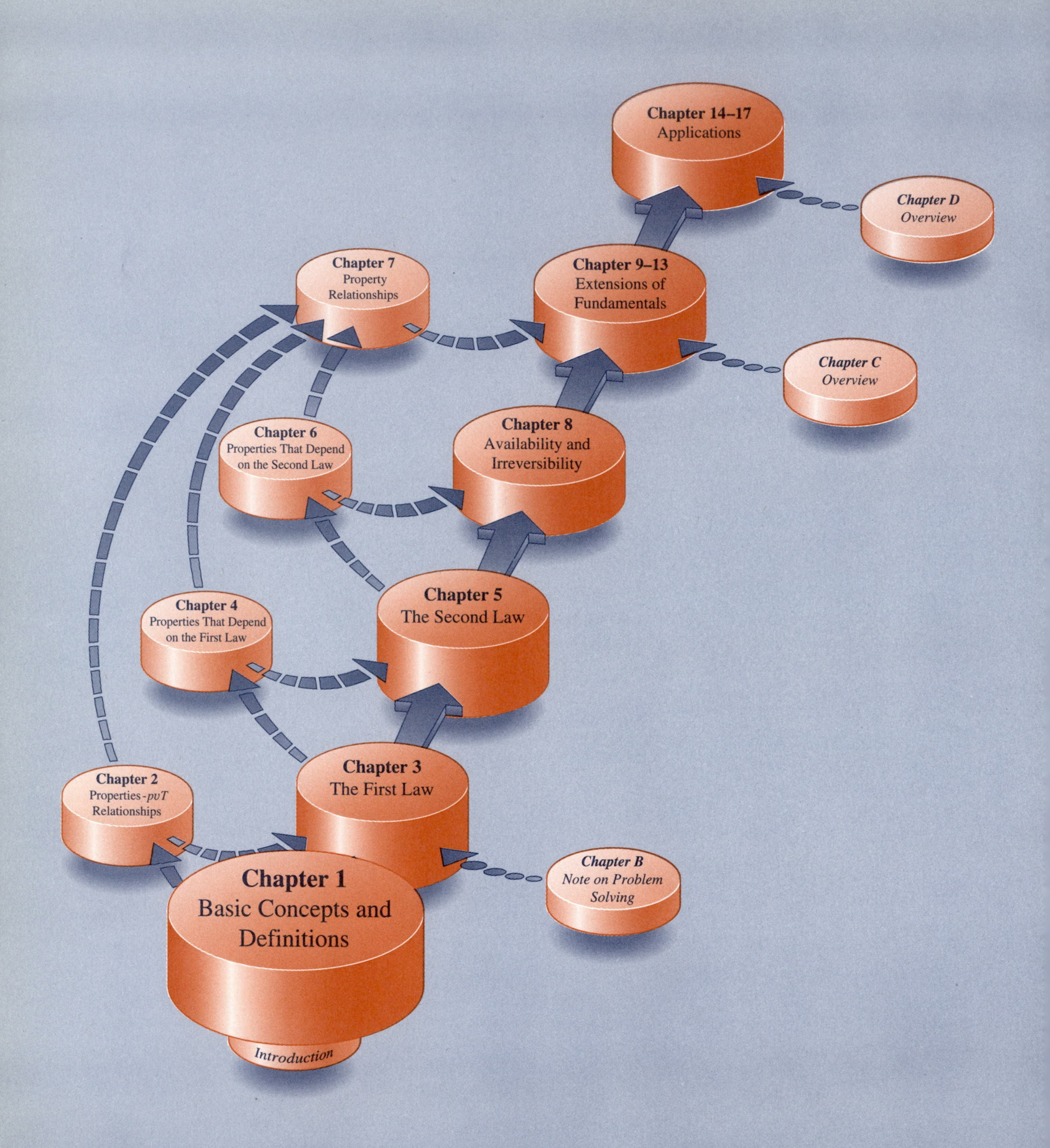

Chapter 14–17
Applications
Chapter D
Overview
Chapter 7
Property
Relationships
Chapter 9–13
Extensions of
Fundamentals
Chapter C
Overview
Chapter 6
Properties That Depend
on the Second Law
Chapter 8
Availability and
Irreversibility
Chapter 4
Properties That Depend
on the First Law
Chapter 5
The Second Law
Chapter 2
Properties-pvT
Relationships
Chapter 3
The First Law
Chapter B
Note on Problem
Solving
Chapter 1
Basic Concepts and
Definitions
Introduction

Chapter 1

Basic Concepts and Definitions

The study of any science begins with definitions of words used to state the fundamental principles. A sound knowledge of these definitions at the outset, just like an understanding of the rules at the start of a baseball game, will prevent many later misunderstandings. In this chapter we build a vocabulary for thermodynamics wherein each new term relies on the definitions of terms already defined or terms that we accept as verbally undefined. A widely accepted tenet of the philosophy of science is that some terms cannot be defined by simpler terms and must be accepted as verbally undefined (see, for example, Reference 1·1 or 1·2). In this book, terms that we arbitrarily accept as verbally undefined are time, length, temperature, mass, and force.

After studying this chapter, you should know all of the definitions required to understand the first law of thermodynamics. Every term defined in this chapter will be used repeatedly throughout the book.

1·1 Thermodynamic Systems

Thermodynamic system

A **thermodynamic system** is defined as any quantity of matter or any region of space to which we direct attention for purposes of analysis. The quantity of matter or region of space must be within a prescribed boundary. This boundary may be deformable or rigid; it may even be imaginary. Everything outside a system boundary is referred to as the **surroundings**. Usually, the term surroundings is restricted to those things outside the system that in some way influence the system.

Surroundings

Closed system

If a system is defined as a *particular quantity of matter*, then the system always contains the same matter, no matter crosses the system boundary, and the mass of the system is constant. Such a system is called a **closed system**. Examples of closed systems are shown in Fig. 1·1: the gas inside a closed

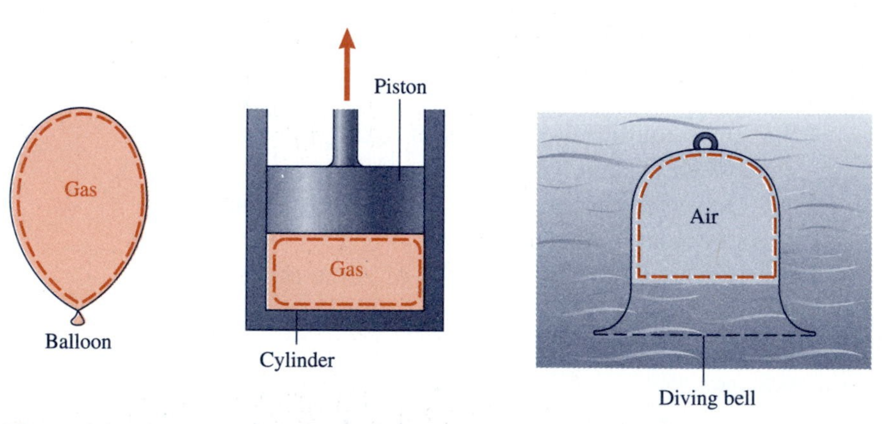

Figure 1·1 Examples of closed systems.

balloon, a gas trapped in a cylinder by a movable piston, and the air in a diving bell. A broken line shows the boundary of each system. In each of these cases, the boundary is deformable, but no matter crosses it.

A special case of a closed system is an *isolated system.* An **isolated system** is a system that in no way interacts with its surroundings. At first, it might seem that a system that does not interact with the surroundings would be of no interest. However, if two or more systems interact with only each other, the combination of these systems can be surrounded by a boundary to define an isolated system that can be useful in analyses.

An isolated system is a special case of a closed system.

An **open system** is defined as a *region of space within a prescribed boundary* that matter may cross. Figure 1·2 shows some open systems with the system boundary in each case shown as a broken line. The balloon (*a*) may have gas entering or leaving at the open lower end. Notice that in this case the open system boundary is deformable. The jet engine (*b*) has air and fuel entering and exhaust gases leaving. For some calculations of jet engine performance, the system selected for study might be only a combustion chamber (*c*). The window air conditioner (*d*) has fluids crossing the boundary at five different sections: indoor air entering, outdoor air entering, conditioned air leaving, warm air discharging to the atmosphere, and liquid water leaving.

Open system
Some authors call an open system a *control volume* and its boundary a *control surface.* These two terms are completely interchangeable with *open system* and *boundary.*

Systems used in thermodynamics range from tiny particles or bubbles to complete complex chemical processing plants or to large regions of the

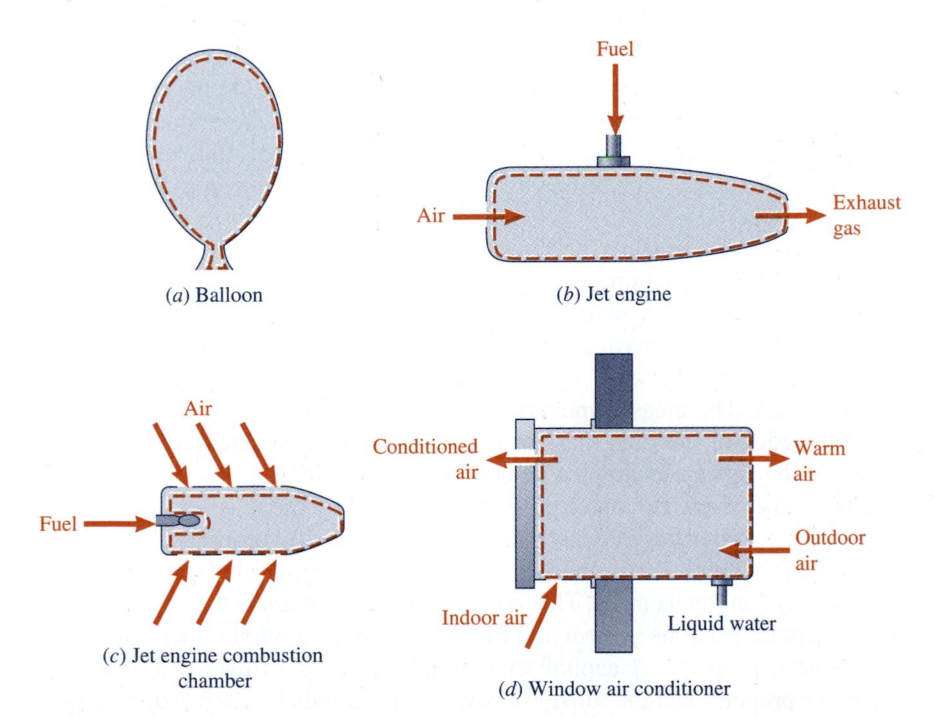

Figure 1·2 Examples of open systems.

In virtually all thermodynamic analyses, the first question you should ask is, What kind of system is involved?

earth's atmosphere. In virtually all thermodynamic analyses, the first step is to answer the question, what kind of system is involved? Even for the same analysis, different systems can be defined, and some will be more convenient than others. In some cases, we can analyze the same phenomenon by means of either a closed system or an open system. Starting a thermodynamics problem by identifying the system is similar to starting a mechanics problem by drawing the free-body diagram.

1·2 Properties, States, and Processes

A property is any observable characteristic of a system.

Properties. A **property** is any observable characteristic of a system. Examples of properties are pressure, temperature, modulus of elasticity, volume, and dynamic viscosity. Any combination of observable characteristics, such as the product of pressure and temperature, is also a property, and such properties can be thought of as *indirectly* observable characteristics of a system. Any number of such properties can be defined, but only a few are useful. Still other properties that cannot be *directly* observed can be defined by means of the laws of thermodynamics. Two such properties, internal energy and entropy, will be introduced in Chapters 4 and 6, respectively.

If a homogeneous system is divided into two parts, the mass of the whole system is equal to the sum of the masses of the two parts. The volume of the whole is also equal to the sum of the volumes of the parts. On the other hand, the temperature of the whole is not equal to the sum of the temperatures of the parts. In fact, the temperature, pressure, and density of the whole are the same as of the parts. This brings us to the distinction between *extensive* and *intensive* properties.

Extensive property

Intensive property—one that has the same value for any part of the system.

If the value of a property of a system equals the sum of the values for the parts of the system, that property is an **extensive** property. Mass, volume, and several other properties (energy, enthalpy, entropy) that will be introduced later are *extensive* properties. By contrast, an **intensive** property is one that has the same value for any part of a homogeneous system as it does for the whole system. The measurement of an intensive property can be made without knowledge of the total mass or extent of the system. Pressure, temperature, and density are examples of intensive properties.

Specific property

If the value of any extensive property is divided by the mass of the system, the resulting property is intensive and is called a **specific** property. For example, specific volume is obtained by dividing the volume (an extensive property) of a system by its mass. This ratio of volume to mass is the same for any part of a homogeneous system and for the system as a whole; therefore, it is an intensive property. A capital letter is usually used as the symbol for an extensive property and the same lowercase letter stands for the corresponding specific property. Thus V is used for volume and v is used for specific volume, and

$$v = \frac{V}{m}$$

where m is used for mass (an exception to the convention of using capital letters for extensive properties).

State

States. The **state** or condition of a system is specified by the values of its properties. If a system has the same values for all of its properties at two different times, the system is in identical states at those two times. Usually, we need to know only a few properties to specify a state completely. Precisely how many properties are required to specify the state of a system depends on the complexity of the system.

Equilibrium state

A system is in an **equilibrium state** (or in **equilibrium**) if no changes can occur in the state of the system without the aid of an external stimulus. We can test if a system is in equilibrium by isolating it and observing whether any change in its state occurs. We shall see that the temperature must be the same throughout a system in equilibrium. Otherwise, when the system is isolated, there would be a change in the temperature distribution. Also, there can be no eddying motions of a fluid in a system in equilibrium. If there were, when we isolate the system such motions would eventually cease; thus, the state of the system would have changed without any external stimulus. A system in equilibrium must be homogeneous or must consist of a finite number of homogeneous parts in contact. Such homogeneity is not sufficient to ensure equilibrium, however. For example, a system composed of iron, water vapor, and air at room temperature consists of a finite number of homogeneous parts in contact. This system is not in equilibrium, though, because even with no interaction with the surroundings, it will change state through the oxidation of the iron.

Notice that the properties of a system, such as pressure and temperature, can each be represented by a single number only if the system is in a state of equilibrium. For this reason, equilibrium states are much easier to specify than nonequilibrium states. Elementary thermodynamics is concerned chiefly with equilibrium states.

The state of a closed system is relatively easy to specify. However, we may also, in the case of an open system, speak of the state of the fluid passing through the system at any section instead of the state of the system itself. For example, Fig. 1·3 shows the nozzle of a rocket. As the propellent gas flows from section a to d, its properties continually change: pressure decreases, temperature decreases, density decreases, and velocity increases. Also, the properties at any section may change with time. For example, the pressure at section a decreases with time. Still, the gas at section a at any instant has a measurable pressure, temperature, and so on. We assume that the relation-

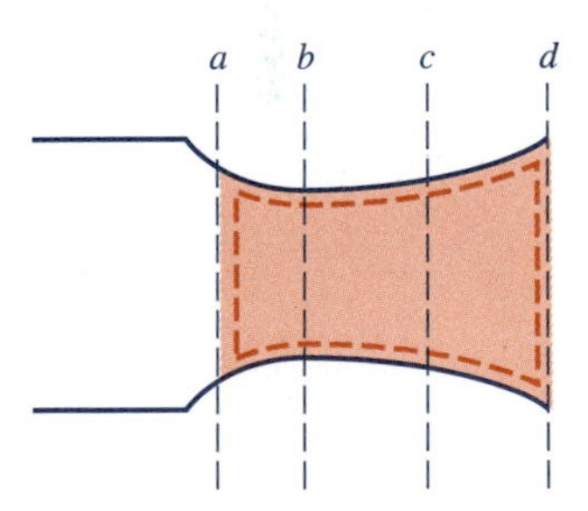

Figure 1·3 Open system: rocket nozzle.

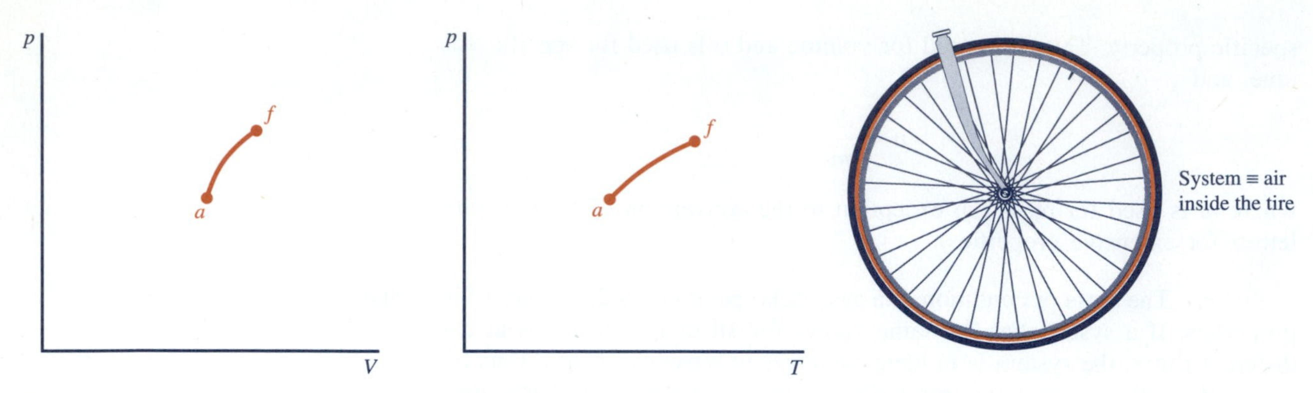

Figure 1·4 An expansion process of a gas, process *a–f*.

ships among the properties of the gas at this section at a given instant are the same as they would be if the gas were in equilibrium in a closed system at the same pressure, temperature, and so on. Thus we consider the fluid flowing within or across the boundaries of an open system to be passing through a series of equilibrium states.

Process
Path

Processes. A change of a system from one state to another is called a **process**. The **path** of the process is the series of states through which the system passes during the process. Figure 1·4 shows a bicycle tire. The system here is a closed system composed of only the air in the tire. The air is initially at a state *a* with pressure, temperature, and volume of p_a, T_a, and V_a. Then, because of the action of the sun, the flexing of the tire as it rolls, and the action of brakes, the air is transformed to a state *f* at a higher pressure, a higher temperature, and a slightly higher volume. States *a* and *f* are shown on pressure–temperature (pT) and pressure–volume (pV) diagrams.

Knowledge of only the end states, *a* and *f*, of the process tells us nothing about the states the system passes through during the process. Figure 1·5 shows three different processes between these same end states. For each process, the system passes through a different series of states between *a* and *f*, or is said to follow a different path. Many useful calculations can be made by knowing only the end states of processes, but for other calculations of engineering interest the path must be known.

Quasiequilibrium processes are infinitesimally close to equilibrium.

If during a process a system passes through a series of only equilibrium states, so that the system at any instant is in equilibrium or infinitesimally close to being in equilibrium, the process is called a **quasiequilibrium** process. A quasiequilibrium process is an idealization similar to a point mass, weightless cord, frictionless pulley, or homogeneous beam. Like these other idealizations, it is quite useful in modeling actual systems.

For a closed system, quasiequilibrium requires that there be no frictional effects and that properties be uniform throughout the system at any instant.

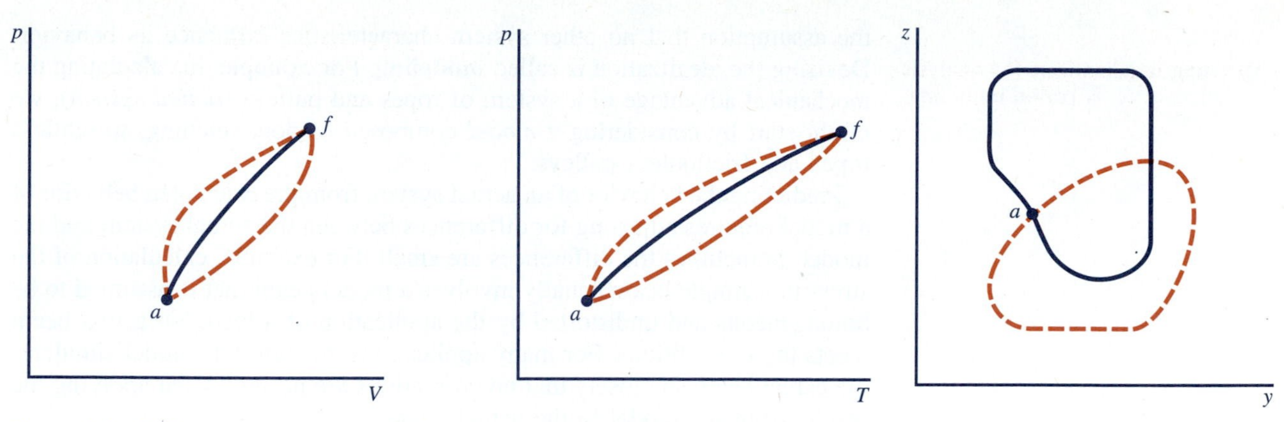

Figure 1·5 Different processes or paths between the same two end states.

Figure 1·6 Cyclic processes or cycles.

(An exception to the uniformity of properties is mentioned in Section 1·5·2.) For an open system, quasiequilibrium requires that there be no friction, but properties may vary from point to point throughout the system at any instant, as pointed out in connection with Fig. 1·3.

If a system undergoes a series of processes and is returned to its original state, it is said to have undergone a **cycle** or **cyclic process**. Figure 1·6 shows two cyclic processes starting and ending at state a. In this diagram, the coordinates y and z can be any two properties. The properties of the system vary during a cycle, but at the completion of a cycle all properties have been restored to their initial values. In other words, the net change in any property is zero for any cycle. This is concisely stated by

Cycle or cyclic process

$$\oint dx = 0 \tag{1·1}$$

where x is any property, and the symbol $\oint$ indicates integration around a cycle.

1·3 Modeling of Systems and Processes

In solving physical problems, we usually focus our attention not on the actual system but rather on some idealized system that is similar to, but simpler than, the actual system. An actual system has many characteristics. Some of these are highly pertinent to the behavior under study, many are immaterial, and some may have slight or unknown influence. The **model** or **idealized system** is defined in terms of only those characteristics that have a major influence on the actual system's behavior. The model is then analyzed under

Model or idealized system

Devising idealizations for analysis is called modeling.

the assumption that no other system characteristics influence its behavior. Devising the idealization is called **modeling**. For example, in calculating the mechanical advantage of a system of ropes and pulleys (*actual system*), we might start by considering a *model* composed of nonstretching, weightless ropes and frictionless pulleys.

Predicting the behavior of an actual system from the calculated behavior of a model requires adjusting for differences between the actual system and the model. Sometimes the differences are small. For example, calculation of the stress in a simple beam usually involves a model beam that is assumed to be homogeneous and undistorted by the application of a load. No actual beam meets these conditions. For many applications, though, this model simulates the actual beam so closely that no corrections are needed when applying the results from the model to the actual beam.

As another example, consider the calculation of the time required for an object to fall from the ceiling to the floor of a room. In modeling the falling process, we might neglect the frictional resistance of the air. If the object is a golf ball or tennis ball, the difference between the time we calculate and the measured time is small. That is, for most purposes the model is satisfactory. If the object is a table tennis ball, the difference is larger, but perhaps unimportant. However, for a falling feather a model in which air friction is neglected would be virtually useless.

It is important to define clearly the models you use to represent any actual system. As pointed out above, one step is to define the thermodynamic system you will analyze. After defining the system, you further define the model by placing restrictions on the system behavior. Examples would be the absence of friction or the absence of heat transfer.

Modeling is an important skill of an engineer. It involves balancing considerations of the extent and accuracy of knowledge about the actual system or process, the purposes for the model, the desired accuracy of results or predictions, and the time and other resources that can be committed to the analysis. Expert modeling requires judgment based on extensive knowledge and imagination. Development of such judgment may require years of experience, but checking on any model's accuracy is an important aspect of sound engineering at any level. It is a habit worthy of early development.

Macroscopic and microscopic models

Macroscopic and Microscopic Models. If all possible physical measurements on a system indicate that it is in the same state it was in at some previous time, we refer to the two states as identical states. This does not mean that we believe each molecule to have the same location and velocity that it previously had. It means only that on a **macroscopic** level the states are identical; on the **microscopic** level we make no direct measurements and consequently have no evidence for conclusions regarding individual molecules.

Classical thermodynamics deals with the macroscopic model only.

The macroscopic approach to thermodynamics is often called **classical thermodynamics**. Engineering thermodynamics uses the classical model. We occasionally use a molecular picture to enhance our understanding of some phenomenon, but it must be clearly understood that the laws of thermodynamics are based solely on macroscopic observations and in no way depend on molecular theory.

Microscopic approaches seek to explain physical phenomena on the basis of molecular behavior. **Kinetic theory** applies the laws of mechanics to individual molecules. Another approach, **statistical thermodynamics**, applies probability considerations to the very large numbers of molecules that comprise any macroscopic quantity of matter.

Kinetic theory
Statistical thermodynamics

1·4 Dimensions and Units

Although at first it may appear surprising, measurements of all physical quantities can be expressed in terms of time, length, mass, temperature, electric current, luminous intensity, and quantity of matter. These seven quantities are called the **primary dimensions**. Other physical quantities have dimensions that can be stated in terms of the primary dimensions. For example, the dimensions of velocity are length/time [L/t], of acceleration are length/time^2 [L/t^2], and of density are mass/length^3 [M/L^3]. The dimensions of force can be derived from Newton's second law,

Primary dimensions: time, length, mass, temperature, electric current, luminous intensity, quantity of matter

$$F = \frac{d}{dt}(mV) \qquad (1\cdot2)$$

as mass·length/time^2 [ML/t^2]. Then the dimensions of pressure, force per unit area, are equivalent to mass/length·time^2 [M/Lt^2].

The selection of primary dimensions is arbitrary. Electric charge might be selected in place of electric current; force might be selected in place of mass. In the field of mechanics, if mass is selected as a primary dimension, the system of dimensions is called an MLt system and the dimension of force is derived from Newton's second law as [ML/t^2]. If force is selected as a primary dimension, the system is called an FLt system, and the dimension of mass is derived from Newton's second law as [Ft^2/L]. Either of these methods is acceptable; there is no redundancy and hence no confusion. However, in engineering there is a widely used system of dimensions that uses both mass and force as primary. This is called an FMLt system. Since one of these four dimensions can be related to the other three by Newton's second law (Eq. 1·2), the system is *overdetermined.* Therefore, in some FMLt system equations a *dimensional constant* is required. This is discussed in Appendix I.

It is important to remember that any equation expressing a physical relationship must have consistency or homogeneity of dimensions. In fact, checking equations for dimensional homogeneity is an important step in problem solving. For example, under certain conditions the equation for the mass rate of flow ($\dot{m}$) of a fluid is

$$\dot{m} = \rho A V$$

where ρ is density, A is an area, and V is a velocity. The dimensional consistency of such an equation is shown by substituting for each quantity its dimensions:

$$\left[\frac{\mathrm{M}}{\mathrm{t}}\right] = \left[\frac{\mathrm{M}}{\mathrm{L}^3}\right][\mathrm{L}^2]\left[\frac{\mathrm{L}}{\mathrm{t}}\right]$$

(Notice that the same symbol may be used for a physical quantity that has a numerical value and for a dimension. For example, T is used for temperature as a variable in an equation and T is used for the dimension of temperature. This causes no confusion because dimensions are always shown in brackets as above.)

Units

Units are specified magnitudes of dimensions that are used for measurement purposes. Some units of length are meter, foot, and inch. Some units of mass are kilogram, pound mass, and ton. Several *systems of units* are in use. Each is based on a selected set of primary dimensions. In other words, an MLt system of dimensions and an FLt system of dimensions would involve different sets of units. The two systems of units used in this book are the International System (SI) and a system that we call the English system. Many variants of the English system are used and many names are applied to them (see Appendix I). Some problems at the ends of chapters involve mixed units, just as engineering practice often does.

Conversion factors are used to convert from one set of units to another.

Conversion factors are used to convert numerical values from one set of units to another. For example, 1 foot = 0.3048 meter, so a length, L_1, of 2.71 feet can be converted into meters as

$$L_1 = 2.71 \text{ ft } (0.3048 \text{ m/ft}) = 0.826 \text{ m}$$

The conversion factor is

$$0.3048 \text{ m/ft} = 1$$

Conversion factors are always dimensionless and have a numerical value of 1.

Notice that a conversion factor is always dimensionless, since it is a ratio of two units having the same dimensions, and always has a numerical value of 1.

Careful attention to dimensions and units will help you avoid countless errors. Any practicing engineer can confirm this. Therefore, time invested in mastering their fundamentals will pay continuing dividends.

1·4·1 International System of Units (SI)

The International System of units (SI, from Système Internationale) is described by strict definitions, rules, and conventions. It is based on the primary dimensions of mass, length, time, temperature, electric current, luminous intensity, and amount of substance. For each of these quantities, a base unit is specified, as shown in Table 1·1. The prefixes used for multiples of units are shown in Table 1·2.

For mechanics, SI is based on an MLt system. As mentioned earlier, in an MLt system, force is a derived dimension defined by Newton's second law:

$$F = \frac{d}{dt}(mV) \qquad (1\cdot 2)$$

$$[\mathrm{F}] = \frac{1}{\mathrm{t}}[\mathrm{M}]\left[\frac{\mathrm{L}}{\mathrm{t}}\right] = [\mathrm{M}]\left[\frac{\mathrm{L}}{\mathrm{t}^2}\right]$$

The units of force are similarly derived as

$$1 \text{ newton} = (1 \text{ kg})(1 \text{ m/s}^2) = 1 \text{ kg}\cdot\text{m/s}^2$$

1·4·2 English System of Units

The English system of units has evolved gradually and, unlike SI, is not based on a single set of standards. Because it evolved through practice, the original English system of units is based on an FMLt system of dimensions. Force is not derived through Newton's second law; instead, force and mass are independently defined. However, since it is possible to derive force or mass through Newton's second law of motion, this system of dimensions is *overdetermined.* Therefore, in an FMLt system,

The FMLt system of units is overdetermined.

$$[\mathrm{F}] \neq [\mathrm{ML/t^2}]$$

as would be true in an MLt system. Newton's second law must be written as

$$F = \frac{1}{g_c}\frac{d}{dt}(mV) \qquad (1\cdot 3)$$

where g_c is a *dimensional constant* (see Appendix I).

One variant of the English system of units is an FLt system with base units

Table 1·1 SI Base Units

Quantity	*Name*	*Symbol*
Length	meter	m
Mass	kilogram	kg
Time	second	s
Electric current	ampere	A
Temperature	kelvin	K
Amount of substance	mole	mol
Luminous intensity	candela	cd

Table 1·2 SI Prefixes

Factor	*Prefix*	*Symbol*
10^{12}	tera	T
10^{9}	giga	G
10^{6}	mega	M
10^{3}	kilo	k
10^{2}	hecto	h
10^{1}	deka	da
10^{-1}	deci	d
10^{-2}	centi	c
10^{-3}	milli	m
10^{-6}	micro	μ
10^{-9}	nano	n
10^{-12}	pico	p
10^{-15}	femto	f
10^{-18}	atto	a

of pound force, foot, and second and a derived unit of mass, the slug. Newton's second law provides the definition of the unit of mass,

$$m = \frac{F}{a} \qquad (1\cdot4)$$

so that

$$1 \text{ slug} = (1 \text{ lbf})/(1 \text{ ft/s}^2) = 1 \text{ lbf}\cdot\text{s}^2/\text{ft}$$

Many engineers use this system, but in numerical calculations still use the pound mass as the mass unit by using the conversion factor.

$$1 \text{ slug} = 32.174 \text{ lbm} \quad \text{or} \quad 32.174 \text{ lbm/slug} = 1$$

because many useful and needed data tables use the pound mass. Another conversion factor that is helpful when using the pound mass as a unit of mass in an FLt system is

$$32.174 \frac{\text{ft}^2/\text{s}^2}{\text{ft}\cdot\text{lbf/lbm}} = 1 \quad \text{or} \quad 32.174 \frac{\text{lbm}\cdot\text{ft}}{\text{lbf}\cdot\text{s}^2} = 1$$

Test yourself: What is the difference between these conversion factors and g_c, the dimensional constant?

Notice that this *is* a conversion factor—dimensionless and with a numerical value of 1—and not a dimensional constant like g_c, which is used only in FMLt systems of dimensions.

All equations in this book are written for MLt or FLt systems and are therefore dimensionally homogeneous with no dimensional constants needed. Therefore, the dimensional constant g_c does not appear in the equations in this book. In addition, each equation is valid for numerical calculations in *any consistent set of units.* By a consistent set of units we mean a set that defines the force unit in terms of mass, length, and time units or the mass unit in terms of force, length, and time units (see Table 1·3). Often we obtain data in units that do not form a consistent set, so conversion factors must be used.

Table 1·3 Consistent Sets of Units

	Mass	*Length*	*Time*	*Force*
1	kilogram	meter	second	newton
2	pound mass	foot	second	poundal
3	slug	foot	second	pound force
4	gram	centimeter	second	dyne

1·5 Some Directly Observable Properties

Making calculations for virtually any engineering system requires knowledge of properties. We introduce now some properties that are independent of the laws of thermodynamics. Others that can be defined only on the basis of one or more laws of thermodynamics will be introduced in later chapters.

1·5·1 Density and Specific Volume

Density (ρ) is defined as the mass of a substance divided by the volume the substance occupies or the mass per unit volume:

Density

$$\rho \equiv \frac{\text{mass}}{\text{volume}} = \frac{m}{V}$$

Specific volume (v) is defined as the volume per unit mass or the reciprocal of density:

Specific volume

$$v \equiv \frac{V}{m} = \frac{1}{\rho}$$

In some fields, **specific weight** (γ) is useful. It is defined as the weight of a substance divided by its volume, or the weight per unit volume:

Specific weight

$$\gamma \equiv \frac{\text{weight}}{\text{volume}} = \frac{w}{V}$$

Weight is the force of gravity on a substance. It depends on both the mass of the substance and the gravitational field strength. If the only force acting on a body is its weight, the resulting acceleration is the gravitational acceleration g, and Newton's second law of motion,

Weight

$$F = ma \tag{1·4}$$

becomes

$$\text{weight} = mg \tag{1·5}$$

This equation is the basic relation between weight, mass, and gravitational acceleration. If each side of this equation is divided by the volume of the body under consideration, the basic relation between specific weight, density, and gravitational acceleration is obtained:

$$\gamma = \rho g \tag{1·6}$$

Specific gravity is defined as the ratio of the density of a substance to some standard density. Notice that *specific gravity*, as defined here, is a misnomer, because it is unaffected by gravity. The standard density used with solids and liquids is often that of water at one atmosphere and some specified

Specific gravity

temperature such as 0°C, 4°C (the temperature of maximum density for water at a pressure of one atmosphere), or 20°C. Since specific gravity is a dimensionless ratio, its numerical value is independent of any system of units.

1·5·2 Pressure

Pressure

Pressure is defined as the normal force exerted by a system on a unit area of its boundary. The pressure may vary from place to place on the system boundary, even when the system is in equilibrium. For example, consider a system consisting of a fluid (either gas or liquid) in a closed tank. A simple force balance on the fluid shows that the pressure increases toward the bottom of the tank as a result of the weight of the fluid.

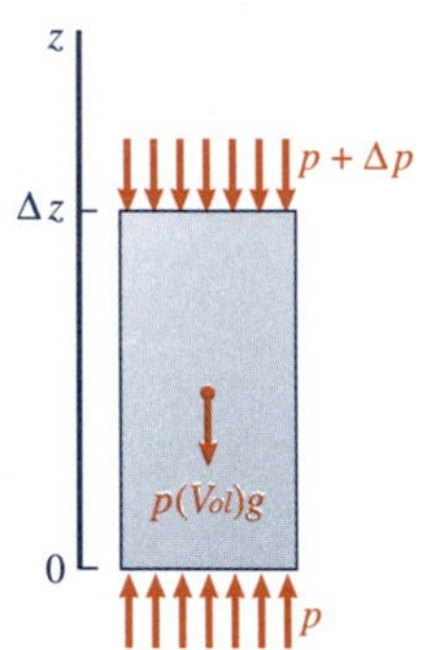

$$0 = \Sigma F_z$$
$$0 = pA - (p + \Delta p)\,A - \rho\,(Vol)g$$
$$\Delta pA = -\,\rho\,(Vol)g = -\,\rho(A\,\Delta z)g$$
$$\Delta p = -\,\rho g\,\Delta z$$

For a fluid in static equilibrium, the relationship between pressure and elevation within the fluid is given by the *basic equation of fluid statics*:

$$dp = -\gamma\, dz \tag{1·7}$$

where γ is the specific weight of the fluid and z is the elevation. The minus sign results from the convention of measuring z positively upward. Thus, as z increases in a fluid, p decreases. This equation can be derived from a force balance on an element of fluid in static equilibrium. In general, the specific weight is a function of pressure and temperature. However, for liquids, which are only slightly compressible, the specific weight can be assumed constant with respect to pressure, so that the basic equation can be integrated to give

$$\Delta p = -\gamma\,\Delta z \quad \text{and} \quad \Delta p = -\rho g\,\Delta z \tag{1·8a}$$

A manometer is a simple and widely used instrument that indicates a pressure difference by balancing a measurable length of fluid column against the pressure difference (see Fig. 1·7). As a result of the use of manometers and the direct proportionality between pressures and manometric fluid heights, pressures are often expressed in units such as inches of mercury, inches of water, or millimeters of mercury. Therefore, Eq. 1·8*a* is often written as

$$\Delta p = \gamma h \quad \text{and} \quad \Delta p = \rho g h \tag{1·8b}$$

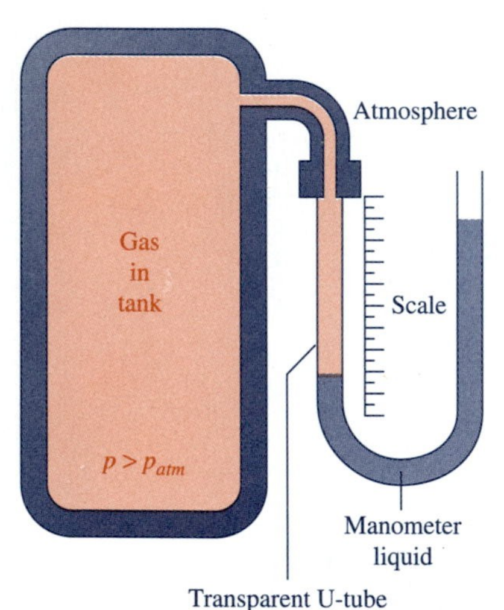

Figure 1·7 A simple U-tube manometer used to measure the pressure in a tank.

Vacuum readings, no matter how they are obtained, are usually expressed in millimeters or inches of mercury. A pressure that is less than atmospheric pressure by 140 mm of mercury is spoken of as a vacuum of 140 mm mercury and is written as 140 mm Hg vac.

Because the density of any manometric liquid varies with temperature, the same pressure difference results in different manometer deflections at different temperatures. Consequently, a *conventional millimeter of mercury* and a *conventional inch of mercury*, based on the density of mercury at 0°C or 32 F and a standard acceleration of gravity, are defined by

$$1 \text{ mm Hg} = 0.1333 \text{ kPa} \quad \text{and} \quad 1 \text{ in. Hg} = 0.4912 \text{ psi}$$

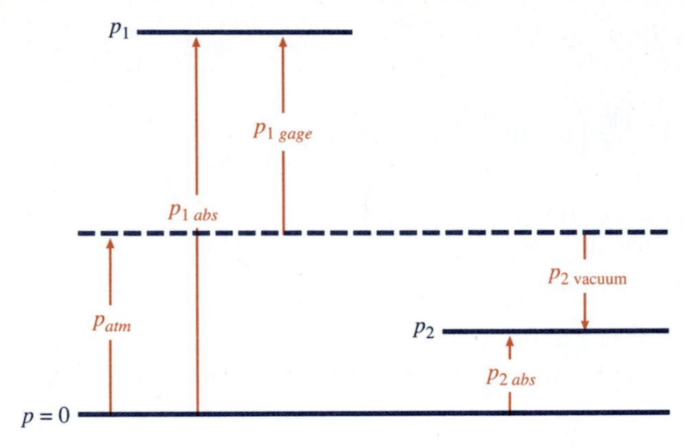

Figure 1·8 Absolute pressure, atmospheric pressure, gage pressure, and vacuum relationships.

Most pressure-measuring instruments measure the difference between the pressure of a fluid and the pressure of the atmosphere. This pressure *difference* is called **gage pressure**. The absolute pressure of the fluid is then obtained by the relation

Gage pressure is the difference between the pressure of a fluid and the pressure of the atmosphere.

$$p_{abs} = p_{atm} + p_{gage}$$

For a pressure lower than atmospheric pressure, the gage pressure is negative and the term **vacuum** is applied to the gage pressure. For example, a gage pressure of -40 kPa is spoken of as a *vacuum* of 40 kPa. The relationships among absolute pressure, gage pressure, atmospheric (or barometric) pressure, and vacuum are shown graphically in Fig. 1·8.

Vacuum

Example 1·1 Manometer

PROBLEM STATEMENT

Determine the pressure difference indicated by a laboratory manometer deflection of (*a*) 26.8 cm of mercury, and (*b*) 13.9 in. of water. The local acceleration of gravity is 9.780 m/s^2 or 32.09 ft/s^2.

SOLUTION

In both parts *a* and *b*, assume that the temperature of the laboratory manometer is room temperature, say, 20°C or 68 F.

(*a*) From tables of properties of mercury, at 20°C the density of mercury is 13,550 kg/m^3. Using the basic equation of fluid statics yields

$$\Delta p = \rho g h = 13{,}550 \frac{\text{kg}}{\text{m}^3}\left(9.780\frac{\text{m}}{\text{s}^2}\right) 26.8\text{ cm}\left(\frac{\text{m}}{100\text{ cm}}\right)$$

$$= 35{,}500\frac{\text{kg}}{\text{m}\cdot\text{s}^2} = 35{,}500\frac{\text{N}}{\text{m}^2} = 35{,}500\text{ Pa}$$

$$= 35.5\text{ kPa}$$

(*b*) We calculate the pressure difference in lbf/in^2 or psi. For water at room temperature, the density as found in many property tables is 62.3 lbm/ft^3. To use a consistent set of units (pound force, slug, foot, second), we convert the units of the density:

$$\rho = 62.3 \frac{\text{lbm}}{\text{ft}^3}\left(\frac{\text{slug}}{32.174\text{ lbm}}\right) = 1.936\text{ slug/ft}^3$$

Then

$$\Delta p = \rho g h = 1.936 \frac{\text{slug}}{\text{ft}^3}\left(32.09\frac{\text{ft}}{\text{s}^2}\right) 13.9\text{ in.}\left(\frac{\text{ft}}{12\text{ in.}}\right)$$

$$= 72.0\frac{\text{slug}\cdot\text{ft}}{\text{s}^2\cdot\text{ft}^2} = 72.0\frac{\text{lbf}}{\text{ft}^2} = 0.500\frac{\text{lbf}}{\text{in}^2} = 0.500\text{ psi}$$

1·5·3 Temperature

Temperature

The familiar sense perception of hot and cold are qualitative indications of the **temperature** of a body, but numerical values cannot be accurately assigned to various temperatures on the basis of physiological sensations. Fortunately, when the temperature of a body changes, several other properties also change. Any of these temperature-dependent properties might be used as an indirect measurement of temperature. For example, both the volume and the electrical resistance of a bar of steel increase as the steel gets hotter. You can probably think of many other temperature-dependent properties of materials, and several different ones are actually used in the measurement of temperature.

Sometimes it is tempting to try to define a term that has been accepted as undefined. For example, temperature is sometimes defined in terms of molecular activity. Such a definition may have some value, but it is ripe with potential pitfalls: At the same temperature, steel and water do not have the same molecular activity.

If a hot body and a cold body are brought into contact with each other while isolated from all other bodies, the hot body becomes colder and the cold body becomes hotter. (There are cases in which the temperature of only one of the bodies changes.) Finally, all changes in the properties of the bodies cease. The bodies are then at the same temperature and are said to be in **thermal equilibrium** with each other. It should be noted that such *equality of temperature* is possible even though no other properties have equal values.

Thermal equilibrium

A thermometer is a body with a readily measurable property that is a function of temperature. In a mercury-in-glass thermometer, the volume of the mercury depends on its temperature. In a resistance thermometer, the electrical resistance of the thermometer element is a temperature-dependent property. For a thermometer to indicate the temperature of another body, the thermometer and the other body must be in contact with each other long enough and must be sufficiently isolated from other bodies so that they will attain thermal equilibrium with each other. The temperature of the thermometer is then the temperature of the other body.

A numerical scale of temperature permits temperatures to be specified quantitatively. One way to establish a temperature scale is first to assign numerical values to certain accurately reproducible temperatures. Two that have been used in the past are called the *ice point* and the *steam point.* The **ice point** is defined as the equilibrium temperature of a mixture of ice and air-saturated liquid water at a pressure of one atmosphere (101.325 kPa or 14.696 psia). The **steam point** is defined as the equilibrium temperature of pure liquid water in contact with its vapor at one atmosphere. These two temperatures are used as reference temperatures because they are accurately reproducible in any laboratory. On the **Celsius scale**, the ice point is assigned

Ice point

Steam point

Anders Celsius (1701–1744), a Swedish astronomer, presented a paper before the Swedish Academy of Sciences in 1742 describing the centigrade thermometer scale with 100 degrees between the ice and steam points, with the ice point designated as the zero of the scale. The present Celsius scale is very nearly the same as the original centigrade scale proposed by Celsius.

the value 0 and the steam point is assigned the value 100. On the **Fahrenheit scale**, the values 32 and 212 are assigned.

Gabriel Daniel Fahrenheit (1686–1736), a German instrument maker who lived mostly in England and Holland, was the first to use mercury-in-glass thermometers. Alcohol and linseed oil had earlier been used as thermometric fluids. Fahrenheit's scale was a modification of one proposed by Sir Isaac Newton with 0 for the ice point and 12 for normal human body temperature. Fahrenheit lowered the zero of the scale to the temperature of a salt–ice mixture and made the degree smaller so that body temperature was 96. His measurements showed the ice and steam points to be at 32 and 212, respectively, based on the reference points at 0 and 96. Subsequently, the 32 and 212 were adopted as reference points, and refinements in thermometers have revealed that the salt–ice mixture minimum temperature and normal body temperature are not exactly 0 and 96 on the present Fahrenheit scale.

We establish a mercury-in-glass thermometer scale by first bringing such a thermometer to the ice point and marking the position of the mercury surface in the stem with a 0. We then bring the thermometer to the temperature of the steam point and mark the position of the mercury surface 100. The stem is then divided into 100 equal parts or degrees. Similarly, we can establish a

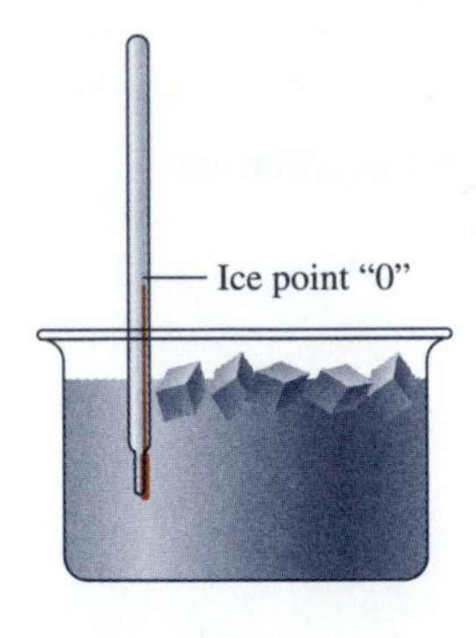

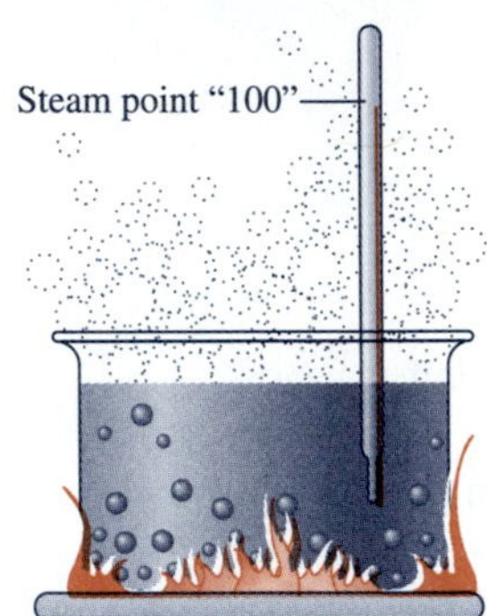

resistance thermometer scale by first noting the electrical resistance of a wire at the ice point and at the steam point and then making a straight-line plot of temperature versus resistance. We can follow the same procedure with a thermocouple to obtain a straight-line plot (based on two points) of temperature versus emf (electromotive force) for a fixed cold-junction temperature. If these three thermometers are placed together in a fluid at the ice point, all will read 0. If they are placed together in a fluid at the steam point, all will read 100. Agreement at these two points results, of course, from calibrating the thermometers at these two temperatures.

If, however, the three thermometers are placed in the same fluid and read 50, 49, and 53, which one is ''correct''? We cannot answer this question from the data given. There is no reason to establish any one as a standard in preference to the others. Thus a shortcoming of any temperature scale defined in terms of the physical properties of a single substance is apparent. Chapter 5 shows how it is possible to establish a *thermodynamic scale of temperature* which is independent of the properties of any substance.

Until then, for practical purposes, we need to define a numerical temperature scale. We can improve on scales that depend on the physical properties of a single substance by using a constant-volume gas thermometer. Constant-volume gas thermometers utilize the variation of pressure with temperature of a fixed volume of gas. This is, of course, a physical property dependence; however, constant-volume gas thermometer results are independent of the gas used.

A constant-volume gas thermometer consists of a gas contained in a constant-volume vessel that is provided with a means for measuring the gas pressure. The gas pressure varies with temperature, as shown in Fig. 1·9. The equation of the line shown is

$$T = T_0 + \left(\frac{T_s - T_i}{p_s - p_i}\right) p$$

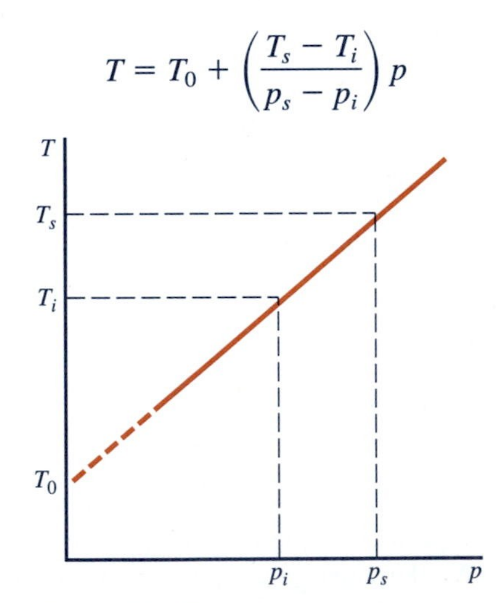

Figure 1·9 Temperature–pressure relationship of a constant-volume gas thermometer.

where the subscripts i, s, and 0 denote, respectively, conditions at the ice point, the steam point, and the temperature corresponding to an extrapolated gas pressure of zero. Experiments show that at low pressures the ratio p_s/p_i approaches the same value, 1.3661, for all gases. This value and the assignment of Celsius scale values of $T_i = 0°C$ and $T_s = 100°C$ show that $T_0 = -273.15°C$.

The **absolute Celsius gas thermometer scale**, defined by $T_0 = 0$ and $(T_s - T_i) = 100$ in the equation above, is very close to the *Kelvin scale*, a thermodynamic temperature scale, which is defined in Chapter 5.

The absolute Celsius gas thermometer scale and the Kelvin scale are very nearly the same.

Despite its simplicity in principle, a constant-volume gas thermometer is difficult to use. Also, even though only two fixed points define a temperature scale, for practical purposes in calibrating thermometers, additional fixed points across a wider temperature range are needed. Therefore, the International Temperature Scale, ITS-90, adopted by the International Committee on Weights and Measures specifies (1) numerical values for several easily reproducible temperatures (fixed points) over a wide temperature range and (2) temperature-measuring devices and interpolation methods in various temperature ranges. ITS-90 agrees closely with the thermodynamic temperature scale discussed in Chapter 5.

Four temperature scales are encountered in engineering practice: Celsius, Kelvin, Fahrenheit, and Rankine. The units on these scales are the kelvin (K), the degree Celsius (°C), degree Fahrenheit (F), and degree Rankine (R). The relationships among them are shown in Fig. 1·10 and are given by the following equations in which temperatures on the various scales are distinguished by subscripts:

William John MacQuorn Rankine (1820–1872), Scottish engineer and professor of civil engineering at the University of Glasgow, made several outstanding contributions to thermodynamics and its applications. He was the author of several books and many papers on thermodynamics, mechanics, canals, steam engines, and water supply systems.

SI calls for the symbol °C to be used instead of C only to avoid confusion with the symbol for coulomb.

$$T_K = T_C + 273.15 = \frac{5}{9} T_R$$

$$T_R = T_F + 459.67 = \frac{9}{5} T_K$$

The values 273.15 and 459.67 are often replaced by the approximate values 273 and 460.

1·6 Point and Path Functions

In preceding sections, we have discussed properties. Later in this chapter, we introduce work and heat, which are interactions between systems. To under-

Properties like pressure, temperature, entropy, and enthalpy are point functions; work and heat are path functions.

stand the differences between properties (like pressure, temperature, and specific volume) and quantities like work and heat, we must understand the difference between point and path functions. Properties are point functions; work and heat are path functions.

If x is a function of two independent variables, y and z, expressed by the notation

$$x = f(y, z)$$

Point function

Exact differential

then x is called a **point function**, because at each point on a plane of yz coordinates there is a discrete value of x. The differential dx of a point function x is an **exact differential**, and

$$dx = \left(\frac{\partial x}{\partial y}\right)_z dy + \left(\frac{\partial x}{\partial z}\right)_y dz$$

This may be written as

$$dx = M\,dy + N\,dz$$

where

$$M = \left(\frac{\partial x}{\partial y}\right)_z \qquad N = \left(\frac{\partial x}{\partial z}\right)_y$$

Then

$$\frac{\partial M}{\partial z} = \frac{\partial}{\partial z}\left(\frac{\partial x}{\partial y}\right) = \frac{\partial^2 x}{\partial z\,\partial y} \qquad \frac{\partial N}{\partial y} = \frac{\partial}{\partial y}\left(\frac{\partial x}{\partial z}\right) = \frac{\partial^2 x}{\partial y\,\partial z}$$

and consequently, since the order of differentiation is immaterial,

$$\frac{\partial M}{\partial z} = \frac{\partial N}{\partial y}$$

In fact, this provides a test for exactness. *A differential in the form $dx = M\,dy + N\,dz$ is an exact differential* (*hence is the differential of a point function*) *if and only if*

$$\frac{\partial M}{\partial z} = \frac{\partial N}{\partial y}$$

A proof of this relation can be found in any elementary textbook on differential equations. For an exact differential dx

$$\int_1^2 dx = x_2 - x_1$$

where $x_2 = f(y_2, z_2)$ and $x_1 = f(y_1, z_1)$. The value of the integral is independent of the path followed on yz coordinates in going from (y_1, z_1) to (y_2, z_2). It would be the same for any of the paths shown in Fig. 1·11.

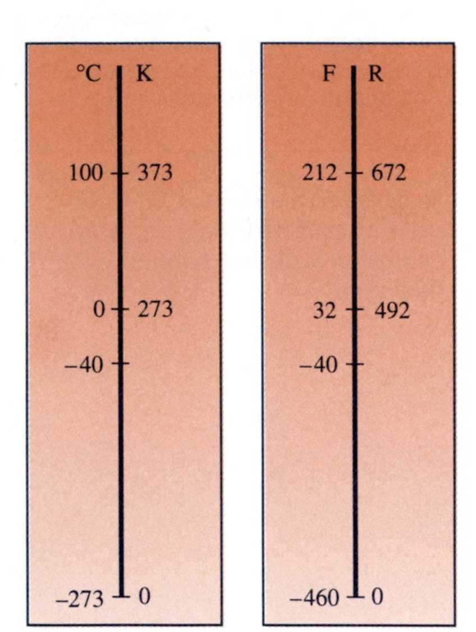

Figure 1·10 Temperature scale relationships. (−40°C and −40 F are marked because they are useful in a quick method for converting temperatures between Celsius and Fahrenheit scales. See Problem 1·30.)

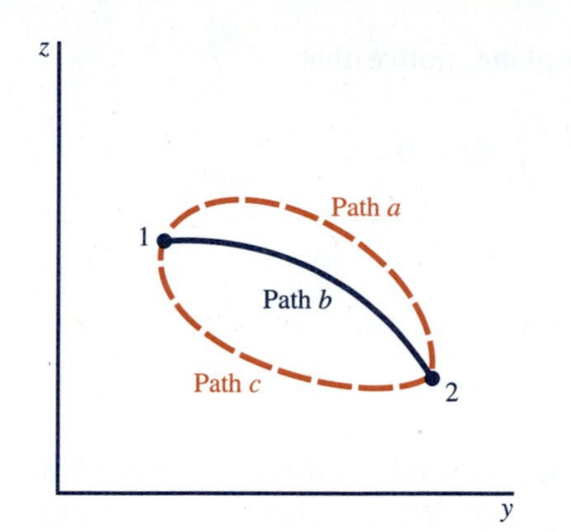

Figure 1·11 Multiple paths between points 1 and 2.

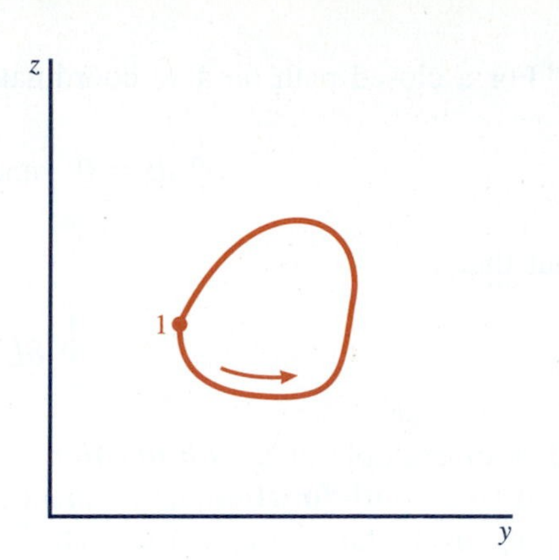

Figure 1·12 A cyclic process or cycle.

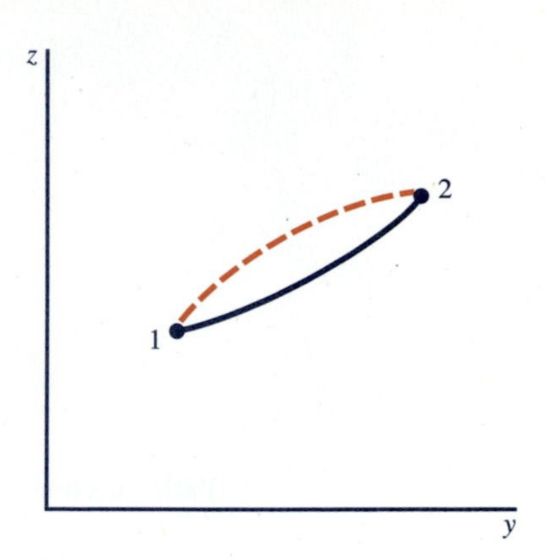

Figure 1·13 Path function: lengths of lines between points 1 and 2.

For any closed path or cycle as shown in Fig. 1·12, the initial and end states (or points on a property diagram) are identical; so

$$\oint dx = 0 \tag{1·1}$$

All thermodynamic properties are point functions. Therefore, the cyclic integral of property is always zero. Also, if for *any* cyclic path $\oint dx = 0$, x must be a property. (We will use this as a test to see if certain quantities are properties.)

The cyclic integral of a point function, and hence of any thermodynamic property, is always zero.

Some quantities related to processes shown on yz diagrams may not be functions of y and z and hence may not be point functions; this is true for work and heat. For example, the length of a line connecting two points, 1 and 2, on a yz coordinate plane (Fig. 1·13) is a path function. Let the length of some path connecting 1 and 2 be L. There is no value of L for point 1 or point 2, nor is there a single value for points 1 and 2 together. There is, in general, a different value of L for each of the many possible paths between 1 and 2. For any small segment of one of these paths, the limiting relationship is

$$(\delta L)^2 = (dz)^2 + (dy)^2$$

and for the entire path,

$$L_{1-2} = \int_1^2 \delta L = \int_1^2 \sqrt{1 + (dy/dz)^2}\, dz$$

This cannot be evaluated unless the relationship between y and z is known, that is, unless the path is specified. Knowledge of the end points alone is insufficient, because L is a path function and $L \neq f(z, y)$.

For a closed path on a yz coordinate plane, notice that

$$\oint dz = 0 \quad \text{and} \quad \oint dy = 0$$

but that

$$\oint \delta L \neq 0$$

L is an example of a *path function.*

Path function

If G is a **path function**, it is a quantity that depends on the path followed in going from state 1 (y_1, z_1) to state 2 (y_2, z_2), and *no* relation of the form $G = f(y, z)$ exists, because specifying a value of y and a value of z does not determine a value of G. The notation G_1 or G_2 should not be used, since this implies that there is a particular value of G at state 1 or state 2, but this is not true. The value of G corresponding to a particular path between states 1 and 2 is therefore not spoken of as the "change" in G, but simply as the value of G for that particular path.

δ = a very small amount of

d = the difference between two values very close together

This value of G is equal to the sum of the G values for any number of segments into which the path may be divided. If these segments are made smaller and smaller, the limiting value of G of one segment is δG. The symbol δ can be interpreted as "a very small amount of," whereas a corresponding loose interpretation of the differential operator d would be "the difference between two values very close together."

For any specific path between states 1 and 2, we may write

$$\int_1^2 \delta G = G_{1-2}$$

Again, for a path function

$$\int_1^2 \delta G \neq G_2 - G_1$$

because (1) a path function cannot be evaluated in terms of end states alone, and (2) there are no values such as G_1 and G_2 that can be assigned to states 1 and 2.

There may be a relationship of the form

$$\delta G = M'\,dy + N'\,dz$$

but, since G is a path function, δG is an *inexact differential* and

$$\frac{\partial M'}{\partial z} \neq \frac{\partial N'}{\partial y}$$

Under special conditions, the cyclic integral of a path function *may* be zero, but in general

$$\oint \delta G \neq 0$$

The cyclic integral of a path function is in general not equal to zero.

As we have seen, thermodynamic properties are point functions. In later sections of this chapter, we will see that work and heat, interactions between systems, are path functions.

Example 1·2 Point Functions

PROBLEM STATEMENT

For some substances, $pv = BT$, where p, v, T, and B are pressure, specific volume, temperature, and a constant, respectively. The quantities s and r are given by

$$\delta s \quad \text{or} \quad ds = \frac{c\,dT}{T} - \frac{v\,dp}{T}$$

and

$$\delta r \quad \text{or} \quad dr = \frac{g\,dT}{T} + \frac{p\,dv}{v}$$

where c and g are constants. Determine whether each of these quantities is a point function.

SOLUTION

Since, in each case, the differential is equal to an expression of the form of $M\,dy + N\,dz$, we need only use the test for exactness:

$$\frac{\partial M}{\partial z} = \frac{\partial N}{\partial y}$$

In the case of δs or ds, $M = c/T$, $N = -v/T$, $z = p$, and $y = T$. Then for $pv = BT$

$$\left[\frac{\partial(c/T)}{\partial p}\right]_T = \left[\frac{\partial(-v/T)}{\partial T}\right]_p$$

The derivative on the left-hand side is obviously zero. The one on the right-hand side cannot be evaluated immediately because v is a function of p and T. Therefore, we make a substitution from the original equation:

$$0 = \left[\frac{\partial(-B/p)}{\partial T}\right]_p$$

$$0 = 0$$

Hence, s is a point function. In the case of r, we have

$$\left[\frac{\partial(g/T)}{\partial v}\right]_T = \left[\frac{\partial(p/v)}{\partial T}\right]_v$$

$$0 \neq \frac{B}{v^2}$$

Hence r is not a point function.

Example 1·3

PROBLEM STATEMENT

Evaluate $\int_A^B dR$ over each of the two paths, (1) $y = x^2$ and (2) $y = 3x - 2$, connecting points A (1, 1) and B (2, 4) for each case:

$$(a)\ R = \int (x\,dy + y\,dx)$$

$$(b)\ R = \int (x\,dy - y\,dx)$$

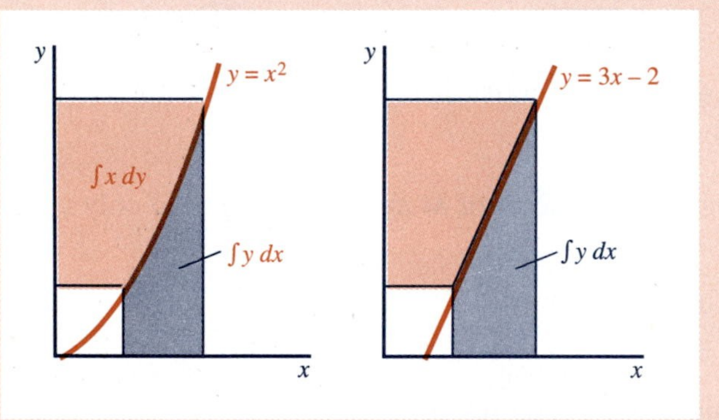

SOLUTION

(*a*) For path 1 ($y = x^2$),

$$\int_A^B dR = \int x\,dy + \int y\,dx = \int 2x^2\,dx + \int x^2\,dx$$

$$= 3\int x^2\,dx = x^3\Big]_{x=1}^{x=2} = 8 - 1 = 7$$

For path 2 ($y = 3x - 2$),

$$\int_A^B dR = \int x\,dy + \int y\,dx = \int 3x\,dx + \int (3x - 2)\,dx$$

$$= \int (6x - 2)\,dx = 3x^2 - 2x\Big]_{x=1}^{x=2} = 8 - 1 = 7$$

The same result for both paths should have been expected because application of the test for exactness shows that $x\,dy + y\,dx$ is exact; hence $R = f(x, y)$ and R has a particular value for each value of x and y. It is readily seen by inspection that $x\,dy + y\,dx = d(xy)$; therefore, $R = xy$, and the simplest way to evaluate $\int_A^B dR$ is by

$$\int_A^B dR = R_B - R_A = x_B y_B - x_A y_A = 2(4) - 1(1) = 7$$

No knowledge of any particular path connecting A and B is necessary because R is a point function.

This particular function allows a simple graphical explanation of its independence of the path chosen between A and B. In each figure, the area between the curve and the x axis represents $\int y\,dx$. This area is not the same for the two paths. The same is true of the areas between the curves and the y axes that represent $\int x\,dy$. However, the *sum* of the two areas on one figure is the same as on the other figure, so that $\int x\,dy + \int y\,dx$ has the same value for either path.

(*b*) For the path 1 ($y = x^2$),

$$\int_A^B dR = \int x\,dy - \int y\,dx = \int 2x^2\,dx - \int x^2\,dx$$

$$= \int x^2\,dx = \frac{x^3}{3}\Big]_{x=1}^{x=2} = \frac{8}{3} - \frac{1}{3} = \frac{7}{3}$$

For the path 2 ($y = 3x - 2$),

$$\int_A^B dR = \int x\,dy - \int y\,dx = \int 3x\,dx - \int (3x - 2)\,dx$$

$$= \int 2\,dx = 2x\Big]_{x=1}^{x=2} = 4 - 2 = 2$$

Different results for the two paths should have been expected, since application of the test for exactness to $R = \int (x\,dy - y\,dx)$ shows that $R \neq f(x, y)$. The differential of R should then be δR instead of dR. On the figures above it is apparent that the *difference* in areas $\int x\,dy$ and $\int y\,dx$ does depend on the path followed between A and B.

1·7 Conservation of Mass

Since a closed system is defined as a particular quantity of matter, the system always contains the same matter and no matter crosses the boundary. Therefore, the mass of the system is constant. For any two states, 1 and 2,

Conservation of mass for a closed system

$$m_1 = m_2$$

This is a statement of the principle of **conservation of mass for a closed system**.

Matter may cross the boundary of an open system, so the amount of mass within the system boundary may change. Conservation of mass requires that

$$\begin{Bmatrix} \text{Increase of} \\ \text{mass within the} \\ \text{system} \end{Bmatrix} = \begin{Bmatrix} \text{Net amount of mass} \\ \text{crossing the boundary} \\ \text{into the system} \end{Bmatrix} \tag{1·9}$$

or, on a time rate basis,

$$\begin{Bmatrix} \text{Rate of} \\ \text{increase of mass} \\ \text{within system} \end{Bmatrix} = \begin{Bmatrix} \text{Net rate of} \\ \text{mass influx} \\ \text{across boundary} \end{Bmatrix} \tag{1·10}$$

Since density may vary throughout an open system, the total mass within the system at any instant is

Derivation of terms in the time rate conservation of mass equation

$$m = \int \rho \, dVol$$

and the time rate of change of mass within the system is

We use the symbol *Vol* for volume when there might be some confusion with *V* for velocity.

$$\begin{Bmatrix} \text{Rate of} \\ \text{increase of mass} \\ \text{within system} \end{Bmatrix} = \frac{dm}{dt} = \frac{d}{dt} \int \rho \, dVol$$

The rate of mass flux across any section of a system boundary can be expressed in terms of the density and velocity at that section. Figure 1·14 shows fluid crossing a section of a system boundary at velocity V which is normal to the area of the boundary. (This derivation can easily be generalized to cover cases where the velocity is not normal to the surface.) During time interval Δt, a volume of fluid ΔVol passes through an element of area ΔA. This volume is $\Delta L \, \Delta A$, where L is normal to the surface ΔA. The mass that crosses the boundary during Δt is then $\rho \, \Delta Vol$, and the rate at which mass crosses the boundary through element dA is $\rho \, \Delta L \, \Delta A / \Delta t$ or $\rho V \, \Delta A$, where V is the velocity normal to the area ΔA. Integrating over the area of this section where matter is crossing the boundary of the open system, the mass rate of flow, $\dot{m}$, entering is

$$\dot{m} = \int \rho V \, dA$$

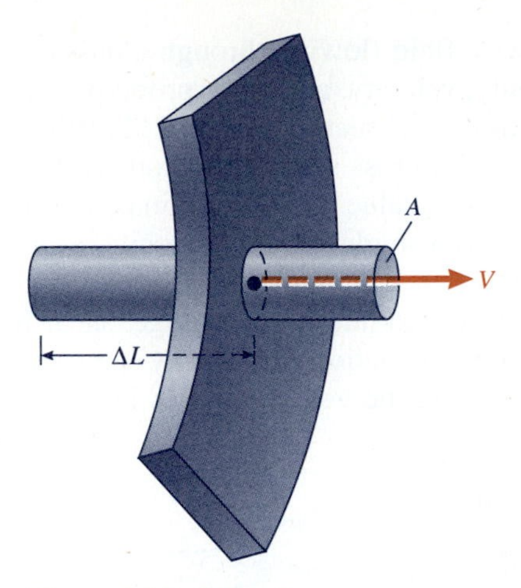

Figure 1·14 Mass flux across a system boundary.

Matter may enter or leave the open system through several openings, so the net rate of mass transfer across the boundary is the summation of quantities for all openings. This can be expressed concisely using vector notation as is often done in fluid mechanics, but we will simply show the net rate of mass transfer as

$$\begin{Bmatrix} \text{Net rate of} \\ \text{mass influx} \\ \text{across boundary} \end{Bmatrix} = \sum_{\text{inlets}} \int \rho V\, dA - \sum_{\text{exits}} \int \rho V\, dA$$

Thus, Eq. 1·10 can be expressed as

$$\frac{d}{dt}\int \rho\, d\mathit{Vol} = \sum_{\text{inlets}} \int \rho V\, dA - \sum_{\text{exits}} \int \rho V\, dA \qquad (1\cdot11)$$

Continuity equation

This result of the conservation of mass principle is called the **continuity equation**.

Fluid properties, such as density and temperature, and flow properties, such as velocity and acceleration, usually vary from point to point in a flow field at any instant. Such flows are said to be *three dimensional* because the properties depend on three spatial coordinates. Flows in which the properties vary with only two spatial coordinates are called *two-dimensional* flows. Many flows can be successfully modeled for some purposes as *one-dimensional* flows in which properties vary with only one coordinate, usually the coordinate in the direction of flow.

Figure 1·15*a* shows a fluid flowing through a passage from a section 1 to a section 2. The density, velocity, and other properties are uniform across the passage cross section at each section, 1 and 2, but they may change in the direction of flow from one cross section to another. Also, the flow direction may change, Still, for each value of the coordinate defined as distance in the direction of flow, there is a single value of velocity, a single value of density, and so on, so the flow is one dimensional.

One-dimensional flow

In the actual flow of a fluid through a passage, shear forces of the passage wall on the fluid cause a variation in velocity across any cross section, as shown in Fig. 1·15*b*, where the velocity varies in the x and y directions but

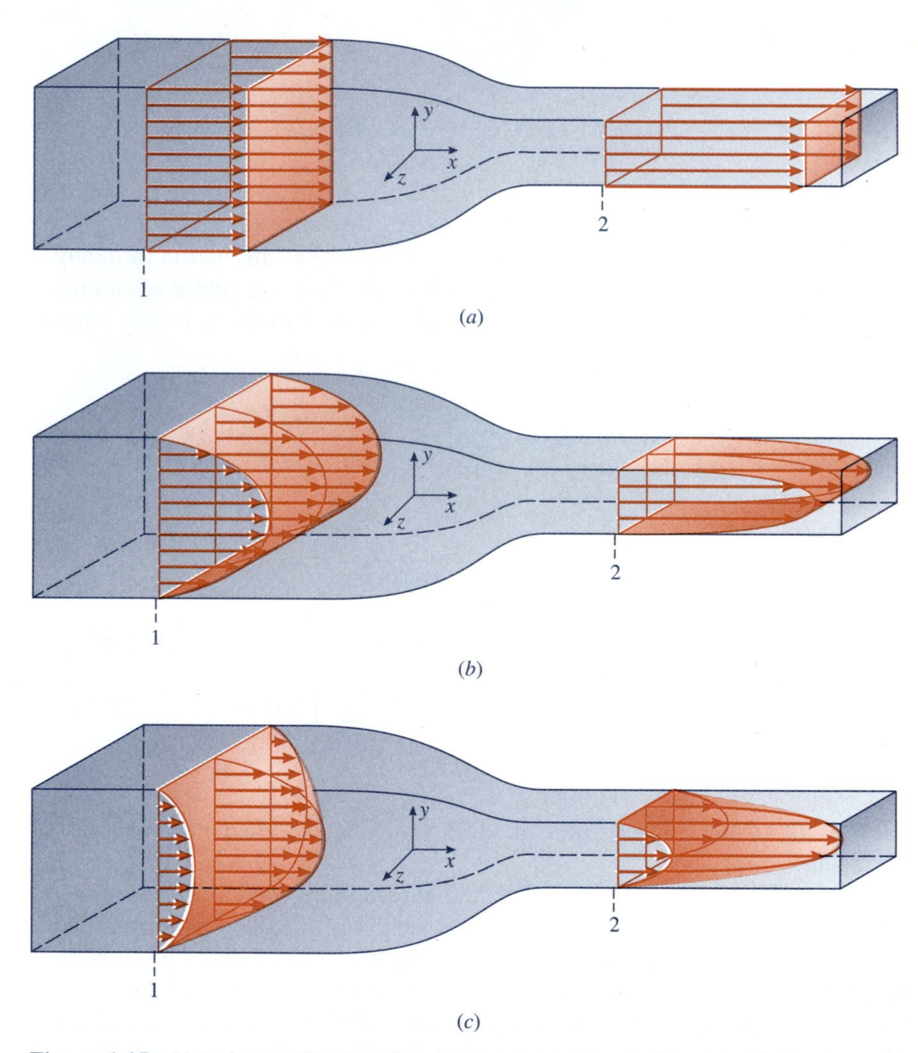

Figure 1·15 Velocity profiles for flow in a passage: (*a*) one-dimensional flow, (*b*) two-dimensional flow, (*c*) three-dimensional flow.

not in the z direction. This is a two-dimensional flow. [If the passage is circular and the flow is axially symmetric, the flow is two dimensional because the velocity (and any other property) distribution can be completely described in terms of two spatial coordinates: distance along the passage and radius from the passage centerline. If the flow is unsymmetrical, it is three dimensional.] Figure 1·15*c* shows a three-dimensional flow. The flow around an airplane wing, a turbine blade, or almost any other object is three dimensional. Similarly, the flow of an actual fluid through any passage is always two or three dimensional.

Two-dimensional flow

Three-dimensional flow

To simplify analyses, we often model a three- or two-dimensional flow through any cross-sectional area A as one dimensional by using an average value for the density, which seldom varies greatly across the section, and calculating an average velocity normal to the area as

$$V = \frac{\dot{m}}{\rho A} \tag{1·12}$$

where $\dot{m}$ is the mass rate of flow or the mass of the fluid flowing through the area per unit time, ρ is the average density across the section, and A is the area of the section, all at the same instant. For one-dimensional flow at each inlet and exit, Eq. 1·11 for the rate of change of mass within a system at any instant becomes

$$\frac{dm}{dt} = \sum_{\text{inlets}} \rho VA - \sum_{\text{exits}} \rho VA$$

For flows with large velocity variations across the stream or with high swirl components, one-dimensional modeling may introduce unacceptable errors. (Included among the problems at the end of Chapter 3 are some calculations of the errors involved in modeling some two-dimensional flows as one dimensional. An important part of effective modeling is the estimation of errors.)

1·8 Steady Flow

The flow through an open system is **steady flow** (and the system is often called a **steady-flow system**) if all properties at each point within the system remain constant with respect to time. Property values vary from one point to another within a steady-flow system, but at each point they are constant with time. From this definition it follows that if the flow is steady, the following must be true:

Steady flow

1. The properties of the fluids crossing the boundary remain constant at each point on the boundary.

2. The flow rate at each section where matter crosses the boundary is constant. (The flow rate cannot change as long as all properties, including velocity, at each point remain constant.)
3. The rate of mass flow into the system equals the rate of mass flow out. Consequently, the amount of mass within the system is constant.
4. The volume of the system remains constant. The system boundary must be rigid.
5. All interactions with the surroundings occur at a constant rate.

These conditions are not only necessary, they are sufficient to insure steady flow. All of these conditions are stated in terms of observations that can be made at the boundary of the system, so no knowledge of events inside a system is needed to determine if steady flow prevails. If any of these conditions is not met, the flow is *unsteady* or *transient.*

Unsteady or transient flow

For the case of steady flow, the mass within the system is constant, so Eq. 1·11, the *continuity equation*, becomes

Continuity equation for steady flow

$$\sum_{\text{inlets}} \int \rho V\, dA = \sum_{\text{exits}} \int \rho V\, dA$$

This is the **continuity equation for steady flow**. For one-dimensional flow this becomes

$$\sum_{\text{inlets}} \rho VA = \sum_{\text{exits}} \rho VA \tag{1·13}$$

For the frequently encountered case of a single inlet and a single outlet,

$$\dot{m} = \rho_1 V_1 A_1 = \rho_2 V_2 A_2 \tag{1·14}$$

Example 1·4

PROBLEM STATEMENT

Nitrogen flows steadily through a converging tube. Near the inlet the nitrogen is at 10.0 psia, 40 F, with a density of 0.0523 lbm/ft^3 and has a velocity of 2000 fps (feet per second) through a cross-sectional area of 0.40 sq ft. At a section farther downstream, the calculated density is 0.124 lbm/ft^3 and the calculated velocity is 1200 fps. Determine the cross-sectional area at the downstream section.

SOLUTION

Analysis: A sketch is first made, and the data are placed on it.

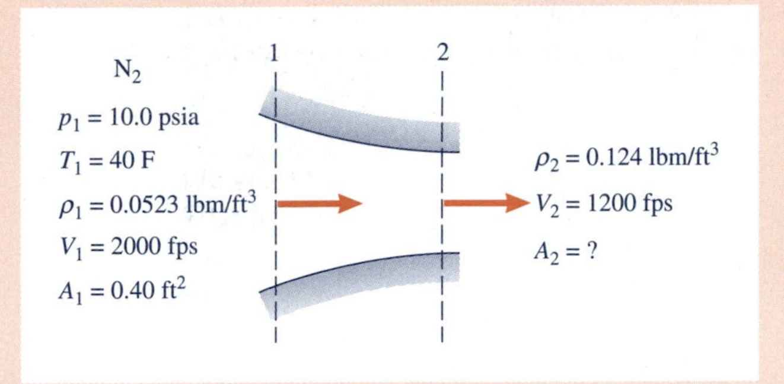

Conservation of mass for this open system under steady-flow conditions is expressed by the continuity equation:

$$\rho_1 V_1 A_1 = \rho_2 V_2 A_2 \tag{1·14}$$

and inspection shows that we have sufficient data to solve this equation for A_2. For this simple calculation, an estimate of the result can be made directly from the single equation involved. (That the velocity is decreased as the area is reduced may be surprising, but in Chapter 13 it is shown that this is a characteristic of the supersonic flow of gases under some conditions.)

Calculations: The continuity equation for steady flow,

$$\rho_1 A_1 V_1 = \rho_2 A_2 V_2$$

applies. Rearranging and substituting numerical values gives

$$A_2 = A_1 \frac{\rho_1 V_1}{\rho_2 V_2} = 0.40 \frac{0.0523(2000)}{0.124(1200)} = 0.28 \text{ ft}^2$$

Example 1·5 Air-Conditioning Unit

PROBLEM STATEMENT

Atmospheric air at 30°C with a density of 1.12 kg/m^3 enters an air-conditioning unit through a cross-sectional area of 0.060 m^2 with a velocity of 10.0 m/s. Water is removed from the air and leaves the air-conditioning system as a liquid at a rate of 45.5 kg/h. Air leaving the unit is at 15°C with a density of 1.21 kg/m^3. The discharge cross-sectional area is 0.075 m^2. Determine the velocity of the air leaving.

SOLUTION

Analysis: We first make a sketch of the system and place data values on the sketch. The problem statement suggests that the flow is steady, since it gives single values of properties at inlet and exits, including a flow rate of the liquid water leaving. Further, it is reasonable for such a system to operate in a steady-flow mode. Therefore, we model this system as a steady-flow open system, as shown in the figure.

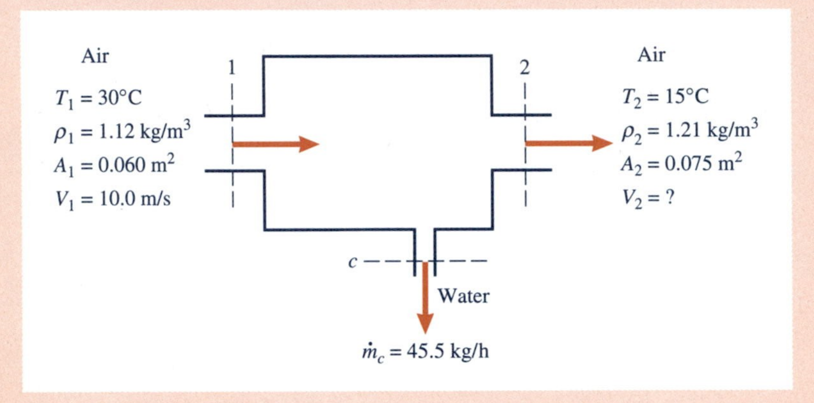

The air exit velocity can be found from

$$\dot{m}_2 = \rho_2 A_2 V_2$$

if $\dot{m}_2$ can be found. The principle involved is the conservation of mass, so a mass balance in the form of the continuity equation for steady flow is

$$\dot{m}_1 = \dot{m}_2 + \dot{m}_c$$
$$\rho_1 A_1 V_1 = \rho_2 A_2 V_2 + \dot{m}_c$$

Values are known for all quantities except V_2, so we can solve for V_2.

Before solving, it is well to estimate the result. The mass leaving at $\dot{m}_2$ is less than the mass entering at 1 by the amount of water removed from the air. However, the mass of water in humid air is much less than the mass of air, so that $\dot{m}_2$ is only slightly less than $\dot{m}_1$. The density is about 10 percent higher and the area is 25 percent larger at the exit than at the inlet, so we expect the velocity at the outlet to be slightly lower at the exit than at the inlet, perhaps 6 or 7 or 8 m/s. (At this stage you may have little confidence in your ability to estimate results, but you will become more skilled with practice. Even at this early stage, estimating results is valuable, since if your calculated result is of the order of, say, either 1 m/s or 100 m/s, you will know that you have made an error.)

Calculations: The continuity equation for steady flow applied to this system is

$$\dot{m}_1 = \dot{m}_2 + \dot{m}_c$$

or

$$\rho_1 A_1 V_1 = \rho_2 A_2 V_2 + \dot{m}_c$$

Rearranging and substituting numerical values yields

$$V_2 = \frac{1}{\rho_2 A_2}(\rho_1 A_1 V_1 - \dot{m}_c)$$

$$= \frac{1}{1.21\ \text{kg/m}^3(0.075\ \text{m}^2)}\left[1.12\frac{\text{kg}}{\text{m}^3}(0.060\ \text{m}^2)10.0\frac{\text{m}}{\text{s}} - 45.5\frac{\text{kg}}{\text{h}}\left(\frac{\text{h}}{3600\ \text{s}}\right)\right]$$

$$= 7.3\ \text{m/s}$$

This result is within the range of our estimated result.

The first step in analyzing an open system is to decide whether the flow is steady. In many actual systems the flow is steady. In others the flow is transient (unsteady) but can, for many purposes, be successfully modeled as steady flow to simplify analyses. For example, consider the flow through a reciprocating compressor as shown in Fig. 1·16. The flow within the cylinder is surely not steady. In fact, during part of the compressor operating cycle, gas is being drawn into the cylinder; during another part gas is being discharged, and during still another part, both valves are closed so that the gas

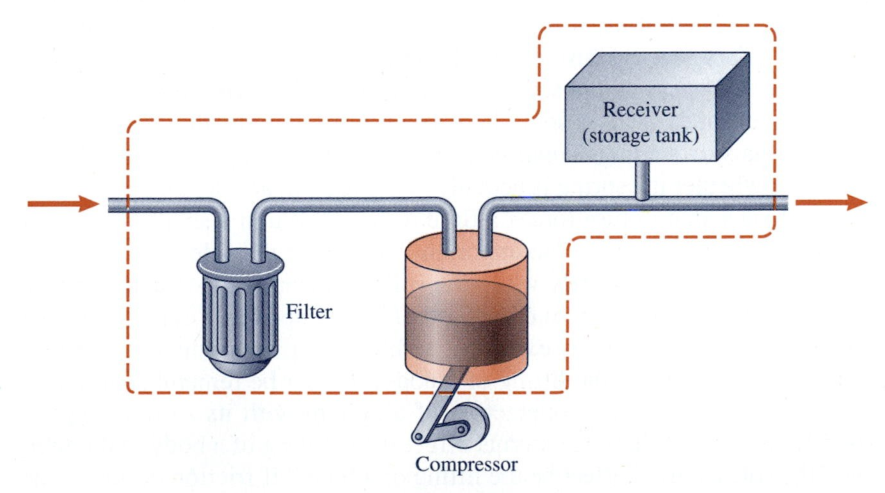

Figure 1·16 Flow through a reciprocating compressor.

within the cylinder is a closed system. For the design of the valves and valve passages the compressor must be modeled as a transient system. However, for other purposes, time-averaged values of properties at various sections can be used in modeling the compressor as a steady-flow system. If the open system is defined as shown by the dotted red line, so that the system includes the inlet filter and the receiver (or storage tank), the fluctuations in flow at the inlet and exit are reduced and a steady-flow model may adequately simulate the actual situation.

Recall that in Section 1·2 we defined a quasiequilibrium process for closed and open systems. For an open system quasiequilibrium means that at any instant the fluid at each point in the system is in a state that would be an equilibrium state if the fluid at that point were isolated. That is, relationships among the properties of the flowing fluid are the same as they would be if the fluid were in an equilibrium state in a closed system. In addition, there must be no frictional effects.

1·9 Work

Work is an interaction between a system and its surroundings. **Work** is done by a system on its surroundings if the sole external effect of the interaction could be the lifting of a body. The magnitude of work is the product of the weight of the body and the distance it is lifted.

This definition points out that work involves both a system and its surroundings. It tells how to identify and measure the work done *by* a system. Work done *on* a system must be defined as work done *by* some other system. This roundabout method is necessary because a definition of work as ''work is done on a system if the sole external effect could be the fall of a body'' is *not valid.* (This fact is illustrated in Example 1·10.)

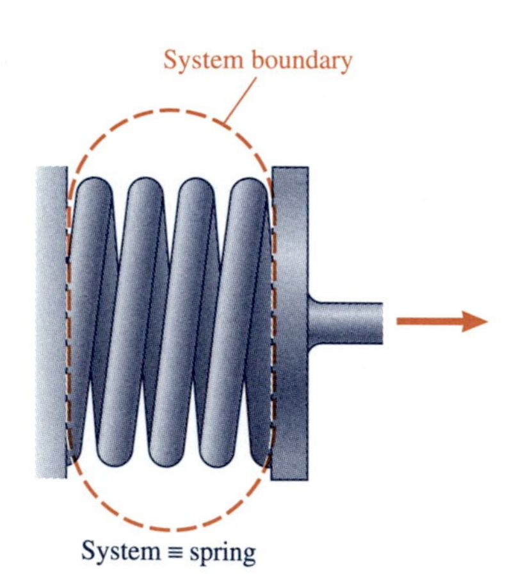

Figure 1·17 A spring doing work on the surroundings.

Consider a system composed of a compressed coil spring (Fig. 1·17). As the spring expands against some part of its surroundings, the action of the spring on its surroundings *could* be reduced to the lifting of a body. It does not matter whether the spring is actually being used to accelerate a body or to push a plunger that in turn forces a fluid to flow through a small opening or to cause some other effect. The important fact is that the sole external effect *could* be the lifting of a body while the spring undergoes the same process.

Figure 1·18 shows a system composed of gas trapped in a cylinder behind a movable piston. If the gas expands, pushing the piston outward, the sole external effect *could* be the lifting of a body. It must be remembered that in deciding whether a certain interaction of a system with its surroundings is work, we ask, not if the sole external effect *is* the lifting of a body, but rather, *could* the sole external effect be the lifting of a body? If friction occurs in the surroundings, there may actually be effects other than the lifting of a body. For example, the temperature of some part of the surroundings may increase. However, if friction is reduced, the limiting case in which the sole external effect is the lifting of a body is approached. The limiting condition of no friction *in the surroundings* that we might use in identifying and measuring work does not restrict us to the consideration of systems that involve no friction. We simply consider, as one possibility, the limiting case of no friction in the surroundings.

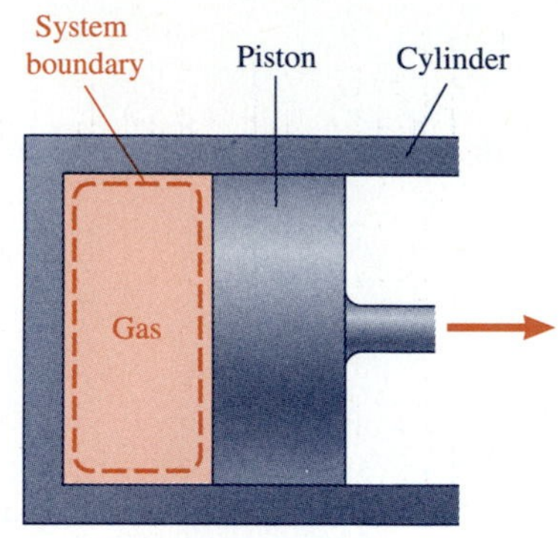

Figure 1·18 A gas doing work on the surroundings.

In the case of both the expanding spring and the expanding gas, work is done by the action of a force exerted by the system on a moving boundary of the system. This kind of work is called *mechanical* work. In both cases, the force is a varying force. However, we calculate the magnitude of work in terms of a constant force (the weight of the body that *could* be lifted) times a distance. We can always devise systems of levers, gears, pulleys, and so forth to convert the varying force into a constant force that can be used to lift a body. It is simpler, though, to express the work in terms of force and displacement at the system boundary.

If we refer to Fig. 1·19, the work done to lift a body of weight w an infinitesimal distance dz can be expressed in terms of the force F on the system boundary as it moves an infinitesimal distance ds, where F is in the direction of ds:

$$\delta W = \text{weight } dz = F\,ds$$

If we integrate the above expression, we find that the general equation for mechanical work,

$$W = \int F\,ds \tag{1·15}$$

General equation for mechanical work

involves values only at the system boundary.

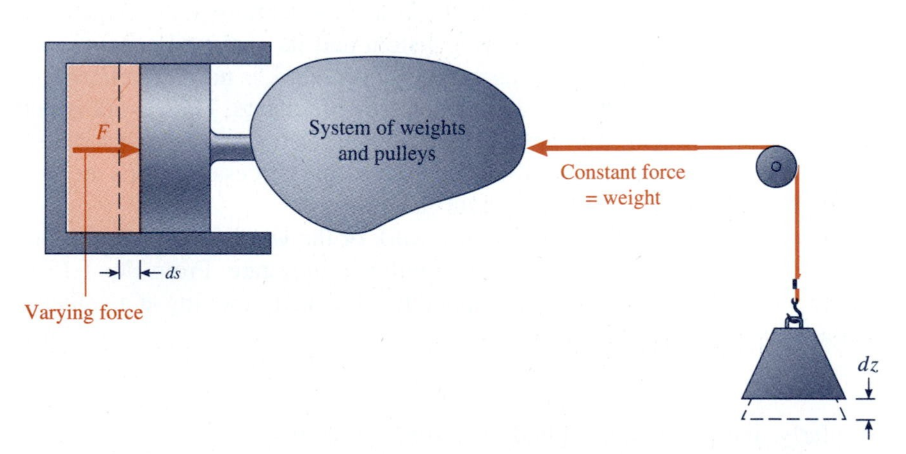

Figure 1·19 Relating the variable force on a system boundary to the constant force to lift a weight.

Many systems do work by exerting a torque on a rotating shaft. The torque may vary as the shaft rotates, and if this is the case, then

Work by torque exerted on a shaft

$$W = \int T \, d\theta$$

where the torque T is the component of torque in the direction of the angular displacement. Note that the motion of the system boundary in this case is not in a direction normal to the boundary surface.

We have now expressed the magnitude of mechanical work in terms of quantities at the system boundary instead of as the product of the weight of a body and the distance it could be lifted.

Sign convention: Work done by a system is positive; work done on a system is negative.

A widespread convention is that work done *by* a system is expressed by *positive* numbers and work done *on* a system is expressed by *negative* numbers. Of course, since work is an interaction between systems, work done by one system must be done on some other system, so the same work is positive with respect to one system and negative with respect to another. When system A does 1000 J of work on system B, we express this as

$$W_A = 1000 \text{ J} \quad \text{or} \quad W_B = -1000 \text{ J}$$

In this book, the work of a system is occasionally specified as work_{in} (meaning work done on the system) to avoid dependence on the sign convention. Where the subscript indicating direction is not used, however, the usual convention is followed: The term work without a subscript indicating direction means work done by the system to which the term is applied. As an illustration, if 1000 J of work is done on a system, we can write

$$W = -1000 \text{ J}$$
$$W_{\text{in}} = 1000 \text{ J}$$

A symbol W_{out} would be redundant, since W means work done by a system.

Even though we speak of the work of a system, it is important to remember that work is an interaction between a system and its surroundings.

The state of a system changes as work is done. Work is not a characteristic that can be observed while a system is in a particular state. Thus, work is not a property of a system; it is not a *point function.* Between two given states, the amount of work done depends on the path of the process between the two states. For this reason, work is a *path function.*

Power

Power is defined as the rate of doing work or the work per unit time. The symbol $\dot{W}$ is used, where the overdot signifies a time rate. From Eq. 1·15 it can be seen that when a force F is applied to a boundary moving at a velocity V, the power output of a system is

$$\dot{W} = FV$$

Similarly, the power transmitted by a rotating shaft is

$$\dot{W} = T\omega$$

where T is torque and ω is angular speed of the shaft.

Calculations of steady-flow systems often involve the work per unit mass, w, which is related to the power and the mass rate of flow, $\dot{m}$, by

$$\dot{W} = \dot{m}w \tag{1·16}$$

The dimensions of power are obviously FL/t or ML^2/t^3, and typical units are kJ/s, N·m/s, watt (W), kilowatt (kW), ft·lbf/s, B/s, and horsepower (hp).

Next, we discuss the calculation of work for some quasiequilibrium processes.

1·9·1 Work of a Quasiequilibrium Process of a Simple Compressible Closed System

A simple compressible substance involves no effects of electricity, magnetism, anisotropic stress, or surface tension. (An incompressible fluid is a special limiting case in which the density is constant.) A closed system that is composed of a simple compressible substance and involves no effects of motion or gravity is called a **simple compressible closed system**. You may well ask how a system that involves no gravity or motion can be a useful model. The answer is that in many situations encountered in practice the effects of gravity and motion are negligible compared with other effects. For example, if we are told that air at 300 kPa, 30°C, is stored in a 1-m-diameter spherical tank, we generally do not ask where in the tank the air pressure is 300 kPa. The reason we do not ask is that the difference in pressure between air at the highest point and the lowest point in the tank is only about 0.3 kPa, and for most purposes this slight variation in pressure caused by gravity can be neglected. Similarly, the effects of motion such as the acceleration of part of the system by means of a moving piston are often negligible.

Simple compressible closed system

Recall from Section 1·2 that a *quasiequilibrium process* is one during which a system remains in equilibrium or infinitesimally close to being in equilibrium. During a quasiequilibrium process, properties must be uniform throughout a closed system and there can be no frictional effects.

Derivation of work of a simple compressible closed system

To develop an expression for the work of a quasiequilibrium process of a simple compressible closed system, consider a gas trapped in a cylinder and expanding against a piston, as shown in Fig. 1·20. The gas expands from an initial state 1 to a final state 2. The piston moves slowly so that the effects of motion on the system are negligible. The pressure is uniform throughout the system at any stage of the expansion, but the value of this uniform pressure changes as the expansion proceeds. At any stage of the expansion, the force on the piston is the product of the pressure of the gas and the area of the piston. Since this force acts in the direction of motion of the piston, the work done by the gas on the piston while the piston moves a distance ds is

$$\delta W = F\,ds = pA\,ds$$

Since $A\,ds$ is the volume increase dV of the system as the piston travels the distance ds,

$$\delta W = p\,dV$$

and the total amount of work done by the gas on the piston as the gas expands from state 1 to state 2 is

Work for a simple compressible closed system

$$W_{1-2} = \int_1^2 p\, dV \qquad (1\cdot 17a)$$

The area beneath a curve on pressure–volume coordinates is $\int p\, dV$. Therefore, the work of the process described above is represented by an area on a plot of the system pressure versus its volume (pV diagram) as shown in the lower part of Fig. 1·20. The crosshatched area represents the work done by the gas as its volume increases by an amount dV. The total area beneath curve 1–2 is $\int_1^2 p\, dV$ and represents the work done by the system as it passes from state 1 to state 2.

Frequently, we are interested not in the total work of a system but in the work per unit mass in the system, so Eq. 1·17*a* becomes

$$w_{1-2} = \int_1^2 p\, dv \qquad (1\cdot 17b)$$

Equation 1·17 will be used frequently. Remember that it is valid only for a quasiequilibrium process of a simple compressible closed system.

A simple compressible closed system can follow any one of many different paths as it changes from one state to another by means of quasiequilibrium processes. On a pV diagram, in general, the area beneath each path differs from that beneath other paths. This result is as expected, because work is a path function and its value therefore depends on the path of the process between any two states and not just on the end states.

Work done as the system expands is work done by the system, because the sole external effect could be the lifting of a body. When work is done by a system, $\int p\, dV$ is positive. When a system is compressed work is done *on* the system and $\int p\, dV$ is negative. Thus the equation

$$W = \int p\, dV \qquad (1\cdot 17a)$$

agrees with the sign convention of work out of (or by) the system as positive and work into (or on) the system as negative.

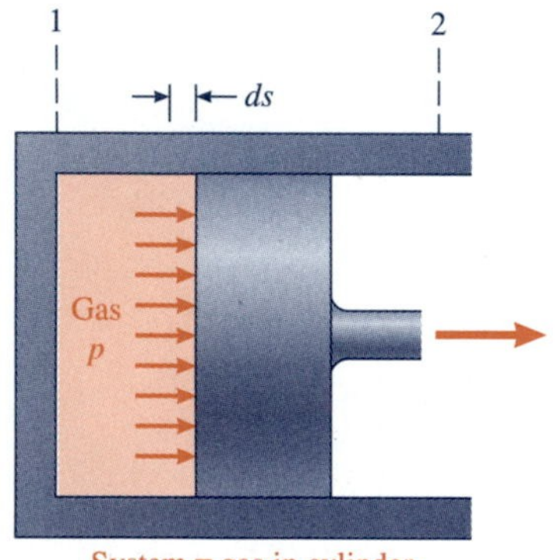

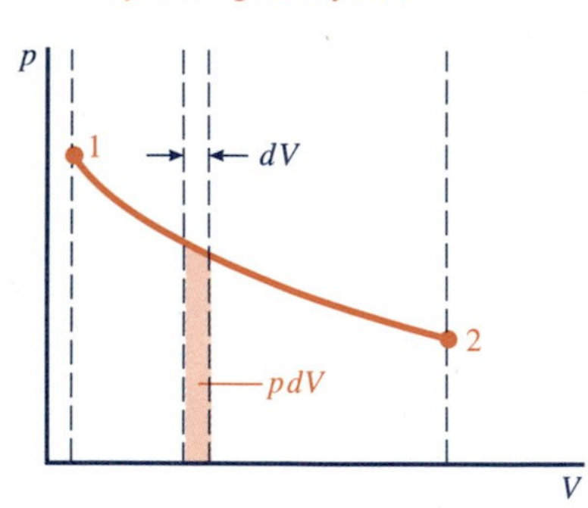

Figure 1·20 A quasiequilibrium process of a closed system.

The Work of a Cyclic Quasiequilibrium Process of a Closed System. A closed system may pass by means of quasiequilibrium processes from a state 1 to a state 2 along a path such as 1–*a*–2 in Fig. 1·21. It may then complete a cycle by returning to state 1. During the return process, the system may pass through different states from those it passed through during the initial processes 1–*a*–2. Let one of the states passed through during the return process be state *b*, as shown in Fig. 1·21.

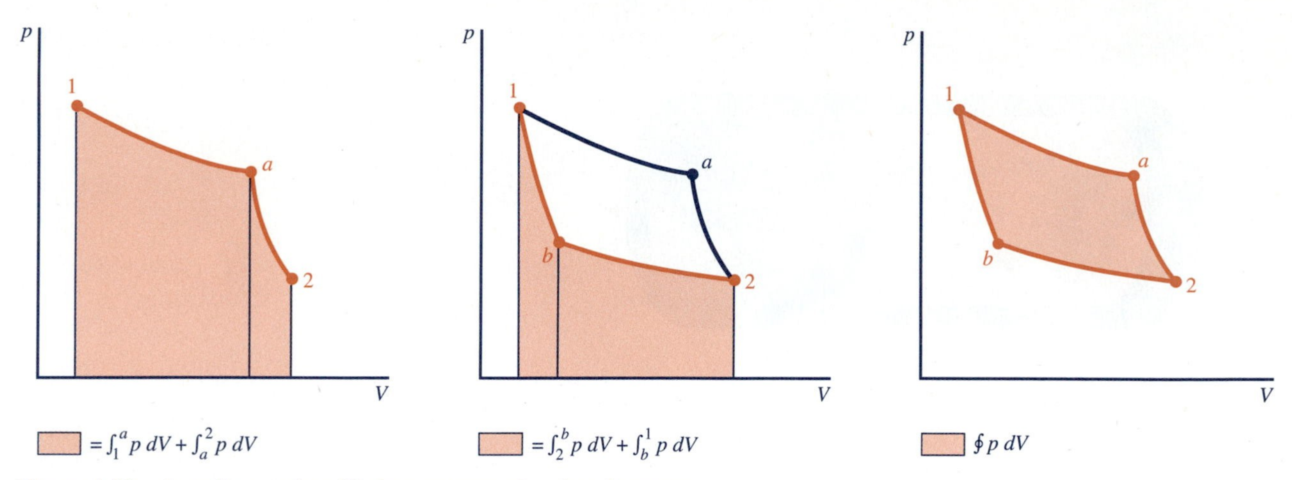

Figure 1·21 A cyclic quasiequilibrium process of a closed system.

During some of the processes of the cycle the system does work on its surroundings; during other processes work is done on the system. The net work of the system for the cycle is the sum of the work of all the processes making up the cycle. For the cycle made up of four processes,

$$\left\{\begin{array}{c}\text{Work of closed system}\\ \text{during quasiequilibrium cycle}\end{array}\right\} = \int_1^a p\,dV + \int_a^2 p\,dV + \int_2^b p\,dV + \int_b^1 p\,dV$$

$$= \oint p\,dV$$

where the symbol $\oint$ denotes integration along a closed or cyclic path.

The work done by the system during each process is represented by the area beneath the *pV* plot of the process, and Fig. 1·21 shows that the net work of the cycle is represented by the area enclosed within the diagram. Notice that the net work of a cycle is generally not zero, even though (by definition of a cycle) the system is returned to its initial state. This is another reminder that work is a path function and cannot be evaluated from only the end states of a process.

Example 1·6

PROBLEM STATEMENT

Consider a system composed of a gas contained in the cylinder, as shown in the figure. The fluid expands from a volume of 1.40 to 1.60 cu ft while the pressure remains constant at 100 psia and the paddle wheel does 3600 ft·lbf of work on the system. Calculate (*a*) the work done by the system on the piston and (*b*) the net amount of work done on or by the system.

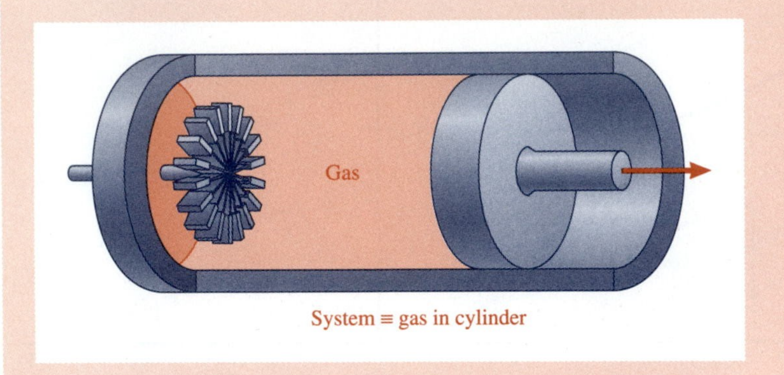

SOLUTION

(*a*) We are dealing with a closed system. The process is obviously not a quasiequilibrium process because fluid shear forces (friction) and eddying motions are involved in the interaction between the paddle wheel and the gas (system). However, *if the action of the paddle wheel is such that the pressure at the piston face is uniform and of known value at each state of the process*, the work done on the piston (or in other words the work associated with a change in volume of the system) can be calculated by

$$W = \int p\, dV$$

For the special case of a constant pressure, this relation becomes

$$W = p\int_1^2 dV = p(V_2 - V_1)$$

$$= 100\frac{\text{lbf}}{\text{in}^2}(144)\frac{\text{in}^2}{\text{ft}^2}(1.60 - 1.40\ \text{ft}^3) = 2880\ \text{ft·lbf}$$

(*b*) Since the paddle-wheel work is work done on the system, we write in accordance with the usual sign convention, $W = -3600$ ft·lbf. The net work of the system (meaning work done *by* the system in accordance with the convention) is

$$W_{\text{net}} = W_{\text{piston}} + W_{\text{paddle}}$$

$$= 2800 + (-3600) = -720\ \text{ft·lbf}$$

$$W_{\text{net, in}} = 720\ \text{ft·lbf}$$

Recall that these results are based on a model in which the pressure on the piston face is 100 psia throughout the process.

Example 1·7

PROBLEM STATEMENT

Carbon dioxide is compressed in a cylinder in an equilibrium process from initial conditions of $p_i = 100$ kPa, $V_i = 0.0040\ \text{m}^3$, to a final pressure of $p_f = 500$ kPa. During the process $pV^{1.22} = \text{constant}$. Compute the work.

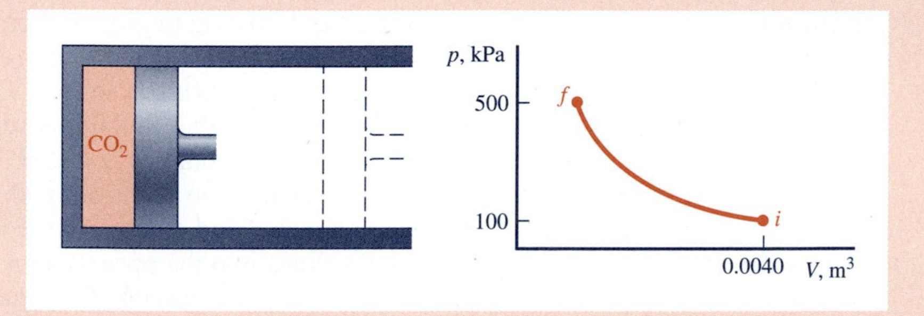

SOLUTION

A sketch of the system and a pV diagram are shown above. We are dealing with

1. A closed system composed of the carbon dioxide.
2. A quasiequilibrium process.

For these conditions, $W = \int p\,dV$, and since we know a relationship between p and V, namely $pV^{1.22} = \text{constant} = C$, we can integrate this expression:

$$W = \int_i^f p\,dV = \int_i^f \frac{C}{V^{1.22}}\,dV = C\int_i^f V^{-1.22}\,dV = \frac{C}{-0.22}\left(V_f^{-0.22} - V_i^{-0.22}\right)$$

At this point one might solve $p_i V_i^{1.22} = p_f V_f^{1.22} = C$ for C and V_f. However, as a general guideline, expressions that include dimensional quantities with fractional exponents can be simplified, so we proceed:

$$W = \frac{p_i V_i^{1.22} V_i^{-0.22}}{-0.22}\left[\left(\frac{V_f}{V_i}\right)^{-0.22} - 1\right] = \frac{p_i V_i}{-0.22}\left[\left(\frac{p_i}{p_f}\right)^{(1/1.22)(-0.22)} - 1\right]$$

$$= -\frac{p_i V_i}{0.22}\left[\left(\frac{p_f}{p_i}\right)^{0.180} - 1\right] = -\frac{100\ \text{kPa}(0.0040\ \text{m}^3)}{0.22}\left[\left(\frac{500}{100}\right)^{0.180} - 1\right]$$

$$= -0.612\ \text{kPa·m}^3 = -0.612\ \text{kN·m} = -0.612\ \text{kJ}$$

The work quantity is negative as we expect for a compression that involves work *input* to the system.

1·9·2 Work of a Quasiequilibrium Process of a Simple Compressible Substance in Steady Flow

Equation 1·17, which was developed in Section 1·9·1 for the work of a quasiequilibrium process of a simple compressible closed system, is highly valuable. Because engineers frequently deal with steady-flow systems, we need an analogous expression for the work of a quasiequilibrium process of a simple compressible substance flowing through a steady-flow open system.

Derivation of work of a simple compressible substance in steady flow

The first step is to obtain an expression for the net force on an open system in terms of the change in momentum of the fluid flowing through it. For this purpose we will consider an open system as shown bounded by the broken line in Fig. 1·22*a* with fluid entering at section 1 and leaving at section 2. We also consider a closed system as shown in Fig. 1·22*b* that initially at time t coincides with the open system and is composed of volumes A and B. The closed system must always include the same matter, so it moves with the flow and at a later time $t + \Delta t$ occupies volumes B and C. During the time interval Δt, an amount of fluid of volume A flows into the open system and fluid of volume C leaves. The open system is always composed of volumes A and B.

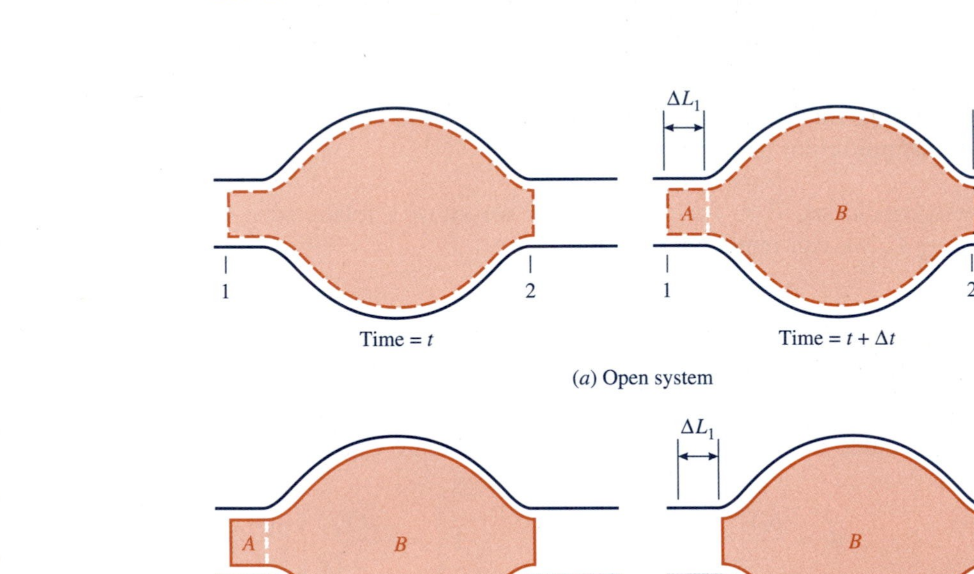

Figure 1·22 Steady-flow open system with a closed system passing through it. The broken red line is the open system boundary. At time t the broken red line is also the boundary of the closed system. At time $t + \Delta t$ the closed system contains volumes B and C.

We can apply Newton's second law.

$$\sum F = \frac{d}{dt}(mV) \tag{1·2}$$

where ΣF is the sum of all forces on the closed system. This is, of course, a vector equation, but for our purposes here we need not emphasize its vector character. Therefore, for our present purposes the open system may be restricted to colinear inlet and exit. Since the velocity varies throughout the closed system,

$$\sum F = \frac{d}{dt}\int_{\substack{closed\\system}} V\,dm \tag{1·18}$$

From the definition of a derivative, this can be expressed as

$$\sum F = \lim_{\Delta t \to 0} \frac{\left(\int_{cl\ sys} V\,dm\right)_{t+\Delta t} - \left(\int_{cl\ sys} V\,dm\right)_t}{\Delta t} \tag{a}$$

Referring again to Fig. 1·22, we note that

$$\begin{aligned}\left(\int_{cl\ sys} V\,dm\right)_{t+\Delta t} &= \left(\int_B V\,dm + \int_C V\,dm\right)_{t+\Delta t}\\ &= \left(\int_{open\ sys} V\,dm - \int_A V\,dm + \int_C V\,dm\right)_{t+\Delta t}\end{aligned} \tag{b}$$

and

$$\left(\int_{cl\ sys} V\,dm\right)_t = \left(\int_{open\ sys} V\,dm\right)_t \tag{c}$$

Substituting b and c into a yields

$$\sum F = \lim_{\Delta t \to 0} \frac{\left(\int_{open\ sys} V\,dm\right)_{t+\Delta t} - \left(\int_A V\,dm\right)_{t+\Delta t} + \left(\int_C V\,dm\right)_{t+\Delta t} - \left(\int_{open\ sys} V\,dm\right)_t}{\Delta t} \tag{d}$$

We now add the restriction of *steady flow*. For steady flow, all properties at each point in an open system remain constant with time, so the first and last terms in the numerator of Eq. d are equal in magnitude. Therefore,

$$\sum F = \lim_{\Delta t \to 0} \frac{\left(\int_C V\,dm\right)_{t+\Delta t} - \left(\int_A V\,dm\right)_{t+\Delta t}}{\Delta t} \tag{e}$$

Since the limit of a sum equals the sum of the limits,

$$\sum F = \lim_{\Delta t \to 0} \frac{\left(\int_C V\,dm\right)_{t+\Delta t}}{\Delta t} - \lim_{\Delta t \to 0} \frac{\left(\int_A V\,dm\right)_{t+\Delta t}}{\Delta t} \qquad (f)$$

(Term 1) (Term 2)

Let us now restrict the flow across section 2 to *one-dimensional* flow with V perpendicular to A_2. The first term on the right-hand side of Eq. f can be simplified as follows:

$$\text{Term 1} = \lim_{\Delta t \to 0} \frac{\left(\int_C V\,dm\right)_{t+\Delta t}}{\Delta t} = \lim_{\Delta t \to 0} \frac{\left(\int_{A_2} V\rho\,\Delta L_2\,dA\right)_{t+\Delta t}}{\Delta t}$$

where $dm = \rho\,dVol = \rho\,\Delta L_2\,dA$. Integration is over A_2, since all of the fluid in volume C at time $t + \Delta t$ passes through area A_2 during the time interval Δt. Further, remembering that the flow is steady and one dimensional and then passing to the limit, we find that

$$\text{Term 1} = \lim_{\Delta t \to 0} \frac{(\rho V_2\,\Delta L_2 A_2)_{t+\Delta t}}{\Delta t} = V_2 \dot{m}_2$$

In a similar manner the second term of Eq. f becomes

$$\text{Term 2} = \lim_{\Delta t \to 0} \frac{\left(\int_A V\,dm\right)_{t+\Delta t}}{\Delta t} = V_1 \dot{m}_1$$

For steady flow with a single inlet and single exit, $\dot{m}_1 = \dot{m}_2 = \dot{m}$. Equation f then becomes

$$\sum F = \dot{m}(V_2 - V_1) \qquad (1\cdot19)$$

This is a valuable equation. It relates the summation of the forces on an open system to properties at the inlet and outlet for one-dimensional steady flow of a fluid. It is a vector equation, although, as noted above, for our present purposes a special case of colinear V_1 and V_2 is adequate. We will use this equation now in deriving an expression for the work of a quasiequilibrium process of a simple compressible substance in steady flow.

Figure 1·23 shows a small steady-flow system that is part of a larger system. The length ΔL is small enough so that the variations of p, A, and V along ΔL are very nearly linear. This does not mean that the p, A, and V vary linearly with L, but only that ΔL is so small that a linear relation is a good approximation for the short distance ΔL. This approximation becomes better as ΔL becomes smaller, and later in the derivation an exact expression will be

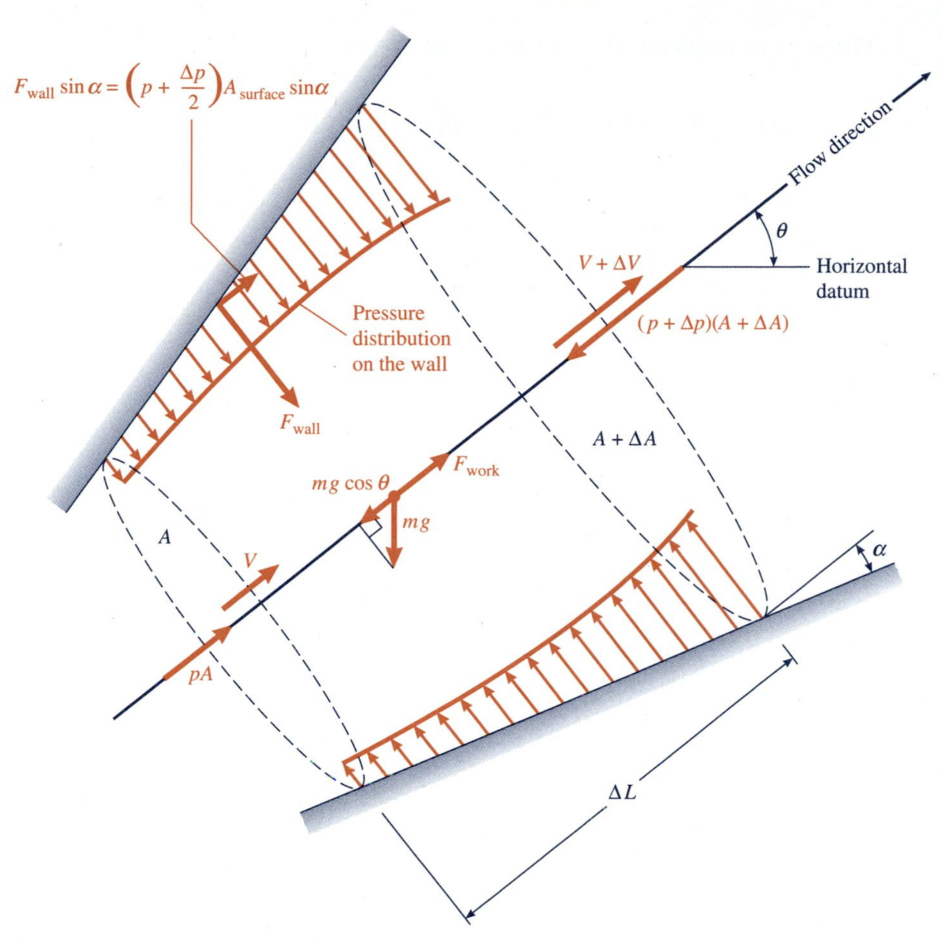

Figure 1·23 A small steady-flow system that is part of a larger steady-flow system with one-dimensional, quasiequilibrium flow of a simple compressible substance.

obtained by letting ΔL approach the differential length dL. The forces acting *on the open system* in a direction parallel to the direction of flow are

1. The force of the adjacent fluid on the inlet area, pA.
2. The force of the adjacent fluid on the exit area, $(p + \Delta p)(A + \Delta A)$.
3. The component of the weight of the element, weight $\cos\theta = mg \cos\theta$.
4. The component of the normal wall force F_{wall} in the direction of flow, $p_{avg}A_{surface} \sin\alpha$. Since the element is assumed to be small enough so that p varies almost linearly along ΔL, the average pressure is the arithmetic mean of p and $(p + \Delta p)$, and the normal wall force component in the direction of flow is $(p + \Delta p/2)A_{surface} \sin\alpha$ or $(p + \Delta p/2)\Delta A$.
5. The force F_{work} by which work is done on the fluid. This force is applied to the fluid by means of some impeller that is not shown in the figure.

The sum or resultant of these forces is

$$\sum F = pA - (p + \Delta p)(A + \Delta A) - \rho\left(A + \frac{\Delta A}{2}\right)\Delta L g \cos\theta + \left(p + \frac{\Delta p}{2}\right)\Delta A + F_{\text{work}}$$

$$= -A\,\Delta p - \frac{\Delta p\,\Delta A}{2} - \rho A\,\Delta L g \cos\theta + \rho g \frac{\Delta A\,\Delta L}{2}\cos\theta + F_{\text{work}}$$

Substituting this expression for the sum of the external forces in the flow direction into Eq. 1·19 and neglecting terms of the order $\Delta p\,\Delta A$ and $\Delta A\,\Delta L$ gives

$$-A\,\Delta p - \rho A\,\Delta L g \cos\theta + F_{\text{work}} = \dot{m}(V + \Delta V - V) \qquad (a)$$

Rearranging yields

$$F_{\text{work}} = A\,\Delta p + \dot{m}\,\Delta V + \rho A\,\Delta L g \cos\theta$$

where F_{work} is in the direction of flow. This force acting on the fluid as it moves the distance ΔL does work *on* the fluid, so that the amount of work is

$$W_{\text{in}} = F_{\text{work}}\,\Delta L = A\,\Delta p\,\Delta L + \rho A V\,\Delta V\,\Delta L + \rho A\,\Delta L g \cos\theta\,\Delta L$$

Divide by the mass of fluid in the open system, $\rho A\,\Delta L$, to obtain the work input per unit of mass,

$$w_{\text{in}} = \frac{\Delta p}{\rho} + V\,\Delta V + g\cos\theta\,\Delta L = v\,\Delta p + V\,\Delta V + g\cos\theta\,\Delta L$$

Now, if ΔL is made to approach dL, then the other differences also approach differentials, and the work (per unit mass) done on the fluid in the distance dL is

$$\delta w_{\text{in}} = v\,dp + V\,dV + g\,dz \qquad (1\cdot 20a)$$

or for flow between sections a finite distance apart

Work for a simple compressible substance in one-dimensional steady flow

$$w_{\text{in}} = \int v\,dp + \Delta\left(\frac{V^2}{2}\right) + g\,\Delta z \qquad (1\cdot 20b)$$

$$w = -\int v\,dp - \Delta\left(\frac{V^2}{2}\right) - g\,\Delta z \qquad (1\cdot 20c)$$

This relation gives the *work of a quasiequilibrium process of a simple compressible substance in one-dimensional steady flow.* (Actually, the derivation requires that the flow is one dimensional at inlet and exit, not throughout the open system.) Also, Eq. 1·20 applies to the special case of an incompressible substance. Notice that it was derived from the principles of mechanics just as the analogous expression, Eq. 1·17, for the work of a quasiequilibrium process of a simple compressible closed system was derived. These two expressions, Eq. 1·17 and Eq. 1·20, for work under specified conditions will be used often in modeling actual processes.

On a pv diagram, the term $\int v\,dp$ is represented by the area between the quasiequilibrium process path and the p axis (see Fig. 1·24).

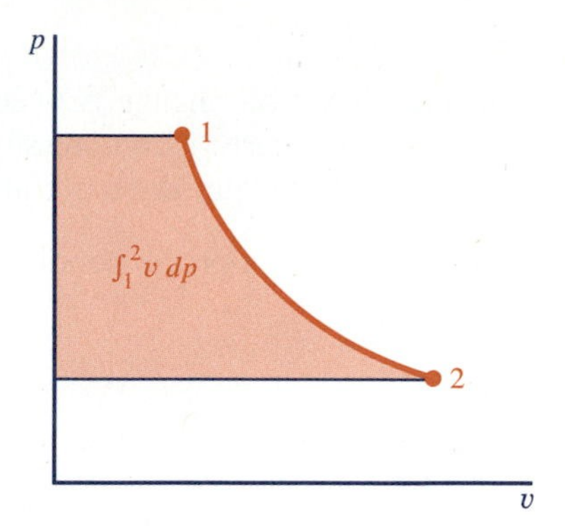

Figure 1·24 $\int v\,dp$ on a pv diagram.

Example 1·8

PROBLEM STATEMENT

It is desired to produce a circular jet of water 1.50 cm in diameter at atmospheric pressure with a velocity of 40.0 m/s. This is to be done by having water from a pipeline, where its velocity is quite low, flow through a nozzle to form the jet. If the expansion is modeled as a quasiequilibrium process, what must be the pressure in the pipeline?

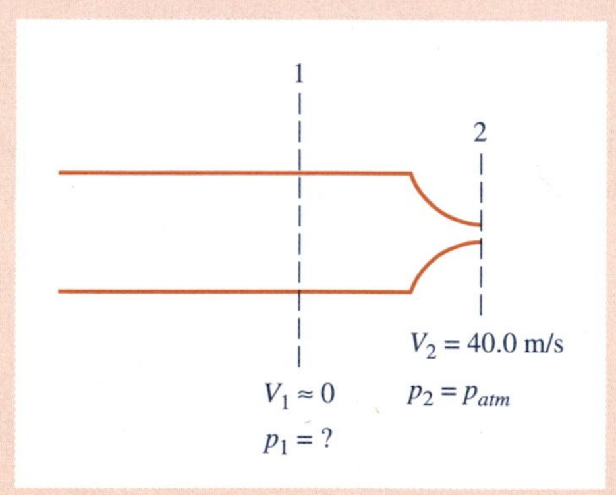

SOLUTION

Analysis: We assume that a steady flow is desired, since no specification of a time interval is given. Then, for the quasiequilibrium process, we can apply the relationship

$$w = -\int_1^2 v\,dp - \frac{V_2^2 - V_1^2}{2} - g(z_2 - z_1) \qquad (1{\cdot}20)$$

We note that

1. No work is done, $w = 0$.
2. $V_1 \ll V_2$, so we take $V_1 = 0$.

3. The elevation change between the pipeline and the jet is likely to be small, so we take $z_2 = z_1$.
4. Water is nearly incompressible, so we take v = constant and can then integrate $\int v\,dp$.
5. The jet diameter does not affect the required pressure.

Calculations: For the conditions listed in the analysis of this quasiequilibrium steady-flow process, the expression

$$w = -\int v\,dp - \frac{V_2^2 - V_1^2}{2} - g(z_2 - z_1)$$

simplifies to

$$(p_1 - p_2)v = \frac{V_2^2}{2}$$

$$p_1 = p_2 + \frac{\rho V^2}{2}$$

For water at room temperature, ρ is approximately 1000 kg/m^3. Substituting this value and other known values into the equation gives

$$p_1 = p_{atm} + 1000\,\frac{\text{kg}}{\text{m}^3}\,\frac{(40.0)^2}{2}\,\frac{\text{m}^2}{\text{s}^2}$$

$$p_1 = p_{atm} + 800{,}000\,\frac{\text{N}}{\text{m}^2} = p_{atm} + 800\text{ kPa}$$

so the pressure in the pipeline must be 800 kPa above atmospheric pressure.

Example 1·9

PROBLEM STATEMENT

A turbine operating on a steady flow of nitrogen is to produce 0.800 kW of power by expanding nitrogen from 300 kPa, 350 K (inlet specific volume of 0.346 m^3/kg), to 120 kPa. For preliminary design, the inlet velocity is assumed to be 30 m/s, the exit velocity is assumed to be 50 m/s, and the expansion will be considered as a quasiequilibrium process such that $pv^{1.40}$ = constant. Determine the required flow rate.

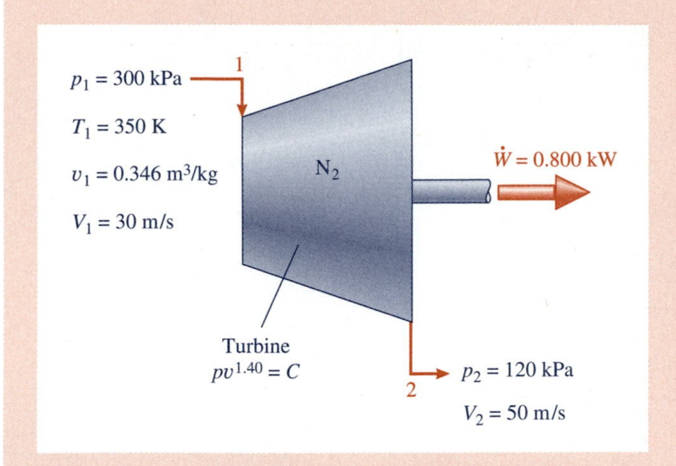

SOLUTION

Analysis: The system is an open system under steady-flow conditions. We further assume that the flow at inlet and exit is one dimensional, and this is implied by the specification of a single number for velocity at each of these sections.

If the work per unit mass can be determined, the flow rate can be obtained from the expression

$$\dot{W} = \dot{m}w \tag{1·16}$$

For a steady-flow quasiequilibrium process with one-dimensional flow at inlet and exit, we can determine work from

$$w = -\int_1^2 v\,dp - \frac{V_2^2 - V_1^2}{2} - g(z_2 - z_1) \tag{1·20}$$

since we have a relationship between v and p for the process, $pv^{1.4} = C$. We will assume that the change in elevation is negligible.

Calculations:

$$\begin{aligned}
w &= -\int_1^2 v\,dp - \frac{V_2^2 - V_1^2}{2} - g(z_2 - z_1) \\
&= -\int_1^2 \left(\frac{C}{p}\right)^{1/1.40} dp - \frac{V_2^2 - V_1^2}{2} + 0 = -C^{1/1.40}\int_1^2 p^{-1/1.4}\,dp - \frac{V_2^2 - V_1^2}{2} \\
&= -p_1^{1/1.40}v_1\left(\frac{1.40}{1.40 - 1}\right)\left(p_2^{1-(1/1.40)} - p_1^{1-(1/1.40)}\right) - \frac{V_2^2 - V_1^2}{2} \\
&= -\frac{1.4}{0.4}p_1v_1\left[\left(\frac{p_2}{p_1}\right)^{1-(1/1.40)} - 1\right] - \frac{V_2^2 - V_1^2}{2}
\end{aligned}$$

$$w = -\frac{1.4}{0.4}(300\text{ kPa})1000\frac{\text{N}}{\text{m}^2\cdot\text{kPa}}\left(0.346\,\frac{\text{m}^3}{\text{kg}}\right)\left[\left(\frac{120}{300}\right)^{0.286} - 1\right] - \frac{(50)^2 - (30)^2}{2}\,\frac{\text{m}^2}{\text{s}^2}$$

$$= 82{,}950\,\frac{\text{N}\cdot\text{m}}{\text{kg}} = 83.0\,\frac{\text{kJ}}{\text{kg}}$$

The required flow rate can now be determined:

$$\dot{m} = \frac{\dot{W}}{w} = \frac{0.800\text{ kW}}{83.0\text{ kJ/kg}} = 0.00964\text{ kg/s}$$

Comment: At this stage of your study of thermodynamics, you may not be able to estimate the required flow rate accurately. However, if you were now to calculate the area of the turbine inlet for this flow and found it to be, say, 10 m^2 or 1 mm^2, you would find such a result to be unreasonable. Calculating A_1 from $A_1 = \dot{m}v_1/V_1$ yields an area of approximately 1 cm^2. This is not unreasonable, so it does not suggest any error in our calculation.

Additional Comment: In making numerical calculations as above, we usually carry numbers within a calculator or computer through the entire solution, even though we may write down intermediate values that are rounded to an appropriate number of significant digits. This is why the results obtained may not agree exactly with those obtained by inserting the rounded intermediate values into equations.

1·9·3 Work of Quasiequilibrium Processes of Some Other Systems

In the two preceding subsections, we have developed expressions for the work of quasiequilibrium processes for two cases: simple compressible closed systems and open systems with steady one-dimensional flow of a simple compressible substance. Obviously, many other kinds of systems involve work, so we mention here that for many quasiequilibrium work processes there are other expressions for work and that there is also an interesting similarity among these expressions.

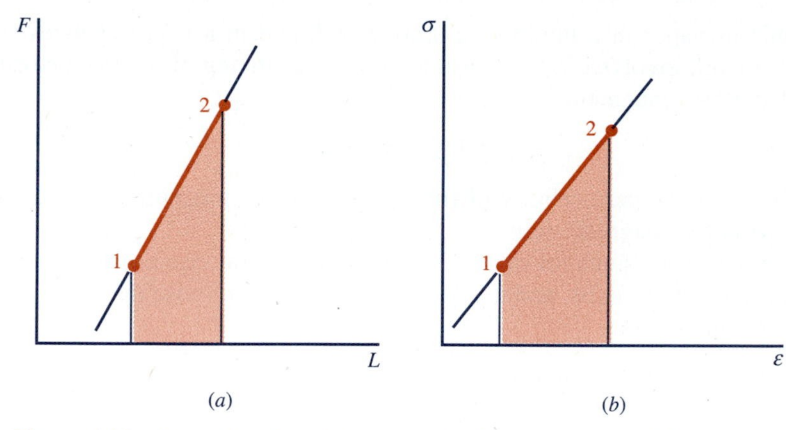

Figure 1·25 Force–length and stress–strain diagrams for stretching an elastic wire.

The work done by an elastic wire is given in differential form by

$$\delta W = -F\,dL \quad \text{or} \quad \delta W = -\sigma(\text{volume})\,d\varepsilon$$

Work done by an elastic wire

where F is the tensile force, σ is the tensile stress, and ε is the unit strain (dL/L). The work done by a wire that is stretched from state 1 to state 2 is the negative of the area below the FL diagram, Fig. 1·25*a*. The negative of the area below the corresponding stress–strain diagram, Fig. 1·25*b*, is the work per unit volume. For stretching the wire, the work is negative because work is done *on* the system as it goes from state 1 to state 2.

The work done by a surface film as it changes area in a quasiequilibrium manner is given by

$$\delta W = -S\,dA$$

Work done by a surface film

where S is the surface tension and A is the area of the film. Two simple configurations for stretching a surface film are shown in Fig. 1·26. Of greater practical significance in some engineering applications is the work done as a

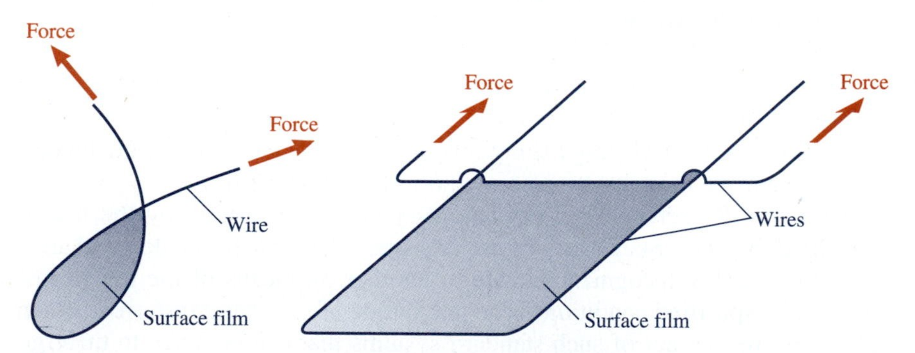

Figure 1·26 Two configurations for stretching a surface film.

bubble of vapor in a liquid or a droplet of liquid in a vapor changes size.

The work involved in the quasiequilibrium changing of the magnetization of a uniform magnetic field is given by

Work done by quasiequilibrium changes in a magnetic field

$$\delta W = -\mu_0 H\, dM$$

where μ_0 is the permeability of free space, H is the magnetic field strength, and M is the magnetization.

Similar expressions for other kinds of work of quasiequilibrium processes of closed systems exist, and it is significant that in each case the general form of the equation is

General form of the equation describing work

$$\delta W = \text{(intensive property)}\ d\text{(extensive property)}$$

1·10 Heat

As discussed in Section 1·6, if two bodies at different temperatures are brought into contact with each other while isolated from all other bodies, they interact with each other so that the temperature of one or both will change until both bodies are at the same temperature. This *interaction* between the bodies or systems is the result of only the temperature difference between them and is called *heat.*

Heat is an interaction between a system and its surroundings which is caused by a difference in temperature between the system and its surroundings.

Conventionally, we say that heat is added to a body which becomes hotter or is taken from a body which becomes colder. (We will encounter exceptions to this when work is also involved, however.) The definition above provides a means of *recognizing* the interaction called heat, and particularly provides a means of distinguishing this interaction from the one called work. In addition, we must have some method of *measuring* heat.

> Unfortunately, conventional expressions such as "heat is added" and "heat is transferred," which stem from the now discredited caloric theory, may give the erroneous impression that heat is a substance. No such difficulty is encountered with work, the other interaction between systems.

Heat transfer across a system boundary can be measured by the use of *standard systems.* Standard systems are ones that can be made to change from one readily recognizable state to another by means of the transfer of heat under specified conditions. The magnitude of the heat transferred is then given by the number of such standard systems that can be made to undergo the specified state changes as a result of the interaction. This is analogous to counting the number of standard weights that can be lifted a specific distance to measure work. For example, suppose we wish to know how much heat is transferred from a block of steel that cools from the temperature of boiling water at one atmosphere (100°C) to the temperature of freezing water at one atmosphere (0°C). The standard systems we use might be 1-kg blocks of ice at 0°C. The steel block can be cooled with one such block of ice after another so that each block of ice is melted but the temperature of the resulting liquid is not changed. The number of such standard systems (blocks of ice) required is a measure of the amount of heat removed from the steel block. Note that the conditions under which the standard systems interact with the steel block must be specified. The blocks of ice, for example, can be made to change state even without heat transfer, so the state changes are related to heat transfer only if they are carried out under the same specified conditions in each instance. The important conclusion is that it is possible to describe certain operations by which heat can be measured; hence, heat can be defined operationally.

The standard systems and state changes that have been used historically to measure heat have almost always been quantities of water caused to undergo specified temperature changes. The original definition of the units of heat, the calorie and the British thermal unit (Btu, abbreviated further as B), were based on specified masses of water at specified temperatures caused to undergo a temperature change of one degree on a particular scale of temperature while under a constant pressure of 1 atm. The calorie and the Btu are no longer defined in this manner. They are defined in terms of the joule (= 1 N·m), as will be shown in Chapter 3.

Sign convention: Heat added to a system is positive; heat taken from a system is negative.

The sign convention of heat is that heat added *to* a system is expressed by *positive* numbers and heat taken *from* a system is expressed by *negative* numbers. Suppose that 100 units of heat are transferred from system A to system B. Following the convention described, we express this as

$$Q_A = -100 \text{ units} \quad \text{or} \quad Q_B = 100 \text{ units}$$

In this book, direction subscripts are used for heat just as for work. Where the subscript indicating direction is not used, the usual convention is followed: Q (without a subscript) stands for the net amount of heat *added* to the system to which the term is applied. As an illustration, suppose that heat is transferred from a system in the amount of 100 units. We can express this in the following ways:

$$Q = -100 \text{ units}$$
$$Q_{\text{out}} = 100 \text{ units}$$

A symbol Q_{in} would be redundant, since Q means heat added to a system.

Heat, like work, is an interaction between systems. The amount of heat transferred depends on how the system was changed from one state to another. Heat, like work, is a *path function.* Thus the amount of heat transferred to or from a system during a process cannot be determined from the end states alone. It is not a property of a system.

In everyday conversation where there is no concern with measurements or with consistency among definitions one hears statements about "the heat in a fuel" or "the heat in a radiator." The definition of heat used in thermodynamics makes such statements meaningless.

The rate at which heat is transferred is represented by the symbol $\dot{Q}$, where the overdot again signifies a time rate.

Adiabatic

The word **adiabatic** means *without heat transfer.* We speak of *adiabatic processes* and *adiabatic systems.*

The first law of thermodynamics expresses a relationship between heat and work and also relates heat and work to state changes of a system. The second law of thermodynamics (Chapter 5) limits the relationship between heat and work and points out differences in their possible effects. Before getting to the first and second laws, it is well to review the essential difference between heat and work which is established by their definitions: Heat is an interaction caused by a temperature difference between a system and its surroundings; work is done by a system if the sole effect of the system on its surroundings could be reduced to the lifting of a body.

Consider a system consisting of air and an externally driven paddle wheel in a thermally insulated vessel. The system and the surroundings are initially at the same temperature. If the paddle wheel is turned by an external motor,

> At this stage you may ask if thermodynamics is concerned only with peculiar systems that a practicing engineer is unlikely to encounter. The reply is that the systems used in proofs and explanations are occasionally unusual because, for the illustration of principles, it is desirable to use systems that involve only one interaction or effect at a time. Sometimes this calls for a highly simplified system (like the paddle-wheel system here), and sometimes it calls for an elaborately controlled system (like the standard systems mentioned earlier in this section, where pressure and temperature were both controlled and interactions with systems other than the steel block were prevented). These systems are chosen to illustrate some principle or definition; in turn, the application of the principle or definition to some more common or more complex system is more readily understood.

the temperature of the system increases. The interaction between the system and its surroundings is the result of the torque exerted on the rotating shaft and is not the result of a temperature difference between the system and its surroundings, because (1) the system and surroundings were at the same temperature at the beginning of the process and (2) the vessel is thermally insulated to prevent any interaction resulting from a temperature difference. Thus the interaction is work, not heat. It is unfortunate and misleading that in everyday language it might be said that the gas was "heated by the paddle wheel" or "heated by friction."

Example 1·10

PROBLEM STATEMENT

Referring to the figure, we see that the air, oil, ice, and liquid water within the outer container are all initially at a temperature of 0°C. The outer container is thermally insulated so that no heat can be transferred through it. The weight is allowed to fall, turning the paddle wheel in the oil by means of the pulley. After some time has elapsed, it is observed that the air, oil, ice, and liquid water are again at 0°C but that some of the ice in the inner container has melted. Identify the work done on or by each system and the heat transferred to or from each of the systems that are identified as follows: System A is everything within the outer container, system B is everything within the inner container, and system C is everything within the outer container except system B. (System A = system B + system C.)

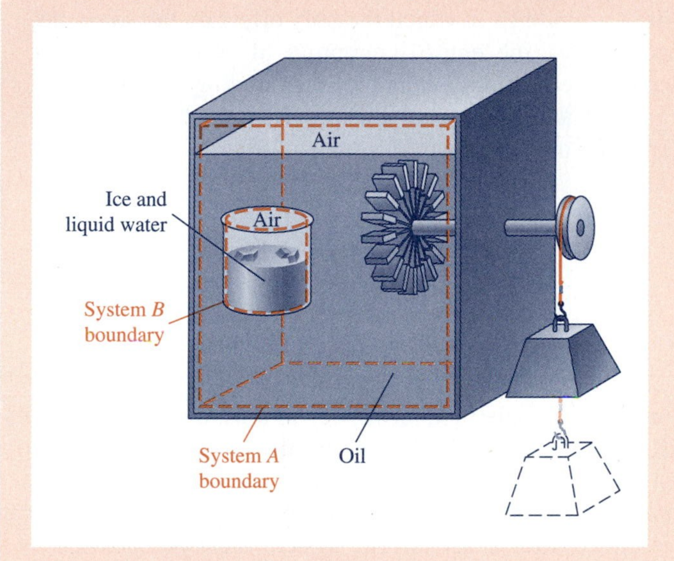

SOLUTION

Because of the paddle-wheel action, as the weight falls the temperature of the oil (and the air above the oil) increases. The resulting temperature difference between the oil and the contents of the inner container (system B) causes a transfer of heat from the oil into the inner container. This transfer of heat melts some of the ice and causes the temperature of the oil and air to return to 0°C. The interactions involving the various systems are as follows.

System B: Heat—There is a transfer of heat into system B from system C inasmuch as there is an interaction that results from a temperature difference. Work—No mechanical work is done on or by system B because its interaction with its surroundings involves no motion of the system boundary.

System A: Heat—No heat is transferred to or from system A because the system is thermally insulated from its surroundings. (Something occurs *within* system A as a result of a temperature difference *within* the system, but with regard to system A this is not heat transfer because heat is an interaction *between a system and its surroundings.*) Work—Work is done on system A by means of the shaft that turns the paddle wheel. Notice that as far as the action of the weight and pulley system external to system A is concerned, the action within system A could be reduced to the lifting of a weight (by removing the oil and replacing the paddle wheel by a weight and pulley arrangement).

System C: Heat—Heat is transferred from system C to system B. No other heat is transferred to or from system C because that part of its boundary which is not common to system B is thermally insulated. Work—Work is done on system C by means of the shaft that turns the paddle wheel.

As an additional point of interest, consider system B and its surroundings. After the process described is completed, the temperature of everything outside system B is the same as at the beginning of the process, so the sole effect on the surroundings of system B is the fall of the weight. In Section 1·10 it was pointed out, however, that an interaction between a system and its surroundings in which the only effect on the surroundings is the *fall* of a body is not necessarily work. In fact, the entire interaction of system B with its surroundings is the result of a temperature difference between the system and its surroundings (even though the temperature difference is zero at both the beginning and the end of the process); so the interaction between system B and its surroundings is only a transfer of heat.

1·11 Summary

This chapter has introduced undefined terms, definitions, and important relationships that will be used repeatedly throughout the remainder of the book. Learning these is essential for a sound understanding of thermodynamics.

A *thermodynamic system* is any quantity of matter or region of space to which attention is directed for analysis. The system must be enclosed by a prescribed boundary that may be rigid or deformable. If no matter crosses the boundary, the system is a *closed system.* If matter does cross the system boundary, the system is an *open system.* Everything outside the system boundary is called the *surroundings.* An *isolated system* is a closed system that in no way interacts with its surroundings.

A *property* is an observable characteristic of a system while the system is in any one condition or *state.* An *intensive property* is one that has the same value for all parts of a homogeneous system and for the entire system. Examples are pressure, temperature, and density. An *extensive property* is one that has a value for the entire system equal to the sum of the values for the parts of the system. Examples are mass and volume. A *specific* property is an extensive property of a system divided by the mass of the system. An example is specific volume.

A *process* of a system is a change of that system from one state to another. The *path* of the process is the series of states the system passes through during the change. A *cycle* or *cyclic process* is a process or series of processes that returns the system to the state it was in before the process began. The net change in any property of a system that executes a cycle is zero; that is, if x is a property

$$\oint dx = 0 \tag{1·1}$$

A system is in an *equilibrium state* or in *equilibrium* if no changes can occur in the state of the system without an external stimulus. If during a process a system passes through a series of only equilibrium states so that the system at any instant is in equilibrium or infinitesimally close to being in equilibrium, the process is called a *quasiequilibrium* process. During a quasiequilibrium process, there must be no frictional effects. (Other characteristics will be discussed later.)

Any actual system has many characteristics. Some of these have great influence on the system's behavior, some have little or no effect, and some have unknown effects. We therefore usually devise for study an *idealized* system or *model* that in key respects is much simpler than an actual system but simulates its behavior. In a similar manner, we deal with *idealized* processes or *models* that simulate actual processes. Devising the idealizations or *modeling* is an important skill in engineering.

For a fluid in static equilibrium, the relationship between pressure and elevation is the basic equation of fluid statics:

$$dp = -\gamma\, dz \tag{1·7}$$

where γ is the specific weight of the fluid and z is the elevation. For an incompressible fluid, the magnitude of the pressure difference over an elevation difference of h is given by

$$\Delta p = \gamma h \quad \text{and} \quad \Delta p = \rho g h \tag{1·8}$$

For any thermodynamic system, the conservation of mass principle can be expressed by the *continuity equation*:

$$\frac{d}{dt}\int \rho\, dVol = \sum_{\text{inlets}} \int \rho V\, dA - \sum_{\text{exits}} \int \rho V\, dA \tag{1·11}$$

Steady flow through an open system exists when all properties at each point throughout the system remain constant with respect to time. For the special case of *one-dimensional steady flow with a single inlet and single outlet*, the continuity equation becomes

$$\dot{m} = \rho_1 V_1 A_1 = \rho_2 V_2 A_2 \tag{1·14}$$

Work is an interaction between a system and its surroundings. Work is done by a system on its surroundings if the sole external effect of the interaction could be the lifting of a body. The magnitude of work is the product of the weight of the body and the distance it is lifted.

The work of a process depends on the path of the process, not just on the end states, so work is a *path function*, not a *point function* as any property is.

The sign convention for work is that work done by a body is represented by positive numbers and work done on a body is represented by negative numbers.

Power $\dot{W}$ is the rate of doing work. For steady-flow systems,

$$\dot{W} = \dot{m}w \tag{1·16}$$

Mechanical work is work that is done by the action of a force exerted on a moving boundary of a system. A general equation for mechanical work is

$$W = \int F\,ds \tag{1·15}$$

where F is the force in the direction of the displacement s.

Two important expressions for work of quasiequilibrium processes under certain conditions are the following: For work of a *quasiequilibrium process of a simple compressible closed system*,

$$W_{1-2} = \int_1^2 p\,dV \qquad w_{1-2} = \int_1^2 p\,dv \tag{1·17}$$

For work of a *quasiequilibrium process of a simple compressible substance in one-dimensional steady flow*,

$$w = -\int v\,dp - \Delta\left(\frac{V^2}{2}\right) - g\,\Delta z \tag{1·20}$$

Heat is an interaction between a system and its surroundings caused by a difference in temperature between the system and the surroundings. Heat can be measured in terms of specified changes of state of certain standard systems. Heat, like work, is a *path function.*

The sign convention is that heat added to a system is designated by positive numbers, heat rejected is designated by negative numbers. Notice that this is opposite to the sign convention for work.

References

1·1 Bridgman, P. W., *The Logic of Modern Physics*, Macmillan, New York, 1927. (Chapter I treats definitions of terms in an absorbing manner.)

1·2 Rapoport, Anatol, *Operational Philosophy*, Harper, New York, 1957. (Chapter 1, entitled "The Problem of Definition: What is X?" is a delightful, informative essay.)

Problems

1·1 For analyzing each of the listed events, specify the kind of system you would use and sketch the system boundary, showing any flows of matter:
(*a*) Inflating a bicycle tire with a hand pump.
(*b*) Cooling the power supply of a desktop computer.
(*c*) The change in size of a toy balloon left in the sun.
(*d*) The change in air pressure within a sealed package as temperature changes.
(*e*) Cooling an electric power distribution transformer by spraying water on it.

1·2 For analyzing each of the listed events, specify the kind of system you would use and sketch the system boundary, showing any flows of matter:
(*a*) Pumping water from a lake to an overhead storage tank.
(*b*) A gas in a cylinder expanding against a piston to compress a coil spring.
(*c*) The cooling of a disk brake as an automobile is stopped.
(*d*) Mixing of foods by an electric mixer.
(*e*) Heating water in a household gas-fired water heater.

1·3 The specific volume of a gas is 1.5 m^3/kg. What volume does 7.5 kg occupy? What is the density of the gas?

1·4 The density of a gas is 0.28 lbm/ft^3. What volume does 12.0 lbm occupy? What is the specific volume of the gas?

1·5 A tank with a volume of 0.85 m^3 contains 16 kg of gas. Determine the specific volume. Calculate the final specific volume and the final density of the gas remaining in the tank after 13 kg of gas escapes.

1·6 A tank with a volume of 6.0 cu ft contains 0.40 lbm of gas. Determine the specific volume and the density. Calculate the final specific volume and the final density of the gas in the tank after 1.85 lbm of additional gas is pumped into the tank.

1·7 A vacuum gage reads 60.2 cm of mercury. Determine the absolute pressure in kPa if the barometer reads 75.7 cm of mercury.

1·8 A vacuum gage reads 26 in. of mercury when barometric pressure is 29.5 in. of mercury. What is the indicated pressure in psia?

1·9 A pressure gage connected to a tank reads 380 kPa while the barometer reads 745 mm Hg. What is the absolute pressure in the tank?

1·10 Determine the pressure on the hull of a submarine submerged 40 m below sea level. The specific gravity of the seawater is 1.025. The barometric pressure is 75.0 cm of mercury.

1·11 Determine the pressure, in psia, 60 ft below the surface of the sea. The specific gravity of the water is 1.022. The barometer reading is 29.2 inches of mercury.

1·12 Create a pressure versus depth chart for a diver. (You must make some decisions. State your assumptions.)

1·13 At the inlet and exhaust of a turbine the absolute steam pressures are 6000 kN/m^2 and 4.0 cm Hg, respectively. Barometric pressure is 75 cm Hg. Calculate the gage pressure for the entering steam and the vacuum gage pressure for the exhaust steam.

1·14 Air flowing through a duct passes through a bank of filters. The pressure drop across the filter bank is measured by a manometer filled with water as shown in the figure. The manometer deflection is 17.2 cm. What is the pressure drop in kPa? In psi?

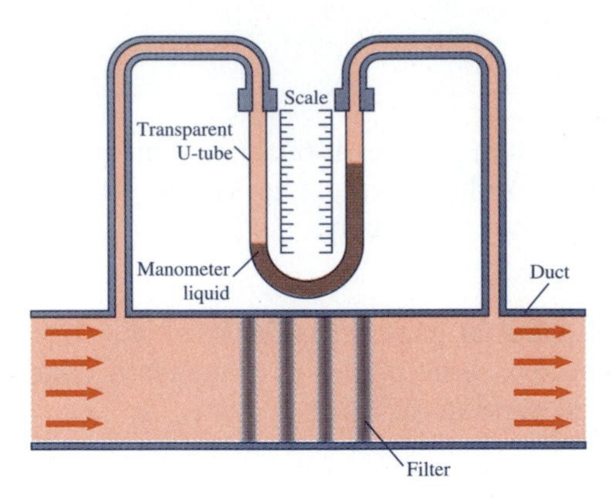

Problem 1·14

1·15 Determine the pressure difference between water lines *A* and *B* as indicated by the manometer shown in the figure.

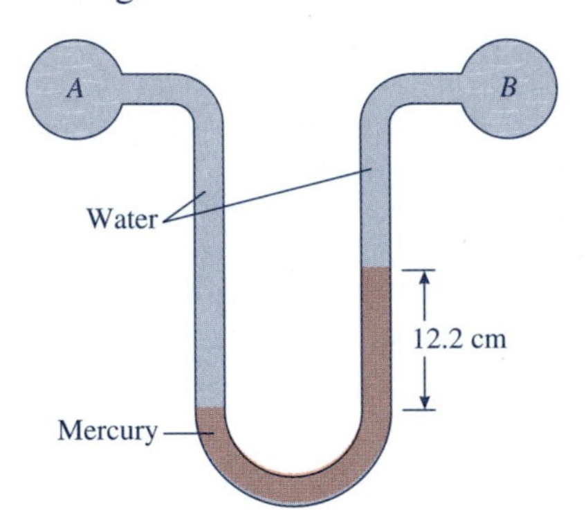

Problem 1·15

1·16 Determine the pressure difference between water lines *A* and *B* as indicated by the manometer shown in the figure.

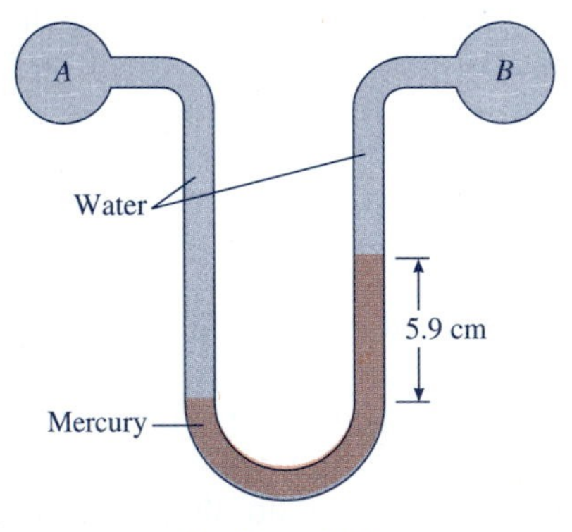

Problem 1·16

1·17 The flow rate of air through a pipe is measured by means of an orifice as shown in the figure. If the manometer is filled with an oil with a specific gravity of 0.88 and the manometer deflection is 8.9 cm, what is the pressure drop across the orifice in kPa? In psi?

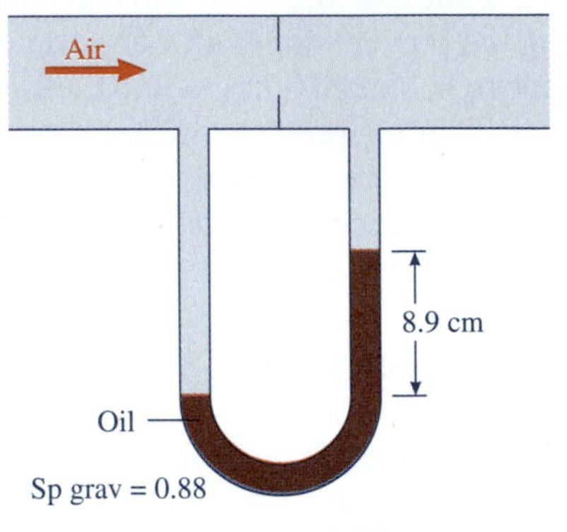

Problem 1·17

1·18 A man is 10 feet below the surface of a swimming pool. How much force does the water exert on one square foot of his chest? Justify your use of either absolute or gage pressure in this calculation.

1·19 A mercury manometer at sea level and one at an altitude of 4000 m have the same deflection. The temperatures are the same. Explain which one indicates the greater pressure difference.

1·20 In the "standard atmosphere" that is used in many aeronautical applications, atmospheric pressure is 61.7 kPa at an altitude of 4000 m and 57.8 kPa at an altitude of 4500 m. What is the average density of air between these altitudes?

1·21 A precision mercury manometer measuring a pressure difference has a deflection of 424.4 mm. If the mercury temperature is 29°C and the acceleration of gravity at the manometer location is 978.6 cm/s^2, how much error would be involved in using the conventional millimeter of mercury in converting the manometer deflection to units of kPa?

1·22 The gas in a vertical cylinder supports the piston as shown. The mass of the piston is 400 kg; the area of the cylinder cross section is 0.0276 m². Atmospheric pressure is 96 kPa. Calculate the pressure of the gas. (Indicate, of course, whether you are stating absolute or gage pressure.) What assumptions do you make?

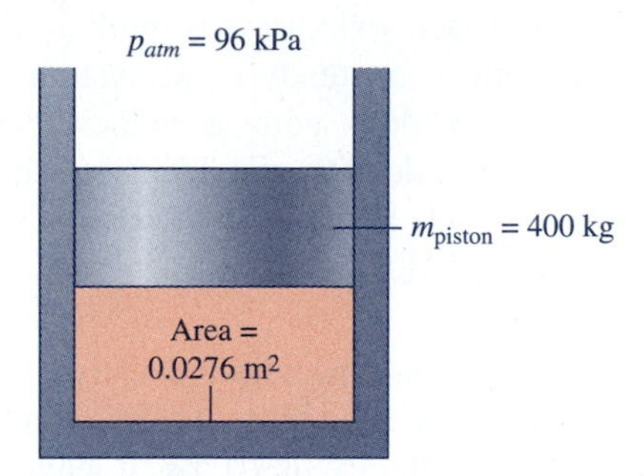

Problem 1·22

1·23 What is the normal range of barometric pressures where you live? What percentage variation does this amount to?

1·24 One occasionally hears statements such as, "Today it is 97 F in the shade." What temperature is being given? How would you answer the question, "What is it in the sun?" Explain what temperature a mercury-in-glass thermometer measures when it is held in the shade of trees and when it is held in sunlight.

1·25 What is the normal range of ambient temperatures where you live? What percentage variation of absolute temperature does this amount to?

1·26 Refer to Fig. 1·9. Verify that substituting the values given for the ratio p_s/p_i and for the Celsius temperatures at the ice and steam points into the equation of the line gives a value of −273.15°C for the temperature at zero pressure in a constant-volume gas thermometer.

1·27 Convert the following Celsius temperatures to Fahrenheit: (*a*) 0, (*b*) −40, (*c*) 25, (*d*) 130, (*e*) 600.

1·28 Convert the following Fahrenheit temperatures to Celsius: (*a*) −70, (*b*) 0, (*c*) 40, (*d*) 600, (*e*) 2400.

1·29 On the Réaumur temperature scale the ice point is zero and the steam point is 80. What is the Réaumur temperature of absolute zero?

1·30 Determine the validity of the following procedures for converting between Celsius and Fahrenheit temperature scales.

(*a*) Add 40 to the temperature on one of these scales, multiply by $\frac{5}{9}$ or $\frac{9}{5}$, and subtract 40 to obtain the temperature on the other scale.

(*b*) Double the Celsius temperature scale value, subtract 10 percent of that result, and add 32 to obtain the Fahrenheit temperature.

(*c*) Subtract 32 from the Fahrenheit temperature, halve this amount, and add one-ninth of the result to obtain the Celsius temperature.

1·31 Three thermodynamic quantities X, Y, and Z are defined as $X = \int (p\,dv + v\,dp)$; $Y = \int (p\,dv - v\,dp)$, and $Z = \int (R\,dT + p\,dv)$, where $R = pv/T =$ constant. Which of these quantities X, Y, and Z are properties?

1·32 Evaluate F, G, and H as defined by

$$F = \int (x^2\,dy + y\,dx)$$

$$G = \int [(x - y)\,dy + (y^2 - 2x)\,dx]$$

$$H = \int [2y(x - 2y)\,dx + x(x - 8y)\,dy]$$

over each of the following paths from $x = 0, y = 0$ to $x = 1, y = 2$: (*a*) $y = 2x$, (*b*) $y = 2x^2$, and (*c*) $y = 2x^{1/2}$.

1·33 Evaluate I, J, and K as defined by

$$I = \int (xy\,dx + xy\,dy)$$

$$J = \int [(x + y - 20)\,dx - (x - 2y)\,dy]$$

$$K = \int 2y\,dx$$

over each of the following paths from $x = 20, y = 20$ to $x = 40, y = 60$: (*a*) $y = 2x - 20$, (*b*) $y = 0.025x^2 + 0.5x$, and (*c*) $y = 60 - 0.10(x - 40)^2$.

1·34 Consider a liquid in a circular pipe of radius r_0. At one instant, the velocity is a maximum u_{max} at the pipe axis and zero at the pipe wall. Determine the ratio V/u_{max}, where V is the average velocity at a given section, if the velocity u at any radius r is given by $u = u_{max}(1 - r/r_0)^2$.

1·35 Consider a liquid flowing steadily in a circular pipe of radius r_0. Determine the ratio V/u_{max}, where V is the average velocity across the pipe and u_{max} is the maximum velocity, if the velocity u at any radius is given by $u = u_{max}[1 - (r/r_0)^2]$.

1·36 A gas with a density of 0.232 kg/m^3 flows at a velocity of 280 m/s through a cross-sectional area of 0.00850 m^2. Determine the mass rate of flow.

1·37 Ammonia with a specific volume of 1.940 m^3/kg flows at a rate of 3.30 kg/s through a tube with a cross-sectional area of 0.010 m^2. Determine the average velocity across the tube cross section.

1·38 Methane with a specific volume of 0.848 m^3/kg has an average velocity of 60.8 m/s across a flow area of 0.0150 m^2. Determine the mass rate of flow.

1·39 Air entering a diesel engine passes through a filter that prevents dust from reaching the engine. The pressure drop through the filter affects engine performance. Evaluate the idea of modeling the filter as a steady-flow system.

1·40 A type of air conditioner used in hot dry climates draws in outdoor air and sprays a very fine mist of water into it. The water evaporates into the air, lowering its temperature and increasing its humidity before it leaves the air conditioner. Evaluate the idea of modeling the air conditioner as a steady-flow system.

1·41 A lawn mower is driven by a one-cylinder gasoline engine. Evaluate the idea of modeling the muffler of the engine as a steady-flow system.

1·42 Liquid sulfur dioxide with a density of 1394 kg/m^3 enters an expansion valve at a rate of 40.0 kg/h with a velocity of 0.38 m/s. The sulfur dioxide partially vaporizes as it passes through the valve so that it leaves with a density of 11.9 kg/m^3. Inlet and outlet areas are equal, and the flow is steady. Calculate (*a*) the outlet area and (*b*) the outlet velocity.

1·43 Furnace exhaust gas with a density of 0.80 kg/m^3 enters a steady-flow system through a 60-cm-diameter duct with a velocity of 11.0 m/s. It leaves with a density of 1.15 kg/m^3 through a 20-cm-diameter duct. Determine the outlet velocity and the mass rate of flow.

1·44 A gas mixture with a density of 0.095 lbm/cu ft enters a steady-flow system through a 14-in.-diameter duct with a velocity of 10 fps. It leaves with a density of 0.23 lbm/cu ft through a 4-in.-diameter duct. Determine the outlet velocity and the mass rate of flow.

1·45 Methane with a density of 0.52 kg/m^3 enters a burner with a velocity of 30 m/s through a cross-sectional area of 8.2 cm^2. Air with a density of 1.20 kg/m^3 enters at a mass flow rate 21 times that of the methane. Exhaust gas with a density of 0.680 kg/m^3 leaves through a cross-sectional area of 196 cm^2. Flow is steady. Determine the exhaust velocity.

1·46 Air with a density of 0.042 lbm/ft^3 enters an aircraft jet engine at a velocity of 400 mph through a cross-sectional area of 9.20 sq ft. For cabin cooling, 6.0 lbm/s of air is drawn off part way through the compressor. Fuel is then added to the air at an air/fuel mass ratio of 38 and burned. The exhaust gases will have a density of 0.019 lbm/ft^3 and a velocity of 1180 fps. Determine the exhaust cross-sectional area.

1·47 A mine hoist raises a cage having a mass of 750 kg at a rate of 50 m/s. What power is developed by the hoist? (This calculation requires you to make some assumptions.)

1·48 How much power is required to propel a jet aircraft in steady level flight at 1350 km/h if the thrust required is 400 kN?

1·49 An aircraft with a mass of 22,000 kg flies at a speed of 500 km/h while climbing at a rate of 800 m/min. The engines produce a thrust of 320 kN in the direction of flight. Calculate the power required.

1·50 A thrust of 75,000 lbf is required to propel a jet aircraft in level flight at 560 mph. How much power is developed?

1·51 Calculate the power developed by a locomotive that draws a train at 50 km/h up a grade that rises 30 cm vertically in each 30 m measured horizontally (a 1.0-percent grade). The total mass of the locomotive and cars is 4600 tonnes. If the frictional resistance is 30 N/t, what power is developed by the engine?

1·52 Determine the power delivered by a locomotive that draws a train at 80 km/h down a 0.20-percent grade. The total mass of the locomotive and cars is 2800 tonnes. Frictional resistance is 45 N/t.

1·53 A 1000-ton train rolls steadily at 60 mph on a level track. Frictional resistance is 15 lbf/ton. Determine (*a*) the force required and (*b*) the power (in hp) developed by the locomotive.

1·54 A locomotive draws a train at 4 mph up a grade that rises 1 ft vertically in each 100 ft measured horizontally. The total weight of the locomotive and cars is 4000 tons. Frictional resistance is 4 lbf/ton. Calculate the power developed by the locomotive.

1·55 A tank 4 m long, 3 m wide, and 2 m deep is half full of water. Calculate the work required to raise all the water over the top edge of the tank.

1·56 A tank 10 ft long, 7 ft wide, and 6 ft deep is half full of water. Calculate the work (in foot-pounds) required to raise all the water over the top edge of the tank.

1·57 A 2-in.-diameter hole is to be drilled into the earth to a depth of 300 ft. Determine the work required to raise the earth material to the surface if the average weight of 1 cu ft is 112 lbf.

1·58 A 10-hp electric motor has nothing attached to its shaft. The motor is operating at 3600 rpm and is connected to a 115-V power line. What is the power output?

1·59 A coil spring stands on end. Its linear spring constant is 0.98 N/mm. An object with a mass of 1.10 kg is placed on the spring so that it is compressed to a new equilibrium position. How much work is done on or by the spring as it is compressed? How much work is done on or by a similar spring with the same spring constant if it is hanging from a hook and an object of 1.10 kg mass is attached to the lower end of the spring, thus stretching it to a new equilibrium position?

1·60 Although dp/dt is the rate of change of pressure with time, power, $\dot{W} = dW/dt$, cannot be defined as the rate of change of work with time. Explain.

1·61 Derive an expression for the work done in the elastic stretching of a steel wire in terms of modulus of elasticity and strain.

1·62 Determine the amount of work required to stretch a steel wire 1.00 mm in diameter and initially 400 mm long to a length of 410 mm at room temperature. (The modulus of elasticity is 200×10^3 MPa.)

1·63 When a tugboat tows a barge at a velocity of 5.0 km/h, the tension in the tow cable is 145,000 N. If the power required to tow the barge varies with velocity cubed, determine the tension in the cable at a towing velocity of 6.5 km/h.

1·64 Calculate the work done on a body of 350 kg mass to accelerate it from a velocity of 40 m/s to 100 m/s in (*a*) 10 s, (*b*) 1 min.

1·65 Calculate the work done on a body with a mass of 4.2 slugs to accelerate it from a velocity of 200 to 600 fps in (*a*) 15 s, (*b*) 1 min.

1·66 Calculate the work required to lift a 25-kg body from an elevation of 208 m above mean sea level to an elevation 80 m higher in (*a*) 2 min, (*b*) 10 min.

1·67 Calculate the work required to lift a 90-slug body from an elevation of 640 ft above mean sea level to an elevation 60 ft higher in (*a*) 2 min, (*b*) 10 min.

1·68 Assuming that the gravitational acceleration g varies with altitude h above the earth's surface according to the relationship

$$g = \frac{a}{(b + h)^2}$$

where a and b are constants, sketch a curve of the weight of a given object versus h, and write an equation for the work done in lifting an object of mass m to an elevation h. Write the equation in terms of m, h, a, and b.

1·69 Calculate the work performed when a gas expands from a volume of 0.080 m^3 to 0.190 m^3 under a constant pressure of 850 kPa. Sketch a pressure–volume diagram for the process.

1·70 Calculate the work performed when a gas expands from a volume of 0.809 cu ft to 1.80 cu ft under a constant pressure of 120 psia. Sketch a pressure–volume diagram for the process.

1·71 A vertical U-tube has a circular cross section with a diameter of 1.0 cm. One leg of the U-tube is capped and the other is open to the atmosphere. The two ends are at the same elevation. Mercury in the U-tube traps nitrogen in the closed leg so that the mercury level is 30 cm below the capped end. In the other leg, the mercury surface is at the open end, so increasing the pressure of the nitrogen causes mercury to overflow. Atmospheric pressure is 95.0 kPa. The nitrogen is heated so that it expands until the level of the mercury in that leg is lowered to 40 cm below the capped end. Calculate the work done by the nitrogen as it expands. (You will undoubtedly neglect the effect of the change in mercury temperature near the nitrogen. If this effect were considered, would the calculated work be greater, smaller, or the same?)

1·72 A rigid steel tank contains 0.40 m^3 of oxygen at a pressure of 1200 kPa. If the oxygen is cooled so that its pressure is reduced to 900 kPa, how much work is performed? Draw a pressure–volume diagram for the process.

1·73 A toy balloon filled with a light hydrocarbon gas floats against the ceiling of a room. If the gas cools and the balloon contracts, is any work done on or by the gas? Draw a pressure–volume diagram for the cooling process.

1·74 A gas is trapped by a frictionless piston in a vertical cylinder as shown in the figure. The piston is supported solely by the gas pressure. The gas pressure is initially 250 kPa. The gas is then heated so that the piston is lifted 10 cm. Calculate the work done by the gas. How much of this work is done to lift the piston and how much is done to push back the atmosphere? What is the mass of the piston?

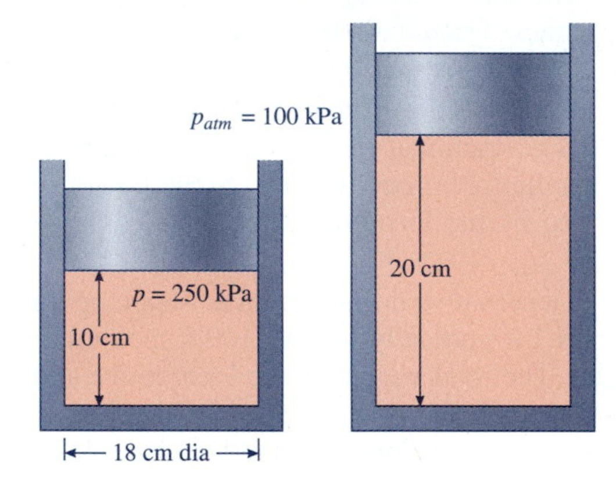

Problem 1·74

1·75 In a quasiequilibrium process in a closed system, a gas expands from a volume of 0.150 m^3 and a pressure of 120 kPa to a volume of 0.25 m^3 in such a manner that $p(V + 0.030)$ = constant, where V is in m^3. Calculate the work.

1·76 In a quasiequilibrium process in a closed system, a gas is compressed from a volume of 4.0 cu ft and a pressure of 15 psia to a volume of 3.2 cu ft in such a manner that $p(V + 0.030)$ = constant, where V is in cu ft. Calculate the work.

1·77 In a closed system, a quasiequilibrium process occurs for which the pressure changes according to the equation $p = 300V + 1000$, with the pressure in psfa and the volume in cubic feet. The volume changes from 6.0 to 4.0 cu ft. Calculate the work.

1·78 Air is compressed slowly and frictionlessly in a cylinder according to the relationship $pV^{1.4} = C$ from an initial volume of 0.8 m^3 and pressure of 100 kN/m^2 to a final volume of 0.4 m^3. The symbol C represents a constant. Calculate the work.

1·79 Air is compressed slowly and frictionlessly in a cylinder according to the relationship $pV^{1.4} = C$ from an initial volume of 2 cu ft and pressure of 30 psia to a final volume of 1.2 cu ft. The symbol C represents a constant. Determine the work.

1·80 A gas expands according to the law $pv^a = C$, where a and C are constants. Derive a general equation for the work done when the gas expands from an initial absolute pressure and volume of p_1 and V_1 to a final absolute pressure and volume of p_2 and V_2 (*a*) in a closed system, (*b*) in a steady-flow system.

1·81 A spherical balloon has a diameter D_1 when the pressure of the gas inside is p_1 and atmospheric pressure is p_0. The gas is heated, causing the balloon diameter to increase to D_2. The pressure of the gas is proportional to the balloon diameter. How much work is done by the gas? How much work is done on the atmosphere? Express your answers in terms of p_1, D_1, and the ratio D_2/D_1.

1·82 A vapor in a cylinder is compressed frictionlessly from 20 to 60 psia in such a manner that $pv =$ constant. The initial specific volume is 25.0 ft^3/lbm. Sketch the process on pv coordinates, and calculate the work.

1·83 A gas in a cylinder behind a piston undergoes a quasiequilibrium process that follows the relationship $(V - 0.2)10^5 = (p - 100)^2$. Initially, the gas is at 200 kPa and has a volume of 0.30 m^3. The final pressure is 400 kPa. Sketch a pV diagram of the process and determine the work.

1·84 A gas in a closed system expands in a quasiequilibrium process from a volume of 0.20 m^3 at 200 kPa, 80°C to a volume of 0.60 m^3 in such a manner that $p = 225 - 125V$, with p in kPa and V in m^3. How much work is done?

1·85 Give a specific example of (*a*) a closed-system process in which $\int p\,dv = 0$ and work $\neq 0$, and (*b*) a steady-flow process in which work $= 0$ and $\int v\,dp \neq 0$.

1·86 If the gears of the counter mechanism of a nutating disk or ''wobble-plate'' liquid meter become hard to turn, will the pressure drop across the meter be affected for a given flow rate?

1·87 Air is compressed in a frictionless steady-flow process from 70 kN/m^2, -29°C and a specific volume of 1 m^3/kg to 100 kN/m^2 in such a manner that $p(v + 0.5) =$ constant, where v is in m^3/kg. Inlet velocity is negligibly small and discharge velocity is 100 m/s. The flow rate is 0.20 kg/s. Calculate the work required per kilogram of air.

1·88 Carbon dioxide is compressed frictionlessly in a steady-flow process in such a manner that $pv^{1/2} =$ constant from 100 to 500 kPa. The initial specific volume is 0.560 m^3/kg. Calculate the work per kilogram of gas.

1·89 Air is compressed in a frictionless steady-flow process from 10 psia, 60 F and a specific volume of 19.25 cu ft/lbm to 15 psia in such a manner that $p(v + 4) =$ constant, where v is in cu ft/lbm. Inlet velocity is negligibly small and discharge velocity is 350 fps. The flow rate is 0.280 lbm/s. Calculate the work required per pound of air.

1·90 Refer to Fig. 1·23 and the derivation of Eq. 1·20 in Section 1·9. The pressure and cross-sectional area at one end of the fluid element are p and A; those at the other end are $p + \Delta p$ and $A + \Delta A$. Carry out the derivation by using p and A as the pressure and area near the center of the element. Then, at the ends of the element the pressure will be $p + \Delta p/2$ and $p - \Delta p/2$, and so on.

1·91 Under the assumption of steady frictionless flow, calculate the power to pump 0.00094 m^3/s of water ($\rho = 999.1\ kg/m^3$) from a sump into an open tank as shown in the figure. The pipe discharging into the overflow tank has an inside diameter of 2.10 cm.

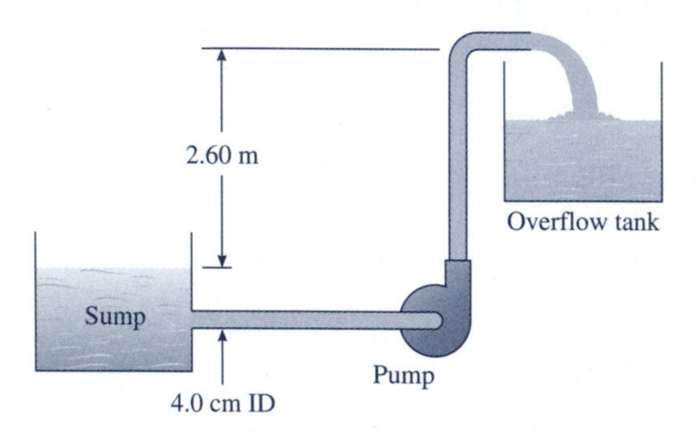

Problem 1·91

1·92 Under the assumption of steady frictionless flow, calculate the power to pump 14.0 gpm of water from a sump into an open tank as shown in the figure. A $\frac{1}{2}$-in. pipe (inside diameter of 0.618 in.) discharges into the overflow tank.

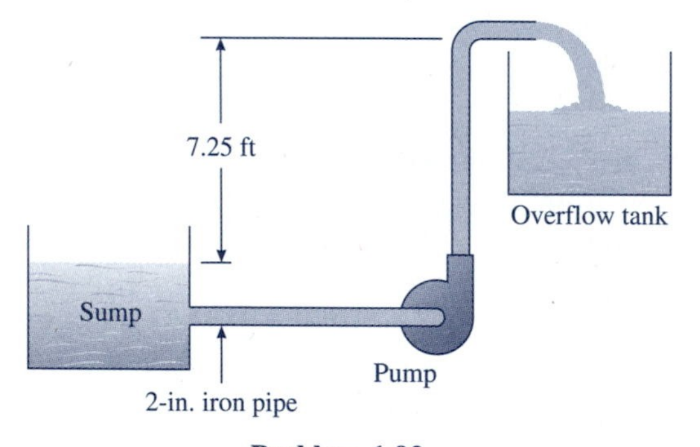

Problem 1·92

1·93 Water flows through the venturi meter shown in the figure. The pressure difference between the upstream, section 1, and the throat section 2 is indicated by the deflection of the manometer that contains mercury and, of course, water. Determine the flow rate.

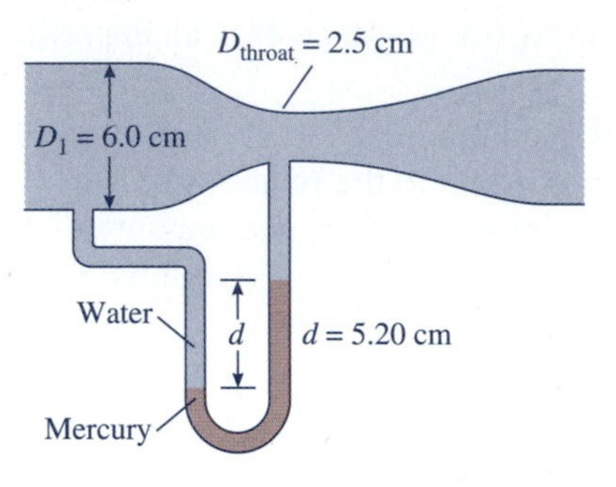

Problem 1·93

1·94 At the end of a pipe, water is to pass through a converging nozzle to form a jet as shown in the figure. Assuming frictionless flow, predict the reading of the pressure gage shown.

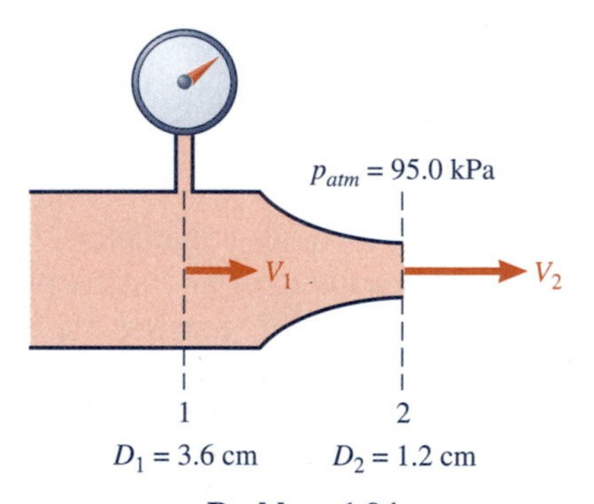

Problem 1·94

1·95 Nitrogen is compressed steadily from one atmosphere to four atmospheres in a quasiequilibrium process for which $pv^{1.34}$ = constant. Nitrogen entering the compressor has negligible velocity and a density of 1.118 kg/m^3. Nitrogen leaving the compressor has a velocity of 95 m/s. Calculate the work per kilogram of nitrogen.

1·96 Methane enters a compressor at 14.0 psia with a very low velocity and is compressed to 55 psia. The entering methane has a density of 0.040 lbm/ft^3. The methane leaves the compressor with a velocity of 250 fps. The flow rate is 0.38 lbm/s. Assuming a quasiequilibrium process in which $pv^{1.33}$ = constant, calculate the work done on the methane in B/lbm.

1·97 Air entering an aircraft jet engine is slowed before entering the first row of compressor blades as shown in the figure. At the inlet, section 1, the pressure is 70 kPa, the density is 0.930 kg/m³, and the velocity is 180 m/s. The cross-sectional area at section 1 is 0.500 m²; at section 2, 0.570 m². Assuming the flow to be steady and frictionless with $pv^{1.4}$ = constant, determine the pressure at section 2.

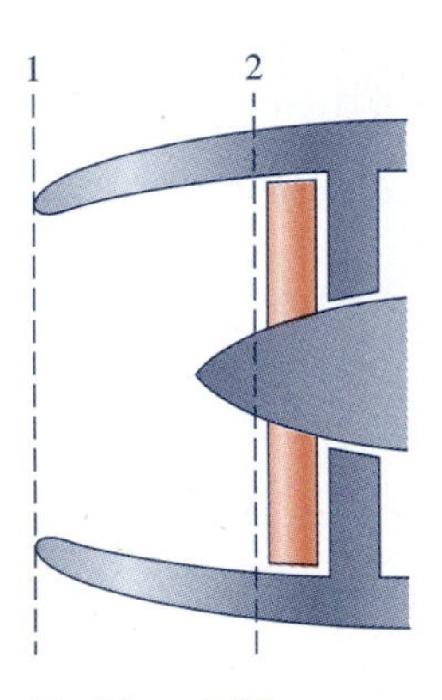

Problem 1·97

1·98 A liquid flows steadily through a horizontal converging tube that is a truncated circular cone with an included angle of θ. At the inlet, the diameter is D_1 and the pressure and average velocity are p_1 and V_1. Determine the pressure p at any section a distance L downstream from the inlet in terms of p_1, V_1, θ, L, and the liquid density.

1·99 Water is to be aerated by one of the two methods shown in the figure. Just above the surface of a pond a pipe with an inside diameter of 5.0 cm ends. Either a diverging section, with a serrated top edge that causes the overflowing water to be separated into many streams to increase the surface area as the water falls back into the pond, or a converging nozzle section will be attached to the pipe. For each case, calculate the pressure at the gage, assuming steady, quasiequilibrium flow at a rate of 6.0 kg/s. How high will the water issuing from the nozzle rise, assuming friction can be neglected?

1·100 In everyday conversation the noun *heat* and the verb *to heat* are often used with meanings different from those used in engineering thermodynamics. One type of misuse leads to confusion between an interaction between systems and a property. The other type of misuse leads to a confusion between different types of interactions between systems. Give several examples of each type of misuse.

1·101 For each process and each system listed, indicate whether the work is greater than, less than, or equal to zero, and do the same for heat, observing the usual sign convention for work and heat:

(*a*) A coil spring stands on end on a rigid surface. A weight is placed on the spring, compressing it. Consider as the system (1) the weight, (2) the spring.

(*b*) A paddle wheel turned by a motor stirs a liquid in an insulated vessel. Consider as the system (1) the liquid, (2) the paddle wheel.

(*c*) A gas in an insulated cylinder is compressed so that its pressure and temperature both increase. The system is the gas.

(*d*) A steel wire is bent back and forth until it becomes hot to the touch. The system is the wire.

(*e*) Carbon dioxide is compressed in a water-cooled compressor. Steady flow prevails. The system is (1) the carbon dioxide, (2) the cooling water, (3) a section of the compressor cylinder wall.

1·102 For each process and each system listed, indicate whether the work is greater than, less than, or equal to zero, and do the same for heat, observing the usual sign conventions for work and heat:

(*a*) An automobile stands with bright sunlight on its left side. The temperature of the air in the tires is thereby increased. The system is (1) the air in the tires, (2) the automobile.

(*b*) A household refrigerator is driven by an electric

motor. Steady-flow conditions prevail for several hours. The system is (1) the motor, (2) the entire contents of the refrigerator cabinet.

(*c*) An aircraft engine drives a propeller through a speed-reducing gearbox. Operation is steady. Power delivered to the propeller is 96 percent of the power delivered to the gearbox. The system is (1) the gearbox, (2) the propeller.

(*d*) Two people play racquetball on an enclosed court for an hour. The system is (1) the air in the room, (2) one of the people.

(*e*) A toy pinwheel turns in a breeze. The system is (1) the pinwheel, (2) the air that passes through the pinwheel.

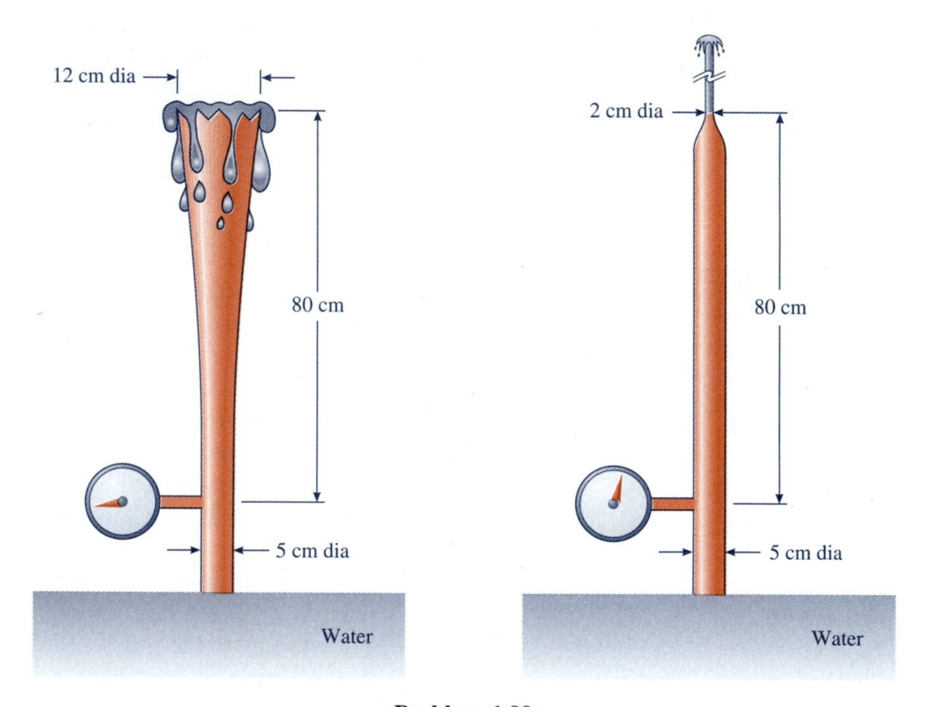

Problem 1·99

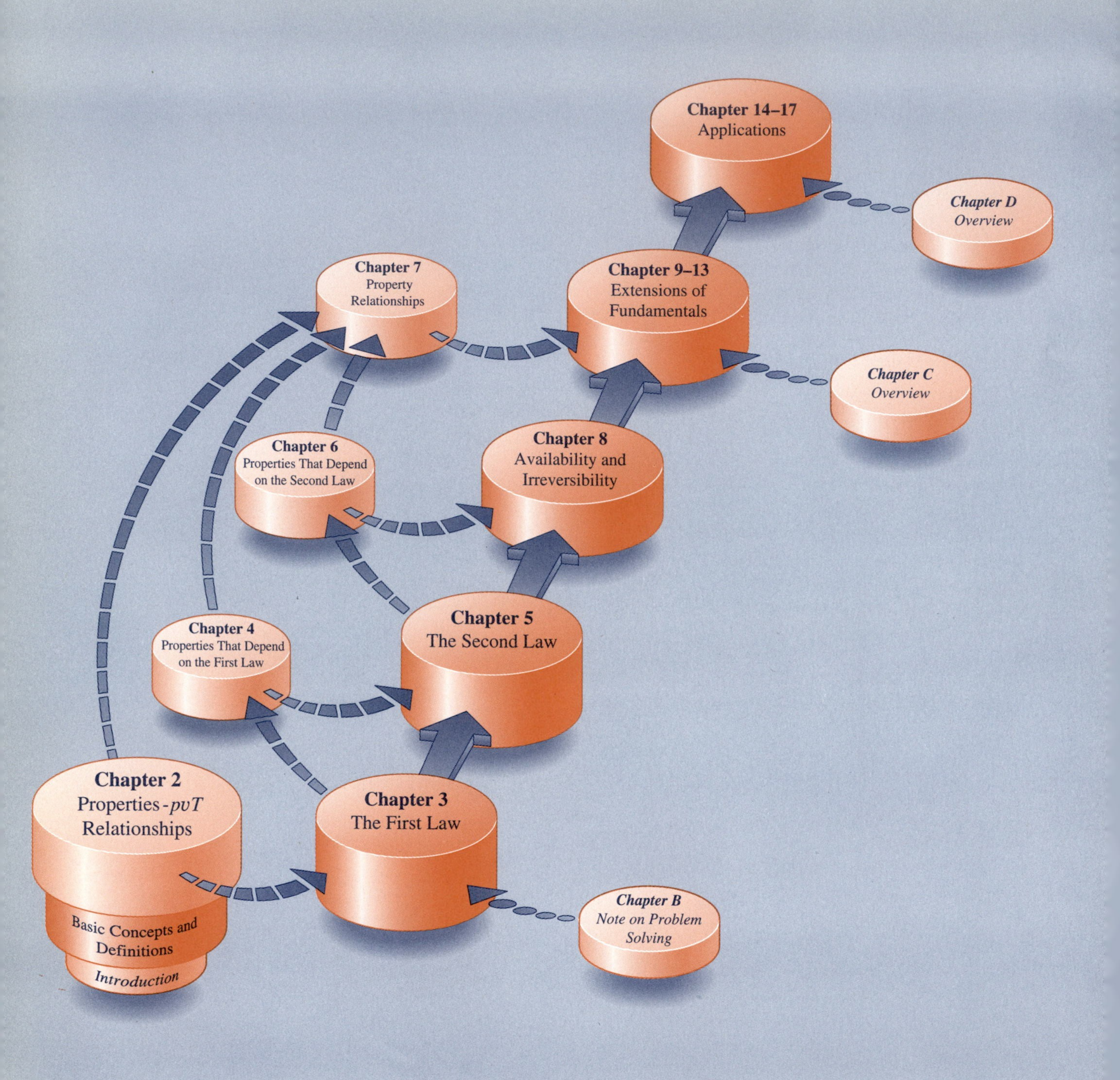

Chapter 14–17
Applications
Chapter D
Overview
Chapter 7
Property
Relationships
Chapter 9–13
Extensions of
Fundamentals
Chapter C
Overview
Chapter 6
Properties That Depend
on the Second Law
Chapter 8
Availability and
Irreversibility
Chapter 5
The Second Law
Chapter 4
Properties That Depend
on the First Law
Chapter 2
Properties-pvT
Relationships
Basic Concepts and
Definitions
Introduction
Chapter 3
The First Law
Chapter B
Note on Problem
Solving

Chapter 2

Properties—pvT Relationships

In the preceding chapter, we established several important definitions. We could now proceed to the first of the broad principles of thermodynamics, the first law, which requires no knowledge of the properties of substances. Instead, we will first present some physical property relationships that do not depend on the first law. We take this approach so that immediately after introducing the first law in Chapter 3 we can solve some problems involving more than just the first law. Our choices would be severely limited without some property data or relationships. Therefore, this first chapter on physical properties treats relationships among pressure, temperature, and specific volume (or density) for various substances.

2·1 Phases of Substances

At room conditions, copper is a solid, water is a liquid, and oxygen is a gas, yet we know that each of these substances can exist as a solid, liquid, or gas at other pressures and temperatures. That is, each one can exist in different *phases*.

Phase

A **phase** is any homogeneous part of a system that is physically distinct. Each phase in a system is separated from other phases by interfaces called *phase boundaries*. In a system consisting of an ice cube in liquid water, there are two phases, solid and liquid, and the phase boundary is the surface of the ice cube. The ice is homogeneous and the liquid is homogeneous. The pressure and temperature are the same for the solid and the liquid, but the two phases are distinct because their densities (as well as other properties) are different.

The phases we first think of are solid, liquid, and gas. A gas mixture is always a single phase, because the molecules of the constituent gases intermingle completely throughout the entire volume of the gas and make the mixture homogeneous. We cannot distinguish among the constituents. There can, however, be more than one solid or liquid phase of a substance. For example, at room temperature, sulfur exists in three solid phases known as monoclinic, rhombic, and amorphous forms. They all have the same chemical composition of pure sulfur, but they differ in crystal structure. Even at the same pressure and temperature they differ in other physical properties. You are undoubtedly also familiar with two solid forms of carbon: graphite and diamond.

A system composed of liquid water with a layer of oil on top of it would clearly involve two phases with a planar phase boundary separating the two. If the liquids are thoroughly agitated so that the oil is distributed in tiny globules throughout the water, the system is still composed of the same two phases, even though the oil phase is distributed into separate globules. The phase boundary then consists of many near-spherical bounding surfaces of the oil globules. A mixture of water and alcohol would be a single phase because these liquids are miscible and form a homogeneous mixture.

2·1·1 Molecular Model of Solids, Liquids, and Gases

Recall that classical thermodynamics is a macroscopic model of nature. In other words, the study of physical properties in classical thermodynamics does not involve considerations of molecular structure and behavior. Nevertheless, some understanding of molecular phenomena can be helpful, so, let us now consider very briefly the characteristics of solids, liquids, and gases from the molecular point of view.

In crystalline solids, the units of the crystal structure are commonly atoms or ions rather than molecules, although in certain types of crystals molecules maintain their identity. In this discussion of solids, however, we use the word molecule to simplify comparison with liquids and gases. The molecules of a solid are very close together and are arranged in a three-dimensional pattern that is repeated with minor irregularities throughout the solid. In noncrystalline solids, such as glasses, the immediate neighbors of any molecule are generally arranged in a fixed pattern, but the regularity of the pattern does not extend throughout the solid. Crystalline solids possess long-range order as well as short-range order, whereas noncrystalline solids possess only short-range order. In all solids the molecules are so close together that the forces of molecules on each other are large. (These are repulsive forces at very close spacing and attractive forces when there is a greater separation between molecules.) These forces tend to keep the molecules in fixed positions relative to each other and give solids their resistance to deformation. The molecules of a solid continually oscillate or vibrate about their equilibrium positions, but the amplitude of this motion is small, so that a molecule is never far from its equilibrium position. The velocity of the molecules during the oscillation depends on the temperature of the solid: the higher the temperature, the higher the velocity. As the temperature of a solid is increased, the molecular velocities increase until the attractive forces are partially overcome. Then groups of molecules shift position relative to other such groups and the solid becomes a liquid.

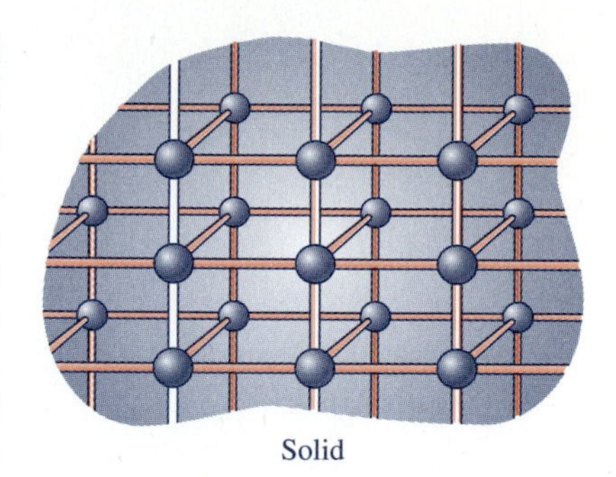

Solid

In a liquid, the molecular spacing is of the same order of magnitude as in a solid. The molecules are not held in relatively fixed positions throughout the entire phase, but small groups of molecules retain a semblance of crystalline order. That is, each molecule retains an orientation with respect to some of its neighbors that it had in the solid phase. These small groups of molecules in which remnants of crystalline structure persist are not firmly held in position relative to neighboring groups. For this reason, a liquid in static equilibrium cannot withstand a shearing stress.

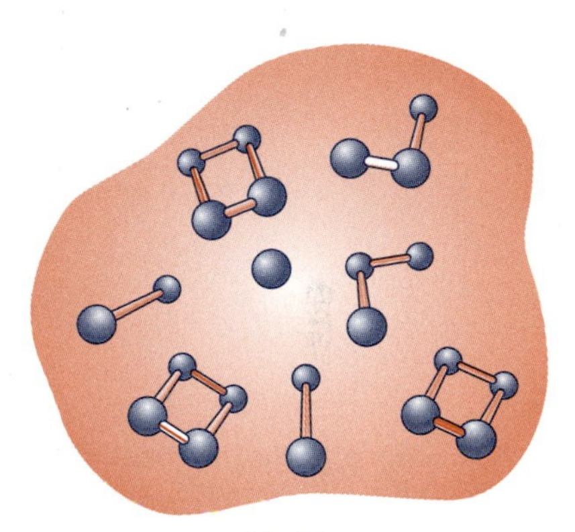

Liquid

The distances between molecules in a liquid are generally slightly greater than the distances between molecules in the solid phase of the same substance. A notable exception is water. Water, unlike most substances, expands on freezing, so the molecular spacing is slightly greater in the solid phase than in the liquid phase.

A gas is composed of molecules that are relatively far apart. There is no regularity or permanence in their arrangement in space, for gas molecules are continually in motion, colliding with each other and rebounding to travel in new directions. For gases of low density, collisions are the only interactions of consequence between molecules, because they are so widely separated that intermolecular forces are small. For gases of high density, intermolecular forces may be significant.

In air at normal room conditions, there is a large range of molecular velocities, but an average value is about 500 m/s. On the average, a molecule of oxygen or nitrogen in such air travels about 0.00006 mm between collisions and experiences about 8 billion collisions per second. One cubic centimeter

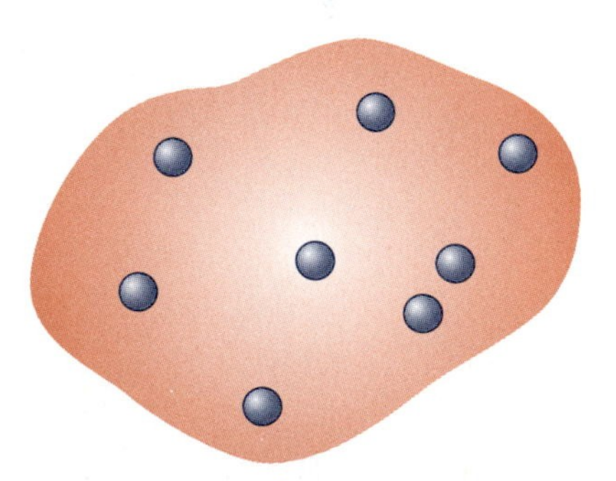

Gas

of normal room air contains about 2.5×10^{19} molecules; but, if this gives the impression that the molecules are jammed together, you should note that the molecules themselves occupy only about 1/1000 of the volume that the gas fills.

2·1·2 Pure Substances

Pure substance

A **pure substance** is a substance that is chemically homogeneous and fixed in chemical composition. A system composed of liquid and vapor phases of H_2O is a pure substance. Even if some of the liquid is vaporized during a process, the system will still be chemically homogeneous and unchanged in chemical composition. It is not *physically* homogeneous, of course, because the liquid density is usually much greater than the vapor density.

A mixture may be a pure substance. For example, air is a mixture of (principally) oxygen and nitrogen. In its gaseous phase it may be heated, cooled, compressed, or expanded with no change in its chemical composition. If, however, it is cooled until part of it liquefies, it is no longer a pure substance, because the liquid will contain a higher fraction of nitrogen than the original mixture does. Atmospheric air also contains water vapor, and if the air is cooled to the extent that some of the water vapor condenses, it is no longer a pure substance: the chemical composition of the condensate is different from that of the still gaseous part of the system. Additionally, after some moisture is removed, the gaseous part has a chemical composition slightly different from its original composition. A mixture of oxygen and carbon monoxide is a pure substance as long as it remains fixed in composition. If some of the CO combines with some of the O_2 to form CO_2, the system does not involve a pure substance during the process because its chemical composition is changing. A system cannot be treated as a pure substance during any process that involves a chemical reaction.

A substance cannot be treated as pure if it undergoes a chemical reaction.

The state of a pure substance at rest is usually specified completely by the values of two independent intensive properties, provided that there are no electric, magnetic, solid distortion, or surface tension effects. For example, if the pressure and temperature of air are specified, then the values of all other properties of the air, such as density, viscosity, and thermal conductivity, are fixed. If the pressure and density of air are specified, then the temperature and all other properties have fixed values. Care must be exercised to see that the two intensive properties selected to specify a state are *independent*. Two properties are independent if either one can be varied throughout a range of values while the other remains constant.

2·1·3 Equilibrium of Phases of a Pure Substance

For any one phase of a substance, the pressure and temperature can be varied independently over wide ranges and hence are independent properties. However, *when two phases of a pure substance coexist in equilibrium, there is a fixed relationship between their pressure and temperature.* At any given pressure there is but one temperature at which two specific phases will exist together in equilibrium. Conversely, the two phases can coexist in equilibrium at a given temperature only at a particular pressure. In this case, pressure and temperature are not independent properties. Specifying both will *not* determine the state of the system. More information is required.

Multiple Phases. The conditions under which two or more phases of a pure substance can exist together in equilibrium are called the **saturation conditions**. The pressure and temperature are called the **saturation pressure** and the **saturation temperature**. It is important to note that specifying the pressure and the condition of saturation determines the temperature, and vice versa. Any phase of a substance existing under such conditions is called a **saturated phase**. For example, liquid water and water vapor in equilibrium with each other are spoken of as *saturated liquid* and *saturated vapor*. Ice in equilibrium with liquid or vapor or both is spoken of as a *saturated solid*.

Saturation conditions

Saturation pressure and temperature

Saturated phase

The presence of two or more phases is not required in order to have a saturated phase. A phase is saturated, even if it exists alone, if it is at a pressure and a temperature under which two or more phases *could* exist together in equilibrium. For example, liquid water at its boiling temperature is a saturated phase.

As an illustration of the equilibrium of two phases of a pure substance, consider the familiar substance *water*. Liquid water (a single phase) can exist in equilibrium under a pressure of 1 atm at any temperature ranging from 0°C to 100°C. However, it cannot exist at a temperature higher than 100°C while under this pressure. Water vapor (a single phase) can exist under a pressure of 1 atm only at temperatures of 100°C and higher. The only temperature at which either the liquid phase or the vapor phase or both can exist under a pressure of 1 atm is 100°C. Therefore, 100°C is the saturation temperature of water at 1 atm (101.325 kPa). Likewise, 1 atm is said to be the saturation pressure of water at 100°C. If the temperature of liquid in a system composed only of water is slowly increased while the pressure is held constant at 1 atm, no vapor will form until the temperature reaches 100°C. Then, no matter how much vapor is formed, as long as both liquid and vapor are present in equilibrium and the pressure is held constant, the temperature will be 100°C. Not until all of the liquid has changed to vapor can the temperature rise above the saturation temperature of 100°C. If the pressure is not 101.325 kPa, then vaporization will occur at a different temperature. For liquid–vapor equilibrium of any substance, the higher the pressure, the higher the saturation temperature.

The relationship between liquid–vapor saturation pressures and temperatures is shown for several substances in Fig. 2·1. These curves are called liquid–vapor saturation curves, vaporization curves, or condensation curves. The upper and lower ends of these curves, known respectively as the critical states and the triple-phase states of the substances, will be discussed later.

The temperature at which *any* two phases of a pure substance can coexist in equilibrium depends on the pressure. This applies to solid–liquid and

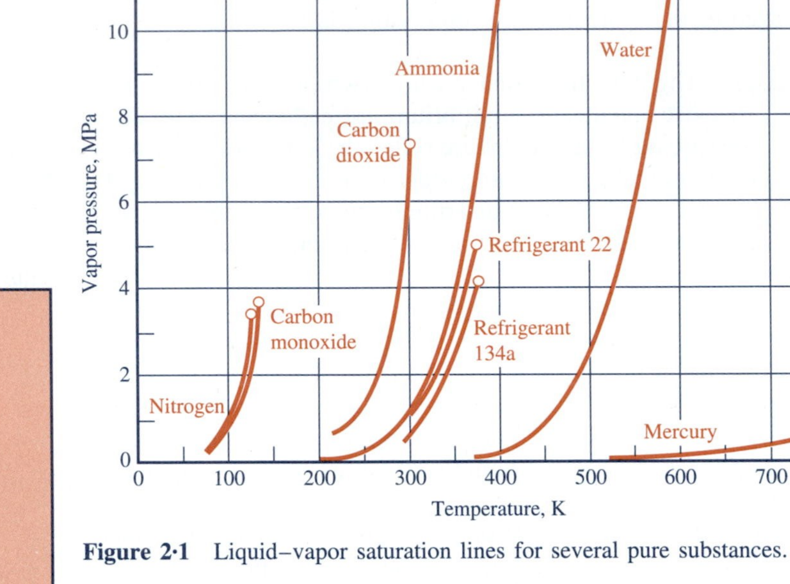

Figure 2·1 Liquid–vapor saturation lines for several pure substances.

Figure 2·2 Solid–liquid saturation line (or melting or freezing line) for a substance that contracts on freezing.

solid–vapor equilibrium as well as to the liquid–vapor equilibrium already discussed. Thus the melting or freezing temperature of a substance depends on the pressure, although the variation of this temperature with pressure is usually small. For most substances, the freezing temperature increases as the pressure increases, so that the solid–liquid saturation curve or melting line appears as in Fig. 2·2. For water and other substances that expand on freezing, increasing the pressure lowers the freezing temperature so the melting (or freezing) line appears as in Fig. 2·3. A familiar demonstration of this phenomenon involves resting a wire weighted at both ends across a block of ice. The increased pressure caused by the wire lowers the freezing point slightly, causing the ice directly below the wire to melt. The liquid water formed passes above the wire and freezes because the pressure on it is again atmospheric. In this manner it is possible to pass a wire completely through a block of ice.

Figure 2·4 shows pressure versus freezing temperature for water. At any point on the line, the liquid and solid may exist together in equilibrium. The area to the left of the line represents the solid, and that to the right the liquid. For example, at a pressure of 40 MPa and a temperature of $-5°C$ water can exist only in the solid form, ice. At 40 MPa and $-2°C$, only liquid water may exist.

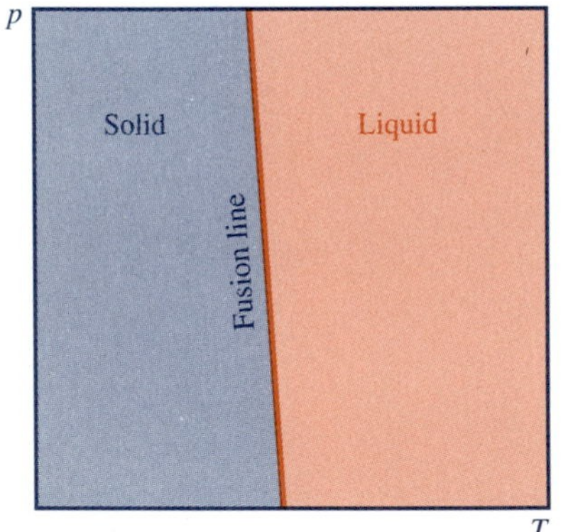

Figure 2·3 Solid–liquid saturation line (or melting or freezing line) for a substance that expands on freezing.

For most substances, there is some pressure below which liquid cannot exist. Below this pressure the solid and vapor phases can coexist in equilibrium, and the relation between pressure and temperature for these solid–

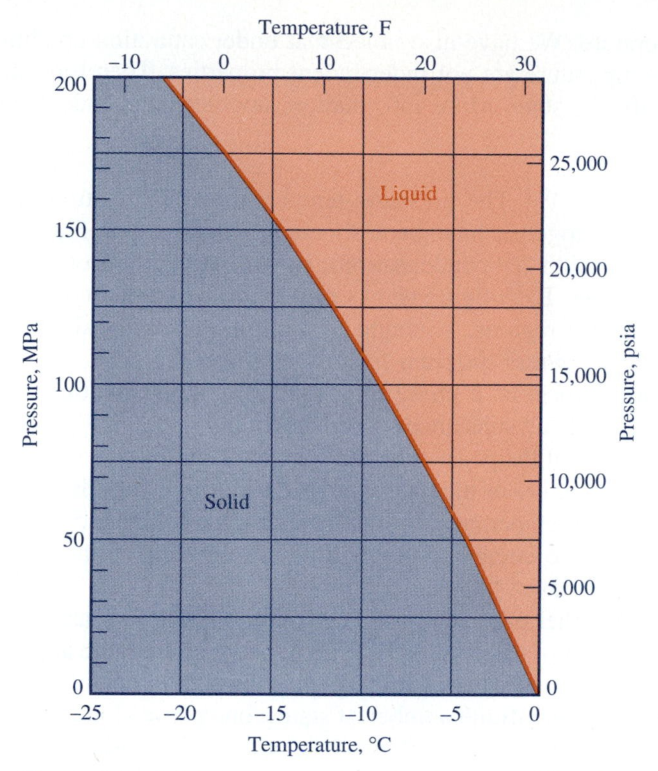

Figure 2·4 Solid–liquid saturation line for water. (Data from N. E. Dorsey, *Properties of Ordinary Water Substance*, Reinhold, New York, 1940.)

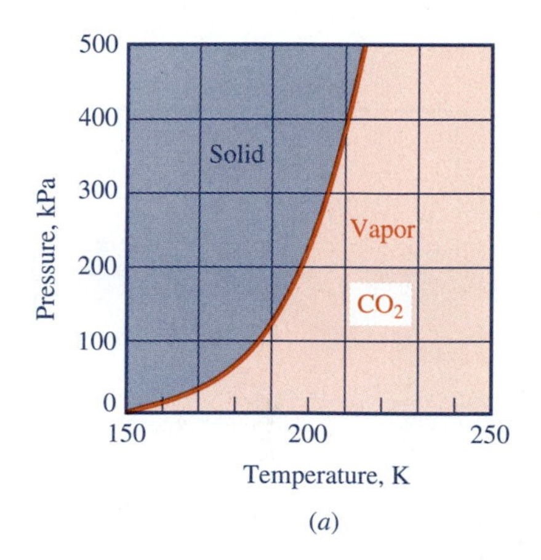

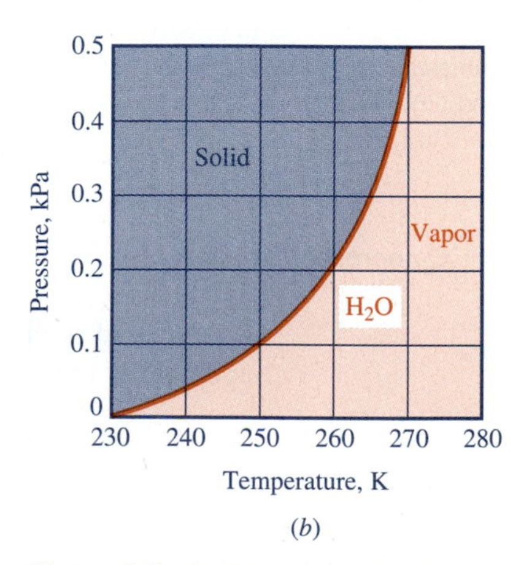

Figure 2·5 Solid–vapor saturation lines (sublimation lines) for (*a*) carbon dioxide and (*b*) water. Notice that the pressure ranges differ by a factor of 1000.

vapor saturation states is fixed for each substance. The transformation from a solid to a vapor is known as **sublimation**; so the solid–vapor saturation line on a pT diagram is often called a **sublimation line**. At any combination of pressure and temperature on this line, saturated solid, saturated vapor, or a mixture of the two can exist. For all substances, the higher the pressure, the higher the sublimation temperature. Figure 2·5 shows sublimation lines for carbon dioxide and water.

A familiar example of sublimation is the transformation of solid carbon dioxide (dry ice) to a gas instead of to a liquid. As another example, ice disappears from a sidewalk by the process of sublimation while the temperature remains well below 0°C. (In each of these cases, the presence of atmospheric air means that we are not dealing with phase equilibrium of a pure substance, but the quantitative differences are small.)

We have seen that different phases of a pure substance can exist together in equilibrium only under certain conditions. We have discussed liquid–vapor, solid–liquid, and solid–vapor equilibrium and have noted that each of these pairs of phases can exist together only at certain combinations of pressure

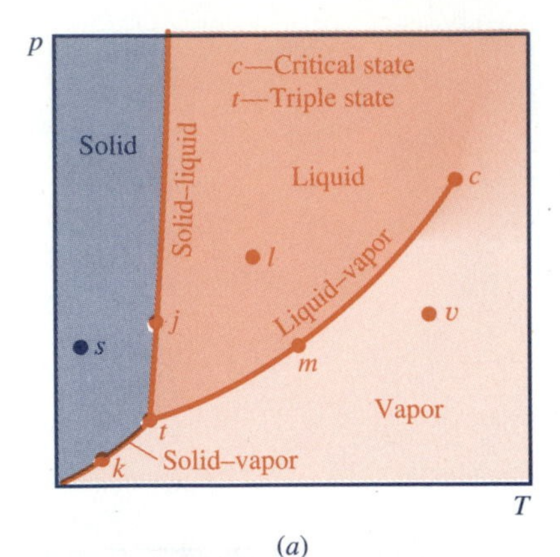

(a)

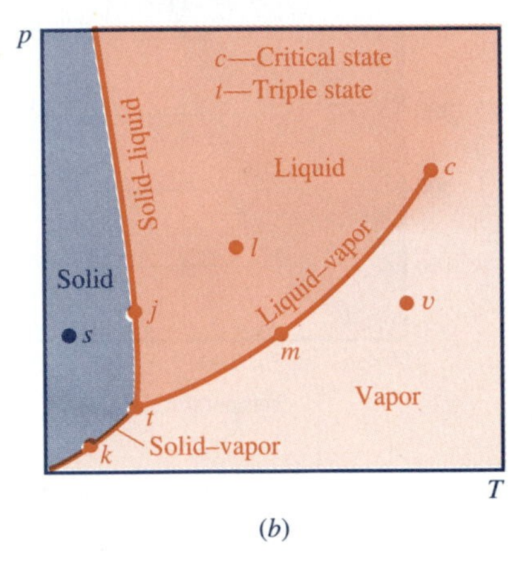

(b)

Figure 2·6 Phase diagrams for (*a*) a substance that contracts on freezing and (*b*) a substance that expands on freezing.

and temperature. We have also noted that under saturation conditions, pressure and temperature are not independent properties. Therefore, they do not fully specify the state. More information, another independent property, is needed.

Phase Diagrams. The pressure–temperature (pT) relationships for two phases existing together have been shown schematically in Figs. 2·1 through 2·5 as vaporization (or condensation), melting (or fusion or freezing), and sublimation lines. Each figure shows one of these two-phase lines separating two single-phase regions. A single pT diagram showing all three of these lines is called a *phase diagram* for a pure substance.

The general form of a phase diagram for a substance that contracts on freezing, as nearly all substances do, is shown in Fig. 2·6*a*. Figure 2·6*b* shows the general form of the phase diagram for a substance like water that expands on freezing. On either of the phase diagrams in Fig. 2·6, each *area* represents the pressure and temperature ranges in which a single phase can exist. Each *line* represents the combinations of pressure and temperature under which two phases may exist together.

In Fig. 2·6, either part *a* or part *b*, each of the points l, v, and s represents a single state. Each of these states is a single phase so pressure and temperature are independent variables and fully specify the state. Each of the points m, j, and k represents an infinite number of states: one phase alone, the other phase alone, or any ratio of the two phases in a mixture. This illustrates that whenever two phases exist in equilibrium, p and T are not independent so are insufficient to fully specify the state. The state of *each phase* within the mixture, however, is specified by either p or T and the knowledge that another phase is or could be present.

Compressed or subcooled liquid

Liquid existing at a temperature lower than its saturation temperature, state d in Fig. 2·7*a*, or, in other words, at a pressure higher than its saturation pressure, as shown in Fig 2·7*b*, is called a **compressed liquid** or a **subcooled liquid** to distinguish it from a saturated liquid. Thus all points in the liquid region that are not on the liquid–vapor or liquid–solid saturation line represent states of compressed liquid. Water flowing from a drinking fountain and mercury in an open cup at room temperature are both compressed liquids. In a like manner, a solid at a temperature below its saturation temperature is called a **compressed solid**, and a vapor at a temperature above its saturation temperature is called a **superheated vapor**. These modifiers (compressed, subcooled, superheated) are generally used only when their omission might reasonably mislead a reader into inferring the existence of saturation conditions.

Compressed solid
Superheated vapor

The triple point: where all three phases of a substance coexist

The intersection of the vaporization line, melting line, and sublimation line of a phase (pT) diagram represents the conditions under which three phases can coexist in equilibrium and is called the **triple-phase state**, **triple-phase point**, or **triple point**. These conditions are represented by a point only on a pressure–temperature diagram; on other property diagrams they are represented by a line or an area. If a substance can exist in more than three phases, it has more than one triple point. For example, at high pressures several solid phases of water different from common ice have been observed, so water has several (at least seven, in fact) triple points. References to *the* triple point of water always pertain to the state at which liquid, solid, and vapor can exist. Triple-point data for several substances are given in Table 2·1, and more data are given in Appendix Table A·1. Phase diagrams for carbon dioxide and water in the vicinity of their triple points are shown in Figs. 2·8 and 2·9.

If three phases of a pure substance exist together in a system, only one value of pressure and only one value of temperature are possible. The state of *each phase* present in the system is fixed. However, to determine the state *of the system* at the triple-point pressure and temperature, more information is required, namely, the proportions of the phases present. For example, if only 1 percent of the mass is in the vapor phase, the specific volume of the system is much less than if 20 percent of the mass is vapor.

In a similar manner, each point on the vaporization (or condensation) line fully defines the state of *each phase* present. However, the state of the *system* is not determined because each point represents an infinite number of mixture states ranging from all liquid and no vapor to no liquid and all vapor. Consequently, each point on the vaporization line on a pT diagram represents states with a range of mixture specific volumes.

The critical point: where the liquid and vapor phases of a substance are indistinguishable

As pressure and temperature increase, the specific volume of saturated liquid increases and that of saturated vapor decreases, so that at the end of the vaporization line, point c on Fig. 2·6, they are equal, and liquid and vapor are indistinguishable from each other. Point c is called the **critical state** or **critical point**. The **critical pressure** and the **critical temperature** are the highest

Critical pressure and temperature

Table 2·1 Triple-Point and Critical-Point Data

	Triple Point				*Critical Point*			
	Pressure		*Temperature*		*Pressure*		*Temperature*	
Substance	kPa	psia	°C	F	MPa	psia	°C	F
Ammonia	6.1	0.89	−78	−108	11.3	1640	132	270
Carbon dioxide	517	75	−57	− 70	7.39	1071	31	88
Helium	5.1	0.75	−271	−456	0.23	33	−268	−450
Hydrogen	7.0	1.0	−259	−434	1.3	188	−213	−400
Mercury	0.165×10^{-6}	23.9×10^{-9}	−39	−38	18.2	2650	899	1650
Nitrogen	12.5	1.81	−210	−346	3.4	493	−147	−233
Oxygen	0.15	0.022	−219	−361	5.0	731	−119	−182
Refrigerant 22					4.99	723	96	205
Refrigerant 134a					4.07	590	101	214
Water	0.611	0.0886	0.01	32.02	22.1	3206	374	705

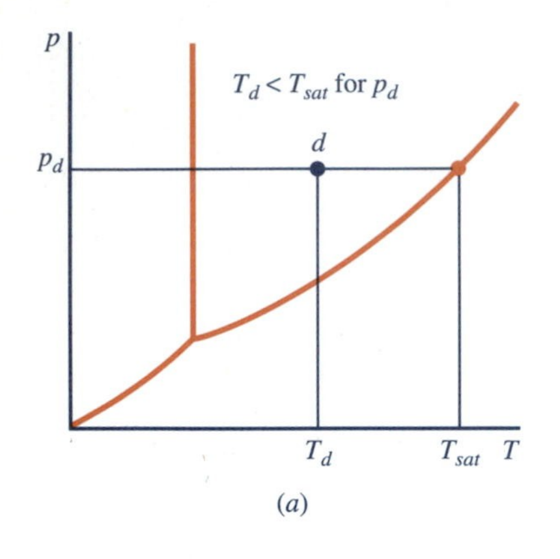

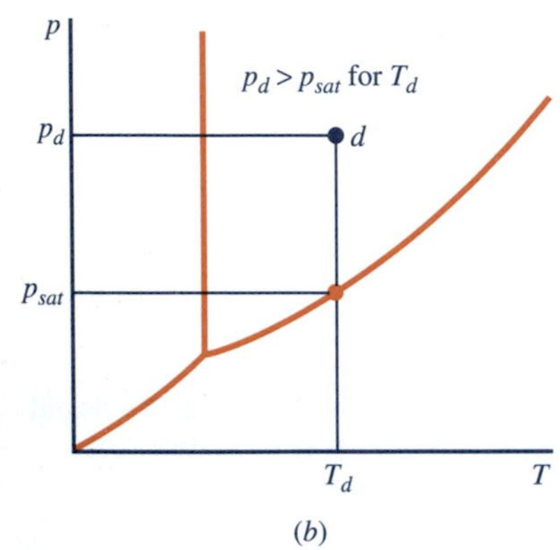

Figure 2·7 A compressed liquid state described in two ways: (*a*) Liquid at a temperature lower than its saturation temperature and (*b*) liquid at a pressure greater than its saturation pressure.

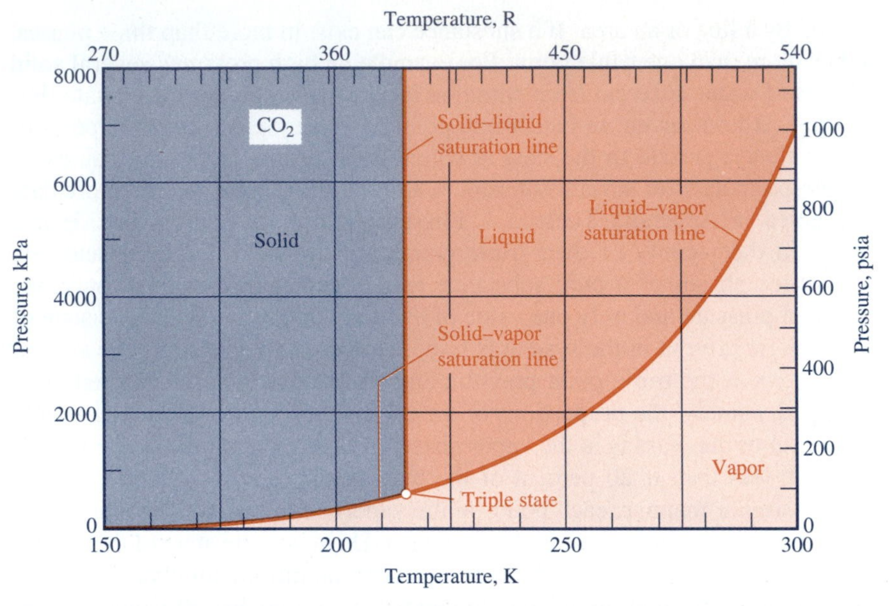

Figure 2·8 Phase diagram for carbon dioxide in the vicinity of the triple point.

pressure and temperature at which distinguishable liquid and vapor phases can exist together in equilibrium. We will return to a discussion of the critical point later in this section. Table 2·1 gives critical point data for a few substances, and more data are provided in Appendix Table A·1.

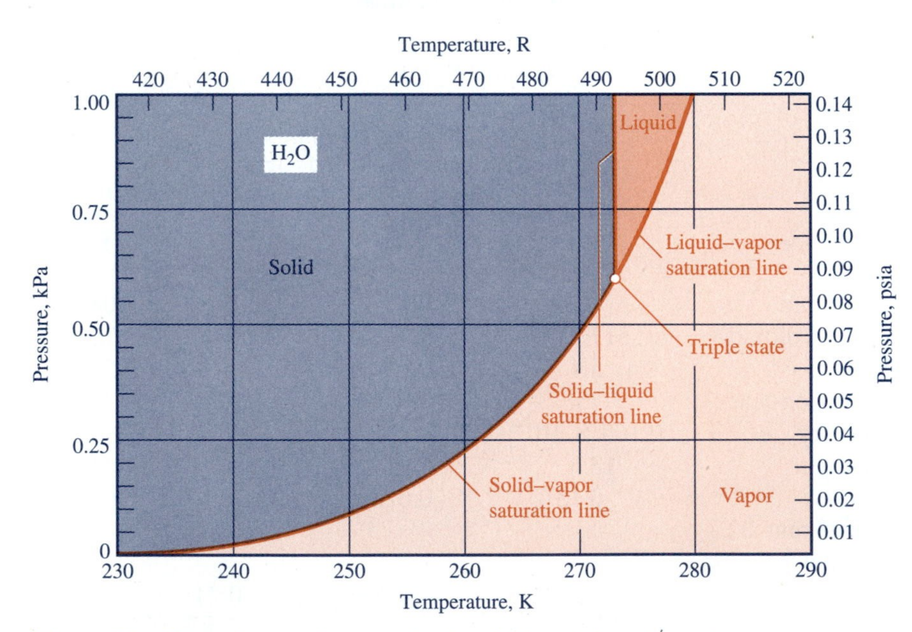

Figure 2·9 Phase diagram for water in the vicinity of the triple point.

The pT diagrams show the conditions under which a single phase, two phases, or three phases of a pure substance can exist in equilibrium. Other property diagrams are useful for other purposes.

Other Property Diagrams. A pressure–specific volume (pv) diagram of a pure substance that contracts on freezing is shown in Fig. 2·10*a* beside a pT diagram to the same pressure scale, Fig 2·10*b*. All the regions shown on the pv diagram appear also on the pT diagram. Line a–c, called the **saturated liquid line**, is a plot of the specific volume of saturated liquid versus pressure. As the pressure increases (and consequently the saturation temperature increases), the specific volume of saturated liquid increases slightly. The volume scale of Fig. 2·10*a* has been distorted to magnify this increase. The region immediately to the left of the saturated liquid line represents compressed or subcooled liquid states.

Saturated liquid line

The line c–l, called the **saturated vapor line**, is a plot of the specific volume of saturated vapor versus pressure. As the pressure increases, the specific volume of saturated vapor decreases. The region immediately to the right of the saturated vapor line represents superheated vapor states.

Saturated vapor line

Solid, liquid, and vapor states are represented by points in three different regions on a pv diagram just as they are on a pT diagram. However, mixtures of two phases are represented by *areas* on a pv diagram while they are represented by *lines* on a pT diagram. Mixtures of three phases are represented by a line on a pv diagram and by a *point* on a pT diagram. We will

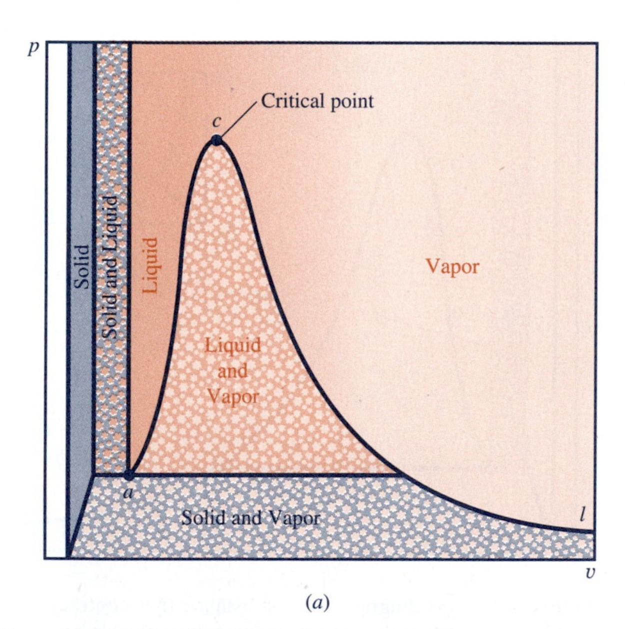

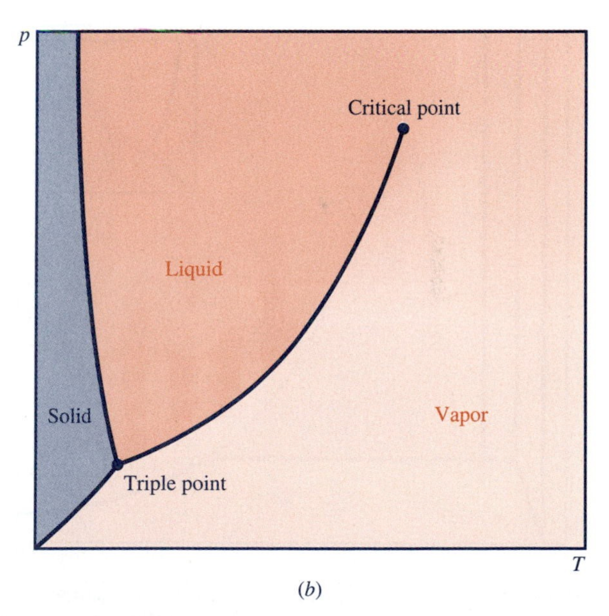

Figure 2·10 (*a*) pv diagram (with v scale distorted) and (*b*) pT diagram, both for a substance that contracts on freezing.

now look at various parts of the *pv* diagram, starting with liquid and vapor regions as highlighted in Fig. 2·11.

The area beneath the dome formed by the saturated liquid line *a–c* and the part of the saturated vapor line *c–l* down to line *a–b* represents *mixtures* of liquid and vapor. All the states represented by this *area* on the *pv* diagram are represented by the *line* on a *pT* diagram between the triple point and the critical point.

In Fig. 2·12, saturated solid states at pressures higher than the triple-phase pressure are represented by points along the line *i–m*, called the saturated solid line. Liquid at the freezing temperature is represented by points along the line *a–n*. In accordance with the definition of saturation conditions, liquid at the freezing temperature could be called saturated liquid except that this would lead to confusion with liquid at the boiling temperature. Hence the line *a–n* is called the freezing liquid line. Mixtures of solid and liquid are represented by points in the region bounded by lines *a–n*, *i–m*, and *i–a*. All points within this region and on its boundaries on the *pv* diagram lie on the freezing line on the *pT* diagram. Line *o–w* represents states of minimum specific volume for the solid at various pressures, that is, the solid can be compressed no further.

Below the triple-phase pressure, liquid cannot exist. Figure 2·13 highlights the region beneath line *o–i–a–b–l* that represents mixtures of solid and vapor. The states in this region of a *pv* diagram lie on the sublimation line of a *pT* diagram. Line *o–i* represents saturated solid states and line *b–l* repre-

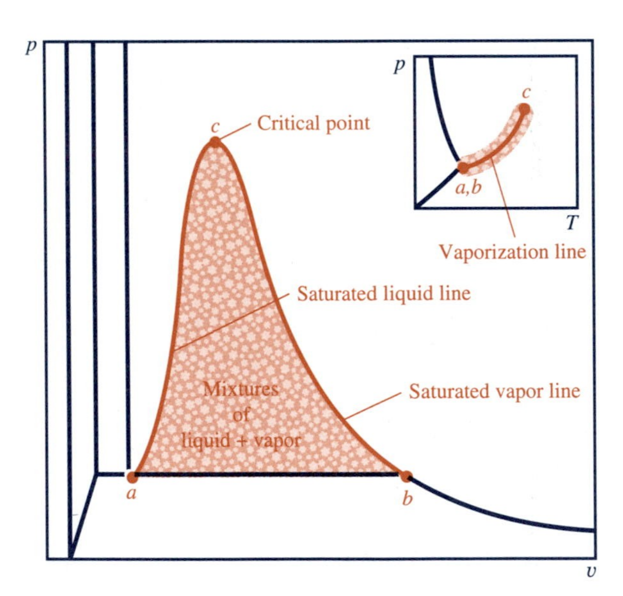

Figure 2·11 *pv* diagram for a substance that contracts on freezing, with liquid–vapor region highlighted.

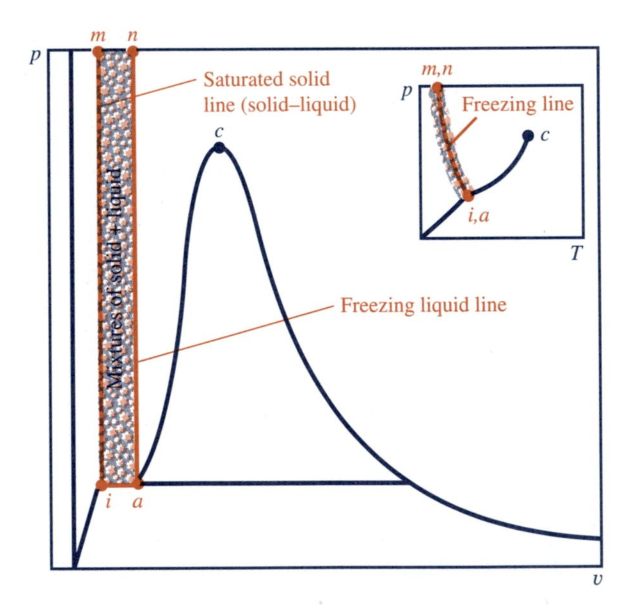

Figure 2·12 *pv* diagram for a substance that contracts on freezing, with solid–liquid region highlighted.

sents saturated vapor states. The line i–a–b on the pv diagram represents the states in which liquid, solid, and vapor can exist together. Notice that the triple state, which is represented by a point on a pT diagram, is represented by a line on a pv diagram. On some other property diagrams, it is represented by an area.

Saturated liquid at the triple-phase pressure is represented by point a in Fig. 2·13, and point b represents saturated vapor. The saturated solid at the same pressure (and temperature) is represented by point i. Points on the triple-phase line between i and a represent mixtures of solid and liquid, solid and vapor, or solid–liquid–vapor mixtures. Points between a and b represent solid–vapor, liquid–vapor, or solid–liquid–vapor mixtures.

The broken lines in Fig. 2·14 are lines of constant temperature. In all two-phase regions constant-temperature lines coincide with constant-pressure lines.

It must be kept in mind that in each pv diagram shown here the specific volume scale is greatly distorted. For all substances the change in volume between points a and b on Fig. 2·13 is many times greater than that between i and a; and the critical specific volume is only slightly greater than v_a or v_i. To emphasize this point, Figs. 2·15 and 2·16 show with true linear scales partial pv diagrams for carbon dioxide and water.

Two-dimensional property diagrams suffice for many thermodynamic analyses, but a better picture of the relationship among pressure, specific volume, and temperature of a substance is given by a three-dimensional pvT

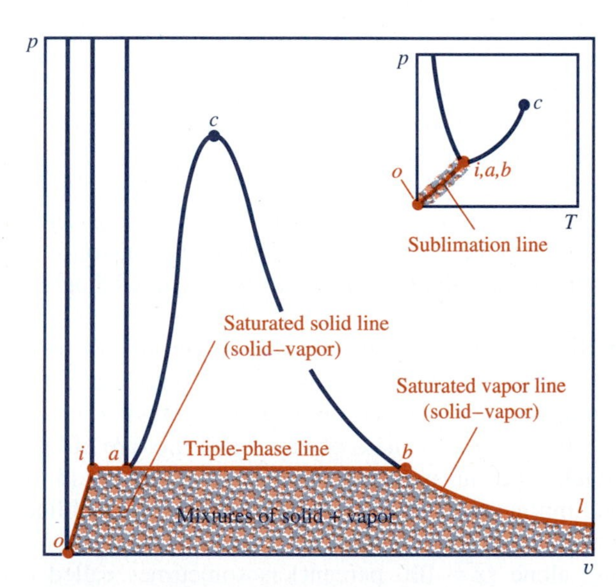

Figure 2·13 *pv* diagram for a substance that contracts on freezing, with solid–vapor region highlighted.

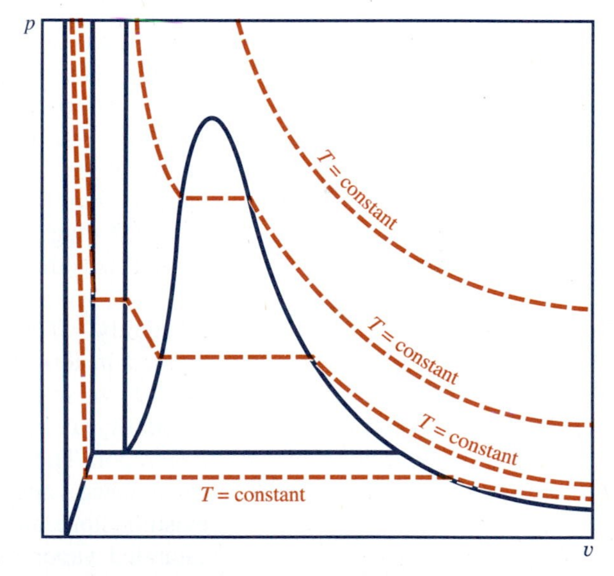

Figure 2·14 *pv* diagram for a substance that contracts on freezing, showing constant-temperature lines.

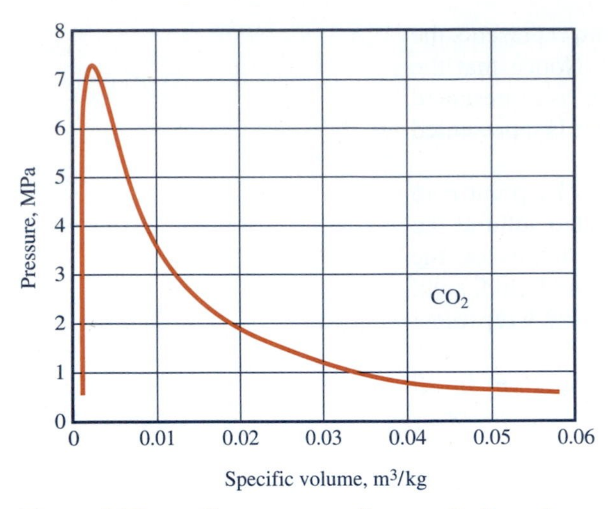

Figure 2·15 *pv* diagram to true linear scale for carbon dioxide.

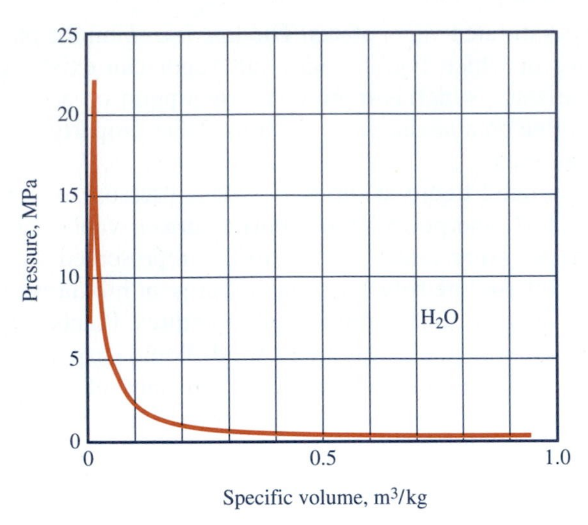

Figure 2·16 *pv* diagram to true linear scale for water.

diagram like Fig. 2·17. All equilibrium states of the substance are represented by points on the *pvT* surface. Figure 2·17 shows the *pv* and *pT* diagrams, which have been discussed above, as well as a *Tv* diagram, as projections of the *pvT* surface. Points representing mixtures of two phases all lie in surfaces that have elements perpendicular to the *pT* plane. That is why two-phase regions, which are areas on the *pv* diagram, project into lines on the *pT* diagram. Various surfaces other than the *pvT* surface are useful in some types of thermodynamic analyses.

For a substance that expands on freezing, various regions of the *pvT* diagram overlap on a *pv* projection. Inspection of Figs. 2·18 and 2·19 reveals that point *r* of the *pv* diagram can represent solid (r_1), solid–liquid (r_2), or a liquid–vapor mixture (r_3). Point *s* can represent a solid, a solid–liquid mixture, or a liquid. However, these regions and states are unambiguously shown on a *pvT* diagram, since each point on the *pvT* surface represents only one state. This is the advantage of a *pvT* diagram over two-dimensional diagrams.

Quality is the fraction by mass of vapor in a mixture of liquid and vapor.

Quality. The properties of a mixture of two phases can always be expressed in terms of the properties of the two individual phases and their proportions in the mixture. Their proportions are usually expressed in terms of **quality**, x, defined as the fraction by mass of vapor in a mixture of liquid and vapor. The term quality has no meaning for a compressed liquid or a superheated vapor. The limiting values of quality are 0 for a saturated liquid existing alone and 1.0 or 100 percent for a saturated vapor existing alone. A saturated vapor existing alone ($x = 100$ percent) is sometimes called *dry saturated vapor*. A mixture of liquid and vapor is called *wet vapor*.

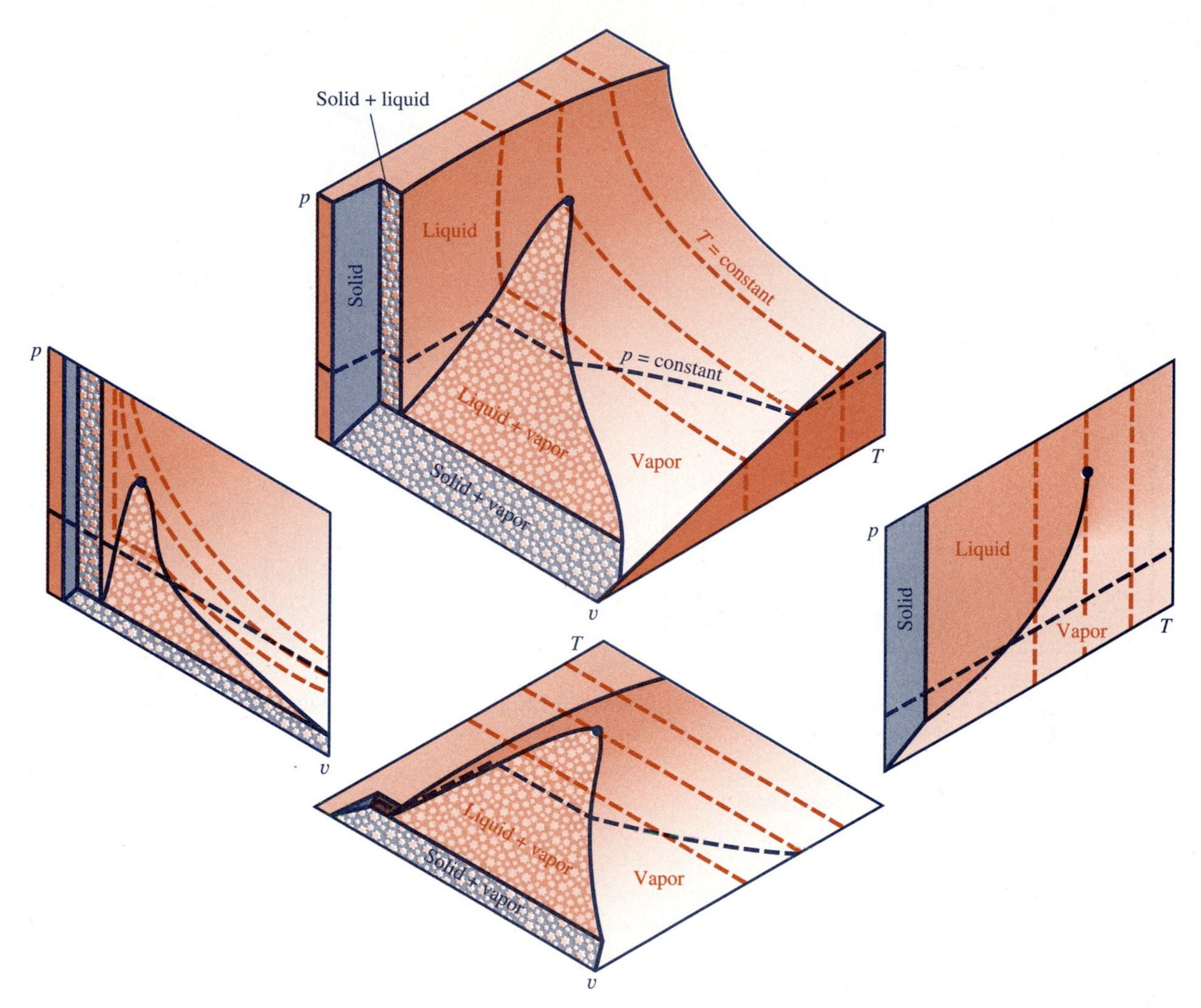

Figure 2·17 *pvT* surface for a substance that contracts on freezing (v scale distorted).

Consider a liquid–vapor mixture, state z, at the pressure of states f and g in Fig. 2·20. The liquid in the mixture is represented by point f; its specific volume is v_f. The vapor in the mixture is represented by point g; its specific volume is v_g. The mixture has a specific volume v_z that is greater than v_f but less than v_g, and v_z can be expressed in terms of v_f, v_g, and x. The volume of the mixture is the sum of the volume of the liquid and of the vapor because the liquid, although it may be dispersed in very small drops, occupies a volume from which the vapor is excluded. Thus, letting m_L and m_V denote the mass of liquid and of vapor, respectively, the specific volume of the

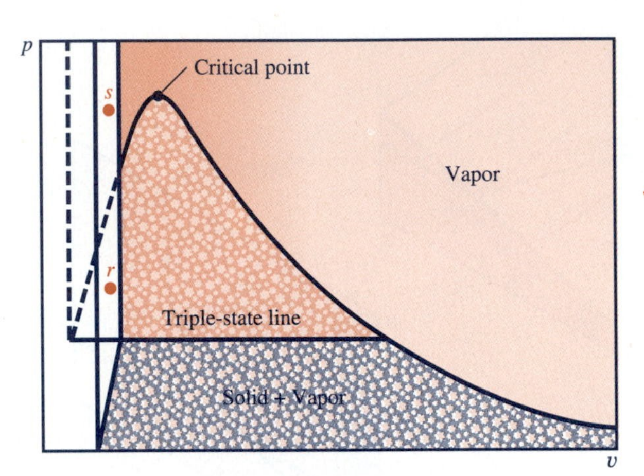

Figure 2·18 *pv* diagram for a substance that, like water, expands on freezing (*v* scale distorted).

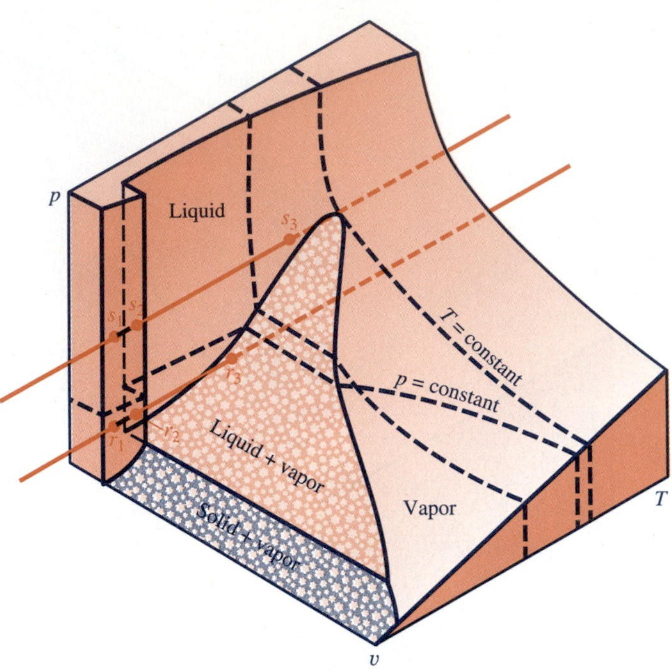

Figure 2·19 *pvT* surface for a substance that, like water, expands on freezing (*v* scale distorted).

mixture is

$$v_z = \frac{V}{m} = \frac{V_L + V_V}{m_L + m_V} = \frac{m_L v_f + m_V v_g}{m_L + m_V}$$

$$= \frac{m_L}{m_L + m_V} v_f + \frac{m_v}{m_L + m_V} v_g$$

Using the definition of quality x,

$$x \equiv \frac{m_V}{m_L + m_V} \quad \text{and} \quad \frac{m_L}{m_L + m_V} = 1 - x$$

gives

$$v_z = (1 - x_z)v_f + x_z v_g \tag{2·1a}$$

$$v_z = v_f + x_z(v_g - v_f) \tag{2·1b}$$

The difference $(v_g - v_f)$ is denoted by the symbol v_{fg}, so that

$$v_z = v_f + x_z v_{fg} \tag{2·1c}$$

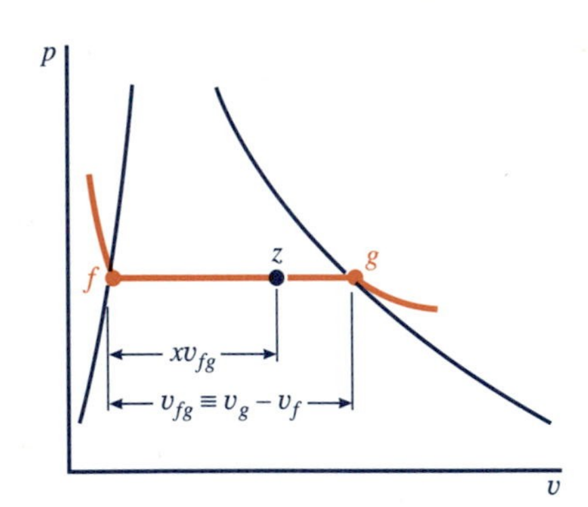

Figure 2·20 Defining quality of a liquid–vapor mixture.

Therefore, when $x = 50$ percent, the point representing the state of the mixture is midway between points f and g; when $x = 20$ percent, the point is one fifth of the distance along f–g from point f, and so forth. A physical interpretation of Eq. 2·1*b* or 2·1*c* is that the volume of 1 kg of mixture is the volume of 1 kg of saturated liquid plus the increase in volume during the vaporization of x kilograms of substance.

The Critical State. The critical state, the limiting state for the existence of saturated liquid and vapor, has already been introduced. The *critical pressure* is the highest pressure under which distinguishable liquid and vapor phases can exist in equilibrium. The *critical temperature* is the highest temperature at which distinguishable liquid and vapor phases can exist in equilibrium. As the critical state is approached from lower pressures and temperatures, the properties of saturated liquid and saturated vapor approach each other. Therefore, properties such as v_{fg} reach the limiting value of zero at the critical point.

Since the properties, especially density, of liquid and vapor phases of a substance under most conditions are so different, it is sometimes difficult to visualize conditions under which the two phases become indistinguishable from each other. It is often helpful to consider the following experiment.

If a rigid transparent vessel is filled with a liquid–vapor mixture of a substance in a state represented by point a on Fig. 2·21, all the liquid will in time settle to the bottom, and a meniscus can be seen separating the two phases. If the mixture is then heated, the pressure and temperature will increase. Inspection of Fig. 2·21 shows that the fraction of liquid, as well as the liquid specific volume, will increase, causing the meniscus to rise until it reaches the top of the vessel at point b. During further heating to point m, the vessel contains liquid only. If the vessel had originally been filled with a mixture represented by point d, heating would have caused the evaporation of the liquid, and the meniscus would have fallen until the last trace of liquid in

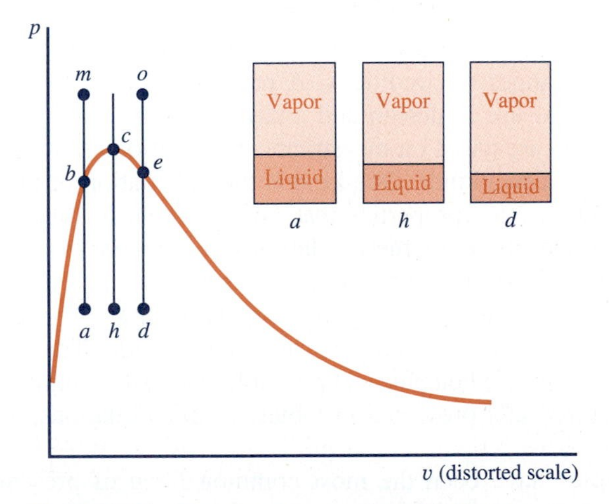

Figure 2·21 Experiments near the critical state.

the bottom was vaporized as the substance reached the state represented by point e. Further heating to point o would superheat the vapor.

For the most interesting case, the vessel is initially filled with a mixture at state h such that the mixture specific volume v_h is equal to the specific volume at the critical state. As the mixture is heated, it approaches the critical state with the liquid becoming less dense and the vapor becoming more dense. As the critical state is approached, the meniscus fades away because there are no longer two phases present. The single phase that remains may be considered as either a liquid or a vapor, and there is fully as much reason for one designation as for the other.

It appears that the amounts of liquid and vapor initially placed in the vessel must be very carefully controlled (in other words, point h must be carefully selected) to ensure that constant-volume heating will cause the meniscus to disappear somewhere near the center of the tube instead of rising to the top as the entire sample is liquefied or falling to the bottom as the entire sample is vaporized. Actually, experiments show that the meniscus between liquid and vapor phases disappears not only at the critical state but in a region around it where the two phase densities are nearly equal. This means that the disappearance of the meniscus can be demonstrated without making the substance pass exactly through its critical state. Conversely, the critical state cannot be precisely identified by the disappearance of the meniscus. These facts were not anticipated before the performance of such experiments. In fact, the investigation of properties near the critical state offers several eloquent examples of the importance of careful experimentation in conjunction with analytical studies.

2·2 Property Data

Engineers need data for many substances, and various applications require data over wide ranges of conditions. A recurring task is to obtain accurate property values to use in design and analysis.

Published data are based on measurements at various states supplemented by calculations of two kinds. One kind is the calculation of properties from relationships with other properties that may be more readily or more accurately measurable. Some of these relationships are discussed in Chapter 7 following the study of the second law of thermodynamics. The second kind of calculation is the interpolation of property values for states other than those states for which measurements have been made. These interpolation calculations are founded on physical principles as well as numerical methods.

Data are stored in many forms. Choose the one that is best for you.

Data are stored and presented in tables, charts, equations, and computer software.

Printed tables have been the most common form of presenting data. A wide range of states can be accommodated and, if tabular increments are small, users must do only a minimum amount of interpolation between tabular values. The limits of validity of a table are clear: the table simply ends. For many substances over a wide range of states, tables are the most widely available form of data. We will see later that for some kinds of calculations, charts are more convenient than tables.

If relatively simple equations adequately represent property relationships, the equations are more convenient than tables and have the advantage that they can be used directly in building mathematical models. In such cases, we usually require for each substance (and perhaps for certain ranges of states) a small number of constants or coefficients in a data base. A drawback of equations for physical property data is that they can readily be applied outside their ranges of validity.

Property data are stored in computer software for two kinds of retrieval. One form is *lookup* software where the user enters certain properties and the computer returns other property values. The user then copies or enters these values into calculations just as values from a printed table are entered. The advantage of lookup software over a printed table is that interpolation is done by the computer. The other form is *online* software that supplies property values directly into the user's application program. No copying or reentering is involved. Online software is more convenient in all respects and reduces chances for error. Both kinds of property data software can be based on either discrete tabulated values or on the original data correlations used for producing tables. Again, we must be careful not to apply computer programs outside their ranges of validity, although it is possible to include error-trapping to reduce this danger.

The software accompanying this book shows how useful online data can be.

Tables that give properties of two or more phases of a substance are nearly always divided into tables of saturation states, that is, states where two or more phases can exist together in equilibrium, and single-phase states. Figure 2·22 shows *pv* and *Tv* diagrams for ammonia. A table of the saturation states

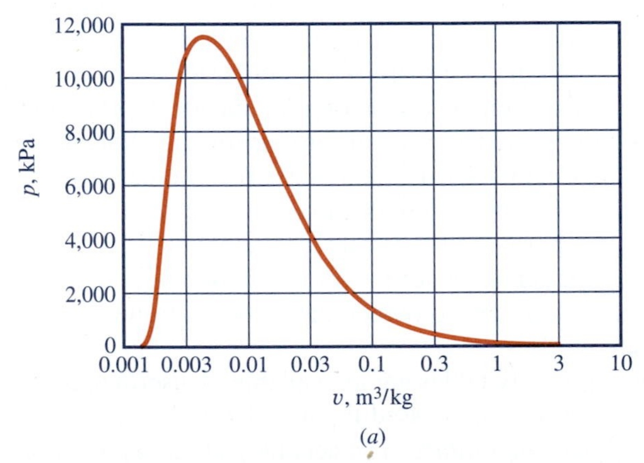

(*a*)

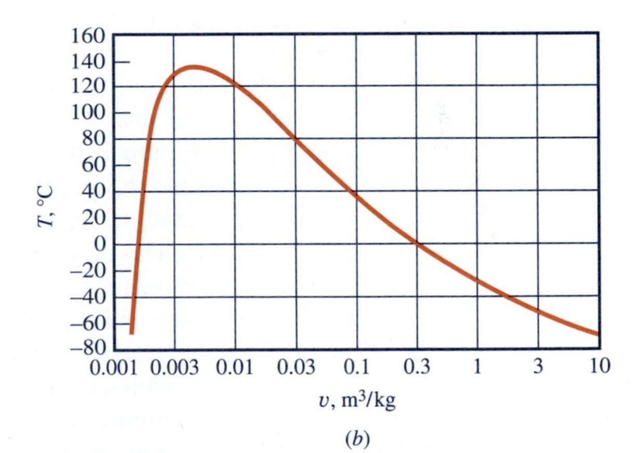

(*b*)

Figure 2·22 *pv* and *Tv* diagrams for ammonia.

Ammonia: Liquid–Vapor Saturation, Temperature Table

Temp. °C	Pressure kPa	v_f m³/kg	v_g m³/kg	h_f kJ/kg	h_{fg} kJ/kg	h_g kJ/kg	s_f kJ/kg·K	s_{fg} kJ/kg·K	s_g kJ/kg·K
–50	40.9	0.001 424	2.6270	123.4	1416.3	1539.7	0.5908	6.3468	6.9376
–45	54.5	0.001 437	2.0064	145.2	1402.7	1547.9	0.6873	6.1483	6.8356
–40	71.7	0.001 450	1.5526	167.1	1388.8	1555.9	0.7824	5.9566	6.7390
–35	93.1	0.001 463	1.2161	189.2	1374.5	1563.7	0.8758	5.7714	6.6472
–30	119.5	0.001 476	0.9634	211.4	1359.7	1571.1	0.9680	5.5920	6.5600
–25	151.5	0.001 490	0.7712	233.7	1344.6	1578.3	1.0586	5.4184	6.4770
–20	190.2	0.001 504	0.6233	256.2	1329.0	1585.1	1.1481	5.2496	6.3977
–15	236.3	0.001 519	0.5084	278.8	1312.9	1591.7	1.2362	5.0858	6.3220
–10	290.8	0.001 534	0.4181	301.6	1296.4	1597.9	1.3232	4.9263	6.2495
–5	354.9	0.001 550	0.3465	324.4	1279.4	1603.8	1.4090	4.7710	6.1800

Figure 2·23 A sample of ammonia tables—saturation states.

of ammonia gives data only along the line *a–c–b*. Figure 2·23 is a small section of such a table where the argument is temperature. That is, integral (whole number) values of temperature are given in the first column. The second column gives the corresponding saturation pressures. The third and fourth columns give the specific volumes of saturated liquid v_f, and the specific volumes of saturated vapor v_g. Many tables also include data for $v_{fg} \equiv v_g - v_f$. Properties of mixtures, such as for state z in Fig. 2·20, are never tabulated; they are calculated from the quality and the properties of the constituent phases by means of equations such as $v_z = v_f + x_z v_{fg}$. This is true for v and also for other properties, such as internal energy (u), enthalpy (h), and entropy (s), which are introduced in subsequent chapters.

Saturation state tables are single-argument tables because, as we have mentioned before, either pressure or temperature is sufficient to determine the properties of the saturated phases. Properties of superheated vapor and compressed liquid are presented in double-argument tables, because specifying the state of a single-phase fluid requires two independent properties. Figure 2·24 is part of a superheated vapor table for ammonia. Sometimes superheated vapor states and compressed liquid states are presented in the same table. The steam tables of Table A·6 in the appendix use such a format.

The property tables in this book (and most other textbooks) are abridged from source tables that extend across wider ranges of pressure and temperature and have smaller table increments to reduce the amount of interpolation required in most applications.

To approximate compressed liquid values, use the properties of the saturated liquid at the *temperature* of interest.

Often compressed liquid property tables are unavailable. A useful approximation is to take the properties of compressed liquid to be the same as those of saturated liquid *at the same temperature*. The accuracy of this approximation is part of the reason why there is little motivation to publish compressed liquid property tables.

Ammonia: Superheated Vapor

p = 400 kPa (T_{sat} = –1.89°C)			Temp.	p = 450 kPa (T_{sat} = 1.25°C)			Temp.	p = 500 kPa (T_{sat} = 4.13°C)		
v, m³/kg	h, kJ/kg	s, kJ/kg·K	°C	v, m³/kg	h, kJ/kg	s, kJ/kg·K	°C	v, m³/kg	h, kJ/kg	s, kJ/kg·K
0.3094	1607.3	6.1382	T_{sat}	0.2767	1610.6	6.0969	T_{sat}	0.2503	1613.6	6.0598
0.3127	1612.5	6.1653	0				0			
0.3267	1636.9	6.2485	10	0.2883	1633.2	6.1826	10	0.2575	1629.6	6.1233
0.3406	1661.1	6.3293	20	0.3008	1657.8	6.2635	20	0.2690	1654.5	6.2043
0.3544	1685.3	6.4076	30	0.3132	1682.3	6.3420	30	0.2803	1679.3	6.2829
0.3680	1709.2	6.4836	40	0.3255	1706.6	6.4182	40	0.2915	1703.9	6.3593
0.3815	1733.1	6.5574	50	0.3377	1703.7	6.4922	50	0.3026	1728.3	6.4336
0.3949	1756.8	6.6289	60	0.3497	1754.6	6.5642	60	0.3135	1752.5	6.5059
0.4081	1780.4	6.6984	70	0.3616	1778.4	6.6342	70	0.3243	1776.5	6.5764
0.4212	1803.9	6.7659	80	0.3733	1802.1	6.7023	80	0.3350	1800.3	6.6451
0.4342	1827.2	6.8315	90	0.3849	1825.6	6.7687	90	0.3455	1824.0	6.7122

Figure 2·24 A sample of ammonia tables—superheated vapor states.

Various formats are used for property tables, but adapting to new formats is easy if you keep in mind that

1. *Saturation* tables are single-argument (pressure or temperature) tables that give only the properties of the two saturated phases that can exist together. They give properties only along saturation lines.
2. *Superheated* tables or *compressed*-liquid tables are two-argument tables that give properties throughout a single-phase region.

Several examples of the use of property tables will now be given, using the abbreviated tables included in the appendix. It is suggested that you go through the example problems to see how tables are used and then use the software included with this book to solve these same example problems. The tables include properties other than pressure, temperature, and specific volume (or density), but in keeping with the topic of this chapter, we use them now only for pvT relationships. Later we will use the same tables for other properties.

Example 2·1 Use of Property Tables

PROBLEM STATEMENT

In a household refrigerator, refrigerant R134a evaporates at −28°C inside the evaporator tubes that surround the freezer compartment. It condenses at 40°C in the condenser tubes on the rear of the refrigerator. Determine the pressure inside the evaporator and inside the condenser. Also, if saturated liquid enters the evaporator and saturated vapor leaves it, what is the increase in specific volume of the R134a?

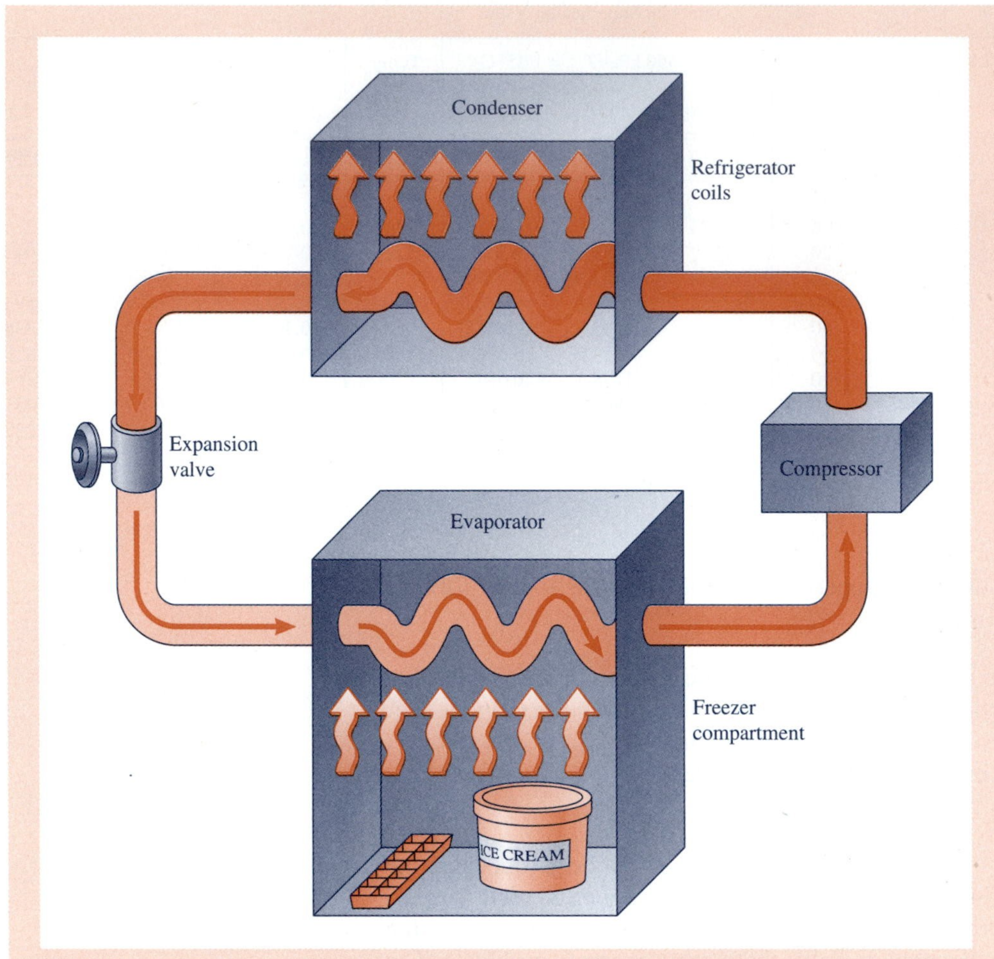

SOLUTION

We go to the R134a saturation table, Table A·10·1, and entering the temperatures −28 and 40°C we see that the saturation pressures are 92.62 and 1017.6 kPa, respectively.

Taking values of v_f and v_g from the tables at −28°C, we find the increase in specific volume in the evaporator to be

$$\Delta v = v_g - v_f = 0.2069 - 0.0007265 = 0.2062 \text{ m}^3/\text{kg}$$

Example 2·2 Use of Property Tables

PROBLEM STATEMENT

Five kilograms of steam at 200 kPa occupies a volume of (*a*) 2.60 m^3, (*b*) 5.20 m^3. Determine the temperature in each case and the quality for any two-phase state.

SOLUTION

(*a*) We do not know what phase the steam is in, whether it is superheated or wet, that is, containing both liquid and vapor. To determine this, we calculate the specific volume

$$v = \frac{\text{volume}}{\text{mass}} = \frac{2.60}{5} = 0.52 \text{ m}^3\text{/kg}$$

and then compare this with the values given in Table A·6·2 at 200 kPa for saturated liquid ($v_f =$ 0.00106049 m³/kg) and for saturated vapor (v_g = 0.8858 m³/kg). We see that $v_f < v < v_g$, so the steam is a wet vapor ($0 < x < 100$ percent) and its temperature must be the saturation temperature for 200 kPa: 120.24°C. The quality is

$$x = \frac{v - v_f}{v_g - v_f} = \frac{0.52 - 0.0016049}{0.8858 - 0.0016049} = 0.586$$

(*b*) For this state, the specific volume is

$$v = \frac{\text{volume}}{\text{mass}} = \frac{5.20}{5} = 1.04 \text{ m}^3\text{/kg}$$

This is greater than the specific volume of saturated vapor v_g at 200 kPa, so the steam is superheated. After entering the compressed water and superheated steam table (Table A·6·3) at 200 kPa and v = 1.04 m³/kg, interpolation gives T = 183.5°C.

Example 2·3 Use of Property Tables

PROBLEM STATEMENT

As part of a test of a vacuum-canning machine, a vessel with a volume of 0.0011 m³ is filled with dry saturated steam at an atmospheric pressure of 95 kPa. It is then sealed and cooled to room temperature, 25°C. If the vessel volume is unchanged, what is the final pressure in the vessel? How much of the steam condenses?

SOLUTION

We sketch a *pv* diagram and place the initial state, state 1, on it. The final state is at the same volume (and the same specific volume) but at a lower temperature. Consequently, we see from the *pv* diagram that it will be a liquid–vapor mixture. For a temperature of 25°C, the saturation steam tables (Table A·6·1) show that the pressure must be 3.17 kPa. Since the volume of the vessel does not change,

$$v_2 = v_1 = v_{g1} = 1.7776 \text{ m}^3\text{/kg}$$

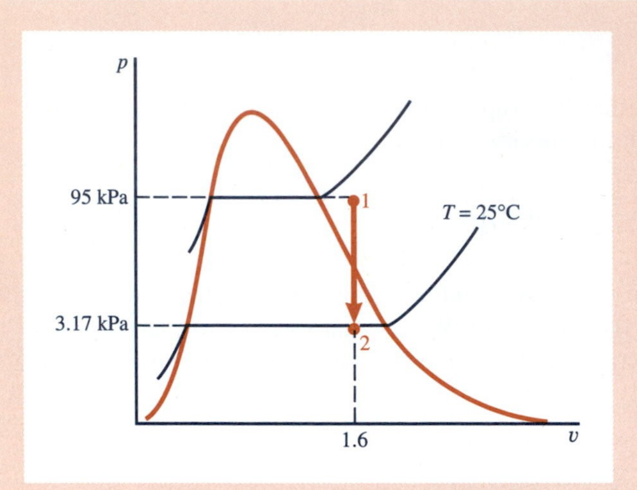

For the final state, the quality x_2 is the fraction of the mixture that is vapor, so the fraction of the steam that condenses is

$$1 - x = 1 - \frac{v_2 - v_{f2}}{v_{g2} - v_{f2}} = 1 - \frac{1.7766 - 0.001003}{43.139 - 0.001003} = 0.959$$

Comment: This value indicates that our *pv* sketch is quite distorted.

Example 2·4 Use of Property Tables

PROBLEM STATEMENT

Two refrigerants, R134a and ammonia, are being considered for a refrigeration plant that will need evaporator and condenser temperatures of −16 and 35°C. Compare the maximum pressures of the two refrigerants in this application. Also compare the specific volumes of dry saturated vapor leaving the evaporator.

SOLUTION

The maximum pressure occurs in the condenser where the temperature is a maximum and will be the saturation pressure for that temperature. From Tables A·10 and A·8 for the two refrigerants, the following values are obtained:

	R134a	Ammonia
Saturation pressure at 35°C, in kPa	888	1350
v_g at −16°C, in m³/kg	0.1255	0.529

Comment: Calculations like these are important in selecting refrigerants. Generally, the higher the required pressure, the higher the cost of equipment. The specific volume and other factors determine the size of the equipment and the amount of work required to compress the refrigerant.

Example 2·5 Use of Property Tables

PROBLEM STATEMENT

Fill in the blanks in the following table for the properties of water.

	p, kPa	T, °C	x, %	v, m³/kg
(*a*)	300	200		
(*b*)	300		65	
(*c*)		200		0.1050
(*d*)	10,000			0.0584
(*e*)	20,000	120		
(*f*)	101.325	100		
(*g*)		120		0.620

SOLUTION

Values found in the steam tables are inserted into the table in *italics.* In a few cases, calculations similar to those in earlier examples were made. *M* denotes *meaningless* and *I* denotes *indeterminate.*

	p, kPa	T, °C	x, %	v, m³/kg
(*a*)	300	200	*M*	*0.7163*
(*b*)	300	*133.6*	65	*0.394*
(*c*)	*1553.7*	200	*0.823*	0.1050
(*d*)	10,000	*1000*	*M*	0.0584
(*e*)	20,000	120	*M*	*0.0010513*
(*f*)	101.325	100	*I*	*I*
(*g*)	*198.48*	120	*0.695*	0.620

Explanation:

(*a*) For 300 kPa, the saturation table (Table A·6·2) shows that the saturation temperature is 133.6°C. The specified temperature is higher than this, so the water must be a superheated vapor, and v can be read directly from the superheated vapor table, Table A·6·3.

(*b*) The quality is specified, so the water is a mixture of liquid and vapor. From the saturation table at 300 kPa, we read $v_f = 0.001073$ m³/kg and $v_g = 0.6056$ m³/kg, $v = (1 - x)v_f + xv_g = (1 - 0.65) \times 0.001073 + 0.65(0.6056) = 0.394$ m³/kg. We also read the saturation temperature, $T = 133.6$°C.

(*c*) For 200°C, the saturation table shows that $v_g = 0.1273\ m^3/kg$. The specified value of v is less than v_g, so the water must be a mixture of liquid and vapor. Therefore, the pressure is the saturation pressure for 200°C and the quality is calculated as $x = (v - v_f)/(v_g - v_f) = (0.1050 - 0.0011564)/(0.1273 - 0.0011564) = 0.823$.

(*d*) For a pressure of 10,000 kPa, the saturation table shows that $v_g = 0.01802\ m^3/kg$. The specified specific volume is greater than this value, so the water must be a superheated vapor. For superheated vapor at 10,000 kPa, we read down the specific volume column and see that the value of 0.0584 m^3/kg corresponds to a temperature of 1000°C.

(*e*) For a pressure of 20,000 kPa, the saturation table shows that the saturation temperature is 365.8°C. The specified temperature is well below this, so the phase of this state is compressed liquid. The value for v is read directly from the compressed liquid table, Table A·6·3.

(*f*) For a temperature of 100°C, we look into the steam tables to see if the saturation pressure is higher or lower than 101.325 kPa and we see that it is exactly this value. Therefore, the steam is saturated and the state is not fully specified, since we do not have two *independent* intensive properties. Consequently, the other properties are indeterminate.

(*g*) We enter the saturated steam table at 120°C to see if the value of $v = 0.620\ m^3/kg$ is more than v_g or less than v_f. It turns out to be between these two values, so the state is one of a liquid–vapor mixture. We calculate the quality of the state by the equation, $v = v_f + x(v_g - v_f)$.

Example 2·6 Use of Property Tables

PROBLEM STATEMENT

Fill in the blanks in the following table for the properties of R134a.

	p, psia	T, F	x, %	v, ft^3/lbm
(*a*)	120	140		
(*b*)	20		35	
(*c*)		100		0.290
(*d*)	200			0.250
(*e*)	120	70		

SOLUTION

Values found in the R134a tables are inserted into the table in *italics.* In a few cases, calculations similar to those in earlier examples were made. *M* denotes *meaningless* and *I* denotes *indeterminate.*

	p, psia	T, F	x, %	v, ft^3/lbm
(*a*)	120	140	*M*	*0.462*
(*b*)	20	*−2.39*	35	*0.804*
(*c*)	*139.0*	100	*84.5*	0.290
(*d*)	200	*144*	*M*	0.250
(*e*)	120	70	*M*	*0.0131*

Explanation:

(*a*) For 120 psia, the saturation table (Table A·10·2E) shows that the saturation temperature is 90.5 F. The specified temperature is higher than this, so the R134a must be a superheated vapor, and v can be read directly from the superheated vapor table, Table A·10·3E.

(*b*) The quality is specified, so the R134a is a mixture of liquid and vapor. From the saturation table at 20 psia, we read $v_f = 0.01186$ ft^3/lbm and $v_g = 2.2765$ ft^3/lbm, $v = (1 - x)v_f + xv_g = (1 - 0.35) \times 0.01186 + 0.35(2.2765) = 0.804$ ft^3/lbm.

(*c*) For 100 F, the saturation table shows that $v_g = 0.3408$ ft^3/lbm. The specified value of v is less than v_g, so the water must be a mixture of liquid and vapor. Therefore, the pressure is the saturation pressure for 100 F and the quality is calculated as $x = (v - v_f)/(v_g - v_f) = (0.290 - 0.01390)/(0.3408 - 0.01390) = 0.845$.

(*d*) For a pressure of 200 psia, the saturation table shows that $v_g = 0.2303$ ft^3/lbm. The specified specific volume is greater than this value, so the water must be a superheated vapor. For superheated vapor at 200 psia, we interpolate in the specific volume column and see that the value of 0.250 ft^3/lbm corresponds to a temperature of 144 F.

(*e*) For a pressure of 120 psia, the saturation table shows that the saturation temperature is 90.46 F. The specified temperature is below this, so the phase of this state is compressed liquid. If we have no compressed liquid table at hand, we use the approximation that v of a compressed liquid is nearly equal to v_f *at the same temperature.*

2·3 Equations of State for Gases

Property tables may provide pvT (and other) data for two or three phases. Simple equations that represent pvT relationships in all phases for any substance do not exist. However, the gaseous phase, especially at lower pressures and higher temperatures, can often be modeled by a simple equation, the *ideal-gas equation of state*, or more complex equations of state that are presented in Chapter 7.

2·3·1 The Ideal-Gas Equation of State

Ideal gas

An **ideal gas** is defined as a substance for which the equation of state is

$$pv = \frac{\overline{R}}{M}T \tag{2·2}$$

Universal gas constant

where $\overline{R}$ is called the **universal gas constant** and M is the molar mass of the substance. The value of $\overline{R}$ is

$$\overline{R} = 8.31441 \text{ kJ/kmol·K} = 1.9859 \text{ B/lbmol·R} = 1545.3 \text{ ft·lbf/lbmol·R}$$

The quantity $\overline{R}/M$ is called the *gas constant* R for a particular gas, so that the ideal-gas equation becomes

$$pv = RT \tag{2·3}$$

Molar mass **is frequently called *molecular weight*. This latter term is, of course, a misnomer for this quantity that is independent of gravitational effects.**

where R is the gas constant for that gas. R values for several gases are given in Table 2·2, and a more extensive list is provided in Appendix Table A·1.

Alternative forms of the ideal-gas equation of state are

Alternative forms of the ideal-gas equation of state

$$pv = RT \qquad pV = N\overline{R}T$$

$$pV = mRT \qquad pV = \frac{m\overline{R}T}{M}$$

$$p = \rho RT \qquad pV = NMRT$$

where N is the number of moles of the gas, and the other symbols are as previously defined. An ideal gas is a hypothetical substance. We define it as

Table 2·2 *R* Values for Several Gases

	Molar Mass, M	Gas Constant, R	
Gas	kg/kmol *or* lbm/lbmol	kJ/kg·K	ft·lbf/lbm·R
Air	28.97	0.287	53.3
Argon, Ar	39.94	0.208	38.7
Carbon dioxide, CO_2	44.01	0.189	35.1
Helium, He	4.004	2.077	386.0
Hydrogen, H_2	2.016	4.124	766.0
Methane, CH_4	16.04	0.518	96.3
Nitrogen, N_2	28.01	0.297	55.2
Octane, C_8H_{18}	114.23	0.0728	13.5
Oxygen, O_2	32.00	0.260	48.3
Steam, H_2O	18.02	0.461	85.7

a substance that follows the equation of state $pv = RT$. Historically, the concept of an ideal gas resulted from the work of Boyle, Charles, and Gay-Lussac, all of whom measured properties of real gases. Their work and other later studies indicate that for real gases at low pressure and relatively high temperature the equation of state is very nearly $pv = RT$. A pvT surface for an ideal gas is shown in Fig. 2·25.

Another important characteristic of an ideal gas will be shown on the basis of experiments in Chapter 4 and derived from basic principles in Chapter 7. We mention it here as a preview that modeling a substance as an ideal gas provides more than just a simple equation of state (pvT equation).

2·3·2 Applying the Ideal-Gas Equation of State to Real Gases

Assessing the accuracy of the ideal-gas equation of state

How accurately does the ideal-gas equation, $pv = RT$, model a real gas? This is an important question. Using such a simple equation of state is convenient, but we must know how much accuracy we sacrifice for the convenience. A qualitative answer is that usually $pv = RT$ is increasingly accurate for real gases as

1. Molar mass decreases.
2. Pressure decreases.
3. Temperature increases.

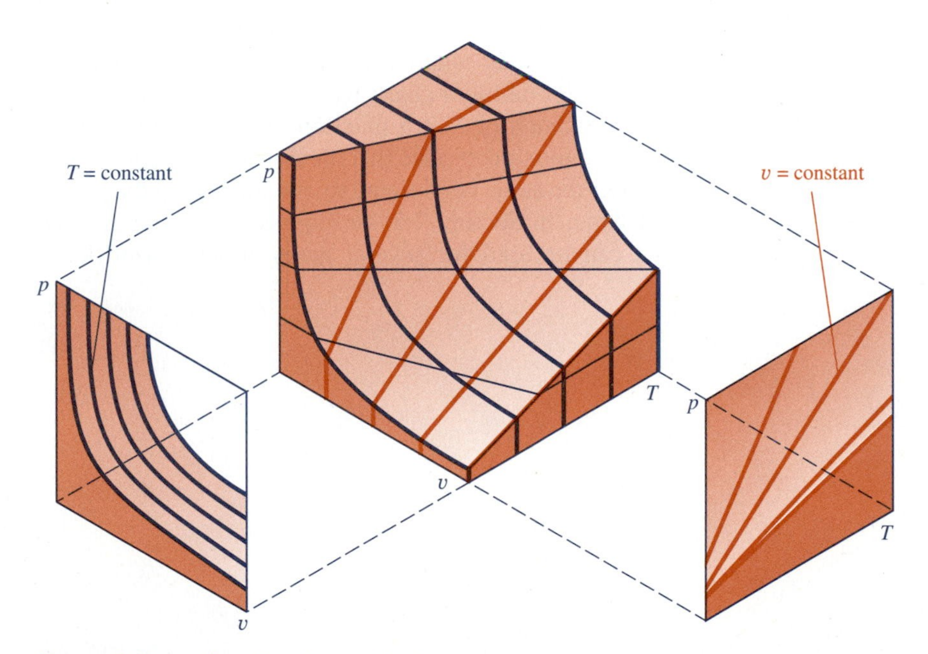

Figure 2·25 pvT surface for an ideal gas.

This is a statement of a general trend, but it cannot be relied on in all ranges of pressure and temperature. We can say only that $pv = RT$ represents the characteristics of many real gases accurately enough for many engineering calculations as long as the gases are not simultaneously at pressures above their critical pressures and temperatures below their critical temperatures. For example, for air at room temperature, the error in $pv = RT$ is less than 1 percent for pressures as high as 2.7 MPa (nearly 400 psia). For air at 1 atm, the error in $pv = RT$ is less than 1 percent for temperatures as low as −130°C. For hydrogen at 1 atm, the error in $pv = RT$ is less than 1 percent for temperatures as low as −220°C.

In modeling real gases as ideal gases, be aware that uncertainty regarding the actual composition of gases involved in engineering calculations may limit accuracy. For example, the composition of atmospheric air varies with location and time even beyond the expected changes in humidity (water vapor content) that are discussed in Chapter 9. Commercially available gases, which are nominally pure, usually contain traces of other gases. Consequently, the number of significant digits you may find in tables for molar masses or gas constants may not be justified throughout many calculations.

A quantitative answer to the question regarding the accuracy of $pv = RT$ for real gases is given by the *compressibility factors*. For real gases, the ideal-gas equation of state can be modified to

$$pv = ZRT \tag{2·4}$$

Compressibility factor

where Z, the **compressibility factor**, is an empirically determined function of p and T for each gas. The compressibility factor can be thought of as a "correction factor" to the ideal-gas equation of state when it is applied to a real gas. Compressibility factor tables or charts have been published for some gases, and Fig. 2·26 shows charts of Z as a function of p and T to the same scale for oxygen and carbon dioxide. Notice that the Z values are quite different for the two gases under the same conditions, so one must have the Z chart or table for the gas at hand. In Section 7·5 we discuss a *generalized compressibility factor* that can be used for several gases at some sacrifice in accuracy.

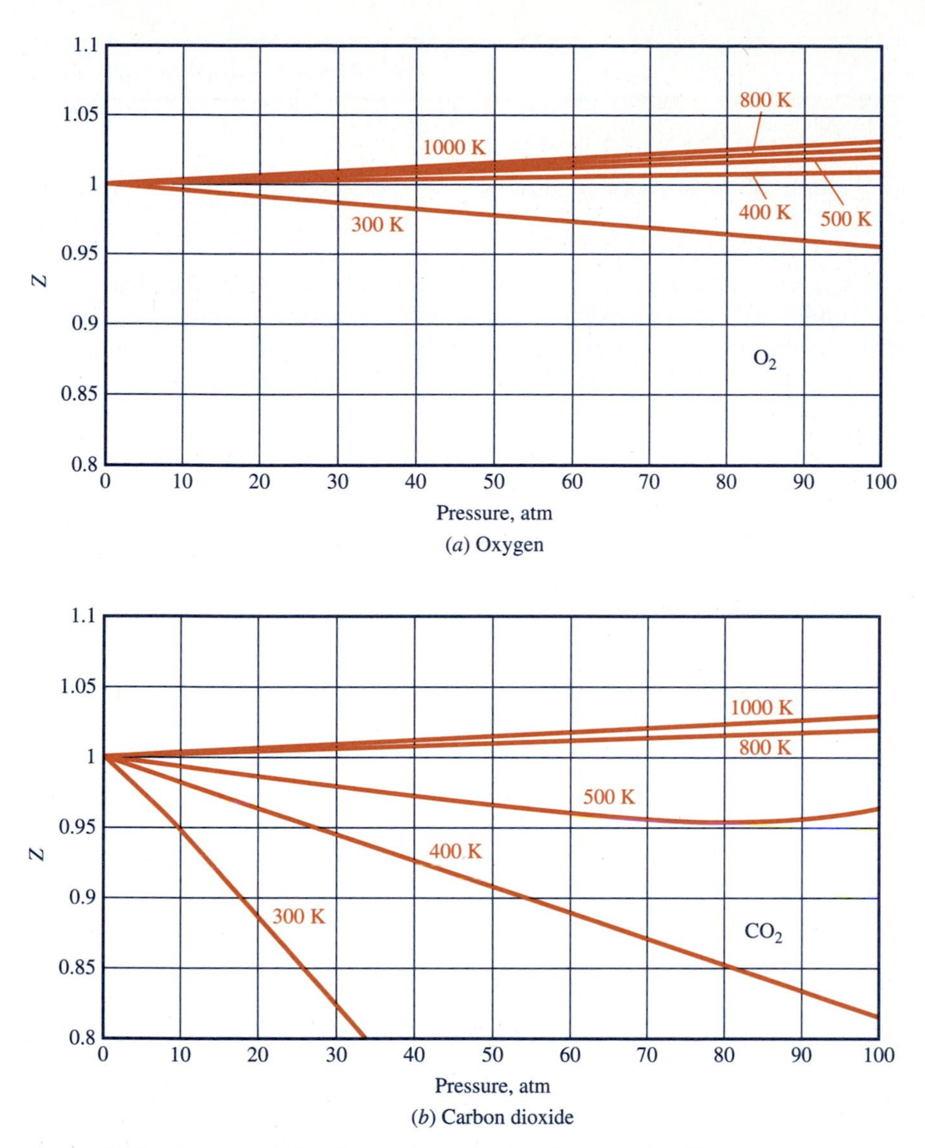

Figure 2·26 Compressibility factors for oxygen and carbon dioxide.

Example 2·7 Ideal-Gas Equation of State

PROBLEM STATEMENT

Determine the density and specific volume of air at room conditions.

SOLUTION

Analysis: We will model the air as an ideal-gas; that is, we will assume that its behavior fits the ideal-gas equation of state, $pv = RT$. ''Room conditions'' are not specified, so let us consider for locations near sea level that barometric pressure might vary from 97 to slightly over 101 kPa (or, say, 14.2 to 14.7 psia) and that room temperature might vary from 15 to 32°C (or, say, from 60 to 90 F). Minimum density occurs with minimum pressure and maximum temperature; maximum density occurs with maximum pressure and minimum temperature, so we will calculate both.

Calculations: Using the maximum pressure and minimum temperature, we find

$$\rho = \frac{p}{RT} = \frac{101 \text{ kPa}}{0.287 \dfrac{\text{kJ}}{\text{kg}\cdot\text{K}}(288 \text{ K})} = 1.22 \text{ kg/m}^3$$

$$v = \frac{1}{\rho} = 0.82 \text{ m}^3\text{/kg}$$

For the minimum pressure and maximum temperature,

$$\rho = \frac{p}{RT} = \frac{97}{0.287(305)} = 1.11 \text{ kg/m}^3$$

$$v = \frac{1}{\rho} = 0.90 \text{ m}^3\text{/kg}$$

Comment: It is well for an engineer to remember typical numerical values of some quantities, such as the density or specific volume of room temperature air. (Density is of the order of 1 kg/m^3 and specific volume is of the order of 1 m^3/kg. In English units, densities are of the order of 0.07 to 0.08 lbm/ft^3—0.002 to 0.0025 slug/ft^3—and specific volumes are of the order of 13 to 14 ft^3/lbm.) These values are convenient in making preliminary calculations and in checking the reasonableness of various calculations.

Example 2·8 Ideal-Gas Equation of State

PROBLEM STATEMENT

Determine the mass of helium at 600 kPa, 40°C, to fill a tank with a volume of 35 m^3.

SOLUTION

Analysis: We will model the helium as an ideal gas. (This is reasonable because even though the pressure is above the critical pressure of helium, the temperature is far above the critical temperature, and helium has a molar mass of only 4 kg/kmol.) Let us first estimate the mass of helium on the basis of estimating the mass of *air* that would be held under the specified conditions. We recall that air at room conditions has a density of 1.1 to 1.2 kg/m^3, so the mass of room air to fill 35 m^3 would be about 40 kg. But the pressure in the tank will be about six times atmospheric pressure, so the mass of air at this higher pressure would be six times as much or approximately 240 kg. The temperature is only slightly higher than a typical room temperature (ratio of 313/293 of kelvin temperatures), so let us adjust 240 kg to 225 kg. Finally, helium has a molar mass about one-seventh (4.003/28.97) that of air, so the mass of helium would be one-seventh as much as the mass of air, or a little over 30 kg.

Calculations:

$$m = \frac{pV}{RT} = \frac{MpV}{\overline{R}T} = \frac{4.003\text{ kg/kmol}}{8.31441\text{ kJ/kmol·K}}\,\frac{600\text{ kPa}(35\text{ m}^3)}{(273 + 40)\text{K}} = 32.3\text{ kg}$$

Comment: The calculated result is close to our estimate, so we regard the result as reasonable. With practice, you can make such estimates quickly and accurately, and doing so can help detect many calculation errors.

Example 2·9 Ideal-Gas Equation of State

PROBLEM STATEMENT

Acetylene, C_2H_2 (also called ethyne), is to be compressed steadily at a rate of 5.80 lbmol/h from 14.3 psia, 125 F, to 70 psia. The discharge temperature is estimated to be 320 F. If the inlet velocity is not to exceed 40 fps and the discharge velocity is not to exceed 80 fps, determine the minimum cross-sectional area at the inlet and at the outlet.

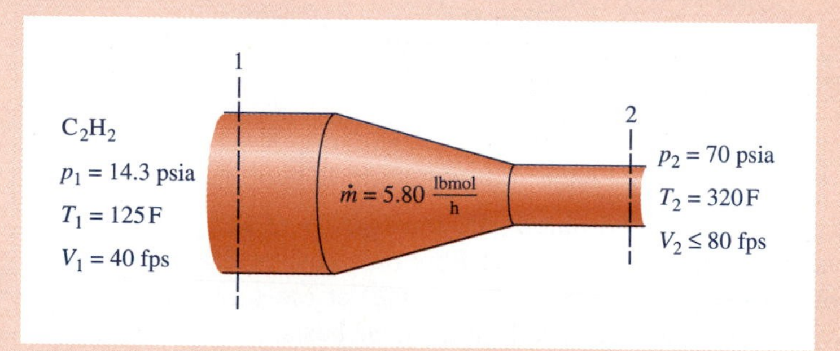

SOLUTION

Analysis: We first make a sketch of the system showing the given conditions. We model the flow at inlet (section 1) and outlet (section 2) as one dimensional. At each section, the minimum cross-sectional area corresponds to the maximum velocity, so we determine the areas from the continuity equation:

$$A_{min} = \frac{\dot{m}}{\rho V_{max}} = \frac{\dot{N}M}{\rho V_{max}} \tag{a}$$

where $\dot{N}$ is the molar flow rate (lbmol/h) and M is the molar mass (lbm/lbmol).

At each section, $\dot{m}$ and V_{max} are known and, modeling the ethyne as an ideal gas, we can calculate ρ by

$$\rho = \frac{p}{RT} = \frac{Mp}{\overline{R}T} \tag{b}$$

Combining Eqs. a and b gives

$$A_{min} = \frac{\dot{N}\overline{R}T}{pV_{max}} \tag{c}$$

and for both sections values of all quantities on the right-hand side of Eq. c are known. (In passing, we note that Eq. c is reasonable physically: Increasing the molar flow rate or the temperature would require a larger area and increasing the pressure or the velocity would require a smaller area.)

Calculations: Following the analysis,

$$A_{1,min} = \frac{\dot{N}\overline{R}T_1}{p_1 V_{1,max}}$$

$$= \frac{5.80\dfrac{\text{lbmol}}{\text{h}}\left(1545\dfrac{\text{ft·lbf}}{\text{lbm·R}}\right)585\text{ R}}{3600\dfrac{\text{s}}{\text{h}}\left(14.2\dfrac{\text{lbf}}{\text{in}^2}\right)144\dfrac{\text{in}^2}{\text{ft}^2}\left(40\dfrac{\text{ft}}{\text{s}}\right)}$$

$$= 0.0178 \text{ ft}^2 = 2.56 \text{ in}^2$$

$$A_{2,min} = \frac{\dot{N}\overline{R}T_2}{p_2 V_{2,max}} = \frac{5.80\dfrac{\text{lbmol}}{\text{h}}\left(1545\dfrac{\text{ft·lbf}}{\text{lbm·R}}\right)780}{3600\dfrac{\text{s}}{\text{h}}\left(70\dfrac{\text{lbf}}{\text{in}^2}\right)144\dfrac{\text{in}^2}{\text{ft}^2}\left(80\dfrac{\text{ft}}{\text{s}}\right)}$$

$$= 0.00241 \text{ ft}^2 = 0.346 \text{ in}^2$$

Comment: Notice that we made no use of either the gas constant or the molar mass of acetylene. In fact, the solution is independent of the gas involved. This is true because the flow rate was specified on a molar basis, and all ideal gases have the same molar density (moles per unit volume) at the same pressure and temperature. In the chemical industry, flow rates are often specified on a molar basis.

2·3·3 The Ideal-Gas Temperature Scale

Constant-volume gas thermometers (see Section 1·5·3) using different gases agree with each other more and more closely as the pressure at any one temperature is reduced. Also, as the pressure of a gas approaches zero, the gas behaves more and more like an ideal gas. Thus the gas thermometer scale defined by extrapolating real gas behavior to the condition of zero pressure is called the ideal-gas temperature scale. The ideal-gas temperature scale defined in this manner does not depend on the properties of any one gas but on the properties of gases in general. The defining equation then is

$$\frac{T_2}{T_1} = \left(\frac{p_2}{p_1}\right)_{v=\text{constant},\ p_1 \to 0}$$

At this point it still appears that the ideal-gas temperature scale, since it depends on physical properties of some substance or substances, is no better than one based on the expansion of mercury or on the change in electrical resistance of some material. We will show in Chapter 5, however, that the ideal-gas scale is equivalent to one that is entirely independent of physical properties.

2·4 Equations of State for Liquids and Solids

pvT relationships are not as important for liquids as for gases simply because the variation of specific volume with temperature is much smaller in liquids than in gases, and the variation of specific volume with pressure is smaller still. This can be seen by a study of the compressed liquid table for any substance. Variations of other properties with temperature (and to a smaller extent with pressure) are significant, and some of these are discussed in Chapter 7. Some empirical *pvT* relationships have been developed for liquids, and even some of the equations of state introduced in Section 7·6 for gases have been applied to liquids, often with different constants.

Liquids and gases when at rest are isotropic and can sustain no shearing stresses. Consequently, pressure and temperature are sufficient to specify a state. Solids, however, are markedly different from liquids and gases. The state of a solid is not specified by only temperature and density. The state also involves stress, and stress in a solid is not a scalar as pressure is in a liquid or gas at rest. Consequently, there are no *pvT* correlations for solids. More complex relationships involving additional properties are required. (An equation for the volume expansivity of a solid as a function of temperature relates v to T under specified stress conditions, but this is not equivalent to a *pvT* relationship.) Thermodynamics is important in correlating solid proper-

Some *pvT* relations exist for liquids, but there are none for solids.

ties and predicting the behavior of solids, but thermodynamic analyses of solids do not involve pvT relationships.

2·5 Summary

A *phase* is any homogeneous part of a system that is physically distinct. Each phase in a system is separated from other phases by *phase boundaries*. In any system, there can be only one gas phase, but two or more liquid or solid phases may exist.

A *pure substance* is chemically homogeneous and fixed in chemical composition. In the absence of electric, magnetic, solid distortion, and surface tension effects, the state of a pure substance at rest is usually specified by two independent variables.

The pressure and temperature of any one phase of a pure substance can be varied independently over wide ranges, but when two or more phases exist together in equilibrium, there is a fixed relationship between the pressure and temperature. Conditions under which two or more phases can exist together are called *saturation conditions*. A pressure–temperature diagram showing saturation conditions is called a *phase diagram*. The areas between the saturation lines on a phase diagram represent single phases.

For a mixture of liquid and vapor, the *quality* x is defined as the fraction by mass of vapor in the mixture. The specific volume of the mixture v is given by

$$v = v_f + x(v_g - v_f) \tag{2·1}$$

where v_f is the specific volume of saturated liquid and v_g is the specific volume of saturated vapor. Other specific properties of mixtures can be calculated in a similar manner.

Conditions under which three phases of a pure substance can exist together in equilibrium are known as the *triple-phase state* or *triple-point* conditions.

As pressure is increased, the evaporation or boiling temperature of a pure substance increases. Also, the specific volume of the saturated liquid increases and the specific volume of the saturated vapor decreases until the two are equal at the *critical point*. At pressures higher than the critical pressure and temperatures higher than the critical temperature, there is no distinction between liquid and vapor.

Property data are stored in tables, charts, and computer software, and for limited ranges of conditions, relationships among properties are modeled by equations ranging from quite simple ones like the *ideal-gas equation of state* to highly complex equations.

An *ideal gas* is defined as a substance for which the equation of state is

$$pv = RT \tag{2·3}$$

where p is pressure, v is specific volume, T is temperature, and R is the *gas constant* for the gas. For any gas,

$$R = \frac{\overline{R}}{M}$$

where $\overline{R}$ is the *universal gas constant* and M is the molar mass of the gas. The value of $\overline{R}$ is

$$\overline{R} = 8.31441 \text{ kJ/kmol·K} = 1.9859 \text{ B/lbmol·R} = 1545.3 \text{ ft·lbf/lbmol·R}$$

For real gases, the ideal-gas equation of state is sufficiently accurate for many purposes at lower pressures, at higher temperatures, and for lower-molar-mass gases.

Another pvT relation for real gases is

$$pv = ZRT \tag{2·4}$$

where Z, the *compressibility factor*, which can be thought of as a ''correction factor'' to the ideal-gas equation of state, is a function of p and T for the gas at hand.

References and Suggested Reading

2·1 *Chemical Engineers' Handbook*, Robert H. Perry and Cecil H. Chilton, eds., 5th ed., McGraw-Hill, New York, 1973.

2·2 *Handbook of Chemistry and Physics*, CRC Press, annual editions.

2·3 *Handbook of Tables for Applied Engineering Science*, Ray E. Bolz and George L. Tuve, eds., The Chemical Rubber Company, Cleveland, Ohio, 1970.

2·4 *Mechanical Engineers' Handbook*, Myer Kutz, ed., Wiley, New York, 1986.

2·5 *Marks' Standard Handbook for Mechanical Engineers*, Eugene A. Avallone and Theodore Baumeister III, eds., 9th ed., McGraw-Hill, New York, 1991.

Sources of Property Data

2·6 *JANAF Thermochemical Tables*, 3rd ed., published by the American Chemical Society and the American Institute of Physics for the National Bureau of Standards, 1986. (Also, *Journal of Physical and Chemical Reference Data*, Vol. 14, 1985, Supplement No. 1.)

2·7 *Tables of Thermal Properties of Gases*, National Bureau of Standards Circular No. 564, 1955. (Although this reference is old, it is still accurate. Notice that these tables are for real gases and include the effect of pressure even for pressures that might often be considered as low.)

2·8 Hilsenrath, Joseph H., et al., *Tables of Thermodynamic and Transport Properties*, Pergamon, Elmsford, N.Y., 1960.

2·9 Keenan, Joseph H., Jing Chao, and Joseph Kaye, *Gas Tables*, 2nd ed., Wiley, New York, 1980 (English units), 1983 (SI units).

2·10 Grigull, Ulrich, Johannes Straub, and Peter Schiebener, eds., *Steam Tables in SI-Units*, 2nd revised ed., Springer-Verlag, New York, 1984.

2·11 Haar, Lester, John S. Gallagher, and George S. Kell, *NBS/NRC Steam Tables*, Hemisphere, New York, 1984.

2·12 Keenan, Joseph H., Frederick G. Keyes, Philip G. Hill, and Joan G. Moore, *Steam Tables*, Wiley, New York, 1969 (English units), 1978 (SI units).

2·13 Haar, Lester, and John S. Gallagher, "Thermodynamic Properties of Ammonia," *Journal of Physical and Chemical Reference Data*, Vol. 2, No. 3, 1978, pp. 635–792.

2·14 Vargaftik, N. B., *Tables on the Thermophysical Properties of Liquids and Gases in Normal and Dissociated States*, 2nd ed., Wiley, New York, 1975.

2·15 *Thermodynamic Properties: GASPROPS, REFRIG, STEAMCALC*, Software Systems Corp., Wiley, New York, 1985. (One of several packages of personal computer software for properties.)

Problems

2·1 Which of the systems described below are composed of pure substances during the described process? (The containers and partitions are not parts of the systems.)

(*a*) A tank contains oxygen and nitrogen on opposite sides of a partition. The partition is removed (or punctured).

(*b*) A tank contains nitrogen at 130 kPa on one side of a partition and nitrogen at 100 kPa on the other side. The partition is removed.

(*c*) Air is contained in a tank. Also within the tank is a covered dish containing water. The cover is removed from the dish.

(*d*) A tank contains liquid water and steam. The tank is cooled so that some of the steam condenses.

2·2 Which of the systems described below are composed of pure substances during the described processes? (The containers and partitions are not parts of the systems.)

(*a*) A tank contains air, water vapor, and iron filings. Solid iron oxide is formed very slowly.

(*b*) A tank contains ice and water vapor. Some of the ice sublimes, but no liquid water is formed.

(*c*) A tank contains liquid water on one side of a partition and steam at a lower pressure on the other side. The partition is ruptured.

(*d*) Air flows over very cold tubes and some of it is condensed. The resulting liquid is slightly richer in nitrogen than the gas.

2·3 In Section 2·1 it is stated that in a system composed of small globules of oil dispersed in liquid water, the phase boundaries are "near spherical" surfaces. Explain why they are not spherical.

2·4 An open vessel of liquid water is in a closed room, and the air in the room contains water vapor. This system can exist in equilibrium at a fixed pressure and several different temperatures. Explain why this does not conflict with the statement that two phases of a pure substance can coexist in equilibrium at a given pressure at only one temperature.

2·5 From Fig. 2·4 show that the difference between the triple-point temperature for water and the ice-point temperature is 0.01 degrees C. (This cannot be a precise calculation in view of the scale of the figure.)

2·6 At room temperature, approximately what pressure is required to liquefy (*a*) carbon dioxide, (*b*) ammonia, (*c*) refrigerant R134a?

2·7 At what temperature does nitrogen boil under a pressure of (*a*) 1000 kPa, (*b*) 2000 kPa, (*c*) 3000 kPa?

2·8 What are the liquefaction temperatures for nitrogen under a pressure of (*a*) 100 psia, (*b*) 200 psia, (*c*) 400 psia?

2·9 A tank contains liquid nitrogen. To maintain a temperature of 100 K, to what pressure must the pressure regulator be set?

2·10 What is the saturation temperature for refrigerant R22 for a pressure of (*a*) 3000 kPa, (*b*) 300 psia?

2·11 Under what pressure(s) can ice and liquid water at −10°C exist together in equilibrium?

2·12 At what temperature(s) can ice and liquid water exist together in equilibrium under a pressure of 7000 psia?

2·13 What pressure is required to melt ice at −12°C?

2·14 At what temperature would water at a pressure of 10,000 psia freeze?

2·15 Is it possible to maintain ice at 45 MPa and −3°C?

2·16 Can ice exist at 5000 psia, 25 F?

2·17 It has been suggested that the liquid–vapor saturation curve of a pure substance can be represented by an equation of the form $p = p_0 e^{-c/T}$, where T is absolute temperature and p_0 and c are constants. Comment on how well such an equation fits the data for water.

2·18 Name a substance other than water that expands on freezing. In what applications is this characteristic important?

2·19 On a phase (pT) diagram, sketch some lines of constant density in each of the single-phase regions. Justify the usual signs and relative magnitudes of the slopes $(\partial p/\partial T)_v$ and $(\partial v/\partial T)_p$.

2·20 Sketch a phase (pT) diagram for water and sketch on this diagram a few lines of constant specific volume. Make a magnified sketch of the region near the triple-phase point to show that for a pressure around 1 atm, the liquid has a maximum density at approximately 4°C (39 F). (Insufficient data are available in this book to make an accurate plot, but you can make one that is qualitatively correct.)

2·21 On a pv diagram, how many states of a system can be represented by a single point on the triple-phase line?

2·22 Sketch a Tv diagram for water and identify the various lines and regions on it. Show constant-pressure lines for pressures below the triple-phase value, between the triple-phase and critical values, and above the critical value.

2·23 A sealed vessel containing a carbon dioxide block is opened to the atmosphere. What change does the block undergo? What will be the surface temperature of the block?

2·24 Carbon dioxide gas initially at 10°C is cooled at a constant pressure of 100 kPa to −100°C. Describe the phase change that occurs.

2·25 For each of the five states of carbon dioxide, indicate the phase(s) of the substance by *S*, *L*, *V*, (solid, liquid, vapor) or a combination:

State	1	2	3	4	5
p, kPa	1000	1000	1000	100	6000
T, °C	50	5	−60	−60	25
Phase					

2·26 For each of the five states of carbon dioxide, indicate the phase(s) of the substance by *S*, *L*, *V*, (solid, liquid, vapor) or a combination:

State	1	2	3	4	5
p, psia	14.7	14.7	35	100	6000
T, F	20	230	45	250	23
Phase					

2·27 Comment on the definition of critical temperature as the temperature above which the liquid phase of a substance cannot exist.

2·28 In the first paragraph of Section 2·2, it is stated that in a system composed of liquid and vapor phases the density of one phase is usually much greater than that of the other. Explain the need for the word "usually."

2·29 You are given two intensive properties of a pure substance and asked to find several other intensive properties. Give an example of a case in which this cannot be done because a value is (*a*) indeterminate, (*b*) meaningless.

2·30 For each of the five states of water, fill in the blanks in the following table with the property values or with *M* or *I* (for meaningless or indeterminate):

State	1	2	3	4	5
p, kPa	200	300	2000		5000
T, °C		133.56	300	150	100
x, %	80				
v, m³/kg				0.361	

2·31 For each of the five states of water, fill in the blanks in the following table with the property values or with *M* or *I* (for meaningless or indeterminate):

State	1	2	3	4	5
p, psia	14.696	200	300		10
T, F			120	300	200
x, %		92			
v, ft³/lbm	2.880			0.310	

2·32 For each of the five states of ammonia, fill in the blanks in the following table with the property values or with *M* or *I* (for meaningless or indeterminate):

State	1	2	3	4	5
p, kPa		270		2033	400
T, °C	−10	0	130	50	
x, %	40				
v, m³/kg			0.390		0.251

2·33 For each of the five states of ammonia, fill in the blanks in the following table with the property values or with *M* or *I* (for meaningless or indeterminate):

State	1	2	3	4	5
p, psia		50		30	180
T, F	60	60	−10		
x, %			34	100	
v, ft³/lbm	0.240				0.105

2·34 For each of the five states of refrigerant R134a, fill in the blanks in the following table with the property values or with *M* or *I* (for meaningless or indeterminate):

State	1	2	3	4	5
p, kPa		270		250	400
T, °C	−10	0	70		
x, %	40				
v, m³/kg			0.00750	0.090	0.025

2·35 For each of the five states of refrigerant R134a, fill in the blanks in the following table with the property values or with *M* or *I* (for meaningless or indeterminate):

State	1	2	3	4	5
p, psia		50		30	180
T, F	60	60	−10		
x, %			34	100	
v, ft³/lbm	0.240				0.105

2·36 Potassium at 1 atm boils at approximately 760°C. The specific volumes of the saturated liquid and vapor are 0.0015 and 1.991 m^3/kg, respectively. Compute the specific volume of a mixture of 70 percent quality.

2·37 A closed tank contains 0.50 m^3 of dry saturated steam at a gage pressure of 249 kPa. (*a*) What is its temperature? (*b*) How many kilograms of steam does the tank contain? Barometric pressure equals 101 kPa.

2·38 A tank having a volume of 0.050 m^3 contains 80 percent saturated vapor and 20 percent saturated liquid water by volume at a temperature of 25°C. If the liquid and vapor are agitated until thoroughly mixed, what would be the quality of the mixture?

2·39 Cesium at 100 psia has a saturation temperature of 1740 F. Specific volumes of saturated liquid and saturated vapor are 0.00991 and 1.597 ft^3/lbm, respectively. Determine the specific volume of a mixture of 60 percent quality.

2·40 Water at 60 F stands 6 cm deep in an open container with an inside diameter of 2 cm. During a 96-h period, the water level drops 2 mm as a result of evaporation. Determine the net average number of molecules leaving per second during the 4-day period. (In this chapter, we have not mentioned Avogadro's number, 6.022×10^{23}, but you have encountered it in other courses.)

2·41 Pure water at room temperature is stored in a bottle, with the water surface standing in the bottle neck which has a diameter of 20 mm. During a period of 24 h, water evaporates so that the liquid level drops by 0.10 mm. Determine the average rate (molecules/s) at which molecules leave the liquid phase. (In this chapter, we have not mentioned Avogadro's number, 6.022×10^{23}, but you have encountered it in other courses.)

Problem 2·41

2·42 A rigid container with a volume of 0.170 m^3 is initially filled with steam at 180 kPa, 340°C. It is cooled to 90°C.

(*a*) At what temperature does a phase change start to occur?
(*b*) What is the final pressure?
(*c*) What mass fraction of the water is liquid in the final state?
(*d*) Calculate the work done during the cooling process.

2·43 A method of obtaining a low pressure in a sealed vessel is to fill the vessel with steam before sealing, seal, and then condense the steam. A vessel having a volume of 0.15 m^3 is filled with dry saturated steam at 101 kPa. The vessel is sealed and then chilled to 30°C.

(*a*) What is the final pressure?
(*b*) Sketch *pv* and *pT* diagrams of the process, showing saturation lines on each diagram.

2·44 The water level in a storage tank 9 ft high is regulated by a float-controlled valve. The water level is indicated by a simple gage glass 8 ft long on the side of the tank. When storing cold water, the gage glass shows that the water level is maintained 8 ft above the tank bottom by the float-controlled valve, but when hot (200 F) water is in the tank, the gage shows that the same control device maintains the water level 3 in. lower. How do you account for this?

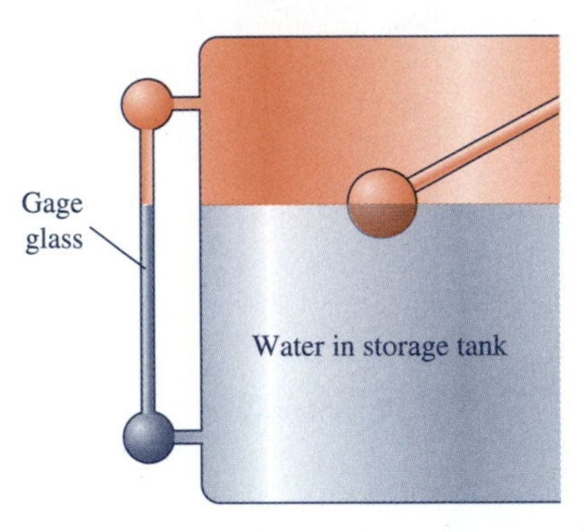

Problem 2·44

2·45 Ammonia, initially dry and saturated at 900 kPa, expands in a closed system to 600 kPa, 15°C. The expansion is a quasiequilibrium process that can be represented by a straight line on a *pv* diagram. Calculate the work.

2·46 Ammonia, initially dry and saturated at 200 kPa, is compressed in a steady-flow system to 500 kPa, 70°C. The compression is a quasiequilibrium process that can be represented by a straight line on a *pv* diagram. The flow rate is 0.116 kg/s. Calculate the power input.

2·47 Steam, initially dry and saturated at 200 psia, expands in a closed system to 50 psia, 300 F. The expansion is a quasiequilibrium process that can be represented by a straight line on a *pv* diagram. Calculate the work.

2·48 Ammonia enters a compressor at 200 kPa, 95 percent quality, with a velocity of 15 m/s through a cross-sectional area of 10.0 cm^2. It is discharged at 580 kPa, 59°C, through a cross-sectional area of 3.65 cm^2. The flow is steady. Determine the mass rate of flow and the discharge velocity.

2·49 Refrigerant R134a enters a compressor at 200 kPa, 93 percent quality, with a velocity of 22 m/s through a cross-sectional area of 12.0 cm^2. It is discharged at 600 kPa, 65°C, through a cross-sectional area of 5.25 cm^2. Determine the mass rate of flow and the discharge velocity.

2·50 Water enters one end of a long coiled tube at 5000 kPa, 220°C, at a steady rate of 2.20 kg/s. It is heated as it passes through the tube so that it leaves as superheated steam at 4200 kPa, 480°C. The inside diameter of the tube is 3.12 cm. Determine the velocity of the steam leaving.

2·51 A steam turbine exhausts at 5 cm of mercury with a quality of 91 percent. The flow rate is 140,000 kg/h. What cross-sectional area is needed at the exhaust flange if the velocity is not to exceed 40 m/s?

2·52 Given an ideal gas, is x a property if $x = (c/T)\,dT + (p/T)\,dv$ and c is a constant?

2·53 A spherical balloon 3.2 m in diameter contains hydrogen at 25°C and 100 kPa absolute. Compute the mass of hydrogen in the balloon.

2·54 What volume must be provided in a methane tank to hold 10 kg of methane at 1500 kPa and 30°C? If the temperature drops to 0°C, what will be the pressure?

2·55 The density of a gas is 1.15 kg/m^3 at 0°C and 100 kPa. Using the ideal-gas equation of state, compute the molar mass.

2·56 Determine the volume required to store 25 kg of ethene at 320 kPa and 35°C. If the temperature rises to 65°C, what will be the pressure in the storage tank?

2·57 A leaking tank initially contains 50 kg of helium at 900 kPa and 50°C. Determine the mass of helium that leaks out if the final pressure is 400 kPa and the final temperature is 25°C.

2·58 Compute the volume occupied by 1 kmol of CO at 101.3 kPa and 80°C.

2·59 Hydrogen gas is to be stored in a cylindrical tank having an internal diameter of 0.2 m and a length of 0.6 m. If the maximum allowable pressure and temperature are 2 MPa and 60°C, how many moles of hydrogen can be stored at the maximum temperature and pressure?

2·60 Air at 25 kPa gage, 25°C, is contained in a tank that has a volume of 0.05 m^3. Barometric pressure is 90 kPa, and the local acceleration of gravity is 9.61 m/s^2. Calculate the weight of air in the tank.

2·61 Air at 200 mm of mercury vacuum, 25°C, is contained in a tank that has a volume of 0.040 m^3. Barometric pressure is 690 mm of mercury, and the local acceleration of gravity is 9.75 m/s^2. Calculate the mass and the weight of air in the tank.

2·62 An elevator shaft in a building is 100 m high. The lower end of the shaft is open so that the air pressure inside the shaft is the same as outside. The outdoor temperature is −5°C and the air inside the shaft is at 25°C. Barometric pressure is 101.35 kPa. At the top of the shaft, what is the difference between the air pressure inside the shaft and outside the building? State your assumptions clearly.

2·63 A spherical balloon 48 ft in diameter contains helium at 80 F and 29.0 in. of mercury absolute. Compute the mass of helium in the balloon.

2·64 What volume must be provided in an air receiver to hold 20 lbm of air if the pressure is 250 psia and the temperature is 100 F? If the temperature drops to 32 F, what will be the pressure?

2·65 The density of a gas is 0.07704 lbm/cu ft at 32 F and 14.7 psia. Using the ideal-gas equation of state compute the molar mass.

2·66 Determine the volume required to store 50 lbm of air at 250 psia and 70 F. If the temperature rises to 175 F, what will be the pressure in the storage tank?

2·67 A closed tank contains 100 lbm of air at 140 psia, 125 F. How much air leaks out if the final pressure is 60 psia and the final temperature is 80 F?

2·68 Compute the volume occupied by 1 lbmol of CO at 14.7 psia and 190 F.

2·69 Hydrogen gas is to be stored in a cylindrical tank having an internal diameter of 8 in. and a length of 2 ft. If the maximum allowable pressure and temperature are 300 psia and 140 F, how many moles of hydrogen can be stored at the maximum temperature and pressure?

2·70 Air in a tire is initially at 380 kPa, 20°C, when the tire volume is 0.120 m^3. As the tire is warmed by the sun, the pressure becomes 450 kPa and the tire volume increases by 5 percent. Determine the final temperature and the mass of air in the tire.

2·71 An airplane has two tires that are inflated to 1500 kPa. Each tire has a volume of 0.021 m^3. How much would the weight of the airplane be reduced by using helium instead of air in the tires? What other considerations are involved in deciding whether to use helium?

2·72 An air bubble with a volume of 0.15 m^3 trapped in some debris 30 m below the surface of a lake is dislodged and rises to the surface. What is the volume of the bubble as it reaches the surface? If one of your assumptions is not met, would the volume actually be larger or smaller than the value you have calculated?

2·73 An air bubble trapped at the bottom of an open standpipe filled with water to a depth of 15 m is released. Describe the motion of the water surface as the bubble rises and escapes into the atmosphere.

2·74 Consider a closed rigid vessel 33.9 ft high (inside dimension) that is completely filled with water except for a small bubble of air that is held at the bottom. (The bubble might be held by means of an inverted cup, for example.) The pressure at the top of the water column is 1 atm, so the pressure of the bubble is 2 atm. The bubble is then released. After the bubble reaches the top of the vessel, what is the pressure of the water at the bottom? If two or more equal-size bubbles were initially at the bottom of the vessel, how would the pressure at the bottom vary if they were released one at a time?

2·75 How much air leaks into or from a building when air temperature inside changes from 15 to 25°C while the pressure remains essentially constant? Express your answer in terms of the building volume. Would subtracting the furniture volume from the building volume improve the value significantly?

2·76 A rigid vessel with a volume of 0.182 m^3 is divided into two equal volumes by a partition. Initially, hydrogen at 70.00 kPa, 20°C, is held on one side of the partition and the other side of the tank is evacuated. After 6 h it is discovered that some hydrogen has passed through the partition so that the pressure on the first side has dropped to 69.90 kPa with no change in temperature. Determine the average number of molecules per second passing through the partition during the 6-h period. (In this chapter, we have not mentioned Avogadro's number, but you have encountered it in other courses.)

2·77 A vessel with a volume of 0.32 m^3 containing air at 120.0 kPa and 12°C, was sealed 4200 years ago and placed in a tomb in Egypt. Since that time enough air has leaked from the vessel to cause the pressure to drop to 119.9 kPa with no change in temperature. Determine the average rate (molecules per second) at which air has left the vessel during the 4200 years. (In this chapter, we have not mentioned Avogadro's number, but you have encountered it in other courses.)

2·78 One-tenth pound of air is contained in a cylinder at 15 psia and 40 F. This air is to be compressed to twice its initial pressure and one-half its initial volume. Calculate (*a*) the net work done on or by the gas, (*b*) the net heat added to or taken from the gas, and (*c*) the net change in internal energy if (1) the pressure is first doubled at constant volume and the volume is then halved at constant pressure, and (2) the volume is first halved at constant pressure and the pressure is then doubled at constant volume.

2·79 Air trapped in a cylinder expands frictionlessly against a piston so that pV = constant. Initially the air is at 400 kN/m^2, 4°C, and occupies a volume of 0.02 m^3. The local value of g is 9.51 m/s^2.

(*a*) To what pressure must the air expand in order to perform 8100 J of work?
(*b*) What is the mass of air in the system?

2·80 Carbon dioxide is expanded frictionlessly in a closed system so that $pV^{1.5}$ = constant from initial conditions of 275 kPa, 170°C, V_1 = 0.06 m^3 to a final volume of V_2 = 0.12 m^3. Calculate (*a*) the work done and (*b*) the heat transferred.

2·81 Air trapped in a cylinder expands frictionlessly against a piston so that pV = constant. Initially the air is at 60 psia, 40 F, and occupies a volume of 0.50 cu ft. The local value of g is 31.8 ft/s^2.

(*a*) To what pressure must the air expand in order to perform 6000 ft·lbf of work?
(*b*) What is the mass of air in the system?

2·82 Carbon dioxide is expanded frictionlessly in a closed system so that $pV^{1.30}$ = constant from initial conditions of 40 psia, 340 F, V_1 = 2.0 cu ft to a final volume of V_2 = 4.0 cu ft. Calculate (*a*) the work done and (*b*) the heat transferred.

2·83 Methane enters a machine at 95 kPa, 35°C, with a velocity of 9.0 m/s through a cross-sectional area of 0.040 m^2. It leaves the machine at 220 kPa, 90°C, through a cross-sectional area of 0.015 m^2. Heat removed from the methane flowing through the machine amounts to 60.4 kJ/kg. The flow is steady. Determine the outlet velocity.

2·84 Methane enters a machine at 14.0 psia, 95°F, with a velocity of 30 fps through a cross-sectional area of 0.4 sq ft. It leaves the machine at 32.0 psia, 200 F, through a cross-sectional area of 0.15 sq ft. Heat removed from the methane flowing through the machine amounts to 26.1 B/lbm. The flow is steady. Determine the outlet velocity.

2·85 The density of CO_2 at 8000 kPa and 125°C is measured and found to be 133 kg/m^3. Determine the compressibility factor for CO_2 at this state.

2·86 The measured density of oxygen at 40 atm and −53°C is 77.7 kg/m^3. Determine the compressibility factor for oxygen at this state.

2·87 How much error is involved in calculating the density of oxygen at 30 atm and room temperature by the ideal-gas equation of state?

2·88 How much error is involved in calculating the density of carbon dioxide at 30 atm and room temperature by the ideal-gas equation of state?

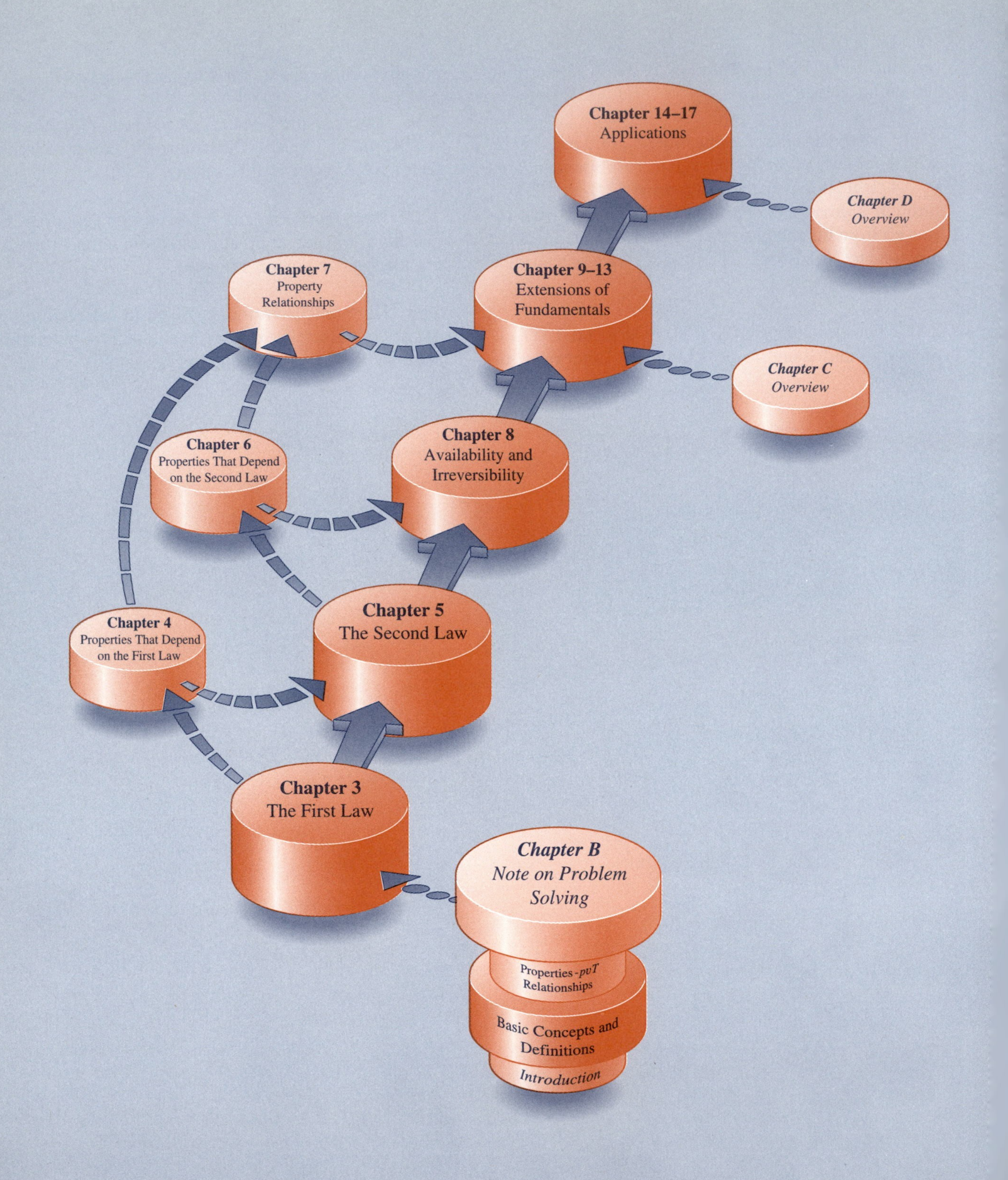
Chapter 14–17
Applications
Chapter D
Overview
Chapter 7
Property
Relationships
Chapter 9–13
Extensions of
Fundamentals
Chapter C
Overview
Chapter 6
Properties That Depend
on the Second Law
Chapter 8
Availability and
Irreversibility
Chapter 4
Properties That Depend
on the First Law
Chapter 5
The Second Law
Chapter 3
The First Law
Chapter B
Note on Problem
Solving
Properties - pvT
Relationships
Basic Concepts and
Definitions
Introduction

Chapter B

Note on Problem Solving

The variety of work in engineering is limitless. This variety is one of the reasons engineering work is challenging and offers great personal satisfaction. Engineers are problem solvers, and engineering calls for many different approaches to problem solving.

A common type of engineering problem involves various physical principles and knowledge that are applied to predict the behavior of a system. The system may be an existing one or it may be one that is conceived during a design, that is, one that has never existed.

A problem that involves only one physical principle is called simple. All seven of the example problem solutions of Chapter 1 are simple problems. Engineering problems are usually complex. By complex we mean that several principles are involved. The principles may need to be applied sequentially or simultaneously. Data may have to be obtained from one or more sources and processed before it can be used. Since several steps are involved in the solution of the problem, you need to plan your solution approach before starting calculations.

No single approach, strictly followed, works for all problems. Engineers have many problem-solving tools. They choose an appropriate mixture of them for work on any one design, analysis, or investigation. Through experience you will develop the judgment to select the best tools for any job. Nonetheless, some general guidelines for problem solving are consistently useful and should be kept in mind. They can be used as needed when you are solving problems and will help you to avoid errors and keep from getting "stuck."

The guidelines presented here may be more extensive than necessary at this early stage of your study of thermodynamics. You will not exercise some of them until you solve the more complex problems in later chapters. If we waited until later to introduce them, however, you might miss their benefits in forming good problem-solving habits. Therefore, follow them as much as you can now. Later in your study, review this chapter. At that time its value is likely to be more apparent.

Your abilities in problem solving will grow only with practice. They will grow most rapidly if your practice is on problems of increasing challenge that continually use the principles, information, and methodologies you are learning.

Understand the Problem

The first step in solving a problem is obvious. Make sure that you understand the problem. What is wanted? If you are to design a new or improved product, the first step is to understand the customer's needs and desires. At various stages as a design evolves, analysis of alternative designs and predictions of behavior are required. Another engineering job is to analyze an existing system or device to explain or predict its behavior under various conditions. In all these cases, the first step is to define the problem. What is the situation? Does it make physical sense? Does it appear determinate? That is, as the problem is defined, do the conditions and constraints actually *determine* the results or are there other pertinent factors that must be considered? These questions must be answered first. A labeled sketch is usually helpful at this stage.

More Detailed Understanding

After a general understanding of the problem is established, the next step is to fill in the specifics that define the problem. Some questions to ask yourself are

- *What kind of system is involved?* In answering this question, describe the system in sufficient detail so that principles involved can be applied to it. This must always be done. What assumptions are made in modeling the system?
- *What substances are involved?* Are there equations of state? Are property data for the substance available? What assumptions need to be made in modeling the substance?
- *What features, characteristics, and data are known?* Are all the data available from the sources at hand (databases, handbooks, reference works)? What is the quality of the data available? Are better data needed? Is a trip to the library necessary? Is information from a manufacturer needed?
- *What kind of processes are involved?* Are they equilibrium processes? What constraints are there? Again, what assumptions are made?
- *What are the expected ranges of the results?*
- *What order of accuracy is wanted in the result?* A rough approximation that might be called a "first-order" result? A more accurate or second-order result? A high-order result of great accuracy? All these different orders of results have important roles in engineering.

Nearly always, a sketch labeled with the answers to the questions above should be made.

In engineering design practice, additional questions are often productive: Are these constraints the right ones? Are they really necessary? If one or more were removed, would better solutions result? (Better might mean quicker, more accurate, more robust, cheaper, etc.)

Analysis Starts

Once the problem is fully understood, analysis starts, and you should ask questions such as

- *What principles are involved?*
- *What principles or formulations involve the quantities sought?*
- *Do we have sufficient knowledge to use them? If not, what other quantities must we obtain?*
- *What principles or formulations involve these quantities?*
- *Do we have sufficient knowledge to use these additional principles or formulations? If not, what other quantities must we still determine?*
- *Are limiting case analyses possible and helpful?*

Continue in these directions until you can calculate the quantities needed in the formulations that involve the quantities you were originally looking for.

Before starting numerical calculations, always estimate the results. At first this is difficult, but with practice your ability will increase remarkably.

When Difficulties Arise: What To Do When You Are Stuck

When you are making no progress on a solution, ask yourself, "Why am I stuck?" or "What is it that I need to start moving again?" This question appears almost trivial at first, but it often elicits an answer that breaks the impasse. Perhaps you need a certain quantity. Look again to see if you already have a value for it. Then ask more questions of yourself: What are the various ways I could determine it? Can I assume a value? Can I use limiting case values? 0? 1? Further questions to ask are: Can I solve this problem by applying entirely different principles? What principles apply that I have not yet used?

Review the problem statement, your sketch and definition of the system, and your list of assumptions to see that these are all reasonable and consistent. Do the equations you have written apply to the situation at hand? If so, are they correctly written? These questions will not resolve every difficulty you encounter, but they will help in many cases.

Numerical Calculations

Delay numerical calculations as long as possible. Finish the analysis first so that you will have a complete outline of the numerical calculations and confidence that the problem can be solved. If you want to make numerical calculations for various ranges of input quantities, completing the entire analysis first is clearly advantageous. Of course, there are exceptions; there are times when for good reasons you may wish to complete the numerical calculations for one part of a problem before completing the symbolic solution. (For example, you may wish to check the validity of an early assumption.)

Decide how to do the numerical calculations: longhand, using a calculator and perhaps some graphical methods? By computer? If you will use a computer, will you use an equation solver, spreadsheet, special program, or a program you will write yourself? Part of engineering judgment is knowing when and how best to use a computer. You can best develop this judgment through guided practice.

Beside estimating overall results as suggested earlier, form the rewarding habit of estimating from the input numbers the results of individual numerical calculations. This habit will detect many errors before they cost you much time.

Finally

Is the result reasonable? Is the answer robust? In engineering practice, we often must assess how sensitive the result is to the assumptions made and to any parameters used. It is well to form early the habit of doing this. Check some details: Are the units correct? Is the sign correct? Is this really what I was trying to determine? Then, have I presented the solution in a form that is easy for a reader to understand?

Some Comments on Problem Solving in the Academic Environment

The fundamentals of problem solving are the same in practice and in academia. However, in academia one of your foremost goals is to develop your problem-solving abilities, and you are doing this while learning principles, information, and procedures that are new to you. You are probably studying several courses simultaneously and often feeling that you do not have enough time for all the courses. Therefore, some additional pointers may be helpful. Some of these you may resist, but we advise you to try them. Many students have found them valuable, and not only in thermodynamics.

As you start each problem, record the time. Record the time again as you complete each problem. Also, recording the time in the margin at intermediate points during a long problem solution will help you to learn where your time goes while solving problems. Always keep some time pressure on yourself as you solve homework problems. This will help you to focus on the problem at hand and filter out distractions. By all means, stop to reflect on critical points in your problem solution and try to clear up troublesome points as they occur. However, while you are making calculations, writing down solution steps, extracting tabular data, or performing similar operations, keep moving! Developing the habit of keeping time pressure on yourself will pay dividends. Among other benefits, this will help when you are solving problems under stressful conditions, as happens in every engineer's career.

Avoid the practice of making numerical calculations first on scratch paper or in the margin. Try to complete the entire analysis of a complex problem before starting calculations; but once you have the problem completely analyzed, the solution you start writing should be your final one, presented in a form that will be clear to any reader.

After completing each problem spend a few additional minutes to enhance the return from your time and effort by asking yourself some questions: What does this problem solution show me? What have I learned? What caused difficulty? What principles are involved? Could I now solve a similar problem in less time?

If some parts of the problem solution have caused you difficulty, try to determine why this occurred. If it appears that there are things you should learn or abilities you should develop, start working on them right now. Do not just make a note to work on them in the future. (You may not have 2 hours right now, but spend at least 15 minutes.)

Some Important Points in Summary

There is no one procedure (or methodology) that is followed in every instance. Instead, we need a general approach, many tools, plus alertness and flexibility. We must continually ask questions about what we are doing and what we are dealing with. Some questions may be useless at one stage of a problem solution and highly fruitful at another stage.

Do not assume from this presentation that problem

solving always proceeds in a simple sequential manner. There is often some backtracking, jumping ahead, or side trips. Whatever it takes to make progress toward an optimum solution is to be used. Even expert problem solvers make mistakes, but their continual checking and questioning allows them to detect mistakes early and correct them.

Most of all: *Keep moving*. Make every action and every time investment count. Do not misunderstand this admonition, though. Occasional reflection on what has already been accomplished, evaluation of results, and consideration of alternative paths at various stages of problem solving are necessary and productive.

This note on problem solving, unlike other chapters in this book, includes no list of problems, because we do not want to emphasize problem solving per se separately from thermodynamics. However, as you develop the methods presented here you will find that they are general and apply in other areas as well as in thermodynamics.

Two example problem solutions follow. The second problem requires results for a range of specified values and an output plot, so a computer solution is indicated. Various equation solver software packages are available commercially, and each has its own strengths. We show solutions using two widely used packages.

Example 1 Heat Exchanger

PROBLEM STATEMENT

Methane, CH_4, enters a heat exchanger as shown in the figure at 95 kPa, 20°C, with a velocity of 28.0 m/s. The methane leaves at 90 kPa and a temperature 65 Celsius degrees higher than the inlet temperature. The inlet and outlet cross-sectional areas for the methane flow are 0.80 m² and 0.94 m², respectively. The exhaust gases that flow around the tubes carrying the methane enter at 370°C and leave at 322°C. Determine the exit velocity of the methane.

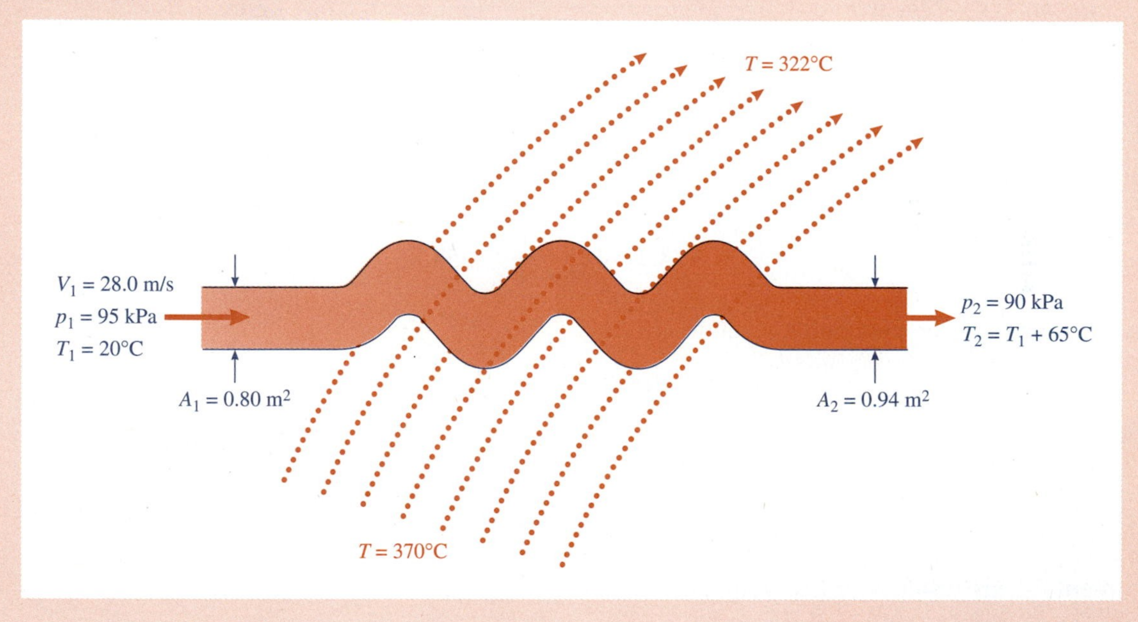

SOLUTION

Analysis: Since we are looking for the exit velocity, the continuity equation (or principle of mass conservation) comes to mind. Single values of various properties are given and there is no indication that the flow is unsteady; therefore, we assume that it is steady. A single value of each property is given at the inlet and at the outlet, so we will model the flow as one-dimensional. For steady, one-dimensional flow, the continuity equation is

$$\rho_1 A_1 V_1 = \rho_2 A_2 V_2$$

Both areas and V_1 are known, and we are seeking V_2, so we must find values of the density at inlet and outlet. Since pressure and temperature at each section are known, we can determine the densities from the ideal-gas equation of state. At a pressure of approximately one atmosphere and the temperature given, the ideal-gas equation of state should accurately model methane, which has a molar mass of only 16 kg/kmol. Thus we can calculate both densities to use in the continuity equation.

Before starting the calculations, let us estimate the result. If nothing else were changed between inlet and outlet, the larger outlet area would cause a decrease in velocity in the ratio $A_1/A_2 = 0.80/0.94$. The air density decreases through the heat exchanger by virtue of both the slight decrease in pressure and the increase in temperature, and this decrease in density tends to increase the velocity. Judging the magnitudes of these effects, we estimate that the outlet velocity will not differ much from the inlet velocity. That is, we expect it to be neither less than 20 m/s nor more than 40 m/s.

Calculations: We start by calculating the density at inlet and outlet by the ideal-gas equation of state,

$$\rho_1 = \frac{p_1}{RT_1} = \frac{p_1 M}{\bar{R}T_1} = \frac{95 \text{ kPa}\left(16\dfrac{\text{kg}}{\text{kmol}}\right)}{8.31441\dfrac{\text{kJ}}{\text{kmol·K}}(293 \text{ K})} = 0.624 \text{ kg/m}^3$$

and with $T_2 = T_1 + 65 = 20 + 65 = 85°\text{C}$,

$$\rho_2 = \frac{p_2}{RT_2} = \frac{p_2 M}{\bar{R}T_2} = \frac{90 \text{ kPa}\left(16\dfrac{\text{kg}}{\text{kmol}}\right)}{8.31441\dfrac{\text{kJ}}{\text{kmol·K}}(358 \text{ K})} = 0.484 \text{ kg/m}^3$$

Then we rearrange the continuity equation,

$$\rho_1 A_1 V_1 = \rho_2 A_2 V_2$$

to find

$$V_2 = V_1\frac{\rho_1 A_1}{\rho_2 A_2} = 28.0\frac{0.624(0.80)}{0.484(0.94)} = 30.7 \text{ m/s}$$

Comment: This result is in line with our estimate.

This is not a difficult problem, yet it is *complex* in that it involves two ideas: the continuity equation and the ideal-gas equation of state.

Some data mentioned in the statement of the problem are not needed. This is often the case in problems in this book. In practice, of course, usually nothing is "given." The engineer must decide what data to measure or otherwise determine before starting calculations.

Example 2 Compressor

PROBLEM STATEMENT

A compressor with a very large inlet area and a discharge area of 880 mm^2 is to be used to compress air from 100 kPa, 20°C, to a pressure between 120 kPa and 240 kPa at a rate of 0.0460 kg/s. Assuming that the compression is a quasiequilibrium process in which pv^n = constant with $n = 1.32$, plot the power input versus discharge pressure for the range of 120 to 240 kPa.

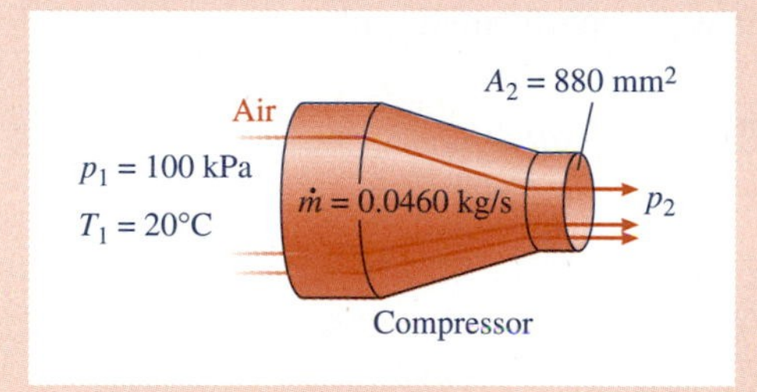

SOLUTION

Analysis: We will use a computer to save time because the calculations are to be done for several values of the discharge pressure and because a curve is to be plotted.

The single value given for flow rate and the single values for various properties imply that the flow is steady, so we define the system as a steady-flow system and assume one-dimensional flow at the inlet and outlet. We can calculate the power input from

$$\dot{W}_{in} = \dot{m}w_{in}$$

and the flow rate is specified. Since the compression is a quasiequilibrium steady-flow process, the work input is given by

$$w_{in} = \int_1^2 v\,dp + \frac{V_2^2 - V_1^2}{2}$$

and we can integrate the first term on the right side of the equation because a relationship between p and v

is specified. $V_1 \approx 0$ because the inlet area is very large. To obtain V_2, we will use the continuity equation

$$V_2 = \frac{\dot{m}v_2}{A_2}$$

The only quantity on the right side that we do not have a value for is v_2, and we can get this from the relationship

$$p_1 v_1^n = p_2 v_2^n = c$$

which we can use if we have an independent method of calculating v_1. The pressure and temperature ranges are such that we can model the gas by the ideal-gas equation of state, so we can calculate v_1 by

$$v_1 = \frac{RT_1}{p_1}$$

While making this analysis, we can enter the equations into the equation solver. We also describe each equation as we enter it. These annotations help readers understand the solution. (We may find them helpful ourselves if we return to the solution at a later time.)

Calculations: First, the integration needed in the expression for w_{in} is performed. $pv^n = c$, so

$$\int_1^2 v\,dp = \int_1^2 \left(\frac{c}{p}\right)^{1/n} dp = \int_1^2 (p_1 v_1^n)^{1/n} p^{-1/n}\,dp$$

$$= \frac{1}{p_1^n v_1}\left(\frac{n}{n-1}\right)[(p_2^{(n-1)/n} - p_1^{(n-1)/n})]$$

$$= p_1 v_1\left(\frac{n}{n-1}\right)\left[\left(\frac{p_2}{p_1}\right)^{(n-1)/n} - 1\right]$$

$$= RT_1\left(\frac{n}{n-1}\right)\left[\left(\frac{p_2}{p_1}\right)^{(n-1)/n} - 1\right]$$

so that

$$w_{in} = RT_1\left(\frac{n}{n-1}\right)\left[\left(\frac{p_2}{p_1}\right)^{(n-1)/n} - 1\right] + \frac{V_2^2}{2}$$

This equation is added to those entered into the equation solver during the analysis. Notice that there are five equations and five unknowns: $\dot{W}_{in}$, w_{in}, V_2, v_2, and v_1. (We show here results from two equation solvers, TK Solver and Mathcad. Both of these can perform numerical integration, but for this example we are entering only algebraic equations into the equation solvers.)

Completing the Computer Model for TK Solver: After entering the equations on the Rule Sheet, we enter on the Variable Sheet the appropriate units and the input values. For the discharge pressure, p_2, we enter a list of values ranging from 120 to 240 kPa. Other variables that change values as p_2 is changed are also marked by an L in the status column to show that they will each have a list of values.

```
═══════════════════════ RULE SHEET ══════════════════════════════════════
S Rule──────────────────────────────────────────────────────────────────
  ;───────────────── Problem Solving Example 2 ──────────────────────────
  ;Model the process as one-dimensional steady flow.
  Wdotin = mdot * win           ;Relation among power, work, and flow rate
  mdot = A2 * V2 / v2           ;Continuity equation
  p1 * v1^n = p2 * v2^n         ;Specified pv relation for the process
  p1 * v1 = R * T1              ;Ideal gas equation of state for inlet gas
  win = R * T1 *(n/(n-1))*((p2/p1)^((n-1)/n) - 1) + V2^2/2    ;Expression for
                                ;work of steady-flow, quasiequilibrium process
                                ;that is obtained from win = INT(vdp) + V2^2/2
                                ;Calculation units are m, s, kg, K, J, Pa.

═══════════════════════ VARIABLE SHEET ══════════════════════════════════
St Input──── Name─── Output─── Unit────── Comment────────────────────────
                                ────────── Problem Solving Example 2 [PSEX2.TK]
L            Wdotin  1.38      kJ/s       Power input in kJ/s or kW
   0.046     mdot              kg/s       Mass flow rate
L            win     30.1      kJ/kg      Work in per unit mass
L            v2      0.652     m^3/kg     Specific volume at outlet
   880       A2                mm^2       Cross-sectional area at outlet
   100       p1                kPa        Pressure at inlet
             v1      0.841     m^3/kg     Specific volume at inlet
   1.32      n                            Exponent in pv relation
L  140       p2                kPa        Pressure at outlet
   0.287     R                 kJ/(kg*K)  Gas constant for air
   20        T1                C          Temperature at inlet
L            V2      34.1      m/s        Velocity at outlet
```

Results of Example Problem 2

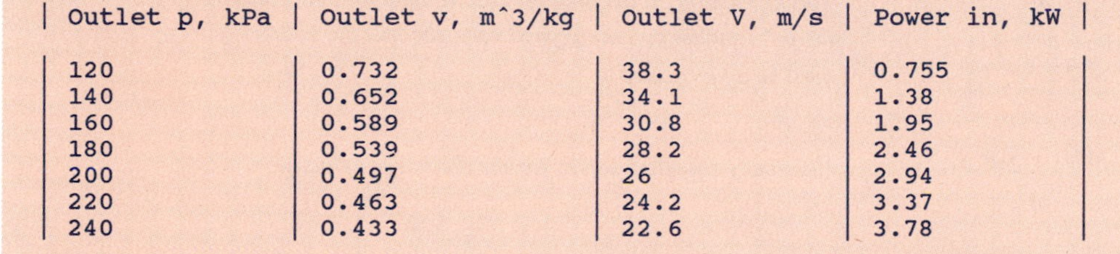

Outlet p, kPa	Outlet v, m^3/kg	Outlet V, m/s	Power in, kW
120	0.732	38.3	0.755
140	0.652	34.1	1.38
160	0.589	30.8	1.95
180	0.539	28.2	2.46
200	0.497	26	2.94
220	0.463	24.2	3.37
240	0.433	22.6	3.78

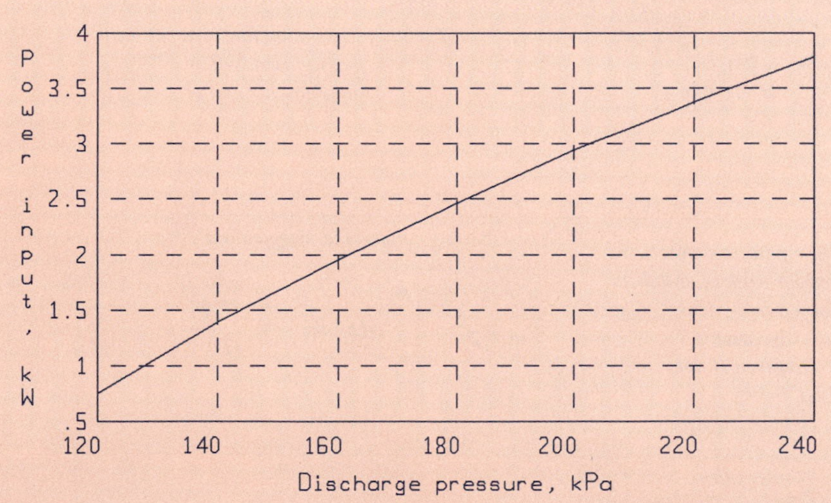

Example 2 TK Solver solution. (*TK Solver is a registered trademark of Universal Technical Systems, Inc., Rockford, Illinois.*)

Mathcad® 5.0 Solution

Given:

$p_1 := 100 \cdot \text{kPa}$	*Pressure at inlet*	Additional Definitions
$T_1 := (273.15 + 20) \cdot \text{K}$	*Temperature at inlet*	$\text{kPa} \equiv 1000 \cdot \text{Pa}$
$mdot := 0.0460 \cdot \frac{\text{kg}}{\text{sec}}$	*Mass flow rate*	$\text{kJ} \equiv 1000 \cdot \text{joule}$
		$R := 0.287 \cdot \frac{\text{kJ}}{\text{kg} \cdot \text{K}}$
$A_2 := 880 \cdot \text{mm}^2$	*Cross-sectional area at discharge*	
$n := 1.32$	*Exponent in pv relation*	
$p2_{min} := 120 \cdot \text{kPa}$	*Range of final pressures*	
$p2_{max} := 240 \cdot \text{kPa}$		

Solution:

$$v_1 := R \cdot \frac{T_1}{p_1}$$

The ideal gas equation of state, solved for the initial specific volume

$$v_2(p_2) := \left(\frac{p_1}{p_2}\right)^{\frac{1}{n}} \cdot v_1$$

This expression for the final specific volume follows from the pv^n condition. Since p_2 will vary, the specific volume must be a function of it.

$$V_2(p_2) := \frac{mdot \cdot v_2(p_2)}{A_2}$$

Continuity equation, solved for the discharge velocity

$$w_{in}(p_2) := R \cdot T_1 \cdot \frac{n}{n-1} \cdot \left[\left(\frac{p_2}{p_1}\right)^{\frac{n-1}{n}} - 1\right] + \frac{V_2(p_2)^2}{2}$$

Expression for work per unit mass of steady-flow, quasiequilibrium process

---

Alternatively, the equation shown to the right can be used for work in. (Mathcad can either perform a numerical integration or the algebraic manipulation.) However, if this expression is used, then $v_2(p_2)$ must be changed to v(p).

The following equation is "toggled out":

$$w_{in}(p_2) := \int_{p_1}^{p_2} v(p)\, dp + \frac{V_2(p_2)^2}{2}$$

$$Wdot_{in}(p_2) := mdot \cdot w_{in}(p_2)$$

Relation among power, work, and flow rate

Create array of discharge pressures

$$i := 0..12$$

$$p2_i := \frac{p2_{max} - p2_{min}}{12} \cdot i + p2_{min}$$

Compute and display results

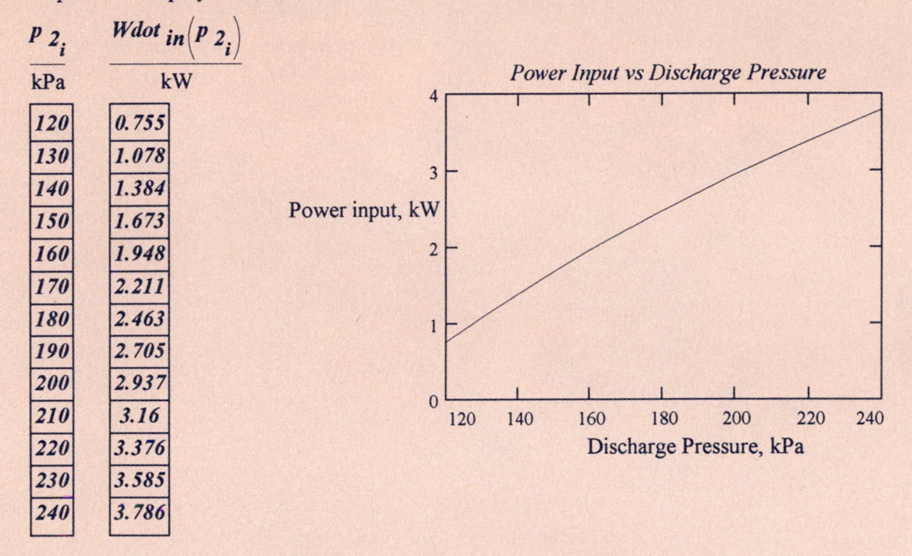

$\frac{p2_i}{\text{kPa}}$	$\frac{Wdot_{in}(p2_i)}{\text{kW}}$
120	0.755
130	1.078
140	1.384
150	1.673
160	1.948
170	2.211
180	2.463
190	2.705
200	2.937
210	3.16
220	3.376
230	3.585
240	3.786

Example 2 Mathcad solution. (*Mathcad is a registered trademark of MathSoft, Inc., Cambridge, Massachusetts. Used by permission of MathSoft, Inc.*)

Completing the Computer Model for Mathcad: We enter the given quantities and the equations in the sequence shown so that the equations can be solved from top to bottom. Then we create an array of discharge pressures covering the range from 120 to 240 kPa.

Results: The results for each equation solver are shown in a plot and also in a table. The computer equation solver prepared both of these automatically.

Comment: This kind of calculation where results are needed for a series of input values is frequently encountered in engineering design. Often results are needed for ranges of two or more input variables, so computer use is essential to avoid the tedium of repetitive calculations and preparation of output displays.

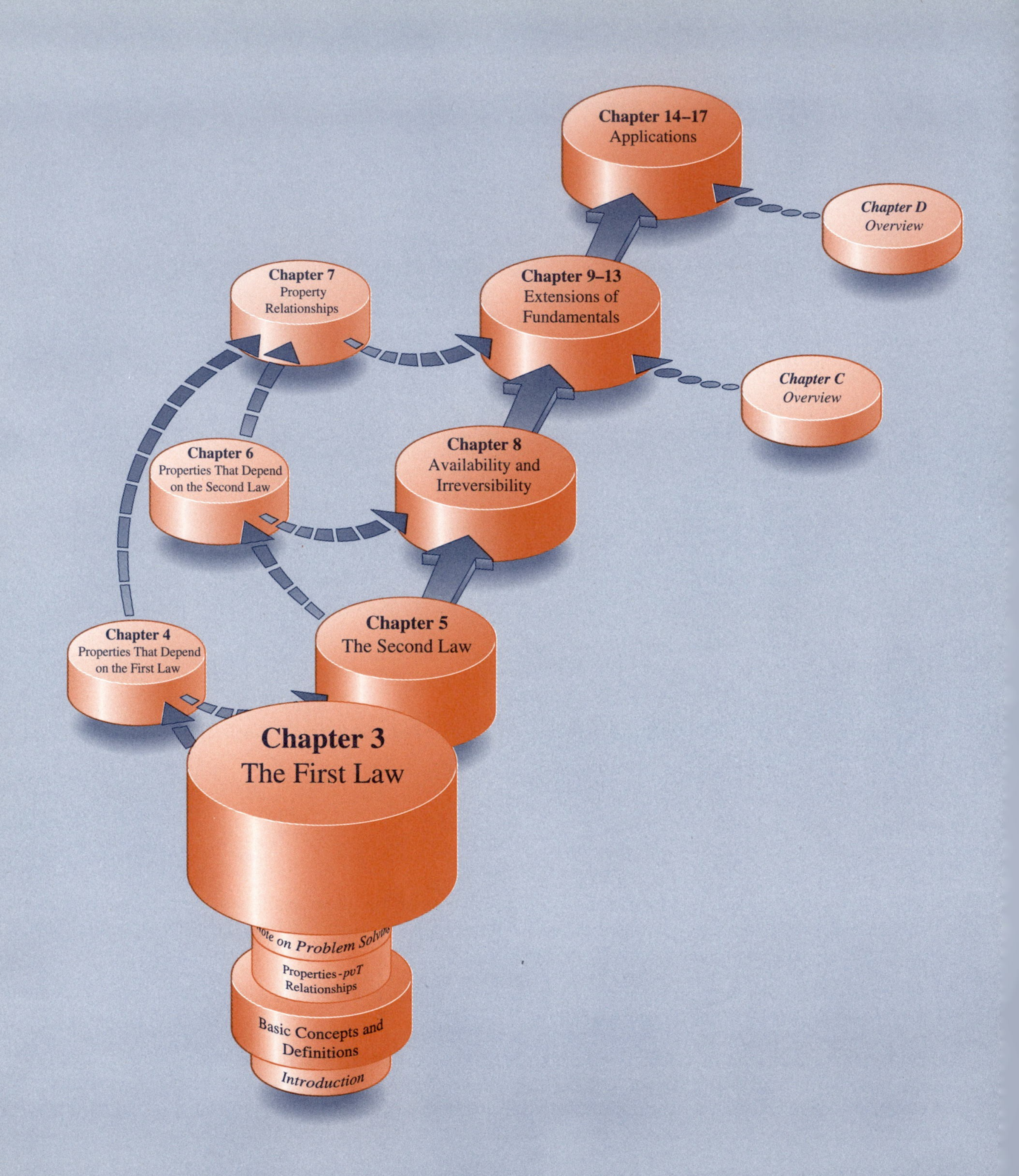

Chapter 14–17
Applications
Chapter D
Overview
Chapter 7
Property
Relationships
Chapter 9–13
Extensions of
Fundamentals
Chapter C
Overview
Chapter 6
Properties That Depend
on the Second Law
Chapter 8
Availability and
Irreversibility
Chapter 4
Properties That Depend
on the First Law
Chapter 5
The Second Law
Chapter 3
The First Law
on Problem Solv
Properties-*pvT*
Relationships
Basic Concepts and
Definitions
Introduction

Chapter 3

The First Law of Thermodynamics

In this chapter we discuss the relationship between work and heat. This relationship leads to the first law of thermodynamics and the definition of energy, concepts that are useful in a wide range of engineering analyses. After studying this chapter you will be able to relate work, heat, and the changes in properties of substances for various processes in open and closed systems.

3·1 The First Law for Cyclic Processes

Drill bits, saw blades, and bicycle tire pumps become hot during use. This is obviously not from heat transfer because the objects become warmer than the surroundings. They get hotter when work is done on them. However, these same effects *could* be caused by heat transfer. This suggests that some effects can be caused equivalently by heat or work and that there is some relationship between heat and work. This is a scientific hypothesis.

If you were to design some experiments to test this hypothesis, how might you proceed? You might start by measuring heat and work of various systems as they change state. However, you could not determine a relationship between work and heat directly, because the system is also changing state. To remove this effect, you might consider only cyclic processes. Then there would be no net change in the state of the system.

Consider a closed system that can do work on the surroundings or have work done on it, and that can have heat transfer both to and from the surroundings. This would be a useful device for testing the hypothesis. Figure 3·1 shows such a system. There are no limitations on the kinds of processes that may occur as long as the system can be returned to its initial state. Confirming when a cycle has occurred will, of course, require some measurements within the system. However, all the necessary measurements of work and heat can be made entirely outside the system. For example, the standard systems used to measure heat might be quantities of water brought into contact with the system. Thus, if the system rejects heat, the water temperature would rise. Conversely, if heat is transferred to the system, the temperature of the water would drop. Likewise, work might be measured by counting the number of standard weights lifted a specified distance.

Such measurements of heat and work for cycles of closed systems result in a plot like that of Fig. 3·2. This plot shows that the net work of the cycle, $\oint \delta W$, is proportional to the net heat transfer of the cycle, $\oint \delta Q$. *Proportional!* There is indeed a relationship between work and heat, and the hypothesis has been supported.

It is this relationship that forms the basis of the first law of thermodynamics.

The direct proportionality shown in Fig. 3·2 holds for any arbitrary units of heat and work. Historically, different units were used. However, the proportionality that is shown by experiments means that heat and work can be expressed in the same units. Figure 3·3 is a replot of Fig. 3·2 with Q and W measured in the same units. We conclude that for any closed system,

$$\oint \delta Q = \oint \delta W \tag{3·1}$$

For any cycle of a closed system, the net heat transfer equals the net work.

The first law of thermodynamics

This is a statement of **the first law of thermodynamics**. It applies to any closed system that undergoes a cycle; there are no restrictions or limitations.

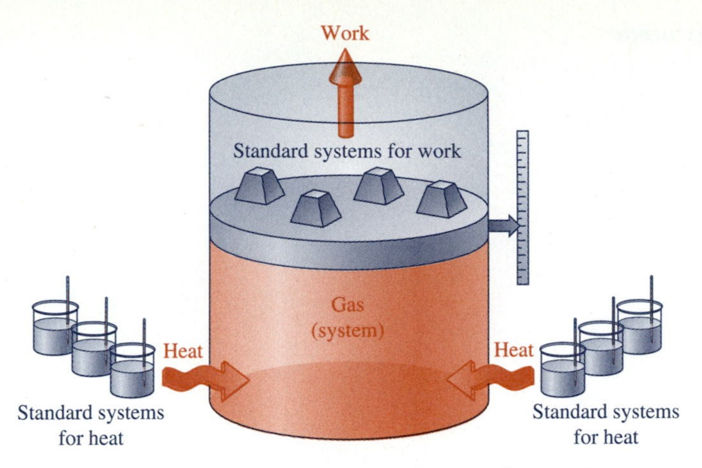

Figure 3·1 Closed system that can undergo various cycles.

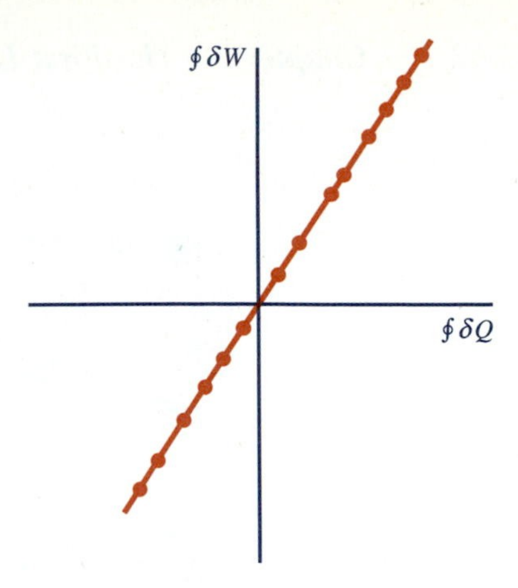

Figure 3·2 Experiments show that the net work of any cycle is proportional to the net heat transfer.

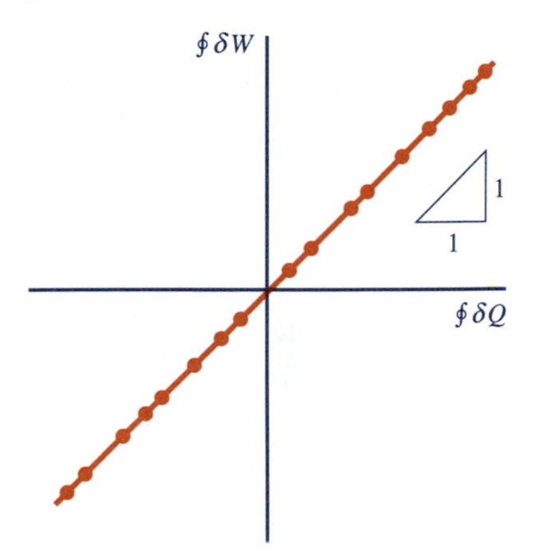

Figure 3·3 Heat and work can be measured in the same units, and for any cycle, measurements show that $\oint \delta Q = \oint \delta W$.

The first law of thermodynamics is supported by a vast number of experiments; no exceptions have been observed. Consequently, we accept it as a law of nature. We deduce many results from it and use it in innumerable applications. Notice that it is an empirical law, based on induction from experimental observations. It has not been proven.

The relationship between heat and work has been shown using only cyclic processes. Using cycles eliminates any possibility that the relationship is influenced by net changes in the system. In the next section, we see what happens when the process is not cyclic.

Example 3·1

PROBLEM STATEMENT

A gas held in a cylinder behind a piston as shown in the figure undergoes a cycle composed of three processes:

Process 1–2: a compression without heat transfer. For this process, $W_{1-2} = -58$ kJ.
Process 2–3: an expansion during which heat is added and work is done on the piston. $Q_{2-3} = 253$ kJ; $W_{2-3} = 95$ kJ.
Process 3–1: a cooling of the gas while the piston is stationary to bring the gas to its initial state.

Determine the net heat transfer and the heat transfer of process 3–1.

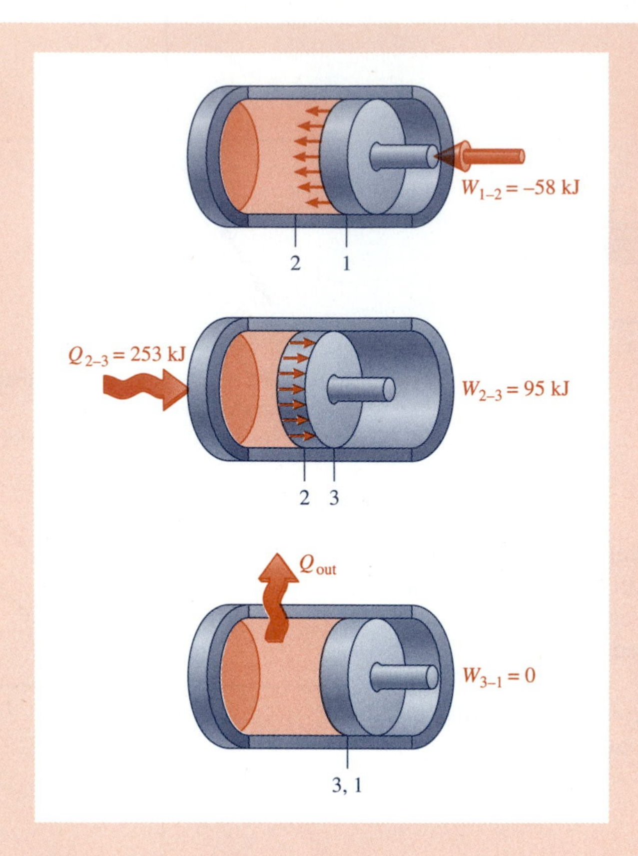

SOLUTION

For a cyclic process, the first law tells us that the net heat transfer equals the net work done, so we will first calculate the work of the cycle.

During process 3–1, the piston is stationary, so no work is done. Consequently, the net work of the cycle is

$$\oint \delta W = W_{1-2} + W_{2-3} + W_{3-1} = -58 + 95 + 0 = 37 \text{ kJ}$$

Then, by the first law,

$$\oint \delta Q = \oint \delta W = 37 \text{ kJ}$$

The net heat transfer is made up of the heat transfer of the three processes:

$$\oint \delta Q = Q_{1-2} + Q_{2-3} + Q_{3-1}$$

so

$$Q_{3-1} = \oint \delta Q - Q_{1-2} - Q_{2-3} = 37 - 0 - 253 = -216 \text{ kJ}$$

The negative sign indicates that 216 kJ of heat is transferred *from* the system to the surroundings during process 3–1.

Example 3·2 Heat Pump

PROBLEM STATEMENT

A heat pump is used for heating a building. It picks up heat from well water at 10°C and discharges heat to a building to maintain it at 20°C. Over a period of 14 days, a meter shows that 1490 kW·h of electrical work was put into the heat pump. For an average supply of heat to the building of 120,000 kJ/h, how much heat was removed from the well water during this time period? (This kind of calculation is involved in evaluating the suitability of a well for supplying heat to a heat pump.)

SOLUTION

We first define the system for study as the heat pump, including the refrigerant that flows through the heat exchangers to absorb heat from the well water and give up heat to the building. No matter crosses the system boundary, so it is a closed system. We make a sketch of the system and show the data on it.

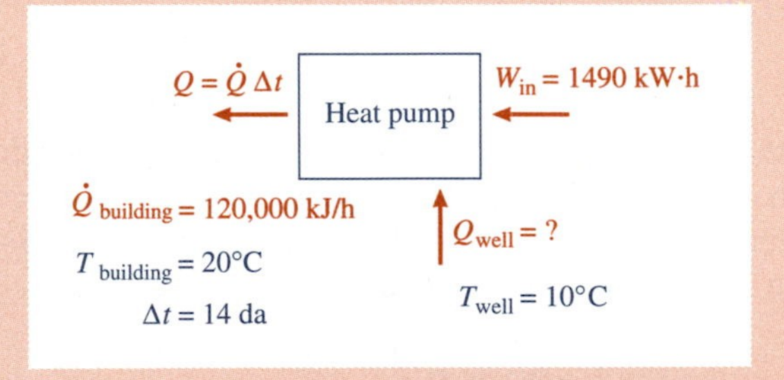

At the end of the 14-day period the system is in the same state it was in at the beginning, so the system has undergone a cycle (or, actually, many cycles) during this time interval. The first law tells us that

$$\oint \delta Q = \oint \delta W$$

Also, assuming that there is no heat transfer between the system and anything other than the well water and the building,

$$\oint \delta Q = Q_{\text{well}} + Q_{\text{building}}$$

so,

$$
\begin{aligned}
Q_{\text{well}} &= \oint \delta Q - Q_{\text{building}} = \oint \delta W - Q_{\text{building}} \\
&= \oint \delta W - \dot{Q}_{\text{building}}\,\Delta t \\
&= -1490\ \text{kW·h}(3600\ \text{s/h}) - (-120{,}000\ \text{kJ/h})14\ \text{days}(24\ \text{h/day}) \\
&= 35{,}000{,}000\ \text{kJ}
\end{aligned}
$$

(The positive sign indicates that heat is added *to* the heat pump from the well. Similarly, $\dot{Q}_{\text{building}} = -120{,}000$ kJ/h indicates that heat transfer is *from* the heat pump into the building.)

3·2 The First Law for Noncyclic Processes: Defining Stored Energy

Now we consider the first law for noncyclic processes.

The first law of thermodynamics as stated above applies only to cyclic processes of closed systems. We now extend the application to noncyclic processes. In doing this, we define a valuable property, stored energy.

Equation 3·1, the first law, can be rearranged as

$$\oint (\delta Q - \delta W) = 0 \tag{3·1}$$

which allows us to show that the quantity $(\delta Q - \delta W)$ is a property. Consider a closed system that is changed from a state 1 to a state 2 by some process shown as path A on a property diagram such as the pV diagram of Fig. 3·4. The system is returned to state 1 via path B. The first law states that

$$\oint_{1-A-2-B-1} (\delta Q - \delta W) = 0$$

or

$$\int_{1-A}^{2} (\delta Q - \delta W) + \int_{2-B}^{1} (\delta Q - \delta W) = 0 \tag{a}$$

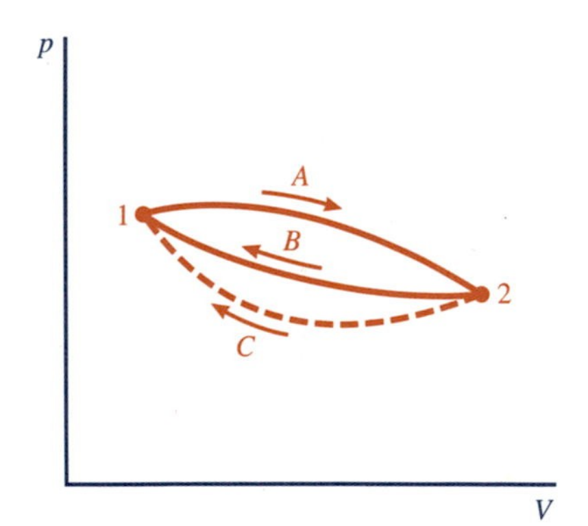

Figure 3·4 Different paths for the same state change of a closed system.

If C is any other path by which the system could be restored from state 2 to state 1, then it is likewise true that

$$\int_{1-A}^{2} (\delta Q - \delta W) + \int_{2-C}^{1} (\delta Q - \delta W) = 0 \tag{b}$$

Comparing Eqs. *a* and *b*, we see that

$$\int_{2-B}^{1} (\delta Q - \delta W) = \int_{2-C}^{1} (\delta Q - \delta W)$$

Since B and C were any two paths between states 2 and 1, it follows that the value of $\int(\delta Q - \delta W)$ is the same for *all* paths between the two states. In other words, the value of $\int(\delta Q - \delta W)$ or $(Q - W)$ depends only on the end states of a process. *Thus* $\int(\delta Q - \delta W)$ *or* $(Q - W)$ *is a property*. This property is called **stored energy** and is denoted by the symbol E. Thus

Stored energy

$$\Delta E \equiv \int (\delta Q - \delta W) \qquad (3\cdot 2a)$$

$$\Delta E \equiv Q - W \qquad (3\cdot 2b)$$

The defining equation for stored energy also serves as a useful statement of the first law.

This defining equation for ΔE can be regarded as a useful statement of the first law.

Since E is a property, its differential is exact, and the differential form of the defining equation is

$$dE \equiv \delta Q - \delta W \qquad (3\cdot 2c)$$

The definition of stored energy given by Eqs. 3·2 is an *operational* definition. Equation 3·2*b*, for example, states that the change in the stored energy of a closed system during any process is obtained by subtracting the net amount of work done by the system from the net amount of heat added to the system. Although stored energy has been defined using a closed system, we shall see later that stored energy is a useful property of open systems as well.

Only the *change* in the value of E between two states can be evaluated by means of Eqs. 3·2. In fact, thermodynamics provides no information about absolute values of E for any system. Fortunately, it is only the change in E that is important in engineering problems. Consequently, a value of $E = 0$ can be assigned to any particular state of a system. Then, from measurements of heat and work during a process that transforms the system to another state, the value of E at the other state can be determined. Since the state for which $E = 0$ is selected arbitrarily, some states will have negative values of E. Many different states of a closed system may have the same value of E, just as many states have the same pressure.

You should note that this definition is precise and different in kind from the definition you might find in a small dictionary. In scientific work where precise meanings are essential, important terms such as energy must be either defined operationally or explicitly accepted as verbally undefined.

Since heat and work are related to the stored energy of a system, they are considered forms of *energy in transition* or *transitional energy*. Thus *energy* refers to both stored energy (a property of a system) and to heat and work (interactions between systems). This is why we always call E *stored* energy. Some books call it internal energy, but we will reserve this name for one particular form of stored energy in conformity with many reference works and published tables of thermodynamic properties.

Since heat and work are forms of energy in transition, Eq. 3·2 becomes a statement of the *conservation of energy principle*: The increase in the energy stored in a system is equal to the net transfer of energy into the system. This is equivalent to the statement that energy can be neither created nor destroyed.

For an isolated system, both work and heat must be zero, so that Eqs. 3·2 reduce to

The stored energy of an isolated system is constant.

$$\Delta E_{\text{isolated system}} = 0 \tag{3·3}$$

That is, the stored energy of an isolated system must remain constant. This does not mean that there can be no change of state of an isolated system; it does mean that an isolated system can exist only in those states that have the same stored energy as the initial state.

In many analyses it is convenient to work not with the total work, heat, and stored energy of a system but instead with these quantities on a per unit mass basis. Thus, dividing Eqs. 3·1, 3·2*b*, and 3·3 by the mass of the system gives

$$\oint (\delta q - \delta w) = 0 \tag{3·1}$$

$$\Delta e \equiv q - w \tag{3·2b}$$

$$\Delta e_{\text{isolated system}} = 0 \tag{3·3}$$

Before looking further into the nature of E, let us summarize the preceding discussion of the first law of thermodynamics:

1. The first law is a generalization based on the results of many experiments. It cannot be deduced from any other physical principles; it is entirely empirical.
2. As a result of the first law, energy can be operationally defined in terms of heat and work, which are themselves operationally defined.
3. The first law can be expressed by any one of the following equations or equivalent word statements:

$$\oint (\delta Q - \delta W) = 0 \tag{3·1}$$

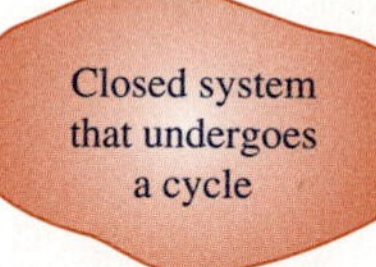

$\oint \delta Q = \oint \delta W$

Whenever a closed system executes a cycle, the net work done by the system is equal to the net heat transfer to the system. (This is the statement closest to the experimental results that support the first law.)

$$\Delta E = Q - W \tag{3·2b}$$

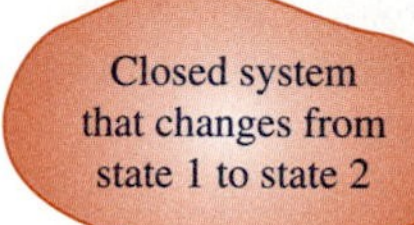

$Q_{1-2} = E_2 - E_1 + W_{1-2}$

The net heat added to a closed system minus the net work done by the system is equal to the increase in the stored energy of the system. (This is the statement most convenient for application to many engineering problems. It defines stored energy.)

$$\Delta E_{\text{isolated system}} = 0 \tag{3·3}$$

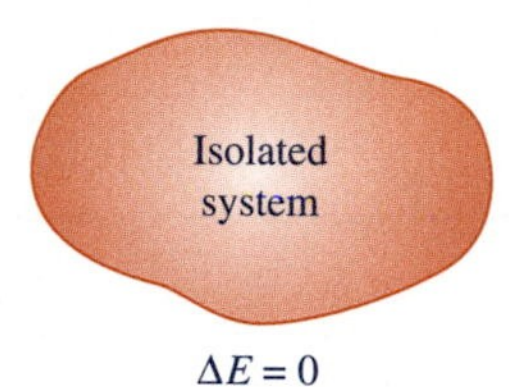

$\Delta E = 0$

The stored energy of an isolated system remains constant. (In a few engineering problems this is a convenient statement for direct application.)

4. One form of the first law is the conservation of energy principle: Energy can be neither created nor destroyed, although it can be stored in various forms and can be transferred from one system to another.

Example 3·3

PROBLEM STATEMENT

A system is composed of gas contained in a cylinder fitted with a piston. The gas expands from a state 1 for which $E_1 = 70$ kJ to a state 2 for which $E_2 = -20$ kJ. During the expansion, the gas does 60 kJ of work on the surroundings. Determine the amount of heat transferred to or from the system during the process.

SOLUTION

We first make a sketch of the system and show the data on it. Applying the first law in the form

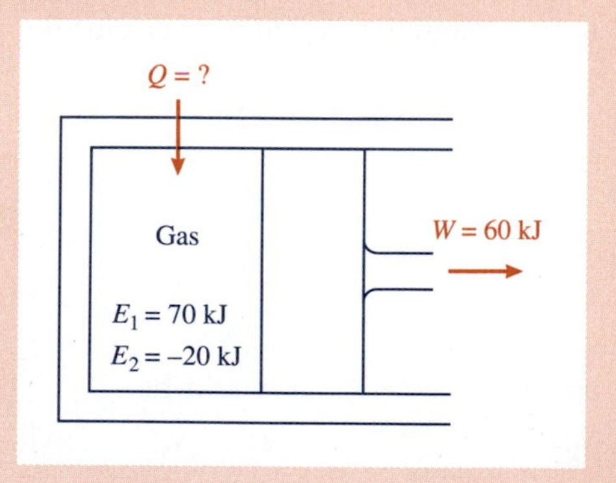

$$Q = E_2 - E_1 + W$$

to this system yields

$$Q = -20 - 70 + 60 = -30 \text{ kJ}$$

Since Q without a subscript stands for heat added to a system, the minus sign of the result indicates that 30 kJ of heat is transferred from the system. The result can be written also as $Q_{out} = 30$ kJ.

Notice that in this process heat and work are both taken from the system. The total decrease in the stored energy of the system is equal to the sum of the energy removed as heat and as work.

Example 3·4

PROBLEM STATEMENT

Determine the final stored energy of a mass of water that has an initial stored energy, $E = 20$ kJ, and then undergoes a process during which 10 kJ of work is done on the water and 3 kJ of heat is removed from it.

SOLUTION

We first make a sketch of the closed system composed of the water and show the data on it. Applying the first

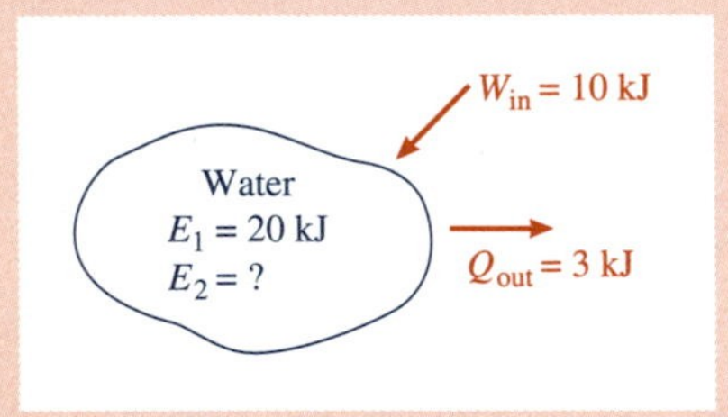

law to the system, the final amount of stored energy is equal to the initial amount of stored energy plus the energy added as work minus the energy removed as heat:

$$\begin{aligned} E_2 &= E_1 + W_{in} - Q_{out} \\ &= 20 + 10 - 3 = 27 \text{ kJ} \end{aligned}$$

If the first law is applied as

$$E_2 = E_1 - W + Q$$

the substitution of numerical values (recalling that Q denotes Q_{in} and W denotes W_{out}) would be

$$E_2 = 20 - (-10) + (-3) = 27 \text{ kJ}$$

3·3 The Nature of Stored Energy

Stored energy E is defined by

$$\Delta E \equiv Q - W \tag{3·2}$$

Many engineering problems involve the use of Eq. 3·2 or one of the many special relations that can be derived from it. Often the value of either Q or W is specified for a process and we want to find out how much of the other form of transitional energy is required to cause a particular change of state. If we could determine ΔE independently in terms of other property changes, we could use the first law to determine either work or heat with knowledge of the other. This would increase the utility of the first law.

In certain cases we can readily correlate E with other properties. We now investigate some of these cases.

Potential Energy. Consider a system composed of a body that experiences an elevation change. No other property of the body changes (see Fig. 3·5). If this change in elevation is brought about without a transfer of heat, then the change in stored energy is equal to the work done on the body.

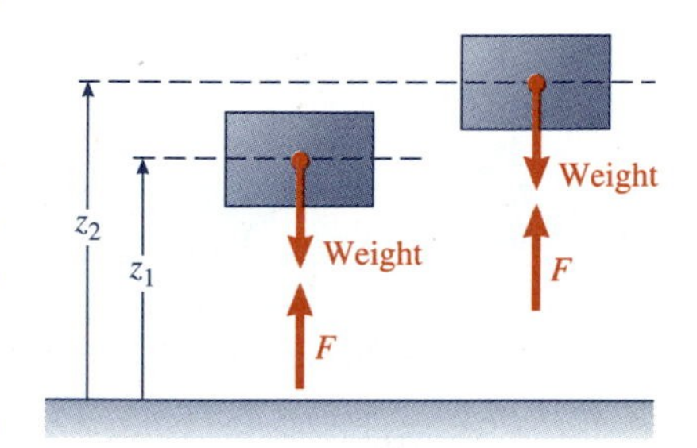

Figure 3·5 Work is done on a body to lift it from elevation z_1 to elevation z_2.

$$\Delta E_{\text{change in elevation}} = W_{\text{on body}} = \int_1^2 F\,dz \tag{3·4}$$

The force required to lift the body is equal to the weight of the body, so the preceding equation becomes

$$\Delta E_{\text{change in elevation}} = \int_1^2 F\,dz = \int_1^2 (\text{weight})\,dz = \int_1^2 mg\,dz \tag{3·5}$$

where m is the mass of the body and g is the acceleration of gravity. Usually, the variation in g with elevation is negligible, so that

$$\Delta E_{\text{change in elevation}} = mg\int_1^2 dz = mg(z_2 - z_1) \tag{3·6}$$

Experiments show that even for complex processes involving changes in other properties, if the elevation of the system changes, the effect on the stored energy is always partly attributable to the elevation change. This form of the stored energy that is a function of elevation alone is called *potential energy*, *PE*. Thus

Potential energy is a form of stored energy.

$$\Delta PE = PE_2 - PE_1 = (\text{weight})(z_2 - z_1) = mg(z_2 - z_1) \tag{3·7a}$$

We may arbitrarily select the elevation where the potential energy of a body is zero. If we assign $PE_0 = 0$, at an elevation z_0 and PE is the potential energy at any other elevation, then the equation

$$PE - PE_0 = mg(z - z_0)$$

reduces to

$$PE = mgz = (\text{weight})z \tag{3·7b}$$

Potential energy

where z is measured from the datum plane at an elevation of z_0. Thus **potential energy** is often defined as the energy stored in a system as a result of its location in the earth's gravitational field, and its magnitude is the product of the weight of the system and the distance between the system center of gravity and some arbitrary horizontal datum plane.

Strictly speaking, in the earth's gravitational field, potential energy is stored in the system composed of the earth and the relatively small body under study. Throughout this book, "potential energy of a body" will mean "potential energy of the system composed of the earth and the body."

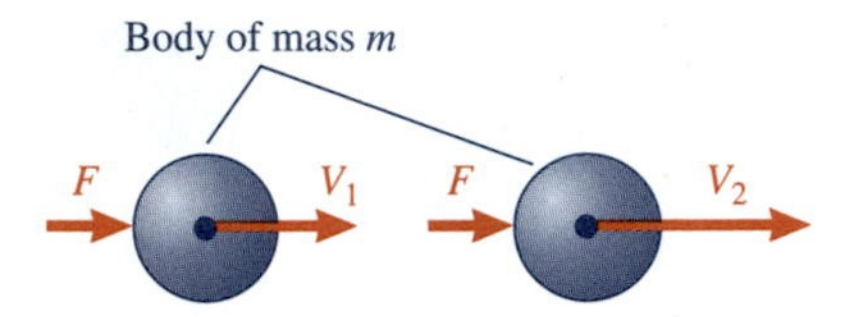

Figure 3·6 Work is done on a body to accelerate it from velocity V_1 to velocity V_2.

Kinetic Energy. Consider a system composed of a body that experiences a change in velocity. Its elevation, temperature, volume, and all other properties remain constant (see Fig. 3·6). If such a change in velocity is brought about without heat transfer, application of the first law shows that the change in stored energy is equal to the work done on the body to accelerate it:

$$\Delta E_{\text{change in velocity}} = W_{\text{on body}} = \int_1^2 F\,dL = \int_1^2 ma\,dL$$

Noting that $a = dV/dt$ and $V = dL/dt$, this becomes

$$\Delta E_{\text{change in velocity}} = m\int_1^2 \frac{dV}{dt}dL = m\int_1^2 V\,dV = \frac{m}{2}(V_2^2 - V_1^2)$$

Experiments show that whenever a system undergoes a process involving a change in velocity, part of the change in stored energy is always attributable to the change in velocity alone, regardless of what other properties may change. This part of the stored energy is called *kinetic energy*, *KE*. Thus,

$$\Delta KE = KE_2 - KE_1 = m\left(\frac{V_2^2 - V_1^2}{2}\right) \tag{3·8a}$$

It is customary and convenient to assign a value of zero kinetic energy to a system with zero velocity relative to some arbitrary inertial frame of reference. When this is done, the kinetic energy of the system is

$$KE = \frac{mV^2}{2} \tag{3·8b}$$

Kinetic energy

where V is the velocity of the system relative to some specified frame of reference. Thus **kinetic energy** is defined as the energy stored in a system by virtue of the motion of the system, and its magnitude is given by $mV^2/2$. If not all parts of the system have the same velocity, the kinetic energy is given by

$$KE = \frac{1}{2} \int V^2 \, dm \tag{3·8c}$$

where V is the velocity of any elemental mass dm, and the integration must be carried out over the entire mass of the system. It can be shown that the kinetic energy of a solid body rotating about a fixed axis is $I\omega^2/2$, where ω is the angular velocity and I is the moment of inertia of the body about the axis of rotation.

The kinetic energy of a solid body rotating about a fixed axis is a form of stored energy.

Other Forms of Stored Energy. Using procedures similar to those used to identify potential and kinetic energies, we can define several other forms of energy related to magnetic fields, electrical effects, solid distortion effects, and surface tension effects. However, even in the absence of these effects, a difference in the stored energy of a system can still be discerned by measurements of heat and work. This form of stored energy is called **internal energy** and is represented by the symbol U. Thus

Internal energy

$$U \equiv E - KE - PE - \text{(magnetic energy)} - \text{(electric energy)} - \text{(surface energy)} - \text{(solid distortion energy)}$$

The ease with which internal energy is correlated with other properties of a system depends on the nature of the system. We discuss this point in Chapters 4 and 7.

Although the principles of classical thermodynamics are independent of any assumptions regarding the structure of matter, it is often helpful to think of internal energy, in the absence of chemical or nuclear reactions, as the summation of the kinetic and potential energies of the molecules of a substance. The kinetic energy of molecules is associated with translational, rotational, and vibratory motions of molecules. These molecular motions increase as the temperature increases. Molecular potential energy is related to the forces between molecules. These forces are large in a solid where molecules are close together, smaller in a liquid, and very small in a gas where molecules are separated from each other by distances that are large with respect to molecular dimensions. Even though the force decreases with increased spacing, the integral of force times distance, which is the work required to separate the molecules or the molecular potential energy, continually increases during the transition from a solid to a liquid to a gas. Therefore, molecular potential energy is highest for gases and lowest for solids.

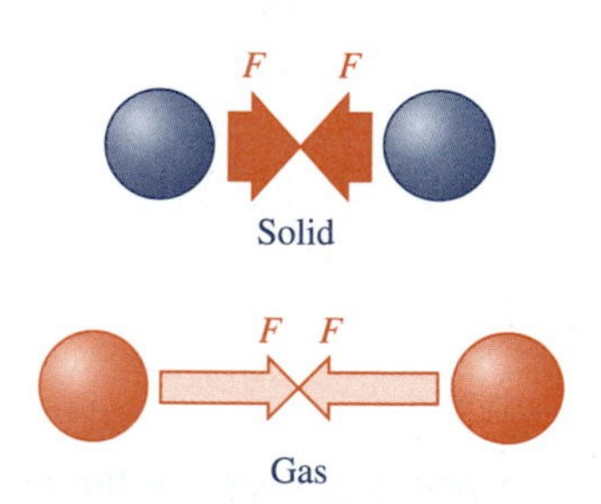

Molecular potential energy is the work required to separate molecules.

In a process that involves a chemical reaction, the change in the internal energy of a system is related to changes in the internal structure of molecules, that is, to changes on the atomic level rather than on the molecular level. Nuclear reactions involve changes on the subatomic level, that is, changes within the atoms of a substance. Measurements of only heat and work provide no means of distinguishing among the molecular, atomic, and nuclear levels of energy storage, so all are included under the one category of internal energy.

In connection with nuclear reactions and the conversion of matter to energy, the conservation of energy and conservation of mass principles must be generalized to treat matter as a form of energy. This subject is not covered in this book. Note, however, that the first law of thermodynamics as stated here in Eq. 3·1 does hold in any case.

Thermodynamics provides information on only changes in internal energy. It is impossible to determine absolute values of internal energy from measurements of heat and work. Just as the potential energy of a body is assigned the value of zero at some convenient horizontal datum plane, we often assign the internal energy of a substance the value of zero at some arbitrary reference state.

Introducing internal energy allows us to write an equation for the total change in the stored energy of a system:

$$\Delta E = \Delta U + \Delta PE + \Delta KE + \Delta(\text{magnetic energy}) + \Delta(\text{electric energy}) + \Delta(\text{surface energy}) + \Delta(\text{solid distortion energy}) \qquad (3\cdot 9a)$$

In the absence of electric, magnetic, and surface tension effects, this equation becomes

$$\Delta E = \Delta U + \Delta KE + \Delta PE \qquad (3\cdot 9b)$$

If we assign a value of $E = 0$ to a reference state where $U = 0$, $V = 0$, and $z = 0$, then E can be evaluated at any other state by

$$E = U + \frac{mV^2}{2} + mgz \qquad (3\cdot 9c)$$

provided, of course, that all parts of the system have the same velocity and z is the distance of the center of gravity from the datum plane. Per unit mass, this is written

$$e = u + \frac{V^2}{2} + gz \qquad (3\cdot 9d)$$

3·4 The First Law Applied to Closed Systems

Some engineers use the term "stationary system" or "nonflow system" to designate a closed system that is at rest or in a state of uniform horizontal motion.

Usually, engineering systems analyzed as closed systems do not move. Therefore, in this book when the term *closed system* is used and there is no indication to the contrary, it will be assumed that effects of gravity and motion are negligible. This means that there is no change in the kinetic energy or the potential energy of a closed system. If electric, magnetic, solid distortion, and surface tension effects are also negligible, the system is called a **simple compressible system** and the total change in the stored energy of the system is the change in internal energy U, or $\Delta E = \Delta U$. Thus the first law applied to a closed system under these conditions becomes

Simple compressible system

$$Q - W = \Delta U$$
$$Q - W = U_2 - U_1 \qquad (3\cdot10)$$

The first law applied to a simple compressible system

where the subscripts 1 and 2 refer respectively to the states at the beginning and the end of a process. Notice that Eq. 3·10 is a special case of Eq. 3·2.

Example 3·5

PROBLEM STATEMENT

A quantity of gas in a closed system expands, performing 34.0 kJ of work while its internal energy decreases by 27.0 kJ. Determine the heat transfer of the process.

SOLUTION

We first make a sketch of the closed system and show data on it. (We do not know that the system is a

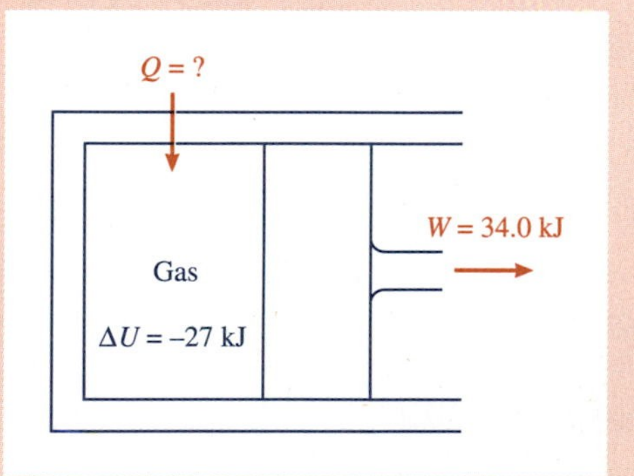

cylinder fitted with a piston, but such a model suffices.) For a closed system, we apply the first law in the form

$$Q = \Delta U + W \qquad (3\cdot10)$$

and substitute numerical values to obtain

$$Q = -27.0 + 34.0 = 7.0 \text{ kJ}$$

The positive result indicates that heat is added to the system. If a negative value were obtained for Q, the conclusion would be that heat is removed from the system during the process.

Example 3·6

PROBLEM STATEMENT

Radon gas initially at 65 kPa, 200°C, is to be heated in a closed, rigid container until it is at 400°C. The mass of the radon is 0.393 kg. A table of properties shows that at 200°C, the internal energy of radon is 26.6 kJ/kg; at 400°C it is 37.8 kJ/kg. Determine the amount of heat required.

SOLUTION

The system is a closed system composed of the radon gas. We make a sketch and place the data on it. Since

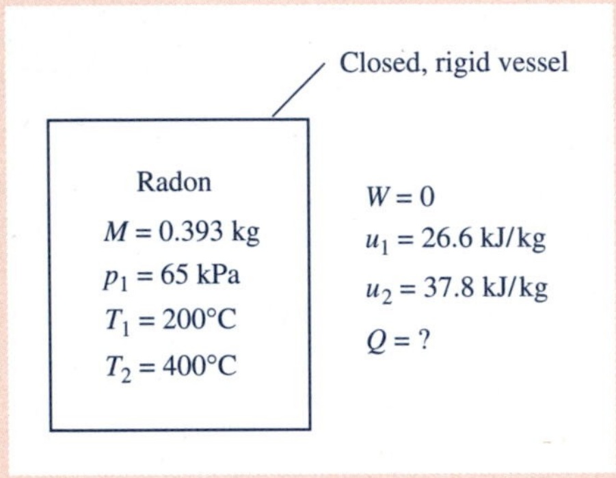

the radon is in a rigid container, no work is done. The first law applied to the closed system is then

$$\begin{aligned} Q &= U_2 - U_1 + W = m(u_2 - u_1) + W \\ &= 0.393 \text{ kg } (37.8 - 26.6) \text{ kJ/kg} + 0 \\ &= 4.4 \text{ kJ} \end{aligned}$$

Example 3·7

PROBLEM STATEMENT

Neon is held in a cylinder fitted with a piston so that the gas can be slowly and frictionlessly compressed from an initial pressure of 14.3 psia to a final pressure of 44.3 psia. The initial temperature and volume are

75 F and 1.60 cu ft, and the compression process is such that $pv^{1.3}$ = constant. The internal energy of neon in ft·lbf/lbm is given by $u = 1.50pv$, where p is in psfa (pounds force per square foot absolute) and v is in ft^3/lbm. Determine the heat transfer of the compression process.

SOLUTION

Analysis: The system is a closed system composed of the neon in the cylinder. A sketch of the system and a pV diagram of the compression process are made.

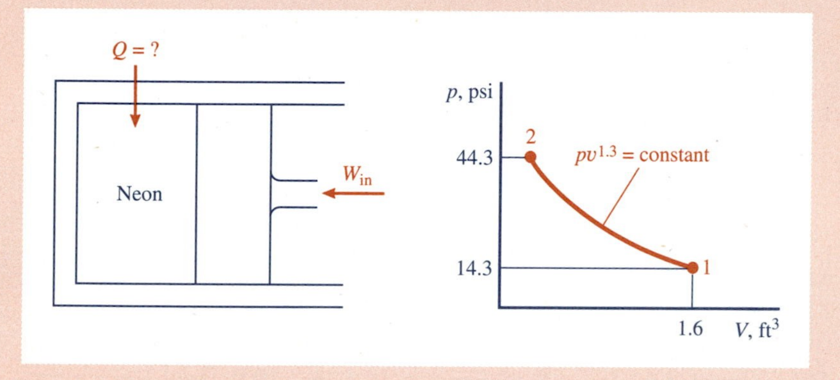

The heat transfer can be found from the first law for a closed system if other quantities in the first law statement can be determined. We write out the first law equation and show how the needed quantities can be determined. Since the compression process is slow and frictionless, we assume that it is a quasiequilibrium process so that work can be evaluated by $\int p\,dv$.

$$Q = U_2 - U_1 + W = m(u_2 - u_1 + w)$$

$$m = \frac{p_1 V_1}{RT_1} \qquad u_2 - u_1 = 1.50(p_2 v_2 - p_1 v_1) \qquad w = \int_1^2 p\,dv$$

Can be integrated because the pv relationship is known: $pv^{1.3} = p_1 v_1^{1.3} = p_2 v_2^{1.3}$

$$R = \frac{\overline{R}}{M} \qquad v_2 = \left(\frac{p_1}{p_2}\right)^{1/1.3} v_1 \longrightarrow v_1 = \frac{V_1}{m}$$

In this analysis, we started writing the first law equation because it contains Q, the quantity we seek. However, that equation contained some quantities that we did not have numerical values for, so we wrote other equations for those quantities. We continued this process until we had equations containing only known quantities. At this point we see that the problem is solvable, and the analysis provides a guide as to which equations we should write first in the solution. Starting the calculations by writing an equation for which we have no numerical values would be wasteful.

Calculations: m is found from the ideal-gas equation of state:

$$m = \frac{p_1 V_1}{RT_1} = \frac{p_1 V_1 M}{\bar{R} T_1}$$
$$= \frac{14.3 \text{ lbf}}{\text{in}^2}\left(\frac{144 \text{ in}^2}{\text{ft}^2}\right)\left(\frac{\text{lbmol·R}}{1545 \text{ ft·lbf}}\right)\frac{(1.60 \text{ ft}^3)}{(535 \text{ R})}\frac{20.18 \text{ lbm}}{\text{lbmol}}$$
$$= 0.0804 \text{ lbm}$$

v_1 is found from the ratio V_1/m:

$$v_1 = \frac{V_1}{m} = \frac{1.60 \text{ ft}^3}{0.0804 \text{ lbm}} = 19.9 \text{ ft}^3/\text{lbm}$$

v_2 is found from $pv^{1.3} = \text{constant} = p_1 v_1^{1.3} = p_2 v_2^{1.3}$ as

$$v_2 = v_1\left(\frac{p_1}{p_2}\right)^{1/1.3} = 19.9\left(\frac{14.3}{44.3}\right)^{0.769} = 8.34 \text{ ft}^3/\text{lbm}$$

Then

$$u_2 - u_1 = 1.50(p_2 v_2 - p_1 v_1) = 1.50\left(\frac{144 \text{ in}^2}{\text{ft}^2}\right)\left[44.3\frac{\text{lbf}}{\text{in}^2}\left(8.34\frac{\text{ft}^3}{\text{lbm}}\right) - 14.3\frac{\text{lbf}}{\text{in}^2}\left(19.9\frac{\text{ft}^3}{\text{lbm}}\right)\right]\frac{\text{B}}{778 \text{ ft·lbf}}$$
$$= 23.6 \text{ B/lbm}$$

The work input for this quasiequilibrium process of a closed system is

$$w = \int_1^2 p\, dv = \int_1^2 \frac{\text{constant}}{v^{1.3}}\, dv$$
$$= \int_1^2 \frac{p_1 v_1^{1.3}}{v^{1.3}}\, dv = p_1 v_1^{1.3}\left(\frac{1}{-0.3}\right)\left(\frac{1}{v_2^{0.3}} - \frac{1}{v_1^{0.3}}\right)$$
$$= -p_1 v_1\left(\frac{1}{0.3}\right)\left[\left(\frac{v_1}{v_2}\right)^{0.3} - 1\right] = -14.3\frac{\text{lbf}}{\text{in}^2}\left(144\frac{\text{in}^2}{\text{ft}^2}\right)\left(19.9\frac{\text{ft}^3}{\text{lbm}}\right)\frac{1}{0.3}\left[\left(\frac{19.9}{8.34}\right)^{0.3} - 1\right]$$
$$= -40{,}720 \text{ ft·lbf/lbm} = -52.3 \text{ B/lbm}$$

Substituting the values of Δu and w into the first law formulation,

$$Q = m(u_2 - u_1 + w)$$

gives

$$Q = 0.0804 \text{ lbm}(23.6 - 52.3)\frac{\text{B}}{\text{lbm}} = -2.31 \text{ B}$$

The minus sign indicates that heat was removed from the neon during the compression; that is,

$$Q_{\text{out}} = -Q = -(-2.31) = 2.31 \text{ B}$$

Comment: It is well to look back on our solution to notice that this is a complex problem. It involves several principles or independent ideas: (1) the first law, (2) the ideal-gas equation of state, (3) the expression for work of a quasiequilibrium process of a simple compressible system, (4) the pv relationship specified for the process, and (5) the relationship between u and pv specified for the neon. Clearly, for such a problem it is well to write out the complete analysis before starting numerical calculations.

3·5 The First Law Applied to Open Systems

In engineering, open systems are encountered much more frequently than closed systems. The following section shows how the first law may be applied to an open system. Section 3·5·2 then treats the important special case of steady flow, and Section 3·5·3 presents several examples of calculations for open systems.

3·5·1 Open Systems—General

Derivation of the first law for an open system

For deriving a general expression of the first law for an open system, consider a closed system as it flows through an open system as shown in Fig. 3·7. Figure 3·7*a* shows the closed system (solid-line boundary) and the open system (broken-line boundary) at time t; Fig. 3·7*b* shows the same two systems at time $t + \Delta t$. The open system boundary (as well as the closed system

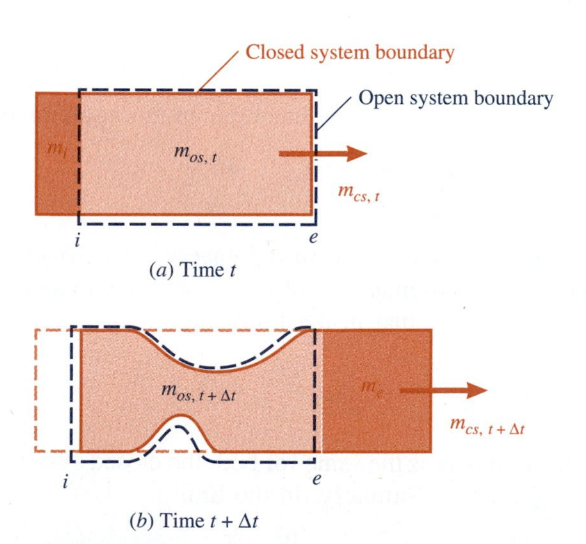

Figure 3·7 A closed system passing through an open system: (*a*) at time t; (*b*) at time $t + \Delta t$. Notice that the shape of the open system may change as well as that of the closed system.

boundary) may be flexible, so during the time interval Δt, parts of the boundary may move.

At time t, the mass of the closed system $m_{cs,t}$ is the sum of the mass in the open system at that instant $m_{os,t}$ and the mass designated as m_i in Fig. 3·7*a*. At time $t + \Delta t$ the mass of the closed system is the sum of the mass in the open system at that instant $m_{os,t+\Delta t}$ and the mass m_e shown in Fig. 3·7*b*. The closed system always contains the same matter, so its mass is constant:

$$m_{cs,t} = m_{cs,t+\Delta t}$$

$$m_{os,t} + m_i = m_{os,t+\Delta t} + m_e$$

There is no requirement that m_{os} be constant or that $m_i = m_e$.

Applying the first law (Eq. 3·2) to the closed system over the time interval Δt gives

$$E_{t+\Delta t} - E_t = Q - W \tag{3·2}$$

The stored energy terms for the closed system can be evaluated as

$$E_t = E_{os,t} + m_i e_i$$

$$E_{t+\Delta t} = E_{os,t+\Delta t} + m_e e_e$$

where e_i and e_e are assumed to be uniform throughout the masses m_i and m_e. Substituting these expressions for E_t and $E_{t+\Delta t}$ into the first law expression for the closed system and rearranging gives

$$E_{os,t+\Delta t} - E_{os,t} = Q_{cs} - W_{cs} + m_i e_i - m_e e_e$$

Dividing this equation by the time interval Δt gives

$$\frac{E_{os,t+\Delta t} - E_{os,t}}{\Delta t} = \frac{Q_{cs}}{\Delta t} - \frac{W_{cs}}{\Delta t} + \frac{m_i}{\Delta t} e_i - \frac{m_e}{\Delta t} e_e \tag{a}$$

In the limit as Δt approaches zero, the left-hand side of this equation becomes

$$\lim_{\Delta t \to 0} \frac{E_{os,t+\Delta t} - E_{os,t}}{\Delta t} = \frac{dE_{os}}{dt} \tag{b}$$

which is the time rate of change of the stored energy of the open system.

As Δt approaches zero, the boundaries of the closed system and the open system approach each other so that in the limit

$$\lim_{\Delta t \to 0} \frac{Q_{cs}}{\Delta t} = \lim_{\Delta t \to 0} \frac{Q_{os}}{\Delta t} = \dot{Q} \tag{c}$$

That is, the rate of heat transfer is the same for both the closed system and the open system, $\dot{Q}_{cs} = \dot{Q}_{os} = \dot{Q}$. Similarly, in the limit

$$\lim_{\Delta t \to 0} \frac{W_{cs}}{\Delta t} = \dot{W}_{cs} = \lim_{\Delta t \to 0} \frac{W_{os}}{\Delta t} = \dot{W}_{os} = \dot{W} \tag{d}$$

In the limit as Δt approaches zero, the last two terms in Eq. *a* become

$$\lim_{\Delta t \to 0} \frac{m_i}{\Delta t} e_i = \dot{m}_i e_i \qquad (e)$$

$$\lim_{\Delta t \to 0} \frac{m_e}{\Delta t} e_e = \dot{m}_e e_e \qquad (f)$$

where $\dot{m}_i$ and $\dot{m}_e$ are the mass rates of flow at the inlet and exit sections. We earlier assumed that properties such as e_i and e_e were uniform throughout m_i and m_e. In the limit, this is equivalent to a restriction to one-dimensional flow at the inlet and the outlet. (Later we will see that this restriction can be removed.)

Substituting from Eqs. *b* through *f* into Eq. *a* gives in the limit for the open system

$$\frac{dE}{dt} = \dot{Q} - \dot{W}_{\text{total}} + \dot{m}_i e_i - \dot{m}_e e_e \qquad (3\cdot 11)$$

where the subscript is added to $\dot{W}$ to emphasize that this term includes all forms of work, including what is called *flow work.*

Wherever matter crosses the boundary of a system, work is done on or by the system. Therefore, the total work for an open system is usually separated into two parts: (1) the work required to push a fluid into or out of the system, called flow work, and (2) all other forms of work.

To derive an expression for the work required to push a fluid into or out of a system, consider a system as shown in Fig. 3·8 with fluid entering at section *i* and leaving at section *e*. In particular, consider the process by which a volume V_i of fluid is pushed into the system at section *i*. The fluid element of volume V_i has a cross-sectional area A_i and a length L_i as it crosses the system boundary at section *i*. The force acting on this element to push it across the system boundary is $F_i = p_i A_i$. This force acts through a distance L_i so that the work done in pushing the element across the system boundary is

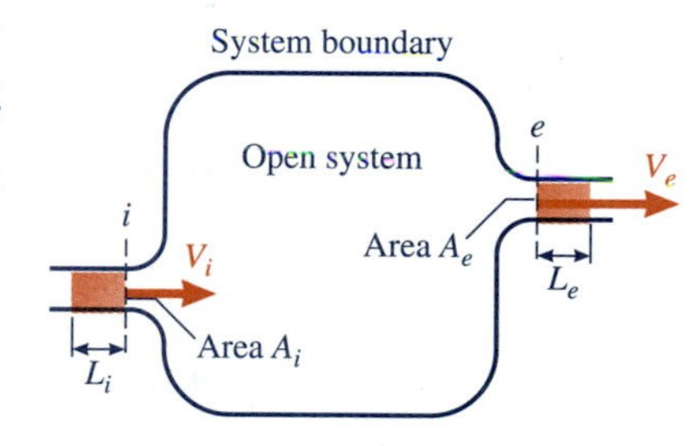

Figure 3·8 Flow work of fluid entering and leaving an open system.

$$\text{Work} = F_i L_i = p_i A_i L_i = p_i V_i$$

This is work done on the system by the fluid outside the system, because the effect within the system could be reduced entirely to the lifting of a weight. In a similar manner, the work performed by the system to push an element of fluid out at section *e* is

$$\text{Work} = p_e V_e$$

This work that must be done in causing fluid to flow into or from a system is called **flow work**.

Flow work

The flow work per unit mass crossing the boundary of a system is pv. The time rates of flow work at the inlet and the outlet are given by $-\dot{m}_i p_i v_i$ and $\dot{m}_e p_e v_e$. The negative sign for the flow work at the inlet is in accordance with the sign convention that work done *on* a system is negative.

Thus, with flow work identified separately from other kinds of work, the $\dot{W}_{\text{total}}$ term in Eq. 3·11 becomes

$$\dot{W}_{\text{total}} = \dot{W} + \dot{m}_e p_e v_e - \dot{m}_i p_i v_i$$

where $\dot{W}$ is the rate of work done by an open system by means of impellers, pistons, other moving boundaries, or any other means, excluding only the work done where matter crosses the system boundary.

Since the work (or rate of doing work) term in first law equations for open systems is usually separated into two parts, the term *work* without modifiers is *always* understood to mean all forms of work except flow work. The complete two-word name is *always* used when referring to flow work. Therefore, Eq. 3·11 becomes

$$\frac{dE}{dt} = \dot{Q} - \dot{W} + \dot{m}_i p_i v_i - \dot{m}_e p_e v_e + \dot{m}_i e_i - \dot{m}_e e_e \qquad (3\cdot 12a)$$

or

$$\frac{dE}{dt} = \dot{Q} - \dot{W} + \dot{m}_i(e_i + p_i v_i) - \dot{m}_e(e_e + p_e v_e) \qquad (3\cdot 12b)$$

This equation is the first law of thermodynamics expressed on a time rate basis for an open system with one-dimensional flow at inlet and outlet. It expresses the time rate of change of energy stored in the open system in terms of the heat transfer to the system, the work done by the system, the flow work, and the stored energy e carried by matter entering and leaving the system.

Equation 3·12 can be generalized to cover cases of multidimensional flow at the inlet and outlet. If there is more than one inlet and outlet, Eq. 3·12 can be written as

$$\frac{dE}{dt} = \dot{Q} - \dot{W} + \sum_{\text{inlets}} \dot{m}(e + pv) - \sum_{\text{outlets}} \dot{m}(e + pv) \qquad (3\cdot 12c)$$

In many applications, effects of electricity, magnetism, and surface tension are negligible so that, as mentioned in Section 3·3,

$$e = u + \frac{V^2}{2} + gz \qquad (3\cdot 9d)$$

Then Eq. 3·12 becomes

General expression for the first law for an open system

$$\frac{dE}{dt} = \dot{Q} - \dot{W} + \dot{m}_i\left(u_i + p_i v_i + \frac{V_i^2}{2} + gz_i\right) - \dot{m}_e\left(u_e + p_e v_e + \frac{V_e^2}{2} + gz_e\right) \qquad (3\cdot 12d)$$

If the open system itself is at rest,

$$\frac{dE}{dt} = \frac{dU}{dt}$$

Then Eq. 3·12 becomes

The first law for an open system at rest

$$\frac{dU}{dt} = \dot{Q} - \dot{W} + \dot{m}_i\left(u_i + p_i v_i + \frac{V_i^2}{2} + gz_i\right) - \dot{m}_e\left(u_e + p_e v_e + \frac{V_e^2}{2} + gz_e\right) \quad (3\cdot12e)$$

Enthalpy

Because the combination of properties $u + pv$ occurs frequently in thermodynamics, it has been given a name, *enthalpy*. **Enthalpy**, H or h, is defined as

$$H \equiv U + pV$$

(where V is volume) or, per unit mass,

$$h \equiv u + pv$$

It should be noted that u represents a form of stored energy, but pv does not; therefore, their sum is not a form of stored energy. It will be seen later that in certain applications enthalpy may be treated as energy, but this should not obscure the fact that *enthalpy is simply a useful property defined as a combination of other properties and is not a form of energy.*

In most thermodynamic applications, only *changes* in enthalpy are important.

In engineering calculations, enthalpy is used more frequently than internal energy. Consequently, tables of thermodynamic properties always include enthalpy and sometimes include internal energy also. Since we cannot obtain absolute values of internal energy, we cannot obtain absolute values of enthalpy. Fortunately, only changes in enthalpy are important.

Introducing enthalpy into Eq. 3·12*d* gives

$$\frac{dU}{dt} = \dot{Q} - \dot{W} + \dot{m}_i\left(h_i + \frac{V_i^2}{2} + gz_i\right) - \dot{m}_e\left(h_e + \frac{V_e^2}{2} + gz_e\right) \quad (3\cdot12e)$$

Various other forms of Eq. 3·12 can be developed. For example, for an incremental change in state of an open system associated with incremental masses δm_i entering and δm_e leaving, Eq. 3·12*b* becomes

$$dE = \delta Q - \delta W + (e_i + p_i v_i)\,\delta m_i - (e_e + p_e v_e)\,\delta m_e \quad (3\cdot13)$$

For an open system that changes from a state 1 to a state 2, we have

$$E_2 - E_1 = Q - W + \int_i (e + pv)\,\delta m - \int_e (e + pv)\,\delta m \quad (3\cdot14)$$

where E_1 and E_2 are the initial and final stored energies of the system, Q and W are the heat and work during the process, and the two integral terms account for both the stored energy carried across the system boundary and the flow work of the entering and leaving streams.

You can readily develop other forms of the first law to apply to open systems as needed. Rather than memorize any special form, you should learn and understand the general first law statement or energy balance:

The general first law statement is an energy balance.

$$\begin{pmatrix}\text{Net}\\\text{increase}\\\text{in stored}\\\text{energy of}\\\text{system}\end{pmatrix} = \begin{pmatrix}\text{Net amount of}\\\text{energy added to}\\\text{system as heat}\\\text{and all forms of}\\\text{work}\end{pmatrix} + \begin{pmatrix}\text{Stored}\\\text{energy}\\\text{of matter}\\\text{entering}\\\text{system}\end{pmatrix} - \begin{pmatrix}\text{Stored}\\\text{energy}\\\text{of matter}\\\text{leaving}\\\text{system}\end{pmatrix} \qquad (3{\cdot}15)$$

Remember that the work term for an open system can be separated into two parts: (1) flow work, and (2) all other forms of work. An alternative form for Eq. 3·15 is

$$\begin{pmatrix}\text{Net}\\\text{increase}\\\text{in stored}\\\text{energy of}\\\text{system}\end{pmatrix} = \begin{pmatrix}\text{Heat}\\\text{added}\\\text{to}\\\text{system}\end{pmatrix} + \begin{pmatrix}\text{All}\\\text{forms}\\\text{of work}\\\text{done by}\\\text{system}\end{pmatrix} - \begin{pmatrix}\text{Stored}\\\text{energy}\\\text{of matter}\\\text{leaving}\\\text{system}\end{pmatrix} + \begin{pmatrix}\text{Stored}\\\text{energy}\\\text{of matter}\\\text{entering}\\\text{system}\end{pmatrix}$$

In the general case of an open system, the properties of the matter crossing the system boundary may vary. If the variations of p and v *across a stream entering or leaving* the system are known, the flow-work term for that section can be evaluated as $\int_0^m pv\,\delta m$, where m is the total amount of mass that crosses the boundary there. Alternatively, if the variations of p, v, and the mass rate of flow *with time* are known, it is more convenient to write the flow-work term for one section as $\int_{t_1}^{t_2} \dot{m}pv\,dt$, where t_1 and t_2 are the times at which the process starts and ends. In a similar manner, the stored energy of fluid crossing the system boundary can be evaluated as $\int_0^m e\,\delta m$ or as $\int_{t_1}^{t_2} \dot{m}e\,dt$. Furthermore, if the power input or output of the system is known as a function of time, W may be evaluated as $\int_{t_1}^{t_2} \dot{W}\,dt$. Thus, the terms in Eq. 3·15 can be evaluated as follows for a process that changes the open system from state 1 to state 2:

Formulations of the terms in an energy balance

$$\begin{pmatrix}\text{Net increase in stored}\\\text{energy of system}\end{pmatrix} = E_2 - E_1$$

$$\begin{pmatrix}\text{Net amount of}\\\text{energy added to}\\\text{system as heat}\\\text{and work}\end{pmatrix} = \begin{cases} Q - W + \sum\limits_{\text{inlets}} \int pv\,\delta m - \sum\limits_{\text{exits}} \int pv\,\delta m \\ Q - W + \sum\limits_{\text{inlets}} \int \dot{m}pv\,dt - \sum\limits_{\text{exits}} \int \dot{m}pv\,dt \\ Q - \int \dot{W}\,dt + \sum\limits_{\text{inlets}} \int \dot{m}pv\,dt - \sum\limits_{\text{exits}} \int \dot{m}pv\,dt \end{cases}$$

$$\begin{pmatrix}\text{Stored energy of matter}\\ \text{entering system}\end{pmatrix} = \sum_{\text{inlets}} \int e\,\delta m = \sum_{\text{inlets}} \int \dot{m}e\,dt$$

$$\begin{pmatrix}\text{Stored energy of matter}\\ \text{leaving system}\end{pmatrix} = \sum_{\text{exits}} \int e\,\delta m = \sum_{\text{exits}} \int \dot{m}e\,dt$$

The closed-system energy balance can be considered a special case of the open-system energy balance. To illustrate this, notice that Eq. 3·13

$$\delta Q - \delta W + (e_i + p_i v_i)\,\delta m_i - (e_e + p_e v_e)\,\delta m_e = dE \qquad (3\cdot13)$$

for the special case of a *closed system*, where $\delta m_i = \delta m_e = 0$, becomes

$$\delta Q - \delta W = dE$$

which was introduced in Section 3·2. This serves as a check. However, we cannot *derive* the closed-system equation from the open-system equation, since this would reverse the logical structure we have been following. The first law is an inductive conclusion based on observations on *closed systems*. It can be extended to open systems. However, starting from the open-system result and deducing the closed-system energy balance is a circular development. It works out, of course, but it proves nothing.

In case you are tempted at this point to memorize some of the many energy balance equations that have been displayed in the last three sections, we repeat for emphasis: *Learn and understand the first-law relationship or energy balance:*

$$\begin{pmatrix}\text{Net}\\ \text{increase}\\ \text{in stored}\\ \text{energy of}\\ \text{system}\end{pmatrix} = \begin{pmatrix}\text{Net amount of}\\ \text{energy added to}\\ \text{system as heat}\\ \text{and all forms of}\\ \text{work}\end{pmatrix} + \begin{pmatrix}\text{Stored}\\ \text{energy}\\ \text{of matter}\\ \text{entering}\\ \text{system}\end{pmatrix} - \begin{pmatrix}\text{Stored}\\ \text{energy}\\ \text{of matter}\\ \text{leaving}\\ \text{system}\end{pmatrix} \qquad (3\cdot15)$$

General statement of the first law

From this, develop any other first law equations you may need for particular applications.

3·5·2 Open Systems—Steady Flow

Steady flow, as defined in Section 1·9, requires that all properties at each point in a system remain constant with respect to time. This means that the mass rate of flow at each section where matter crosses the system boundary remains constant, the net mass flow across the boundary is zero, the system volume remains constant, and all interactions with the surroundings occur at a constant rate. Under these conditions, we can simplify the various formulations for the first law applied to open systems given in Section 3·5·1.

Since for steady flow the net increase in stored energy of the system is zero, the first law relationship for open systems,

$$\begin{pmatrix}\text{Net}\\\text{increase}\\\text{in stored}\\\text{energy of}\\\text{system}\end{pmatrix}=\begin{pmatrix}\text{Net amount of}\\\text{energy added to}\\\text{system as heat}\\\text{and all forms}\\\text{of work}\end{pmatrix}+\begin{pmatrix}\text{Stored}\\\text{energy}\\\text{of matter}\\\text{entering}\\\text{system}\end{pmatrix}-\begin{pmatrix}\text{Stored}\\\text{energy}\\\text{of matter}\\\text{leaving}\\\text{system}\end{pmatrix}\qquad(3{\cdot}15)$$

becomes

A general expression for the first law applied to an open, steady-flow system

$$\begin{pmatrix}\text{Net amount of}\\\text{energy added to}\\\text{system as heat}\\\text{and all forms}\\\text{of work}\end{pmatrix}=\begin{pmatrix}\text{Stored}\\\text{energy}\\\text{of matter}\\\text{leaving}\\\text{system}\end{pmatrix}-\begin{pmatrix}\text{Stored}\\\text{energy}\\\text{of matter}\\\text{entering}\\\text{system}\end{pmatrix}\qquad(3.16)$$

For steady flow, the first law equation for an open system with one-dimensional flow through more than one inlet and outlet, Eq. 3·12*c*, becomes

$$\dot{Q}-\dot{W}=\sum_{\text{outlets}}\dot{m}(e+pv)-\sum_{\text{inlets}}\dot{m}(e+pv)\qquad(3{\cdot}17a)$$

Where effects of electricity, magnetism, and surface tension are negligible,

$$e=u+\frac{V^2}{2}+gz\qquad(3{\cdot}9d)$$

and if there is only one inlet and one outlet, steady flow requires that $\dot{m}_e = \dot{m}_i = \dot{m}$ and Eq. 3·17*a* (or Eq. 3·12*d*) becomes

$$\dot{Q}-\dot{W}=\dot{m}\left(u_e+\frac{V_e^2}{2}+gz_e+p_ev_e\right)-\dot{m}\left(u_i+\frac{V_i^2}{2}+gz_i+p_iv_i\right)\qquad(3{\cdot}17b)$$

or

$$\dot{Q}-\dot{W}=\dot{m}\left(h_e+\frac{V_e^2}{2}+gz_e\right)-\dot{m}\left(h_i+\frac{V_i^2}{2}+gz_i\right)\qquad(3{\cdot}17c)$$

Dividing this equation by the mass rate of flow gives the first law equation on a unit mass basis for a steady-flow system with one inlet and one outlet and one-dimensional flow:

The first law for steady-flow systems

$$q-w=h_e-h_i+\frac{V_e^2-V_i^2}{2}+g(z_e-z_i)\qquad(3{\cdot}18)$$

This is the most frequently used form of the first law equation for steady-flow systems because many steady-flow systems have only one inlet and one outlet, one-dimensional flow is a common assumption, effects of electricity, magnetism, and surface tension are often negligible, and property values are usually available on a per unit mass basis. Figure 3·9 shows a typical diagram for the analysis of such a steady-flow system.

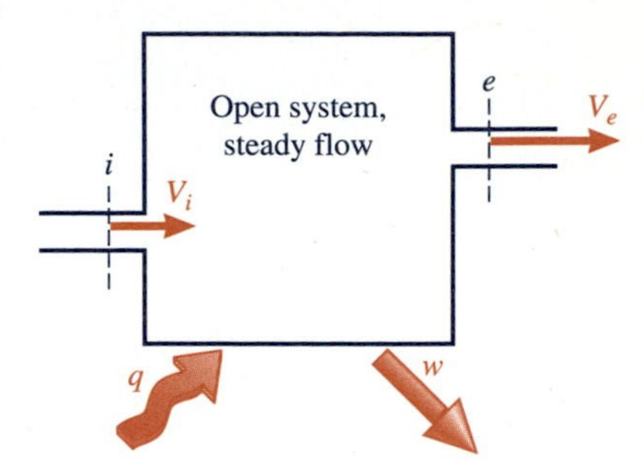

Figure 3·9 A common steady-flow system.

As mentioned at the end of Section 1·8, in many important actual systems, the flow is steady. Also, for many purposes unsteady flows can be adequately modeled as steady if the state of the system and properties of matter crossing the boundaries of the system periodically return to the same values and properly averaged values are used.

Cases of two-dimensional and three-dimensional flows crossing a system boundary can be handled by integrating property values across the flow area.

Many applications call for simplifying Eq. 3·18, as you will see in the following example problems. For example, heat exchangers, devices for adding or removing heat from a fluid, involve no work. If the fluid is a gas, potential energy changes are negligible. Often the velocity changes little from the inlet to the outlet, so kinetic energy changes are negligible.

As a fluid is being compressed or expanded in a compressor, engine, or turbine, there may be some heat transfer, but it is usually much smaller than the work and often it is negligible. In fact, even when thermodynamic considerations make simultaneous work and heat transfer attractive, the fluid is often made to flow through separate devices in sequence, with one device involving only work and the other involving only heat transfer. The reason is that design constraints for optimum heat transfer and for optimum work conflict with each other. For example, optimum heat transfer calls for a flow passage with a maximum ratio of surface area to volume, and high turbulence and mixing are often desired. The opposite conditions are desired for the flow passage through an engine, turbine, or compressor.

For a gas or vapor flowing through any system, the change in potential energy is usually negligible. For the flow through a nozzle or diffuser, no work is done, because there is no moving shaft or moving system boundary, and usually the heat transfer is negligibly small, so

The first law for flow through a nozzle or a diffuser

$$h_i + \frac{V_i^2}{2} = h_e + \frac{V_e^2}{2}$$

For the flow through a partially closed valve or a porous plug of material, no work is done, and if there is no heat transfer and no change in kinetic energy, the process is called a **throttling process**. (Such a process occurs in the capillary tube of a household refrigerator.)

Throttling process

In a throttling process, the first law shows that the enthalpy at the inlet and the outlet is the same.

$$h_e = h_i$$

This does not mean that enthalpy is constant, but only that the enthalpy is the same at inlet and outlet. Many more special forms of Eq. 3·18 will arise as you apply the first law in the modeling of various steady-flow systems.

3·5·3 Applications of the First Law to Open Systems

The following example problems are all different, yet they provide but a small sample of applications of the first law. They involve steady flow and transient flow, single fluid streams and multiple streams, adiabatic processes and processes with heat transfer, and both compressible and incompressible fluids. Some of the example problems are simple, involving only one principle, and some are complex. Study them as samples of how you can apply the general principles to various situations, not as solution techniques to be remembered until a suitable problem arises.

Example 3·8 Dehumidifier

PROBLEM STATEMENT

One part of an air-conditioning system is a dehumidifier. Very warm atmospheric air containing water vapor enters the dehumidifier with an enthalpy of 90.0 kJ/kg at a rate of 210 kg/h. Heat is removed from the air as it passes over a bank of tubes through which cold water flows. Atmospheric moisture that condenses on the tubes drains from the dehumidifier with an enthalpy of 34.0 kJ/kg at a rate of 4.0 kg/h. Air leaving has an enthalpy of 23.8 kJ/kg. Velocities through the dehumidifier are quite low. Determine the rate of heat removal from the air stream through the dehumidifier.

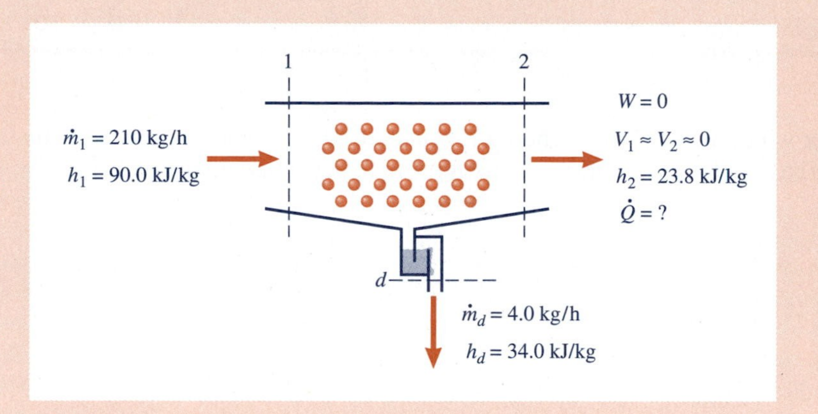

SOLUTION

Analysis: We define the system as the dehumidifier space that the air and atmospheric moisture flow through. A sketch of the system is made and we place data on it. We assume that the flow is steady. This is a reasonable assumption for such a device and it is also implied by the specification of flow rates. Since we are to calculate a heat transfer rate, we think immediately of the first law in a formulation such as

$$\dot{Q} - \dot{W} = \sum_{\text{exits}} \dot{m}\left(h + \frac{V^2}{2} + gz\right) - \sum_{\text{inlets}} \dot{m}\left(h + \frac{V^2}{2} + gz\right)$$

There is obviously no work done in the dehumidifier, for there is no shaft or other means of applying a moving force. It is clear from the diagram that potential energy changes are negligible, and the specification of low velocities makes kinetic energy changes negligible. Therefore, the first law can be simplified and the mass rates of flow can be determined using the conservation of mass principle.

Calculations: Conservation of mass for a steady-flow system tells us that

$$\sum_{\text{exits}} \dot{m} = \sum_{\text{inlets}} \dot{m}$$

or

$$\dot{m}_2 + \dot{m}_d = \dot{m}_1$$
$$\dot{m}_2 = \dot{m}_1 - \dot{m}_d = 210 - 4.0 = 206 \text{ kg/h}$$

The first law applied to this steady-flow system simplifies to

$$\begin{aligned}\dot{Q} &= \dot{m}_2 h_2 + \dot{m}_d h_d - \dot{m}_1 h_1 \\ &= 206(23.8) + 4(34.0) - 210(90.0) = -13{,}900 \text{ kJ/h}\end{aligned}$$

Example 3·9 Wind Turbine

PROBLEM STATEMENT

A simple model for a wind turbine (or windmill) is shown where the wind speed is V_1 and the speed of the air in the wake of the turbine is V_2. The air that passes through the turbine is decelerated so that the wake

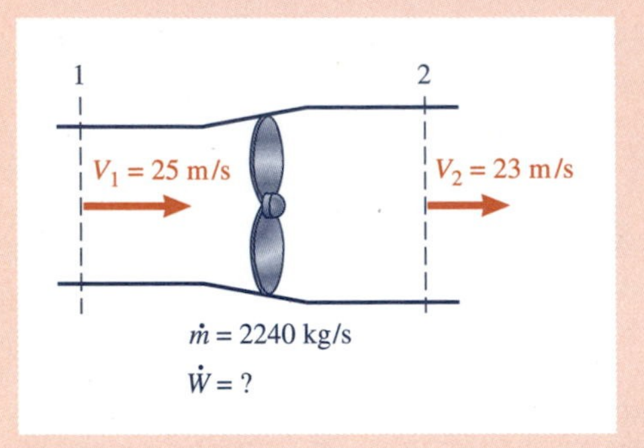

diameter is greater than the turbine rotor diameter, which is in turn greater than the diameter of the imaginary tube of upstream air that passes through the rotor. There is no change in pressure or temperature of the air. For a 10-m-diameter wind turbine that passes 2240 kg/s of air in a 25-m/s wind, the average wake velocity is 23 m/s. At what rate is energy being taken from the air (in kW)?

SOLUTION

Analysis: We define the open system here as having an imaginary boundary that encloses all the air that flows through the turbine rotor. We assume steady flow and note that there is no heat transfer because the air in the system and that in the surroundings are at the same temperature. The air has the same pressure and temperature at sections 1 and 2, so it is in the same state and has the same enthalpy. The potential energy does not change. Therefore, the first law equation for steady flow can be greatly simplified to yield the work done by the air on the wind turbine rotor.

Calculations: The first law equation for steady flow through an open system with one-dimensional flow at the single inlet and the single outlet

$$\dot{Q} - \dot{W} = \dot{m}\left[h_2 - h_1 + \frac{V_2^2 - V_1^2}{2} + g(z_2 - z_1)\right]$$

simplifies to

$$\dot{W} = \dot{m}\left(\frac{V_1^2 - V_2^2}{2}\right)$$

$$= 2240\frac{\text{kg}}{\text{s}}\left(\frac{25^2 - 23^2}{2}\frac{\text{m}^2}{\text{s}^2}\right)$$

$$= 107{,}500\frac{\text{kg}}{\text{s}}\frac{\text{m}^2}{\text{s}^2} = 107{,}500 \text{ J/s}$$

$$= 107.5 \text{ kW}$$

Comments: In the next-to-last step, we wrote out the units to show the substitution from the relation 1 J = 1 kg·m^2/s^2. Repeated use of the kinetic energy in the first law equation reminds us that the units m^2/s^2 are equivalent to J/kg. (Similarly, in the English system, the units ft^2/s^2 are equivalent to ft·lbf/slug.)

The result, 107.5 kW, is the amount of power given up by the air under the assumptions we have made in this simple model. This is more than the power that can be delivered by the wind turbine. *Part* of the difference stems from the assumption of one-dimensional flow downstream of the rotor. In the actual system (not the idealized one we have used), swirl components of the velocity and eddying motions in the wake of the rotor are significant. Still, the "first-order analysis" we have made here is often useful.

Example 3·10 Water Pump

PROBLEM STATEMENT

Water flows at a very low velocity in a pipeline at 30 psig, 60 F. Water is to be drawn from this line at a steady rate of 25 gpm by a pump and discharged through a pipe 25 ft higher where the pressure is 60 psig. The discharge line has an inside diameter of 1.05 inches. *Assuming there is no heat transfer and no change in internal energy* of the water, determine the amount of power delivered to the water by the pump.

SOLUTION

Analysis: This is a steady-flow situation. We sketch the system to include the pump and the piping con-

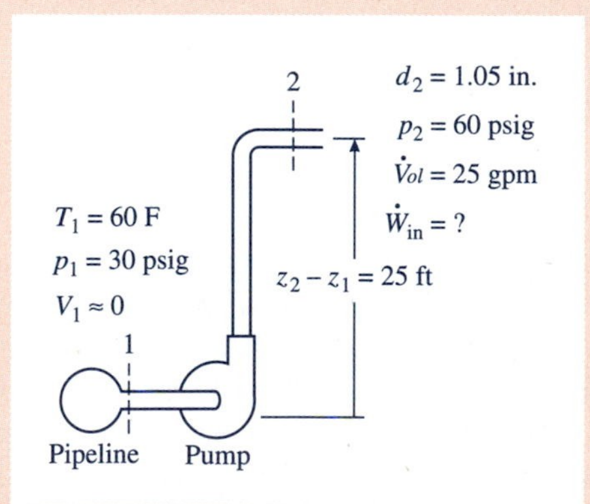

nected to it. The power delivered to the water is the product of the mass rate of flow and the work done on the water per unit mass. The mass rate of flow is simply the volume rate of flow, which is specified, multiplied by the density. Since water under these conditions is nearly incompressible, the density is the same at inlet and outlet and the volume rate of flow is also constant from inlet to outlet.

The work per unit mass can be found from the first law, and the various quantities in the equation can be evaluated as shown, where a check mark indicates a known value (and v_1 and v_2 are replaced by v for the incompressible fluid)

$$w_{in} = \overbrace{u_2 - u_1}^{\checkmark} + \overset{\checkmark}{p_2}\overset{\checkmark}{v} - \overset{\checkmark}{p_1}\overset{\checkmark}{v} + \frac{V_2^2 - \overset{\checkmark}{V_1^2}}{2} + g\left(\overbrace{z_2 - z_1}^{\checkmark}\right) - \overset{\checkmark}{q}$$

$$V_2 = \frac{\dot{m}}{\overset{\checkmark}{\rho} A_2} \qquad \dot{m} = \overset{\checkmark}{\rho}\, \dot{Vol}$$

$$A_2 = \frac{\pi}{4}\overset{\checkmark}{d_2^2}$$

Only gage pressures are specified, but only a pressure difference is required, so there is no need for an assumption regarding atmospheric pressure. This analysis shows that we can calculate the power input and also suggests a sequence for numerical calculations (such as Table A·2).

Calculations: Using a density value of water from steam tables or elsewhere (such as Table A·5) gives

$$\dot{m} = \rho\,\dot{Vol} = 62.4\frac{\text{lbm}}{\text{ft}^3}\left(25\frac{\text{gal}}{\text{min}}\right)231\frac{\text{in}^3}{\text{gal}}\left(\frac{1}{1728}\frac{\text{ft}^3}{\text{in}^3}\right)\frac{1}{60}\frac{\text{min}}{\text{s}} = 3.48\frac{\text{lbm}}{\text{s}}$$

$$V_2 = \frac{\dot{m}}{\rho A_2} = \frac{\dot{m}}{\rho}\left(\frac{4}{\pi d^2}\right) = 3.48\frac{\text{lbm}}{\text{s}}\left(\frac{1}{62.4}\frac{\text{ft}^3}{\text{lbm}}\right)\frac{4}{\pi(1.05)^2\ \text{in}^2}\left(144\frac{\text{in}^2}{\text{ft}^2}\right)$$

$$= 9.26\ \text{ft/s}$$

$$w_{in} = u_2 - u_1 + p_2 v - p_1 v + \frac{V_2^2 - V_1^2}{2} + g(z_2 - z_1) - q$$

If we consider the very low inlet velocity as negligible and substitute the other zero values, we get

$$w_{in} = \frac{p_2 - p_1}{\rho} + \frac{V_2^2}{2} + g(z_2 - z_1)$$

This equation is valid for any consistent set of units (see Section 1·4). Since we are using mixed units (ft, s, and lbf in pressure, and lbm in density), some conversion factors are likely to be required. If we use pressure in lbf/ft^2 and density in lbm/ft^3, the first term on the right-hand side of the equation above has units of ft·lbf/lbm, energy per unit mass. Each of the other two terms, however, has units of ft^2/s^2. Since 1 lbf = 1 slug·ft/s^2, the units ft^2/s^2 are equivalent to ft·lbf/slug, also energy per unit mass. Adding the terms in the equation requires that they be in the same units, so we use a conversion factor with the kinetic and potential energy terms to get them into ft·lbf/lbm (see Section 1·4·2). Also, since only a pressure difference is involved in the first law equation, we can use gage pressures just as well as absolute pressures. Thus, substituting numerical values into the first law equation (and assuming the local acceleration of gravity g is 32.2 ft/s^2), we get

$$w_{in} = (60 - 30)\frac{\text{lbf}}{\text{in}^2}\left(\frac{1}{62.4}\frac{\text{ft}^3}{\text{lbm}}\right)144\frac{\text{in}^2}{\text{ft}^2} + \frac{(9.26)^2}{2}\frac{\text{ft}^2}{\text{s}^2}\left(\frac{1}{32.174}\frac{\text{lbf}\cdot\text{s}^2}{\text{lbm}\cdot\text{ft}}\right)$$

$$+ 32.2\frac{\text{ft}^2}{\text{s}^2}(25\ \text{ft})\frac{1}{32.174}\frac{\text{lbf}\cdot\text{s}^2}{\text{lbm}\cdot\text{ft}}$$

$$= 95.6 \text{ ft}\cdot\text{lbf/lbm}$$

$$\dot{W}_{\text{in}} = \dot{m}w_{\text{in}} = 3.48\frac{\text{lbm}}{\text{s}}\left(95.6\frac{\text{ft}\cdot\text{lbf}}{\text{lbm}}\right)\frac{1}{550}\frac{\text{s}\cdot\text{hp}}{\text{ft}\cdot\text{lbf}}$$

$$= 0.60 \text{ hp}$$

Comment: This result is for a model with no heat transfer and no change in the internal energy of the water. Actually, these terms are not both zero and consequently the power requirement is greater than we have calculated under these assumptions. (More accurate modeling of such systems is treated in Chapter 14.)

Example 3·11 Refrigeration Condenser

PROBLEM STATEMENT

In a household refrigerator, refrigerant in the vapor phase enters the heat exchanger (condenser) on the back of the cabinet, where heat is removed to the room air, so that the refrigerant condenses to a liquid. Refrigerant properties at the condenser inlet are $u_1 = 85.2$ B/lbm, $h_1 = 93.5$ B/lbm, $v_1 = 0.298$ ft^3/lbm and at the outlet are $u_2 = 26.0$ B/lbm, $h_2 = 26.3$ B/lbm, $v_2 = 0.0123$ ft^3/lbm. If the flow rate of refrigerant is 42.0 lbm/h, at what rate is heat transferred to the room air that flows over the condenser?

SOLUTION

Analysis: The system here is an open system: the condenser. Consider the flow as steady, as suggested by

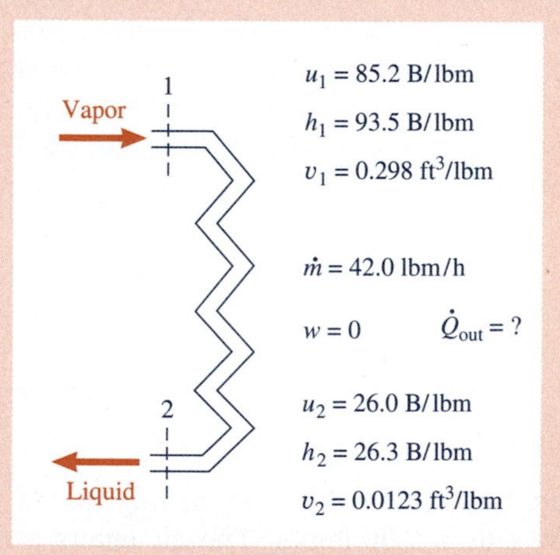

the specified flow rate. This is clearly a first law application, since we are looking for a rate of energy exchange. The heat transfer rate *to* the room air is equal to the heat transfer rate *from* the refrigerant. Let us determine the heat transfer per unit mass of refrigerant and then multiply by the mass rate of flow to determine $\dot{Q}$. We have

$$q_{out} = -q = h_1 - h_2 + \frac{V_1^2 - V_2^2}{2} + g(z_1 - z_2) - w$$

The system is a heat exchanger with rigid boundaries and no means of doing work, so the work term is zero. We have no information on velocities of the refrigerant, but it is likely that the velocity of the liquid leaving is very low, since its density is much greater than the density of the incoming vapor. The vapor velocity at inlet is likely to be low because designing the refrigerator with a high vapor velocity would cause excessive frictional pressure drop through the condenser. We will tentatively assume that the kinetic energy change is negligible and will review this assumption after making calculations. The potential energy change is also very slight, because even if the elevation change were 5 ft, the potential energy change would be approximately 5 ft·lbf/lbm or 0.006 B/lbm, which is negligible in comparison with the change in enthalpy. The first law equation then reduces to

$$q_{out} = h_1 - h_2 = 93.5 - 26.3 = 67.2 \text{ B/lbm}$$

$$\dot{Q}_{out} = \dot{m}q_{out} = 42.0(67.2) = 2820 \text{ B/h}$$

Comment: This value is based on the assumption of negligible kinetic energy at the inlet. Let us now check this assumption. If we evaluate $V_1^2/2$, we see that even for a value of V as high as 70 ft/s, this term has a value of only 0.1 B/lbm, which for most purposes is negligible in comparison with the value of 67.2 B/lbm calculated for q_{out}.

Is the result reasonable? The rate of heat rejection from a refrigerator operating steadily must equal the sum of the power input and the rate of any other energy input. Therefore, the power input to this refrigerator must be less than 2820 B/h or 1.1 hp. Household refrigerators are usually driven by motors rated at between 0.25 and 1 hp, so the result is in a reasonable range.

Finally, you may have noticed that the statement of the problem included some data that were not needed in the solution. Problems that arise in engineering practice do not carry sharply defined lists of "given" or "known" quantities. Sometimes nothing is "given" and a significant step may be the determination of what is known and what can be assumed. Of course, not all that is known or assumable is useful. Consequently, the problem statements in this book occasionally include information that is not needed in the solution.

Example 3·12 Heat Exchanger

PROBLEM STATEMENT

A large engine oil cooler cools lubricating oil at atmospheric pressure from 240 to 150 F by having the oil flow across a bank of tubes through which air flows. The oil flow rate is 0.30 lbm/s. The air enters at atmospheric pressure, 80 F, flows with little change in pressure, and is discharged at a temperature not to exceed 140 F. The heat exchanger is well insulated. Determine the minimum flow rate of the air. (Enthalpy values for the oil are 198 and 102 B/lbm at inlet and outlet, respectively. For the air, enthalpies at inlet and outlet are 129.1 and 143.5 B/lbm, respectively.)

SOLUTION

Analysis: If we defined one open system as the oil passage through the heat exchanger and another as the air passage through the heat exchanger, then there would be heat transfer between the two systems. We could apply the first law to the oil passage to calculate $\dot{Q}$ for the oil. Then we could use the first law applied to a system defined as the air passage through the heat exchanger and use in it $\dot{Q}_{air} = -\dot{Q}_{oil}$. However, if we define the open system for study as the entire heat exchanger, which is well insulated, there is no heat transfer with the surroundings. (The heat transfer from the oil to the cooling air occurs within the system, not across a system boundary, so it does not enter the calculations using the first law applied to the entire heat exchanger.) We make a sketch of the system, noting that work and heat are both zero. Changes in kinetic and

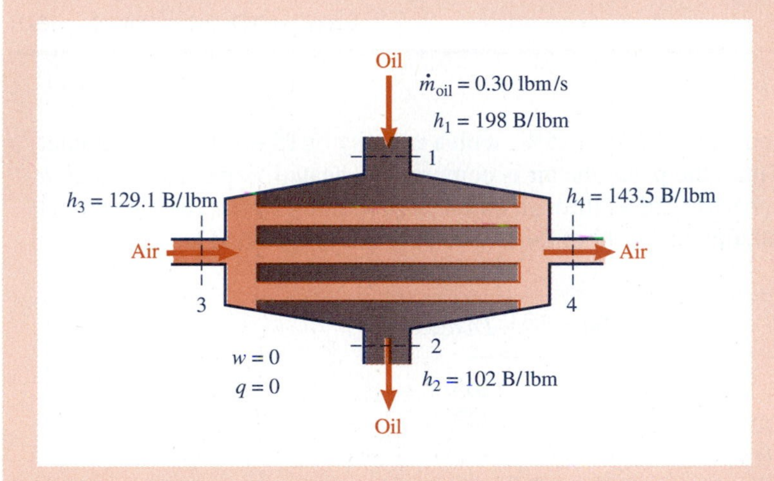

potential energy are assumed to be negligible, so the first law equation

$$\dot{Q} - \dot{W} = \sum_{\text{exits}} \dot{m}\left(h + \frac{V^2}{2} + gz\right) - \sum_{\text{inlets}} \dot{m}\left(h + \frac{V^2}{2} + gz\right)$$

can be simplified to

$$\sum_{\text{exits}} \dot{m}h = \sum_{\text{inlets}} \dot{m}h$$

and solved for the air flow rate.

Calculations: The first law for this steady-flow system is

$$\dot{m}_{oil}h_2 + \dot{m}_{air}h_4 = \dot{m}_{oil}h_1 + \dot{m}_{air}h_3$$

$$\dot{m}_{air}(h_4 - h_3) = \dot{m}_{oil}(h_1 - h_2)$$

$$\dot{m}_{air} = \dot{m}_{oil}\frac{h_1 - h_2}{h_3 - h_4} = 0.30\frac{198 - 102}{143.5 - 129.1} = 2.0 \text{ lbm/s}$$

Comment: Calculating a typical atmospheric air density from the ideal-gas equation of state shows that this is approximately 25 cu ft of air per second or 1500 cu ft per minute. Clearly, the engine for which this oil cooler is used is much larger than a typical automobile engine. (If you have no basis for estimating air flow rates, you might start by noting that a household kitchen exhaust fan capacity is typically of the order of 200 cu ft per minute.)

Example 3·13 Gas Turbine Power Plant

PROBLEM STATEMENT

A gas turbine power plant takes in atmospheric air at 95 kPa, 15°C, with a velocity of 15 m/s through an inlet having a cross-sectional area of 1.28 m^2. Within the plant, the air is compressed, heated, expanded through a turbine, and then discharged into the atmosphere at 205°C through an exhaust opening with a cross-sectional area of 0.71 m^2. The internal energy and enthalpy of the air at the two temperatures specified are as follows:

	Inlet	*Discharge*
Temperature, °C	15	205
Internal energy, kJ/kg	205.7	343.5
Enthalpy, kJ/kg	288.4	480.7

The power output of the plant is 2040 kW. Determine the net amount of heat added to the air in kJ/kg.

SOLUTION

Analysis: The plant description and data indicate that the plant is a steady-flow system with a single inlet and a single outlet. A sketch of the system is made and data are entered on the sketch. Since we are looking for an energy quantity, we write a first law statement or energy balance for the system:

$$q = w + h_2 - h_1 + \frac{V_2^2 - V_1^2}{2} + g(z_2 - z_1) \tag{a}$$

(We could use an energy balance with Δu and Δpv as separate terms instead of Δh, but using Δh will reduce the amount of numerical calculations needed.) Now we must evaluate each of the terms on the right-hand side of the equation to obtain the value of q. The four terms will be considered in the order in which they appear in the energy balance above.

The work output (in kJ/kg) can be found from the power output and the mass rate of flow:

$$w = \frac{\dot{W}}{\dot{m}} \tag{b}$$

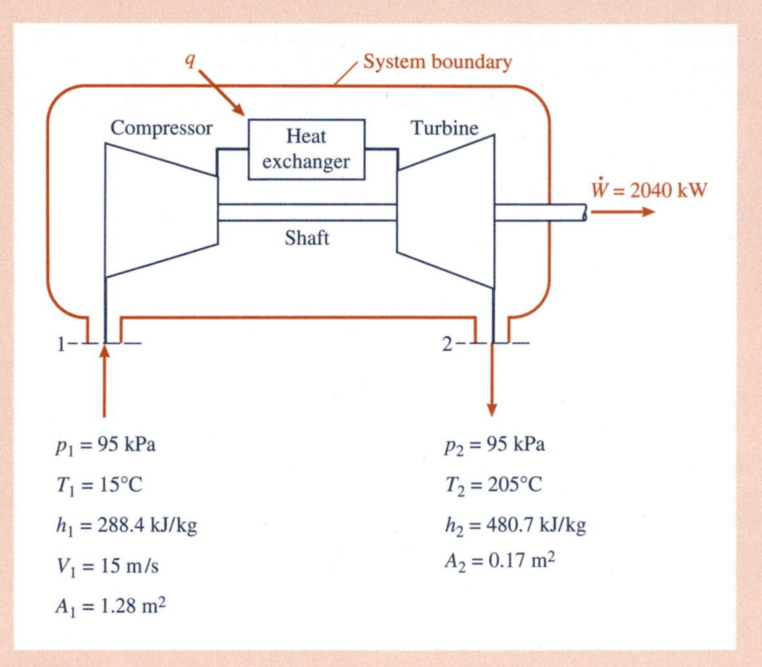

The power output $\dot{W}$ is specified, but the mass rate of flow must be calculated by

$$\dot{m} = \rho_1 A_1 V_1 \qquad (c)$$

A_1 and V_1 are specified, but we do not know the density ρ_1. It is a function of pressure and temperature as given by the ideal-gas equation of state:

$$\rho_1 = \frac{p_1}{RT_1} \qquad (d)$$

and p_1 and T_1 are specified. R for air is, of course, known. Thus Eqs. *b*, *c*, and *d* together provide the value of w for use in Eq. *a*. Enthalpy values are given in the problem statement.

The change in kinetic energy requires knowledge of both V_1 and V_2. V_1 is specified, but the value of V_2 must be obtained. Since A_1 is specified, and ρ_2 can be calculated just as ρ_1 was from known values of p_2 and T_2, we can use the continuity equation to find V_2:

$$\rho_2 A_2 V_2 = \rho_1 A_1 V_1 = \dot{m}$$

The change in potential energy can be found only if we have some information about the elevations of the inlet and outlet, and we do not have this information. However, if we notice the magnitudes of the other terms in the energy balance and then notice that an elevation change of 102 m is required to cause a change in potential energy of 1 kJ/kg (for $g = 9.81$ m/s^2), we should not hesitate to assume that the change in potential energy is negligibly small. (An elevation difference of 10 m between inlet and outlet of a gas-turbine plant would be unusually great, and it would correspond to a potential energy change of only 0.1 kJ/kg.)

Thus our analysis of the problem shows us (1) that we can solve the problem, and (2) the procedure to use in our solution.

An engineer tackling a problem usually analyzes it thoroughly in some manner such as presented above. For an easy problem like this, the analysis would not be written out as it is here, and it would be made so rapidly that one would hardly be aware of it. The analysis might be outlined in writing as shown below:

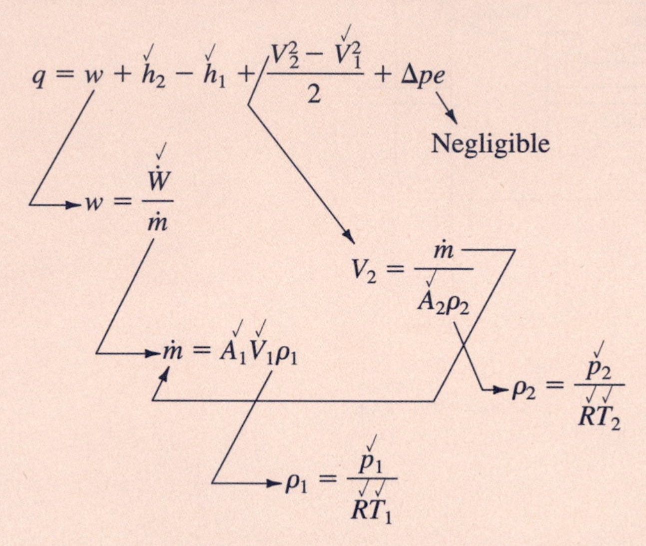

The check marks (√) over certain symbols indicate that the numerical values of those quantities are known. This form of analysis provides a clear outline of the solution. It also suggests the sequence of writing equations for numerical calculations in order to minimize backtracking and excursions to obtain needed values.

Calculations:

$$\rho_1 = \frac{p_1}{RT_1} = \frac{95}{0.287(288)} = 1.15 \text{ kg/m}^3$$

$$\rho_2 = \frac{p_2}{RT_2} = \frac{95}{0.287(478)} = 0.692 \text{ kg/m}^3$$

$$\dot{m} = A_1V_1\rho_1 = 1.28(15)1.15 = 22.1 \text{ kg/s}$$

$$w = \frac{\dot{W}}{\dot{m}} = \frac{2040}{22.1} = 92.4 \text{ kJ/kg}$$

$$V_2 = \frac{\dot{m}}{A_2\rho_2} = \frac{22.1}{0.71(0.692)} = 44.9 \text{ m/s}$$

$$q = w + h_2 - h_1 + \frac{V_2^2 - V_1^2}{2}$$

$$= 92.4\frac{\text{kJ}}{\text{kg}} + (480.7 - 288.4)\frac{\text{kJ}}{\text{kg}} + \frac{(44.9)^2 - (15)^2}{2}\frac{\text{J}}{\text{kg}}\left(\frac{\text{kJ}}{1000 \text{ J}}\right)$$

$$= 286 \text{ kJ/kg}$$

ALTERNATIVE SOLUTION

Problems of this kind can be solved very quickly using a computer with a program for solving simultaneous equations. Of course, we have shown above that this example problem can be solved once for a single set of values quickly by longhand. However, if many such problems are to be solved or some specified quantities are to cover a range of values, a computer solution is far more convenient. Therefore, we present an example computer solution of this problem.

```
==================== RULE SHEET ==========================================
S Rule-------------------------------------------------------------------
  ;---------------- Example 3·13 -----------------------------------------
  ;Open system, steady flow (1-D, one inlet, one outlet).  Fluid is air
  ;in p and T range where ideal-gas equation of state applies.

  q = h2 - h1 + (V2^2 - V1^2)/2 + delpe + w    ; First law, steady flow
  Wdot = mdot * w                               ;Power, work, flow rate relation
  mdot = A2 * V2/v2                             ;Continuity equation
  mdot = A1 * V1/v1                             ;     "         "
  p1 * v1 = R * T1                              ;Ideal-gas equation of state at state 1
  p2 * v2 = R * T2                              ; "    "    "    "    "    "  state 2
```

TK Solver Rule Sheet showing equations entered in arbitrary order.

```
=================== VARIABLE SHEET =========================================
St Input---- Name--- Output--- Unit----- Comment---------------------------
                                --------- * E3-13.TK ----- Example 3·13 -----
              q        286       kJ/kg     Heat transfer per unit mass
   480.7      h2                 kJ/kg     Enthalpy at outlet
   288.4      h1                 kJ/kg        "     "  inlet
              V2       44.9      m/s       Velocity at outlet
   15         V1                 m/s          "     "  inlet
   0          delpe              kJ/kg     Change in potential energy
              w        92.5      kJ/kg     Work per unit mass
   2040       Wdot               kW        Power
              mdot     22.1      kg/s      Mass rate of flow
   0.71       A2                 m^2       Cross-sectional area of outlet
              v2       1.44      m^3/kg    Specific volume at outlet
   1.28       A1                 m^2       Cross-sectional area of inlet
              v1       0.871     m^3/kg    Specific volume at inlet
   95         p1                 kPa       Pressure at inlet
   0.287      R                  kJ/(kg*K) Gas constant for air
   15         T1                 C         Temperature at inlet
   95         p2                 kPa       Pressure at outlet
   205        T2                 C         Temperature at outlet
```

TK Solver Variable Sheet showing input and output variable values.

Mathcad® 4.0 Solution

Given:

Properties at state 1

$p_1 := 95 \cdot kPa$ — *Pressure*

$T_1 := (273.15 + 15) \cdot K$ — *Temperature*

$h_1 := 288.4 \cdot \frac{kJ}{kg}$ — *Enthalpy*

$A_1 := 1.28 \cdot m^2$ — *Area*

$V_1 := 15 \cdot \frac{m}{s}$ — *Velocity*

Properties at state 2

$p_2 := 95 \cdot kPa$

$T_2 := (273.15 + 205) \cdot K$

$h_2 := 480.7 \cdot \frac{kJ}{kg}$

$A_2 := 0.71 \cdot m^2$

Additional information

$Wdot := 2040 \cdot kW$ — *Power*

$\Delta pe := 0 \cdot \frac{kJ}{kg}$ — *Change in potential energy*

Additional Definitions

$kPa \equiv 1000 \cdot Pa$

$kJ \equiv 1000 \cdot joule$

$s \equiv 1 \cdot sec$

$R := 0.287 \cdot \frac{kJ}{kg \cdot K}$

Solution:

$v_1 := R \cdot \frac{T_1}{p_1}$ — $v_1 = 0.871 \cdot \frac{m^3}{kg}$

$v_2 := R \cdot \frac{T_2}{p_2}$ — $v_2 = 1.44 \cdot \frac{m^3}{kg}$

The ideal gas equation of state applied at states 1 and 2, solved for the initial specific volume

$mdot := A_1 \cdot \frac{V_1}{v_1}$ — *Continuity equation, at state 1* — $mdot = 22.1 \cdot \frac{kg}{s}$

$V_2 := \frac{mdot \cdot v_2}{A_2}$ — *Continuity equation, solved for the velocity at state 2* — $V_2 = 44.9 \cdot \frac{m}{s}$

$w := \frac{Wdot}{mdot}$ — *Power, work, flow rate relation* — $w = 92.5 \cdot \frac{kJ}{kg}$

$q := h_2 - h_1 + \frac{V_2^2 - V_1^2}{2} + \Delta pe + w$ — *First law, steady flow* — $q = 286 \cdot \frac{kJ}{kg}$

Mathcad solution. (*Mathcad is a registered trademark of MathSoft, Inc., Cambridge, Massachusetts. Used by permission of MathSoft, Inc.*)

Three principles are involved in the solution of this problem: the first law, conservation of mass, and the ideal-gas equation of state. In addition, we use the relationship among mass rate of flow, work per unit mass, and power. Application of these principles to the steady-flow system described in the problem statement involves six equations. We enter these six equations into an equation-solving program. We show here solutions using both TK Solver and Mathcad.

Comment: We repeat for emphasis that solving one problem like this requires no computer and is perhaps just as well done longhand. However, time is well invested in learning to use the computer for relatively simple solutions before moving on to more complex problems where computer use is essential.

Example 3·14 Gas Flow Through a Nozzle

PROBLEM STATEMENT

Helium at 300 kPa, 60°C, enters a nozzle with negligible velocity and expands steadily without heat transfer in a quasiequilibrium manner to 120 kPa. The process is such that $pv^{1.67}$ = constant. Calculate the exit velocity.

SOLUTION

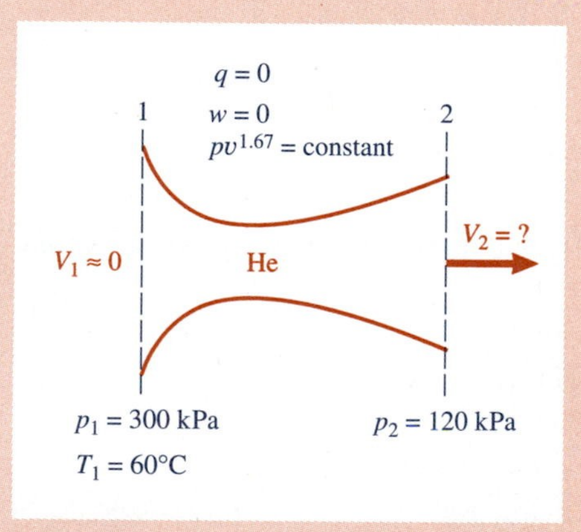

Analysis: Since we are looking for a velocity, we might think first of the continuity equation, but we have no information regarding the flow cross-sectional areas (except that the inlet area must be very large to make the velocity negligible), so the continuity equation will not help. We turn next to the first law for steady flow:

$$q - w = h_2 - h_1 + \frac{V_2^2 - V_1^2}{2} + g(z_2 - z_1)$$

where $q = 0$, no work is done in a nozzle, V_1 is negligible, and the potential energy change for the gas flowing through the nozzle is negligible. There appears to be no way of determining h_2 and h_1 (unless one has already studied Chapter 4!), so we appear to be "stuck." The first law equation or energy balance written above is valid, but we do not have enough information to solve it for V_2. Therefore, we immediately seek some other means of solving the problem. Perhaps we can find w_{in} by means of the expression that is based on the principles of mechanics (Section 1·10·2):

$$w_{in} = \int_1^2 v\,dp + \frac{V_2^2 - V_1^2}{2} + g(z_2 - z_1)$$

We have noted before that we can eliminate the work and potential energy terms. Since we have a functional relationship between p and v, we can integrate $\int v\,dp$, and so we will be able to determine V_2. (The unsuccessful attempts to use the continuity equation and an energy balance are shown above to illustrate that no harm is done by making such false starts as long as we quickly recognize that we are not on the right track and immediately look for another method of solution.)

Calculations:

$$w_{in} = \int_1^2 v\,dp + \frac{V_2^2 - V_1^2}{2} + g(z_2 - z_1)$$

Dropping the terms that are negligible and using $pv^{1.67} = pv^n = c$, we find

$$\frac{V_2^2}{2} = -\int_1^2 v\,dp = -\int_1^2 \left(\frac{c}{p}\right)^{1/n} dp = -c^{1/n}\int_1^2 p^{-1/n}\,dp$$

$$= -\frac{(p_1 v_1^n)^{1/n}}{\left(1 - \dfrac{1}{n}\right)}[p_2^{1-1/n} - p_1^{1-1/n}] = -\frac{p_1^{1/n} p_1^{1-1/n} v_1}{\left(\dfrac{n-1}{n}\right)}\left[\left(\frac{p_2}{p_1}\right)^{(n-1)/n} - 1\right]$$

$$\frac{V_2^2}{2} = -\left(\frac{n}{n-1}\right)p_1 v_1\left[\left(\frac{p_2}{p_1}\right)^{(n-1)/n} - 1\right]$$

Under the specified conditions, helium is well modeled by the ideal-gas equation of state, so we substitute from $p_1 v_1 = RT_1$ and solve:

$$V_2 = \sqrt{-2\left(\frac{n}{n-1}\right)RT_1\left[\left(\frac{p_2}{p_1}\right)^{(n-1)/n} - 1\right]}$$

$$= \sqrt{\frac{-2(1.67)}{1.67 - 1}\left(2.007\frac{\text{kJ}}{\text{kg·K}}\right)333\text{ K}\left(1000\frac{\text{J}}{\text{kJ}}\right)\left[\left(\frac{120}{300}\right)^{(1.67-1)/1.67} - 1\right]}$$

$$= 1010\text{ m/s}$$

Comment: The conversion factor of 1000 J/kJ is required because to obtain velocity in m/s, the term under the radical must have units of J/kg instead of kJ/kg.

Further comment: Notice that although early in the solution we considered using the first law, we did not use it in our final solution.

Example 3·15 First Law, Open System

PROBLEM STATEMENT

Machines shipped from a manufacturing plant are filled with dry nitrogen and sealed to prevent oxidation of internal parts. The nitrogen is delivered to the filling sites in steel bottles that have an internal volume of 0.120 m^3. The bottles are filled from a pipeline carrying dry nitrogen at 800 kPa, 35°C. A bottle to be filled from the line typically contains nitrogen at 150 kPa, 20°C. The bottle is connected to the pipeline through a valve and the valve is opened and held open until the pressure in the bottle equals that in the pipeline. Properties of nitrogen are related by $h = 1.40u = 1.04T = 3.50pv$, with h and u in kJ/kg, T in K, p in kPa, and v in m^3/kg. (You can recognize that the last part of this equation, $1.04T = 3.50pv$, is a statement of the ideal-gas equation of state.) Assuming that there is no heat transfer between the nitrogen and any part of the surroundings, that conditions throughout the bottle are uniform at any instant, and that changes in kinetic and potential energy are negligible, determine the final mass and temperature of nitrogen in the bottle.

SOLUTION

Analysis: We first make a schematic diagram of the bottle connected to the pipeline and then place data on

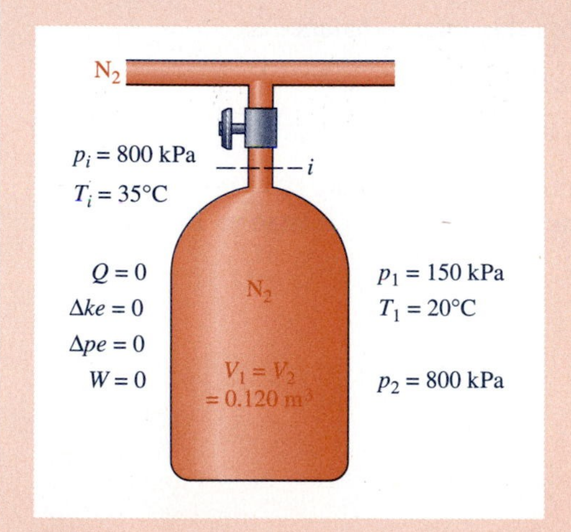

the diagram and in the table below. We define the system as the region inside the bottle and note that the properties of nitrogen entering the system are constant and equal to those of the nitrogen in the pipeline.

	N_2 initially in the bottle	N_2 flowing into the bottle	N_2 finally in the bottle
Pressure, kPa	$p_1 = 150$	$p_i = 800$	$p_2 = 800$
Volume, m^3	$V_1 = 0.120$	—	$V_2 = 0.120$
Mass, kg	$m_1 = p_1V_1/RT_1$	$m_i = m_2 - m_1$	m_2
Temperature, °C	$T_1 = 20$	$T_i = 35$	T_2

We are looking for both the mass and temperature of the nitrogen finally in the bottle. These are related by the expression

$$1.04T_2 = 3.50p_2v_2 = 3.50\,\frac{p_2V_2}{m_2}$$

We have values for p_2 and V_2, so we have two unknowns, m_2 and T_2, but only one equation. The first law provides an equation that relates quantities such as mass, internal energy, and enthalpy. Another equation involving some of these quantities is $h = 1.4u = 1.04T$. Let us then start with the first law.

Calculations: For this open system involving no heat transfer, no work, and no matter leaving the system, we write the first law as

$$\begin{pmatrix}\text{Stored energy}\\ \text{of N}_2\text{ finally}\\ \text{in the bottle,}\\ \text{state 2}\end{pmatrix} = \begin{pmatrix}\text{Stored energy}\\ \text{of N}_2\text{ initially}\\ \text{in the bottle,}\\ \text{state 1}\end{pmatrix} + \begin{pmatrix}\text{Stored energy}\\ \text{of N}_2\text{ that}\\ \text{flows into the}\\ \text{bottle at state } i\end{pmatrix} + \begin{pmatrix}\text{Flow work}\\ \text{of N}_2\text{ that}\\ \text{flows into the}\\ \text{bottle at state } i\end{pmatrix}$$

Neglecting the kinetic and potential energy terms, we can express this energy balance as

$$\begin{aligned} m_2u_2 &= m_1u_1 + m_i(u_i + p_iv_i) \\ &= m_1u_1 + (m_2 - m_1)h_i \qquad (a) \end{aligned}$$

The initial mass in the bottle m_1 can be obtained from the ideal-gas equation of state, $p_1V_1 = m_1RT_1$, so the first law equation we have written involves two unknowns: m_2 and u_2. Here m_2 is related to T_2 and known quantities by the equation

$$1.40u_2 = 1.04T_2 = 3.50p_2v_2 = 3.50\frac{p_2V_2}{m_2}$$

$$u_2 = 2.50\frac{p_2V_2}{m_2}$$

Similarly,

$$u_1 = 2.50\frac{p_1V_1}{m_1}$$

Also,

$$h_i = 1.04T_i$$

Substituting these values into the first law equation

$$m_2u_2 = m_1u_1 + (m_2 - m_1)h_i$$

gives

$$m_2\left(2.50\frac{p_2V_2}{m_2}\right) = m_1\left(2.50\frac{p_1V_1}{m_1}\right) + (m_2 - m_1)1.04T_i$$

Rearranging and noting that $V_2 = V_1$ yields

$$2.50V_1(p_2 - p_1) = (m_2 - m_1)1.04T_i$$

$$m_2 = m_1 + \frac{2.50V_1}{1.04T_i}(p_2 - p_1) = \frac{p_1V_1}{RT_1} + \frac{2.50V_1}{1.04T_i}(p_2 - p_1)$$

$$= \frac{150(0.120)}{0.297(293)} + \frac{2.50(0.120)}{1.04(308)}(800 - 150) = 0.816 \text{ kg}$$

The final temperature can be obtained from the ideal-gas equation of state:

$$T_2 = \frac{p_2V_2}{m_2R} = \frac{800(0.120)}{0.816(0.297)} = 396 \text{ K} = 123°\text{C}$$

Comment: The final temperature of nitrogen in the bottle is higher than either the initial temperature or the temperature of nitrogen in the pipeline. (In fact, the steel bottles, if not insulated, would be dangerous to touch.) This is because the nitrogen initially in the bottle has been compressed by the nitrogen entering and all of the nitrogen finally in the bottle has stored energy which includes the flow work done on the open system by the nitrogen entering.

SOLUTION WITH A SLIGHTLY DIFFERENT APPROACH

Some people prefer to start with the first law equation in differential form. The analysis is the same as given in the solution above, but we start with the first law (for the case of $Q = 0$, $W = 0$, and no changes in kinetic or potential energy) in the form:

$$\begin{pmatrix}\text{Stored energy and}\\ \text{flow work added to}\\ \text{system by entering}\\ \text{mass } \delta m\end{pmatrix} = \begin{pmatrix}\text{Increase in stored}\\ \text{energy of system}\\ \text{while mass}\\ \delta m \text{ enters}\end{pmatrix}$$

$$(u_i + p_iv_i)\ \delta m_i = dU$$

$$h_i\ \delta m_i = dU$$

If m is the mass in the tank, dm is the change in mass in the tank. Since nitrogen enters at only section i, $\delta m_i = dm$:

$$h_i\ dm = dU$$

Since $u = 2.50pv$, then $U = mu = 2.50mpv = 2.50pV$, and $dU = 2.50(V\,dp + p\,dV) = 2.50V\,dp$, since V is constant. Thus,

$$h_i\ dm = 2.50V\,dp$$

Integrating, noting that h_i is constant, gives

$$h_i(m_2 - m_1) = 2.50V_1(p_2 - p_1)$$

$$m_2 = m_1 + \frac{2.50V_1}{1.04T_i}(p_2 - p_1) = \frac{p_1V_1}{RT_1} + \frac{2.50V_1}{1.04T_i}(p_2 - p_1)$$

Substitution of numerical values gives $m_2 = 0.816$ kg, and T_2 can then be found as in the first solution.

Example 3·16 First Law, Open System

PROBLEM STATEMENT

A balloon is to be partially filled at atmospheric pressure with helium from a well-insulated storage tank that has a volume of 6.20 m^3. At the time of filling the balloon, atmospheric pressure is 95.0 kPa. The helium in the tank is initially at a gage pressure of 500 kPa and 20°C before the valve on the tank is opened. Assuming that conditions are uniform throughout the tank at any instant and that heat transfer is negligible, determine the mass of helium that has been put into the balloon when a pressure gage on the storage tank reads 300 kPa. For helium under these conditions, $h = 1.667u = 5.193T$, with h and u in kJ/kg and T in K.

SOLUTION

Analysis: We first make a sketch of the storage tank. The mass of helium that has been put into the balloon

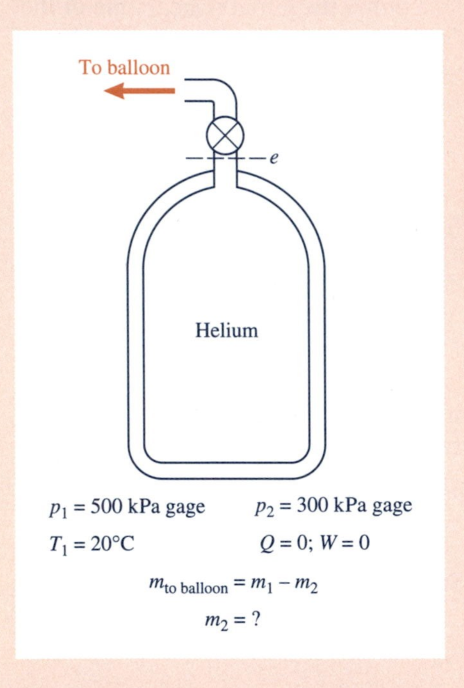

is the difference between the masses initially and finally in the storage tank. The initial mass in the tank can be calculated from the ideal-gas equation of state, since pressure, temperature, and volume are known. We cannot calculate the final mass in the tank in this manner because the final temperature is unknown, so we have two unknowns, m_2 and T_2, in the ideal-gas equation of state. We see in the problem statement that T is related to u, and another relationship involving m_2, T_2, and u_2 is the first law applied to the open system defined as the helium within the tank at any instant.

Section e where the helium leaves the system is just upstream of the valve, so the kinetic energy at e is very small. Also, the properties at e at any instant are equal to the properties of helium inside the tank. Since the properties of helium within the tank (and therefore at section e) vary with time, we will use a differential form of the first law.

Calculations: As an infinitesimal mass of helium δm_e flows from the system, the mass of helium within the system m changes by an amount dm, and $\delta m = -dm$. Since $Q = 0$ and $W = 0$,

$$\begin{pmatrix}\text{Decrease in stored energy} \\ \text{of helium in tank}\end{pmatrix} = \begin{pmatrix}\text{Stored energy of} \\ \delta m \text{ leaving tank}\end{pmatrix} + \begin{pmatrix}\text{Flow work of} \\ \delta m \text{ leaving tank}\end{pmatrix}$$

$$-dU = e\,\delta m + pv\,\delta m = (e + pv)\,\delta m_e$$

where a property without a subscript is a property of the helium in the tank. Potential energy changes can be neglected. As noted above, section e is just upstream of the small valve opening, so the kinetic energy at the exit is negligibly small, $e_e = u_e$, and the last equation becomes

$$-dU = (u_e + p_e v_e)\,\delta m_e$$

Since the properties of the helium leaving are the same as those of the helium in the tank, $u_e = u$, $p_e = p$, and $v_e = v$. Recalling also that $\delta m_e = -dm$, we have

$$-dU = -(u + pv)\,dm$$
$$d(mu) = (u + pv)\,dm$$
$$m\,du + u\,dm = (u + pv)\,dm$$
$$m\,du = pv\,dm$$

Since internal energy can be expressed in terms of T as $1.667u = 5.193T$ or $u = 3.115T$, so that $du = 3.115\,dT$, the energy balance becomes

$$3.115m\,dT = pv\,dm$$

From the ideal-gas equation of state, $T = pv/R$, so making this substitution gives

$$\frac{3.115m}{R}\,d(pv) = pv\,dm$$

$$\frac{dm}{m} = \frac{3.115}{R}\,\frac{d(pv)}{pv}$$

Integrating between states 1 and 2 yields

$$\ln\frac{m_2}{m_1} = \frac{3.115}{R}\ln\left(\frac{p_2v_2}{p_1v_1}\right)$$

$$\frac{m_2}{m_1} = \left(\frac{p_2v_2}{p_1v_1}\right)^{3.115/R} = \left(\frac{p_2V_2m_1}{p_1V_1m_2}\right)^{3.115/R}$$

We now substitute the numerical value of R (2.077 kJ/kg·K) to simplify the exponents in the following steps, and we note that the tank volume is constant, $V_2 = V_1 = V$, so

$$\frac{m_2}{m_1} = \left(\frac{p_2}{p_1}\right)^{1.5/2.5} = \left(\frac{p_2}{p_1}\right)^{0.600}$$

$$m_2 = m_1\left(\frac{p_2}{p_1}\right)^{0.600}$$

The mass of helium that has been put into the balloon equals the decrease in the mass of helium in the storage tank:

$$m_{\text{to balloon}} = m_1 - m_2 = m_1 - m_1\left(\frac{p_2}{p_1}\right)^{0.600} = m_1\left[1 - \left(\frac{p_2}{p_1}\right)^{0.600}\right]$$

Evaluating m_1 by the ideal-gas equation of state, and recalling that absolute pressure is the sum of gage pressure and atmospheric pressure, we have

$$m_{\text{to balloon}} = \frac{p_1 V}{RT_1}\left[1 - \left(\frac{p_2}{p_1}\right)^{0.600}\right]$$

$$= \frac{595(6.20)}{2.077(293)}\left[1 - \left(\frac{395}{595}\right)^{0.600}\right] = 1.32\text{ kg}$$

3·6 Using the First Law to Assess Performance

From the first law we conclude that energy is conserved. Energy can be converted from one form to another, but it cannot be created or destroyed. The first law therefore establishes an "energy accounting scheme" that is useful in assessing the performance of any energy conversion system or device. Such assessment is needed for the proper selection or operation of existing equipment and for choosing among alternatives in the design of new equipment.

Assessing the performance of a system requires first specifying its purpose. Engines, refrigerators, and heat pumps serve different purposes, so they have different performance indexes.

Engines: Thermal Efficiency. The purpose of an *engine* is to convert energy supplied to it as heat or as stored energy in a fuel into work. Optimum performance is the production of a required amount of work using a minimum of energy input. A heat engine receives heat and produces work while executing a cycle. A heat engine may be as simple as a gas confined within a piston–cylinder assembly, as shown in Fig. 3·10*a*, or as complex as an entire power plant, Fig. 3·10*b*.

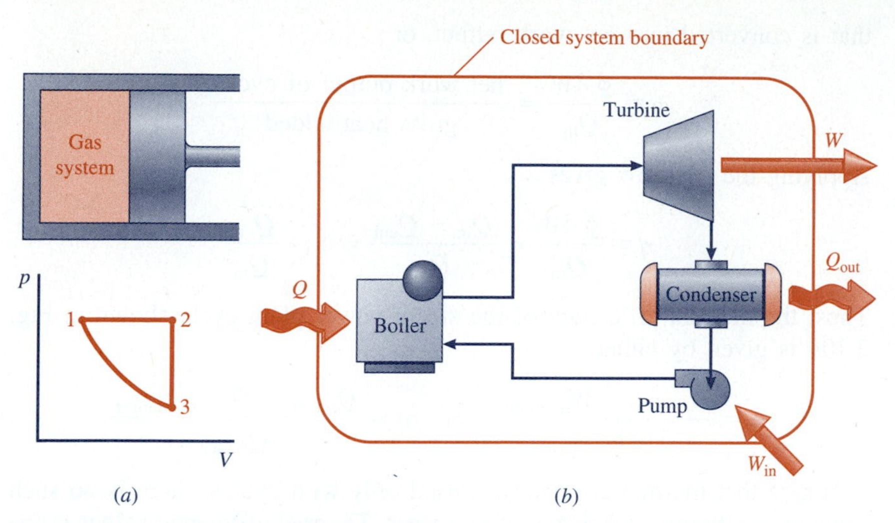

Figure 3·10 Heat engines. Energy input is heat.

In Fig. 3·10*a*, as the gas is heated, it expands to do work on the piston, process 1–2. In process 2–3, the gas is cooled while the piston is stationary and no work is done. Then the piston is moved inward, doing work on the gas while no heat is transferred, process 3–1. A cycle is thus completed with the net effect that heat has been converted to work.

In the power plant shown in Fig. 3·10*b*, heat is added to water in the boiler to generate steam that then expands adiabatically through the turbine, doing work. The steam flows from the turbine into the condenser. In the condenser, heat is removed from the steam to condense it. The liquid leaving the condenser enters a pump that pumps it into the boiler to complete a cycle. Work is done on the liquid flowing through the pump. Even though each of the four pieces of equipment has fluid flowing into and out of it and hence must be treated as an open system, the four pieces of equipment and the connecting piping together always contain the same fluid. No matter enters or leaves this larger system, so it can be treated as a closed system.

In each of these heat-engine cycles some heat is rejected from the system. By the first law,

$$\oint \delta W = \oint \delta Q = Q_{in} - Q_{out}$$

where Q_{in} and Q_{out} are respectively the *gross* amount of heat added and the *gross* amount of heat rejected. Since in the cycles described Q_{out} is not zero, the net work done by the system is less than the gross amount of heat added. That is, not all the heat added is converted into work. **Thermal efficiency** is defined as that fraction of the gross heat input to a heat engine during a cycle

Thermal efficiency

that is converted into net work output, or

$$\eta \equiv \frac{\oint \delta W}{Q_{\text{in}}} = \frac{\text{net work output of cycle}}{\text{gross heat added}}$$

Applying the first law gives

$$\eta \equiv \frac{\oint \delta W}{Q_{\text{in}}} = \frac{Q_{\text{in}} - Q_{\text{out}}}{Q_{\text{in}}} = 1 - \frac{Q_{\text{out}}}{Q_{\text{in}}}$$

Thus, the thermal efficiency of the steam power plant cycle shown in Fig. 3·10*b* is given by either

$$\eta = \frac{W_{\text{turbine}} - W_{\text{in, pump}}}{Q_{\text{boiler}}} \quad \text{or} \quad \eta = \frac{Q_{\text{boiler}} - Q_{\text{out, condenser}}}{Q_{\text{boiler}}}$$

Notice that thermal efficiency is used only with cycles; there is no such thing as the thermal efficiency of a process. Thermal efficiency values range from 0 up to but not including 1 (or from 0 to less than 100 percent). The maximum value is limited by the second law of thermodynamics, as will be explained in Chapter 5.

The heat supplied to a power plant as shown in Fig. 3·10*b* may come from a nuclear reactor or from the combustion of a fuel in a furnace, as shown in Fig. 3·11 where the system boundary encloses the source of heat to the steam generator. This is an open system. Fuel and air enter the system; exhaust gases and perhaps solid refuse leave it. The energy input to this larger system, and entire power plant, is the stored energy of the entering fuel and air, not heat transfer. The performance index for such a plant is not thermal efficiency as we have defined it above, because there is no heat input to the

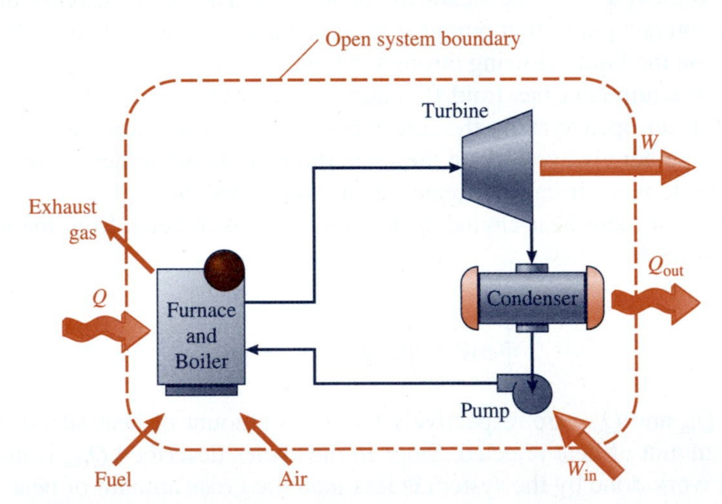

Figure 3·11 Power plant. Energy input is the energy of fuel and air entering.

system. For various kinds of power plants, different indexes are defined, and some of these are introduced later in this book.

Refrigerators: Coefficient of Performance. The purpose of a *refrigerator* is to remove heat from a region at a lower-than-ambient temperature. Doing this requires energy input, usually in the form of work, from elsewhere. Optimum performance is the removal of a required amount of heat using a minimum of energy input to the refrigerator.

A widely used type of refrigerator called a *vapor-compression refrigerator* is shown in Fig. 3·12. The refrigerant in a vapor phase is compressed by the compressor. In this process, work is done on the refrigerant and some heat may be removed. The refrigerant then flows into the condenser where heat is removed to condense it to a liquid. The liquid flows through an expansion valve (or a capillary tube) into the evaporator where the pressure and temperature are low. The evaporating refrigerant absorbs heat from the low-temperature region in the surroundings. Refrigerant flows through each of the four named components, but those components and the piping connecting them always contain the same refrigerant, so together they constitute a closed system.

The refrigerator operates cyclically, so there is no change in its stored energy. Applying the first law, we have

$$W_{\text{in}} + Q_{\text{in}} = Q_{\text{out, condenser}} + Q_{\text{out, compressor}}$$

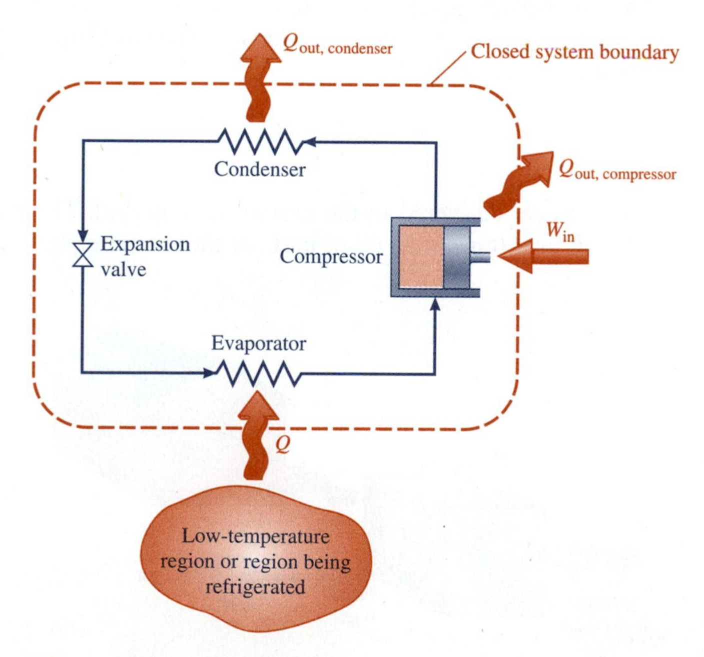

Figure 3·12 Refrigerator.

where we want to maximize the heat removed from the low-temperature region, Q_{in}, and minimize the work input. The *coefficient of performance for a refrigerator* is defined as

The coefficient of performance for a refrigerator

$$\beta_R \equiv \frac{Q_{in,L}}{W_{in}}$$

where $Q_{in,L}$ is the amount of heat removed from the region being refrigerated. (There may be stray heat transfer into the refrigerator from other sources, so $Q_{in,L}$ is not necessarily the total heat transfer into the system.)

The maximum value of the coefficient of performance for a refrigerator is limited by the second law of thermodynamics.

Refrigerator coefficients of performance range from slightly more than 0 to values well above 1. For ice-making machines and household refrigerators, coefficients of performance of 4 to 5 are typical. Maximum values under various conditions are limited by the second law of thermodynamics.

Heat Pumps: Coefficient of Performance. The purpose of a *heat pump* is to supply heat to a region by taking heat from a lower-temperature region. This requires an energy input, usually in the form of work (see Fig. 3·13). Optimum performance is achieved as the amount of work required for a specified amount of heat delivered to the higher-temperature region is minimized.

Obviously, the same equipment can serve as either a heat pump or a refrigerator. The difference is only in purpose: the refrigerator's purpose is to remove heat from a lower-temperature region; the heat pump's purpose is to provide heat to a higher-temperature region. Because the *purposes* are different, different indexes of performance are called for. The *coefficient of performance of a heat pump* is defined as

The coefficient of performance for a heat pump

$$\beta_{HP} \equiv \frac{Q_{out,H}}{W_{in}}$$

where $Q_{out,H}$ is the heat delivered to the region to be heated. There may be stray or intentional heat transfer to other parts of the surroundings, so $Q_{out,H}$

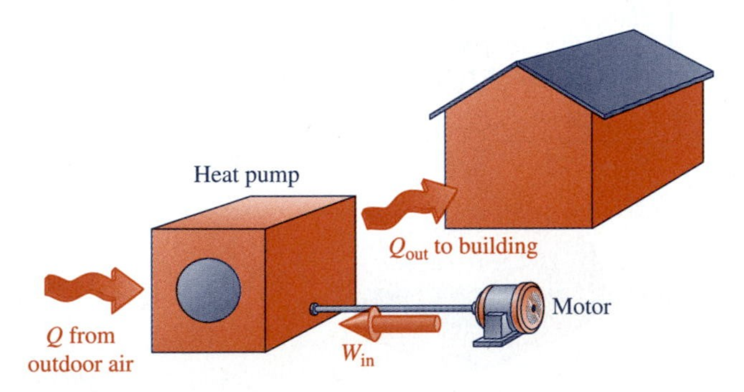

Figure 3·13 Heat pump.

may not be the total heat transfer from the heat pump. Notice that for both refrigerators and heat pumps, the coefficient of performance is defined as a ratio of the heat transfer desired to the net work input required. Also notice that coefficients of performance, like thermal efficiency, are defined only for cycles, not for individual processes. The coefficient of performance for a heat pump may range from slightly above 0 to well above 1.

Heat pump coefficients of performance may be greater than 1.

Limitations of the First Law. The indexes of performance discussed above are all made possible by energy accounting based on the first law. However, the maximum possible values of these indexes under various conditions cannot be predicted by the first law alone. An independent principle, the second law of thermodynamics, is required to answer the questions: How much of the heat available at a given temperature for a heat engine can be converted into work? and What is the minimum amount of work input required to transfer a given amount of heat from a lower-temperature region to a higher-temperature region? Consequently, we will treat the second law before considering various applications of thermodynamics.

In the design of equipment and entire plants, efficiency and coefficients of performance are always important, but maximizing them is often not the most important objective. Trade-offs with other considerations such as cost, manufacturability, size, weight, environmental impact, reliability, serviceability, and delivery time must be made. Making trade-offs wisely, however, requires accurate predictions of thermodynamic performance.

3·7 Historical Note on the First Law

The historical approach is usually not the best method of study for a person seeking a working knowledge of a science, because the development of a science rarely proceeds in an orderly, logical manner. The history of a science is usually the story of a few giant strides and many smaller steps in the right direction interspersed with, and nearly obscured by, aimless groping, mistakes, confusion, misunderstandings, and trips down attractive byroads that turn out to be dead ends. Consequently, the study of "original sources" is often more confusing than helpful. After mastering the fundamentals of the subject, however, you may find the early work fascinating and enlightening, not just for historical interest, but as an illustration of the way science and engineering develop. This is especially true in thermodynamics.

Thermodynamics has a colorful history.

A science begins to develop as a result of either (1) practical needs for the understanding of certain phenomena so that the behavior of systems can be predicted or controlled or (2) curiosity. The science grows as hypotheses are formulated, tested, modified, and accepted or discarded. At the same time, the vocabulary of the science is continually modified for convenience and clarity. Often even the definitions of basic terms are changed.

The concept of energy used in mechanics in the seventeenth century was restricted to what we now call kinetic energy, potential energy, and work. The seventeenth-century energy analyses in mechanics were of great value whenever frictional effects could be ignored, but they could not cope with any process in which there was a frictional effect because the relationship involved in the conversion of work to heat was unknown.

The development of the steam engine introduced a number of practical problems regarding the inverse conversion: heat to work. In 1698, Thomas Savery built a steam engine for pumping water. Thomas Newcomen built a much-improved engine around 1705, and for three-quarters of a century Newcomen engines were widely used. They were succeeded by the engines of James Watt, who patented the separate condenser in 1769 as the first of

James Watt (1730–1819), Scottish engineer, was trained in his youth as an instrument maker. While serving in this capacity at the University of Glasgow he became interested in improving the performance of steam engines. His interest intensified when he was called on to repair a model of a Newcomen engine. His first patent covered not only a separate condenser but also the condenser air pump, insulation of the engine cylinder, steam jacketing of the cylinder, and the use of steam on both sides of the piston. During the period of his early work on steam engines, Watt also engaged in engineering work pertaining to canals, harbors, and bridges. James Watt and Matthew Boulton, a progressive manufacturer and businessman, formed a happy partnership that fully utilized the talents of both men. Engines built by Boulton & Watt, Birmingham, played an important role in the industrial growth of Great Britain during the nineteenth century.

many improvements. The practical need to predict and evaluate the performance of these early steam engines stimulated interest in the relationship between work and heat, two concepts that had developed independently. The idea of work had been developed through mechanics. Heat was the subject of an entirely different "science," and the caloric theory, which postulated that heat was an indestructible fluid, was widely accepted.

Although several people had suggested that the caloric theory was unsound and that heat might be produced without limit by means of work, the first experimental attack on the caloric theory was made by Count Rumford (Benjamin Thompson) beginning in 1787. While directing the manufacture of

Benjamin Thompson (1753–1814) was born in Woburn, Massachusetts. At the age of 14 he is reported to have calculated a solar eclipse with an error of only a few seconds. At 19 he married a wealthy widow and immediately became prominent in Boston social circles. He remained loyal to the Crown when the American Revolution began and in 1776 prudently left Boston for London to become active in the British civil service. Five years later he was involved in a plot to sell British navy information to the French. For 11 years he was in Munich, holding various positions, including minister of war and minister of police in the Bavarian government. He was made a count of the Holy Roman Empire and selected the title of Rumford from the name of an American township. His cannon-boring experiments were made while he was in Bavaria. To Rumford's credit are also prison reforms, stove and fireplace improvements, and numerous other items. He was a founder of the Royal Institution of London. To his discredit are several instances of fantastic duplicity and intrigue. His second wife was the wealthy widow of Antoine Lavoisier.

cannon, Rumford observed that a large amount of heat was evolved from a cannon as it was bored, and that the amount of heat was not directly related to the amount of metal removed or to the size of the chips produced, as would be required by the caloric theory. Experiments showed that more heat was evolved and more work was required when a dull boring tool was used instead of a sharp one. These and other observations convinced Rumford that the caloric theory was unsound, but his views were not widely accepted. He reported his cannon-boring experiments in 1798 and his demonstration that bodies do not change weight when they are heated or cooled in 1799.

Sir Humphry Davy reported in 1799 further evidence of the conversion of work to heat, which he supported by experiments such as the rubbing together of two pieces of ice. Still the caloric theory was not generally abandoned, although more people began to question it. Marc Séguin, a French engineer, stated the theory of the equivalence of heat and work in 1839 and performed some experiments to support it, but his experiments were not convincing. Julius Robert Mayer, a German physician, stated the theory independently in 1842 and calculated the mechanical equivalent of heat (the relationship between the foot-pound and the Btu) from data on the specific heat of gases. The equivalence of heat and work was finally placed on a sound experimental basis by James Prescott Joule, who between 1840 and

James Prescott Joule (1818–1889) was born into a wealthy family and inherited a large brewery in Manchester, England. His financial independence enabled him to devote his life to scientific research, chiefly in the field of electricity and thermodynamics. He had a much stronger interest in accurate measurements than most of the scientists of his time. His work in thermodynamics included that on the equivalence of work and heat and also contributions on the physical properties of gases.

1849 carefully measured the mechanical equivalent of heat by several methods. His results were published in 1843 and 1849. The law of conservation of energy gained widespread acceptance after the publication in 1848 of a paper in which Hermann Helmholtz, then a surgeon in the Prussian army, clearly showed the applications of the law in several scientific fields. It is noteworthy that Mayer and Helmholtz both had great difficulty publishing their papers in scientific journals because their ideas were so much at odds with the generally accepted caloric theory.

The original papers and some histories of thermodynamics show that much confusion was caused by the dual use of the term heat: (1) to designate an interaction between bodies and (2) to designate "something" stored in a body. (Expressions such as "the heat in a body" are frequently found.) The concepts of heat and temperature were sometimes not clearly differentiated. Sometimes the wrong question was asked. There was confusion between (1) the question of what relation heat has to other effects or how it is measured and (2) the nebulous question of what heat *is*. An operational approach tells us that the first question is a fruitful one but that the second is not.

3·8 Summary

Heat and work were defined operationally and independently in Chapter 1. The relationship between work and heat is expressed by the first law of thermodynamics, which can be formulated in many ways. For any closed system undergoing a cycle,

$$\oint \delta Q = \oint \delta W \tag{3·1}$$

From this statement of the first law it follows that for any path between two states of a closed system, the quantity $(\delta Q - \delta W)$ or $(Q - W)$ has the same value. This quantity is therefore a property. We call it *stored energy*, and its definition

$$\Delta E \equiv Q - W \tag{3·2b}$$

is a useful statement of *the first law of thermodynamics for a closed system.*

Both heat and work are forms of energy in transition as distinct from stored energy. Thus the first law is a statement of the *conservation of energy principle*.

Energy can be stored in many forms such as potential, kinetic, electrical, magnetic, surface, and other energies.

Potential energy is the energy stored in a system as a result of its location in a gravitational field, and its magnitude is given by

$$PE = (\text{weight})z \tag{3·7b}$$

where z is the distance of the system's center of gravity from some arbitrary horizontal datum plane.

Kinetic energy is the energy stored in a system by virtue of the motion of the system, and its magnitude is given by

$$KE = \frac{mV^2}{2} \tag{3·8}$$

where m is the mass of the system and V is its velocity relative to an arbitrary frame of reference. If not all parts of the system have the same velocity, a more general expression must be used (see Eq. 3·8*c*).

Internal energy U is the form of energy stored in a system that is independent of gravity, motion, electricity, magnetism, and surface tension. The internal energy of a substance is related to the potential and kinetic energies of the molecules of the substance and to the internal structure of the molecules. In the absence of motion, gravity, electricity, magnetism, and surface tension, $U = E$. If motion and gravity are considered, the stored energy of a system is

$$E = U + \frac{mV^2}{2} + mgz \tag{3·9}$$

For a simple compressible closed system, the first law statement becomes

$$Q - W = \Delta U \qquad (3\cdot10)$$

For open systems, the general statement of the first law is

$$\begin{pmatrix}\text{Net}\\ \text{increase}\\ \text{in stored}\\ \text{energy of}\\ \text{system}\end{pmatrix} = \begin{pmatrix}\text{Net amount of}\\ \text{energy added to}\\ \text{system as heat}\\ \text{and all forms of}\\ \text{work}\end{pmatrix} + \begin{pmatrix}\text{Stored}\\ \text{energy}\\ \text{of matter}\\ \text{entering}\\ \text{system}\end{pmatrix} - \begin{pmatrix}\text{Stored}\\ \text{energy}\\ \text{of matter}\\ \text{leaving}\\ \text{system}\end{pmatrix} \qquad (3\cdot15)$$

Whenever matter crosses the boundary of a system, flow work is done. At any section where matter crosses a system boundary the *flow work* per unit mass is

$$\text{flow work per unit mass} = pv$$

Flow work does not apply a force to a moving system boundary or a torque to a rotating shaft, so it is treated separately from all other kinds of work. Therefore, the symbol W means all work other than flow work, and the energy balance for an open system can be formulated as

$$\frac{dE}{dt} = \dot{Q} - \dot{W} + \sum_{\text{inlets}} \dot{m}(e + pv) - \sum_{\text{outlets}} \dot{m}(e + pv) \qquad (3\cdot12c)$$

For a simple compressible fluid,

$$\frac{dU}{dt} = \dot{Q} - \dot{W} + \dot{m}_i\left(u_i + p_i v_i + \frac{V_i^2}{2} + gz_i\right) - \dot{m}_e\left(u_e + p_e v_e + \frac{V_e^2}{2} + gz_e\right) \qquad (3\cdot12d)$$

or, using *enthalpy*, which is defined as $h \equiv u + pv$,

$$\frac{dU}{dt} = \dot{Q} - \dot{W} + \dot{m}_i\left(h_i + \frac{V_i^2}{2} + gz_i\right) - \dot{m}_e\left(h_e + \frac{V_e^2}{2} + gz_e\right) \qquad (3\cdot12e)$$

For the frequently encountered case of steady flow, the left-hand side of each of the above three equations is zero.

Thermal efficiency, a performance parameter that applies to heat engine cycles, is defined as

$$\eta \equiv \frac{\oint \delta W}{Q_{\text{in}}} = \frac{\text{net work output of cycle}}{\text{gross heat added during cycle}}$$

Coefficient of performance, a performance parameter that applies to refrigeration and heat pump cycles, is defined for refrigerators as

$$\beta_R \equiv \frac{Q_{\text{in},L}}{W_{\text{in}}} = \frac{\text{heat absorbed from low-temperature region}}{\text{net work input of cycle}}$$

and for heat pumps as

$$\beta_{HP} \equiv \frac{Q_{\text{out},H}}{W_{\text{in}}} = \frac{\text{heat discharged to high-temperature region}}{\text{net work input of cycle}}$$

As you leaf through the pages of this textbook, you may have the impression that many equations are necessary for the study of thermodynamics. Closer inspection, however, shows that most of the equations are simply special forms of a few basic relations. These special forms are included not because you should learn each of them. They are presented only to show you a few of the many special forms that you can derive from the basic relations in order to apply the basic principles to specific systems and processes. The broad scope of thermodynamics is indicated by the large number and great diversity of applications made of its basic principles. Because the different types of systems to which the basic principles of thermodynamics have been applied are so numerous and because there are so many more as yet untried applications, it would be futile to try to learn all the special forms of the basic relations. Instead, an engineer needs a sound understanding of the basic principles, their ranges of application, and their limitations. The best way to achieve this understanding is through practice in the application of the principles to many different situations.

To show how few relations are needed to solve the problems in this chapter, the basic principle formulations are presented below in red. These are often not the most general forms. For example, the conservation of mass and conservation of energy equations for steady-flow and closed systems are actually special cases of the more general equations for an open system. However, the formulations listed here are convenient starting points in many engineering analyses because they can be readily generalized or restricted as the situation demands. In all cases, effects of electricity, magnetism, solid distortion, and surface tension are assumed to be negligible. Remember that learning the restrictions on each equation and the precise meaning of the various terms is just as important as learning the equation itself.

Summary of basic principle formulations: ***Notice not how*** **many** ***there are, but how*** **few** ***there are for a tremendously broad range of applications.***

Basic Principle Formulations

Simple Compressible Closed System:

Conservation of mass: $m_1 = m_2$

Conservation of energy: $Q = U_2 - U_1 + W$ (3·10)

For quasiequilibrium processes, an expression for mechanical work of simple compressible systems:

$$W = \int p\, dV \qquad (3\cdot17)$$

Open System: one inlet, one outlet

Conservation of mass: $m_2 = m_1 + m_i - m_e$

Conservation of energy:

$$\frac{dE}{dt} = \dot{Q} - \dot{W} + \dot{m}_i\left(h_i + \frac{V_i^2}{2} + gz_i\right) - \dot{m}_e\left(h_e + \frac{V_e^2}{2} + gz_e\right) \quad (3\cdot12d)$$

Open System, Steady Flow: one inlet, one outlet

Conservation of mass: $\dot{m}_i = \dot{m}_e = \rho_i A_i V_i = \rho_e A_e V_e$ (3·14)

Conservation of energy:

$$q - w = h_e - h_i + \frac{V_e^2 - V_i^2}{2} + g(z_e - z_i) \quad (3\cdot18)$$

For quasiequilibrium processes of a simple compressible substance an expression for mechanical work per unit mass of fluid:

$$w_{\text{in}} = \int v\,dp + \Delta ke + \Delta pe \quad (1\cdot20)$$

Suggested Reading

3·1 Moran, Michael J., and Howard N. Shapiro, *Fundamentals of Engineering Thermodynamics*, 2nd ed., Wiley, New York, 1992. (Energy is first introduced from the potential and kinetic energy concepts of mechanics, and the first law is stated initially for adiabatic closed systems as $E_2 - E_1 = -W_{\text{adiabatic}}$. The first law is introduced prior to any discussion of pvT relations or tabulated property data for substances. We mentioned this as a possible sequence of topics in the introduction to Chapter 2.)

3·2 Russell, Lynn D., and George A. Adebiyi, *Classical Thermodynamics*, Saunders, Philadelphia, 1993, Chapter 6. (Several statements of the first law are given. You will see that the statement we use is referred to as the *Poincaré statement.*)

3·3 Wark, Kenneth, *Thermodynamics*, 5th ed., McGraw-Hill, New York, 1988, Chapter 2. (Notice that the sign convention for work is different from that used in the present book and by most authors.)

3·4 Fox, Robert W., and Alan T. McDonald, *Introduction to Fluid Mechanics*, 4th ed., Wiley, New York, 1992, Chapter 4. (A general approach for applying conservation of mass and conservation of energy to open systems is presented.)

Problems

3·1 A gearbox operates steadily with a power output of 45.0 kW while the input shaft delivers 48.6 kW to the gearbox. Determine the heat transfer. Explain why "steadily" in this case means that the gearbox operates cyclically.

3·2 A household refrigerator operates steadily over a period of time when the door is not opened. The power input is measured as 0.82 kW. The heat transfer from the condenser coils on the back of the refrigerator is measured as 2540 kJ/h. Determine the rate at which heat is transferred from the room air into the refrigerator. To make a decision on whether the insulation thickness should be increased to reduce this heat transfer from the room, what costs should be compared?

3·3 In Section 3·1 it is stated that the increase in the stored energy of any closed system is equal to the net transfer of energy into the system. It is then stated that this is equivalent to the statement that energy can be neither created nor destroyed. Explain the reasoning for the equivalence.

3·4 Give five examples of systems that undergo changes in state while isolated. (Isolated means that no heat, work, or mass crosses the system boundary.)

3·5 For each of the five cases of processes of a closed system, fill in the blanks where possible:

	Q	W	E_1	E_2	ΔE
(*a*)	35 kJ	15 kJ	180 kJ		
(*b*)	35 kJ	−15 kJ		60 kJ	
(*c*)	25 kJ				15 kJ
(*d*)		50 kJ	80 kJ		−40 kJ
(*e*)	−40 kJ		220 kJ	380 kJ	

3·6 A closed system changes from state A to state B while 55 kJ of heat is added and 70 kJ of work is done. As the system is returned to state A, 35 kJ of work is done on it. Determine the heat transfer of process B–A.

3·7 A closed system undergoes a process during which 600 kJ of heat is added to the system and the stored energy increases by 1800 kJ. Calculate the work.

3·8 A gas in a closed system expands, doing 800 kJ of work while receiving 600 kJ of heat. Calculate the change in stored energy.

3·9 During a process 1–2, 4000 kJ of heat is added to a closed system and the system's stored energy increases by 3000 kJ. During the return process 2–1 that restores the system to state 1, 1000 kJ of work is done on the system. Determine the heat transfer of process 2–1.

3·10 A closed system does 600 kJ of work while 220 kJ of heat is removed. Then the system is restored to its initial state by a process in which 100 kJ of heat is added to it. Determine ΔE for this second process.

3·11 The stored energy of a closed system changes from 250 to 210 kJ/kg while the system performs 20 kJ/kg of work. Compute the heat transfer.

3·12 During the expansion of 85 kg of gas in a closed system, heat is added to the gas in the amount of 1100 kJ. The stored energy decreases by 1800 kJ. Compute the work.

3·13 While 2000 kJ of heat is removed from a closed system, its stored energy decreases by 600 kJ. Determine the work.

3·14 A vapor trapped in a cylinder behind a piston has an initial stored energy of 85.0 kJ. When 15.0 kJ of heat is added, the system does 36.5 kJ of work. The mass of vapor in the system is 0.69 kg. Determine the final stored energy of the system.

3·15 One-half kilogram of gas is held in a rigid tank. An external motor does 50.0 kJ/kg of work on the gas by means of an impeller while the stored energy of the gas increases from 120.0 to 160.0 kJ/kg. Determine the heat transfer in kJ/kg.

3·16 A closed system contains 30 g of a gas. How much heat (in joules) is added to the system to produce 5000 N·m of work while the stored energy decreases by 200 J/g?

3·17 A gas in a cylinder fitted with a piston undergoes a cycle composed of three processes. First, the gas expands at constant pressure with a heat addition of 40.0 kJ and a work output of 10.0 kJ. Then it is cooled at constant volume by a removal of 50 kJ of heat. Finally, an adiabatic process restores the gas to its initial state. Determine (*a*) the work of the adiabatic process and (*b*) the stored energy of the gas at each of the other two states if its stored energy in the initial state is assigned the value of zero.

3·18 A closed system changes from state 1 to state 2 while 25 B of heat is added and 35 B of work is done. As the system is returned to state 1, 20 B of work is done on it. What is the heat transfer during process 2–1?

3·19 The stored energy of a fluid in a closed system changes from 500 to 440 B/lbm while the fluid performs 30,000 ft·lbf/lbm of work. Compute the heat transfer.

3·20 While 30 lbm of gas expands in a closed system, 420 B of heat is added and the stored energy decreases by 1200 B. How much work is done?

3·21 Compute the work of a closed-system process during which 450 B of heat is added and the stored energy of the system increases by 2300 B.

3·22 An expanding gas does 12,000 ft·lbf of work while it receives 19 B of heat. Calculate the change in its stored energy.

3·23 During a process 1–2 of a closed system, 200 B of heat is added to the system and the stored energy of the system increases by 150 B. During the return process 2–1, which restores the system to its initial state, 80 B of work is done on the system. Determine the heat transfer of process 2–1.

3·24 During a certain process a closed system does 30 B of work while 10 B of heat is removed. Then the system is restored to its initial state by means of a process in which 4 B of heat is added to it. Determine ΔE for this second process.

3·25 Three copper blocks in a rigid insulated container are initially at different temperatures. As they exchange heat to come to the same temperature, their stored energies change by -260 kJ, $+140$ kJ, and $+100$ kJ. Nothing else is in the container except air. Determine ΔE of the air.

3·26 Calculate the work required to accelerate a 140-kg motorcycle and 80-kg rider from rest to a speed of 30 m/s on a level highway if there are no frictional effects.

3·27 Calculate the amount of work required to accelerate a 3000-lbm automobile from rest to a speed of 60 mph on a level highway if there are no frictional effects.

3·28 A body with a mass of 20 kg has a potential energy of 400 kJ relative to datum plane *A* and a potential energy of -300 kJ relative to datum plane *B*. The local acceleration of gravity is 9.76 m/s^2. What is the relative location of plane *A* with respect to plane *B*?

3·29 A body with a mass of 40 lbm has a potential energy of 450 ft·lbf relative to datum plane *A* and a potential energy of -320 ft·lbf relative to datum plane *B*. The local acceleration of gravity is 31.5 ft/s^2. What is the relative location of plane *A* with respect to plane *B*?

3·30 Ethane initially at 35 kPa with a volume of 0.120 m^3 is compressed frictionlessly in a cylinder until its volume is halved in such a manner that its pressure and volume are linearly related, $p = a + bV$. The final pressure is 80 kPa, and the change in internal energy of the ethane is 3.22 kJ. Determine the heat transfer.

3·31 Ethene in a closed system is compressed frictionlessly from 95 to 190 kPa in such a manner that pV = constant. Initially, the density is 1.11 kg/m^3 and the volume is 0.045 m^3. During the compression, heat is removed from the ethene in the amount of 2.96 kJ. Determine the change in internal energy of the ethene.

3·32 Neon in a cylinder expands frictionlessly against a piston so that $p^{0.7}V$ = constant from initial conditions of 300 kPa, 90°C, and a volume of 0.024 m^3, to a final pressure of 120 kPa. The internal energy change is −2.59 kJ and the enthalpy change is −4.31 kJ. Determine the heat transfer.

3·33 A cylinder with a volume of 2 cu ft contains 0.5 lbm of a gas at 20 psia and 50 F. The gas is compressed in a quasiequilibrium process to 120 psia in such a manner that p (in psia) = constant − $100V$, with V in cu ft. Its final temperature is 20 F. The internal energy of the gas is given by u (in B/lbm) = $0.20T$, with T in Fahrenheit. Sketch a pV diagram and calculate (*a*) heat transfer and (*b*) enthalpy change per pound.

3·34 In a closed system 0.27 lbm of air expands in a quasiequilibrium process from 30 psia, 140 F, until its volume is doubled. The initial volume is 2.0 cu ft, and the expansion follows the path pV = constant. During the process 28.5 B/lbm of heat is added. Determine the internal energy change of the air in B/lbm.

3·35 Air in a cylinder expands frictionlessly against a piston in such a manner that p^2V = constant. Initially the air is at 30 psia, 140 F, and occupies a volume of 4 cu ft. The final volume is 16 cu ft. Heat is added to the air in the amount of 100 B. Sketch a pV diagram and determine the change in internal energy per pound of air.

3·36 A well-insulated tank filled with neon has an electric resistance heating element in it. Because current from an external storage battery flows through the heating element, the temperature of the gas rises. Determine the algebraic sign (> 0, $= 0$, < 0) of the heat transfer and of the work of a system defined as (*a*) just the neon and (*b*) the neon and the heating element. The heating element has a resistance of 5 Ω. If during the process the potential difference across the element is 12 V and the process occurs in 15 min, determine, if possible, ΔE for the system composed of (*c*) just the neon and (*d*) the neon and the heating element.

3·37 A single-phase electric transformer operates under steady conditions with 440 V across the primary and a current of 8.60 A with unity power factor. The secondary provides 30.0 A at 120 V, unity power factor. Determine the net heat transfer.

3·38 A robot arm is moved by means of an electric motor and a speed-reducing gear train mounted within the arm. During a complete cycle of arm operation, the work output from the gear train is 2.40 kJ. The arm completes seven cycles of operation per minute. The average efficiency of the motor and gear train combination is estimated to be 75 percent. To simulate the heat dissipation required, the motor and gear train are to be replaced by an electric resistance heating element for the purpose of a heat dissipation study. How much energy should be dissipated by the electric resistance heating element?

3·39 Refer to Example 1·1. If the stored energy change of the system during the compression is 0.602 kJ, determine the heat transfer.

3·40 Describe in detail the energy transfers that occur in the operation of a Cartesian diver, the device shown in the figure. When the rubber membrane is pressed downward or lifted, the location of the diver in the flask is changed. Consider as the system (*a*) the entire

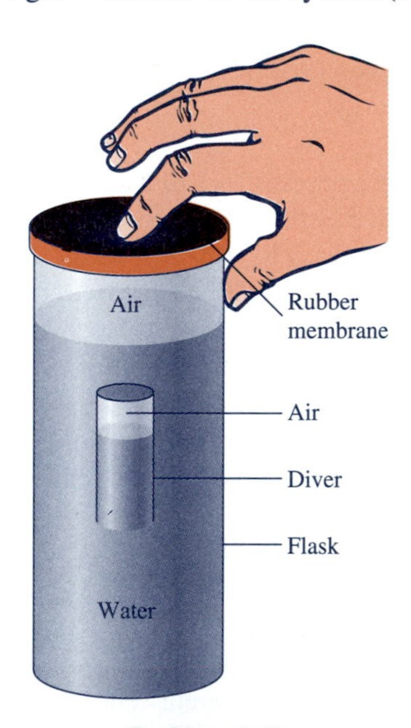

Problem 3·40

device, (*b*) the air within the device, (*c*) the water within the device, (*d*) the air at the top of the device, and (*e*) the air within the diver.

3·41 Consider an air bubble rising through an open tank of water. Does the size of the bubble change? What are the energy transfers across a boundary that encloses only the bubble? What are the energy transfers across a boundary that encloses all the water and the bubble?

3·42 Occasionally, one sees the relation $\delta Q = dU + p\,dV$ referred to as the first law of thermodynamics. Comment on this, stating the restrictions that apply to this equation.

3·43 For a closed system, $\oint \delta Q = \oint \delta W$. Show by an example that this relationship does not hold for an open system.

3·44 Write an energy balance for a body falling freely in a vacuum, considering the body as the system and using a stationary frame of reference. Notice that there is a force on the body and that the body is moving. Does this mean that work is done on or by the body? If so, with what other system is the work exchanged? If not, what is the meaning of the product of the body weight and the distance moved?

3·45 Comment on this proposal for destroying energy: A steel coil spring is compressed, and in this manner more energy is stored in it than in the same spring when uncompressed. The compressed spring is held to a constant length by means of a glass thread tied around it and is then submerged in an acid that dissolves the spring but not the thread. Therefore, it is claimed, the energy used to compress the spring is lost and can never be recovered.

3·46 An aircraft moves in a north-to-south direction at constant speed and constant altitude above sea level. The acceleration of gravity at constant altitude changes with latitude; therefore the weight of the craft itself (excluding the fuel) changes, and consequently its potential energy changes. Explain what is wrong with this reasoning, or explain the energy transformation that is involved.

3·47 Derive an expression for the kinetic energy of a rotating solid body (*a*) from Eq. 3·8*c* and (*b*) from the work required to put the body into motion.

3·48 Derive an expression involving the modulus of elasticity for the change in stored energy of a wire that is stretched adiabatically within its elastic limit. (Be careful with the algebraic sign.)

3·49 Consider two masses in interstellar space. The attractive force between them varies inversely as the square of the distance separating them. How does the potential energy of the two masses vary with that distance?

3·50 Consider a storage battery that has a terminal potential of 12 V while it delivers a current of 10 A for 3 h. The stored energy of the battery decreases by 1205 B. Determine the heat transfer.

3·51 A dc electric motor operates steadily at 800 rpm when the torque applied to its shaft is 96 lbf·in. It draws a current of 50 A at 24 V. Determine the rate of heat transfer.

3·52 Two streams of a liquid flow through identical circular pipes at the same mass flow rate. The velocity of one is uniform across the pipe cross section, and the velocity of the other varies across the pipe cross section in such a manner that the velocity u at any radius r is given by $u/u_{max} = 1 - (r/r_0)^2$, where u_{max} is the velocity at the pipe axis and r_0 is the pipe radius. Determine the ratio of the kinetic energy of the stream with the parabolic velocity distribution to that of the stream with the uniform distribution.

3·53 Seventy kilojoules of work is done by each kilogram of fluid passing through an apparatus under steady-flow conditions. In the inlet pipe, which is located 30 m above the floor, the specific volume is 3 m^3/kg, the pressure is 300 kPa, and the velocity is 50 m/s. In the discharge pipe, which is 15 m below the floor, the specific volume is 9 m^3/kg, the pressure is 60 kPa, and the velocity is 150 m/s. Heat loss from the fluid is 3 kJ/kg. Determine the change in internal energy of the fluid passing through the apparatus.

3·54 The flow rate through a steam nozzle is 450 kg/h. The initial and final pressures are 1400 and 14 kPa. The initial and final velocities are 150 and 1200 m/s. There is no heat transfer. Calculate the change in enthalpy.

3·55 The flow rate through a steam nozzle is 1000 lbm/h. The initial and final pressures are 200 and 2 psia. The initial and final velocities are 500 and 4000 fps. There is no heat transfer. Calculate the change in enthalpy.

3·56 A fluid flows through a turbine at a rate of 2.50 kg/s. The inlet and exit velocities are 30 and 120 m/s, respectively. The initial and final enthalpies are 2930 and 2675 kJ/kg, respectively. Heat loss is 45.0 kJ/s. Compute the power output.

3·57 The flow rate through a turbine is 40,000 lbm/h. The inlet and exit velocities are 6000 and 24,000 fpm, respectively. The initial and final enthalpy values are 1260 and 1000 B/lbm, respectively, and the heat loss amounts to 140,000 B/h. Calculate the power output.

3·58 Air expands through a nozzle from a pressure of 500 kPa to a final pressure of 100 kPa. The enthalpy decreases by 100 kJ/kg. The flow is adiabatic and the inlet velocity is very low. Calculate the exit velocity.

3·59 Air expands through a nozzle from 75 to 15 psia. The enthalpy decreases by 48 B/lbm. The flow is adiabatic and the inlet velocity is very low. Calculate the exit velocity.

3·60 An air compressor takes in air at 100 kPa, 40°C, and discharges it at 690 kPa, 208°C. The initial and final internal energy values for the air are 224 and 346 kJ/kg, respectively. The cooling water around the cylinders removes 70 kJ/kg from the air. Neglecting changes in kinetic and potential energy, calculate the work.

3·61 An air compressor takes in air at a pressure of 14.4 psia, 73 F, and discharges it at 100 psia, 280 F. The initial and final internal energy values for the air are 12 and 47 B/lbm, respectively. The cooling water around the cylinders removes 33 B/lbm of entering air. Neglecting changes in kinetic and potential energy, calculate the work.

3·62 Steam expands adiabatically through a nozzle from 1400 to 14 kPa. The initial and final enthalpy values are 3300 and 2800 kJ/kg, respectively, and the initial velocity is very low. Determine the final velocity.

3·63 Steam expands adiabatically through a nozzle from 200 to 2 psia. The initial and final enthalpy values are 1289 and 955 B/lbm, respectively, and the initial velocity is very low. Determine the final velocity.

3·64 The supply line to a steam radiator is located 2 ft above the discharge line. Dry saturated steam at 15 psia enters the radiator. Its internal energy is 1078 B/lbm. Saturated liquid at 15 psia leaves the radiator with an enthalpy of 181 B/lbm. Neglecting kinetic energy changes, calculate the heat released from the radiator per pound of steam entering.

3·65 An aircraft fuel with a specific gravity of 0.840 is to be pumped at a rate of 5.20 kg/s from a pressure of 40 kPa to a pressure of 800 kPa. The fuel temperature is approximately 5°C. Velocity change through the pump is very low. The required power input to the pump is estimated to be 140 percent of that required to do the pumping frictionlessly. State the assumptions you make and determine the required power input.

3·66 A hydraulic turbine is located slightly lower than the level of the water surface downstream of a dam. The difference in water level between the upstream and downstream sides of the dam is 22.0 m. Water entering the turbine has a velocity of 3.2 m/s, and that leaving has a velocity of 1.6 m/s. The flow rate through the turbine is 1.16 m^3/s. For frictionless flow, determine the turbine power output.

3·67 The flow rate through a hydraulic turbine is 6 cfs. The pressure of the water is 30.6 psig at the inlet, which has a cross-sectional area of 0.50 sq ft. At the outlet, 8 ft below the inlet, the velocity is 2.5 fps and the pressure is 4 psig. Calculate the power imparted to the turbine by the fluid.

3·68 The water level in an open tank is maintained 3.20 m above the centerline of a nozzle through which water flows from the tank into the atmosphere. The exit diameter of the nozzle (and the diameter of the water jet leaving) is 13.0 mm. Determine the flow rate of the water jet. State clearly the assumptions you make and how they affect the calculation.

3·69 Steam flows through a turbine at a rate of 40,000 kg/h with inlet and outlet enthalpy values of 2800 and 2100 kJ/kg, respectively. Inlet and outlet velocities are 30 and 200 m/s, respectively. Heat loss to the surroundings amounts to 300,000 kJ/h. Calculate the power output.

3·70 A gas flows steadily through a turbine at a rate of 1.40 kg/s, entering at 500 kN/m^2, 940 K ($h_1 = 900$ J/g, $u_1 = 724$ J/g), and leaving at 100 kN/m^2, 770 K ($h_2 = 690$ J/g, $u_2 = 546$ J/g, $\rho_2 = 0.70$ kg/m^3) through a cross-sectional area of 0.010 m^2. The inlet velocity is negligibly low. The power output is 240 kW. Calculate the heat transfer in kJ/kg and the molar mass of the gas.

3·71 Air flows at a rate of 0.9 kg/s through a compressor, entering at 100 kPa, 5°C, with a velocity of 60 m/s, and leaving at 200 kPa, 70°C, with a velocity of 120 m/s. The enthalpy of the air increases by 65.3 kJ/kg as it passes through the compressor; its internal energy increases by 46.6 kJ/kg. Heat transferred from the air to the cooling water circulating through the compressor casing amounts to 19 kJ/kg. Calculate the power required by the compressor and the inlet and outlet areas.

3·72 Air flows at a rate of 2.0 lbm/s through a compressor, entering at 15.0 psia, 40 F, with a velocity of 200 fps, and leaving at 30.0 psia, 160 F, with a velocity of 400 fps. The enthalpy of the air increases by 28.8 B/lbm as it passes through the compressor; its internal energy increases by 20.5 B/lbm. Heat transferred from the air to the cooling water circulating through the compressor casing amounts to 8.0 B/lbm. Calculate the power required by the compressor and the inlet and outlet areas.

3·73 Air enters a machine at 100 kPa, 30°C, at low velocity and leaves at 100 kPa, 55°C, with a velocity of 90 m/s through an opening having a cross-sectional area of 0.00090 m^2. The enthalpy of the air increases 25.2 kJ/kg, and its internal energy increases 18.0 kJ/kg. A blower delivers 0.75 kW to the air as it passes through the machine. Calculate the heat added to or removed from the air in kJ/kg.

3·74 Methane flows frictionlessly through a turbine at a rate of 13.2 kg/h, expanding from 375 to 110 kPa in such a manner that $pv^{1.25} = 117$, with p in kPa and v in m^3/kg. The inlet velocity is quite low, but the exit velocity is 60 m/s. The internal energy change of the methane passing through the turbine is -107.6 kJ/kg and the enthalpy change is -139.7 kJ/kg. Determine the heat transfer per kilogram of methane.

3·75 Air enters a wind tunnel at 96.0 kPa, 20°C, with low velocity and then flows steadily through a nozzle into the test section. If the expansion in the nozzle is such that $pv^{1.40}$ = constant, to what pressure must the air be expanded to achieve a velocity of 220 m/s in the test section? Assume that the flow is frictionless.

3·76 A gas flows steadily through a machine while expanding frictionlessly from 60 to 15 psia according to the relation $pv^{1.25} = 32{,}000$ with p in psf and v in cu ft/lbm. The enthalpy decreases by 35 B/lbm, and the change in kinetic energy is negligible. Calculate the heat transfer in B/lbm.

3·77 Air enters a gas turbine at 400 kPa, 370°C ($u =$ 468.1 kJ/kg, $h = 652.7$ kJ/kg), and leaves at 100 kPa.

The net increase in kinetic energy of the air passing through the turbine is 12 kJ/kg. There is no heat transfer. The flow rate is 3.4 kg/s. The power output is 410 kW. Determine the enthalpy of the air at the turbine outlet.

3·78 Air enters a turbine at 60 psia, 700 F (u = 201.6 B/lbm, h = 281.1 B/lbm), and leaves at 15 psia. The net increase in kinetic energy of the air passing through the turbine is 5.0 B/lbm. There is no heat transfer. The flow rate is 15 lbm/s. The power output is 1100 hp. Determine the enthalpy of the air at the turbine outlet.

3·79 Air enters a compressor at 100 kPa, 30°C (h = 303.4 kJ/kg, u = 216.5 kJ/kg), with negligible velocity and is discharged at 490 kPa, 260°C (h = 537.4 kJ/kg, u = 384.4 kJ/kg). The discharge velocity is 150 m/s. The power input is 2400 kW. The flow rate is −9.0 kg/s. Determine the heat transfer in kJ/kg and the discharge area in cm^2.

3·80 Air enters a compressor at 14 psia, 80 F (h = 129 B/lbm, u = 92 B/lbm), with negligible velocity and is discharged at 70 psia, 500 F (h = 231 B/lbm, u = 165 B/lbm). The discharge velocity is 500 fps. The power input is 3200 hp. The flow rate is 20 lbm/s. Determine the heat transfer in B/lbm and the discharge area in cm^2.

3·81 The power output of a steam turbine is 6000 kW when the flow rate is 41,200 kg/h. Barometric pressure is 730 mm of mercury. Determine the amount of heat, in kJ/kg, added to or removed from the steam passing through the turbine if inlet and exhaust conditions are respectively: absolute pressures, 1400 and 7.0 kPa; temperatures, 300 and 39°C; velocities, 60 and 150 m/s; internal energies, 2784.5 and 2338.3 kJ/kg; and enthalpies, 3039.7 and 2476.1 kJ/kg.

3·82 A blower supplying air to a diesel engine draws in air at 90 kPa, 20°C (u = 209.3 kJ/kg), with low velocity at a rate of 12.3 m^3/min. It discharges air into the engine at 106 kPa, 30°C (u = 303.4 kJ/kg), with a velocity of 24 m/s. The flow is adiabatic. Calculate the blower's power consumption and its discharge area.

3·83 A blower draws in air at 14 psia, 81 F (u = 93.1 B/lbm), V = 10 fps, and discharges it at 15 psia, 91 F (u = 94.8 B/lbm), V = 50 fps. Flow is adiabatic and 1500 cfm enters the blower. Calculate the blower's power consumption and its discharge area.

3·84 Carbon monoxide flows steadily through a machine at a rate of 0.40 kg/s, entering at 170 kPa, 17°C (u_1 = 215.1 kJ/kg, h_1 = 301.2 kJ/kg), and leaving at 90 kPa, −7°C (u_2 = 197.2 kJ/kg, h_2 = 276.3 kJ/kg). The inlet velocity is quite small. Discharge is through a cross-sectional area of 0.0070 m^2. Heat transfer into the gas as it passes through the machine is estimated as 1.2 kJ/s. Determine the power output.

3·85 Carbon dioxide flows steadily through a machine at a rate of 0.5 lbm/s, entering with negligible velocity at 20 psia, 80 F (u_1 = 67.7 B/lbm, h_1 = 92.1 B/lbm). Carbon dioxide leaves the machine at 60 psia, 200 F (u_2 = 87.5 B/lbm, h_2 = 117.3 B/lbm), through an opening that has an area of 0.4 sq in. Power input to the machine is 25 hp. Determine the heat transfer per pound of carbon dioxide.

3·86 A piece of electronic equipment that has a steady power consumption of 600 W has outside dimensions of 29 by 26 by 14 cm. The output signals carry a negligible amount of energy. The cooling air blown through the unit can experience an increase in internal energy of not more than 15 kJ/kg and an increase in enthalpy of not more than 21 kJ/kg. At what rate (kg/s) must cooling air be supplied?

3·87 A gaseous hydrocarbon fuel enters a burner at 1 atm, 25°C, with an enthalpy of −3556 kJ/kg at a rate of 18.0 kg/s. Air at 1 atm, 25°C, enters the burner with an enthalpy of 300 kJ/kg at a rate of 391 kg/s. The combustion products leaving the burner at 540°C have an enthalpy of −1693 kJ/kg. Determine the amount of heat transferred from the burner in kJ/kg of fuel.

3·88 A gas enters an insulated steady-flow system at

200 kPa, 27°C, with an enthalpy of 214.1 kJ/kg and a velocity of 70 m/s, at a rate of 0.00230 kg/s. Within the system, the gas stream is separated into two streams, and measurements show that one stream leaves at 100 kPa with an enthalpy of 206.0 kJ/kg and a very low velocity at a rate of 0.00090 kg/s. The other leaving stream also has a very low velocity. What is its enthalpy?

3·89 Two fluids flow through a heat exchanger. Fluid A flows at a rate of 3000 kg/h, entering with an enthalpy of -321.6 kJ/kg and leaving with an enthalpy of -344.0 kJ/kg. Fluid B flows at a rate of 165 kg/h and enters with an enthalpy of 42.0 kJ/kg. All fluid velocities are very low, and it is estimated that 4 percent of the heat removed from the higher temperature fluid A is lost as heat to the surroundings because the insulation of the heat exchanger is not perfect. Determine the enthalpy of fluid B as it leaves the heat exchanger.

3·90 An insulated tank initially contains 0.35 kg of a gas with an internal energy of 220.0 kJ/kg. Additional gas with an internal energy of 260.0 kJ/kg and an enthalpy of 350.0 kJ/kg enters the tank until the total mass of gas contained is 0.90 kg. Calculate the internal energy (kJ/kg) of the gas finally in the tank.

3·91 A well-insulated tank of volume V contains a gas initially at a pressure p_i, with a specific volume of v_i. The internal energy of the gas is given by $u = apv$, where a is a constant. The gas leaks from the tank through a porous plug at a rate that varies with time. Determine the value of a required to have the mass of gas remaining in the tank at any time proportional to the gas pressure raised to the power of 0.80.

3·92 A well-insulated tank initially holds air at a low pressure p_i. Air from the surrounding atmosphere then leaks slowly into the tank through a porous plug. For air, $u = bpv$, where b is a constant. Show that the mass in the tank at any instant is given by $m = m_i(pv + A)/(p_iv_i + A)$, where A is a constant that depends on the conditions of the surrounding atmosphere.

3·93 A gas in a closed tank of volume V is initially at $p = p_i$ and $T = T_i$. A small valve is opened, allowing gas to escape at a rate of $\dot{m} = Kp\sqrt{T}$, where K is a constant and p and T are the pressure and temperature of the gas at any instant. There is no heat transfer. The internal energy of the gas is given by $u = cT$, where c is a constant. Determine the rate of temperature change with time just after the valve is opened in terms of R (the gas constant), c, K, V, and T_i.

3·94 Air with properties such that $h = 1.4u = 1.004T$, where u and h are in kJ/kg and T is in kelvins, is initially trapped in a well-insulated rigid tank. An impeller in the tank is turned by an external motor and air is bled from the tank through a small valve to keep the temperature constant at 40°C. Assume that the pressure and temperature are uniform throughout the tank at any instant. Determine the mass rate of flow of air from the tank at an instant when the pressure is 100 kPa and the power input is 0.037 kW.

3·95 An insulated tank holds air initially at 140 kPa, 15°C. An electrical heating coil in the tank steadily dissipates 100 W to the air in the tank while a pressure-regulating valve bleeds air from the tank to the atmosphere to maintain the pressure constant in the tank. The tank volume is 0.110 m^3. For air, $h = 1.4u = 1.04T$, where h and u are in kJ/kg and T is in kelvins. Determine the temperature of the air in the tank 2 min after the initial conditions.

3·96 An incompressible liquid enters a device at a constant temperature T_1 and flows through the device at a constant mass flow rate $\dot{m}$. During the initial operating time period between $\tau = 0$ and $\tau = a$, the heat added to the liquid as it passes through the device is expressed as $\dot{Q} = \dot{Q}_0 + b\tau$, where $\dot{Q}_0$ and b are constants. The volume of liquid in the device remains constant. No work is done. The energy stored in the liquid within the device at any instant is $E = E_1 + FT_2$, where E_1 and F are constants and T_2 is the temperature of the liquid leaving at that instant. The enthalpy and internal energy per unit mass of liquid can be expressed as $h = CT$ and $u = DT$, where C and D

are constants. Derive an expression for T_2 at any instant τ, with $0 < \tau < a$.

3·97 A small leak develops in a 20-cu-ft insulated tank which initially contains air at 40 psia. What fraction of the air escapes by the time the pressure drops to 20 psia? (For air, $h = 1.4u = 0.24T$, where h and u are in B/lbm and T is in Rankine.)

3·98 A 4-cu-ft tank is to be filled from a compressed air line that carries air at 100 psia, 200 F. Initially the tank contains air at room conditions, 14.7 psia, 80 F. Determine the final mass of air in the tank. Assume that the filling process is adiabatic and that for air, $h = 1.4u = 0.24T$, where h and u are in B/lbm and T is in Rankine. List all the assumptions you make as part of your solution.

3·99 One part of a refrigeration system is a compressor that takes in refrigerant at 125 kPa, -10°C ($v_1 = 1.0059$ m^3/kg, $u_1 = 1324.1$ kJ/kg, $h_1 = 1449.8$ kJ/kg), with a velocity of 50 m/s and discharges it at 1000 kPa, 70°C ($v_2 = 0.1570$ m^3/kg, $u_2 = 1433.3$ kJ/kg, $h_2 = 1590.3$ kJ/kg), with a velocity of 105 m/s. The power input to the compressor is 2.2 kW. Determine the heat transfer per kilogram of refrigerant for a flow rate of 0.014 kg/s.

3·100 Carbon monoxide is to be compressed steadily from 100 kPa, 110°C, to 650 kPa at a rate of 74 kg/h. On the basis of data from similar compressors, it is estimated that the heat removal rate from the gas being compressed is 15 percent of the power input. The exit temperature is 365°C. Enthalpies at inlet and exit are 398 and 671 kJ/kg, respectively. Inlet velocity is quite low. The exit cross-sectional area is 1.71 cm^2. Determine the power input.

3·101 Nitrogen is to be compressed steadily from 15 psia, 300 F, to 100 psia at a rate of 170 lbm/h. On the basis of data from similar compressors, it is estimated that the heat removal rate from the gas being compressed is 17 percent of the power input. The exit temperature is 700 F. Enthalpies at inlet and exit are 2081.9 and 4865.2 B/lbm, respectively. Inlet velocity is quite low. The exit cross-sectional area is 0.262 in^2. Determine the power input.

3·102 Nitrogen is heated while it flows steadily through a constant-area tube. At the inlet the nitrogen is at 200 kPa, 100°C, and has a velocity of 175 m/s. It is heated to 300°C. At the outlet the pressure is 115 kPa. At the inlet, $u_1 = 277$ kJ/kg and $h_1 = 387$ kJ/kg; at the exit, $u_2 = 428$ kJ/kg and $h_2 = 598$ kJ/kg. Determine the heat transfer per kilogram of nitrogen.

3·103 Nitrogen flows steadily and adiabatically through a long pipe. At one section where the nitrogen is at 250 kPa, 150°C, the velocity is 225 m/s. Because of friction, the pressure drops in the direction of flow. For nitrogen under these conditions, $u = 0.743T$ and $h = 1.04T$. Calculate the velocity and temperature at a section farther downstream where the pressure is 140 kPa.

3·104 For the conditions of Example 3·16 make a plot of mass of helium left in the tank versus pressure gage reading.

3·105 A rule of thumb for wind-powered generators is that the power output is proportional to the cube of the wind velocity. Explain the basis for this rule.

3·106 Prove that in a frictionless steady-flow process of an incompressible fluid with no work done, $q - \Delta u = 0$. Does this mean that such a process is adiabatic?

3·107 In a household refrigerator, energy is supplied by the electric power supply lines and energy is removed from the condenser to the surrounding room air (see Example 3·10). List all the other ways that energy may enter or leave the system defined as the entire refrigerator.

3·108 A 50,000-lbm aircraft with a landing speed of 150 knots is to be stopped by the arresting gear of an aircraft carrier within 400 ft of the point of touchdown. The energy of the aircraft is to be absorbed by a hydraulic oil and then dissipated to the surrounding atmosphere. Each B of energy absorbed by the hydraulic oil causes its temperature to rise 0.5 F. The oil should experience a temperature rise of not more than 100 degrees F. How many pounds of oil are needed to absorb the energy from a single landing aircraft? How would the amount of oil required be changed if the landing length were reduced to 300 ft? Would it be advisable to capture this energy to reduce the amount of fuel needed to power the ship instead of dissipating it as heat to the atmosphere? Explain your reasoning.

3·109 Describe in detail the energy transfers that occur in the operation of a portable electric resistance heater, including a fan, used to heat rooms.

3·110 Describe in detail the energy transfers that occur in the starting of an automobile engine.

3·111 Describe in detail the energy transfers that occur when an automobile is braked to a stop.

3·112 Describe in detail the energy transfers that occur in the operation of an electric drill boring a hole in steel.

3·113 An automobile travels steadily on a smooth, level highway. List in sequence all the energy conversions that occur, starting with the chemical energy stored in the fuel. Under steady conditions, what fraction of the fuel energy goes to the atmosphere? (In one respect, conditions are not steady: The amount of fuel in the fuel tank decreases, so the total mass of the vehicle changes.)

3·114 Answer the questions of Problem 3·113 for the case of a constant-slope highway with travel direction (*a*) upward, (*b*) downward.

3·115 An automobile starts from rest and accelerates to a normal driving speed on a level highway. List in sequence all the energy conversions that occur, starting with the chemical energy stored in the fuel, during the acceleration phase of the trip.

3·116 Comment on the following definition of thermal efficiency: Thermal efficiency is the ratio of the energy outputs from a system to the energy inputs to it.

3·117 For an actual refrigeration unit removing an amount of heat Q_L from a low-temperature region and discharging an amount of heat Q_H to a high-temperature region, the coefficient of performance is defined as $\beta_R \equiv Q_L/W_{in}$. Explain why it would not be satisfactory to define the coefficient of performance as $\beta_R \equiv Q_L/(Q_H - Q_L)$.

3·118 A power plant receives 300 GJ/h from burning fuel. Energy leaving with exhaust gases amounts to 50 GJ/h more than the energy introduced by air entering, cooling water absorbs 200 GJ/h, and heat loss to the surrounding atmosphere amounts to 10 GJ/h. Determine the plant thermal efficiency.

3·119 A closed system undergoes a cycle during which it receives 3050 kJ from a fluid at a very high temperature, rejects 550 kJ to part of the surroundings at 120°C, and also rejects heat to part of the surroundings at 17°C. The work output of the cycle is 680 kJ. Determine the thermal efficiency of the cycle.

3·120 A refrigerator removes heat steadily from a food compartment at 6°C at a rate of 800 kJ/min while its power input is 3.05 kW. Cooling water removes heat from the refrigerating unit at a rate of 870 kJ/min, and some heat is transferred from the refrigerating unit to the atmosphere. Determine (*a*) the heat loss to the atmosphere and (*b*) the coefficient of performance.

3·121 When a household refrigerator is running, the driving electric motor requires a power input of 450 W. If it runs one third of the time and its coefficient of performance is 2.7, at what rate is heat discharged to the room that contains the refrigerator?

3·122 A building heated by electric resistance heaters requires an average of 21 kW during one winter month. If the building were to be heated by means of a heat pump that absorbs heat from a solar-heated pond at a temperature of 7°C and has a coefficient of performance of 3.8, (*a*) what would the power requirement be and (*b*) at what rate would heat be withdrawn from the pond?

3·123 A rule of thumb for the power required for some kinds of refrigeration systems is "one horsepower per ton," meaning that for each ton of refrigeration (defined as a refrigeration rate of 200 B/min), the required power output is about 1 hp. What is the corresponding coefficient of performance?

3·124 Sometimes efficiency and coefficient of performance are called simply "performance indexes" with a general definition of

$$\text{Performance index} = \frac{\text{desired output}}{\text{input that costs}}$$

Comment on the validity, generality, and usefulness of this definition.

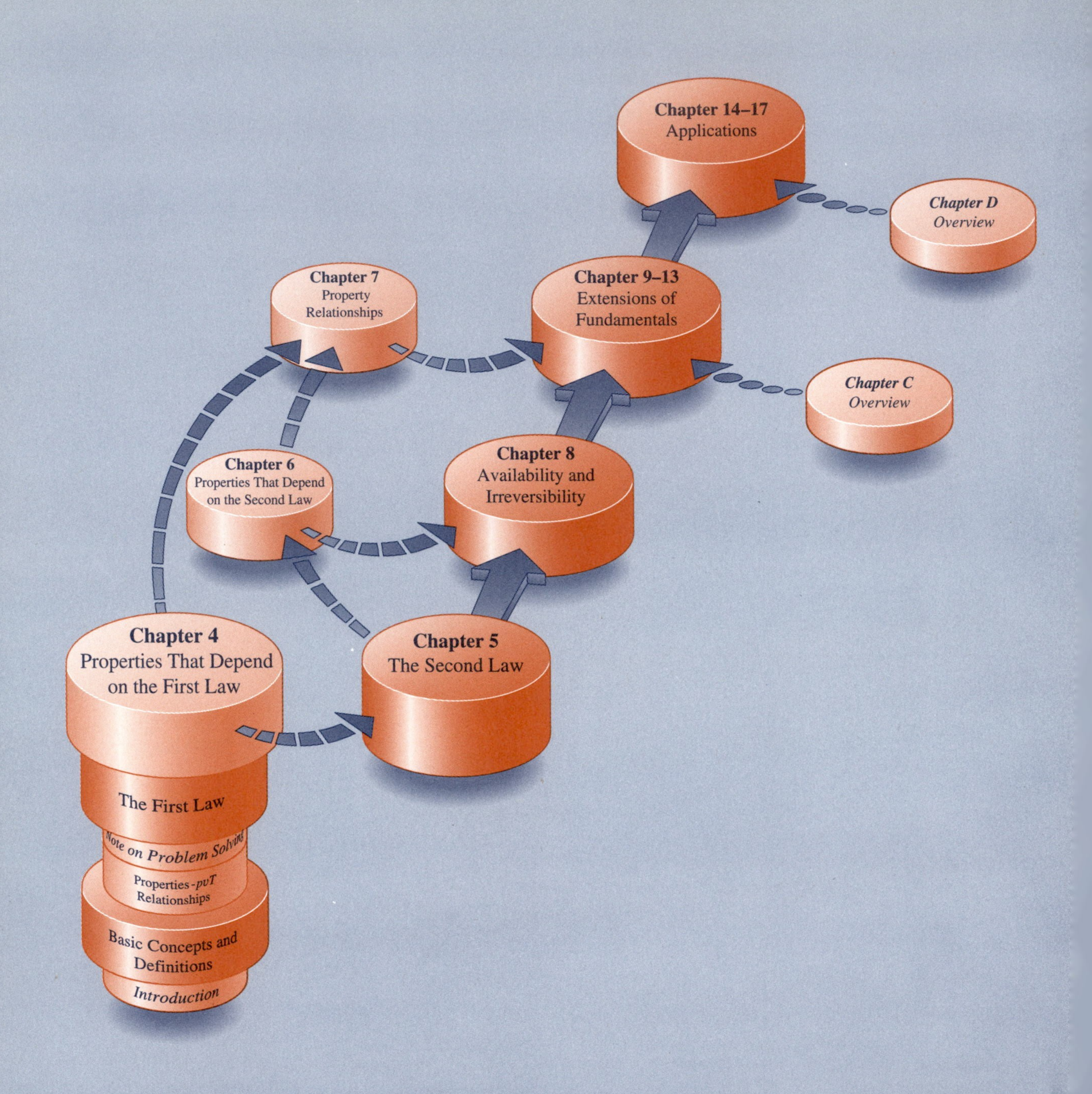
Chapter 14–17
Applications
Chapter D
Overview
Chapter 7
Property
Relationships
Chapter 9–13
Extensions of
Fundamentals
Chapter C
Overview
Chapter 6
Properties That Depend
on the Second Law
Chapter 8
Availability and
Irreversibility
Chapter 4
Properties That Depend
on the First Law
Chapter 5
The Second Law
The First Law
Note on Problem Solving
Properties-pvT
Relationships
Basic Concepts and
Definitions
Introduction

Chapter 4

Properties That Depend on the First Law

Chapter 2 covered relationships among the pressure, specific volume, and temperature (p, v, and T) of substances for which tabular data are available and for the special case of ideal gases. The first law allows us to define additional properties: internal energy, enthalpy, and specific heats. We now study relationships among these properties and p, v, and T.

4·1 Internal Energy and Enthalpy of Substances

Usually, for any phase of a pure substance, two independent intensive properties define the state. Thus, the internal energy and enthalpy of a substance

> If it is *usually* true that two independent intensive properties define the state, when is it not true? Exceptions are rare, and a general statement to cover all cases is beyond the scope of this book. Partial answers to the question are given by two statements found in many thermodynamics books: the phase rule and the state principle. These are outwardly simple but require extensive knowledge in application. Exceptions to the statement, *for any phase of a pure substance, two independent intensive properties define the state*, are likely to be the same cases where the phase rule or the state principle is intricate to apply. The phase rule is introduced in Chapter 10.

are related to other properties so that typically

$$u = f(v, T) \qquad u = f(p, T) \qquad h = f(p, T) \qquad h = f(p, \rho)$$

These relationships are seldom simple, just as the pvT relationships for substances are seldom simple (Chapter 2).

The pT diagrams of Figs. 4·1 and 4·2 show that for water and ammonia, enthalpy is a strong function of temperature but a weak function of pressure; that is, the change in enthalpy with temperature is much greater than its change with pressure. In fact, at low pressures constant enthalpy lines approach constant temperature lines. For vapors, enthalpy depends strongly on both temperature and pressure except at low pressures. Figures 4·3 and 4·4 are plotted with logarithmic pressure scales to magnify the low-pressure parts of the pT diagrams of Figs. 4·1 and 4·2. They show that at low pressures, pressure has so little influence on enthalpy that constant-enthalpy lines approach lines of constant temperature.

Two different substances, water and ammonia, have been used for these diagrams. Nonetheless, the plots are quite similar in form. This is the case for most other substances, so that our conclusions regarding the relationship of enthalpy to other properties are somewhat general. Similar plots can be made with lines of constant internal energy. (For liquids, the difference between enthalpy and internal energy, $h - u = pv$, is always small because v for liquids is very small.)

These same general conclusions may be deduced from ph diagrams, a kind of diagram used frequently in the refrigeration field. Figures 4·5 and 4·6 are ph diagrams for water and ammonia, respectively. The pressure scales of these diagrams are logarithmic, so the coincidence of constant-temperature lines and constant-enthalpy lines at low pressures is more evident.

In Chapter 2, we saw that published tables list properties of single phases and of saturation states. In Section 2·3, specific volume v was the property used to illustrate the structure of property tables. Values of internal energy u and enthalpy h are also tabulated for single phases and for saturation states.

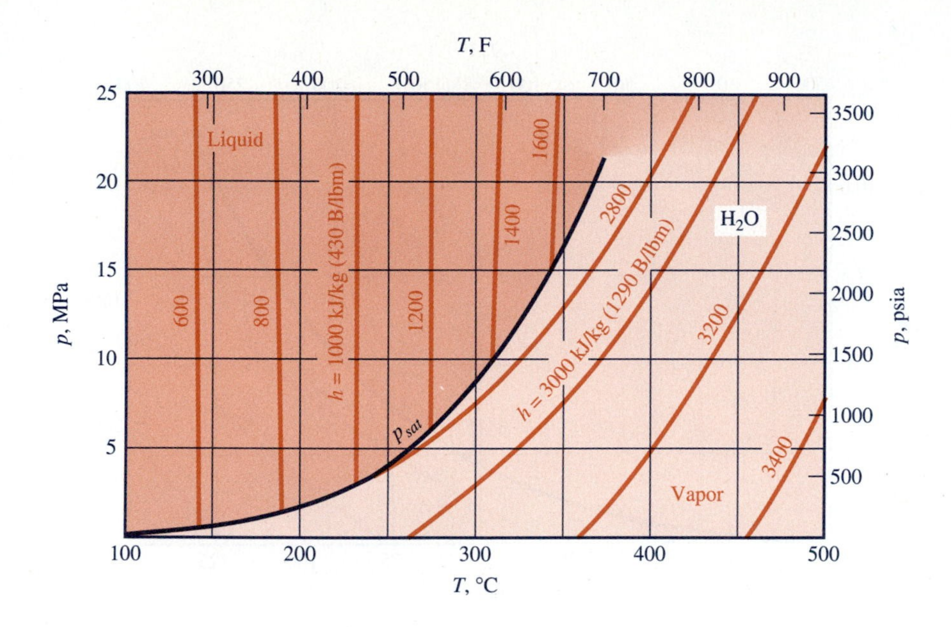

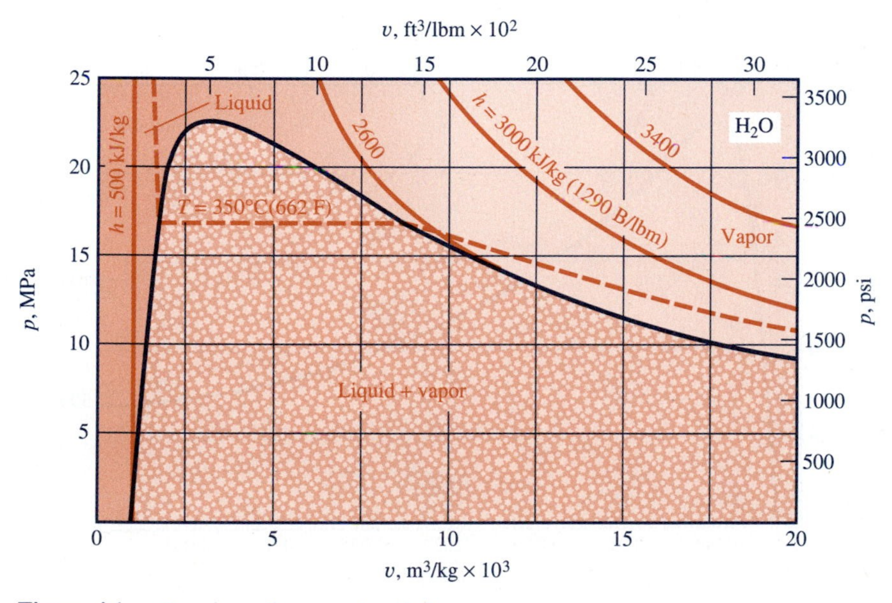

Figure 4·1 pT and pv diagrams for H_2O.

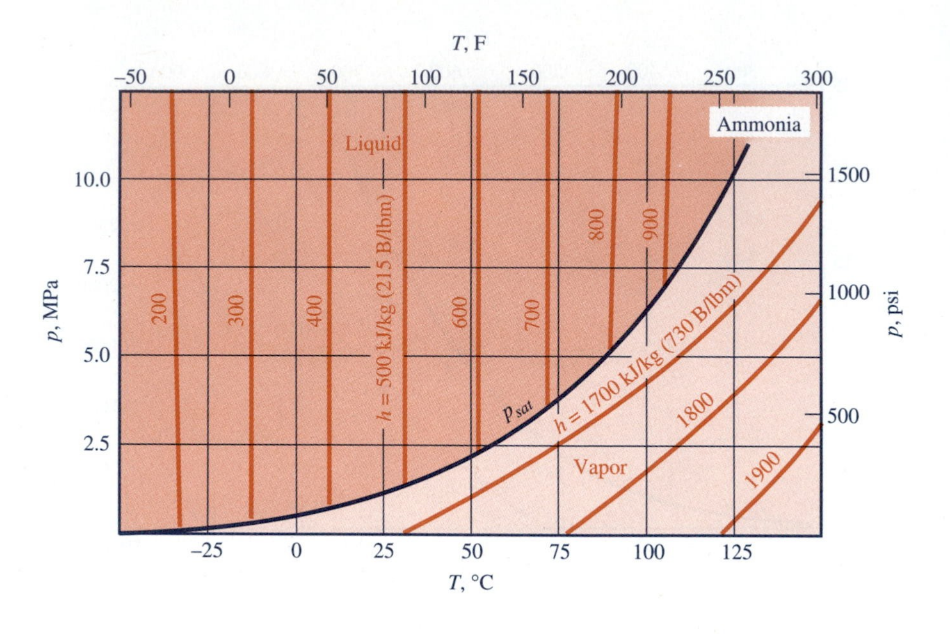

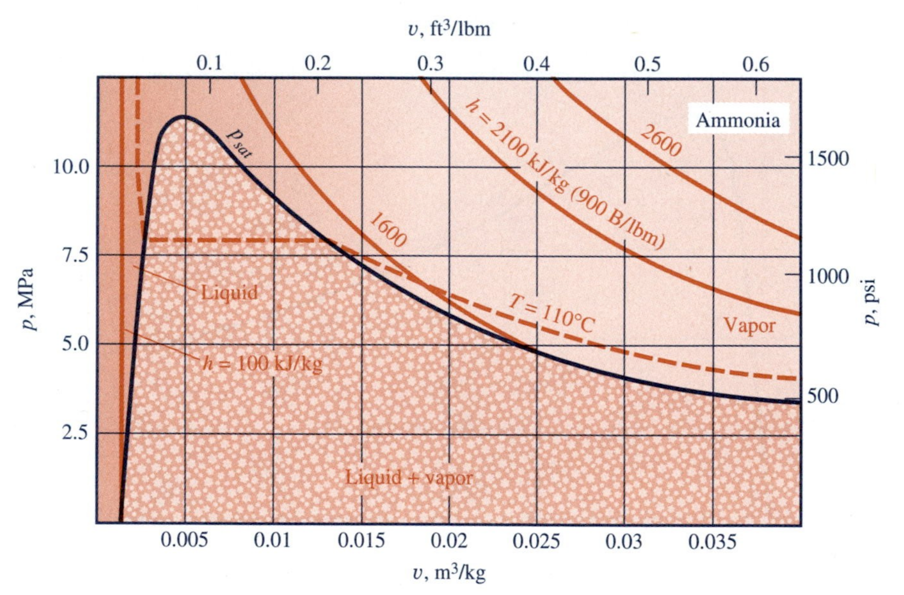

Figure 4·2 pT and pv diagrams for ammonia.

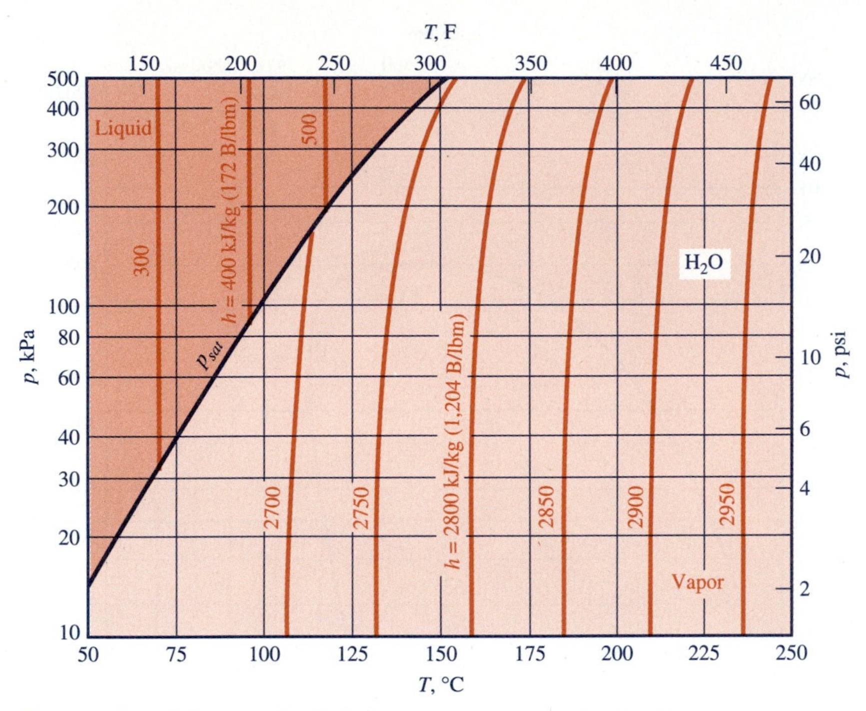

Figure 4·3 pT diagram for H_2O, low-pressure range, showing how constant-enthalpy lines approach constant-temperature lines at low pressure.

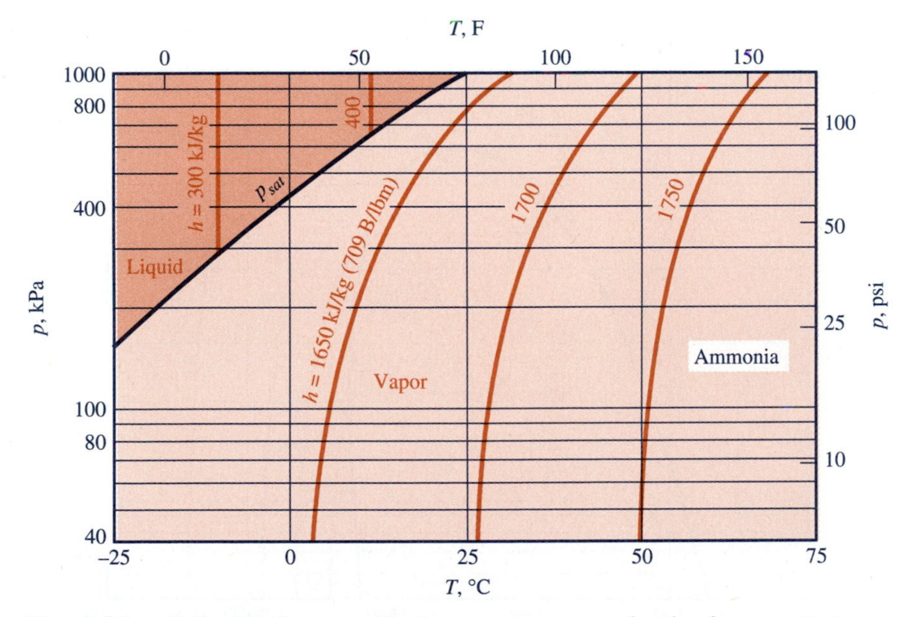

Figure 4·4 pT diagram for ammonia, low-pressure range, showing how constant-enthalpy lines approach constant-temperature lines at low pressure.

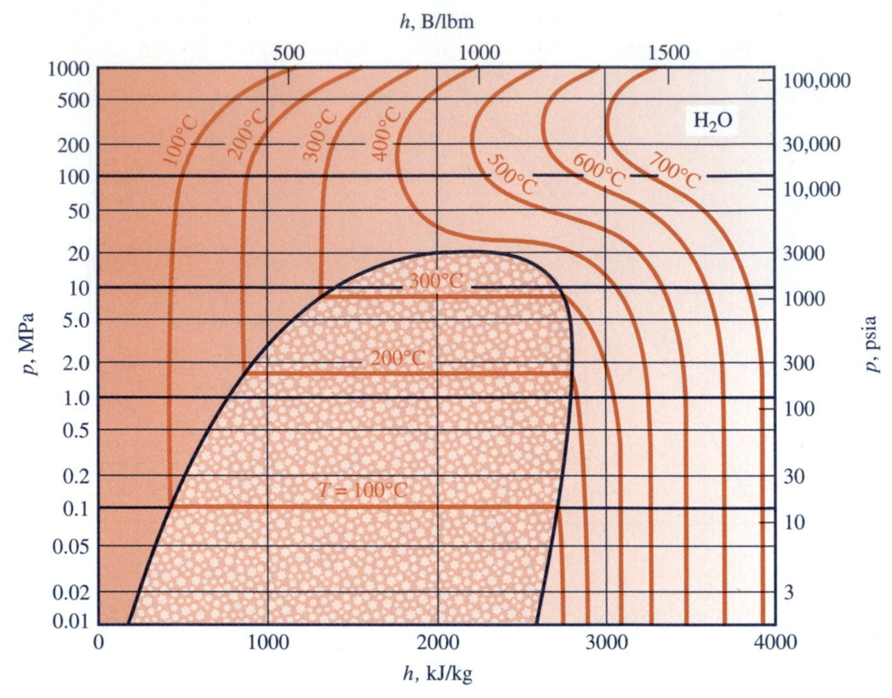

Figure 4·5 *ph* diagram for H_2O.

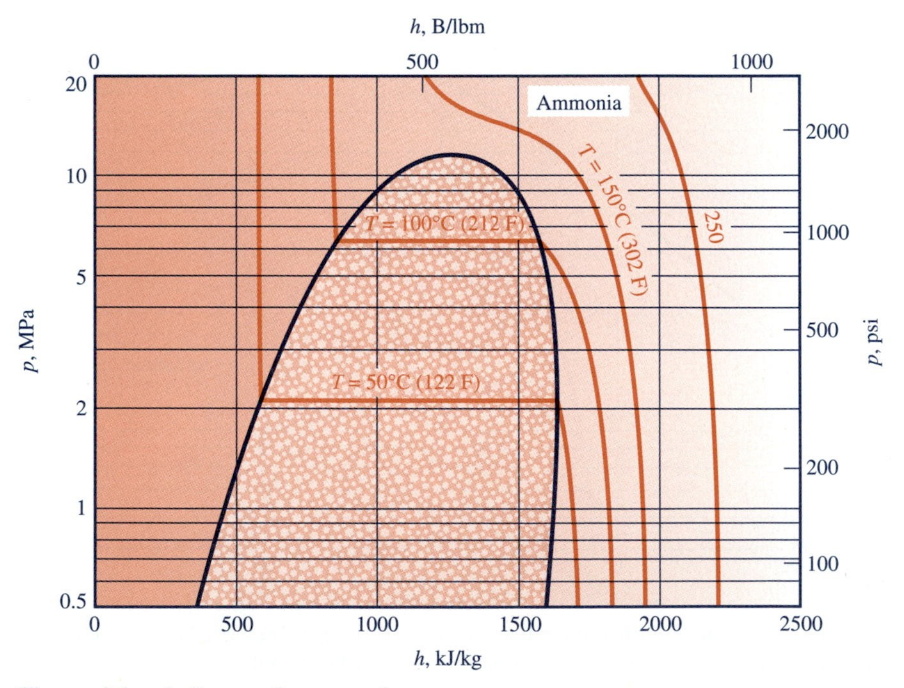

Figure 4·6 *ph* diagram for ammonia.

The state at which $h = 0$ or $u = 0$ is selected arbitrarily when tables are compiled, so that different tables for the same substance are likely to have different values of h or u at any specified state. For any two states, the differences in h or u values must be the same for all tables.

4·1·1 Latent Heats

Figures 4·1 and 4·2 show that saturated liquid and saturated vapor at given pressure or temperature have markedly different values of enthalpy. The enthalpy of saturated liquid and of saturated vapor are designated by h_f and h_g, respectively, and their difference is

$$h_{fg} \equiv h_g - h_f$$

This quantity has been called the **latent heat of vaporization** because, as we will show, under certain conditions h_{fg} equals the amount of heat required to vaporize a unit mass of liquid and the heat added causes no change in temperature. (The lack of a temperature change led to the idea that the heat is "hidden" or "latent.")

Latent heat of vaporization

In a closed system initially composed of a saturated liquid, if heat is added in a quasiequilibrium process while the pressure is held constant, the amount of heat added per unit mass of fluid is

$$\begin{aligned} q &= u_2 - u_1 + w \\ &= u_2 - u_1 + \int_1^2 p\,dv = u_2 - u_1 + p(v_2 - v_1) \\ &= u_2 + p_2 v_2 - (u_1 + p_1 v_1) \\ &= h_2 - h_1 \end{aligned}$$

If the initial state is saturated liquid and the final state is saturated vapor, then $h_1 = h_f$ and $h_2 = h_g$ and

$$q = h_g - h_f = h_{fg}$$

Thus h_{fg} is equal to the heat added to evaporate a unit mass of saturated liquid to saturated vapor in a closed system in a constant-pressure quasiequilibrium process.

A physical interpretation of h_{fg} for closed systems

A similar conclusion applies to a steady-flow system. If we apply the first law to a quasiequilibrium steady-flow, constant-pressure process with no work and no changes in potential or kinetic energy, we see that the heat added to evaporate saturated liquid to a saturated vapor under those conditions is also h_{fg}.

A physical interpretation of h_{fg} for steady-flow systems

These first law analyses show that h_{fg} is equal to the heat added only under special conditions. In fact, the vaporization can be caused with no heat transfer at all by doing work on the fluid. Therefore, the name *latent heat of vaporization* is a misnomer, but it is widely used.

In tables of thermodynamic properties, h_{fg} is often tabulated as well as h_f and h_g. Sometimes, u_f, u_g, and u_{fg} are tabulated although they can be readily calculated by

$$u_f = h_f - pv_f \quad \text{and} \quad u_g = h_g - pv_g$$

and

$$u_{fg} = h_{fg} - pv_{fg}$$

Let us now look at the physical picture of heating and vaporization. Consider a closed system containing initially a compressed liquid. If heat is added to the system in a quasiequilibrium constant-pressure process, the temperature rises as shown in either part of Fig. 4·7 until the liquid reaches its saturation temperature, state 2. The heat added in process 1–2 is called *sensible* heat because it changes the temperature of the system. Once state 2 is reached, no further energy can be added to the liquid without causing a change in phase to begin. Further heat addition at constant pressure initiates evaporation of the liquid. As long as the pressure is held constant, the temperature remains constant, so this is *latent* heating. After all of the liquid has been evaporated, the system is in state 3. Further heat addition at constant pressure increases the temperature of the vapor, so this process involves, like process 1–2, *sensible* heating because the temperature of the substance increases as heat is added.

Sensible heating

Latent heating

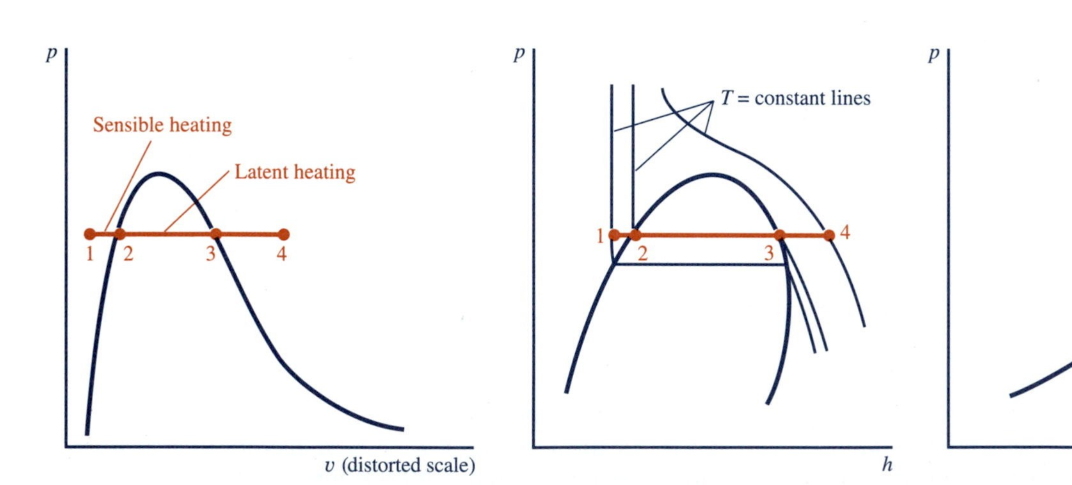

Figure 4·7 Heating a substance in a quasiequilibrium constant-pressure process of a closed system.

Example 4·1 Use of Property Tables

PROBLEM STATEMENT

Fill in the blanks in the following table for the properties of water.

	p, kPa	T, °C	x, %	v, m³/kg	h, kJ/kg
(*a*)	10,000	500			
(*b*)	400	143.6			
(*c*)		160	91.0		
(*d*)	5,000	180			
(*e*)		100			1,100.0
(*f*)				0.452	2,990.0

SOLUTION

Values found in the steam tables are inserted into the table in *italics.* In some cases, calculations similar to those in Chapter 2, Examples 2·1 through 2·5, were made. *M* denotes *meaningless* and *I* denotes *indeterminate*.

	p, kPa	T, °C	x, %	v, m³/kg	h, kJ/kg
(*a*)	10,000	500	*M*	*0.03279*	*3,374.4*
(*b*)	400	143.6	*I*	*I*	*I*
(*c*)	*617.7*	160	91.0	*0.2796*	*2,570.6*
(*d*)	5,000	180	*M*	*1.131*	*767.6*
(*e*)	*101.3*	100	*30.2*	*0.506*	1,100.0
(*f*)	*540*	*265*	*M*	0.452	2,990.0

Explanation:

(*a*) For 10,000 kPa, the saturation table for water (Table A·6·2) shows that the saturation temperature is 311°C. The specified temperature is higher than this, so the water must be a superheated vapor, and v and h can be read directly from the superheated vapor table. Point a on the ph diagram shows the location of this state.

(*b*) For 400 kPa, the saturation table for water (Table A·6·2) shows that the saturation temperature is 143.6°C, the same as the specified value. Therefore, the water is saturated so that p and T are dependent of each other and do not fully specify the state. It could be a saturated liquid, a saturated vapor, or a mixture of the two in any proportion. Since the proportions are unknown, x, v, and h cannot be determined.

(*c*) The quality is specified, so the state is one of a liquid–vapor mixture. The saturation pressure can be read directly from the saturation table (Table A·6·1). Then v and h are determined from $v = v_f + x(v_g - v_f)$ and $h = h_f + x(h_g - h_f)$.

(*d*) For 5000 kPa, the saturation temperature (from Table A·6·2) is 263.98°C. The specified temperature is lower than this, so we are dealing with a compressed liquid. The values of v and h can be read directly from the compressed liquid table that is part of Table A·6·3.

(*e*) At 100°C, $h_g = 2675.7$ kJ/kg. The specified value for h is lower than this but higher than the value for h_f, so the state is one of a liquid–vapor mixture. The quality can be determined from $h = h_f + x(h_g - h_f)$. Then v can be calculated from $v = v_f + x(v_g - v_f)$.

(*f*) Here neither pressure nor temperature is specified so the direct approaches as used in the above cases cannot be used. One must search the tables for the state with the specified values of v and h. Charts A·7·1 and A·7·3, on which some lines of constant p or constant v as well as lines of constant h are shown, can be quite helpful in finding the p and T ranges in which the state lies.

Locating the state points on a property diagram is often helpful. The five determinate points are shown here on a *ph* diagram.

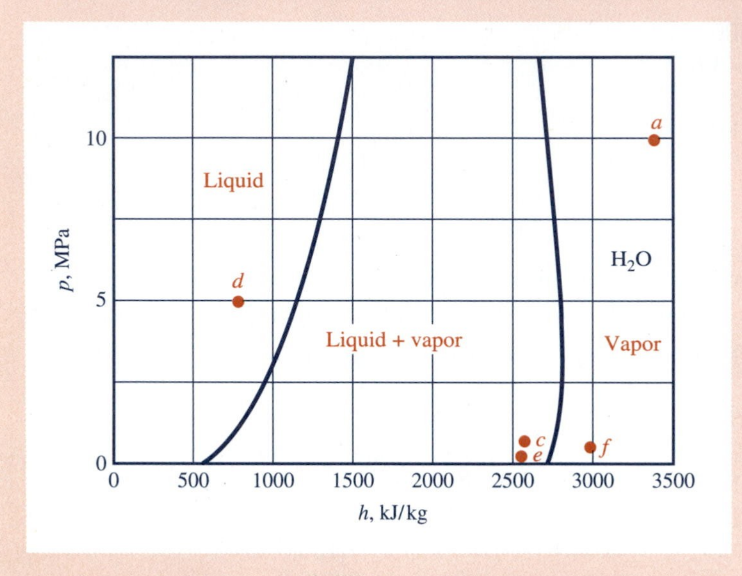

Comment: These values can also be obtained from the software model STEAM included with this book. The STEAM model also displays the state points on *pv* and *ph* diagrams (as well as on another diagram that will be introduced in Chapter 6).

Example 4·2 Use of Property Tables

PROBLEM STATEMENT

Fill in the blanks in the following table for the properties of R134a.

	p, psi	T, F	x, %	v, ft³/lbm	h, B/lbm
(*a*)	120	140			
(*b*)		−10	95		
(*c*)		0			32.4
(*d*)	120	80			
(*e*)	30				28.0

SOLUTION

Values found in the R134a tables are inserted into the table in *italics.* In some cases, calculations similar to those in Chapter 2, Examples 2·1 through 2·5, were made. *M* denotes *meaningless* and *I* denotes *indeterminate*.

	p, psi	T, F	x, %	v, ft³/lbm	h, B/lbm
(*a*)	120	140	*M*	*0.4610*	*126.6*
(*b*)	*16.67*	−10	95	*2.558*	*95.7*
(*c*)	*21.2*	0	*23.0*	*0.503*	32.4
(*d*)	120	80	*M*	*0.01334*	*37.3*
(*e*)	30	*15.4*	*13.3*	*0.216*	28.0

Explanation:

(*a*) For 120 psia, the saturation table for R134a (Table A·10·2E) shows that the saturation temperature is 90.54 F. The specified temperature is higher than this, so the R134a must be a superheated vapor, and v and h can be read directly from the superheated vapor table. Point a on the ph diagram shows the location of this state.

(*b*) The quality is specified, so the state is one of a liquid–vapor mixture. The saturation pressure can be read directly from the saturation table. Then v and h are determined from $v = v_f + x(v_g - v_f)$ and $h = h_f + x(h_g - h_f)$.

(*c*) At 0 F, $h_g = 101.75$ B/lbm. The specified value for h is lower than this but higher than the value for h_f (11.63 B/lbm), so the state is one of a liquid–vapor mixture. The quality can be determined from $h = h_f + x(h_g - h_f)$. Then v can be calculated from $v = v_f + x(v_g - v_f)$.

(*d*) For 120 psia, the saturation temperature, from Table A·10·2E, is 90.54 F. The specified temperature is lower than this, so we are dealing with a compressed liquid. If compressed liquid tables are not at hand, we use the approximation that properties of a slightly compressed liquid are equal to those of saturated liquid *at the same temperature*. Thus we take $v = v_f$ at 80 F and $h \approx h_f$ at 80 F.

(*e*) At 30 psia, $h_f = 16.3$ B/lbm and $h_g = 103.96$ B/lbm. The specified value for h is between these two values, so the state is one of a liquid–vapor mixture. The quality can be determined from $h = h_f + x(h_g - h_f)$. Then v can be calculated from $v = v_f + x(v_g - v_f)$.

The five states, a through e, are shown here on two ph diagrams, one with a linear pressure scale and one with a logarithmic pressure scale.

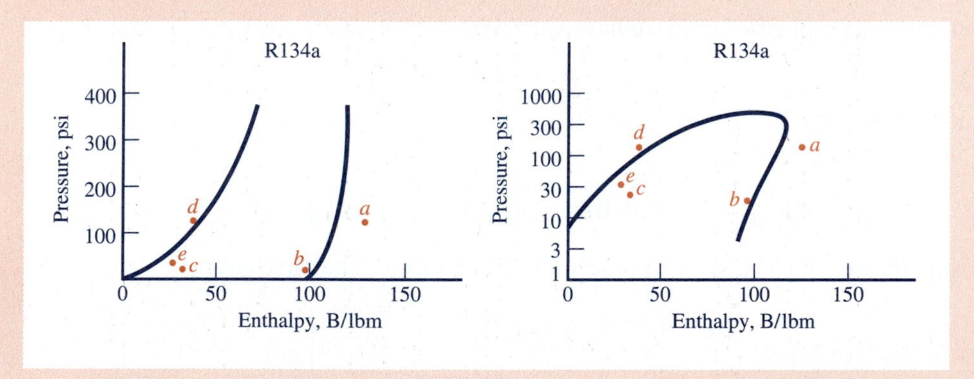

Example 4·3 Refrigeration System

PROBLEM STATEMENT

In a refrigeration system using R134a, refrigerant enters the compressor at 200 kPa, $x = 0.94$, and is discharged at 1000 kPa, 43°C. The flow rate is 1.9 kg/s. During the compression, heat transfer from the refrigerant is estimated to be 10 percent of the work input. Determine the power input.

SOLUTION

Analysis: We first make a sketch of the open system and place data values on it. The single values of flow rate and inlet and outlet properties suggest that the flow is steady, so we will assume that it is. No data are given that permit us to determine velocities, so we will assume that the change in velocity (and therefore of kinetic energy) between inlet and outlet is negligible. The change in potential energy is also assumed to be negligible.

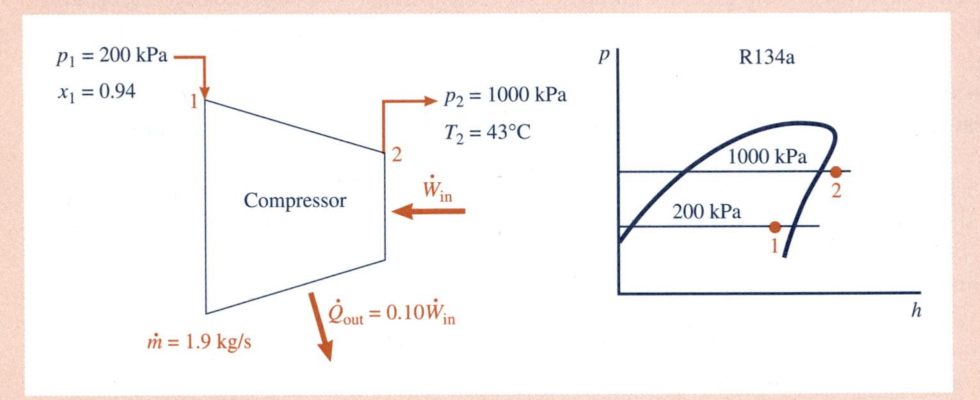

Since the flow rate is known, if we calculate the work per unit mass, we can immediately calculate the power input. The work per unit mass can be obtained from the first law, which with no changes in kinetic and potential energies is

$$w_{in} = h_2 - h_1 + q_{out}$$

An estimate of q_{out} in terms of w_{in} is specified, and we have sufficient data to determine the properties h_1 and h_2, so we proceed.

Calculations: From the pressure and the quality of the R134a at the inlet, we can use the saturated R134a table, Table A·10·2, to determine the inlet enthalpy:

$$h_1 = h_{f1} + x_1 h_{fg1} = 38.22 + 0.94(206.3) = 232.2 \text{ kJ/kg}$$

From Table A·10·3 for superheated R134a, we can read $h_2 = 275.5$ kJ/kg. The first law is

$$w_{in} = h_2 - h_1 + q_{out}$$

and with the specified estimate of q_{out}

$$w_{in} = h_2 - h_1 + 0.1w_{in}$$
$$= \frac{h_2 - h_1}{0.9} = \frac{275.5 - 232.2}{0.9} = 48.1 \text{ kJ/kg}$$

The power input is

$$\dot{W}_{in} = \dot{m}w_{in} = 1.9\frac{\text{kg}}{\text{s}}\left(48.1\frac{\text{kJ}}{\text{kg}}\right) = 91.4\frac{\text{kJ}}{\text{s}} = 91.4 \text{ kW}$$

Comment: The power input value shows that this refrigeration system is much larger than a household refrigerator, although the temperature limits of the system are similar. Further attention shows that the mass flow rate of 1.9 kg/s far exceeds that of a household refrigerator.

Example 4·4

PROBLEM STATEMENT

A vessel with a volume of 0.0011 m^3 is filled with dry sa, urated steam at a pressure of 95 kPa. It is then sealed and cooled to 25°C. Determine the amount of heat that must be removed.

SOLUTION

Analysis: Heat is removed from the steam only after the vessel is sealed, so we are dealing with a closed system. The volume of the system is constant, so when we sketch a *pv* diagram for the process we locate the initial state 1 on the saturation line and the final state 2 at the same specific volume but at a lower pressure, because lowering the temperature lowers the pressure of a mixture of saturated phases.

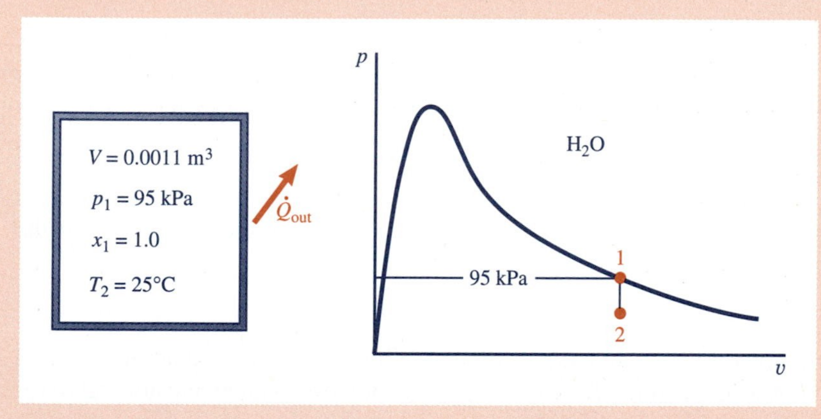

No work is done, so the first law for the closed system is

$$Q_{out} = U_1 - U_2 = m(u_1 - u_2)$$

State 1 is specified as a dry saturated vapor at a given pressure, and state 2 is specified by its temperature and the fact that $v_2 = v_1$, so we can find the values of the u's and solve the problem.

Calculations: Finding u_2 requires finding the quality x_2, so we find this from

$$v_2 = v_{f2} + x_2 v_{fg2}$$

$$x_2 = \frac{v_2 - v_{f2}}{v_{fg2}} = \frac{v_1 - v_{f2}}{v_{g2} - v_{f2}} = \frac{1.7776 - 0.00100}{43.357 - 0.0100} = 0.041$$

Then

$$u_1 = u_{g1}$$
$$u_2 = u_{f2} + x_2 u_{fg2}$$

If the tables display no u values, we must determine them from $u = h - pv$. Therefore,

$$u_1 = u_{g1} = h_{g1} - p_1 v_{g1} = 2672.9 - 95(1.7776) = 2504.0 \text{ kJ/kg}$$
$$u_2 = h_2 - p_2 v_2 = (h_{f2} + x_2 h_{fg2}) - p_2 v_2$$
$$= 104.75 + 0.041(2441.5) - 3.1691(1.7776) = 199.2 \text{ kJ/kg}$$

Then the first law is

$$Q_{out} = U_1 - U_2 = m(u_1 - u_2) = \frac{V}{v}(u_1 - u_2)$$

$$= \frac{0.0011 \text{ m}^3}{1.7776 \text{ m}^3\text{/kg}}(2504.0 - 199.2) \text{ kJ/kg} = 1.43 \text{ kJ}$$

Example 4·5 Expansion Valve in a Refrigeration System

PROBLEM STATEMENT

In a refrigeration system using Refrigerant 134a, saturated liquid at 65 F enters an expansion valve and is throttled to a pressure corresponding to an evaporator temperature of −10 F. Determine the quality of the refrigerant entering the evaporator.

SOLUTION

Analysis: We first make a flow diagram and place data values on it. We also sketch a *ph* diagram, at least showing the saturation line and state 1. Assume that the flow is steady. The process in an expansion valve (or

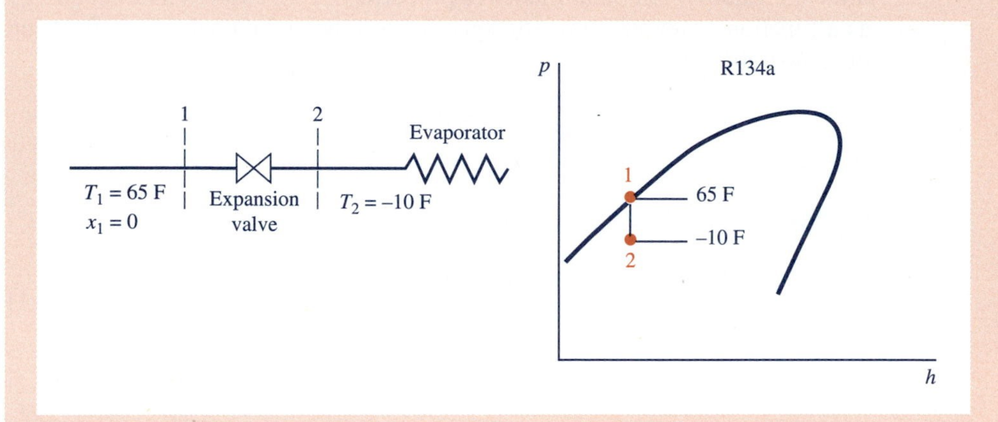

throttling valve) is one with no work done and no heat transfer. Potential energy changes are negligible. If we also assume that kinetic energy changes are negligible, the first law reduces to

$$h_2 = h_1$$

Therefore, for state 2 as the refrigerant leaves the valve we know the values of two properties, T and h. Consequently, we can readily determine the quality. We now add state 2 to the *ph* diagram.

Calculations: The R134a at state 1 is a saturated liquid, and the R134a tables at 65 F show $h_f = 33.0$ B/lbm. The R134a at state 2 is a liquid–vapor mixture, so

$$h_1 = h_2 = h_{f2} + x_2(h_{g2} - h_{f2})$$

Solving for x_2 and obtaining values for h_{f2} and h_{g2} from Table A·10·1E, we find

$$x_2 = \frac{h_2 - h_{f2}}{h_{g2} - h_{f2}} = \frac{33.0 - 9.0}{101.7 - 9.0} = 0.259$$

Comment: This means that 25.9 percent of the refrigerant entering the evaporator is vapor. The remaining 74.1 percent is a liquid that will evaporate as it absorbs heat from the refrigerated space around the evaporator tubes.

4·1·2 Specific Heats

Two useful thermodynamic properties, c_p and c_v, will now be introduced because we use them in the evaluation of u and h. They were originally called specific heats, because they were first defined as the amount of heat added to a substance to cause a unit increase in its temperature under special, carefully controlled conditions. They are actually much more widely useful than this early definition suggests, but the name specific heat has persisted. Often they are simply called "c sub p" and "c sub v."

As mentioned in Section 4·1, for any phase of a pure substance, two independent intensive properties usually define the state and consequently determine the values of all other intensive properties. If we express internal energy as a function of temperature and specific volume and enthalpy as a function of temperature and pressure,

$$u = f(T, v)$$
$$h = f(T, p)$$

we have for the differentials du and dh

$$du = \left(\frac{\partial u}{\partial T}\right)_v dT + \left(\frac{\partial u}{\partial v}\right)_T dv \tag{4·1}$$

$$dh = \left(\frac{\partial h}{\partial T}\right)_p dT + \left(\frac{\partial h}{\partial p}\right)_T dp \tag{4·2}$$

Two of the partial derivatives in these expressions are the definitions of c_v and c_p:

Definitions of specific heats, c_p and c_v

$$c_v \equiv \left(\frac{\partial u}{\partial T}\right)_v \tag{4·3}$$

$$c_p \equiv \left(\frac{\partial h}{\partial T}\right)_p \tag{4·4}$$

These two quantities are properties of any phase of a pure substance. (For a two- or three-phase mixture, p and T are not independent, so c_v and c_p cannot be evaluated.)

Figure 4·8 shows that c_p values for liquid and vapor water vary with pressure and temperature. The same is true for other substances. Data on c_p for various substances are given in the appendix (Tables A·3 and A·5, Chart A·4), and in Chapter 7 we discuss general property relationships that involve c_v and c_p. Section 4·5 treats the special case of specific heats of ideal gases.

For an incompressible fluid, v is constant. No energy can be added to such a fluid in a closed system by increasing the pressure, and

$$c_v = \left(\frac{\partial u}{\partial T}\right)_v = \frac{du}{dT}$$

A change in enthalpy is related to a change in internal energy by

$$dh = du + d(pv) = c_v\, dT + v\, dp$$

Specific heat of an incompressible substance

Differentiating this expression with respect to temperature at constant pressure shows that

$$c_p = c_v = c$$

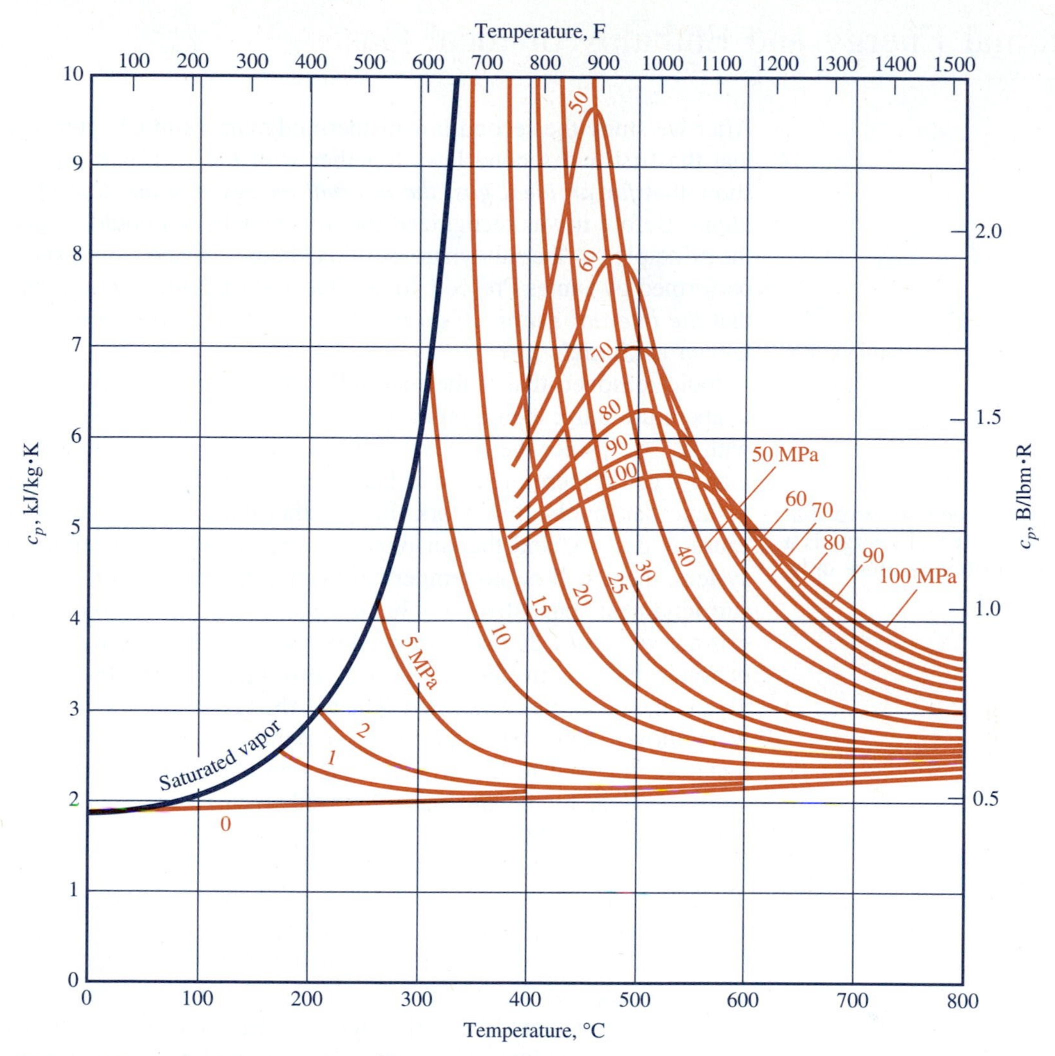

Figure 4·8 c_p for liquid and vapor water (from Ref. 4·16).

where c is called simply the *specific heat* of the incompressible fluid. Then

$$du = c\,dT$$

$$dh = du + v\,dp$$

$$\Delta h = \Delta u + v\,\Delta p$$

For liquids, v is so small that one often sees $\Delta h \approx \Delta u$.

For liquids, $\Delta h \approx \Delta u$

Because the definition of specific heat involves a temperature derivative, the units can be interchangeably kJ/kg·°C or kJ/kg·K. With T [K] = T [°C] + 273.15, we see that dT [K] = dT [°C]. The size of the unit is the same on the Kelvin and Celsius scales.

4·2 Internal Energy and Enthalpy of Ideal Gases

After we study the second law of thermodynamics in Chapter 5, we will see that the first and second laws together with the ideal-gas equation of state show that *for an ideal gas, the internal energy is a function of temperature alone!* Before it was recognized that this conclusion could be deduced from the principles of thermodynamics, it had been reached from experiments first performed by James Prescott Joule. Because of Joule's work, the conclusion that *the internal energy of an ideal gas is a function of temperature only* is known as **Joule's law**.

Joule's law

Joule reasoned that if the internal energy of a substance is a function of temperature only, then a process with $q = 0$ and $w = 0$ would be a process with $\Delta T = 0$, since the first law would show $\Delta u = 0$. The interdependence of u and T could therefore be checked experimentally by means of an adiabatic system that involved no work. Perfect thermal insulation is not needed, because if $\Delta T = 0$ and the surroundings are at the initial temperature of the system, there will be no temperature difference between the system and the surroundings; hence there will be no heat transfer. Thus, to check his hypothesis that $u = f(T)$, Joule submerged two tanks connected by a valve in a tank of water that was insulated from the surroundings. One tank was filled with air and the other was evacuated. The air, the tanks, and the surrounding water were allowed to come to the same temperature. Then the valve between the tanks was opened to let air pass from one tank to the other. Of course, no work was done by the air. Joule observed no change in the temperature of the water bath. From this he concluded that no heat was transferred to or from the air; hence, since there was also no work done, the internal energy of the air did not change. Since there was no heat exchange between the air and the water, the temperature of the air must have remained constant, even though the volume changed. Thus Joule concluded that $(\partial u/\partial v)_T = 0$; that is, the internal energy of the gas is a function of temperature only.

Joule's experiments showed that for a gas, internal energy is a function of temperature only.

If in an experiment like Joule's the temperature of the water changes, does this prove that the internal energy of the gas under study is not a function of temperature only? Yes, it does so conclusively. Suppose that the water temperature rises. Then there must have been a heat transfer to the water from the gas, and the gas temperature must have risen above that of the water during the expansion process. Since heat was removed from the gas and no work was done, the internal energy of the gas must have decreased. This expansion process is shown as process 1–2 on the uv and Tv diagrams of Fig. 4·9. To restore the gas that finally fills both tanks to its initial temperature, heat must be removed. In this constant-volume process, shown as 2–3 in Fig. 4·9, no work is done as heat is removed, and so the internal energy of the gas decreases. Thus, when the gas has been restored to its initial temperature ($T_3 = T_1$), its internal energy is lower than its initial value ($u_3 < u_1$). Clearly, the internal energy of a gas that behaves in this manner is not a function of temperature only.

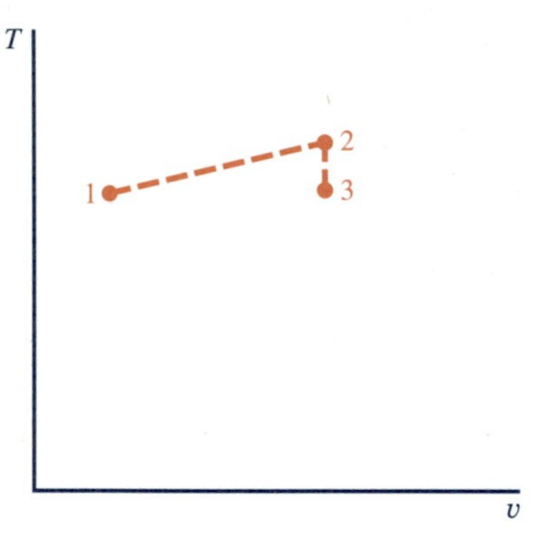

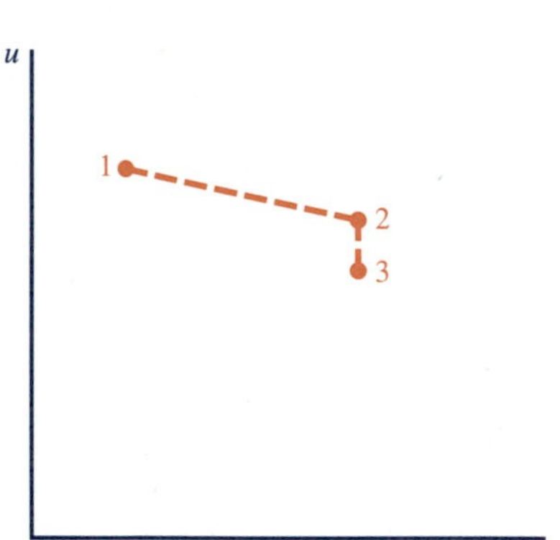

Figure 4·9 Reasoning about Joule's experiment.

For *real* gases internal energy is not a function of temperature only, but Joule did not detect a temperature change of the water, because for the pressure and temperature ranges of his experiments, air behaves very nearly as an ideal gas and also because the mass of the water was so great compared to that of the air that the temperature rise of the water was too small to be detected by his instruments. Joule later ran other experiments with different equipment to show that for real gases, internal energy is not a function of temperature alone. However, in the ranges of pressure and temperature where $pv = RT$ is sufficiently accurate, Joule's law holds. In other words, it is valid for ideal gases only.

For real gases, internal energy is not a function of temperature only, but Joule originally was not able to detect this.

From Joule's law it is easy to establish the relationship between u and T for ideal gases. For a gas, T and v are certainly independent properties, so we may write, as before,

Joule's law can be used to establish the relationship between *u* and *T* for ideal gases.

$$u = f(T, v)$$

Then

$$du = \left(\frac{\partial u}{\partial T}\right)_v dT + \left(\frac{\partial u}{\partial v}\right)_T dv$$

Recalling the definition of c_v (Eq. 4·3), we can write

$$du = c_v\, dT + \left(\frac{\partial u}{\partial v}\right)_T dv$$

However, Joule's law states that $(\partial u/\partial v)_T = 0$ for an ideal gas, so that

For an ideal gas, internal energy is a function of temperature only.

$$du = c_v\, dT \tag{4·5}$$

for *any process* of an ideal gas.

For any finite change of state,

$$\Delta u = \int c_v\, dT \tag{4·6}$$

Change in internal energy for an ideal gas

To emphasize that this relationship holds for *all processes* of an ideal gas, let us look at the *pv* diagram for an ideal gas shown in Fig. 4·10. For the equation of state $pv = RT$, the constant-temperature lines are hyperbolas. In accordance with Joule's law, the constant-temperature lines are also lines of constant internal energy. Therefore, $u_a = u_b = u_c = u_d$, and

$$u_a - u_1 = u_b - u_1 = u_c - u_1 = u_d - u_1 = \int_1^2 c_v\, dT$$

No matter what path is followed by an ideal gas between a state 1 and any state at a temperature T_2, the change in internal energy per unit mass is given by $\int_1^2 c_v\, dT$. This point troubles many students, because they see c_v, which we (unfortunately) call specific heat at constant volume, and then conclude *incorrectly* that Eq. 4·6 holds only for constant-volume processes.

We can also show that the enthalpy of an ideal gas is a function of temperature only. By definition, $h \equiv u + pv$, and for an ideal gas this becomes

For an ideal gas, enthalpy is a function of temperature only.

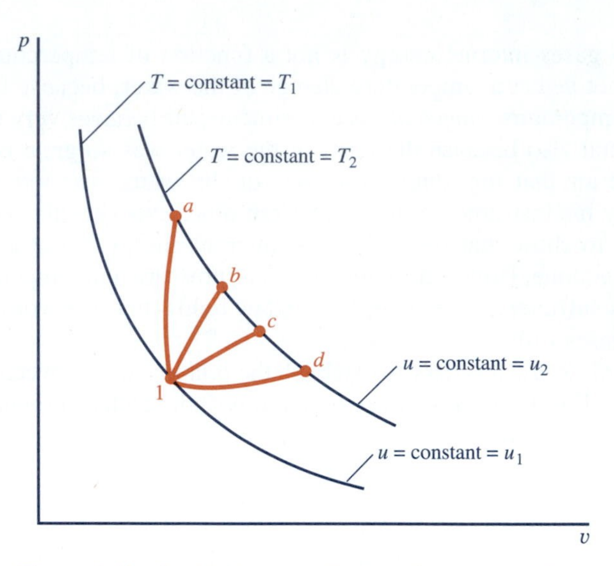

Figure 4·10 Joule's law: pv diagram showing that Δu of an ideal gas is the same for all processes having the same ΔT.

$h = u + RT$. Both terms on the right-hand side of this equation depend solely on temperature, so h is a function of temperature only, $h = f(T)$. To establish the relationship between h and T for an ideal gas, let us start by considering that for any pure substance (again in the absence of certain effects) in a state where pressure and temperature are independent properties

$$h = f(p, T)$$

and

$$dh = \left(\frac{\partial h}{\partial T}\right)_p dT + \left(\frac{\partial h}{\partial p}\right)_T dp \tag{4·2}$$

Recalling the definition of c_p, we can write this as

$$dh = c_p\, dT + \left(\frac{\partial h}{\partial p}\right)_T dp$$

For an ideal gas, we have already shown that h is a function of T only, so $(\partial h/\partial p)_T = 0$, and the last expression for dh reduces to

$$dh = c_p\, dT \tag{4·7}$$

which holds for *any process* of an ideal gas. Since h is a function of T only, then c_p must be a function of T only. For any finite change of state,

Change in enthalpy for an ideal gas

$$\Delta h = \int c_p\, dT \tag{4·8}$$

The change in enthalpy for an ideal gas is expressed in terms of c_p, but it is not restricted to a constant-pressure process.

The quantities $\int c_v\, dT$ and $\int c_p\, dT$ that appear in Eqs. 4·6 and 4·8 can be evaluated by several methods, some of which are mentioned in the following section.

In closing, we repeat for emphasis that *for an ideal gas*

$$du = c_v \, dT \qquad (4\cdot5)$$

and

General equations for the internal energy and enthalpy of an ideal gas

$$dh = c_p \, dT \qquad (4\cdot7)$$

for all processes and u and h are functions of temperature only.

4·2·1 Specific Heats of Ideal Gases

The preceding section pointed out that c_p and c_v for ideal gases are functions of temperature only. We will now show that their difference, $c_p - c_v$, is constant.

For an ideal gas the definition of enthalpy, $h \equiv u + pv$, gives us

$$h = u + RT$$

and, since R is constant,

$$dh = du + R \, dT$$

For any process of an ideal gas $dh = c_p \, dT$ and $du = c_v \, dT$, so that

$$c_p \, dT = c_v \, dT + R \, dT$$

$$c_p = c_v + R$$

$$c_p - c_v = R \qquad (4\cdot9)$$

For ideal gases, the difference between c_p and c_v is constant.

The ratio of the specific heats, c_p/c_v, is often useful and is designated by k:

$$k \equiv \frac{c_p}{c_v}$$

Ratio of specific heats (Some authors use γ instead of k.)

Combining Eq. 4·9 and the definition of k gives

$$c_p = \frac{Rk}{k - 1} \quad \text{and} \quad c_v = \frac{R}{k - 1}$$

As pressure decreases, the behavior of real gases approaches that of ideal gases, so the specific heats of real gases measured at very low pressures are called the ideal-gas specific heats or the zero-pressure specific heats. The symbols c_{p0} and c_{v0} are often used to indicate that the values concerned are those for zero pressure. When dealing with only ideal gases, we omit the 0 subscripts, even though the c_p and c_v values we use for an ideal gas are the c_{p0} and c_{v0} values for the corresponding real gas.

When dealing with ideal gases we omit subscripts referring to the effect of pressure on specific heats.

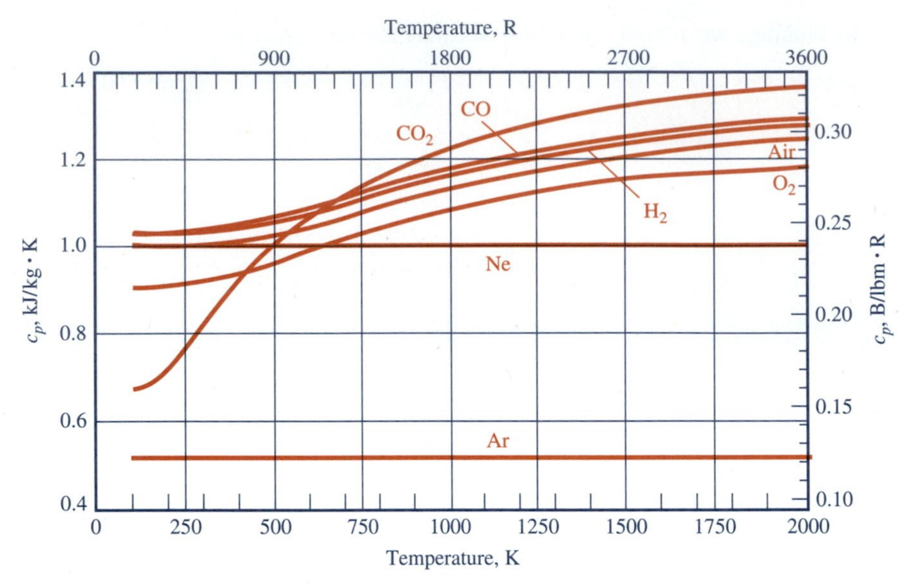

Figure 4·11 c_p for several ideal gases.

Figure 4·11 shows c_p of several ideal gases as a function of temperature. Also, c_v can be obtained from the relationship, $c_v = c_p - R$. These data are shown in more detail in Chart A·4 and Table A·3 in the appendix. Table A·1 gives for several gases the polynomial coefficients for equations of the form

$$\frac{c_p}{R} = \frac{\bar{c}_p}{\bar{R}} = a + bT + cT^2 + dT^3 + eT^4 \tag{4·10}$$

that fit the data.

Plotting the data of Fig. 4·11 in terms of molar specific heat or in terms of ratios, c_p/R or $\bar{c}_p/\bar{R}$, collapses the data significantly (see Fig. 4·12). Molar specific heats are related to specific heats on a mass basis by the relationships,

Molar specific heats. An overbar indicates a molar quantity.

$$\bar{c}_p = Mc_p \quad \text{and} \quad \bar{c}_v = Mc_v$$

where M is the molar mass.

This can be proven from kinetic theory. See Section 4·2·4.

Figures 4·11 and 4·12 show, among other things, that all the monatomic gases have constant specific heats and the molar specific heats are the same for all of them. For all other gases shown, specific heats increase monotonically with temperature. At low temperatures, the molar specific heats of all the diatomic gases shown are the same.

The specific heat ratio is either constant or decreases monotonically with temperature for all gases.

Figure 4·13 shows the variation of k ($\equiv c_p/c_v$) with temperature. In accordance with the c_p variations shown in Fig. 4·11, we see that k for monatomic gases is constant and that k for all other gases decreases monotonically with temperature.

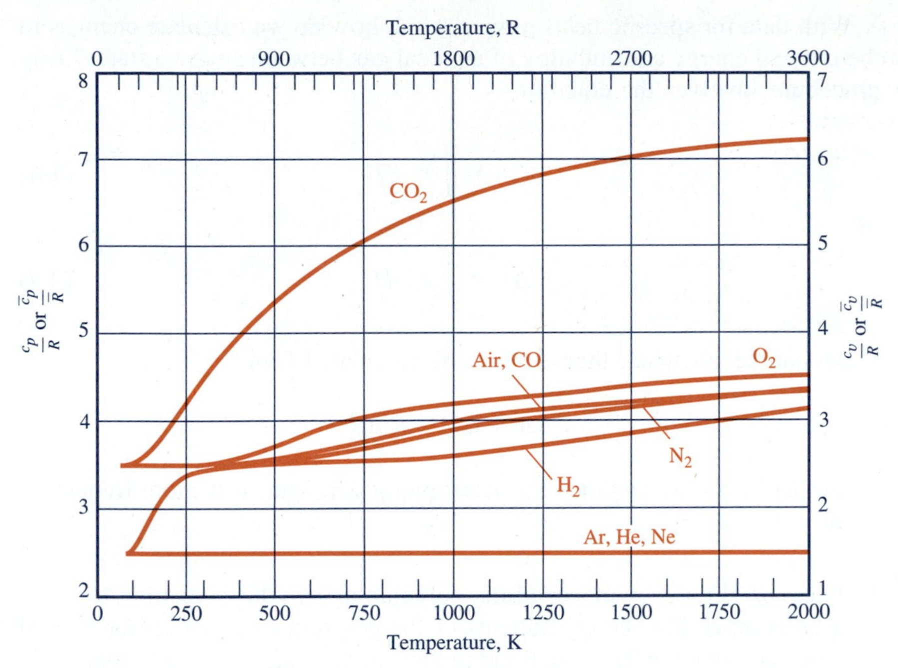

Figure 4·12 c_p/R or $\bar{c}_p/\bar{R}$ for several ideal gases.

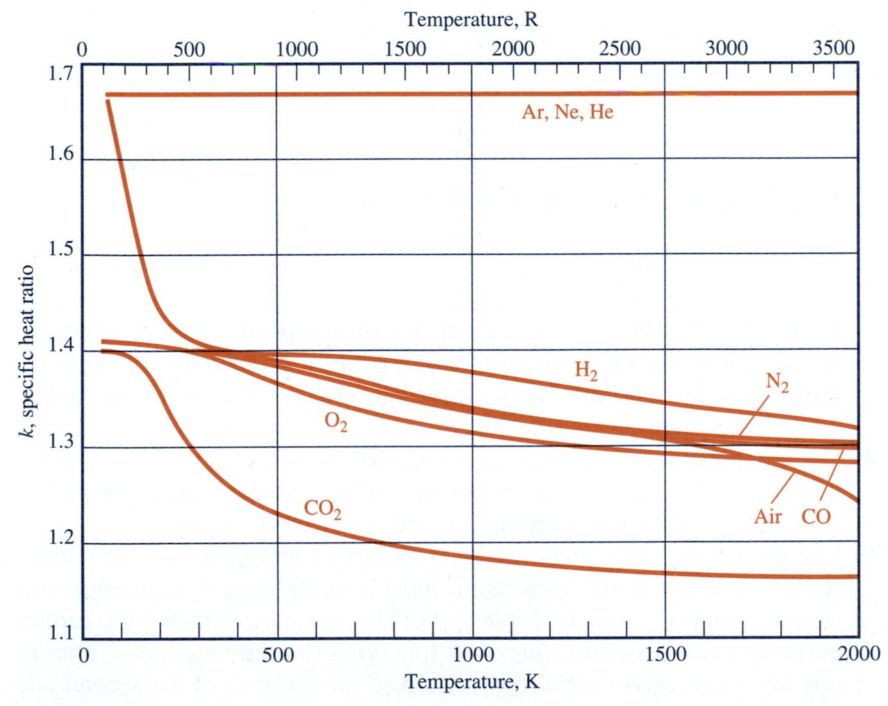

Figure 4·13 $k = c_p/c_v$ for several ideal gases.

With data for specific heats now in hand, how do we calculate changes in the internal energy and enthalpy of an ideal gas between any two states? Any procedure involves the equation

$$\Delta u = \int c_v \, dT \tag{4·6}$$

or

$$\Delta h = \int c_p \, dT \tag{4·8}$$

If Δh can be obtained, then Δu can be calculated from

$$\Delta u = \Delta h - R\,\Delta T$$

Various methods for determining the change in enthalpy for an ideal gas.

so we list here only methods for determining Δh, since it is more frequently used:

1. Integrate Eq. 4·8 by means of the polynomial (or other) equation for c_p as a function of T. (The coefficients of the polynomial equation for several gases are given in Table A·1 and in the software accompanying this book. Coefficients for over 400 gases are given in Reference 4·14.)
2. Integrate the $c_p T$ relationship graphically as shown in Fig. 4·14, where the area beneath the $c_p T$ curve between T_1 and T_2 is $h_2 - h_1$.
3. Use mean values of specific heats defined by

$$c_{p,\text{mean}} \equiv \frac{\int_1^2 c_p \, dT}{T_2 - T_1} \quad \text{and} \quad c_{v,\text{mean}} \equiv \frac{\int_1^2 c_v \, dT}{T_2 - T_1}$$

so that for any process of an ideal gas

$$\Delta h = c_{p,\text{mean}}\,\Delta T \quad \text{and} \quad \Delta u = c_{v,\text{mean}}\,\Delta T$$

as illustrated in Fig. 4·15. Evaluating a mean specific heat requires the specification of a temperature interval, so a mean specific heat is not a property in the thermodynamic sense because its value does not depend only on the state of the system.

4. Use computer software (such as model IGPROP included with this book) for online retrieval of h, u, and other temperature-dependent properties. This method obviates interpolation and copying of values.
5. Use published tables such as those shown in abridged form in Tables A·12, A·15, and A·16. Such tables have been developed from equations similar to those given in Table A·1. (The use of gas tables for further purposes is discussed in Chapter 6 following the introduction of a property known as *entropy*. Entropy is defined on the basis of the second law of thermodynamics, which is introduced in the next chapter.)

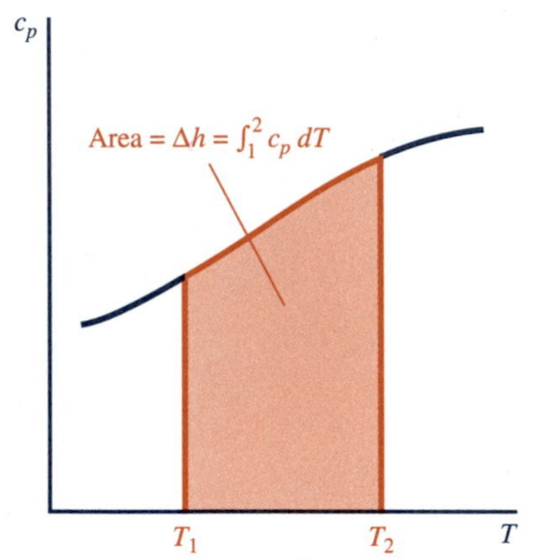

Figure 4·14 Δh of an ideal gas.

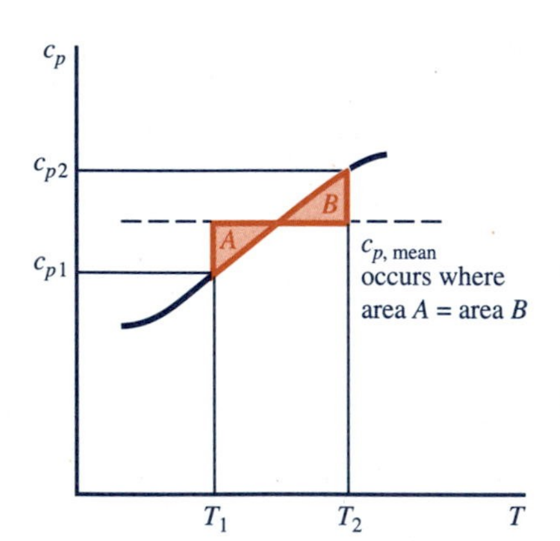

Figure 4·15 Determination of mean specific heat.

Example 4·6 Cooling of Carbon Dioxide

PROBLEM STATEMENT

Determine the rate of heat removal required to cool carbon dioxide in a steady-flow, constant-pressure, quasiequilibrium process from 700 to 40°C at a rate of 45 kg/h. The pressure of CO_2 passing through the heat exchanger is 110 kPa.

SOLUTION

Analysis: We first make a sketch of the steady-flow system and put the data on it. We can determine the rate of heat transfer from

$$\dot{Q} = \dot{m}q$$

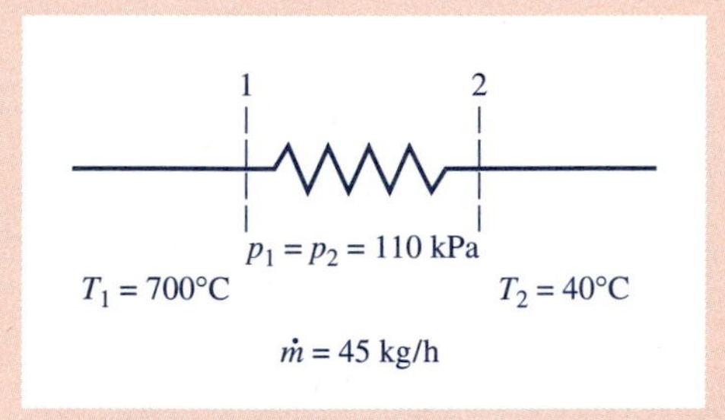

The value of $\dot{m}$ is known and we can find q from the first law:

$$q = h_2 - h_1 + \frac{V_2^2 - V_1^2}{2} + g(z_2 - z_1) + w$$

No work is done in a heat exchanger and the change in potential energy is surely negligible. We have no information on velocities or flow areas, so we will make the calculation assuming the change in kinetic energy is negligible. Thus the first law becomes

$$q = h_2 - h_1$$

and we need only to evaluate the enthalpy change. This can be done by several methods.

Calculation of $h_2 - h_1$ based on integration of $\int c_p\, dT$: We can integrate $\int c_p\, dT$ by using the polynomial expression for T as shown by Eq. 4·10 and in Table A·1 as

$$q = h_2 - h_1 = \int_1^2 c_p\, dT = \int_1^2 R(a + bT + cT^2 + dT^3 + eT^4)\, dT$$

$$= R\left[a(T_2 - T_1) + \frac{b}{2}(T_2^2 - T_1^2) + \frac{c}{3}(T_2^3 - T_1^3) + \frac{d}{4}(T_2^4 - T_1^4) + \frac{e}{5}(T_2^5 - T_1^5)\right]$$

Substituting values of $T_1 = 973$ K, $T_2 = 313$ K, and from Table A·1 the low-temperature-range values $a = 2.227$, $b = 9.991 \times 10^{-3}\ \text{K}^{-1}$, $c = -9.802 \times 10^{-6}\ \text{K}^{-2}$, $d = 5.397 \times 10^{-9}\ \text{K}^{-3}$, and $e = -1.281 \times 10^{-12}\ \text{K}^{-4}$ into this equation gives

$$q = -713 \text{ kJ/kg}$$

Calculation of $h_2 - h_1$ based on mean value of c_p: Estimating a mean specific heat for CO_2 between 313 and 973 K from Fig. 4·10 as 1.08 kJ/kg·K, we have

$$q = h_2 - h_1 = c_{p,\text{mean}}(T_2 - T_1) = 1.08(313 - 973) = -713 \text{ kJ/kg}$$

Calculation of $h_2 - h_1$ based on tabular values of h: Linearly interpolating in Table A·15 gives for $T_1 = 973$ K, $h_1 = 938.8$ kJ/kg, and for $T_1 = 313$ K, $h_2 = 225.9$ kJ/kg so that

$$q = h_2 - h_1 = 225.9 - 938.8 = -712.9 \text{ kJ/kg}$$

Calculation of $h_2 - h_1$ using an online computer data model for values of h: In this case, we enter the equation,

$$q = h_2 - h_1$$

We load the appropriate model for ideal-gas properties, enter the known values for T_1 and T_2, and solve to get the result displayed on both the TK Solver and Mathcad printouts, $q = -713$ kJ/kg.

```
═══════════════════ RULE SHEET ══════════════════════════════════
S Rule────────────────────────────────────────────────────────────
  ;────────────── Example 4·6 ─────────────────────────────────────
  q = h2 - h1       ;First law applied to the heat exchanger modeled as a
                    ;steady-flow system with no change in potential and
                    ;kinetic energies

  ;────────────── Ideal gas properties ─────────────────── IGPROP.TK
  call IGPR (gas, T1, h1, phi1, pr1)
  call IGPR (gas, T2, h2, phi2, pr2)

═══════════════════ VARIABLE SHEET ══════════════════════════════
St Input─── Name─── Output─── Unit─── Comment────────────────────
                              ─────── * E4·6.TK ─── Example 4·6 ──
            h1      938.4     kJ/kg   Enthalpy, state 1
            h2      225.4     kJ/kg   Enthalpy, state 2
            q       -713      kJ/kg   Heat

                              ─ Model: * IGPROP.TK ─ Ideal gas properties ─
   'CO2     gas                       * Dive (>) for list of input choices
   973      T1                K       Temperature, state 1
   313      T2                K       Temperature, state 2
```

TK Solver solution

Mathcad® 5.0 Solution

Background relationship to calculate the enthalpy of CO_2:

Constants for CO_2 :

$a := 2.227$

$b := 9.991 \cdot 10^{-3} \cdot K^{-1}$

$c := -9.802 \cdot 10^{-6} \cdot K^{-2}$

$d := 5.397 \cdot 10^{-9} \cdot K^{3}$

$e := -1.281 \cdot 10^{-12} \cdot K^{-4}$

$R_{CO2} := 0.189 \cdot \frac{kJ}{kg \cdot K}$

Additional Definitions

$kPa \equiv 1000 \cdot Pa$

$kJ \equiv 1000 \cdot joule$

$s \equiv 1 \cdot sec$

$h \equiv 1 \cdot hr$

Enthalpy as a function of temperature:

$$h_{CO2}(T) := R_{CO2} \cdot \left(a \cdot T + \frac{b}{2} \cdot T^2 + \frac{c}{3} \cdot T^3 + \frac{d}{4} \cdot T^4 + \frac{e}{5} \cdot T^5 \right)$$

Solution:

Note: Only the region below this line would appear for the online solution.

Given:

$T_1 := 973 \cdot K \quad T_2 := 313 \cdot K$

Heat required to cool carbon dioxide:

$q := h_{CO2}(T_2) - h_{CO2}(T_1)$

$q = -713 \cdot \frac{kJ}{kg}$

Mathcad solution. (*Mathcad is a registered trademark of MathSoft, Inc., Cambridge, Massachusetts. Used by permission of MathSoft, Inc.*)

Calculation of $\dot{Q}$: Using the given mass rate of flow and the value of q obtained by any one of the methods shown, we find that the rate of heat transfer is

$$\dot{Q} = \dot{m}q = 45(-713) = -32{,}100 \text{ kJ/h}$$

Comment: In this case, the four methods for calculating $h_2 - h_1$ give identical results. Often, however, linear interpolation in tables and the difficulty in estimating the mean value of c_p from a chart like Fig. 4·10 or A·4 introduce inaccuracies.

Example 4·7 Jet Engine Compressor

PROBLEM STATEMENT

Under certain operating conditions, the compressor of a small aircraft jet engine takes in air at 95 kPa, 25°C, at a rate of 300 m³/min at a very low velocity. At the compressor discharge, the air is at 200 kPa, 120°C. The discharge cross-sectional area is 0.028 m². Determine the power required to drive the compressor.

SOLUTION

Analysis: We first make a sketch of the system and put the data on it. Since a flow rate is specified and single values of properties are given at the inlet and the outlet, we assume that the flow is steady. We can calculate the power input from

$$\dot{W}_{in} = \dot{m} w_{in}$$

and we can calculate w_{in} from the first law:

$$w_{in} = h_2 - h_1 + \frac{V_2^2 - V_1^2}{2} + g(z_2 - z_1) - q$$

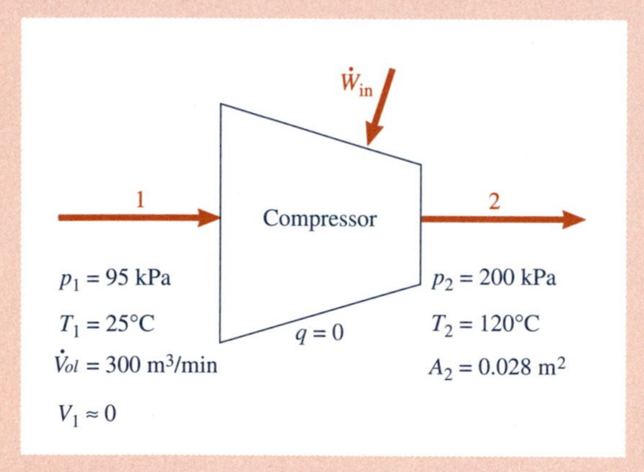

For a jet engine compressor the change in potential energy is negligible. It is reasonable to model the process as adiabatic ($q = 0$) because the flow passage is short, its ratio of surface area to volume is low, the mass rate of flow per cross-sectional area is high so that the heat transfer per unit mass would be small, and the temperature difference between the air passing through the compressor and the surroundings is relatively small.

We can find the enthalpy difference by any of the methods used in Example 4·6. V_2 can be found from

$$V_2 = \frac{\dot{m} v_2}{A_2}$$

v_2 needed here can be obtained from the ideal-gas equation of state, which is quite accurate for air at approximately 2 atm, 120°C,

$$v_2 = \frac{RT_2}{p_2}$$

Thus we see that we can determine w_{in}.

The flow rate $\dot{m}$ can be obtained from

$$\dot{m} = \frac{\dot{Vol}_1}{v_1}$$

where $\dot{Vol}_1$ is the volume rate of flow at the inlet that is specified. v_1 can be obtained from the ideal-gas equation of state as v_2 was.

Now we see that we can solve the problem. If we are to do the solution sequentially (as in a longhand solution), we will start with the last equation and go in reverse order through the equations displayed above.

Calculations (longhand as one alternative): Following the procedure outlined in the analysis, we obtain

$$v_1 = \frac{RT_1}{p_1} = \frac{0.287 \dfrac{\text{kJ}}{\text{kg·K}}(25 + 273.15)\text{ K}}{95 \text{ kPa}} = 0.901 \text{ m}^3\text{/kg}$$

$$v_2 = \frac{RT_2}{p_2} = \frac{0.287(120 + 273.15)}{200} = 0.564 \text{ m}^3\text{/kg}$$

$$\dot{m} = \frac{\dot{Vol}_1}{v_1} = \frac{300 \text{ m}^3\text{/min}}{60 \text{ s/min}(0.901 \text{ m}^3\text{/kg})} = 5.55 \text{ kg/s}$$

$$V_2 = \frac{\dot{m}v_2}{A_2} = \frac{5.55(0.564)}{0.028} = 111.8 \text{ m/s}$$

Then, obtaining enthalpy values from the air table, Table A·12, we substitute numerical values into the first law equation:

$$\begin{aligned} w_{in} &= h_2 - h_1 + \frac{V_2^2 - V_1^2}{2} \\ &= 396.1 - 300.4 + \frac{(111.8)^2 - 0 \text{ m}^2\text{/s}^2}{2(1000 \text{ m}^2\text{·kg/kJ·s}^2)} \\ &= 101.9 \text{ kJ/kg} \end{aligned}$$

$$\dot{W}_{in} = \dot{m}w_{in} = 5.55(101.9) = 566 \text{ kJ/s} = 566 \text{ kW}$$

Calculations (computer solution as another alternative): For this problem, a computer solution is unnecessary. However, solving some short problems using a computer is good preparation for more extensive calculations and for when a problem needs to be solved for an array of input variables. Even for a problem such as this one, an advantage of using a computer is that frequently used equations, with annotations, can be stored and retrieved to avoid rewriting. Thus you can spend more of your time analyzing the problem and less rewriting common equations.

The analysis of this problem shows that we will use

1. The first law
2. The relationship among work per unit mass, power, and mass rate of flow
3. The continuity equation
4. The ideal-gas equation of state
5. A relationship between enthalpy and temperature for air

The equations for items 1 through 4 are well known, so you can readily enter them into an equation-solving program. In this case, we save time by calling these equations from basic equation models that have already been prepared and stored in the software accompanying this book. Model FLSF (**F**irst **L**aw **S**teady **F**low) includes the first law equation for steady one-dimensional flow with one inlet and one outlet, the continuity equation for these conditions, and the relationship among $\dot{m}$, $\dot{W}$, and w. Model EQS-IG contains the ideal-gas equation of state. A relationship between h and T is contained in model AIRPROP. After loading these three models and removing any lines that do not apply to this problem, we have the equations, with annotations, on the Rule sheet as shown in the figure. All variables are automatically displayed on the Variable sheet, so we enter all known values as inputs and solve for the results shown on the Variable sheet.

```
====================== RULE SHEET ==========================================
S Rule--------------------------------------------------------------------
  ;----------------- Example 4·7 ------------------------------------------
  ;----------------- First law, steady flow ------------------ FLSF.TK
  q=delh + (V2^2-V1^2)/2 + delpe + w      ;First law, steady flow
  delh = h2 - h1
  Wdot=mdot*w                             ;Power, work, flow rate relation
  mdot = A2 * V2/v2                       ;Continuity equation
  mdot = vrf1/v1    ;Relating mass rate of flow to volume rate of flow

  ;----------------- Equation of state, ideal gas ------------ EQS-IG.TK
  p1 * v1 = R * T1                        ;Equation of state at inlet
  p2 * v2 = R * T2                        ;Equation of state at outlet

  ;----------------- Air properties, for air as an ideal gas ------- AIRPROP.TK
  call IGPR (gas, T1, h1, s01, pr1)       ;s0 and pr are extraneous here
  call IGPR (gas, T2, h2, s02, pr2)
```

```
═══════════════════════════ VARIABLE SHEET ═══════════════════════════════
St Input──── Name─── Output─── Unit────── Comment──────────────────────────
                               ────────── * E4-7.TK ──── Example 4·7 ─────
                               ── Model: * FLSF.TK ── First law, steady flow ──
   0         q                 kJ/kg      Heat transfer per unit mass
             delh     95.6     kJ/kg      Change in enthalpy
             V2       111.8    m/s        Velocity at outlet
   0         V1                m/s        Velocity at inlet
   0         delpe             kJ/kg      Change in potential energy
             w        -101.9   kJ/kg      Work per unit mass
             h2       396.1    kJ/kg      Enthalpy at outlet
             h1       300.4    kJ/kg      Enthalpy at inlet
             Wdot     -565.7   kJ/s       Power
             mdot     5.551    kg/s       Mass rate of flow
   0.028     A2                m^2        Cross-sectional area of outlet
   95        p1                kPa        Pressure at inlet
             v1       0.9007   m^3/kg     Specific volume at inlet
             v2       0.5642   m^3/kg     Specific volume at outlet
   200       p2                kPa        Pressure at outlet
   300       vrf1              m^3/min    Volume rate of flow at inlet
                               ── Model: * EQS-IG.TK ── Equation of state ────
   0.287     R                 kJ/(kg*K)  Gas constant for the gas
   25        T1                C          Temperature at inlet
   120       T2                C          Temperature at outlet
                               ── Model: * AIRPROP.TK ─── Air properties ────
```

Comment: The value for power (or $\dot{W}$) is negative, indicating that work is done *on* the air in the compressor. Is the value reasonable? At this stage, you may not be able to make an accurate estimate of the power required to drive such a compressor, but you can establish at least some wide limits. Start by noting that this is a *small* jet engine. If the exit area is circular, its diameter is only about the length of your hand, so this is not a jet engine for an airliner where the compressor power might be tens of thousands of kilowatts. At the other extreme, a compressor taking in 300 m^3/min of air and doubling its pressure obviously requires more power than a small lawn mower engine produces—perhaps 2 or 3 kW. Limits of 3 and 10,000 kW may seem absurdly broad, but they show how you can start to make estimates of magnitudes.

4·2·2 Ideal Gases with Constant Specific Heats—A Special Case

For ideal gases, c_p and c_v usually vary with temperature, although the difference $c_p - c_v = R$ is constant. There are two cases in which constant specific heats are used:

For monatomic gases, specific heats are constant.

1. For monatomic gases, specific heats are constant as indicated by Figs. 4·10–4·12.
2. For other gases in limited temperature ranges, specific heats are nearly constant. For example, as the data in the appendix show, between 0 and 100°C, the variation in c_p of air, nitrogen, and carbon monoxide is less than 1 percent, and for oxygen and hydrogen it is less than 2 percent. (For this temperature range c_p of carbon dioxide, however, changes by more than 11 percent, and for hydrocarbon gases the variation may be much greater.) Air-conditioning systems and many other gas-handling systems operate within limited temperature ranges, so using constant specific heats is accurate and simplifies some calculations.

Even for cases when the variation of specific heats with temperature is significant, constant specific heats may be used for ''scoping'' calculations, rough estimates, or calculations where only a first order of accuracy (i.e., a low accuracy) is needed. Wisely deciding what order of accuracy is needed or justified in a calculation may require well-developed engineering judgment. We will return to this point in later chapters.

When the specific heats of an ideal gas are constant, several relationships are simplified. For example,

$$\Delta u = \int c_v \, dT \quad \text{and} \quad \Delta h = \int c_p \, dT \tag{4·6, 4·8}$$

are simplified to

Simplified equations for Δu and Δh when specific heats are constant.

$$\Delta u = c_v \, \Delta T \quad \text{and} \quad \Delta h = c_p \, \Delta T \tag{4·11, 4·12}$$

4·2·3 Quasiequilibrium Adiabatic Process of Ideal Gases with Constant Specific Heats

The quasiequilibrium adiabatic process is frequently encountered in thermodynamic analyses. Therefore, it is worthwhile to establish a pv relationship for this process for use in integrating $\int p\, dv$ and $\int v\, dp$. The pv relationship can also be combined with the equation of state to give pT and vT relationships that are useful. These steps are easily performed for an ideal gas with constant specific heats, so we will now derive the pv relationship for a *quasiequilibrium adiabatic process of an ideal gas with constant specific heats.*

For a closed system the first law may be written

$$\delta q = du + \delta w$$

Derivation of the pv relationship for a quasiequilibrium adiabatic process of an ideal gas with constant specific heats

If the process is *adiabatic*, this becomes

$$0 = du + \delta w$$

and if it is also a *quasiequilibrium process of a simple compressible substance*, we have

$$0 = du + p\,dv$$

If the substance is an *ideal* gas,

$$0 = c_v\,dT + p\,dv$$

To get a pv relation, we must eliminate T. We can do this by use of the ideal-gas equation of state, which gives $T = pv/R$ and $dT = (p\,dv + v\,dp)/R$. Thus, we have

$$0 = \frac{c_v}{R}(p\,dv + v\,dp) + p\,dv$$

that we rearrange as follows:

$$0 = v\,dp + \left(\frac{R}{c_v} + 1\right)p\,dv = v\,dp + \left(\frac{R + c_v}{c_v}\right)p\,dv$$

$$0 = v\,dp + \frac{c_p}{c_v}p\,dv = v\,dp + kp\,dv$$

$$0 = \frac{dp}{p} + k\frac{dv}{v}$$

We have not yet imposed the restriction of constant specific heats, but now we do so in order to integrate the last expression. Recall that the difference between c_p and c_v is constant (and equal to R), so that a constant k implies constant values of c_p and c_v. Integration gives

pv relationship for a quasiequilibrium adiabatic process of an ideal gas with constant specific heats

$$pv^k = \text{constant} \tag{4·13a}$$

This equation holds for a *quasiequilibrium adiabatic process of an ideal gas with constant specific heats.* For such a process between states 1 and 2

$$p_1v_1^k = p_2v_2^k \tag{4·13b}$$

Although the derivation above was based on a closed system, the pv relation is the same if the quasiequilibrium adiabatic process occurs in an open system. Now we develop the corresponding pT and vT relationships. From the ideal-gas equation of state we have

$$\frac{p_1v_1}{T_1} = \frac{p_2v_2}{T_2}$$

which when combined with Eq. 4·13*b* gives

Useful relationships for a quasiequilibrium adiabatic process of an ideal gas with constant specific heats

$$\frac{T_2}{T_1} = \left(\frac{p_2}{p_1}\right)^{(k-1)/k} = \left(\frac{v_1}{v_2}\right)^{k-1} \tag{4·13c}$$

Equations 4·13 are *pv*, *pT*, and *Tv* relations that apply only to a particular type of process of an ideal gas with constant specific heats; they are not equations of state. They involve only two properties at a time. Since for a pure substance there are generally two *independent* properties, an equation of state must involve three properties.

Other Polytropic Processes. For some quasiequilibrium processes of an ideal gas the *pv* relation is

$$pv^n = \text{constant} \tag{4·14a}$$

where *n* is a constant for each process. A process represented by such an equation is called a **polytropic** process and *n* is called the *polytropic exponent* for the process; *n* is *not* a property of the gas. Combining Eq. 4·14*a* with the equation of state gives

$$\frac{T_2}{T_1} = \left(\frac{p_2}{p_1}\right)^{(n-1)/n} = \left(\frac{v_1}{v_2}\right)^{n-1} \tag{4·14b}$$

Polytropic processes

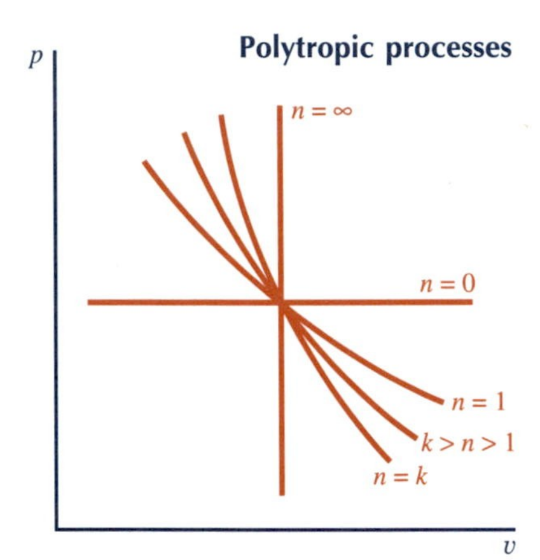

Figure 4·16 Polytropic processes of an ideal gas.

Notice that the quasiequilibrium constant-pressure, constant-volume, constant-temperature, and adiabatic processes are special cases of polytropic processes for which the values of *n* are 0, ∞, 1, and *k*, respectively. (If the reason for $n = \infty$ in the constant-volume case is not apparent, consider that when $n = \infty$, $p^{1/\infty}v = \text{constant}$; hence $v = \text{constant}$.) Various polytropic processes are shown on a *pv* diagram in Fig. 4·16. During a polytropic expansion for which $k > n > 1$, heat is added to the gas but its temperature decreases because the work done by the system exceeds the heat added.

Example 4·8 Compression of Neon

PROBLEM STATEMENT

Neon in a closed system is to be compressed adiabatically in a quasiequilibrium process from 96 kPa, 20°C, to 396 kPa. The initial volume is 0.240 m^3. Determine (*a*) the final volume, (*b*) the final temperature, and (*c*) the work.

SOLUTION

Analysis: We first make a labeled sketch of the system and a property (*pv*) diagram. Since the compression is a quasiequilibrium adiabatic process of an ideal gas with constant specific heats, we can calculate the final

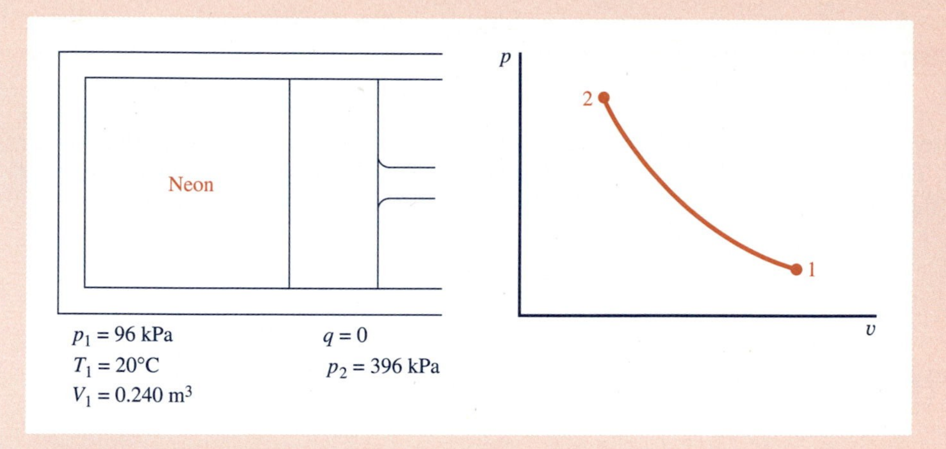

volume and temperature from equations such as pv^k = constant (or pV^k = constant) and the ideal-gas equation of state. We can then use the first law to determine the work, because $q = 0$ and the internal energy change can be calculated from the temperature change.

Solution: The final volume is calculated from the pv (or pV) relationship for a quasiequilibrium adiabatic process of an ideal gas with constant specific heats:

(*a*)

$$p_1 V_1^k = p_2 V_2^k \qquad (A)$$

$$V_2 = V_1 \left(\frac{p_1}{p_2}\right)^{1/k} = 0.240 \left(\frac{96}{396}\right)^{1/1.667} = 0.1026 \text{ m}^3$$

(*b*) Eliminating V's by combining Eq. *A* and the equation

$$\frac{p_1 V_1}{T_1} = \frac{p_2 V_2}{T_2}$$

gives

$$T_2 = T_1 \left(\frac{p_2}{p_1}\right)^{(k-1)/k} = 293 \left(\frac{396}{96}\right)^{(1.667-1)/1.667} = 516.5 \text{ K} = 243°\text{C}$$

(*c*)

$$\begin{aligned} W_{\text{in}} &= U_2 - U_1 - Q = m(u_2 - u_1) - 0 \\ &= mc_v(T_2 - T_1) = \frac{p_1 V_1}{RT_1} c_v (T_2 - T_1) \\ &= \frac{p_1 V_1}{R} c_v \left(\frac{T_2}{T_1} - 1\right) = \frac{96(0.240)}{0.412} 0.619 \left(\frac{516.6}{293} - 1\right) \\ &= 26.4 \text{ kJ} \end{aligned}$$

where we used the ideal-gas equation of state to find the mass and obtained values of R and c_v from Table A·1.

Example 4·9 Steady-Flow Compression of Nitrogen

PROBLEM STATEMENT

Nitrogen is compressed steadily at a rate of 0.46 kg/s from 100 kPa, 20°C, to a final pressure that varies between 160 and 320 kPa. The compression is polytropic with a polytropic exponent of 1.32. Determine (*a*) the final temperature, (*b*) the work per kilogram, and (*c*) the heat transfer per kilogram as functions of the final pressure.

SOLUTION

Analysis: We first make a labeled sketch of the system and a *pv* diagram. In the pressure range specified, nitrogen can be modeled as an ideal gas. We will assume that the temperature range is small enough that we can use constant specific heats, and we will check this assumption after getting the results.

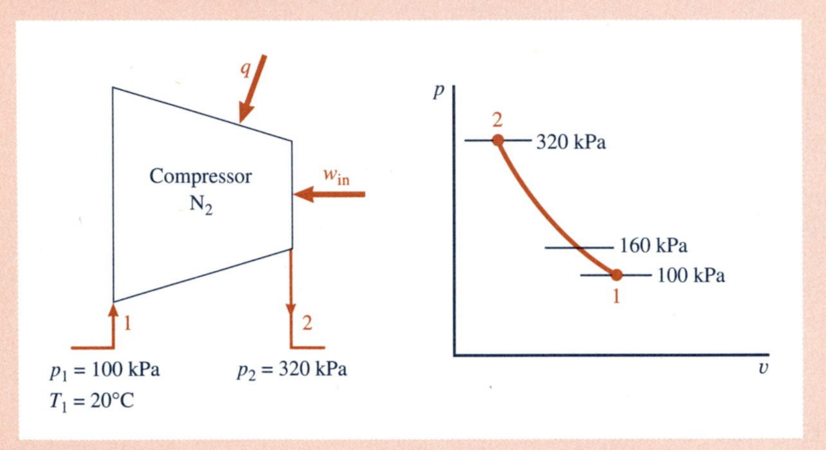

By combining the ideal-gas equation of state with the *pv* relationship for a polytropic process, $pv^n =$ constant, we can solve for the final temperature,

$$T_2 = T_1\left(\frac{p_2}{p_1}\right)^{(n-1)/n}$$

The work for this steady-flow quasiequilibrium process can be obtained from $w = -\int v\,dp - \Delta ke - \Delta pe$. We assume that the changes in kinetic and potential energies are negligible. We can perform the integration because we have a relationship between p and v:

$$w_{\text{in}} = \int_1^2 v\,dp = \int_1^2 v_1\left(\frac{p_1}{p}\right)^{1/n} dp = v_1 p_1{}^{1/n}\int_1^2 p^{-1/n}\,dp$$

$$= p_1^{1/n} v_1 \frac{n}{n-1}[p_2^{(n-1)/n} - p_1^{(n-1)/n}] = p_1 v_1 \frac{n}{n-1}\left[\left(\frac{p_2}{p_1}\right)^{(n-1)/n} - 1\right]$$

$$= RT_1 \frac{n}{n-1}\left[\left(\frac{p_2}{p_1}\right)^{(n-1)/n} - 1\right]$$

We can then find q values from the first law,

$$q = h_2 - h_1 + w$$

because we will have calculated values of w and can calculate enthalpy changes from $\Delta h = c_p \, \Delta T$. We can quickly make these calculations longhand for any given discharge pressure, but since we want results for several values of discharge pressure, we will use a computer equation-solving program.

Solution: We enter all of the equations mentioned in the analysis above on the Rule sheet as shown in the figure. On the Variable sheet we indicate that p_2, T_2, w, and q all will have lists of values. We enter for the list of pressures values of 160, 200, 240, 280, and 320 kPa. We present the results in both a table and graphs.

```
=================== RULE SHEET ==========================
S Rule------------------------------------------------------
  ;------------------ Example 4·9 -------------------------
  q = delh - win                ;First law for steady flow, delke = delpe = 0
  delh = cp * (T2 - T1)         ;Enthalpy change for ideal gas with
                                ;          constant specific heats
  T2/T1=(p2/p1)^((n-1)/n)       ;Expression from combining pv^n = C and pv/T = R
  win = R * T1 * (n/(n-1))*((p2/p1)^((n-1)/n) - 1)      ;Integral of vdp
```

```
=================== VARIABLE SHEET ======================
St Input---- Name--- Output--- Unit------ Comment-----------------------
                               ---------- * E4-9.TK ----- Example 4·9 ----
L            T2      55.4      C          Discharge temperature
   20        T1                C          Inlet temperature
L  160       p2                kPa        Discharge pressure
   100       p1                kPa        Inlet pressure
   1.32      n                            Polytropic exponent
L            win     43.3      kJ/kg      Work input
   0.297     R                 kJ/(kg*K)  Gas constant
L            delh    36.8      kJ/kg      Enthalpy change
   1.04      cp                kJ/(kg*K)  Specific heat at constant pressure
L            q       -6.51     kJ/kg      Heat transfer
```

p2, kPa	T2, °C	win, kJ/kg	q, kJ/kg
160	55.4	43.3	-6.51
200	73.6	65.7	-9.88
240	89.3	84.9	-12.8
280	103	102	-15.3
320	115	117	-17.6

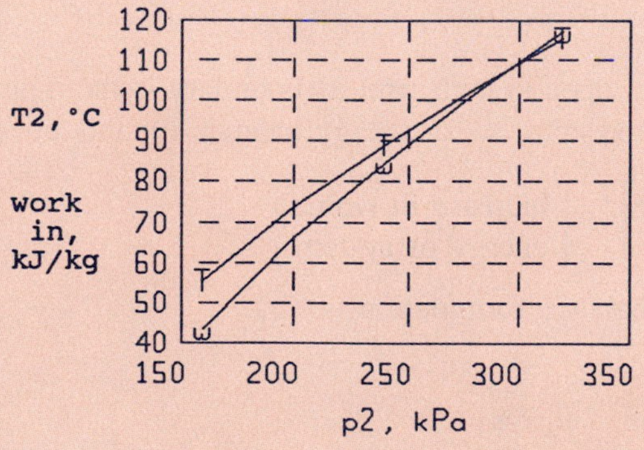

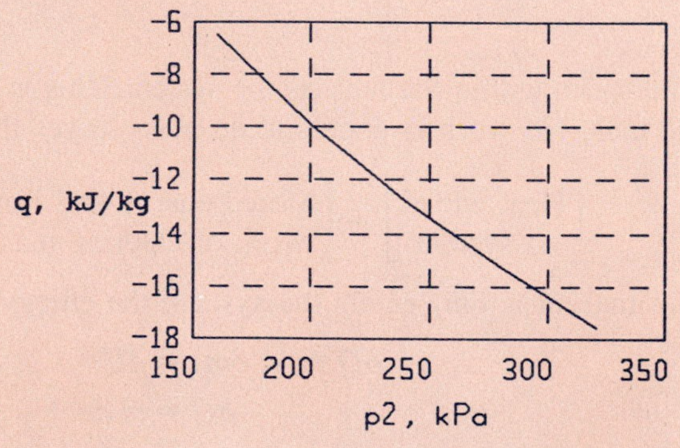

T2, °C

work in, kJ/kg

120 110 100 90 80 70 60 50 40

150 200 250 300 350

p2, kPa

q, kJ/kg

-6 -8 -10 -12 -14 -16 -18

150 200 250 300 350

p2, kPa

Comment: We note that the maximum value of T_2 is 115°C. The data in the appendix (Table A·3 or Chart A·4) for specific heat of nitrogen show that in the temperature range involved here the assumption of constant specific heats is reasonable.

Example 4·10 Isothermal Tank Filling

PROBLEM STATEMENT

A 60-cu-ft tank contains carbon monoxide initially at 20 psia, 100 F. A compressor delivers carbon monoxide to the tank, raising the pressure from 20 to 40 psia. As the compressor discharge pressure increases, so does the discharge temperature. The discharge temperature and pressure are related by $T = 80p^{0.25}$, where T is in Rankine and p is in psfa. Determine the heat transfer necessary to maintain the temperature of the carbon monoxide in the tank at 100 F.

SOLUTION

First a sketch of the compressor and tank is made, and then we define the following symbols:

- Subscript i denotes properties of the CO initially in the tank.
- Subscript f denotes properties of the CO finally in the tank.
- Subscript 1 denotes properties of the CO entering the tank.
- m, p, T, U, etc. = the mass, pressure, temperature, internal energy, etc. of CO in the tank at any instant.
- m_1 = the mass of CO that has already entered the tank $m_1 = m - m_i$.

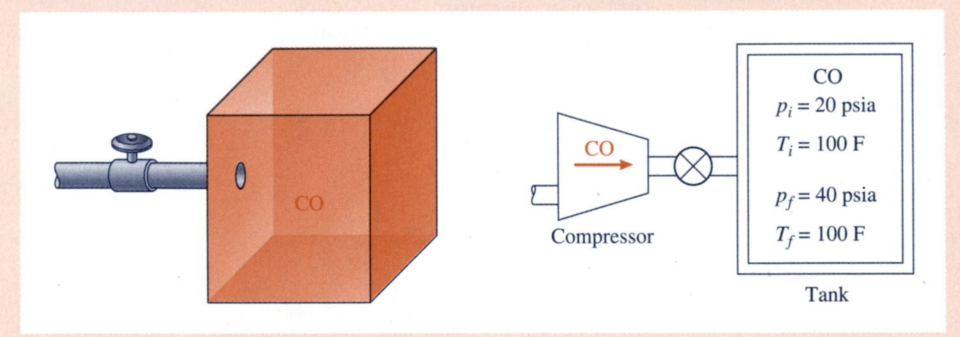

Consider the space within the tank as the system. This is an open system, and its boundary is the inner surface of the tank. No work is done and no mass leaves the system; so an energy balance for this open system is

$$\begin{bmatrix}\text{Heat added}\\ \text{to system}\end{bmatrix} + \begin{bmatrix}\text{Stored energy and flow}\\ \text{work of entering mass}\end{bmatrix} = \begin{bmatrix}\text{Increase in stored}\\ \text{energy of system}\end{bmatrix}$$

As an infinitesimal mass δm_1 enters the system, the energy balance formulation is

$$\delta Q + h_1\,\delta m_1 = dU$$

$$\delta Q = m\,du + u\,dm - h_1\,\delta m_1$$

Since the temperature of the CO in the tank is held constant, u is constant, and therefore $m\,du = 0$:

$$\delta Q = u\,dm - h_1\,\delta m_1$$

We must evaluate u and h_1.

Let us first see if the temperature range is narrow enough for specific heats to be constant. The compressor discharge (i.e., tank entrance) temperatures for the 20- and 40-psia pressure limits are

$$T_{1,\,20\text{ psia}} = 80p_1^{0.25} = 80[20(144)]^{0.25} = 586\text{ R} = 126\text{ F}$$
$$T_{1,\,40\text{ psia}} = 80p_1^{0.25} = 80[40(144)]^{0.25} = 697\text{ R} = 237\text{ F}$$

Therefore, the CO temperature varies from 100 to 237 F. In this range we see from Table A·3 and Chart A·4 that we can treat the specific heats as constant. (We will use $c_p = 0.250$ B/lbm·R and $c_v = 0.179$ B/lbm·R.) Thus we have $u - u_0 = c_v(T - T_0)$ and $h - h_0 = c_p(T - T_0)$. For this problem, let $u_0 = 0$ and $h_0 = 0$ when $T_0 = 0$, so that we have $u = c_vT$ and $h = c_pT$. The energy balance now becomes

$$\begin{aligned}\delta Q &= c_vT\,dm - c_pT_1\,\delta m_1\\ &= c_vT\,dm - c_p(80)p_1^{0.25}\,\delta m_1\end{aligned}$$

But we know that

$$p_1 = p = \frac{mRT}{V}$$

and

$$\delta m_1 = d(m - m_i) = dm = d\left(\frac{pV}{RT}\right) = \frac{V}{RT}\,dp$$

Making these two substitutions in the energy balance above gives

$$\delta Q = c_v\frac{V}{R}\,dp - 80c_pp^{0.25}\frac{V}{RT}\,dp$$

$$\begin{aligned}Q &= \frac{V}{R}\left(c_v\int_i^f dp - \frac{80c_p}{T}\int_i^f p^{0.25}\,dp\right)\\ &= \frac{V}{R}\left[c_v(p_f - p_i) - \frac{80c_p}{1.25T}(p_f^{1.25} - p_i^{1.25})\right]\\ &= \frac{60\text{ ft}^3}{55.2\,\dfrac{\text{ft·lbf}}{\text{lbm·R}}}\left\{0.179\,\frac{\text{B}}{\text{lbm·R}}\left(144\,\frac{\text{in}^2}{\text{ft}^2}\right)\left(40 - 20\,\frac{\text{lbf}}{\text{in}^2}\right)\right.\\ &\quad\left. - \frac{80\,\dfrac{\text{R}}{(\text{lbf/ft}^2)^{0.25}}\left(0.25\,\dfrac{\text{B}}{\text{lbm·R}}\right)}{1.25(560\text{ R})}\left[144^{1.25}(40^{1.25} - 20^{1.25})\left(\frac{\text{lbf}}{\text{ft}^2}\right)^{1.25}\right]\right\}\\ &= -343\text{ B}\end{aligned}$$

The minus sign indicates that heat must be removed from the CO.

4·2·4 Kinetic Theory and Statistical Mechanics Explanations

Although classical thermodynamics does not depend on assumptions regarding the microscopic structure of matter, a molecular description will help you understand why the internal energy of an ideal gas is a function of temperature only.

An ideal gas is sometimes described microscopically as a gas in which there are no forces among the molecules. Before discussing molecules, however, let us first consider an analogy. Consider a body that has a mass of 1 kg. At some locations on the surface of the earth this body weighs 9.8 N, and the work required to lift it 1 m is 9.8 N·m. Thus, as the body is moved 1 m farther from the earth, its potential energy increases by 9.8 N·m, or 9.8 J. This same body when at an elevation of 100 km above the surface of the earth weighs only about 9.5 N, so only 9.5 N·m of work is required to move it 1 m farther from the earth. As the body moves farther from the earth, less work is required to move it through each meter of travel because the gravitational force continues to decrease. Eventually, a negligible amount of work will be required to move the body farther from the earth. Then the body can be moved through space with only negligible changes in its potential energy.

Now consider the molecules of an ideal gas. It is the sum of their potential and kinetic energies that comprises the internal energy of the gas. Like the body and the earth we have just discussed, the molecules of an ideal gas are so far apart that the forces among them are negligibly small. The distances between molecules are so great that even if the distances are halved or doubled, there is no appreciable change in the potential energy of the molecules. Thus the internal energy of an ideal gas is independent of the gas volume. When the gas temperature is changed, however, the velocity and the kinetic energy of the molecules change, so the internal energy is a function of temperature.

From kinetic theory we can derive the ideal-gas equation of state by applying principles of mechanics to individual molecules. We assume that the molecules are spaced so widely apart that they exert no forces on each other except when they collide. The potential energy of the molecules remains unchanged. Only their kinetic energy changes.

As noted earlier, Figs. 4·11 and 4·12 show that the specific heats of monatomic gases are constant and for other gases specific heats increase monotonically with temperature. This is an important observation, and kinetic theory provides at least a qualitative explanation.

To determine specific heats we assume that all the internal energy of a gas is in the kinetic energy of the molecules. Calculating the kinetic energy of molecules is based on two ideas from kinetic theory: the concept of degrees of freedom and the equipartition of energy principle. The number of degrees of freedom f refers to the number of independent quantities that must be specified to determine the energy of a molecule. For the purposes of this discussion, we consider only translational, rotational, and vibrational degrees of freedom. For example, a molecule constrained to move in only the x

direction has one degree of freedom because if V_x is specified then the energy can be calculated as $mV_x^2/2$. If the molecule is free to move in any direction, then V_x, V_y, and V_z must be specified, and the molecule has three degrees of freedom. If the moment of inertia of the molecule about the x axis is not zero and the molecule rotates about this axis, then there is rotational kinetic energy and the angular velocity must be specified, adding another degree of freedom. Rotation about other axes adds other degrees of freedom, and vibration of the atoms within a molecule adds still others.

Statistical mechanics' equipartition of energy principle states that when only translational, rotational, and vibrational degrees of freedom are considered, the total energy of molecules is divided equally among their degrees of freedom and the internal energy is given by

General equations for specific heat derived from kinetic theory in terms of the number of degrees of freedom.

$$\bar{u} = \frac{f}{2}\bar{R}T$$

Thus

$$\bar{c}_v = \left(\frac{\partial \bar{u}}{\partial T}\right)_v = \frac{f}{2}\bar{R}$$

$$\bar{c}_p = \bar{c}_v + \bar{R} = \frac{f}{2}\bar{R} + \bar{R} = \frac{f+2}{2}\bar{R}$$

and

$$k = \frac{\bar{c}_p}{\bar{c}_v} = \frac{f+2}{f}$$

Monatomic molecule:

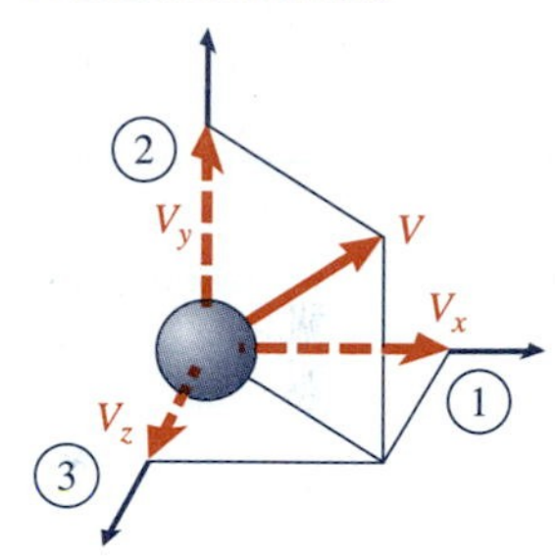

For a monatomic gas, kinetic energy involves only the three linear velocity components along three spatial axes, and on this basis we can calculate that

$$\bar{c}_v = \frac{3}{2}\bar{R} \qquad \bar{c}_p = \frac{5}{2}\bar{R} \qquad k = \frac{5}{3}$$

for a monatomic gas (see Ref. 4·3 or 4·4).

Consider now a diatomic molecule composed of two point masses. In addition to translational kinetic energy, it may possess rotational kinetic energy by virtue of rotation about the two mutually perpendicular axes that are perpendicular to the line joining the two atoms. Energy of rotation about the axis connecting the atoms is zero because the moment of inertia about this axis is zero for point particles. The atoms may also vibrate along the line connecting them. This involves two additional degrees of freedom because both potential and kinetic energy are involved. (The atoms within a molecule exert forces on each other even though the intermolecular forces are negligible.) Both position and velocity of the atoms must be specified in order to determine the vibrational energy. Thus a diatomic molecule may have seven degrees of freedom: three associated with translation, two with rotation, and two with vibration. However, statistical mechanical reasoning suggests that vibration occurs only at higher temperatures, so at normal room temperature, for example, a diatomic gas usually has only five degrees of freedom. At very

Diatomic molecule:

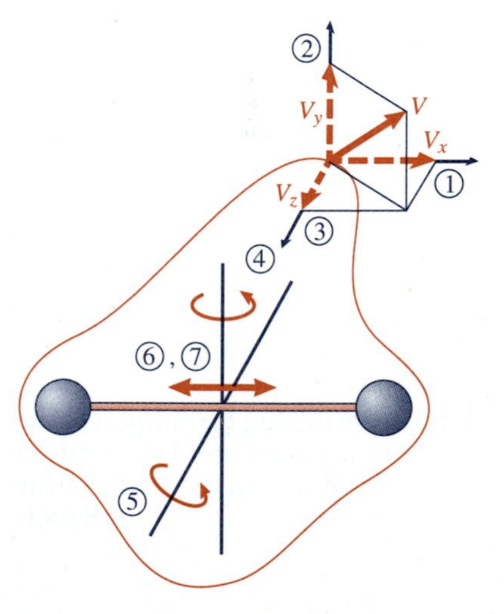

low temperatures rotation ceases, so $f = 3$. Thus, for a diatomic gas, kinetic theory leads us to expect

Kinetic theory result for the specific heat of a diatomic gas

At low temperatures: $\bar{c}_v = \frac{3}{2}\bar{R}$ $\quad \bar{c}_p = \frac{5}{2}\bar{R}$ $\quad k = \frac{5}{3}$

At intermediate temperatures: $\bar{c}_v = \frac{5}{2}\bar{R}$ $\quad \bar{c}_p = \frac{7}{2}\bar{R}$ $\quad k = \frac{7}{5}$

At high temperatures: $\bar{c}_v = \frac{7}{2}\bar{R}$ $\quad \bar{c}_p = \frac{9}{2}\bar{R}$ $\quad k = \frac{9}{7}$

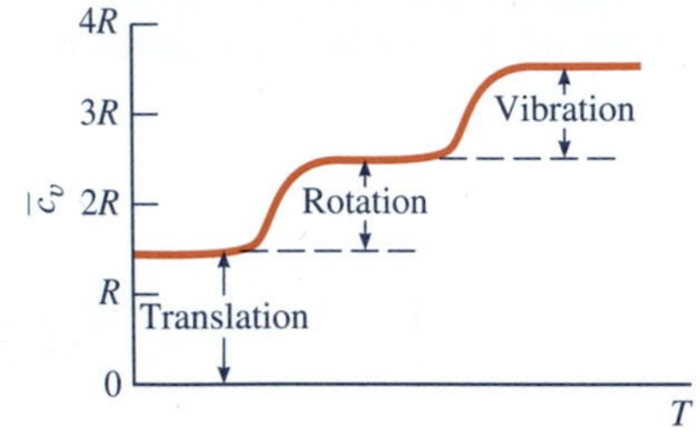

Figure 4·17 c_v variation with temperature according to the kinetic theory model of an ideal gas.

Such variation of $\bar{c}_v$ is shown in Fig. 4·17. It has been proposed that the reason for the gradual change shown in Fig. 4·17 is that, as the temperature increases, not all the molecules begin to rotate or vibrate at the same temperature. As more molecules begin to rotate and then to vibrate, specific heat increases gradually. Quantum mechanics provides a more complete and rigorous explanation.

The simplified approach presented above has numerous shortcomings. It does not explain the specific heat variation of gases with more complex molecules. Even for monatomic and diatomic gases, the explanation given above is somewhat inaccurate because the model is oversimplified. Overall, quantum mechanics provides a much more accurate model.

4·3 Sources of Property Data

The variety of substances encountered in engineering practice is unlimited, so an engineer must be able to find reliable property data on many, some common, some uncommon. We list here several sources:

- For widely used substances, textbooks like this one give abbreviated property data as well as references to more extensive data.
- Handbooks (Refs. 4·5 through 4·9) give data on a wider variety of substances but often for limited data ranges. They also provide references to more extensive data for each substance.
- Collections of thermophysical property data for various substances can be found in any technical library. References 4·10 through 4·19 are samples of such collections, including some that are available in computer programs. When searching in a library, use, in addition to the name of the substance, key words such as *data*, *properties*, *physical*, and *thermophysical.*

When searching for property data in a library, use key words in addition to the name of the substance.

- Some organizations have extensive continuing projects for periodically publishing property data. First-time users can be overwhelmed by the apparent complexity of some of these publications, but practice makes them simple to use. These publications often provide valuable information on data sources and sometimes evaluate the data. Evaluations are important when data from different sources disagree. Several such organizations are listed in the Sources of Property Data at the end of this chapter.

Data from the sources listed above are increasingly available in computer program format. Automatic interpolation and unit conversions are usually included in the computer programs, so they are often preferred over printed tables. Most useful are computer programs that provide the property data in *online* mode so that the user can import values without copying or reentering them.

The property data provided in the software that accompanies this text is all in *online* format.

4·4 Summary

For any phase of a pure substance, two independent intensive properties usually define the state. Under such conditions, the internal energy and the enthalpy are functions of two independent variables. The functional relationship is typically complex, so for most substances u and h values are obtained from correlations of properties based on experimental data.

For most substances, thermodynamic property data are published for saturated liquid, saturated vapor, and superheated vapor. Occasionally, compressed liquid data are included.

Where compressed liquid data are not available, a useful approximation is that *the properties of a compressed liquid are equal to the properties of saturated liquid at the same temperature.*

The difference between the enthalpy of saturated vapor and that of saturated liquid at the same pressure (and temperature), $h_g - h_f = h_{fg}$, is called the *latent heat of vaporization.*

Heating of a saturated mixture at constant pressure to cause phase change is referred to as *latent* heating because the temperature does not change. Heating of a single phase is called *sensible* heating because the temperature does change.

Specific heats are properties defined as

$$c_v \equiv \left(\frac{\partial u}{\partial T}\right)_v \quad \text{and} \quad c_p \equiv \left(\frac{\partial h}{\partial T}\right)_p \qquad (4\cdot3,\ 4\cdot4)$$

Generally, specific heats of a substance vary with both temperature and pressure.

For an ideal gas, internal energy and enthalpy are functions of temperature only:

$$u = u(T) \quad \text{and} \quad h = h(T)$$

This statement is known as Joule's law. It follows that *for any process of an ideal gas,*

$$\Delta u = \int c_v \, dT \quad \text{and} \quad \Delta h = \int c_p \, dT \qquad (4\cdot6,\ 4\cdot8)$$

Two important relations involving specific heats of ideal gases are

$$c_p - c_v = R \qquad (4\cdot9)$$

and

$$k \equiv \frac{c_p}{c_v}$$

For the *special case of ideal gases with constant specific heats,*

$$\Delta u = c_v \, \Delta T \quad \text{and} \quad \Delta h = c_p \, \Delta T \qquad (4\cdot11,\ 4\cdot12)$$

For the *further special case* of a *quasiequilibrium adiabatic process* of an *ideal gas* with *constant specific heats,*

$$pv^k = \text{constant} \qquad (4\cdot13a)$$

and

$$\frac{T_2}{T_1} = \left(\frac{p_2}{p_1}\right)^{(k-1)/k} = \left(\frac{v_1}{v_2}\right)^{k-1} \qquad (4\cdot13c)$$

Suggested Reading

4·1 Howell, John R., and Richard O. Buckius, *Fundamentals of Engineering Thermodynamics*, 2nd ed., McGraw-Hill, New York, 1992, Chapter 3. (In this book, the term *control mass* is used for *closed system.*)

4·2 Russell, Lynn D., and George A. Adebiyi, *Classical Thermodynamics*, Saunders, Philadelphia, 1993, Chapter 3. (This book gives extended explanations of some definitions and property relationships.)

4·3 Sears, F. W., Mark W. Zemansky, and H. D. Young, *University Physics*, 7th ed., Addison-Wesley, Reading, MA, 1987, Chapter 20.

4·4 Giancoli, Douglas C., *Physics for Scientists and Engineers*, 2nd ed., Prentice-Hall, Englewood Cliffs, NJ, 1989. Chapters 19 and 21.

Sources of Property Data

Handbooks:

4·5 *Chemical Engineers' Handbook*, 6th ed., Robert H. Perry and Don W. Green, eds., McGraw-Hill, New York, 1984.

4·6 *Handbook of Chemistry and Physics*, annual editions, CRC Press, Boca Raton, FL.

4·7 *Handbook of Tables for Applied Engineering Science*, Ray E. Bolz and George L. Tuve, eds., The Chemical Rubber Company, Cleveland, OH, 1970.

4·8 *Mechanical Engineers' Handbook,* Myer Kutz, ed., Wiley, New York, 1986.

4·9 *Marks' Standard Handbook for Mechanical Engineers,* 9th ed., Eugene A. Avaflone and Theodore Baumeister III, eds., McGraw-Hill, New York, 1991.

Collections of Thermophysical Property Data for Various Substances:

4·10 *JANAF Thermochemical Tables,* 3rd ed., published by the American Chemical Society and the American Institute of Physics for the National Bureau of Standards, 1986. (Also, *Journal of Physical and Chemical Reference Data*, Vol. 14, 1985, Supplement No. 1.)

4·11 *Tables of Thermal Properties of Gases*, National Bureau of Standards Circular No. 564, 1955. (Although this reference is old, it is still accurate. Notice that these tables are for real gases and include the effect of pressure even for pressures that might often be considered as low.)

4·12 Hilsenrath, Joseph H., et al., *Tables of Thermodynamic and Transport Properties*, Pergamon, Elmsford, NY, 1960.

4·13 Keenan, Joseph H., Jing Chao, and Joseph Kaye, *Gas Tables*, 2nd ed., Wiley, New York, 1980 (English units), 1983 (SI units).

4·14 Gordon, Sanford, and B. J. McBride, "Computer Program for Calculation of Complex Chemical Equilibrium Compositions, Rocket Performance, Incident and Reflected Shocks, and Chapman–Jouget Detonations," *NASA SP-273*, 1971 (and interim revision, March 1976).

4·15 Grigull, Ulrich, Johannes Straub, and Peter Schiebener, eds., *Steam Tables in SI-Units*, 2nd revised ed., Springer, New York, 1984.

4·16 Haar, Lester, John S. Gallagher, and George S. Kell, *NBS/NRC Steam Tables*, Hemisphere, New York, 1984.

4·17 Keenan, Joseph H., Frederick G. Keyes, Philip G. Hill, and Joan G. Moore, *Steam Tables*, Wiley, New York, 1969 (English units), 1978 (SI units).

4·18 Haar, Lester, and John S. Gallagher, "Thermodynamic Properties of Ammonia." *Journal of Physical and Chemical Reference Data*, Vol. 2, No. 3, 1978, pp. 635–792.

4·19 Vargaftik, N. B., *Tables on the Thermophysical Properties of Liquids and Gases in Normal and Dissociated States*, 2nd ed., Wiley, New York, 1975.

4·20 *Thermodynamic Properties: GASPROPS, REFRIG, STEAMCALC*, Software Systems Corp., Wiley, New York, 1985. (One of several packages of personal computer software for properties.)

See also the continuing series of publications on property data from organizations such as the American Society of Heating, Refrigerating, and Air-Conditioning Engineers (ASHRAE), American Petroleum Institute, U.S. National Institute of Standards and Technology, Thermophysical Properties Research Center at Purdue University, and Thermodynamics Research Center at Texas A&M University. Refrigerant property data are published by refrigerant manufacturers.

Problems

4·1 Steam at 260°C contained in a cylinder fitted with a piston initially has a quality of 70 percent and a volume of 0.02 m^3. The steam is expanded at constant temperature until it is dry and saturated (i.e., $x = 100$ percent). Determine (*a*) the mass of steam in the cylinder and (*b*) the work done during the expansion process.

4·2 Dry saturated steam at 6500 kPa is contained in a rigid tank having a volume of 0.20 m^3. How much heat is required (in kJ/kg) to increase the pressure to 7900 kPa?

4·3 A method of obtaining a low pressure in a sealed vessel is to fill the vessel with steam before sealing, seal, and then condense the steam. A vessel having a volume of 0.05 m^3 is filled with dry saturated steam at 101 kPa. The vessel is sealed and then chilled to 30°C. (*a*) What is the final pressure? (*b*) Sketch *pv* and *pT* diagrams of the process, showing saturation lines on each diagram, (*c*) Calculate the amount of heat added to or taken from the steam.

4·4 Two kilograms of steam at 1.0 MPa, 88 percent quality, is heated in a closed-system frictionless process until the temperature is 330°C. Calculate the amount of heat transferred if the process is at (*a*) constant pressure and (*b*) constant volume.

4·5 One-tenth kilogram of dry, saturated steam at 180 kPa is heated in a closed system to a final condition of 300 kPa, 200°C. Work done on the steam during the process amounts to 8.0 kJ/kg. Calculate the amount of heat transferred per kilogram of steam.

4·6 Steam in a closed system is expanded from 520 kPa, 80 percent quality, to 300 kPa, 250°C. During the expansion, 540 kJ/kg of heat is added to the steam. Calculate the work.

4·7 One-tenth kilogram of steam at 100 kPa, 80 percent quality, is contained in a rigid, thermally insulated vessel. A paddle wheel inside the vessel is turned by an external motor until the steam pressure is 120 kPa. Determine the amount of work done on the steam.

4·8 In a closed system, dry saturated steam at 690 kPa is heated in a constant-volume process until its pressure is 1200 kPa. It is then expanded adiabatically to 690 kPa, 370°C, and later cooled to the saturation temperature at constant pressure. Calculate the net work done. State any assumptions made.

4·9 Ninety grams of steam initially dry and saturated at 70 kPa (state 1) is heated at constant pressure until its volume is 0.25 m^3 (state 2). It is then heated at constant volume until it is at 95 kPa (state 3). Calculate the total heat added per kilogram of steam between states 1 and 3. Be sure to state your assumptions.

4·10 (*a*) Solve Problem 4·9 for a quasiequilibrium steady-flow system (v_2 = 2.8 m^3/kg and 2–3 is a constant specific volume process). (*b*) Is the amount of work done the same? Why or why not?

4·11 In a closed system, steam initially at 67 psia, 80 percent quality, expands to 40 psia, 300 F. During the expansion, heat is added to the steam in the amount of 231.0 B/lbm. Calculate the work.

4·12 One-tenth pound of steam at 16.0 psia, 80 percent quality, is contained in a rigid, thermally insulated vessel. A paddle wheel inside the vessel is turned by an external motor until the steam is at 22 psia. Determine the amount of work done on the steam.

4·13 Two-tenths pound of steam initially dry and saturated at 10 psia (state 1) is heated at constant pressure until its volume is 9.0 ft^3 (state 2). It is then heated at constant volume until it is at 14.0 psia (state 3). What is the total heat added per pound of steam between states 1 and 3?

4·14 (*a*) Solve Problem 4·13 for a frictionless steady-flow system (v_2 = 44.99 ft^3/lbm and process 2–3 is at constant specific volume). (*b*) Is the amount of work done the same? Why or why not?

4·15 Determine the required heat transfer rate for a steady-flow process that produces 80,000 lbm/h of steam at 500 psia, 800 F, from entering water at 530 psia, 100 F.

4·16 Steam flows steadily through a turbine at a rate of 900 kg/h, entering at 700 kPa, 200°C, with negligible velocity and leaving at 7 kPa, 92 percent quality, through an opening of 0.031 m^2 cross-sectional area. The turbine shaft speed is 3000 rpm and the power output is 125 kW. Calculate the heat transfer in kJ/kg.

4·17 Steam enters a turbine at 250 kPa, 150°C, with negligible velocity and is exhausted at 10 kPa with a velocity of 150 m/s. The flow rate is 18,000 kg/h and the turbine power output is 1500 kW. The flow is adiabatic. Determine the quality (if wet) or temperature (if superheated) of the exhaust steam.

4·18 What throttle temperature is required for a steam turbine that is to develop 9000 kW from a flow rate of 39,000 kg/h if the steam enters at 4100 kPa and leaves at 30 mm of mercury absolute containing 10 percent moisture? Assume negligible heat transfer and kinetic energy change.

4·19 Steam enters a turbine at 2800 kPa, 310°C, and leaves at 60 mm of mercury (h = 2550 kJ/kg). The flow rate is 35,000 kg/h. If the inlet opening has a cross-sectional area of 0.02 m^2, what exhaust opening area is required in order to have the change in kinetic energy between inlet and exhaust not more than 1 kJ/kg?

4·20 Saturated liquid water at 1250 kPa is throttled at a steady rate of 140 kg/h to a pressure of 100 kPa. What fraction of the water evaporates during the throttling process? Calculate the density of the fluid leaving the throttle valve.

4·21 Steam flows at a rate of 1.54 kg/s through a

tube that has a constant cross-sectional area of 45 cm^2. The steam enters at 480 kPa, 90 percent quality, with a velocity of 120 m/s, and leaves at 4600 kPa, 420°C. Calculate the amount of heat added per kilogram of steam.

4·22 A nozzle discharges 3.30 kg/s of steam at 175 kPa, 170°C, through an exit area of 0.01 m^2. Assuming that the flow is adiabatic and that the kinetic energy of the steam at inlet is negligibly small, determine the inlet temperature if the inlet pressure is 300 kPa.

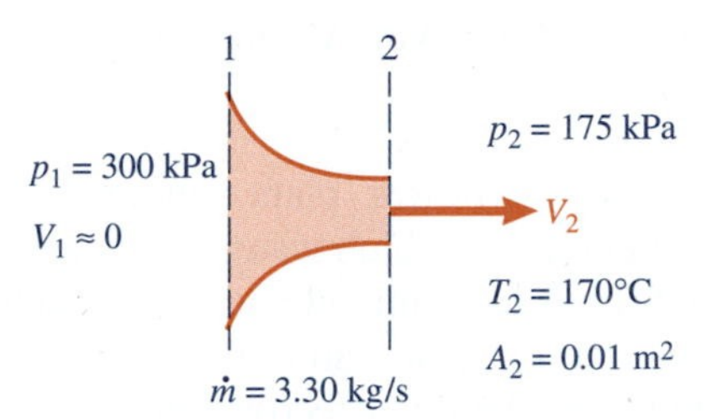

Problem 4·22

4·23 Steam at 125 kPa containing 60 percent moisture enters a condenser through a flow area of 0.014 m^2 with a velocity of 200 m/s. The condensate leaves at 125 kPa, 40°C. Calculate the heat transfer in kJ/h.

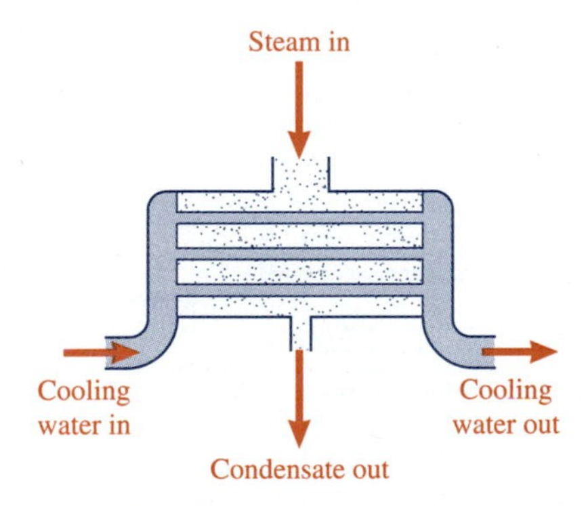

Problems 4·23 and 4·28

4·24 Steam flows steadily through a turbine at a rate of 2000 lbm/h, entering at 100 psia, 400 F, with negligible velocity and leaving at 1 psia, 90 percent quality, through an opening of 0.333 ft^2 cross-sectional area. The turbine shaft speed is 7000 rpm and the power output is 150 hp. Calculate the heat transfer in B/lbm.

4·25 Calculate the throttle temperature required for a steam turbine that is to develop 10,000 kW from a flow rate of 95,000 lbm/h if the steam enters at 600 psia and leaves at 1.0 in. of mercury absolute containing 10 percent moisture. Assume that there is negligible heat transfer and negligible change in kinetic energy.

4·26 Saturated liquid water at 200 psia is throttled at a steady rate of 300 lbm/h to atmospheric pressure. What fraction of the water vaporizes during the throttling process, and what is the density of the H_2O leaving the throttle valve?

4·27 Steam flows at a rate of 3.59 lbm/s through a tube that has a constant cross-sectional area of 0.05 ft^2. The steam enters at 70 psia, 90 percent quality, with a velocity of 400 ft/s, and leaves at 65 psia, 800 F. Calculate the amount of heat added per pound of steam.

4·28 Steam at 16 psia containing 60 percent moisture enters a condenser through a flow area of 0.15 ft^2 with a velocity of 700 ft/s (see figure with Problem 4·23). The condensate leaves at 16 psia, 100 F. Calculate the heat transfer in B/h.

4·29 A two-phase mixture of ammonia at −10°C has an enthalpy of 690.5 kJ/kg. Determine its internal energy.

4·30 Ammonia enters an adiabatic refrigerator compressor at 100 kPa, 95 percent quality, and leaves at 1000 kPa, 100°C. Power delivered to the ammonia by

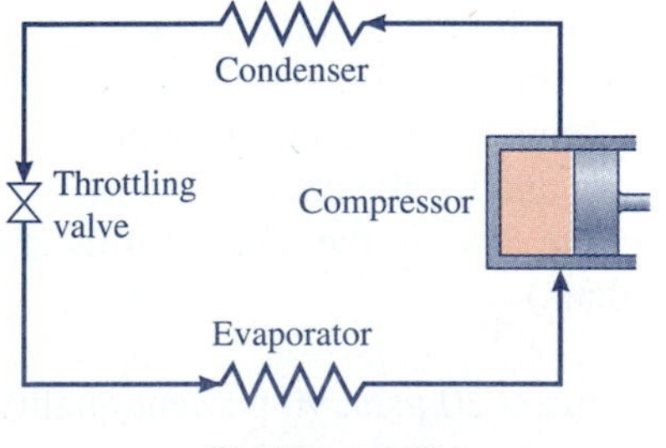

Problem 4·30

the compressor is 25.0 kW. Determine the ammonia flow rate.

4·31 In a refrigeration system ammonia is throttled from 1500 kPa, 30°C, to 200 kPa (see figure with Problem 4·30). Determine the quality of the ammonia leaving the throttling valve.

4·32 Ammonia enters a refrigeration system evaporator at 150 kPa, 20 percent quality, and leaves at 150 kPa, −20°C (see figure with Problem 4·30). The flow rate is 0.5 kg/s. Determine the amount of heat transfer in kJ/kg.

4·33 Refrigerant 134a is compressed at a rate of 72 kg/h from 100 kPa, 90 percent quality, to 800 kPa (see figure with Problem 4·30). Power input is 1.05 kW. Find the discharge temperature if the heat transfer rate from the compressor is 5.90 kJ/min.

4·34 Refrigerant 134a is compressed adiabatically from 125 kPa, 90 percent quality, to 1000 kPa, 60°C (see figure with Problem 4·30). Power delivered to the refrigerant is 22.0 kW. Determine the refrigerant flow rate.

4·35 Refrigerant 134a is throttled from 1200 kPa, 35°C, to 400 kPa (see figure with Problem 4·30). Determine the quality of the refrigerant leaving the throttling valve.

4·36 Refrigerant 134a enters a refrigeration system evaporator at 200 kPa, 20 percent quality, and leaves at 200 kPa, 0°C (see figure with Problem 4·30). The flow rate is 0.122 kg/s. Determine the heat transfer in kJ/kg.

4·37 Carbon dioxide enters a heat exchanger at −40°C, 30 percent quality, and leaves as dry saturated vapor at the same temperature. Calculate (*a*) the heat added in kJ/kg and (*b*) the change in internal energy in kJ/kg. (See the list at the end of this chapter for sources of data.)

4·38 Ammonia at 20 psia, 90 percent quality, enters a compressor at a rate of 12 lbm/min. Power input to the compressor is 46.7 hp, and heat is removed from the ammonia during compression at a rate of 330 B/min. Discharge pressure is 200 psia. Determine the discharge temperature.

4·39 An insulated tank with a volume of 10 m^3 is completely evacuated except for a sealed vial of water. The liquid water in the vial is initially at 100°C. The vial is suddenly broken, and the system is allowed to come to equilibrium. Determine the final pressure in the tank and the fraction of the water that evaporates for a vial volume of (*a*) 0.0600 m^3, (*b*) 0.0060 m^3.

4·40 In a steam power plant, an open feedwater heater is a vessel into which water from various places in the plant flows to be mixed with steam so that water at near saturation temperature leaves to go to the steam generator. An open feedwater heater operating at 110 kPa receives the following measured water flows: 30,000 kg/h at 40°C, 2500 kg/h at 80°C, and 800 kg/h at 20°C. Also, saturated liquid at 200 kPa is throttled into the heater at a rate of 2000 kg/h. Steam enters the heater at 110 kPa containing 2 percent moisture. Water leaving the heater is at a temperature 2°C below the saturation temperature and its measured flow rate is 39,000 kg/h. The heater is insulated, but

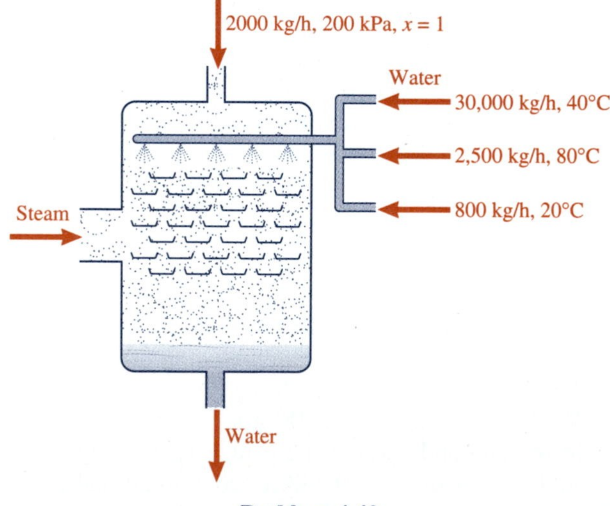

Problem 4·40

there is a question as to whether the insulation is fully effective. What conclusion can you draw?

4·41 In a refrigeration system, 680 kg/h of refrigerant 134a at 800 kPa, 40°C, enters a condenser and liquid at a temperature 5 Celsius degrees lower than the saturation temperature leaves. The cooling medium is water entering at 13°C. It is estimated that heat transfer from the condenser to the surrounding atmosphere amounts to 4 percent of the energy transferred from the refrigerant. What is the required cooling water flow rate if its exit temperature is not to exceed 25°C?

4·42 A vessel has a volume of 1.45 m^3. Initially, 80 percent of its volume is filled with saturated liquid ammonia at −25°C and the rest is filled with ammonia vapor. Liquid leaves through a small valve while heat is added to maintain the temperature at −25°C. The final mass of liquid in the vessel is one-third of the initial mass of liquid. Determine the amount of heat that must be added.

4·43 A well-insulated vessel has a volume of 1.45 m^3. Initially, 75 percent of its volume is filled with saturated liquid ammonia at −20°C and the rest is filled with saturated vapor. Vapor is allowed to escape through a small valve. Plot pressure in the vessel as a function of the fraction of the vessel filled with liquid.

4·44 Refrigerant 134a vapor initially saturated at 10°C is held in an insulated tank having a volume of 0.30 m^3. A 200-W electric heating element in the tank is activated. A relief valve at the top of the tank maintains the system at constant pressure. Assuming that conditions are uniform throughout the tank at any instant, determine the time required for the refrigerant to reach a temperature of 70°C. (A stepwise solution is needed for an accurate result. Using a mean value of outlet stream properties gives an error of less than 3 percent.)

4·45 A vessel of 5.6 m^3 volume contains steam initially saturated at 800 kPa. Heat is added at a constant rate of 5 kJ/s while an automatic valve allows steam to leave at a constant rate of 0.02 kg/s. What is the state of the steam remaining in the tank 5 min after the initial condition? (Use a mean value for any property of the vapor leaving.)

4·46 An insulated tank with a volume of 45.0 ft^3 is initially filled with liquid water at 60 F. Saturated steam at 1 atm is then bled into the vessel to heat the water while a pressure relief valve allows liquid to leave to maintain the pressure constant at 1 atm. Determine the amount of steam that must be added to raise the water temperature to 180 F.

4·47 An insulated tank with a volume of 20.0 ft^3 contains dry saturated ammonia vapor at 100 psig. It is connected to a large line in which ammonia at 200 psig, 140 F, is flowing. What is the final temperature of the ammonia in the tank for an adiabatic filling process?

4·48 Steam initially saturated at 212 F is held in an insulated tank having a volume of 8.30 ft^3. An electric heating element in the tank adds heat at a constant rate of 0.50 kW. Steam is bled from the tank to hold the pressure constant. Assuming that conditions are uniform throughout the tank at any instant, determine the time required for the steam to reach a temperature of 300 F. (A stepwise solution is needed for an accurate result. Using a mean value of outlet stream properties gives only a small error.)

4·49 A vessel of 29.3 ft^3 volume contains refrigerant 134a vapor initially saturated at 78 psia. Heat is added at a constant rate of 5.0 kW while an automatic valve allows R134a to leave at a constant rate of 0.05 lbm/s. After 5 minutes, what is the state of the remaining R134a? (Use a mean value for any property of the vapor leaving.)

4·50 A steam space heater as shown in the figure is to provide 10,000 B/h to room air. Dry saturated steam will enter the heater and the thermostatic trap will permit only liquid water at a temperature of 140 F to leave. For use in a design study, prepare a chart showing the required steam flow rate as a function of the steam pressure.

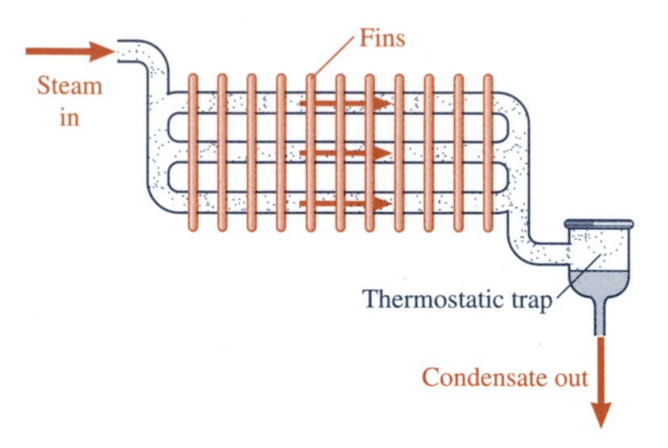

Problem 4·50

4·51 An electrical heating pad used for sore muscles and joints has a power consumption of 40 W. To provide the same heating effect for two hours, what size hot water bottle made of rubber would be required? Assume that no effective heating occurs when the water temperature is less than 50°C (122 F). The maximum temperature depends on available water supplies and safety considerations. Make a plot of water volume as a function of initial temperature.

4·52 A manufacturing plant has one process that requires 3.5×10^6 B/h at 350 F and another that requires 2.15×10^6 B/h at 240 F. It is proposed that steam be generated at 150 psig and supplied to the heating system as shown in the diagram. The traps allow only liquid, no vapor, to pass. Determine the flow rates through the two pressure-reducing valves.

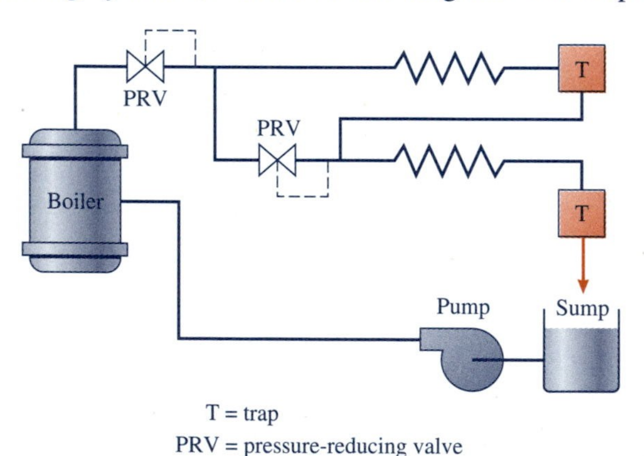

Problem 4·52

4·53 Ice at 0°C on the outside of a tube is melted to liquid at 0°C by a transfer of heat from water flowing inside the tube. Water enters the tube at 40°C and leaves at 25°C. Latent heat of fusion of water at 0°C is 333.4 kJ/kg. (*a*) How much warm water is needed per kilogram of ice? (*b*) Show how the latent heat of fusion given above can be verified from data given in the steam tables.

4·54 Ice at 32 F on the outside of a tube is melted to liquid at 32 F by a transfer of heat from water flowing inside the tube. Water enters the tube at 100 F and leaves at 80 F. Latent heat of fusion of water at 32 F is 143.3 B/lbm. (*a*) How much warm water is needed per pound of ice? (*b*) Show how the latent heat of fusion given above can be verified from data given in the steam tables.

4·55 From Fig. 4·2, explain how c_p of liquid ammonia varies with temperature at a constant pressure. How does it vary with pressure at a constant temperature?

4·56 During a test of an automobile engine, the amount of heat transferred to the cooling water is determined by measuring the amount of cooling water flowing through the engine and its temperature rise. The inlet and outlet temperatures are measured at the points labeled T_i and T_o on the diagram. The flow rate is measured by the precision flow meter FM. All the piping is well insulated. After a test is completed, it is

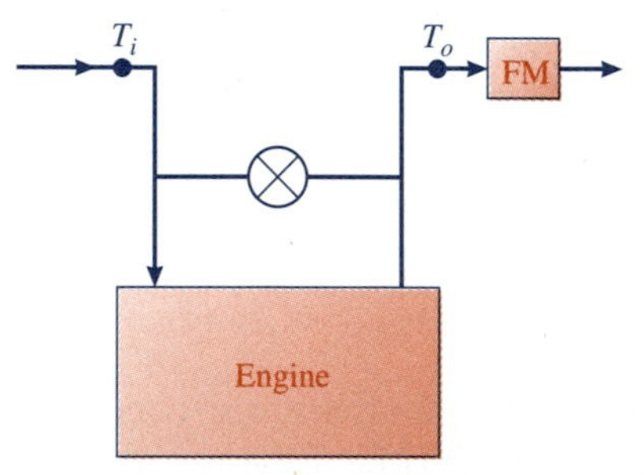

Problem 4·56

discovered that bypass valve B was partially open during the test. Explain why the data cannot be used or might still be used. (If the data can be used, what practical advantages would there be to the arrangement shown instead of a simple inline throttling valve?)

4·57 Calculate $(\partial u/\partial v)_T$ for steam at 10 MPa, 400°C.

4·58 Calculate $(\partial u/\partial v)_T$ for steam at 100 psia, 350 F.

4·59 For steam at 2700 kPa, 250°C, determine from the steam tables approximate values of $(\partial h/\partial T)_p$, $(\partial h/\partial T)_v$, and $(\partial p/\partial T)_v$. Why are the values you obtain approximate? Are these point functions or path functions? Is dh/dT a point or path function?

4·60 Refer to Example 3·8. If cooling water for the dehumidifier is available at 12°C and is to be discharged at a temperature not higher than 20°C, what is the minimum flow rate of water required?

4·61 Hot bricks have been used as "foot warmers" in beds and also to keep bread warm between the time it is cooked and when it is served. Compare the energy storage characteristics of bricks with those of water for the same mass and for the same volume. Assume that the maximum temperature with either medium is 200 F.

4·62 A swimming pool 75 ft long and 45 ft wide has an average depth of 6 ft. It is filled with water at 55 F. To warm the water, some water is circulated through a gas-fired heater capable of adding 80,000 B/h to the water. How long does it take to raise the pool temperature 10 Fahrenheit degrees? To avoid injury, the temperature of the water reentering the pool must not exceed 130 F. What specification should be established for the flow rate of the circulating water?

4·63 A drinking glass is to be filled with water initially at 20°C, but first enough 30-cc ice cubes at −4°C will be placed in the glass to ensure that the water is chilled to 0°C. What is the minimum number of ice cubes required?

4·64 In the discussion of Joule's experiment in the early part of Section 4·2, see the response to the question, "If in an experiment like Joule's the temperature of the water changes, does this prove that the internal energy of the gas under study is not a function of temperature only?" Answer this question by supposing that the water temperature decreases instead of rising as is supposed in Section 4·2.

4·65 For an ideal gas sketch a pv diagram showing lines of constant T, a Tv diagram showing lines of constant p, and a pT diagram showing lines of constant v. Derive an expression for the slope of each line.

4·66 Prove that for an ideal gas

$$du = \frac{1}{k-1}d(pv) \quad \text{and} \quad dh = \frac{k}{k-1}d(pv)$$

4·67 Refer to Problem 3·91. What is the value of k for the gas?

4·68 A closed, rigid, insulated container is occupied by an ideal gas at 50 kPa, 25°C. Work is done on the gas by an external motor through an impeller mounted inside the container until the pressure is 100 kPa. For this gas in the temperature range involved, $c_p = 0.84 + 0.00075T$, where T is in kelvins and c_p is in kJ/kg·K. Calculate (*a*) the enthalpy change of the gas, and (*b*) the work done on the gas.

4·69 The equation for c_p as a function of T displayed in Table A·1 has five terms. Comment on the suggestion that only the first two or three terms might be used as an approximation.

4·70 Ethene (ethylene) is heated under a constant pressure of 1 atm from 20 to 500°C. Compute the heat added.

4·71 Two-tenths kilogram of carbon dioxide is heated from 100 to 2000°C at 1 atm. How much heat is required?

4·72 Determine the amount of heat that must be added to heat 0.2 lbm of carbon dioxide from 100 to 4000 F under a constant pressure of 14.7 psia.

4·73 In calculations for ideal gases, one occasionally sees enthalpy and internal energy evaluated as $h = c_pT$ and $u = c_vT$. Comment on this practice, its soundness, generality, restrictions, and underlying assumptions.

4·74 An ideal gas at 8 psia, 50 F, fills a closed, rigid, thermally insulated container. An impeller inside the container is turned by an external motor until the pressure is 14 psia. For this gas in the temperature range involved, $c_p = 0.2 + 0.0001T$ and $c_v = 0.15 + 0.0001T$, where T is in Rankine and c_p and c_v are in B/lbm·R. Calculate, in B/lbm, (*a*) the enthalpy change of the gas and (*b*) the work done on the gas.

4·75 Using the data of Table A·1, plot curves of k versus temperature in the range of 300 to 1500 K for (*a*) propane, (*b*) ethene (ethylene), and (*c*) methane.

4·76 Helium enters a gas turbine at 625 kPa, 200°C, and exits at 100 kPa, 25°C. Heat loss during the process amounts to 35 kJ/kg. The flow rate is 6350 kg/h. Calculate the power output if the change in kinetic energy is negligible.

4·77 Refer to the first example problem in Chapter B, Note on Problem Solving, and calculate the heat added to the methane in kJ/kg.

4·78 Refer to the second example problem in Chapter B, Note on Problem Solving, and make a plot of the rate of heat transfer (in kJ/h) as a function of the discharge pressure.

4·79 Two kilograms of a gas with $c_v = 0.75$ kJ/kg·K expands adiabatically in a closed system and performs 10 kJ of work. The final temperature is 90°C. Compute the initial temperature.

4·80 A gas flows adiabatically through a nozzle from a pressure of 1200 kPa to a final pressure of 120 kPa. The initial and final specific volume values are 1.186 and 6.289 m^3/kg, respectively. If the initial and final temperatures are 345 and 183 K, respectively, compute the final velocity. Assume that the specific heat at constant volume is 10.0 kJ/kg·K and that the inlet velocity is negligible.

4·81 Three pounds of a certain gas with $c_v = 0.18$ B/lbm·F expands adiabatically in a closed system and in so doing performs 10,000 ft·lbf of work. The final temperature is 200 F. Compute the initial temperature.

4·82 To compress 5 lbm of an ideal gas with $c_v = 0.169$ B/lbm·R, it is required that 25,000 ft·lbf of work be supplied while 25 B of heat is removed. Compute the temperature change.

4·83 A gas flows through a nozzle from a pressure of 180 psia to a final pressure of 20 psia. The initial and final specific volume values are 1.277 and 6.09 cu ft/lbm, respectively. If the initial and final temperatures are 621 and 329 R respectively, compute the final velocity. Neglect the initial velocity and the heat-loss terms. Assume that the specific heat at constant volume is 0.169 B/lbm·R.

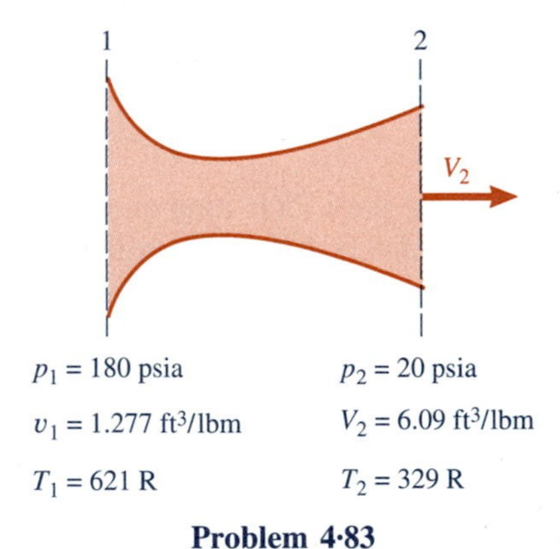

Problem 4·83

4·84 A piece of electronic equipment fits into a case of dimensions 9 × 16 × 22 cm. Steady-state power consumption is 600 W. The output signals carry a negligible amount of energy. It is required that the equipment temperature remain below 60°C. Either forced air or circulating water may be used as a cooling medium. Determine both the required mass flow rate (kg/s) and the required volume flow rate (m^3/s) if (*a*) the coolant is air entering at 1 atm, 20°C, with an allowable air temperature rise of 30 Celsius degrees, (*b*) the coolant is water entering at 1 atm, 20°C, with an allowable water temperature rise of 35 Celsius degrees.

4·85 A single-phase electric transformer operates steadily with a primary voltage of 2300 V and its secondary at 230 V. The rated power output is 750 kW, and the efficiency (the ratio of output power to input power) is 97.0 percent. If the transformer is cooled only by atmospheric air blown over it, what flow rate (kg/s) is required for an entry air temperature of 30°C and an air temperature rise not to exceed 20 Celsius degrees?

4·86 A Hilsch tube or Ranque tube is a short length of insulated pipe, open at one end and capped at the other, with a small opening at the center of the cap. A gas is injected tangentially into the pipe at some point along its length, and even though there are no moving parts, the gas streams leaving at the two ends are observed to be at different temperatures. If air enters such a device at 300 kPa, 40°C, at a rate of 0.100 kg/s and air leaves one end at 100 kPa, 50°C, at a rate of 0.060 kg/s, what is the temperature of the air leaving the other end?

4·87 Twenty liters of nitrogen trapped in a cylinder at 100 kPa, 5°C, is compressed until its volume is 0.01 m^3 and its pressure is 175 kPa. Calculate the following quantities: (*a*) total internal energy change of the nitrogen, (*b*) net heat added to the nitrogen during the process, and (*c*) net work done by the nitrogen during the process. (If there is insufficient information for the calculation of any item, state what additional information is necessary for its determination.)

4·88 In a closed system, 2 lbm of air is heated at constant pressure from 30 psia, 40 F, to 140 F. Because of a frictional effect, the work output is only 10 B. Calculate the amount of heat added to the air.

4·89 One cubic foot of air trapped in a cylinder at 15 psia, 40 F, is compressed until its volume is 0.50 cu ft and its pressure is 25 psia. Calculate the quantities listed below (if there is insufficient information for the calculation of any item, state what additional information is necessary for its determination): (*a*) total internal energy change of the air, (*b*) net heat added to the air during the process, and (*c*) net work done by the air during the process.

4·90 Propane enters a steady-flow system at 175 kPa, 200°C, with a velocity of 60 m/s through a cross-sectional area of 0.05 m^2. It leaves at 80 kPa, 150°C, at the same velocity. The system delivers 60 kW to the surroundings. Calculate the heat transfer in kJ/kg.

4·91 Air enters a gas turbine power plant at 101.35 kPa, 15°C, at a rate of 1150 m^3/min. It is compressed to 410 kPa, heated, and then expanded through a turbine and exhausted at 101.35 kPa, 260°C. The net power output of the plant is 3 MW. Neglecting kinetic energy changes, calculate the net amount of heat added to the air in kJ/kg.

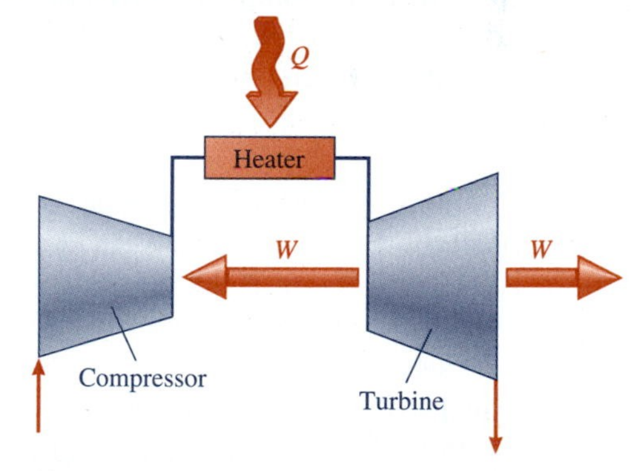

Problem 4·91

4·92 Ethylene at 5°C, 100 kPa, is compressed isothermally to 480 kPa at a rate of 0.95 kg/s by an ideal compressor. The kinetic energy of the air passing through the compressor increases by 11 kJ/kg. Heat removed from the air amounts to 128 kJ/kg. Determine the power input to the compressor.

4·93 Helium is drawn into a multistage centrifugal compressor at 95 kPa, 5°C, with negligible velocity and is discharged at 175 kPa, 150°C, with a velocity of 150 m/s through a cross-sectional area of 0.020 m^2. Power input to the compressor is 1000 kW. Determine the heat transfer in kJ/kg.

4·94 In a gas-turbine power plant, air is taken in at 14.7 psia, 77 F, at a rate of 38,000 cfm, compressed to 60 psia, heated, and then expanded through a turbine and exhausted at 14.7 psia, 500 F (see figure with Problem 4·91). If the net output of the plant is 3860 hp, calculate the net amount of heat added to the air in B/lbm. Neglect changes in kinetic energy.

4·95 A compressor takes in 12,000 cfm of air at 14.0 psia, 80 F, with negligible velocity and discharges air at 30.0 psia, 240 F, through an opening that has a cross-sectional area of 0.25 sq ft. Heat is removed from the air being compressed at a rate of 60 B/s. Determine the power input to the compressor.

4·96 Air enters a compressor at 100 kN/m^2, 4°C, with a velocity of 150 m/s through a cross-sectional area of 0.060 m^2. The air is compressed frictionlessly, steadily, and adiabatically to twice the original pressure and the discharge velocity is very low. Find the power input.

4·97 Two kilograms of carbon monoxide is compressed frictionlessly and adiabatically in a closed system from 100 kPa, 20°C, to 200 kPa. Compute the work and the changes in internal energy and enthalpy.

4·98 Refrigerant 134a enters the condensing coils of a refrigerator at 35°C, 93 percent quality, and leaves as saturated liquid at a rate of 0.50 kg/min. Calculate the required volume flow rate of air (in m^3/s) to condense the refrigerant if the atmospheric air blown over the condenser is supplied at 20°C and leaves at 32°C.

4·99 Partially vaporized water at 110°C with a quality of 12 percent leaves a solar collector at a rate of 0.105 kg/s. Find the flow rate of atmospheric air at 15°C required to cool the water to 90°C at constant pressure if the temperature rise of the cooling air is limited to 25 Celsius degrees.

4·100 Refrigerant 134a enters the condensing coils of a refrigerator at 100 F, 91 percent quality, and leaves as saturated liquid at a rate of 0.90 lbm/min. Determine the volume flow rate of air (in cfm) required to condense the refrigerant if room air is blown over the condenser and its temperature rise is limited to 10 F.

4·101 A tank of volume V contains an ideal gas initially at p_i and T_i. The gas leaks out of the tank through a small opening until the pressure drops to p_f. Heat is added to the gas to keep the temperature constant. Neglect kinetic energy changes. Determine the amount of heat that must be added. Express your answer in terms of quantities included in the following list: V, p_i, p_f, T_i, m_i, m_f.

4·102 An ideal gas in a well-insulated tank leaks out to the atmosphere through a porous plug. Write the first law as it applies to this system in terms of nothing more than the pressure, temperature, and mass of gas in the tank and specific heats and gas constant of the gas.

4·103 Air initially at p_i and T_i is held in an insulated tank. The pressure p_i is less than the pressure of the surrounding atmosphere, p_0. A small leak develops in the tank. Determine the temperature in the tank when the air pressure in the tank has reached p_0.

4·104 A gas accumulator is a device that stores varying amounts of gas to reduce flow rate variations in parts of a gas distribution system. A vertical cylinder with a weight-loaded piston supported by the gas in the cylinder is a constant-pressure accumulator. Consider such an accumulator that receives an ideal gas at a constant rate $\dot{m}_1$ and at constant pressure and temperature p_1 and T_1 and has an outward flow rate given by $\dot{m}_2 = \dot{m}_0 \sin \omega\tau$, where ω is a constant and τ is time. The accumulator is well insulated and it operates continuously under the given conditions. Assume that the gas is well mixed in the accumulator so that the properties of the gas leaving are essentially the same as those of the gas throughout the accumulator. Determine the outlet temperature as a function of the specified conditions and time.

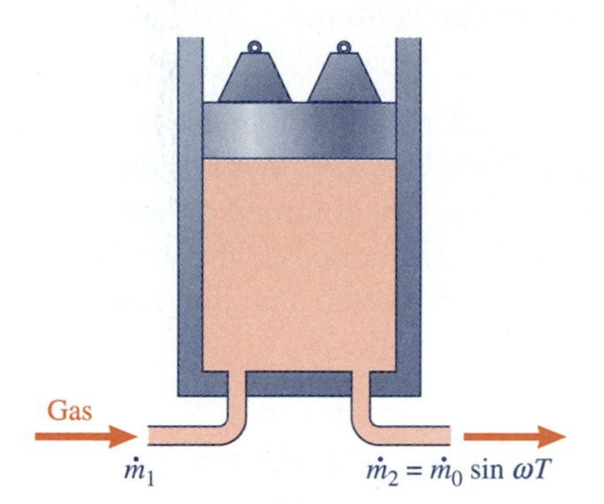

Problem 4·104

4·105 Air is drawn from a large test chamber by a vacuum pump at a variable mass rate of flow while a small amount of atmospheric air leaks into the chamber at a constant rate. The velocities of both entering and leaving air are very low. The surroundings are at a higher temperature than the air in the chamber. Write an energy balance equation for the system composed of the air in the test chamber, and simplify the equation as far as possible.

4·106 A tank of 100 cu ft volume contains air initially at 100 psia, 100 F. Heat is added at a constant rate of 5 B/s while an automatic valve allows air to leave the tank at a constant rate of 0.05 lbm/s. What is the temperature of the air in the tank 5 min after the initial condition?

4·107 Refer to Problem 4·106. Starting from the initial conditions, how long will it take for the air in the tank to reach 350 F?

4·108 Refer to Problem 4·106. Starting from the initial conditions, what is the pressure of the air in the tank after 10 min?

4·109 A computer with a power consumption of 160 W is to be cooled by forced air convection. Cooling air leaving should have a temperature not greater than 30 Celsius degrees above room temperature and a velocity not greater than 5 m/s. All the cooling air leaves through a common opening. For use in a design study, make a chart of discharge area required versus temperature rise for various values of exit velocity.

4·110 Make a chart or charts like that required in Problem 4·109 but suitable for use with power consumptions in the range of 160 to 1000 W.

4·111 An electric-resistance space heater is to operate on a 120-volt ac power supply and have a power consumption of 1200 W. If the fan incorporated into the heater is to provide a velocity not exceeding 10 m/s, determine the cross-sectional flow area required for various temperature rises between 20 and 50 Celsius degrees.

4·112 A gear box operates steadily with input and output shaft speeds of 3600 rpm and 240 rpm, respectively, and input and output shaft torques of 52,000 and 748,000 lbf·ft. For a design study, plot the volume rate of flow of cooling air required to cool the gear box versus the cooling air temperature rise for various ambient temperatures. Do the same for water as a coolant, except that the water supply temperature is fixed at 15°C (59 F). Also note the following constraint: No part of the gear box casing should have a temperature exceeding 85°C (185 F).

4·113 A room 18 ft × 14 ft × 7.5 ft high contains four metal benches weighing 150 lbf each and 8 metal stools weighing 10 lbf each. To maintain a constant air temperature of 65 F in the room requires a heat input rate of 2.3 kW from electric-resistance heaters to balance the heat loss to the surroundings. When a power outage is expected, it is decided to bring in 5-gallon cans of hot water. How many such cans of water at what temperature would be required to assure that the room temperature will not drop below 65 F for at least 2 hours? Express the results of your calculations in a plot.

4·114 Make as many measurements as you can on an electric hand-held hair dryer, and make a quantitative energy balance to the greatest extent that you can.

4·115 From data in this book and elsewhere, what general conclusions can you make about the relative magnitudes of the specific heats of solid, liquid, and gaseous phases of a substance?

4·116 From data in this book and elsewhere, what general conclusions can you draw regarding a correlation between specific heats and molar masses of gases?

4·117 Discussions in this chapter of variations of specific heats of ideal gases lead to the conclusion that c_p cannot decrease with increasing temperature. Yet, Reference 4·11 shows that c_p/R decreases from 3.5146 at 200 K to 3.5082 at 290 K, and R surely is constant. Resolve this paradox.

4·118 An "aerosol" can of whipped cream (or shaving cream) uses air as a propellant. What initial pressure is required to ensure that essentially all of the product can be expelled? (This requires numerous decisions and assumptions.)

4·119 Refer to Example 1·7. Determine the stored energy change of the system during the compression and the heat transfer.

4·120 Refer to Example 1·9. Determine the heat transfer and the exhaust temperature.

4·121 A piston–cylinder arrangement contains 0.10 kg of argon at 200 kPa, 15°C. The argon expands frictionlessly and isothermally until its pressure is 100 kPa. Calculate the heat transfer.

4·122 One kilogram of oxygen is compressed isothermally in a closed system from 100 kPa and 25°C to 300 kPa. Calculate (*a*) work, (*b*) heat transfer, and (*c*) change in internal energy.

4·123 Work is done on helium in a closed rigid tank by means of a paddle wheel turned by an external motor. The tank contains 0.25 m^3 of helium initially at 100 kPa and 5°C. 25 kJ of heat is added to the helium and the final pressure in the tank is 200 kPa. Sketch a *pv* diagram of the process, and calculate the amount of work done on the helium.

4·124 A closed rigid tank contains 0.14 m^3 of air initially at 100 kN/m^2, 4°C. Work is done on the air by means of a paddle wheel turned by an external motor, and 5 kcal of heat is added to the air. The final pressure is 200 kN/m^2. Sketch a *pv* diagram of the process and calculate the work.

4·125 Two pounds of air is compressed isothermally in a closed system from 15 psia, 40 F to 45 psia. Calculate (*a*) work, (*b*) heat transfer, and (*c*) change in internal energy.

4·126 An ideal gas is heated from state 1 to state 2 at constant pressure and is then further heated from state 2 to state 3 at constant volume. Sketch for this sequence of processes six property diagrams: *pv*, *Tv*, *pT*, *uv*, *ph*, and *uT*, where in each case the first variable is to be the ordinate.

4·127 An ideal gas is compressed frictionlessly and adiabatically from state 1 to state 2. It is then expanded frictionlessly and isothermally from state 2 to state 3, and $v_3 = v_1$. Sketch for this sequence of processes six property diagrams: *pv*, *Tv*, *pT*, *uv*, *ph*, and *uT* where in each case the first variable is to be the ordinate.

4·128 Nitrogen in a closed system is heated at a constant pressure of 1 atm until its volume has increased by 30 percent. Sketch the following diagrams showing the process: *pv*, *pT*, *pu*, *ph*, *vT*, *vu*, *vh*, *Tu*, and *uh*. The first variable listed is to be plotted on the vertical axis.

4·129 In a closed system, a gas is expanded frictionlessly at constant pressure until its volume has increased by one-fifth. What fraction of the heat added would be converted into work if the gas were (*a*) air, (*b*) ethane, or (*c*) helium?

4·130 One-tenth kilogram of an ideal gas with a molar mass of 40 expands frictionlessly at constant pressure in a closed system from 100 kPa, 60°C, to 160°C when 5 kJ of heat is added. Determine c_v.

4·131 A 22.0-cu-ft rigid tank is filled with an ideal gas at 3000 psia, 100 F. The addition of 13.7 B of heat raises the gas temperature by 40 F. If the constant-pressure specific heat of the gas is 0.238 B/lbm·R, what is the value of k for the gas?

4·132 In a closed system, 1.25 kg of ethane is heated at constant pressure from 200 kPa, 0°C, to 60°C. The work output is only 18.2 kJ. Calculate the amount of heat added to the gas.

4·133 One-hundredth kilogram of air at 300 kPa, 10°C, is trapped inside a vertical cylinder that is fitted at the top with a weighted piston so that the pressure of the air is held constant. There is no heat transfer. A paddle wheel in the cylinder is turned until the volume of the air has increased by 20 percent. Determine (*a*) the net amount of work done on the air and (*b*) the amount of work done on the air by the paddle wheel.

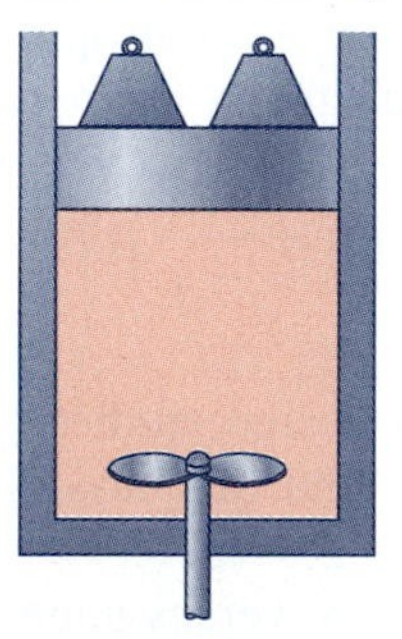

Problem 4·133

4·134 One-tenth cubic meter of methane is contained in a cylinder at 100 kPa, 10°C. It is compressed to twice its initial pressure and one-half its initial volume. Calculate (*a*) the net work done on or by the gas, (*b*) the net heat added to or taken from the gas, and (*c*) the net change in internal energy if (1) the pressure is first doubled at constant volume and the volume is then halved at constant pressure, (2) the constant-pressure process precedes the constant-volume process.

4·135 Starting with the first law as applied to a steady-flow process, prove that pv^k = constant for a frictionless adiabatic process of an ideal gas with constant specific heats.

4·136 Determine the heat transfer during a polytropic process in which the temperature of octane in a closed system changes from 115 to 5°C and the octane does 45 kJ of work per kilogram.

4·137 One cubic meter of gas at 170°C expands polytropically in a cylinder until the temperature is 5°C and the volume is 4 m^3. Determine the value of n.

4·138 Air enters a nozzle at 690 kN/m^2, 170°C, with negligible velocity. It expands adiabatically without friction as it flows through the nozzle and leaves at 345 kN/m^2. The nozzle exit area is 5.20 cm^2. Calculate the exit velocity.

4·139 Oxygen in a closed system expands frictionlessly and adiabatically from 200 kN/m^2, 170°C, to 100 kN/m^2. Calculate the work done per kilogram of oxygen.

4·140 Air at 275 kPa, 60°C, enters a nozzle with negligible velocity and expands adiabatically and in a quasiequilibrium manner to 140 kPa. The outlet area of the nozzle is 16.4 cm^2. Calculate (*a*) the outlet velocity and (*b*) the value of $\int v\,dp$ for this process.

4·141 An ideal reciprocating compressor draws in 5.7 m^3/min of air at 100 kPa, 5°C, and compresses it polytropically with n = 1.35 to 500 kPa. The air is cooled by cooling water which flows at a rate of 1.63 kg/min and undergoes a temperature rise of 5 Celsius degrees. Calculate the compressor power requirement.

4·142 Air is compressed in a frictionless, steady-flow process from 70 kN/m^2, 15°C, to 100 kN/m^2 in such a manner that $p(v + 0.2)$ = constant, where v is in m^3/kg. Inlet velocity is negligibly small and discharge velocity is 100 m/s. Calculate the heat transfer.

4·143 Air flows steadily and frictionlessly through a compressor at a rate of 1.5 kg/s. At inlet the air is at 70 kPa, 5°C, and at outlet the air is at 140 kPa, 120°C. Between inlet and outlet of the system, the air is compressed in such a manner that pv^2 = constant. The inlet area is very large and the outlet area is 45 cm^2. Sketch the process on *pv* and *Tv* coordinates, and calculate the heat transfer in kJ/kg.

4·144 One kilogram of argon that under a pressure of 1.4 MPa occupies a volume of 0.06 m^3 expands at constant temperature until the volume is doubled. It is then compressed at constant pressure to a volume of 0.06 m^3, after which its pressure is raised to 1.4 MPa at constant volume. Draw approximately to scale the *pv* diagram for the processes of this closed system. Determine the net change in internal energy for the processes involved. Calculate the net work done and the net heat flow during these processes.

4·145 One kilogram of octane is compressed frictionlessly and adiabatically from 100 kPa and 0.40 m^3 to 0.20 m^3, then expanded at constant pressure to 0.4 m^3, and finally cooled at constant volume until the initial pressure is reached. Sketch the processes on a *pv* diagram approximately to scale. Shade the area representing the work for the constant-pressure process. Determine the work done, the heat added, and the change in internal energy for the adiabatic process.

4·146 Six pounds of nitrogen expands frictionlessly and adiabatically in a closed system from 400 psia, 100 F, to half of the initial pressure. Compute the change in internal energy, the change in enthalpy, the work, and the heat transfer.

4·147 Two pounds of air is compressed frictionlessly and adiabatically in a closed system from 15 psia, 70 F, to 150 psia. Compute the change in internal energy, the change in enthalpy, and the work.

4·148 Air enters a nozzle at 100 psia, 340 F, with negligible velocity. It expands adiabatically without friction as it flows through the nozzle and leaves at 50 psia. The nozzle exit area is 0.80 sq in. Calculate the exit velocity.

4·149 Calculate the power required to compress air frictionlessly and adiabatically from 12 psia, 40 F, to 24 psia. Average discharge velocity is 600 fps through a cross-sectional area of 0.5 sq ft, and the inlet velocity is negligibly small.

4·150 An ideal reciprocating compressor draws in 200 cfm of air at 15 psia, 40 F, and compresses it polytropically with $n = 1.35$. The compression ratio is 5:1. Cooling water that removes heat from the air flows at a rate of 10.2 lbm/min and undergoes a temperature rise of 12 Fahrenheit degrees. Calculate the compressor power requirement.

4·151 Air is compressed in a frictionless steady-flow process from 10 psia, 60 F, to 15 psia in such a manner that $p(v + 5)$ = constant, where v is in cu ft/lbm. If the inlet velocity is negligibly small and the discharge velocity is 250 fps, what is the heat transfer?

4·152 Show that for a polytropic (quasiequilibrium) process of an ideal gas in a closed system

$$q = c_v\left(\frac{k - n}{n - 1}\right)(T_2 - T_1)$$

4·153 For use in design studies, a chart that shows temperature at the end of quasiequilibrium adiabatic compression of air is needed. The chart should show discharge temperature versus gage discharge pressure for three values of inlet temperature typical of construction sites in North America. Use a range of gage discharge pressures from 0 to 15 atm. Use constant specific heats. (In later chapters this problem is restated to include additional effects.)

4·154 An ideal gas is to be compressed frictionlessly and adiabatically from initial atmospheric conditions of p_1 and T_1 to a higher pressure p_2. Plot curves of T_2 versus p_2/p_1 for various constant values of k of common gases at room temperature.

4·155 An ideal gas is to be expanded frictionlessly and adiabatically from initial atmospheric conditions of p_1 and T_1 to a lower pressure p_2. Plot curves of T_2 versus p_2/p_1 for various constant values of k of common gases for $T_1 = 25°C$.

4·156 A cylindrical tank open at the bottom is mostly submerged in water, open end down, with the axis of the cylinder vertical. The pressure inside the cylinder is greater than atmospheric pressure p_0. A small valve in the top of the tank is opened, allowing air to escape and the water level inside the tank to rise. It is desired to determine the temperature of the air in the tank at any instant, assuming that the process is adiabatic. Write the differential equation for this system in terms of physical constants and p, V, T, and m, where these symbols stand for the pressure, volume, temperature, and mass of the air inside the tank at any instant.

4·157 Derive an expression for the amount of heat that must be added to an ideal gas in a tank of volume V in order to increase its temperature from T_i to T_f while gas is bled from the tank to hold the pressure constant.

4·158 A cylinder with its axis vertical and the cylinder head at the top is fitted with a piston. Initially the piston is at the top of the cylinder with negligible clearance volume. The piston is held in position by the atmospheric pressure acting behind it. A small hole in the cylinder head is opened to let air from the atmosphere flow into the cylinder, allowing the piston to drop slowly while the pressure within the cylinder remains constant. The piston and cylinder are made of thermally insulating material. When the piston comes to rest against stops, the final temperature and volume of air in the cylinder are designated by T_f and V_f. Atmospheric temperature and pressure are T_0 and p_0. Determine T_f.

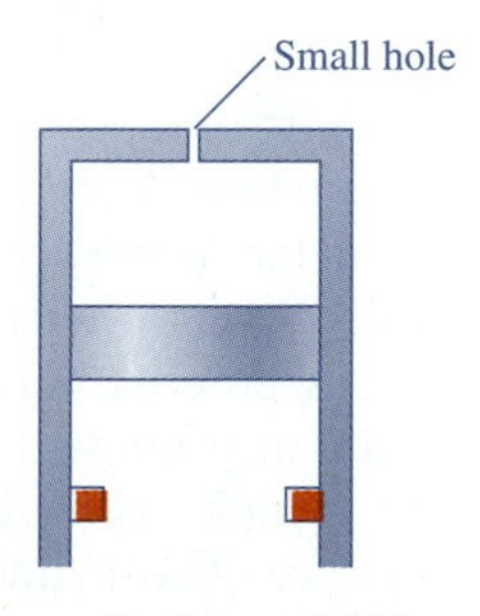

Problem 4·158

4·159 Solve Problem 4·158 with an initial clearance volume V_i containing a mass of air m_i at $p_i < p_0$, and $T_i = T_0$.

4·160 An arresting device is a large cylinder fitted with a piston and initially containing air at atmospheric pressure and temperature, p_0, and T_0. When the piston is struck, it immediately has a velocity V_i and then decelerates at a rate $dV/dt = Ap$ while air escapes to the atmosphere through a small opening at a rate $\dot{m} = Bp^{0.5}$, where A and B are constants. Neglecting heat transfer, derive an expression for the temperature of the air in the cylinder as a function of time, V_i, A, B, and the initial properties of the air. Sketch curves of the air pressure and the piston velocity versus time.

4·161 An empty, one-room, well-insulated building has an internal volume of 5500 m^3. A small vent maintains the pressure inside the building equal to the outside atmospheric pressure of 101 kPa. How much heat must be added by a space heater in the building to increase the inside air temperature from 7 to 25°C?

4·162 Acetylene initially at 95 kPa, 40°C, is held in an insulated tank having a volume of 0.30 m^3. An electric heating element within the tank adds heat at a constant rate of 100 W. Acetylene is bled from the tank to hold the pressure constant. Assuming that conditions are uniform throughout the tank at any instant, determine the time required for the acetylene to reach a temperature of 120°C.

4·163 A tank with a volume of 1.2 m^3 is initially evacuated. Atmospheric air seeps into the tank through a porous plug so slowly that there is ample time for heat transfer to keep the temperature inside the tank equal to the atmospheric temperature of 20°C. Finally the pressure inside the tank is equal to the atmospheric pressure of 95 kPa. Determine the amount of heat added to or removed from the air in the tank.

4·164 An insulated tank with a volume of 0.80 m^3 contains oxygen at 100 kPa, 20°C. It is connected to a large line in which oxygen at 1 MPa, 50°C, is flowing. For an adiabatic filling process, find the final temperature of the tank contents.

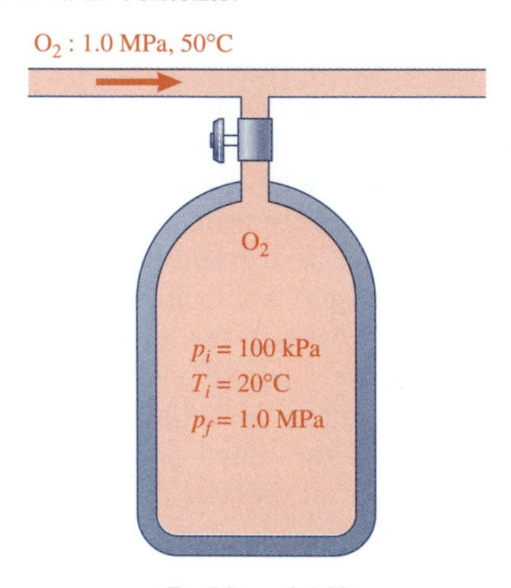

Problem 4·164

4·165 An insulated tank initially contains 10 kg of air at 600 kPa, 100°C. Air is to be bled from the tank to the atmosphere, causing both the pressure and the temperature of the air remaining in the tank to decrease. The bleed valve is to be regulated, however, so that the temperature drops at the constant rate of 0.10 Celsius degrees per second. Assuming that the process is adiabatic and that kinetic energy effects are negligible, determine the mass rate of flow from the tank as the bleeding of air begins.

4·166 An ideal gas escapes from an insulated tank at a constant low rate of $\dot{m}$. Derive expressions for the rate of change of pressure and of temperature of the gas in the tank with respect to time in terms of $\dot{m}$ and the properties of the gas at any instant.

4·167 For the gases Ar, He, CO, N_2, and O_2, make a table comparing the c_v, c_p, and k values at 0°C as obtained from kinetic theory with the actual values at low pressure and 0°C.

4·168 One-tenth of a cubic meter of air at 1 atm, 20°C, is trapped in a cylinder. It is to be compressed in a quasiequilibrium adiabatic process to 5 atm. Plot to scale both a pV diagram and a diagram of piston force versus piston displacement. On each diagram, show the curves for three cases: stroke/bore ratios of 4, 1, and 0.25.

4·169 Solve Problem 4·168 for a quasistatic isothermal process.

4·170 Packages with masses of 80 kg moving at 5 m/s on a conveyor are to be brought to rest so they can be removed from the conveyor by an existing pickup device. The deceleration must not exceed $5g$, where $g = 9.81$ m/s^2. A proposed arresting device is shown in the schematic diagram, where the gas in the cylinder is air initially at room pressure and temperature. Assume that the stroke/bore ratio is to be unity. Determine the size of cylinder needed. Plot force versus displacement.

4·171 What limits the repetition rate of the arresting device of Problem 4·170? What are the good and bad features of the device? What improvements do you suggest?

4·172 Refer to Problem 4·170. Should the vent be opened during operation to limit the deceleration? How would this change the force-displacement curve?

4·173 Refer to Problem 4·170. What would be the effects of increasing the initial pressure of the air in the cylinder?

4·174 Packages with masses of 80 kg moving at 5 m/s on a conveyor are to be brought to rest so they can be removed from the conveyor by an existing pickup device. The deceleration must not exceed $5g$, where $g = 9.81$ m/s^2. A proposed arresting device is shown in the schematic diagram with Problem 4·170, where the gas in the cylinder is air initially at room pressure and temperature. The cylinder length must not exceed 0.40 m. Determine the dimensions of a cylinder that will meet these specifications. Plot force versus displacement during the deceleration for various stroke/bore ratios.

4·175 Packages with masses in the range of 40 kg to 120 kg moving with velocities in the range of 2 to 11 m/s on a conveyor are to be brought to rest so they can be removed from the conveyor by an existing pickup device. The deceleration must not exceed $5g$, where $g = 9.81\ \text{m/s}^2$. A proposed arresting device is shown in the schematic diagram with Problem 4·170, where the gas in the cylinder is air initially at room pressure and temperature. The engineer proposing the device points out that one advantage is that this device restores each package, regardless of its mass or speed, to approximately its position at the time of initial impact. Determine reasonable dimensions of a cylinder that will meet these specifications. Plot force versus displacement during the deceleration. What improvements in the arresting device do you suggest?

4·176 Refer to Problem 4·175. The piston stroke changes with the mass and velocity of the packages arrested. For the results you obtained for Problem 4·175, what is the maximum stroke/bore ratio?

4·177 An alternative proposal for a package-arresting device described in Problem 4·170 is to have a paddle wheel that, through a screw or pulley arrangement, is made to rotate in a tank of oil by the packages being decelerated. Would such a device work? If so, what are its advantages and disadvantages in comparison with the device of Problem 4·170?

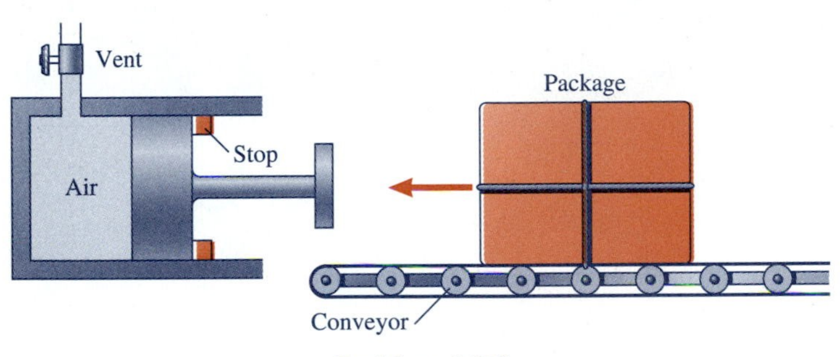

Problem 4·170

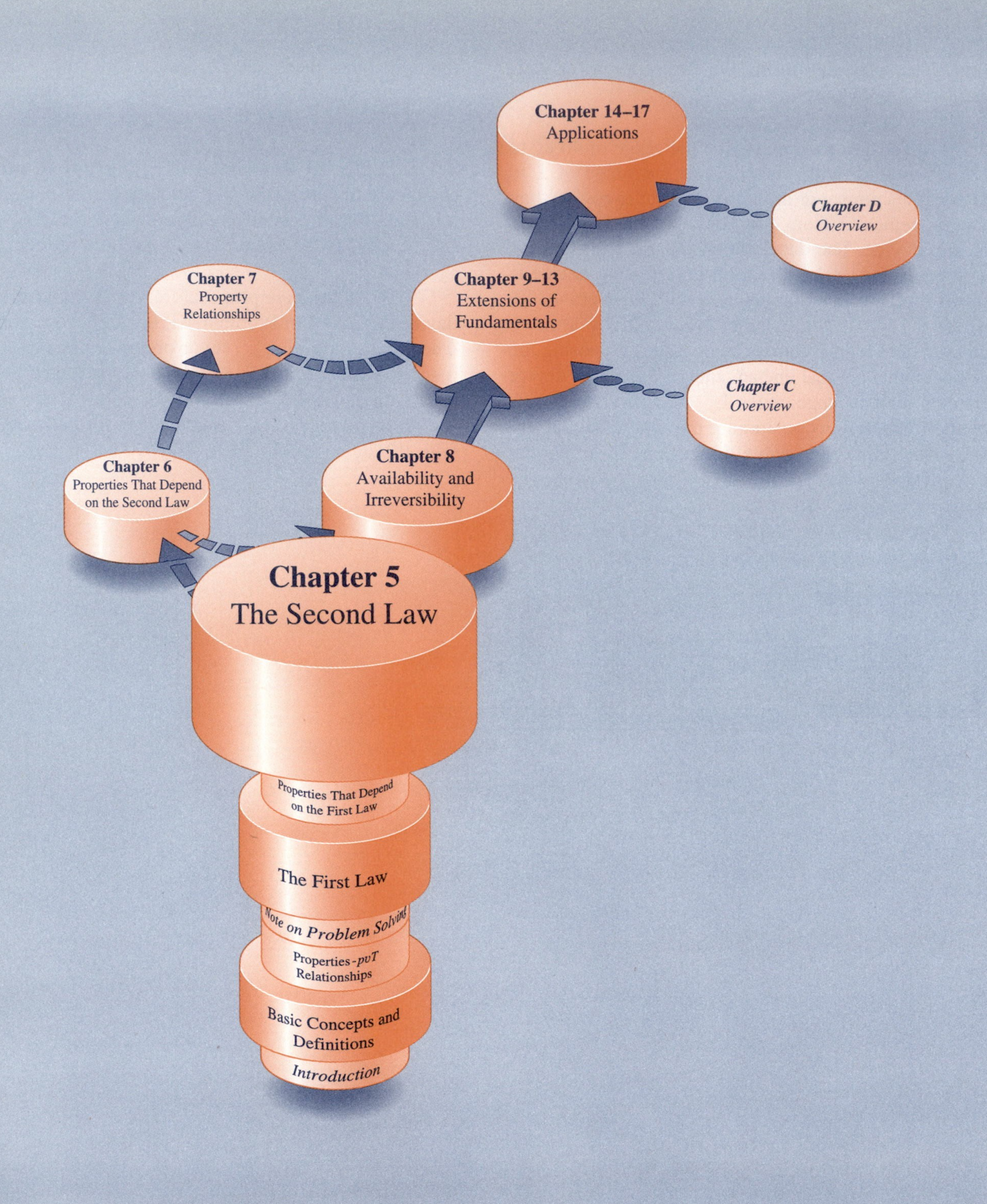
Chapter 14–17
Applications
Chapter D
Overview
Chapter 7
Property
Relationships
Chapter 9–13
Extensions of
Fundamentals
Chapter C
Overview
Chapter 6
Properties That Depend
on the Second Law
Chapter 8
Availability and
Irreversibility
Chapter 5
The Second Law
Properties That Depend
on the First Law
The First Law
Note on Problem Solving
Properties-pvT
Relationships
Basic Concepts and
Definitions
Introduction

Chapter 5

The Second Law of Thermodynamics

In this chapter, we will first discuss the limitations of the first law. As valuable as it is in calculations of energy conversions, the first law cannot tell us whether or not certain energy conversions *can* occur. To determine this, we need a second independent principle, the second law of thermodynamics, which is the main subject of this chapter.

The second law shows that some processes cause changes in a system and its surroundings that can be completely undone; that is, both the system and the surroundings can be returned to their initial states. Other processes cause changes such that the system and all its surroundings can never both be returned to their initial states! These processes are called reversible and irreversible processes, respectively. Engineers are interested in whether processes are reversible or irreversible, because for devices that produce work (like engines and turbines), reversible processes deliver more work than irreversible processes. Also, devices that require work (like compressors, pumps, and refrigerators) require less work input when reversible processes are used instead of irreversible ones.

5·1 Limitations of the First Law; Why We Need a Second Law

The second law allows us to determine the extent of an energy conversion and whether an energy conversion is possible.

The first law is a versatile and valuable tool. However, there are energy conversion phenomena that cannot be explained using the first law alone. The first law expresses the relationship between work and heat, and it allows us to define stored energy. But it does not allow us to predict the extent of an energy conversion nor does it indicate whether a proposed conceived energy conversion process is possible.

Heat cannot be completely converted to work.

As an example, experience shows that at least one type of desirable energy conversion—heat to work—cannot be carried out completely. Furthermore, certain processes that would in no way violate the first law cannot occur. Let us examine these two lessons of experience in the following illustrations.

1. Consider a gasoline engine. Energy stored in the fuel and in the combustion air is delivered to the engine. Energy leaves the engine as work via the drive shaft, as heat, and as stored energy in the exhaust gases. There is also an energy transfer as flow work of the fluids entering and leaving. For economical operation the work output should be as great as possible for a given amount of energy input. Thus the energy leaving the engine in the exhaust gases and as heat should be reduced to a minimum. As far as the first law is concerned, these energy losses could be reduced to zero. Then the work output would equal the energy input. However, all attempts to obtain such performance from an engine have failed. No matter what ingenious accessories have been used, the complete conversion of fuel energy into work by an engine has not been accomplished.

2. A steam power plant affords another example of a limited conversion of energy. Even the best steam power plants require for the production of 100 kJ of work an energy input of about 250 kJ. Thus, for every 100 kJ of work produced, about 150 kJ is rejected to the surroundings as some form of energy other than work.

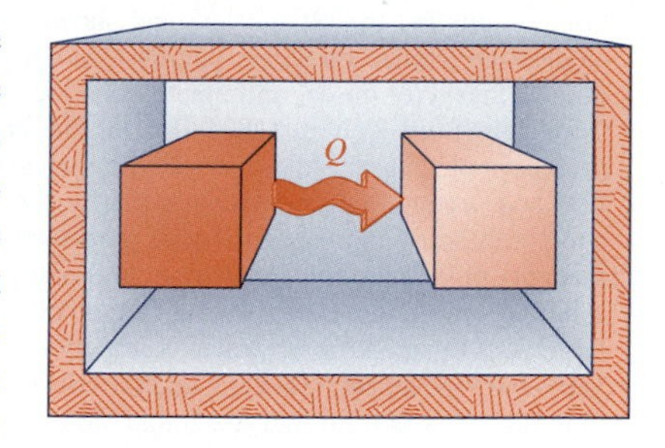

3. Consider now two blocks of copper at different temperatures that are enclosed together in a thermally insulated box. Of course, energy will be transferred from the higher-temperature to the lower-temperature block as heat. If there is nothing else inside the box, the amount of energy lost by the higher-temperature block will be equal to the amount of energy gained by the lower-temperature block. This is in accordance with the first law. The first law, however, would be satisfied also by a process whereby energy was transferred from the lower-temperature block to the higher-temperature block, but this latter process never occurs. Thus, satisfaction of the first law does not ensure that a process can occur.

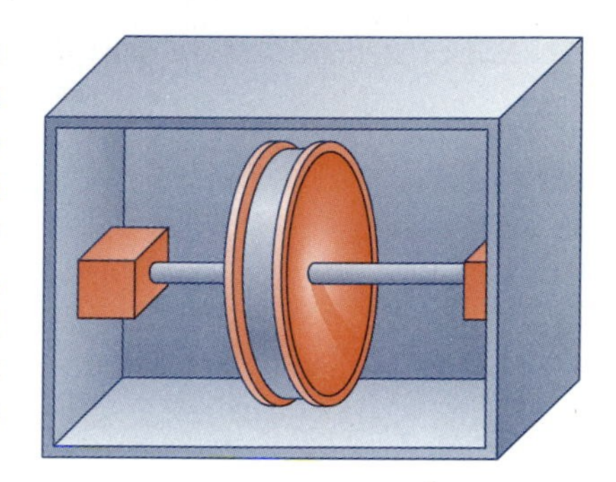

4. As another example, a spinning flywheel mounted on a shaft between two bearings in a vacuum will come to rest as a result of friction in the bearings. In this process the kinetic energy of the flywheel and shaft is reduced to zero while the internal energy of the bearings, the lubricant, and part of the shaft is increased by the same magnitude. Energy is conserved. Energy would also be conserved in the reverse process that would cool the bearings, lubricant, and parts of the shaft and accelerate the flywheel and shaft until their kinetic energy equals the decrease in internal energy of those parts that were cooled. Describing such an event seems absurd, though, because it is common knowledge that this process does not occur.

On the basis of these observations, we naturally wonder if there is some law of nature that agrees with our observations in these simple cases and can guide us in more complex cases where it is not so obvious that certain processes cannot occur. Such a law of nature does exist and is called the second law of thermodynamics.

5·2 The Second Law of Thermodynamics

Much experience, including many deliberate experiments, has led to the second law of thermodynamics. This law has been stated in many different forms, some of which appear at first to bear no relation whatsoever to each other. However, if any one of the statements of the second law is accepted as a postulate, all the other statements can be proved from this starting point. None of the statements, though, can be derived from any other law of nature.

The second law of thermodynamics, like the first law, cannot be derived from any other law of nature.

Two well-known statements of the second law of thermodynamics are the Clausius statement and the Kelvin–Planck statement.

Rudolph Julius Emmanuel Clausius (1822–1888) was a German mathematical physicist. After a study of the work of Sadi Carnot (see Section 5·6) he presented in 1850 a clear general statement of the second law. He applied the second law and showed the value of the property entropy (see Chapter 6) in an exhaustive treatise on steam engines. Although he made significant contributions in the areas of optics, kinetic theory of gases, and electrolysis, he is most famous for his work related to the second law. William Thomson (Lord Kelvin) (1824–1907), one of the outstanding physicists of all time, was for 53 years professor of natural philosophy at the University of Glasgow. In 1851 he presented a paper in which the first and second laws were combined for the first time. In addition to helping firmly establish the first law and formulating the second law of thermodynamics, he published papers on geophysics, electricity, magnetism, telegraphy, navigation, and many other branches of science. He invented many instruments for scientific and engineering work. The encouragement that he gave his students and other scientists played a large part in scientific advances made by others. Max Planck (1858–1947), professor of physics at the University of Berlin, clarified several concepts in thermodynamics but is best known for his work on radiation in which he laid the foundation of the quantum theory.

Clausius statement of the second law

The Clausius statement may be given as follows: **It is impossible for any device to operate in such a manner that it produces no effect other than the transfer of heat from one body to another body at a higher temperature.** This statement warrants close investigation. It does not say that it is impossible to transfer heat from a lower-temperature body to a higher-temperature body. Indeed, this is exactly what a refrigerator does. A refrigerator cannot operate, though, unless it receives an energy input, usually in the form of work. This energy input from the surroundings constitutes an effect other than the transfer of heat from the lower-temperature body to the higher-temperature body.

The Clausius statement is sometimes given as ''Heat cannot of itself pass from a cold to a hot body.'' In the words of Planck, ''As Clausius repeatedly and expressly pointed out, this principle does not merely say that heat does not flow directly from a cold to a hot body . . . but it expressly states that heat can in no way and by no process be transported from a colder to a warmer body without leaving further changes.''

Kelvin–Planck statement of the second law

A statement of the second law that pertains more directly to heat engines is the Kelvin–Planck statement: **It is impossible for any device to operate in a cycle and produce work while exchanging heat only with bodies at a single fixed temperature.** A common version of the Kelvin–Planck statement is, ''It is impossible for any device operating in a cycle to absorb heat from a single reservoir and produce an equivalent amount of work.'' The term *reservoir* refers to an **energy reservoir**, which is defined as a body or system that can absorb or reject a finite amount of energy with no appreciable change in its temperature. In practice, the atmosphere or the water of a river or lake can serve as an energy reservoir. Thousands of kilojoules can be discharged to river water by a power plant without appreciably raising the temperature of the river. A furnace atmosphere that is maintained at a constant temperature by the combustion of fuel serves as an energy reservoir in a steam power plant.

Energy reservoir

The Clausius and Kelvin–Planck statements of the second law are entirely equivalent. This equivalence can be demonstrated by showing that the violation of either statement can result in a violation of the other one. Each case will now be demonstrated.

The Clausius and the Kelvin–Planck statements of the second law are equivalent.

Violation of Clausius Statement ⇒ Violation of Kelvin–Planck Statement. Referring to Fig. 5·1*a*, the device marked "Clausius violator" causes heat $|Q_H|$ to be transferred from the energy reservoir at T_L to the one at the higher temperature T_H without causing any other effects. This kind of device is impossible according to the Clausius statement. We will now show that if it were possible, it would also violate the Kelvin–Planck statement. Let a heat engine operate with heat input $|Q_H|$ from the reservoir at T_H and with an amount of heat $|Q_L|$ rejected to the reservoir at T_L. Experience shows that actual heat engines can operate in this manner. Application of the first law shows that the amount of heat rejected $|Q_L|$ is equal to $(|Q_H| - |W|)$, where W is the work output of the heat engine. Since the amount of heat rejected at T_H by the Clausius violator is equal to that absorbed at the same temperature by the heat engine, the operation of these two devices, bounded by a broken line in Fig. 5·1*a*, produces no change in the reservoir at T_H. In fact, the reservoir could be eliminated by having the Clausius violator transfer heat directly to the heat engine. Thus the system composed of the two devices operates in a cycle, absorbs a net amount of heat $(|Q_H| - |Q_L|)$ from a single reservoir (at T_L), produces work, and produces no other effects. This is a violation of the Kelvin–Planck statement.

We use absolute values of Q and W in this discussion because doing so is convenient. Also, in evaluating engine and refrigerator performance the signs are clear from the context. As an exercise, you might try the same developments using the sign convention strictly.

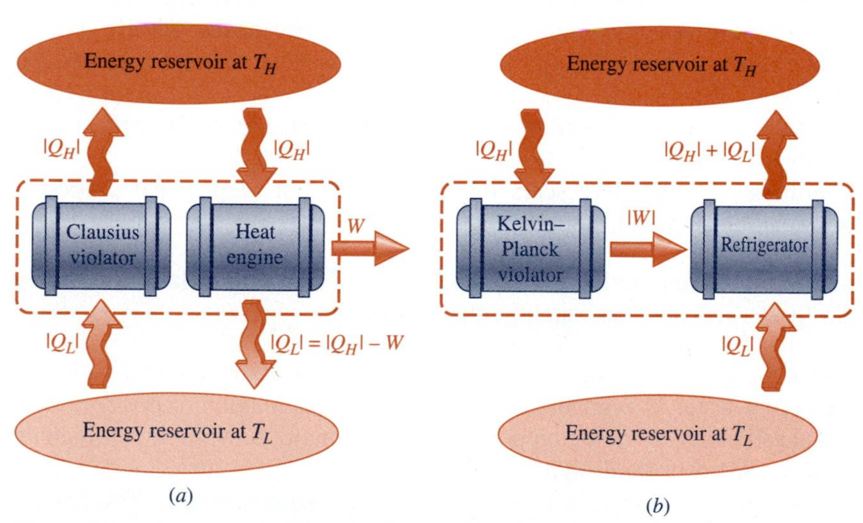

Figure 5·1 Illustration of the equivalence of the Clausius and Kelvin–Planck statements of the second law.

Violation of Kelvin–Planck Statement ⇒ Violation of Clausius Statement. The device labeled "Kelvin–Planck violator" in Fig. 5·1*b* absorbs an amount of heat $|Q_H|$ from the reservoir at T_H, produces an equivalent amount of work $|W|$, and produces no other effects. This kind of device violates the Kelvin–Planck statement. To show that the operation of such a device can result in a violation of the Clausius statement, let a refrigerator be used to transfer heat from the reservoir at T_L to the reservoir at T_H. Such a refrigerator can certainly be operated, but it requires some work input. Let the refrigerator be driven by the Kelvin–Planck violator. Then see what the system composed of both devices does. It operates cyclically, causes heat to be transferred from one reservoir to another one at a higher temperature, and produces no other effects! Thus a violation of the Kelvin–Planck statement results in a violation of the Clausius statement. Our conclusion is that these two statements of the second law of thermodynamics are equivalent.

As yet, all attempts to disprove the second law have failed.

We will deduce several corollaries from the second law in the following chapter. To disprove the second law it is necessary only to disprove any one of its corollaries. As yet, all attempts to do so have failed.

5·3 The Uses of the Second Law

Why is the second law valuable? In later chapters we will see that the second law and its corollaries provide means for

- Determining the maximum possible efficiencies of heat engines
- Determining the maximum coefficients of performance of refrigerators
- Determining whether any particular process we may conceive is possible or not
- Predicting in which direction a chemical reaction or any other process will proceed
- Defining a temperature scale that is independent of physical properties
- Correlating physical properties

What kinds of questions can be answered by means of the second law?

For example, after studying the applications of the second law, we will be able to answer questions such as the following:

1. In a steam power plant, the maximum furnace temperature is 1400°C. Cooling water is available at 15°C. What is the maximum possible thermal efficiency of the steam power plant, no matter how many refinements are included in its design?
2. A refrigerator located in a room where the air temperature is 20°C freezes 5 kg of ice per hour from water initially at 10°C. The refrigerator rejects heat to the air in the room. What is the absolute minimum power requirement of the refrigerator?

3. Air expands adiabatically from 200 kPa, 40°C, to 100 kPa. What is the lowest possible final temperature?
4. Is it possible to compress air adiabatically from 100 kPa, 15°C, to 200 kPa, 30°C? To 200 kPa, 105°C?
5. If carbon dioxide is expanded adiabatically through a nozzle from zero velocity at 40 psia, 100 F, to 15 psia, what is the maximum velocity it can reach?
6. Sixteen kilograms of oxygen and 28 kg of carbon monoxide fill a tank. The mixture is heated until it reaches 250 kPa, 500°C. To what extent does the reaction $CO + \frac{1}{2}O_2 \rightarrow CO_2$ occur?
7. Is there some limiting temperature below which a body cannot be cooled?
8. Steel expands when heated. When it is stretched adiabatically within the elastic limit, does its temperature increase or decrease?

The answers to questions such as these can be obtained from the second law through a process of deductive reasoning. So that we do not have to reason all the way from the second law itself each time we answer such questions, we carry some of the reasoning through once and establish certain useful definitions and corollaries of the second law from which we can then begin our reasoning for particular applications. This chapter answers none of the questions listed above, but it does establish some definitions and corollaries that we need to answer all these questions in later chapters.

5·4 Reversible and Irreversible Processes

A process is **reversible** if, after it has occurred, both the system and the surroundings can by any means whatsoever be returned to their original states. Any other process is **irreversible**.

Reversible processes

Irreversible process

Reversible processes are important because they provide the maximum work from work-producing devices and the minimum work input to devices that absorb work to operate. For these devices and many others, reversible processes are standards of comparison. To determine whether a process is reversible, it is necessary to apply the second law. Use of the second law for this purpose is illustrated in the following section.

Reversible processes are standards of comparison.

5·4·1 Illustrations

We will examine irreversible processes before reversible processes for two reasons. First, it is often easier to show that a process is not reversible than to show that one is. Second, once certain irreversible phenomena are identified, we can often recognize reversible processes simply by the absence of these irreversible phenomena.

It is often easier to show that a process is irreversible than to show that it is reversible. We can identify reversible processes by the absence of irreversible ones.

So, how can we identify irreversible processes? Recall that if a process is reversible, then both the system and the surroundings can be restored to their initial states. However, if a process is irreversible, the reverse process is impossible. Therefore, we can determine if a process is reversible by determining if the reverse process is possible. We can prove that a process is impossible as follows: (1) Assume that the process is possible. (2) Combine this process with other processes, known from experience to be possible, to form a cycle which violates the second law. If such a cycle can be devised, then the assumption of step 1 is false and the process in question is impossible.

Two-step methodology for proving that a process is impossible

Determining if stirring is a reversible process

Consider the process in which a gas in a closed, rigid, thermally insulated tank is stirred by a paddle wheel (Fig. 5·2). The system is the gas within the tank. (Assume there is never more than a negligible amount of heat transfer between the gas and the paddle wheel itself.) Let the paddle wheel be turned by the action of a falling weight that turns a pulley on a shaft. The motion of the paddle wheel is resisted by shearing forces in the gas, and thus work is done on the gas by the paddle wheel. The gas changes from state A to state B. Application of the first law shows that during this process the internal energy of the gas increases. The temperature of the gas increases.

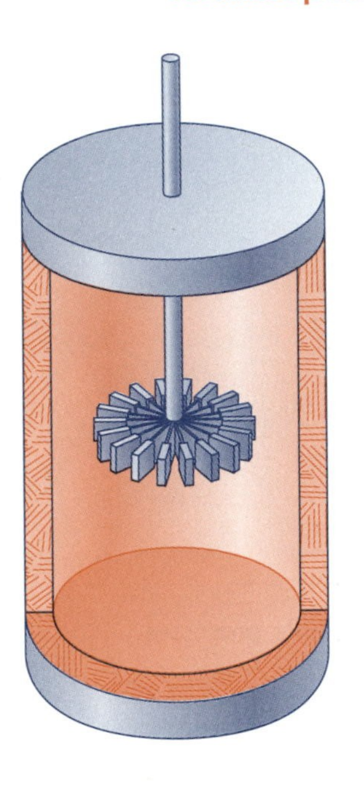

Is this process reversible? If it is reversible, then it is possible to restore both the system and the surroundings to their initial states. That is, there must be some process that results in the weight's being lifted to its initial position while the internal energy of the gas (and hence its temperature) decreases. What constraints are there? No change except the lowering of the weight was made in the surroundings during the original process. Therefore, no change in the surroundings except the lifting of the weight can be made during the reverse process. If this reverse process is possible, then the stirring process is reversible.

Is the reverse process possible? Let us answer this by first *assuming* that the reverse process is possible. Then consider a cycle composed of two processes.

Process 1. The process described above in which the weight is raised as the temperature and internal energy of the gas decrease. (This is the process we have assumed to be possible.) The system changes from an initial state B to state A (see Fig. 5·3).

Process 2. A process in which heat is transferred from some constant-temperature energy reservoir in the surroundings to the gas while the paddle wheel is stationary. (Part of the thermal insulation of the tank must be removed during this process.) This process continues until the gas is brought to its initial temperature T_B. (The energy reservoir must therefore be at a temperature higher than T_B.)

The net results of this cycle are as follows:

1. The system has executed a cycle and returned to its initial state.

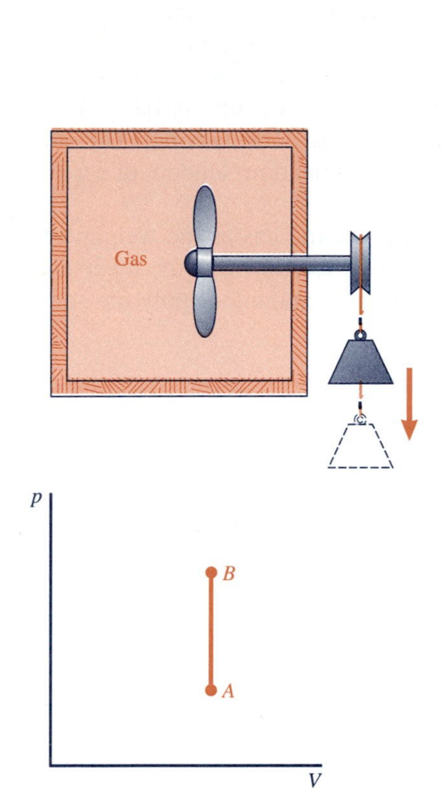

Figure 5·2 Irreversible process. System changes from state A to state B as weight drops and paddle wheel does work on gas.

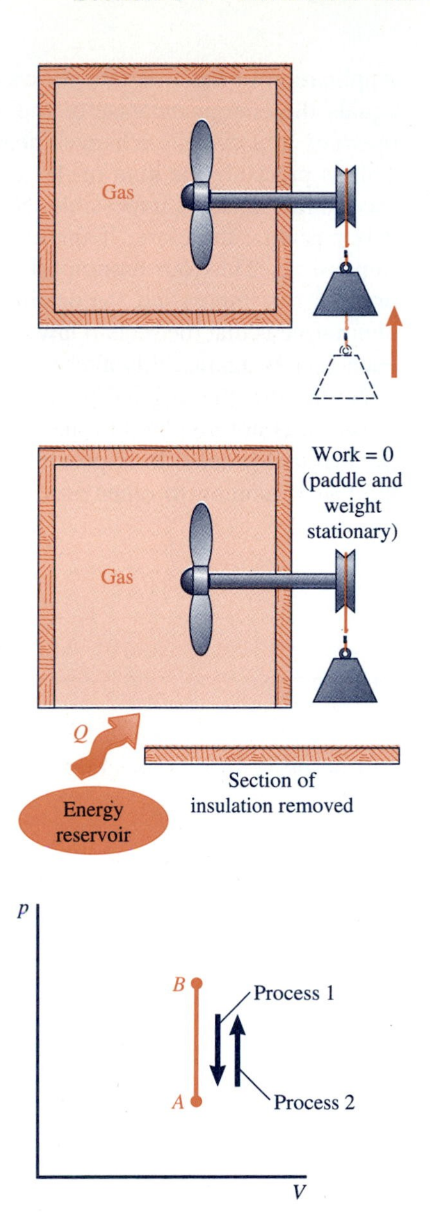

Figure 5·3 Cycle incorporating the assumed-possible reversal of the stirring process.

2. The surroundings have changed in two ways:
(*a*) The weight is at a higher level.
(*b*) The amount of energy stored in the energy reservoir has been decreased.

Application of the first law shows that the energy decrease of the reservoir equals the energy increase of the weight. Thus the system is a device that operates in a cycle, exchanges heat with a single reservoir, and does work. This is precisely the kind of device that the Kelvin–Planck statement of the second law declares impossible. Now it is a matter of experience that process 2 is possible; therefore, if the cycle is impossible, then process 1 must be impossible. Thus our assumption that process 1 is possible is false. Since process 1 is impossible, the original stirring process is irreversible. The only alternative is that the second law is false, and a tremendous amount of experience argues against this alternative.

Stirring is an irreversible process.

Notice that the original process, the stirring of the gas as the weight was lowered, is not involved in the cycle we devised to show that the reverse of the original process is impossible. The following examples apply the same kind of reasoning to other processes.

Example 5·1 Heat Transfer Across a Finite Temperature Difference

PROBLEM STATEMENT

Demonstrate that the transfer of heat across a finite temperature difference is irreversible.

SOLUTION

If the transfer of heat from a body at a temperature T_H to a body at a lower temperature T_L is reversible, then the transfer of heat from the body at T_L to the one at T_H with no other effects is possible. *Assume* that this transfer of heat from T_L to T_H is possible. Then, referring to the figure, construct a cycle as follows:

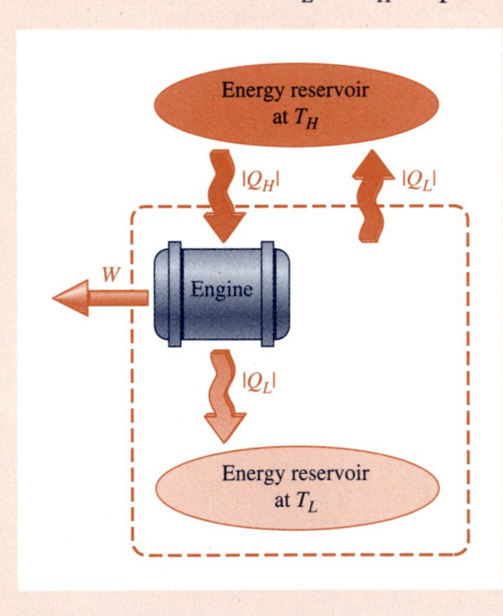

1. Let an amount of heat $|Q_H|$ be transferred from a reservoir at T_H to a heat engine that operates cyclically, producing work and rejecting an amount of heat $|Q_L|$ to a reservoir at a lower temperature T_L. We know from experience that this can be done.
2. Then let an amount of heat $|Q_L|$ be transferred from the reservoir at T_L to the one at T_H in accordance with the assumption that such a process is possible.

Since $|Q_L|$ was added to the low-temperature reservoir in the first process and the same amount of heat was withdrawn from it during the second process, this reservoir executed a cycle. The heat engine also executed a cycle. Therefore the entire system enclosed by the broken line in the diagram (the heat engine and the reservoir at T_L) executed a cycle. During this cycle this composite system produced work while exchanging heat with a single reservoir. Such a cycle violates the second law.

Checking back to see which process of the proposed cycle might actually be impossible and thus prevent the execution of such a cycle, we see that the first process is proved by experience to be possible, so the second process, which we *assumed* to be possible, must actually be impossible. If this process—the transfer of heat from T_L to T_H—is impossible, then the transfer of heat across the finite temperature difference from T_H to T_L must be irreversible.

Example 5·2 Free Expansion of a Gas

PROBLEM STATEMENT

Demonstrate that the free expansion of a gas is an irreversible process. An example of a free or unrestrained expansion of a gas is the following: An insulated tank is separated into two parts by a partition. A gas is held on one side of the partition; the other side of the tank is evacuated. This is state *A* of the system. An opening is then made in the partition and the gas expands to fill the entire tank. This is state *B*. No work is done as the gas expands.

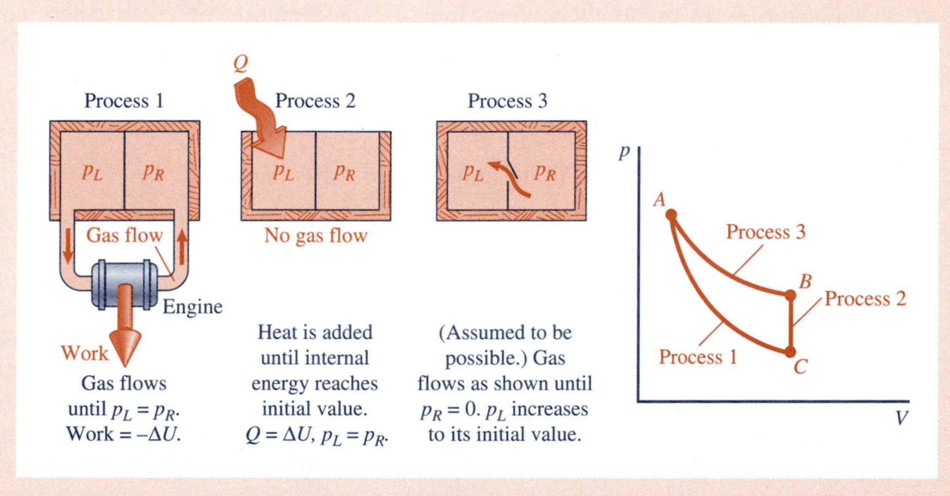

SOLUTION

Consider the gas as the system. First, assume that the reverse process is possible. The reverse process begins with the gas occupying the entire tank and requires the gas on one side of the partition to move against an increasing pressure to the other side of the tank. This must occur with no interaction with the surroundings since the free expansion process occurred with no interaction with the surroundings.

Next consider a cycle of three processes:

Process 1. Starting with the gas all on one side of the partition, state A, let part of it expand through an engine and into the other part of the tank until the pressure is the same on the two sides of the partition. In expanding through the engine, the gas does work so that its internal energy is decreased. Let heat transfer occur between the two parts of the tank so all of the gas comes to the same temperature, state C.

Process 2. Remove part of the tank insulation, and add heat from an external reservoir to the gas until its internal energy is restored to its initial value. The gas is now in state B with $U_B = U_A$. State B is also the state the gas would be in following a free expansion from state A.

Process 3. Starting with the gas in state B, let the reverse of a free expansion occur to restore the system to its initial state A. Thus a cycle is completed.

This cycle results in a production of work while heat is absorbed from a single reservoir. It thereby violates the second law. Inspection of the cycle reveals that processes 1 and 2 are shown by experience to be possible, whereas process 3 is *assumed* to be possible. If the second law is accepted, then the assumption regarding process 3 is false. The reverse of a free expansion is therefore impossible and a free expansion must be irreversible.

5·4·2 Characteristics of Reversible and Irreversible Processes

By reasoning similar to that used in the preceding examples it can be shown that processes involving mixing, inelastic deformation of a substance, and certain other effects are also irreversible. We use this information to conclude that a *reversible process* must involve *no*

Irreversible effects

- Friction
- Heat transfer across a finite temperature difference
- Free expansion
- Mixing
- Inelastic deformation

The test for reversibility involves the application of the second law of thermodynamics.

Various other effects (such as an electric current flow through a resistance) are also irreversible but are not listed here. In all cases the test for reversibility involves the application of the second law of thermodynamics.

At this point you may infer from the examples above that we use a second law formalism to reach conclusions that are obvious from common experience. Indeed, it is well that these simple examples do confirm common experience. In one sense, their doing so makes them good examples. The reasoning that may appear unnecessary here, though, is valuable because it applies to cases where common experience does not provide an obvious conclusion. Some of the questions listed in Section 5·4, in fact, refer to such cases, and we will encounter others in subsequent chapters.

Common features of all reversible processes

Let us now identify some *features that are common to all reversible processes*. Consider first a system composed of a gas trapped in a cylinder fitted with a frictionless gas-tight piston. Let the cylinder and piston be made of a material that is a perfect heat insulator. If the piston is slowly pushed into the cylinder, the pressure and temperature of the gas increase uniformly throughout the gas. A very small decrease in the external force on the piston will permit the gas to expand, and, if the expansion is very slow, the pressure decreases uniformly throughout the system. For each position of the piston, the pressure during the expansion is the same as it was during the compression. Consequently, the work done by the gas during expansion equals the work done on the gas during compression. When the gas reaches its initial volume, the net work is zero. Also, there has been no heat transfer. The surroundings, as well as the system, have therefore been returned to their initial state. Consequently, the very slow frictionless adiabatic process is reversible.

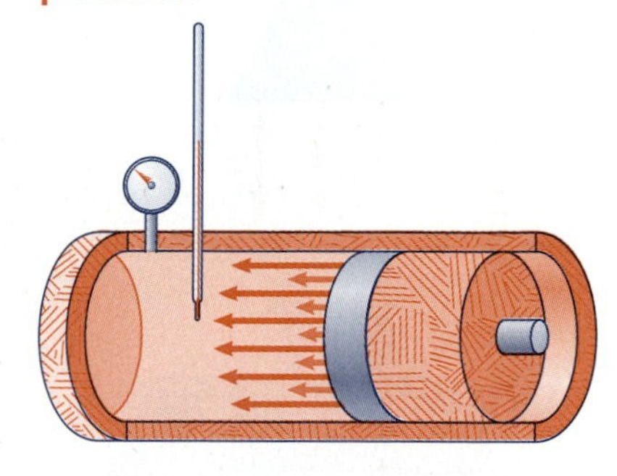

A very slow frictionless adiabatic process is reversible.

In contrast, if the adiabatic compression is performed by a *rapid* inward motion of the piston, the process is not reversible. During the process, the pressure near the piston face is higher than that elsewhere in the cylinder. A pressure wave is initiated, and it travels through the gas until the pressure becomes uniform. Then, even if the gas expands slowly to its initial volume, for each position of the piston, the pressure near the piston face is lower than it was during the compression process. The work done during the expansion is therefore less than that done on the gas during compression. At the end of the expansion process, the stored energy of the system is greater than it was initially; but, since the system volume equals its initial value, the excess stored energy cannot be removed *as work* while the system is restored to its initial state. Let the excess stored energy be removed by a transfer of heat from the system to the surroundings while the piston is stationary. The system has now been returned to its initial state. Turning our attention now to the surroundings, we see that work was taken from the surroundings in order to compress the gas. Perhaps the work was done by the lowering of a weight or the unwinding of a coil spring in the surroundings. Then the system performed less work on the surroundings to raise the weight part of the way to its initial position or to rewind the spring partially. Finally, heat was transferred to the surroundings. For the surroundings to be returned to their initial state, this heat must be converted completely into work to raise the weight or wind the spring without causing any other effects. Any device that could perform this conversion would violate the second law; hence we conclude that the system and the surroundings cannot *both* be restored to their initial states. Therefore, the adiabatic compression of the gas during which the pressure is not uniform throughout the gas is irreversible.

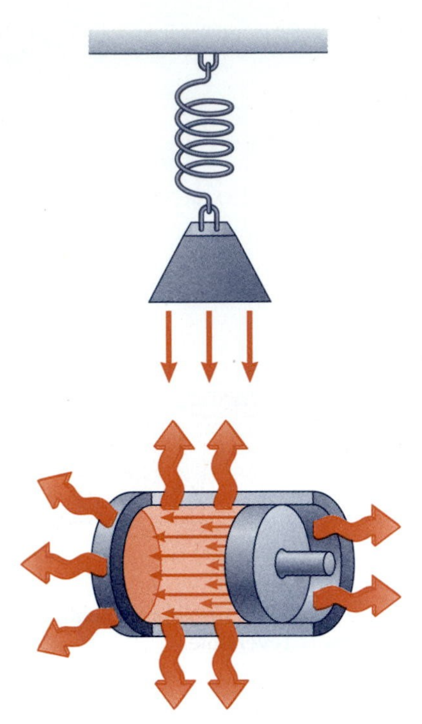

A very slow frictionless isothermal process is reversible.

A reversible process may be approximated by the deflection of a spring. If a very small load is slowly applied to a tension spring, the spring will elongate a minute distance. After elongation, if the load is reduced so that the spring is allowed to contract to its original position, the work performed by the spring as it contracts will be approximately equal to that required to stretch the spring. This process approaches a reversible process only if the force is applied and reduced incrementally; otherwise, unconstrained vibrations and other effects will occur that would make the process irreversible.

Another example of a reversible process is the frictionless isothermal process. Consider a gas in a cylinder having a gas-tight frictionless piston and cylinder walls that are perfect conductors of heat. If the gas is very slowly compressed, its temperature tends to rise; but as soon as the temperature is slightly greater than that of the surroundings, heat transfer from the system occurs to keep the gas temperature constant as work is done on it. If the gas is allowed to expand very slowly, its temperature tends to decrease; but as soon as it is slightly lower than the temperature of the surroundings, heat transfer to the system occurs to keep the gas temperature constant as work is done by the system. Minute temperature differentials are used, so the time required for the expansion or compression to occur is extremely long. This type of process fulfills the requirements of a reversible process as the temperature difference between the system and the surroundings approaches zero. The temperature must be uniform throughout the gas so that only an infinitesimal change in the temperature of the surroundings causes the surroundings to be at a higher or lower temperature than the entire system.

Heat transfer can be reversible even when the temperature of the system varies, provided the temperature of the surroundings also varies.

Can a transfer of heat be reversible if the temperature of the system varies with time? Yes, provided the temperature of the surroundings also varies with time so that the difference in temperature between the system and the surroundings is never more than an infinitesimal amount.

In the design of actual devices, reversible heat transfer is impractical.

The greater the temperature difference between two bodies, the greater the rate of heat transfer between them will be. This is true whether the heat is transferred by conduction, convection, or radiation. As the temperature difference is made smaller, the time required to transfer a given quantity of heat increases. As the temperature difference approaches an infinitesimal value, the time required to transfer any finite amount of heat grows toward an infinite value. Obviously, in the design of heat transfer devices, energy transfer rates approaching zero are unacceptable. Thus, in actual devices, the temperature difference is not infinitesimal and the processes cannot be considered reversible. Reversible heat transfer is the *limiting case* of heat transfer as the temperature difference between two bodies approaches zero.

The steady frictionless adiabatic flow of a fluid through a nozzle is a reversible process.

Another example of a reversible process is the steady frictionless adiabatic flow of a fluid through a nozzle. Application of the first law to such a system shows that, as kinetic energy increases in the direction of flow, the enthalpy decreases. If the nozzle is followed by a frictionless diffuser, as shown in Fig. 5·4, the fluid undergoes an increase in enthalpy and a decrease in kinetic energy between sections 2 and 3, and it can be discharged in a state 3 which

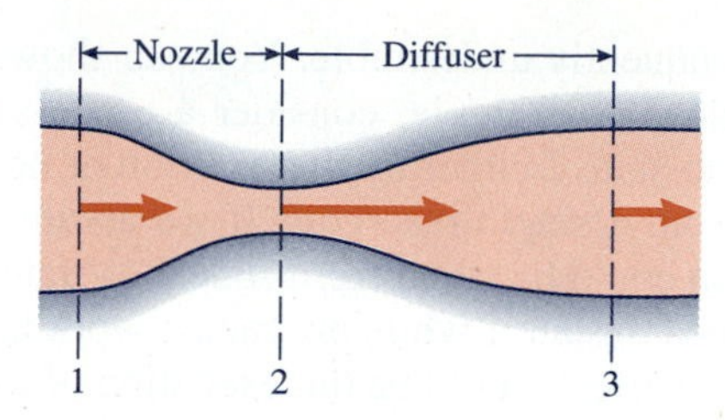

Figure 5·4 Reversible processes in a nozzle and diffuser.

is identical with state 1. Thus, after the nozzle process has occurred, it is possible to restore the fluid flowing and all parts of the surroundings to their initial states.

We have specified that reversible processes must be carried out slowly, without friction, and with uniform properties throughout the system if it is a closed system. These are also the requirements for a quasiequilibrium process as introduced in Section 1·3. If the system is an open one, a reversible process must also meet the requirements of a quasiequilibrium process. Consequently, *any reversible process must be a quasiequilibrium process.*

Any reversible process must be a quasiequilibrium process.

Study of various reversible and irreversible processes such as those just described leads to several conclusions regarding reversible processes:

Characteristics of reversible and irreversible processes

1. A reversible process must be such that, after it has occurred, the system and the surroundings can be made to traverse, in the reverse order, the states they passed through during the original process. All energy transformations of the original process would be reversed in direction but unchanged in form or magnitude.
2. The direction of a reversible process can be changed by making infinitesimal changes in the conditions that control it.
3. During a reversible process, the system and the surroundings must each at all times be in states of equilibrium or infinitesimally close to states of equilibrium; that is, the process must be a quasiequilibrium one.
4. A reversible process must involve no friction, unrestrained expansion, mixing, heat transfer across a finite temperature difference, or inelastic deformation.

A reversible process must meet each of the conditions listed. If any one of the conditions is not met, the process is irreversible. Thus we can apply these conditions to test any process for reversibility without having to go through the process of reasoning all the way from a second law statement.

Work can always be completely converted into heat, but the extent to which heat can be converted into work is limited. Therefore, work is the more valuable form of energy in transition. A process that uses work to produce the same effect that could be produced by the less valuable form of

Work is the more valuable form of energy in transition.

energy, heat, is consequently undesirable. You can show that such a process is always irreversible. For example, consider a process that takes a closed system from state 1 to state 2 either by the addition of heat or by work input. The heat is added from a single reservoir. If we assume that the process is reversible, the system could be taken from state 1 to state 2 by means of heat input and then returned to state 1 while producing work in an amount equal to the heat input. The net effect would be the execution of a cycle by the system while withdrawing heat from a single energy reservoir and producing an equivalent amount of work. This would be a violation of the Kelvin–Planck statement of the second law. Therefore, our assumption must be false and we conclude that a process brought about by work input in place of heat input is irreversible.

Engineers use reversible processes as limiting cases in analyses.

All actual processes are irreversible. Reversible processes do not occur. Nevertheless, reversible processes are extremely useful and serve as standards of comparison since they are often the limiting cases of actual processes. Many actual processes are difficult to analyze completely; therefore, an engineer must frequently base an analysis or design on reversible processes and then adjust the results to apply them to actual processes. This is similar to the use of ideal systems in the analysis or design of actual systems (see Section 1·4). Engineers use reversible processes in the same way that they use point masses, frictionless pulleys, weightless cords, and homogeneous beams: as idealizations to simplify the analysis of actual systems and processes.

The concept of reversible processes is also important because it permits the definition of a highly useful property, entropy, which is introduced in Chapter 6.

5·4·3 Internal and External Reversibility

A process is irreversible if it involves heat transfer across a finite temperature difference between the system and the surroundings. However, the *system* can behave during this irreversible process just as though the heat were being transferred reversibly across an infinitesimal temperature difference. Such a process is said to be **internally reversible** because nothing occurs within the system to make it irreversible, but it is **externally irreversible**.

Internally reversible
Externally irreversible

A process is reversible only if it is both internally and externally reversible.

A process meets our definition of a reversible process only if it is both internally and externally reversible. Frictionless adiabatic and frictionless isothermal processes described in the preceding section are *internally and externally reversible*. A process that involves friction or some other irreversibility within the system and also heat exchange with surroundings at a different temperature is *internally and externally irreversible*.

Internally reversible	*Externally reversible*	*Resulting process is*
√	√	**Reversible**
√	×	**Irreversible**
×	√	**Irreversible**
×	×	**Irreversible**

Some relationships that have been developed for reversible processes involve only properties of the system and consequently are valid for processes that are internally reversible even if they are externally irreversible. As an example, a gas expanding in a quasiequilibrium process in a cylinder against a moving piston may have heat added across a finite temperature difference.

This external irreversibility in no way affects the *pv* relationship, so the relation

$$w = \int p\, dv \qquad (1\cdot17)$$

still holds. The restriction on this equation as well as on the analogous relationship for a steady-flow system,

$$w = -\int v\, dp - \Delta\left(\frac{V^2}{2}\right) - g\,\Delta z \qquad (1\cdot20)$$

has been that the equation holds for quasiequilibrium processes of simple compressible substances. Now, in the light of the current discussion, from here on the restriction "quasiequilibrium" will be replaced by "internally reversible." We could not have used the term *reversible* in earlier chapters, because the definition of the term depends on the second law.

Quasiequilibrium replaced by reversible

One can also conceive processes that are externally reversible but internally irreversible, but they are seldom useful in modeling actual processes.

5·5 Reversible Cycles

A cycle composed entirely of reversible processes is called a reversible cycle. If all of the processes are only internally reversible, the cycle is an internally reversible cycle. As an example of a reversible cycle, consider an ideal gas trapped in a cylinder behind a piston. Let the gas go through the following processes:

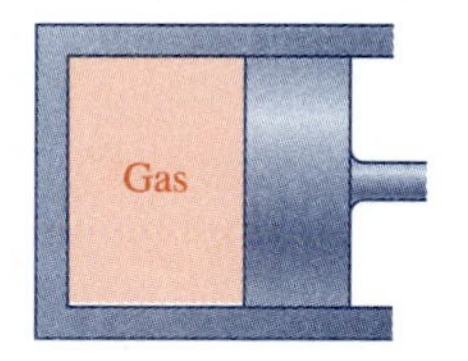

Process	*Description*	*Comment*
1–2	Gas expands reversibly at constant pressure.	During this process, heat is added to the gas and the gas does work on the surroundings.
2–3	Gas expands reversibly and adiabatically to the initial temperature.	During this process, there is no heat transfer and the gas does work on the surroundings.
3–1	Gas is compressed reversibly and isothermally to its initial state.	During this process, work is done on the gas and heat is removed.

This cycle is shown on a *pV* diagram in Fig. 5·5*a*. Since the mean pressure during the expansion is greater than that during the compression, there is a

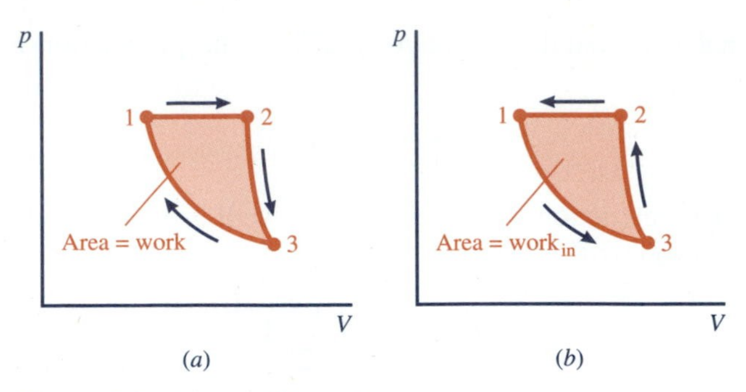

Figure 5·5 A reversible cycle.

net work output from this cycle. This work output is represented by the area within the cycle diagram of Fig. 5·5*a*.

For this cycle to be reversible, each process must be reversible. This means that heat can be transferred only across infinitesimal temperature differences. Consequently, for the constant-pressure process, the temperature of the part of the surroundings exchanging heat with the system must vary so as to be always only infinitesimally different from the system temperature. This is somewhat impractical, even though some actual processes are designed to approach this behavior. If the temperatures of parts of the surroundings are constant, the heat transfer processes are irreversible. However, the irreversibility is external, so the processes of the cycle can all still be internally reversible. The system behavior is the same, so all calculations involving only system properties are the same. For example, we can still calculate work as $\int p\, dV$. For applications to power generation and refrigeration, we will introduce in Chapters 15 and 16 several ideal cycles that are internally reversible.

When reversed, this cycle removes heat from the part of the surroundings that it is in contact with during the isothermal expansion (process 1–3) and discharges heat at higher temperatures during the constant-pressure compression (process 2–1). Thus, it is a refrigeration cycle with a net work input to the system. Figure 5·5*b* shows the *pV* diagram.

Example 5·3 Internally Reversible Cycle

PROBLEM STATEMENT

Helium in a closed system undergoes a cycle composed of the following three internally reversible processes: (1) a constant-pressure expansion at 300 kPa from 20 to 145°C, (2) a constant-volume cooling to 20°C, and (3) an isothermal compression to 300 kPa. Determine (*a*) the energy transfers per kilogram of helium for each process, (*b*) the thermal efficiency of the cycle, and (*c*) the energy transfers for each process when helium undergoes this cycle in reverse.

SOLUTION

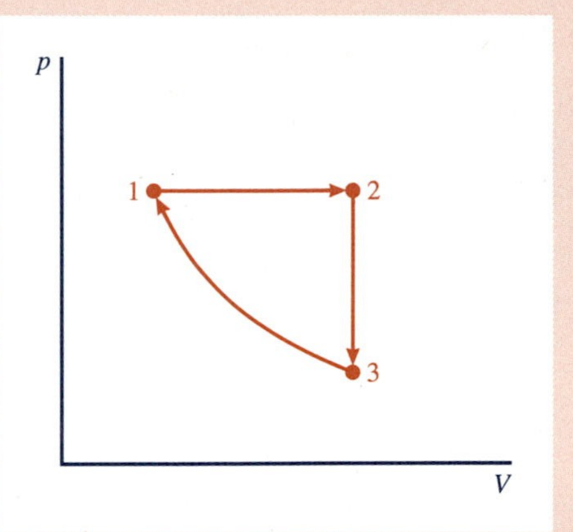

(*a*) Helium under these conditions behaves as an ideal gas. For the reversible constant-pressure expansion,

$$w_{1-2} = \int_1^2 p\,dv = p(v_2 - v_1) = p_2v_2 - p_1v_1 = R(T_2 - T_1)$$
$$= 2.08(145 - 20) = 260 \text{ kJ/kg}$$

For helium c_v is constant, so

$$u_2 - u_1 = c_v(T_2 - T_1) = 3.12(145 - 20) = 390 \text{ kJ/kg}$$

Applying the first law yields

$$q_{1-2} = u_2 - u_1 + w_{1-2} = 390 + 260 = 650 \text{ kJ/kg}$$

For the constant-volume cooling, work = 0 and

$$q_{2-3} = u_3 - u_2 = c_v(T_3 - T_2) = 3.12(20 - 145) = -390 \text{ kJ/kg}$$

For the isothermal compression, since it is an internally reversible process of a simple compressible substance,

$$w_{3-1} = \int_3^1 p\,dv = \int_3^1 \frac{RT}{v}\,dv = RT_1 \ln\frac{v_1}{v_3} = RT_1 \ln\frac{p_3}{p_1} = RT_1 \ln\frac{p_3}{p_2} = RT_1 \ln\frac{T_3}{T_2}$$
$$= 2.08(293) \ln\frac{293}{418} = -216 \text{ kJ/kg}$$

Applying the first law yields

$$q_{3-1} = w_{3-1} + u_1 - u_3 = w_{3-1} + c_v(T_1 - T_3) = -216 + 0 = -216 \text{ kJ/kg}$$

The results can be listed as follows:

Process	q, kJ/kg	w, kJ/kg
1–2	650	260
2–3	−390	0
3–1	−216	−216
Cycle	44	44

(*b*)

$$\eta = \frac{\oint \delta w}{q_{\text{in}}} = \frac{44}{650} = 6.7 \text{ percent}$$

where q_{in} is the *gross* heat added.

(*c*) For the reversed cycle the energy transfers for each process are equal in magnitude but opposite in sign to those of the original cycle:

Process	q, kJ/kg	w, kJ/kg
1–3	216	216
3–2	390	0
2–1	−650	−260
Cycle	−44	−44

We show later that for specified temperature limits, a reversible cycle has the highest efficiency possible.

Just as reversible processes serve as standards of comparison for actual processes, reversible cycles can be benchmarks for actual cycles. In fact, later in this chapter we demonstrate that for specified temperature limits, an actual cycle can never have a higher efficiency than a reversible cycle.

In a reversible cycle (not just an *internally* reversible cycle) that involves heat exchange with two energy reservoirs at fixed temperatures T_H and T_L, heat can be transferred only when the system is at a constant temperature infinitesimally higher or lower than T_H or T_L. In other words, the system can exchange heat only during isothermal processes. Therefore, all processes that are not isothermal must be adiabatic. A cycle that meets these conditions is the Carnot cycle, named for Sadi Carnot, the French engineer who in 1824 first described it. The Carnot cycle is of such great importance that we now describe it in detail.

Nicholas Leonard Sadi Carnot (1796–1832), member of an illustrious family, studied at the École Polytechnique and was an officer in the French Army Engineers. He had broad interests and was an accomplished athlete. The only paper he published during his lifetime, *Reflections on the Motive Power of Heat*, is one of the milestones of scientific thought. He originated the use of cycles in thermodynamic analysis and laid the foundations for the second law by describing and analyzing the Carnot cycle and stating the Carnot principle, which is discussed in Section 5·6.

5·5·1 The Carnot Cycle

The Carnot cycle, since it is reversible, is the limiting case for both an engine and a refrigerator. The Carnot engine is a heat engine that converts heat into work with the highest possible efficiency. The engine operates while exchanging heat with two energy reservoirs and always involves four processes: two reversible isothermal and two reversible adiabatic processes. A Carnot cycle can be executed by many different types of systems. The system can be a liquid, a gas, an electric cell, a soap film, a steel wire, or a rubber band, to name just a few. In any case, a Carnot cycle involves

The Carnot cycle can be used to find the maximum possible efficiency.

1. A system
2. An energy reservoir at some temperature T_H
3. An energy reservoir at some lower temperature T_L
4. Some means of periodically insulating the system from one or both of the reservoirs
5. A part of the surroundings that can both absorb work and do work on the system.

To study the Carnot cycle, consider a system composed of a gas (not necessarily an ideal gas—see Example 5·4 for this special case) held in an insulated cylinder fitted with an insulating piston (Fig. 5·6). The insulation of the cylinder head can be removed and the cylinder can be placed in contact with either energy reservoir.

The cycle begins with the gas in state 1 as shown on the pV diagram labeled "Gas" in Fig. 5·6. The temperature of the gas is T_H (or $T_H - dT$). The four processes that make up the Carnot engine cycle are as follows:

Process 1–2. Heat is to be transferred from the high-temperature reservoir to the gas. The insulation is removed from the cylinder head and the cylinder is placed in contact with the energy reservoir at T_H. The gas then expands very slowly, doing work on the surroundings. As the gas expands, its temperature tends to decrease, but heat is added from the reservoir at T_H to maintain the gas at a temperature of $T_H - dT$ during the reversible isothermal expansion. The gas temperature cannot increase above T_H because heat is being transferred from the reservoir to the gas. The gas temperature is not allowed to decrease below $T_H - dT$ because the objective is to get maximum work from the gas, and we see from the pV diagram that allowing the temperature to fall would reduce the work output of the isothermal expansion. When the gas reaches state 2, the cylinder is removed from contact with the high-temperature reservoir and the insulation is replaced.

Process 2–3. The gas expands reversibly and adiabatically, doing work; its pressure and temperature decrease. (Whenever the gas is at a temperature between $T_H - dT$ and $T_L + dT$ there should be no heat transfer. Heat transfer between a reservoir and the gas at some intermediate temperature would be wasteful, because no work would be produced, al-

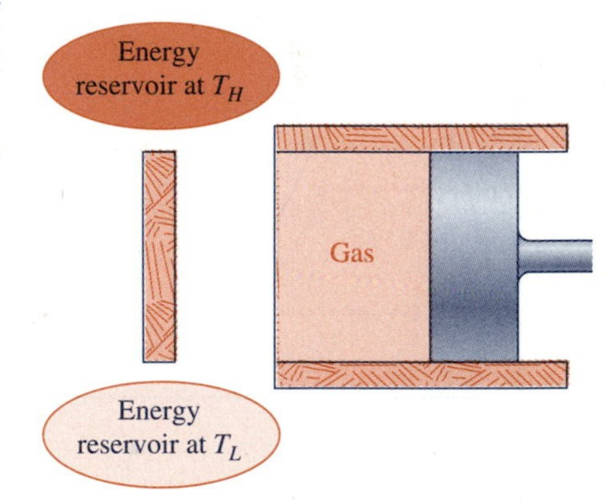

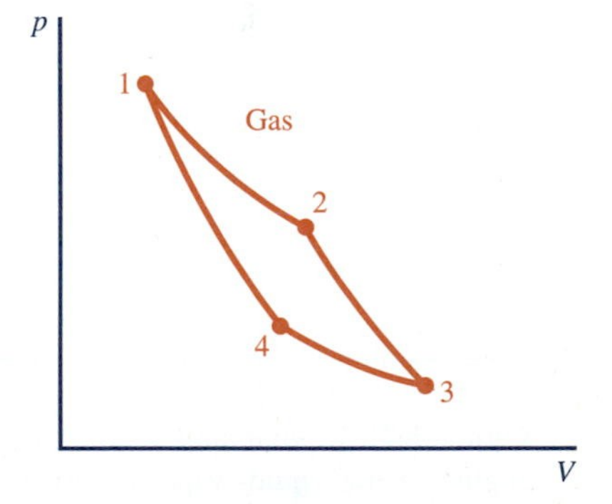

Figure 5·6 A Carnot engine using a gas as working substance.

though we know that some work *could* be produced by interposing another heat engine between the reservoir and the gas in the cylinder.) When the gas temperature reaches $T_L + dT$, the insulation on the cylinder head is removed and the cylinder is placed in contact with the reservoir at T_L.

Process 3–4. The gas is compressed reversibly and isothermally with heat transfer to the low-temperature reservoir to keep the gas temperature at $T_L + dT$. (The work input during this process should be kept to a minimum. If the temperature were allowed to rise, the pressure would also rise, and this would increase the amount of work required to compress the gas.) The reversible isothermal compression is continued until the gas reaches state 4. State 4 was selected so that a reversible adiabatic process can be used to return the system to state 1. When state 4 is reached, the cylinder is removed from contact with the low-temperature reservoir and the insulation is reapplied to the cylinder head.

Process 4–1. The gas is compressed reversibly and adiabatically to the initial state. Work is done on the gas and its temperature increases from $T_L + dT$ to $T_H - dT$.

The area within the pV diagram of Fig. 5·6 represents the net work of this externally reversible cycle. Since the system executes a cycle, the net change in its stored energy is zero. The net effects of this cycle are (1) heat is removed from the energy reservoir at T_H, (2) heat is added to the energy reservoir at T_L, and (3) work is produced. Application of the first law shows that the work produced is equal to the difference between the heat added to the system at T_H (or $T_H - dT$) and the heat rejected at T_L (or $T_L + dT$).

If the working fluid in the cycle just described were a liquid–vapor mixture of some pure substance instead of a gas, the isothermal processes 1–2 and 3–4 would also be constant-pressure processes. During process 1–2, evaporation would occur, and condensation would occur during process 3–4. If all the liquid were evaporated and the vapor became superheated during the isothermal heat addition, only that part of process 1–2 which occurred with both liquid and vapor present would be a constant-pressure process. pv diagrams for these cases are shown in Fig. 5·7. Remember that the heat-addition and heat-rejection processes of the Carnot cycle are always isothermal. For a particular substance, an isothermal process may have constant values of other properties, but this characteristic of the substance is merely incidental to the execution of a Carnot cycle. The Carnot cycle requires that the heat-addition and heat-rejection processes be only reversible and isothermal—nothing else.

Figure 5·8 shows a flow diagram and a pv diagram for a steady-flow Carnot cycle that uses a liquid–vapor mixture as the working substance. In steady-flow systems like this, the ideal reversible processes can be more closely approximated by actual processes than is the case with the closed system engine described above.

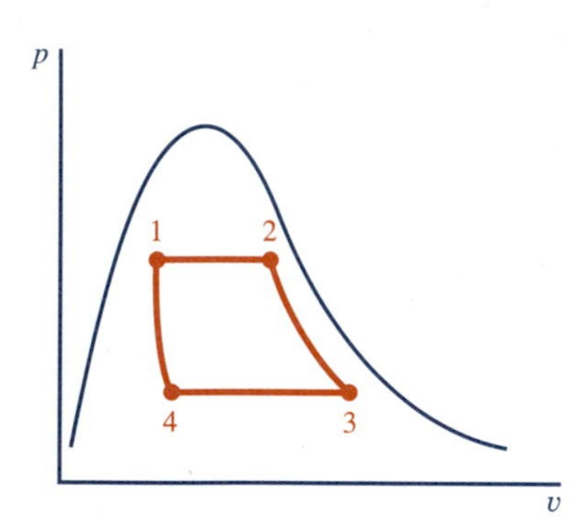

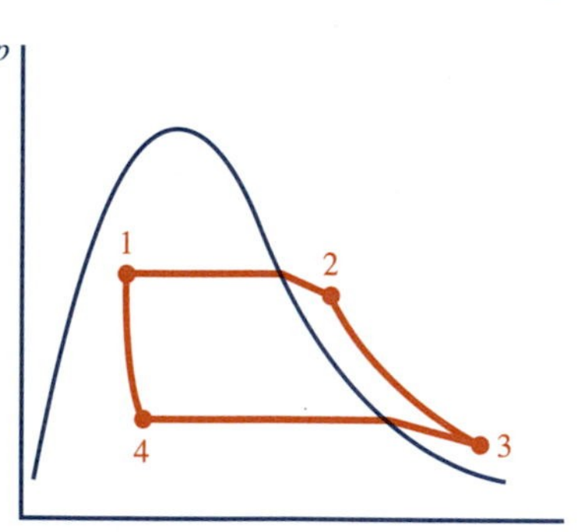

Figure 5·7 *pv* diagrams for Carnot engines using liquid–vapor mixtures as working substance.

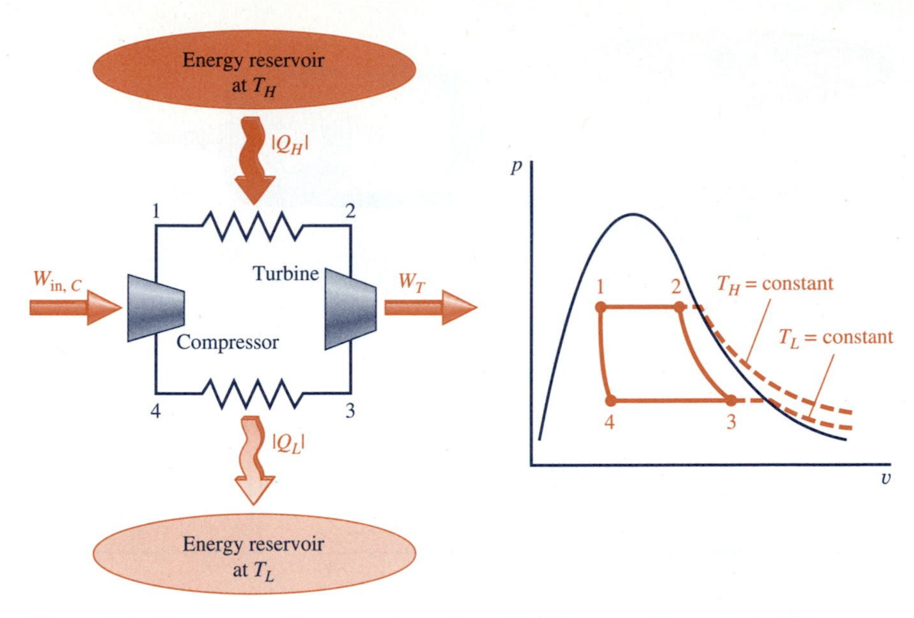

Figure 5·8 Steady-flow Carnot engine using a liquid–vapor mixture as working substance.

We will show later that a Carnot engine has the highest possible efficiency for specified temperature limits.

5·5·2 The Reversed Carnot Cycle

A Carnot engine operated in reverse is a refrigerator or heat pump. It absorbs heat from a low-temperature reservoir and rejects heat to a high-temperature reservoir. To do this, it requires work from the surroundings. (In a household refrigerator, heat is absorbed from the refrigerator compartment, heat is rejected to the surrounding air, and an electric motor provides the work input.)

The working substance of a Carnot refrigerator absorbs heat $|Q_L|$ during an isothermal process at T_L. Its temperature is then increased adiabatically to T_H. During an isothermal process at T_H, heat $|Q_H|$ is rejected to the higher-temperature reservoir (see Fig. 5·9). The cycle is then completed by an adiabatic process that lowers the temperature of the working substance and returns it to its initial state. All processes are of course reversible. You can readily give detailed descriptions of the four processes as we did above for the Carnot engine. Application of the first law shows that

$$\text{Net work}_{\text{in}} = |Q_H| - |Q_L|$$

For given temperature limits, a Carnot refrigerator or heat pump has the highest possible coefficient of performance. However, Carnot engines and

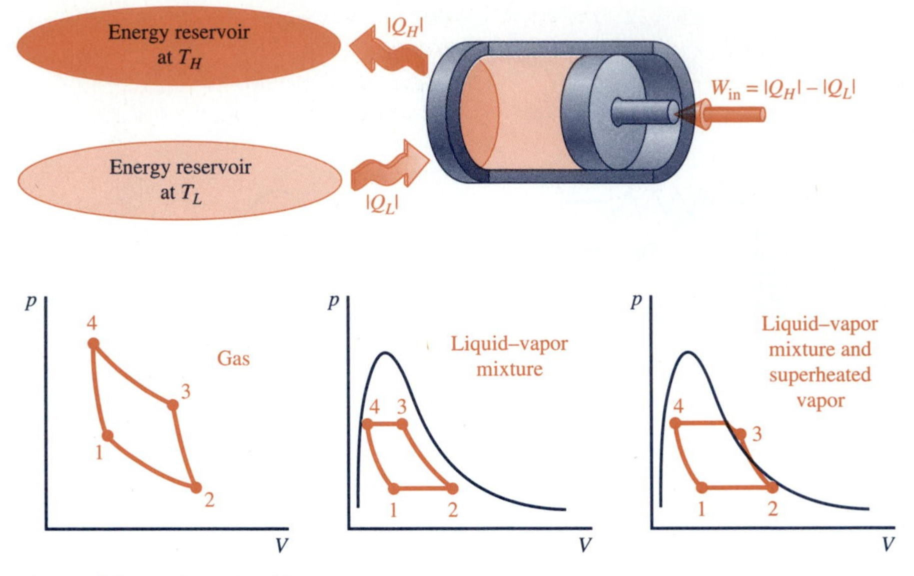

Figure 5·9 A Carnot refrigerator.

refrigerators are not built and used in practice for many reasons, including the impracticability of heat transfer across infinitesimally small temperature differences. Nevertheless, the Carnot cycle as a limiting case and as a standard of comparison is of great utility in engineering work.

Example 5·4 Carnot Engine Using Ideal Gas

PROBLEM STATEMENT

For a Carnot engine using an ideal gas with constant specific heats as working fluid, derive an expression for thermal efficiency in terms of the reservoir temperatures T_H and T_L. (This problem statement implies that the efficiency of the cycle is a function of temperature limits only, but we have not yet proved this. We do so in Section 5·7.)

SOLUTION

First, we make a diagram of the engine and the reservoirs as well as a pV diagram of the ideal-gas working fluid.

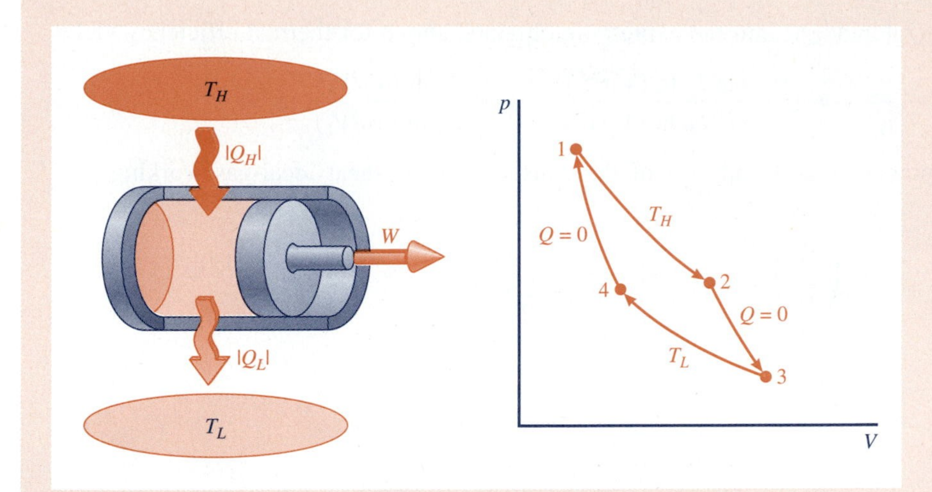

By the definition of thermal efficiency,

$$\eta = \frac{\oint \delta W}{Q_{\text{in}}} = \frac{W}{Q_{\text{in}}}$$

Applying the first law to this closed system that executes a cycle yields

$$\eta = \frac{\oint \delta W}{Q_{\text{in}}} = \frac{\oint \delta Q}{Q_{\text{in}}} = \frac{Q_{\text{in}} - Q_{\text{out}}}{Q_{\text{in}}} = 1 - \frac{Q_{\text{out}}}{Q_{\text{in}}} = 1 - \frac{|Q_L|}{|Q_H|}$$

where Q_{in} or $|Q_H|$ is the total heat transferred from the reservoir at T_H to the engine, and Q_{out} or $|Q_L|$ is the total heat transferred from the engine to the reservoir at T_L. Applying the first law to the system during the reversible isothermal expansion 1–2, we get

$$|Q_H| = U_2 - U_1 + W_{1-2}$$

For an ideal gas, internal energy is a function of temperature only; therefore $\Delta U = 0$ for an isothermal process. Also, for a reversible process of a closed system, $W = \int p\, dV$, so that

$$|Q_H| = 0 + \int_1^2 p\, dV$$

Substituting for p from the ideal-gas equation of state, and noting that m, R, and T are constant for this process, gives

$$|Q_H| = 0 + \int_1^2 \frac{mRT_H}{V}\, dV = mRT_H \int_1^2 \frac{dV}{V} = mRT_H \ln \frac{V_2}{V_1}$$

By the same reasoning,

$$|Q_L| = U_3 - U_4 - W_{3-4} = 0 - \int_3^4 p\, dV = -mRT_L \ln \frac{V_4}{V_3} = mRT_L \ln \frac{V_3}{V_4}$$

Substituting these values for $|Q_H|$ and $|Q_L|$ into the expression obtained above for thermal efficiency yields

$$\eta = 1 - \frac{|Q_L|}{|Q_H|} = 1 - \frac{mRT_L \ln (V_3/V_4)}{mRT_H \ln (V_2/V_1)} = 1 - \frac{T_L \ln (V_3/V_4)}{T_H \ln (V_2/V_1)}$$

For the reversible adiabatic processes 2–3 and 4–1 of the constant-specific-heat ideal-gas working substance,

$$\frac{V_3}{V_2} = \left(\frac{T_2}{T_3}\right)^{1/(k-1)} = \left(\frac{T_H}{T_L}\right)^{1/(k-1)} \quad \text{and} \quad \frac{V_4}{V_1} = \left(\frac{T_1}{T_4}\right)^{1/(k-1)} = \left(\frac{T_H}{T_L}\right)^{1/(k-1)}$$

Thus

$$\frac{V_3}{V_2} = \frac{V_4}{V_1}$$

and

$$\frac{V_3}{V_4} = \frac{V_2}{V_1}$$

so that the expression for thermal efficiency reduces to

$$\eta = 1 - \frac{T_L}{T_H}$$

Note that the thermal efficiency of a Carnot cycle, using an ideal gas as a working fluid, increases as the ratio of T_H to T_L increases. Writing the efficiency expression as

$$\eta = \frac{T_H - T_L}{T_H}$$

shows that, for a given temperature of either reservoir, the efficiency is increased by increasing the temperature difference between the reservoirs.

Example 5·5 Carnot Refrigerator Using Ideal Gas

PROBLEM STATEMENT

Derive an expression, in terms of the reservoir temperatures T_L and T_H, for the coefficient of performance of a Carnot refrigerator using an ideal gas with constant specific heats as working fluid. (As in the preceding example, we have assumed here that the coefficient of performance is a function of temperature only. We will prove this in Section 5·7.)

SOLUTION

First, we make a diagram of the Carnot refrigerator and the reservoirs as well as a pV diagram of the ideal-gas working fluid. By the definition of coefficient of performance for a refrigerator (Section 3·6) and application of the first law,

$$\beta_R = \frac{|Q_L|}{W_{\text{in}}} = \frac{|Q_L|}{|Q_H| - |Q_L|}$$

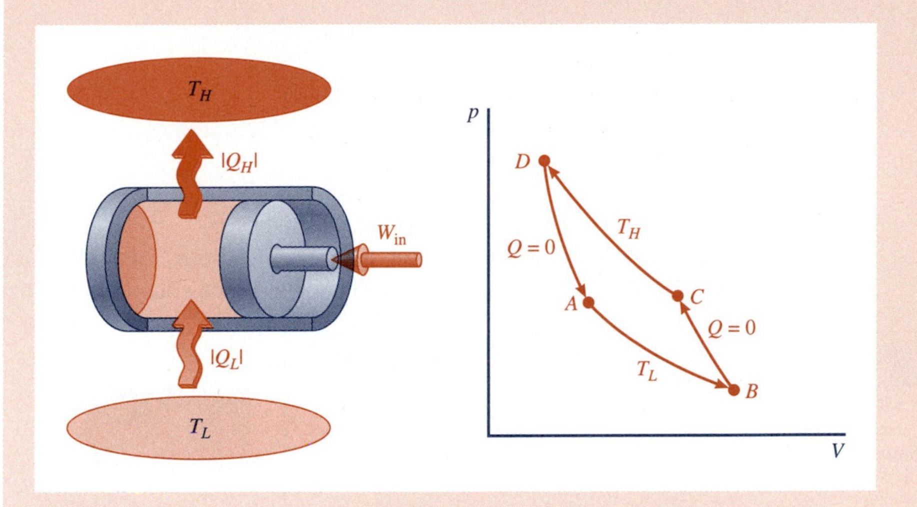

$|Q_L|$ and $|Q_H|$ can be evaluated by the first law, Joule's law, $w = \int p\, dV$, and $pV = mRT$ as in Example 5·4 to give

$$|Q_L| = Q_{A-B} = mRT_L \ln \frac{V_B}{V_A}$$

$$|Q_H| = Q_{\text{out},C-D} = -Q_{C-D} = -mRT_H \ln \frac{V_D}{V_C} = mRT_H \ln \frac{V_C}{V_D}$$

It can also be shown, as in Example 5·3, that

$$\frac{V_B}{V_A} = \frac{V_C}{V_D}$$

so that substitutions in the expression for coefficient of performance give

$$\beta_R = \frac{mRT_L \ln (V_C/V_D)}{mRT_H \ln (V_C/V_D) - mRT_L \ln (V_C/V_D)} = \frac{T_L}{T_H - T_L}$$

Notice that this is *not* the reciprocal of the efficiency expression obtained in Example 5·4. Finally, notice some numerical values of β_R for various reservoir temperatures (recalling that absolute temperatures are used in the expression derived):

T_L, °C	T_H, °C	β_R
−5	30	7.7
−5	20	10.7
−5	10	17.9
−50	20	3.2
−150	20	0.72
−250	20	0.085

5·5·3 Other Reversible Cycles

The Carnot cycle is not the only reversible heat-engine cycle. Two others are known as the Stirling and Ericsson cycles. These warrant consideration be-

The Reverend Robert Stirling (1790–1878) was the first person to propose the use of regeneration in heat-engine cycles. John Ericsson (1803–1889), Swedish–American engineer and inventor, built a steam locomotive, the *Novelty*, that competed with Stephenson's *Rocket* in 1829. Among his inventions were the revolving naval gun turret, the marine screw propeller, and the steam fire engine. He is well known as the designer and builder of the ironclad *Monitor* used by the United States in answer to the Confederate *Merrimac* (or *Virginia*) during the Civil War. Under Ericsson's supervision the *Monitor* was built in only 126 days.

cause they both employ the principle of *regeneration* that is used in many modern steam and gas-turbine power plants and elsewhere.

For reversible cycles, heat can be transferred between the system and fixed-temperature reservoirs only during isothermal processes. However, during nonisothermal processes, it is possible for heat to be transferred between the system and some regenerative energy-storage device that absorbs heat from the system during part of the cycle and returns the same amount of heat to the system during another part of the cycle. Of course, if the regenerator is to be reversible, heat must always be transferred between the system and the regenerator across an infinitesimal temperature difference.

Stirling Cycle. Figure 5·10 shows a schematic diagram of a Stirling engine and a *pv* diagram for an ideal-gas working substance. The engine

The engine shown in Fig. 5·10 is not the same physical arrangement as that proposed by Stirling, but the thermodynamic cycle is the same.

consists of a cylinder with a piston at each end. In the middle of the cylinder, between the pistons, is the regenerator. This can be a plug of wire gauze or a

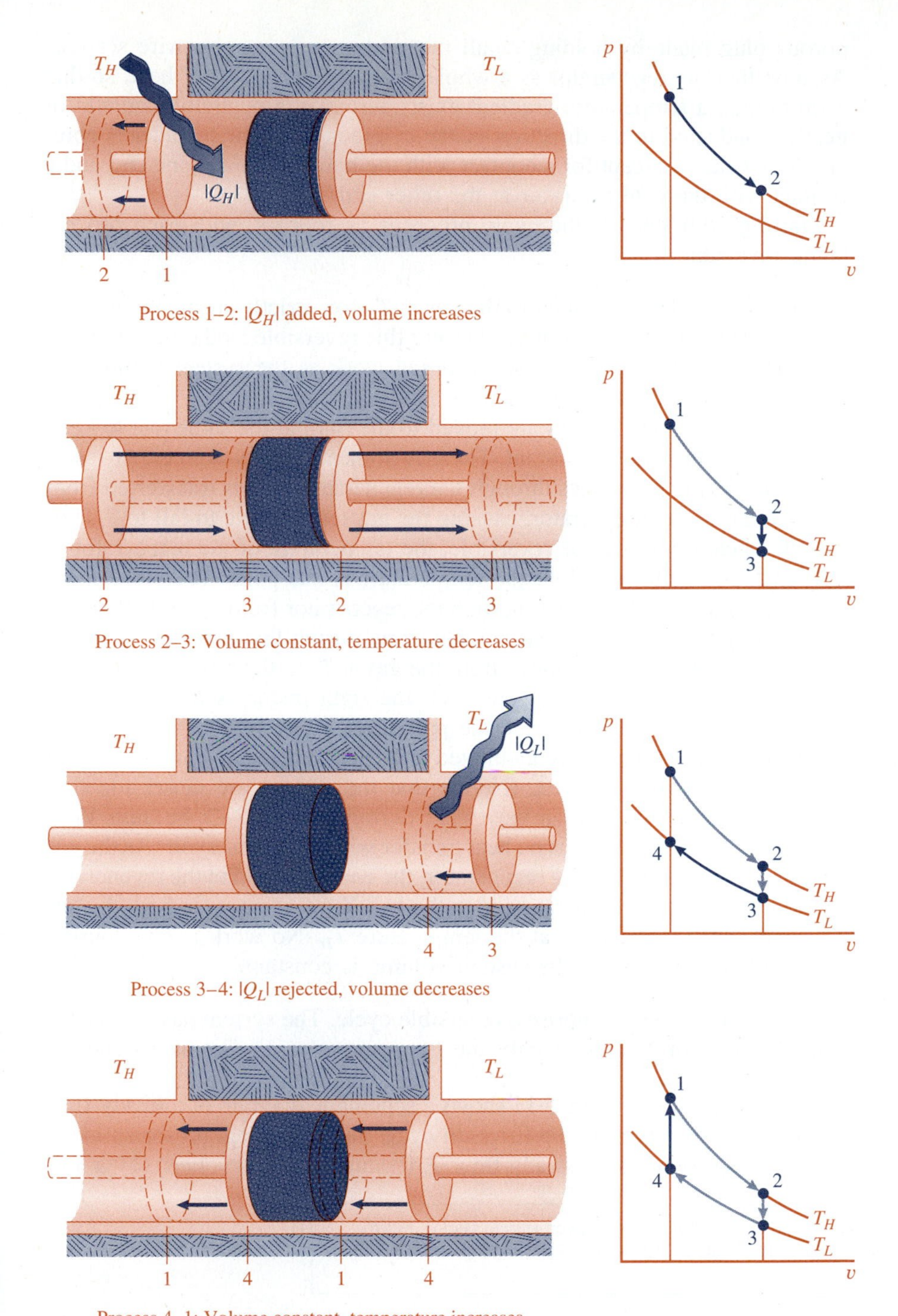

Figure 5·10 Stirling engine.

porous plug made by holding small metal shot between two wire screens. Assume that the regenerator as a whole is a poor conductor of heat, so that even though a temperature gradient exists across it, a negligible amount of heat is conducted in the direction of the cylinder axis. The cylinder is completely insulated except for a contact with the hot reservoir at one end and a contact with the cold reservoir at the other end.

Starting with state 1, the cycle proceeds as follows, with each process being reversible:

Process 1–2. Heat is added to the gas at T_H (or, strictly speaking, at $T_H - dT$) from the reservoir at T_H. During this reversible isothermal process, the left piston moves outward, doing work as the system volume increases and the pressure falls.

Process 2–3. Both pistons are moved to the right at the same rate to keep the system volume constant. There is no heat transfer with either reservoir. As the gas passes through the regenerator, heat is transferred from the gas to the regenerator, causing the gas temperature to fall to T_L. For this heat transfer to be reversible, the temperature of the regenerator at each point must equal the gas temperature at that point. Therefore, there is a temperature gradient through the regenerator from T_H at the left end to T_L at the right end. No work is done during this process.

Process 3–4. Heat is removed from the gas at T_L to the reservoir at T_L. To hold the gas temperature constant, the right piston is moved inward, doing work on the gas, and the pressure increases.

Process 4–1. Both pistons are moved to the left at the same rate to keep the system volume constant. (Notice that the pistons are closer together during this process than they were during process 2–3, because $V_4 - V_1 < V_2 - V_3$.) There is no heat transfer with either reservoir. As the gas passes back through the regenerator, the energy stored in the regenerator during process 2–3 is returned to the gas. Thus, it emerges from the left end of the regenerator at the temperature T_H. No work is done during this process because the system volume is constant.

These four processes comprise a reversible cycle. The system has exchanged a net amount of heat with only the two energy reservoirs at T_H and T_L.

Ericsson Cycle. Figure 5·11 shows a schematic diagram of a steady-flow power plant operating on the Ericsson cycle and a *pv* diagram for an ideal-gas working fluid. The steady-flow Ericsson engine involves (1) a turbine through which the gas expands isothermally, doing work and absorbing heat from an energy reservoir at T_H, (2) a compressor that compresses the gas

> Ericsson's original engine, of which several models have been built, was a closed-system engine, but the ideal thermodynamic cycle is the same for the steady-flow engine. The steady-flow engine is described here in order to introduce another type of regenerator.

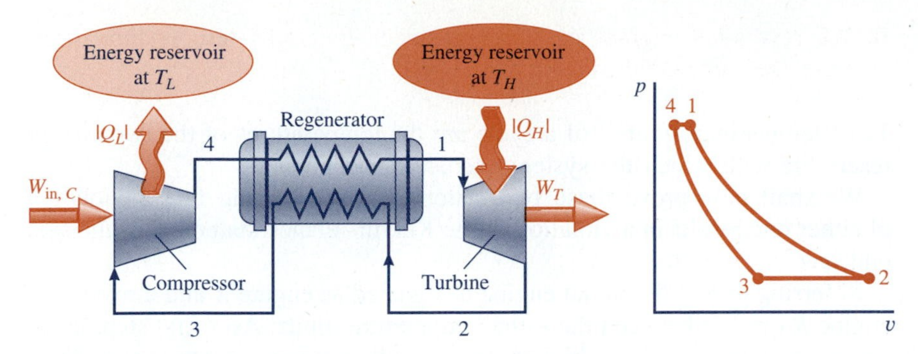

Figure 5·11 Ericsson engine.

isothermally while heat is rejected from the gas at T_L (or $T_L + dT$) to the energy reservoir at T_L, and (3) a counterflow heat exchanger which is used as a regenerator. In the regenerator, gas from the compressor is heated in process 4–1 and the gas from the turbine is cooled in process 2–3. Notice in Fig. 5·11 that the gas entering the left end of the regenerator (state 4) is at T_L, as is the gas that leaves the left end (state 3). Also, both the gas leaving the right end (state 1) and the gas entering the right end (state 2) are at T_H. Throughout the regenerator there is no more than an infinitesimal temperature difference between the two gas streams at any one section, so that the operation of the regenerator is reversible. (Of course, no actual regenerator would be designed for an infinitesimal temperature difference because then a finite amount of heat could not be transferred during a finite period of time. Remember that reversible heat transfer is the *limiting* case as the temperature difference is made smaller and smaller.) All processes of the Ericsson cycle are reversible. The only parts of the surroundings involved in heat transfer are the two energy reservoirs at T_H and T_L.

The Carnot, Stirling, and Ericsson cycles have been introduced here to show how a reversible cycle can be executed. The next chapter introduces some far-reaching consequences of the second law that involve the performance of reversible engines.

5·6 The Carnot Corollaries

Many valuable deductions follow from the second law of thermodynamics. We present now some deductions that pertain to the efficiencies of engines and the coefficients of performance of refrigerators and heat pumps. These are often called the *Carnot corollaries*. The deductions regarding engines are

The Carnot corollaries are also known as the Carnot principle.

I. No engine can be more efficient than a reversible engine operating between the same temperature limits.

Carnot corollaries for engines

II. All reversible engines operating between the same temperature limits have the same efficiency.

The "temperature limits" of a cycle are the temperatures of the two energy reservoirs with which the system exchanges heat.

We shall now prove these two statements by showing that a violation of either one results in a violation of the Kelvin–Planck statement of the second law.

Referring to Fig. 5·12*a*, an engine designated as engine X and a reversible engine R operate between the same temperature limits. As a first step, let us *assume* that engine X has a higher efficiency than the reversible engine. Then for the same amount of heat $|Q_H|$ supplied to each engine, $W_X > W_R$ and $|Q_{LX}| < |Q_{LR}|$. Now let the reversible engine be reversed to operate as a refrigerator, as shown in Fig. 5·12*b*. The reversed engine rejects heat $|Q_H|$ to the energy reservoir at T_H and requires a work input of $|W_R|$. Since $|W_R|$ is less than the work output W_X of engine X, engine X can drive the reversed engine and still deliver work in the amount $(W_X - |W_R|)$ to other parts of the surroundings. The reversed engine is rejecting heat in the amount $|Q_H|$ to the reservoir at T_H, and engine X is absorbing the same amount of heat from this reservoir. Therefore, there is zero net exchange of heat with the reservoir, and the reservoir could in fact be eliminated by having the reversed engine discharge heat directly into engine X. Now look at the system that is made up of engine X and the reversed engine together (enclosed by the broken line in Fig. 5·12*b*). It is operating cyclically, exchanging heat with a single reservoir

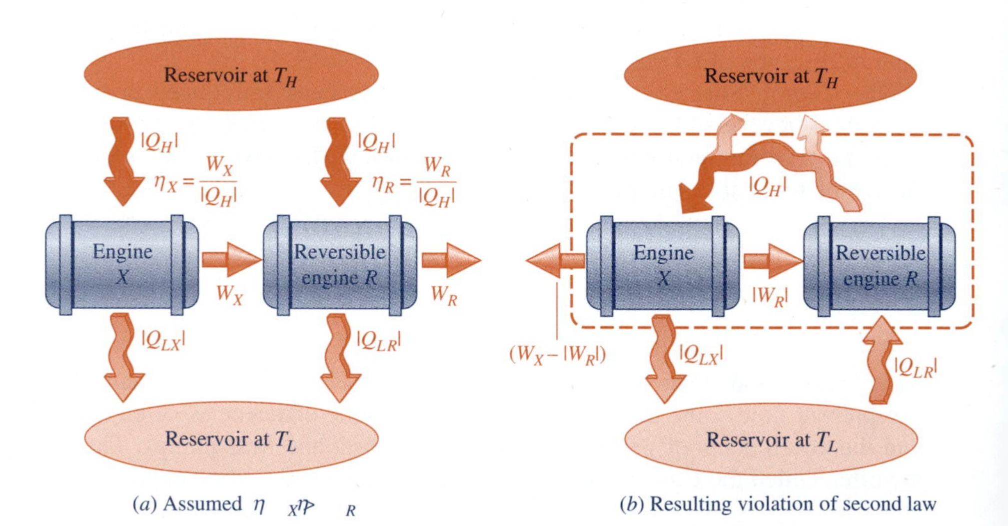

(*a*) Assumed $\eta_X > \eta_R$ (*b*) Resulting violation of second law

Figure 5·12 Proof of the Carnot principle.

(the one at T_L), and producing work.* This is precisely what the Kelvin–

*Notice that the single reservoir could be the atmosphere, the ocean, or the water of a river or lake. The engine and reversed engine together form a device that could draw energy only from one of these sources and convert it into work continuously while producing no other effects. What a marvelous device this would be! Do you believe that such a device is possible? On what is your belief based?

Planck statement of the second law declares to be impossible. Consequently, our assumption that $\eta_X > \eta_R$ must be false. *Conclusion: No engine can be more efficient than a reversible engine operating between the same temperature limits.*

To prove the second of the Carnot corollaries, let both engines in Fig. 5·12*a* be reversible engines. Assume that their efficiencies are different so that their work outputs are different for the same amount of heat input. Then reverse the less efficient engine. The more efficient engine can drive the reversed engine and have some work left over, even though a net amount of heat is drawn from only one reservoir. This is the same violation of the second law that we reached earlier. We must conclude that two reversible engines operating between the same temperature limits cannot have different efficiencies. *Conclusion: All reversible engines operating between the same temperature limits have the same efficiency.* This latter conclusion means that if we can determine the efficiency of a reversible engine operating on any particular substance as a function of its temperature limits, then this same equation applies to all reversible engines.

All reversible engines operating between the same temperature limits have the same efficiency.

Example 5·4 shows that for a Carnot engine using an ideal gas as working substance and operating between reservoir temperatures of T_H and T_L

$$\eta = 1 - \frac{T_L}{T_H}$$

where T_H and T_L are temperatures on the ideal-gas absolute temperature scale. In accordance with the second of the Carnot corollaries, then this equation applies to *any* reversible engine. In the next section, however, we arrive at this equation from only the Carnot corollaries and the definition of a thermodynamic temperature scale with no reference to a calculation based on a particular working substance.

For refrigerators (and heat pumps), the Carnot corollaries are as follows:

Carnot corollaries for refrigerators and heat pumps

I. No refrigerator can have a higher coefficient of performance than a reversible refrigerator operating between the same temperature limits.
II. All reversible refrigerators operating between the same temperature limits have the same coefficient of performance.

The proofs of these corollaries are similar to those used for the corollaries dealing with engines.

The same expression for coefficient of performance as a function of reservoir temperatures holds for all reversible refrigerators. Therefore, the expression derived in Example 5·5 for a refrigerator using an ideal gas with constant specific heats applies to any reversible refrigerator.

5·7 The Thermodynamic Temperature Scale

In Section 1·6·3 we discussed the shortcoming of any temperature scale defined in terms of the physical properties of a substance. Now we know that (1) the efficiency of reversible engines is a function of temperature only, and (2) efficiency involves only heat and work and can therefore be measured independently of properties. In view of these two facts, a temperature scale can be defined by the relationship

$$\eta_{\text{reversible}} \equiv \frac{W}{Q_{\text{in}}} = f(T_H, T_L)$$

The procedure is discussed in the following paragraphs. A temperature scale that is independent of the properties of substances is called a *thermodynamic temperature scale*.

First, let us make clear just what we mean by "defining a temperature scale" and why we must do it. We are not trying to define temperature, because we accept it—along with mass, length, and time—as verbally undefined. When we define a temperature scale, we specify a method for assigning numerical values to various temperatures.

We must define a thermodynamic temperature scale for scientific use.

Since childhood you have been accustomed to reading thermometers—usually mercury-in-glass or alcohol-in-glass—to get numbers for different temperature levels. So, when part way through a thermodynamics course you are told that we must define a temperature scale, your reaction may well be, "What's wrong with the one we've been using?" Let us answer this question partly by referring to the thermometer calibration difficulty mentioned in Section 1·6·3. Even thermometers of the same kind calibrated at two temperatures often disagree with each other at other temperatures. You may say, "Well, why not arbitrarily establish one thermometer as a standard and thereby define a temperature scale?" This is a reasonable suggestion (provided that other people would adopt the same thermometer as a standard), but if we were to use such a scale in scientific work we would be annoyed by some features of it. There may be a question as to whether this scale should be used outside the temperature range in which the standard thermometer itself operates. Also, we will find that certain quantities (such as the efficiency of reversible engines) are functions of temperature only, but the functional relationship is not quite the same for different temperature ranges. Certain other difficulties will also arise. Thus the thermometer scale we use

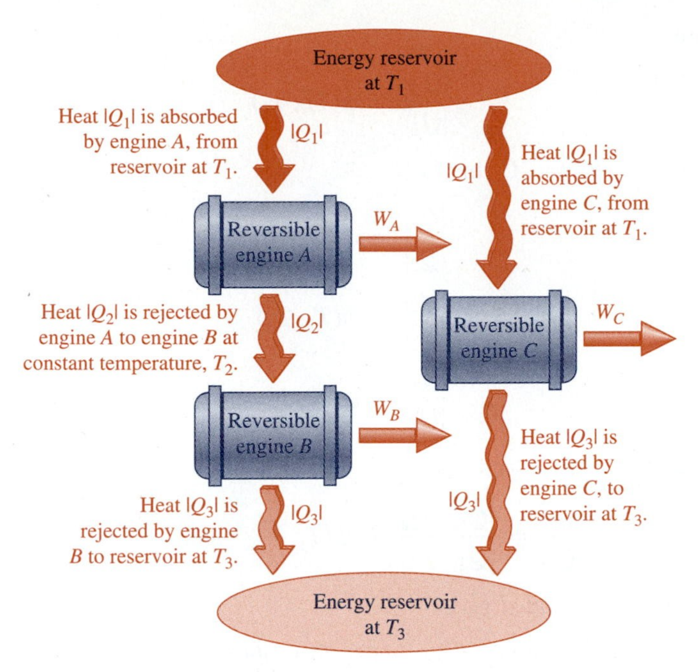

Figure 5·13 Arrangement of reversible engines for proof that $|Q_1|/|Q_3| = f(T_1)/f(T_3)$.

in everyday life will not suffice. It is suitable for indicating when we should wear an overcoat or when an automobile engine is too hot; but for a scientific tool we need some other means of assigning numerical values to temperatures. That is what we mean by saying that we must ''define a temperature scale.''

Defining a thermodynamic temperature scale

Consider three reversible engines operating as shown in Fig. 5·13. Engines A and C each absorb heat in the amount $|Q_1|$ from the reservoir at T_1. Engine C rejects heat $|Q_3|$ to the reservoir at T_3. Engine A rejects heat $|Q_2|$ at a constant temperature T_2 to engine B. Engine B rejects heat to the reservoir at T_3, and the amount rejected must be $|Q_3|$, the same amount that engine C rejects. This must be true because engines A and B taken together constitute a reversible heat engine operating between the same temperature limits as engine C; so engines A and B taken together must have the same efficiency as engine C. Since the heat input to engines A and B combined is the same as the heat input to engine C, the heat rejected must be the same.

By the Carnot corollaries, the efficiency of a reversible engine is a function of its operating temperature limits only:

$$\eta = \frac{W}{Q_{\text{in}}} = \frac{Q_{\text{in}} - Q_{\text{out}}}{Q_{\text{in}}} = 1 - \frac{Q_{\text{out}}}{Q_{\text{in}}} = 1 - \frac{|Q_L|}{|Q_H|} = f(T_H, T_L)$$

Therefore

$$\frac{|Q_H|}{|Q_L|} = \psi(T_H, T_L)$$

For engines *A*, *B*, and *C* this is

$$\frac{|Q_1|}{|Q_2|} = \psi(T_1, T_2) \qquad (a)$$

$$\frac{|Q_2|}{|Q_3|} = \psi(T_2, T_3) \qquad (b)$$

$$\frac{|Q_1|}{|Q_3|} = \psi(T_1, T_3) \qquad (c)$$

The product of Eqs. *a* and *b* is

$$\frac{|Q_1|}{|Q_3|} = \frac{|Q_1|}{|Q_2|}\frac{|Q_2|}{|Q_3|} = \psi(T_1, T_2) \cdot \psi(T_2, T_3) \qquad (d)$$

Combining Eqs. *c* and *d* gives

$$\psi(T_1, T_3) = \psi(T_1, T_2) \cdot \psi(T_2, T_3) \qquad (e)$$

Look closely at Eq. *e*. The left-hand side is a function of T_1 and T_3 only; therefore the right-hand side must be a function of T_1 and T_3 only. The value of T_2 does not affect the value of the product on the right-hand side. This tells us something about the form of the function ψ. To satisfy this condition, the function ψ must have the form

$$\psi(T_1, T_2) = \frac{f(T_1)}{f(T_2)}$$

$$\psi(T_2, T_3) = \frac{f(T_2)}{f(T_3)}$$

Substituting into Eq. *e* gives

$$\psi(T_1, T_3) = \psi(T_1, T_2) \cdot \psi(T_2, T_3) = \frac{f(T_1)}{f(T_2)} \cdot \frac{f(T_2)}{f(T_3)} = \frac{f(T_1)}{f(T_3)}$$

Substituting now into Eq. *c* gives

$$\frac{|Q_1|}{|Q_3|} = \frac{f(T_1)}{f(T_3)}$$

which tells us much more than Eq. *c* does regarding the dependence of $|Q_1|/|Q_3|$ on the reservoir temperatures. This is as far as deductive reasoning will take us. The second law thus requires *only* that

$$\frac{|Q_H|}{|Q_L|} = \frac{f(T_H)}{f(T_L)} \qquad (5\cdot1)$$

for any reversible engine. The function $f(T)$ can be chosen arbitrarily to define a temperature scale that is *independent of the physical properties of any substance*. Such a temperature scale is called a **thermodynamic temperature scale**. As proposed by Lord Kelvin, we can let $f(T) = T$ so that a thermodynamic temperature scale is *defined* by

$$\frac{|Q_H|}{|Q_L|} = \frac{T_H}{T_L} \tag{5·2}$$

Thermodynamic temperature scale

On this scale, the **Kelvin scale**, the ratio of two temperatures is equal to the ratio of the amounts of heat transferred between a reversible heat engine and two reservoirs at these temperatures.

Kelvin temperature scale

Equation 5·2 shows that T_H and T_L must have the same sign. Since we prefer positive numbers for a temperature scale, negative temperatures are precluded from the Kelvin or thermodynamic temperature scale.

Negative temperatures cannot exist on the thermodynamic temperature scale.

To define zero on the Kelvin scale or, in other words, an absolute zero temperature, consider a series of reversible engines, as shown in Fig. 5·14. One engine absorbs heat $|Q_1|$ from the energy reservoir at T_1. It rejects heat $|Q_2|$ at a temperature T_2 to another reversible engine that in turn rejects heat $|Q_3|$ at a temperature T_3 to another reversible engine and so forth. Let the temperatures be selected so that each engine does the same amount of work. Thus

$$W_1 = W_2 = W_3 = \cdots$$

and

$$|Q_1| - |Q_2| = |Q_2| - |Q_3| = |Q_3| - |Q_4| = \cdots$$

and application of the defining equation of the Kelvin scale (Eq. 5·2) shows that

$$T_1 - T_2 = T_2 - T_3 = T_3 - T_4 = \cdots$$

That is, the engines in the series do equal work when the temperature differences across them are equal.

As more engines are added to the series, the total work output increases. By the first law, the total work output cannot exceed $|Q_1|$. In the limiting case where there are enough engines in the series to make the total work output $(W_1 + W_2 + W_3 + \cdots)$ equal to $|Q_1|$, the last engine will reject zero heat. The last engine in the series (or the whole series of engines taken together) is then violating the Kelvin–Planck statement of the second law. Thus the operation of the series of engines with zero heat rejection from the last engine cannot be accomplished, although it can be approached as a limiting case. As the number of engines in the series is increased and the heat rejection of the last engine approaches zero, the temperature at which this heat is rejected also approaches zero on the Kelvin scale in accordance with the defining equation

$$\frac{|Q_H|}{|Q_L|} = \frac{T_H}{T_L}$$

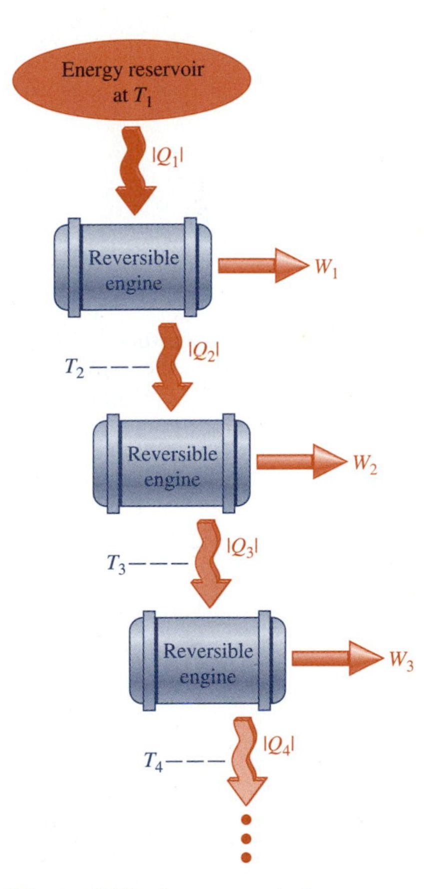

Figure 5·14 Arrangement of reversible engines used in discussing the Kelvin temperature scale.

Absolute zero temperature

Absolute zero temperature can be defined as follows: *If a reversible heat engine operates between two energy reservoirs, absorbing a constant heat input from the hotter reservoir, and the temperature of the colder reservoir is successively lowered, the amount of heat rejected decreases. As the amount of heat rejected approaches zero, the temperature of the colder reservoir approaches absolute zero.* Notice that this definition makes no mention whatsoever of the physical properties of any substance either at absolute zero or at any other temperature. Also, the second law alone does not lead to the conclusion that it is impossible for the temperature of any system to be absolute zero.

The Kelvin or thermodynamic temperature scale is not completely defined until the unit is established. This was done by the international General Conference on Weights and Measures in 1954 and 1967: ''The kelvin, unit of thermodynamic temperature, is the fraction 1/273.16 of the thermodynamic temperature of the triple point of water.'' Thus a numerical value in kelvins for any other temperature is given by

$$T = T_{TP} \frac{|Q|}{|Q_{TP}|} = 273.16 \frac{|Q|}{|Q_{TP}|}$$

where $|Q|$ and $|Q_{TP}|$ are the amounts of heat transferred by a reversible engine operating between a reservoir at T and a reservoir at T_{TP}, the triple-point temperature of water.

The Celsius temperature scale is then defined by

$$T[°C] = T[K] - 273.15$$

The ice-point temperature is 0.01 K lower than the triple-point temperature.

Notice that 0°C is assigned to the *ice-point* temperature, which is approximately 0.01 K lower than the triple-point temperature. (The ice-point temperature is the temperature of a mixture of ice and air-saturated water in equilibrium at 1 atm.) The size of the kelvin was selected to agree very closely with the size of a Celsius degree, but the definition of the kelvin is now independent of any other temperature scales. Other scales are defined in terms of the Kelvin or thermodynamic temperature scale.

Even though the thermodynamic temperature scale is defined in terms of the performance of reversible engines, it is unnecessary to build and operate a reversible engine to determine numerical values on the thermodynamic temperature scale. In fact, we can instead show the relationship between the thermodynamic temperature scale and the ideal-gas temperature scale. We do this next.

From the definition of thermal efficiency and the first law, we can write for any heat engine

$$\eta = \frac{W}{Q_{in}} = \frac{Q_{in} - Q_{out}}{Q_{in}} = 1 - \frac{Q_{out}}{Q_{in}}$$

From the definition of the thermodynamic temperature scale, which depends on the second law, we can write for any reversible heat engine operating between reservoirs at T_H and T_L

$$\frac{Q_{\text{out}}}{Q_{\text{in}}} = \frac{|Q_L|}{|Q_H|} = \frac{T_L}{T_H} \tag{5·2}$$

Therefore, the efficiency of any reversible heat engine operating between reservoirs at T_H and T_L is

$$\eta = 1 - \frac{T_L}{T_H} \tag{5·3}$$

The efficiency of any reversible heat engine operating between reservoirs at T_H and T_L

where T_H and T_L are temperatures on the absolute thermodynamic scale. By similar reasoning it can be shown that the coefficient of performance of a reversible refrigerator is

$$\beta_R = \frac{T_L}{T_H - T_L} \tag{5·4}$$

The coefficient of performance of a reversible refrigerator operating between reservoirs at T_H and T_L

where the temperatures are on the thermodynamic temperature scale. Equations 5·3 and 5·4 are exactly the same as the equations derived in Examples 5·4 and 5·5 in terms of temperatures on the ideal-gas temperature scale. The conclusion is that the ideal-gas temperature scale agrees identically with the thermodynamic temperature scale.

In practice, highly accurate temperature measurements use the International Temperature Scale (ITS), an accurately reproducible scale for laboratory use. The ITS is defined by

1. Assigning values to certain accurately reproducible temperatures such as boiling and melting points
2. Specifying the type of thermometer to be used in each range of the scale
3. Specifying the interpolation formula to be used for each thermometer between the assigned values

The ITS was established to agree with the thermodynamic scale and it does so very closely. In Chapter 7 we will show how temperatures measured by any thermometer can be related to thermodynamic temperature scale values.

Example 5·6

PROBLEM STATEMENT

Determine the heat input to a Carnot engine that operates between 350 and 25°C and produces 100 kJ of work.

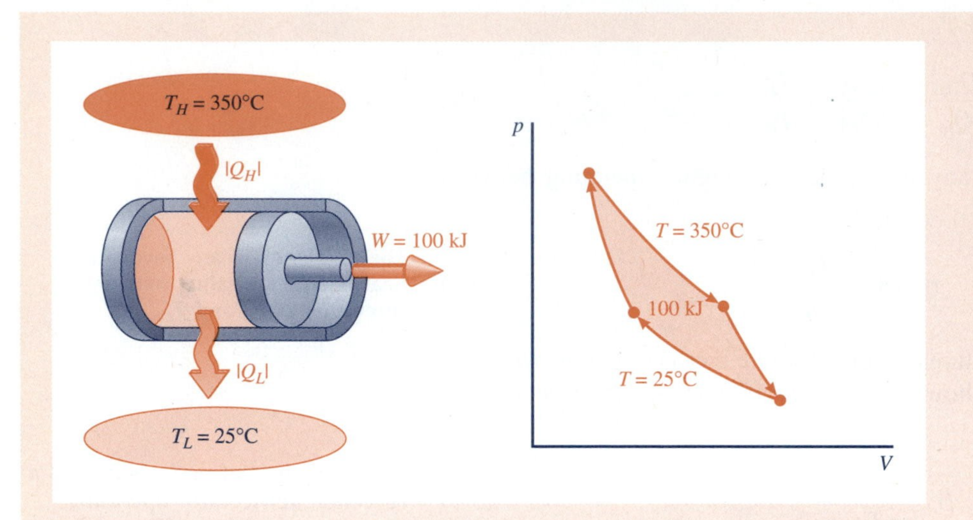

SOLUTION

Applying the first law and the definition of the thermodynamic temperature scale to this engine, we have two simultaneous equations:

$$|Q_H| - |Q_L| = W$$

$$\frac{|Q_H|}{|Q_L|} = \frac{T_H}{T_L}$$

Solving these for $|Q_H|$ yields

$$|Q_H| - |Q_H|\left(\frac{T_L}{T_H}\right) = W$$

$$|Q_H| = \frac{W}{1 - T_L/T_H} = \frac{100}{1 - 298/623} = 192 \text{ kJ}$$

(Notice that the last equation is the same as $|Q_H| = W/\eta$.)

Example 5·7

PROBLEM STATEMENT

Compute the power required to drive a reversible refrigerator if 100 kJ/min is absorbed from the cold region and the isothermal processes occur at 40 and −5°C.

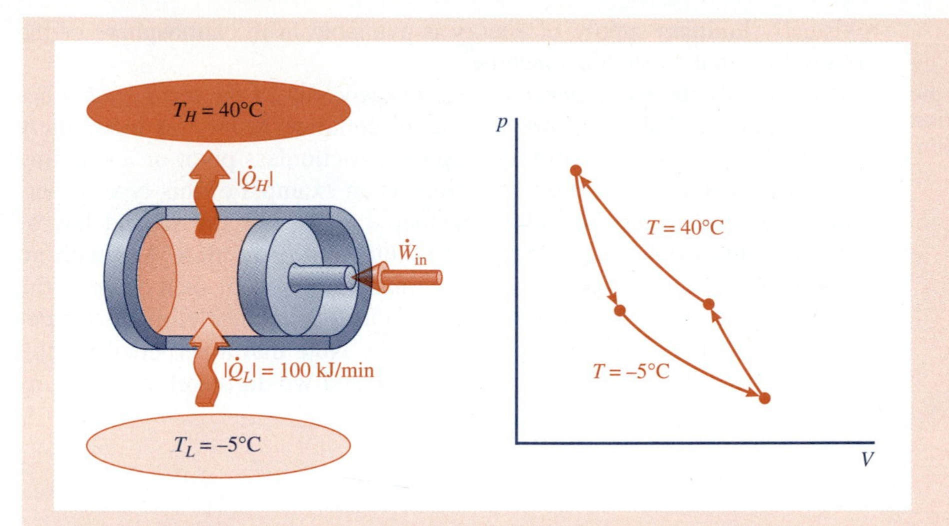

SOLUTION

The power input is the difference between the rate at which heat is rejected and the rate at which heat is absorbed. The rate of heat rejection can be found by

$$|\dot{Q}_H| = |\dot{Q}_L|\frac{T_H}{T_L} = 100(313/268) = 117 \text{ kJ/min}$$

and then

$$\dot{W}_{in} = |\dot{Q}_H| - |\dot{Q}_L| = 117 - 100 = 17 \text{ kJ/min} = 0.28 \text{ kW}$$

5·8 Perpetual-Motion Machines

A perpetual-motion machine of the first kind violates the first law.

Perpetual-motion machines may be classified into three kinds. A *perpetual-motion machine of the first kind* violates the first law by operating in a cycle and producing a greater net work output than the net amount of heat put into the machine. It thus creates energy. Many such machines have been proposed, and several of them have been patented, but none of them has actually operated as a perpetual motion machine of the first kind.

A perpetual-motion machine of the second kind violates the second law.

A *perpetual-motion machine of the second kind* violates the second law by producing work while operating cyclically and exchanging heat only with bodies at a single fixed temperature. Such a device does not violate the first law. No energy is created as would be the case for a perpetual-motion machine of the first kind. Nevertheless, such a device is just as valuable because a virtually limitless supply of energy is available in the atmosphere or the oceans for input to such a machine.

A perpetual-motion machine of the third kind requires the elimination of friction.

Occasionally the name *perpetual-motion machine of the third kind* is applied to devices that, once set in motion, continue in motion indefinitely without slowing down. A spinning top on a frictionless pivot or a spinning flywheel mounted in frictionless bearings is an example of this type of perpetual-motion machine. It violates neither the first nor the second law of thermodynamics but requires only the elimination of friction. Although no one has succeeded in completely eliminating friction from such devices, the extent to which friction can be reduced, short of complete elimination, depends only on the time and money available. Note that a perpetual-motion machine of the third kind produces no work and would therefore not be as valuable as one of the first or second kind.

5·9 Irreversible Processes and Molecular Disorder

Irreversible processes always result in an increase of molecular disorder.

As mentioned earlier, the principles of thermodynamics depend on no assumptions regarding the existence or behavior of molecules. Even so, a molecular picture occasionally supplements our understanding of some effect. As an illustration, we consider briefly a conclusion that can be deduced from the principles of statistical thermodynamics: Irreversible processes always result in an increase of molecular disorder.

Consider the sliding of a block along a horizontal plane. Friction between the block and plane causes the block to stop. Energy is transformed from kinetic energy of the block to internal energy of the block and plane. Near their rubbing surfaces, the block and plane experience an increase in temperature. This temperature increase shows up on the molecular scale as an increase in molecular velocities, and thus as an increase in the kinetic energy of the molecules. The initial kinetic energy of the block can also be considered as kinetic energy of molecules, since all the molecules in the block are initially moving with the block; they have a common translatory motion superimposed on their individual motions. Thus the energy that we speak of as being "transformed" is in the form of kinetic energy of molecules both before and after the transformation occurs. The essential difference is that initially the energy is related to an *ordered* motion of the molecules, whereas finally it is related to a *random* or *disordered* motion. Is the reverse transformation possible? The first law, which requires only conservation of energy, does not prohibit it; but the second law, as discussed earlier in this chapter, leads us to classify the original process as irreversible, so that the reverse transformation is impossible.

As another example, a paddle wheel stirring a gas imposes an ordered motion on the gas molecules, but this ordered motion is rapidly dissipated into a random motion that is observed as an increase in gas temperature. Our discussion of the second law leads us to believe that the reverse process is impossible: the random molecular motions will not spontaneously become ordered in such a manner as to turn the paddle wheel. But, from the molecular point of view, is such an occurrence absolutely impossible? If the molecules are all in random motion, might there be some chance that at some time a sufficient number will be moving in such ways as to turn the paddle wheel? Statistical thermodynamics tells us that such an event is possible, but that the probability of its occurrence is extremely small—so small that for any calculations or predictions we make we can safely consider the occurrence to be impossible. Thus we can say that the ordered molecular state is a state of extremely low probability and that the disordered molecular state is more probable. We can apply similar reasoning from statistical thermodynamics to any irreversible process, and we reach the same conclusion in each instance: *In an irreversible process, any isolated system proceeds toward more probable states, and more probable states are ones of greater molecular disorder.*

In an irreversible process, any isolated system proceeds toward states of greater molecular disorder.

Remember that this conclusion from statistical thermodynamics cannot be reached from, and has no bearing on, the principles of thermodynamics as presented in this book, because this book deals with classical thermodynamics or thermodynamics from the macroscopic point of view. For purposes of engineering analysis and design, classical thermodynamics is at present more useful than statistical thermodynamics. However, statistical thermodynamics is becoming more valuable in explaining certain phenomena and physical properties. Therefore, its importance in engineering is growing.

5·10 Summary

The second law of thermodynamics is a far-reaching principle of nature that has been stated in many forms. For engineers, one of the following two forms is usually the most valuable:

1. The Clausius statement: *It is impossible for any device to operate in such a manner that it produces no effect other than the transfer of heat from one body to another body at a higher temperature.*
2. The Kelvin–Planck statement: *It is impossible for any device to operate in a cycle and produce work while exchanging heat only with bodies at a single fixed temperature.*

These two statements of the second law and many other statements are entirely equivalent in their consequences. If any one of them is taken as a starting point, all of the others can be deduced.

A *reversible process* is a process such that, after it has occurred, both the system and all the surroundings can be returned to the states they were in before the process occurred. Any other process that occurs is *irreversible*.

One way to prove that a process is irreversible is as follows: (1) Assume that the process is reversible so that the reverse process is possible. (2) Combine this reverse process with other processes, known from experience to be possible, to form a cycle that violates the second law. If such a cycle can be devised, then the assumption of step 1 is false, and the process originally considered is irreversible.

Any process is irreversible if it involves friction, heat transfer across a finite temperature difference, free expansion, mixing, or inelastic deformation.

Reversible processes must be executed very slowly to avoid friction and can involve heat transfer only across infinitesimal temperature differences. Also, a reversible process must be such that, after it has occurred, the system and the surroundings can be made to traverse in the reverse order the states they passed through during the original process, with energy transformations of the original process reversed in direction. During a reversible process, the system and the surroundings must each at all times be in states of equilibrium or infinitesimally close to states of equilibrium. The direction of a reversible process can be changed by making infinitesimal changes in the conditions that control it.

Reversible processes actually do not occur in nature. Neither do point masses, frictionless pulleys, weightless cords, and homogeneous beams; but engineers find reversible processes to be just as useful as these other idealizations in analyzing and designing actual systems and processes.

A process is *internally reversible* if all irreversible effects occur outside the system boundary.

All relationships that have been presented as restricted to quasiequilibrium processes are from here on referred to as being restricted to internally reversible processes. (The term *reversible* could not be used earlier, because the concept of reversibility depends on the second law.)

A cycle composed entirely of reversible processes is a *reversible cycle*. An example of a reversible cycle that operates between energy reservoirs at two fixed temperatures is the *Carnot* cycle. Other reversible cycles are the *Stirling* and *Ericsson* cycles.

No engine can be more efficient than a reversible engine operating between the same temperature limits. The "temperature limits" of a cycle are the temperatures of the two energy reservoirs with which the system exchanges heat.

All reversible engines operating between the same temperature limits have the same efficiency.

These two italicized statements are two *Carnot corollaries*. Similar statements apply to reversible refrigerators.

A temperature scale that is entirely independent of the physical properties of any substance can be defined. Such a scale is called a *thermodynamic temperature scale*. The most commonly used thermodynamic temperature scale is one proposed by Kelvin and defined by the relationship

$$\frac{T_H}{T_L} = \frac{|Q_H|}{|Q_L|} \tag{5·3}$$

where T_H and T_L are the temperatures of the energy reservoirs between which a reversible engine operates when it absorbs heat $|Q_H|$ from the high-temperature reservoir and rejects heat $|Q_L|$ to the low-temperature reservoir. The ideal-gas absolute temperature scale agrees identically with the Kelvin thermodynamic temperature scale.

Negative temperatures are impossible on the Kelvin scale.

It must be remembered that the relationship

$$\frac{|Q_H|}{|Q_L|} = \frac{T_H}{T_L} \tag{5·3}$$

applies only to reversible engines operating between reservoirs at T_H and T_L. The efficiency of any externally reversible engine operating between temperatures of T_H and T_L is given by

$$\eta = 1 - \frac{T_L}{T_H} = \frac{T_H - T_L}{T_H}$$

Suggested Reading

5·1 Atkins, P. W., *The Second Law*, Scientific American Books, New York, 1984. (As the preface states, this is a nonmathematical account requiring very little scientific background and "should therefore be accessible to any persistent reader." It is sound and interesting.)

5·2 Black, William Z., and James G. Hartley, *Thermodynamics*, 2nd ed., HarperCollins, New York, 1991, Chapter 5. (This book uses the sign conventions for heat and work throughout the treatment of engines and refrigerators.)

5·3 Howell, John R., and Richard O. Buckius, *Fundamentals of Engineering Thermodynamics*, 2nd ed., McGraw-Hill, New York, 1992, Chapter 5. (This book often uses absolute values of heat in its treatment of engines and refrigerators. Also, this book presents a postulational approach to the second law in addition to the phenomenological approach used in this book and all others listed here.)

5·4 Moran, Michael J., and Howard N. Shapiro, *Fundamentals of Engineering Thermodynamics*, 2nd ed., Wiley, New York, 1992, Chapter 5. (This book and the following one abandon the sign convention for heat during part of the treatment.)

5·5 Van Wylen, Gordon J., Richard E. Sonntag, and Claus Borgnakke, *Fundamentals of Classical Thermodynamics*, 4th ed., Wiley, New York, 1994, Chapter 6.

Problems

5·1 Explain whether each of the following constitutes a violation of the second law: (*a*) An electric motor doing work receives current from a storage battery. The battery and motor exchange heat with only the atmosphere. (*b*) A radiometer (a paddle wheel with each paddle bright and reflective on one side, dull on the other side, with a vertical shaft mounted in an evacuated glass globe) receives heat from the sun. Though there is usually no work output, it is surely possible to arrange a mechanism for having the radiometer do work.

5·2 The solution of Example 5·2 is carried out using the Kelvin–Planck statement of the second law. Carry out the same solution using the Clausius statement.

5·3 Prove from the Clausius statement or the Kelvin–Planck statement the following, which is known as the Carathéodory statement of the second law: In the vicinity of any state of a substance there are some states that cannot be reached (from the state first mentioned) by adiabatic processes alone.

5·4 The first law of thermodynamics has been paraphrased as "You can't get something for nothing" and the second law as "You can't get as much as you thought you could." Other people have suggested "You can't win" and "You can't even break even." Are these apt paraphrases? Explain why or why not.

5·5 Prove whether the following statement is true or false: A closed system cannot execute a cycle while exchanging heat with a single energy reservoir.

5·6 Comment on the following statement: A reversible process, when undone, leaves no history.

5·7 In Section 5·4 it is stated that the frictionless adiabatic and frictionless isothermal processes described earlier in that section are *internally* and *externally reversible.* Could either of them be *internally reversible* and *externally irreversible*? Explain.

5·8 A reversible process must be both internally and externally reversible. A process may be made irreversible by *either* internal or external irreversibilities. Describe a process that meets the following sets of conditions: (*a*) internally reversible and externally irreversible, (*b*) internally irreversible and externally reversible, and (*c*) both internally and externally irreversible.

5·9 One person states that in a reversible isothermal expansion of an ideal gas in a closed system there is no change in internal energy so the heat added equals the work done. Another person states that such a process violates the second law because it amounts to a system absorbing heat from a constant-temperature energy reservoir and producing an equivalent amount of work. Resolve this conflict.

5·10 Oxygen in a closed system expands isothermally from 250 kPa, 80°C, to 110 kPa. The mass of oxygen is 3 kg. Determine the heat transfer if (*a*) the process is reversible, and (*b*) the process is irreversible and the work is 85 percent of that produced by the reversible process.

5·11 Two kilograms of oxygen is to be compressed isothermally in a closed system from 95 kPa, 20°C, to 300 kPa. Determine the heat transfer if (*a*) the process is reversible, and (*b*) the process is irreversible so that 80 percent more work is required than is the case with the reversible process.

5·12 Nitrogen in a closed system is compressed isothermally from 10 N/cm^2, 50°C, to 20 N/cm^2. Calculate the heat transfer in J/g if (*a*) the process is reversible, (*b*) irreversibilities increase the amount of work required to 10 percent more than is required in the reversible process.

5·13 Air in a well-insulated cylinder expands reversibly from initial conditions of 200 kPa, 75°C, to a final pressure of 120 kPa. Determine the final temperature.

5·14 Air enters a turbine at 40 kN/m^2, 900°C, and leaves at 10 kN/m^2. The flow is reversible and adiabatic. Inlet velocity is quite low and outlet velocity is 150 m/s through a cross-sectional area of 0.20 m^2. Calculate the power output in kW.

5·15 Air in a closed system expands isothermally from 45 psia, 120 F, to 15 psia. The mass of air is 1.72 lbm. Determine the amount of heat added if (*a*) the process is reversible and (*b*) irreversibilities are present that reduce the work to 80 percent of that produced by the reversible process.

5·16 Nitrogen in a closed system is compressed isothermally from 14 psia, 100 F, to 35 psia. Calculate the heat transfer per pound if (*a*) the process is reversible and (*b*) irreversibilities are present that increase the amount of work required by 50 percent.

5·17 A Carnot engine uses methane as the working fluid. At the start of the isothermal expansion, the methane is at 500 kPa, 90°C, and occupies a volume of 0.05 m^3. The volume at the end of this process is 0.10 m^3, and the temperature at the end of the adiabatic expansion is 0°C. Determine (*a*) the heat added and (*b*) the heat rejected.

5·18 A reversed Carnot cycle using octane as the working fluid operates between temperature limits of 20 and 200°C. During the isothermal compression, the volume is halved, and the minimum specific volume during the cycle is 0.15 m^3/kg. Determine the coefficient of performance and the amount of heat absorbed from the low-temperature region per kilogram of octane.

5·19 Carbon dioxide expands isothermally in a Carnot engine from $V_1 = 0.03$ m^3 to $V_2 = 0.09$ m^3. During the adiabatic expansion process, the enthalpy of the CO_2 decreases from 200 to 100 kJ. Assuming the enthalpy and internal energy of the CO_2 to be zero at 0 K and specific heats to be constant over the temperature range, determine (*a*) the efficiency of the engine and (*b*) the pressure at the end of the adiabatic expansion.

5·20 Consider a Carnot engine using air as the working substance. At the beginning of the isothermal expansion, the air is at 550 kPa and occupies a volume of 5 liters. The pressure and volume at the end of the adiabatic expansion are 150 kPa and 15 liters. Determine the efficiency of the engine.

5·21 In a Carnot engine using air as the working fluid, the air is at 500 kPa, 95°C, and occupies a volume of 0.2 m^3 at the beginning of the isothermal expansion. The volume doubles during the isothermal expansion. The temperature at the end of the adiabatic expansion is 0°C. Determine (*a*) the heat added and (*b*) the heat rejected.

5·22 A Carnot engine produces 45 kW while operating between temperatures of 200 and 35°C. At what rate (kJ/min) does the engine absorb heat from the source?

5·23 A Carnot engine receives 220,000 kJ/h from a source at 350°C and rejects 130,000 kJ/h to another cold reservoir. What is the temperature of the cold reservoir?

5·24 A Carnot engine receives 400 kJ/s from a source at 500°C and delivers 230 kW of power. Determine (*a*) the efficiency and (*b*) the temperature of the receiver.

5·25 A Carnot engine works between temperature limits of 530 and 25°C. For each 100 kJ absorbed from the source, compute (*a*) the heat rejected and (*b*) the work developed by the engine.

5·26 A Carnot engine receives 8.5 kJ from a source at 260°C and during the cycle performs 4.1 kJ of work. At what temperature is heat rejected?

5·27 A heat engine operating on the Carnot cycle produces 13.5 kJ of work while operating between temperature limits of 260 and 5°C. Determine the efficiency.

5·28 An ice machine consists of a reversed Carnot

engine operating between 0 and 30°C. How many kilograms of ice will it produce per hour from water at 0°C if the power input is 1.5 kW?

5·29 A reversed Carnot engine operates between 30 and −5°C. If 1400 kJ/min is to be removed from the cold region, calculate the power required.

5·30 A reversed Carnot engine operates with temperature limits of −20 and 35°C and rejects 36,000 kJ/h to a receiver at 35°C. Calculate the power required to operate the machine.

5·31 A Carnot refrigerator removes 16,000 kJ/min from a cold storage room at −20°C. Heat is discharged to the atmosphere at 20°C. Determine the power required.

5·32 A Carnot heat pump is used for heating a building. Heat is taken from the outside air at −5°C, and the building is to be maintained at 20°C. To maintain this temperature, 520,000 kJ/h is required for heating. Find (*a*) the heat taken from the outside per hour and (*b*) the power required.

5·33 A reversed Carnot engine operating between −20 and 40°C rejects 160,000 kJ/h to the receiver at 40°C. Compute the power required to operate the machine.

5·34 A reversed Carnot engine absorbs 425 kJ/min from a source at a temperature of −25°C and rejects heat at 32°C. What is the coefficient of performance?

5·35 What is the coefficient of performance of a reversed Carnot engine operating between 5 and 35°C that delivers 105,000 kJ/h to the higher-temperature reservoir?

5·36 Maintaining a building at 22°C requires a heat input of 85,000 kJ/h. How much power is required by a Carnot heat pump if it picks up heat from (*a*) the atmosphere at −5°C, (*b*) well water at 8°C?

5·37 A Carnot refrigerator removes heat at a rate of 40.0 kJ/s from a cold storage room at −20°C. The refrigerator is driven by a Carnot engine that takes heat from a reservoir at 400°C and discharges heat to the atmosphere at 25°C. Determine (*a*) the power required to drive the refrigerator and (*b*) the total rate of heat rejected to the atmosphere by the combined devices.

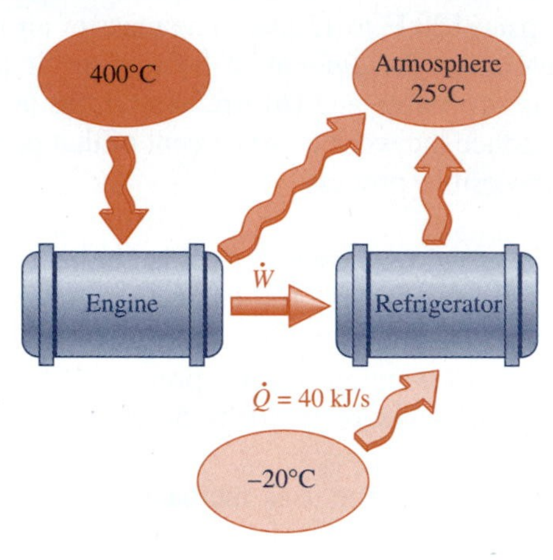

Problem 5·37

5·38 A Carnot engine using wet steam as the working fluid operates between temperature limits of 150 and 35°C. Sketch a *pv* diagram of the cycle. For a work output of 100 kJ, calculate the amount of heat rejected from the working fluid at the lower temperature.

5·39 A Carnot refrigerator using R134a as a working fluid removes 10,000 kJ/h from a region at 5°C. The maximum pressure for the cycle is 700 kPa. Sketch a *pv* diagram of the cycle and calculate the power input to the refrigerator.

5·40 A Carnot refrigerator using wet steam as a working fluid removes 80,000 kJ/h from a region at 5°C. The maximum pressure of the working fluid is 70 kPa. Sketch a *pv* diagram of the cycle and calculate the power input to the refrigerator.

5·41 A Carnot cycle uses 0.1 kg of steam as a working fluid. At the beginning of the isothermal expansion the working fluid is at 1350 kPa, 20 percent quality. At the end of the isothermal expansion, the steam is dry and saturated. During the isothermal compression the steam is at 70 kPa. Calculate (*a*) the efficiency of the cycle and (*b*) the amount of work done per cycle.

5·42 A reversed Carnot cycle uses 0.1 kg of ammonia as a working fluid. At the beginning of the isothermal expansion, the ammonia is at 80 kPa, 30 percent quality; at the end of the isothermal expansion, the ammonia is dry and saturated. Heat is rejected at 20°C. Determine (*a*) the change in internal energy of the ammonia during the isothermal expansion and (*b*) the work input per cycle.

5·43 A closed-system Stirling cycle using helium as the working fluid operates with maximum pressure and temperature of 1200 kPa, 1000°C. The cycle has an overall volume ratio of 4.2 and minimum pressure and temperature of 75 kPa, 61°C. Determine the thermal efficiency of the cycle.

5·44 For an Ericsson cycle as shown in Fig. 5·11, $p_1 = 500$ kPa, $T_1 = 600$°C, $p_3 = 100$ kPa, $T_3 = 60$°C, $\dot{m} = 0.220$ kg/s, and the working fluid is argon. Determine the thermal efficiency.

5·45 Reversing the cycle described in Example 5·3 would make an unsatisfactory refrigerator if the region to be refrigerated is to be held at 23°C and there is no lower-temperature region in the surroundings. Explain this with the help of suitable diagrams. Modify the reversed cycle by adding a single process so that it can keep the refrigerated region at 23°C.

5·46 A reversed Carnot cycle using hydrogen as the working fluid operates between temperature limits of 10 and 200°C. During the isothermal compression the volume is halved, and the minimum specific volume during the cycle is 2 cu ft/lbm. Determine the amount of heat absorbed from the low-temperature region per pound of hydrogen and the coefficient of performance of the cycle as a refrigerator and as a heat pump.

5·47 Consider a Carnot engine using air as a working fluid and having an overall volume ratio of 9. During the isothermal heat rejection which begins with the air at 14.0 psia, 40 F, the volume is decreased to one third of its maximum value. Determine the efficiency of this cycle.

5·48 A Carnot engine develops 80 hp while rejecting 320,000 B/h at 45 F. Compute the temperature of the source or hot body.

5·49 A Carnot engine operates with temperature limits of 1200 F and 70 F. For a power output of 140 hp, calculate the heat supplied, the heat rejected, and the efficiency of the engine.

5·50 A Carnot engine operating with temperature limits of 1000 and 100 F rejects 200 B/min. Calculate the power output.

5·51 The efficiency of a Carnot engine discharging heat to a cooling pond at 80 F is 30 percent. If the cooling pond receives 800 B/min, what is the power output of the engine? What is the temperature of the high-temperature source?

5·52 A Carnot engine containing 8 lbm of air has at the beginning of the expansion stroke a volume of 10 cu ft and a pressure of 220 psia. The exhaust temperature is 40 F. If 8 B of heat is added during the cycle, find (*a*) the efficiency of the engine and (*b*) the work of the cycle.

5·53 Find the efficiency of a heat engine operating on the Carnot cycle and producing 10,000 ft·lbf of work while operating between temperature limits of 900 and 40 F.

5·54 If the work output of a Carnot engine operating between 100 and 600 F is 2200 B, how much heat must be absorbed from the high-temperature reservoir and how much rejected to the low-temperature reservoir?

5·55 A reversed Carnot engine is used as an ice machine: it operates between 32 and 84 F. How many pounds of ice will it produce per hour from water at 32 F if the power input is 2 hp?

5·56 A reversed Carnot engine absorbs 100 B/min from a cold reservoir at −10 F. Find the coefficient of performance if heat is rejected at 90 F if the reversed engine is used as (*a*) a refrigerator, (*b*) a heat pump.

5·57 A reversed Carnot engine operating between 40 and 100 F delivers 100,000 B/h to the higher-temperature reservoir. Compute the coefficient of performance.

5·58 A Carnot engine using wet steam as the working fluid operates between temperature limits of 300 and 100 F. Sketch a *pv* diagram of the cycle. For a work output of 100 B, calculate the amount of heat rejected from the working fluid at the lower temperature.

5·59 A Carnot refrigerator is to remove 400 B/h from a region at −60 F and discharge heat to the atmosphere at 40 F. The Carnot refrigerator is to be driven by a Carnot engine operating between an energy reservoir at 1040 F and the atmosphere at 40 F. How much heat, in B/h, must be supplied to the Carnot engine at 1040 F?

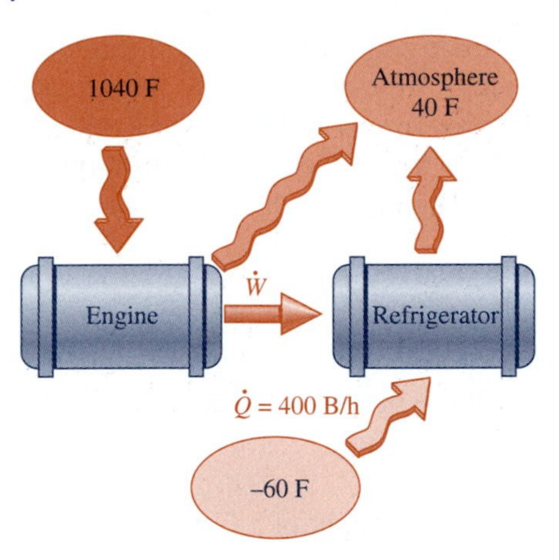

Problem 5·59

5·60 Solve Problem 5·59 with the Carnot engine rejecting heat to the region at −60 F instead of to the atmosphere. The refrigerator must then absorb 400 B/h plus the heat rejected by the engine.

5·61 A reversed Carnot cycle uses 0.2 lbm of steam as working fluid. At the beginning of the isothermal expansion, the steam is at 0.5 psia, 30 percent quality; at the end of the isothermal expansion, the steam is dry and saturated. Heat is rejected at 240 F. Determine (*a*) the change in internal energy of the steam during the isothermal expansion and (*b*) the work input per cycle.

5·62 Write in terms of only the reservoir temperatures, T_H and T_L, expressions for the coefficient of performance of (*a*) a Carnot refrigerator and (*b*) a Carnot heat pump.

5·63 For a constant heat-rejection temperature, make a chart of Carnot cycle efficiency versus heat source temperature.

5·64 Show, on a common set of axes, curves of Carnot refrigerator coefficient of performance and Carnot heat pump coefficient of performance versus low temperature for a fixed heat-rejection temperature.

5·65 If the working fluid of a Carnot cycle is an ideal gas, is the work of the adiabatic compression equal in magnitude to that of the adiabatic expansion? Support your answer.

5·66 For a Carnot heat pump that uses an ideal gas as a working fluid, derive an expression for the coefficient of performance in terms of the reservoir temperatures T_L and T_H. Make a table of β_{HP} values for the same reservoir temperatures as used in the table of Example 5·5.

5·67 The Stirling engine physical arrangement shown in Fig. 5·10 has serious drawbacks as a practical engine, so other physical arrangements are used. For one of the common ones you can find described in the literature, make a diagram like Fig. 5·10, showing the physical configuration at each of the four states of the working fluid, the *pv* diagram, and a description of each of the processes. Explain any terms such as "power piston" or "displacer piston" that you use, and state the chief advantages of this arrangement over the one shown in Fig. 5·10.

5·68 The Ericsson engine shown in Fig. 5·11 involves steady-flow equipment, but this cycle can also be carried out with a piston–cylinder arrangement. Describe one such physical arrangement and explain its operation with the aid of a sketch and a *pv* diagram.

5·69 A closed-system Stirling cycle using helium as working fluid has an overall volume ratio of 3.8 and an overall temperature ratio of 3.0. Maximum pressure and temperature in the cycle are 300 psia and 1100 F. Determine the power output of the cycle for a heat input rate of 55,000 B/h.

5·70 The overall temperature and volume ratios of a closed-system Ericsson cycle are 3.0 and 4.8, respectively. The maximum temperature in the cycle is 1200 F and the maximum pressure is 500 psia. Determine the cycle thermal efficiency.

5·71 What steps would be involved in executing a Carnot cycle using a rubber band as the working substance and water at the ice point and water at the 1 atm boiling point as the energy reservoirs? Sketch force–length and *Ts* diagrams. (If you are unaware of some of the peculiarities of rubber, some tabletop experiments can be quite enlightening regarding properties such as the thermal coefficient of expansion.)

5·72 Referring to Fig. 5·12, demonstrate the Carnot principle by letting one of the initial assumptions be that $Q_{LX} = Q_{LR}$ and proving that such a device would violate the second law.

5·73 Prove that no refrigerator can have a higher coefficient of performance than a reversible refrigerator operating between the same temperature limits.

5·74 Prove that all reversible refrigerators operating between the same temperature limits have the same coefficient of performance.

5·75 It is possible for an actual engine to be more efficient than a Carnot engine. Is this statement true? Explain.

5·76 Lowering the temperature at which a heat engine rejects heat increases its efficiency. A steam power plant usually rejects its heat to water from a river, lake, or ocean; so would it be advisable to cool the water first by means of a refrigeration system driven by the power plant? Explain.

5·77 Answer questions 1 through 5 in Section 5·3.

5·78 Power plants have been proposed that operate on the difference in temperature between water at great depths in the ocean and near the surface. In a location where the surface temperature is 35°C and the temperature at a great depth is 5°C, what is the maximum efficiency of such a power plant?

5·79 Referring to Fig. 5·13, demonstrate that $Q_1/Q_3 = f(T_1)/f(T_3)$ without making reference to any heat engines except *A* and *B*.

5·80 When Sir Isaac Newton established the first numerical temperature scale, would it have been possible to assign numbers so that temperatures of colder bodies were represented by larger numbers? (This means that a man in Chicago stepping outdoors in January might turn up his coat collar and say, ''Brrr, the temperature is really getting up there today.'') If impossible, state why. If possible, do you see any objections?

5·81 Upon learning that a temperature of 0.0014 K has been reached in a cryogenic laboratory, someone states that the scientists practically reached absolute zero. Comment on this statement.

5·82 Write Eq. 5·2 using the sign convention instead of absolute values. Starting then from this equation, show that a negative temperature violates the second law.

5·83 Occasionally someone says, when evaluating an apparent perpetual-motion machine of the first kind, that the machine would work if it were not for friction. Explain why any device that would operate as a perpetual-motion machine of the first kind in the absence of friction would also work as such a machine despite frictional effects.

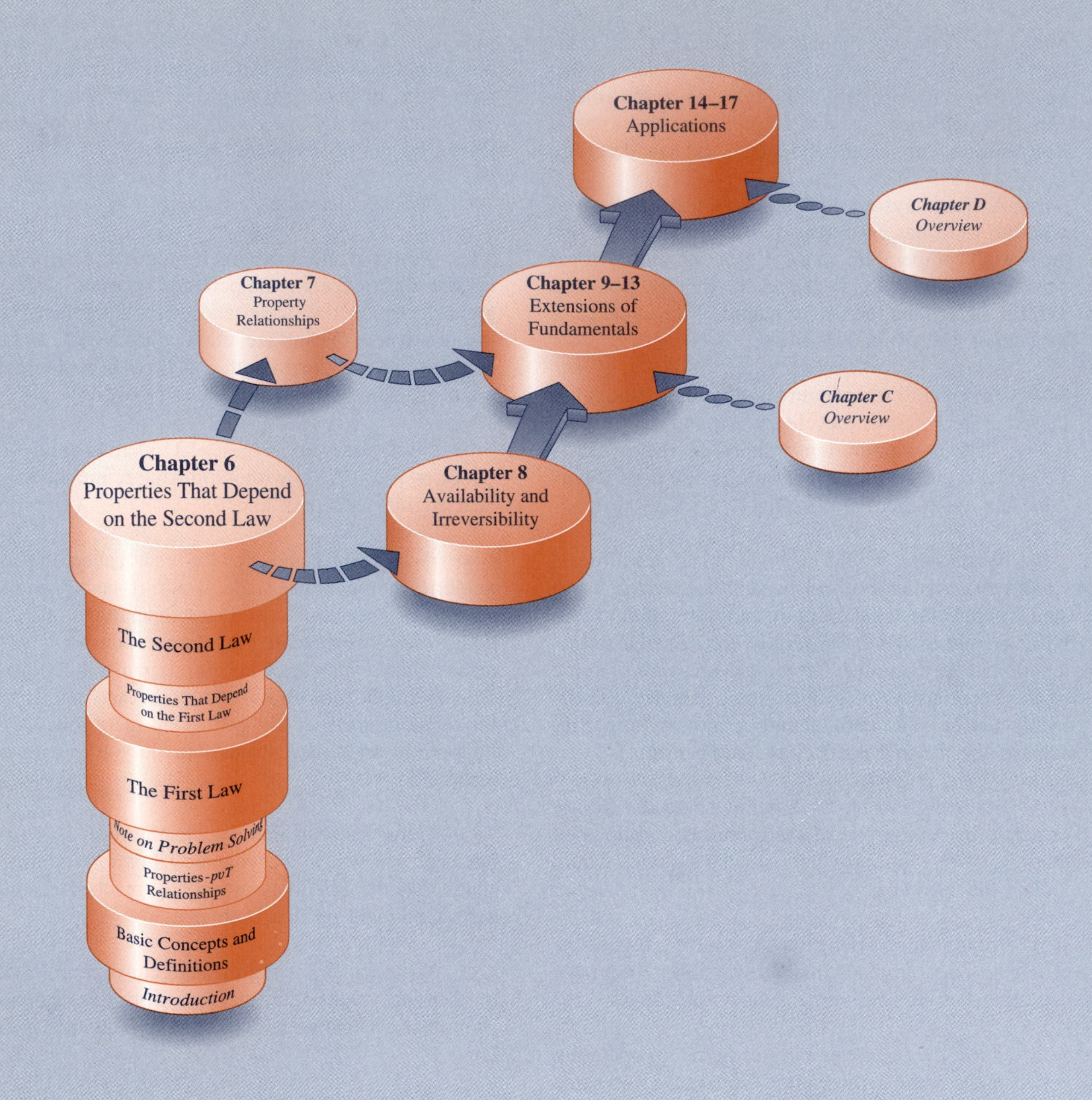
Chapter 14–17
Applications
Chapter D
Overview
Chapter 7
Property
Relationships
Chapter 9–13
Extensions of
Fundamentals
Chapter C
Overview
Chapter 6
Properties That Depend
on the Second Law
Chapter 8
Availability and
Irreversibility
The Second Law
Properties That Depend
on the First Law
The First Law
Note on Problem Solving
Properties-pvT
Relationships
Basic Concepts and
Definitions
Introduction

Chapter 6

Properties That Depend on the Second Law—Entropy and Others

Chapter 3 showed that the first law of thermodynamics leads to the definition of a very useful property: stored energy, E. The term *energy* is used in everyday conversation, often in the same sense as in engineering and science. Consequently, you were familiar with the term before meeting it in the study of thermodynamics, and this familiarity may even have helped you to gain an understanding of the nature of stored energy, which when rigorously defined is based on the first law.

The second law also leads to the definition of a very useful property: entropy, S. Unlike energy, the term *entropy* is not used in everyday conversation. Many introductory physics courses do not mention it. Consequently, entropy is probably an unfamiliar term. Furthermore, it is defined in terms of a calculus operation, and no direct physical picture of it can be given. For these reasons, the property entropy may at first appear somewhat nebulous. To gain an understanding of entropy, you should study its uses and keep asking the questions, "What is it used for?" and "How is it used?" If you are looking for a physical description as an answer, the question "What *is* entropy?" is fruitless. (Recall that the development of the first law was impeded by the question, "What *is* heat?" Only when attention was turned to the question, "How is heat related to work and other effects?" was progress made toward the understanding of heat.) Consequently, this chapter both introduces entropy and discusses several of its uses. We will use the property entropy frequently throughout the rest of this book.

6·1 The Property Entropy

To show that a quantity is a property; show that the cyclic integral is equal to zero.

One way to prove that a quantity is a property is to show that the cyclic integral of the quantity is always zero. Physically, this means that if a system executes a cycle, the quantity always returns to its original value when the system returns to its initial state. The value of the quantity thus depends only on the state of the system. This is true for all properties and for properties only. We will now prove that $\oint (\delta Q/T)_{rev} = 0$ and therefore that $\int (\delta Q/T)_{rev}$ is a property. This property is called entropy, S.

Proof that entropy is a property

As a first step in the proof, we show that any reversible process can be approximated by a series of reversible adiabatic and reversible isothermal processes. For example, the reversible process represented by line 1–2 on the pV diagram of Fig. 6·1 may be approximated by a series of reversible adiabatic and isothermal steps as shown by 1–b, b–c, and c–2, provided that the individual steps are so chosen that the area under curve 1–2 is equal to that under 1–b–c–2. Of course, the greater the number of adiabatic and isothermal steps, the closer the series of lines will approach the original reversible process. For purposes of discussion, the two adiabatic lines and the one isothermal line will suffice. Consider a closed system. Since the areas under the two curves are made equal, $\int p\, dV$ will be the same and the work is the same:

$$W_{1-2} = W_{1-b-c-2}$$

Applying the first law gives

$$Q_{1-2} = U_2 - U_1 + W_{1-2}$$

and

$$Q_{1-b-c-2} = U_2 - U_1 + W_{1-b-c-2}$$

Since the initial and final internal energy (property) values are the same for any path and the steps are selected so that the work terms are equal,

$$Q_{1-2} = Q_{1-b-c-2}$$

In other words, the heat transferred during the reversible process 1–2 is the same as that transferred during the isothermal change b–c because no heat is transferred during the adiabatic steps 1–b and 2–c. This is significant because *it is now always possible to replace a reversible process by a series of reversible adiabatic and isothermal processes so that the internal energy change, the heat transferred, and the work performed are the same.*

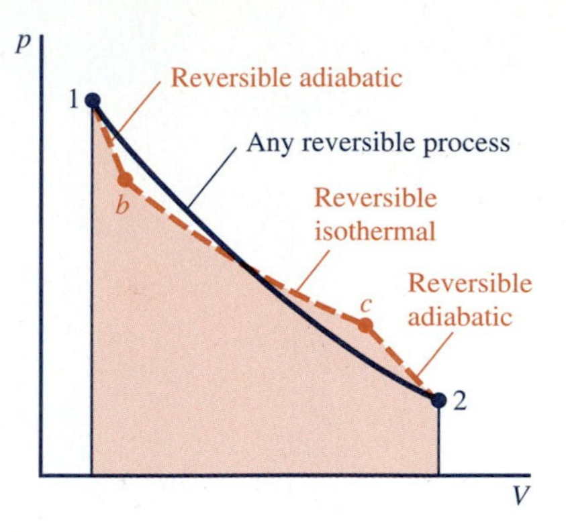

Figure 6·1 Simulating any reversible process by a series of reversible adiabatics and reversible isothermals.

A system at state 1 as shown in Fig. 6·2*a* is assumed first to undergo a reversible process 1–2–3 and then to proceed along the reversible path 3–4–1 to the initial state 1. These processes form a cycle. We can replace the original processes by a series of reversible adiabatic and isothermal processes as shown in Fig. 6·2*b* in a manner similar to that previously described. These various adiabatic (solid) and isothermal (dashed) lines may then be connected to represent a number of Carnot cycles, as shown in Fig. 6·2*c*. For the Carnot cycles a–b–c–d and e–f–g–h, the second law leads to

$$\frac{|Q_{H,a-b}|}{T_{H,a-b}} = \frac{|Q_{L,c-d}|}{T_{L,c-d}}$$

$$\frac{|Q_{H,e-f}|}{T_{H,e-f}} = \frac{|Q_{L,g-h}|}{T_{L,g-h}}$$

where the T's are the temperatures of the system during the isothermal pro-

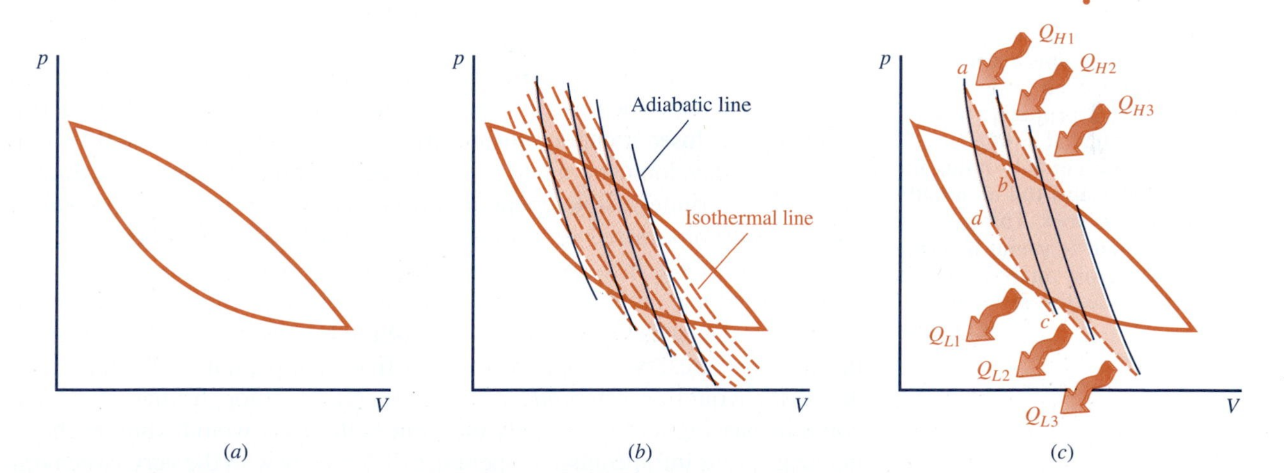

Figure 6·2 Simulating a reversible cycle by a series of reversible adiabatics and reversible isothermals.

cesses. To simplify previous developments, only absolute values of Q_H and Q_L were used. The sign convention of positive for heat absorbed and negative for heat rejected converts the preceding equations to

$$\frac{Q_{H,a-b}}{T_{H,a-b}} + \frac{Q_{L,c-d}}{T_{L,c-d}} = 0$$

$$\frac{Q_{H,e-f}}{T_{H,e-f}} + \frac{Q_{L,g-h}}{T_{L,g-h}} = 0$$

Therefore, the summation of the ratios of heat transfer to absolute temperature for *all* the Carnot cycles equals zero, or

$$\left(\frac{Q_{H,a-b}}{T_{H,a-b}} + \frac{Q_{L,c-d}}{T_{L,c-d}}\right) + \left(\frac{Q_{H,e-f}}{T_{H,e-f}} + \frac{Q_{L,e-f}}{T_{L,e-f}}\right) + \cdots = 0$$

This summation expression can be simplified by letting Q represent heat transfer and T the absolute temperature at which the heat is transferred. Hence, for the reversible cycle,

$$\sum \frac{Q}{T} = 0$$

If the number of Carnot cycles is greatly increased, the stepped paths more closely approximate the actual processes. Finally, as the number of Carnot cycles becomes very large, the summation of the Q/T terms becomes equal to the integral of $\delta Q/T$ and the preceding equation becomes

Entropy is a property.

$$\oint \left(\frac{\delta Q}{T}\right)_{rev} = 0 \tag{6·1}$$

From this result *we conclude that* $\int (\delta Q/T)_{rev}$ *is a property.* This means that the value of $\int (\delta Q/T)_{rev}$ for any process depends only on the end states and not on the path followed between the states.

If $\int (\delta Q/T)$ is integrated along *irreversible* paths between two states, it is found that in general a different numerical value is obtained for each path; that is, $\int (\delta Q/T)$ is not a property. In fact, it can be shown that $\oint (\delta Q/T) \leq 0$, where the equality holds for reversible cycles and the inequality for irreversible ones. This general statement is useful in some thermodynamic analyses and is known as the *inequality of Clausius*.

The notation *rev* on the integral sign indicates that the integration must be carried out along some *reversible* path connecting the two states. This restriction requires closer attention. Since Carnot cycles, which are reversible, were used in the development above, it would be necessary to have heat transfer across only infinitesimal temperature differences between the system and the energy reservoirs in the surroundings. Therefore, as the number of Carnot cycles in Fig. 6·2 increases, the number of reservoirs needed at different temperatures increases. If instead of utilizing this large number of reservoirs the system executes the same processes of the Carnot cycles while exchanging heat with reservoirs only at the two fixed temperatures, T_1 and T_3, the processes would be *externally* irreversible. Notice, though, that the system could be passing through exactly the same states as it would while exchanging heat across infinitesimal temperature differences with the very large number of reservoirs. Therefore, if Q is the heat transfer of the system and T is the absolute temperature of the system, the value of the integral $\oint (\delta Q/T)_{rev}$

is zero for the system whether the process is reversible or only internally reversible. Consequently, in evaluating the integral it is necessary only that the process be *internally reversible*, so we can write

$$\oint \left(\frac{\delta Q}{T}\right)_{\substack{int\\rev}} = 0 \qquad (6{\cdot}2)$$

The property $\int (\delta Q/T)_{int\ rev}$ is called entropy, is denoted by the symbol S, and is defined by

$$\Delta S \equiv \int \left(\frac{\delta Q}{T}\right)_{\substack{int\\rev}} \qquad (6{\cdot}3a)$$

The definition of entropy

or, per unit mass,

$$\Delta s \equiv \int \left(\frac{\delta q}{T}\right)_{\substack{int\\rev}} \qquad (6{\cdot}3b)$$

This is an operational definition. It tells us how to obtain numbers for ΔS, even though the operations prescribed are ''pencil-and-paper'' operations. Notice that we actually define the *change in entropy* ΔS instead of entropy S, just as we earlier defined ΔE instead of E. In engineering work it is usually only the *change* in S that is important; therefore a value of $S = 0$ can be assigned to any particular state of a system arbitrarily. Once this is done, the entropy at any other state x is given by

In engineering work, usually only changes in entropy are important.

$$S_x = S_0 + \int_0^x \left(\frac{\delta Q}{T}\right)_{\substack{int\\rev}}$$

Expression for entropy on an arbitrary scale

where state 0 is the one for which $S = S_0$.

Because S is a property, ΔS between two states is the same, no matter what path, reversible or irreversible, is followed as a system changes from one state to the other. Equation 6·3 states, however, that the numerical value for ΔS must be obtained by integrating $\int(\delta Q/T)$ along *some reversible* path. Examples of this calculation are given in the next section.

Notice the parallelism between the definitions of ΔE and ΔS:

As a result of	*It can be shown that*	*So that a property can be defined as*
The first law	$\oint (\delta Q - \delta W) = 0$	$\Delta E \equiv Q - W$
The second law	$\oint \left(\frac{\delta Q}{T}\right)_{\substack{int\\rev}} = 0$	$\Delta S \equiv \int \left(\frac{\delta Q}{T}\right)_{\substack{int\\rev}}$

Parallelism between the definitions of ΔE and ΔS

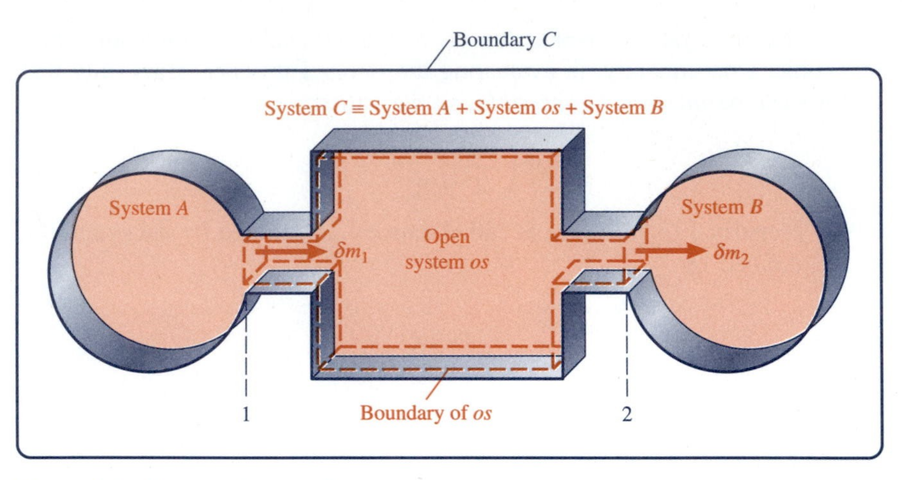

Figure 6·3 Entropy change of an open system.

Determination of a general expression for ΔS for an open system

The definition $\Delta S \equiv \int (\delta Q/T)_{int\ rev}$ holds for any closed system or any fixed quantity of matter, just as the definition $\Delta E \equiv Q - W$ does. To determine a general expression for ΔS of an open system, consider any open system *os* as in Fig. 6·3 (where for convenience only one inlet and one outlet are shown). Let the mass δm_1 entering the open system have a specific entropy s_1. Let this mass come from an open system *A* that might be in the particular form shown in Fig. 6·4. System *A* exchanges no heat with the surroundings, and the only work done on system *A* by the piston is equal to the flow work delivered to the open system as mass δm_1 crosses the boundary. (Notice that this restriction on system *A* does not restrict the generality of the open system's behavior at all.)

As mass δm_1 is transferred from system *A* to the open system *os*, the entropy change of *A* is $-s_1\, \delta m_1$. Let the mass leaving the open system go to a system *B* that behaves in the same manner as system *A*. The entropy change of *B* is then $s_2\, \delta m_2$. Since there is no heat transfer to systems *A* and *B*, the heat transfer to closed system *C*, as defined in Fig. 6·3, is the same as that to open system *os*. Thus, for the closed system within boundary *C*,

$$dS_C = \left(\frac{\delta Q}{T}\right)_{\substack{int\\ rev,C}} = \left(\frac{\delta Q}{T}\right)_{\substack{int\\ rev,os}}$$

where *T* is the temperature of that part of the system to which heat is transferred. (Recall that the temperature throughout an open system is generally not uniform.) Then, since entropy is an extensive property,

$$dS_C = dS_{os} + dS_A + dS_B$$

and

$$dS_{os} = dS_C - dS_A - dS_B$$

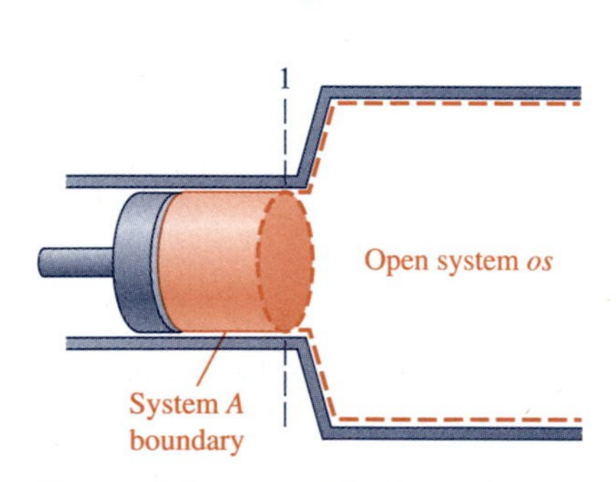

Figure 6·4 A possible form of open system *A*.

$$dS_{os} = \left(\frac{\delta Q}{T}\right)_{\substack{int\\rev,os}} + s_1\,\delta m_1 - s_2\,\delta m_2 \qquad (6\cdot 4)$$

General expression for dS for an open system

This is the general expression for dS of an open system.

For steady flow, the properties at any point within the system remain constant with respect to time, so $dS_{os} = 0$ and $\delta m_1 = \delta m_2$. Then Eq. 6·4 becomes

$$0 = \left(\frac{\delta Q}{T}\right)_{\substack{int\\rev,os}} - (s_2 - s_1)\,\delta m$$

Furthermore, for steady flow, $\int (\delta Q/T)_{int\ rev}$ for the system is the same as $\int (\delta Q/T)_{int\ rev}$ for the mass flowing through the system. The integration can then be performed between the entering and exiting states of the fluid, and we have

$$s_2 - s_1 = \int_1^2 \left(\frac{\delta q}{T}\right)_{\substack{int\\rev}}$$

which agrees with the definition of Δs that holds for any fixed mass, including a mass that is moving through an open system.

Notice that the definition of entropy leads to the conclusion that for either a closed system or a steady-flow system, *for any reversible process*

$$q = \int T\,ds \qquad (6\cdot 5)$$

Heat transfer for any *reversible* process can be represented by an area on a *Ts* diagram.

so that heat transfer is represented by an area on a property diagram with coordinates of T and s. (For an open system, T is the temperature of that part of the system to which heat is transferred.) We will use this conclusion in subsequent chapters, so we will discuss it further in Section 6·5.

6·1·1 Calculation of Entropy Changes from the Definition

Entropy values are included in published tables and computer programs of thermodynamic properties, but often we must calculate entropy changes for various processes. We can calculate ΔS between any two states of a substance directly from the definition by selecting any reversible path connecting the two states and integrating along that path. Examples of this procedure follow.

Example 6·1

PROBLEM STATEMENT

Calculate the change in entropy of neon that is heated reversibly at constant pressure of 120 kPa from 20°C to 110°C in a closed system.

SOLUTION A

Since the mass of neon in the system is unspecified, we will calculate the entropy change per unit mass. Under the low pressure specified, neon can be treated as an ideal gas. Because it is monatomic, its specific

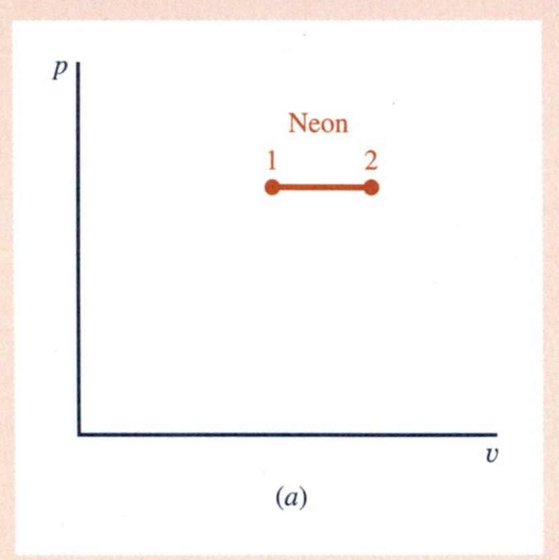

(a)

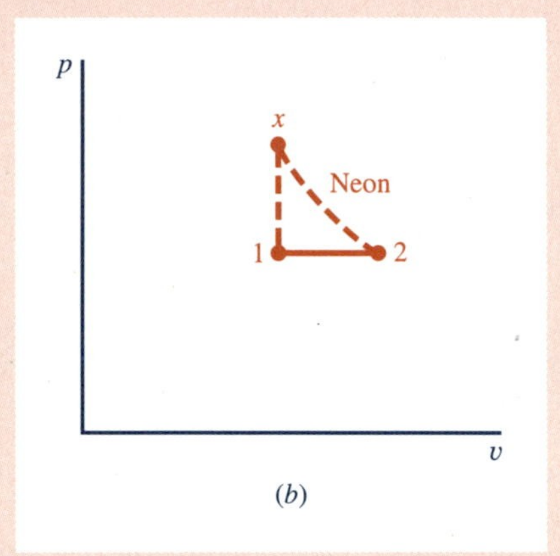

(b)

heats are constant. We first make a *pv* diagram of the process. For a constant-pressure process of an ideal gas, $v_2/v_1 = T_2/T_1 = 383/293 = 1.3$, so we sketch the *pv* diagram approximately to scale. We will calculate Δs from the defining equation

$$\Delta s \equiv \int_1^2 \left(\frac{\delta q}{T}\right)_{\substack{int\\rev}} \tag{6·3}$$

by evaluating the integral along any reversible path between the end states. Let us use the reversible constant-pressure path that the system actually follows. We must find a relationship between δq and dT in order to integrate. This can be done by the following steps in which each step is explained, or the restriction imposed is stated, by the words in parentheses:

$$\begin{aligned}
\delta q &= du + \delta w && \text{(first law, closed system)}\\
&= du + p\,dv && \text{(reversible process)}\\
&= du + d(pv) && \text{(constant pressure)}\\
&= d(u + pv) = dh && \text{(definition of } h\text{)}\\
&= c_p\,dT && \text{(ideal gas)}
\end{aligned}$$

Making this substitution for δq and obtaining the c_p value from Table A·3 gives

$$\Delta s = \int_1^2 \frac{c_p\,dT}{T}$$

$$\Delta s = c_p \ln \frac{T_2}{T_1} = 1.03 \ln \frac{383}{293} = 0.276 \text{ kJ/kg·K}$$

(Notice that the units of c_p may be either kJ/kg·°C or kJ/kg·K, as explained in Section 4·3, but the entropy change units must be kJ/kg·K because the definition of Δs involves *absolute* temperature, not a temperature change in the denominator.)

SOLUTION B

Let us evaluate

$$\Delta s = \int_1^2 \left(\frac{\delta q}{T}\right)_{\substack{int\\rev}}$$

over some reversible path other than the constant-pressure path. (There is no reason to choose any other path in this case except to illustrate that the same result will be obtained.) Let us choose a path as shown on the diagram that consists first of a reversible constant-volume heating from the initial state 1 to the final temperature. Call this state for which $v = v_1$ and $T = T_2$ state x. The second part of the reversible path is a reversible isothermal process from state x to state 2. Since Δs is the same for all paths between states 1 and 2, we have

$$\Delta s = \int_1^2 \left(\frac{\delta q}{T}\right)_{\substack{int\\rev}} = \int_1^x \left(\frac{\delta q}{T}\right)_{\substack{int\\rev}} + \int_x^2 \left(\frac{\delta q}{T}\right)_{\substack{int\\rev}}$$

Applying the first law and noting that $1\text{–}x$ and $x\text{–}2$ are reversible processes, we find

$$\Delta s = \int_1^x \frac{du}{T} + \int_x^2 \frac{du + p\,dv}{T}$$

Since the system is composed of an ideal gas,

$$\Delta s = \int_1^x \frac{c_v\,dT}{T} + \int_x^2 \frac{0 + p\,dv}{T} = \int_1^x \frac{c_v\,dT}{T} + \int_x^2 \frac{R\,dv}{v}$$

With c_v and R constant,

$$\Delta s = c_v \ln \frac{T_x}{T_1} + R \ln \frac{v_2}{v_x}$$

Noting that $T_x = T_2$, $v_x = v_1$, and $v_2/v_1 = T_2/T_1$, we have

$$\Delta s = c_v \ln \frac{T_2}{T_1} + R \ln \frac{T_2}{T_1} = (c_v + R) \ln \frac{T_2}{T_1}$$
$$= c_p \ln \frac{T_2}{T_1}$$

This, as expected, is the same expression that was obtained in solution A by integrating $\int (\delta q/T)$ along the reversible constant-pressure path.

Example 6·2

PROBLEM STATEMENT

Calculate the change in entropy per kilogram of neon heated reversibly at constant pressure of 120 kPa from 20°C to 110°C in a steady-flow system.

SOLUTION

For a pure substance, entropy is a function of any other two properties. Since here the initial pressure and temperature and the final pressure and temperature are the same as in Example 6·1, the entropy change per

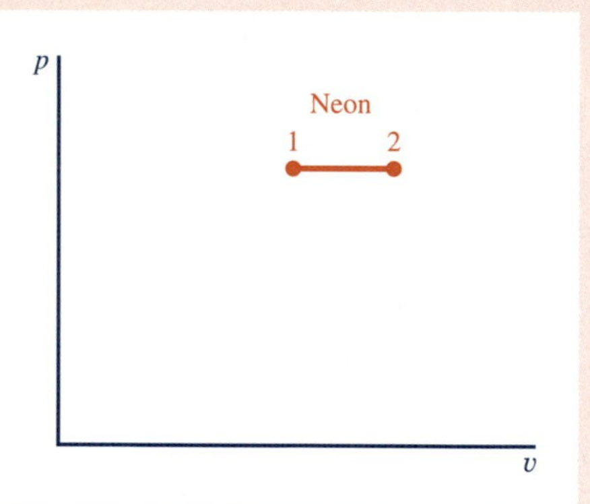

unit of mass must be the same. The fact that the neon in one case is in a closed system and in the other case is in a steady-flow system makes no difference. Had we not realized this, we might have calculated Δs as follows.

The first law applied to a steady-flow system gives an energy balance that, in differential form, is

$$\delta q = dh + dke + dpe + \delta w$$

For a reversible steady-flow process

$$\delta w = -v\,dp - dke - dpe$$

so that

$$\delta q = dh + dke + dpe - v\,dp - dke - dpe$$

and for a constant pressure, $v\,dp = 0$ so that

$$\delta q = dh$$

For a steady-flow system, $dS_{os} = 0$ and $\delta m_1 = \delta m_2$; so the expression for dS of an open system,

$$dS_{os} = \left(\frac{\delta Q}{T}\right)_{\substack{int\\ rev,os}} + s_1\,\delta m_1 - s_2\,\delta m_2 \qquad (6\cdot 4)$$

becomes

$$dS_{os} = 0 = \left(\frac{\delta Q}{T}\right)_{\substack{int\\ rev,os}} + (s_1 - s_2)\,\delta m$$

$$s_2 - s_1 = \int_1^2 \left(\frac{\delta q}{T}\right)_{\substack{int\\ rev}} = \int_1^2 \frac{dh}{T} = \int_1^2 \frac{c_p\,dT}{T}$$

and for a constant c_p of 1.03 kJ/kg·°C (or 1.03 kJ/kg·K)

$$\Delta s = c_p \ln \frac{T_2}{T_1} = 1.03 \ln \frac{383}{293} = 0.276 \text{ kJ/kg·K}$$

(Notice that this is the entropy change of the neon flowing through the system, not that of the system itself. From the definition of steady flow, $\Delta S_{system} = 0$.)

Example 6·3 Calculation of Entropy Change

PROBLEM STATEMENT

Atmospheric air at 14.5 psia, 75 F, enters a compressor and is discharged at 50 psia, 250 F. Determine the change in entropy of the air.

SOLUTION

No information is given on the compression process, but this is of no concern because the entropy change depends only on the end states, and for each of them both p and T are specified. We can calculate Δs for any process by evaluating $\int (\delta q/T)$ along *any reversible* path connecting the same end states. We sketch a pv

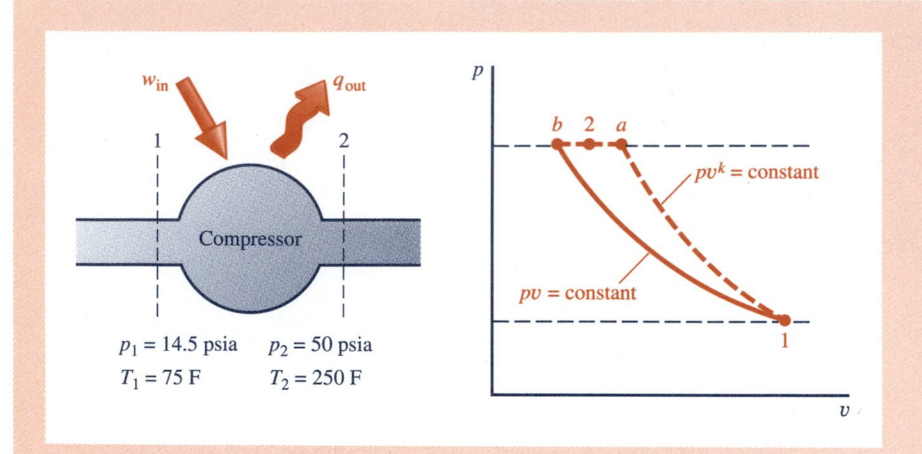

diagram showing two possible *reversible* paths between states 1 and 2:

1–a–2, adiabatic compression to p_2 followed by constant-pressure heating or cooling to T_2
1–b–2, isothermal compression followed by constant-pressure heating to T_2

To sketch the *pv* diagram approximately to scale, we can first calculate

$$\frac{v_2}{v_1} = \frac{p_1 T_2}{p_2 T_1} = \frac{14.5(710)}{50(535)} = 0.385$$

$$\frac{v_a}{v_1} = \left(\frac{p_1}{p_2}\right)^{1/k} = \left(\frac{14.5}{50}\right)^{1/1.40} = 0.412$$

$$\frac{v_b}{v_1} = \frac{p_1}{p_2} = \frac{14.5}{50} = 0.290$$

$$T_a = T_1\left(\frac{p_a}{p_1}\right)^{(k-1)/k} = 535\left(\frac{50}{14.5}\right)^{(1.40-1)/1.40} = 762 \text{ R} = 302 \text{ F}$$

(For the limited temperature range involved, we use constant specific heats.)

Path 1–a–2; a reversible adiabatic process followed by a reversible constant-pressure process:

$$\Delta s = \int_{1-a}^{2} \frac{\delta q}{T} = \int_{1}^{a} \frac{\delta q}{T} + \int_{a}^{2} \frac{\delta q}{T} = 0 + \int_{a}^{2} \frac{dh + dke + dpe + \delta w}{T}$$

For a reversible steady-flow process of a simple compressible substance, $\delta w = -v\,dp - dke - dpe$. Making this substitution, we obtain

$$\Delta s = \int_{a}^{2} \frac{dh - v\,dp}{T} = \frac{c_p\,dT - 0}{T}$$

Over this small temperature range, we can use a constant c_p to obtain

$$\Delta s = c_p \ln \frac{T_2}{T_a} = 0.242 \ln \frac{710}{762} = -0.0171 \text{ B/lbm·R}$$

(Although we show here only an appropriate number of significant digits for each intermediate result, the numerical values have been carried within a calculator throughout the calculations. This makes a difference, especially where exponentials and logarithms are involved.)

Path 1–b–2; a reversible isothermal process followed by a reversible constant-pressure process:

$$\Delta s = \int_1^2 \frac{\delta q}{T} = \int_1^b \frac{\delta q}{T} + \int_b^2 \frac{\delta q}{T} = \frac{q_{1-b}}{T_1} + \int_b^2 \frac{\delta q}{T} \tag{A}$$

For a reversible isothermal steady-flow process of an ideal gas,

$$\begin{aligned} q_{1-b} &= h_b - h_1 + \Delta ke + \Delta pe + w \\ &= 0 + \Delta ke + \Delta pe - \int_1^b v\,dp - \Delta ke - \Delta pe = -\int_1^b v\,dp \\ &= -\int_1^b RT_1 \frac{dp}{p} = -RT_1 \ln \frac{p_b}{p_1} \end{aligned}$$

For a reversible constant-pressure steady-flow process of an ideal gas, we showed in our calculations for the first path

$$\int_b^2 \frac{\delta q}{T} = c_p \ln \frac{T_2}{T_b}$$

Making these substitutions into Eq. A above yields

$$\begin{aligned} \Delta s = s_2 - s_1 &= -R \ln \frac{p_b}{p_1} + c_p \ln \frac{T_2}{T_b} \\ &= -0.0685 \ln \frac{50}{14.5} + 0.241 \ln \frac{710}{535} = -0.0166 \text{ B/lbm·R} \end{aligned}$$

(A different c_p value is used for process b–2 in this path and process a–2 in the earlier path because the temperature ranges are different. See Table A·3E.)

In the examples above, we have calculated entropy changes directly from the definition of entropy. In Section 6·2 we will introduce a more convenient method of calculating entropy changes.

6·2 Useful Relationships Among Properties: The *T ds* Equations

Derivation of the *T ds* equations

By combining the definition of entropy with the first law, we now derive a highly useful relationship that is often used for calculating entropy and for other purposes. Entropy is defined by

$$\Delta s \equiv \int \left(\frac{\delta q}{T}\right)_{\substack{int\\rev}} \tag{6·3}$$

From the first law, for any closed system that passes through equilibrium states and for which the only reversible work equals $\int p\,dv$, $\delta q_{rev} = du + p\,dv$. Therefore,

> The relation $\delta q_{rev} = du + p\,dv$ holds also for a steady-flow system as shown by
>
> $$\delta q = du + d(pv) + dke + dpe + \delta w$$
>
> which for a reversible process is
>
> $$\begin{aligned}\delta q_{rev} &= du + p\,dv + v\,dp + dke + dpe - v\,dp - dke - dpe\\ &= du + p\,dv\end{aligned}$$
>
> Notice, however, that for the steady-flow system $p\,dv \neq \delta w$.

$$\Delta s = \int \frac{du + p\,dv}{T} \tag{6·6a}$$

or

$$ds = \frac{du + p\,dv}{T} \tag{6·6b}$$

If the relationship among p, v, T, and u is known, this expression can be integrated to give the change in entropy between any two equilibrium states. We know that since entropy is a property (and thus a point function), Δs must be the same for any process, reversible or irreversible, between two given states. This, of course, also means that the relation holds for a fluid flowing through an open system as well as for a closed system because entropy is a function of properties such as p, v, T, and u and is independent of the position or state of motion of the system.

The integration of the right-hand side of Eq. 6·3 must be performed for some reversible process between the end states. However, Eq. 6·6 involves only properties, and since the functional relationship among p, v, T, u, and s is independent of any process, Eq. 6·6 is in no way restricted to reversible processes. This equation does require a continuous functional relationship among the properties involved, and this in turn requires that we deal with only equilibrium states. Also, the equation is restricted to pure substances. (If mixing or a chemical reaction occurs, entropy can change even though internal energy and volume are held constant; so obviously the equation does not apply. This point is discussed in Chapters 9 and 11.)

Equation 6·6 can be rearranged as

$$T\,ds = du + p\,dv = du + d(pv) - v\,dp = dh - v\,dp \qquad (6\cdot7)$$

and it will be seen that this is a convenient form.

Conclusion: The relations, commonly called the $T\,ds$ equations,

$$T\,ds = du + p\,dv \qquad (6\cdot7a)$$

$$T\,ds = dh - v\,dp \qquad (6\cdot7b)$$

The *T ds* equations (also called the Gibbs equations)

are valid for any process of a pure substance. (Remember that we have assumed the absence of electricity, magnetism, solid distortion effects, and surface tension. The equations can be broadened to include these other effects, but doing so is beyond the scope of this book.) These equations are highly useful for purposes beyond the calculation of entropy changes.

In the following table, notice the differences among the restrictions that apply to the four equations.

Equations	*Restrictions*
(a) $\delta q = du + \delta w$	This is a statement of the first law, applicable to any simple compressible closed system.
(b) $\delta q = du + p\,dv$	This is a statement of the first law, restricted to reversible processes of a closed system. By comparison with Eq. a, $p\,dv = \delta w$. (This equation can also be applied to reversible processes of a fluid flowing through an open system, but then $p\,dv \neq \delta w$.)
(c) $T\,ds = du + \delta w$	This is a statement combining the first and second laws and is restricted to reversible processes of a closed system. By comparison with Eq. a, $T\,ds = \delta q$.
(d) $T\,ds = du + p\,dv$	This is a relationship among properties (p, v, T, s, u) *valid for all processes between equilibrium states*. It is based on the first and second laws, but in itself it is a statement of neither. In general, $T\,ds \neq \delta q$ and $p\,dv \neq \delta w$. (By comparison with Eq. a it is seen, in fact, that for a closed system, $T\,ds = \delta q$ *only* when $p\,dv = \delta w$, and vice versa.)

To illustrate the difference between Eqs. a and d, consider the system composed of a gas trapped in a closed, rigid, thermally insulated vessel fitted with a paddle wheel, so that work can be done on the gas. Let the paddle wheel be turned by an external motor. Notice that $p\,dv = 0$ but $\delta w \neq 0$, and $\delta q = 0$ but $T\,ds \neq 0$. Equation a, the first law, reduces to

$$0 = du + \delta w$$

Equation d, the $T\,ds$ equation, reduces to

$$T\,ds = du + 0$$

Thus we see how utterly different Eqs. a and d are. Both apply to this irreversible process but, except for du, there is no correspondence between their terms. Equations b and c do not apply because they are restricted to reversible processes.

Learn the *T ds* equations.

Learn the $T\,ds$ equations. They are useful in a number of ways. For one thing, they provide a means of calculating the change in entropy of a system if the relationship among p, v, T, and u (or h) is known. For example, they can be easily integrated for an ideal gas so that it is unnecessary to learn any special equations for the entropy change of an ideal gas. Many other useful results follow from these equations.

Example 6·4

PROBLEM STATEMENT

Determine for an ideal gas the slope of (a) a constant-pressure line and (b) a constant-volume line on a Ts diagram.

SOLUTION

For an ideal gas, the $T\,ds$ equations

$$T\,ds = dh - v\,dp \tag{6·7b}$$

$$T\,ds = du + p\,dv \tag{6·7a}$$

become

$$T\,ds = c_p\,dT - v\,dp$$

$$T\,ds = c_v\,dT + p\,dv$$

For a constant-pressure process, $dp = 0$. For a constant-volume process, $dv = 0$. The desired slopes can consequently be obtained readily as

$$\text{Slope of constant-pressure line} = \left(\frac{\partial T}{\partial s}\right)_p = \frac{T}{c_p}$$

$$\text{Slope of constant-volume line} = \left(\frac{\partial T}{\partial s}\right)_v = \frac{T}{c_v}$$

Since $c_p > c_v$, the slope of a constant-pressure line is less than that of a constant-volume line *at the same temperature*. You can see now that the general form of these lines is as shown in the figure.

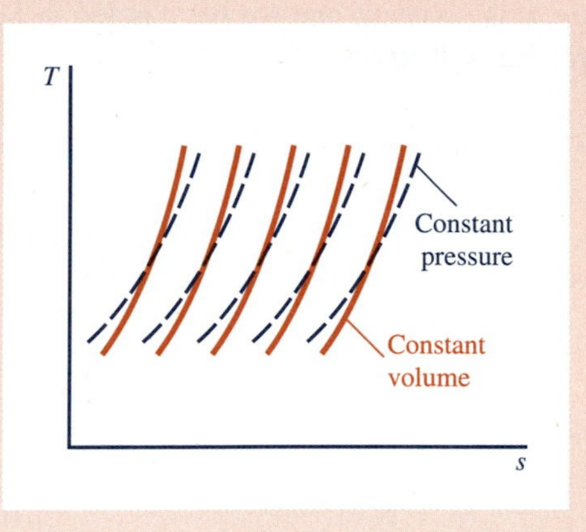

6·2·1 Calculation of Entropy Changes from the *T ds* Equations

The *T ds* equations,

$$T\,ds = du + p\,dv \tag{6·7a}$$

$$T\,ds = dh - v\,dp \tag{6·7b}$$

provide a direct means of calculating entropy changes:

$$s_2 - s_1 = \int_1^2 \frac{du}{T} + \int_1^2 \frac{p\,dv}{T} \tag{6·8a}$$

$$s_2 - s_1 = \int_1^2 \frac{dh}{T} - \int_1^2 \frac{v\,dp}{T} \tag{6·8b}$$

Separate evaluation of the two terms on the right of each equation may require specifying some path between states 1 and 2. However, since the entropy change is always independent of the path between states 1 and 2, the *combination* of the two terms on the right is always path-independent. A pvT relationship is needed for the second term on the right. For most pvT relationships, the value of the integral depends on the path. A uT or hT relationship is needed for the first term on the right, but for most substances u and h are functions of two independent properties, not just of T, so there is no unique uT or hT relationship and the path must be specified. Obviously, if either term is independent of the path, then the other one must be, too, because $(s_2 - s_1)$ is independent of the path.

For some important special cases, u and h are functions of temperature only, so the first term on the right-hand side of Eqs. 6·8 can be evaluated from the end states alone. It follows that the second term can also be evaluated from end states alone. Two of these cases are *incompressible substances* and *ideal gases.*

Entropy change for an incompressible substance

For an *incompressible substance*, v is constant and we have noted in Section 4·3 that $du = c\,dT$. Therefore, Eq. 6·8 gives

$$s_2 - s_1 = c \ln \frac{T_2}{T_1} \tag{6·9}$$

For an *ideal gas*, specific heats are functions of temperature only and the pvT relationship is such that the second term on the right-hand side of Eqs. 6·7 can be readily integrated, so Eqs. 6·8 become

Entropy change for an ideal gas

$$s_2 - s_1 = \int_1^2 \frac{c_v\,dT}{T} + R \ln \frac{v_2}{v_1} \tag{6·10a}$$

$$s_2 - s_1 = \int_1^2 \frac{c_p\,dT}{T} - R \ln \frac{p_2}{p_1} \tag{6·10b}$$

and evaluation of these expressions requires only a functional relationship between c_v or c_p and T.

The integration in Eqs. 6·10 might be time-consuming in some cases, so values of the integral in Eq. 6·10*b* have been published in printed tables and computer software. Notice that entropy is a function of both pressure and temperature, so a double-argument table (like the superheated steam tables) would be required to tabulate values of entropy for an ideal gas. However, Eq. 6·10*b* shows that for an ideal gas, the statement that entropy is a function of temperature and pressure,

$$s = f(T, p)$$

can be specified more closely as

$$s = f(T) + f(p)$$

The part of entropy that is a function of temperature is given the symbol ϕ.

Consequently, $f(T)$ can be listed in a table with temperature as the single argument (just as h and u values are listed). Then, to calculate the entropy change between two states, we determine the $f(T)$ difference from tables to use in Eq. 6·10. We assign the function of temperature the symbol ϕ. Then

$$\phi_2 - \phi_1 = \int_{T_1}^{T_2} \frac{c_p\,dT}{T} \tag{6·11}$$

The value of ϕ at any one temperature can be selected arbitrarily when tables are compiled. Then the entropy change of an ideal gas between any two states 1 and 2 is

$$s_2 - s_1 = \phi_2 - \phi_1 - R \ln \frac{p_2}{p_1} \tag{6·12}$$

[In some books and data tables, the quantity we have defined as ϕ is assigned the symbol $s°$. Sometimes, this is done even when the superscript has been defined as referring to a standard pressure, although ϕ (or $s°$ as used in these books and data tables) is independent of pressure, so that this practice can be confusing. We mention this now because in using ideal-gas property tables, you may find the column of ϕ values, which depend on temperature only, under a heading of $s°$. We will discuss the relationship of ϕ to $s°$ in Chapter 11.]

ϕ is *independent of pressure.*

For an *isentropic process* of an ideal gas we have from Eq. 6·12

$$s_2 - s_1 = 0 = \phi_2 - \phi_1 - R \ln \frac{p_2}{p_1}$$

Rearranging gives

$$\frac{p_2}{p_1} = \exp\left(\frac{\phi_2 - \phi_1}{R}\right) = \frac{\exp(\phi_2/R)}{\exp(\phi_1/R)} = \frac{f(T_2)}{f(T_1)}$$

This new function of temperature is called **relative pressure**, p_r. By definition,

Relative pressure. Notice that p_r is dimensionless and is not a pressure. The term *relative pressure* is consequently somewhat misleading. Also, be careful not to confuse *relative pressure, p_r,* with *reduced pressure, p_R.*

$$p_r \equiv \exp\left(\frac{\phi}{R}\right)$$

Then, for an *isentropic process*, we have

$$\frac{p_2}{p_1} = \frac{p_{r2}}{p_{r1}}$$

Thus a table of p_r versus temperature gives us the pressure–temperature relationship for isentropic processes.

A table of p_r versus T provides the pressure–temperature relationship for isentropic processes.

For any two states at the same entropy the ratio of specific volumes can also be expressed as a ratio of temperature functions. For any two states 1 and 2 of an ideal gas

$$\frac{v_2}{v_1} = \frac{p_1 T_2}{p_2 T_1}$$

If $s_2 = s_1$, then

$$\frac{v_2}{v_1} = \frac{p_1 T_2}{p_2 T_1} = \frac{p_{r1} T_2}{p_{r2} T_1} = \frac{T_2/p_{r2}}{T_1/p_{r1}} = \frac{\text{function of } T_2}{\text{function of } T_1}$$

This defines the **relative specific volume**, $v_r \equiv T/p_r$, so that, for an *isentropic process*,

Relative specific volume

$$\frac{v_2}{v_1} = \frac{v_{r2}}{v_{r1}}$$

We have noted that the selection of a temperature at which the enthalpy or internal energy is zero is arbitrary. Also, in the tabulation of ϕ, p_r, and v_r, various additive terms or multipliers may be introduced that have no effect on the use of the table. They must be considered, however, in verifying tabular values by means of defining equations such as $p_r \equiv \exp(\phi/R)$. These terms or factors should be defined in the documentation that accompanies each table.

For ideal gases, h, p_r, u, v_r, and ϕ are tabulated as a function of temperature only.

A sample of an ideal-gas thermodynamic property table is shown in Table 6·1. Notice that this is a single-argument table. All the properties listed are functions of temperature only. Such tables for several gases are in the appendix, Tables A·12 and A·15.

For *ideal gases with constant specific heats*, Eqs. 6·10 simplify to

Entropy change for an ideal gas with constant specific heats

$$s_2 - s_1 = c_v \ln \frac{T_2}{T_1} + R \ln \frac{v_2}{v_1} \tag{6·13a}$$

$$s_2 - s_1 = c_p \ln \frac{T_2}{T_1} - R \ln \frac{p_2}{p_1} \tag{6·13b}$$

For the special case of isentropic processes of ideal gases with constant specific heats, you can derive from Eqs. 6·13 the pv, pT, and vT relations of Eq. 4·13.

For *real gases*, where u and h are not functions of temperature only and $\int v\, dp/T$ depends on the path between end states, the calculation of entropy changes is considered in Section 7·6.

Another case in which the $T\,ds$ equations (Eqs. 6·7) can be integrated is that of *two-phase mixtures of a pure substance*. For such a mixture, if either the pressure or the temperature is constant, they are both constant, so the

Table 6·1 Properties of Air at Low Pressure

T, K	T, °C	h, kJ/kg	p_r	u, kJ/kg	v_r	ϕ, kJ(kg·K)
1000	726.85	1048.0	116.0	761.0	8.622	7.973
1001	727.85	1049.2	116.4	761.9	8.597	7.974
1002	728.85	1050.3	116.9	762.7	8.572	7.975
1003	729.85	1051.4	117.4	763.6	8.547	7.976
1004	730.85	1052.6	117.8	764.4	8.521	7.977
1005	731.85	1053.7	118.3	765.3	8.496	7.978
1006	732.85	1054.9	118.8	766.1	8.471	7.979
1007	733.85	1056.0	119.2	767.0	8.446	7.981
1008	734.85	1057.1	119.7	767.9	8.421	7.982
1009	735.85	1058.3	120.2	768.7	8.396	7.983

integration is simplified. For example, you can readily verify in any property tables for saturation states that $h_{fg} = Ts_{fg}$, $h_{ig} = Ts_{ig}$, and $h_{if} = Ts_{if}$.

General equations for entropy changes of *any pure substance* are presented in Section 7·3.

Example 6·5 Calculation of Entropy Change

PROBLEM STATEMENT

Steam in a closed system expands reversibly and isothermally from 180°C, $x = 0.65$, to 500 kPa. Sketch the process on a property diagram where the work is represented by an area and on another one where the heat transfer is represented by an area. Calculate the work.

SOLUTION

Analysis: For a reversible process of a closed system, $w = \int p\,dv$, so work is represented by an area on a pv diagram. For any reversible process, $q = \int T\,ds$, so heat is represented by an area on a Ts diagram.

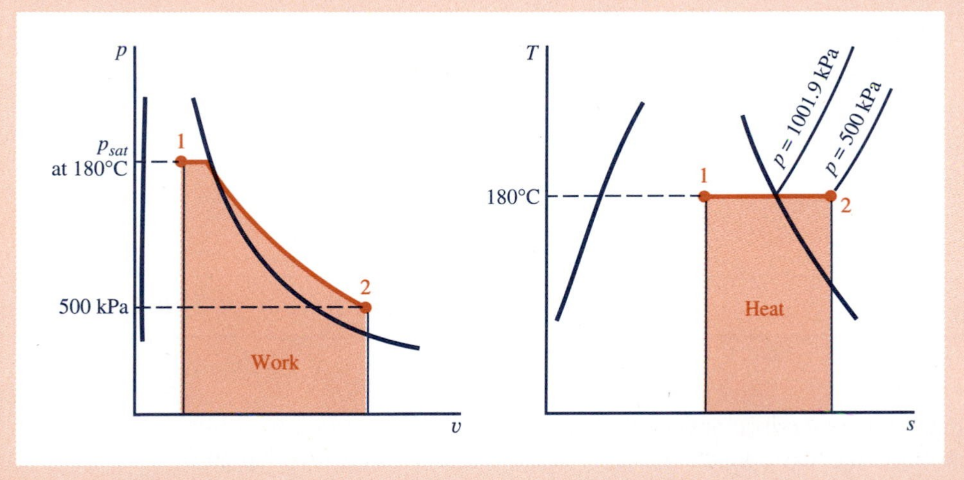

Both the beginning and end states of the process are specified, so we can determine from the steam tables values of u_1 and u_2. Then from the first law,

$$w = u_1 - u_2 + q$$

we can obtain w if q can be determined by another means. For this reversible isothermal process, we can readily integrate $\int T\,ds$ to determine q.

Calculations: First, we will obtain the needed property values from the steam tables:

$$u_1 = h_1 - p_1 v_1 = h_{f1} + x_1 h_{fg1} - p_1(v_{f1} + x_1 v_{fg1})$$
$$= 763.25 + 0.65(2014.6) - 1001.9[0.001127 + 0.65(0.19403 - 0.001127)]$$
$$= 1946.0 \text{ kJ/kg}$$
$$s_1 = s_{f1} + x_1 s_{fg1} = 2.13966 + 0.65(4.4456) = 5.029 \text{ kJ/kg·K}$$
$$u_2 = 2609.5 \text{ kJ/kg} \qquad s_2 = 6.9644 \text{ kJ/kg·K}$$
$$q = \int_1^2 T\,ds = T(s_2 - s_1) = (180 + 273)(6.9644 - 5.029) = 877.0 \text{ kJ/kg}$$
$$w = u_1 - u_2 + q = 1946.0 - 2609.5 + 877.0 = 214 \text{ kJ/kg}$$

Comment: Notice that the internal energy of the system increased during the process, but the heat added was larger than the increase in internal energy, so work was done by the system.

Example 6·6 Calculation of Entropy Change

PROBLEM STATEMENT

Air in a cylinder is compressed reversibly and isothermally from 95 kPa, 25°C, to 290 kPa. The initial volume is 0.162 m^3. Determine the work input, the heat transfer, and the entropy change of the air being compressed.

SOLUTION

Analysis: The system is a closed one composed of the air within the cylinder. We first make a labeled sketch of the system and two property diagrams: pv and Ts. For a reversible process in a closed system, we can calculate work by $W = \int p\,dV$. For an isothermal process of an ideal gas, $\Delta U = 0$, so we can obtain the heat transfer from the first law, $Q = \Delta U + W$. Then ΔS can be obtained from its definition or a $T\,ds$ equation.

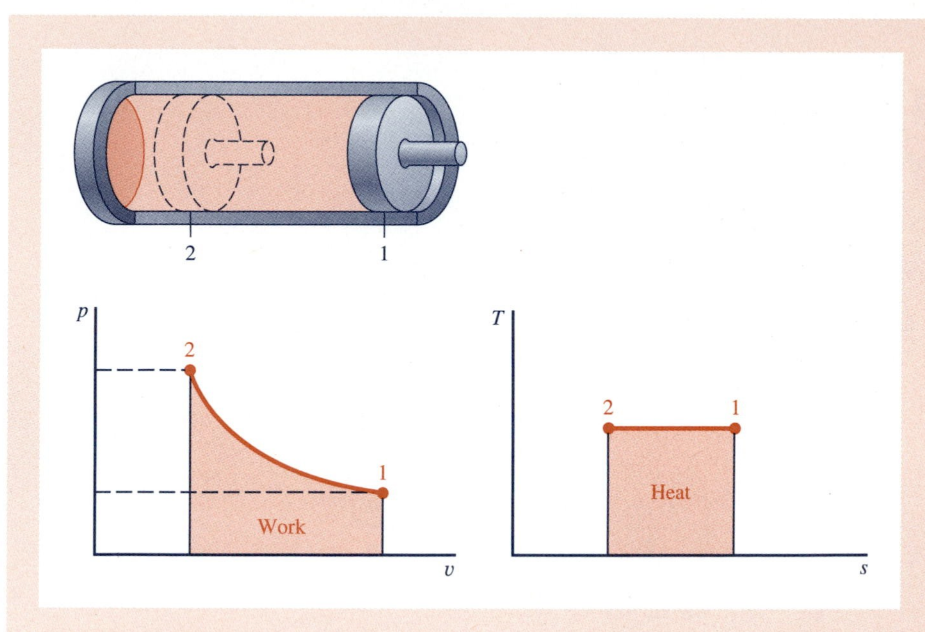

Calculations:

$$W = \int_1^2 p\,dV = \int_1^2 \frac{mRT}{V}\,dV = mRT \ln \frac{V_2}{V_1} = p_1 V_1 \ln \frac{p_1}{p_2}$$

$$= 95(0.162) \ln \frac{95}{290} = -17.2 \text{ kJ}$$

$$Q = U_2 - U_1 + W = 0 - 17.2 = -17.2 \text{ kJ}$$

$$S_2 - S_1 = \int_1^2 \left(\frac{\delta Q}{T}\right)_{\substack{int\\rev}} = \frac{Q}{T} = \frac{-17.2}{25 + 273} = -0.0576 \text{ kJ/K}$$

Example 6·7 Calculation of Entropy Change

PROBLEM STATEMENT

Carbon dioxide is cooled in a heat exchanger from 900 F to 200 F. The inlet pressure is 15.2 psia, and a pressure drop of 1.3 psi is caused by friction as the gas flows through the heat exchanger. Calculate the entropy change of the CO_2 flowing through the heat exchanger.

SOLUTION

Analysis: Carbon dioxide at approximately one atmosphere pressure and temperatures of 200 F and higher

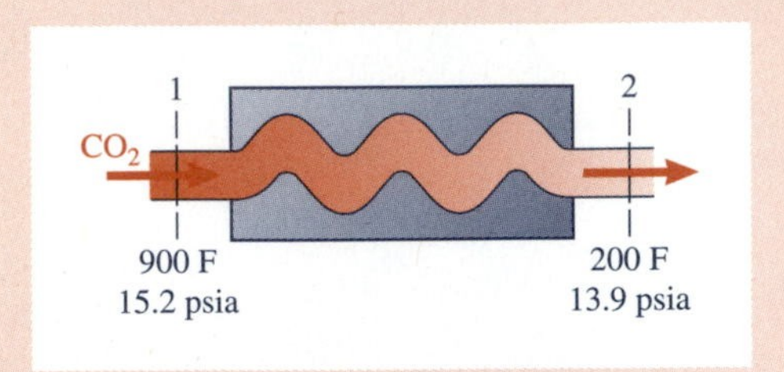

can be accurately modeled as an ideal gas. The specific heats vary significantly over the specified temperature range, so an assumption of constant specific heats is unjustified. Therefore, integrating

$$T\,ds = dh - v\,dp$$

gives

$$s_2 - s_1 = \int_1^2 \frac{dh}{T} - \int_1^2 \frac{v\,dp}{T} = \int_1^2 \frac{c_p\,dT}{T} - R\ln\frac{p_2}{p_1}$$

$$= \phi_2 - \phi_1 - R\ln\frac{p_2}{p_1}$$

Values of ϕ can be obtained from Table A·15E or from the software model for ideal-gas properties to give

$$s_2 - s_1 = 1.2035 - 1.3816 - \frac{35.1}{778}\ln\frac{15.2 - 1.3}{15.2}$$

$$= -0.174\ \text{B/lbm·R}$$

Example 6·8 Calculation of Entropy Change

PROBLEM STATEMENT

Carbon monoxide flows steadily and adiabatically through a turbine. At the inlet, the CO is at 220 kPa, 10°C; at the exhaust, the CO is at 120 kPa, −17°C. Define the system as the entire flow passage within the turbine. Determine the entropy change of (*a*) the CO, (*b*) the system, and (*c*) the surroundings.

SOLUTION

We first make a sketch of the steady-flow system and *pv* and *Ts* diagrams of the process. We will model the CO as an ideal gas, and in this limited temperature range we can use constant specific heats. Therefore, the

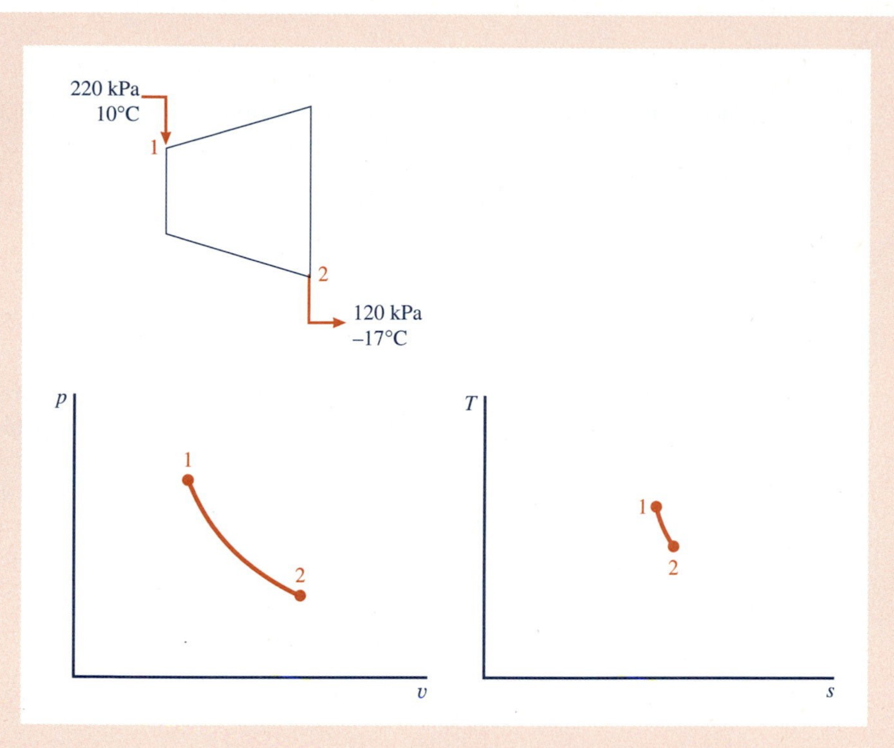

entropy change of the CO can be calculated by integrating one of the $T\,ds$ equations.

(*a*)

$$T\,ds = dh - v\,dp$$

$$s_2 - s_1 = \int_1^2 \frac{dh}{T} - \int_1^2 \frac{v\,dp}{T} = \int_1^2 \frac{c_p\,dT}{T} - R\ln\frac{p_2}{p_1}$$

For constant c_p, this becomes

$$s_2 - s_1 = c_p \ln\frac{T_2}{T_1} - R\ln\frac{p_2}{p_1}$$

$$= 1.04 \ln\frac{273 - 17}{273 + 10} - 0.297\ln\frac{120}{220} = 0.0757 \text{ kJ/kg·K}$$

where the c_p value is from Table A·3.

(*b*) This is a steady-flow system, so the properties at any point within the system remain constant with respect to time. Therefore, the entropy change of the system, as for any steady-flow system, is zero.

(*c*) For the *surroundings* of this steady-flow system, the entropy increases, because each kilogram of CO that leaves the surroundings (i.e., enters the system) has an entropy that is lower than the entropy of CO that enters the surroundings by 0.0757 kJ/kg·K. The entropy change of the surroundings is therefore 0.0757 kJ/kg·K.

6·3 Entropy as a Coordinate

As mentioned at the close of Section 6·1, the definition of entropy, Eq. 6·3, can be rearranged so that *for a reversible process,*

$$Q = \int T\,dS \tag{6·5}$$

for a closed system and

$$q = \int T\,ds \tag{6·5}$$

For any reversible process, the heat transfer can be represented by the area beneath the path on a *TS* diagram.

for either a closed or steady-flow system. Therefore, heat transfer is represented by an area on a *TS* or *Ts* diagram. Figure 6·5 shows a reversible process 1–2 on a *TS* diagram. The area beneath the path is $\int_1^2 T\,dS$, so it represents the heat transferred to the system during the reversible process. If the process is reversed so as to proceed from state 2 to state 1, then the heat transferred to the system is $Q_{2-1} = \int_2^1 T\,dS$. For the same path, $\int_1^2 T\,dS$ and $\int_2^1 T\,dS$ are equal in magnitude but opposite in sign. This is consistent with the sign convention for heat.

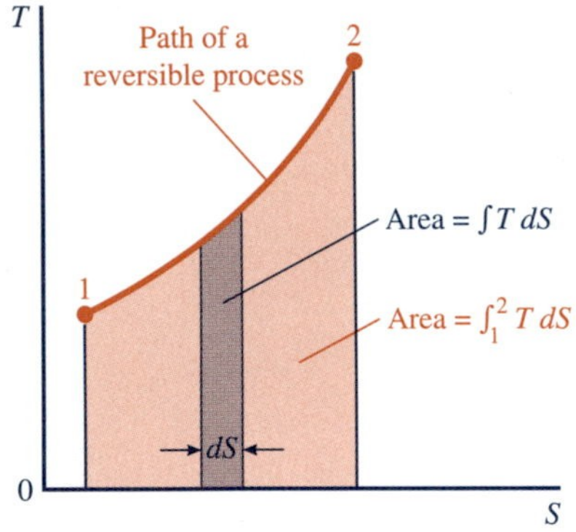

Figure 6·5 *TS* diagram of a reversible process.

The relation

$$q = \int T\,ds \tag{6·5}$$

is in some respects analogous to the relations

$$w = \int p\,dv \tag{1·17}$$

and

$$w + \Delta ke + \Delta pe = -\int v\,dp \tag{1·20}$$

Each of these three equations relates heat or work to an area on a property diagram for reversible processes. Equation 1·17 applies only to a closed system and Eq. 1·20 applies only to a steady-flow system; but Eq. 6·5 applies to either type of system.

Temperature–entropy diagrams are frequently used in the analysis of processes and cycles because for reversible processes areas on the diagrams represent heat transfer.

A Carnot cycle can always be represented by a rectangle on a *TS* diagram.

A Carnot cycle is composed of two reversible isothermal processes and two reversible adiabatic or isentropic processes. Therefore, it is always represented by a rectangle on a *TS* diagram. In Fig. 6·6, process *a–b* is the reversible isothermal heat addition at T_H. The area beneath line *a–b* represents the heat added $|Q_H|$. Since process *b–c* is a reversible adiabatic process, the area beneath line *b–c* must be zero. Process *c–d* is the reversible isothermal rejection of heat $|Q_L|$ to the energy reservoir at T_L. $|Q_L|$ is represented by the area

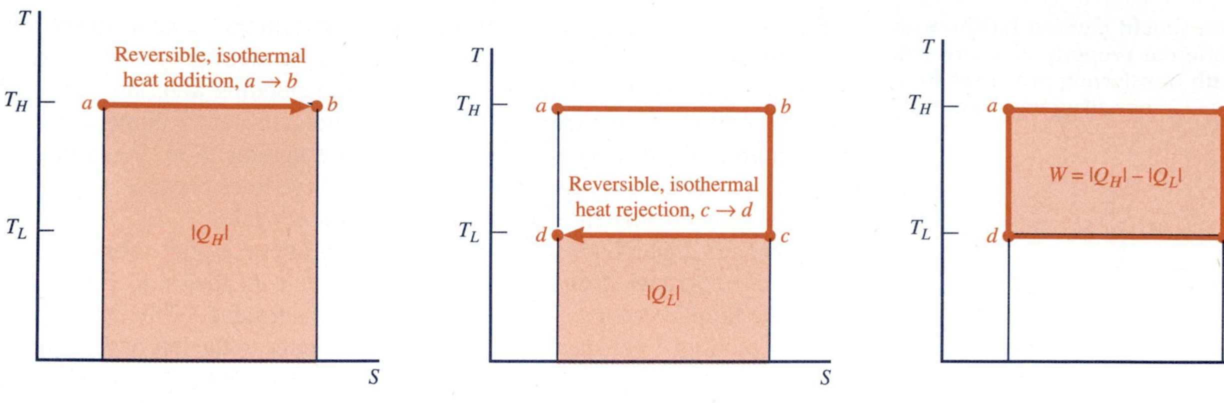

Figure 6·6 *TS* diagram of a Carnot cycle.

beneath line c–d. The cycle is completed by means of the reversible adiabatic process d–a that returns the system to its initial state a.

The area a–b–c–d–a is the difference between the area beneath line a–b (representing $|Q_H|$) and the area beneath line c–d (representing $|Q_L|$), so it represents ($|Q_H| - |Q_L|$) or the work done by the cycle. Areas on a *Ts* diagram represent heat transfer for all reversible processes of a substance, but they represent work only for reversible *cycles*, because for any cycle of a closed system

The area within the rectangle represents the work of the Carnot cycle.

$$\oint \delta W = \oint \delta Q \tag{3·1}$$

It is important to note that the area beneath the path of an irreversible process on a *Ts* diagram has no significance. It does *not* represent heat transfer because

$$Q_{irrev} \neq \int T\,dS$$

It can be shown rigorously that $Q_{irrev} < \int T\,ds$. As an example that $Q_{irrev} \neq \int T\,ds$, consider a gas stirred by a paddle wheel inside a closed, rigid, thermally insulated vessel. The gas goes from state 1 to state 2. A reversible process between the same two end states would be a constant-volume heating, so that $S_2 > S_1$ and $\int_1^2 T\,dS > 0$. Yet the paddle-wheel process occurs in an insulated vessel so that $Q = 0$. Therefore, for this irreversible process $Q < \int T\,dS$.

Irreversible processes are often shown as broken lines on *TS* diagrams (see Fig. 6·7) for two reasons. First, a broken line serves as a reminder that the area beneath an irreversible process path has no significance. Also, the exact path of an irreversible process between two states is frequently unknown or cannot be represented on a property diagram because the system may not be in an equilibrium state at all times.

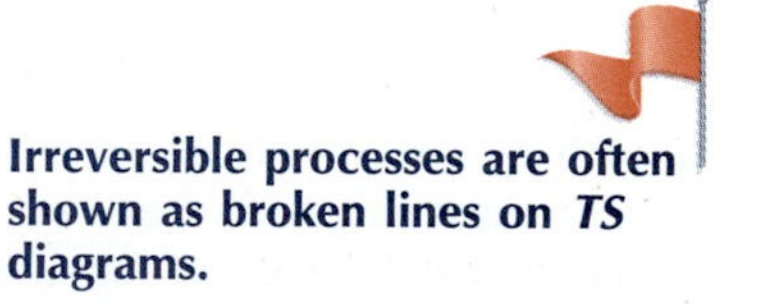

Irreversible processes are often shown as broken lines on *TS* diagrams.

You should develop facility with different property diagrams and with transferring processes from one diagram to another.

Two or more property diagrams are often useful in the same analysis, and you now have the ability to transfer a process or constant-property line from one diagram to another. As an example of corresponding diagrams, Fig. 6·8 shows, for an ideal gas, lines of constant pressure, constant volume, constant temperature, and constant entropy on *pv* and *Ts* diagrams. You should be able to explain the shapes and orientations of all of these lines because we will be using them frequently in analyzing processes and cycles.

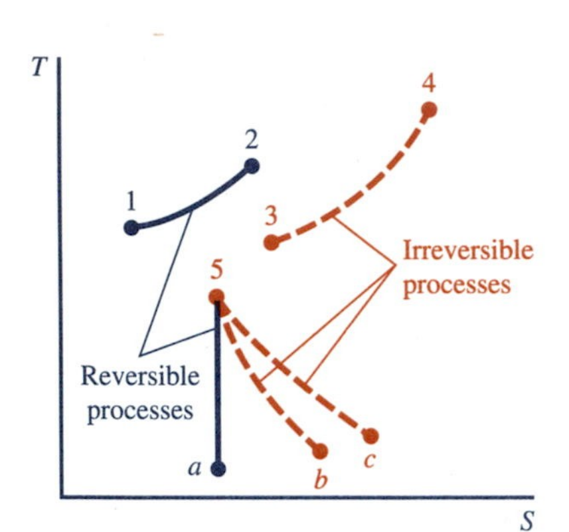

Figure 6·7 Reversible and irreversible processes on a *TS* diagram. Irreversible processes are shown as broken lines.

In Chapters 2 and 4 we showed the saturation lines for pure substances on *pv* and *ph* diagrams. Now we consider them on a *Ts* diagram. Figure 6·9 shows the liquid–vapor saturation line for a pure substance. This plot can be made by plotting T versus s_f for saturated liquid states to the left of the critical state and T versus s_g to the right. Some lines for constant pressure, volume, enthalpy, temperature, and entropy are also shown. Notice the following features:

- Within the two-phase (or "wet" vapor) region, constant-pressure lines coincide with constant-temperature lines.
- Throughout the superheat region, constant-pressure and constant-volume lines have slopes that increase with increasing temperature, and at any state the constant-volume line is steeper than the constant-pressure line.
- At low pressures, the constant-enthalpy lines in the superheated vapor region approach constant-temperature lines as pressure decreases. This means that the vapor behavior is approaching that of an ideal gas as pressure is lowered.

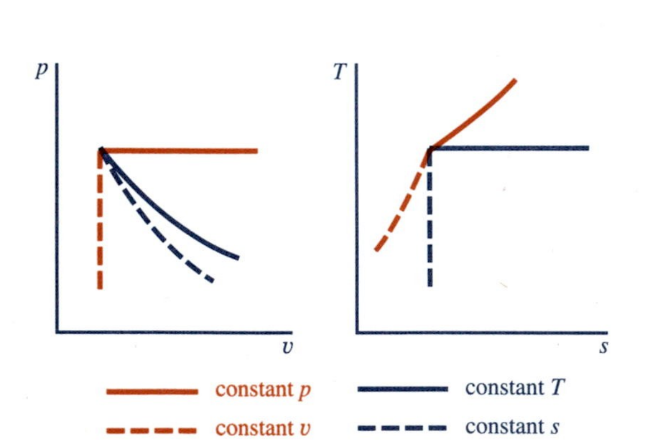

Figure 6·8 Lines of constant pressure, constant volume, constant temperature, and constant entropy shown on *pv* and *Ts* diagrams for an ideal gas.

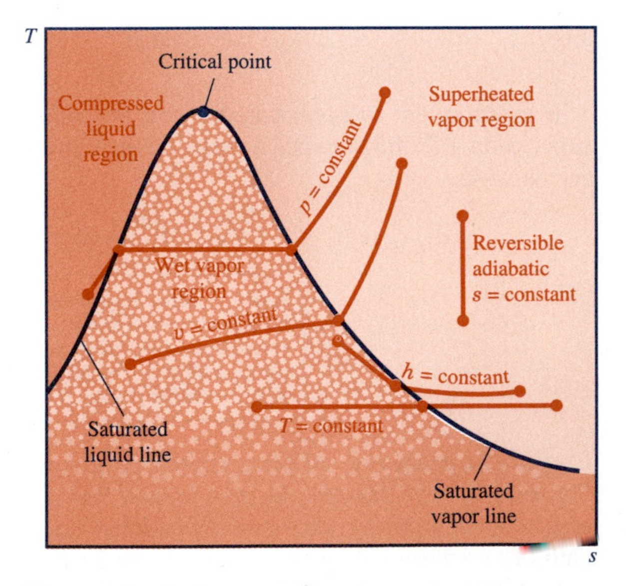

Figure 6·9 *Ts* diagram of two phases, liquid and vapor, of a pure substance.

Observations like these are helpful in making thermodynamic analyses.

Temperature–entropy diagrams for water and air are in the appendix, Charts A·7·1 and A·13. (Note that when air is condensed at constant pressure, the composition of the liquid differs from that of the vapor so that air is no longer a pure substance. Certain features on the *Ts* diagram for air are therefore different from those of a pure substance *Ts* diagram.)

Enthalpy–Entropy (*hs*) Diagrams. Another commonly used diagram is the enthalpy–entropy diagram, often called a Mollier chart (see, for example, Chart A·7·2). A reversible adiabatic process must be a constant-entropy or *isentropic* process; therefore, such a process is represented by a vertical line on any diagram that has entropy on the abscissa. Further, enthalpy is a convenient coordinate because for many processes in steady-flow systems, enthalpy change is equal to the work done, the heat transferred, or the change in kinetic energy. Therefore, many processes of steady-flow systems can be easily represented on an *hs* diagram, and needed property values can be read directly from it. Areas on the *hs* diagram have no significance.

Mollier diagrams are useful because isentropic processes are represented by vertical lines and for many steady-flow processes, the enthalpy change is equal to the work done, the heat transferred, or the change in kinetic energy.

Example 6·9 Refrigeration Compressor

PROBLEM STATEMENT

In an ideal refrigeration system, refrigerant R134a is compressed reversibly and adiabatically at a steady rate of 28.8 lbm/min from 15 psia, 92.0 percent quality, to 90 psia. Kinetic energy changes are negligible. Determine the power required.

SOLUTION

Analysis: For a steady-flow process, $\dot{W} = \dot{m}w$. Work can be found from the first law,

$$w_{in} = h_2 - h_1 + \Delta ke + \Delta pe - q$$

which for an adiabatic process and no change in kinetic or potential energy becomes

$$w_{in} = h_2 - h_1$$

h_1 can be found from $h_1 = h_{f1} + x_1 h_{fg1}$. In addition to the pressure at the discharge, we know that $s_2 = s_1$ for the *reversible adiabatic* process. The *Ts* diagram helps us to see that, if s_2 is greater than s_g at 40 psia, the R134a discharged is superheated, but if s_2 is less than s_g at 200 psia, the R134a discharged is wet. If the discharge is superheated, h_2 can be obtained from the tables by using p_2 and s_2; if the discharge is wet, then

$$h_2 = h_{f2} + x_2 h_{fg2} \quad \text{and} \quad x_2 = \frac{s_2 - s_1}{s_{fg2}}$$

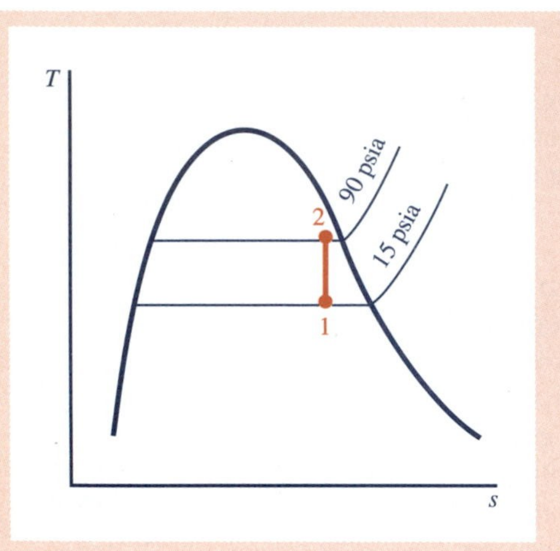

Calculations: Referring to the R134a tables in the appendix, or using software model R134a, we find

$$h_1 = h_{f1} + x_1 h_{fg1} = 7.77 + 0.920(93.27) = 93.6 \text{ B/lbm}$$
$$s_2 = s_1 = s_{f1} + x_1 s_{fg1} = 0.0180 + 0.920(0.2093) = 0.2105 \text{ B/lbm·R}$$

The value of s_2 is less than s_{g2}, so the R134a is wet at discharge:

$$x_2 = \frac{s_2 - s_{f2}}{s_{fg2}} = \frac{0.2105 - 0.0746}{0.1455} = 0.934$$
$$h_2 = h_{f2} + x_2 h_{fg2} = 35.6 + 0.934(77.49) = 108.0 \text{ B/lbm}$$

Applying the first law, recalling that $q = 0$, and assuming that $\Delta ke + \Delta pe = 0$, we find

$$w_{\text{in}} = h_2 - h_1 = 108.0 - 93.6 = 14.4 \text{ B/lbm}$$

and the power input is

$$\dot{W}_{\text{in}} = \dot{m} w_{\text{in}} = 28.8(14.4) = 415 \text{ B/min} = 7.30 \text{ kW}.$$

6·4 The Increase of Entropy Principle

The principle of the increase of entropy

The property entropy provides a means of determining if a process is possible and, if so, whether it is reversible or irreversible. This is accomplished using *the increase of entropy principle*: *the entropy of an isolated system always increases or, in the limiting case of a reversible process, remains constant with respect to time*. In mathematical form,

$$\left(\frac{dS}{dt}\right)_{\text{isolated system}} \geq 0$$

or, with the understanding that time is the independent variable, this statement is usually written

$$\Delta S_{\text{isolated system}} \geq 0 \tag{6·14}$$

The inequality holds for irreversible processes; the equality holds for reversible processes.

Since an isolated system is one that in no way interacts with its surroundings, this principle may appear to be severely restricted in application. However, a judicious selection of boundaries often makes it possible to work with isolated systems. A proof of the increase of entropy principle follows.

Proof of the increase of entropy principle

In Section 6·1 we saw that a system can always be made to go from one state to another by a series of reversible adiabatic and reversible isothermal processes. In fact, two states of a system can always be connected by one reversible adiabatic process and one reversible isothermal process. Let any closed system undergo an adiabatic process from a state 1 to a state 2 as shown in Fig. 6·10. Now the system can be restored to state 1 by means of the following processes.

Process 2–a: A reversible adiabatic process that restores the system to its initial temperature, followed by

Process a–1: A reversible isothermal process that restores the system to its initial state.

In this way the system has executed a cycle. During the cycle it could exchange heat with its surroundings only during the reversible isothermal process a–1. This heat exchange could have been with an energy reservoir at constant temperature. Therefore, in accordance with the second law, there can be no net work output from the system during the cycle; that is, $\oint \delta W \leq 0$. Then applying the first law,

$$\oint \delta Q = \oint \delta W$$

we see that $\oint \delta Q \leq 0$: Either (1) heat was removed from the system during process a–1, or (2) the cycle was completely adiabatic so that state a was identical with state 1. Consequently,

$$S_1 - S_a \leq 0 \tag{a}$$

where the equality holds if the reversible adiabatic process 2–a alone returned the system to state 1 so that state a and state 1 are identical. This could be the case only if the original adiabatic process 1–2 had been reversible. For process 2–a,

$$S_a - S_2 = 0 \tag{b}$$

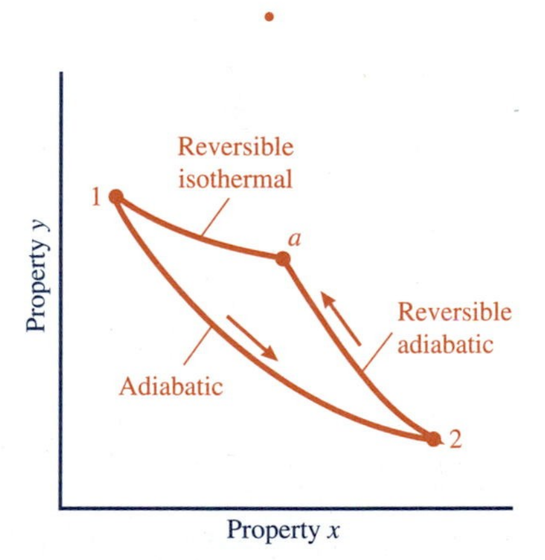

Figure 6·10 Proof of the increase of entropy principle.

For the cycle,

$$\Delta S = (S_2 - S_1) + (S_a - S_2) + (S_1 - S_a) = 0 \qquad (c)$$

Comparison of Eqs. *a*, *b*, and *c* shows that

$$S_2 - S_1 \geq 0$$

where the equality holds only if the adiabatic process 1–2 is reversible. Recalling that process 1–2 was an adiabatic process of any closed system, we have shown that, in general,

The statement of entropy increase for an adiabatic closed system

$$\Delta S_{\text{adiabatic closed system}} \geq 0$$

This is one statement of the principle of the increase of entropy: The entropy of an adiabatic closed system always increases or, in the limiting case of a reversible process, remains constant. The independent variable is understood to be time.

An isolated system has already been defined as a system that in no way interacts with its surroundings. Therefore, every isolated system is an adiabatic closed system, and we conclude that

$$\Delta S_{\text{isolated system}} \geq 0 \qquad (6{\cdot}14)$$

Thus an alternative statement of the principle of the increase of entropy is: The entropy of an isolated system always increases or, in the limiting case of a reversible process, remains constant. The independent variable is understood to be time. This statement is actually less general than the one that refers to an adiabatic closed system, because an isolated system is a special case (work = 0) of an adiabatic closed system. Nevertheless, this is a convenient and widely used form of the increase of entropy principle.

The increase of entropy principle provides another criterion of reversibility in addition to those listed in Section 5·5·3. For any process of an isolated system,

- If the entropy of the system remains constant, the process is reversible.
- If the entropy of the system increases, the process is irreversible.
- The entropy of an isolated system cannot decrease.

Consider the heating of a block of steel from a temperature T_1 to a temperature T_2. Heat is supplied from a constant-temperature reservoir (perhaps steam condensing at constant pressure) at T_H. T_H is greater than T_2. Neither the block nor the reservoir is isolated, but the block and the reservoir together constitute an isolated system. Since the transfer of heat from the reservoir to the block across a finite temperature difference is irreversible, the entropy of

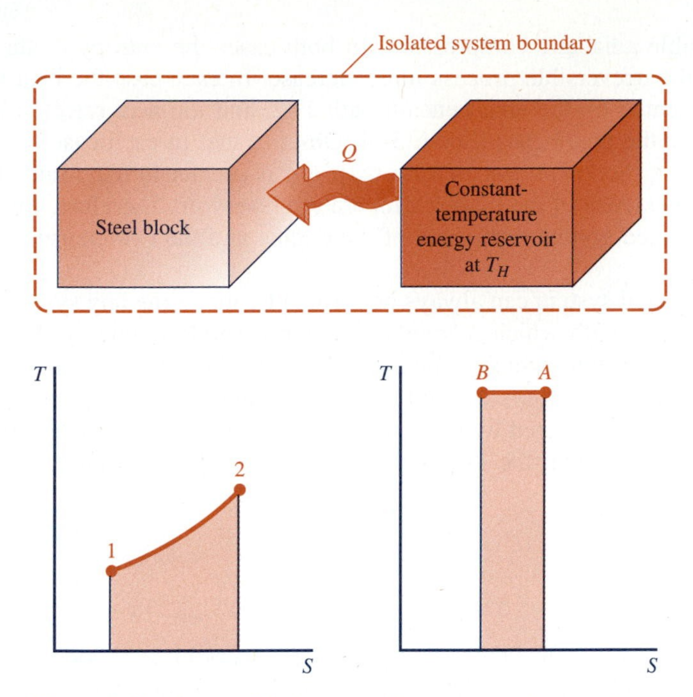

Figure 6·11 Irreversible heat transfer causes an increase in entropy of the isolated system.

the isolated system increases. This can be seen from the *TS* diagrams of Fig. 6·11. Since the heating of the block and the cooling of the reservoir are *internally* reversible, the amount of heat transferred is represented by the area beneath the path on each *TS* diagram. The two areas must be equal; so the lower temperature of the block requires that $|\Delta S_{\text{block}}| > |\Delta S_{\text{reservoir}}|$ and thus $\Delta S_{\text{isolated system}} > 0$.

A Carnot cycle modified by an irreversible adiabatic expansion is shown on a *TS* diagram in Fig. 6·12*a*. Figure 6·12*b* shows one modified by an

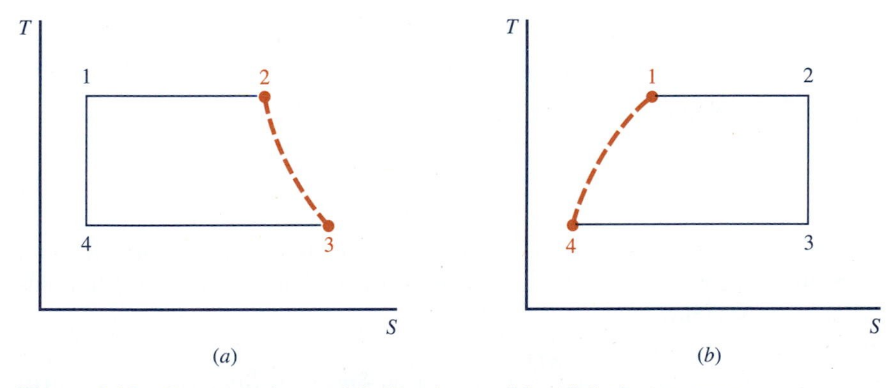

Figure 6·12 Carnot cycles modified by irreversible adiabatic processes.

irreversible adiabatic compression. In both cases the entropy of the system during the irreversible process must increase. In each case the heat supplied is represented by the area beneath path 1–2, and the heat rejected is represented by the area beneath path 3–4. Observe that in each case the entropy *increase* of the lower-temperature reservoir is greater in magnitude than the entropy *decrease* of the higher-temperature reservoir. Therefore, the entropy of an isolated system composed of the engine and the two energy reservoirs increases.

An isolated system can always be formed by drawing the boundary around the system and the surroundings with which it interacts.

An isolated system can always be formed by including any system and the surroundings with which it interacts within a single boundary. Sometimes, the original system, which is then only a part of the isolated system, is called a *subsystem*. Since a system and its surroundings include, by definition, everything that is affected by the process, the combination is sometimes called the *universe*, so that the increase of entropy principle is stated as

$$\Delta S_{\text{universe}} \geq 0 \tag{6·15}$$

where

$$\Delta S_{\text{universe}} = \Delta S_{\text{system}} + \Delta S_{surr}$$

Recall that we define the surroundings as everything outside the system boundary but usually restrict the term to things outside the system that in some way affect the behavior of the system. Consequently, the term *universe* generally refers to everything that is involved in a process and need not include things that have no effect on the process.

The entropy of a system is not restricted; it may increase, decrease, or remain the same.

Notice that no conclusion has been drawn regarding ΔS of systems in general. The entropy of a *system* may increase, decrease, or remain constant. It is only the entropy of the universe or of an isolated system that must increase or, in the case of a reversible process, remain constant.

Open systems occur frequently in practice, so next we show how the increase of entropy principle can be applied to them. An isolated system can always be formed by including within one boundary an open system and its surroundings. Denoting the properties of the open system by the subscript *os*, we then have

$$dS_{\text{isolated system}} = dS_{os} + dS_{surr} \geq 0 \tag{a}$$

As shown in Section 6·1

$$dS_{os} = \left(\frac{\delta Q}{T}\right)_{\substack{int\\rev,os}} + s_1\,\delta m_1 - s_2\,\delta m_2 \tag{b}$$

where δm_1 is the mass entering the open system and δm_2 is the mass leaving it. Similarly, for the surroundings of the open system (which constitute another open system),

$$dS_{surr} = \left(\frac{\delta Q}{T}\right)_{\substack{int\\rev,surr}} + s_2\,\delta m_2 - s_1\,\delta m_1 \tag{c}$$

Substituting from Eqs. *b* and *c* into Eq. *a*, we have *for any open system*,

$$\left(\frac{\delta Q}{T}\right)_{\substack{int\\rev,os}} + \left(\frac{\delta Q}{T}\right)_{\substack{int\\rev,surr}} \geq 0$$

For *steady flow*, $dS_{os} = 0$, so that Eq. *a* becomes

$$dS_{surr} \geq 0$$

or

$$\left(\frac{\delta Q}{T}\right)_{\substack{int\\rev,surr}} + s_2\,\delta m - s_1\,\delta m \geq 0 \qquad (6\cdot16)$$

for the case of a single inlet (section 1) and a single outlet (section 2). A frequently encountered special case in engineering is the *adiabatic* steady-flow process. The increase of entropy principle shows that

$$\Delta s \geq 0$$

where Δs is the change in entropy of the fluid flowing through the system. (Remember that Δs for the steady-flow system itself, which is a region in space in which the properties at each point remain constant, is zero for all processes.) Some possible adiabatic compression and expansion paths for steadily flowing fluids are shown in *Ts* diagrams in Fig. 6·13.

One of the most valuable uses of the increase of entropy principle is in predicting the direction in which a chemical reaction will proceed. A reaction can never proceed in such a direction as to cause the entropy of the universe to decrease. This point is discussed in Chapter 11 in connection with combustion reactions.

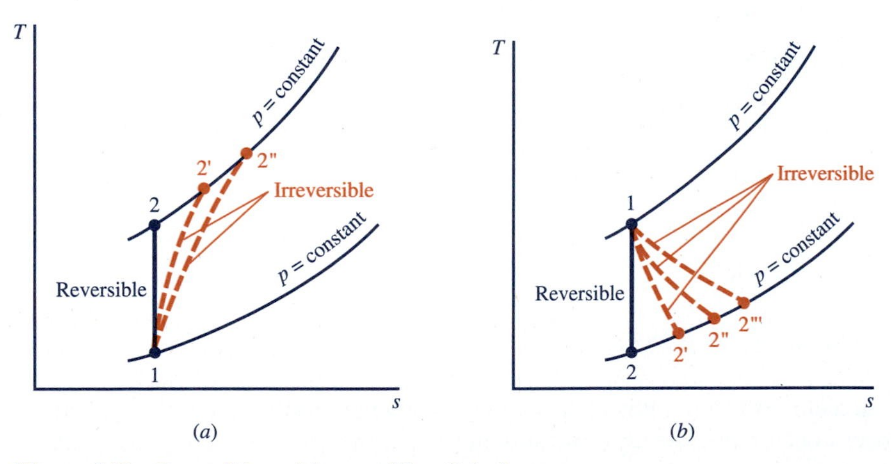

Figure 6·13 Reversible and irreversible adiabatic processes.

Example 6·10 Application of the Increase of Entropy Principle

PROBLEM STATEMENT

Is it possible to compress air adiabatically from 100 kPa, 15°C, to (*a*) 200 kPa, 30°C? (*b*) 200 kPa, 105°C?

SOLUTION

Analysis: Whether we are dealing with a closed system or a steady-flow system, it is impossible for the entropy of a substance undergoing an adiabatic process to decrease. It must increase, or in the limiting case

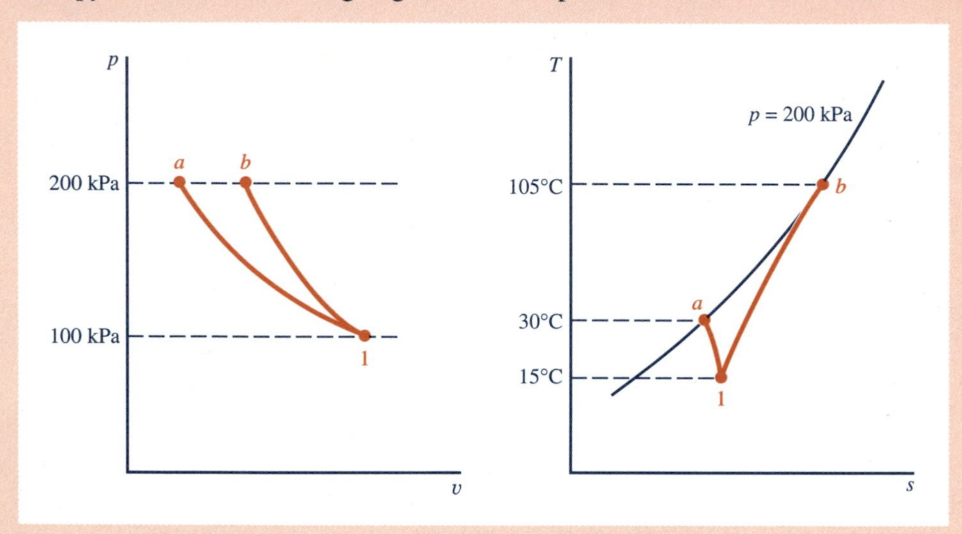

of a *reversible* adiabatic process, remain constant. For the pressure and temperature ranges here, we can model air as an ideal gas. Also, in the specified temperature ranges, variations in specific heats are small, so we will use specific heats that are constant at mean values for each temperature range. We will calculate the entropy change for each case by integrating the $T\,ds$ equation, $T\,ds = dh - v\,dp$.

Calculations: For an ideal gas with constant specific heats, the equation for both cases, a and b, is

$$s_2 - s_1 = \int_1^2 ds = \int_1^2 \frac{dh - v\,dp}{T} = \int_1^2 \frac{c_p\,dT}{T} - \int_1^2 \frac{R\,dp}{p}$$

$$= c_p \ln \frac{T_2}{T_1} - R \ln \frac{p_2}{p_1}$$

Substituting numerical values for each case yields

(*a*) $$s_2 - s_1 = 1.005 \ln \frac{303}{288} - 0.287 \ln \frac{200}{100} = -0.148 \text{ kJ/kg·K}$$

If the adiabatic process occurs in a closed system, this is the entropy change per unit mass of the system and of the universe. If the process occurs in a steady-flow system, this is the entropy change per unit mass of the surroundings and of the universe. A decrease in entropy of the universe is impossible, therefore the process described is impossible.

(*b*)

$$s_2 - s_1 = 1.005 \ln \frac{378}{288} - 0.287 \ln \frac{200}{100} = 0.0742 \text{ kJ/kg·K}$$

Here the entropy of the universe increases, so this process is possible.

Comment: This question is one that was listed in Section 5·3 as answerable by means of the second law.

Example 6·11 Application of Increase of Entropy Principle

PROBLEM STATEMENT

If carbon dioxide is expanded steadily and adiabatically through a nozzle from a very low velocity at 40 psia, 600 F, to 15 psia, what is the maximum velocity it can reach?

SOLUTION

Analysis: At these pressures and temperatures, CO_2 can be modeled as an ideal gas, but of course the

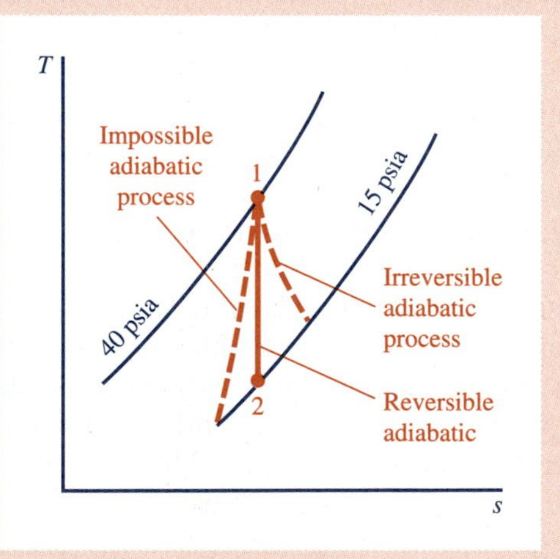

specific heats may vary with temperature. No work is done in a nozzle, so the first law with $q = 0$, $w = 0$, no change in potential energy, and negligible inlet kinetic energy gives

$$\frac{V_2^2}{2} = h_1 - h_2$$

State 1 is fixed, so maximum V_2 will result from a minimum h_2. For an ideal gas, minimum h_2 means minimum T_2, and a Ts diagram shows that minimum T_2 results from a reversible adiabatic or isentropic process. Any lower value of T_2 at the specified pressure would involve a decrease in entropy of the CO_2 and thus of the universe, so it is impossible. We therefore calculate V_2 for an isentropic process. We will determine state 2 by the relationship for isentropic processes of ideal gases, $p_{r2}/p_{r1} = p_2/p_1$.

Calculations: For an isentropic process of an ideal gas,

$$p_{r2} = p_{r1}\left(\frac{p_2}{p_1}\right) = 4.580\left(\frac{15}{40}\right) = 1.717$$

where values are obtained from CO_2 printed tables or the software model for ideal-gas properties. For this value of p_{r2}, we find $h_2 = 169.3$ B/lbm, which gives

$$V_2 = \sqrt{2(h_1 - h_2)}$$

$$= \sqrt{2\left(212.4 - 169.3\ \frac{\text{B}}{\text{lbm}}\right)778\frac{\text{ft·lbf}}{\text{B}}\left(32.174\frac{\text{lbm·ft}}{\text{lbf·s}^2}\right)}$$

$$= 1470 \text{ ft/s}$$

Comment: This question is one that was listed in Section 5·3 as answerable by means of the second law.

Up to this point, our use of the increase of entropy principle has involved only the question of whether the entropy change of the universe (or of an isolated system) for a process is zero or greater than zero, because that tells us whether the process is reversible or irreversible. We have not been concerned with the magnitude of the entropy increase. However, in Chapter 8 and in most of the subsequent chapters, we will make use of this magnitude in calculating a valuable quantity, *irreversibility*, I, which enables us to evaluate quantitatively the effects of irreversible processes.

Sometimes, the increase in entropy of the universe (or of an isolated system) itself is used as a measure of the effects of irreversible processes and is called "entropy production" or "entropy generation." We do not use these terms in this book because, in practice, using irreversibility, I, provides a more direct comparison with energy quantities of interest. This is covered in Chapter 8.

6·5 Helmholtz and Gibbs Functions

Enthalpy was introduced as a derived property defined as the sum of U and pV. Many other derived properties might be defined by arbitrary combinations of other properties. However, doing so makes sense only if such derived properties are useful. Two derived properties that are quite useful in some areas of thermodynamics are the Helmholtz function, A, and the Gibbs function, G, which are defined as

Josiah Willard Gibbs (1839–1903) received from Yale University in 1863 the first Ph.D. in engineering awarded in America. After further study in Europe he became professor of mathematical physics at Yale and held that position until his death. He made significant contributions in several fields, but a single paper, "On the Equilibrium of Heterogeneous Substances," places him in the first rank among scientists. Because the value of this contribution can be appreciated only by people with knowledge of thermodynamics, the name of this man who was one of the outstanding scientists of all time is virtually unknown to the general public.

Helmholtz and Gibbs functions

$$A \equiv U - TS \quad \text{or} \quad a \equiv u - Ts \tag{6·17}$$

$$G \equiv H - TS \quad \text{or} \quad g \equiv h - Ts \tag{6·18}$$

Consider a closed system that is initially and finally at the temperature T_0 of the atmosphere around it and exchanges heat only with the atmosphere. (During the process the system temperature may differ from T_0. For example, the system might be compressed or expanded adiabatically so that its temperature changes and then exchange heat with the atmosphere until its temperature is again T_0.) It can be shown that the maximum amount of heat that can be transferred to the system is $T_0(S_2 - S_1)$, so that the maximum work output of the system is given by

$$\begin{aligned} W_{max} &= U_1 - U_2 + Q_{max} = U_1 - U_2 + T_0(S_2 - S_1) \\ &= U_1 - T_0S_1 - (U_2 - T_0S_2) = A_1 - A_2 \end{aligned}$$

Some uses of the Helmholtz and Gibbs functions

Conclusion: If a closed system passes from one state to another state at the same temperature while exchanging heat only with surrounding atmosphere at that temperature ($T_2 = T_1 = T_0$), *the maximum work of the process is equal to the decrease in A of the system.* This conclusion is not restricted to systems composed of pure substances, so it may be applied, for example, to a process involving a chemical reaction.

Part of the work done by a closed system may be done in pushing back the atmosphere if the system expands. This part of the work is $p_0(V_2 - V_1)$, where p_0 is atmospheric pressure and V_1 and V_2 are the initial and final volumes of the system. If *useful work* means the work done in excess of that required to push back the atmosphere, then

Useful work is that done in excess of the work required to push back the atmosphere.

$$W_{useful} = W - p_0(V_2 - V_1)$$

The maximum useful work if the system exchanges heat with only the surroundings at T_0 is

$$\begin{aligned} W_{max\ useful} &= U_1 - U_2 + T_0(S_2 - S_1) - p_0(V_2 - V_1) \\ &= U_1 + p_0V_1 - T_0S_1 - (U_2 + p_0V_2 - T_0S_2) \end{aligned}$$

and, if $p_2 = p_1 = p_0$ and $T_2 = T_1 = T_0$, then

$$W_{max\ useful} = H_1 - T_0S_1 - (H_2 - T_0S_2) = G_1 - G_2$$

Conclusion: If a closed system passes between two states, each at the pressure and temperature of the surrounding atmosphere, and exchanges heat with the atmosphere only, the maximum useful work (that is, work done on systems other than the atmosphere) of the process is equal to the decrease in G of the system. These conditions are met by a system undergoing a chemical reaction in an open vessel. The system need not be composed of a pure substance.

Consider a steady-flow process for which $\Delta ke = \Delta pe = 0$, $T_1 = T_2 = T_0$, and the system exchanges heat with only the surrounding atmosphere at T_0. Under such conditions, the maximum work per unit mass output is

$$\begin{aligned} w_{max} &= h_1 - h_2 + q_{max} = h_1 - h_2 + T_0(s_2 - s_1) \\ &= h_1 - T_0s_1 - (h_2 - T_0s_2) = g_1 - g_2 \end{aligned}$$

Conclusion: If a substance enters and leaves a steady-flow system at the temperature of the surrounding atmosphere and exchanges heat with only the atmosphere, the maximum work per unit mass that can be produced is equal to the decrease in g of the substance.

As an example of the use of the Gibbs function, suppose that we determine the G value for the air and fuel entering a gasoline engine and for the products of combustion after they are cooled to atmospheric temperature. The difference between these values is the maximum amount of work that can conceivably be obtained from this particular flow of air, fuel, and products.

A and G have been introduced here simply as examples of useful properties that, like entropy, can be defined as a consequence of the second law. We will use them in Chapters 7, 12, and 17.

6·6 Uses of Entropy

The introduction to this chapter pointed out that searching for a general physical picture of entropy is fruitless. Instead of trying to answer the question "What *is* entropy?," you should concentrate on the questions "What is it used for?" and "How is it used?" Let us review some of the uses of entropy that have already been discussed.

Reviewing uses of entropy already discussed

1. *Entropy is used as a convenient coordinate.* The ideal process in many machines and flow passages is a constant-entropy process. Also, on a temperature-entropy diagram the area beneath the path of a reversible process represents heat transfer. By leafing through the remaining pages of this book, you will see how frequently diagrams with entropy as a coordinate are used.

2. *Entropy is used in the calculation of heat transfer in reversible processes,* $Q_{rev} = \int T\,dS$. This is of limited practical value because in most cases where $\int T\,dS$ can be integrated, sufficient data are available for the calculation of Q by means of the first law.
3. *Entropy is used to determine whether a process is reversible, irreversible, or impossible.* This use is based on the increase of entropy principle, $\Delta S_{\text{isolated system}} \geq 0$, where the equality holds only for reversible processes.
4. *Entropy is used in the correlation of physical property data.* This use is based on the relations $T\,ds = du + p\,dv = dh - v\,dp$ and on the definitions of a and g. Further illustrations appear in Chapter 7.

Several other uses of entropy will be introduced in later chapters.

6·7 Entropy and Probability; Entropy in Other Fields

Section 5·9 stated a conclusion from statistical thermodynamics: In an irreversible process an isolated system proceeds toward more probable states or states of greater disorder. In this chapter we have seen that a conclusion of classical thermodynamics is that during an irreversible process an isolated system proceeds toward states of greater entropy. A natural question is whether there is any direct relationship between probability and entropy. The answer to this question is yes. Before looking at this relationship, let us see what is meant by the probability of a state.

The most frequently used illustration begins with the consideration of a box that is divided into two equal parts by a half-partition (see Fig. 6·14). A marble is in the box, but the box has been shaken so that the marble is as likely to be on one side of the partition as on the other. There are two possible configurations or arrangements. The probability that the marble is in side X of the box is $\frac{1}{2}$. The probability that the marble is in side Y of the box is also $\frac{1}{2}$. (Since it is a certainty that the marble is somewhere in the box, the probability that it is in either side X or side Y is 1.) Now consider the case of two marbles, called A and B, in the box. There are four possible, and equally probable, arrangements, as shown in Table A. The probability of each ar-

Table A

Arrangement	*Marbles in X*	*Marbles in Y*
1	*A* and *B*	None
2	*A*	*B*
3	*B*	*A*
4	None	*A* and *B*

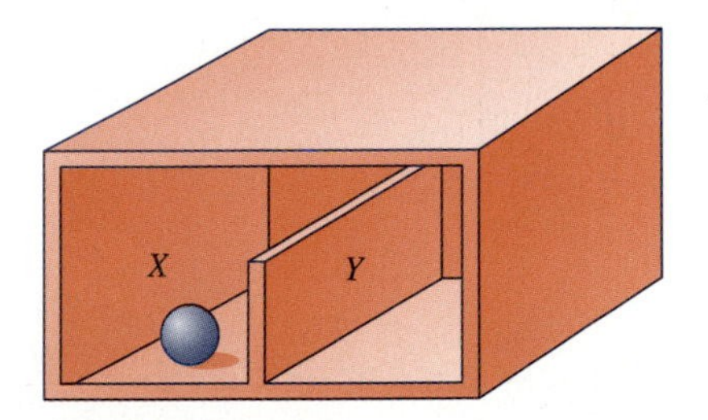

Figure 6·14 Marble in a half-partitioned box: a system that can exist in either of two states.

Table B

State	*Number of Marbles in X*	*Number of Marbles in Y*	*Probability*
a	2	0	$\frac{1}{4}$
b	1	1	$\frac{1}{2}$
c	0	2	$\frac{1}{4}$

rangement is $\frac{1}{4}$ because there are four possible arrangements. If the marbles cannot be distinguished from each other, arrangements 2 and 3 are identical. We say that arrangements 2 and 3 produce the same state: the state in which there is one marble in each side of the box. Therefore, only three states are possible, as shown in Table B.

State b has a probability of $\frac{1}{2}$ because two of the four possible arrangements result in state b. The probability of state a is $\frac{1}{4}$ because only one of the four possible arrangements results in this state; the same is true of state c. Since there are no states other than a, b, or c possible, the sum of the probabilities of these three states is 1. For four marbles in the box there are 16 possible arrangements which can produce five different states, and their probabilities can be calculated as shown in Table C.

If we go to a more general case of n marbles, the probability of having all n marbles in X is $(\frac{1}{2})^n$. In other words, the state in which all marbles are in one side of the box is a state of low probability. The probability of having r marbles out of n in side X (that is, the number of combinations of n things taken r at a time, divided by the total number of arrangements) is

$$\frac{n(n-1)(n-2)\cdots(n-r+1)}{r!(2)^n}$$

Table C

Number of Marbles in X	*Number of Marbles in Y*	*Probability*
4	0	$\frac{1}{16}$
3	1	$\frac{1}{4}$
2	2	$\frac{3}{8}$
1	3	$\frac{1}{4}$
0	4	$\frac{1}{16}$

Let us now extend our consideration from a few marbles to many molecules of a gas that are free to move from one side of the box to the other. The opening in the partition can be very small in comparison with the size of the box and still permit enormous numbers of molecules to pass back and forth. It is assumed that any molecule is equally likely to be in side X or side Y. The probabilities of uniform and nearly uniform distributions of n molecules are shown in Table 6·2. It is seen that even for as few as 1000 molecules the probability of an appreciable departure from an equal distribution is very small. (Note that one cubic millimeter of air at normal room conditions contains more than 2×10^{16} molecules.) Any state in which the molecules are unequally distributed is a state of lower probability than one in which they are more nearly equally distributed. Having thus looked into the meaning of the probability of a state, we still face the question, "How is this related to entropy?"

In seeking the relationship between entropy and probability, let us first recall that entropy is an extensive property, so that if systems A and B are combined to form system C, then

$$S_C = S_A + S_B$$

Table 6·2 Probabilities of Some Distributions of Molecules

		Number of Combinations in Which Number of Molecules (r) in X Is			*Probability of Combinations in Which Number of Molecules in X Is*		
Number of Molecules, n	*Number of Combinations of Molecules, 2^n*	$0.5n$	$0.49n$ *to* $0.51n$	$0.45n$ *to* $0.55n$	$0.5n$	$0.49n$ *to* $0.51n$	$0.45n$ *to* $0.55n$
2	4	2			0.500		
4	16	6			0.375		
6	64	20			0.313		
8	256	70			0.273		
10	1024	252			0.246		
20	1.049×10^6	0.1848×10^6		0.5207×10^6	0.176		0.496
40	1.100×10^{12}	0.1378×10^{12}		0.6272×10^{12}	0.125		0.570
100	1.2677×10^{30}	0.1009×10^{30}	0.2987×10^{30}	0.9238×10^{30}	0.0796	0.236	0.729
200	1.6069×10^{60}	0.0950×10^{60}	0.4439×10^{60}	1.3862×10^{60}	0.0563	0.276	0.863
1000	1.0715×10^{301}	0.0270×10^{301}	0.5282×10^{301}	1.0700×10^{301}	0.0252	0.493	0.9986
2000	1.1481×10^{602}	0.0205×10^{602}	0.7357×10^{602}	1.148×10^{602}	0.0178	0.641	0.999993

On the other hand, the probability of the state of the combined system is

$$P_C = P_A P_B$$

A relationship between probability and entropy. This equation can also be obtained through statistical mechanics. That method also allows C to be evaluated from other physical properties.

These relationships are satisfied by letting

$$S = C \ln P$$

because then

$$S_C = S_A + S_B = C(\ln P_A + \ln P_B) = C \ln P_A P_B = C \ln P_C$$

Thus the change in entropy of a system that passes from a state 1 to a state 2 is

$$S_2 - S_1 = C \ln \frac{P_2}{P_1}$$

where P_2 and P_1 are the probabilities of the final and initial states, respectively, and C is a constant.

As an example, consider an ideal gas that undergoes a free expansion in a closed system until its volume is doubled, $V_2 = 2V_1$ (see Fig. 6·15). From the reasoning above, the probability that all n molecules of the gas would be in the half of the enclosure called V_1 just after the partition is opened is $(\frac{1}{2})^n$. This is the probability of the initial state:

$$P_1 = \left(\frac{1}{2}\right)^n$$

For a final equilibrium state, the distribution of molecules throughout the total volume V_2 is very nearly uniform, so that the probability of the final state is very nearly 1:

$$P_2 = (1)^n = 1$$

The constant C for this case is $N\bar{R}/n$, where N is the number of moles of gas, $\bar{R}$ is the universal gas constant, and n is the number of molecules in the system. Then

$$S_2 - S_1 = \frac{N\bar{R}}{n} \ln \frac{1}{(1/2)^n} = \frac{N\bar{R}}{n} \ln (2)^n = N\bar{R} \ln 2 = N\bar{R} \ln \frac{V_2}{V_1}$$

and this is the same expression for ΔS that we obtain from classical thermodynamics for the free expansion. A conclusion is that more probable states are states of higher entropy.

When a solid is melted or a liquid is vaporized, there is a change toward a less ordered molecular state or an increase in the "randomness" of molecular motions and configurations. Such processes serve as additional illustrations of the relationship between entropy increase and the increase in molecular disorder.

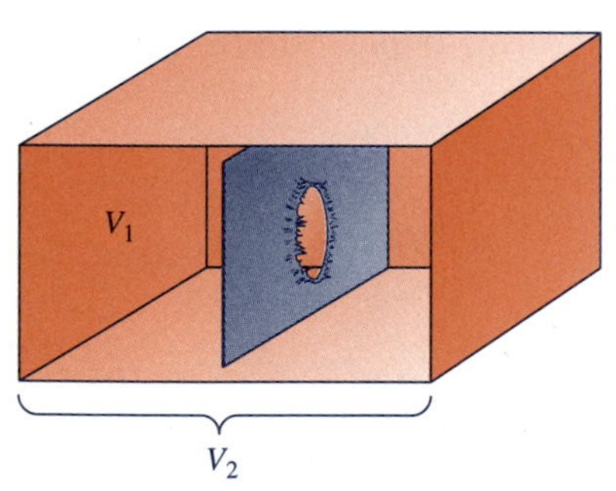

Figure 6·15 Gas is initially in volume V_1. Partition is punctured to allow free expansion to final volume of V_2.

In addition to this relationship between entropy and disorder, there are relationships between entropy and other quantities that instead of being conserved tend to increase or decrease. Indeed, the term *entropy*, which was first coined in classical thermodynamics, has been introduced into several fields. One such field is information theory. One might intuitively expect some kind of inverse relationship between entropy and information, because if a message with a certain amount of information is put into a communication system, the message that comes out of the system can carry at most the same amount of information, but all events that occur in any real communication system tend to cause a loss of information.

The conclusion that concepts can be transferred from one field to another is important. Even more important, however, is the conclusion that the use of entropy in classical thermodynamics, statistical mechanics, information theory, and other fields does not generate any conflict among the fields, but rather strengthens and promotes progress in all of them.

6·8 Summary

From the second law it can be shown that

$$\Delta S \equiv \int \left(\frac{\delta Q}{T}\right)_{\substack{int\\rev}} \tag{6·3}$$

is a property. This property is called entropy. It is actually the *change* in entropy that is defined by Eq. 6·3, but in engineering work we are usually concerned only with entropy changes rather than with absolute values.

There is a strong parallelism between the definitions of ΔS from the second law and ΔE from the first law. ΔS or Δs between any two states of a system can be calculated by integrating $\int (\delta Q/T)$ or $\int (\delta q/T)$ along any reversible path between the two states. It is usually easier, however, to calculate Δs from the equations

$$T\,ds = du + p\,dv \tag{6·7a}$$

$$T\,ds = dh - v\,dp \tag{6·7b}$$

or

$$s_2 - s_1 = \int_1^2 \frac{du}{T} + \int_1^2 \frac{p\,dv}{T} \tag{6·8a}$$

$$s_2 - s_1 = \int_1^2 \frac{dh}{T} - \int \frac{v\,dp}{T} \tag{6·8b}$$

These can be developed from the first and second laws for any pure substance changing from one equilibrium state to another in the absence of electrical, magnetic, solid deformation, and surface tension effects.

For an ideal gas, specific heats are functions of temperature only and the second term on the right of Eq. 6·8 can be readily integrated, so we have

$$s_2 - s_1 = \int_1^2 \frac{c_p\, dT}{T} - R \ln \frac{p_2}{p_1} \tag{6·10b}$$

The integral defines ϕ, the part of entropy that is a function of temperature only, so that

$$\phi_2 - \phi_1 \equiv \int_1^2 \frac{c_p\, dT}{T}$$

and then

$$s_2 - s_1 = \phi_2 - \phi_1 - R \ln \frac{p_2}{p_1} \tag{6·12}$$

ϕ is a function of temperature only, so it can be tabulated in ideal-gas tables which are single-argument tables. Other tabulated functions of temperature are the relative pressure, $p_r \equiv e^{\phi/R}$, and the relative specific volume, v_r, defined as follows: For an isentropic process of an ideal gas,

$$\frac{p_{r2}}{p_{r1}} = \frac{p_2}{p_1} \quad \text{and} \quad \frac{v_{r2}}{v_{r1}} = \frac{v_2}{v_1}$$

The heat transfer of any reversible process of either a closed system or a steady-flow system is given by

$$q = \int T\, ds$$

Consequently, the heat transfer of any reversible process is represented as an area on a *Ts* diagram. Entropy is also useful as a coordinate in conjunction with other properties besides temperature.

Whether a process is reversible, irreversible, or even possible can be determined by the *increase of entropy principle*:

$$\Delta S_{\text{isolated system}} \geq 0 \tag{6·14}$$

or

$$\Delta S_{\text{universe}} \geq 0 \tag{6·15}$$

in which the equality applies to reversible processes. Application of this principle to a steady-flow system results in

$$\left(\frac{\delta Q}{T}\right)_{\substack{int \\ rev,surr}} + \int_{\text{out}} s\, \delta m - \int_{\text{in}} s\, \delta m \geq 0 \tag{6·16}$$

Two properties that are quite useful, particularly in processes involving chemical reactions, are the Helmholtz function, A, defined as

$$A \equiv U - TS \quad \text{or} \quad a \equiv u - Ts \tag{6·17}$$

and the Gibbs function, G, defined as

$$G \equiv H - TS \quad \text{or} \quad g \equiv h - Ts \tag{6·18}$$

If a closed system passes from one state to another state at the same temperature while exchanging heat with only the surrounding atmosphere at that temperature ($T_2 = T_1 = T_0$), the maximum work of the process is equal to the decrease in A of the system. This conclusion is not restricted to systems composed of pure substances, so it may be applied, for example, to a process involving a chemical reaction. If a closed system passes between two states, in each of which it is at the pressure and temperature of the surrounding atmosphere, and exchanges heat only with the atmosphere, the maximum useful work (i.e., work done on systems other than the atmosphere) of the process is equal to the decrease in G of the system. If a substance enters and leaves a steady-flow system at the temperature of the surrounding atmosphere and exchanges heat with only the atmosphere, the maximum work per unit mass that can be produced is equal to the decrease in g of the substance.

In addition to its uses in classical thermodynamics, which are listed in this chapter, entropy is valuable in several other fields.

Suggested Reading

6·1 Atkins, P. W., *The Second Law*, Scientific American Books, New York, 1984, Chapter 2. (We again suggest this nonmathematical account as being interesting and informative to someone who has studied the mathematical basis.)

6·2 Keenan, Joseph H., *Thermodynamics*, Wiley, New York, 1941, Chapters VI–VIII. (This reference is older than most that we list. It is a classic work that has influenced most of the thermodynamics books written since it appeared. Now that you have studied the second law and some of its consequences, a reading of this work can be quite rewarding.)

Problems

6·1 A closed system can go from one specified state to another specified state by means of several different adiabatic paths. Is the work the same for all paths? Explain.

6·2 Is $\int_{rev} (\delta w/p)$ a property?

6·3 Prove that the minimum work input to a refrigerator that cools a rigid body from atmospheric temperature T_0 to a lower temperature T_L is $W_{in} = T_0(S_0 - S_L) + U_L - U_0$, where S and U are properties of the body being cooled.

6·4 Prove that for any process of a closed system $\Delta s \geq \int (\delta q/T)$.

6·5 A closed system composed of five kilograms of carbon monoxide undergoes a reversible, isentropic expansion from 300 kPa, 50°C, to 150 kPa. Calculate the work done per kilogram of carbon monoxide.

6·6 Nitrogen flowing through a compressor at a rate of 5 kg/s is compressed isentropically from 100 kPa, 5°C, to 200 kPa. The kinetic energy of the nitrogen is increased by 7.0 kJ/kg. Calculate the power input to the nitrogen.

6·7 In a reversible process of a closed system, the heat added is given by $Q = aT + bT^2$, where $a =$ 200 J/K and $b = 3.20$ J/K^2. Calculate the entropy change of the system as its temperature changes from 50°C to 200°C.

6·8 During a constant-temperature process at 130°C, the entropy of 3 kg of ethane decreases by 2.7 kJ/K. For an initial pressure of 800 kPa, compute (*a*) the work done and (*b*) the final volume.

6·9 One-tenth kilogram of octane is heated at constant volume from 60°C, 300 kPa, to 280°C, compressed reversibly and adiabatically to 1.4 MPa, and then cooled under constant-volume conditions to 110°C. Determine the change in entropy of the octane for the individual steps and for the complete series of steps.

6·10 Two kilograms of carbon dioxide at a pressure of 1400 kPa with a volume of 0.10 m^3 undergoes a constant-pressure change until the volume is doubled. The pressure is then reduced by 50 percent under constant-volume conditions, after which the gas is compressed isothermally to its original volume. Compute the total change in entropy of the system for the three steps.

6·11 One-tenth kilogram of propane at 100 kPa, 60°C, is contained in a closed, rigid, thermally insulated vessel. A paddle wheel inside the vessel is turned by an external motor until the pressure is 130 kPa. Sketch *pv* and *Ts* diagrams for this process, and calculate the change in entropy of the propane.

6·12 Methane initially at 110 kPa, 5°C, has its volume doubled while its temperature rises to 90°C. Calculate the change in enthalpy, the change in internal energy, the change in entropy, the work done, and the heat transferred during the process. If it is impossible to calculate any of these quantities, explain why.

6·13 One-twentieth kilogram of acetylene in a closed system is compressed irreversibly from 100 kPa, 5°C, to 200 kPa. During the process 1.8 kJ of heat is removed from the acetylene, and the work done on the acetylene amounts to 2.7 kJ. Determine the entropy change of the acetylene.

6·14 Air with a low velocity enters a machine at 95 kPa, 17°C, and leaves at 240 kPa, 120°C. In case *A*, the discharge velocity is low; in case *B* the discharge velocity is 40 m/s. Calculate the entropy change of the air in each case.

6·15 Air flowing through a compressor at a rate of 12 lbm/s is compressed isentropically from 15 psia, 40 F, to 30 psia. The kinetic energy of the air is increased by 3.0 B/lbm. Calculate the power input to the air.

6·16 One pound of air is heated at constant volume from 140 F, 50 psia, to 540 F, compressed reversibly and adiabatically to 200 psia, and then cooled under constant-volume conditions to 240 F. Determine the change in entropy of the air for the individual steps and for the complete series of steps.

6·17 One-half pound of air at 15 psia, 140 F, is contained in a closed rigid thermally insulated vessel. A paddle wheel inside the vessel is turned by an external motor until the pressure is 20 psia. Sketch *pv* and *Ts* diagrams for this process, and calculate the change in entropy of the air.

6·18 Helium initially at 30 psia, 40 F, experiences an increase in temperature to 200 F while its volume doubles. Calculate the change in enthalpy, the change in internal energy, the change in entropy, the work done, and the heat transferred during the process. If it is impossible to calculate any of these quantities, explain why.

6·19 One-tenth pound of oxygen in a closed system is compressed irreversibly from 15 psia, 40 F, to 30 psia. During the process 1.7 B of heat is removed from the oxygen, and the work done on the oxygen amounts to 2.6 B. Determine the entropy change of the oxygen.

6·20 State the conditions under which each of the following equations is true:

$$\oint T\,ds = -\oint v\,dp$$

$$\oint T\,ds = \oint p\,dv$$

6·21 A heat engine utilizing air as the working fluid completes its cycle in three steps as follows: process 1–2, adiabatic compression from $T_1 = 40°C$ to $T_2 = 150°C$; process 2–3, isothermal expansion at 150°C; process 3–1, dT/ds = constant. Assuming that each step in the cycle is reversible, calculate the efficiency of the engine.

6·22 Air expands irreversibly in a cylinder from $p_1 = 280$ kPa, $T_1 = 60°C$, to $p_2 = 140$ kPa. The air does work in the amount of 30 kJ/kg and 14 kJ/kg of heat is removed from the air during the expansion. The initial volume is 0.00878 m^3. Calculate the entropy change of the air per kilogram.

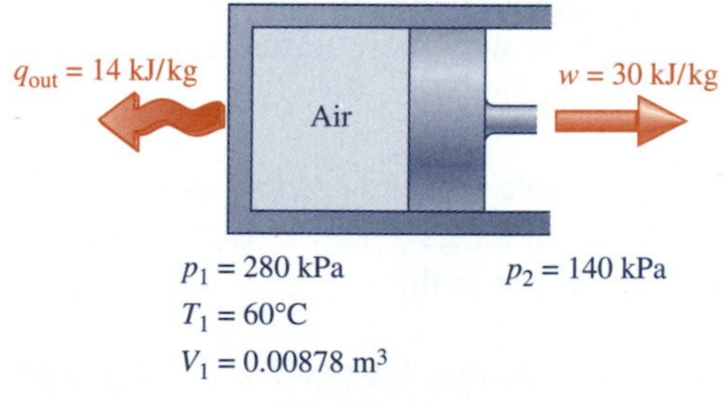

Problem 6·22

6·23 Air initially at 4 bar, 60°C, expands reversibly in a piston–cylinder arrangement. It expands to 200 kPa in an isothermal process and then expands adiabatically from $p_2 = 200$ kPa to $p_3 = 100$ kPa. Sketch the processes on *pv* and *Ts* coordinates and calculate for the entire expansion from state 1 to state 3: (*a*) q, (*b*) Δu, (*c*) Δs.

6·24 Heat is transferred from a very large mass of water at 90°C to 0.1 kg of oxygen that expands irreversibly in a cylinder fitted with a piston from 400 kPa, 60°C, to 150 kPa, 35°C. The oxygen does 4.00 kJ of work. Calculate (*a*) the heat transfer to the oxygen and (*b*) the entropy change of the universe.

6·25 Nitrogen in a closed system is heated reversibly from 105 kPa, 5°C, to 125 kPa, 115°C, so that dT/ds = constant. (*a*) Calculate the work done per pound of nitrogen. (*b*) Sketch a large accurate *Ts* diagram for the process, showing lines of constant pressure for 105 and 125 kPa. Show also some constant-volume lines. (*c*) Sketch a large accurate *pv* diagram for the process. (Support your sketches by some calculations proving that you have shown the correct shapes or slopes.)

6·26 A gas with a molar mass of 47.9 kg/kmol that is to be used in many calculations during an engineering design project has c_p versus T values as shown in the table. For the pressure range involved, the gas can be modeled as an ideal gas. Make a table with T as argument showing values of h, u, ϕ, and p_r. For ϕ at 298.15°C use 5.633 kJ/kg·K; for h, use 302.7 kJ/kg.

T, K	c_p, kJ/kg·K
298.15	1.487
300	1.492
400	1.707
500	1.871
600	1.995
700	2.090
800	2.167
900	2.230
1000	2.285

6·27 A gas with a molar mass of 44.0 lbm/lbmol that is to be used in many calculations during an engineering design project has c_p versus T values as shown in the table. For the pressure range involved, the gas can be modeled as an ideal gas. Make a table with T as argument showing values of h, u, ϕ, and p_r. For ϕ at 77 F use 1.16 B/lbm·R; for h, use 54.0 B/lbm.

T, F	c_p, B/lbm·R
77	0.133
100	0.141
200	0.175
300	0.202
400	0.224
500	0.242
600	0.257
700	0.269
800	0.279
900	0.288
1000	0.295

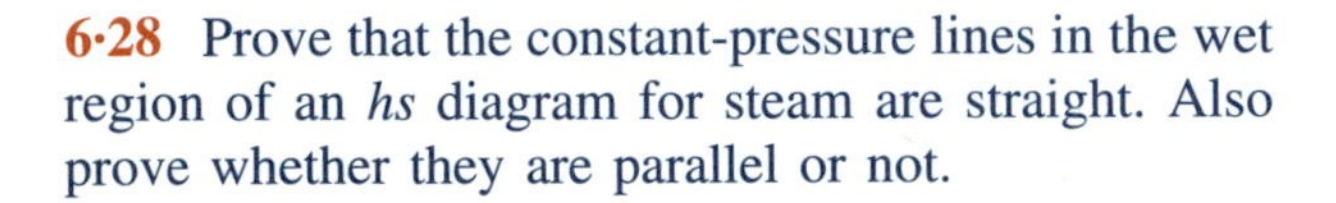

6·28 Prove that the constant-pressure lines in the wet region of an *hs* diagram for steam are straight. Also prove whether they are parallel or not.

6·29 In Example 6·4 it is shown that for an ideal gas $(\partial T/\partial s)_p = T/c_p$ and $(\partial T/\partial s)_v = T/c_v$. Show that these relations are true for any pure substance.

6·30 Which one of the *Ts* diagrams shown here for an ideal gas properly shows two constant-pressure lines? Support your answer, and draw two constant-volume lines on a similar diagram.

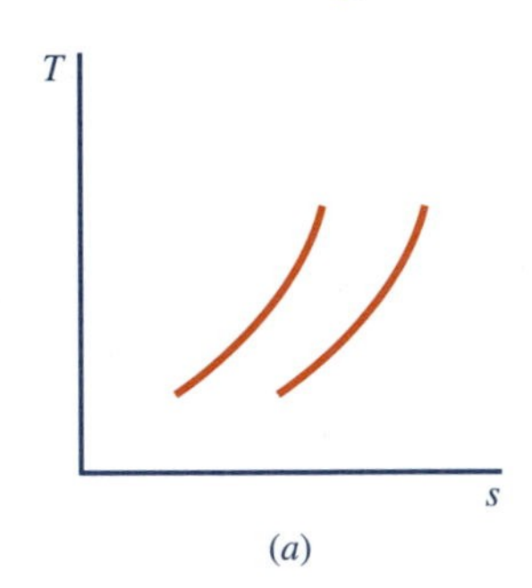

(*a*)

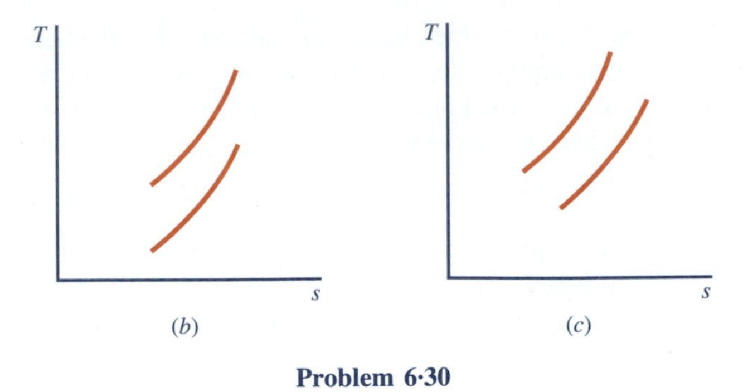

(*b*) (*c*)

Problem 6·30

6·31 Derive the following expressions for an ideal gas with constant specific heats:

$$s_2 - s_1 = c_p \ln \frac{T_2/T_1}{(p_2/p_1)^{(k-1)/k}}$$

$$s_2 - s_1 = c_v \ln \frac{p_2}{p_1} + c_p \ln \frac{v_2}{v_1}$$

6·32 (*a*) In a closed system, a gas undergoes a cycle made up of the following processes: 1–2, reversible isothermal compression; 2–3, reversible constant-volume heating; 3–4, reversible constant-pressure expansion; 4–1, reversible adiabatic expansion. Sketch *pv* and *Ts* diagrams, and state whether each of the following quantities is positive, zero, negative, or indeterminate in sign: $\oint \delta W$, $\oint \delta Q$, $\oint dS$, $\oint dU$, $\oint dH$. (*b*) Repeat part *a* with process 4–1 changed to an irreversible adiabatic expansion.

6·33 Demonstrate by means of a *ph* diagram that the first term on the right-hand side of Eq. 6·8*b*, $\int_1^2 dh/T$, usually depends on the path between the two end states. Using a *pv* diagram, do the same for the second term. Also, show why these terms are independent of the path for an ideal gas.

6·34 Demonstrate that if $\int_1^2 v\, dp/T$ is independent of the path between states 1 and 2, then $\int_1^2 dh/T$ is also independent of the path.

6·35 Demonstrate that $\int_1^2 v\, dp/T$ is usually dependent on the path of the process between states 1 and 2.

6·36 Near the definition of ϕ in Eq. 6·11, the statement is made that the value of ϕ at any one temperature can be selected arbitrarily when tables are compiled. Refer to the "Sources of Data and Calculation Methods" section of *Gas Tables*, by Keenan, Chao, and Kaye (Ref. 4·13) and write a statement of how this fact was used in compiling the tables for air in that work.

6·37 Early in Section 6·2·1 is the statement that each of the two terms on the right-hand side of Eq. 6·8 usually depends on the path of a process, even though the combination of the two terms is path-independent. Give a numerical example of this point for the case of steam.

6·38 For use in design studies, a chart that shows temperature at the end of quasiequilibrium adiabatic compression of air is needed. The chart should plot discharge temperature versus gage discharge pressure for three values of inlet temperature typical of construction sites in North America. Use a range of gage discharge pressures from 0 to 15 atm. (This problem appeared in Chapter 4 to be solved using constant specific heats. In Chapter 9 it appears again to include effects of variable humidity.)

6·39 Starting with the air table values at 300 K and using the specific-heat relation given in Table A·1, calculate the values of ϕ and h at 1000 K. Compare your results with the tabular values.

6·40 During a constant-pressure heating process, 260 kJ/kg of heat is added to hydrogen at 1 atm, 50°C. Determine the final temperature.

6·41 During a constant-pressure heating process, 260 kJ/kg of heat is added to propane at 1 atm, 50°C. Determine the final temperature.

6·42 One-half kilogram of dry air expands isentropically until the final specific volume is six times the initial value. If the temperature at the start of the process is 1400 K, compute (*a*) the final temperature, (*b*) the final enthalpy, and (*c*) the entropy change.

6·43 Air at 800 kPa, 650°C, enters a nozzle with negligible velocity and expands isentropically to 350 kPa. Calculate the exit velocity.

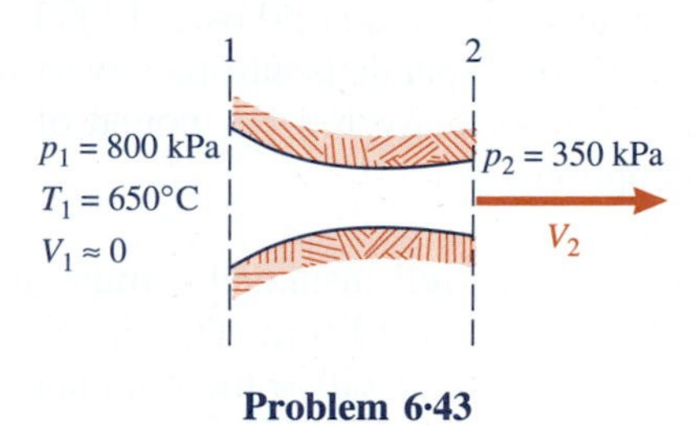

Problem 6·43

6·44 Air enters a gas turbine at 725 kPa, 650°C, and expands adiabatically to 100 kPa, 315°C. Neglecting changes in kinetic energy, calculate the work done per kilogram of air. Is the process reversible?

6·45 In a closed system, air is compressed isentropically from 95 kPa, 150°C, to 400 kPa and is then heated reversibly at constant volume until its pressure is 550 kPa. Using air tables, calculate the heat added to the air in kJ/kg.

6·46 Air enters a well-insulated turbine at 500 kPa, 900°C, and leaves at 100 kPa. The flow is steady at a rate of 1.08 kg/s. State either the minimum possible exhaust temperature or the additional information required to determine it.

6·47 Starting with the air table values at 600 R and using the specific-heat relation given in Table A·1E, calculate the values of ϕ and h at 2000 R. Compare your results with the tabular values.

6·48 An insulated tank having a volume of 5000 cu ft contains dry air initially at 60 psia, 1000 R. Air flows out through a small nozzle. Determine the state of the air in the tank after one-half the original mass has left. State any assumptions that you make.

6·49 During a constant-pressure heating process, 122 B/lbm of heat is added to air initially at 1 atm, 140 F. Determine the final temperature.

6·50 Air at 75 psia, 1420 F, enters a nozzle with negligible velocity and expands isentropically to 15 psia. Calculate the exit velocity.

6·51 Air enters a turbine at 50 psia, 1200 F, at a rate of 8.10 lbm/s and expands isentropically to 14.5 psia. Calculate the work delivered per pound of air. State your assumptions.

6·52 Air enters a well-insulated turbine at 70 psia, 1600 F, and exhausts at 14 psia. The flow is steady at a rate of 5.8 lbm/s. State either the minimum possible exhaust temperature or the additional information required in order to determine it.

6·53 In an isentropic process, steam flows through a turbine from 1.4 MPa, 250°C, to an exhaust pressure of 50 mm Hg abs. The flow rate is 4500 kg/h, and kinetic energy changes are negligible. Find the power output.

6·54 Calculate the work required per kilogram of R134a that is compressed reversibly and adiabatically in a steady-flow system from 700 kPa, 50°C, until its density is doubled. Kinetic energy changes are negligible.

6·55 Steam is supplied to a turbine at 1.4 MPa, 300°C, and exhausted to the atmosphere. To what pressure must the incoming steam be throttled to reduce the work per kilogram to two-thirds of that obtained without throttling? Assume that the flow through the turbine is reversible and adiabatic in either case and that the exhaust pressure is not changed.

6·56 In a closed system, steam expands reversibly from a dry saturated condition at 100 kPa to a dry saturated condition at 70 kPa in such a manner that the process is represented by a straight line on a *Ts* diagram. Determine the work per kilogram of steam.

6·57 Dry saturated steam at 800 kPa enters a steady-flow system and is expanded reversibly and isothermally to 500 kPa. There is no change in kinetic energy. The flow rate is 1.0 kg/s. Sketch the process on a *Ts* diagram, and calculate the amount of work done per kilogram of steam.

6·58 In a closed system, dry saturated steam at 700 kPa is heated in a constant-volume process to 1.2 MPa. It is then expanded isothermally to the original pressure and finally cooled to the saturation temperature at that pressure. If all processes are reversible, what is the net work?

6·59 Dry saturated steam initially at 100 kPa, 99.6°C, is compressed reversibly in a cylinder so that $T = 151.7(s - 5.79)^2$, where T is in kelvins and s is in J/g·K. The final state is 400 kPa, 400°C. Calculate (*a*) the heat transfer in J/g and (*b*) the work in J/g.

6·60 A system undergoes a reversible process such that the process on a *TS* diagram is a straight line. Some property values are $T_1 = 60°C$, $U_1 = 170$ kJ, $H_1 = 220$ kJ, $S_1 = 0.2300$ kJ/K, $T_2 = 170°C$, $U_2 = 190$ kJ, $H_2 = 247$ kJ, and $S_2 = 0.3000$ kJ/K. Calculate the work of the process.

6·61 A Carnot engine operates between a source at 150°C and a sink at 50°C. If 200 kJ is transferred to the sink each minute, compute the changes in entropy per unit time for the system and for the universe for the isothermal expansion.

6·62 A reversed Carnot engine operates between temperature limits of 20°C and 50°C. Sketch the cycle on a *Ts* diagram and indicate the areas that represent the heat transferred from the reservoir at 20°C and the work input. If 100 kJ is absorbed from the reservoir at 20°C, compute the change in entropy of the system for the isothermal compression process.

6·63 An engine operates on a slightly modified Carnot cycle between constant temperature limits of 250°C and 30°C. During the isothermal expansion the entropy of the system changes 1.3 kJ/K for 2 kg of the working substance. A slight increase in entropy of 0.05 kJ/K occurs during the adiabatic expansion because of fluid friction within the engine. Sketch the cycle on a *Ts* diagram. Compute the heat added, heat rejected, work performed, and efficiency of the cycle. Also, determine the heat added, heat rejected, work performed, and efficiency on the basis that the adiabatic expansion is reversible.

6·64 A refrigerating machine patterned after a reversed Carnot cycle operates between rooms at −20°C and 30°C. Each hour, 12,500 kJ is to be removed from the cold room. One expansion process is an irreversible adiabatic process that causes an entropy increase for the system of 10 percent of the total entropy change for the reversible isothermal expansion. Draw a *Ts* diagram and show the area that represents the heat absorbed at the lower temperature. Calculate the power required.

6·65 A Carnot engine operates between temperature limits of 280°C and 5°C. A total of 100 kJ of heat is transferred from the hot reservoir to the working fluid. Calculate (*a*) the total work of the engine, (*b*) the entropy change of the reservoir at 280°C, and (*c*) the entropy change of the reservoir at 5°C.

6·66 Fifty kilograms of water at 5°C and 80 kg of water at 80°C are mixed in an insulated vessel at a constant pressure of 1 atm. Determine the entropy change of the system composed of the 130 kg of water.

6·67 Air in a closed system is heated reversibly from 100 kPa, 7°C, to 119.7 kPa, 119°C, so that dT/ds is constant.

(*a*) Calculate the work done per kilogram of air.
(*b*) Sketch a large, accurate *Ts* diagram for the process, showing lines of constant pressure for 100 and 119.7 kPa. Show also some constant-volume lines.
(*c*) Sketch a large, accurate *pv* diagram for the process. (Support your sketches by calculations proving that you have shown the correct shapes or slopes.)

6·68 A rigid, insulated vessel with a volume of 0.55 m^3 initially contains saturated refrigerant 134a at 25°C. A valve on the vessel is opened to allow the refrigerant to escape to the atmosphere. Determine the mass and the state of the refrigerant remaining in the vessel when the pressure inside the vessel has reached equilibrium with that of the surrounding atmosphere.

6·69 Calculate the work per pound required to compress steam reversibly and adiabatically in a steady-flow system from 100 psia, 330 F, until its density is doubled. Kinetic energy changes are negligible.

6·70 Steam is supplied to a turbine at 200 psia, 600 F, and exhausted at 14.7 psia. To what pressure must the incoming steam be throttled to reduce the work per pound to two-thirds of that obtained without throttling? Assume that the flow through the turbine is reversible adiabatic in either case and that the exhaust pressure is unchanged.

6·71 In a reversible isothermal process, 100 B/lbm of heat is added to a closed system initially composed of saturated liquid water at 400 F. Determine the work done during this process.

6·72 Dry saturated steam at 118 psia enters a steady-flow system and is expanded reversibly and isothermally to 74 psia. There is no change in kinetic energy. The flow rate is 2.0 lbm/s. Sketch the process on a *Ts* diagram, and calculate the amount of work done per pound of steam.

6·73 In a closed system, dry saturated steam at 100 psia is heated in a constant-volume process until its pressure is 200 psia. It is then expanded isothermally to 100 psia and finally cooled to the saturation temperature at constant pressure. All processes are reversible. Calculate the net work done.

6·74 A Carnot engine operates between a source at 800 F and a receiver at 100 F. If 200 B is transferred to the receiver each minute, compute the changes in entropy per unit time for the system and the universe for the isothermal expansion.

6·75 A reversed Carnot engine operates between temperature limits of 70 F and 120 F, and 100 B is absorbed from the cold reservoir. Sketch the cycle on a *Ts* diagram and indicate the areas that represent the heat transferred from the reservoir at 70 F and the work input. Also calculate the change in entropy of the system for the isothermal compression process.

6·76 An engine operates on a Carnot cycle except that, because of fluid friction, the entropy increases slightly during the adiabatic expansion. The engine operates between constant temperature limits of 240 F and 100 F, and 50 B/min is added during the isothermal expansion. The increase in entropy for the working fluid during the adiabatic expansion is 5 percent of the entropy change for the isothermal expansion process. Sketch the cycle on a *Ts* diagram and find the power output of the engine.

6·77 An ideal gas undergoes the following four reversible processes: It is heated at constant volume from state 1 to state 2, expanded isothermally to state 3, expanded adiabatically to state 4, which is at the same pressure as state 1, and then restored to state 1 by a constant-pressure process.

(*a*) Sketch *pv* and *Ts* diagrams of the cycle.

(*b*) State whether each of the following quantities is greater than zero, less than zero, equal to zero, or of indeterminate sign: $\oint \delta q$, $\oint \delta w$, $\oint du$, $\oint ds$.

6·78 A vapor in a closed system passes through a cycle made up of three processes: isentropic expansion 1–2, constant-pressure cooling 2–3, and constant-volume heating 3–1. Sketch *pv* and *Ts* diagrams for the cycle if state 1 is superheated vapor and (*a*) states 2 and 3 are also superheated, (*b*) states 2 and 3 are "wet vapor" states.

6·79 An ideal gas in a closed system undergoes a cycle made up of three processes: Process 1–2 is a reversible constant-volume process that starts at p_1 and T_1 and goes to $p_2 = 1.5p_1$; process 2–3 is an irreversible adiabatic process with an end state of $p_3 = p_1$ and $T_3 = 1.2T_1$; and process 3–1 is a reversible constant-pressure process that returns the system to state 1. Sketch to scale *pv*, *Ts*, *vs*, and *uv* diagrams of the cycle.

6·80 A fluid passing through a steady-flow system can be heated reversibly from state 1 to state 2 in either of two ways. In one case, $T = a + bs$; in the other case, $T = c + es^2$, where a, b, c, and e are constants. In which case is the heat transfer greater?

6·81 Sketch *Ts*, *pv*, *Tv*, *ph*, *hs*, and *hv* diagrams of a Carnot cycle using as a working fluid (*a*) an ideal gas, (*b*) a liquid–vapor mixture.

6·82 An ideal gas undergoes a cycle made up of three processes: Process 1–2 is a constant-pressure cooling; process 2–3 is a constant-volume heating until $T_3 = T_1$; and process 3–1 can be represented as a straight line on a *pv* diagram. Sketch a *Ts* diagram. Number the states and show clearly the shapes of the process lines.

6·83 An ideal gas undergoes a cycle made up of three processes: Process 1–2 is an isentropic expansion: process 2–3 is an isothermal compression until $p_3 = p_1$; and process 3–1 can be represented by a straight line on a *Ts* diagram. Sketch a *pv* diagram. Number the states and show clearly the shapes of the process lines.

6·84 Show that h_{fg} for a pure substance can be represented by an area on a *Ts* diagram.

6·85 Shown are two *ph* diagrams for processes of a refrigerant. Sketch the corresponding *Ts* diagrams.

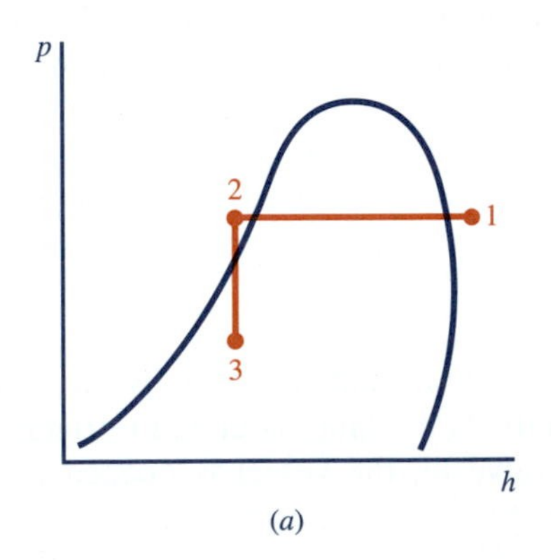

(*a*)

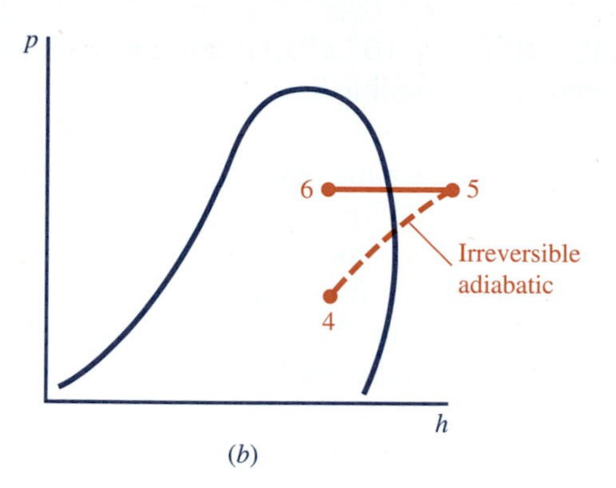

(*b*)

Problem 6·85

6·86 Shown are two *Ts* diagrams for processes of a refrigerant. Sketch the corresponding *pv* and *ph* diagrams.

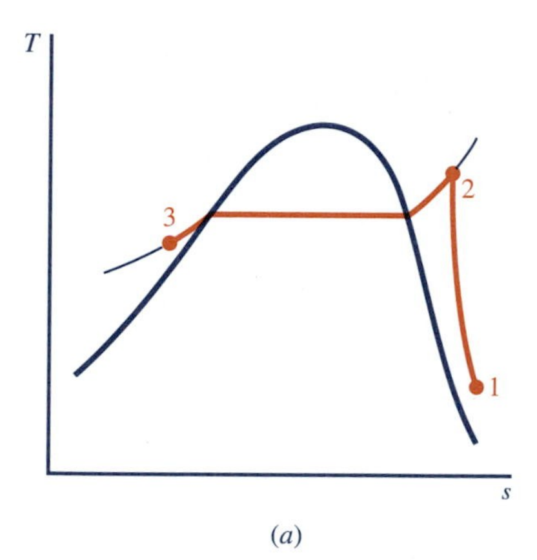

(*a*)

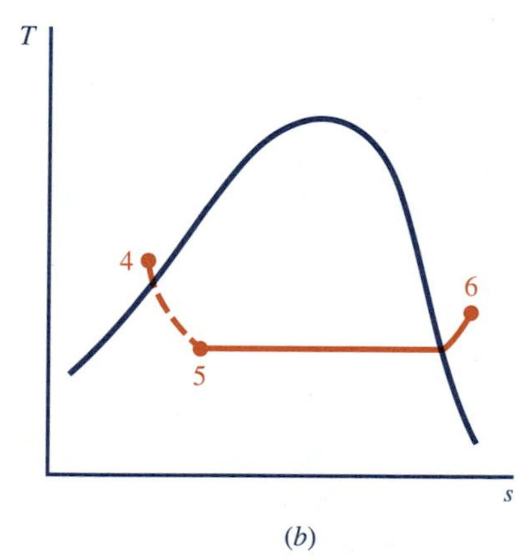

(*b*)

Problem 6·86

6·87 Sketch *pv* and *ph* diagrams corresponding to the *Ts* diagram of Fig. 6·9 and number the states in the same manner.

6·88 Is an adiabatic expansion of steam from 400 kPa, 150°C, to 101.3 kPa, 94 percent quality, reversible, irreversible, or impossible? State the quality ranges for reversible, for irreversible, and for impossible adiabatic expansions between the specified initial conditions and the specified final pressure.

6·89 Dry saturated SO_2 at -25°C is compressed in a steady-flow process to 4 bar with negligible heat transfer. What is the minimum final temperature? Is it the same for a closed-system adiabatic compression?

6·90 Is an adiabatic expansion of oxygen from 250 kPa, 60°C, to 125 kPa, 25°C, possible? Prove your answer. What is the lowest possible final temperature for an adiabatic expansion of oxygen from the specified initial conditions to 125 kPa? What is the maximum possible final temperature?

6·91 A turbine operating on nitrogen has inlet conditions of 200 kPa, 90°C, and exhausts to the atmosphere at 100 kPa. If the expansion is adiabatic and $\Delta ke = 0$, is an exhaust temperature of 5°C possible? If kinetic energy changes cannot be neglected, is this exhaust temperature possible?

6·92 One-fourth kilogram of carbon monoxide in a closed system expands from 250 kPa, 120°C, to 125 kPa, 25°C, producing 8.0 kJ of work. During this process, the entropy of the surroundings increases by 0.0324 kJ/K. (*a*) Calculate the heat transfer. (*b*) Is the process reversible? Prove your answer.

6·93 Air enters a turbine at 300 kPa, 120°C, and leaves at 100 kPa, 50°C. The flow rate is 1.4 kg/s, and the power output is 80 kW. Some heat is transferred to the surrounding atmosphere at 5°C. Calculate, per kilogram of air passing through the turbine, (*a*) the heat transfer, (*b*) the entropy change of the universe.

6·94 Air expands steadily through a turbine from 200 kPa, 60°C, to 95 kPa, 15°C. For each kilogram of air passing through the turbine the entropy of the surroundings decreases by 0.04 kJ/K. Is the expansion as described reversible, irreversible, or impossible? Support your answer.

6·95 Air is compressed irreversibly from 105 kPa, 5°C, to 210 kPa, 96°C.
(*a*) Calculate Δs of the air.
(*b*) Give all the information you can on the numerical value of the entropy change of the surroundings per kilogram of air.

6·96 Air expands reversibly and adiabatically through a turbine from 2 bar, 115°C, to 70 kPa. Inlet and exhaust velocities are both 60 m/s. The inlet cross-sectional area is 35 cm^2.
(*a*) Sketch *pv* and *Ts* diagrams of the process, indicating any areas or distances that represent heat or work.
(*b*) Calculate the power output in kW.

6·97 Two kilograms of steam in a closed system expands reversibly from a dry saturated condition at 3.45 MPa to 140 kPa, 150°C, so that the process can be represented by a straight line on a *Ts* diagram. The process is reversible. The lowest temperature in the surroundings is 5°C. Sketch the *Ts* diagram and calculate (*a*) the work done on or by the steam, (*b*) the entropy change of the surroundings.

6·98 Air enters a compressor from the atmosphere at 85 kPa, 5°C, and leaves at 170 kPa, 60°C. The flow rate is 4.5 kg/s. Kinetic energy changes are negligible. During the compression, heat is transferred from the air to a coolant, and for each pound of air compressed the entropy of the coolant increases by 0.020 kJ/K. Show whether or not the compressor operates reversibly.

6·99 Air is to be compressed adiabatically from 90 kPa, 20°C, to 400 kPa at a rate of 0.26 kg/min. Give all the information you can regarding the outlet temperature.

6·100 For the steady adiabatic expansion of air from 300 kPa, 40°C, to 105 kPa, what range of exhaust temperatures is possible?

6·101 Is an adiabatic expansion of steam from 60 psia, 300 F, to 14.7 psia, 94 percent quality, reversible, irreversible, or impossible? State the quality limits for reversible, for irreversible, and for impossible adiabatic expansions between the specified initial state and the specified final pressure.

6·102 Dry saturated sulfur dioxide at −10 F is compressed adiabatically to 60 psia in a steady-flow process. What is the minimum final temperature? Is it the same for a closed-system adiabatic compression?

6·103 Two pounds of air in a closed system expands from 40 psia, 240 F, to 20 psia, 80 F, producing 32.0 B of work. During this process, the entropy of the surroundings increases by 0.0456 B/R.
(*a*) Calculate the heat transfer.
(*b*) Prove whether or not the process is reversible.

6·104 Assume that air enters a turbine at 30 psia, 240 F, and is exhausted at 15 psia, 160 F. The flow rate is 0.18 lbm/s, and the flow is adiabatic.
(*a*) Calculate Δs.
(*b*) Sketch the process on *pv* and *Ts* coordinates.
(*c*) Is the process possible?

6·105 Consider the compression process described in Example 6·3. Is it possible? Reversible? Adiabatic?

6·106 Is it possible for a system to undergo an isentropic process that is not reversible and adiabatic? Explain your answer. Would your answer be different if the system is isolated?

6·107 Consider a closed system that is initially and finally at the temperature of the surrounding atmosphere, T_0, and exchanges heat with only the atmosphere. Show that the maximum amount of heat that can be transferred to the system is $T_0(S_2 - S_1)$, where S_2 and S_1 are the final and initial entropies of the system.

6·108 The following is a statement of the second law known as the Carathéodory statement: In the vicinity of any equilibrium state of a closed system there are some states that cannot be reached by adiabatic processes. Demonstrate that this statement is equivalent to the Kelvin–Planck or Clausius statement.

6·109 Consider a closed system that is initially and finally at the temperature T_0 of the atmosphere and exchanges heat with only the atmosphere. During the process the system temperature may differ from T_0. For example, the system might be compressed in an adiabatic process so that its temperature rises and then returned to a state of equilibrium with the atmosphere through heat transfer. Prove that the maximum amount of heat that can be transferred to the system is given by $T_0(S_2 - S_1)$.

6·110 See Problem 4·86 dealing with a Hilsch tube or Ranque tube. Apply the second law to see that the described process is possible. If it is possible, is it reversible?

6·111 For any two phases of a pure substance existing together in equilibrium, the Gibbs functions must be the same. Verify that ammonia table values for a few cases of liquid–vapor equilibrium meet this condition.

6·112 Prove that for a reversible isothermal process of a closed system, the work equals the decrease in the Helmholtz function.

6·113 Prove that for a reversible isothermal process of a steady-flow system with no change in kinetic or potential energy, the work equals the decrease in the Gibbs function.

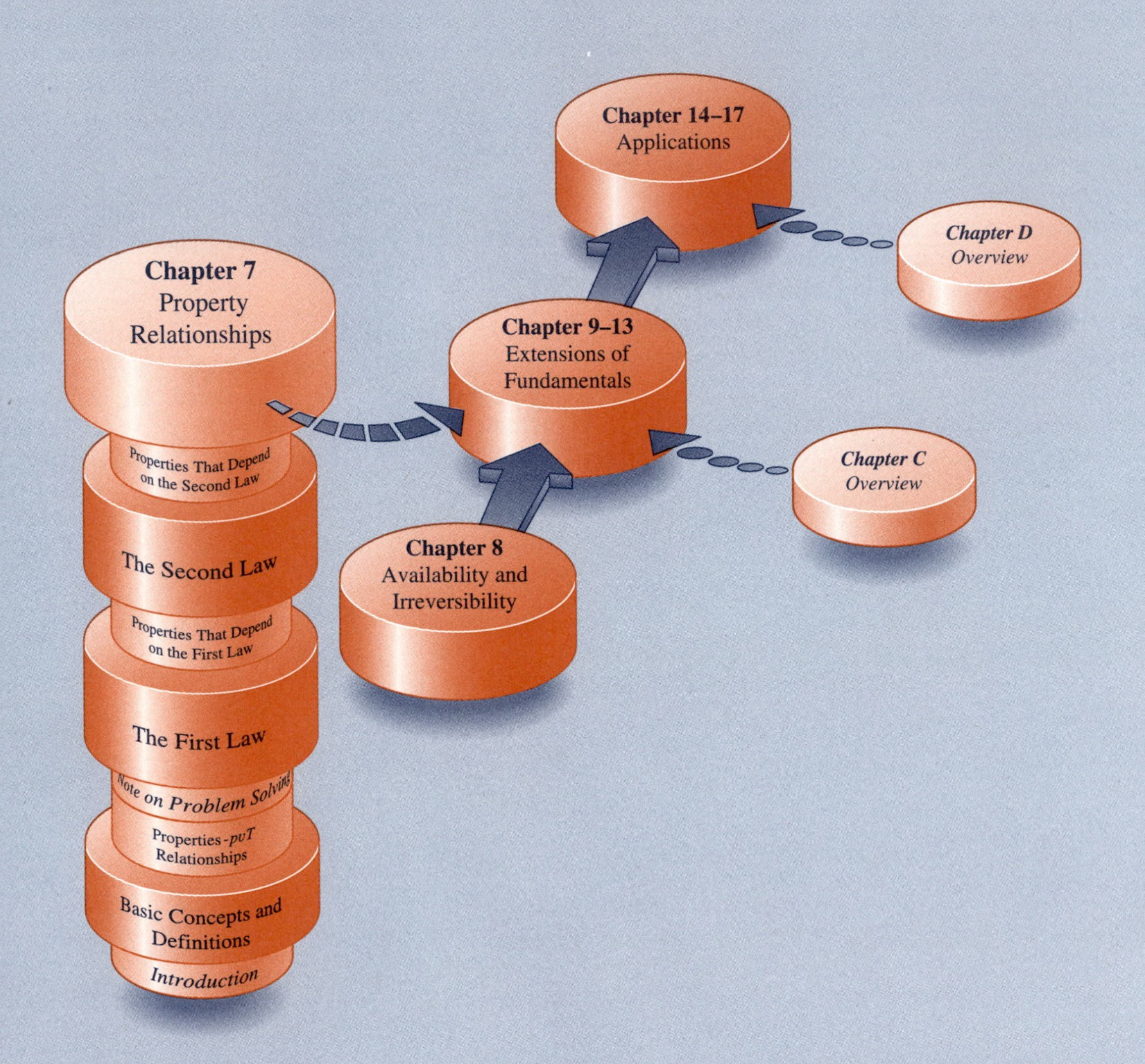
Chapter 14–17
Applications
Chapter D
Overview
Chapter 7
Property
Relationships
Chapter 9–13
Extensions of
Fundamentals
Chapter C
Overview
Properties That Depend
on the Second Law
Chapter 8
Availability and
Irreversibility
The Second Law
Properties That Depend
on the First Law
The First Law
Note on Problem Solving
Properties-*pvT*
Relationships
Basic Concepts and
Definitions
Introduction

Chapter 7

Property Relationships

Before introducing the first law, we discussed relationships among the directly measurable properties p, v (or ρ), and T: pressure, specific volume (or density), and temperature. Then, on the basis of the first law, we defined two more properties, u and h, internal energy and enthalpy, that expanded the range of calculations we could make. The second law led to three additional properties, s, a, and g: entropy, Helmholtz function, and Gibbs function.

Now, on the basis of both the first and second laws of thermodynamics, we are ready to look into relationships among these properties and various others of practical importance. In this chapter we discuss a few of these relationships as samples of the many that engineers find useful in various fields of work.

Engineers select fluids on the basis of their properties.

These property relationships are important for several reasons. First, just as engineers select structural materials on the basis of properties such as strength, ductility, and corrosion resistance and then build these properties into structural design models, they select fluids for various applications on the basis of other properties. The thermophysical properties of fluids vary widely, though, and the ways in which they vary can be as important as the property values themselves. Consequently, knowledge of property relationships is needed for accurate modeling of thermodynamic systems.

Sometimes, the data an engineer needs for a particular substance are not available. Then a knowledge of general property relationships may permit the calculation of some properties from others. When properties outside the range of published values are needed, extrapolations are most reliable if they are based on knowledge of general property relationships.

Property relations are essential in organizing experimental programs to determine property values.

If the scarcity of data on some substance is so restrictive that experimental measurements are required, general property relationships are essential in organizing the experimental program to get the needed information with the least expenditure of time and money. In fact, the cost of preparing the extensive tables of properties that are now available for many substances would be prohibitive if general property relationships did not make it possible to gain much information from a small amount of data.

Liquids and even solids in addition to gases can be simple compressible substances.

This chapter deals with properties of pure, simple compressible substances. Throughout the literature of thermodynamics, *pure substance* often means *pure, simple compressible substance*, and this sense is used throughout this chapter. Notice that liquids and even solids in the absence of anisotropic stress, although they are much less compressible than gases, can also be simple compressible substances. Mixtures of variable composition are treated in Chapters 9, 10, 11, and 12.

7·1 The Maxwell Equations

The Maxwell equations can easily be derived from the $T\,ds$ equations.

The Maxwell equations relate entropy to the three directly measurable properties, p, v, and T, for pure simple compressible substances. You can easily derive them by starting with a $T\,ds$ equation,

$$T\,ds = du + p\,dv \tag{6·7}$$

and then simply substituting into Eq. 6·7 the definitions of h, a, and g:

$$h \equiv u + pv$$

$$a \equiv u - Ts \tag{6·18}$$

$$g \equiv h - Ts \tag{6·19}$$

to obtain

$$du = T\,ds - p\,dv \tag{6·7}$$

$$dh = T\,ds + v\,dp \tag{6·7}$$

$$dh = -p\,dv + s\,dT \tag{7·1}$$

$$dg = v\,dp - s\,dT \tag{7·2}$$

Then, since these are differentials of properties, they are exact differentials, so we can immediately write, as shown in Section 1·6,

$$\left(\frac{\partial T}{\partial v}\right)_s = -\left(\frac{\partial p}{\partial s}\right)_v \tag{7·3}$$

$$\left(\frac{\partial T}{\partial p}\right)_s = \left(\frac{\partial v}{\partial s}\right)_p \tag{7·4}$$

$$\left(\frac{\partial p}{\partial T}\right)_v = \left(\frac{\partial s}{\partial v}\right)_T \tag{7·5}$$

$$\left(\frac{\partial v}{\partial T}\right)_p = -\left(\frac{\partial s}{\partial p}\right)_T \tag{7·6}$$

The Maxwell equations

These four equations are called the **Maxwell equations**. Notice that each one involves entropy, s, and all three of the properties p, v, and T.

Several other useful relationships among properties can be derived from Eqs. 6·7, 7·1, and 7·2. For example, taking u as a function of s and v, we have, as pointed out in Section 1·6,

$$du = \left(\frac{\partial u}{\partial s}\right)_v ds + \left(\frac{\partial u}{\partial v}\right)_s dv$$

Comparing this with Eq. 6·7,

$$du = T\,ds - p\,dv \tag{6·7}$$

and noting that ds and dv are independent, we can equate the coefficients of ds and dv to give

$$\left(\frac{\partial u}{\partial s}\right)_v = T \qquad \left(\frac{\partial u}{\partial v}\right)_s = -p \tag{7·7}$$

In a similar manner, we can write expressions for dh, da, and dg for $h = f(s, p)$, $a = f(v, T)$, and $g = f(p, T)$ and compare them with Eqs. 6·7, 7·1, and 7·2, respectively, to obtain

$$\left(\frac{\partial h}{\partial s}\right)_p = T \qquad \left(\frac{\partial h}{\partial p}\right)_s = v \tag{7·8}$$

$$\left(\frac{\partial a}{\partial v}\right)_T = -p \qquad \left(\frac{\partial a}{\partial T}\right)_v = -s \tag{7·9}$$

$$\left(\frac{\partial g}{\partial p}\right)_T = v \qquad \left(\frac{\partial g}{\partial T}\right)_p = -s \tag{7·10}$$

Notice again how readily these relationships can be derived from the $T\,ds$ equations. Many other relations among the eight variables p, v, T, s, h, u, a, and g can be derived from those already shown. Some of these relations will be used later in this chapter.

7·2 Some Applications of the Maxwell Equations

Some interesting conclusions can be drawn directly from the Maxwell equations and Eqs. 7·7 to 7·10. For example, from the second of Eqs. 7·8, that is,

James Clerk Maxwell (1831–1879), a member of a wealthy Scottish family, received his formal education at the University of Edinburgh and Cambridge University. He was professor of physics and astronomy at King's College, London (1860–1868), and in 1871 became the first professor of experimental physics at Cambridge University. He made scientific contributions in many areas, including mechanics, optics, electricity and magnetism, kinetic theory of gases, and thermodynamics.

$$\left(\frac{\partial h}{\partial p}\right)_s = v \tag{7·8}$$

A conclusion from the Maxwell equations

we conclude that since v is always positive, in any isentropic process the enthalpy of a substance decreases if the pressure decreases. Notice also that for isentropic processes of any system,

$$\Delta h = \int v\,dp \qquad (s = \text{constant, only})$$

As another example, the first of Eqs. 7·7,

$$\left(\frac{\partial u}{\partial s}\right)_v = T \tag{7·7}$$

can be written as

$$\left(\frac{\partial u}{\partial T}\right)_v \left(\frac{\partial T}{\partial s}\right)_v = T$$

Recalling the definition of c_v, we have

$$c_v\left(\frac{\partial T}{\partial s}\right)_v = T$$

or

$$\left(\frac{\partial T}{\partial s}\right)_v = \frac{T}{c_v} \qquad (7{\cdot}11)$$

Thus the slope of a constant-volume line on a Ts diagram is equal to T/c_v for any substance. Example 6·4 shows this to be true for an ideal gas, but here we see that the conclusion is independent of any equation of state.

Another conclusion from the Maxwell equations

The Maxwell equations provide a simple proof that the specific heats of an ideal gas are functions of temperature only. The third Maxwell equation is

Proof that the specific heats of an ideal gas are functions of temperature only

$$\left(\frac{\partial s}{\partial v}\right)_T = \left(\frac{\partial p}{\partial T}\right)_v \qquad (7{\cdot}5)$$

and we have just shown that

$$\left(\frac{\partial s}{\partial T}\right)_v = \frac{c_v}{T} \qquad (7{\cdot}11)$$

As long as s is a continuous function of T and v, and the derivatives are also continuous, the order of differentiation is immaterial, and the ''mixed'' second-order partial derivatives are equal,

$$\left[\frac{\partial}{\partial v}\left(\frac{\partial s}{\partial T}\right)_v\right]_T = \left[\frac{\partial}{\partial T}\left(\frac{\partial s}{\partial v}\right)_T\right]_v \qquad (a)$$

This is usually written

$$\frac{\partial^2 s}{\partial v \partial T} = \frac{\partial^2 s}{\partial T\, \partial v}$$

Substituting from Eqs. 7·5 and 7·11 into Eq. a gives

$$\left[\frac{\partial}{\partial v}\left(\frac{c_v}{T}\right)\right]_T = \left[\frac{\partial}{\partial T}\left(\frac{\partial p}{\partial T}\right)_v\right]_v$$

$$\left(\frac{\partial c_v}{\partial v}\right)_T = T\left(\frac{\partial^2 p}{\partial T^2}\right)_v$$

For an ideal gas, $pv = RT$, and therefore $(\partial^2 p/\partial T^2)_v = 0$. Thus

$$\left(\frac{\partial c_v}{\partial v}\right)_T = 0$$

which indicates that c_v is a function of temperature only.

For an ideal gas, $c_v = f(T)$.

The Maxwell equations allow us to evaluate entropy from data on other properties.

Let us now see how the Maxwell equations help in the evaluation of entropy from data on the directly measurable properties pressure, specific volume, temperature, and specific heats. For a pure substance entropy is a function of any two of the properties: p and v, p and T, or v and T. Two partial derivatives involving s are associated with each of the three pairs of measurable properties, making a total of six:

$$\underset{\mathbf{1}}{\left(\frac{\partial s}{\partial p}\right)_v} \quad \underset{\mathbf{2}}{\left(\frac{\partial s}{\partial v}\right)_p} \quad \underset{\mathbf{3}}{\left(\frac{\partial s}{\partial p}\right)_T} \quad \underset{\mathbf{4}}{\left(\frac{\partial s}{\partial T}\right)_p} \quad \underset{\mathbf{5}}{\left(\frac{\partial s}{\partial v}\right)_T} \quad \underset{\mathbf{6}}{\left(\frac{\partial s}{\partial T}\right)_v}$$

If we can express each of these derivatives in terms of p, v, T, and other readily measurable properties, we will have complete information on the entropy of any substance for which we have data on the other properties. The third and fifth derivatives can be evaluated from pvT data by means of two Maxwell equations alone:

$$\mathbf{3}: \left(\frac{\partial s}{\partial p}\right)_T = -\left(\frac{\partial v}{\partial T}\right)_p \qquad \mathbf{5}: \left(\frac{\partial s}{\partial v}\right)_T = \left(\frac{\partial p}{\partial T}\right)_v$$

The last derivative can be evaluated as

$$\mathbf{6}: \left(\frac{\partial s}{\partial T}\right)_v = \frac{c_v}{T} \tag{7·11}$$

as shown above; and it can be shown in similar fashion (starting from the first of Eqs. 7·8) that the fourth derivative can be evaluated as

$$\mathbf{4}: \left(\frac{\partial s}{\partial T}\right)_p = \frac{c_p}{T} \tag{7·12}$$

The two remaining derivatives can be expressed as

$$\mathbf{1}: \left(\frac{\partial s}{\partial p}\right)_v = \left(\frac{\partial s}{\partial T}\right)_v \left(\frac{\partial T}{\partial p}\right)_v = \frac{c_v}{T}\left(\frac{\partial T}{\partial p}\right)_v$$

and

$$\mathbf{2}: \left(\frac{\partial s}{\partial v}\right)_p = \left(\frac{\partial s}{\partial T}\right)_p \left(\frac{\partial T}{\partial v}\right)_p = \frac{c_p}{T}\left(\frac{\partial T}{\partial v}\right)_p$$

Thus we have obtained from the Maxwell equations a complete description of the entropy variation of a pure substance in terms of data on p, v, T, c_p, and c_v.

Notice that only *derivatives* of entropy have been expressed in terms of pvT and specific-heat data, so that entropy values can be obtained only by integration. The integration introduces arbitrary functions or "constants of integration" that cannot be evaluated from pvT and specific-heat data alone. The same difficulty is experienced in trying to evaluate u, h, a, or g from pvT and specific-heat data alone.

Entropy can be evaluated from pvT and specific-heat data only by integration. This introduces arbitrary constants.

7·2·1 Characteristic Functions

Inspection of Eqs. 7·7 through 7·10 shows that at least some properties can be completely evaluated by differentiation alone, and thus no arbitrary functions will be involved if we start with data on certain groups of properties. For example, if we have complete uvs data, we determine p and T by

$$p = -\left(\frac{\partial u}{\partial v}\right)_s \qquad T = \left(\frac{\partial u}{\partial s}\right)_v \tag{7·7}$$

Then

$$h \equiv u + pv = u - \left(\frac{\partial u}{\partial v}\right)_s v$$

$$a \equiv u - Ts = u - \left(\frac{\partial u}{\partial s}\right)_v s$$

$$g \equiv h - Ts = u - \left(\frac{\partial u}{\partial v}\right)_s v - \left(\frac{\partial u}{\partial s}\right)_v s$$

Thus we can evaluate p, T, h, a, and g completely from uvs data alone. The uvs relation for a substance

$$f(u, v, s) = 0$$

is called a **characteristic function**. A characteristic function is one from which all properties of a substance can be obtained by differentiation alone, so that no arbitrary functions that require supplementary data for their evaluation are introduced. Other characteristic functions are

Characteristic function

$$f(h, p, s) = 0$$
$$f(a, v, T) = 0$$
$$f(g, p, T) = 0$$

Unfortunately, the formulation of the properties of a substance cannot be started from a characteristic function because in none of them are all three properties directly measurable. As noted earlier, $f(p, v, T)$ is not a characteristic function. The usefulness of the characteristic functions is that once any one of them is formulated for a substance, all properties can then be determined without the use of additional data. This procedure was used to generate the steam tables in this book.

7·2·2 The Clapeyron Equation

Derivation of the Clapeyron equation

As one more example of the use of the Maxwell equations to correlate physical properties, we now derive the *Clapeyron equation* for property relationships during phase changes of a pure substance. In general, $p = f(T, v)$ and therefore

$$dp = \left(\frac{\partial p}{\partial T}\right)_v dT + \left(\frac{\partial p}{\partial v}\right)_T dv$$

and

$$\frac{dp}{dT} = \left(\frac{\partial p}{\partial T}\right)_v + \left(\frac{\partial p}{\partial v}\right)_T \frac{dv}{dT}$$

For two phases of a pure substance existing together in equilibrium, pressure and temperature are dependent only on each other. Therefore, $(\partial p/\partial v)_T = 0$, so that the general expression reduces to

$$\frac{dp}{dT} = \left(\frac{\partial p}{\partial T}\right)_v$$

Also, if s' and v' represent the properties of one phase and s'' and v'' those of the other phase, then

$$\left(\frac{\partial s}{\partial v}\right)_T = \frac{s'' - s'}{v'' - v'}$$

Thus the third Maxwell equation

$$\left(\frac{\partial p}{\partial T}\right)_v = \left(\frac{\partial s}{\partial v}\right)_T \tag{7·5}$$

becomes, for two phases of a pure substance coexisting in equilibrium,

$$\frac{dp}{dT} = \frac{s'' - s'}{v'' - v'} \tag{b}$$

The equation

$$T\,ds = dh - v\,dp \tag{6·7}$$

for the case of a phase change at constant pressure (and, consequently, at constant temperature) becomes

$$T(s'' - s') = (h'' - h') \tag{c}$$

The quantity $(h'' - h')$ is the latent heat of the phase transformation. Combining Eqs. *b* and *c* yields

$$\frac{dp}{dT} = \frac{h'' - h'}{T(v'' - v')} \qquad (7\cdot 13)$$

Clapeyron equation

This equation is called the **Clapeyron equation**. It relates three readily measurable properties: the slope of the saturation pressure–temperature line, the latent heat, and the change in volume during a phase transformation. Thus it can be used to check the consistency of measurements or to obtain information on one of these properties from data on the other two.

We have noticed earlier (Section 2·2) that the freezing temperature of water decreases as the pressure increases. We have also noted the unusual behavior of water in expanding as it freezes. Now we see that these two characteristics are related by the Clapeyron equation as it applies to the liquid–solid transformation:

$$\frac{dp}{dT} = \frac{h_f - h_i}{T(v_f - v_i)} = \frac{h_{if}}{Tv_{if}}$$

Since $h_{if} > 0$ and for water $v_{if} < 0$, the slope of the fusion line of a phase (pT) diagram must be negative. We already knew this about water from independent observations, but the Clapeyron equation now permits us to generalize: The freezing temperature of *any* substance that expands on freezing is lowered as the pressure is increased.

A general conclusion from the Clapeyron equation

Because the latent heats h_{fg} and h_{ig} and the volume changes v_{fg} and v_{ig} are all positive for all substances, the sublimation and vaporization curves have positive slopes on the phase diagrams of all substances.

Example 7·1 Application of the Clapeyron Equation

PROBLEM STATEMENT

Show that on a phase diagram for water the sublimation line and the vaporization line have different slopes at the triple point (i.e., that in the figure line a–c is not simply a continuation of line b–a).

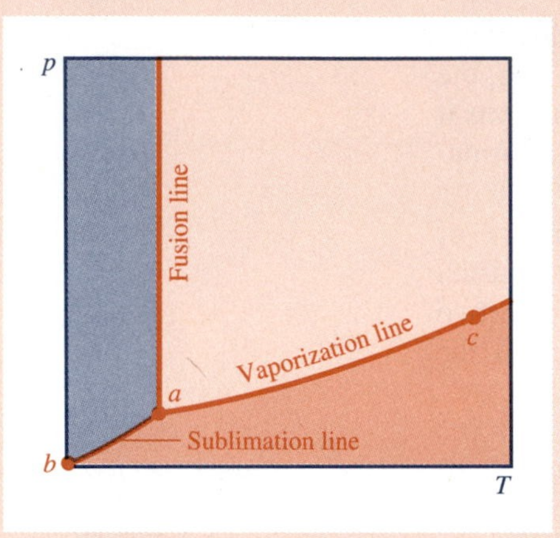

SOLUTION

We must prove that dp/dT has different values at the triple point for lines b–a and a–c. The difference between the slopes can be evaluated by means of the Clapeyron equation as follows:

$$\left(\frac{dp}{dT}\right)_{b-a} - \left(\frac{dp}{dT}\right)_{a-c} = \frac{s_{ig}}{v_{ig}} - \frac{s_{fg}}{v_{fg}}$$

At the triple state, $s_{ig} = s_{if} + s_{fg}$ and $v_{ig} = v_{if} + v_{fg}$, so that the difference in slopes is

$$\left(\frac{dp}{dT}\right)_{b-a} - \left(\frac{dp}{dT}\right)_{a-c} = \frac{(s_{if} + s_{fg})v_{fg} - s_{fg}(v_{if} + v_{fg})}{v_{fg}(v_{if} + v_{fg})}$$

$$= \frac{s_{if}v_{fg} - s_{fg}v_{if}}{v_{fg}(v_{if} + v_{fg})}$$

Since v_{if} is very small compared with v_{fg}, while s_{if} is smaller than s_{fg} to a lesser degree, this reduces to approximately

$$\left(\frac{dp}{dT}\right)_{b-a} - \left(\frac{dp}{dT}\right)_{a-c} \approx \frac{s_{if}}{v_{fg}}$$

Both quantities on the right-hand side are positive, so the slope of the sublimation line is greater than that of the vaporization line at the triple point.

7·3 General Equations for Internal Energy, Enthalpy, and Entropy in Terms of p, v, T, and Specific Heats

The entropy of a pure substance may be expressed as a function of T and v, $s = f(T, v)$, so that

$$ds = \left(\frac{\partial s}{\partial T}\right)_v dT + \left(\frac{\partial s}{\partial v}\right)_T dv$$

From Section 7·1,

$$\left(\frac{\partial s}{\partial T}\right)_v = \frac{c_v}{T} \qquad (7\cdot11)$$

and the third Maxwell equation is

$$\left(\frac{\partial s}{\partial v}\right)_T = \left(\frac{\partial p}{\partial T}\right)_v \qquad (7\cdot5)$$

Making these two substitutions in the expression for ds gives

$$ds = c_v \frac{dT}{T} + \left(\frac{\partial p}{\partial T}\right)_v dv \qquad (7\cdot14)$$

By starting with $s = f(T, p)$, it can be shown in a similar manner that

General equations for entropy

$$ds = c_p \frac{dT}{T} - \left(\frac{\partial v}{\partial T}\right)_p dp \qquad (7\cdot15)$$

Equations 7·14 and 7·15 are general equations for the entropy change of a pure substance. They can be integrated if specific-heat temperature and pvT data are available. They are also useful in obtaining general expressions for du and dh, as will now be shown.

For any process between equilibrium states of a pure substance,

$$du = T\,ds - p\,dv \qquad (6\cdot7)$$

Substituting into this equation the value of ds from Eq. 7·14, we have

General equation for internal energy

$$du = c_v\,dT + \left[T\left(\frac{\partial p}{\partial T}\right)_v - p\right] dv \qquad (7\cdot16)$$

Substituting the value of ds from Eq. 7·15 into

$$dh = T\,ds + v\,dp \qquad (6\cdot7)$$

gives

General equation for enthalpy

$$dh = c_p\,dT - \left[T\left(\frac{\partial v}{\partial T}\right)_p - v\right] dp \qquad (7\cdot17)$$

Equations 7·16 and 7·17 are general equations for du and dh of a pure substance. They too can be integrated if specific-heat temperature and pvT data are available. Notice that integration of Eqs. 7·14, 7·15, 7·16, and 7·17 can give only *changes* in s, u, and h. Absolute values cannot be obtained from pvT and specific-heat data alone.

All properties can be determined from measurements of only a few quantities.

These general equations make possible the generation of extensive and accurate property tables from relatively few measurements. As an illustration of this point, consider Fig. 7·1, which is a skeleton Ts diagram for some substance. All properties (h, u, s, g, etc.) on line 1–2–3–4–5 can be determined from measurements of only the following quantities:

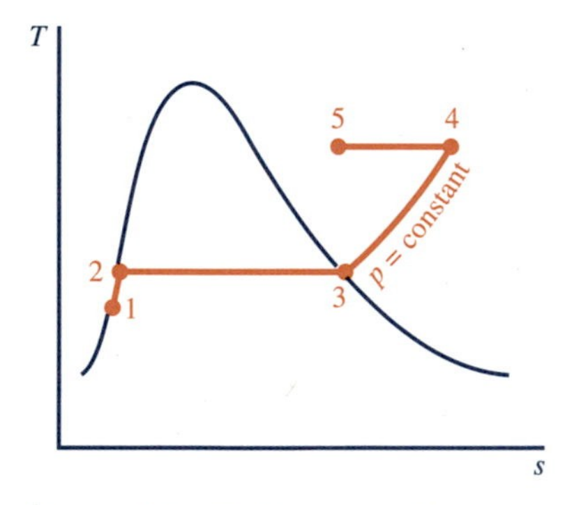

Figure 7·1 *Ts* diagram to illustrate property determination.

1. c_p of the liquid between states 1 and 2, where 1 is a base state at which values of h and s are assigned arbitrarily
2. The saturation temperature and h_{fg} for the pressure of states 2, 3, and 4
3. c_p of the vapor along the constant-pressure line from 3 to 4
4. pvT data in the region around the constant-temperature line between 4 and 5

If the base state 1 is a saturated liquid state, the process 1–2 can be taken as an isentropic process from p_1 to p_2 followed by a constant-pressure process to state 2. For all liquids at low pressures, Δh and ΔT for the isentropic process are very small. For the constant-pressure part of the process, Eqs.

7·17 and 7·15 reduce to

$$\Delta h = \int_{T_1}^{T_2} c_p \, dT$$

$$\Delta s = \int_{T_1}^{T_2} c_p \frac{dT}{T}$$

From the measured values of the saturation temperature and h_{fg} at p_2, s_{fg} can be determined from a $T\,ds$ equation such as Eq. 6·7. For the constant-pressure process 3–4, Δh and Δs can again be obtained from Eqs. 7·17 and 7·15, because the c_p data along this constant-pressure line make numerical integration possible. For the constant-temperature process 4–5, Eqs. 7·17 and 7·15 simplify to

$$\Delta h = -\int_{p_4}^{p_5} \left[T\left(\frac{\partial v}{\partial T}\right)_p - v \right] dp$$

$$\Delta s = -\int_{p_4}^{p_5} \left(\frac{\partial v}{\partial T}\right)_p dp$$

The partial derivatives in these equations can be evaluated numerically or graphically from the pvT data in the vicinity of process 4–5.

The result of these calculations is that all properties at a state 5 can be determined from properties at a state 1 and a small amount of experimental data. The approach outlined here is a simplified one. Much more extensive calculations would be made to ensure accuracy and consistency in published tables. For example, several paths might be used between states 1 and 5. These additional paths would require measurements of saturation temperature and h_{fg} at various pressures. If v_{fg} is also measured, the Clapeyron equation provides a check on the consistency of these measurements.

Published tables of properties usually include a listing of data sources and an explanation of the methods used in generating the table. Although these explanations may be at a level beyond that of an introductory course in thermodynamics, a perusal of them shows the key role of the general equations as presented in this chapter.

Example 7·2

PROBLEM STATEMENT

Develop an expression for the change in internal energy of a gas which follows the equation of state

$$p = \frac{RT}{v - b} - \frac{a}{v^2}$$

SOLUTION

Differentiating the equation of state gives

$$\left(\frac{\partial p}{\partial T}\right)_v = \frac{R}{v-b}$$

Substituting this value into the general expression for du,

$$du = c_v\,dT + \left[T\left(\frac{\partial p}{\partial T}\right)_v - p\right]dv \tag{7·16}$$

gives

$$du = c_v\,dT + \left[\frac{RT}{v-b} - p\right]dv$$

Substituting $[RT/(v-b) - a/v^2]$ for p gives

$$du = c_v\,dT + \frac{a}{v^2}\,dv$$

Integrating yields

$$u_2 - u_1 = \int_1^2 c_v\,dT - a\left(\frac{1}{v_2} - \frac{1}{v_1}\right)$$

Knowledge of the c_vT relation is needed for the integration of $\int c_v\,dT$.

7·4 Specific Heat Relations

The preceding section presents general equations for internal energy, enthalpy, and entropy changes in terms of p, v, T, and specific heats. pvT relationships were presented earlier and will be revisited in Section 7·6·1, so we now derive some general relationships for specific heats. We start by deriving equations for the difference between specific heats $(c_p - c_v)$ and for the ratio of specific heats c_p/c_v.

Derivation of some equations involving specific heats

Equating the two general equations for ds (Eqs. 7·14 and 7·15) gives

$$c_v\frac{dT}{T} + \left(\frac{\partial p}{\partial T}\right)_v dv = c_p\frac{dT}{T} - \left(\frac{\partial v}{\partial T}\right)_p dp$$

Solving for dT, we find

$$dT = \frac{T(\partial p/\partial T)_v}{c_p - c_v}\,dv + \frac{T(\partial v/\partial T)_p}{c_p - c_v}\,dp$$

The temperature may be expressed as a function of v and p. Hence,

$$dT = \left(\frac{\partial T}{\partial v}\right)_p dv + \left(\frac{\partial T}{\partial p}\right)_v dp$$

Equating the coefficients in the last two equations results in the two identical relations,

$$\left(\frac{\partial T}{\partial v}\right)_p = \frac{T(\partial p/\partial T)_v}{c_p - c_v} \quad \text{and} \quad \left(\frac{\partial T}{\partial p}\right)_v = \frac{T(\partial v/\partial T)_p}{c_p - c_v}$$

Thus

$$c_p - c_v = T\left(\frac{\partial p}{\partial T}\right)_v \left(\frac{\partial v}{\partial T}\right)_p \tag{7·18}$$

The difference between c_p and c_v can thus be determined for any substance if the equation of state (pvT relation) is known or if the two partial derivatives can be measured. For liquids and solids, it is difficult to measure $(\partial p/\partial T)_v$, so a different form of Eq. 7·18 is desired.

The general relation (see Appendix II)

$$\left(\frac{\partial x}{\partial y}\right)_z \left(\frac{\partial y}{\partial z}\right)_x \left(\frac{\partial z}{\partial x}\right)_y = -1$$

when applied to $f(p, v, T)$ is

$$\left(\frac{\partial p}{\partial T}\right)_v \left(\frac{\partial T}{\partial v}\right)_p \left(\frac{\partial v}{\partial p}\right)_T = -1$$

or

$$\left(\frac{\partial p}{\partial T}\right)_v = -\left(\frac{\partial p}{\partial v}\right)_T \left(\frac{\partial v}{\partial T}\right)_p$$

Making this substitution in Eq. 7·18, we find

$$c_p - c_v = -T\left(\frac{\partial p}{\partial v}\right)_T \left(\frac{\partial v}{\partial T}\right)_p^2 \tag{7·19}$$

The coefficient of volume expansion β and the isothermal compressibility κ_T are defined as

$$\beta \equiv \frac{1}{v}\left(\frac{\partial v}{\partial T}\right)_p \quad \text{and} \quad \kappa_T \equiv -\frac{1}{v}\left(\frac{\partial v}{\partial p}\right)_T$$

Making these substitutions in Eq. 7·19 gives

$$c_p - c_v = \frac{Tv\beta^2}{\kappa_T} \tag{7·19}$$

The difference between specific heats

T, v, and κ_T are always positive, and β^2 is either positive or zero; so c_p can never be less than c_v. When $\beta = 0$ (as for water at 1 atm and about 4°C), $c_p = c_v$. For solids and liquids, c_v is difficult to measure, so Eq. 7·19 is used to calculate c_v from c_p and the other properties.

To obtain a general expression for the ratio of the specific heats, c_p/c_v, let us combine

$$\frac{c_p}{T} = \left(\frac{\partial s}{\partial T}\right)_p \tag{7·12}$$

and

$$\frac{c_v}{T} = \left(\frac{\partial s}{\partial T}\right)_v \tag{7·11}$$

to give

$$k = \frac{c_p}{c_v} = \frac{(\partial s/\partial T)_p}{(\partial s/\partial T)_v}$$

To obtain derivatives that appear in the Maxwell equations, expand both the numerator and the denominator to give

$$k = \frac{c_p}{c_v} = \frac{(\partial s/\partial v)_p(\partial v/\partial T)_p}{(\partial s/\partial p)_v(\partial p/\partial T)_v}$$

Each of the four partial derivatives here appears in one of the Maxwell equations. Substituting for each one from the proper Maxwell equation gives

$$k = \frac{c_p}{c_v} = \frac{-(\partial p/\partial T)_s(\partial s/\partial p)_T}{-(\partial v/\partial T)_s(\partial s/\partial v)_T}$$

$$= \left(\frac{\partial p}{\partial v}\right)_s\left(\frac{\partial v}{\partial p}\right)_T \tag{7·20}$$

If we define isentropic compressibility κ_s by

$$\kappa_s \equiv -\frac{1}{v}\left(\frac{\partial v}{\partial p}\right)_s$$

we have

The ratio of specific heats

$$k = \frac{c_p}{c_v} = \frac{\kappa_T}{\kappa_s} \tag{7·21}$$

This relationship holds for gases, liquids, and solids.

A scale showing specific heats of various common solids, liquids, and gases at room conditions is displayed in the margin of this page. Notice that usually, but not always, specific heats are highest for gases and lowest for solids. Notice also that c_p of water is much higher than that of most other liquids. General trends and relative values like these can be readily found in handbooks and other publications. Engineers who also have such information as part of their general knowledge find it helpful in quickly making estimates and judging the reasonableness of numerical results.

General trends of specific heats with temperature

Figures 4·11 and 4·12 for gases and Fig. 7·3 for solids show that specific heats usually remain constant or increase with increasing temperature. This is usually true for liquids, too. In the vicinity of the critical state, however, very high values of c_p occur, as shown for air and water in Figs. 7·2 and 4·8, respectively, so these general statements do not hold in the critical state region.

There are several methods for approximating specific heats of substances from other data. A knowledge of some of these is often helpful. In Section 4·5 it was shown that for monatomic and diatomic gases that can be treated as ideal gases the $\bar{c}_p$ values are approximately $5\bar{R}/2$ and $7\bar{R}/2$, respectively, and the $\bar{c}_v$ values are approximately $3\bar{R}/2$ and $5\bar{R}/2$, respectively.

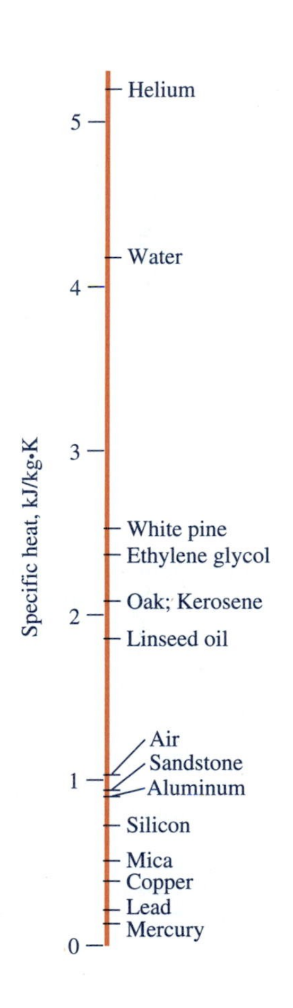

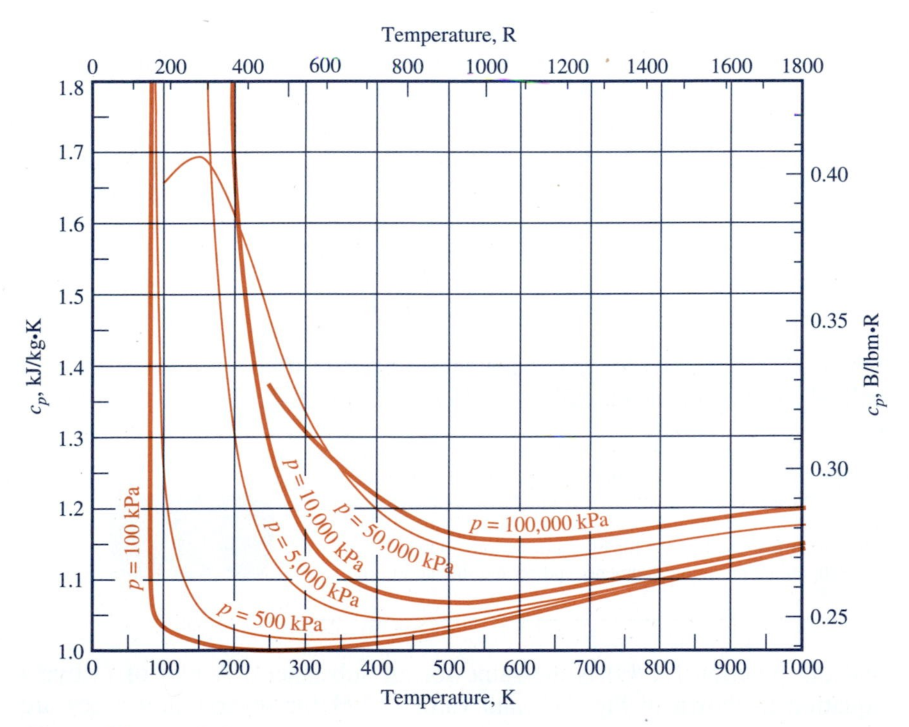

Figure 7·2 c_p of air.

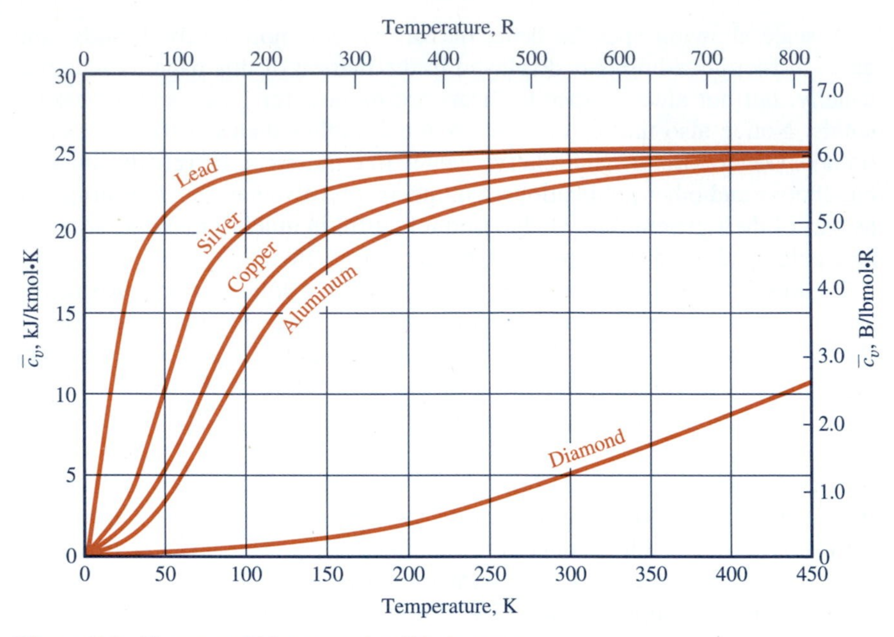

Figure 7·3 $\bar{c}_v$ versus T for several solid elements.

Figure 7·3 shows $\bar{c}_v$ versus temperature for several solid substances. Two things are readily noticed: (1) As temperature increases, $\bar{c}_v$ approaches a value of about 25 kJ/kmol·K (6 B/lbmol·R) for all the substances, and (2) the curves are approximately the same for all substances except for a temperature scale factor, although for diamond this is not apparent in Fig. 7·3. A valuable generalization was provided by Peter Debye, who by use of quantum statistical mechanics showed that for all isotropic solids $\bar{c}_v$ is given very nearly by

Debye's generalization for the specific heat of isotropic solids

$$\bar{c}_v = 3\bar{R}f\left(\frac{T}{\Theta}\right)$$

where $\bar{R}$ is the universal gas constant (Yes, $\bar{R}$ is used in connection with solids!), Θ is a constant, called the *Debye temperature*, for each substance,

Peter Joseph William Debye (1884–1966) was born in the Netherlands, earned an electrical engineering degree at the Technical University of Aachen, and completed his doctoral studies in physics at the University of Munich. He served on university faculties in Germany, Switzerland, the Netherlands, and the United States. In 1936 he was awarded the Nobel Prize for Chemistry "for his contributions to our knowledge of molecular structure through his investigations on dipole moments and on the diffraction of X rays and electrons in gases."

and the function $f(T/\Theta)$ is the same for all substances. A plot of Debye's equation is shown in Fig. 7·4, and values of Θ for several substances are given in Table 7·1.

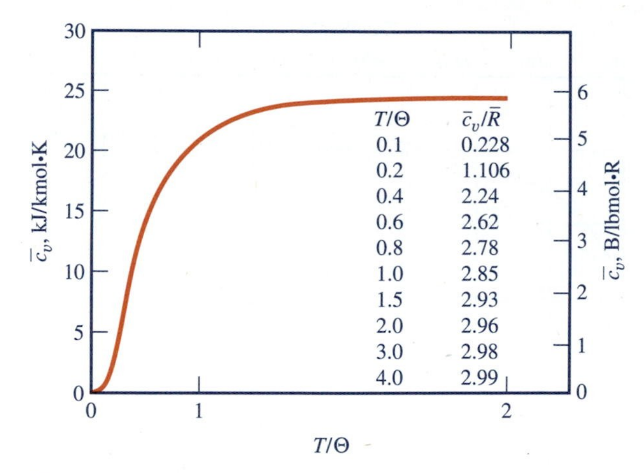

Figure 7·4 Debye's equation.

Table 7·1 Debye Temperatures, Θ

Substance	Θ, K	Θ, R
Aluminum	428	770
Cadmium	209	376
Calcium	230	414
Copper	343	617
Diamond	2230	4014
Graphite	420	756
Iron	467	841
Lead	105	189
Mercury	72	130
Silver	225	405
Sodium	158	284
Zinc	327	589

Source: American Institute of Physics Handbook, 3rd ed., McGraw-Hill, New York, 1972, pp. 4-115, 4-116.

For most substances there is disagreement among published values for the Debye temperatures. One reason is that the Debye temperature is not a directly measurable property but is a parameter that fits experimental values of specific heats to the Debye function. If the fit is made for different temperature ranges, the Debye temperature will be slightly different (or for some substances, such as cadmium, greatly different).

Notice that, as T/Θ increases, $\bar{c}_v$ approaches $3\bar{R}$ (or approximately 25 kJ/kmol·K). As T/Θ approaches zero, Debye's equation becomes

$$c_v \propto T^3$$

and this is often called the **Debye T^3 law**. For very high values of T/Θ, $\bar{c}_v$ of some solids is higher than the $3\bar{R}$ given by Debye's equation. Some isotropic solids have c_v values well over $3\bar{R}$ even at fairly low temperatures. Examples are ice with $\bar{c}_v = 4.5\bar{R}$ at 0°C and the element barium with $\bar{c}_v = 4.7\bar{R}$ at 20°C. Thus it is apparent that Debye's equation must be applied with caution. Nevertheless, it is a convenient means of estimating $\bar{c}_v$ values, and it is an important step in the development of a complete theory of the solid phase.

Debye's T^3 law

Remember that Debye's equation pertains to $\bar{c}_v$, not $\bar{c}_p$. For solids and liquids, data referred to simply as "specific-heat" values are usually $\bar{c}_p$ (or c_p) values. The difference between $\bar{c}_p$ and $\bar{c}_v$ increases as temperature increases, as shown by Eq. 7·19,

For solids, the difference between the specific heats increases with increasing temperature.

$$\bar{c}_p - \bar{c}_v = \frac{T\bar{v}\beta^2}{\kappa_T} \qquad \text{(7·19, molar basis)}$$

Figure 7·5 shows the magnitude of this difference over a wide range of temperature for one solid.

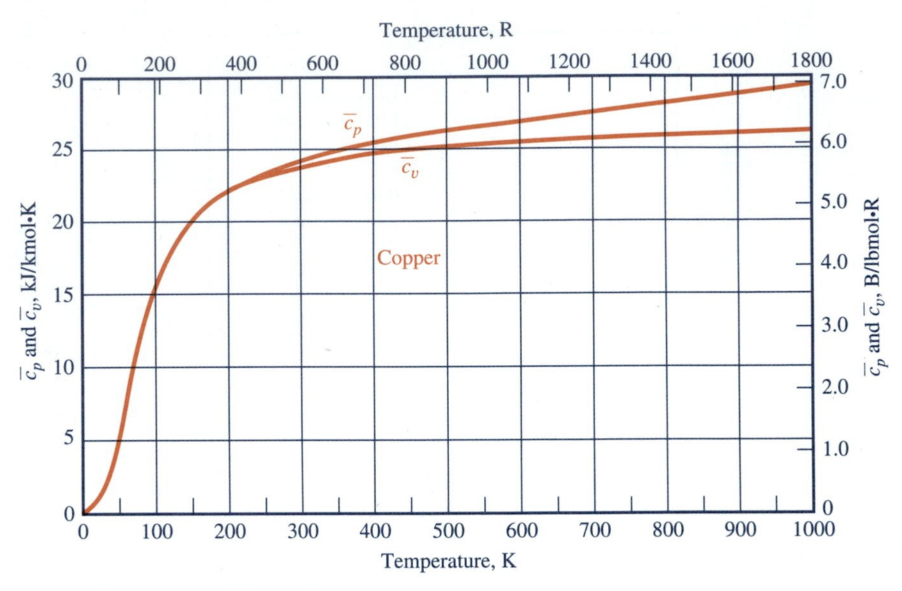

Figure 7·5 $\overline{c}_p$ and $\overline{c}_v$ for copper.

7·5 The Joule–Thomson Coefficient

A useful property of substances that, like specific heats, is defined as a partial derivative is the **Joule–Thomson coefficient**, μ:

Joule–Thomson coefficient

$$\mu \equiv \left(\frac{\partial T}{\partial p}\right)_h$$

It is important partly because it can be measured more accurately and more readily than can certain other useful properties that are directly related to it.

Measuring the Joule–Thomson coefficient

The Joule–Thomson coefficient of a liquid or gas is measured by allowing the fluid to expand steadily through a porous plug as shown in Fig. 7·6. The plug and adjacent piping are thermally insulated, so $q = 0$. No work is done, and the changes in potential and kinetic energy can be made negligibly small. The first law as applied to any steady-flow system,

$$q + h_1 + \frac{V_1^2}{2} + gz_1 = h_2 + \frac{V_2^2}{2} + gz_2 + w \tag{3·18}$$

becomes for this system

$$h_1 = h_2$$

The inlet pressure and temperature (p_1 and T_1) are held constant. The downstream pressure is held at several different values successively, and at each of these T_2 is measured. Then T_2 is plotted against p_2 as in Fig. 7·7*a*. Each plotted point represents a state for which the enthalpy is equal to h_1. If enough measurements are made, we can pass a curve through the points and with reasonable confidence label that curve a constant-enthalpy line (Fig. 7·7*b*). The slope of this line at any point is $(\partial T/\partial p)_h$, the Joule–Thomson coefficient. To obtain the Joule–Thomson coefficient at various pressures and temperatures, it is necessary to use several combinations of initial pressure and temperature, each combination providing a different constant h line, as shown in Fig. 7·8. Values of the Joule–Thomson coefficient for several gases are given in Fig. 7·9.

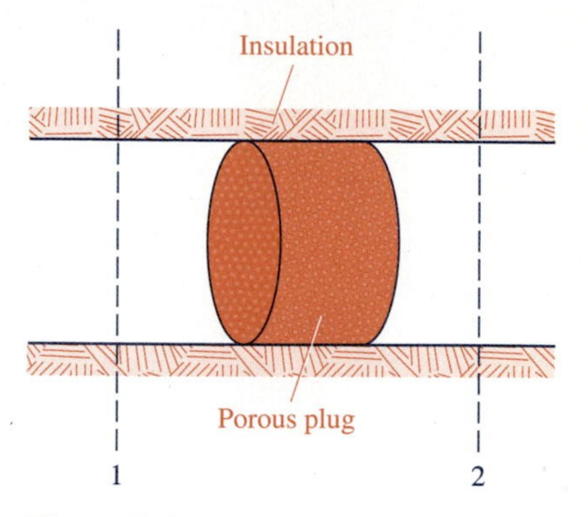

Figure 7·6 Apparatus for measuring Joule–Thomson coefficient.

Inversion line

The broken line in Fig. 7·8 passes through the maximum temperature points of the constant-enthalpy lines. It is called the **inversion line**. To the left of this line, the Joule–Thomson coefficient is positive; to the right of the inversion line, the Joule–Thomson coefficient is negative. Expansions that occur to the left of the inversion line between states of equal enthalpy (such as from *m* to *b*) result in a decrease in temperature, while expansions (such as *a* to *m*) occurring to the right of the inversion line result in a temperature rise.

A Joule–Thomson expansion used for refrigeration

The Joule–Thomson expansion can be used for refrigeration. Figure 7·8 shows that this is possible only when the initial temperature is lower than the maximum inversion temperature or the temperature at which the upper part of the inversion line cuts the axis. For example, starting from point *a*, a gas will experience an overall drop in temperature if its pressure is lowered sufficiently—say to point *c*. A gas initially at point *r* or *s* cannot be cooled by a Joule–Thomson expansion (i.e., throttling), no matter how low the down-

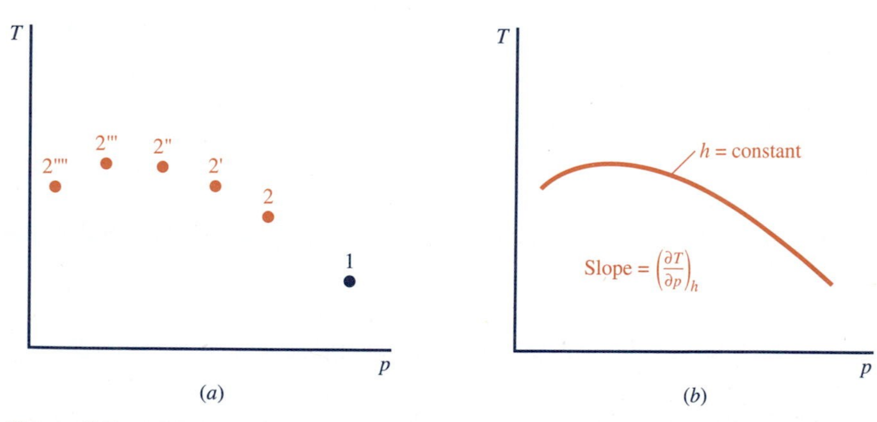

Figure 7·7 (*a*) Joule–Thomson expansion data points and (*b*) the resulting constant-enthalpy line, the slope of which is the Joule–Thomson coefficient.

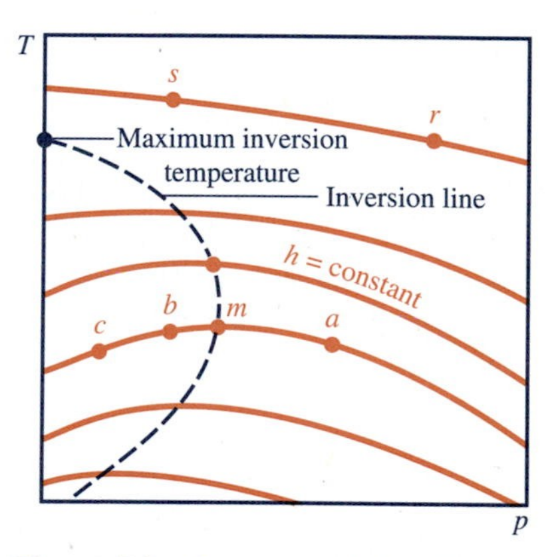

Figure 7·8 Constant-enthalpy lines and the inversion line for a substance.

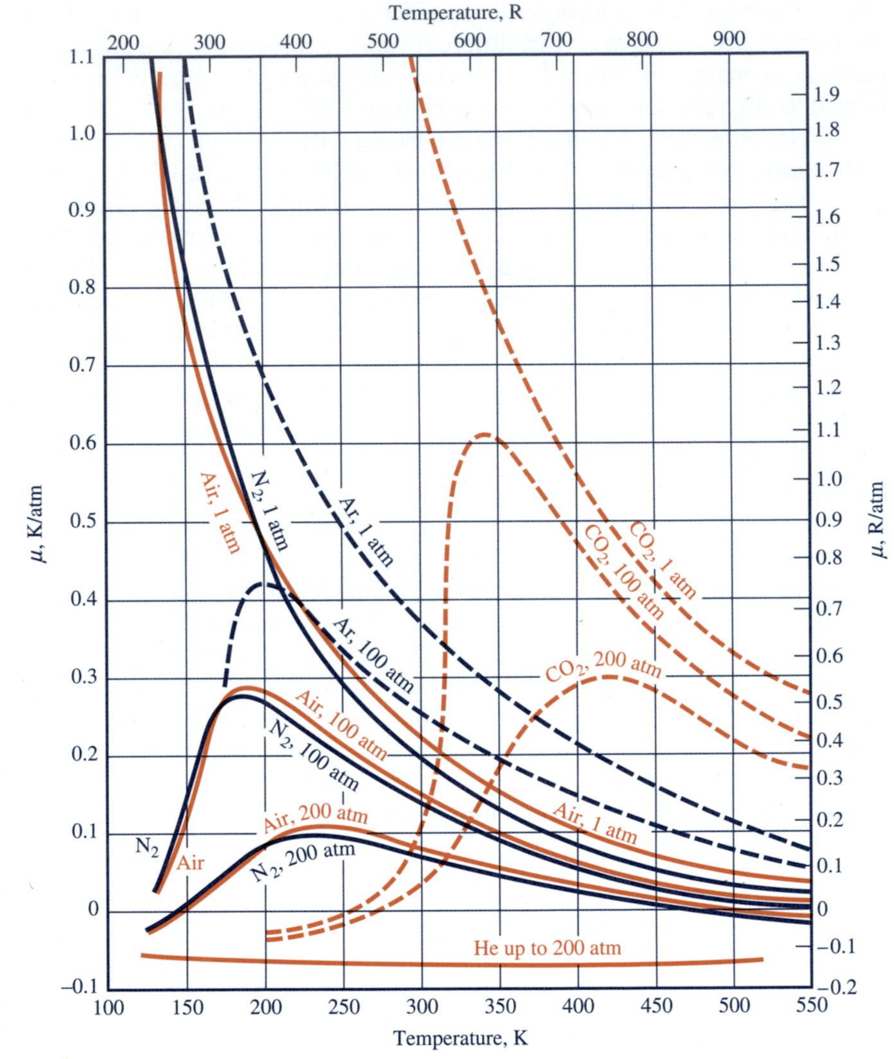

Figure 7·9 Joule–Thomson coefficients.

Table 7·2 Maximum Inversion Temperatures

Gas	Maximum Inversion Temperature K	R
Air	659	1186
Argon	780	1404
Hydrogen	205	369
Nitrogen	621	1118
Oxygen	764	1375

Source: American Institute of Physics Handbook, 3rd ed., McGraw-Hill, New York, 1972.

stream pressure is. The maximum inversion temperatures for several gases are given in Table 7·2.

7·5·1 The Joule–Thomson Coefficient in Terms of pvT and c_p Data

The Joule–Thomson coefficient of a substance can be calculated from the pvT relation and data on c_p. We have seen that a general equation for enthalpy change of a pure substance is

$$dh = c_p\, dT + \left[v - T\left(\frac{\partial v}{\partial T}\right)_p\right] dp \tag{7·17}$$

or

$$dT = \frac{dh}{c_p} + \frac{1}{c_p}\left[T\left(\frac{\partial v}{\partial T}\right)_p - v\right] dp \tag{d}$$

Since $T = f(h, p)$, the differential dT is

$$dT = \left(\frac{\partial T}{\partial h}\right)_p dh + \left(\frac{\partial T}{\partial p}\right)_h dp \tag{e}$$

Because h and p can be varied independently, the corresponding coefficients in Eqs. *d* and *e* must be equal. Equating the coefficients of dp, we get

$$\mu = \left(\frac{\partial T}{\partial p}\right)_h = \frac{1}{c_p}\left[T\left(\frac{\partial v}{\partial T}\right)_p - v\right] \tag{7·22a}$$

By inspection this can be written also as

$$\mu = \frac{T^2}{c_p}\left[\frac{\partial(v/T)}{\partial T}\right]_p \tag{7·22b}$$

The Joule–Thomson coefficient in terms of pvT and c_p

Equations 7·22 give μ in terms of the equation of state (pvT relation) and c_p.

7·5·2 The Constant-Temperature Coefficient

Another coefficient that is useful and fairly easy to measure is the **constant-temperature coefficient** c_T defined by

$$c_T \equiv \left(\frac{\partial h}{\partial p}\right)_T$$

The constant-temperature coefficient

For its measurement, a fluid is expanded slowly through a porous plug and heat is added or removed (by means of an electric heating coil or a flow of coolant) to maintain the outlet temperature equal to the inlet temperature. The amount of heat required to hold the temperature constant is measured so that enthalpy changes can be computed from the first law.

The general relationship among the partial derivatives of x, y, and z when the three are functionally related (see Appendix II) is

$$\left(\frac{\partial x}{\partial y}\right)_z\left(\frac{\partial y}{\partial z}\right)_x\left(\frac{\partial z}{\partial x}\right)_y = -1$$

Applying this to T, p, and h, we see the relationship among μ, c_T, and c_p:

$$\left(\frac{\partial T}{\partial p}\right)_h \left(\frac{\partial p}{\partial h}\right)_T \left(\frac{\partial h}{\partial T}\right)_p = -1$$

$$\frac{\mu c_p}{c_T} = -1$$

A convenient relationship for checking data

$$\mu = -\frac{c_T}{c_p} \tag{7·23}$$

This is a convenient relationship for checking the consistency of experimental data on these properties.

Equations 7·22 and 7·23 show that for an ideal gas, $\mu = 0$ and $c_T = 0$.

7·5·3 Relating Thermometer Readings to Thermodynamic Temperature Scales

Another use of the Joule–Thomson coefficient is to relate readings of actual thermometers or other temperature scales to the thermodynamic temperature scale. We have defined the thermodynamic temperature scale in terms of the performance of reversible engines, and doing so eliminated any dependence on the properties of any substance. But reversible engines are an idealization that cannot be used in an actual thermometer. The problem is to find a means of relating temperatures on actual, practical thermometers to those on the thermodynamic scale.

The temperature T in Eq. 7·22*a*,

$$\left(\frac{\partial T}{\partial p}\right)_h = \frac{1}{c_p}\left[T\left(\frac{\partial v}{\partial T}\right)_p - v\right] \tag{7·22a}$$

is the thermodynamic temperature. Let θ be the temperature indicated by some thermometer such as a real-gas thermometer or a mercury-in-glass thermometer. T and θ are functionally related; for each value of one there is a unique value of the other. To make the derivatives in Eq. 7·22*a* involve θ instead of T, we note that

$$\left(\frac{\partial T}{\partial p}\right)_h = \left(\frac{\partial \theta}{\partial p}\right)_h \frac{dT}{d\theta}$$

$$\frac{1}{c_p} = \left(\frac{\partial T}{\partial h}\right)_p = \left(\frac{\partial \theta}{\partial h}\right)_p \frac{dT}{d\theta}$$

$$\left(\frac{\partial v}{\partial T}\right)_p = \left(\frac{\partial v}{\partial \theta}\right)_p \frac{d\theta}{dT}$$

Making these substitutions in Eq. 7·22*a* yields

$$\left(\frac{\partial \theta}{\partial p}\right)_h \frac{dT}{d\theta} = \left(\frac{\partial \theta}{\partial h}\right)_p \frac{dT}{d\theta}\left[T\left(\frac{\partial v}{\partial \theta}\right)_p \frac{d\theta}{dT} - v\right]$$

and rearranging gives

$$\frac{dT}{T} = \frac{(\partial v/\partial \theta)_p \, d\theta}{(\partial \theta/\partial p)_h(\partial h/\partial \theta)_p + v}$$

Letting T_0 and θ_0 be corresponding values of T and θ, we have

An equation relating thermodynamic temperatures to thermometer readings

$$\int_{T_0}^{T} \frac{dT}{T} = \ln \frac{T}{T_0} = \int_{\theta_0}^{\theta} \frac{(\partial v/\partial \theta)_p \, d\theta}{(\partial \theta/\partial p)_h(\partial h/\partial \theta)_p + v}$$

All terms in the integrand of the right-hand side of this equation can be measured, and the integral is a function of T (and hence θ) only. Therefore, the right-hand side can be integrated numerically or graphically to provide a relationship between θ and T, and this is just what we are looking for. Notice that we placed no restrictions on the substance used for the measurement of the quantities in the integrand. Also notice that the three partial derivatives, $(\partial v/\partial \theta)_p$, $(\partial \theta/\partial p)_h$, and $(\partial h/\partial \theta)_p$, are, respectively, the product of the coefficient of thermal expansion and the specific volume, the Joule–Thomson coefficient, and the specific heat at constant pressure, all defined in terms of θ instead of T. By measuring these three properties of a substance over a range of temperature, we are able to relate our actual thermometer readings to thermodynamic temperature scale values.

7·6 Real Gases

Chapter 2 dealt with *pvT* relationships of substances for which tabular data are available and for the special case of ideal gases. Chapter 4 then dealt with internal energy, enthalpy, and specific heats of the same substances. Now we consider real gases for which $pv = RT$ is inaccurate and for which extensive tables of properties are unavailable. For real gases, u, h, and specific heats are usually functions of both pressure and temperature, so relationships for these quantities as well as for s, a, and g are not as simple as they are for ideal gases. This section presents property relationships and some data for real gases.

We consider four equations of state for real gases:
- **the van der Waals equation**
- **the Redlich–Kwong/Redlich–Kwong–Soave equations**
- **the Benedict–Webb–Rubin equation**
- **the virial equations**

7·6·1 Equations of State for Gases

For real gases, many equations of state have been proposed. All are more complex than $pv = RT$, because the increased accuracy over a wide range of states requires increased complexity. We will mention here only four of the many that have been published.

The van der Waals Equation. The van der Waals equation of state is of pedagogical and historical interest. It is based on physical reasoning and was the first of many attempts to model the pvT behavior of real gases more accurately than the ideal-gas equation.

The ideal-gas equation of state can be derived from kinetic theory. Among the assumptions made in the derivation are that there are no attractive forces between molecules and that the volume of molecules themselves is negligible compared to the total volume occupied by the gas. These two assumptions become unreasonable as the density increases. In 1873, J. D. van der Waals presented an equation based on his reasoning that $p\bar{v} = \bar{R}T$ would be improved by replacing $\bar{v}$ by the volume per mole of the space *between* the molecules, $\bar{v} - b$, where b is the volume per mole of the molecules. Since b is the volume per mole of the molecules if they were all crowded together, it is of the same order of magnitude as the molar specific volume of the liquid phase of the substance. This modification of the ideal-gas equation of state gives $p(\bar{v} - b) = \bar{R}T$. Van der Waals further reasoned that the pressure of a real gas is less than that of an ideal gas at the same temperature and density because of the attractive forces between molecules, and that this reduction in pressure is proportional to the square of the density or to $(1/\bar{v})^2$. Thus the van der Waals equation of state in molar form is

van der Waals based his equation of state on physical reasoning.

$$p = \frac{\bar{R}T}{\bar{v} - b} - \frac{a}{\bar{v}^2} \tag{7·24}$$

or

The van der Waals equation of state

$$\left(p + \frac{a}{\bar{v}^2}\right)(\bar{v} - b) = \bar{R}T$$

Notice that the effect of each modification is greatest for states of high density or low $\bar{v}$.

The constants a and b can be evaluated from the critical pressure and critical temperature. This means of evaluation is based on the observation that on a pv plot the van der Waals equation for temperatures lower than the critical temperature has isotherms with maxima and minima, unlike the behavior of any real substance. As temperature increases, the $\bar{v}$ values at these maxima and minima approach each other. The isotherm on which they coincide has neither a maximum nor a minimum but an inflection point where, of course, $(\partial p/\partial \bar{v})_T = 0$. This is characteristic of the isotherm passing through the critical state of a substance, so the critical point properties can be related to the van der Waals constants as

The van der Waals constants can be related to the properties at the critical state.

$$\bar{v}_c = 3b \qquad p_c = \frac{a}{27b^2} \qquad T_c = \frac{8a}{27\bar{R}b}$$

Thus, once the critical pressure and temperature of a substance are known, a and b can be calculated. Values obtained from critical state properties may give inaccurate results for states far removed from the critical state, so sometimes values of a and b obtained from experimental data in various ranges of pressure and temperature are used. The values of the van der Waals constants given in Table A·1 are from critical state data. Care must be exercised to ensure that homogeneous units are used.

The van der Waals equation is a two-constant equation of state. Several other two-constant equations of state for gases have been published and represent the pvT behavior of real gases better than the ideal-gas equation of state. However, as you might expect, more complex equations are required for still greater accuracy over wide ranges of pressure and temperature.

The Redlich–Kwong and Redlich–Kwong–Soave Equations. In 1949, Redlich and Kwong presented a two-constant equation of state that is more accurate than the van der Waals equation over a wide range:

$$p = \frac{\overline{R}T}{(\overline{v} - b)} - \frac{a}{\overline{v}(\overline{v} + b)T^{0.5}} \tag{7·25}$$

Redlich–Kwong equation of state

(The a and b in the Redlich–Kwong equation are not the same as those in the van der Waals equation.) Values of the two empirical constants can be obtained from critical-state data as

$$a = 0.427\frac{\overline{R}^2T_c^{2.5}}{p_c} \quad \text{and} \quad b = 0.0866\frac{\overline{R}T_c}{p_c}$$

The numerical coefficients, 0.427 and 0.0866, in these expressions are dimensionless, so they can be used with any consistent set of units.

An improvement over the Redlich–Kwong equation of state called the *Redlich–Kwong–Soave* equation is a three-constant equation:

$$p = \frac{\overline{R}T}{(\overline{v} - b)} - \frac{a}{\overline{v}(\overline{v} + b)T^{0.5}}\left\{1 + m\left[1 - \left(\frac{T}{T_c}\right)^{0.5}\right]\right\}^2 \tag{7·26}$$

Redlich–Kwong–Soave equation of state

where m is given by

$$m = 0.48 + 1.574\omega - 0.176\omega^2$$

and ω, called the *acentric factor*, is a constant for each gas. Acentric factor values are given in Table A·1.

The Benedict–Webb–Rubin Equation of State. The Benedict–Webb–Rubin equation is an eight-constant equation that was developed for light hydrocarbons and mixtures of light hydrocarbons:

$$p = \frac{RT}{v} + \left(B_0RT - A_0 - \frac{C_0}{T^2}\right)\frac{1}{v^2} + (bRT - a)\frac{1}{v^3} + \frac{a\alpha}{v^6} + \frac{c}{v^3T^2}\left(1 + \frac{\gamma}{v^2}\right)e^{-\gamma/v^2} \tag{7·27}$$

Benedict–Webb–Rubin equation of state

The constants of this equation for several gases are given in Table A·2.

Virial Equations of State. It is often convenient to have an equation of state in the form

$$\frac{pv}{RT} = A_0 + A_1 p + A_2 p^2 + A_3 p^3 + \cdots \tag{7·28}$$

Virial equations of state

or

$$\frac{pv}{RT} = B_0 + \frac{B_1}{v} + \frac{B_2}{v^2} + \frac{B_3}{v^3} + \cdots$$

where the A's and B's are functions of temperature only. These equations are called **virial equations**, and the A's and B's are called *virial coefficients*. Several other forms of virial equations are also used. One advantage of virial equations of state is that it is relatively easy to determine the virial coefficients from experimental pvT data.

General Comments on Equations of State for Gases. At low densities, all equations of state for real gases should approach the ideal-gas equation of state, and the samples shown here all do so. Some of the equations of state for real gases apply to liquid states, as well, but with less accuracy even when constants established from liquid data are used.

Beware of spurious roots.

Many empirical equations of state have multiple roots. Usually, only one root is physically significant, and you must be careful to distinguish this root from the physically spurious ones. Estimates of the result based on the ideal-gas equation of state or available data for other states can help in evaluating solutions.

Some general statements regarding the relative accuracies of various equations can be made, but they should be followed with caution. A sample comparison of the results of some equations of state with the accurate and extensive data from the steam tables is shown in Fig. 7·10. Do not conclude, however, that the relative accuracies of these equations of state are the same for all substances.

Engineers must learn to balance accuracy and calculation simplicity. The software accompanying this book makes calculations using complex equations of state easy and rapid.

Making hand calculations of these complex equations of state is surely unappealing. In fact, if the calculation had to be done by hand, one would be tempted to sacrifice accuracy by using a simpler equation of state. Part of an engineer's skill is knowing how and when to make such trade-offs. Fortunately, we do not have to make complex calculations by hand. The software that accompanies this book includes models for the van der Waals, Redlich–Kwong, Redlich–Kwong–Soave, and Benedict–Webb–Rubin equations of state. In each model, the constants for several gases are included in online form, so that they can be used in an application with no copying from tables.

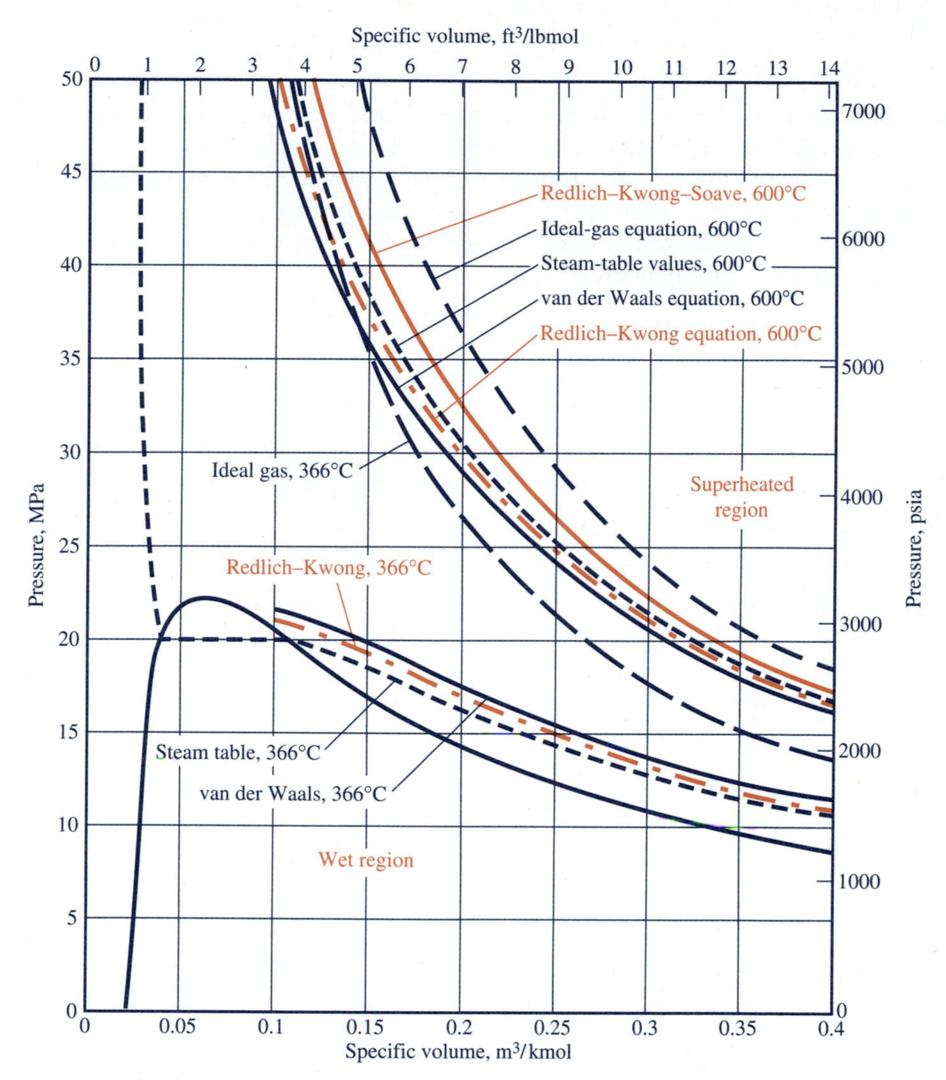

Figure 7·10 Comparison of four equations of state with steam-table data.

Example 7·3 Real-Gas pvT Equations of State

PROBLEM STATEMENT

Methane, CH_4, is held at 70.0 MPa, −38.8°C, in a tank having a volume of 2.20 m^3. Determine the mass by means of the (*a*) van der Waals, (*b*) Redlich–Kwong, (*c*) Redlich–Kwong–Soave, and (*d*) Benedict–Webb–Rubin equations of state.

SOLUTION

Analysis: Since we know the volume of the tank, we can obtain the mass of methane if we find the specific volume. None of the equations of state specified is readily solved explicitly for the specific volume, so we turn to a computer solution. (If we were forced to solve the problem without a computer, we would assume values for v and solve for corresponding values of p. After two or three such calculations, a plot of p versus v would guide us toward the v value that satisfies the equation for the specified p value.) If a guess value of v is needed, we use the value from the ideal-gas equation of state,

$$v = \frac{RT}{p} = \frac{0.5183(234.4)}{70{,}000} = 0.00174 \text{ m}^3/\text{kg}$$

Calculations: Using an equation solver, TK Solver, we enter the equation, $v = V/m$, and call each equation of state to the Rule sheet, as shown in the figure for the Redlich–Kwong–Soave equation. The variables appear on the Variable sheet where we enter all known values. The figure shows the Variable sheet after solving. The results for all four equations of state are tabulated:

```
════════════════════ RULE SHEET ════════════════════════════════
S Rule──────────────────────────────────────────────────────────
  ;──────────────── Example 7·3 ─────────────────────────────────
  m = V/v
  ;──────────────── Equation of state, Redlich-Kwong-Soave ───── EQS-RKS.TK
  call RKS (p,T;v)

════════════════════ VARIABLE SHEET ════════════════════════════
St Input──── Name─── Output─── Unit───── Comment───────────────────
                               ───────── * E7-3.TK ──── Example 7·3 ────
                m     626       kg        Mass of methane
   2.2          V               m^3       Volume of tank
                               ── Model:  * EQS-RKS ── Equation of state, RKS ──
   'CH4         gas                       * Dive (>) for list of input choices
                v     0.00352   m^3/kg    Specific volume of methane
   70000        p               kPa       Pressure of methane
   -38.8        T               C         Temperature of methane
```

Equation of State	v, m³/kg	m, kg
van der Waals	0.00352	625
Redlich–Kwong	0.00276	798
Redlich–Kwong–Soave	0.00351	627
Benedict–Webb–Rubin	0.00283	777

Use of software accompanying this book: Models for these four equations of state are included in the software accompanying this book. Each model has the equation of state already entered in a function that we can call to the Rule sheet as needed. For several gases, the constants are tabulated and can be called online as a function of the variable "gas." All variables, including the name of the gas, are listed on the Variable sheet. However, quantities such as the constants in the equations of state and the critical pressure and temperature that the user does not need to display because they are internal to the solution are not shown on the Variable sheet. (They can be displayed there if desired.) We simply enter known values in the appropriate places on the Variable sheet and enter the Solve command to obtain the solutions.

Comment: The purpose of this example is to show the kind of variation one should expect from different equations of state. In this instance, the experimental value of v is 0.00278 m^3/kg (Reference 7·3), so that *in this case* the Redlich–Kwong and Benedict–Webb–Rubin equations are the most accurate. They give results within 2 percent of the experimental value.

As mentioned earlier, these equations of state may have multiple roots. Comparing multiple solution values with the approximate value obtained from the ideal gas equation of state usually indicates which solution is the physically significant one.

7·6·2 The Hypothesis of Corresponding States and Generalized Compressibility Factors

Section 2·3·2 pointed out that the ideal-gas equation of state accurately models real gases only in certain ranges of pressure and temperature. For use outside these ranges, it can be modified to

$$pv = ZRT \tag{2·4}$$

Compressibility factor

where Z, the compressibility factor for the gas, is a function of p and T that can be presented in tables or charts like Fig. 2·26. A drawback of using compressibility factors is that for many gases compressibility factor data over wide ranges of pressure and temperature are unavailable.

It would be convenient if one compressibility-factor chart could be used for several gases. Two methods of doing this are used. One is by means of an approximation sometimes call the **hypothesis of corresponding states**: If any two gases have equal values for the ratio of pressure to critical pressure and equal values for the ratio of temperature to critical temperature, then the ratio of specific volume to critical specific volume is the same for the two gases. The ratios of pressure, temperature, and specific volume to the corresponding critical values are called **reduced coordinates** or **reduced properties**. Reduced pressure, reduced temperature, and reduced specific volume are defined by

The hypothesis of corresponding states

$$p_R \equiv \frac{p}{p_c} \qquad T_R \equiv \frac{T}{T_c} \qquad v_R \equiv \frac{v}{v_c}$$

Reduced coordinates or reduced properties

where the subscript R denotes a reduced property and the subscript c denotes a property at the critical state. In terms of these symbols, the hypothesis of corresponding states says that for all gases

$$v_R = f(p_R, T_R) \qquad (f)$$

and the function is the same for all gases.

If this approximation were accurate, a diagram of p_R versus T_R with v_R as a parameter plotted from data on any gas would apply to all others. Actually, the hypothesis of corresponding states is not sufficiently accurate over a wide range for much engineering work. To illustrate the inaccuracy, notice that the left-hand side of Eq. *f* can be written as

$$v_R = \frac{v}{v_c} = \frac{ZRTp_c}{pZ_cRT_c} = \frac{Z}{Z_c}\left(\frac{T_R}{p_R}\right)$$

so that

$$\frac{Z}{Z_c}\frac{T_R}{p_R} = f(p_R, T_R)$$

and

$$\frac{Z}{Z_c} = F(p_R, T_R)$$

Thus it follows that at the same values of p_R and T_R, all gases have the same value of Z/Z_c. However, at very low values of p_R, $Z = 1$ for all gases; so Z_c must be the same for all gases if the hypothesis stated by Eq. *f* holds. Experiments show that this is not the case. For example, Z_c is 0.230 for water, 0.275 for carbon dioxide, and 0.305 for hydrogen. Consequently, this approximation is inaccurate at low pressures. Further investigation shows that it is most useful in the vicinity of the critical state.

Another approach is based on the observation that at the same reduced pressure and reduced temperature all gases have approximately the same compressibility factor, except near the critical state. Thus

$$Z = f(p_R, T_R) \qquad (g)$$

where the function is the same for all gases. (This is also sometimes called the hypothesis of corresponding states, but Eqs. *f* and *g* are not the same.) Obviously, this does not hold for the critical state, where $p_R = 1$ and $T_R = 1$, because we have seen above that Z_c is not the same for all gases. For states well removed from the critical state, however, this **generalized compressibility factor** gives more accurate results than Eq. *f*.

Generalized compressibility factor

Generalized compressibility factor charts, with Z plotted against p_R with T_R as a parameter, are found in many books. In Fig. 7·11 we show a plot of Z as a function of p_R and T_R for CO_2 and O_2. You can see that the lines of Fig.

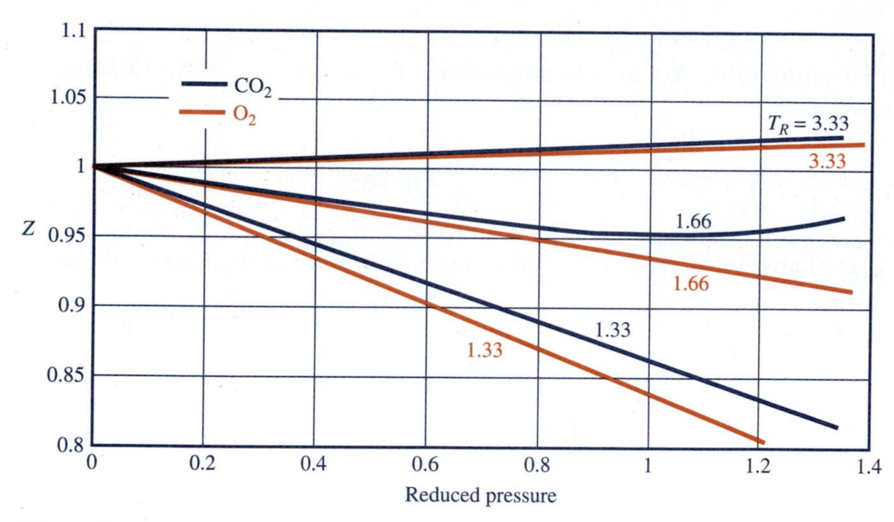

Figure 7·11 Compressibility factors for oxygen and carbon dioxide as functions of reduced coordinates. Compare with Fig. 2·26.

2·26*a* and *b* are brought much closer together by plotting against p_R and T_R, but still they do not coincide. This is a reminder that generalized compressibility factors are approximations.

An improvement based on the correlation of data for many gases over wide ranges of pressure and temperature (Reference 7·4), is to take

$$Z = Z^{(0)} + \omega Z^{(1)} \tag{7·29}$$

where $Z^{(0)}$ and $Z^{(1)}$ are functions of p_R and T_R, and ω, the **acentric factor**, depends on the gas. Tables and charts for $Z^{(0)}$ and $Z^{(1)}$ are in the appendix. Values of p_c, T_c, and ω for several gases are listed in Table A·1.

Acentric factor

Example 7·4 Use of Generalized Compressibility Factor

PROBLEM STATEMENT

Methane, CH_4, is held at 50.0 MPa, 0°C, in a tank having a volume of 2.20 m^3. By means of a generalized compressibility factor, determine the mass of methane in the tank.

SOLUTION

Since the solution method is specified, we will use

$$m = \frac{pV}{ZRT}$$

To determine the value of Z, we first go to Table A·1 to determine values for methane: $p_c = 5.117$ MPa, $T_c = 283$ K, and $\omega = 0.089$. Then we have

$$p_R = \frac{p}{p_c} = \frac{50.0}{5.117} = 9.77 \qquad T_R = \frac{T}{T_c} = \frac{273}{283} = 0.965$$

Entering Tables A·23·1 and A·23·2 or Charts A·24·1 and A·24·2 with these values, we interpolate to obtain

$$Z^{(0)} = 1.273 \qquad Z^{(1)} = -0.315$$

Then

$$Z = Z^{(0)} + \omega Z^{(1)} = 1.273 + 0.089(-0.315) = 1.25$$

$$m = \frac{pV}{ZRT} = \frac{50.0\ \text{MPa}\left(1000\ \dfrac{\text{kPa}}{\text{MPa}}\right)2.2\ \text{m}^3}{1.25\left(0.518\ \dfrac{\text{kJ}}{\text{kg·K}}\right)273\ \text{K}} = 624\ \text{kg}$$

Comment: The solution above involves double interpolation in the tables for $Z^{(0)}$ and $Z^{(1)}$. This is best done by a computer. The software accompanying this book includes a model for the Lee–Kesler Z values. This model uses Table A·23 and a double-interpolation procedure. It can be merged with a model that provides p_c, T_c, ω, and R for the gas and with another model in which the user enters the equations $m = pV/ZRT$ and $Z = Z^{(0)} + \omega Z^{(1)}$. We enter the known values for p, T, and V and solve.

7·6·3 Enthalpy and Entropy of Real Gases

The generalized compressibility factors can be used to approximate enthalpy and entropy changes.

If we have relationships among p, v, and T and specific heats for a pure simple compressible substance, we can calculate enthalpy and entropy changes from Eqs. 7·17, 7·14, and 7·15. In the absence of an equation of state (pvT equation), we can use generalized compressibility factors to approximate the pvT relationship. We now show how the generalized compressibility factors can be used to approximate enthalpy and entropy changes. Just as a compressibility factor is a modification of the simple ideal-gas pvT relationship, the enthalpy and entropy calculation methods we now develop for real gases are extensions of the simple Δh and Δs relationships for an ideal gas.

Consider a process 1–2 of a real gas as shown on the Ts and ph diagrams of Fig. 7·12. At state 1, $p_R > 1$. At state 2, $T_R < 1$ and p_R is less than 1 but not very low. Consequently, it is unlikely that the ideal-gas equation of state would apply to these states. For pvT relationships at each state, we might use the generalized compressibility factor as described in Section 7·6·2. How can we calculate Δh and Δs?

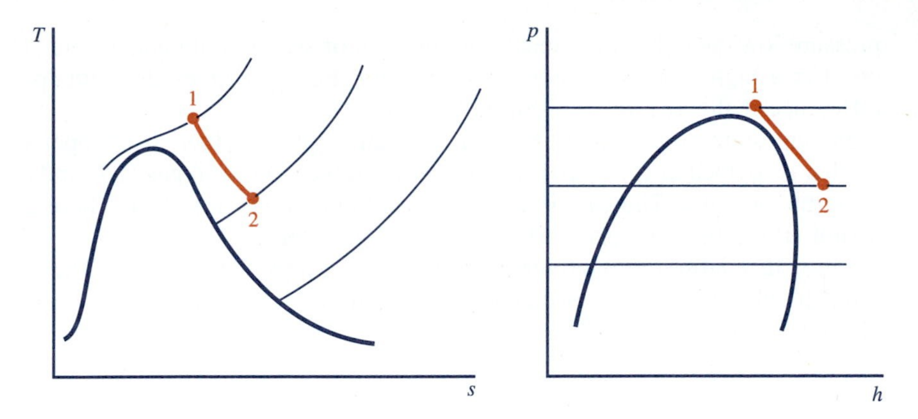

Figure 7·12 Process 1–2 of a real gas shown on *Ts* and *ph* diagrams.

For any gas, the change in enthalpy is given by

$$h_2 - h_1 = \int_1^2 c_p\, dT - \int_1^2 \left[T\left(\frac{\partial v}{\partial T}\right)_p - v \right] dp \qquad (7\cdot17)$$

The second term on the right-hand side can be evaluated from pvT data; the first one requires a knowledge of the c_pT relationship along the process path from state 1 to state 2. For many gases, the pvT data needed might be approximated by a generalized compressibility factor relationship using reduced coordinates, but data on c_pT relationships for gases at high pressures are scarce.

Since the change of any property between two states is the same for any process between them, a fruitful approach is to evaluate the enthalpy change by using the path 1–1′–2′–2 shown in Fig. 7·13. This path consists of a constant-temperature process from state 1 to a state 1′, a constant-pressure process from state 1′ to a state 2′, and a constant-temperature process from state 2′ to state 2. States 1′ and 2′ are at such a low pressure that the ideal-gas equation of state applies. Thus we evaluate the enthalpy change between states 1 and 2 as

$$h_2 - h_1 = (h_2 - h_{2'}) + (h_{2'} - h_{1'}) + (h_{1'} - h_1) \qquad (h)$$

Each of the three terms on the right-hand side can be evaluated by means of Eq. 7·17. Notice that for each of the three processes, one term on the right side of Eq. 7·17 is zero.

For process 1′–2′, the pressure is constant at a low value, so Eq. 7·17 gives

$$h_{2'} - h_{1'} = \int_{1'}^{2'} c_p\, dT$$

which can be evaluated from the ideal-gas c_pT relation. Processes 1–1′ and 2′–2 are both constant-temperature processes between a state (1′ or 2′) at a

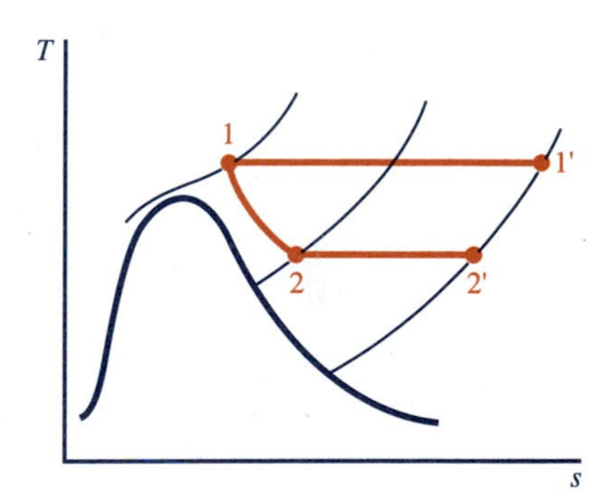

Figure 7·13 Another path (1–1′–2′–2) between states 1 and 2 for evaluating $h_2 - h_1$. States 1′ and 2′ are ideal-gas states.

pressure low enough for the ideal-gas equation of state to hold and a state (1 or 2) at a higher pressure. For either process, Eq. 7·17 gives the enthalpy difference if the pvT relationship is known.

For any state x of a real gas, for which the enthalpy is h_x, there is a property h_x^* that is defined as the enthalpy of the gas if it were an ideal gas at T_x and a very low pressure. Thus, h_x^* is a property of the real gas in state x. There is a similarly defined property u^* for any real-gas state.

Using this definition and referring to Fig. 7·13, notice that $h_1^* = h_{1'}$. Therefore, the change in enthalpy between two real-gas states can be written in general as

$$h_2 - h_1 = (h_2 - h_2^*) + (h_2^* - h_1^*) + (h_1^* - h_1) \tag{i}$$

At this point you may ask why the two forms of the expression for $h_2 - h_1$ are presented. Why not always use Eq. h based on a graphical explanation? The answer is that for enthalpy changes this would indeed be satisfactory; however, when we get to entropy changes, we will see that no simple graphical representation is possible, and we would like to have general expressions for Δh and for Δs that are similar in form. Equation i is in that form.

Enthalpy departure or enthalpy residual

The difference between the enthalpy of any state and the enthalpy of an ideal gas at the same temperature is known as the **enthalpy departure** *or* **enthalpy residual**.

For any state, the enthalpy departure ($\equiv h^* - h$) is given by Eq. 7·17 as

$$h^* - h = \int_p^{p_0} \left[v - T\left(\frac{\partial v}{\partial T}\right)_p \right] dp$$

where p_0 is the pressure of the ideal-gas state. In the absence of any other pvT relationship for the gas at hand, we use a compressibility factor so that in this equation we make the substitution $v = ZRT/p$ to obtain

$$h^* - h = -R \int_p^{p_0} \frac{T^2}{p} \left(\frac{\partial Z}{\partial T}\right)_p dp$$

If ZpT data are available for the gas, this expression can be evaluated graphically or numerically. If ZpT data are not available, we can use the approximation of a generalized compressibility factor or Zp_RT_R chart. Substituting $p = p_Rp_c$ and $T = T_RT_c$ into the equation results in

$$\frac{h^* - h}{RT_c} = \int_0^{p_R} T_R^2 \left(\frac{\partial Z}{\partial T_R}\right)_{p_R} d \ln p_R$$

The integral can be evaluated numerically or graphically from a table of generalized compressibility factors. This has been done on the basis of the Lee–Kesler tables and charts for $Z^{(0)}$ and $Z^{(1)}$ (Tables A·23·1 and A·23·2 and Charts A·24·1 and A·24·2). The resulting *generalized enthalpy departure tables and charts* (Tables A·23·3 and A·23·4 and Charts A·24·3 and A·24·4) give the components of the expression

$$\frac{h^* - h}{RT_c} = \left(\frac{h^* - h}{RT_c}\right)^{(0)} + \omega\left(\frac{h^* - h}{RT_c}\right)^{(1)}$$

Generalized enthalpy departure tables and charts provide the components to calculate the enthalpy change for real gases.

Remember that these tables and charts are based on generalized compressibility factors and correlation of data from many gases, so high accuracy is not to be expected. Calculations requiring great accuracy must be based on experimental data and data correlations for the specific gases under consideration.

Generalized entropy departure tables and charts can be developed in a similar manner, starting with the general equation for entropy change:

$$s_2 - s_1 = \int_1^2 c_p \frac{dT}{T} - \int_1^2 \left(\frac{\partial v}{\partial T}\right)_p dp \tag{7·15}$$

One difference between developing a generalized departure chart for entropy and developing one for enthalpy is that for an ideal gas, entropy is not a function of temperature only as enthalpy is. Furthermore, for an ideal gas $s \to \infty$ as $p \to 0$ at constant T. (Sketch a few constant-pressure lines on a Ts diagram to illustrate this.) The difficulty this introduces can be avoided by using some arbitrary low pressure, p_0, as a standard; then it is seen that the value of the arbitrary low pressure does not affect the results. Applying Eq. 7·15 to a constant-temperature process between the state of the real gas (at p, T) and a state (at p_0, T) at a low enough pressure for the gas to behave as an ideal gas gives

$$s_{p,T} - s^*_{p_0,T} = -\int_{p_0}^{p} \left(\frac{\partial v}{\partial T}\right)_p dp \tag{j}$$

where the asterisk indicates an ideal-gas property.

If the gas could exist as an ideal gas at state p, T, the change in entropy between the state at p_0, T and that at p, T would be

$$s^*_{p,T} - s^*_{p_0,T} = -\int_{p_0}^{p} \left(\frac{\partial v}{\partial T}\right)_p dp = -R\int_{p_0}^{p} \frac{dp}{p} \tag{k}$$

Subtracting Eq. j from Eq. k gives

$$s^*_{p,T} - s_{p,T} = -\int_{p_0}^{p} \left[\frac{R}{p} - \left(\frac{\partial v}{\partial T}\right)_p\right] dp$$

which is the difference in entropy between the hypothetical state of the gas as an ideal gas at p and T and the real gas state at the same pressure and temperature. This is the **entropy departure** or **entropy residual**. Again, as in the case of enthalpy departure, the entropy departure tables and charts give

Entropy departure or entropy residual

Generalized entropy departure tables and charts provide the components to calculate the entropy change for real gases.

the components of the expression

$$\frac{s^* - s}{R} = \left(\frac{s^* - s}{R}\right)^{(0)} + \omega\left(\frac{s^* - s}{R}\right)^{(1)}$$

The calculation of entropy change of a real gas between states 1 and 2 is carried out by

$$s_2 - s_1 = (s_2 - s_2^*) + (s_2^* - s_1^*) + (s_1^* - s_1) \qquad (l)$$

The first term on the right-hand side is the entropy difference between the real gas and the hypothetical ideal gas, both at p_2 and T_2. The second term is the entropy difference of the hypothetical ideal gas between states 1* and 2*. The third term is the entropy difference between the real gas and the hypothetical ideal gas, both at p_1 and T_1.

Notice that s_x^* is defined as the entropy of a gas at state x *if it were an ideal gas.* h_x^* was defined as the enthalpy of the real gas at T_x and a very low pressure, but since the enthalpy of an ideal gas is independent of pressure, the following would be an almost identical definition: h_x^* is the enthalpy of a gas at state x *if it were an ideal gas.* The difference between the definitions of enthalpy departure and of entropy departure is that for the enthalpy departure, the ideal-gas state can be shown on a property diagram for the real gas as in Fig. 7·13 because there is a region on the diagram where the real gas behaves as an ideal gas. For the entropy departure, no such hypothetical ideal-gas state at the pressure and temperature of state x can be shown on a real-gas property diagram. For example, no diagram similar to Fig. 7·13 can be created to accompany Eq. *l*.

High accuracy should not be expected for enthalpy and entropy changes calculated using the departure tables and charts.

Again, remember that the enthalpy and entropy departure tables and charts are based on generalized compressibility factors and correlation of data from many gases, so high accuracy is not to be expected.

Example 7·5 High-Pressure Expansion of Methane

PROBLEM STATEMENT

Methane expands through a well-insulated turbine at a rate of 0.087 kg/s from 10 MPa, 27°C, to 2.0 MPa, −45°C. Kinetic energy changes are negligible. Determine the power output and the change in entropy of the methane flowing through the turbine.

SOLUTION

Analysis: We first sketch a flow diagram and a Ts diagram. We will apply the first law to calculate the work per unit mass, so we need to find the change in enthalpy. At the specified conditions, methane is unlikely to

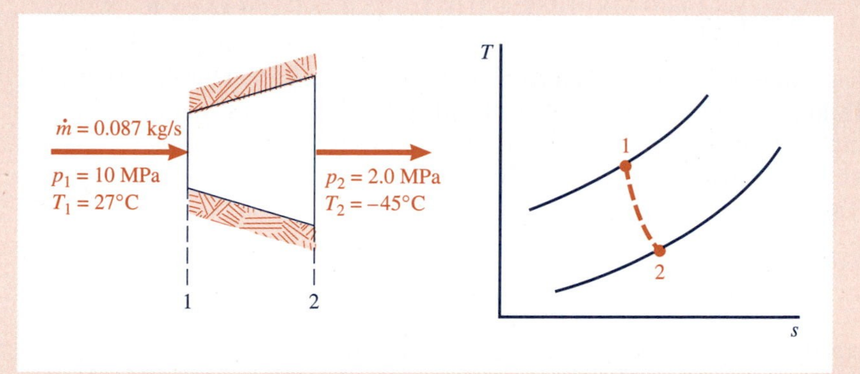

behave as an ideal gas. To check this assumption, we obtain values of p_c = 4.266 MPa, T_c = 370 K, and ω (from Table A·1) for methane and then calculate p_R and T_R for each state. Then we obtain values of $Z = Z^{(0)} + \omega Z^{(1)}$ by using Charts A·24·1 and A·24·2 or Tables A·23·1 and A·23·2 or the software model for Lee–Kesler Z values. The results are

$$p_{R1} = 2.17 \quad T_{R1} = 1.57 \quad Z_1^{(0)} = 0.855 \quad Z_1^{(1)} = 0.186 \quad Z_1 = 0.857$$

$$p_{R2} = 0.43 \quad T_{R2} = 1.2 \quad Z_2^{(0)} = 0.914 \quad Z_2^{(1)} = 0.021 \quad Z_2 = 0.914$$

The departures shown from the ideal-gas value of $Z = 1$ confirms that we cannot treat the propane as an ideal gas. Therefore, in the absence of propane property tables, we will use the generalized enthalpy and entropy charts in the appendix or the Lee–Kesler model that includes h and s departures in addition to Z values.

Calculations: We must calculate both the change in enthalpy (to use in the first law) and the change in entropy, so we write the expressions for these quantities as

$$h_2 - h_1 = (h_2 - h_2^*) + (h_2^* - h_1^*) + (h_1^* - h_1) \tag{a}$$

$$s_2 - s_1 = (s_2 - s_2^*) + (s_2^* - s_1^*) + (s_1^* - s_1) \tag{b}$$

To obtain the values of the enthalpy and entropy departures (the first and third terms on the right-hand side of these equations), we go first to either the tables or charts or the software model LKZHS and obtain values for the enthalpy and entropy departure components based on p_R and T_R values for each state to give

$$h_2 - h_2^* = -RT_c\left[\left(\frac{h_2^* - h_2}{RT_c}\right)^{(0)} + \omega\left(\frac{h_2^* - h_2}{RT_c}\right)^{(1)}\right] = -0.518(191)[0.330 + 0.011(0.17)] = -32.8 \text{ kJ/kg}$$

$$h_1^* - h_1 = RT_c\left[\left(\frac{h_1^* - h_1}{RT_c}\right)^{(0)} + \omega\left(\frac{h_1^* - h_1}{RT_c}\right)^{(1)}\right] = 0.518(191)[1.016 + 0.011(-0.06)] = 100.5 \text{ kJ/kg}$$

$$s_2 - s_2^* = -R\left[\left(\frac{s_2^* - s_2}{R}\right)^{(0)} + \omega\left(\frac{s_2^* - s_2}{R}\right)^{(1)}\right] = -0.518[0.192 + 0.011(0.16)] = -0.100 \text{ kJ/kg·K}$$

$$s_1^* - s_1 = R\left[\left(\frac{s_1^* - s_1}{R}\right)^{(0)} + \omega\left(\frac{s_1^* - s_1}{R}\right)^{(1)}\right] = 0.518[0.488 + 0.011(0.15)] = 0.254 \text{ kJ/kg·K}$$

The middle term on the right-hand side of Eq. *a* is the enthalpy difference for an ideal gas between 27°C and −45°C. We can obtain this from the software model IGPROP (Ideal Gas Properties) or we can use an average specific heat by calculating specific heats at the two end temperatures from the equation for specific heat given in Table A·1. The latter approach gives $c_{p1} = 2.226$ kJ/kg·K and $c_{p2} = 2.075$ kJ/kg·K for an average value of 2.15 kJ/kg·K. Then Eq. *a* becomes

$$\begin{aligned} h_2 - h_1 &= (h_2 - h_2^*) + c_{p,av}(T_2 - T_1) + (h_1^* - h_1) \\ &= -32.8 + 2.15(-45 - 27) + 100.5 = -87.1 \text{ kJ/kg} \end{aligned}$$

The first law for this adiabatic steady-flow process with no change in kinetic energy gives

$$w = h_1 - h_2 = 87.1 \text{ kJ/kg}$$

and the power output is

$$\dot{W} = \dot{m}w = 0.087(87.1) = 7.6 \text{ kW}$$

The middle term of Eq. *b* is the entropy change of an ideal gas between states 1 and 2, so from a $T\,ds$ equation we have

$$s_2^* - s_1^* = \int_{1*}^{2*} ds = \int_{1*}^{2*} \frac{dh - v\,dp}{T} = \int_{T_1}^{T_2} \frac{c_p\,dT}{T} - R \ln \frac{p_2}{p_1}$$

Again using the average specific heat, we have

$$\begin{aligned} s_2 - s_1 &= (s_2 - s_2^*) + c_{p,av} \ln \frac{T_2}{T_1} - R \ln \frac{p_2}{p_1} + (s_1^* - s_1) \\ &= -0.100 + 2.15 \ln \frac{228}{300} - 0.518 \ln \frac{2}{10} + 0.254 = 0.40 \text{ kJ/kg·K} \end{aligned}$$

Comment: In view of the limited accuracy of the generalized enthalpy and entropy departure tables and charts, it is well to present the final results with not more than two significant digits. More significant digits would imply an accuracy that is not to be expected.

7·7 Summary

This chapter has presented property relationships for pure, simple compressible substances. These relationships are based on both the first and second laws of thermodynamics. Among these relationships are the Maxwell equations, which relate entropy to the directly measurable properties pressure, temperature, and specific volume:

$$\left(\frac{\partial T}{\partial v}\right)_s = -\left(\frac{\partial p}{\partial s}\right)_v \tag{7·3}$$

$$\left(\frac{\partial T}{\partial p}\right)_s = \left(\frac{\partial v}{\partial s}\right)_p \tag{7·4}$$

$$\left(\frac{\partial p}{\partial T}\right)_v = \left(\frac{\partial s}{\partial v}\right)_T \tag{7·5}$$

$$\left(\frac{\partial v}{\partial T}\right)_p = -\left(\frac{\partial s}{\partial p}\right)_T \tag{7·6}$$

The Maxwell equations, as well as many other useful relationships, can be derived quickly from the four equations

$$du = T\,ds - p\,dv \tag{6·7}$$

$$dh = T\,ds + v\,dp \tag{6·7}$$

$$da = -p\,dv - s\,dT \tag{7·1}$$

$$dg = v\,dp - s\,dT \tag{7·2}$$

One conclusion we reach from a study of property relationships is that if we can determine the relationships within certain groups of properties, all other properties can then be obtained by differentiation alone, so that no additional data are required to evaluate constants of integration. These groups are called *characteristic functions*, and they are (u, v, s), (h, p, s), (a, v, T), and (g, p, T).

General expressions for entropy, internal energy, and enthalpy in terms of pvT and specific-heat data can be derived as

$$ds = c_v \frac{dT}{T} + \left(\frac{\partial p}{\partial T}\right)_v dv = c_p \frac{dT}{T} - \left(\frac{\partial v}{\partial T}\right)_p dp \tag{7·14, 7·15}$$

$$du = c_v\,dT + \left[T\left(\frac{\partial p}{\partial T}\right)_v - p\right] dv \tag{7·16}$$

$$dh = c_p\,dT - \left[T\left(\frac{\partial v}{\partial T}\right)_p - v\right] dp \tag{7·17}$$

General relations involving c_p and c_v are

$$c_p - c_v = T\left(\frac{\partial p}{\partial T}\right)_v \left(\frac{\partial v}{\partial T}\right)_p \tag{7·18}$$

$$c_p - c_v = -T\left(\frac{\partial p}{\partial v}\right)_T \left(\frac{\partial v}{\partial T}\right)_p^2 = \frac{Tv\beta^2}{\kappa_T} \tag{7·19}$$

$$k = \frac{c_p}{c_v} = \left(\frac{\partial p}{\partial v}\right)_s \left(\frac{\partial v}{\partial p}\right)_T = \frac{\kappa_T}{\kappa_s} \tag{7·20, 7·21}$$

where β is the coefficient of volume expansion and κ_T and κ_s are respectively the isothermal compressibility and the isentropic compressibility.

The Joule–Thomson coefficient,

$$\mu \equiv \left(\frac{\partial T}{\partial p}\right)_h$$

is useful in itself and also because it can be measured more accurately and more readily than certain other useful properties that are directly related to it. It can also be related to pvT and specific-heat data.

One of the uses of the Joule–Thomson coefficient is to make possible the measurement of thermodynamic temperatures by real thermometers, even though thermodynamic temperature scales are defined in terms of idealizations such as reversible engines.

Some of the many equations of state that have been proposed for gases have been presented in this chapter, and we have seen that from pvT equations plus specific-heat data we can calculate internal energy, enthalpy, entropy, and other quantities.

Compressibility factors for modifying the ideal-gas equation of state to fit real-gas behavior were introduced in Chapter 2. We noted that a drawback of compressibility factors is that we need ZpT data for each gas, and such data are not generally available. In the absence of such data, we can use generalized compressibility factors. These are based on the observation that at the same reduced pressure and reduced temperature, all gases have approximately the same compressibility factor. Reduced pressure, p_R, and reduced temperature, T_R, are defined as

$$p_R \equiv \frac{p}{p_c} \qquad T_R \equiv \frac{T}{T_c}$$

where the subscript R denotes a reduced property and the subscript c denotes a property at the critical state.

Tables and charts of generalized compressibility factors as functions of p_R and T_R have been developed by correlating data on many gases. We present those developed by Lee and Kesler where

$$Z = Z^{(0)} + \omega Z^{(1)} \tag{7·29}$$

Separate tables and charts are published for $Z^{(0)}$ and $Z^{(1)}$, and ω is the acentric factor, which is a characteristic of each gas.

For real gases, enthalpy and entropy differences can be calculated approximately by means of generalized *departure tables or charts* that give the differences between properties of the real gas and properties of an ideal gas under similar conditions.

References

7·1 Daubert, Thomas E., *Chemical Engineering Thermodynamics*, McGraw-Hill, New York, 1985. (Chapter 2 on pvT properties of fluids is brief and informative, providing both background and some extensions of the material presented in this book.)

7·2 Haar, Lester, John S. Gallagher, and George S. Kell, *NBS/NRC Steam Tables*, Hemisphere, New York, 1984. (The appendix presents an equation for a in terms of ρ and T. In other words, a *characteristic function* $f(a, \rho, T)$ or $f(a, v, T)$ is given, and then it is shown that other properties are determined from this equation by differentiation alone.)

7·3 Vennix, Alan J., Thomas W. Leland, Jr., and Riki Kobayashi, "Low-Temperature Volumetric Properties of Methane," *Journal of Chemical and Engineering Data*, Vol. 15, No. 2, 1970, pp. 238–243. (This is the source of the experimental values mentioned in Example 7·3.)

7·4 Lee, Byung Ik, and Michael G. Kesler, "A Generalized Thermodynamic Correlation Based on Three-Parameter Corresponding States," *AIChE Journal*, Vol. 21, No. 3, May 1975, pp. 510–527.

Problems

7·1 For a single phase of a pure substance, sketch lines of constant entropy and constant temperature on pv coordinates and prove that the relative slopes you show are correct.

7·2 Determine the numerical value of the slope of a constant-pressure line on an hs diagram for air at 2 atm, 35°C (95°F). If you were you to find this slope for liquid water at the same pressure and temperature, would it be greater than, less than, or the same as that for air? Explain your reasoning.

7·3 Determine the value of $(\partial s/\partial v)_T$ for ethene at 150 kPa, 20°C.

7·4 Determine $(\partial p/\partial T)_s$ at 100 kPa, 200°C, (*a*) for nitrogen and (*b*) approximately for refrigerant R134a.

7·5 Determine the value of $(\partial s/\partial v)_T$ for air at 10 psia, 40 F.

7·6 Determine $(\partial p/\partial T)_s$ at 11 psia, 480 F, (*a*) for air and (*b*) approximately for steam.

7·7 Show from the Maxwell equations that for an ideal gas $(\partial u/\partial v)_T = 0$.

7·8 Show that the latent heat of vaporization may be expressed as

$$h_{fg} = T\int_{v_f}^{v_g}\left(\frac{\partial p}{\partial T}\right)_v dv$$

7·9 Assuming that a vapor in equilibrium with a liquid at a certain temperature and pressure behaves as an ideal gas, develop the following relation on the basis of the Clapeyron equation:

$$\frac{d(\ln p)}{dT} = \frac{h_{fg}}{RT^2}$$

7·10 Determine the change in atmospheric pressure with altitude near sea level. Using this value and values of T_{sat}, h_{fg}, and v_{fg}, for water at a single pressure, determine the change in the water boiling temperature with altitude near sea level.

7·11 From the data below on superheated ammonia, determine for ammonia at 600 kPa, 20°C, (*a*) c_p and (*b*) the value of Y in the following approximate expression for enthalpy change: $\Delta h = c_p\,\Delta T + Y\,\Delta p$.

Pressure, kPa		*Temperature*, °C 15	20	25
500	h	1642.1	1654.6	1667.0 kJ/kg
	v	0.2633	0.2690	0.2747 m³/kg
600	h	1635.2	1647.9	1661.0
	v	0.2163	0.2212	0.2261
700	h	1628.1	1641.2	1654.2
	v	0.1828	0.1871	0.1914

7·12 From the data given below compute the value of h_{fg} for steam at 0.70 MPa, and check the results with the steam-table value.

p Saturation, MPa	*T Saturation*, °C	*Specific Volume*, m³/kg $v_f \times 10^3$	$v_g \times 10^3$
0.60	158.85	1.1006	315.7
0.70	164.97	1.1080	272.8
0.80	170.43	1.1148	240.4

7·13 From the following data determine the value for the latent heat of vaporization of refrigerant R134a at 240 K:

T Saturation, K	*p Saturation*, kPa	*Specific Volume*, m³/kg $v_f \times 10^3$	$v_g \times 10^3$
250	115.53	0.731	167.972
240	72.40	0.718	260.807
230	43.23	0.705	422.982

7·14 From the data of Fig. 2·4 and the compressed-liquid and saturated-solid data of Table A·6, estimate the latent heat of fusion of water at 140 MPa. State clearly any assumptions or approximations that you make.

7·15 By means of the Clapeyron equation and the data of Table A·6·4, determine the saturation pressure for water vapor at −42°C.

7·16 Compute the amount that the melting-point temperature of ice is lowered for an increase in pressure of 1 atm. The specific volumes of ice and water at 0°C are 0.001091 and 0.001000 m³/kg, respectively.

7·17 Ice and water at 0.6109 kPa and 0.01°C have the following properties:

Phase	*Enthalpy*, kJ/kg	*Specific Volume*, m³/kg
Liquid	0.02	0.001000
Solid	−333.41	0.001091

Assuming that the enthalpy and specific volume are independent of the pressure, compute the melting temperature or equilibrium temperature at 1000 atm.

7·18 From the following data determine the value for the latent heat of vaporization of refrigerant R134a at −20 F.

T Saturation, F	*p Saturation*, psia	*Specific Volume*, cu ft/lbm v_f	v_g
−10	16.62	0.0117	2.712
−20	12.89	0.0116	3.447
−30	9.85	0.0115	4.437

7·19 By means of the Clapeyron equation and the data of Table A·8·1 or A·8·2, determine the saturation pressure for ammonia at −55 F.

7·20 An approximate expression for the variation of saturation pressure with temperature is $p = Ae^{-b/T}$, where A and b are constants. Show that this expression can be derived from the Clapeyron equation by making certain assumptions (such as the independence of h_{fg} and pressure).

7·21 The velocity of sound c in a medium is given by

$$c = \sqrt{\left(\frac{\partial p}{\partial \rho}\right)_s}$$

Find an expression for the velocity of sound in terms of such quantities as p, u, T, R, k, for (*a*) an ideal gas and (*b*) an incompressible liquid.

7·22 For an ideal gas show that the coefficient of volume expansion is equal to the reciprocal of the absolute temperature.

7·23 Prove that for an ideal gas the isothermal compressibility, $-\dfrac{1}{v}(\partial v/\partial p)_T$, is equal to the inverse of the pressure of the gas.

7·24 Derive an expression for the change in enthalpy of a gas that follows the equation of state $p(v - b) = RT$.

7·25 From Eq. 7·18 determine the value of $(c_p - c_v)$ for an ideal gas.

7·26 A certain gas follows the equation of state $p(v - b) = RT$. Show that for a reversible adiabatic process of this gas

$$T(v - b)^{R/c_v} = \text{constant}$$

7·27 Derive the following expression from fundamental relations:

$$\left(\frac{\partial u}{\partial v}\right)_T = T\left(\frac{\partial p}{\partial T}\right)_v - p$$

7·28 Develop the following relation for the Helmholtz function, a:

$$a = u + T\left(\frac{\partial a}{\partial T}\right)_v$$

7·29 Derive the following from fundamental relations:

$$\left(\frac{\partial h}{\partial p}\right)_T = v - T\left(\frac{\partial v}{\partial T}\right)_p$$

7·30 Prove the following equality:

$$\left(\frac{\partial c_v}{\partial v}\right)_T = T\left(\frac{\partial^2 p}{\partial T^2}\right)_v$$

7·31 Derive the relation

$$\left(\frac{\partial^2 a}{\partial T^2}\right)_v = \frac{c_v}{T}$$

7·32 A certain gas follows the equation of state $p(\bar{v} - b) = \bar{R}T$. Determine the coefficient of volume expansion at 27°C for this gas, assuming that under the conditions stated, v equals 1500 cm^3/mol and b equals 20 cm^3/mol.

7·33 Determine from steam-table data the value of β for water at (*a*) 70 kPa, 150°C, (*b*) 20.5 MPa, 400°C, and (*c*) 20.5 MPa, 150°C.

7·34 Determine the Joule–Thomson coefficient of refrigerant R134a at (*a*) 0.04 MPa, −20°C, and (*b*) 1.6 MPa, 90°C.

7·35 Determine the Joule–Thomson coefficient of water at (*a*) 70 kPa, 150°C, and (*b*) 20.5 MPa, 400°C.

7·36 Determine the Joule–Thomson coefficient of water at (*a*) 10 psia, 300 F, and (*b*) 3000 psia, 750 F.

7·37 Near the end of Section 2·1·1, it is said that the state of a pure substance is *usually* specified by the values of two independent intensive properties. Explain how exceptions to this statement are shown by (*a*) Fig. 7·8 and (*b*) Chart A·7·3.

7·38 In Section 7·6 it is stated that the constant b in the van der Waals equation of state is of the same order of magnitude as the specific volume of the liquid phase of a substance. For any four of the substances of Table A·2, report on the validity of this statement.

7·39 Develop the following form of the van der Waals equation in terms of the reduced coordinates:

$$\left(p_R + \frac{3}{v_R^2}\right)\left(v_R - \frac{1}{3}\right) = \frac{8}{3}T_R$$

7·40 By means of the van der Waals equation of state, plot for nitrogen on pv coordinates two isotherms: one for 0°C and one for 100°C, both in a pressure range extending to 300 atm. At 100, 200, and 300 atm show the points that would be obtained by the ideal-gas equation of state and by means of compressibility factors.

7·41 Derive an expression in terms of the van der Waals constants, other properties of the gas, and the pressure limits, for the work of a reversible isothermal expansion of a van der Waals gas for (*a*) a closed system and (*b*) a steady-flow system with no changes in kinetic or potential energy.

7·42 An expression is needed for the work of a reversible isothermal expansion of a Redlich–Kwong gas in terms of the Redlich–Kwong constants, other properties of the gas, and the pressure limits of the process. Derive this expression for (*a*) a closed system and (*b*) a steady-flow system with no changes in kinetic or potential energy.

7·43 Determine $(\partial c_p/\partial p)_T$ for a van der Waals gas.

7·44 Establish a relation for the change in enthalpy of a gas that follows the van der Waals equation of state.

7·45 Determine an expression for the change in entropy of a van der Waals gas.

7·46 Determine the Joule–Thomson coefficient of a van der Waals gas in terms of a, b, $\overline{R}$, T, $\overline{v}$, and c_p.

7·47 Prove that for a van der Waals gas the inversion temperature is given by

$$T = \frac{2a}{b\overline{R}}\left(1 - \frac{b}{\overline{v}}\right)^2$$

7·48 Show that the van der Waals equation of state gives a value for the compressibility factor at the critical state of $Z_c = \frac{3}{8}$. (You can do this by noting that on a pv diagram a constant-temperature line has a zero slope at the critical point.) Compare this with actual values of Z_c for some real gases.

7·49 Compute the specific volume of steam at 1.0 MPa, 400°C, by means of (*a*) the ideal-gas equation of state, (*b*) a compressibility factor, (*c*) the Redlich–Kwong equation of state. Compare these results to the steam-table value and give the percent error for each method.

7·50 Compute the specific volume of propane at 4.25 MPa, 244°C, by means of (*a*) a generalized compressibility factor, (*b*) the Redlich–Kwong equation.

7·51 Compute the specific volume of nitrogen at 500 atm, 200°C, by means of (*a*) the ideal-gas equation of state, (*b*) the van der Waals equation, (*c*) the Redlich–Kwong equation, and (*d*) reduced coordinates.

7·52 A tank with a total volume of 3.14 m^3 contains 3 moles of sulfur dioxide at 65°C. Compute the pressure by means of (*a*) the ideal-gas equation and (*b*) the van der Waals equation.

7·53 Using the Redlich–Kwong–Soave equation, compute the pressure of a quantity of air compressed to a volume of 0.0002 m^3 at a temperature of 5°C. Before compression, the air occupied a volume of 0.03 m^3 at 20°C and 101.3 kPa.

7·54 Compute the specific volume of steam at 30 MPa, 500°C, using the generalized compressibility factor chart and compare it with the corresponding steam-table value.

7·55 Determine the density of helium at 3.0 MPa, −220°C.

7·56 Compute the specific volume of nitrogen at 1000 atm, 400 F, by means of (*a*) the ideal-gas equation of state, (*b*) the van der Waals equation, (*c*) the Redlich–Kwong equation, (*d*) the Redlich–Kwong–Soave equation, and (*e*) reduced coordinates.

7·57 Compute the pressure of neon for a temperature of 100 F and a specific volume of 0.15 cu ft/lbm by means of the Redlich–Kwong–Soave equation.

7·58 Using the Redlich–Kwong–Soave equation, compute the pressure of a quantity of air compressed to a volume of 0.008 cu ft at a temperature of 40 F. Before compression, the air occupied a volume of 1 cu ft at 70 F and 14.7 psia.

7·59 Compute the compressibility factor for steam at 2000 psia, 900 F.

7·60 A closed, rigid, two-gallon tank contains nitrogen at 3000 psia, 392 F. Calculate the nitrogen pressure that results when the gas in the tank is cooled to 32 F.

7·61 Determine the density of carbon monoxide at 1500 psia, −300 F.

7·62 Determine the power required to compress ethylene steadily, reversibly, and isothermally at a rate of 0.31 kg/s from 10 MPa, 10°C, to 40 MPa. Determine also the rate of heat removal and the size of the discharge line if the mean velocity at outlet is not to exceed 20 m/s.

7·63 Oxygen in a closed system expands reversibly and adiabatically from 50 MPa, 100°C, to 15 MPa. Calculate the work done per kilogram of oxygen.

7·64 Air at 30 MPa is to be cooled at constant pressure at a rate of 20 g/s from 17°C to −5°C by flowing over the evaporator coils of a refrigerator. Refrigerant R134a is supplied to the evaporator as saturated liquid at −10°C. If the temperature of the refrigerant exiting the evaporator is not to exceed 10°C, find the minimum flow rate of refrigerant needed.

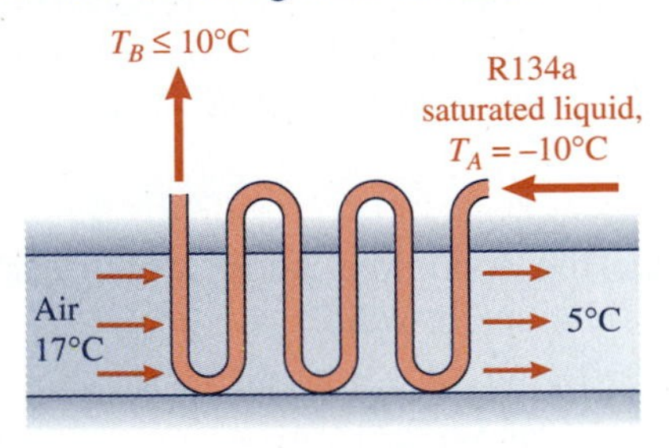

Problem 7·64

7·65 Air enters a compressor at 3.0 MPa, 40°C, with a velocity of 100 m/s through a cross-sectional area of 4.2 cm². The air is compressed steadily and adiabatically to 9.0 MPa. The discharge velocity is very low. The power input is 30 percent greater than that required for reversible adiabatic compression. Determine the power input and the discharge temperature.

7·66 Air in a vessel of 0.50 m³ volume is initially at 25 MPa, 20°C. Heat is added to raise the temperature of the air in the vessel to 120°C while air is bled from the vessel through a regulating valve to hold the pressure constant. Determine the heat transfer.

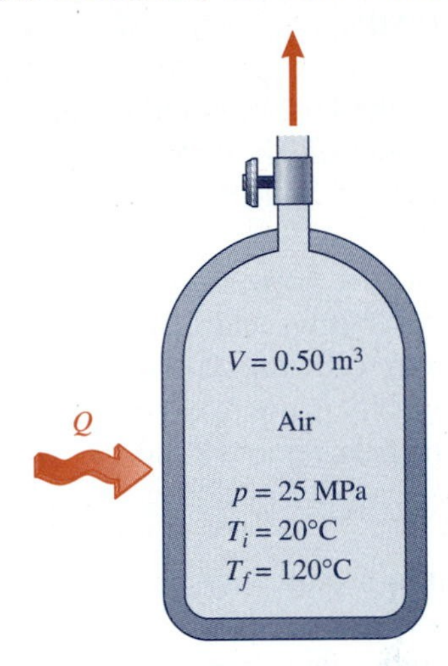

Problem 7·66

7·67 Carbon dioxide in a vessel of 0.50 m³ volume is initially at 50 MPa, 40°C. Gas is bled from the vessel to hold the pressure constant while heat is added to raise the temperature of the gas remaining in the vessel to 130°C. Determine the heat transfer.

7·68 An insulated vessel with a volume of 0.15 m³ initially contains air at atmospheric pressure, 20°C. It is connected to a line through which air at 30 MPa,

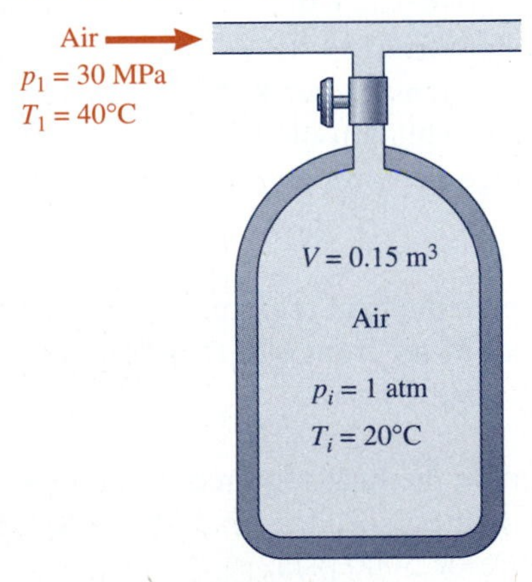

Problem 7·68

40°C, flows. If the filling process is adiabatic, determine the final temperature of the air in the vessel.

7·69 Compute the heat required to raise the temperature of 0.5 kg of air from 300°C to 500°C at a constant pressure of 70 MPa.

7·70 Compute the heat required to increase the temperature of 1 mole of air from 150°C to 250°C at a constant pressure of 25 MPa.

7·71 Determine the power required to compress ethane steadily, reversibly, and isothermally at a rate of 0.45 lbm/s from 100 atm, 200 F, to 400 atm. If the mean outlet velocity is not to exceed 50 fps, determine also the rate of heat removal and the size of the discharge line.

7·72 Oxygen in a closed system expands reversibly and adiabatically from 7500 psia, 240 F, to 1500 psia. Calculate the work done per pound of oxygen.

7·73 Oxygen enters a compressor at 800 psia, 140 F, with a velocity of 200 ft/s through a cross-sectional area of 2.12 sq. in. In a steady and adiabatic process, the oxygen is compressed to 3000 psia. Assume that the outlet velocity is negligible and that the power input is 30 percent greater than that required for reversible adiabatic compression. Determine the power input and the discharge temperature.

7·74 Nitrogen expands reversibly and isothermally in a closed system from 14 MPa, 35°C, to 1 MPa. Determine the heat transfer per kilogram by using (*a*) the van der Waals equation of state and the general equation for change in a property, (*b*) generalized property charts.

7·75 Compute the heat required to raise the temperature of 1 lbm of air from 600 to 1000 F at a constant pressure of 10,000 psia.

7·76 Compute the heat required to increase the temperature of 1 lbmol of air from 300 to 700 F at a constant pressure of 5000 psia.

7·77 The compressibility factor has been defined as the ratio of the specific volume of a real gas to the specific volume predicted by the ideal-gas equation of state. Does this agree with the definition presented in Section 7·6?

7·78 Consider a Carnot engine using as a working material (*a*) a steel wire, (*b*) a rubber band. Sketch and label corresponding points and processes on force–length and temperature–entropy diagrams for these engines. (*Suggestion*: Some simple experiments with a rubber band show that rubber and steel have quite different coefficients of thermal expansion. Further interesting conclusions result from applying the equations of Appendix II as we have done in this chapter to relate properties.)

7·79 Near the end of Section 2·2, it is stated that in room air the oxygen and nitrogen molecules occupy about 1/1000th of the gas volume. How does this value relate to the value of b, the ''co-volume'' for oxygen or nitrogen in the van der Waals equation?

7·80 Section 1·3 pointed out that properties may be classified as three types: (1) those that are directly observable, (2) those that can be defined by means of the laws of thermodynamics, and (3) those that are defined as combinations of other properties. Categorize each property in the following list of quantities accordingly: p, T, u, v, h, s, Q, R, A, U, W, m, S, c_p, V, G.

7·81 State the conditions under which each of the following relations is valid:

(*a*) $\text{work} = \int p\, dV$
(*b*) $h_2 - h_1 = c_p(T_2 - T_1)$
(*c*) $q_{in} + \text{work}_{in} = \Delta h + \Delta ke$
(*d*) $c_p - c_v = R$
(*e*) $\rho_1 A_1 V_1 = \rho_2 A_2 V_2$
(*f*) $\text{work}_{in} = \int v\, dp + \Delta ke + \Delta pe$
(*g*) $du = c_v\, dT$
(*h*) $h = u + pv$
(*i*) $Q = U_2 - U_1 + \text{work}$
(*j*) $pv^k = \text{constant}$

7·82 State the conditions under which each of the following relations is valid:
(*a*) $du = T\,ds - p\,dv$
(*b*) $\left(\frac{\partial h}{\partial s}\right)_p = T$
(*c*) $\Delta S_{\text{system}} = 0$
(*d*) $T\,ds = dh - v\,dp$
(*e*) $\Delta S_{\text{isolated system}} = 0$
(*f*) $\Delta S = \int \frac{\delta Q}{T}$
(*g*) $\Delta S_{\text{universe}} = 0$
(*h*) $\left(\frac{\partial T}{\partial s}\right)_v = \frac{T}{c_v}$
(*i*) $\frac{Q_1}{Q_2} = \frac{T_1}{T_2}$
(*j*) $\oint dx = 0$

7·83 State in not more than two pages which parts of Chapters 1 through 7 you believe you have a sound understanding of and which parts you are weakest in, giving as well as you can the reasons why any parts have caused difficulty.

7·84 Show from the Maxwell equations that since water at 1 atm has a maximum density at around 4°C, entropy is independent of pressure at this temperature and that consequently, constant-pressure lines in the liquid region of a *Ts* diagram in the vicinity of 1 atm coincide with each other and cross the saturation line around 4°C.

7·85 The table shows specific heat at constant volume versus *T* data for a solid element. Determine the Debye temperature for this element.

T, K	$\bar{c}_v$, kJ/(kmol·K)
50	11.4
100	20.4
150	22.7
200	23.6
250	24.0
300	24.3
350	24.5

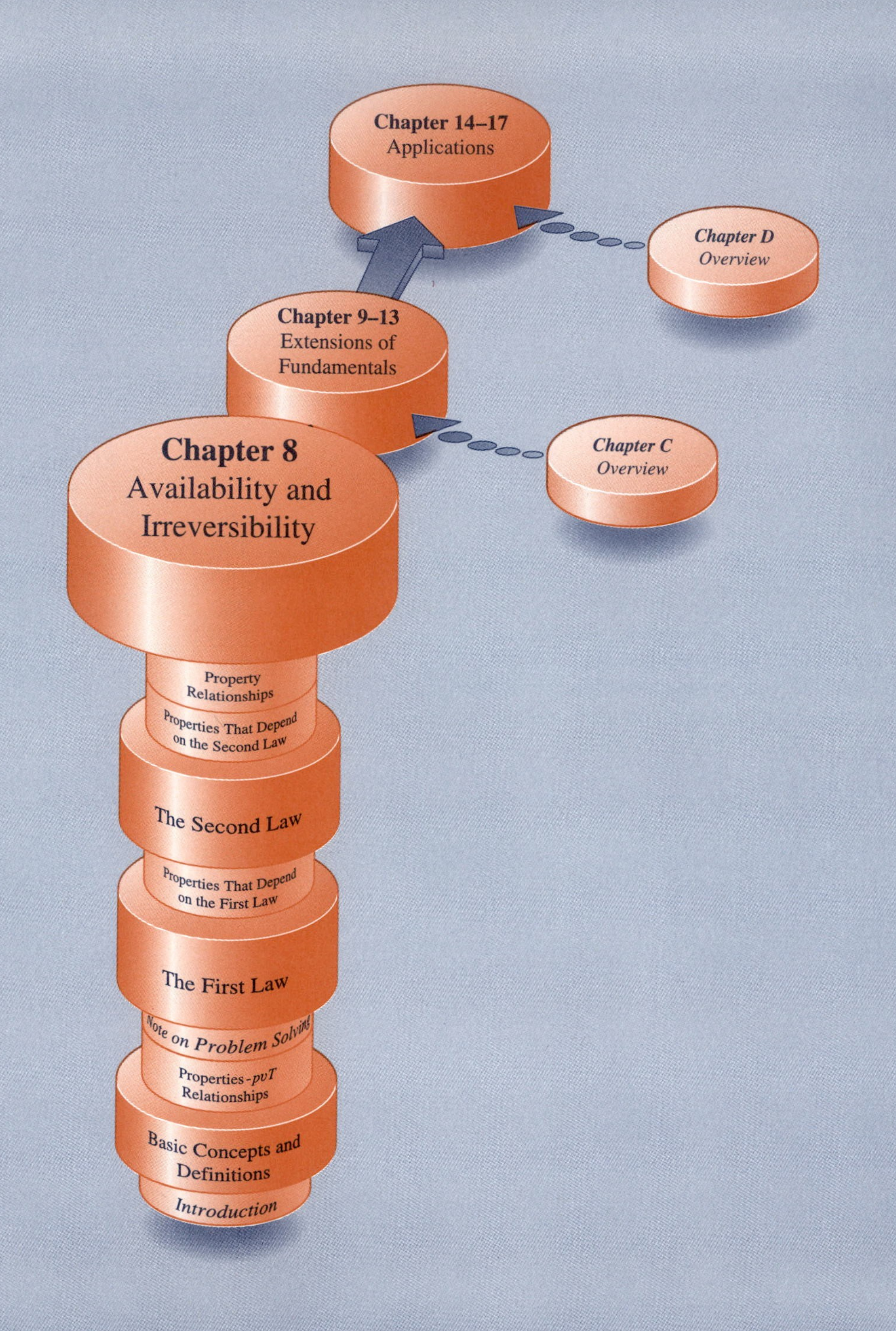

Chapter 14–17
Applications
Chapter D
Overview
Chapter 9–13
Extensions of
Fundamentals
Chapter C
Overview
Chapter 8
Availability and
Irreversibility
Property
Relationships
Properties That Depend
on the Second Law
The Second Law
Properties That Depend
on the First Law
The First Law
Note on Problem Solving
Properties-pvT
Relationships
Basic Concepts and
Definitions
Introduction

Chapter 8

Availability and Irreversibility

Even if a conclusion is supported by reasoning from basic principles and is widely accepted, if it does not agree with the intuition you have developed, you should investigate to see why it does not. Each time you resolve such a disagreement, you will most likely enhance your understanding.

From your study of the first law and properties of gases, you know that 1 kilogram of air at 100 kPa, 40°C, contains the *same amount of stored energy* as 1 kilogram of air at 200 kPa, and the same temperature. Still, you may be slightly uncomfortable with this conclusion. It appears that the air at the higher pressure can do more work, even if it has the same stored energy.

Consequently, the higher-pressure air would be more valuable. Let us see if we can quantify this difference which lies in the ability of a system to do work.

The question, "How much work can a system do?" is a worthwhile question. A more fruitful question is, "How much work can a system do as it changes from one specified state to another?" This chapter sharpens these questions further and answers them by introducing quantities such as *availability* and *available* and *unavailable energy*. It also shows how to quantify the effect of irreversible processes by use of the quantity *irreversibility*.

To reach the valuable simple equations for availability and irreversibility we must first derive an expression for *maximum work*. Although in engineering practice we seldom need to calculate numerical values of maximum work, it serves as the foundation for defining availability which we frequently need to calculate. The powerful results make worthwhile the following few pages deriving expressions for maximum work.

8·1 Maximum Work

As we will see later in this chapter, calculations of availability, available energy, and irreversibility all involve an *atmosphere*, so we first define this term. We have defined a *system* as a region in space within prescribed boundaries. The *surroundings* comprise everything outside the system boundaries. For almost all thermodynamic systems one part of the surroundings is an

Atmosphere

atmosphere of uniform pressure and temperature. This atmosphere is so large in comparison with the system that its pressure and temperature are not influenced by any process of the system; but the atmosphere does influence the behavior of the system. In Fig. 8·1, for example, the system and surroundings are defined as

System:	The gas in the cylinder
Surroundings:	The coil spring, the piston, the cylinder, the energy reservoir at T_R, and the atmosphere

The system exchanges heat with both the energy reservoir at T_R and with the atmosphere. Whenever the system changes volume, it exchanges work with both the spring and the atmosphere, since the force of the piston on the system comprises both the spring force and the force resulting from atmospheric pressure, p_0. Therefore, the total work done by the system is equal to that delivered to parts of the surroundings other than the atmosphere plus that done on the atmosphere.

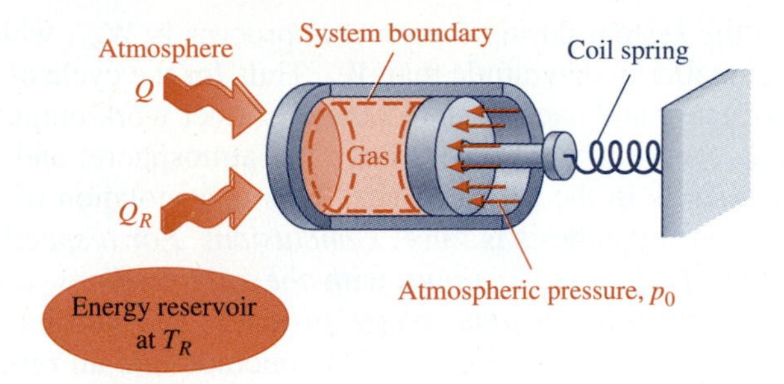

Figure 8·1 System and surroundings including the atmosphere.

8·1·1 Maximum Work for a System Exchanging Heat with Only the Atmosphere

Let us determine the maximum amount of work that can be done by a system that goes from one specified state to another while exchanging heat with only the atmosphere. To be general, consider an open system, as shown in Fig. 8·2, that for some infinitesimal change has a mass δm_1 entering and a mass δm_2 leaving. (The special cases of a closed system and of a steady-flow system can be developed later from this general approach.) The first law applied to this system for an infinitesimal change is

$$\delta Q - \delta W + (e_1 + p_1 v_1)\,\delta m_1 - (e_2 + p_2 v_2)\,\delta m_2 = dE \qquad (3\cdot 13)$$

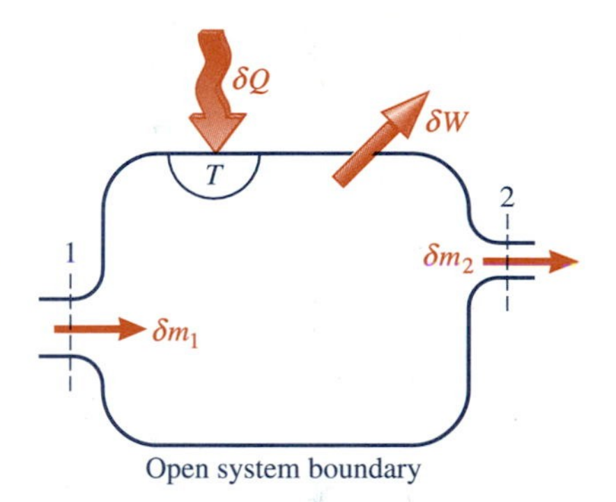

Figure 8·2 Open system that exchanges heat with only the atmosphere.

For specified end states and specified flows into and from the system, the first law can give us *only the difference* ($Q - W$), and many different values of Q and W are possible. We must use the second law to answer the question: For specified end states, specified flows into and from the system, and heat exchange with only the atmosphere, what is the maximum amount of work attainable? We will permit the use of auxiliary devices in producing this maximum work only if they operate cyclically so that they experience no permanent change.

We find as we study this problem that the maximum work can be done by means of an externally reversible process, as shown by the following reasoning: Consider a system as shown in Fig. 8·2 that exchanges heat with only the atmosphere, has specified amounts of mass in specified states crossing its boundaries, and passes from an initial state i to a final state f. *Suppose* that under these conditions some process A can occur to produce more work output W_A than some reversible process R between the same end states. (This is our hypothesis.) Let this process A occur, Fig. 8·3a. Then return the system to state i by the reverse of the reversible process R, Fig. 8·3b. During this reverse process, parts of the surroundings that exchanged mass with the system during process A are also returned to their initial states. The amount of

Maximum work is produced only by reversible processes.

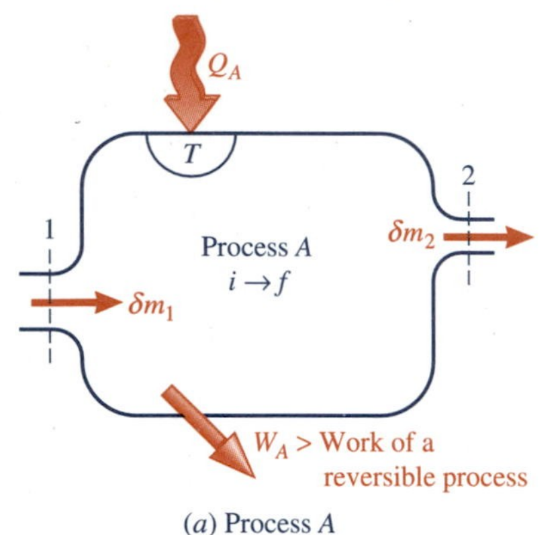

(*a*) Process *A*

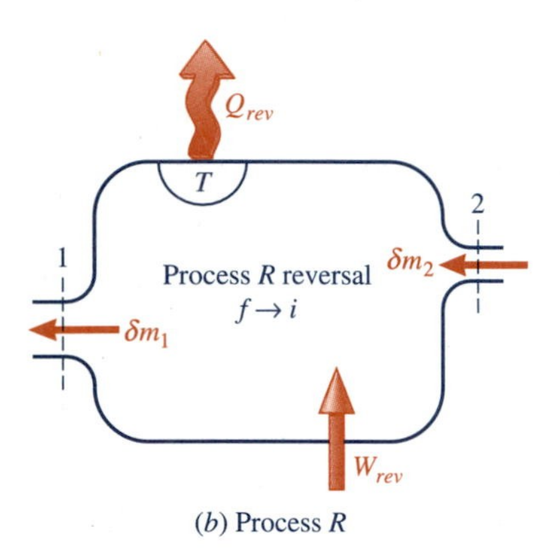

(*b*) Process *R*

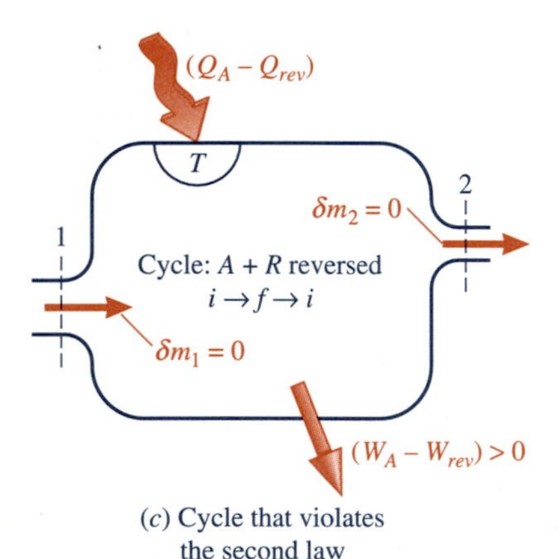

(*c*) Cycle that violates the second law

Figure 8·3 Proof that maximum work results from reversible processes.

work done on the system during the reverse process is W_{rev}, which (by our hypothesis) is smaller in magnitude than W_A. Thus for the cycle of the system from state i to state f and back to state i there is a net work output of $(W_A - W_{rev})$, heat has been exchanged with only the atmosphere, and there have been no other effects in the surroundings. This is a violation of the second law. Therefore, our hypothesis is false. *Conclusion*: *For a specified change in state and specified mass exchange with the surroundings, a system exchanging heat with only the atmosphere produces the maximum work by means of a reversible process.* We can also conclude that all reversible processes conducted under the same conditions result in the same work output.

For the system shown in Fig. 8·2, let T be the temperature of that part of the system where heat enters or leaves. (Remember that for an open system in general, T may vary from point to point throughout the system.) T_0 is the temperature of the atmosphere, also called the *ambient* temperature. Heat cannot be transferred reversibly directly between the system and the atmosphere because in general there is a finite difference between T and T_0, but heat can be transferred reversibly by means of a reversible engine (such as a Carnot engine) interposed between the system and the atmosphere. This arrangement is shown in Fig. 8·4. Let δQ be the heat added to the system. The heat entering the engine from the atmosphere is δQ_0. For the reversible engine

$$\frac{\delta Q}{T} = \frac{\delta Q_0}{T_0}$$

and

$$\delta W_{eng} = \delta Q_0 - \delta Q = \left(\frac{T_0}{T} - 1\right) \delta Q$$

(Notice that $\delta W_{eng} \geq 0$ under all conditions. When $T < T_0$, heat enters the system so that δQ is positive. When $T > T_0$, heat leaves the system so that δQ is negative.) The maximum amount of work that can be obtained then is

$$\begin{aligned} \delta W_{max} &= \delta W_{sys} + \delta W_{eng} \\ &= \delta W_{sys} + \left(\frac{T_0}{T} - 1\right) \delta Q \\ &= (\delta W_{sys} - \delta Q) + T_0 \frac{\delta Q}{T} \end{aligned} \qquad (a)$$

Remember that δQ is transferred reversibly because for maximum work the system must undergo a reversible process. By the first law, for the open system

$$\delta W - \delta Q = -dE_{os} + (e_1 + p_1 v_1)\, \delta m_1 - (e_2 + p_2 v_2)\, \delta m_2 \qquad (3 \cdot 13)$$

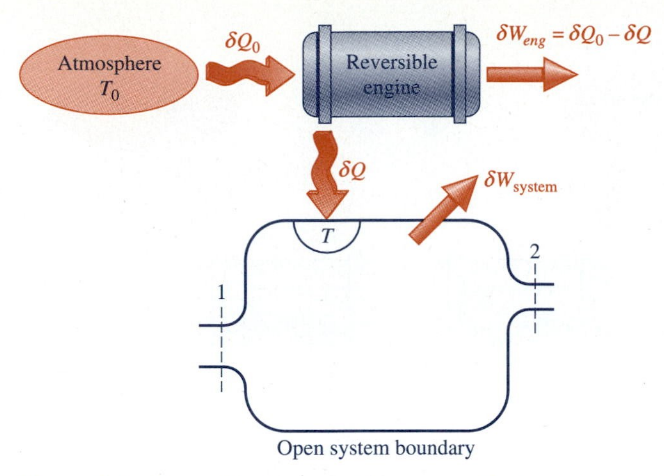

Figure 8·4 Reversible heat transfer across a finite temperature difference between a system and the atmosphere.

Also, for an open system we have shown that

$$\left(\frac{\delta Q}{T}\right)_{\substack{int\\rev}} = dS_{os} - s_1\delta m_1 + s_2\delta m_2 \tag{6·4}$$

Making these two substitutions from Eqs. 3·13 and 6·4 into Eq. *a* for δW_{max} and collecting terms, we have

$$\delta W_{max} = -d(E - T_0S)_{os} + (e_1 + p_1v_1 - T_0s_1)\,\delta m_1 - (e_2 + p_2v_2 - T_0s_2)\,\delta m_2 \tag{8·1}$$

Maximum work for system exchanging heat with only the atmosphere

This is the maximum work for a system exchanging heat with only the atmosphere while undergoing an infinitesimal change in state.

8·1·2 Maximum Work for a System Exchanging Heat with the Atmosphere and a Reservoir at T_R

Now consider the case in which the system exchanges heat with some reservoir at a temperature T_R as well as with the atmosphere. Let the reservoir exchange no mass with the surroundings. Applying Eq. 8·1 to the *combination of the open system and the reservoir*, as shown in Fig. 8·5, and denoting properties of the reservoir by the subscript *R*, we find

$$\delta W_{max} = -d(E - T_0S)_{os} - d(E - T_0S)_R + (e_1 + p_1v_1 - T_0s_1)\,\delta m_1 - (e_2 + p_2v_2 - T_0s_2)\,\delta m_2 \tag{b}$$

We usually assume that an energy reservoir undergoes only internally revers-

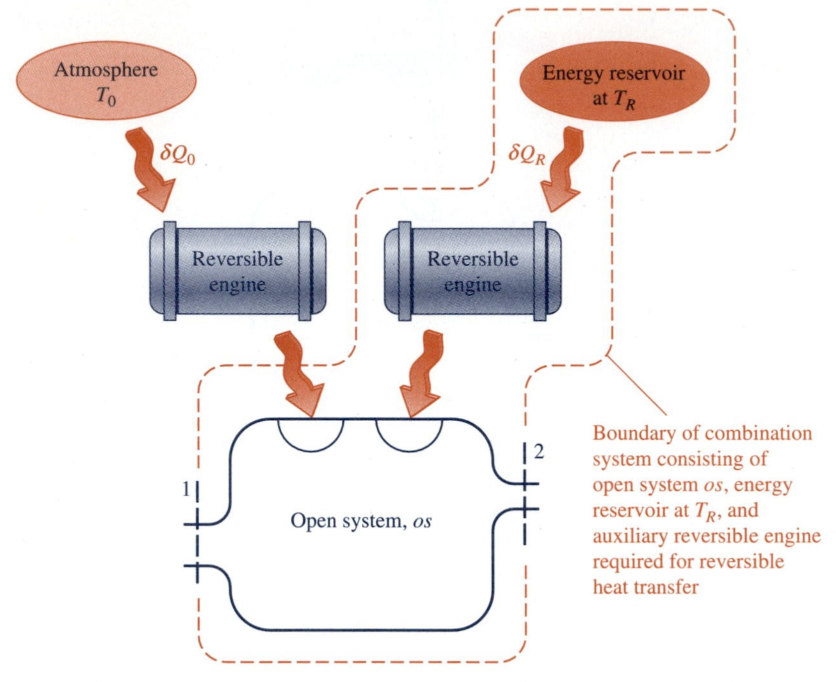

Figure 8·5 System exchanging heat with the atmosphere and also with a reservoir at temperature T_R.

ible changes and does no work. Then the first law applied to the reservoir, with δQ_R being the heat transferred *to* the system and *from* the reservoir, is

$$\delta Q_R = -dE_R \tag{c}$$

Then we have for the second term on the right-hand side of Eq. *b*

$$-d(E - T_0 S)_R = \delta Q_R + T_0\, dS_R \tag{d}$$

But for a reversible process of the reservoir

The minus sign is required because δQ_R is heat transferred *to* the system *from* the reservoir.

$$dS_R = -\frac{\delta Q_R}{T_R}$$

Then Eq. *d* becomes

$$-dE_R + T_0\, dS_R = \delta Q_R - T_0 \frac{\delta Q_R}{T_R} = \delta Q_R\left(1 - \frac{T_0}{T_R}\right)$$

and substitution of this into Eq. *b* followed by rearrangement gives

Maximum work for system exchanging heat with the atmosphere and an energy reservoir at T_R

$$\delta W_{max} = -d(E - T_0 S)_{os} + (e_1 + p_1 v_1 - T_0 s_1)\,\delta m_1 - (e_2 + p_2 v_2 - T_0 s_2)\,\delta m_2 + \delta Q_R\left(1 - \frac{T_0}{T_R}\right) \tag{8·2}$$

This is the expression for the maximum work that can be obtained from an open system that exchanges heat with only the atmosphere at T_0 and an energy reservoir at T_R.

Equation 8·2 can be readily simplified for the case of a closed system or the case of steady flow.

Special Case of a Closed System. For a *closed system*, $\delta m_1 = \delta m_2 = 0$, and

$$\delta W_{max} = -d(E - T_0 S) + \delta Q_R \left(1 - \frac{T_0}{T_R}\right) \qquad (8\text{·}3a)$$

For a change from state i to state f,

$$W_{max} = (E_i - T_0 S_i) - (E_f - T_0 S_f) + Q_R \left(1 - \frac{T_0}{T_R}\right) \qquad (8\text{·}3b)$$

Maximum work, closed system

Special Case of a Steady-Flow System. Under *steady-flow* conditions, $dE_{os} = 0$, $dS_{os} = 0$, and $\delta m_1 = \delta m_2 = \delta m$. Then

$$\delta W_{max} = [(e_1 + p_1 v_1 - T_0 s_1) - (e_2 + p_2 v_2 - T_0 s_2)]\, \delta m + \delta Q_R \left(1 - \frac{T_0}{T_R}\right) \qquad (8\text{·}4a)$$

Per unit mass of fluid entering at section 1 and leaving at section 2,

$$w_{max} = (e_1 + p_1 v_1 - T_0 s_1) - (e_2 + p_2 v_2 - T_0 s_2) + q_R \left(1 - \frac{T_0}{T_R}\right) \qquad (8\text{·}4b)$$

Recall that in the absence of electrical, magnetic, solid distortion, and surface tension effects, we can evaluate e as

$$e = u + \frac{V^2}{2} + gz$$

Therefore, Eq. 8·4b becomes

$$w_{max} = \left(h_1 + \frac{V_1^2}{2} + gz_1 - T_0 s_1\right) - \left(h_2 + \frac{V_2^2}{2} + gz_2 - T_0 s_2\right) + q_R \left(1 - \frac{T_0}{T_R}\right) \qquad (8\text{·}4c)$$

Maximum work, steady-flow system

Example 8·1 Maximum Work

PROBLEM STATEMENT

A tank with a volume of 1.60 m^3 is evacuated. Atmospheric air outside the tank is at 95 kPa, 15°C. What is the maximum possible amount of work that can be done by allowing atmospheric air to enter the tank until the tank contents come to equilibrium with the atmosphere?

SOLUTION

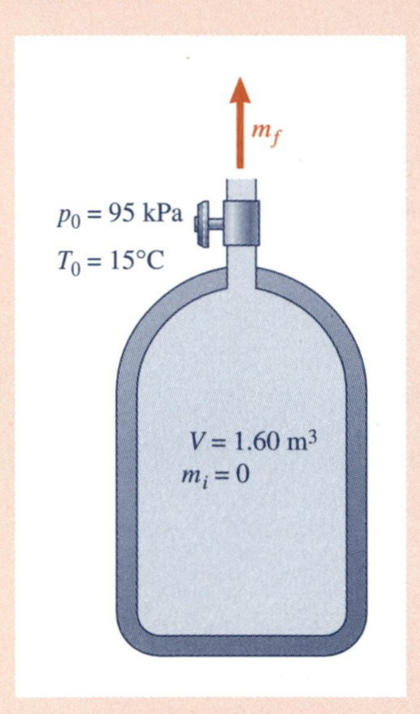

For any open system that exchanges heat with only the atmosphere,

$$\delta W_{max} = -d(E - T_0S)_{os} + (e_1 + p_1v_1 - T_0s_1)\,\delta m_1 - (e_2 + p_2v_2 - T_0s_2)\,\delta m_2 \tag{8·1}$$

In this case, $\delta m_2 = 0$. All the air that enters the tank crosses the open-system boundary at T_0 and p_0 with negligible kinetic energy, so that the properties of the mass δm_1 are constant and equal to those of the atmosphere. (If the kinetic energy of the mass crossing the system boundary is appreciable, then the pressure and temperature of it are lower than p_0 and T_0. Application of the first law to fluid flowing between some stagnant part of the atmosphere and the system boundary then shows that $h_0 = h_1 + V_1^2/2$, so that any kinetic energy possessed by the fluid at section 1 is balanced by a decrease in h_1 below h_0.) Thus the equation above can be readily integrated to give

$$W_{max} = (E_i - T_0S_i) - (E_f - T_0S_f) + (e_0 + p_0v_0 - T_0s_0)(m_f - m_i)$$

where the subscripts i and f denote initial and final conditions in the tank. E_i, S_i, and m_i are each zero because there is initially nothing in the tank; so we can simplify the last equation to

$$\begin{aligned} W_{max} &= 0 - (E_f - T_0S_f) + (e_0 + p_0v_0 - T_0s_0)m_f \\ &= -e_fm_f + T_0s_fm_f + e_0m_f + p_0v_0m_f - T_0s_0m_f \end{aligned}$$

The gas finally in the tank comes to equilibrium with the atmosphere and is at rest; so $p_f = p_0$, $T_f = T_0$ and therefore $e_f = e_0$, $s_f = s_0$, and $v_f = v_0$:

$$W_{max} = p_0 v_0 m_f = p_0 V_{\text{tank}} = 95(1.60) = 152 \text{ kJ}$$

(Notice that the mechanism for producing this work is entirely unspecified. No matter what devices are used, however, we can obtain no more than 152 kJ of work from the interaction of this system and the atmosphere.)

Example 8·2 Maximum Work

PROBLEM STATEMENT

Steam at 1000 kPa, 275°C, enters a steady-flow system with negligible velocity and leaves at 100 kPa, 120°C, with a velocity of 160 m/s. The flow rate is 9500 kg/h. Heat is exchanged only with the surrounding atmosphere at 15°C. Determine the maximum possible power output.

SOLUTION

For a steady-flow system exchanging heat with only the atmosphere, the maximum work is

$$w_{max} = h_1 + \frac{V_1^2}{2} + gz_1 - T_0 s_1 - \left(h_2 + \frac{V_2^2}{2} + gz_2 - T_0 s_2\right) \tag{8·4c}$$

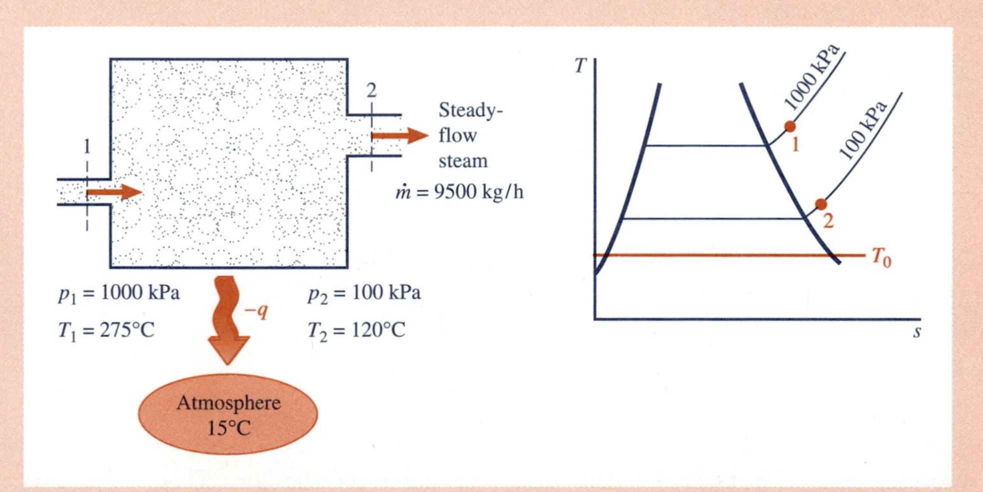

In this case, changes in potential energy can be neglected, and initial kinetic energy is negligible. The equation above becomes

$$w_{max} = (h_1 - T_0 s_1) - \left(h_2 + \frac{V_2^2}{2} - T_0 s_2\right) = h_1 - h_2 - T_0(s_1 - s_2) - \frac{V_2^2}{2}$$

Obtaining property values from the steam tables we get

$$w_{max} = 2996.6 - 2716.3 - 288(7.0257 - 7.4656) - \frac{(160)^2}{2(1000)} = 394 \text{ kJ/kg}$$

The maximum possible power output is then

$$\dot{W}_{max} = \dot{m}w_{max} = \frac{9500(394)}{3600} = 1040 \text{ kW}$$

(Notice again that the mechanism for producing this much power is unspecified, but, no matter what devices are used, this power output cannot be exceeded. An actual machine operating with the specified end conditions will produce less power because of irreversible effects.)

Example 8·3 Minimum Work for a Compression

PROBLEM STATEMENT

Atmospheric air is compressed steadily from 14.0 psia, 60 F, to 70.0 psia, 240 F, by a compressor that is cooled only by atmospheric air. Neglect kinetic energy changes, and determine the minimum work required per pound of air.

SOLUTION

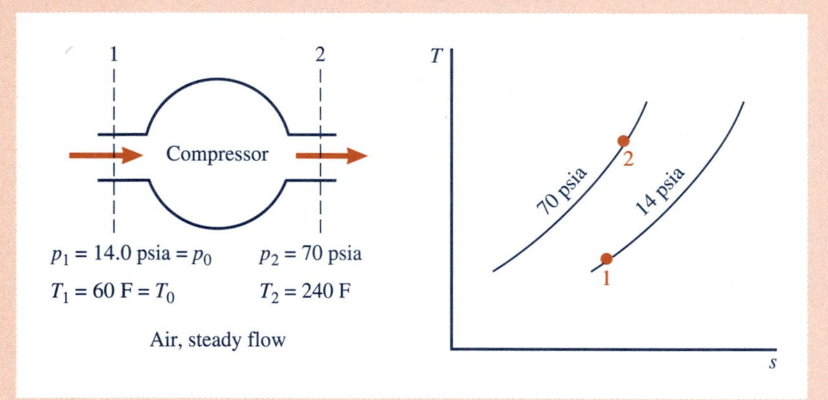

For this compression process, $w_{max} < 0$ because work must be done *on* the air. Then $-w_{max} =$ minimum w_{in}. For example, if we find that $-w_{max} = 10$ units, this means that $-w = 11$ units is possible but $-w = 9$ units is impossible; the minimum w_{in} is 10 units. We calculate w_{max} for this steady-flow process (with $\Delta ke = \Delta pe = 0$) by

$$\begin{aligned} w_{max} &= (h_1 - T_0 s_1) - (h_2 - T_0 s_2) \\ &= h_1 - h_2 - T_0(s_1 - s_2) \end{aligned} \qquad \text{(8·4c, simplified)}$$

Treating air as an ideal gas with constant specific heats (see Table A·3) and using the $T\,ds$ equation, $T\,ds = dh - v\,dp$, we get

$$w_{max} = c_p(T_1 - T_2) - T_0 \int_2^1 ds$$

$$= c_p(T_1 - T_2) - T_0 \left(\int_2^1 \frac{dh}{T} - \int_2^1 \frac{v\,dp}{T} \right)$$

$$= c_p(T_1 - T_2) - T_0 \left(\int_2^1 c_p \frac{dT}{T} - R \int_2^1 \frac{dp}{p} \right)$$

$$= c_p(T_1 - T_2) - T_0 \left(c_p \ln \frac{T_1}{T_2} - R \ln \frac{p_1}{p_2} \right)$$

$$= 0.241(60 - 240) - 520 \left(0.241 \ln \frac{520}{700} - \frac{53.3}{778} \ln \frac{14}{70} \right) = -63.5 \text{ B/lbm}$$

Since $w_{max} = -63.5$ B/lbm, the minimum work input is 63.5 B/lbm.

8·2 Availability

In the preceding section we showed how to calculate the maximum work obtainable from a system that goes from one specified state to another while exchanging heat with only the atmosphere and other constant-temperature energy reservoirs. As noted earlier, some of this work may be done on the atmosphere and therefore serves no useful purpose.

Useful work is defined as the work done by a system exclusive of that done on the atmosphere. Whenever the volume of the system increases, work is done pushing back the atmosphere. Thus, if p_0 is the uniform and constant pressure of the atmosphere and V_i and V_f are the initial and final volumes of a system, then

Useful work

$$W_{useful} = W - p_0(V_f - V_i)$$

Notice that if $V_f < V_i$, then $W_{useful} > W$, where W is the work done by the system.

An important question is the following: For any given initial state of a system that exchanges heat with only the atmosphere, what final state is required for the maximum useful work to be produced by the combination of the system and the atmosphere? The answer is that the maximum useful work is done when the system final state has pressure and temperature equal to those of the atmosphere. This answer is supported by the following reasoning: Let the atmosphere be at a pressure p_0 and a temperature T_0. If the system temperature T is different from T_0, work can always be produced by transferring heat from either the system or the atmosphere, whichever is at the higher temperature, to a heat engine that operates cyclically, produces work, and rejects heat to the lower-temperature region, either the atmosphere or the system. The cases of $T > T_0$ and $T < T_0$ are both shown schematically in Fig. 8·6. In both cases, the temperature of the system approaches T_0 as work is done. When $T = T_0$, no work can be obtained in this manner. However, even when $T = T_0$, work can be obtained if $p \neq p_0$ by letting the system change in volume until $p = p_0$. When $T = T_0$ and $p = p_0$, no work can be produced by interaction of the system and the atmosphere. They are in equilibrium with each other, and the system is said to be in the **dead state**. (The stored energy of the system in the dead state is designated as E_0 and is not necessarily zero.) Therefore, *the maximum useful work can be produced by means of a process that carries the system to the dead state.* This fact leads to the definition of *availability*.

Dead state

Availability

The **availability** of a system is defined as the *maximum useful work that can be obtained from the system–atmosphere combination as the system goes from a given state to the dead state while exchanging heat with only the*

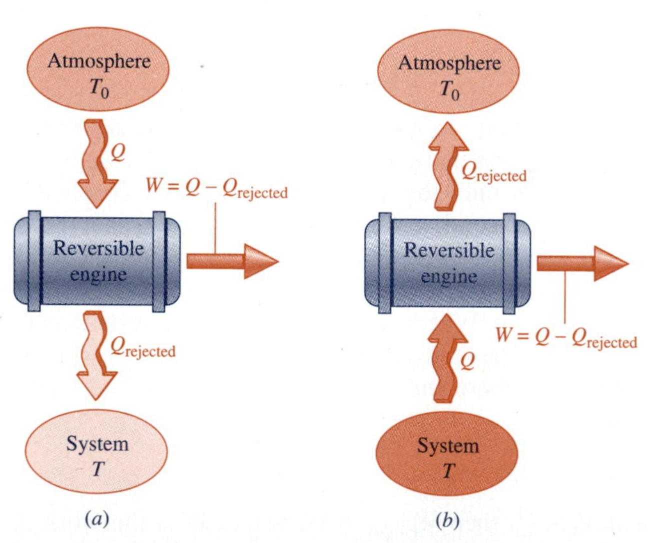

Figure 8·6 Obtaining work by reversible transfer of heat between a system and the atmosphere whether $T > T_0$ or $T < T_0$.

atmosphere. Notice that availability is not a property of the system alone; its value depends on p_0 and T_0 as well as on the properties of the system. It would be more precise to speak of the availability of a system–atmosphere combination, but common usage is to refer simply to the availability of the system.

We have been referring to T_0 and p_0 as the temperature and pressure of the surrounding atmosphere. For some applications, however, a power plant or refrigeration plant rejects heat to a large body of water—a lake, river, or ocean—instead of to the surrounding atmosphere, and in such cases T_0 should be the temperature of the water. T_0 is often called the **sink temperature** because an energy reservoir such as the atmosphere or the water of a river or lake that is used for absorbing heat from other systems is called a sink. T_0 is also referred to as the *lowest temperature in the surroundings.* This actually means the lowest temperature of an energy reservoir to which appreciable quantities of heat may be rejected.

Sink temperature = temperature of the atmosphere = ambient temperature = lowest temperature in the surroundings

In the preceding section we showed how to calculate maximum work for *any* two end states and heat exchange with only the atmosphere. Therefore, we can readily calculate *availability* by letting the final state be the dead state and subtracting the work done by the system on the atmosphere.

Availability of a closed system

Closed Systems. The availability of a *closed system* is given the symbol Φ (and $\phi \equiv \Phi/m$) and is given by

$$\begin{aligned}\Phi_1 &\equiv W_{max\ useful,1-0}\\ &= W_{max,1-0} - p_0(V_0 - V_1)\\ &= (E_1 - T_0S_1) - (E_0 - T_0S_0) - p_0(V_0 - V_1) \qquad (8\text{·}5a)\\ &= (E_1 + p_0V_1 - T_0S_1) - (E_0 + p_0V_0 - T_0S_0)\end{aligned}$$

where the subscript 0 denotes properties of the system when in the dead state. Since useful work can always be obtained if a closed system is in a state other than the dead state, $\Phi \geq 0$ *for all states.* Notice that the maximum useful work obtainable from a closed system that exchanges heat with only the atmosphere as it passes from any state 1 to any state 2 is given by the decrease in Φ:

$$\begin{aligned}W_{max\ useful} &= -\Delta\Phi = \Phi_1 - \Phi_2\\ &= (E_1 + p_0V_1 - T_0S_1) - (E_2 + p_0V_2 - T_0S_2) \qquad (8\text{·}5b)\end{aligned}$$

If the system also receives heat Q_R from an energy reservoir at T_R, we have, from Eq. 8·3,

$$W_{max\ useful} = -\Delta\Phi + Q_R\left(1 - \frac{T_0}{T_R}\right) \qquad (8\text{·}5c)$$

Example 8·4 Availability of Gas Stored in Tanks

PROBLEM STATEMENT

A tank holds 1 kg of air at 100 kPa, 40°C, and another tank holds 1 kg of air at 200 kPa, 40°C. The atmosphere is at 100 kPa, 20°C. In which tank is the stored energy greater? Determine the availability of the air in each tank. (This is the case mentioned in the introduction to this chapter.)

SOLUTION

The stored energy, which is the same as the internal energy for a gas stored in a tank, is the same in each tank, because at the specified pressure and temperature air can be accurately modeled as an ideal gas, and the internal energy of an ideal gas is a function of temperature only. The availability is not the same, however, as the following calculations show.

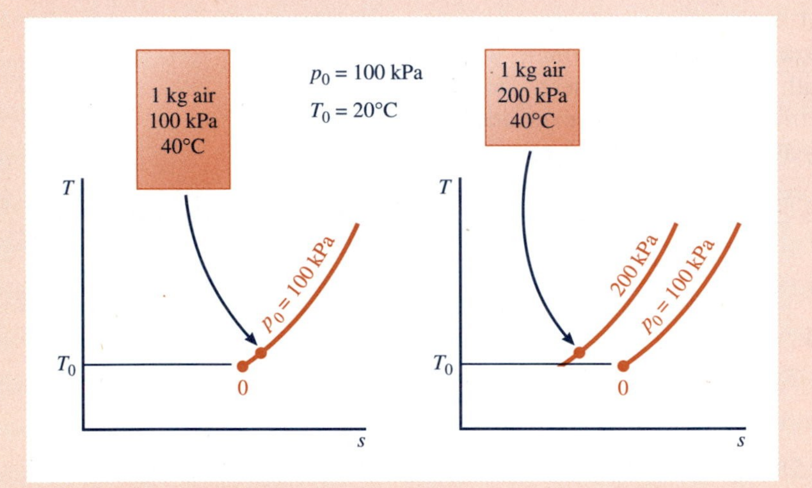

Air in the First Tank: With properties of gas in the tank shown with no subscripts, the availability is

$$\begin{aligned}\Phi &= E + p_0V - T_0S - (E_0 + p_0V_0 - T_0S_0)\\ &= m(e + p_0v - T_0s) - m(e_0 + p_0v_0 - T_0s_0)\\ &= m[u - u_0 + p_0(v - v_0) - T_0(s - s_0)]\end{aligned}$$

For the limited temperature range involved, we will use constant specific heats in evaluating $u - u_0$ and integrating a $T\,ds$ equation to get $s - s_0$. We can then express the relationship for Φ in terms of m, p, p_0, T, and T_0:

$$\begin{aligned}\Phi &= m\left[c_v(T - T_0) + p_0\left(\frac{RT}{p} - \frac{RT_0}{p_0}\right) - T_0\left(c_p \ln\frac{T}{T_0} - R\ln\frac{p}{p_0}\right)\right]\\ &= m\left[c_v(T - T_0) + R\left(\frac{p_0}{p}T - T_0\right) - T_0c_p\ln\frac{T}{T_0} + RT_0\ln\frac{p}{p_0}\right]\end{aligned}$$

$$= mRT_0\left[\frac{c_v}{R}\left(\frac{T}{T_0} - 1\right) + \frac{p_0}{p}\frac{T}{T_0} - 1 - \frac{c_p}{R}\ln\frac{T}{T_0} + \ln\frac{p}{p_0}\right]$$

$$= 1(0.287)293\left[\frac{0.717}{0.287}\left(\frac{313}{293} - 1\right) + \frac{100}{100}\left(\frac{313}{293}\right) - 1 - \frac{1.004}{0.287}\ln\frac{313}{293} + 0\right]$$

$$= 0.66 \text{ kJ}$$

Air in the Second Tank: Substituting values of $p = 200$ kPa, $T = 40°\text{C} = 313$ K into the equation derived above for Φ yields

$$\Phi = mRT_0\left[\frac{c_v}{R}\left(\frac{T}{T_0} - 1\right) + \frac{p_0}{p}\frac{T}{T_0} - 1 - \frac{c_p}{R}\ln\frac{T}{T_0} + \ln\frac{p}{p_0}\right]$$

$$= 1(0.287)293\left[\frac{0.717}{0.287}\left(\frac{313}{293} - 1\right) + \frac{100}{200}\left(\frac{313}{293}\right) - 1 - \frac{1.004}{0.287}\ln\frac{313}{293} + \ln\frac{200}{100}\right]$$

$$= 14.0 \text{ kJ}$$

Comment: The results show that although the energy of the gas in the two tanks is the same, the amount of work that can be obtained as the system goes to the temperature and pressure of the atmosphere while exchanging heat with only the atmosphere is much greater for the higher-pressure air.

Steady-Flow Systems. We turn now to the availability of a *steadily flowing* fluid. Since the volume of a steady-flow system does not change, no work is done on or by the atmosphere (except flow work as described in Section 3·5), and there is no distinction between work and useful work. We call the availability of the steadily flowing fluid *stream availability* Y (and $y = Y/m$), to distinguish it from the availability of a closed system Φ (or ϕ). **Stream availability** is defined as the maximum work that can be obtained as the fluid goes reversibly to the dead state while exchanging heat with only the atmosphere. In the dead state the fluid is at p_0 and T_0 and is also at rest, $V_0 = 0$. In the absence of electrical, magnetic, solid distortion, and surface tension effects, we obtain from Eq. 8·4*c*

Stream availability

$$y_1 = w_{max,1-0} = \left(h_1 + \frac{V_1^2}{2} + gz_1 - T_0 s_1\right) - (h_0 + gz_0 - T_0 s_0) \quad (8{\cdot}6a)$$

In availability calculations it is often convenient to use the *Darrieus function*, which is defined as $b \equiv h - T_0 s$. For any given temperature of the atmosphere, b is a property of a substance. Thus

$$y_1 = \left(b_1 + \frac{V_1^2}{2} + gz_1\right) - (b_0 + gz_0) \quad (8{\cdot}6b)$$

and, for a change from state 1 to state 2

$$w_{max} = -\Delta y = y_1 - y_2 = \left(b_1 + \frac{V_1^2}{2} + gz_1\right) - \left(b_2 + \frac{V_2^2}{2} + gz_2\right) \quad (8\cdot6c)$$

Stream availability is sometimes called simply the availability of a flowing fluid. This practice can lead to errors and should be used only with caution.

For a steadily flowing fluid that goes from one state to another while exchanging heat with only the atmosphere, the maximum work obtainable per unit mass is $-\Delta y$. If the fluid also receives heat in the amount q_R per unit mass from an energy reservoir at T_R, the maximum work that can be obtained per unit mass is

$$w_{max} = -\Delta y + \left(1 - \frac{T_0}{T_R}\right)q_R \quad (8\cdot6d)$$

Example 8·5 Availability in a Steam Power Plant

PROBLEM STATEMENT

Steam enters a well-insulated turbine at 800 kPa, 300°C, with negligible velocity and leaves at 30 mm Hg, 94 percent quality, with a velocity of 150 m/s. The flow rate is 54,000 kg/h. Atmospheric pressure is 730 mm Hg. The plant rejects heat to lake water at 15°C. Determine (*a*) the power output, (*b*) the maximum power for the given end states, and (*c*) the maximum amount of power that could be obtained from the exhaust steam.

SOLUTION

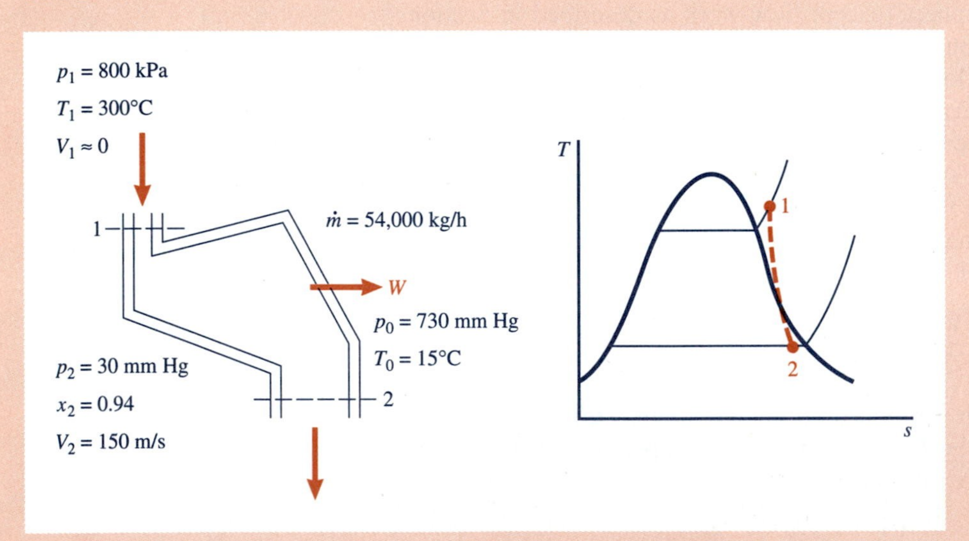

Analysis: We first sketch a flow diagram and a *Ts* diagram. Since the turbine is described as well insulated, we assume adiabatic flow. Then, from the specified end states we can calculate the work by the first law and combine the work with the specified flow rate to determine the power output. The maximum power is the product of the flow rate and the maximum work, and the maximum work is the decrease in stream availability between inlet and exhaust. The maximum power that can be obtained from the exhaust is proportional to the exhaust stream availability or the maximum work that could be obtained by having the steam go from the exhaust state to the dead state at p_0 and T_0.

Calculations: Obviously, for the first law equation and the availability equations we will need h and s values at the inlet (state 1), the exhaust (state 2), and the dead state (state 0), so we obtain those first from the steam tables or the computer program STEAM. (Actually, if we are going to solve the problem by means of a computer program that can call the steam properties from STEAM as needed, we do not need to list the property values explicitly.)

$$h_1 = 3055.9 \text{ kJ/kg} \qquad s_1 = 7.2319 \text{ kJ/kg·K}$$

The exhaust pressure is

$$p = \gamma\,\Delta z = \rho_{Hg} g\,\Delta z_{Hg} = 1000(13.6)\frac{\text{kg}}{\text{m}^3}\left(9.81\frac{\text{m}}{\text{s}^2}\right)\frac{30 \text{ mm}}{1000 \text{ mm/m}} = 4000 \text{ Pa}$$
$$= 4.0 \text{ kPa}$$
$$h_2 = h_{f2} + x_2 h_{fg2} = 121.28 + 0.94(2432.2) = 2407.5 \text{ kJ/kg}$$
$$s_2 = s_{f2} + x_2 s_{fg2} = 0.42203 + 0.94(8.0505) = 7.9895 \text{kJ/kg·K}$$

Water in the dead state of 730 mm Hg, 15°C, is compressed liquid. For such a low pressure, the enthalpy and entropy of compressed liquid are very nearly equal to those of saturated liquid at the same temperature:

$$h_0 = h_{f,15^\circ\text{C}} = 62.93 \text{ kJ/kg}$$
$$s_0 = s_{f,15^\circ\text{C}} = 0.2243 \text{ kJ/kg·K}$$

You need not make this approximation if you use the computer model STEAM. It gives values of $h_0 =$ 62.91 kJ/kg and $s_0 = 0.2200$ kJ/kg·K for the compressed liquid. You can see that the approximation is accurate.

(*a*) The first law applied to the turbine for adiabatic steady flow gives

$$w = h_1 - h_2 - \frac{V_2^2}{2} = 3055.9 - 2407.5 - \frac{150^2}{2(1000)} = 637 \text{ kJ/kg}$$

$$\dot{W} = \dot{m}w = \frac{54{,}000(637)}{3600} = 9560 \text{ kW}$$

(*b*) The maximum work for the specified end states is

$$w_{max} = h_1 - T_0 s_1 - \left(h_2 + \frac{V_2^2}{2} - T_0 s_2\right)$$
$$= 3055.9 - 288(7.2319) - \left[2407.5 + \frac{150^2}{2(1000)} - 288(7.9895)\right]$$
$$= 855 \text{ kJ/kg}$$

Then

$$\dot{W}_{max} = \dot{m}w_{max} = \frac{54{,}000(855)}{3600} = 12{,}830 \text{ kW}$$

(*c*) The stream availability of the exhaust steam is

$$y_2 = h_2 + \frac{V_2^2}{2} - T_0 s_2 - (h_0 - T_0 s_0)$$

$$= 2407.5 + \frac{150^2}{2(1000)} - 288(7.9895) - [62.9 - 288(0.2243)] = 119 \text{ kJ/kg}$$

$$\dot{W}_{max,2-0} = \dot{m}y_2 = \frac{54{,}000(119)}{3600} = 1790 \text{ kW}$$

Comment: In part *b*, w_{max} could have been calculated for the adiabatic process as $w_{max} = y_1 - y_2$.

8·3 Available and Unavailable Energy

The second law limits the conversion of energy from one form to another. Work can always be converted completely into heat, but heat cannot be converted completely into work. The extent to which a quantity of energy can be converted into work is important, so we define *available energy* and *unavailable energy*.

Available energy, also called *exergy*

Available energy is energy that can by some means be converted into useful work. This applies to both energy in transition and all forms of stored energy. Kinetic energy, potential energy, and work itself are all available energy. Internal energy and heat may be partially available energy, but they cannot be entirely available energy. Available energy is also called *exergy*.

Unavailable energy

Unavailable energy is energy that can by no means be converted into work. Some energy is entirely unavailable. For example, the energy of the atmosphere is huge in amount, but no fraction of it can be converted into work by any means.

The discussions of maximum work and availability have shown that the maximum conversion of energy into work is achieved through reversible processes. Consequently, unavailable energy is sometimes defined as energy that cannot be converted into work *even by reversible engines*.

All energy, stored or transitional, can be classified as follows:

Energy that can be completely converted into work	Available energy
Energy that can be partially converted into work	A combination of available and unavailable energy
Energy that *cannot* be converted into work	Unavailable energy

In terms of available and unavailable energy, the first law might be stated as follows:

First law: Energy consists of available energy and unavailable energy, and the sum of the available energy and the unavailable energy of the universe remains constant.

The second law might be stated in any of the following forms, among others (see Reference 8·1):

Second law: It is impossible to convert unavailable energy into available energy.
Second law: The unavailable energy of the universe continually increases or, in the case of reversible processes only, remains constant.
Second law: The available energy of the universe continually decreases or, in the case of reversible processes only, remains constant.

Each of these three statements can be deduced from either of the others or from any of the other second law statements. Notice that each of them limits the direction or extent of the energy conversion that the first law states is possible. (It is well to repeat here that by *universe* we mean a system plus its surroundings, where the term *surroundings* is restricted to those things outside a system that in some way influence the system.)

In nonscientific literature, the terms *energy waste* and *energy loss* are occasionally used. In many such cases it would be more accurate to refer to the waste or loss of *available energy*.

8·3·1 Available and Unavailable Energy of a System and the Surrounding Atmosphere

In the preceding section, we discussed availability (ϕ) and stream availability (y). These are not properties of a system but of a system–atmosphere combination, and they indicate the maximum useful work that can be obtained from processes that change the system to the pressure and temperature of the atmosphere. Consequently, ϕ and y are used to determine the fraction of *stored energy* that is available energy.

Turning to *transitional energy*, we see that all work is available energy but that only a part of heat is available energy. We now see how that part can be calculated.

8·3·2 Available and Unavailable Parts of Heat Transfer

As an introduction to the available and unavailable parts of heat transfer, let us consider two questions.

We confine our attention to devices that operate cyclically in order to make certain that no part of the work produced results from the depletion of the stored energy of the devices.

First, suppose an energy reservoir at 600 K gives up 1000 kJ of heat. If the surrounding atmosphere, which is the coldest energy reservoir in the surroundings, is at 300 K, how much of this heat can be converted into work by devices that are not permanently changed, that is, that operate cyclically? We answer this question by noting that the maximum work can be obtained by the use of a reversible engine (such as a Carnot, Stirling, or Ericsson engine) that receives the 1000 kJ at 600 K and rejects heat at the temperature of the atmosphere, 300 K, as shown in Fig. 8·7*a*. The efficiency of such an engine is $1 - (T_{low}/T_{high}) = 1 - (300/600) = 50$ percent, so the answer to the question is that 500 kJ of the 1000 kJ given up by the reservoir can be converted into work.

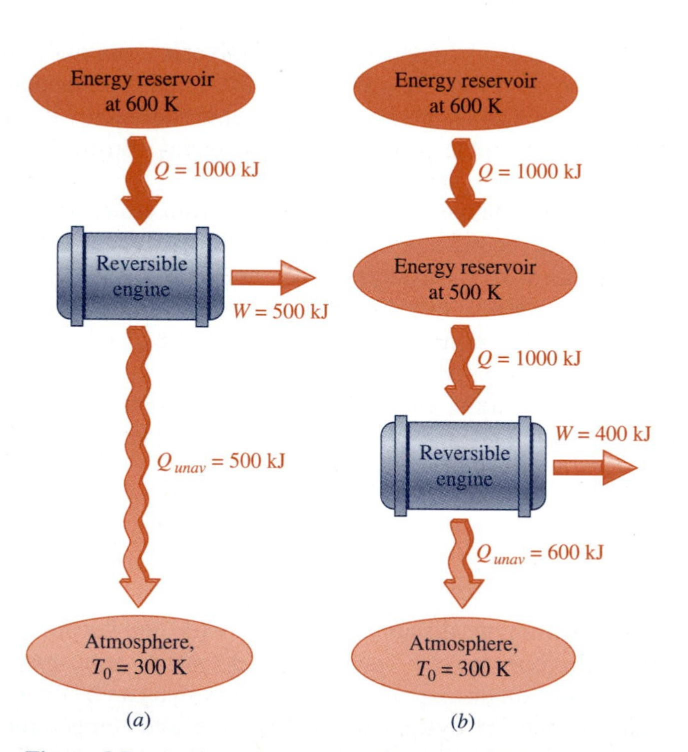

Figure 8·7 As the same amount of heat is added at lower temperatures, the amount of work obtainable decreases.

For the second question, suppose that the 1000 kJ is transferred from the reservoir at 600 K to a reservoir at 500 K. Then, after this transfer of heat has been made, and with the atmosphere still at 300 K, how much of the 1000 kJ can be converted into work by cyclically operating devices? In answering this question we first note that the 1000 kJ stored in the reservoir at 500 K can be supplied to a reversible engine as in Fig. 8·7*b* at 500 K but at no higher temperature. The efficiency of such an engine is $1 - (T_{\text{low}}/T_{\text{high}}) = 1 - (300/500) = 40$ percent; so the answer to this second question is that a maximum of 400 kJ of the 1000 kJ transferred to the 500 K reservoir can be converted into work. Yet this same 1000 kJ as removed from the reservoir at 600 K could have been converted to the extent of 500 kJ into work if it had not first been transferred to the reservoir at 500 K.

Such questions are important in engineering. For another example of the type of question we want to answer, consider a steam power plant as shown in Fig. 8·8. For each 1000 kJ of energy input to the plant, suppose that 250 kJ is converted into work, 600 kJ is rejected as heat to the cooling water, and 150 kJ is carried away by the stack gases. Which loss is more serious—the 600 kJ to the cooling water or the 150 kJ in the stack gases? Obviously, one is four times bigger than the other in terms of energy quantities, but is it four times as valuable? It certainly is not. The reason for this is that heat is rejected to the cooling water from the power plant at a temperature just slightly higher than the cooling water temperature. Therefore, a reversible engine absorbing this heat might have an efficiency on the order of only 3 percent so that even a reversible engine could produce less than 20 kJ of work from the 600 kJ of heat. The stack gases leave at a much higher temperature, so a reversible engine absorbing heat from them might convert as much as one-third of the stack-gas energy, or 50 kJ, into work. So in the evaluation of energy rejected by a system the question is not simply how much is rejected but how much of that rejected could be converted into work?

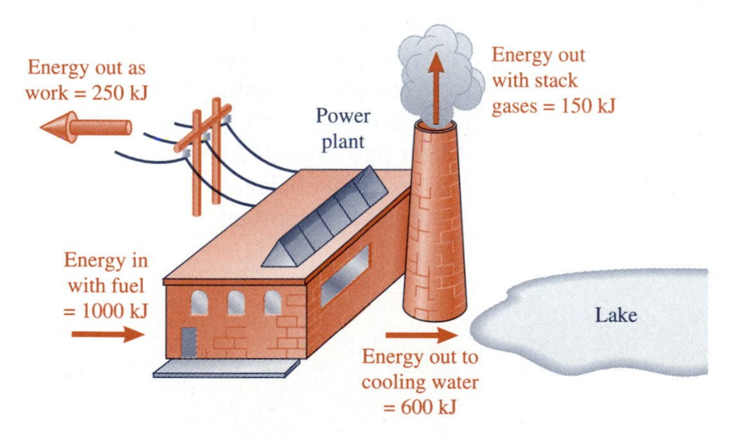

Figure 8·8 Energy transfers of a steam power plant.

Unavailable energy = energy that cannot be converted into work *even by reversible engines*

The available part of the heat added to or taken from a system is that part that could be converted into work by reversible engines. (This is the maximum amount of work that could be obtained by any means whatsoever, because for specified temperature limits no engine can be more efficient than reversible engines.) *That part of heat added to or taken from a system that could not be converted into work even by reversible engines is called the unavailable part.* The symbols Q_{av} and Q_{unav} are used for the available and unavailable parts and Q for the total amount of heat, so $Q = Q_{av} + Q_{unav}$.

For a Carnot engine that receives heat Q_H at T_H and rejects heat at T_0, the lowest temperature in the surroundings, the available and unavailable parts of Q_H are the same as the work done and the heat rejected.

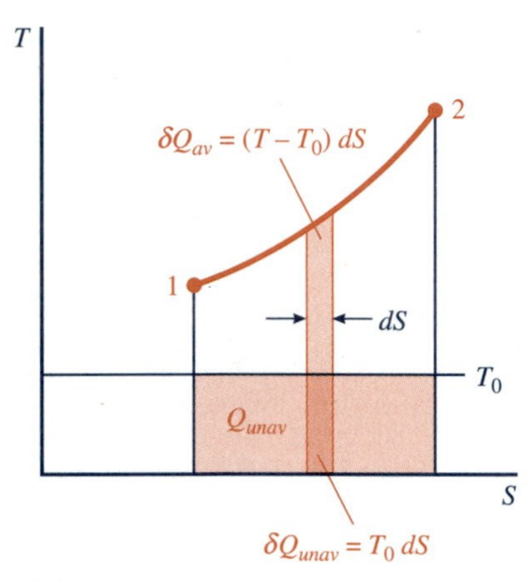

Figure 8·9 Available and unavailable parts of heat.

If heat Q is added to a system in a reversible process such as 1–2 in Fig. 8·9, we can determine the available and unavailable parts by imagining that each small fraction of the heat δQ is added instead to a Carnot (or other reversible) engine at some temperature T. T varies from T_1 to T_2, but each Carnot engine receives so little heat that it operates between essentially constant temperature limits of T and T_0. The available and unavailable portions of δQ are represented by the areas $(T - T_0)\, dS$ and $T_0\, dS$ in Fig. 8·9 for any one of these engines. To add all of the heat Q to Carnot engines at the same temperatures at which the original system absorbs heat, an infinite number of Carnot engines is required because only an infinitesimal amount of heat can be added to each engine at constant temperature. The available and unavailable parts of the heat Q are respectively the sums of δQ_{av} and δQ_{unav} for all the engines. Thus

$$Q_{unav,1-2} = \int_1^2 T_0\, dS = T_0(S_2 - S_1) = T_0\, \Delta S \tag{8·7}$$

$$Q_{av,1-2} = \int_1^2 (T - T_0)\, dS = \int_1^2 T\, dS - T_0(S_2 - S_1) = \int_1^2 T\, dS - Q_{unav}$$

Remember that heat transfer can be represented by an area on a TS diagram only for internally reversible processes. Under some conditions, the area $T_0\, \Delta S$ has physical significance even for irreversible processes.

The first law is the basis for ''energy accounting'' procedures in which heat, work, and stored energy are compared and balanced. The second law makes possible the division of heat into available and unavailable parts and thus adds valuable information to an energy balance. It shows that something more than just the quantity of energy is important.

Example 8·6 Available Part of Heat Transfer

PROBLEM STATEMENT

Consider the transfer of 1000 kJ of heat from a reservoir at 1200 K to 5 kg of a gas initially at 200 kPa, 600 K, in a closed tank. For the gas, $c_v = 0.800$ kJ/kg·K throughout the temperature range involved. The lowest temperature in the surroundings is 300 K. Determine how much of the heat removed from the reservoir is available and unavailable and how much of that absorbed by the gas is available and unavailable.

SOLUTION

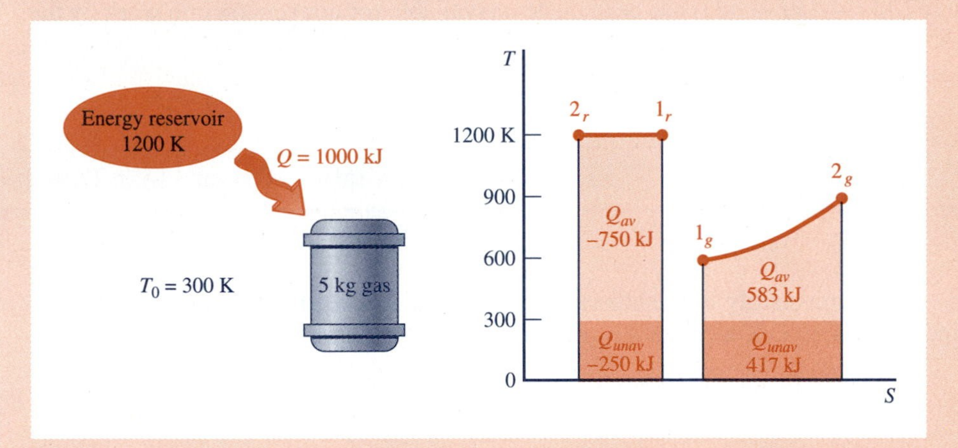

We first sketch a *TS* diagram for the reservoir and for the gas. Since for each system (reservoir and gas) the process can be internally reversible, the area beneath the *TS* diagram path, 1_r–2_r equals the area beneath 1_g–2_g because in each case the area represents the 1000 kJ of heat transferred. To find T_{g2}, we write

$$Q = \Delta U_g = mc_v(T_{g2} - T_{g1})$$

$$T_{g2} = T_{g1} + \frac{Q}{mc_v} = 600 + \frac{1000}{5(0.800)} = 850 \text{ K}$$

Then

$$\Delta S_g = \int_1^2 \frac{\delta Q}{T} = \int_1^2 \frac{dU}{T} = \int_1^2 \frac{mc_v\, dT}{T} = mc_v \ln \frac{T_2}{T_1}$$

$$= 5(0.800) \ln \frac{850}{600} = 1.39 \text{ kJ/K}$$

$$\Delta S_r = \int_1^2 \frac{\delta Q}{T} = \frac{Q}{T} = \frac{-1000}{1200} = -0.833 \text{ kJ/K}$$

Then

$$Q_{unav,g} = T_0\,\Delta S_g = 300(1.39) = 417 \text{ kJ}$$
$$Q_{av,g} = Q - Q_{unav,g} = 1000 - 417 = 583 \text{ kJ}$$
$$Q_{unav,r} = T_0\,\Delta S_r = 300(-0.833) = -250 \text{ kJ}$$
$$Q_{av,r} = Q - Q_{unav,r} = -1000 - (-250) = -750 \text{ kJ}$$

Comment: The last two values are negative because they are fractions of the heat *removed from* the reservoir. Notice that, as the 1000 kJ of heat was given up by the reservoir at 1200 K, 750 kJ of it could have been converted into work; but, as the heat was absorbed by the gas, only 583 kJ of it could have been converted into work. Thus 167 kJ (= 750 − 583) has been made unavailable for conversion into work by the irreversible transfer of heat across a finite temperature difference.

Example 8·7 Available Part of Heat Transfer

PROBLEM STATEMENT

Steam in a closed system expands reversibly and isothermally from dry saturated conditions at 180°C to 500 kPa. The ambient temperature is 20°C. Calculate the available part of the heat transfer to the system and the work.

SOLUTION

Analysis: Since the mass or volume of the system is unspecified, we will make the calculations per unit mass of steam. We can always calculate the unavailable part of the heat transfer by $q_{unav} = T_0\,\Delta s$ and then subtract this from q to obtain q_{av}. The work can be obtained by the first law, $w = u_1 - u_2 + q$.

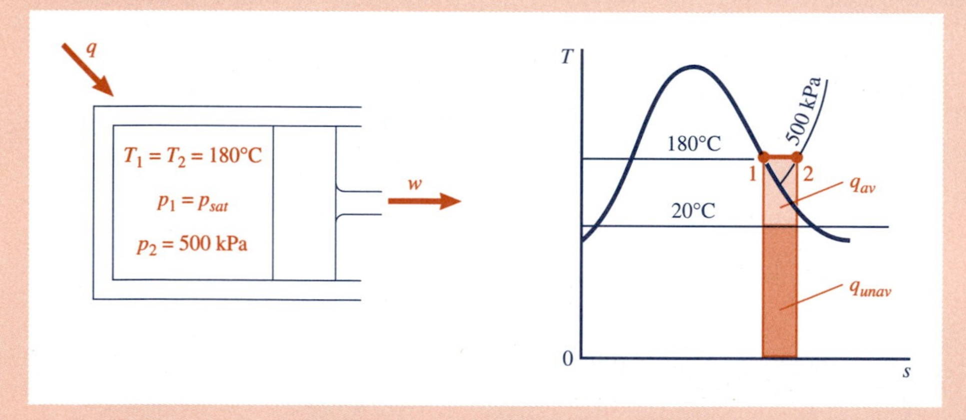

Calculations: Both end states are fully specified, so we can obtain values of u and s from the steam tables or from the computer model for steam properties:

$T_1 = 180°C$	$T_2 = 180°C$
$x_1 = 1$	$p_2 = 500$ kPa
$u_1 = 2583.4$ kJ/kg	$u_2 = 2609.5$ kJ/kg
$s_1 = 6.5853$ kJ/kg·K	$s_2 = 6.9644$ kJ/kg·K

where u values were obtained from $u = h - pv$.

Since the process is reversible,

$$q = \int_1^2 T\,ds$$

and, since it is isothermal, integration gives

$$q = \int_1^2 T\,ds = T(s_2 - s_1) = (180 + 273)(6.9644 - 6.5853) = 171.8 \text{ kJ/kg}$$

The unavailable part of this heat is given, as always, by $q_{unav} = T_0\,\Delta s$:

$$q_{unav} = T_0(s_2 - s_1) = 293(6.9644 - 6.5893) = 111.1 \text{ kJ/kg}$$

The available part is then

$$q_{av} = q - q_{unav} = 171.8 - 111.1 = 60.7 \text{ kJ/kg}$$

Using the first law, we calculate the work as

$$w = u_1 - u_2 + q = 2583.4 - 2609.5 + 171.8 = 145.7 \text{ kJ/kg}$$

Comment: We have calculated that the available part of the heat added to the steam is 60.7 kJ/kg, but the work done during the process is 145.7 kJ/kg. Is there any conflict here? No. During the process, the state of the system changed, so the work is even greater than the total heat added. If the system were to be returned to its initial state, the maximum work that could be obtained from the system during the cycle could not exceed 60.7 kJ/kg. Remember, we say that heat is converted into work only when the system is returned to its initial state so that changes in stored energy are not involved.

8·4 Irreversibility

A reversible process is defined as a process such that, after it has occurred, both the system and all the surroundings can be returned to their initial states. An irreversible process always produces some effects that cannot be undone, so it is impossible to restore *both* the system and all the surroundings to their initial states after an irreversible process has occurred. The total energy of the system and its surroundings remains constant (in accordance with the first law), but one result of an irreversible process is a decrease in the amount of energy that can be converted into work. This decrease can be quantified and is called *irreversibility*. Consider a system that exchanges heat with only the surrounding atmosphere. For a process in which no work is done, the decrease in the amount of useful work attainable is simply the decrease in availability. If, however, some work is done during the process, then some of the decrease in availability can be accounted for by the work actually done. With this line of reasoning in mind, let us define the **irreversibility I of any process** as

Defining equation for *irreversibility*. Note that we use *irreversibility* as a quantity, not just as a qualitative characteristic of a process, such as friction or unrestrained expansion.

$$I \equiv W_{max} - W \tag{8·8a}$$

where W is the work actually produced. During the process, heat may be exchanged with various energy reservoirs as well as with the atmosphere. W_{max} is the maximum work for a process that (1) occurs between the same end states as the actual process, (2) involves the same amounts of heat exchange with the various energy reservoirs other than the atmosphere, and (3) involves heat exchange with the atmosphere also, if necessary. Since the work done on the atmosphere, $p_0(V_f - V_i)$, depends only on the end states of a process, it is the same for the maximum work process and for the actual process. Therefore, if we subtract this quantity from each term on the right-hand side of Eq. 8·8a, we have

$$I = W_{max\ useful} - W_{useful} \tag{8·8b}$$

Notice that the definition of irreversibility I given above is *not* restricted to systems that exchange heat with only the atmosphere. Following the usual convention, the symbol i is used for irreversibility per unit mass, I/m.

Irreversibility is not a property. It is not a characteristic of a system. For any process, however, it can be evaluated from end states of the system *and the surroundings* as we shall now demonstrate.

Consider an open system as shown in Fig. 8·10. The properties of the system are denoted by the subscript os, δm_1 is the mass entering the system, δm_2 is the mass leaving the system, and δQ_R is the amount of heat transferred to the system from an energy reservoir at a temperature T_R. For such a system, an expression for δW_{max} is

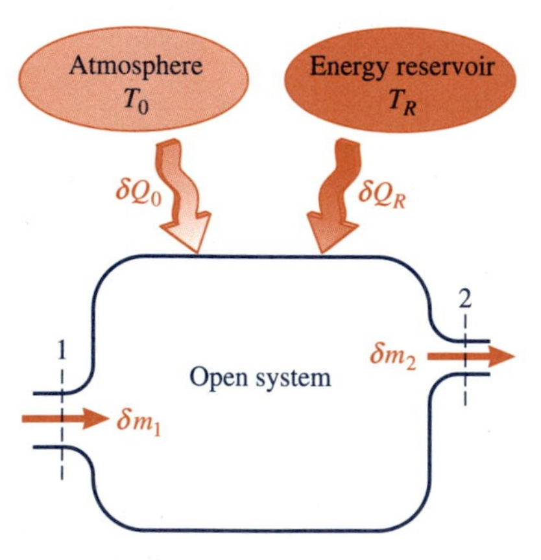

Figure 8·10 Open system exchanging heat with the atmosphere and with a reservoir at T_R.

$$\delta W_{max} = -d(E - T_0S)_{os} + (e_1 + p_1v_1 - T_0s_1)\,\delta m_1 - (e_2 + p_2v_2 - T_0s_2)\,\delta m_2 + \delta Q_R\left(1 - \frac{T_0}{T_R}\right) \tag{8·2}$$

and application of the first law gives the following expression for δW:

$$\delta W = \delta Q - dE + (e_1 + p_1v_1)\,\delta m_1 - (e_2 + p_2v_2)\,\delta m_2 \tag{3·13}$$

Making these two substitutions in the expression above for I (Eq. 8·8a) gives

$$\delta I = \delta W_{max} - \delta W = -T_0s_1\,\delta m_1 + T_0s_2\,\delta m_2 + \delta Q_R - T_0\frac{\delta Q_R}{T_R} - \delta Q \tag{a}$$

Since δQ is the *total* heat transferred to the system and δQ_R is the heat transferred to the system *from the energy reservoir,*

$$\delta Q - \delta Q_R = \delta Q_0$$

where δQ_0 is the heat transferred to the system *from the atmosphere.* As long as the atmosphere experiences no internal irreversibilities (mixing, friction, turbulence, etc.), $-\delta Q_0 = T_0 dS_{atm}$. (The minus sign is needed because δQ_0 is heat transferred *from* the atmosphere to the system.) In the same manner, $\delta Q_R/T_R = dS_R$. Making these substitutions in Eq. *a* gives

$$\delta I = T_0(dS_{os} - s_1\,\delta m_1 + s_2\,\delta m_2 + dS_R + dS_{atm}) \qquad (b)$$

This equation can be readily simplified for the cases of closed systems ($\delta m_1 = \delta m_2 = 0$) and steady-flow systems ($dS_{os} = 0$; $\delta m_1 = \delta m_2$). The quantity ($s_2\,\delta m_2 - s_1\,\delta m_1$) is the entropy change of those parts of the surroundings that exchange mass with the open system. Therefore, the last four terms in the parentheses of Eq. *b* are dS of the surroundings. (Those parts of the surroundings that exchange work with the system can operate reversibly and adiabatically, and hence for them $dS = 0$. Of course, those parts of the surroundings *may* operate irreversibly, but such irreversibility in no way influences the behavior of the system, atmosphere, or energy reservoir.) Thus Eq. *b* can be rewritten as

$$\delta I = T_0(dS_{os} + dS_{surr}) \qquad (8\cdot9a)$$

and, since T_0 = constant,

$$I = T_0(\Delta S_{os} + \Delta S_{surr}) \qquad (8\cdot9b)$$

$$I = T_0\,\Delta S_{\text{universe}} \qquad (8\cdot9c)$$

$$I = T_0\,\Delta S_{\text{isolated system}} \qquad (8\cdot9d)$$

An important, widely useful equation for irreversibility that has no restrictions

Recall again that by *universe* we mean a system plus all parts of the surroundings that in some way influence the system. The last form of Eq. 8·9 is also quite general because an isolated system can always be formed by including within a single boundary everything that affects a system during a process.

Equation 8·9 is usually much easier to apply than Eq. 8·8, and it is one of the most useful of all thermodynamic equations. One reason for its usefulness is that it requires the knowledge of only end states and the heat transfer with the atmosphere and other energy reservoirs.

From Eq. 8·9 and the increase of entropy principle, we conclude that, for all processes,

$$I \geq 0$$

where the equality holds only for reversible processes.

For a *reversible process*, $W = W_{max}$, and both the system and the surroundings can be returned to their initial states. Therefore, there is no decrease in the amount of energy that can be converted into work, and it is reasonable to have $I = 0$.

For any *irreversible process*, $W < W_{max}$; so even without reference to Eq. 8·9 and the increase of entropy principle we see that $I > 0$. An irreversible process cannot be completely undone; it always results in some permanent change in the system or the surroundings. Before an irreversible process occurs, a system and its surroundings are in states such that a certain amount of their energy can be converted into work. After the irreversible process occurs, the system and surroundings are in states such that a smaller fraction of their energy can be converted into work; and this decrease in the amount of energy (stored in the system and surroundings) that can be converted into work is equal to I.

I = decrease in the amount of energy that can be converted into work

Consider a closed system that changes from state 1 to state 2 while exchanging heat with only the atmosphere at T_0. The maximum useful work is given by

$$W_{max\ useful} = \Phi_1 - \Phi_2 \tag{8·5b}$$

so that

$$I = W_{max\ useful} - W_{useful} = \Phi_1 - \Phi_2 - W_{useful}$$

By definition, Φ_1 is the maximum amount of useful work that can be obtained by the interaction of the system initially at state 1 and the atmosphere. Φ_2 has the same significance for state 2. If useful work is done by the system during the process, it could lift a body or compress a spring so that energy in the amount W_{useful} is stored outside the system and can be reconverted into work. Thus, after the process has occurred, the total amount of useful work still obtainable is $(\Phi_2 + W_{useful})$. Thus

$$\begin{aligned} I &= \Phi_1 - (\Phi_2 + W_{useful}) \\ &= \begin{bmatrix}\text{Maximum useful work}\\ \text{initially obtainable}\end{bmatrix} - \left\{\begin{bmatrix}\text{Maximum useful work}\\ \text{finally obtainable}\end{bmatrix}\right. \\ &\qquad \left. + \begin{bmatrix}\text{Useful work done}\\ \text{during the process}\end{bmatrix}\right\} \\ &= \text{Decrease in maximum useful work obtainable} \end{aligned}$$

For the case where work is done on the system, W_{useful} is negative and Φ increases; therefore, the above result still holds. It can readily be shown that this interpretation of I is valid for both open and closed systems and for cases of heat exchange with energy reservoirs other than the atmosphere.

I has several other valuable interpretations. It is sometimes called energy made unavailable because it is the increase in unavailable energy caused by a process. It is also the decrease in available energy or the decrease in exergy, sometimes called the *exergy loss*. I has also been called the *degradation* (Reference 8·1), a highly descriptive name, because it is the amount of energy degraded from the valuable *available* form to the less valuable *unavailable form.*

Since $I = T_0 \, \Delta S_{univ}$ is a measure of the decrease of available energy for a process and T_0 is constant, then of course ΔS_{univ} is itself a measure of the degradation of energy. Consequently, some people simply use the increase of entropy of the universe, ΔS_{univ}, to evaluate the effects of irreversible processes or to compare alternative processes. They sometimes call ΔS_{univ} the *entropy production* or *entropy generation* (although entropy is a property of a system or a substance and not something that is *produced* or that exists by itself).

One advantage of using irreversibility, I, is that its value is usually of the same order as energy quantities involved in the same analysis, whereas numerical values of entropy production (ΔS_{univ}) are not so readily checked for reasonableness.

***I* is usually of the same order as energy quantities involved in a process.**

Continually estimating or predicting results of engineering calculations is good practice. Doing so can facilitate the calculations and also reveal calculation errors. However, estimating numerical values for entropy changes may be difficult for anyone without considerable experience in such calculations, so notice that the general equation $I = T_0 \, \Delta S_{univ}$ is useful for this purpose. As examples, in a compression process, the irreversibility usually does not exceed the work input; in a heat exchanger process, the irreversibility usually does not exceed the total heat transfer; and in a turbine or engine the irreversibility is unlikely to be of a greater order of magnitude than the work output. Other limits of irreversibility can be reasoned for various other processes. Then the ratio I/T_0 provides a limit to the possible entropy change of the universe for the process.

At this point, you may be inclined to believe that reducing irreversibility or losses in availability is always desirable. In practice, however, other goals may conflict. For example, for a given amount of heat transferred, the irreversibility is lower when the temperature difference across which heat is transferred is smaller. However, smaller temperature differences require larger heat transfer areas or greater time for a given amount of heat transfer so that heat exchangers and other equipment become more costly to build, install, and maintain. Engineers frequently must optimize designs among such conflicting constraints.

In engineering, *all* the constraints matter.

Example 8·8 — Irreversibility

PROBLEM STATEMENT

An insulated tank contains 0.60 kg of air initially at 200 kPa, 20°C. An impeller inside the tank is turned by an external motor until the pressure is 230 kPa. Ambient conditions are 95 kPa, 20°C. Determine the irreversibility of the process.

SOLUTION

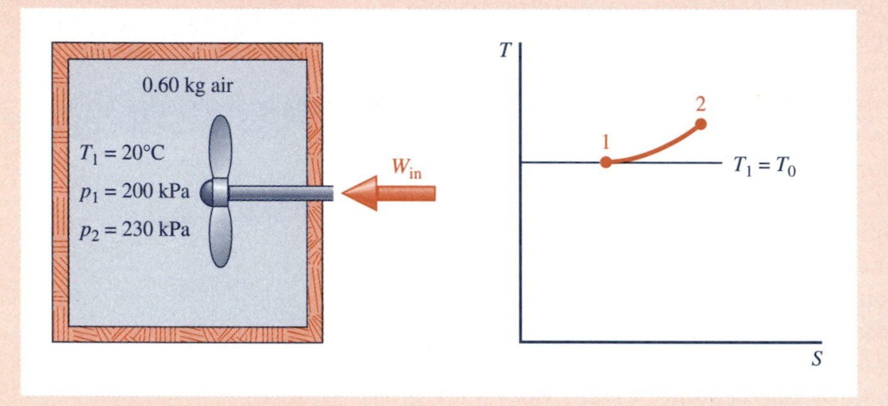

$$I = T_0\, \Delta S_{univ} = T_0(\Delta S_{sys} + \Delta S_{surr})$$

Since there is no change in entropy *of the surroundings* for this adiabatic process of a closed system,

$$I = T_0\, \Delta S_{sys} = T_0 m \int_1^2 ds = T_0 m \int_1^2 \left(\frac{du}{T} + p\frac{dv}{T} \right)$$

Under the conditions specified, air follows the ideal-gas equation of state. We must check to see if using a constant-specific-heat value is satisfactory. The final temperature for this constant-volume process is

$$T_2 = T_1 \frac{p_2}{p_2} = 293\left(\frac{230}{200}\right) = 337 \text{ K}$$

Table A·4 shows that over this temperature range c_v is nearly constant and that a suitable mean value is 0.719 kJ/kg·K.

Therefore,

$$I = T_0 m \int_1^2 \frac{c_v\, dT}{T} + 0 = T_0 m c_v \ln\frac{T_2}{T_1} = T_0 m c_v \ln\frac{p_2}{p_1}$$

$$= 293(0.6)0.719 \ln\frac{230}{200} = 17.7 \text{ kJ}$$

ALTERNATIVE SOLUTION

For this adiabatic process of a closed system, $W_{max} = (E_1 - T_0S_1) - (E_2 - T_0S_2)$, and application of the first law gives $W = E_1 - E_2$; so

$$I = W_{max} - W$$
$$= (E_1 - T_0S_1) - (E_2 - T_0S_2) - (E_1 - E_2) = T_0(S_2 - S_1)$$

From this point the solution is the same as the one presented above.

Discussion: The problem was solved without calculating $\Delta\Phi$ or W_{useful}. For discussion purposes, let us calculate these.

$$W_{useful} = W - p_0(V_2 - V_1) = E_1 - E_2 + Q - p_0(V_2 - V_1)$$
$$= U_1 - U_2 + 0 - 0 = mc_v(T_1 - T_2) = 0.6(0.719)(293 - 337)$$
$$= -19.0 \text{ kJ}$$
$$\Delta\Phi = \Phi_2 - \Phi_1 = (E_2 + P_0V_2 - T_0S_2) - (E_1 + P_0V_1 - T_0S_1)$$
$$= U_2 - U_1 - T_0(S_2 - S_1)$$
$$= 19.0 - 17.7 = 1.3 \text{ kJ}$$

Now we see that by means of the impeller 19.0 kJ of work was done on the system but the availability of the system (and atmosphere) increased by only 1.3 kJ. That is, only 1.3 kJ more of useful work can be obtained from a process that starts at state 2 than can be obtained from one that starts at state 1. After the process, the surroundings can supply 19.0 kJ less work than before; so the net *decrease* in the amount of work obtainable is (19.0 − 1.7 = 17.7 kJ), and this is the irreversibility of the process.

Example 8·9 Irreversibility

PROBLEM STATEMENT

Water at 12 psig flowing through the tubes of a shell-and-tube heat exchanger at a rate of 40 gallons per minute is heated from 60 F to 160 F by means of steam that enters at 3 psig, 92 percent quality. Condensate leaves the heat exchanger at 140 F. The temperature of a large water supply is 60 F. Determine, per pound of water heated, the irreversibility of the process.

SOLUTION

We first make a labeled flow diagram and a *Ts* diagram. We assume that the flow is steady. We observe that pressures are given as gage pressures but that atmospheric pressure is not given. We will assume p_{atm} = 14.5 psia. We will also assume that there are no stray heat losses; that is, all the heat given up by the steam is

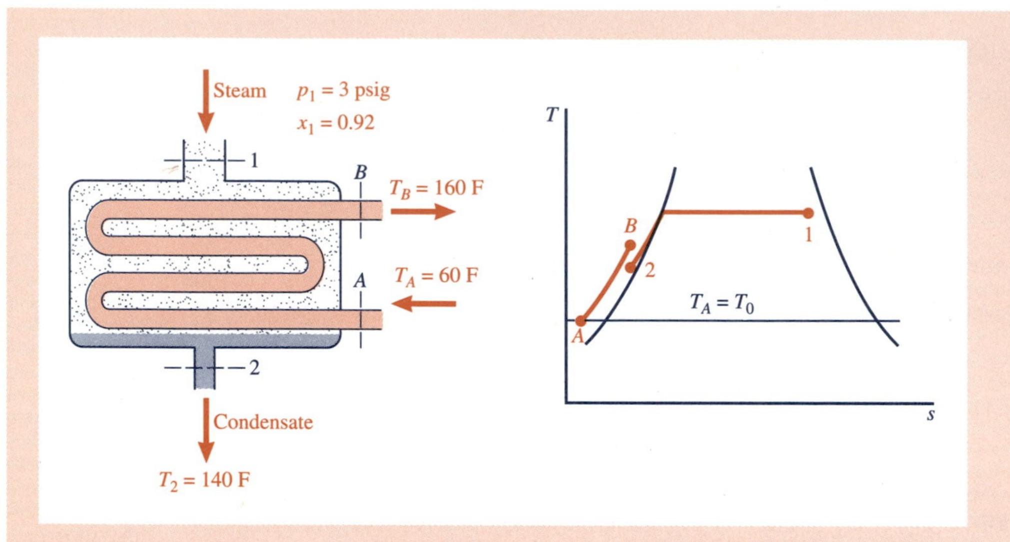

transferred to the water. To find the ratio of steam flow rate to the flow rate of the water being heated, we use the first law for the steady-flow system:

$$\dot{m}_w(h_B - h_A) = \dot{m}_s(h_1 - h_2)$$

$$\frac{\dot{m}_s}{\dot{m}_w} = \frac{h_B - h_A}{h_1 - h_2} = \frac{127.9 - 28.0}{1076.5 - 107.9} = 0.103\,\frac{\text{lbm steam}}{\text{lbm water}}$$

where the enthalpy values are from the steam tables (using $h_1 = h_{f1} + x_1 h_{fg1}$) or computer model STEAM. The irreversibility rate is

$$\begin{aligned} \dot{I} &= T_0\,\Delta\dot{S}_{univ} = T_0(\Delta\dot{S}_{sys} + \Delta\dot{S}_{surr}) \\ &= T_0[0 + \dot{m}_w(s_B - s_A) + \dot{m}_s(s_2 - s_1)] \\ i &= \frac{\dot{I}}{\dot{m}_w} = T_0\left[s_B - s_A + \frac{\dot{m}_s}{\dot{m}_w}(s_2 - s_1)\right] \\ &= 520[0.23095 - 0.0545 + 0.103(0.19853 - 1.6292)] = 15.1 \text{ B/lbm water} \end{aligned}$$

Example 8·10 Irreversibility in a Steam Turbine

PROBLEM STATEMENT

A well-insulated turbine takes in steam at 800 kPa, 300°C, with negligible velocity and exhausts steam at 30 mm Hg, 94 percent quality, with a velocity of 150 m/s. The flow rate is 54,000 kg/h. Atmospheric pressure is 730 mm Hg. The plant rejects heat to lake water at 15°C. Determine the irreversibility per kilogram of steam.

SOLUTION

Analysis: We first sketch a flow diagram and a *Ts* diagram. Since the flow is steady, $\Delta s_{os} = 0$ and the expression for irreversibility,

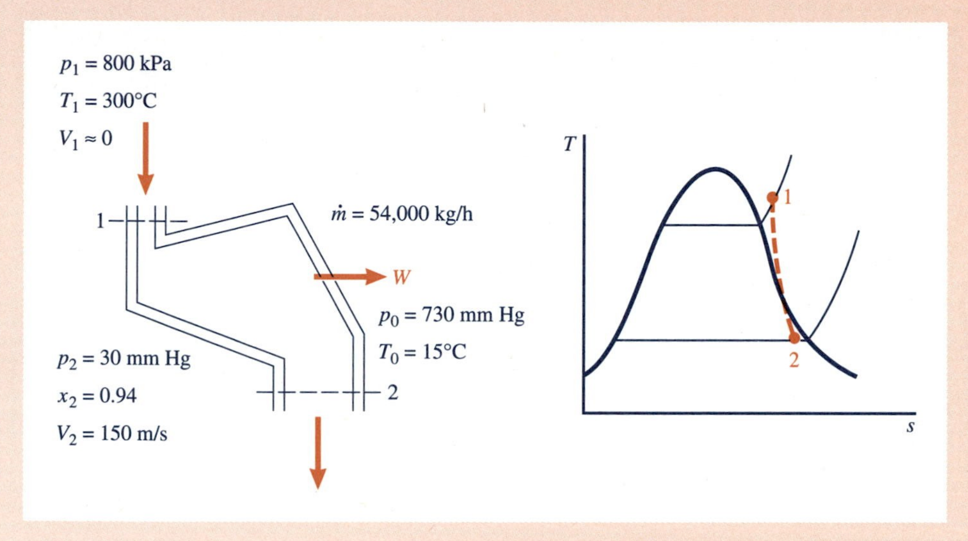

$$i = T_0\, \Delta s_{univ} = T_0(\Delta s_{os} + \Delta s_{surr})$$

becomes for the adiabatic process

$$i = T_0(s_2 - s_1)$$

Taking the entropy values from the steam tables or the computer model STEAM (as in Example 8·5) gives

$$i = 288(7.989 - 7.2319) = 218 \text{ kJ/kg}$$

$$\dot{I} = \dot{m}i = \frac{54{,}000(218.0)}{3600} = 3270 \text{ kW}$$

Here we see that the time rate of irreversibility is the same as the difference between $\dot{W}_{max}$ and $\dot{W}$ calculated in Example 8·5: (12,830 − 9560) = 3270 kW.

Example 8·11 Irreversibility in a Refrigeration System

PROBLEM STATEMENT

In a refrigeration system, saturated liquid R134a enters a capillary tube (a long, small-bore tube) at 600 kPa and leaves at 100 kPa. The flow is steady. The geometries of the fittings on the ends of the tube are such that there is no change in kinetic energy. While passing through the tube, the refrigerant receives 7.0 kJ/kg of heat from an energy reservoir at 0°C. The atmosphere is at 15°C. Determine the irreversibility of the process.

SOLUTION

We first sketch the flow system and a *Ts* diagram.

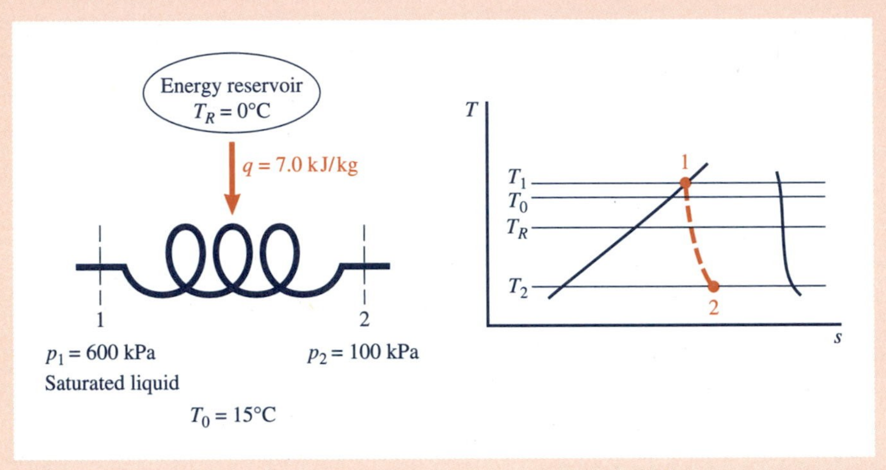

Then

$$i = \frac{I}{m} = \frac{T_0}{m}(\Delta S_{os} + \Delta S_{surr})$$

For any steady-flow system, $\Delta S_{os} = 0$. The entropy of the surroundings changes by an amount $m(s_2 - s_1)$ because of the mass transfer between the system and the surroundings and also by an amount $-Q_R/T_R$, where Q_R is heat transfer *to* the system *from* the reservoir at T_R. Thus

$$i = \frac{T_0}{m}\left[0 + m(s_2 - s_1) - \frac{Q_R}{T_R}\right] = T_0\left(s_2 - s_1 - \frac{q_R}{T_R}\right)$$

s_1 can be read directly from R134a tables or a computer model as s_f at 600 kPa. We know the pressure at the outlet, but we must know one other independent property to determine the state and the value of s_2. (Since at the outlet the R134a is a liquid–vapor mixture, we can find its temperature immediately from the saturated R134a table, but we must still find another property since p and T are not independent for two-phase mixtures.) By application of the first law we can find h_2 because the first law applied to a steady-flow system,

$$h_2 = h_1 + q - w - \Delta ke - \Delta pe$$

in this case reduces to

$$h_2 = h_1 + q = 81.32 + 7.0 = 88.3 \text{ kJ/kg}$$

Now we can find the quality at the outlet,

$$x_2 = \frac{h_2 - h_f}{h_{fg}} = \frac{88.3 - 17.15}{217.4} = 0.327$$

and

$$s_2 = s_f + x_2 s_{fg} = 0.0716 + 0.327(0.8807) = 0.360 \text{ kJ/kg·K}$$

We can finally calculate i by substituting numerical values:

$$i = T_0\left(s_2 - s_1 - \frac{q_R}{T_R}\right) = 288\left(0.360 - 0.3073 - \frac{7.0}{273}\right) = 7.8 \text{ kJ/kg}$$

8·5 Availability Accounting and Available Energy Accounting

The first law is the basis for energy accounting, which provides much valuable information. Still, we often want to know more information. We want to know how much of the energy can be converted into work or we want to know how much a certain process reduces the amount of work obtainable from a system. Availability or available energy accounting that is made possible by the second law provides such information to supplement that provided by the first law. The following example illustrates the calculation of this additional information, and in subsequent chapters several other examples are presented.

There are many ways to present availability or available energy accounting; be alert to finding the most effective ways for any particular application.

Example 8·12 Energy, Availability, and Available Energy Accounting

PROBLEM STATEMENT

A steady-flow power plant using air as a working medium is shown in the flow diagram and the *Ts* diagram. The compressor and the turbine operate adiabatically. Kinetic energy changes are negligible. Pressures and temperatures are shown on the diagrams. The air entering at state 1 is at atmospheric conditions. Assuming that the heat addition process is reversible, make an (*a*) energy accounting, (*b*) availability accounting, and (*c*) available energy accounting.

SOLUTION

For the temperature range involved, c_p varies significantly, so we will use the air tables (Table A·12) or the computer model for air properties.

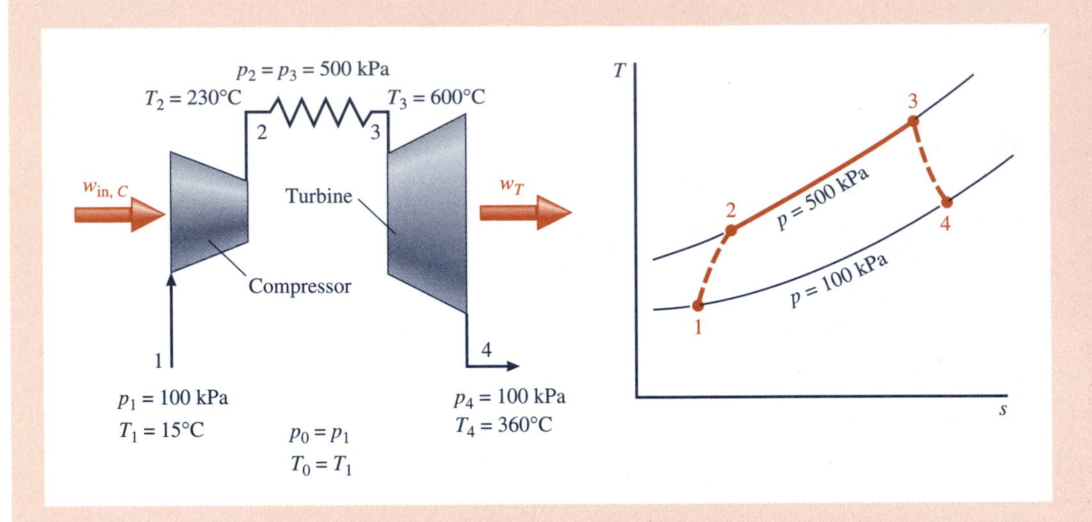

(*a*) Application of the first law to each of the three pieces of equipment gives

$$w_{in,C} = h_2 - h_1 = 508.3 - 290.4 = 217.9 \text{ kJ/kg}$$
$$q = h_3 - h_2 = 904.9 - 508.3 = 396.6 \text{ kJ/kg}$$
$$w_T = h_3 - h_4 = 904.9 - 644.0 = 260.9 \text{ kJ/kg}$$

Since this is an open system, an energy accounting must include the energy entering with the air at the inlet and leaving with the air at the exhaust. At each section, the energy is $u + pv$ or h. We do not have absolute values for internal energy and enthalpy, so we must use values at each section based on the same datum. This can be done by using the air table (or computer model AIRPROP) values. This provides an energy balance (or energy accounting) for the plant as follows:

Energy In		*Energy Out*	
With entering fluid	$h_1 = 290.4$ kJ/kg	With leaving fluid	$h_4 = 644.0$ kJ/kg
Compressor work$_{in}$	$w_{in} = 217.9$	Turbine work	$w = 260.9$
Heat	$q = 396.6$		
Total	$= 904.9$ kJ/kg	Total	$= 904.9$ kJ/kg

This might be shown graphically as in Diagram *A*. If part of the turbine work is used to drive the compressor directly (as is the case in actual gas turbine power plants), then the net work of the plant is $w_T - w_{in,C}$, and the energy accounting can be shown as in Diagram *B*.

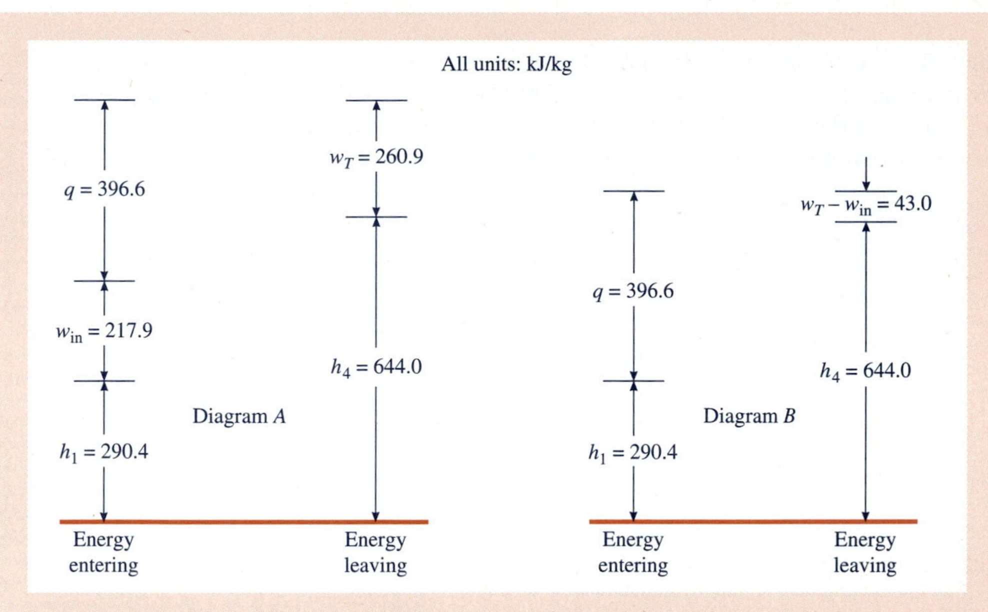

If the energy datum used is the state of the atmospheric air entering the plant, the energy entering with the air would be zero and the energy leaving with the exhaust air is $(h_4 - h_1) = 644.0 - 290.4 = 353.6$ kJ/kg. The energy accounting table is then

Energy In		*Energy Out*	
With entering fluid	$h_1 - h_0 = 0$ kJ/kg	With leaving fluid	$h_4 - h_0 = 353.6$ kJ/kg
Compressor work$_{in}$	$w_{in} = 217.9$	Turbine work	$w = 260.9$
Heat	$q = 396.6$		
Total	$= 614.5$ kJ/kg	Total	$= 614.5$ kJ/kg

The energy accounting is shown graphically in Diagrams *C* and *D*.

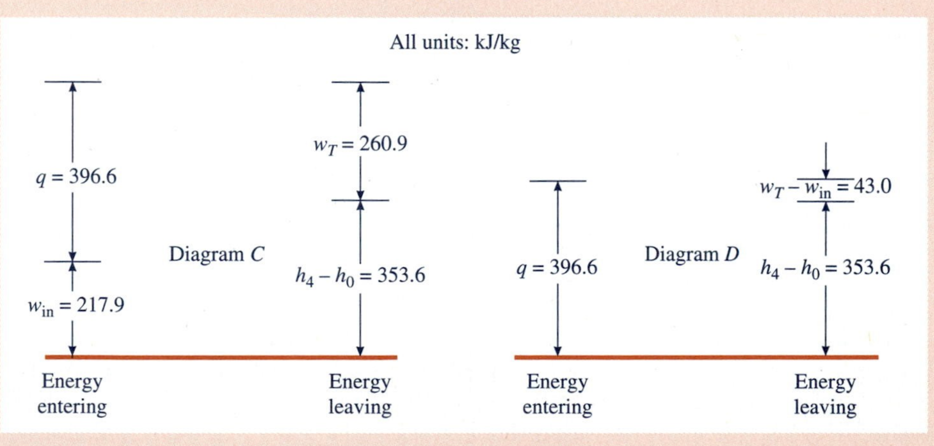

Diagrams *A*, *B*, *C*, and *D* are different ways to show the same energy accounting.

(*b*) For availability accounting, the scale of availability is fixed because, by definition, the availability or stream availability of a fluid in the dead state (100 kPa, 15°C, here) is zero. That is, we do not have a choice of datum for availability as we do for internal energy or enthalpy.

For each state,

$$y = h - T_0 s - (h_0 - T_0 s_0) = h - h_0 - T_0(s - s_0)$$

and

$$s - s_0 = \phi - \phi_0 - R \ln \frac{p}{p_0}$$

Substituting values for each state from the air tables or using the computer model gives

$$y_1 = 0 \qquad y_2 = 188.4 \text{ kJ/kg} \qquad y_3 = 415.4 \text{ kJ/kg} \qquad y_4 = 121.9 \text{ kJ/kg}$$

For the remainder of the solution we will need the entropy changes for the three processes that occur within the plant, so we calculate them now:

$$s_2 - s_1 = \phi_2 - \phi_1 - R \ln \frac{p_2}{p_1} = 7.231 - 6.667 - 0.287 \ln \frac{500}{100} = 0.1023 \text{ kJ/kg·K}$$

$$s_3 - s_2 = \phi_3 - \phi_2 - R \ln \frac{p_3}{p_2} = 7.820 - 7.231 - 0 = 0.589 \text{ kJ/kg·K}$$

$$s_4 - s_3 = \phi_4 - \phi_3 - R \ln \frac{p_4}{p_3} = 7.471 - 7.820 - 0.287 \ln \frac{100}{500} = 0.113 \text{ kJ/kg·K}$$

For each of the three processes, $i = T_0 \, \Delta s_{univ}$, and the values are

$$i_{1-2} = 29.5 \text{ kJ/kg} \qquad i_{2-3} = 0 \qquad i_{3-4} = 32.6 \text{ kJ/kg}$$

where $i_{2-3} = 0$ because the heat transfer process is specified as reversible.

Then an availability accounting can take the following form:

State	*Process*	*y*, kJ/kg	*q*, kJ/kg	*w*, kJ/kg	Δ*y*, kJ/kg	*i*, kJ/kg
1		0				
	1–2		0	−217.9	188.4	29.5
2		188.4				
	2–3		396.6	0	227.0	0
3		415.4				
	3–4		0	260.9	−293.5	32.6
4		121.9				

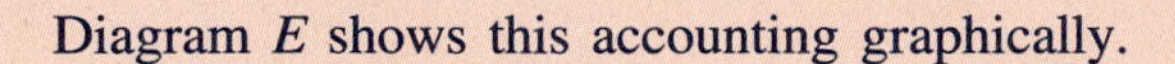

Diagram E shows this accounting graphically.

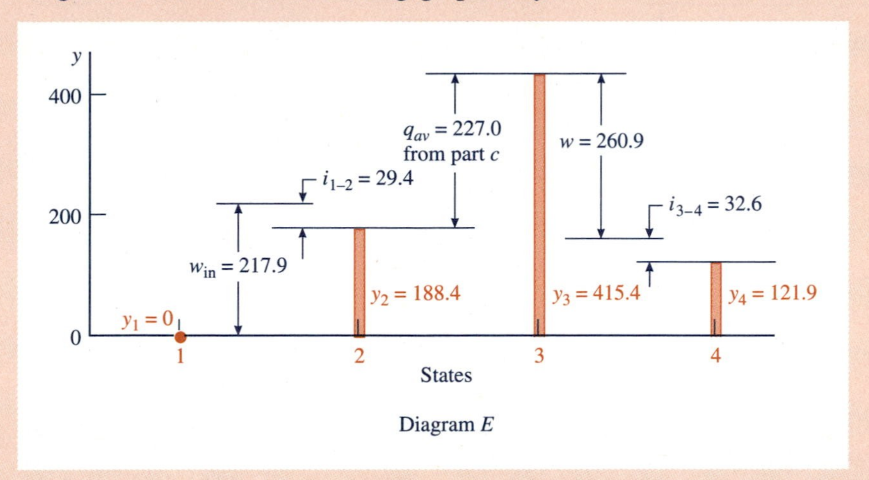

Diagram E

Comment: Notice that in process 1–2, the work input was 217.9 kJ/kg, but the stream availability increased by only 188.4 kJ/kg because the irreversibility value of 29.5 kJ/kg indicates that this amount of energy was made unavailable by the process. Likewise, during the turbine expansion, process 3–4, the stream availability decreased by 293.5 kJ/kg, but the work output was only 260.9 kJ/kg because the process was irreversible. Notice that the availability of the air in state 4 indicates that 121.9 kJ/kg of work could be obtained from the exhaust gas stream, and this is more than the net work output of the plant (260.9 − 217.9 = 43.0 kJ/kg). The 227.0 kJ/kg increase in availability in the heat exchanger was disposed of as follows:

Work	43.0 kJ/kg
Leaving stream availability	121.9
Compressor irreversibility	29.5
Turbine irreversibility	32.6
Total	227.0 kJ/kg

(*c*) For the available energy accounting, let us close the cycle of the power plant by imagining that the air leaving the turbine is cooled in a heat exchanger and then reenters the compressor at state 1. Then we are dealing with a cycle as shown on the *Ts* diagram. The cycle occurs within a closed system, so the only energy transfers are the heat added (q_{2-3}), the net work output ($w_T - w_{in,C}$ or $w_{3-4} + w_{1-2}$), and the heat rejected to the surroundings (q_{4-1}).

The available parts of the two transfers of heat are calculated by

$$q_{av,2-3} = q_{2-3} - q_{unav,2-3} = q_{2-3} - T_0(s_3 - s_2) = 396.6 - 288(0.589) = 227.0 \text{ kJ/kg}$$
$$q_{av,4-1} = q_{4-1} - q_{unav,4-1} = q_{4-1} - T_0(s_1 - s_4) = -353.6 - 288(-0.804) = 121.9 \text{ kJ/kg}$$

where the value of $(s_1 - s_4)$ is obtained as the negative of the sum $(s_2 - s_1) + (s_3 - s_2) + (s_4 - s_3)$. The results are shown in Diagram F.

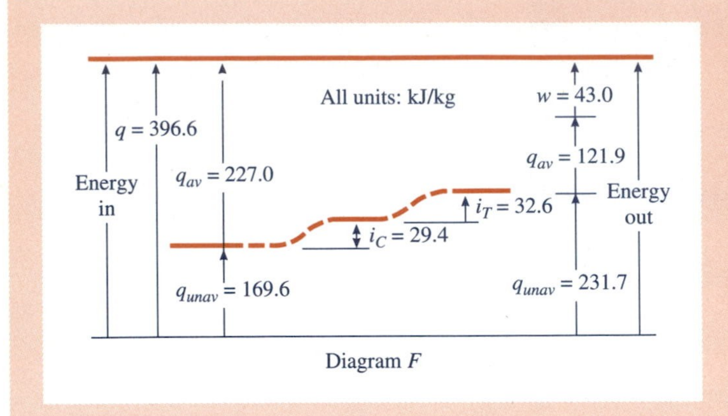

Diagram *F*

Comment: Diagram *F* shows that of the 396.6 kJ/kg of heat added, 169.6 kJ/kg was unavailable energy that could never be converted to work. The irreversibility of the compressor and turbine processes increased the unavailable energy so that, of the heat rejected, 353.6 kJ/kg, the unavailable part was 231.7 kJ/kg and the available part was 121.9 kJ/kg.

General Comment: Energy, availability, and unavailable energy accounting results have been shown for this simple power plant. As the systems studied become more complex, the value of these different accounting schemes becomes more apparent, especially for identifying places within the systems where improvement is most needed.

8·6 Summary

This chapter has treated the questions of

- how much work can be obtained from a system initially in a given state,
- how much of a given quantity of heat can be converted into work, and
- how much the capability of producing work is reduced by irreversible effects during a process.

Answering these questions involves *availability* (ϕ), *stream availability* (y), *unavailable part of heat transfer* (q_{unav}), and *irreversibility* (i), none of which is a property. All these quantities involve, in addition to a system, an *atmosphere* defined as a *part of the surroundings* at uniform pressure (p_0) and temperature (T_0) that is so large that processes of a system do not affect the pressure or temperature of the atmosphere. The atmosphere is an energy reservoir: It can absorb or give up large quantities of energy without experiencing a temperature change. For many analyses, the atmosphere is the atmosphere that surrounds us, the atmosphere that we breathe. Sometimes, though, the system we are dealing with may exchange heat with a large body of water as an energy reservoir. Then for T_0 we use the temperature of the body of water. T_0 is sometimes called the *sink temperature*.

For an open system that exchanges heat with only the atmosphere at T_0 and another energy reservoir at T_R, the maximum work from the combination of system and surroundings for any infinitesimal change in the system is

$$\delta W_{max} = -d(E - T_0 S)_{os} + (e_1 + p_1 v_1 - T_0 s_1)\,\delta m_1 - (e_2 + p_2 v_2 - T_0 s_2)\,\delta m_2 + \delta Q_R\left(1 - \frac{T_0}{T_R}\right) \qquad (8\cdot 2)$$

where δm_1 is the mass entering, δm_2 is the mass leaving, δQ_R is the heat transferred to the system from the reservoir at T_R, and the subscript *os* designates properties of the system. This equation simplifies for a *closed system* as follows:

$$\delta W_{max} = -d(E - T_0 S) + \delta Q_R\left(1 - \frac{T_0}{T_R}\right)$$

and for a *steady-flow system* as follows:

$$\delta W_{max} = [(e_1 + p_1 v_1 - T_0 s_1) - (e_2 + p_2 v_2 - T_0 s_2)]\,\delta m + \delta Q_R\left(1 - \frac{T_0}{T_R}\right) \qquad (8\cdot 4a)$$

Useful work is defined as the work done by a system exclusive of that done on the atmosphere. If p_0 is the pressure of the atmosphere and V_i and V_f are the initial and final volumes of a system, then

$$W_{useful} = W - p_0(V_f - V_i)$$

The greatest possible amount of work can be obtained from a system that exchanges heat with only the atmosphere if the system goes to a final state in which it is in equilibrium with the atmosphere. The pressure and temperature of the system will then be p_0 and T_0, and the system is said to be in the *dead state* because no work can be produced.

The *availability* of a system in any state is the maximum useful work that can be obtained as the system goes from that state to the dead state while exchanging heat with only the atmosphere. Availability is not a property of the system alone; its value depends on p_0 and T_0 as well as on the properties of the system.

The availability of a *closed system*, Φ (and $\phi = \Phi/m$), is given by

$$\Phi_1 \equiv W_{max\ useful,1-0} = (E_1 + p_0 V_1 - T_0 S_1) - (E_0 + p_0 V_0 - T_0 S_0) \qquad (8\cdot 5a)$$

where the subscript 0 denotes properties of the system when in the dead state. The maximum useful work obtainable from a closed system that exchanges heat with only the atmosphere as it passes from state 1 to state 2 is given by the decrease in Φ:

$$W_{max\ useful} = -\Delta\Phi = \Phi_1 - \Phi_2$$
$$= (E_1 + p_0V_1 - T_0S_1) - (E_2 + p_0V_2 - T_0S_2) \quad (8\cdot5b)$$

Since useful work can always be obtained if a closed system is in a state other than the dead state, $\Phi \geq 0$ for all states. If the system also receives heat Q_R from an energy reservoir at T_R, then

$$W_{max\ useful} = -\Delta\Phi + Q_R\left(1 - \frac{T_0}{T_R}\right) \quad (8\cdot5c)$$

The availability of a *steadily flowing* fluid is called *stream availability* and is denoted by the symbol Y (and $y = Y/m$). It is defined as the maximum work that can be obtained as the fluid goes to the dead state while exchanging heat with only the atmosphere. In the dead state the fluid is at p_0 and T_0 and is also at rest, $V_0 = 0$. For a steadily flowing fluid in a state 1, in the absence of electrical, magnetic, solid distortion, and surface tension effects, we have

$$y_1 = \left(h_1 + \frac{V_1^2}{2} + gz_1 - T_0s_1\right) - (h_0 + gz_0 - T_0s_0) \quad (8\cdot6a)$$

The *Darrieus function* is defined as $b \equiv h - T_0s$. *For any given temperature of the atmosphere, b* is a property of a substance. The use of b shortens the preceding equation to

$$y_1 = \left(b_1 + \frac{V_1^2}{2} + gz_1\right) - (b_0 + gz_0) \quad (8\cdot6b)$$

and, for steady flow from state 1 to state 2 and heat exchange with only the atmosphere,

$$w_{max} = -\Delta y = y_1 - y_2 = \left(b_1 + \frac{V_1^2}{2} + gz_1\right) - \left(b_2 + \frac{V_2^2}{2} + gz_2\right) \quad (8\cdot6c)$$

Stream availability, y, is often called simply the availability of a flowing fluid, but remember that for a fluid that is part of a flowing stream the pertinent quantity is y and not ϕ.

For a steadily flowing fluid that goes from one state to another while exchanging heat with only the atmosphere, the maximum work obtainable per unit mass is $-\Delta y$. If the fluid also receives heat q_R per unit mass from an energy reservoir at T_R, the maximum work that can be obtained per unit mass is

$$w_{max} = -\Delta y + q_R\left(1 - \frac{T_0}{T_R}\right) \quad (8\cdot6d)$$

All energy, whether stored or transitional, can be divided into two parts:

available energy (also called *exergy*) and *unavailable energy*. Available energy is energy that can by some means be converted into work. Unavailable energy is energy that cannot be converted into work, even by reversible engines. Unavailable energy cannot be converted into available energy. An equivalent to the increase of entropy principle is that the unavailable energy of the universe continually increases or, in the limit of a reversible process, remains constant.

The *irreversibility*, I, of any process is defined as

$$I \equiv W_{max} - W = W_{max\ useful} - W_{useful} \tag{8·8}$$

W in this equation is the work actually produced. During the process, heat may be exchanged with various energy reservoirs as well as with the atmosphere. W_{max} is the maximum work for a process that (1) is between the same end states as the actual process, (2) involves the same amounts of heat exchange with the various energy reservoirs other than the atmosphere, and (3) involves heat exchange with the atmosphere also if necessary. For any process,

$$I = T_0(\Delta S_{os} + \Delta S_{surr}) = T_0\ \Delta S_{isolated\ sys} = T_0\ \Delta S_{univ} \tag{8·9}$$

For all processes, $I \geq 0$, with the equality holding for reversible processes. I can be interpreted as the decrease in available energy caused by a process.

Availability accounting provides information that supplements that obtained from energy accounting. It involves the determination of availability changes for each of a series of processes. Comparison of the work done with the availability change gives a measure of the degree to which each process approaches the ideal. *Available energy accounting* is similar. Both indicate where efforts spent in improving performance are likely to be most fruitful.

References and Suggested Reading

8·1 Dixon, John R., *Thermodynamics I: An Introduction to Energy*, Prentice Hall, Englewood Cliffs, NJ, 1975. (For irreversibility, this book uses the term *degradation*, which is more descriptive and has other advantages.)

8·2 Moran, Michael J., and Howard N. Shapiro, *Fundamentals of Engineering Thermodynamics*, 2nd ed., Wiley, New York, 1992, Chapter 7.

8·3 Moran, Michael J., *Availability Analysis*, Prentice Hall, Englewood Cliffs, NJ, 1982. (This book provides greater detail on several aspects of availability.)

Problems

8·1 One-twentieth kilogram of acetylene in a closed system expands from 200 kPa, 60°C, to 100 kPa, 35°C, while receiving 1.0 kJ of heat from a reservoir at 120°C. The surrounding atmosphere is at 95 kPa, 25°C. Determine the maximum work. How much of this work would be done on the atmosphere?

8·2 Saturated liquid refrigerant R134a at 1.2 MPa is throttled to 200 kPa. Determine the maximum work if the sink temperature (see the first paragraph of Section 8·6) is 11°C.

8·3 A tank having a volume of 2.0 m^3 initially contains air at atmospheric conditions, 1 atm, 17°C. The pressure in the tank is to be reduced to 20 Pa to test a piece of equipment in the tank. What is the minimum work required to lower the pressure?

8·4 The air pressure in a tank is to be maintained at 20 Pa while a test is conducted. Leakage of atmospheric air into the tank occurs at a rate of 0.10 liter per hour. Determine the minimum power required to maintain the test pressure in the tank.

8·5 Describe a means for obtaining from the system described in Example 8·1 the maximum work calculated.

8·6 Solve Example 8·2 for a range of T_0 values from −20°C to 40°C and present your results in a plot.

8·7 For atmospheric conditions of 1 atm, 0°C, does a closed system consisting of an ideal gas at 1 atm have a higher availability when it is at −20°C or when it is at 20°C?

8·8 For atmospheric conditions of 1 atm, 0°C, does a closed system consisting of an ideal gas at 0°C have a higher availability when it is at 0.5 atm or when it is at 2 atm?

8·9 Calculate the availability of a 20-kg block of ice at 0°C if the surrounding atmosphere is at 95 kPa, 15°C.

8·10 Find the availability of air in a 300-liter tank if the tank is pressurized to 700 kPa while the surrounding atmosphere is at 95 kPa. Both the air in the tank and the atmosphere are at 20°C. Is this the same as the maximum work that can be obtained from expansion of the air into the surrounding atmosphere?

8·11 A tank with a volume of 0.3 m^3 contains air at 95 kPa, 260°C. The surrounding atmosphere is at 95 kPa, 20°C. Determine the availability of the air in the tank.

8·12 Solve Problem 8·11 with the air in the tank initially at 14 kPa, 20°C.

8·13 Solve Problem 8·11 with the air in the tank initially at 95 kPa, −70°C.

8·14 Two moles of ethane in a closed system expands from 200 kPa, 60°C, to 100 kPa, 35°C, while receiving heat from a reservoir at 120°C. Determine the maximum amount of heat that can be transferred from the reservoir to the ethane under these conditions.

8·15 Solve Example 8·1 if the tank initially contains air at 35 kPa, 20°C.

8·16 Solve Example 8·1 if the tank initially contains air at 35 kPa, −20°C.

8·17 Air initially at atmospheric conditions of 100 kPa, 20°C, is held in a rigid, well-insulated tank that has a volume of 2.80 m^3. A paddle wheel in the tank is driven by an external motor until the temperature of the air in the tank rises to 313°C. (*a*) Determine the work done on the air in kJ/kg. (*b*) After the process has occurred, what fraction of the energy gained by the system during the process could possibly be reconverted to work?

8·18 For atmospheric conditions of 1 atm, 0°C, does a steadily flowing ideal gas at 1 atm have a higher stream availability when it is at −20°C or when it is at 20°C?

8·19 For atmospheric conditions of 1 atm, 0°C, does a steadily flowing ideal gas at 0°C have a higher availability when it is at 0.5 atm or when it is at 2 atm?

8·20 Air enters a compressor at atmospheric conditions of 100 kPa, 5°C, and is compressed to 200 kPa, 75°C. The flow rate is 6.0 kg/s. Determine, per kilogram of air, the stream availability of the entering air, the stream availability of the leaving air, and the minimum work input. What is the physical meaning of the minimum work input? Does it refer to an adiabatic compressor? To a water-cooled compressor?

8·21 Helium enters an actual turbine at 300 kPa, 200°C, and expands to 100 kPa, 150°C. Heat transfer to the atmosphere at 101.3 kPa, 25°C, amounts to 7.0 kJ/kg. Calculate the entering stream availability, the leaving stream availability, and the maximum work. What is the physical meaning of the maximum work? Does it refer to an adiabatic turbine? To one with a heat loss of 7.0 kJ/kg?

8·22 Air in a closed system is initially at 300 kPa, 20°C. It expands reversibly and adiabatically until its pressure is equal to the pressure of the atmosphere, 100 kPa. Atmospheric temperature is 20°C. Determine the availability at the beginning and end of the process.

8·23 Air flows steadily, reversibly, and adiabatically through a turbine with no change in kinetic energy. At inlet the air is at 300 kPa, 20°C. The exhaust pressure is 100 kPa. The surrounding atmosphere is at 100 kPa, 20°C. Determine the stream availability at inlet and exhaust.

8·24 Steam enters a radiant superheater at 7 MPa, 370°C, with a velocity of 33 m/s and leaves at 6.5 MPa, 480°C, with a velocity of 45 m/s. Heat is transferred to the steam from a furnace at 1400°C. The atmosphere is at 100 kPa, 15°C. Determine the maximum work for this process. Devise a means for producing this amount of work without changing the process end states.

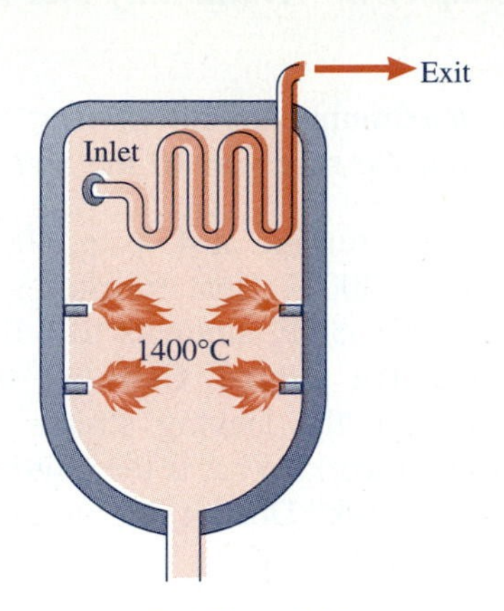

Problem 8·24

8·25 Dry saturated steam at 150°C is held in a rigid, well-insulated tank that has a volume of 0.35 m^3. A paddle wheel in the tank is driven by an external motor until the steam is at 1.4 MPa, 540°C. Atmospheric conditions are 1 bar, 25°C. (*a*) Calculate the work input in kJ/kg. (*b*) After the process has occurred, how much of the energy put in as work during the process could possibly be reconverted to work?

8·26 One-tenth pound of air in a closed system expands from 30 psia, 140 F, to 15 psia, 100 F, while receiving 1.0 B of heat from a reservoir at 250 F. The surrounding atmosphere is at 14.0 psia, 80 F. Determine the maximum work. How much of this work would be done on the atmosphere?

8·27 Saturated liquid ammonia at 170 psia is throttled to 30 psia. Determine the maximum work if the sink temperature (see the first paragraph of Section 8·6) is 65 F.

8·28 Calculate the availability of a 50-lbm block of ice at 32 F if the surrounding atmosphere is at 14.0 psia, 60 F.

8·29 A tank with a volume of 10 cu ft contains air at 14.0 psia, 500 F. What is the availability of the tank contents if the surrounding atmosphere is at the same pressure but only 70 F?

8·30 Two pounds of air in a closed system expands from 30 psia, 140 F, to 15 psia, 100 F, while receiving heat from a reservoir at 250 F. Under these conditions, what is the maximum amount of heat that can be transferred from the reservoir to the air?

8·31 Air enters a turbine at 45 psia, 400 F, and expands to 15 psia, 300 F. Heat transfer to the atmosphere at 14.1 psia, 80 F, amounts to 3 B/lbm of air passing through the turbine. Calculate the entering stream availability, the leaving stream availability, and the maximum work. What is the physical meaning of the maximum work? Does it refer to this turbine or to an adiabatic one?

8·32 In a steam power plant, the heat rejected to the condenser cooling water may be much greater than the energy lost up the stack. Why is it that the stack loss is of greater concern; that is, why are greater efforts made to reduce it rather than the condenser loss?

8·33 For a closed system, the first law gives $\oint \delta Q = \oint \delta W$. Comment on the following equation for a closed system that is proposed as an analogous relation given by the second law:

$$\oint \delta Q_{av} = \oint \delta W + T_0 \, \Delta S_{surr}$$

8·34 Explain whether or not the following statements are true: (*a*) All reversible processes between the same two end states produce the same amount of work. (*b*) All reversible adiabatic processes between the same two end states produce the same amount of work. (*c*) All reversible processes between the same two end states produce the same amount of work if the system exchanges heat with only a constant-temperature reservoir.

8·35 Is it possible for the availability of a closed system to be negative? Explain your answer, giving reasons or examples.

8·36 Give an example of an ideal gas in a closed system that passes through a series of states so that the energy of the system is continually decreasing but the availability is increasing.

8·37 State whether the following statement is true or false and give a proof of your conclusion: The availability of a closed system can be increased only by doing work on the system or by transferring heat to it from a body at a temperature other than T_0.

8·38 State whether the following statement is true or false and give a proof of your conclusion: The stream availability y of a steadily flowing fluid can be increased only by doing work on the fluid or by transferring heat to it from a body at a temperature other than T_0.

8·39 An ideal gas in a closed system is at a pressure $p_1 > p_0$ and a temperature $T_1 = T_0$. It expands adiabatically to p_0 while doing work. Is there an increase, decrease, no change, or an indeterminate change in its (*a*) internal energy, (*b*) availability?

8·40 An ideal gas flows steadily at a pressure $p_1 > p_0$ and a temperature $T_1 = T_0$. It expands adiabatically to p_0 while doing work. Is there an increase, decrease, no change, or an indeterminate change in its (*a*) internal energy, (*b*) enthalpy, (*c*) kinetic energy, (*d*) stream availability?

8·41 An ideal gas flows steadily. Illustrate how this gas can pass through a series of states in which its energy decreases while its stream availability increases.

8·42 Heat is transferred from a furnace at a constant temperature of 1400°C to water that is evaporating at a constant temperature of 250°C. The lowest temperature in the surroundings is 30°C. For each megajoule

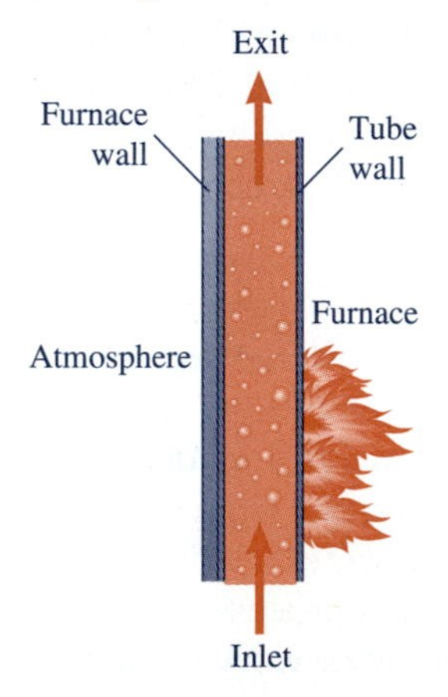

Problems 8·42 and 8·49

of energy transferred, how much would be available if it were absorbed by a fluid at the furnace temperature? How much of the energy is available as it is received by the water at 250°C? Show these quantities as areas on a *TS* diagram.

8·43 Consider the transfer of heat from an energy reservoir at 250°C to 2.5 kg of air initially at 100 kPa, 60°C, trapped in a closed, rigid tank. Heat is transferred until the temperature of the air is 170°C. The temperature of the surroundings is 5°C. (*a*) How much heat is transferred? (*b*) How much of the energy removed from the reservoir is available energy? How much is unavailable? (*c*) How much of the energy added to the air in the tank is available energy? How much is unavailable?

8·44 A closed, rigid container holds dry saturated refrigerant R134a at 800 kPa. Heat is added until the pressure becomes 1000 kPa. How much of the heat added is available energy if the sink temperature is 25°C?

8·45 A system composed of three kilograms of oxygen is cooled at a constant pressure of 140 kPa from 150 to 35°C. The sink temperature is 35°C. All the heat removed is dissipated to the surroundings. How much of the heat removed from the oxygen is available energy? Calculate the entropy change of the system and of the universe. By how much does the total energy of the universe change?

8·46 A closed, rigid tank with a volume of 0.1 m^3 contains air initially at 140 kPa, 60°C. Heat is added reversibly until the pressure becomes 280 kPa. Atmospheric temperature is 15°C. Determine the fraction of the heat added that is unavailable energy.

8·47 In a steady-flow reversible process with no change in potential or kinetic energy, a gas is cooled at a constant pressure of 140 kPa from 170 to 60°C. The sink temperature is −10°C. For this gas, in the temperature range involved, $c_p = 0.8 + 0.0002T$ and $c_v = 0.6 + 0.002T$, where T is in kelvins and c_p and c_v are in kJ/kg·K. Sketch a *Ts* diagram and calculate, in kJ/kg, (*a*) the amount to heat removed from the gas, (*b*) the entropy change of the gas, and (*c*) the amount of heat removed from the gas that is unavailable energy.

8·48 Nitrogen flowing steadily over the tubes of a heat exchanger enters at 105 kPa, 200°C, and is cooled to 150°C. Nitrogen flows through the tubes at the same rate, entering at 400 kPa, 65°C, and absorbing all the heat given up by the hotter nitrogen. Neglect all pressure drops in the heat exchanger. The temperature of the surrounding atmosphere is 25°C. Calculate (*a*) the fraction of the heat given up by the hotter stream that is available energy, and (*b*) the fraction of the heat absorbed by the cooler stream that is available energy.

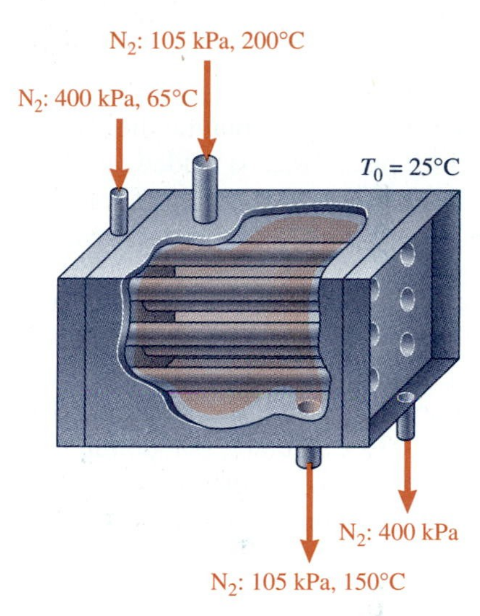

Problem 8·48

8·49 Heat is transferred from a furnace at a constant temperature of 2540 F to water that is evaporating at a constant temperature of 540 F. See figure. The lowest temperature in the surroundings is 90 F. For each 1000 B transferred, how much of the energy would be available if it were absorbed by a fluid at 2540 F? How much of the energy is available as it is received by the water at 540 F? Show these quantities as areas on a *TS* diagram.

8·50 Consider the transfer of heat from an energy reservoir at 540 F to 5 lbm of air initially at 15 psia, 140 F, trapped in a closed, rigid tank. Heat is transferred until the temperature of the air is 340 F. The temperature of the surroundings is 40 F. (*a*) How much heat is transferred? (*b*) How much of the energy removed from the reservoir is available energy? How much is unavailable? (*c*) How much of the energy added to the air in the tank is available energy? How much is unavailable?

8·51 For each 1000 B supplied to a power plant, 600 B is rejected to the atmosphere at 80 F from the working fluid that is condensed at a constant temperature of 90 F, and 200 B of heat is rejected to the atmosphere from working fluid at 800 F. The work output is 200 B. Determine the maximum amount of work that might be obtained from (*a*) the heat rejected from the fluid at 90 F and (*b*) the heat rejected from the fluid at 800 F.

8·52 A closed, rigid container holds dry saturated steam at 17.0 psia. Heat is added until the pressure becomes 32.0 psia. How much of the heat added is available energy if the sink temperature is 32 F?

8·53 Air flows reversibly and at constant pressure through a steady-flow system at a rate of 0.988 lbm/s, entering at 30 psia, 100 F, and leaving at 300 F. The lowest temperature in the surroundings (the sink temperature) is 10 F. Calculate, per pound of air, (*a*) the heat added to the air, (*b*) the work done on or by the air, and (*c*) the amount of heat added to the air that is available energy.

8·54 An energy reservoir at 840°C supplies heat to steam flowing through a heat exchanger at a rate of 900 kg/h. Steam enters the device dry and saturated at 5.2 MPa and leaves at 5.1 MPa, 360°C. Atmospheric temperature is 5°C. Determine the amount of energy made unavailable by this process in kJ/h.

8·55 Steam enters a turbine at 700 kPa, 200°C, and leaves at 7 kPa. The expansion is adiabatic. The flow rate is 9000 kg/h. The power output is 1.1 MW. Condenser cooling water is available from a large river at 5°C. Per kilogram of steam, how much energy is made unavailable by the operation of the turbine?

8·56 Air expands through a turbine from 400 kPa, 500°C, to 100 kPa, 340°C, while it loses 11 kJ/kg of heat to the atmosphere at 101.3 kPa, 20°C. Assuming that changes in kinetic energy are insignificant, calculate for this process (*a*) the decrease in stream availability, (*b*) the maximum work, and (*c*) the irreversibility.

8·57 One-fourth kilogram of air, initially at 400 kPa, 60°C, is expanded adiabatically in a closed system until its volume is doubled and its temperature equals that of the surrounding atmosphere, 5°C. Calculate for this process (*a*) the maximum work, (*b*) the change in availability, and (*c*) the irreversibility.

8·58 Exhaust steam enters a condenser at 7 kPa, 90 percent quality, (state 1) and is condensed to a saturated liquid (state 2) by cooling water that enters at 15°C (state *A*) and leaves at 20°C (state *B*). The steam flow rate is 1.5 kg/s. The sink temperature is 15°C.

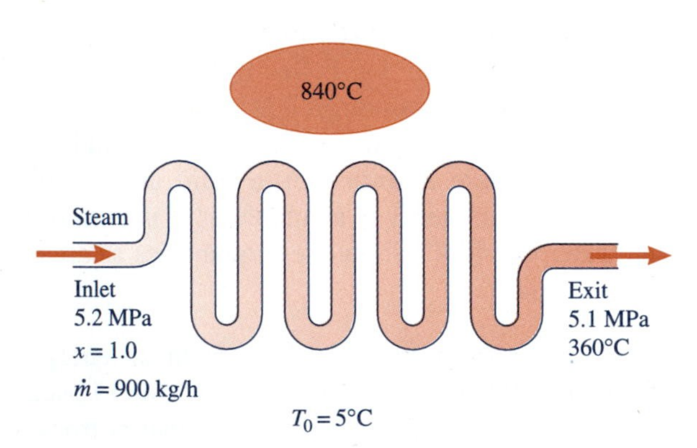

Problem 8·54

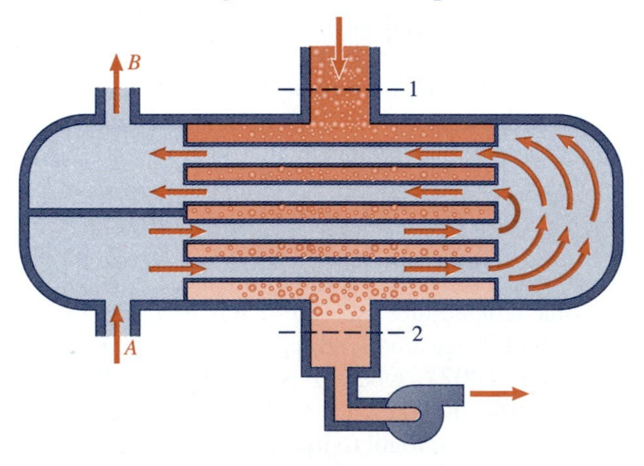

Problems 8·58 and 8·77

Sketch a *Ts* diagram for each of the fluids and calculate (*a*) the maximum work and (*b*) the irreversibility.

8·59 A closed, rigid, thermally insulated tank contains 2 kg of methane at 100 kPa, 60°C. A paddle wheel inside the tank is turned by an external motor until the methane is at 170°C. The sink temperature is 15°C. Calculate the irreversibility of the process.

8·60 The "hot side" air of an air-to-air heat exchanger enters the device at 300°C, 1 atm, and the "cold side" air enters at 60°C, 1.7 MPa, and leaves at 115°C. Frictional pressure losses are negligible, and the flow rate on each side is 4.5 kg/s. Calculate (*a*) the irreversibility of the process per kilogram of "hot side" air for a sink temperature of 15°C, and (*b*) the entropy change of the universe resulting from an hour's operation of the heat exchanger.

8·61 Water at 90°C is sprayed into steam at 700 kPa, 270°C, that flows at a rate of 9000 kg/h to desuperheat it to dry saturated steam at 700 kPa. The spray chamber in which this occurs is called a desuperheater. The sink temperature is 15°C. Calculate, per kilogram of steam entering the desuperheater, (*a*) the amount of water needed, and (*b*) the irreversibility of the desuperheating process. Devise a means of carrying out this process reversibly.

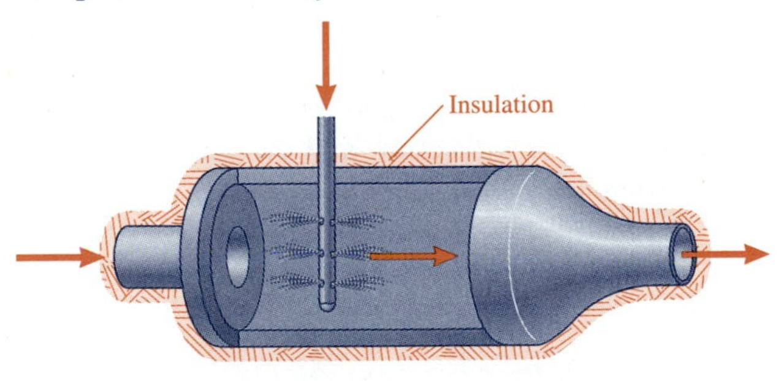

Problems 8·61 and 8·78

8·62 Steam that is dry and saturated at 2.0 MPa is throttled through a well-insulated valve to 600 kPa. The flow rate is 360 kg/h and kinetic energy changes are insignificant. For a sink temperature of 5°C, sketch a *Ts* diagram and calculate the irreversibility of the process.

8·63 Air is compressed adiabatically from 100 kPa, 5°C, to 300 kPa at a rate of 0.910 kg/s. Kinetic energy changes are negligible. Work input is 140 kJ/kg. (*a*) Calculate the irreversibility of the process per kilogram. (*b*) Explain the physical significance of the answer to *a*.

8·64 A steam-heated water heater heats 2.3 kg/min of water at a constant pressure of 95 kPa. The water is heated from 15°C to 70°C by means of steam that enters the heater at 95 kPa, 80 percent quality, and leaves as saturated liquid at nearly the same pressure. The surrounding atmosphere is at 95 kPa, 15°C. Calculate, per kilogram of water heated, the irreversibility of the process.

8·65 Refer to Example 8·2. Calculate the irreversibility of the process if it is possible to do so. If it is impossible to do so, state why.

8·66 Refer to Example 8·3. Calculate the irreversibility of the process if it is possible to do so. If it is impossible to do so, state why.

8·67 A household refrigerator has two compartments, one that is maintained at −15°C and the other that is maintained at 5°C. A container holding 350 grams of orange juice concentrate has been stored in the lower-temperature compartment. The ambient temperature is 20°C. Making reasonable assumptions regarding physical properties, calculate (*a*) the amount of energy made unavailable by taking the sealed container from the lower-temperature compartment into the surrounding room and allowing it to come to room temperature; and (*b*) the amount of energy made unavailable by first moving the sealed container into the 5°C compartment and letting it come to equilibrium and then removing it to the surrounding room and allowing it to come to equilibrium there.

8·68 Air flows adiabatically through a turbine, entering at 400 kPa, 450°C. The exhaust conditions are 115 kPa, 320°C. In one configuration, the turbine has an exhaust velocity of 160 m/s; in another configuration, it is 20 m/s. Determine the irreversibility in each case.

8·69 For the two cases specified in Problem 8·68 compare the work.

8·70 For the two cases specified in Problem 8·68 compare the reversible work.

8·71 Refer to Example 1 in *Chapter B, Note on Problem Solving*. Assume that the methane enters at atmospheric temperature and calculate the irreversibility per kilogram of methane. (Some other assumptions are necessary, so state them clearly.)

8·72 Refer to Example 2 in *Chapter B, Note on Problem Solving*, and make a plot of the irreversibility rate (in kJ/h) as a function of the discharge pressure. Assume that the heat removed from the gas being compressed goes to the atmosphere. State clearly the other assumptions that you make.

8·73 A compressor draws in air from the surrounding atmosphere at 100 kPa, 5°C, and discharges it at 390 kPa. The flow rate is 2.3 kg/s, power input to the compressor is 450 kW, and during compression heat is removed from the air being compressed to the surrounding atmosphere in the amount of 25.0 kJ/kg. Determine the entropy change of the universe and the irreversibility of the compression process, both per kilogram of air compressed.

8·74 Measurements on a compressed air turbine operating steadily show that air enters at 140 kPa, 60°C, with a velocity of 30 m/s through a cross-sectional area of 20 sq cm and leaves at 70 kPa, 25°C, with the same velocity. The power output is 2.35 kW. The turbine casing is uninsulated. The surrounding atmosphere is at 70 kPa, 5°C. Calculate the entropy change of the universe and the irreversibility of the process, both per pound of air.

8·75 Heat is transferred from an energy reservoir at 1540 F to steam flowing at a rate of 2000 lbm/h. Steam enters the heat exchanger dry and saturated at 750 psia and leaves at 740 psia, 680 F. The lowest temperature in the surroundings is 40 F. Determine the amount of energy made unavailable by this process in B/h.

8·76 Air expands through a turbine from 60 psia, 940 F, to 15 psia, 640 F, while heat in the amount of 5 B/lbm is lost to the atmosphere at 14.7 psia, 70 F. Assuming kinetic energy changes during the process are insignificant, find for this process (*a*) the decrease in stream availability, (*b*) the maximum work, (*c*) the entropy change of the universe, and (*d*) the irreversibility.

8·77 Steam enters a condenser at 1 psia, 90 percent quality (state 1), and is condensed to a saturated liquid (state 2) by cooling water that enters at 60 F (state *A*) and leaves at 70 F (state *B*). See figure. The steam flow rate is 200 lbm/min. The sink temperature is 60 F. Sketch a *Ts* diagram for each of the fluids. Calculate (*a*) the maximum work, (*b*) the entropy change of the universe, and (*c*) the irreversibility.

8·78 Steam at 100 psia, 520 F, flowing at a rate of 20,000 lbm/h is desuperheated through the addition of a fine spray of water at 200 F. See figure. The result is dry saturated steam at the same pressure. The spray chamber in which this process occurs is called a desuperheater. Assume that atmospheric temperature is 60 F. Calculate, per pound of steam entering the desuperheater, (*a*) the amount of water needed and (*b*) the irreversibility of the desuperheating process. Devise a means of carrying out this process reversibly.

8·79 Eight hundred lbm/h of dry, saturated steam at 300 psia is throttled to 81 psia. There is no significant change in kinetic energy, and the sink temperature is 40 F. Sketch a *Ts* diagram, and calculate, in B/lbm, the irreversibility of the process.

8·80 Under what conditions can the irreversibility of a process be negative? Give reason(s) for your answer.

8·81 Refer to Problem 8·67. Explain why there would or would not be a saving of energy required to operate the household refrigerator if procedure *a* or *b* were used regularly instead of the other procedure.

8·82 Refer to Example 8·7 and make the investigation that is mentioned in the comment at the end.

8·83 The following sentence appears in Section 8·4: "In a compression process, the irreversibility usually does not exceed the work input; in a heat exchanger process, the irreversibility usually does not exceed the total heat transfer; and in a turbine or engine, the irreversibility is unlikely to be of a greater order of magnitude than the work output." Explain why the words *usually* and *is unlikely to be* are needed. That is, describe some exceptions.

8·84 An engine cycle consists of four processes of air in a closed system:

Process 1–2. An internally reversible constant-pressure expansion with $p_1 = p_2 = 500$ kPa, $T_1 = 220°C$, $T_2 = 600°C$.

Process 2–3. An irreversible adiabatic expansion to 100 kPa, $T_3 = 380°C$.

Process 3–4. An internally reversible constant-pressure compression with $p_3 = p_4 = 100$ kPa, $T_4 = 20°C$.

Process 4–1. An irreversible adiabatic compression to the initial state.

The cycle receives heat from an energy reservoir at 600°C and rejects heat to an energy reservoir at 20°C. Make a diagram that shows to scale the irreversibilities of the four processes and shows how they add to form the total increase in unavailable energy.

8·85 Atmospheric air at 100 kPa, 20°C, enters a filter in which it undergoes a throttling process that lowers its pressure to 90 kPa before entering a compressor where it is compressed adiabatically to 400 kPa, 210°C. It then flows through an aftercooler where it is cooled to 50°C by heat transfer to the atmosphere. The flow is steady. Make a diagram showing to scale the irreversibilities of the individual processes and how they relate to changes in stream availability.

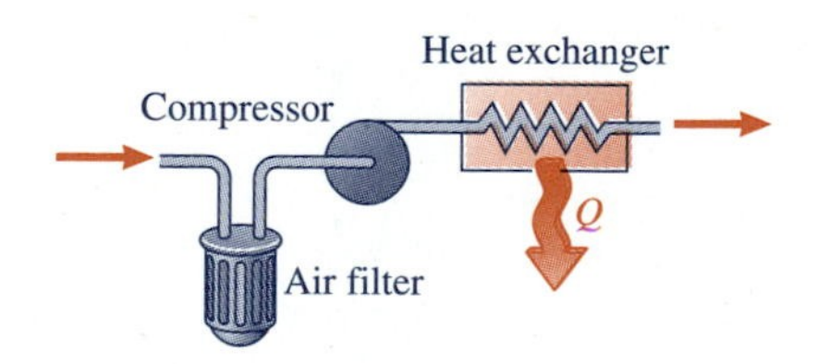

Problems 8·85 and 8·87

8·86 An engine cycle consists of four processes of air in a closed system:

Process 1–2. An internally reversible constant-pressure expansion with $p_1 = p_2 = 70$ psia, $T_1 = 430$ F, $T_2 = 1100$ F.

Process 2–3. An irreversible adiabatic expansion to 14 psia, $T_3 = 700$ F.

Process 3–4. An internally reversible constant-pressure compression with $p_3 = p_4 = 14$ psia, $T_4 = 70$ F.

Process 4–1. An irreversible adiabatic compression to the initial state.

The cycle receives heat from an energy reservoir at 1100 F and rejects heat to an energy reservoir at 50 F. Make a diagram that shows to scale the irreversibilities of the four processes and shows how they add to form the total increase in unavailable energy.

8·87 Atmospheric air at 14.5 psia, 60 F, enters a filter in which it undergoes a throttling process that lowers its pressure to 14.2 psia before entering a compressor where it is compressed adiabatically to 60 psia, 400 F. It then flows through an aftercooler where it is cooled to 120 F by transferring heat to the atmosphere. See figure. The flow is steady. Make a diagram that shows to scale the irreversibility of each process and the relationships of irreversibility values to changes in stream availability.

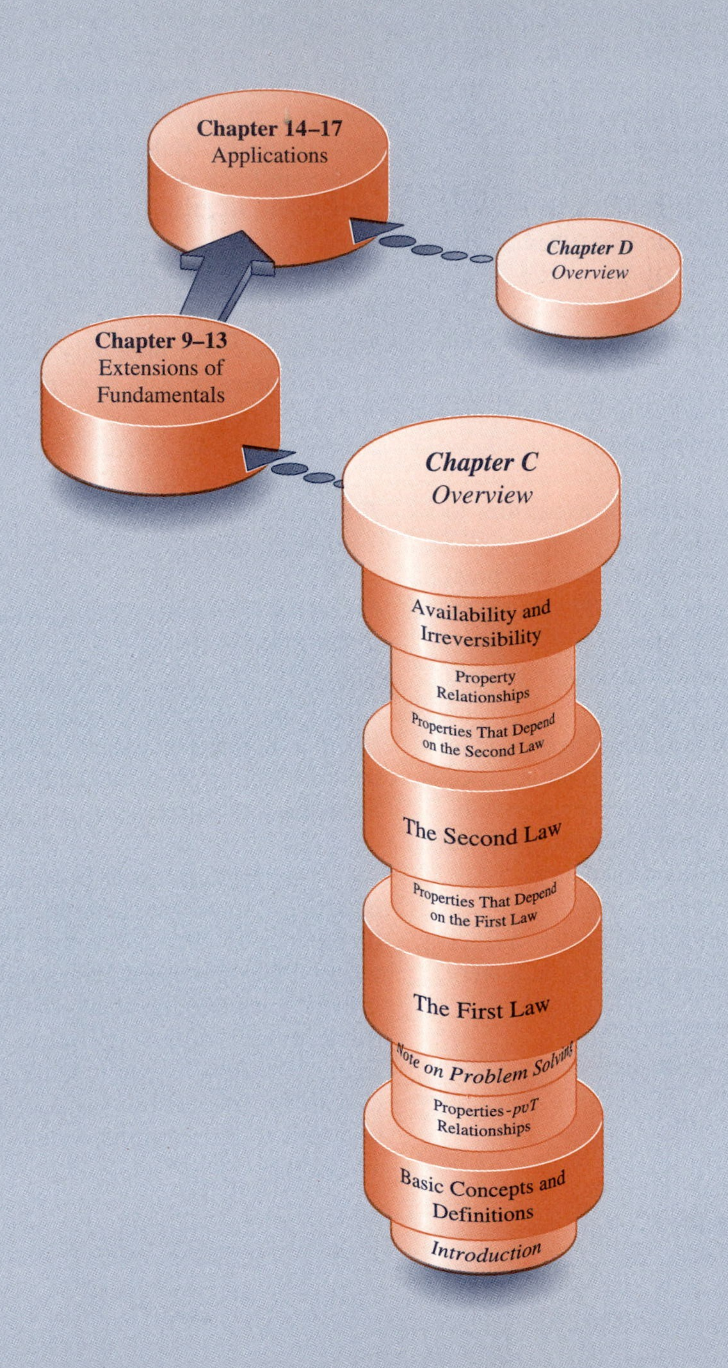

Chapter 14–17
Applications
Chapter D
Overview
Chapter 9–13
Extensions of
Fundamentals
Chapter C
Overview
Availability and
Irreversibility
Property
Relationships
Properties That Depend
on the Second Law
The Second Law
Properties That Depend
on the First Law
The First Law
Note on Problem Solving
Properties-pvT
Relationships
Basic Concepts and
Definitions
Introduction

Chapter C

Overview—Looking Back, Looking Ahead

We have now covered the fundamentals of thermodynamics: the first law, the second law, and some properties and property relationships.

The First Law. The first law is an empirical relationship between heat and work that led to the definition of stored energy. The energy stored in a system can be in various forms: kinetic, potential, internal, magnetic, electrical, and surface energy. The first law enables us to calculate the conversion of energy from one form to another and the transfer of energy to or from a system as heat or work. It is sometimes called the conservation of energy principle. The first law formulations developed in Chapter 3 can be applied to an endless variety of systems and processes. Even with no further enhancements, the first law as presented here is a vital tool in many engineering problems.

The Second Law. The second law is an empirical conclusion used to determine the possibility or impossibility of processes. There are many statements of the second law, but all state that certain processes are impossible. Another way of saying this is that some processes produce effects that can never be completely undone; such processes are called irreversible. The second law leads to the definition of entropy. By now, the parallelism between the definitions of energy and of entropy should be clear. We use entropy or one of the properties derived from entropy when determining if a process is reversible or even possible. The concepts of availability and irreversibility allow us to quantify the effects of irreversible processes in decreasing the amount of energy available for doing work.

Physical Properties. You can look back to confirm that the properties and property relationships introduced in Chapters 1 and 2 were independent of the first law. Then, the properties u and h introduced in Chapter 3 were defined on the basis of the first law. After we defined the properties s, a, and g on the basis of the second law, we used both the first and second laws to establish many property relationships. This interplay among principles and property relationships continues throughout thermodynamics. This is highlighted in the panorama on the facing page.

In addition to learning the principles and property relationships, if you have solved a variety of the problems that are a part of each chapter, you have improved your problem-solving abilities. Some complete problems you solved in Chapters 1 and 2 became just small parts of more complex and extensive problems in later chapters.

What can you do now with this knowledge and ability? You have already seen that you can solve application problems involving a variety of systems and processes, calculating work, heat, and property changes. You can now solve most of the application problems in later chapters on compressible flow (Chapter 13), compression and expansion processes (Chapter 14), and power generation (Chapter 15) as well as many of the problems in the chapter on refrigeration (Chapter 16). To handle many other applications, you need to study extensions of the fundamentals. Therefore, the next four chapters cover topics that are important for direct application and also as foundations for many more applications. Gas and gas–vapor mixtures as covered in Chapter 9 are frequently encountered in engineering. Perhaps the most common gas–vapor mixture is that of air and water vapor that we call atmospheric air. Humidity in some cases has effects that appear disproportionate to the small amount of water vapor present. Binary mixtures of substances are discussed in Chapter 10. Binary mixtures are relevant because a small amount of gas dissolved in a liquid can have significant impact, and the operation of devices such as absorption refrigeration systems that are part of many large air-conditioning systems can be understood only on the basis of binary mixture behavior. Chapter 11 shows how the first and second laws and various property relationships are applied to chemically reacting mixtures, and then Chapter 12 on thermochemical equilibrium illustrates how the second law is used to determine if certain reactions are possible. These determinations cannot be made on the basis of intuition or any principles other than the second law. The next four chapters, then, are important both for direct application and for supporting some of the applications introduced in subsequent chapters.

This is an excellent time for you to review the fundamentals of thermodynamics. A suggested approach:

- Review the table of contents, either as it appears at the front of the book, in the panorama, or at the beginning of each chapter.
- Review the introductory and summary sections of each chapter.
- Review *Chapter B* on problem solving.
- Review the problem solutions you have completed. Notice the principles and property relationships you have used. Also, notice any changes in your solution techniques that have occurred since you started solving thermodynamic problems.

Investing some time in review at this stage is well worth the effort.

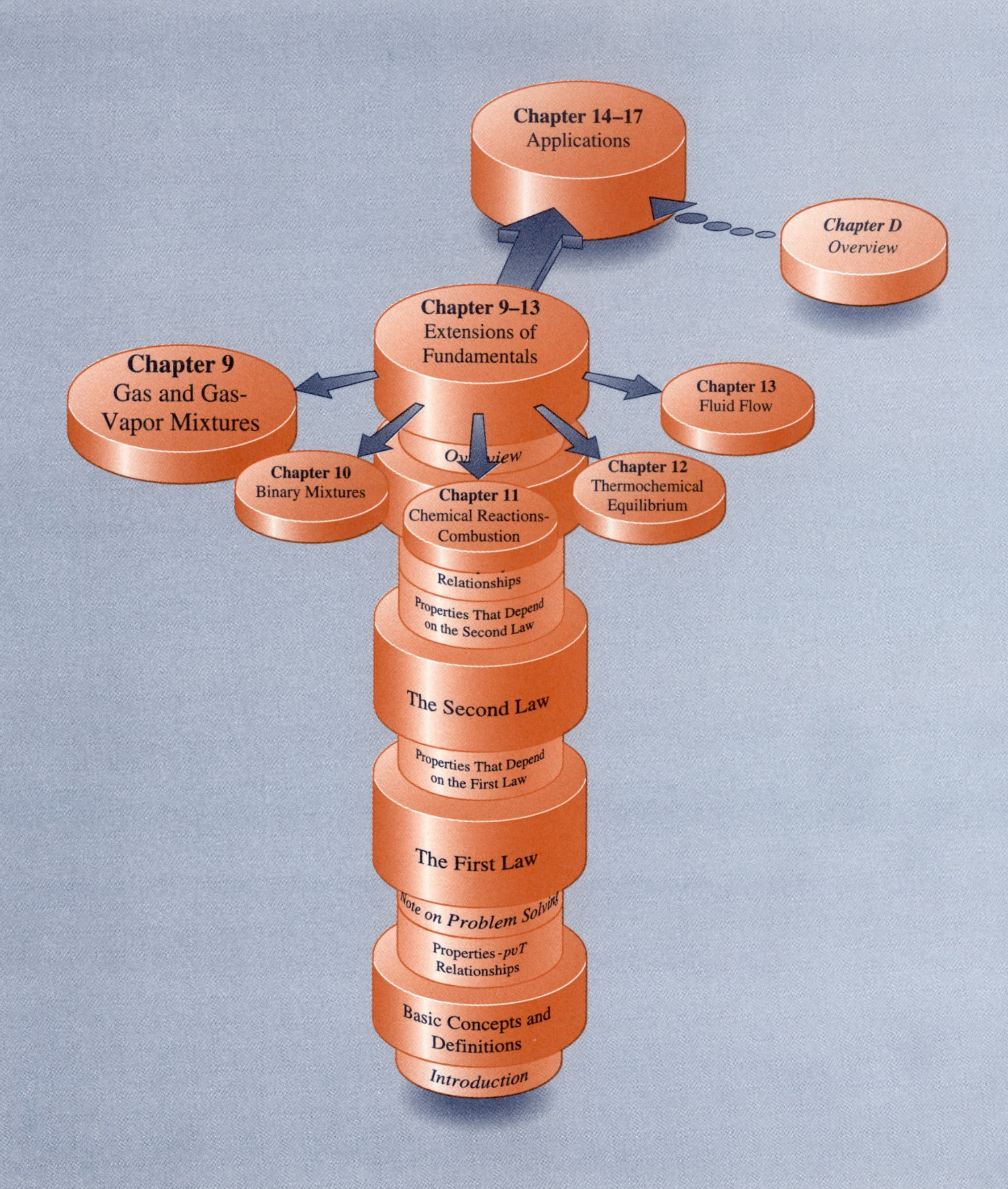

Chapter 14–17
Applications
Chapter D
Overview
Chapter 9–13
Extensions of
Fundamentals
Chapter 9
Gas and Gas-
Vapor Mixtures
Chapter 13
Fluid Flow
Chapter 10
Binary Mixtures
Chapter 11
Chemical Reactions-
Combustion
Chapter 12
Thermochemical
Equilibrium
Relationships
Properties That Depend
on the Second Law
The Second Law
Properties That Depend
on the First Law
The First Law
Note on Problem Solving
Properties-pvT
Relationships
Basic Concepts and
Definitions
Introduction

Chapter 9

Gas and Gas–Vapor Mixtures

A pure substance is defined as a substance that is homogeneous and unchanging in chemical composition. Homogeneous mixtures of gases that do not react with each other are therefore pure substances, and the properties of such mixtures can be determined just like the properties of any other pure substance. The properties of common mixtures such as air and certain combustion products have been tabulated or fit by equations, but since an unlimited number of mixtures is possible, properties of all of them cannot be tabulated.

Because engineers often deal with mixtures, we must be able to calculate the properties of any mixture from the properties of its components. This chapter discusses such calculations, first for gas mixtures and then for gas–vapor mixtures.

9·1 Mixtures of Gases

9·1·1 Analyses of Mixtures

If a gas mixture consists of gases A, B, C, and so on, the mass of the mixture is the sum of the masses of the component gases:

$$m_m = m_A + m_B + m_C + \cdots = \sum m_i \qquad (a)$$

where the subscript m refers to the mixture and the subscript i refers to the ith component. The mass fraction (or concentration) of any component i is defined as

Mass fraction

$$x_i = \frac{m_i}{m_m}, \qquad x_A = \frac{m_A}{m_m}, \qquad \text{etc.}$$

Mass analysis

A **mass analysis** (sometimes called a gravimetric analysis) is expressed in terms of the mass fractions, and

$$1 = x_A + x_B + x_C + \cdots = \sum x_i$$

The total number of moles of a mixture is the sum of the number of moles of its components:

$$N_m \equiv N_A + N_B + N_C + \cdots = \sum N_i$$

The mole fraction y is defined as

Mole fraction

$$y_i \equiv \frac{N_i}{N_m}, \qquad y_A \equiv \frac{N_A}{N_m}, \qquad \text{etc.}$$

Molar analysis

A **molar analysis** is expressed in terms of the mole fractions, and

$$1 = y_A + y_B + y_C + \cdots = \sum y_i$$

The number of moles N, the mass m, and the molar mass M of a component (subscript i) and of an entire mixture (subscript m) are related by

$$m_i = N_i M_i \qquad (b)$$
$$m_m = N_m M_m$$

where M_m is the molar mass of the mixture. Substituting from Eq. *b* into Eq. *a* gives

$$m_m = \sum m_i = \sum N_i M_i$$

and

$$M_m = \frac{m_m}{N_m} = \frac{\Sigma N_i M_i}{N_m} = \sum y_i M_i$$

From the equations above, a useful relationship for conversions of mixture analyses is

$$y_i = x_i \frac{M_m}{M_i}$$

9·1·2 Partial Pressure and Partial Volume

The **partial pressure** p_i of a component i in a gas mixture is defined as

$$p_i \equiv y_i p_m, \qquad p_A \equiv y_A p_m, \qquad \text{etc.}$$

Partial pressure

where y is the mole fraction. From this definition, the sum of the partial pressures of the components of a gas mixture equals the mixture pressure

$$\sum p_i = \sum y_i p_m = p_m \sum y_i = p_m$$

This applies to any gas mixture, whether it is an ideal gas or not.

The **partial volume** V_i of a component i in a gas mixture is defined as

$$V_i \equiv y_i V_m, \qquad V_A \equiv y_A V_m, \qquad \text{etc.}$$

Partial volume

The sum of the partial volumes of the components of a gas mixture equals the volume of the mixture:

$$\sum V_i = \sum y_i V_m = V_m \sum y_i = V_m$$

The partial volume is of course not the actual volume of a component as it exists in the mixture because each component fills the entire volume of the vessel that holds the mixture.

The definitions of partial pressure and partial volume are general and hold for all mixtures. We will see that for mixtures of ideal gases, partial pressure and partial volume have physical significance.

Two different models of gas mixtures, the Dalton model and the Amagat model, are in use. We use both in this book. In our discussions of mixtures of ideal gases, we confine our attention to cases where the mixture itself is an ideal gas. Keep in mind, however, that a mixture of ideal gases is not always itself an ideal gas. How closely the mixture conforms to the ideal-gas equation of state depends on the mixture pressure, temperature, and molar mass.

9·1·3 The Dalton Model

For a mixture of ideal gases that is also an ideal gas, the mixture pressure is

$$p_m = \frac{N_m \bar{R} T_m}{V_m} = \frac{(N_A + N_B + \cdots)\bar{R}T_m}{V_m} = \frac{N_A \bar{R} T_m}{V_m} + \frac{N_B \bar{R} T_m}{V_m} + \cdots$$

$$= p_A(N_A, T_m, V_m) + p_B(N_B, T_m, V_m) + \cdots = \sum p_i(N_i, T_m, V_m)$$

where $p_A(N_A, T_m, V_m)$ is the pressure of N_A moles of component A at the temperature T_m and the volume V_m. Restated, the Dalton model says

The Dalton model is an additive-pressure model.

The pressure of a mixture of ideal gases equals the sum of the pressures of its components if each existed alone at the temperature and volume of the mixture.

We have shown the Dalton model to hold for ideal-gas mixtures, but it holds approximately for real-gas mixtures even in some pressure and temperature ranges where $pv = RT$ is inaccurate.

Combining the ideal-gas equation of state for p_A, the equation above for p_m, and the definition of partial pressure gives

Partial pressure is defined for the general case as

$$p_A = y_A p_m$$

Only for the case of an ideal gas is p_A equal to the pressure of N_A moles of component A evaluated at the mixture volume and temperature.

$$p_A(N_A, T_m, V_m)_{\text{ideal gas}} = \frac{N_A \bar{R} T_m}{V_m} = \frac{N_A}{N_m} p_m = y_A p_m = p_A$$

That is, in a mixture of ideal gases, the partial pressure of each component equals the pressure that component would exert if it existed alone at the temperature and volume of the mixture. Another description of the Dalton model is as follows:

An alternative statement of the Dalton model

In an ideal-gas mixture of ideal gases, each component behaves in all respects as though it existed alone at the temperature of the mixture and its partial pressure or, equivalently, at the temperature and volume of the mixture.

See Fig. 9·1. Therefore, the internal energy and entropy of an ideal-gas mixture are equal respectively to the sums of the internal energies and entropies

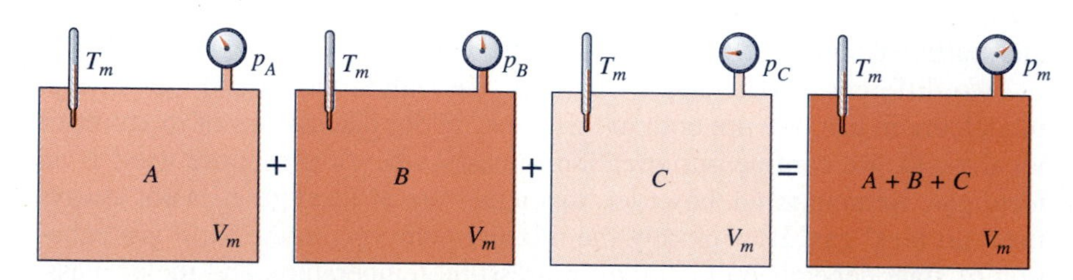

Figure 9·1 The Dalton model.

of the components if each existed alone at the temperature and volume of the mixture or, equivalently, at the temperature of the mixture and the component partial pressure.

9·1·4 The Amagat Model

If a mixture of ideal gases is also an ideal gas, then for a mixture of ideal gases A, B, C, and so on,

$$V_m = \frac{N_m \overline{R} T_m}{p_m} = \frac{(N_A + N_B + \cdots)\overline{R} T_m}{p_m} = \frac{N_A \overline{R} T_m}{p_m} + \frac{N_B \overline{R} T_m}{p_m} + \cdots$$

$$= V_A(N_A, T_m, p_m) + V_B(N_B, T_m, p_m) + \cdots = \sum V_i(N_i, T_m, p_m)$$

where $V_A(N_A, T_m, p_m)$ is the volume N_A moles of component A at the temperature T_m and the pressure p_m. Thus the Amagat model says

The volume of a mixture of ideal gases equals the sum of the volumes of its components if each existed alone at the temperature and pressure of the mixture.

The Amagat model is an additive-volume model.

See Fig. 9·2. Like the Dalton model, the Amagat model is accurate only for ideal gases, but holds approximately for real-gas mixtures even in some ranges of pressure and temperature where $pv = RT$ is inaccurate.

For ideal-gas mixtures, volumetric analyses are frequently used. The **volume fraction** is defined as

Volume fraction

$$\text{Volume fraction of } A \equiv \frac{V_A(N_A, T_m, p_m)}{V_m}$$

$$= \frac{\text{Volume of } A \text{ existing alone at } T_m, p_m}{\text{Volume of the mixture at } T_m, p_m}$$

The volume fraction is *not* defined as the ratio of a component volume to the mixture volume, because this ratio is always unity. This is true because in the actual mixture, each component occupies a volume equal to the mixture volume. In other words, it exists, perfectly mixed, throughout the entire mix-

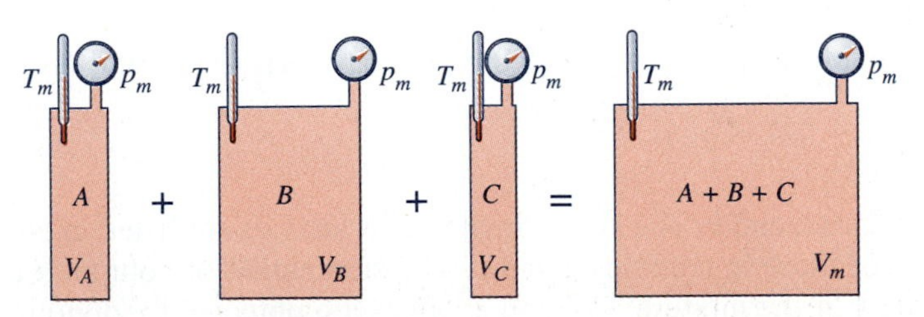

Figure 9·2 The Amagat model.

In practice, it is useful to know that when the temperature of a real-gas mixture is well above the critical temperature of all its components, the Amagat model is usually more accurate than the Dalton model.

ture volume. The partial volume is simply a construct that is useful for gas analysis. Notice also that we define volume fraction or volumetric analysis only for mixtures of *ideal gases*, because only for ideal gases is the Amagat or additive volume model accurate. The volume fraction of a component in an ideal-gas mixture equals its mole fraction, as can be shown by

$$\frac{V_A(N_A, T_m, p_m)}{V_m} = \frac{N_A\overline{R}T_m}{p_m}\left(\frac{p_m}{N_m\overline{R}T_m}\right) = \frac{N_A}{N_m} = y_A$$

and the volume of an ideal-gas mixture component, if it existed alone at p_m and T_m, equals the partial volume of the component in the mixture:

$$V_A(N_A, T_m, p_m)_{\text{ideal gas}} = y_A V_m = V_A$$

The equality of volume fraction and mole fraction in an ideal-gas mixture enables us to write the units of volume fraction as moles of component per mole of mixture, and doing so simplifies the conversion between volumetric and mass analyses. Such conversions must be made because gas mixtures are often analyzed on a volumetric basis, but a mass analysis is generally more useful in relating properties of a mixture to the properties of its components. Conversion from one basis to the other is illustrated in the two examples that follow. Notice that the pressure and temperature of the mixture have no bearing on the conversion. An ideal-gas mixture can be heated, cooled, compressed, or expanded, and its volumetric analysis remains constant as long as its mass analysis does. Two suggestions on making the conversions are (1) use a tabular form if there are more than two components, and (2) write down the units at the head of each column, and observe them carefully.

The Dalton model is more widely used than the Amagat model, but each has advantages, so they often are used together. For example, volumetric analyses, which are based on the Amagat model, are often used in connection with calculations based largely on the Dalton model.

Example 9·1 Analysis Conversion: Volumetric to Mass

PROBLEM STATEMENT

A gas has the following volumetric analysis in percentages: H_2, 46.0; CO, 10.5; CH_4, 31.0; and N_2, 12.5. Determine the mass analysis.

SOLUTION

In the table below, the given data are in columns *a* and *b*. The approximate molar masses are listed in column *c*. The values in column *d* are the products of those in columns *b* and *c*. The summation of column *d* is the mass of 1 mole of mixture or the molar mass of the mixture. Column *e* values are obtained by dividing the column *d* values by the column *d* total.

a Component, i	b Volumetric Analysis, y_i (kmol/kmol of Mixture)	c Molar Mass, M_i (kg/kmol)	d $y_i M_i$ (kg/kmol of Mixture)	e Mass Analysis, $y_i M_i / M_m$ (kg/kg of Mixture)
H_2	0.46	2	0.92	0.075
CO	0.105	28	2.94	0.239
CH_4	0.310	16	4.96	0.402
N_2	0.125	28	3.50	0.284
	1.00		12.32 = M_m	1.000

Example 9·2 Analysis Conversion: Mass to Volumetric

PROBLEM STATEMENT

A gas mixture has the following mass analysis in percentages: H_2, 10; CO, 60; and CO_2, 30. Determine the volumetric analysis.

SOLUTION

Computer spreadsheets are well suited to such calculations and are especially convenient when several cases are to be considered. We show here one spreadsheet solution using approximate molar masses and another

a Component i	b Mass analysis, c_i kg/kg of mixture	c Molar mass, M_i kg/kmol	d c_i/M_i kmol/kg of mixture	e Volumetric analysis, y_i kmol/kmol of mixture
H_2	0.100	2	0.0500	0.639
CO	0.600	28	0.0214	0.274
CO_2	0.300	44	0.0068	0.087
	1.000		0.0782	1.000

a Component i	b Mass analysis, c_i kg/kg of mixture	c Molar mass, M_i kg/kmol	d c_i/M_i kmol/kg of mixture	e Volumetric analysis, y_i kmol/kmol of mixture
H_2	0.100	2.016	0.0496	0.637
CO	0.600	28.010	0.0214	0.275
CO_2	0.300	44.010	0.0068	0.088
	1.000		0.0778	1.000

Excel® spreadsheet solution. Excel is a registered trademark of Microsoft Corporation.

one using precise values. It would be straightforward to add other gases or to compute other quantities of interest.

Comment: Notice that H_2 was only 10 percent of the mixture by mass but a major fraction of the mixture by volume. This is reasonable because H_2 is the least dense component. This kind of check on reasonableness of the result should be made on all mixture analysis conversions.

9·1·5 Properties of Ideal-Gas Mixtures Based on the Dalton Model

To discuss the properties of ideal-gas mixtures, consider a mixture of three ideal gases, *A*, *B*, and *C*. (The results can readily be generalized to a mixture of any number of components.) The properties of such a mixture in terms of the properties of its components are discussed here.

Temperature. For any uniform mixture the temperature is the same for each component and for the mixture:

$$T_m = T_A = T_B = T_C$$

(This is also true for the Amagat model.)

Mass, Number of Moles, and Molar Mass. The mass, the number of moles, and the molar mass of a mixture are shown in Section 9·1 to be given by

$$m_m = m_A + m_B + m_C$$
$$N_m = N_A + N_B + N_C$$
$$M_m = y_A M_A + y_B M_B + y_C M_C$$

These relationships hold for all mixtures, not just for ideal gases. (They also hold for the Amagat model.)

Pressure. Using the Dalton model for an ideal-gas mixture, we showed that the partial pressure equals the component pressure at the temperature and volume of the mixture. Since the sum of the partial pressures must equal the mixture pressure, it follows that the sum of the component pressures at the volume and temperature of the mixture equals the mixture pressure.

The Dalton concept, that each component behaves in all respects as though it existed alone at its partial pressure and the temperature of the mixture, is in keeping with the molecular model that shows the pressure of an ideal gas to be the result of gas molecules bombarding the vessel walls. From this point of view it is easy to separate the pressure of a mixture into parts, each attributable to the bombardment of the vessel walls by the molecules of one component. It is impossible to measure directly the pressure of just one component of a mixture, but nevertheless it is often convenient to treat the partial pressure of a component in an ideal-gas mixture as the pressure exerted by that component *as it exists in the mixture.*

Volume. The volume of each component of a gas mixture is the same as the volume of the mixture because the molecules of each component are free to move throughout the entire space occupied by the mixture.

$$V_A(N_A, p_A, T_m) = V_B(N_B, p_B, T_m) = \cdots = V_m$$

Here, $V_A(N_A, p_A, T_m)$ is the volume of component A as it exists in the mixture, that is, at partial pressure p_A and mixture temperature T_m.

Internal Energy, Enthalpy, Entropy. For a mixture of ideal gases, the Dalton model leads to

$$U_m = U_A(N_A, T_m) + U_B(N_B, T_m) + U_C(N_C, T_m)$$
$$H_m = H_A(N_A, T_m) + H_B(N_B, T_m) + H_C(N_C, T_m)$$
$$S_m = S_A(N_A, p_A, T_m) + S_B(N_B, p_B, T_m) + S_C(N_C, p_C, T_m)$$

and similar expressions for other properties such as A_m and G_m. In these equations, the component properties must be evaluated as though each component existed alone at the temperature and volume of the mixture or at its partial pressure and the mixture temperature. The internal energy and enthalpy of an ideal gas are functions of temperature only, and the only temperature we use in evaluating properties of a mixture or its components is the mixture temperature T_m. However, the entropy of an ideal gas is a function of two properties, so the component entropies must be evaluated at the mixture temperature and the component partial pressure, or, equivalently, at the temperature and volume of the mixture. See again Fig. 9·1 illustrating the Dalton model. Per unit mass these expressions become

$$u_m \equiv \frac{U_m}{m_m} = \frac{U_A + U_B + U_C}{m_m} = \frac{m_A u_A + m_B u_B + m_C u_C}{m_m}$$

$$h_m \equiv \frac{H_m}{m_m} = \frac{m_A h_A + m_B h_B + m_C h_C}{m_m}$$

$$s_m = \frac{S_m}{m_m} = \frac{m_A s_A + m_B s_B + m_C s_C}{m_m}$$

Specific Heats, Gas Constant. Since

$$u_m = \frac{m_A}{m_m} u_A + \frac{m_B}{m_m} u_B + \frac{m_C}{m_m} u_C$$

c_v of a mixture is given by

$$c_{vm} \equiv \left(\frac{\partial u_m}{\partial T}\right)_v = \frac{m_A}{m_m}\left(\frac{\partial u_A}{\partial T}\right)_v + \frac{m_B}{m_m}\left(\frac{\partial u_B}{\partial T}\right)_v + \frac{m_C}{m_m}\left(\frac{\partial u_C}{\partial T}\right)_v$$

$$= \frac{m_A c_{vA} + m_B c_{vB} + m_C c_{vC}}{m_m} = \frac{m_A}{m_m} c_{vA} + \frac{m_B}{m_m} c_{vB} + \frac{m_C}{m_m} c_{vC}$$

$$= x_A c_{vA} + x_B c_{vB} + x_C c_{vC}$$

In a similar manner,

$$c_{pm} = \frac{m_A}{m_m} c_{pA} + \frac{m_B}{m_m} c_{pB} + \frac{m_C}{m_m} c_{pC}$$

and

$$R_m = \frac{m_A}{m_m} R_A + \frac{m_B}{m_m} R_B + \frac{m_C}{m_m} R_C$$

The gas constant of the mixture can also be obtained by

$$R_m = \frac{\overline{R}}{M_m}$$

Example 9·3 Properties of Ideal-Gas Mixtures

PROBLEM STATEMENT

A gas mixture at 100 kPa, 25°C, has a mass analysis of 20 percent hydrogen, 30 percent nitrogen, and 50 percent oxygen. Determine (*a*) the partial pressures of the components and (*b*) the constant-pressure specific heat, c_p.

SOLUTION

(*a*) To determine the partial pressures we first convert the mass analysis to a volumetric or molar analysis:

Component, i	*Mass Analysis, x_i* (kg/kg of Mixture)	*Molar Mass, M_i* (kg/kmol)	*x_i/M_i* (kmol/kg of Mixture)	*Volumetric Analysis, y_i* (kmol/kmol of Mixture)	*Partial Pressure, $y_i p_m$* (kPa)
H_2	0.20	2	0.100	0.792	79.2
N_2	0.30	28	0.0107	0.085	8.5
O_2	0.50	32	0.0156	0.123	12.3
			0.1263	1.000	100.0 = p_m

The partial pressures are calculated as $p_{H_2} = y_{H_2} p_m$, and so on.

(*b*) c_p of the mixture is the weighted average of the c_p's that are obtained from Table A·1:

$$c_{pm} = \sum \frac{m_i}{m_m} c_{p,i} = 0.20(14.3) + 0.30(1.04) + 0.50(0.919) = 3.63 \text{ kJ/kg·K}$$

Example 9·4 Process of an Ideal-Gas Mixture

PROBLEM STATEMENT

A mixture with a molar composition of 70 percent helium and 30 percent oxygen is compressed reversibly and adiabatically from 14.0 psia, 50 F, to 45 psia. Determine (*a*) the final temperature, (*b*) the work per pound of mixture, and (*c*) the entropy change of each component per pound of mixture.

SOLUTION

Analysis: In the pressure and temperature ranges involved, we can model the gas mixture as an ideal gas. The specific heats of helium, a monatomic gas, are constant, and the specific heat variation of oxygen is likely to be small for the limited temperature range resulting from a pressure ratio of approximately three. Therefore, we assume that the mixture can be modeled as an ideal gas with constant specific heats.

For a reversible adiabatic process of an ideal gas with constant specific heats, we can obtain the final temperature from the pT relationship obtained by combining the ideal-gas equation of state and $pv^k =$ constant.

Work can be obtained from the first law since we can evaluate the change in internal energy from the end state temperatures.

The entropy change for each component can be obtained by integrating one of the $T\,ds$ equations, since each component acts as though it exists alone at its partial pressure and the temperature of the mixture.

(*a*) To determine the specific heats of the mixture, we first determine the mass analysis:

Component	*Volumetric Analysis* (lbmol/lbmol of Mixture)	*Molar Mass* (lbm/lbmol)	(lbm/lbmol of Mixture)	*Mass Analysis* (lbm/lbm of Mixture)
He	0.70	4	2.80	0.226
O_2	0.30	32	9.60	0.774
			$12.40 = M_m$	

Then the mixture specific heats, using component specific heats from Table A·1E or A·3E or Chart A·4 at an estimated mean temperature of 150 F, are given by

$$c_{pm} = \frac{m_{He}}{m_m} c_{p,He} + \frac{m_{O_2}}{m_m} c_{p,O_2} = 0.226(1.24) + 0.774(0.222) = 0.452 \text{ B/lbm·R}$$

$$c_{vm} = \frac{m_{He}}{m_m} c_{v,He} + \frac{m_{O_2}}{m_m} c_{p,O_2} = 0.226(0.745) + 0.774(0.160) = 0.292 \text{ B/lbm·R}$$

and

$$k_m = \frac{c_{pm}}{c_{vm}} = \frac{0.452}{0.292} = 1.548$$

Since we are assuming that the specific heats are constant, the final temperature for the reversible adiabatic process is

$$T_2 = T_1\left(\frac{p_2}{p_1}\right)^{(k_m-1)/k_m} = 510\left(\frac{45}{14}\right)^{0.548/1.548} = 771 \text{ R} = 311 \text{ F}$$

(*b*) Applying the first law to the closed system for this adiabatic process, we have

$$w_{in} = u_{m2} - u_{m1} - q = c_{vm}(T_2 - T_1) - 0$$
$$= 0.292(771 - 510) = 76.2 \text{ B/lbm}$$

(*c*) For this reversible adiabatic process the entropy of the mixture must remain constant. The entropy of each component may change, but the sum of the entropy changes of the two components must be zero. For any process of an ideal gas,

$$\Delta s = \int_1^2 ds = \int_1^2 \frac{dh}{T} - \int_1^2 \frac{v\,dp}{T} = \int_1^2 \frac{c_p\,dT}{T} - R\int_1^2 \frac{dp}{p}$$

and if c_p is constant, the entropy change for each component i is

$$\Delta s_i = c_{pi} \ln \frac{T_2}{T_1} - R_i \ln \frac{p_{i2}}{p_{i1}}$$

For each component in the mixture, the pressures to be used are the partial pressures, but note that for each gas,

$$\frac{p_{i2}}{p_{i1}} = \frac{y_{i2}p_{m2}}{y_{i1}p_{m1}} = \frac{p_{m2}}{p_{m1}}$$

since $y_2 = y_1$. Thus, applying the equation for Δs to the helium, we have

$$\Delta s_{He} = 1.24 \ln \frac{771}{510} - \frac{386 \dfrac{\text{ft·lbf}}{\text{lbm·R}}}{778 \dfrac{\text{ft·lbf}}{\text{B}}} \ln \frac{45}{14} = -0.0668 \text{ B/lbm·R}$$

Per pound of mixture,

$$\frac{\Delta S_{He}}{m_m} = \frac{m_{He}\Delta s_{He}}{m_m} = 0.226(-0.0668) = -0.0151 \text{ B/lbm mixture·R}$$

For the oxygen,

$$\Delta s_{O_2} = 0.222 \ln \frac{771}{510} - \frac{48.3}{778} \ln \frac{45}{14} = 0.0193 \text{ B/lbm·R}$$

and, per pound of mixture,

$$\frac{\Delta S_{O_2}}{m_m} = \frac{m_{O_2}\Delta s_{O_2}}{m_m} = 0.774(0.0193) = 0.0149 \text{ B/lbm mixture·R}$$

Within the limits of accuracy of the calculations shown, these two Δs values per pound of mixture for the helium and for the oxygen are equal in magnitude although opposite in sign. It is actually unnecessary to calculate both Δs values in this manner, except as a check on the computations, because we know that their sum is zero.

9·2 Mixing of Ideal Gases

In the preceding two sections, we discussed mixtures of ideal gases and we related properties of mixtures to properties of the component ideal gases. We dealt with *states of mixtures*. Now we turn to the *process of mixing*. We relate the states of the components before mixing to the states of both the mixture and the individual components after mixing occurs.

The usual problem is to determine the properties of a mixture formed by mixing components of known properties. No new principles are involved. We simply apply the first law and the conservation of mass principle to a conveniently selected system. For example, consider the adiabatic mixing of three gases, A, B, and C, at different pressures and temperatures in a closed system of fixed volume. The gases might be initially in three tanks connected by piping, or they might be in three parts of a tank separated by partitions as in Fig. 9·3. If the partitions are ruptured or removed or the valves are opened, the three gases will form a mixture that has a mass and a volume given by

$$m_m = m_A + m_B + m_C$$
$$V_m = V_A + V_B + V_C$$

where V_A, V_B, and V_C are the volumes of the components before mixing. The mixing process was specified as adiabatic and there is no work done; so the internal energy of the system remains constant and

$$U_m = U_A + U_B + U_C$$

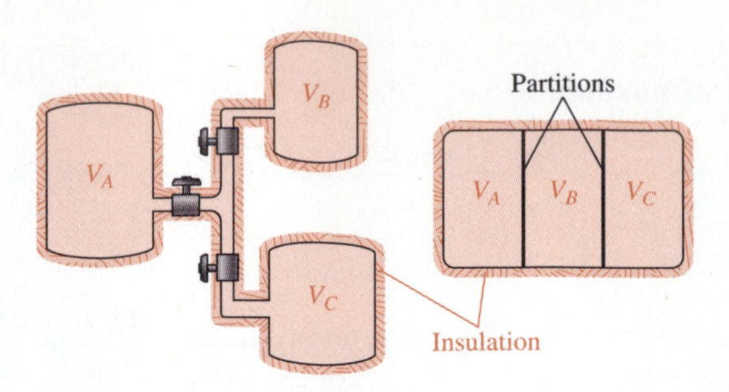

Figure 9·3 Three ideal gases in adiabatic systems before mixing.

where U_A, U_B, and U_C are the internal energies of the components before mixing. The internal energy of the mixture is also equal to the sum of the internal energies of the components after mixing, but the internal energy of each component is generally not the same before and after mixing. Since the internal energy of the entire system remains constant, the sum of the internal energy changes of the components is zero:

$$\Delta U = \Delta U_A + \Delta U_B + \Delta U_C = 0$$

or

$$\Delta U = m_A\, \Delta u_A + m_B\, \Delta u_B + m_C\, \Delta u_C = 0$$

For any ideal gas, internal energy u is a function of temperature only, so the equation above can be written as

$$\Delta U = m_A[u_A(T_m) - u_A(T_A)] + m_B[u_B(T_m) - u_B(T_B)] + m_C[u_C(T_m) - u_C(T_C)] = 0$$

If the uT relation for each gas is entered into an equation solver, this equation can be solved for the mixture temperature. For the special case of c_v values that are constant for each component throughout the temperature range, this equation can be reduced to

$$T_m = \frac{m_A c_{vA} T_A + m_B c_{vB} T_B + m_C c_{vC} T_C}{m_A c_{vA} + m_B c_{vB} + m_C c_{vC}}$$

The derivation of this equation involves no assumption regarding a temperature at which $U = 0$, nor was it stipulated that $U = 0$ at the same temperature for all components.

After the temperature of the mixture has been determined, the pressure can be calculated from

$$p_m = \frac{m_m R_m T_m}{V_m}$$

R_m can be determined from the analysis of the mixture.

Since the mixing process we are considering is irreversible and adiabatic, the entropy of the system must increase. The entropy of the mixture, while equal to the sum of the entropies of the components as they exist in the mixture, is greater than the sum of the entropies of the components before mixing. The entropy change of the entire system is

$$\Delta S = \Delta S_A + \Delta S_B + \Delta S_C > 0$$

and the entropy change for each component can be calculated as though each component existed alone and expanded from its initial conditions to the mixture temperature and volume, its final pressure being therefore its partial pressure in the mixture.

We have illustrated here that no new principles or techniques are involved in the determination of the properties of an ideal-gas mixture formed by mixing components of known properties in a closed, rigid, adiabatic system. *Nonadiabatic mixing* and *mixing in open systems*, both steady flow and transient flow, can also be analyzed by means of the principles that have already been introduced.

Example 9·5 Mixing

PROBLEM STATEMENT

Methane at 100 kPa, 15°C, enters an insulated mixing chamber at a rate of 1.08 kg/s. It is mixed with air at 100 kPa, 160°C, in an air/methane mass ratio of 17.0. The flow is steady and kinetic energy changes are negligible. Ambient pressure and temperature are 100 kPa, 15°C. Determine (*a*) the temperature of the mixture leaving the chamber and (*b*) the irreversibility of the mixing per kilogram of methane.

SOLUTION

Analysis: To find the state of the mixture leaving the mixing chamber, state 3, we will apply the first law, noting that the mixing is adiabatic, kinetic energy changes are negligible, there is no work done in a mixing chamber, and the flow is steady. This will give us values for enthalpy changes across the mixing process, and from the enthalpy changes we can determine the temperature changes. To calculate the irreversibility, we need to calculate the entropy change across the mixing process, and we can do this by calculating the entropy change for each component stream and adding them.

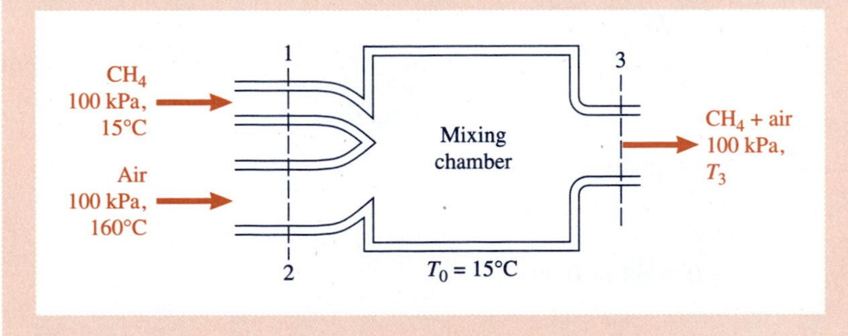

Calculations: (*a*) The first law applied to the mixing chamber under the restrictions listed in the analysis is

$$\dot{m}_{CH_4}h_{CH_4,1} + \dot{m}_{air}h_{air,2} = (\dot{m}_{CH_4} + \dot{m}_{air})h_{m,3} = \dot{m}_{CH_4}h_{CH_4,3} + \dot{m}_{air}h_{air,3}$$

Rearranging gives

$$\dot{m}_{CH_4}(h_3 - h_1)_{CH_4} + \dot{m}_{air}(h_3 - h_2)_{air} = 0$$

Let us use constant specific heats to relate enthalpy changes to temperature changes. Then

$$\dot{m}_{CH_4}c_{p,CH_4}(T_3 - T_1) + \dot{m}_{air}c_{p,air}(T_3 - T_2) = 0$$

$$T_3 = \frac{\dot{m}_{CH_4}c_{p,CH_4}T_1 + \dot{m}_{air}c_{p,air}T_2}{\dot{m}_{CH_4}c_{p,CH_4} + \dot{m}_{air}c_{p,air}}$$

The temperature of the mixture leaving the mixing chamber will be between 160°C and 15°C and, in view of the air/methane mass ratio of 17, closer to 160°C. Therefore, we select from Table A·3 mean values of $c_{p,CH_4} = 2.37$ kJ/kg·K and $c_{p,air} = 1.02$ kJ/kg·K. Making substitutions into the equation for T_3 gives

$$T_3 = \frac{1.08(2.37)288 + 17(1.08)1.02(433)}{1.08(2.37) + 17(1.08)1.02} = 416\text{ K} = 142°\text{C}$$

(*b*) The irreversibility is given by

$$I = T_0\,\Delta S = T_0(m_{CH_4}\,\Delta s_{CH_4} + m_{air}\,\Delta s_{air})$$

$$i = T_0\left(\Delta s_{CH_4} + \frac{m_{air}}{m_{CH_4}}\,\Delta s_{air}\right)$$

For each component, assuming that c_p values are constant,

$$\Delta s = \int ds = \int \frac{dh}{T} - \int \frac{v\,dp}{T} = \int \frac{c_p\,dT}{T} - R\int \frac{dP}{P} = c_p \ln \frac{T_m}{T_{initial}} - R \ln \frac{p_{final}}{p_{initial}}$$

The initial pressure of each component is specified, and it is equal to the final mixture pressure. The final pressure of each component is its partial pressure. Therefore, for each component

$$\Delta s = c_p \ln \frac{T_3}{T_{initial}} - R \ln$$

where

$$y_{CH_4} = \frac{N_{CH_4}}{N} = \frac{m_{CH_4}M_{CH_4}}{m_{CH_4}M_{CH_4} + m_{air}M_{air}} = \frac{1}{1 + \dfrac{m_{air}M_{air}}{m_{CH_4}M_{CH_4}}}$$

$$= \frac{1}{1 + 17\dfrac{16}{28.97}} = 0.0963$$

$$y_{air} = 1 - y_{CH_4} = 1 - 0.0963 = 0.9037$$

Substituting into the equation for i gives

$$i = T_0\left[c_{p,\mathrm{CH_4}} \ln\frac{T_3}{T_1} - R_{\mathrm{CH_4}} \ln y_{\mathrm{CH_4}} + \frac{M_{\mathrm{air}}}{M_{\mathrm{CH_4}}}\left(c_{p,\mathrm{air}} \ln\frac{T_3}{T_2} - R_{\mathrm{air}} \ln y_{\mathrm{air}}\right)\right]$$

$$i = 288\left[2.37 \ln\frac{416}{288} - 0.519 \ln 0.0963 + 17\left(1.02 \ln\frac{416}{433} - 0.287 \ln 0.9037\right)\right]$$

$$= 537 \text{ kJ/kg}$$

9·3 Mixtures of Real Gases

For real gases that do not follow the ideal-gas equation of state, we have no simple pvT relationships and no simple expressions for numerical calculations of internal energy, enthalpy, and entropy even for individual gases. Therefore, we have no reliable general method for calculating the properties of real-gas mixtures. Still, the need for such calculations arises, so various approximations have been proposed.

One method for pvT calculations is to assume that the additive pressure (Dalton) model holds so that

$$p_m = p_A(N_A, T_m, V_m) + p_B(N_B, T_m, V_m) + \cdots$$

In this equation $p_A(N_A, T_m, V_m)$, $p_B(N_B, T_m, V_m)$, . . . denote the pressures that would be exerted by individual components if they existed alone at the temperature and volume of the mixture. These are not partial pressures, because partial pressure is defined by $p_i \equiv y_i p_m$, and $p_i = p_i(N_i, T_m, V_m)$ only for ideal-gas mixtures. By selecting a suitable equation of state it is possible to determine the pressure of each component in a mixture if it existed alone at T_m and V_m. We then add these to determine the mixture pressure.

Another method is to assume that the additive volume (Amagat) model holds so that

$$V_m = V_A(N_A, p_m, T_m) + V_B(N_B, p_m, T_m) + \cdots$$

In this equation $V_A(N_A, p_m, T_m)$ and $V_B(N_B, p_m, T_m)$ denote the volumes of the individual components if they existed alone at the pressure and temperature of the mixture. These are not partial volumes, because partial volume is defined by $V_i \equiv y_i V_m$, but $V_i = V_i(N_i, p_m, T_m)$ only for ideal-gas mixtures. To get the additive volume formulation in terms of molar specific volumes, divide the form given above by the number of moles in the mixture and use the definition of mole fraction, $y_i \equiv N_i/N_m$, to give

$$\bar{v}_m = \frac{V_m}{N_m} = \frac{V_A(N_A, p_m, T_m)}{N_m} + \frac{V_B(N_B, p_m, T_m)}{N_m} + \cdots$$

$$= \frac{y_A V_A(N_A, p_m, T_m)}{N_A} + \frac{y_B V_B(N_B, p_m, T_m)}{N_B} + \cdots$$

$$= y_A \bar{v}_A(p_m, T_m) + y_B \bar{v}_B(p_m, T_m) + \cdots$$

By selecting a suitable equation of state it is possible to determine the molar specific volume of each component if it existed alone at p_m and T_m, but an iterative method may be required if the equations of state are not in an explicit volume form. (For ideal gases, the molar specific volume is the same for all components if they exist alone at the same pressure and temperature, but remember that this is not true for real gases.)

A third method is to use a compressibility factor and the relationship

$$p_m V_m = Z_m N_m \bar{R} T_m$$

Because compressibility factors for mixtures are generally not available, determining Z_m is a challenge. An approximation that has often proved satisfactory is

$$Z_m = y_A Z_A + y_B Z_B + \cdots$$

where y_A, y_B, . . . are the mole fractions of the components and Z_A, Z_B, . . . are their compressibility factors. If the compressibility factors are evaluated at the pressure and temperature of the mixture, the equation for Z_m reduces to the additive volume model.

This is a convenient procedure, because charts of Z as a function of p_R and T_R are available. In determining p_R for each component, use $p_R = p_m/p_c$, not the ratio of the *partial* pressure of the component to its critical pressure.

The changes in enthalpy and entropy for a real-gas mixture can be calculated from

$$\Delta h_m = x_A \, \Delta h_A + x_B \, \Delta h_B + \cdots = \sum x_i \, \Delta h_i$$

or

$$\Delta \bar{h}_m = y_A \, \Delta \bar{h}_A + y_B \, \Delta \bar{h}_B + \cdots = \sum y_i \, \Delta \bar{h}_i$$

and

$$\Delta s_m = x_A \, \Delta s_A + x_B \, \Delta s_B + \cdots = \sum x_i \, \Delta s_i$$

or

$$\Delta \bar{s}_m = y_A \, \Delta \bar{s}_A + y_B \, \Delta \bar{s}_B + \cdots = \sum y_i \, \Delta \bar{s}_i$$

when Δh_i and Δs_i for each component are determined for real gases by methods such as the one shown in Section 7·6·3. The inaccuracies inherent in the use of generalized departure charts that are in turn based on generalized compressibility factors must be recognized.

9·4 Mixtures of Ideal Gases and Vapors

In many engineering systems, such as air-conditioning units, dryers, and humidifiers, the operating fluid is a gas–vapor mixture. In discussing mixtures of gases and vapors, let us call a gas at a temperature lower than its critical temperature a vapor. Thus a vapor can be liquefied by increasing its pressure at constant temperature. This introduces an important consideration in analyzing gas–vapor mixtures that was not present with gas mixtures: The maximum pressure of a vapor in a mixture depends on the temperature of the mixture. To illustrate this, consider a mixture of nitrogen and oxygen at a mixture pressure of 100 kPa and 40°C. The mole fraction of each component can vary from 0 to 1, and the corresponding partial pressure can vary from 0 to 100 kPa. Each component and the mixture can be accurately modeled as an ideal gas. In contrast, consider a mixture of nitrogen and water vapor at 40°C and a mixture pressure of 100 kPa. Each component and the mixture can still be modeled as an ideal gas. However, the steam tables show that at 40°C the maximum pressure under which water vapor can exist is 7.381 kPa.

The maximum pressure of a vapor in a gas–vapor mixture depends on the mixture temperature.

To be precise, 7.381 kPa is the saturation pressure for the pure substance water at 40°C. When nitrogen is present with water liquid and water vapor at 40°C, the equilibrium pressure of the water vapor is slightly different. The difference is inconsequential in this application.

Consequently, the range of composition of this mixture is strictly limited because the mole fraction of water vapor in the mixture cannot exceed 0.07381. Furthermore, the pressure and temperature of the nitrogen–oxygen mixture can be varied over wide ranges without affecting the composition of the mixture; but increasing the pressure or decreasing the temperature of the nitrogen–water-vapor mixture even slightly may cause some of the water vapor to condense, thereby changing the composition of the gas–vapor mixture. The following examples show the application of the principles involved.

Example 9·6 Gas–Vapor Mixture

PROBLEM STATEMENT

A mixture of 0.020 kg of saturated water vapor and 0.50 kg of air is contained in a tank at a temperature of 60°C. Determine the total pressure of the mixture and the volume of the tank.

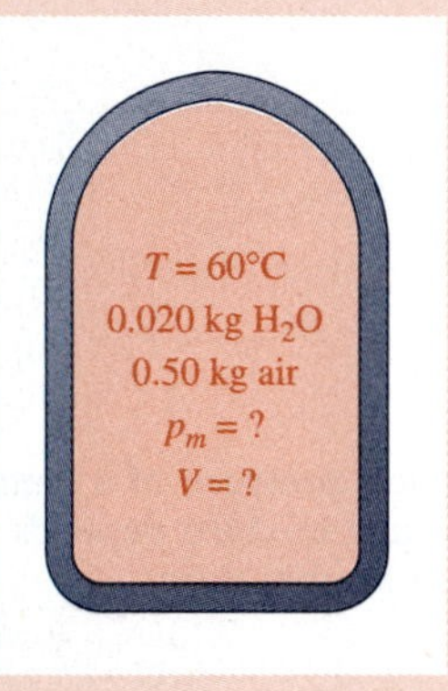

SOLUTION

From the steam tables, the saturation pressure and saturation specific volume for water vapor at 60°C are

$$p_v = 19.9 \text{ kPa}$$
$$v_g = 7.674 \text{ m}^3/\text{kg}$$

The volume occupied by the water vapor or the volume of the tank is

$$V = m_v v_g = 0.020(7.674) = 0.1535 \text{ m}^3$$

This is also the volume occupied by the air. If in accordance with the Dalton model the air behaves as though it existed alone at the temperature and volume of the mixture, the pressure of the air may be determined from the ideal-gas equation of state:

$$p_a = \frac{m_a R_a T}{V} = \frac{0.50(0.287)333}{0.1535} = 311.3 \text{ kPa}$$

Again from the Dalton model, the pressure of the mixture equals the sum of the pressures exerted by the water vapor and the air:

$$p_m = p_v + p_a = 19.9 + 311.3 = 331 \text{ kPa}$$

Example 9·7 Gas–Vapor Mixture

PROBLEM STATEMENT

One kilogram of water vapor and 2.0 kg of air are held in a cylinder that has a volume of 1.109 m^3. If the temperature of the mixture is 120°C, determine the mixture pressure. Determine also the pressure to which the mixture could be compressed isothermally before it ceases to be a pure substance.

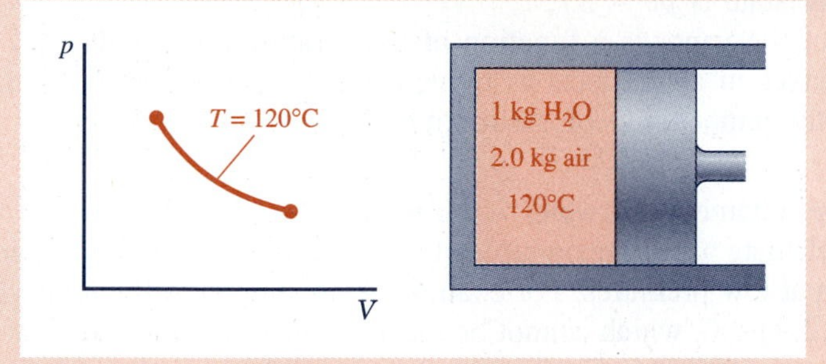

SOLUTION

From the superheated steam table, the pressure of steam at a temperature of 120°C and a specific volume of 1.109 m^3/kg is 160 kPa. The air pressure may be determined from the ideal-gas equation of state:

$$p_a = \frac{m_a R_a T}{V} = \frac{2.0(0.287)393}{1.109} = 203 \text{ kPa}$$

The total pressure of the mixture is, therefore,

$$p_m = p_a + p_v = 203 + 160 = 363 \text{ kPa}$$

As the mixture is compressed isothermally at 120°C, the vapor pressure increases until it reaches its saturation pressure, 198.5 kPa. Vapor at no higher pressure can exist at 120°C, so condensation begins. The gas–vapor mixture composition changes, so it is no longer a pure substance. Until condensation begins, the composition of the mixture is fixed, so the ratio of vapor partial pressure to mixture pressure is constant and

$$\left(\frac{p_v}{p_m}\right)_{\substack{\text{incipient}\\ \text{condensation}}} = \left(\frac{p_v}{p_m}\right)_{\text{initial}}$$

$$p_{m,\substack{\text{incipient}\\ \text{condensation}}} = \frac{p_{sat}}{(p_v/p_m)_{\text{initial}}} = \frac{198.5}{160/363}$$

$$= 450 \text{ kPa}$$

9·4·1 Atmospheric Air

Engineers deal with many different gas–vapor mixtures, but the one that receives the most attention is **atmospheric air**, a mixture of air and water vapor. In most applications involving atmospheric air, the temperatures and therefore the maximum vapor partial pressures are low enough that the vapor can be modeled as an ideal gas. That is,

- the equation of state is $pv = RT$,
- the enthalpy of the vapor is a function of temperature only, and
- the vapor behaves in all respects as though it existed alone at its partial pressure and the temperature of the mixture.

To support the statement that water vapor in atmospheric air follows $pv = RT$, you can calculate from steam-table data some compressibility factor values for steam at low pressures. For example, even at a partial pressure as high as 20 kPa (2.9 psia), which cannot be reached unless the mixture is at a temperature of over 60°C (140 F), the compressibility factor is 0.996^+. For vapor pressures usually encountered in atmospheric air, the compressibility factor is even closer to unity.

That the enthalpy of water vapor at low pressure depends only on temperature can be verified by means of a Ts (or ph or hs) diagram for steam such as Chart A·7. For superheated steam at low pressures the constant-enthalpy lines coincide with constant-temperature lines. Thus no matter what the pressure of water vapor in atmospheric air is, its enthalpy can be read from the superheated vapor tables at the lowest pressure entry in the table and the atmospheric air temperature. Some tables include a separate table for low pressures. For temperatures lower than the lowest temperature entry in the superheated steam or low-pressure steam table, the enthalpy of water vapor at any pressure equals very nearly the enthalpy of saturated vapor at the same temperature.

The enthalpy of low-pressure superheated steam equals very nearly the enthalpy of saturated steam at the same temperature.

For an ideal-gas mixture the partial pressure of each component equals the pressure that that component would exert if it existed alone at the temperature and volume of the mixture. For this reason we often refer to the partial pressure of water vapor in atmospheric air simply as *the pressure of the vapor* in the air, since the vapor behaves in all respects as though it were alone at this pressure and the mixture temperature. (As mentioned earlier, the presence of air affects the equilibrium temperature of liquid and vapor water at a given pressure, making it slightly different from that given in the steam tables for water as a pure substance, but this effect is negligible with atmospheric air.)

Dry air

The air component of atmospheric air is often referred to as the **dry air** to distinguish it from the mixture. Thus atmospheric air is composed of dry air plus water vapor. In weather and air-conditioning applications the temperature range is usually so limited that the c_p of dry air can be considered constant at 1.005 kJ/kg·K (0.240 B/lbm·R).

9·4·2 Relative Humidity and Humidity Ratio

Two terms frequently used in dealing with mixtures of air and water vapor are *relative humidity* and *humidity ratio*. It is important to learn the definitions of these terms and the relationship between them.

Relative humidity ϕ is defined as the ratio of the pressure (i.e., partial pressure) of the vapor in the mixture to the saturation pressure of the vapor at the temperature of the mixture. Let the pressure of the vapor in the mixture be designated by p_v, and let the vapor saturation pressure at the mixture temperature be designated by p_g. (The subscript g has been adopted because it is used in the steam tables to refer to saturated vapor.) Then the relative humidity is defined as

$$\phi \equiv \frac{p_v}{p_g} \tag{9·1}$$

Relative humidity

If the water vapor in the mixture and saturated water vapor at the same temperature can be modeled as ideal gases, relative humidity can be expressed in other forms by substituting for each pressure its equivalent RT/v. Thus

$$\phi \equiv \frac{p_v}{p_g} = \frac{RTv_g}{v_v RT} = \frac{v_g}{v_v} = \frac{\rho_v}{\rho_g} \tag{9·1}$$

The specific volume of the vapor in atmospheric air can thus be determined from the mixture temperature and the relative humidity,

$$v_v = \frac{v_g}{\phi} \tag{9·2}$$

Notice that relative humidity pertains only to the *vapor* in atmospheric air. It is independent of the pressure and density of the dry air in the mixture. It is independent of the barometric pressure.

Atmospheric air that contains saturated water vapor has a relative humidity of 1.0 or 100 percent and is called *saturated air*, although it is only the water vapor in the air which is in a saturation state.

A *Ts* diagram for water vapor in atmospheric air is shown in Fig. 9·4. If point 1 represents the state of the water vapor as it exists in the mixture, relative humidity is the ratio of p_1 to p_2, where state 2 is a saturation state at the mixture temperature.

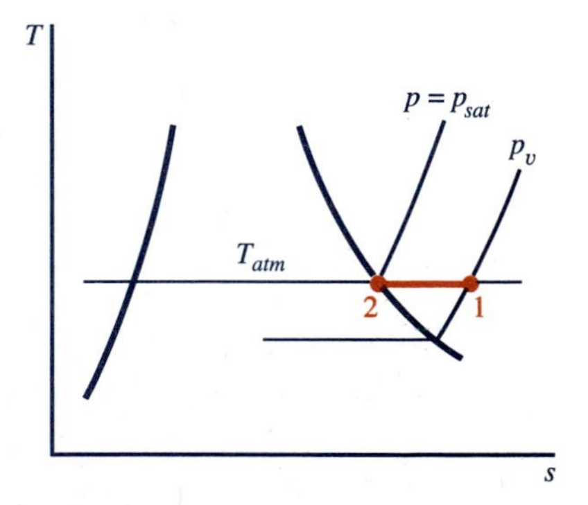

Figure 9·4 *Ts* diagram for the water vapor in atmospheric air.

Humidity ratio ω is defined as the ratio of the mass of vapor in atmospheric air to the mass of dry air. Notice that it is not the same as the mass fraction of water vapor in the mixture. Using the subscript v for vapor and a for dry air, we get

Humidity ratio

$$\omega \equiv \frac{m_v}{m_a} \tag{9·3}$$

and, since the volume is the same for both components of the mixture,

$$\omega \equiv \frac{m_v}{m_a} = \frac{\rho_v}{\rho_a} = \frac{v_a}{v_v} \tag{9·3}$$

If both the air and the water vapor behave as ideal gases, we can express the humidity ratio as

$$\omega \equiv \frac{m_v}{m_a} = \frac{p_v V R_a T}{R_v T p_a V} = \frac{p_v R_a}{p_a R_v} = \frac{p_v R_a}{(p_m - p_v) R_v}$$

Solving for p_v gives

$$p_v = \frac{\omega R_v p_m}{\omega R_v + R_a}$$

This relation is also useful in dealing with combustion products at low pressures, R_a being taken as the gas constant of the dry products of combustion.

The relationship between humidity ratio and relative humidity is obtained by combining Eqs. 9·2 and 9·3:

Relationship between humidity ratio and relative humidity

$$\omega = \frac{v_a}{v_g} \phi \tag{9·4}$$

Example 9·8

PROBLEM STATEMENT

Atmospheric air at 95 kPa, 30°C has a relative humidity of 70 percent. Determine the humidity ratio.

SOLUTION

From the steam tables, the saturation pressure for water vapor at 30°C is 4.246 kPa. The partial pressure of the water vapor in the mixture is then

$$p_v = \phi(p_g) = 0.70(4.246) = 2.97 \text{ kPa}$$

Since the total atmospheric pressure is the sum of the partial pressures, the air pressure is

$$p_a = p_m - p_v = 95 - 2.97 = 92.0 \text{ kPa}$$

The specific volume of the air may be found from the ideal-gas equation

$$v_a = \frac{R_a T}{p_a} = \frac{0.287(303)}{92.0} = 0.945 \text{ m}^3/\text{kg}$$

The value for v_g, taken from the steam tables, is 32.90 m³/kg. The humidity ratio is then

$$\omega = \frac{v_a}{v_g}\phi = \frac{0.945(0.70)}{32.90} = 0.0201 \frac{\text{kg water vapor}}{\text{kg dry air}}$$

9·4·3 Temperatures Used in Psychrometry

Psychrometry is the measurement and analysis of moist atmospheric air. Four temperatures used in psychrometry are dry-bulb temperature, dew-point temperature, adiabatic saturation temperature, and wet-bulb temperature.

A unit of mass used in psychrometry is the grain: 7000 grains = 1 lbm.

Dry-Bulb Temperature. The dry-bulb temperature is simply the temperature of the mixture as it would be measured by any of several types of ordinary thermometers placed in the mixture. Care must be exercised in measuring the temperature of atmospheric air to avoid errors caused by radiant heat transfer between the thermometer and its surroundings. Therefore, where temperature differences exist, shielded thermometer elements should be used. The term *dry-bulb* temperature is used to distinguish the temperature of the mixture from the temperature reading obtained from a thermometer that has its temperature-sensitive element wrapped in water-soaked gauze, the *wet-bulb* temperature.

Dew-Point Temperature. The dew-point temperature of an air–vapor mixture is defined as the saturation temperature of the vapor corresponding to its partial pressure in the mixture. It is thus the temperature at which condensation begins if the mixture is cooled at constant pressure.

A simple laboratory determination of dew-point temperature consists of partially filling a metal cup with water, adding ice, and then stirring while observing the water temperature as it is lowered. The temperature at which moisture begins to collect on the outside of the cup is approximately the dew-point temperature of the room air. The water temperature and the tem-

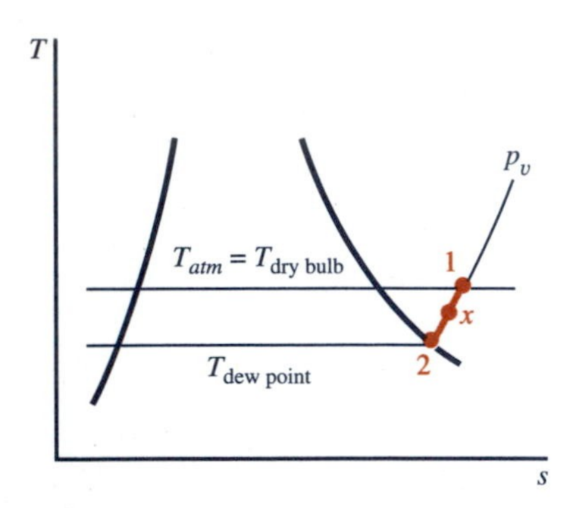

Figure 9·5 T_2 is the dew-point temperature for states 1, 2, and x.

perature of the air in contact with the cup are not exactly the same, and this is one source of error in this method.

If point 1 in Fig. 9·5 represents the water vapor in atmospheric air, the dew point is the temperature at point 2 or the saturation temperature corresponding to the vapor pressure. States 1, 2, and x as well as any other states with the same vapor pressure have the same dew point. For air at 100 percent relative humidity (saturated air) the dew-point temperature equals the dry-bulb temperature.

Example 9·9

PROBLEM STATEMENT

Determine the dew point of atmospheric air at 95 kPa, 30°C, 70 percent relative humidity.

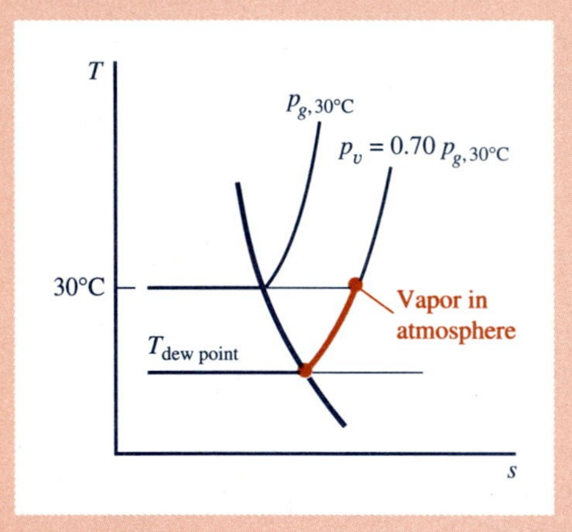

SOLUTION

From the steam tables, the saturation pressure for water vapor at 30°C is 4.246 kPa. The partial pressure of the water vapor in the mixture is then

$$p_v = \phi(p_g) = 0.70(4.246) = 2.97 \text{ kPa}$$

The dew point is the saturation temperature corresponding to this pressure. This is found from the steam tables to be approximately 24°C.

Adiabatic Saturation Temperature. In atmospheric air with a relative humidity of less than 100 percent, the water vapor is at a pressure lower than its saturation pressure. Therefore, if this air is placed in contact with liquid water, some of the water evaporates into the air. The humidity ratio of the air increases. If the evaporation occurs in a thermally insulated container, the temperature of the air decreases because at least part of the latent heat of vaporization of the water that evaporates comes from the air. The lower the initial humidity ratio, the greater the amount of evaporation and the greater the temperature decrease; so we have here the basis of an indirect measurement of humidity ratio.

The adiabatic saturation temperature of atmospheric air is defined as the temperature that results from adiabatically evaporating water into the atmospheric air in steady flow until it becomes saturated, the water being supplied at the final temperature of the mixture. At first it appears that this definition is circular because to determine the adiabatic saturation temperature we must supply water that is at that temperature. Actually the definition is operational and sufficient, and the adiabatic saturation temperature can be found by the following operations: (1) Add water at any temperature adiabatically to steadily flowing atmospheric air until it becomes saturated. (2) Measure the temperature of the saturated air. (3) Change the temperature of the water being added to equal that of the saturated air as measured in step 2. (4) Repeat steps 2 and 3 until the temperature of the saturated air equals that of the water being added. This is the adiabatic saturation temperature of the atmospheric air.

To see how a measurement of the adiabatic saturation temperature can be used to determine the humidity ratio of atmospheric air, consider the steady-flow system shown in Fig. 9·6. Air of unknown humidity ratio ω_1 enters at section 1. The air leaving at section 2 is saturated, and, since the water added is at the same temperature, this is the adiabatic saturation temperature. The total or mixture pressure is constant throughout the system. The mixture pressure and the temperatures at sections 1 and 2 can be measured.

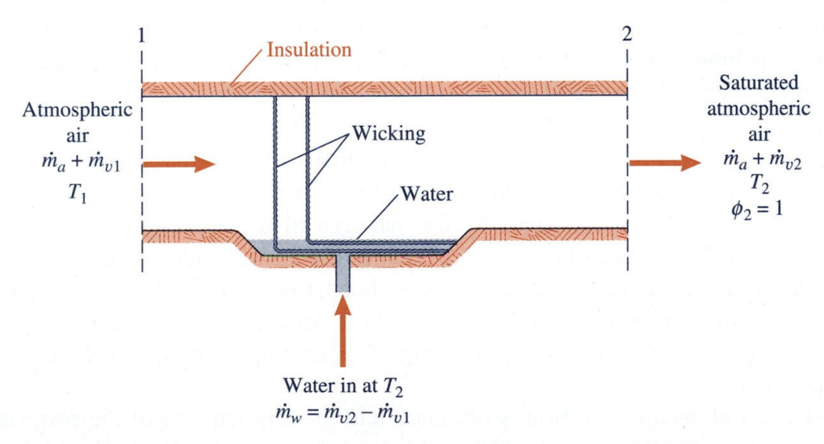

Figure 9·6 Steady-flow system for determining adiabatic saturation temperature.

No work is done, the process is adiabatic, and changes in kinetic and potential energy are negligible. Therefore, applying the first law to this steady-flow system we get

$$m_a h_{a1} + m_{v1} h_{v1} + (m_{v2} - m_{v1}) h_{f2} = m_a h_{a2} + m_{v2} h_{v2}$$

In this equation, $(m_{v2} - m_{v1})$ is the amount of water added, and the enthalpy of the water is written as h_{f2} because the water is introduced at the temperature T_2. Dividing by the mass of air m_a, and noting that $h_{v1} = h_{g1}$ very nearly, as explained in Section 9·4·1, and $h_{v2} = h_{g2}$ exactly because the vapor at section 2 is saturated, we have

$$h_{a1} + \omega_1 h_{g1} + (\omega_2 - \omega_1) h_{f2} = h_{a2} + \omega_2 h_{g2}$$

$$\omega_1 = \frac{h_{a2} - h_{a1} + \omega_2 (h_{g2} - h_{f2})}{h_{g1} - h_{f2}}$$

$$= \frac{c_{pa}(T_2 - T_1) + \omega_2 h_{fg2}}{h_{g1} - h_{f2}}$$

The expression can be evaluated if T_1, T_2, and p_m are measured. At section 2, the air is saturated with water vapor so that $p_{v2} = p_{g2}$ and $\phi_2 = 1.0$. Thus

$$\omega_2 = \frac{v_{a2}}{v_{g2}} = \frac{R_a T_2}{(p_m - p_{g2}) v_{g2}}$$

and ω_2 is a function of only p_m and T_2.

The states of the water vapor in the mixture during the adiabatic saturation process are shown on a Ts diagram in Fig. 9·7. During the process the vapor pressure increases and the temperature decreases; so the adiabatic saturation temperature is higher than the dew-point and lower than the dry-bulb temperature. For the limiting case of a saturated mixture, the dry-bulb, dew-point, and adiabatic saturation temperatures are the same.

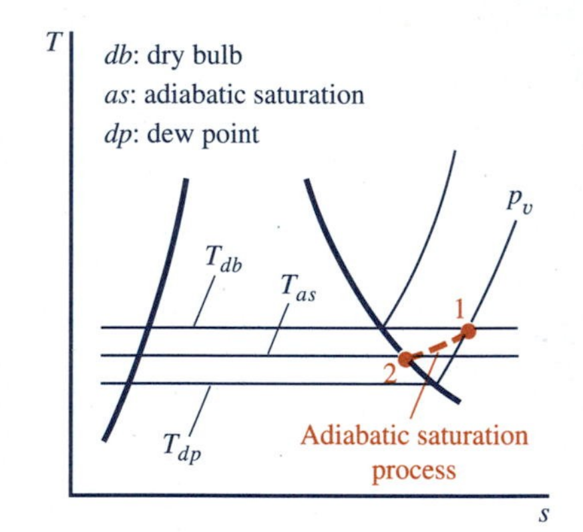

Figure 9·7 Adiabatic saturation.

We have described a method of determining the humidity ratio of atmospheric air from temperature and barometric-pressure measurements. Actually, it is inconvenient to saturate atmospheric air by the procedure described, so other methods are used.

Wet-Bulb Temperature. To avoid the difficulty of adiabatically saturating a sample of atmospheric air, the wet-bulb temperature analysis has been devised. This procedure involves the passage of an unsaturated air–vapor mixture over a wetted surface until a condition of dynamic equilibrium has been attained. When this condition has been reached, the heat transferred from the air–vapor stream to the liquid film to evaporate part of it is equal to the energy carried from the liquid film to the air–vapor stream by the diffusing vapor.

The equilibrium condition is obtained and the temperature of the resulting air–vapor mixture is measured by means of a thermometer, the bulb of which is covered with gauze soaked in water. A schematic diagram is shown in Fig. 9·8. The flow of atmospheric air is provided by a fan or by mounting the thermometer in a holder with a swivel handle so that it can be rotated or whirled through the air. The thermometer reading is called the wet-bulb temperature.

The relationship between the wet-bulb temperature and the adiabatic saturation temperature for any gas–vapor mixture depends on the heat-transfer and diffusion characteristics of the mixture. It happens that for air–water-vapor mixtures in the normal pressure and temperature range of atmospheric air the wet-bulb temperature measured by the usual type of instrument very nearly equals the adiabatic saturation temperature. This is simply fortuitous, and the equality does not hold for most gas–vapor mixtures. For example, in the air–vapor mixtures in oil-storage tanks and in alcohol–air mixtures, the difference between the wet-bulb temperature and the adiabatic saturation temperature may be quite large, and serious errors would follow from the assumption that they are equal.

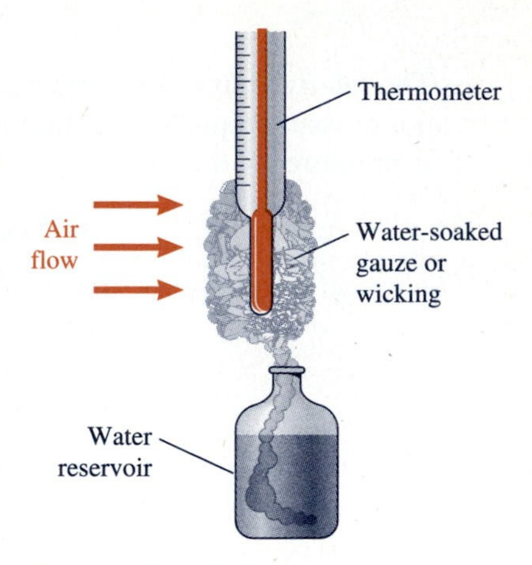

Figure 9·8 Wet-bulb thermometer.

Example 9·10

PROBLEM STATEMENT

Determine the humidity ratio of air of a barometric pressure of 95.0 kPa that has a dry-bulb temperature of 30°C and a wet-bulb temperature of 20°C.

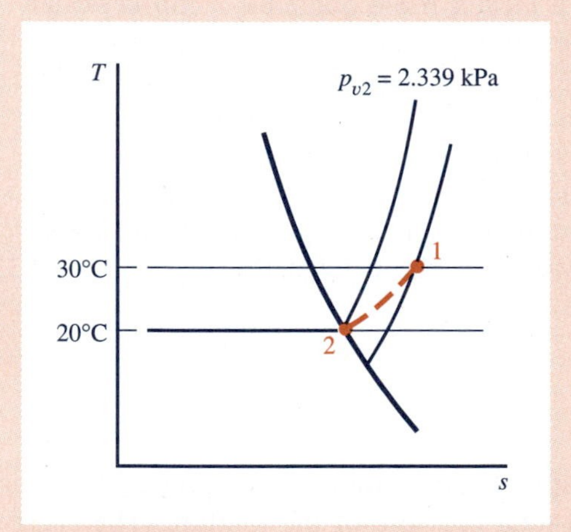

SOLUTION

We first assume that the wet-bulb temperature equals the adiabatic saturation temperature. We then apply the first law to an adiabatic saturation process that proceeds from the specified state to the saturated state at 20°C. Call these two states 1 and 2, respectively.

The relative humidity at state 2 is 100 percent, since complete saturation is assumed. For this condition the vapor pressure equals the saturation pressure at 20°C, which is found from the steam tables to be 2.339 kPa. The pressure of the dry air is then

$$p_{a2} = p_m - p_{v2} = 95.0 - 2.339 = 92.66 \text{ kPa}$$

The specific volume of dry air is

$$v_{a2} = \frac{R_a T_2}{p_{a2}} = \frac{0.287(293)}{92.66} = 0.908 \text{ m}^3\text{/kg}$$

and the humidity ratio at state 2 is

$$\omega_2 = \frac{v_{a2}}{v_{g2}} \phi = \frac{0.908(1.0)}{57.78} = 0.0157 \text{ kg/kg dry air}$$

Application of the first law to the adiabatic saturation process leads (as is shown earlier in this section) to

$$\omega_1 = \frac{c_{pa}(T_2 - T_1) + \omega_2 h_{fg2}}{h_{g1} - h_{f2}}$$

$$\omega_1 = \frac{1.005(20 - 30) + 0.0157(2453.4)}{(2555.3 - 83.8)}$$

$$= 0.0115 \text{ kg/kg dry air}$$

The temperature definitions are all based on equilibrium conditions. Notice, however, that a long time may be required for a system to reach equilibrium. For example, if a closed room contains a cup of water, equilibrium conditions are not reached until all of the water in the cup has evaporated or until the relative humidity is 100 percent with some liquid remaining in the cup. Likewise, the humidity at the seashore is not always 100 percent, and the relative humidity during a rain shower may be less than 100 percent. Someone observing the flow of air through a duct that contains water sprays and has excess water dripping from the walls and structural members might infer that the air emerging has a relative humidity of 100 percent, but actually, because of incomplete mixing or inadequate time for evaporation of the liquid water into the air stream, the leaving air relative humidity may be significantly lower than 100 percent.

9·4·4 Psychrometric Charts

A psychrometric chart for atmospheric air is a plot of the properties of air–water-vapor mixtures at a fixed total pressure of the mixture. The chart is usually based on a pressure of one standard atmosphere, but charts for other barometric pressures are available. The specific properties are expressed not per unit mass of mixture but per unit mass of dry air in the mixture.

A skeleton outline of a psychrometric chart is shown in Fig. 9·9. (Complete charts are included in the appendix.) The horizontal and vertical axes represent dry-bulb temperature and humidity ratio, respectively.)

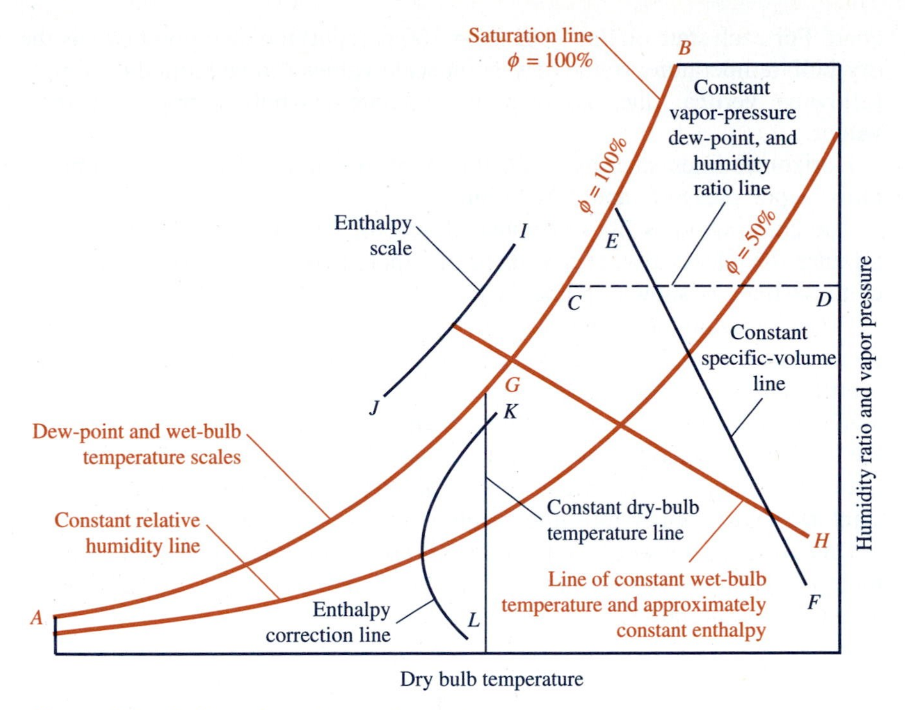

Figure 9·9 Outline of psychrometric chart.

The line A–B is the saturation line and represents states for which the relative humidity is 100 percent. This line and other constant ϕ lines can be readily plotted because for states of constant relative humidity the humidity ratio is a function of only the mixture pressure and the dry-bulb temperature as shown by

$$\omega = \frac{v_a}{v_g}\phi = \frac{R_a T}{p_a v_g}\phi = \frac{R_a T}{(p_m - p_v)v_g}\phi = \frac{R_a T}{(p_m - \phi p_g)v_g}\phi$$

If the air and water vapor both behave as ideal gases and the mixture pressure is constant, the vapor pressure and the humidity ratio are related by

$$\omega \equiv \frac{m_v}{m_a} = \frac{p_v V R_a T}{p_a V R_v T} = \frac{p_v R_a}{(p_m - p_v)R_v}$$

Thus a vapor-pressure scale can be constructed along with the humidity ratio scale on the vertical axis. The humidity ratio scale is usually selected as a linear scale, so the vapor-pressure scale is nonlinear.

The dew point of atmospheric air is a function of the vapor pressure only; so a dew-point temperature scale can also be constructed along the vertical axis. To facilitate reading of the chart, the dew-point temperature scale is laid out along the saturation line A–B instead of as another scale at one side of the chart. For each state on line *A–B* (ϕ = 100 percent) the dew-point equals the dry-bulb temperature; so the dew-point scale values can be plotted by simply following vertical lines up from the various dry-bulb temperature scale values.

Horizontal lines such as *C–D*, therefore, are lines of constant humidity ratio, vapor pressure, and dew point.

The chart also has lines of constant specific volume given as volume of mixture (or, identically, of dry air or of vapor) per unit mass of dry air. One of these lines is shown as line *E–F*.

Consider again the adiabatic saturation process discussed in the preceding section. Every state that air passes through during the adiabatic saturation process has the same adiabatic saturation temperature. Thus the path of an adiabatic saturation process, along which the vapor pressure increases and the dry-bulb temperature decreases, is a line of constant adiabatic saturation temperature and for the special case of air–water-vapor mixtures it is a constant wet-bulb temperature line. Such a line is shown as *G–H*.

Lines of constant wet-bulb temperature are approximately lines of constant mixture enthalpy. To demonstrate this, we first write an energy balance for an adiabatic saturation process:

$$h_{a1} + \omega_1 h_{v1} + (\omega_2 - \omega_1)h_{f2} = h_{a2} + \omega_2 h_{v2}$$

where the subscript 1 refers to any section in the adiabatic saturation flow path and the subscript 2 refers to the fully saturated state. If we define the mixture enthalpy as $h_m \equiv h_a + \omega h_v$, noting that this is the enthalpy *of the mixture* expressed *per unit mass of dry air*, the energy balance becomes

$$h_{m1} - \omega_1 h_{f2} = h_{m2} - \omega_2 h_{f2}$$

Since subscript 1 refers to *any* section in the adiabatic saturation flow path, this last equation can be written as

$$h_m - \omega h_{f2} = \text{constant}$$

Since ωh_f evaluated at the adiabatic saturation temperature is always very small in comparison with h_m, we have

$$h_m \approx \text{constant}$$

for the adiabatic saturation process. Thus, a constant wet-bulb temperature line (which for atmospheric air is very nearly a constant adiabatic saturation temperature line) such as *G–H* is approximately a constant-enthalpy line, so the psychrometric chart carries an enthalpy scale that is read against the constant wet-bulb temperature lines. For accurate determination of the enthalpy of an unsaturated air–vapor mixture, a correction must be applied to the scale value. If we do not neglect the ωh_{f2} term, the enthalpy of the mixture at any state is

$$h_m = h_{m2} - (\omega_2 - \omega)h_{f2} = h_{m2} + \text{correction}$$

where the subscript 2 refers to the adiabatic saturation end state. The correction term, $-(\omega_2 - \omega)h_{f2}$, which literally corrects for the assumption that the mixture enthalpy is constant along a constant wet-bulb temperature line, is a function of humidity ratio and wet-bulb temperature only, and so it can be plotted on the chart. K–L is such a line of constant enthalpy correction. The enthalpy of any mixture then is the enthalpy at the corresponding adiabatic saturation state plus the correction read from a curve such as K–L passing through the mixture state on the chart. (Several other methods are used to show the correction on psychrometric charts, and the correction is usually so small that many charts do not show it at all.)

The same calculations can be made by means of the equations presented here, by the psychrometric chart, by the software accompanying this book, or by other software. Sometimes one method has advantages over the others. Sometimes they are used jointly. For example, rough or scoping calculations can be made by means of a psychrometric chart and followed by more precise calculations using another method. Rapid checking of results obtained by another method can be performed by use of the chart. Some facility with each of the methods is useful.

Example 9·11 Psychrometric Chart

PROBLEM STATEMENT

For atmospheric air with dry-bulb and wet-bulb temperatures of 90 and 70 F, respectively, determine by the use of the psychrometric chart (*a*) the humidity ratio, (*b*) the relative humidity, (*c*) the dew-point temperature, (*d*) the pressure of the water vapor, (*e*) the volume per pound of dry air, and (*f*) the enthalpy per pound of dry air.

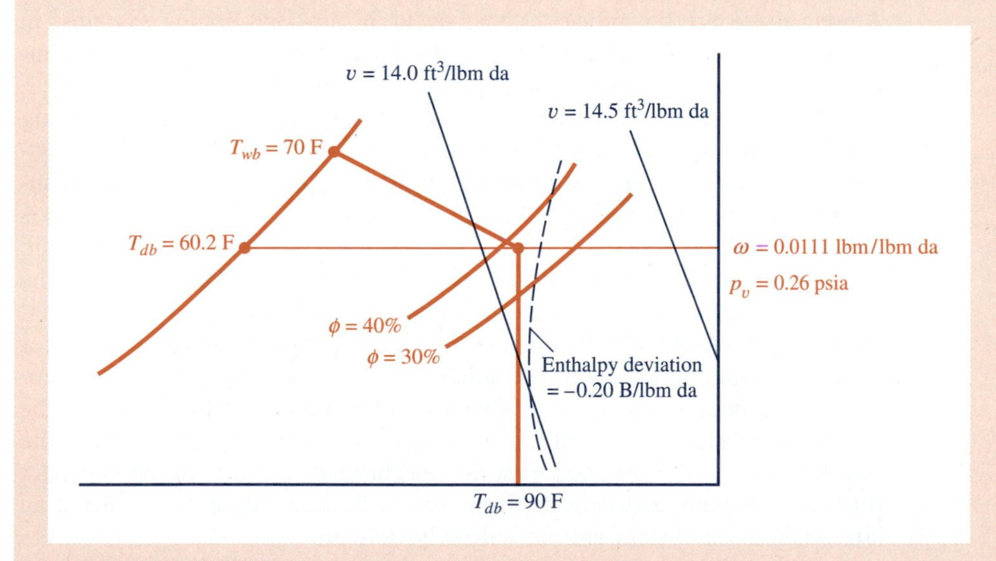

SOLUTION

The particular point on the chart that designates the condition of the atmospheric air is located by finding the intersection of the vertical 90 F dry-bulb temperature line and the sloping 70 F wet-bulb temperature line.

(*a*) The humidity ratio is found by moving along a horizontal line to the proper scale on the diagram. The approximate value is 0.0111 pound of water vapor per pound of dry air.
(*b*) The relative humidity is estimated by observing the location of the point between two of the curved relative humidity lines. The approximate value is 37 percent.
(*c*) The dew-point temperature is found by moving along a horizontal line through the point until the dew-point temperature scale is reached. The approximate value is 60.2 F.
(*d*) The partial pressure of the water vapor is found by following a horizontal line until the vapor-pressure scale is reached. The value is 0.26 psia.
(*e*) The volume per pound of dry air is estimated by observing the location of the point between two of the constant-volume lines. The approximate values is 14.1 cu ft/lbm dry air.
(*f*) The enthalpy may be found by first following the sloping 70 F wet-bulb temperature line until it intersects the enthalpy scale. The enthalpy value is 34.1 B/lbm for saturated air at 70 F wet-bulb temperature. For 90 F dry-bulb and 70 F wet-bulb temperatures, the enthalpy correction is approximately -0.2 B/lbm, so the mixture enthalpy is $34.1 - 0.2 = 33.9$ B/lbm.

9·5 Processes of Mixtures of Ideal Gases and Vapors

Processes of gas–vapor mixtures encountered in engineering are innumerable. Many of these occur under conditions of constant mixture pressure. Among these are the processes that occur in air-conditioning systems, in evaporative cooling, and in drying. In the processes of compressors and engines, of course, the change in mixture pressure may be the dominant factor.

Even within a single field such as air conditioning, the variations among processes encountered are endless. The major functions of an air-conditioning system may be combinations of heating, cooling, humidification, dehumidification, and removal of pollen, dust, and fumes. Energy conservation is always important. So are the initial cost and maintenance costs of equipment. These many considerations call for many balances among factors such as the ratio of recirculated air to fresh air in the air-conditioning system. Further, these considerations always involve varying conditions. An auditorium, for example, must provide comfort conditions when it is occupied by 1000 people or 100 people and for a wide range of outdoor conditions. Even the comfort conditions for people in buildings are known to vary widely. The *ASHRAE comfort zones* shown on the psychrometric charts in the appendix illustrate this point and indicate that even the desired output conditions of an air-conditioning system are not narrowly defined.

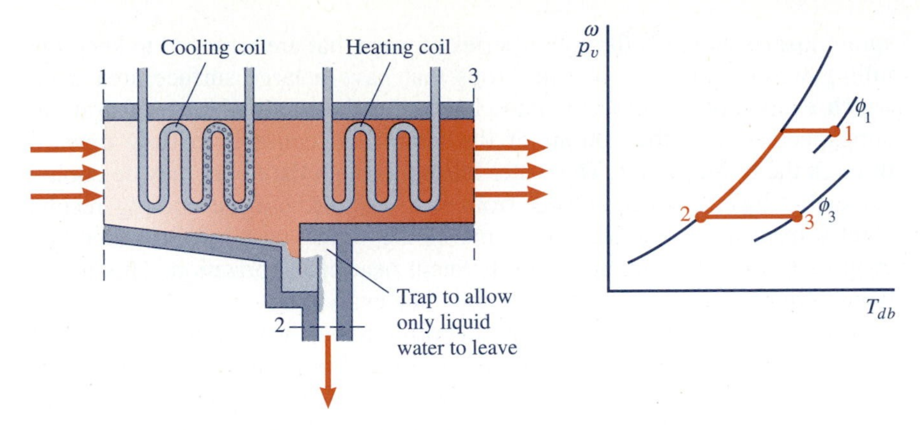

Figure 9·10 Dehumidification and reheating.

One common type of air-conditioning system that cools and dehumidifies air is shown in Fig. 9·10. The air is first cooled to a temperature lower than its initial dew point to cause some of the water vapor to condense. The air leaving the cooler is saturated at a low temperature and is unsuitable for use in rooms occupied by people because it is clammy. It must be heated or mixed with warmer air to produce a condition that is normally regarded as comfortable. The dehumidification and reheating processes are shown on a psychrometric chart in Fig. 9·10. Notice that the temperature to which the mixture must be cooled is the dew point corresponding to the desired final condition.

Where sufficient cooling water cannot be obtained from a river, lake, or ocean, it is necessary to cool water by means of a cooling tower or spray pond so that it can be used over and over again to remove heat from buildings, power plants, refrigerators, and other equipment. Figure 9·11 is a schematic diagram of an induced draft cooling tower. The warm water that is to be cooled is introduced at the top of the tower through distributing troughs or

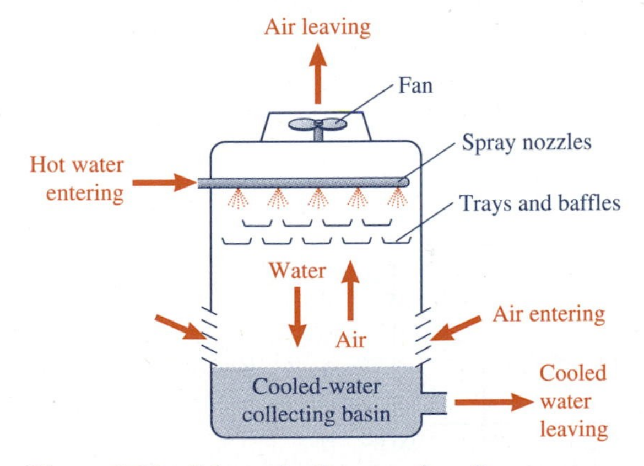

Figure 9·11 Schematic diagram of cooling tower.

spray nozzles and falls through a series of trays that are arranged to keep the falling water broken up in fine drops that have a large surface area from which evaporation can occur. Fans located at the top of the tower draw in atmospheric air at the bottom of the tower and cause it to flow upward through the falling water. The water falling through the tower may be cooled somewhat by a transfer of heat from it to the air, but the cooling results chiefly from the evaporation of some of the water, because the water that evaporates must be supplied with its latent heat of vaporization. This is obtained chiefly from the water that does not evaporate.

In making an energy balance on a cooling tower, we can usually neglect the heat transfer to the fluids within the tower from the surrounding atmosphere. Also, the work of the fans is negligible in comparison with the other energy quantities involved. The following example illustrates the application of the first law (energy balance) and the law of conservation of mass (mass balance) to a cooling tower.

Analysis of weather involves gradients in atmospheric pressure and temperature, buoyancy, and large-scale fluid motions related to terrain and the effects of large solar-heated areas, large cool areas, and bodies of water. Many weather phenomena can be explained by the behavior of air–vapor mixtures as described here. Numerous other phenomena, including aircraft so-called vapor trails, wingtip vortex trails, the fogging of windows and eyeglasses, and the water contamination of partially filled vented gasoline tanks, can likewise be explained.

Example 9·12 Air Conditioner

PROBLEM STATEMENT

Outdoor air at 90 F, 80 percent relative humidity, is to be cooled and dehumidified so that air at 70 F, 45 percent relative humidity, is produced. This is to be done by passing the air through a cooler with provision for removing condensate and then reheating the air, all under a constant pressure 14.2 psia. Determine, per pound of dry air entering, (*a*) the mass of water removed, (*b*) the heat removed from the air in the cooler, and (*c*) the heat added to the air in the reheater.

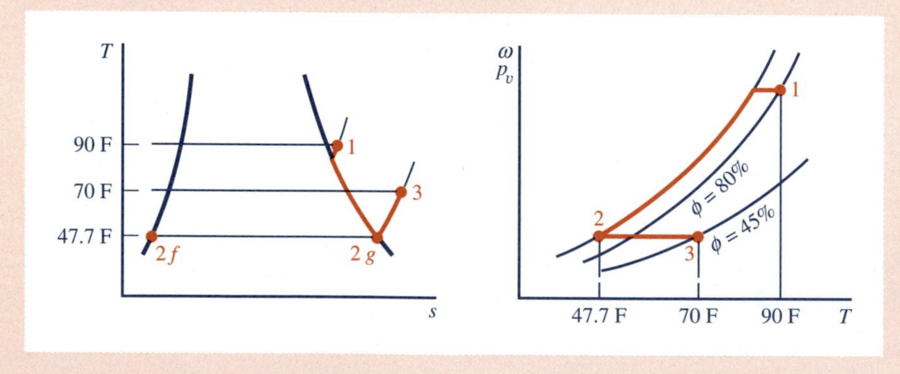

SOLUTION

Refer to Fig. 9·10 for a sketch of the system. The first step in the solution is to determine the humidity ratios for the initial and final conditions:

Initial Condition, State 1	*Final Condition, State* 3
$p_{v1} = \phi_1 p_{g1} = 0.80(0.698) = 0.558$ psia	$p_{v3} = 0.45(0.363) = 0.163$ psia
$p_{a1} = p_m - p_{v1} = 14.2 - 0.558 = 13.64$ psia	$p_{a3} = 14.2 - 0.163 = 14.04$ psia
$v_{a1} = \dfrac{R_a T_1}{p_{a1}} = \dfrac{53.3(550)}{13.64(144)} = 14.92$ cu ft/lbm	$v_{a3} = \dfrac{53.3(530)}{14.04(144)} = 13.97$ cu ft/lbm
$\omega_1 = \dfrac{v_{a1}}{v_{g1}}\phi_1 = \dfrac{14.92(0.80)}{466.6} = 0.0255$ lbm/lbm	$\omega_3 = \dfrac{13.97(0.45)}{866.1} = 0.00726$ lbm/lbm

(*a*) The mass of water removed is

$$\frac{m_w}{m_{da}} = \omega_1 - \omega_3 = 0.0255 - 0.00726 = 0.0182 \text{ lbm/lbm dry air}$$

(*b*) The mixture leaving the cooler will be saturated and must have the same humidity ratio as the mixture in the desired final condition. Thus the temperature at the cooler outlet must be the dew point corresponding to the final condition. This is the saturation temperature corresponding to the final vapor pressure of 0.163 psia and is approximately 47.7 F. The cooler must, therefore, cool the air from 90 F to 47.7 F. During the latter part of this cooling process (while the mixture is at temperatures lower than the dew point of the initial state) water is removed. Application of the first law to the cooler in which work = 0, $\Delta ke = 0$, and the flow is steady gives

$$q_{\text{out}} = h_{a1} + \omega_1 h_{v1} - h_{a2} - \omega_2 h_{v2} - (\omega_1 - \omega_2)h_{f2}$$

We assume that all the condensate leaves at the temperature T_2. Noting that $\omega_2 = \omega_3$, $h_v = h_g$ at the same temperature, and $\Delta h_a = c_{pa}\,\Delta T$, we have

$$\begin{aligned} q_{\text{out}} &= c_{pa}(T_1 - T_2) + \omega_1 h_{g1} - \omega_2 h_{g2} - (\omega_1 - \omega_2)h_{f2} \\ &= 0.24(90 - 47.7) + 0.0255(1100.3) - 0.00726(1081.9) - 0.0182(15.8) \\ &= 30.1 \text{ B/lbm dry air} \end{aligned}$$

(*c*) Application of the first law in similar fashion to the heater gives

$$\begin{aligned} q &= h_{a3} + \omega_3 h_{v3} - h_{a2} - \omega_2 h_{v2} \\ &= c_p(T_3 - T_2) + \omega_3(h_{g3} - h_{g2}) \\ &= 0.24(70 - 47.7) + 0.00726(1091.7 - 1081.9) \\ &= 5.4 \text{ B/lbm dry air} \end{aligned}$$

Comment: This problem can be solved quickly and with only a small loss in accuracy by obtaining properties from the psychrometric chart in the appendix, although the barometric pressure here is 14.2 psia instead of 14.7 psia on which the chart is based. Alternatively, properties can be obtained from the model for psychrometric properties in the software accompanying this book.

Example 9·13 Cooling Tower

PROBLEM STATEMENT

A cooling tower cools water from 45°C to 25°C. Water enters the tower at a rate of 110,000 kg/h. The air entering the tower is at 20°C with a relative humidity of 55 percent; the air leaving is at 40°C with a relative humidity of 95 percent. The air enters at the bottom of the tower and flows upward. Barometric pressure is 92.0 kPa. Determine (*a*) the required flow rate of atmospheric air in kg/h and (*b*) the amount of water lost by evaporation per hour.

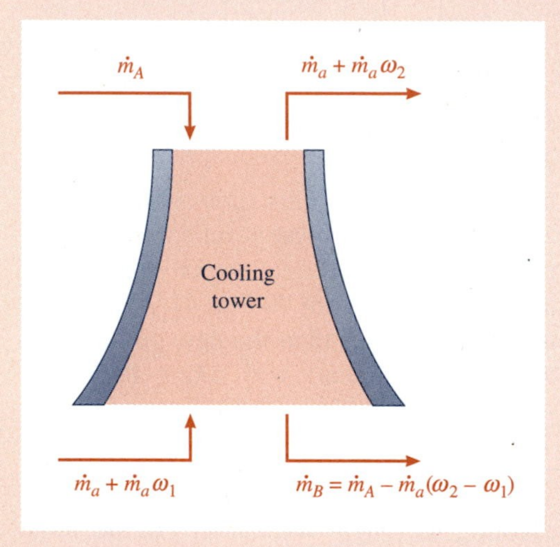

SOLUTION

First a sketch is made of the steady-flow system, the cooling tower. Assume that there is no heat exchange with the surroundings. Designate the air inlet as section 1, the air outlet as section 2, the water inlet as section *A*, and the water outlet as section *B*. The properties of the entering and leaving air streams are calculated as in Example 9·12 with the following results:

Entering Air, State 1	*Leaving Air, State* 2
$p_{v1} = 1.286$ kPa	$p_{v2} = 7.012$ kPa
$p_{a1} = 90.7$ kPa	$p_{a2} = 85.0$ kPa
$\omega_1 = 0.00822$ kg/kg dry air	$\omega_2 = 0.0541$ kg/kg dry air

(*a*) Let $\dot{m}_a$ be the mass flow rate of dry air required. $\dot{m}_{a1} = \dot{m}_{a2} = \dot{m}_a$. Applying the first law to the cooling tower, assuming that $q = 0$, $w = 0$, and $\Delta ke = 0$, we have

$$\dot{m}_a h_{a1} + \dot{m}_{v1} h_{v1} + \dot{m}_A h_A = \dot{m}_a h_{a2} + \dot{m}_{v2} h_{v2} + \dot{m}_B h_B$$

However,

$$\dot{m}_{v1} = \omega_1 \dot{m}_a$$
$$\dot{m}_{v2} = \omega_2 \dot{m}_a$$
$$\dot{m}_B = \dot{m}_A - \dot{m}_a(\omega_2 - \omega_1)$$

so that the energy balance becomes

$$\dot{m}_a h_{a1} + \dot{m}_a \omega_1 h_{v1} + \dot{m}_A h_A = \dot{m}_a h_{a2} + \dot{m}_a \omega_2 h_{v2} + [\dot{m}_A - \dot{m}_a(\omega_2 - \omega_1)] h_B$$

$$\dot{m}_a = \frac{\dot{m}_A(h_A - h_B)}{h_{a2} - h_{a1} + \omega_2 h_{v2} - \omega_1 h_{v1} - (\omega_2 - \omega_1) h_B}$$
$$= \frac{\dot{m}_A(h_A - h_B)}{c_{pa}(T_2 - T_1) + \omega_2 h_{g2} - \omega_1 h_{g1} - (\omega_2 - \omega_1) h_B}$$
$$= \frac{110{,}000(188.42 - 104.76)}{1.005(40 - 20) + 0.0541(2573.4) - 0.00822(2537.2) - (0.0541 - 0.00822)104.76}$$
$$= 68{,}850 \text{ kg dry air/h}$$

(*b*) The amount of water evaporated is

$$\dot{m}_{\text{evaporated}} = \dot{m}_a \omega_2 - \dot{m}_a \omega_1 = 68{,}850(0.0541 - 0.00822) = 3160 \text{ kg/h}$$

9·6 Summary

The mass of a mixture is equal to the sum of the masses of its components:

$$m_m = m_A + m_B + m_C + \cdots$$

The mass fraction (or concentration) of any component i is defined as

$$x_i = \frac{m_i}{m_m}, \qquad x_A = \frac{m_A}{m_m}, \qquad \text{etc.}$$

The mole fraction is defined as

$$y_i \equiv \frac{N_i}{N_m}, \qquad y_A \equiv \frac{N_A}{N_m}, \qquad \text{etc.}$$

The partial pressure p_i of a component i in a gas mixture is defined as

$$p_i \equiv y_i p_m, \qquad p_A \equiv y_A p_m, \qquad \text{etc.}$$

The partial volume V_i of a component i in a gas mixture is defined as

$$V_i \equiv y_i V_m, \qquad V_A \equiv y_A V_m, \qquad \text{etc.}$$

Dalton's model or the additive pressure model states that the pressure of a mixture of ideal gases equals the sum of the pressures of its components if each existed alone at the temperature and volume of the mixture. Another statement of this law is that, in a mixture of ideal gases, each component behaves in all respects as though it existed alone at the temperature and volume of the mixture. It follows that in a mixture of ideal gases the partial pressure of each component is equal to the pressure that the component would exert if it existed alone at the temperature and volume of the mixture.

Amagat's model or the additive volume model states that the volume of a mixture of ideal gases equals the sum of the volumes of its components if each existed alone at the temperature and pressure of the mixture. It follows that in a mixture of ideal gases the partial volume of each component is equal to the volume that the component would occupy if it existed alone at the pressure and temperature of the mixture.

The additive pressure and additive volume models hold strictly only for ideal-gas mixtures, but they hold approximately for real-gas mixtures even in some ranges of pressure and temperature where $pv = RT$ is inaccurate.

Properties of ideal-gas mixtures can be obtained accurately and properties of real-gas mixtures can be obtained approximately in terms of component properties by use of the models described above.

In many gas–vapor mixtures all components can be treated as ideal gases, but an additional fact must be considered: The maximum pressure of a vapor in a mixture depends on the temperature, because it cannot be higher than the saturation pressure corresponding to the mixture temperature.

In *atmospheric air* the water vapor can be treated as an ideal gas; so it follows the equation of state $pv = RT$, and its enthalpy is a function of temperature only. Thus the enthalpy of water vapor in atmospheric air can be read from the steam tables by reading the enthalpy of saturated vapor at the same temperature.

Relative humidity ϕ is defined as the ratio of the pressure (i.e., partial pressure) of the vapor in atmospheric air to the saturation pressure of the vapor at the temperature of the mixture:

$$\phi \equiv \frac{p_v}{p_g} \tag{9·1}$$

Humidity ratio is defined as the ratio of the mass of vapor in atmospheric air to the mass of dry air:

$$\omega \equiv \frac{m_v}{m_a} \tag{9·3}$$

The relationship between relative humidity and humidity ratio for gas–vapor mixtures in which each component behaves as an ideal gas is

$$\omega = \frac{v_a}{v_g} \phi \tag{9·4}$$

The *dew-point temperature* of a gas–vapor mixture is defined as the saturation temperature corresponding to the partial pressure of the vapor in the mixture.

The *adiabatic saturation temperature* of a gas–vapor mixture is defined as the temperature that results from adiabatically adding water to the mixture in steady flow until it becomes saturated, the water being supplied at the final temperature of the mixture.

The *wet-bulb temperature* is the temperature measured by a thermometer wrapped in gauze soaked in water and placed in a stream of atmospheric air. For atmospheric air it happens that the wet-bulb temperature is very close to the adiabatic saturation temperature. This approximation does not hold for other gas–vapor mixtures.

Charts of atmospheric air properties called *psychrometric charts* are useful in reducing the time required for various calculations.

References and Suggested Reading

9·1 Daubert, T. E., *Chemical Engineering Thermodynamics*, McGraw-Hill, New York, 1985, Chapter 6.

9·2 Reid, Robert C., John M. Prausnitz, and Bruce E. Poling, *The Properties of Gases and Liquids*, McGraw-Hill, New York, 1987.

9·3 *ASHRAE Handbook*, American Society of Heating, Refrigerating and Air-Conditioning Engineers, Inc. (This book is published in four volumes: *Fundamentals*, *Systems*, *Equipment*, and *Applications*, with one volume published per year. It is an invaluable source of property data and other information.)

9·4 McQuiston, Faye C., and Jerald D. Parker, *Heating, Ventilation, and Air Conditioning: Analysis and Design*, 4th ed., Wiley, New York, 1994.

Problems

9·1 Prove the following statement: The density of an ideal-gas mixture always equals the sum of the densities of its constituents.

9·2 A mixture consisting of 40 percent oxygen and 60 percent nitrogen by volume is cooled under constant-volume conditions from 1 atm, 85°C, to a final temperature of 10°C. Compute the partial pressures of the constituents and the volumetric analysis at the final temperature.

9·3 A mixture at 20°C consists of 2 parts by mass carbon dioxide, 4 parts hydrogen, and 6 parts nitrogen. Compute the apparent molar mass and the gas constant for the mixture.

9·4 A gas mixture's volumetric analysis shows that it consists of 20 percent carbon dioxide and 5 percent carbon monoxide, while the balance is nitrogen. For a mixture temperature and pressure of 45°C, 500 kPa, find (*a*) the partial pressures of the constituents, (*b*) the apparent molar mass, and (*c*) the apparent gas constant.

9·5 Sampling the flue gases from a blast furnace reveals the following volumetric analysis: 3 percent hydrogen, 25 percent carbon monoxide, 12 percent carbon dioxide, and 60 percent nitrogen. Compute (*a*) the mass analysis, (*b*) the partial pressures, and (*c*) the specific heats for the mixture at 25°C. The total pressure is 1 atm.

9·6 A mixture at 1 bar, 0°C is composed of the following gases: 3 parts oxygen, 2 parts carbon monoxide, 0.5 part hydrogen, 1 part helium, 3 parts nitrogen, and 0.5 part argon, all by mass. Compute the specific heat values and the partial pressures.

9·7 A rigid, 200-liter tank contains oxygen at 30 kPa vacuum, 35°C. Carbon monoxide is forced into the tank until the mixture is at 200 kPa, and then the contents are allowed to cool to their original temperature. Compute both the mass and volumetric analysis of this final mixture.

9·8 A mixture consists of half nitrogen, half carbon dioxide, by mass. If the mixture is compressed from 2 atm, 5°C, to 5 atm, 170°C, what is the entropy change of the CO_2?

9·9 A gas mixture is made up of 60 percent air and 40 percent argon on a volumetric basis. The mixture is compressed from 100 kPa, 20°C, to 200 kPa, 120°C. Determine the entropy change of the gas in kJ/kg·K.

9·10 A tank containing 0.4 m^3 of helium at 700 kPa and 10°C is connected by means of a pipe and valve system to a 0.3-m^3 tank containing nitrogen at 350 kPa and 65°C. If no heat losses occur, compute the resulting pressure and temperature after the valve has been opened and mixing occurs.

9·11 Four kilograms of nitrogen at 12 bar, 10°C, and 2.5 kg of oxygen at 3 bar, 95°C, are mixed adiabatically with no change in the total volume. Determine the mixture pressure and temperature and the irreversibility of the process.

9·12 A system consisting of three tanks, all interconnected by pipes and valves, contains nitrogen, oxygen, and argon. One tank contains 6 kg of nitrogen at 750 kPa, 25°C; another 3 kg of argon at 500 kPa, 50°C; and the third 6 kg of oxygen at 250 kPa, 75°C. After the valves are opened and the system is allowed to come to equilibrium, compute for the mixture (*a*) the temperature, (*b*) the apparent molar mass, (*c*) the constant-pressure specific heat, (*d*) the gas constant, (*e*) the pressure, and (*f*) the partial pressure of each constituent.

9·13 Determine the minimum work input required per kilogram of air to separate air at 1 atm, 15°C, into nitrogen and oxygen, each at 1 atm, 15°C.

9·14 One tank contains 3.76 kmol of nitrogen at 140 kPa, 35°C, and another tank contains 1 kmol of oxygen at the same temperature and pressure. A valve connecting the tanks is opened, and the gases are allowed to mix adiabatically and come to a state of equilibrium. If the lowest temperature in the surroundings is 5°C, determine the irreversibility of the process. Explain the physical significance of this value.

9·15 Propane at 700 kPa, 35°C, flows steadily through a pipeline. A tank of 0.06 m^3 volume that contains air at 101.3 kPa, 15°C, is connected to the pipeline, and propane is allowed to flow into the tank until the total pressure in the tank is 350 kPa. Assuming that no air leaves the tank and that the filling is adiabatic, determine the mass of propane that flows into the tank.

9·16 Propane at 700 kPa, 35°C, flows steadily through a pipeline. A rigid, 60-liter tank initially containing air at 1 atm, 15°C, is connected to the pipeline, and propane is allowed to flow into the tank until the total pressure in the tank is 350 kPa. Assuming that no air leaves the tank, determine the heat transfer that is required to maintain the tank contents at its original temperature.

9·17 Helium at 350 kPa, 50°C, flows through an insulated pipeline. A well-insulated tank 0.06 m^3 in volume containing air at 101.3 kPa, 15°C, is connected to the helium line through a small valve that is opened until the pressure in the tank becomes 200 kPa; the valve is then closed. Assuming that no air flows from the tank and that the process is adiabatic, determine the final temperature of the mixture in the tank.

9·18 A tank is divided by a partition. Initially, 1 kmol of oxygen at 200 kPa, 25°C, is on one side of the partition and 2 moles of nitrogen at the same pressure and temperature is on the other side. The partition is ruptured so that the gases mix. Determine the irreversibility of the mixing.

9·19 Hydrogen and oxygen, each at 100 kPa, 20°C, flow steadily into an adiabatic, constant-pressure mixing chamber and leave as a mixture of stoichiometric proportions. The temperature of the surrounding atmosphere is 20°C. Calculate (*a*) the partial pressure of hydrogen in the mixture and (*b*) the irreversibility of the mixing process per pound of hydrogen.

9·20 The mass fraction of each constituent in a mixture of N_2 and CO_2 is 50 percent. The mixture is compressed from 15 psia, 40 F, to 75 psia, 310 F. Calculate the entropy change of the CO_2.

9·21 Three cubic feet of methane at 300 psia, 100 F, is mixed with 7 cu ft of oxygen at 100 psia, 40 F, by opening a valve between the two tanks. Calculate the heat transfer if the final temperature is 100 F.

9·22 A tank containing 2 lbm of methane at 20 psia, 50 F, and a tank holding 4 lbm of oxygen at 100 psia, 20 F, are connected through a valve. The valve is opened and the gases mix adiabatically. Atmospheric temperature is 50 F. Determine (*a*) the mixture pressure, (*b*) the mixture temperature, (*c*) the volumetric analysis of the mixture, and (*d*) the irreversibility of the process.

9·23 A system is composed of three tanks containing nitrogen, oxygen, and carbon dioxide, all interconnected by pipes and valves. One tank contains 4 lbm of nitrogen at 90 psia, 100 F; another 2 lbm of carbon dioxide at 70 psia, 125 F; and the third 6 lbm of oxygen at 45 psia, 200 F. If the valves are opened and the contents completely mixed, compute for the mixture (*a*) the temperature, (*b*) the apparent molar mass, (*c*) the constant-pressure specific heat, (*d*) the gas constant, (*e*) the pressure, and (*f*) the partial pressure of each constituent.

9·24 Determine the minimum work input required per pound of air to separate air at 1 atm, 60 F, into nitrogen and oxygen, each at 1 atm, 60 F.

9·25 Argon at 100 psia, 100 F, flows steadily through a pipeline. A tank of 2 cu ft volume that contains air at 14.7 psia, 77 F, is connected to the pipeline, and argon is allowed to flow into the tank until the total pressure in the tank is 50 psia. Assuming that no air leaves the tank and that the filling is adiabatic, determine the mass of argon that flows into the tank.

9·26 Argon at 100 psia, 100 F, flows steadily through a pipeline. A tank of 2 cu ft volume that contains air at 14.7 psia, 77 F, is connected to the pipeline, and argon is allowed to flow into the tank until the total pressure in the tank is 50 psia. Assuming that no air leaves the tank, what is the total amount of heat transfer required to maintain the contents of the tank at the original temperature?

9·27 A typical breathing mixture used for divers contains 20 percent oxygen by volume, 15 percent nitrogen, and the balance helium. Another contains the same fractions of oxygen and nitrogen, but the balance is a mixture of helium and neon in the same volumetric ratio they have in air (3.47 mol Ne/mol He). How do the densities and the c_p values of these breathing mixtures compare with those of air?

9·28 Prove that for the adiabatic mixing of ideal gases initially at the same pressure and temperature in a rigid tank the entropy change depends only on the number of moles of constituents and not on what the constituents are.

9·29 State whether or not the following is true and prove your answer: Adiabatic mixing in a constant-volume system of two ideal gases with different k values results in a change in enthalpy but no change in internal energy.

9·30 A gas mixture that is 55 percent and 45 percent by mass of nitrogen and carbon dioxide, respectively, occupies 4.00 m^3 at 100 kPa, 20°C. It is compressed to a final volume of 0.0130 m^3 and a temperature of 35°C. Determine the final pressure using (*a*) the Redlich–Kwong equation of state, (*b*) compressibility factors.

9·31 A 50 percent molar mixture of ethane and propane occupies 4.00 m^3 at 100 kPa, 20°C. It is compressed to a final volume of 0.0130 m^3 and a temperature of 35°C. Determine the final pressure using (*a*) the Redlich–Kwong equation of state, (*b*) compressibility factors.

9·32 Calculate the change in enthalpy and the change in entropy of the gas compressed in Problem 9·30.

9·33 Calculate the change in enthalpy and the change in entropy of the gas compressed in Problem 9·31.

9·34 An insulated tank contains dry air and a small covered vessel of water. The entire contents of the tank are at a uniform temperature. The cover of the vessel is removed so that water evaporates into the air. When the system reaches a state of equilibrium, all parts of the system are again at a uniform temperature. For each of the following properties of the entire system, state whether it has increased, decreased, remained constant, or varied indeterminately: (*a*) temperature, (*b*) internal energy, (*c*) entropy.

9·35 A rigid tank contains 0.4 m^3 of oxygen at 1 bar, 25°C, and no other gas or vapor. A small covered jar inside the tank contains liquid water at the same temperature. The jar is uncovered and some of the liquid evaporates until equilibrium is reached at 25°C, with some heat exchange taking place with the surroundings. (*a*) What is the final tank pressure? (*b*) Calculate the mass of water that has evaporated. (*c*) Indicate whether each of the following quantities increases, decreases, remains constant, or varies indeterminately: (1) entropy of the oxygen, (2) entropy of the H_2O in the system, (3) entropy of the surroundings.

9·36 If the initial relative humidity is 65 percent, how much moisture is removed from moist air at 35°C and 95 kPa when cooled to 5°C? The total pressure remains constant throughout this cooling process.

9·37 A mixture consisting of 0.2 lbm of dry saturated steam and 0.01 lbm of air is contained in a tank at a temperature of 200 F. Compute the total pressure in the tank and the total volume occupied by the mixture.

9·38 Particulate matter is sometimes separated from air by means of cyclone separators or a series of baffles that suddenly changes air flow direction. Comment on the efficacy of these devices for separating atmospheric moisture from air.

9·39 A room of 250 m^3 volume contains dry air at 1 atm, 25°C. How many 250-mL tumblers of water must be evaporated to result in 50 percent relative humidity in the room? What changes will occur in the pressure and temperature of the room during this process if the room is sealed and insulated?

9·40 A sample of air at 50°C has a relative humidity of 100 percent. What is its dew-point temperature? What mass of liquid water per kilogram of dry air will result if the mixture is cooled to 10°C at a constant pressure of 90 kPa?

9·41 Three hundred kilograms per hour of moist air at 30°C and 40 percent relative humidity is to be cooled to 10°C at a constant atmospheric pressure of 101.325 kPa. How much heat must be removed per hour in order for this process to occur?

9·42 Air at 10°C, 90 percent relative humidity, and air at 30°C, 90 percent relative humidity, are mixed steadily and adiabatically. The mass flow rate of the colder stream is twice that of the other stream. What is the resulting state?

9·43 Atmospheric air initially at 160 kPa, 60°C, 100 percent relative humidity, undergoes a reversible adiabatic expansion to 1 atm. Give all the information you can about the entropy change of (*a*) the air, (*b*) the water vapor, (*c*) the mixture, and (*d*) the surroundings.

9·44 Atmospheric air initially at 25 psia, 140 F, 100 percent relative humidity, undergoes a reversible adiabatic expansion to 14.67 psia. Give all the information you can about the entropy change of (*a*) the air, (*b*) the water vapor, (*c*) the mixture, and (*d*) the surroundings.

9·45 Solve Problem 4·109 on the cooling of a computer for an ambient relative humidity of 100 percent.

9·46 Dry air at 30°C has $c_p = 1.005$ kJ/kg·K. How much different is the value for air at 30°C, 95 percent relative humidity?

9·47 A pitot tube used to measure air speeds as high as 100 m/s is calibrated using dry air. Prepare a chart showing how much error in the velocity value results from using this calibration for the pitot tube when it is used in air at 35°C, relative humidity of 100 percent. Make it clear whether the indicated value is too high or too low.

9·48 How much would a washcloth hanging in a room that has a volume of 18.0 m^3 raise the relative humidity in the room if the washcloth is (*a*) soaking wet, (*b*) wrung out? Express your answers by means of a plot involving the initial temperature and relative humidity. You may wish to express your answer in terms of the final relative humidity instead of the increase in relative humidity. You will probably need to perform some experiments.

9·49 A stream of air at 1 atm, 30°C, 80 percent relative humidity, flows at a rate of 28.0 kg/min. It is proposed to mix air at 1 atm, 15°C, with this stream steadily in an insulated mixing chamber so that the resulting temperature is 25°C. What mass rate of flow of the cooler air is required and what will be the relative humidity of the air leaving?

9·50 Determine the irreversibility of the mixing in Problem 9·49.

9·51 How much moisture is removed from moist air initially at 80 F, 14 psia, 60 percent relative humidity, when it is cooled to 40 F? The total pressure remains constant throughout this cooling process.

9·52 Six hundred pounds per hour of moist air at 90 F and relative humidity of 40 percent is to be cooled to 50 F at constant atmospheric pressure. Compute the heat removed per hour. The barometric pressure is 14.696 psia.

9·53 Air is compressed steadily from atmospheric conditions of 14.0 psia, 60 F, 100 percent relative humidity, to 70.0 psia, 240 F, by a compressor that is cooled by only atmospheric air. Neglect kinetic energy changes and determine the minimum work required per pound of dry air. Compare the result with that of Example 8·3.

9·54 An air stream at 90 F, 80 percent relative humidity, is mixed steadily and adiabatically with a second air stream at 50 F and 45 percent relative humidity. The mass rate of flow of the second stream is twice that of the first stream. The pressure is uniform throughout the mixing zone at 1 atm. The temperature of the surrounding atmosphere is 90 F. Calculate the resulting state and the irreversibility.

9·55 An engineer might speak of the "moisture in the atmosphere" or the "moisture in a steam-turbine exhaust." Explain the difference in usage of the term *moisture.*

9·56 Sketch a *Ts* diagram for the water vapor in atmospheric air, showing a few lines of constant relative humidity (i.e., loci of vapor states for which the relative humidity is the same).

9·57 Is it possible to mix, adiabatically and at constant pressure, two air streams, each with a relative humidity of less than 100 percent, and produce condensation in the mixing chamber? Explain.

9·58 What is the dew-point temperature of atmospheric air at 25°C, 1 atm, and 50 percent relative humidity? What are the partial pressures of the constituents?

9·59 A sample of atmospheric air has a pressure and temperature of 101 kPa, 40°C. If the relative humidity is 60 percent, what is the humidity ratio for this sample?

9·60 A 35-liter glass tank contains atmospheric air. When the temperature of the contents is 35°C, its pressure is 140 kPa. If the air is cooled to 15°C, condensation on the glass begins. Use this information to calculate the mass of water in the tank.

9·61 (*a*) Determine the dew point of the products of the following reaction if they are at 100 kPa, 260°C:

$$CH_4 + 2O_2 + 7.52N_2 \rightarrow CO_2 + 2H_2O + 7.52N_2$$

(*b*) Determine the dew point of the products of burning methane with the stoichiometric amount of oxygen with no nitrogen present and the products at 100 kPa, 260°C.

9·62 Determine the humidity ratio of moist air at 14.5 psia, 90 F, for a relative humidity of 60 percent and a barometric pressure of (*a*) 12.5 psia, (*b*) 14.5 psia.

9·63 For a mixture pressure of 11.7 psia, what is the mass of water vapor per pound of dry air for atmospheric air at 100 F and 0.50 relative humidity? What is the dew-point temperature?

9·64 Air at 90 F has a relative humidity of 100 percent. What is the dew-point temperature? What mass of water per pound of dry air will be removed if the mixture is cooled to 10 F at a constant pressure of 12.5 psia?

9·65 The products of the reaction shown are at 20 psia, 450 F. Determine the dew point of the products.

$$H_2 + O_2 + 3.76N_2 \rightarrow H_2O + \tfrac{1}{2}O_2 + 3.76N_2$$

9·66 The percentage composition by mass for a gas–vapor mixture is nitrogen 70, carbon monoxide 4, carbon dioxide 22, water 4. Compute the dew point of the mixture for a total pressure of 1 atm.

9·67 Which of the following has the greater density: dry air or air with a high humidity ratio? Does this explain why fog collects in valleys and low places?

9·68 Under what conditions does a water pipe or a glass of water "sweat"? Under what conditions may water from the atmosphere contaminate the gasoline in an automobile tank?

9·69 Why do building and automobile windows "fog" or "frost"? If a double layer of glass is used, should the air space between them be vented to the inside or the outside?

9·70 What is dew? Under what conditions does it form? Under what conditions does frost form?

9·71 Under what conditions is a person's breath visible?

9·72 A slice of toast is taken from the toaster and placed on a dry plate. A minute later the toast is removed from the plate, and the plate is seen to be wet where the toast was lying. Explain why this is so.

9·73 Under what conditions do eyeglasses fog when the wearer enters a heated building? Is this the same effect as the fogging of a mirror in a shower room?

9·74 Under jungle conditions, machinery that is stored outdoors under draped tarpaulins often collects puddles of water, even though the tarpaulins are impermeable to water. Explain this phenomenon.

9·75 A method for determining dew point is to fill a thin-walled shiny metal cup with water at room temperature and then add ice cubes to the water while stirring it with a thermometer. Read the thermometer periodically and note the reading when condensation on the outside of the cup is first detected. Take this thermometer reading as the dew point. Comment on the basis for this method. What assumptions are involved? What errors are possible? For each error you list, indicate whether it would cause the reading to be too high or too low.

9·76 A prospective purchaser of equipment has written into the specifications: During operation, no visible vapor shall appear at the vent. Comment on this specification. (Is it possible, impossible, easy, or difficult to meet? If you were submitting a bid to supply the equipment, would you comment on this specification?)

9·77 Explain why the annual rainfall in Seattle is much greater than it is in Spokane. Name some other regions where the same pattern of rainfall variation, prevailing wind, and topography cause similar results.

9·78 A sweater that has been stored in a closet is placed in a clear plastic bag that is then securely closed. After the bag is exposed to sunlight for an hour, drops of liquid are observed on its inner surface. Explain this phenomenon and describe experiments that would adequately test your explanation.

9·79 A shower bath that is used by people for several hours a day has both supply pipes, hot and cold, exposed to splashing of the shower water. After a few months of use, one of the pipes has more scale or deposit on its outer surface than the other supply pipe has. Which pipe has this larger amount of scale? Explain this phenomenon and describe experiments that would adequately test your explanation.

9·80 Some kinds of thermal insulation used in the walls of buildings lose their effectiveness when they become moist. Such insulation often has a moisture-blocking film ("vapor barrier") on one side. For installation in a building near London, should the insulation be installed with the vapor barrier toward the interior or toward the outdoors? Explain.

9·81 A sample of atmospheric air has wet- and dry-bulb temperatures of 20°C and 25°C, respectively. Determine (*a*) the vapor partial pressure, (*b*) the relative humidity, (*c*) the humidity ratio, (*d*) the mixture specific volume in cubic meters per kilogram of dry air, and (*e*) the mixture enthalpy per kilogram of dry air. Use the psychrometric chart. Assume barometric pressure of 1 atm.

9·82 For atmospheric air with a relative humidity of 30 percent and a dry-bulb temperature of 35°C, determine by means of the psychrometric chart the humidity ratio and the partial pressure of the water vapor. Assume standard atmospheric pressure.

9·83 If moist air contains 12 grams of water vapor per kilogram of dry air and its relative humidity is 60 percent, determine from the psychrometric chart the dry-bulb temperature and the vapor partial pressure.

9·84 For a barometric pressure of 1 atm, fill in the blanks for the six states in the table below.

State	T, °C	T_{wb}, °C	ϕ, %	ω, kg/kg da	T_{dp}, °C
1	25		40		
2	20	15			
3	20				10
4	30			0.020	
5		20			20
6			60	0.010	

9·85 Locate on the skeleton psychrometric chart five states: (1) 30°C and 80 percent relative humidity, (2) 20°C and 60 percent relative humidity, (3) 20°C and wet-bulb temperature of 15°C, (4) 30°C and a wet-bulb temperature of 20°C, (5) 30°C and a dew point of 15°C.

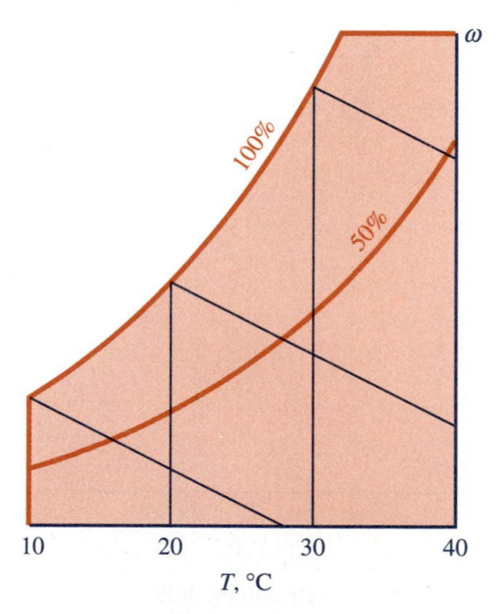

Problem 9·85

9·86 A sample of moist air has wet- and dry-bulb temperatures of 70 F and 80 F, respectively. Determine (*a*) the partial pressure of the water vapor, (*b*) the relative humidity, (*c*) the mass of water vapor per pound of dry air, (*d*) the mixture specific volume in cubic feet per pound of dry air, and (*e*) the mixture enthalpy per pound of dry air. Use the psychrometric chart. Assume standard atmospheric pressure.

9·87 If the relative humidity is 30 percent and the dry-bulb temperature is 100 F, determine by means of the psychrometric chart the humidity ratio and the pressure of the water vapor for atmospheric air. Atmospheric pressure equals 14.696 psia.

9·88 For a barometric pressure of 1 atm, fill in the blanks for the six states in the table below.

State	T, F	T_{wb}, F	ϕ, %	ω, lbm/lbm da	T_{dp}, F
1	70		30		
2	70	50			
3	70				50
4	80			0.020	
5	75		100		
6			70	0.012	

9·89 Locate on the skeleton psychrometric chart five states: (1) 90 F and 80 percent relative humidity, (2) 70 F and 60 percent relative humidity, (3) 70 F and a wet-bulb temperature of 65 F, (4) 90 F and a wet-bulb temperature of 60 F, and (5) 90 F and a dew point of 60 F.

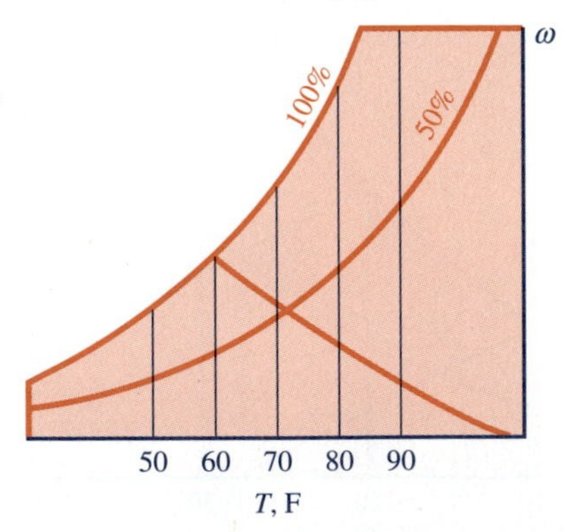

Problem 9·89

9·90 Make a chart that can be entered with atmospheric temperature and the dew-point temperature to give relative humidity. If barometric pressure is a pertinent variable, show lines for 90 kPa and 101.325 kPa.

9·91 Air at 35°C, 40 percent relative humidity, is cooled to 25°C by spraying water at 15°C into it. The mixture pressure remains constant at 101.3 kPa. Assuming that all of the water evaporates and that the mixing occurs in an insulated duct, calculate the mass of water added per kilogram of air.

9·92 Atmospheric air at 101.3 kPa, 20°C, 60 percent relative humidity, is to be used to cool a transformer. The transformer dissipates heat at a rate of 210 MJ/h. To increase the cooling effect per kilogram of dry air and thereby decrease the amount of air needed, the air is first adiabatically saturated with water and then passed over the transformer. If the temperature of the air leaving the transformer should not exceed 30°C, determine the required flow rate in kilograms of dry air per hour both with and without the adiabatic saturation of the incoming air.

9·93 Air at 25°C and relative humidity of 100 percent is heated at constant pressure to a final temperature of 35°C. Determine (*a*) the dew-point temperature at 25°C, (*b*) the initial partial pressure of the water vapor, (*c*) the initial humidity ratio, and (*d*) the final humidity ratio.

9·94 Outside air at 30°C and 45 percent relative humidity is to be conditioned so that the final temperature and relative humidity are 20°C and 30 percent. If the flow process occurs under constant-pressure conditions, compute (*a*) the quantity of water removed per kilogram of dry air, (*b*) the heat removed in the initial cooling process per kilogram of dry air, and (*c*) the heat added after the initial dehumidification process per kilogram of dry air. Assume standard atmospheric pressure.

9·95 An air-conditioning unit consists of a cooler with a condensate trap followed by a heater. Air at 101.3 kPa enters at 30°C, 80 percent relative humidity, and leaves at 20°C, 49 percent relative humidity. For a flow rate of 500 kg dry air per hour, calculate (*a*) the amount of water removed per hour and (*b*) the heat removed (in kJ/kg dry air) in the cooler.

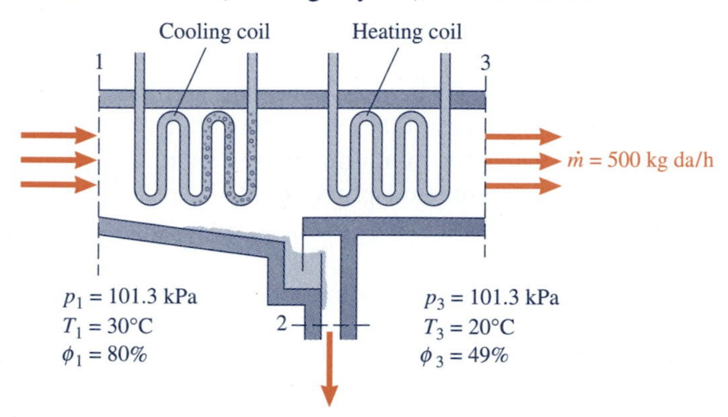

Problem 9·95

9·96 Saturated air at 15°C enters a dehumidifier and passes over the unit's cold coils. The moisture condensed is drained away and then the air passes over the motor, refrigerator, and warm coils. All heat rejected by the device thus goes into reheating the air that emerges at 30°C, 30 percent relative humidity. The flow rate is 5 kg of dry air per minute. For a barometric pressure of 95 kPa, determine the power input to the dehumidifier.

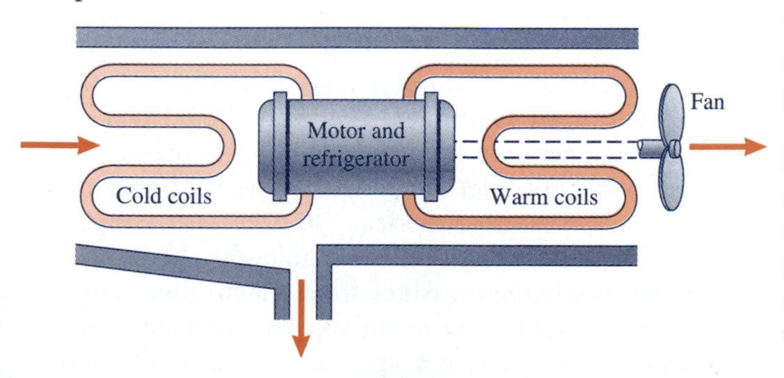

Problem 9·96

9·97 A cooling tower is used to cool an inlet flow of 7500 kg of water per minute from an initial temperature of 40°C to a final temperature of 31°C. Air enters the cooling tower at 25°C and 40 percent relative humidity and leaves as saturated air at 38°C. Compute the mass of inlet air required per minute and the mass of water lost per minute. Atmospheric pressure equals 101.3 kPa.

9·98 Dry air at 5°C enters a heating and humidifying unit and leaves at 25°C, 80 percent relative humidity. Spray water for humidification is supplied to the unit at 15°C. Barometric pressure is 98 kPa. Determine the heat transfer per kilogram of dry air.

9·99 A steady 0.56-kg/s flow of methane is heated at constant pressure as it passes through a duct. The gas enters at 110 kPa, 35°C, and a control system maintains the outlet temperature at 300°C. Changes in velocity are small. (*a*) Determine the rate of heat transfer in kW. (*b*) Determine the rate of heat transfer for the case of methane saturated with water vapor at a mixture pressure of 110 kPa, with the same flow rate of dry methane.

9·100 Compute the amount of heat required, in kilojoules per kilogram of nitrogen, to raise the temperature of water-saturated nitrogen (i.e., $\phi = 1.0$) from 25°C to 600°C in a frictionless steady-flow process under a constant pressure of 1 atm.

9·101 Atmospheric air at 752 mm Hg, 50 percent relative humidity, is compressed isothermally to 155 mm Hg gage. State whether each of the following quantities increases, decreases, remains constant, or varies indeterminately: (*a*) humidity ratio, (*b*) relative humidity, and (*c*) dew point.

9·102 Air at 100 F, 40 percent relative humidity, is cooled to 80 F by spraying water at 60 F into it. Mixture pressure remains constant at 14.7 psia. Assuming that all of the water evaporates and that the mixing occurs in an insulated pipe, calculate the mass of water added per pound of air.

9·103 Saturated air at 80 F is heated at constant pressure to a final temperature of 100 F. Determine (*a*) the dew-point temperature at 80 F, (*b*) the initial water vapor partial pressure, (*c*) the initial humidity ratio, and (*d*) the final humidity ratio.

9·104 Outside air at 90 F, 45 percent relative humidity is to be conditioned so that the final temperature and relative humidity are 70 F and 30 percent. Compute, for an isobaric flow process, (*a*) the quantity of water removed per pound of dry air, (*b*) the heat removed in the initial cooling process per pound of dry air, and (*c*) the heat added per pound of dry air. Assume standard atmospheric pressure.

9·105 An air-conditioning unit consists of a cooler with a condensate trap followed by a heater. Air enters at 90 F, 80 percent relative humidity, and leaves at 70 F, 49 percent relative humidity. This constant-pressure process occurs at 14.67 psia, and the flow rate is 1000 lbm of dry air per hour. Calculate (*a*) the amount of water removed per hour, and (*b*) the heat removed (in B/lbm dry air) in the cooler.

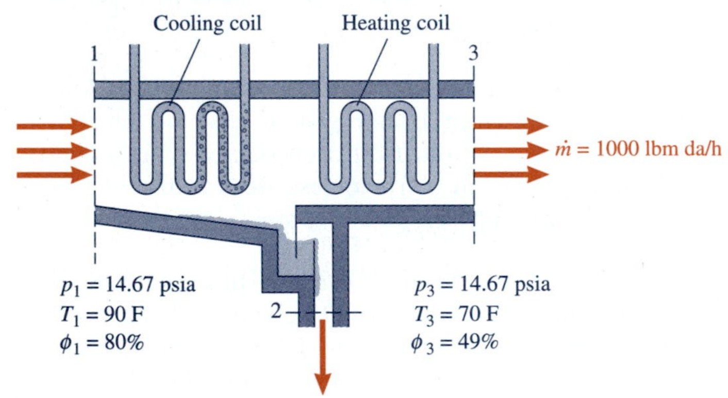

Problem 9·105

9·106 Saturated air at 60 F enters a dehumidifier and first passes over the cold coils of its refrigerating unit. The moisture condensed is drained away and the air then passes over the motor and warm coils of the refrigerating unit. All heat rejected by the refrigerating unit thus goes into reheating the air that emerges at 85 F, 30 percent relative humidity. Barometric pressure is 14.0 psia. Determine the power input to the dehumidifier for a steady flow rate of 10 lbm da/min. (See the figure with Problem 9·96.)

9·107 Water enters a cooling tower at 120 F and leaves at 90 F. The dry-bulb and wet-bulb temperatures of the entering air are 85 F and 70 F, respectively. The air leaving the tower is completely saturated and has a temperature of 105 F. If 100,000 lbm of water enters the tower each hour, compute the mass of air required and the water lost by evaporation. Atmospheric pressure equals 14.696 psia.

9·108 Dry air at 40 F enters a heating and humidifying unit, and leaves at 80 F, 80 percent relative humidity. Spray water for humidification is supplied to the unit at 60 F. Barometric pressure is 14.2 psia. Determine the heat transfer per pound of dry air.

9·109 A mixture of helium and water vapor at 1.30 atm, 30°C, has a relative humidity of 80 percent. (*a*) Calculate the humidity ratio. (*b*) If the pressure is held constant, to what temperature must the mixture be cooled to initiate condensation? (*c*) If the temperature is held constant, to what pressure must the mixture be brought to initiate condensation?

9·110 Explain how it is possible to dehumidify air by passing it through water sprays.

9·111 Consider the design of a cooling tower. What are the relative advantages and disadvantages of a high air velocity? Is it desirable to have the air leaving the tower carry liquid drops?

9·112 For use in design studies, a chart that shows temperature at the end of a reversible adiabatic compression of air is needed. The chart should plot discharge temperature versus gage discharge pressure for three values of inlet temperature typical of construction sites in North America. For each inlet temperature, show the effect of relative humidity. Use a range of gage discharge pressures from 0 atmospheres to 15 atmospheres. (This problem appeared in Chapter 4 to be solved using constant specific heats and in Chapter 6 to be solved for dry air. A variation of this problem appears in Chapter 14 where compressor efficiency is also considered.)

9·113 Three kinds of humidifiers are used to humidify the atmosphere in buildings that house people. In an evaporative humidifier, water is boiled so that vapor enters the atmosphere. In a cold-air evaporator, water is sprayed into the atmosphere in tiny drops. In an ultrasonic humidifier, a small wafer is vibrated at ultrasonic frequencies so that liquid water near the wafer flashes into vapor during part of the oscillation cycle when the pressure near the wafer is very low. Discuss the phenomena involved in each type of humidifier and the relative merits of the three types.

9·114 In hot dry climates, home air conditioning is sometimes provided by simply spraying water into an air stream entering the home. If the desired outlet conditions from the spray unit are temperatures between 20°C and 28°C and relative humidities between 40 percent and 60 percent, what temperature range of entering air at 10 percent relative humidity can be accommodated? Under what conditions would you recommend having some of the inlet air bypass the spray unit and be mixed with air that has passed through it?

9·115 Evaluate the following proposal, specifying the conditions, if any, under which the proposed system is feasible: Hot, humid outdoor air can be dehumidified by spraying cold water into it to chill it to below its dew point so that air leaving the mixing chamber is saturated but at a lower temperature than the incoming air. This air is then mixed with outdoor air to provide air that is within the comfort zone defined as a temperature range of 20°C to 28°C and a relative humidity range of 40 percent to 60 percent.

9·116 Refer to Example 9·12. Suppose that no heat is provided in the heater section of the air conditioner, but instead a quantity of inlet air equal to that which enters the air conditioner is mixed with the air leaving the cooling section. Determine the temperature and relative humidity of the air leaving, assuming that the mixing is complete so that the fluid stream leaving is uniform.

9·117 In a design study for an air conditioner like that shown in Fig. 9·10 and for inlet air conditions as specified in Example 9·12, it is desired to know what final conditions can be achieved by eliminating the heater section and mixing various amounts of inlet air with the air leaving the cooler section. The mixing ratio is defined as the ratio of the mass of bypass air to the mass of air flowing through the cooler section. Plot curves of temperature and relative humidity of leaving air versus mixing ratio in the range of 0 to 3 if the temperature at the end of the cooler is held constant at the value in Example 9·12.

9·118 Solve Problem 9·117 for temperatures at the end of the cooler ranging from 32 F to the value in Example 9·12.

9·119 Refer to Example 9·12. As part of a design study, plot, for the same discharge conditions, two quantities versus the inlet air relative humidity: the temperature at the end of the cooler section and the amount of heat removed in the cooler.

9·120 Refer to Example 9·12. The heat source used in reheating the air being conditioned is outdoor air at 90 F, 80 percent relative humidity. In a design study, it is suggested that instead of removing the condensate from the conditioned air as a liquid, it might be evaporated into the outdoor air passing through the heat exchanger. Evaluate this idea, and if it is feasible, state the requirements such operation would impose.

9·121 A granular product is to be air-dried as it is carried on a conveyor belt. The product enters the dryer at a rate of 2500 kg/h, and 30 percent of this mass is water. In the dryer, at least 90 percent of this water is to be removed. The product is dense and the granular size distribution is relatively uniform. There are no fine particles. For a preliminary design study, variables are the flow rate of air, incoming temperature and humidity, outlet temperature and humidity, and the energy input to the dryer. Experience with similar systems suggests that minimizing energy costs is important. You are to estimate the minimum energy required for drying and the air flow rate. You are also to address the following questions that have been raised: Should the incoming air be dehumidified? Should it be heated? Would spray drying of incoming air be useful? If so, should the spray water be chilled? What are the advantages and disadvantages of having the drying air leave the system with a relative humidity close to 100 percent?

9·122 A piece of equipment is to be cooled by an air stream. For a specified air temperature rise and a specified volume rate of flow, can the cooling be increased by first humidifying or dehumidifying the air?

9·123 The tanks of a portable breathing air supply contain dry air. The air delivered to a user of the supply should have a relative humidity of at least 50 percent, and should be at a temperature not lower than that of the air in the tanks. Devise a system for humidifying the air without chilling it.

9·124 Outdoor air at 90 F and 90 percent relative humidity is to be conditioned to provide 1200 cfm of air at 70 F and 40 percent relative humidity. Cooling water is available at 70 F. A simple reliable system to provide the desired conditioning of the air is desired. Preferred conditions are minimum power input and no heating of air by any means other than outdoor air. Design a system to do this. (Specify the kind of system, flow rates, and power requirement.)

9·125 Outdoor air at 75 F to 95 F with relative humidity in the range of 60 percent to 100 percent is to be conditioned to provide 1200 cfm of air at 70 F and 40 percent relative humidity. Cooling water is available at 70 F. A simple reliable system to provide the desired conditioning of the air is desired. Preferred conditions are minimum power input and no heating of air by any means other than outdoor air. Design a system to do this. (Specify the kind of system, flow rates, and power requirement.)

9·126 Outdoor air is to be dehumidified and cooled as necessary to provide from 1200 to 28,000 cfm of air at 70 F to 75 F and relative humidity between 30 percent and 40 percent to a building. The outdoor air is at 75 F to 95 F with relative humidity in the range of 60 percent to 100 percent, with higher humidity usually occurring with the lower temperatures. Ample cooling water is available at 70 F. A simple reliable system to provide the desired conditioning of the air is desired. Preferred conditions are minimum power input and no heating of air by any means other than outdoor air. Design a system to do this. (Specify the kind of system, flow rates, and maximum power requirement.)

9·127 Air is to be delivered to a test chamber for a spacecraft at 60 kPa, 32°C. Saturated air at 60 kPa and temperatures ranging from 10°C to 30°C is supplied to a heat exchanger that raises the air temperature to 32°C. Neglect frictional pressure drops in the system. For use in a design study, make a plot that shows the final relative humidity as a function of the temperature of air entering the heat exchanger.

9·128 As part of a design study, it is desired to know how low a humidity ratio can be provided by passing moist air at temperatures between 70 F and 100 F through water sprays with water inlet temperature in the range of 55 F to 70 F. Prepare a graphical display that would be useful to an engineer doing the design.

9·129 A building has an air-conditioning system that takes outdoor air, cools it to below the dew point, removes moisture, and then reheats the air to the desired final temperature. Design conditions for the system are outdoor air temperature range of 65 F to 95 F, relative humidity of 50 percent to 100 percent. Discharge air is to be at a temperature between 70 F and 75 F with a relative humidity not higher than 40 percent. An air-conditioning system is to be designed for a nearly identical new building nearby. It has been suggested that passing only some of the incoming air through the entire system while some air bypasses one or more parts of the system and then remixes with the other air could reduce power input, reduce the amount of heat needed, or reduce the size of the refrigerating unit. Make a design study of this suggestion.

9·130 Outdoor air at 0°C is to enter a building and be heated to 20°C by flowing over tubes through which hot furnace gases flow. The air is to be humidified to bring the relative humidity to 50 percent. The air flow rate will be 200 m^3/min. Water is available at 10°C. Sketch the system you propose for the heating and humidification, specifying flow rates and temperatures. (Consider different sequences of the operations. Should the water added to the air be preheated?)

9·131 Outdoor air at temperatures as low as −20°C is to be brought into a building and heated to approximately 20°C by flowing over tubes through which hot furnace gases flow. The air must be humidified to bring the relative humidity to about 50 percent. The air flow rate may be as high as 400 m^3/min. Water is available at 10°C. Sketch the system you propose for the heating and humidification, specifying flow rates and temperatures. (Consider different sequences of the operations. Should the water added to the air be preheated?)

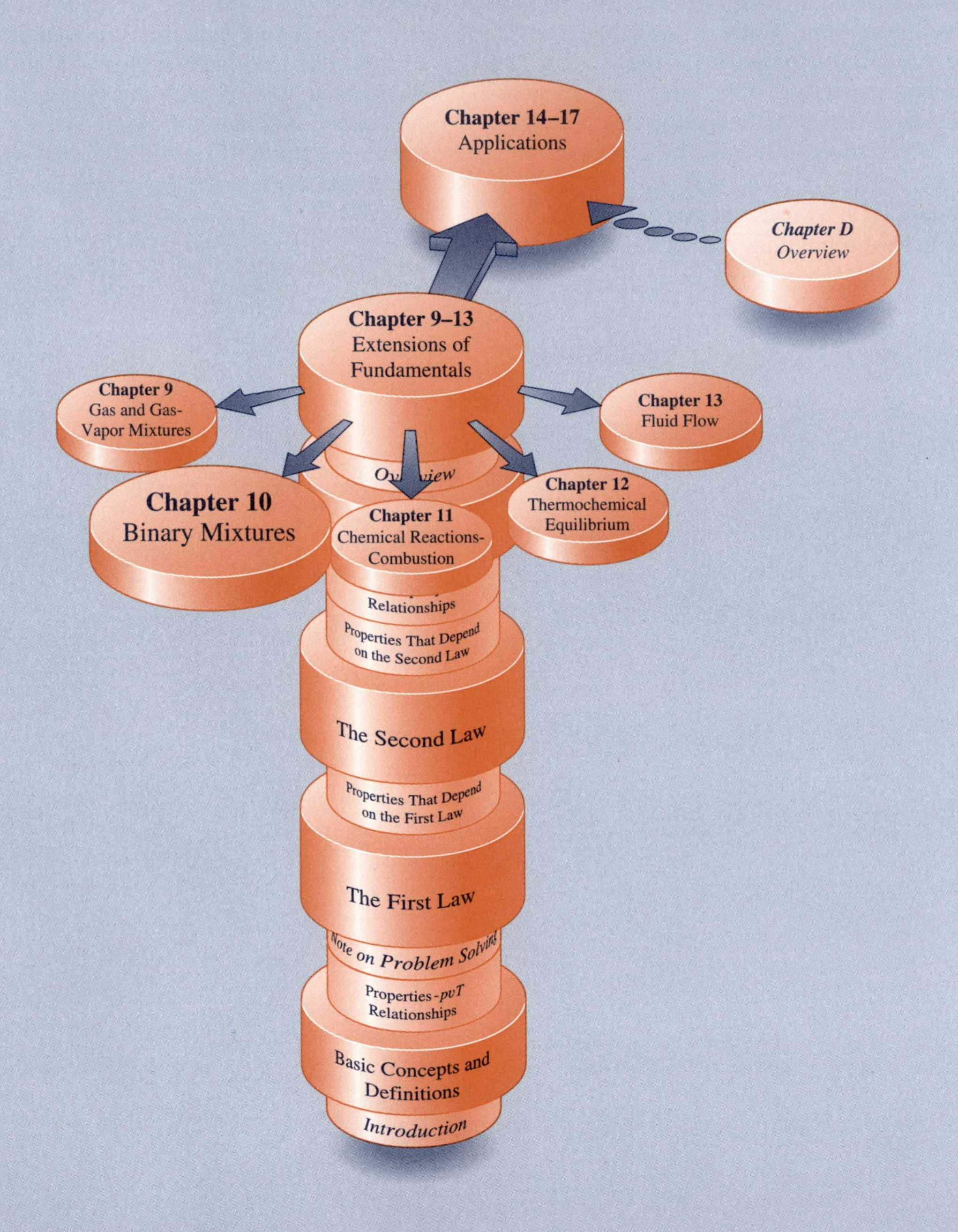

Chapter 14–17
Applications
Chapter D
Overview
Chapter 9–13
Extensions of
Fundamentals
Chapter 9
Gas and Gas-
Vapor Mixtures
Chapter 13
Fluid Flow
Chapter 10
Binary Mixtures
Chapter 11
Chemical Reactions-
Combustion
Chapter 12
Thermochemical
Equilibrium
Relationships
Properties That Depend
on the Second Law
The Second Law
Properties That Depend
on the First Law
The First Law
Note on Problem Solving
Properties-pvT
Relationships
Basic Concepts and
Definitions
Introduction

Chapter 10

Binary Mixtures

Until Chapter 9, we dealt mainly with pure substances whenever we discussed property relationships. In the latter part of Chapter 9, however, we dealt with properties of a nonpure substance: atmospheric air with varying humidity. As long as the chemical composition of the air remains constant, it can be treated as a pure substance. When atmospheric air is dehumidified, however, water vapor is removed, and the overall chemical composition of the air changes. Under such circumstances, we cannot treat air as a pure substance. It is a binary mixture, with dry air and water as the two components. The water may be in only the vapor phase or it may be present in both liquid and vapor phases.

Other mixtures that are nonhomogeneous or of changing composition occur in engineering practice, and in this chapter we treat one important class of these.

10·1 Characteristics of Binary Mixtures

In this chapter we deal with a particular kind of nonpure substance: *binary mixtures*. Binary mixtures contain two substances, both of which may exist in more than one phase. We will use the basic principles of binary mixtures in later chapters to solve specific engineering problems. For example, in Chapter 16, where we discuss absorption refrigeration, an understanding of binary mixtures is vital. Many refrigeration systems use binary mixtures as the working fluid; ammonia and water or lithium–bromide and water are some examples.

There are two important concepts in this chapter. The first concept is

In an equilibrium binary mixture at a given pressure *or* temperature, when the liquid phase and the vapor phase coexist, they have different chemical compositions. Further, the chemical composition of one phase fully determines the chemical composition of the other. We might alternatively express this principle as follows: If the pressure *and* temperature are known for a two-phase equilibrium binary mixture, the chemical compositions in the liquid and vapor phases are different and fully determined.

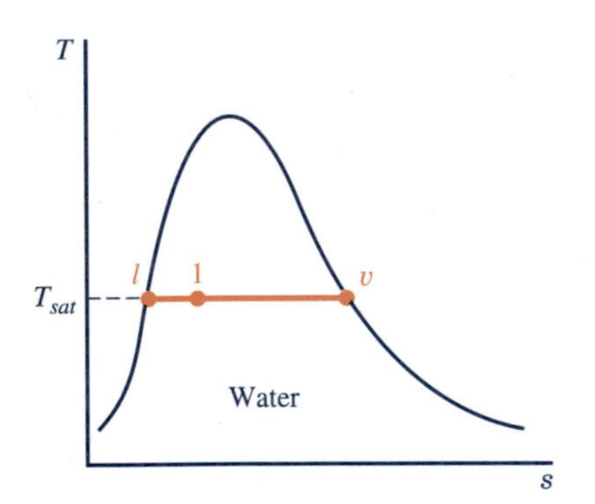

Figure 10·1 *Ts* diagram for water.

This is analogous to observations we made in earlier chapters about the equilibrium compositions of the liquid and vapor phases of a pure substance. Consider steam as an example. In Fig. 10·1, we show a *Ts* diagram for water. Under saturation conditions, if the liquid state l is known, then the vapor state v is also known. If we want to determine the state of the system as shown by point 1 in the figure, however, we require additional information such as the quality. Alternatively, knowledge of either the pressure or the temperature fully determines the state of the liquid and of the vapor. For a more complex system such as a two-phase binary mixture, knowledge of the pressure or temperature alone does not determine the states of both phases. You will see that we need more information to determine the states of the individual phases and of the system in general.

The second concept is

The amount of gas dissolved in a liquid is a function of pressure and temperature and decreases with increasing temperature under most conditions.

You have, no doubt, already observed the effect of temperature on gases dissolved in liquids. A glass of cold water, poured in the evening and left till morning, provides an example of the effect of temperature on the amount of gas that can be held in solution. Through the night as the water temperature rises, small gas bubbles appear on the inside of the glass. The liquid cannot hold as much gas at higher temperatures; therefore, the gas does not remain dissolved.

These are simple statements of the ideas we will discuss in this chapter, but as a preview, they are sufficient.

Some Preliminary Observations on the Phase Rule. Recall our discussion in Chapter 4, Section 4.1, where we state that usually any two *independent* intensive properties of a *pure substance* define the intensive state of the system. Exceptions to this rule are rare. We mentioned in Chapter 4 that the *phase rule* partially addresses the exceptions, but also that the exceptions are likely to occur where the phase rule is difficult to apply.

Complex systems of *nonpure substances* involving several phases pose even greater difficulties for determining the number of *independent* properties required to define the state of the system: the phase rule is difficult to apply, intuition fails us miserably, and we cannot use a simple general rule as we did for pure substances. Therefore, in this chapter we restrict our discussion to binary mixtures. For binary mixtures, the phase rule is usually straightforward to apply. This simple restriction allows us to circumvent much additional effort that, for the applications presented in this book, would provide little return. Yet, we are still able to discuss most aspects of nonpure substances that are of interest in thermodynamic applications. Many references describe the intricacies of the phase rule in depth; we have provided a partial list at the end of this chapter.

Some Definitions. To begin, let us review some pertinent definitions and define some new terms. Two or more properties are *independent* if any one can be varied while any other one is held constant. For example, T and p are independent for an ideal gas, but h and T are not. Under saturation conditions, T and p are not independent for a pure substance. Either the saturation temperature *or* the pressure fully define the state of the liquid and of the vapor.

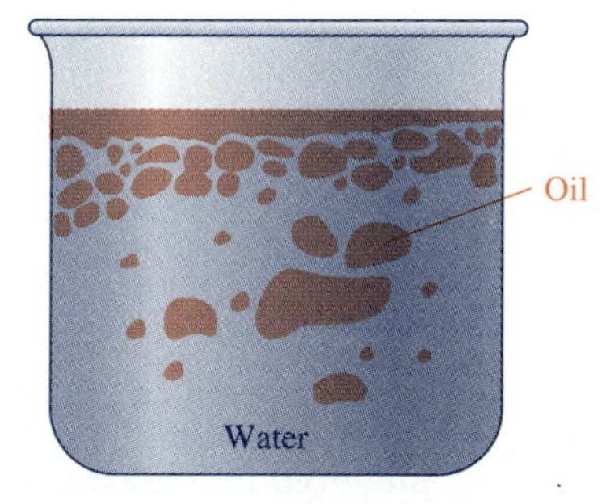

Figure 10·2 Oil droplets form a single phase in water.

As defined in Chapter 2, a *phase* is any homogeneous part of a system that is physically distinct and separated from other parts of the system by bounding surfaces. If all the intensive properties of one part of a system are identical with those of another part, the two parts are a single phase. For example, if many drops of water are present in condensing steam, all of the drops constitute a single liquid phase, even though they are separated from each other. This would also apply to oil agitated in water as shown in Fig. 10·2. In each case, the drops have a common, or homogeneous, chemical composition. Alcohol and water, on the other hand, form a single phase when mixed since no distinct boundaries separate the two components. Mixtures of gases or vapors are always a single phase, as we discussed in Chapter 9.

Recall that if the chemical composition of a gas or gas–vapor mixture is constant, it is a pure substance regardless of how many components are present.

Two liquids are said to be **miscible** if when mixed in any concentration they form a single phase. If when mixed they remain in two phases, each identical in composition before and after mixing, they are said to be **immiscible**. If the mixing results in a single phase for some concentrations and two phases for others, then the liquids are said to be **partially miscible**. It is noteworthy that the miscibility of two liquids may change dramatically with temperature.

Miscibility

Immiscibility

Partial miscibility

An **azeotropic mixture** has either a minimum or maximum boiling temperature or a minimum or a maximum condensation temperature at some intermediate composition of the components. A mixture of water and antifreeze is

Azeotropic mixtures

an example. For *nonazeotropic* mixtures, however, the boiling and condensation temperatures change monotonically from that of one pure component to that of the other in accordance with the mixture composition. For example, for a mixture of ammonia and water, at 1 atmosphere, the boiling temperature of the mixture at 100 percent water is 100°C and at 100 percent ammonia is −33.3°C. At intermediate mixture compositions, the boiling temperature is less than 100°C and greater than −33.3°C.

Solubility is the maximum amount of a substance, the *solute*, that can be dissolved under equilibrium conditions in a liquid, the *solvent*:

Solubility

$$\text{Solubility} = \frac{\text{Maximum mass of solute}}{\text{Mass of solvent}}$$

Alternatively, *solubility* can be expressed on a molar basis as follows:

$$\text{Solubility} = \frac{\text{Maximum moles of solute}}{\text{Moles of solvent}}$$

In either case, *solubility* is a function of temperature and pressure. In practice, gas solubilities are usually small numbers and are measured in units such as parts per million by mass. Note that you must be careful to determine whether data on solubilities are presented as mass or molar ratios.

In the thermodynamic applications presented in this book, we will deal almost exclusively with gas solubility. Note that for a gas, *solubility* refers to the maximum amount of solute that can be *dissolved* in a liquid. Gases dissolved in a liquid do not change the appearance of the liquid. In other words, if gas bubbles are present and visible, the gas is not *dissolved*. In an aquarium, for example, air is bubbled through the water to supply oxygen so that the fish can breathe. The fish cannot use the oxygen in the bubbles themselves, no matter how small. Rather, they use oxygen from the air that is *dissolved* in the water. This oxygen is not visible.

Test yourself: For a given air supply rate, why are small bubbles preferred for aerating an aquarium?

The next few definitions are required to understand the phase rule. For a system, the number of **degrees of freedom**, given the symbol F, represents the number of independent intensive properties. It is also called the **variance**. The **components** C of a system are the chemical species from which the system, in any of its states, can be prepared. However, there is a caveat: To be considered as *components*, the various substances must be capable of *independent variation*. Thus, the number of *components* of a system is the least number of independently variable, chemically distinct species in the system. In complex systems, determining the number of components requires careful analysis because the number of components is not always equal to the number of different species present.

Degrees of freedom

Variance

Components

A system for which there are no independent intensive properties is said to be *invariant*; one with one independent property is called *univariant*; one with two is called *divariant*; and so forth.

The Phase Rule. The phase rule was first published by J. Willard Gibbs in 1876. It is used to determine the maximum number of independent intensive properties or degrees of freedom for a system in equilibrium. This number of properties is necessary and sufficient to specify the state of each phase in the system, but it is insufficient to specify the amount of each phase present. In other words, the number of properties specified by the phase rule is sufficient to determine the intensive but not the extensive state of the system.

For any system in equilibrium that is homogeneous (or composed of a finite number of homogeneous parts in contact) and free of electric, magnetic, gravitational, solid distortion, and surface tension effects, the **phase rule** states that the maximum number of degrees of freedom F is given by

$$F = 2 + C - P \tag{10·1}$$

The phase rule

where C is the number of components and P is the number of phases in the system.

Notice that we present the phase rule here without derivation. Plausible derivations can be based on little more than the principles treated in this textbook; however, a rigorous derivation requires a more extensive background.

The following examples show how the phase rule is applied to determine the degrees of freedom, or the number of independent intensive properties of a system. In each case, the system is in equilibrium and meets the other conditions necessary for application of the phase rule.

Example 10·1

PROBLEM STATEMENT

Consider a system of ice, liquid water, and water vapor. Determine the number of independent intensive properties in this system.

SOLUTION

In this system, there are three phases (solid, liquid, and vapor) and one component (H_2O). Application of the phase rule gives

$$F = 2 + C - P = 2 + 1 - 3 = 0$$

Therefore, there are no independent variables and the system is *invariant*. None of the intensive properties can be varied as long as the three phases exist together in equilibrium. The system is at the triple state, the only state in which these three phases can exist together.

Example 10·2

PROBLEM STATEMENT

Consider a system consisting of ammonia and water in a liquid solution that is in equilibrium with a vapor. Determine the number of degrees of freedom for this binary mixture.

SOLUTION

Because liquid ammonia and water are miscible, there are two phases, liquid and vapor. There are also two components, NH_3 and H_2O (or the two components can also be taken as NH_4OH and H_2O in the liquid phase). The phase rule shows that the system is divariant:

$$F = 2 + C - P = 2 + 2 - 2 = 2$$

This means that specifying the pressure and temperature of this system completely determines the composition of each phase. If the composition of one phase and the pressure are specified, then there is only one possible temperature under which the two phases can exist together, and the composition of the other phase is fully determined.

10·2 *Txy* and *yx* Diagrams

Note that y here is not to be confused with the mole fraction y_i that is introduced in Section 9·1.

Two diagrams are often used to illustrate the states of binary mixtures. The first is the temperature–composition diagram, or the *Txy* diagram, where x and y refer to the mass fractions of the two components in the binary mixture. Figure 10·3 is an example of a *Txy* diagram for two miscible liquids, ammo-

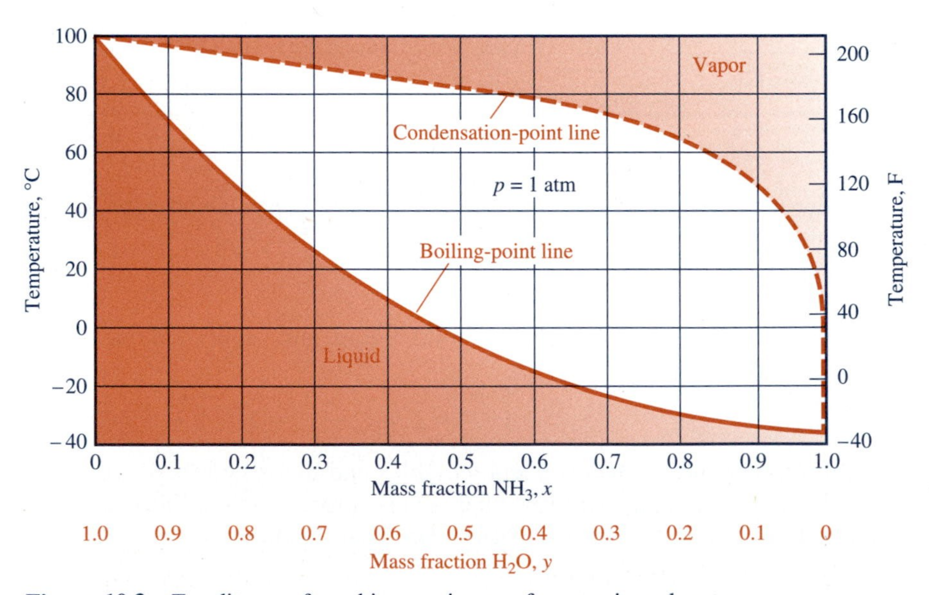

Figure 10·3 *Txy* diagram for a binary mixture of ammonia and water.

nia and water, at a pressure of 1 atm. In Fig. 10·3, x represents the mass fraction of ammonia, y the mass fraction of water, and increasingly intense shading shows increasing mass fraction of water. The solid curve represents the boiling point for any mixture composition. At points above this curve, the liquid will begin to boil and a vapor phase in equilibrium with the liquid phase will be formed. Similarly, the dashed curve represents the condensation point for any mixture composition.

Note that the boiling-point line and the condensation-point line are the same for a pure substance.

Consider a closed system of ammonia and water with composition a, where a represents the mass fraction of ammonia in the overall mixture, initially at temperature T_1 (see Figure 10·4a). At temperature T_2, the maxi-

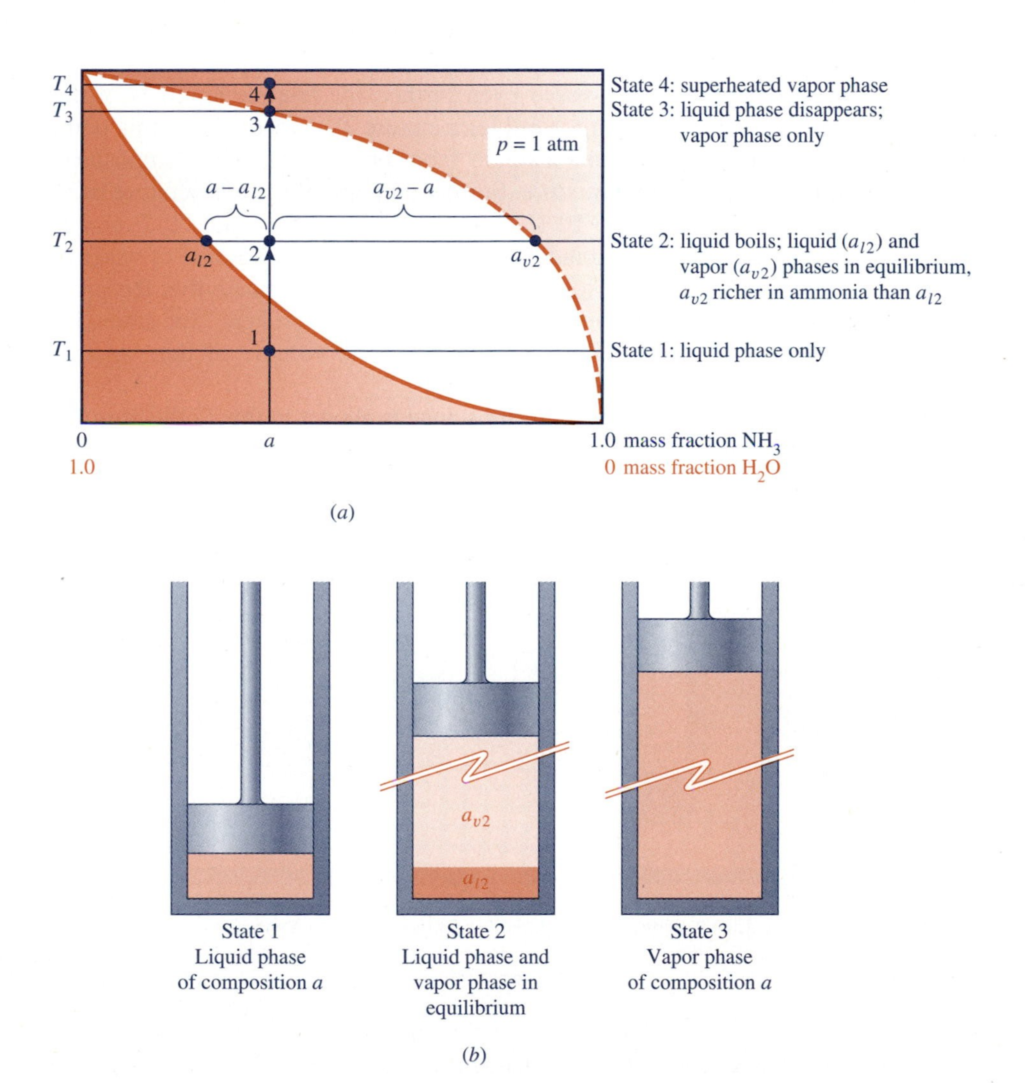

Figure 10·4 (a) *Txy* diagram for ammonia and water describing various states; (b) illustrations of various states of ammonia and water.

mum mass fraction of ammonia possible in the liquid phase is a_{l2}. Since the mass fraction in the closed system is actually a, the liquid boils and vapor is released. The chemical composition of the vapor phase is not equal to that of the liquid phase; rather, the vapor phase, of composition a_{v2}, is richer in ammonia than the liquid phase, of composition a_{l2}. At a temperature equal to T_2, it is not possible for a single phase of a binary mixture of ammonia and water with an overall composition of a to exist. Instead, the overall composition a is maintained by the appropriate ratio of the liquid and vapor phases. For the state labeled 2 in Fig. 10·4*a*, for example, the proportions of the liquid and the vapor are shown by the relative lengths of lines $a_{v2}-a$ and $a-a_{l2}$. If $a_{v2}-a$ and $a-a_{l2}$ are respectively two-thirds and one-third of the total length $a_{l2}-a_{v2}$, then the system is composed of two-thirds liquid and one-third vapor by mass.

As the temperature is increased even further, the relative amounts of the liquid and vapor phases change until temperature T_3 is reached. At this point the liquid phase disappears and a single vapor phase exists with composition equal to that of the original mixture. Further increase of the temperature to T_4 superheats the vapor. Figure 10·4*b* shows sketches of the states 1, 2, and 3. Decreasingly intense shading indicates a decreasing mass fraction of water in the single-phase regions, piston-cylinders are used to indicate that the pressure is constant between states, and the cylinder is "broken" for states involving the vapor phases since in such cases the volume of the vapor is enormously greater than that of the liquid.

Test your understanding: Referring to Fig. 10·4*a*, approximate the maximum temperature at which a single liquid phase of an ammonia–water mixture at composition *a* can exist.

An analogous process for a pure substance is shown in the *Ts* diagram for water, Fig. 10·5*a*. Since the chemical composition of a pure substance does

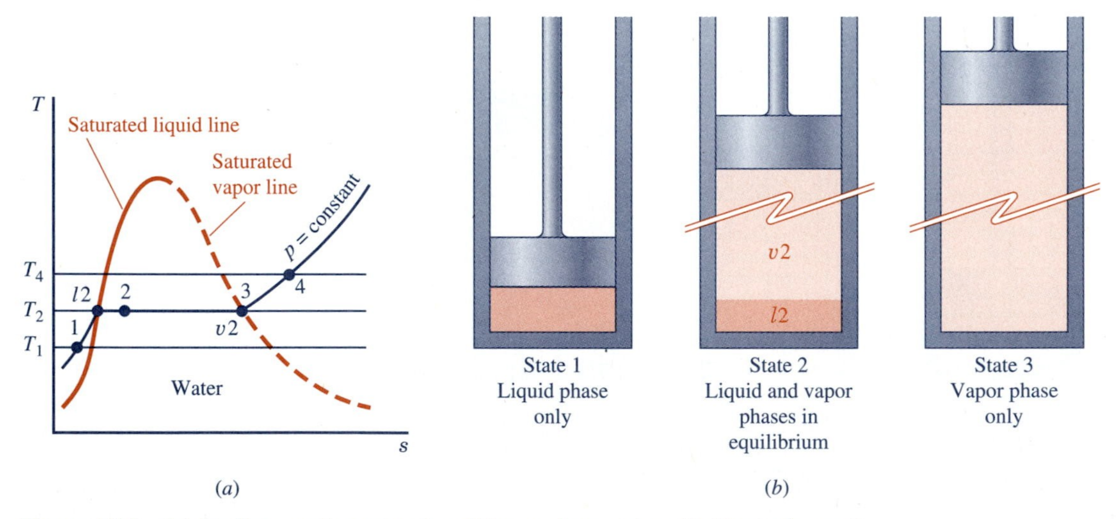

Figure 10·5 (*a*) *Ts* diagram for water describing various states; (*b*) illustrations of various states on *Ts* diagram for water.

not change, we need only ensure that we follow a line of constant pressure. Further, for a pure substance at a given pressure, there is only one saturation temperature. This is unlike a binary mixture, which can have an equilibrium mixture of the vapor and liquid phases at many different temperatures for a given pressure (see Fig. 10·4*a*). In other words, as the ammonia–water system changes from state 2 to state 3, the temperature changes from T_2 to T_3. For pure water, an analogous state change from 2 to 3 occurs at constant temperature (see Fig. 10·5*a*). Despite these differences, this analogy is useful. In Fig. 10·5*a*, at temperature T_1, the system consists of only the liquid phase. When the system reaches a temperature equal to T_2, the liquid boils, thus forming an equilibrium mixture of the liquid phase at state *l*2 and the vapor phase at state *v*2. When the system reaches state 3, the system consists entirely of vapor at state *v*2. At temperature T_4, the vapor is superheated. Figure 10·5*b* shows sketches of the closed system for these states.

As this analogy indicates, the saturated liquid line of the water *Ts* diagram is analogous to the boiling-point line shown in Fig. 10·3 for a binary mixture and, likewise, the saturated vapor line is analogous to the condensation-point line. The area under the saturation dome is analogous to the unshaded area in Fig 10·3.

A *yx* diagram can be plotted from the *Txy* diagram. The equivalent *yx* diagram for an ammonia–water mixture is shown in Fig. 10·6. In the *yx* diagram, the vapor equilibrium region of the *Txy* diagram is mapped to a line. This line represents all mixtures of the liquid and the vapor phases in the same way that the liquid–vapor line in a phase diagram for a pure substance

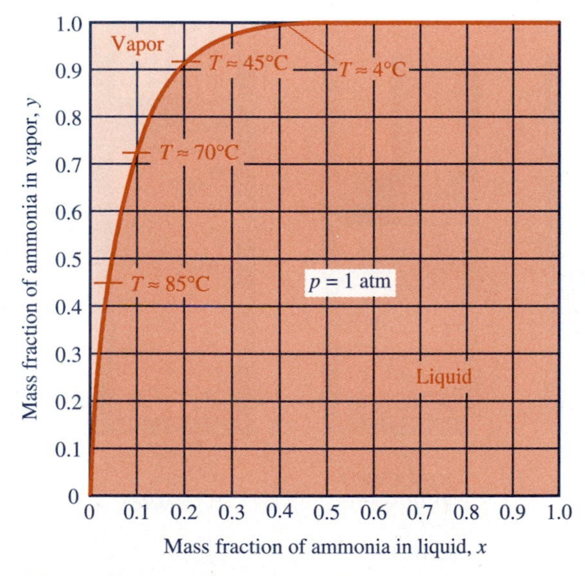

Figure 10·6 *yx* diagram for a binary mixture of ammonia and water.

represents all combinations of liquid and vapor in equilibrium. (See the discussion of Fig. 2·6*a* in Chapter 2). The *yx* diagram is useful for tracing the steps in a distillation or rectification process, as we will see in Section 10·5.

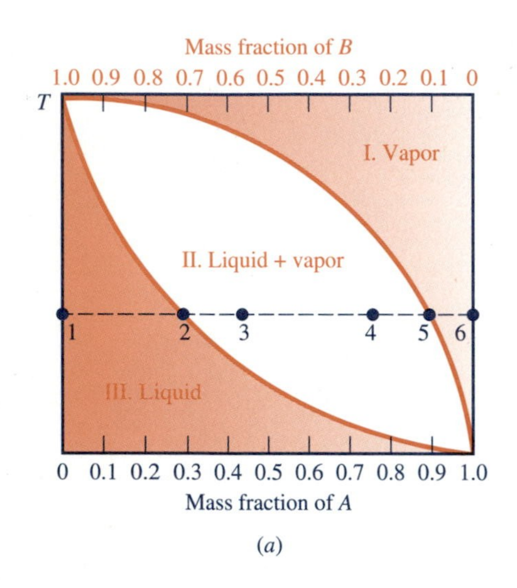

(*a*)

Region	*Description*	*Phase Rule Result*	*Discussion*
I	Vapor consisting of components A and B: one phase is formed.	$F = 3$	Three independent intensive properties are required to define the intensive state of the system. For example, the temperature *and* the composition of the vapor must be specified in addition to the pressure.
II	Mixture of liquid A and liquid B in equilibrium with the vapor: two phases are formed.	$F = 2$	Two independent intensive properties are required. If the pressure is defined (as is usually the case for a *Txy* diagram), then the chemical composition of one phase fully determines the chemical composition of the other. Or the temperature fully determines the chemical composition of both phases.
III	Liquid consisting of components A and B: liquids are miscible, thus, one phase is formed.	$F = 3$	As for Region I, three independent intensive properties are required.

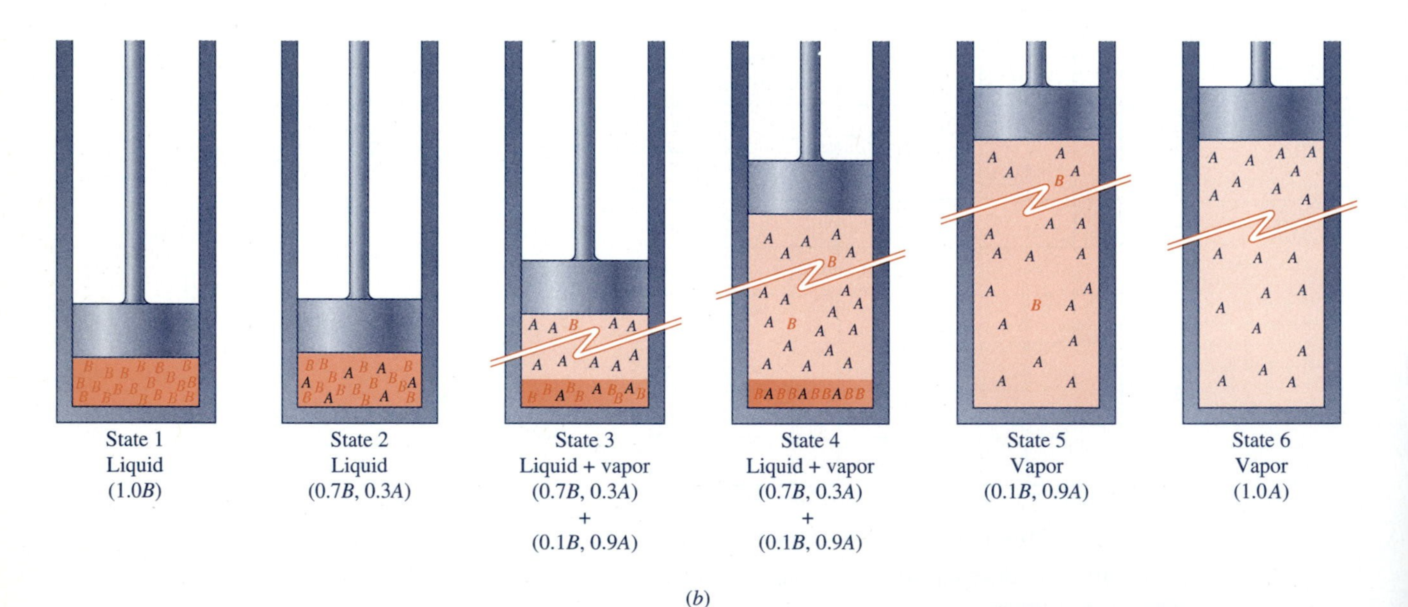

(*b*)

Figure 10·7 (*a*) General *Txy* diagram for two miscible liquids; (*b*) illustrations of various states of two miscible liquids.

10·3 Liquid–Vapor Equilibrium: Miscible Liquids

In this section, we discuss binary mixtures of nonazeotropic miscible liquids. Miscible liquids can have only one liquid phase; mixtures of gases or vapors are always a single phase. Thus, the table accompanying Figure 10·7 shows the various possibilities for the variance of a binary mixture of miscible liquids. As this table indicates, if one phase is present, there are three independent intensive properties. A complete property diagram would therefore be three dimensional. As we discussed above, this difficulty is avoided by using plane diagrams on which two of the properties are plotted for constant values of the third one, usually pressure. At any specified pressure, a binary mixture that consists of two miscible liquids can be described completely by a *Txy* or a *yx* diagram, as was shown above for ammonia and water. Figure 10·7*a* shows a general diagram for two miscible liquids along with some illustrations showing the various compositions possible in each of the regions. In the illustrations of Fig. 10·7*b*, shading indicates an increasing mass fraction of substance *B* in the single-phase regions. As before, piston-cylinders are used to indicate that the pressure is constant between states, and the cylinder is ''broken'' for states involving the vapor phase.

10·4 Liquid–Gas Equilibrium: Solubility

Recall that *solubility* is defined as the maximum amount of a solute that can be dissolved in the solvent. *Gas solubility*, that is, the solubility of a gaseous solute in a liquid solvent, is a function of pressure and temperature. In parts of Chapter 15, Power Systems, and Chapter 16, Refrigeration Systems, this functional relationship is important.

In many systems, gases dissolved in liquids may be highly corrosive. For example, dissolved oxygen in boiler systems causes severe pitting of metal surfaces. Gases dissolved in liquids can also be beneficial, as with chlorine in pools or bromine in spas. In the area of environmental cleanup, aerobic bioremediation of hazardous waste spills is most often limited by the supply of dissolved oxygen to the microbes. Therefore, an understanding of the solubility of gases, whether to avoid damage or to enhance processes, is useful.

In Section 10·5 we will discuss deaerators, which are designed to remove dissolved oxygen.

The solubility of a gas in a liquid can be expressed in terms of **Henry's law**, which can be approximated for many substances with low solubilities as

Henry's law

$$\mathcal{S}_i = \frac{1}{k_i} p_i$$

where $\mathcal{S}_i$ is the solubility of gas i measured in mole fractions, k_i is Henry's law constant for gas i measured in units of pressure per mole fraction, and p_i is the partial pressure of gas i in the gas phase that is in equilibrium with the liquid. The gas phase in equilibrium with the liquid is often referred to as the *head gas*.

Some references call k_i Henry's law constant, others use this name for $1/k_i$. You should check the units to see which is the case.

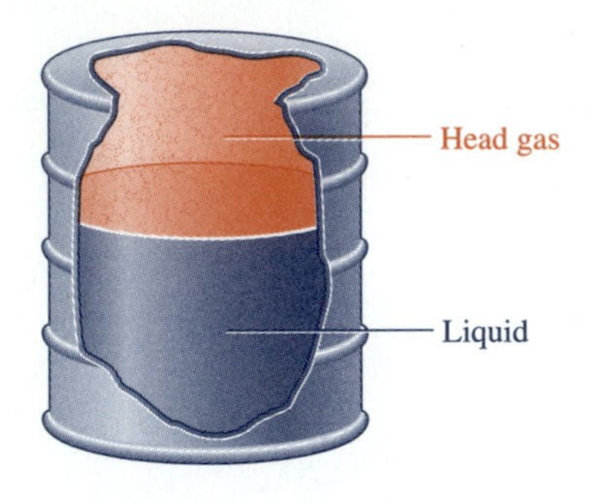

This expression shows that the concentration of the gas in the liquid varies linearly with its partial pressure p_i in the gas phase. Therefore, gases can be removed from the liquid either by reducing the pressure above the liquid or by replacing the head gas with a different chemical species that has a small or zero fraction of the gas we wish to remove from the solution. The former method is sometimes used in boiler systems to remove oxygen from the feedwater and is called vacuum deaeration.

Notice the difference between the effect of temperature on the solubility of solids and gases.

The effect of temperature on the solubility is introduced through the effect on Henry's law constant. Figure 10·8 shows the Henry's law constant as a function of temperature for various gases in water. This figure shows that the solubility of various gases is widely variable (the scale varies over roughly four orders of magnitude). We observe that as temperature increases, solubility decreases. Remember that this conclusion, as well as Henry's law, holds only for gases of low solubility.

10·5 Engineering Systems: Rectifiers and Deaerators

Rectifiers and deaerators are examples of engineering systems that exploit the properties of binary mixtures. Examples of these systems are provided below.

Rectifiers. *Txy* and *yx* diagrams are useful in the study of the separation of volatile miscible liquids into their components. This process is called **rectification**. Figures 10·9*a* and 10·9*b* show a rectification process for an ammonia–water mixture traced on *Txy* and *yx* diagrams with intermediate states *a*–*f* labeled. The mixture of ammonia and water, originally with a small mass fraction of ammonia in the liquid phase, is boiled. The liquid remaining, as boiling progresses, becomes leaner in ammonia, as indicated in Fig. 10·9*b*, and the vapor formed becomes richer in ammonia than the original mixture. If this vapor is separated from the liquid and condensed, the liquid formed from it is therefore richer in ammonia than the original mixture. Notice that on the *yx* diagram, this process is equivalent to moving horizontally to the line of slope equal to 1. The line of slope 1 represents equivalent mass fractions of ammonia in the liquid and vapor phases. Therefore, if we begin with a vapor that has a mass fraction of ammonia equal to 0.3, separate the vapor, and condense it, the liquid now has a mass fraction equal to that of the original vapor, that is, equal to 0.3 (see Fig. 10·9*a*). Note that in the *yx* diagram, unlike the *Txy* diagram, points off the solid line do not represent states of the system. Therefore, the horizontally and vertically drawn lines are only for convenience in tracing the appropriate steps. If the liquid with mass fraction 0.3 is then boiled and the vapor removed and condensed, a liquid even richer in ammonia will be obtained. By continuing the

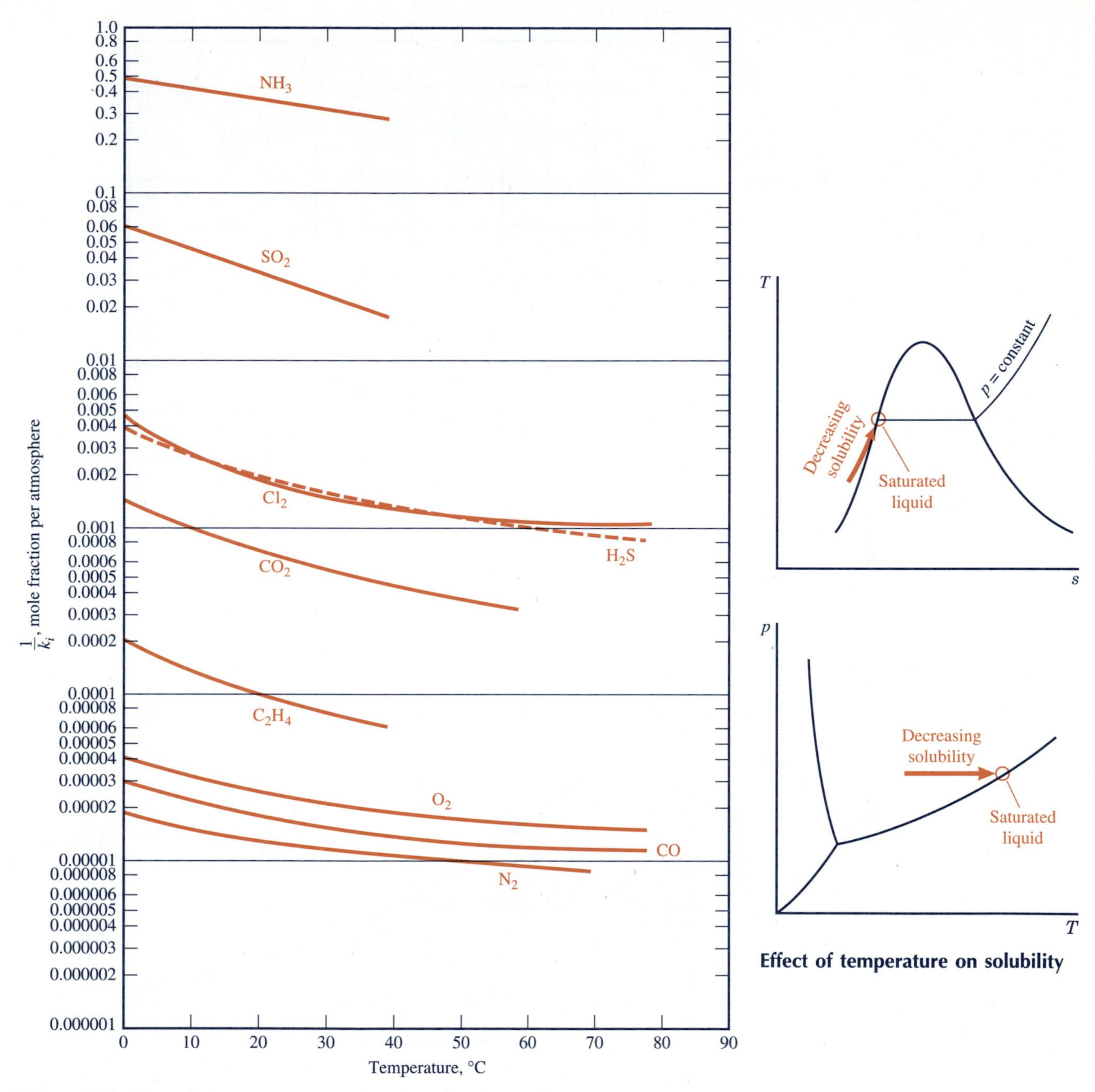

Figure 10·8 Henry's law constant as a function of temperature for various gases in water. (From Ref. 10·2 by permission.)

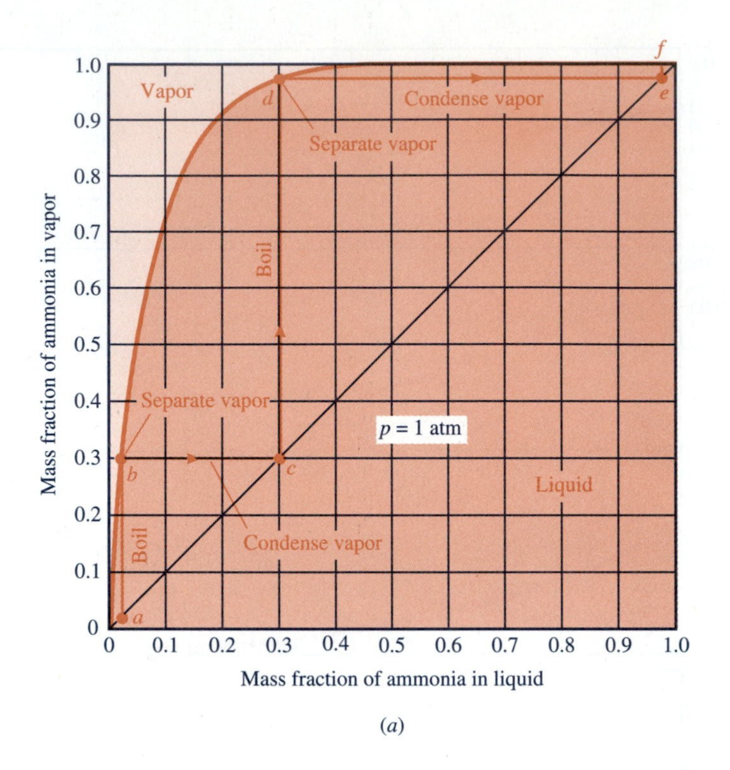

(a)

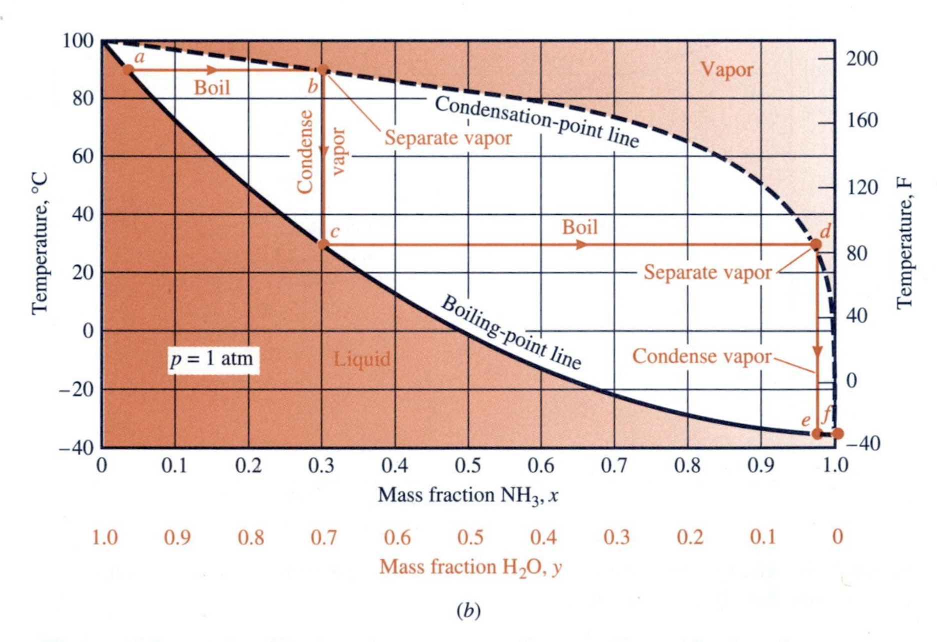

(b)

Figure 10·9 (*a*) Rectification shown on a *yx* diagram; (*b*) rectification shown on a *Txy* diagram.

evaporation, separation, and condensation steps it is possible to obtain complete separation of the ammonia and water.

Figure 10·10 shows schematically a method of partial *rectification.* A binary liquid mixture at its boiling point (state f) enters the rectifier or rectifying tower at section 1 and flows downward over the trays that are arranged to produce a large liquid surface area for heat and mass transfer. Other physical arrangements to promote contact and the attainment of equilibrium are also used, and some of these replace the continuous process by a series of discrete steps. Heat is added at the bottom of the tower, and only liquid is allowed to leave at section 2. The vapor produced from the liquid flows back up the tower to leave at section 3. Heat transferred from the vapor to the counterflowing liquid causes some of the liquid to evaporate and establishes a temperature gradient throughout the tower. Corresponding to the temperature gradient there are concentration gradients in both the liquid and the vapor streams. With reference to the temperature–composition diagram in Fig. 10·10, if a mixture of A and B in state f enters the tower, the vapor driven from it is in state g. This makes the liquid richer in B and raises its boiling point. As it flows downward, its temperature rises, evaporation continues, and the concentration of B increases. The vapor formed lower in the tower is also richer in B. Notice that it is possible to withdraw pure B at section 2, but the vapor leaving at section 3 can be no richer in A than state g. (In an actual rectifier the liquid and vapor at any section are not in equilibrium with each other because heat and mass are being transferred at a finite rate between the phases.)

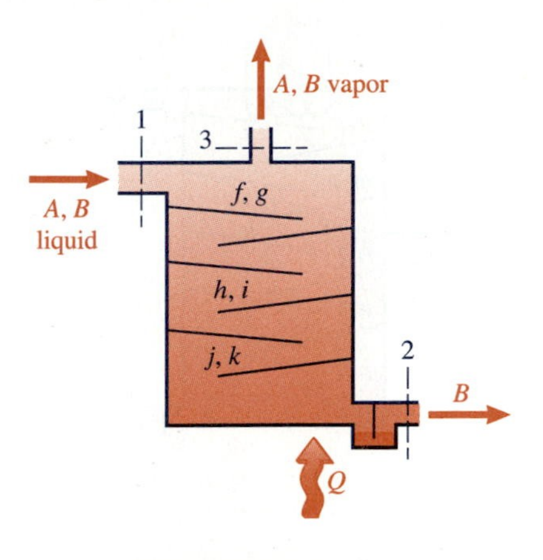

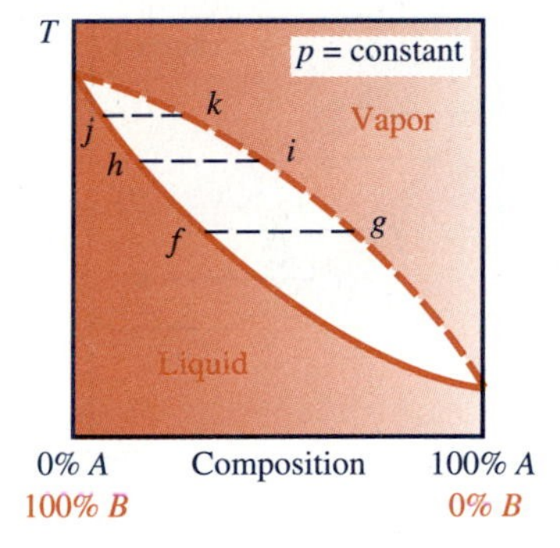

Figure 10·10 Partial rectification.

The maximum yield of pure component B can be determined by means of a mass balance and temperature–composition data. Let x'_A stand for the mass fraction of A in the liquid phase and y'_A stand for the mass fraction of A in the vapor phase. Let $\dot{m}$ represent a total flow rate and $\dot{m}_A$ and $\dot{m}_B$ represent the flow rates of components A and B, respectively. Only B leaves at section 2, so a mass balance is

$$\dot{m}_{B2} = \dot{m}_2 = \dot{m}_1 - \dot{m}_3 = \dot{m}_1 - \frac{\dot{m}_{A3}}{y'_{A3}}$$

All A that enters the system leaves at 3, so $\dot{m}_{A3} = \dot{m}_{A1} = x'_{A1}\dot{m}_1$. Also, the liquid entering at 1 and the vapor leaving at 3 are in contact and are assumed to be in equilibrium with each other; so $y'_{A1} = y'_{A3}$. These substitutions in the mass balance and rearrangement lead to

$$\frac{\dot{m}_{B2}}{\dot{m}_1} = 1 - \frac{x'_{A1}}{y'_{A1}}$$

This is the maximum yield. If less liquid is removed at section 2, the vapor in the tower will be richer in B. Its temperature at each section of the tower will be higher, and this is actually required in order to maintain the required heat transfer from the vapor to the liquid; so the yield given by the equation above is not realized by an actual system.

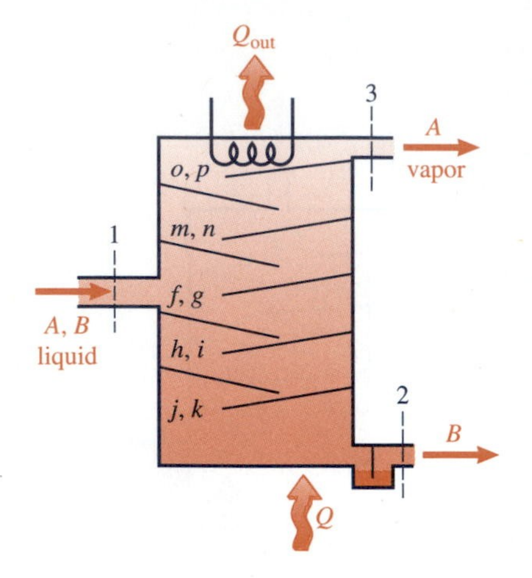

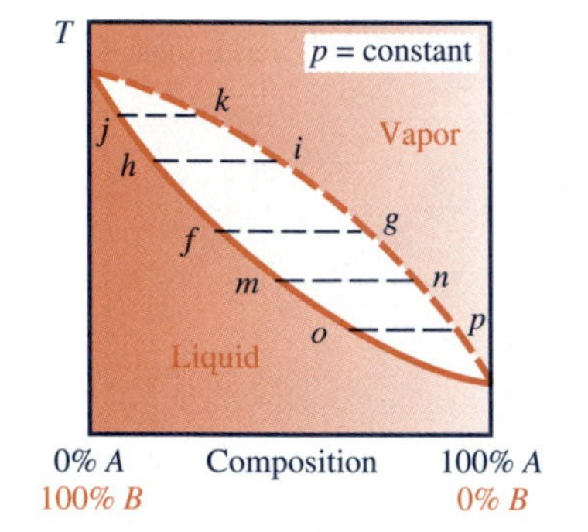

Figure 10·11 Complete rectification.

Complete rectification can be accomplished by means of the system shown schematically in Fig. 10·11. Instead of removing vapor at the section where the liquid enters, the tower is extended above the inlet and a cooler is provided at the top. In the limiting case, pure B liquid leaves at section 2, and pure A vapor leaves at section 3. The condensate from the cooler at the top of the tower flows back down the tower, so that the transfer of heat from the vapor to the liquid and the transfer of mass from the liquid to the vapor occurs throughout the tower. The liquid that flows down the tower and mixes with the incoming stream is called the *reflux*. The mass ratio at the mixing point of reflux to incoming flow is a minimum when these two streams are in the same state as they mix near section 1 and when the vapor flowing up the tower past the mixing point at the inlet is in equilibrium with the liquid.

If the mixture enthalpy data are available, the required heat transfer of a rectification process can be determined by application of the first law.

Deaerators. As we mentioned above, oxygen is strongly corrosive to many metals. For this reason, several methods are used to deaerate feedwater in boiler systems. Some are chemical and rely on the reactivity of various substances, like hydrazine, with the dissolved oxygen. These are not based on solubility; therefore, we will not discuss them. Rather, we will turn our attention to deaeration systems that rely on the principles of solubility to remove oxygen from the water. Figure 10·12 shows a typical steam deaerator as used in power plants. Its purpose is to drive all the dissolved oxygen from the water by bringing the water to its saturation temperature. As shown, the liquid to be deaerated is injected and brought into contact with the steam.

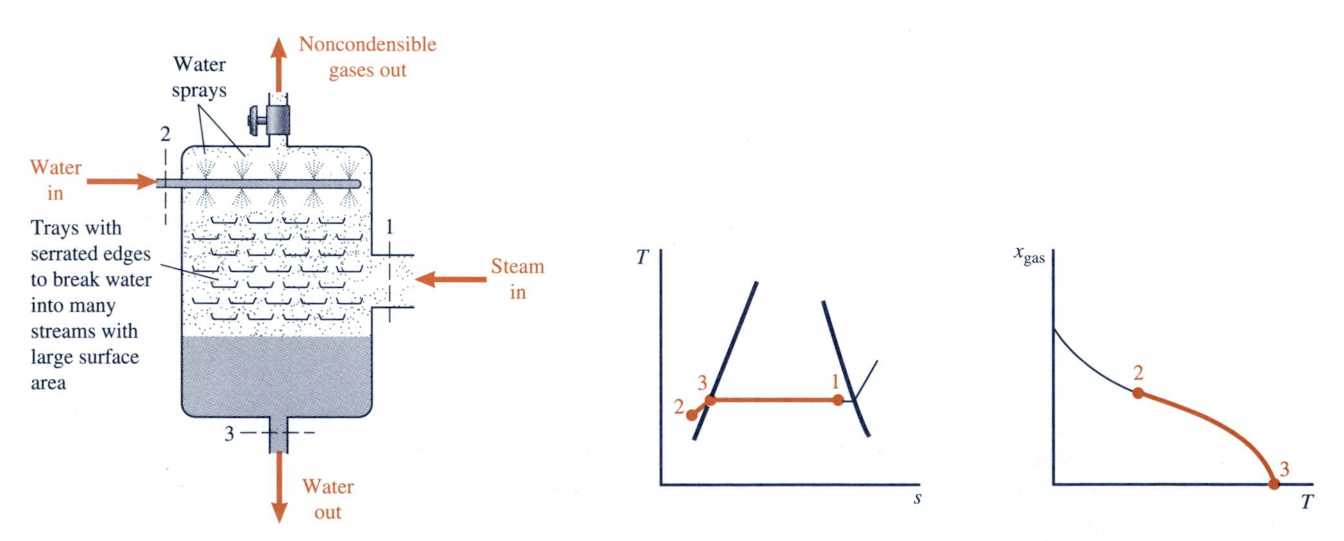

Figure 10·12 Steam deaerator.

Then all the steam is condensed so that liquid water (now deaerated) and the *noncondensible gases* exit the deaerator in separate streams. In practice, a very small amount of the steam may not be condensed and consequently may leave with the gases.

Three factors are important in the design of steam deaerators. First, the design should maximize the surface area of the liquid–vapor interface to increase the rate of heat and mass transfer. Second, the contact time between the liquid and the vapor must be sufficient to allow heat and mass transfer to occur. Third, gases that are removed from the liquid must be removed from the space above the liquid to lower their partial pressures there. All factors are required to ensure that all the liquid is completely deaerated. Recontamination of the deaerated water by the gases must be prevented, and one step to ensure this is to maintain the water temperature very close to the saturation temperature. Figure 10·12 also shows the states of the fluids entering and leaving the deaerator and the oxygen content of the entering and leaving liquid water.

10·6 Summary

If two liquids when mixed in any concentration form a single phase, they are said to be *miscible*. If when mixed they remain in two phases, each identical in composition with one of the liquids before mixing, they are immiscible. If the mixing results in a single phase for some concentrations and two phases for the others, the liquids are said to be partially miscible. Solids, like liquids, show varying degrees of miscibility. Gases are always completely miscible.

The equilibrium composition of the liquid and the vapor phases of a binary mixture in equilibrium are different. Further, the chemical composition of one phase fully determines the chemical composition of the other. We might alternatively express this principle as follows: Specifying the pressure *and* temperature of a two-phase equilibrium binary mixture completely determines the chemical compositions of both the liquid and the vapor phases.

For any system in equilibrium that is homogeneous (or composed of a finite number of homogeneous parts in contact) and free of electric, magnetic, gravitational, solid distortion, and surface tension effects, the *phase rule* states that the maximum number of degrees of freedom F is given by

$$F = 2 + C - P \qquad (10\cdot1)$$

where C is the number of components, and P is the number of phases in the system.

Gas solubility is the maximum amount of a gas that can be dissolved under equilibrium conditions in a liquid; it is usually expressed as a mass ratio. Solubility is a function of temperature and pressure and goes to zero as the liquid approaches liquid–vapor saturation conditions.

The solubility of a gas in a liquid can be expressed in terms of *Henry's law*, which can be approximated for many substances with low solubilities as

$$\mathcal{S}_i = \frac{1}{k_i} p_i$$

where $\mathcal{S}_i$ is the solubility of gas i measured in mole fractions, k_i is Henry's law constant for gas i measured in units of pressure per mole fraction, and p_i is the partial pressure of gas i in the gas phase that is in equilibrium with the liquid.

References

10·1 *Betz Handbook of Industrial Water Conditioning*, 8th ed., Betz Laboratories, Inc., Trevose, Pennsylvania, 1980.

10·2 Henley, Ernest J., and J. D. Seader, *Equilibrium-Stage Separation Operations in Chemical Engineering*, Wiley, New York, 1981.

10·3 Perry, Robert H., and Don W. Green, eds., *Chemical Engineers' Handbook*, 6th ed., McGraw-Hill, New York, 1984. (A source of data on solubility of gases in liquids.)

Problems

10·1 Determine the number of phases for a system consisting of sand and solid salt in the presence of a salt–water solution and vapor.

10·2 A system consists of solid sodium chloride, a solution of sodium chloride, water, ethyl alcohol, and vapor. Determine the number of phases.

10·3 How many components are present in a gas mixture at room temperature and atmospheric pressure consisting of oxygen, hydrogen, and water vapor?

10·4 Determine the number of components and degrees of freedom for a system consisting of CaO, $Ca(OH)_2$, and H_2O at 1 atm, 100° F.

10·5 A system contains ammonia liquid and vapor. How many degrees of freedom are there?

10·6 A system under equilibrium conditions contains a saturated water solution of sodium chloride, excess sodium chloride, ethyl alcohol, and vapor. Determine the number of degrees of freedom.

10·7 For the system in which solid magnesium carbonate dissociates according to the following reaction, determine the number of degrees of freedom: $MgCO_3$(solid) → MgO(solid) + CO_2(gas).

10·8 Derive the following expression for the minimum reflux ratio at the mixing point in a tower like that shown in Fig. 10·11:

$$\frac{\dot{m}_{\text{reflux}}}{\dot{m}_1} = \left(\frac{1 - y'_{A1}}{y'_{A1} - x'_{A1}}\right) x'_{A1}$$

where x'_A is the mass fraction of A in the liquid phase and y'_A is the mass fraction of A in the vapor phase.

10·9 Sketch a *yx* diagram for Fig. 10·10.

10·10 Sketch a *yx* diagram for Fig. 10·11.

10·11 An aqua–ammonia solution at 1 atm, with both liquid and vapor present, has a mass fraction of 30 percent by mass of ammonia in the liquid. For equilibrium conditions, determine (*a*) the temperature and (*b*) the mass fraction of ammonia in the vapor.

10·12 A mixture of 20 percent benzene and 80 percent toluene at 1 atm enters a partial rectifier as shown in Fig. 10·10 at a rate of 6000 lbm/h. Compute the maximum yield of pure toluene.

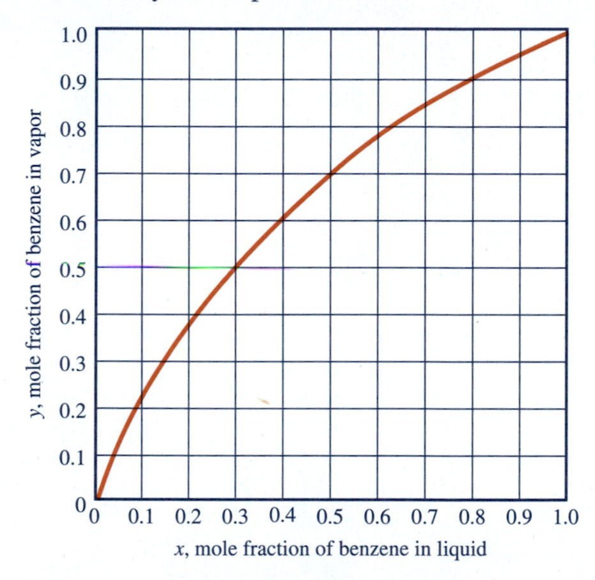

Problems 10·12 and 10·13

10·13 A continuous rectifying column is used to separate completely 12,000 lbm/h of a solution of 60 percent benzene and 40 percent toluene into its components. The pressure is 1 atm. Compute the minimum reflux rate.

10·14 Why does drinking water that has been boiled taste flat? How can its normal taste be restored?

10·15 When a glass of ice water is allowed to stand in a room overnight, bubbles collect on the inner surface of the glass. The composition of these bubbles is not the same as that of room air. Are they richer in nitrogen or richer in oxygen than room air?

10·16 From Fig. 10·8, what can you predict about the composition of the bubbles that collect on the inner surface of a glass that initially is filled with ice water and then allowed to stand for several hours?

10·17 Chlorine is used for the treatment of swimming pool water and bromine is used for the treatment of spa water. From this fact, where would you expect a line for bromine to lie in Fig. 10·8 with respect to the chlorine line?

10·18 Why is chlorine used in swimming pools but bromine is used in spas where the water temperature is higher?

10·19 When the cap is first removed from a bottle of ginger ale, bubbles appear in the liquid and rise through it to escape. Explain this phenomenon. How is it affected by the temperature of the beverage and atmospheric pressure?

10·20 In a steam heating and power plant, water returned from the various heat exchangers and turbines is collected in an open tank. The temperature of the water is usually in a range of 80 F to 140 F. Plot the maximum oxygen content (ppm) of the water in the tank versus temperature.

10·21 A deaerating feedwater heater in a steam power plant operates at 115 kPa. Water at 30°C enters at a rate of 78,000 kg/h. Saturated liquid leaves. Determine the steam flow rate required. Estimate the maximum error in neglecting the energy of the noncondensible gases in the energy balance.

10·22 Explain why the shape of the concentration–temperature curve in Fig. 10·12 is different from the shape of the curves in Fig. 10·8.

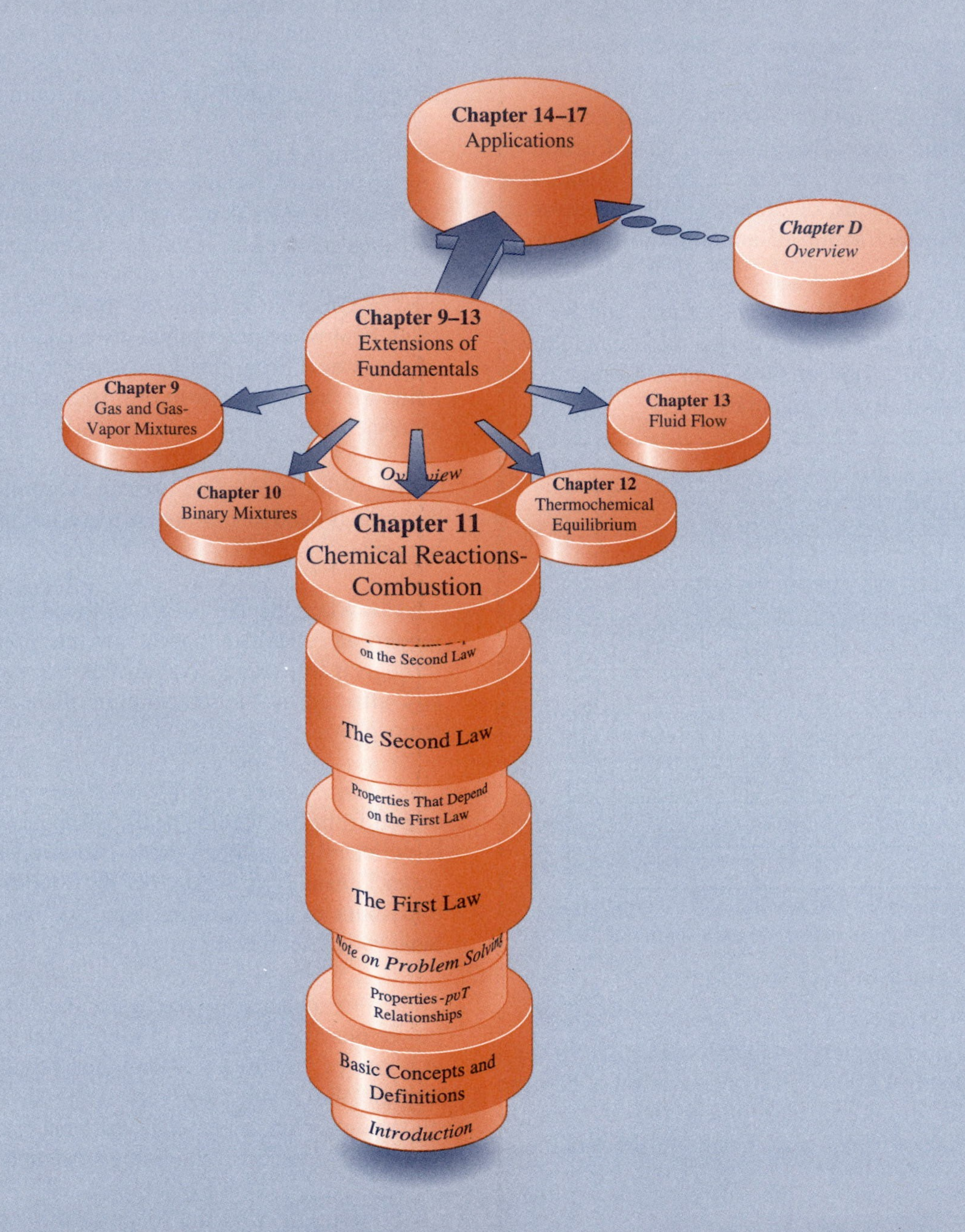
Chapter 14–17
Applications
Chapter D
Overview
Chapter 9–13
Extensions of
Fundamentals
Chapter 9
Gas and Gas-
Vapor Mixtures
Chapter 13
Fluid Flow
Chapter 10
Binary Mixtures
Chapter 12
Thermochemical
Equilibrium
Chapter 11
Chemical Reactions-
Combustion
on the Second Law
The Second Law
Properties That Depend
on the First Law
The First Law
Note on Problem Solving
Properties - pvT
Relationships
Basic Concepts and
Definitions
Introduction

Chapter 11

Chemical Reactions—Combustion

In this chapter we apply thermodynamic principles to chemical reactions. For simplicity, we deal with a particular type of chemical reaction: combustion. Combustion reactions are chosen because of their importance in engineering, but you should keep in mind that the same analysis technique may be applied to other chemical reactions.

Three aspects of chemical reactions are considered in detail in this book:

Aspect of Chemical Reaction	*Purpose*	*Location*
1. The mass balance	To determine the products formed by known reactants or the reactants required to form known products	This chapter, Ch. 11
2. The energy balance or application of the first law	To determine the energy transfers and conversions accompanying a reaction	This chapter, Ch. 11
3. The application of the second law	To determine the extent to which a reaction will proceed and the irreversibility of a reaction	Next chapter, Ch. 12

A fourth important aspect of chemical reactions—the rate at which reactions occur—is treated briefly in Chapter 12.

11·1 A Few Preliminaries

Before discussing the mass balance and the energy balance, we need to look at some fundamentals associated with combustion analysis.

The preceding chapters have mainly treated pure substances and other nonreacting systems. Recall that a pure substance is defined as one that is chemically homogeneous and also chemically *invariant* with respect to time. Therefore, reacting substances are not pure substances. The substances present before a reaction occurs are called the **reactants**, and those present after the reaction has occurred are called the **products**. Both the *reactants* and the *products* are pure substances before or after the reaction, and we calculate their property changes on this basis.

Reactants

Products

11·1·1 The Basic Combustion Reactions

Fuels usually consist of carbon, hydrogen, and sulfur because these constituents *burn* in the presence of oxygen. By *burning* we mean the rapid oxidation of these elements to produce heat.

The basic combustion reactions for these three elements are

$$C + O_2 \rightarrow CO_2$$
$$H_2 + \tfrac{1}{2}O_2 \rightarrow H_2O$$
$$S + O_2 \rightarrow SO_2$$

The above reactions describe what is commonly referred to as **complete combustion**, or complete oxidation, since the products of these reactions, namely, CO_2, H_2O, and SO_2, do not burn further. In other words, they cannot accept additional oxygen atoms in a reaction that produces heat.

Complete combustion

This is in contrast to a reaction such as

$$C + \tfrac{1}{2}O_2 \rightarrow CO$$

wherein the combustion is said to be incomplete because the products are not completely oxidized; that is, CO can be burned further.

The complete combustion of hydrocarbon compounds results in the formation of carbon dioxide and water. Thus

$$C_8H_{18} + 12\tfrac{1}{2}O_2 \rightarrow 8CO_2 + 9H_2O$$

describes the *complete combustion* of *n*-octane, C_8H_{18}. This equation may be interpreted as follows:

$$1 \text{ mole } C_8H_{18} + 12\tfrac{1}{2} \text{ moles } O_2 = 8 \text{ moles } CO_2 + 9 \text{ moles } H_2O$$

where the general term *mole* could be gmol, kmol, lbmol, slugmol, and so on. Alternatively, this equation can be written on a mass basis as

$$114 \text{ kg } C_8H_{18} + 400 \text{ kg } O_2 = 352 \text{ kg } CO_2 + 162 \text{ kg } H_2O$$

where any unit of mass could be used.

Notice that the total mass is the same on both sides of the equation. Also, the total mass of *each* chemical element in the reactants is preserved in the products. By contrast, the total number of moles on each side of the equation need not be the same.

The masses in the last equation are based on approximate molar masses of 12 for carbon and 2 for hydrogen. The more accurate values of 12.011 for the naturally occurring mixture of carbon isotopes and 2.016 for hydrogen (H_2) should be used when greater precision is required.

11·1·2 The Composition of Dry Air

The oxygen for most combustion reactions comes from air. Thus, before we can compute combustion mass balances, we must know the chemical composition of air.

The composition of dry air is given by the following mole fractions: 0.7809 N_2, 0.2095 O_2, 0.0093 Ar (argon), and 0.0003 CO_2. The molar mass of this mixture is 28.967 kg/kmol. For nearly all combustion calculations we can treat the argon and carbon dioxide as additional nitrogen because they are inert and appear in only small quantities. Using this assumption, one mole of air is composed of 0.79 mole N_2 and 0.21 mole O_2. From this molar composition, the mass composition can be derived. Any of the forms shown below can be used to remember the composition of air:

The Composition of Dry Air

Molar Basis	*Mass Basis*
0.21 mole O_2 + 0.79 mole N_2 = 1 mole air	0.232 kg O_2 + 0.768 kg N_2 = 1 kg air
1 mole O_2 + 3.76 moles N_2 = 4.76 moles air	1 kg O_2 + 3.31 kg N_2 = 4.31 kg air

Stoichiometric air

Using Air in Combustion Reactions. **Stoichiometric air** is the quantity of air required to burn one unit of fuel completely. All the oxygen supplied in the reactants is used, and no "free" oxygen appears in the products. Under these circumstances, the products of hydrocarbon combustion consist of carbon dioxide, water, sulfur dioxide, and nitrogen.

Excess air

Excess air is air supplied in excess of that necessary to burn the fuel completely. Excess air appears in the products of combustion unchanged; it simply passes through the reaction. The amount of excess air is normally expressed as a percentage of the stoichiometric amount required for complete combustion of the fuel.

In actual combustion processes, many conditions cause excess air to be either necessary or beneficial. In hazardous waste incinerators, for example, toxic organics are to be broken down, by burning, into their benign constituents. If the combustion reactants are not completely oxidized, toxins can be released. In this case, excess air may increase the level of confidence that combustion is nearly complete.

A Note About Nitrogen. Nitrogen can be treated as inert for most purposes. Therefore, when air is used in a combustion reaction, the nitrogen appears in the products of combustion as a diluent only. For this reason, it is often omitted in writing the combustion reactions. However, note that nitrogen must be included when first or second law calculations are applied. Also, at high temperatures nitrogen reactions may be important.

11·2 The Combustion Mass Balance

In this section we discuss mass balances for two types of combustion: ideal and actual. Ideal combustion is always complete; actual combustion is incomplete. Mass balances for ideal combustion calculations are usually done *a priori,* with knowledge of the reactants only. No additional information is required. Mass balances for actual combustion processes, while identical in principle, usually require some data regarding the products and therefore are done *a posteriori.*

a priori: before the fact

a posteriori: after the fact

In either case, the first principle of combustion is the mass balance:

Combustion mass balance

$$\text{Mass of reactants} = \text{Mass of products} \tag{11·1}$$

This single equation is usually not sufficient to determine all the unknowns in a combustion reaction. In such cases, it is necessary to perform a mass balance on *each* of the chemical elements involved in the reaction. This increases the number of equations and in many cases allows us to complete the analysis. In the following sections we use this technique in both ideal and actual combustion analyses.

The following table summarizes the engineering purpose of both ideal and actual combustion mass balances:

Mass Balance	*Engineering Purpose*
Ideal combustion mass balances	• To determine the amount of air (or other oxidizer) required to burn a fuel completely • To determine the composition of the resulting products
Actual combustion mass balances	• To determine, from an analysis of the products of an actual combustion process, the amount of air actually supplied

11·2·1 Ideal Combustion

For a fuel of known composition, we often need to determine the amount of air (or other oxidizer) required to burn the fuel in the ideal case. This would provide a lower bound to the amount of air required in an actual combustion process. The stoichiometric amount of any oxidizer is the minimum amount required. We may also want to determine the composition of the resulting products.

Consider as an example the ideal combustion of a gaseous fuel with the following volumetric analysis:

Constituent	*Volume Fraction, %*
H_2	9
CO	24
CH_4	2
CO_2	6
O_2	3
N_2	56
Total	100

Let the combustion take place with 50 percent excess air.

Recall that for an ideal gas, a volumetric analysis is equivalent to a molar analysis. To minimize the work with fractions, we will use 100 kmol of fuel. The combustion equation can be written as

$$\underbrace{\begin{matrix} 9H_2 \\ 24CO \\ 2CH_4 \\ 6CO_2 \\ 3O_2 \\ 56N_2 \end{matrix}}_{\text{gaseous fuel}} + \underbrace{1.5x[O_2 + 3.76N_2]}_{\text{50\% excess air}} \rightarrow \underbrace{aH_2O + bCO_2 + cO_2 + dN_2}_{\text{products}} \qquad (a)$$

There are five unknowns, so we will need five independent equations for a solution. A total mass balance alone will therefore not suffice. We must invoke mass, atom, or mole balances on the individual elements.

Performing such balances on the above equation, we find that

$$\text{H balance:} \quad 9(2) + 2(4) = 2a$$
$$a = 13$$
$$\text{C balance:} \quad 24 + 2 + 6 = b$$
$$b = 32$$

We can then solve for the amount of stoichiometric oxygen needed for the combustion. In this balance we do not include the cO_2 term. This unused O_2 appears in the products as the result of the *excess* air supplied for combustion, and thus should not appear in the *stoichiometric* oxygen mass balance:

$$\text{O balance:} \quad 24 + 6(2) + 3(2) + 2x = 13 + 32(2)$$
$$x = 17.5$$

We account for the 50 percent excess air by multiplying this number by 1.5 as shown in Eq. *a* above. This fully defines the amount of nitrogen and oxygen that passes through the reaction as follows:

$$\text{O balance:}\quad 24 + 6(2) + 3(2) + 1.5(17.5)2 = 13 + 32(2) + 2c$$
$$c = 8.75$$

$$\text{N balance:}\quad 56(2) + 1.5(17.5)3.76(2) = 2d$$
$$d = 154.7$$

This solution is an application of elementary chemistry and you may not find it necessary to write out the details. Note again that in this analysis, a mass balance that dealt with the reactants and the products, as a whole, would not be sufficient.

The following examples show how this type of analysis is used in engineering applications. The first example involves the gaseous fuel discussed above. The second example pertains to a solid fuel.

Fuel Analyses. Gaseous fuels are usually analyzed on a volumetric basis. Liquid and solid fuels, on the other hand, are analyzed on a mass basis. Engineers use two forms of mass analysis for solid fuels. In a **proximate analysis** the fuel is separated into arbitrary constituents that are designated as moisture, volatile matter, fixed carbon, and ash. The **ultimate analysis** shows the composition of fuel in terms of its chemical constituents, except for ash which is reported as such and consists of various oxides. You should note that regardless of the type of analysis we start with, it is always possible to convert from the mass basis to the molar basis or vice versa. Consequently, if desired, all combustion calculations could be handled in the same way. This is unnecessary, however, and in each of the following examples the calculations are made directly from the fuel analysis as given.

Proximate analysis

Ultimate analysis

Example 11·1 Ideal Combustion

PROBLEM STATEMENT

For the reaction described by equation *a* above, determine (*a*) the volume of air per cubic meter of fuel required (both measured at the same pressure and temperature), (*b*) the volumetric analysis of the dry products of combustion, (*c*) the mass of the total products of combustion per kilogram of fuel, (*d*) the mass of the dry products of combustion per kilogram of fuel, (*e*) the mass of air supplied per kilogram of fuel, and (*f*) the dew point of the products.

SOLUTION

Recall from the above discussion that the combustion equation for 100 kmol of fuel can be written as

$$\underbrace{\begin{matrix} 9H_2 \\ 24CO \\ 2CH_4 \\ 6CO_2 \\ 3O_2 \\ 56N_2 \end{matrix}}_{\text{gaseous fuel}} + \underbrace{1.5x[O_2 + 3.76N_2]}_{\text{50\% excess air}} \rightarrow \underbrace{aH_2O + bCO_2 + cO_2 + dN_2}_{\text{products}}$$

where we determined that $a = 13$, $b = 32$, $c = 8.75$, $d = 154.7$, and $x = 17.5$.

(*a*) The volume of air per cubic meter of fuel for burning is

$$V_a = (1.5)\frac{17.5 \text{ kmol } O_2}{100 \text{ kmol fuel}}\left(\frac{4.76 \text{ kmol air}}{\text{kmol } O_2}\right) = \frac{1.25 \text{ kmol air}}{\text{kmol fuel}}$$

If we assume ideal gas behavior, then at the same pressure and temperature, 1 kmol of air and 1 kmol of fuel, have the same volume. Therefore, 1.25 kmol air/kmol fuel is equivalent to 1.25 m^3 air/m^3 fuel.

(*b*) The volumetric analysis can be obtained as follows, where for a dry analysis we neglect the water vapor formed:

Product	kmol/100 kmol fuel	*Volumetric Analysis* (%)
CO_2	32	16.4
O_2	8.75	4.5
N_2	154.7	79.1
Total	195.5	100

(*c*) The following table shows the computation of the total mass of products per 100 kmol of fuel:

Product	kmol/100 kmol fuel	*Molar Mass* (kg/kmol)	*Mass* (kg/100 kmol fuel)
H_2O	13	18	234
CO_2	32	44	1408
O_2	8.75	32	280
N_2	154.7	28	4332
Total	208.5		6254

Therefore, to determine the mass of total products per kilogram of fuel, we need to know the molar mass of the fuel. If y_i represents the number of moles of fuel constituent i per 100 kmol of fuel and M_i represents the molar mass of constituent i, then

$$\begin{aligned} M_{\text{fuel}} &= \sum_i y_i M_i \\ &= 9(2) + 24(28) + 2(16) + 6(44) + 3(32) + 56(28) \\ &= 2650 \text{ kg/100 kmol fuel} \end{aligned}$$

Then, the mass of products per kilogram of fuel is

$$\frac{m_p}{m_f} = \frac{6254 \text{ kg products}}{100 \text{ kmol fuel}}\left(\frac{100 \text{ kmol fuel}}{2650 \text{ kg fuel}}\right) = 2.36 \text{ kg products/kg fuel}$$

(*d*) The mass of dry combustion products per kilogram of fuel equals the mass of total products minus the mass of water vapor formed:

$$\begin{aligned} \frac{m_{dp}}{m_f} = \frac{m_p}{m_f} - \frac{m_v}{m_f} &= \frac{(6254 - 234) \text{ kg dp}}{100 \text{ kmol fuel}}\left(\frac{100 \text{ kmol fuel}}{2650 \text{ kg fuel}}\right) \\ &= 2.27 \text{ kg dry products/kg fuel} \end{aligned}$$

(*e*) The mass of air supplied per kg fuel can be found by first calculating the number of moles of air supplied per 100 kmol fuel:

$$\frac{N_a}{N_f} = \frac{1.5(x)4.76}{100} = \frac{1.5(17.5)(4.76)}{100} = \frac{125 \text{ kmol air}}{100 \text{ kmol fuel}}$$

Then

$$\frac{m_a}{m_f} = \frac{N_a M_a}{N_f M_f} = \frac{125 \text{ kmol air}}{100 \text{ kmol fuel}}\left(\frac{28.97 \text{ kg air}}{1 \text{ kmol air}}\right)\left(\frac{100 \text{ kmol fuel}}{2650 \text{ kg fuel}}\right) = 1.36 \text{ kg air/kg fuel}$$

A simple mass balance can be used as a check or as an alternative solution:

$$\begin{aligned} \frac{m_a}{m_f} = \frac{m_p}{m_f} - \frac{m_f}{m_f} &= \frac{2.36 \text{ kg products}}{\text{kg fuel}} - 1.0 \\ &= 1.36 \text{ kg air/kg fuel} \end{aligned}$$

(*f*) The dew point of the products is the saturation temperature corresponding to the partial pressure of the water vapor in the products. (We are assuming that the products behave as an ideal gas.) The water vapor partial pressure is

$$p_v = y_v p_m = \frac{13}{208.5}(101.325) = 6.32 \text{ kPa}$$

The corresponding saturation temperature is found from the saturated steam tables or from the computer model for steam properties by entering this pressure and any value for the quality to indicate saturation conditions. The dew point of the products is 37°C.

Example 11·2 — Ideal Combustion

PROBLEM STATEMENT

A coal has the following ultimate analysis in percentages:

Fuel Analysis

Element	*Percent*
C	70
H	5
S	1
O	12
N	2
Ash	10
Total	100

For complete combustion with 20 percent excess air at 14.7 psia, determine (*a*) the mass of air required per pound of fuel, (*b*) the volumetric analysis of the dry products of combustion, and (*c*) the dew point of the products.

SOLUTION

Analysis: Before beginning the solution, it is useful to write an equation describing the combustion. We will write it for 1 lbm of fuel:

$$\text{Mass of reactants} = \text{Mass of products}$$

$$\underbrace{\begin{matrix}0.70\text{C}\\0.05\text{H}\\0.01\text{S}\\0.12\text{O}\\0.02\text{N}\\0.10\text{ ash}\end{matrix}}_{\text{1 lbm fuel}} + \underbrace{1.2x[0.232\text{O}_2 + 0.768\text{N}_2]}_{\text{20\% excess air}} = \underbrace{a\text{H}_2\text{O}}_{\text{water vapor}} + \underbrace{b\text{CO}_2 + c\text{O}_2 + d\text{N}_2 + e\text{SO}_2}_{\text{dry gases}} + \underbrace{0.10\text{ ash}}_{\text{solid refuse}}$$

Calculations:

(*a*) Performing a mass balance on individual elements we find

$$\text{H balance:}\quad 0.05\text{ lbm} = a\left(\frac{2\text{ lbm H}}{18\text{ lbm H}_2\text{O}}\right) \qquad a = 0.45\text{ lbm H}_2\text{O}$$

$$\text{C balance:}\quad 0.70\text{ lbm} = b\left(\frac{12\text{ lbm C}}{44\text{ lbm CO}_2}\right) \qquad b = 2.57\text{ lbm CO}_2$$

$$\text{S balance:}\quad 0.01\text{ lbm} = e\left(\frac{32\text{ lbm S}}{64\text{ lbm SO}_2}\right) \qquad e = 0.02\text{ lbm SO}_2$$

For stoichiometric burning, $c = 0$, and the multiplier of x is 1. Then, an oxygen mass balance is

$$\text{O balance:}\quad 0.12\text{ lbm O} + x\left(\frac{0.232\text{ lbm O}}{\text{lbm air}}\right) = 0.45\left(\frac{16\text{ lbm O}}{18\text{ lbm H}_2\text{O}}\right) + 2.57\left(\frac{32\text{ lbm O}}{44\text{ lbm CO}_2}\right) + 0.02\left(\frac{32\text{ lbm O}}{64\text{ lbm SO}_2}\right)$$

$$x = (0.40 + 1.87 + 0.01 - 0.12)\left(\frac{1\text{ lbm air}}{0.232\text{ lbm O}}\right)$$

$$x = 9.30\text{ lbm air/lbm fuel}$$

For 20 percent excess air, the mass of air required per lbm of fuel is $1.2(x) = 11.16$ lbm. We can solve for c, the mass of free O_2 in the products, by $c = (0.2)9.30(0.232) = 0.43$ lbm excess O_2. Finally, the mass balance for nitrogen is

$$\text{N balance:}\quad 0.02 + 1.2(9.30)(0.768) = d \qquad d = 8.59\text{ lbm N}_2$$

(*b*) The dry products of combustion consist of all of the gaseous products except water vapor. From the masses of products determined above, we can obtain the volumetric analysis as follows:

Constituent	Mass/lbm fuel	*Molar Mass*	Moles/lbm fuel	*Volumetric Analysis*
CO_2	2.57	44	0.0584	0.154
SO_2	0.02	64	0.0003	0.001
O_2	0.43	32	0.0134	0.035
N_2	8.59	28	0.3068	0.810
Total	11.61		0.379	1.0

(*c*) Assuming that all the water formed is in the vapor phase, the number of moles of water vapor per pound of coal is

$$\frac{N_{H_2O}}{m_f} = 0.45\frac{\text{lbm } H_2O}{\text{lbm fuel}}\left(\frac{\text{lbmol}}{18 \text{ lbm } H_2O}\right) = 0.025 \text{ lbmol } H_2O/\text{lbm fuel}$$

so the total number of moles of gaseous products is

$$\frac{N_p}{m_f} = \frac{N_{dp}}{m_f} + \frac{N_{H_2O}}{m_f} = 0.379 + 0.025 = 0.404 \text{ lbmol gaseous products/lbm fuel}$$

The partial pressure of water vapor is thus

$$p_v = y_v p_m = \frac{N_{H_2O}}{N_p} p_m = \left(\frac{0.025}{0.404}(14.7 \text{ psia})\right) = 0.91 \text{ psia}$$

The dew point is the saturation temperature corresponding to this pressure, approximately 99 F.

11·2·2 Actual Combustion

In the preceding examples we calculated the amount of air required and the resulting product analysis for a fuel burned completely with a specified amount of air. This calculation can be made from the fuel analysis alone; no measured data on an actual combustion process are involved. However, actual combustion processes are not always adequately described by ideal combustion predictions. Sometimes an engineer must determine how much air was *actually* supplied in a combustion process. Often it is difficult to measure the air flow into a combustion chamber—whether it is a jet engine, a boiler furnace, or an internal-combustion engine. In these cases an inferential method is used and additional data are required, namely, the chemical analysis of the products.

Several methods of gas analysis are used. In the past, an instrument called an Orsat analyzer was common. Volumetric analyses are still often called Orsat analyses.

The Mass Balance for Actual Combustion. For the purpose of analyzing actual combustion, the mass balance can be expanded as follows. If a fuel is burned in air, as shown in Fig. 11·1, and the total products of combustion contain some solid refuse, say ash, the mass balance might be written as

$$m_f + m_a = m_{dg} + m_v + m_s \tag{11·2a}$$

where the subscripts are defined as follows:

f = fuel

a = air

dg = dry gaseous products

v = water vapor

s = solid material

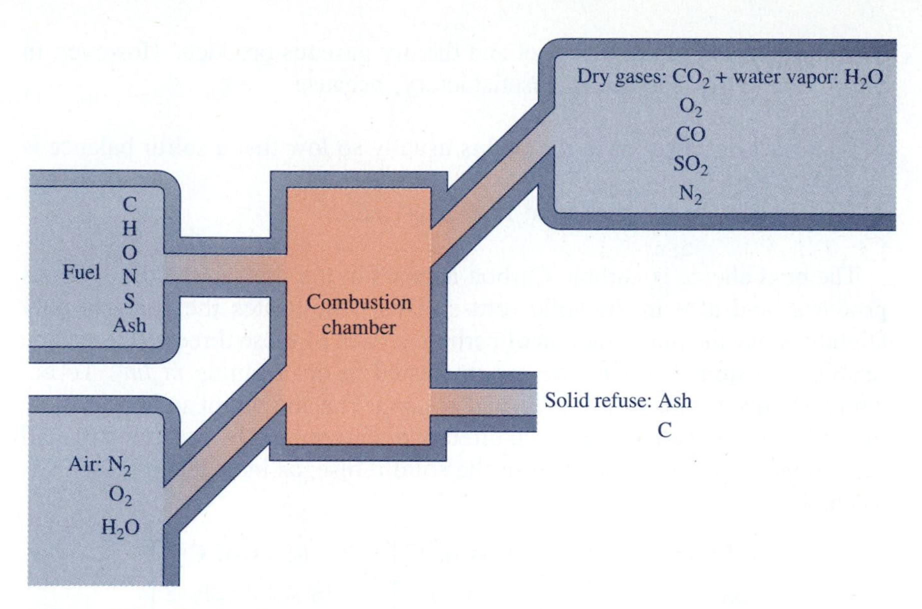

Figure 11·1 Combustion mass accounting.

Usually this analysis is done per unit mass of fuel, therefore, Eq. 11·2*a* becomes

$$1 + \frac{m_a}{m_f} = \frac{m_{dg}}{m_f} + \frac{m_v}{m_f} + \frac{m_s}{m_f} \qquad (11\cdot 2b)$$

Of these masses, m_f and m_s can usually be measured accurately. If we assume that all the hydrogen burns completely, then the ideal hydrogen combustion equation

$$H_2 + \tfrac{1}{2}O_2 \rightarrow H_2O$$

shows that the mass of water in the products is nine times the mass of hydrogen in the fuel; so

$$\frac{m_v}{m_f} = \frac{m_v}{m_H}\frac{m_H}{m_f} = 9\frac{m_H}{m_f}$$

If a significant amount of water vapor enters with the air, it must be added to the water produced from hydrogen in the fuel. Then we are left with two unknowns, m_a and m_{dg}. As was the case for ideal combustion, we can use a mass, atom, or mole balance for each element to supplement the total mass balance. Figure 11·1 shows the constituents in a typical combustion reaction. If we need to determine the mass of air supplied per unit mass of fuel, m_a/m_f, we must first find the mass of dry gaseous products per unit mass of fuel, m_{dg}/m_f. To accomplish this task, we look for an element that appears in both the fuel and the dry gaseous products and only in those two places. If we know the mass fraction of the element in the fuel and also in the dry gaseous product, we can find the mass ratio of dry gaseous products to fuel.

Sulfur appears in only the fuel and the dry gaseous products. However, in practice, a sulfur balance is unsatisfactory, because

1. The fraction of sulfur in the fuel is usually so low that a sulfur balance is inaccurate.
2. SO_2 is usually not measured in exhaust gases.

The next choice is carbon. Carbon appears in the fuel, in the dry gaseous products, and also in the solid refuse. This complicates the analysis only slightly, since the mass fraction of carbon in each of these three places can be readily determined. Again, we are interested in determining m_a/m_f. To accomplish this we must find m_s/m_f and m_{dg}/m_f. The amount of solid refuse per unit mass of fuel can be measured directly or determined by assuming that all the ash in the fuel becomes part of the solid refuse, as indicated in Fig. 11·1. Then,

$$\begin{bmatrix}\text{Mass of C}\\ \text{in dry gases}\end{bmatrix} = \begin{bmatrix}\text{Mass of C}\\ \text{in fuel}\end{bmatrix} - \begin{bmatrix}\text{Mass of C}\\ \text{in solid refuse}\end{bmatrix}$$

This can be expressed per unit mass of the fuel as

$$\left[\frac{\begin{matrix}\text{Mass of C}\\ \text{in dry gases}\end{matrix}}{m_f}\right] = \left[\frac{\begin{matrix}\text{Mass of C}\\ \text{in fuel}\end{matrix}}{m_f}\right] - \left[\frac{\begin{matrix}\text{Mass of C}\\ \text{in solid refuse}\end{matrix}}{m_f}\right]$$

The fuel ultimate analysis, the solid refuse analysis, and the mass of solid refuse per unit mass of fuel together give us the terms on the right-hand side of this equation, and we can then determine the mass of *carbon* in the dry gaseous products per unit mass of fuel, x. If we then determine the mass fraction of *carbon* in the dry gaseous products, y, we can determine m_{dg}/m_f as follows:

$$x = \frac{m_C \text{ in dry gaseous products}}{m_f}$$

$$y = \frac{m_C \text{ in dry gaseous products}}{m_{dg}}$$

$$xm_f = m_C = ym_{dg}$$

and

$$\frac{m_{dg}}{m_f} = \frac{x}{y}$$

Finally, we can use a total mass balance (or a nitrogen or oxygen mass balance) to determine the amount of air supplied per unit mass of fuel.

The following example problem illustrates the mass balance analysis of an actual combustion process.

Example 11·3

PROBLEM STATEMENT

A coal (the same one referred to in Example 11·2) has an ultimate analysis as follows in percentages:

Fuel Analysis

Constituent	*Percent*
C	70
H	5
S	1
O	12
N	2
Ash	10
Total	100

It is burned in a furnace, and the solid refuse is found to contain 33 percent carbon. The volumetric analysis of the dry products of combustion is as follows:

Volumetric Analysis

Constituent	*Percent*
CO_2	14.3
O_2	4.0
CO	1.2

The barometric pressure is 100 kPa. Determine (*a*) the amount of dry air supplied per pound of fuel, (*b*) the percent excess air, and (*c*) the dew point of the products.

SOLUTION

(*a*) We will find the mass of dry gaseous products per pound of fuel and use this information in the total mass balance:

$$m_{\text{entering}} = m_{\text{leaving}}$$

or using Eq. 11·2*b*,

$$1 + \frac{m_a}{m_f} = \frac{m_{dg}}{m_f} + \frac{m_v}{m_f} + \frac{m_s}{m_f}$$

As mentioned earlier, we can use carbon, which appears only in the fuel, dry gaseous products, and refuse, to find the mass of dry gaseous products per pound of fuel:

$$\left[\frac{\text{Mass of C in dry gases}}{m_f}\right] = \left[\frac{\text{Mass of C in fuel}}{m_f}\right] - \left[\frac{\text{Mass of C in solid refuse}}{m_f}\right]$$

Thus

$$\left[\frac{\text{kg C in dry gases}}{\text{kg fuel}}\right] = \left[\frac{\text{kg C in fuel}}{\text{kg fuel}}\right] - \left[\frac{\text{kg C in solid refuse}}{\text{kg refuse}}\right]\left[\frac{\text{kg refuse}}{\text{kg ash}}\right]\left[\frac{\text{kg ash}}{\text{kg fuel}}\right]$$

$$= \left[\frac{0.70}{1}\right] - \left[\frac{0.33}{1}\right]\left[\frac{1}{0.67}\right]\left[\frac{0.10}{1}\right] = \frac{0.65 \text{ kg C in dg}}{\text{kg fuel}}$$

We can then use the results of the volumetric analysis to find the mass of carbon per unit mass of dry gas:

Product	*Mole Fraction* (kmol/100 kmol dg)	*Molar Mass* (kg/kmol)	kg dg/ 100 kmol dg	kg of C in dg/ 100 kmol dg
CO_2	14.3	44	629	171.6
O_2	4.0	32	128	
CO	1.2	28	33.6	14.4
N_2	80.5	28	2254	
Total	100		3045	186.0

Thus,

$$\frac{m_{dg}}{m_f} = 0.65\frac{\text{kg C in dg}}{\text{kg fuel}}\left(\frac{100 \text{ kmol dg}}{186.0 \text{ kg C in dg}}\right)\left(\frac{3045 \text{ kg dg}}{100 \text{ kmol dg}}\right)$$

$$= 10.64\frac{\text{kg dg}}{\text{kg fuel}}$$

Using the total mass balance, we find

$$\frac{m_a}{m_f} = \frac{m_{dg}}{m_f} + \frac{m_v}{m_f} + \frac{m_s}{m_p} - 1$$

$$= 10.64 + \frac{9 \text{ kg water vapor}}{\text{kg } H_2}\left(\frac{0.05 \text{ kg H}}{\text{kg fuel}}\right) + \frac{0.10 \text{ kg ash}}{\text{kg fuel}}\left(\frac{1 \text{ kg refuse}}{0.67 \text{ kg ash}}\right) - 1$$

$$= 10.24 \text{ kg air/kg fuel}$$

(*b*) In Example 11·2 we found that the stoichiometric amount of air required to burn this coal was 9.30 lbm air/lbm fuel. This is equivalent to 9.30 kg air/kg fuel. Therefore, the amount of excess air is

$$\%\text{ Excess air} = \frac{(m_a/m_f) - (m_{a,\text{stoichiometric}}/m_f)}{m_{a,\text{stoichiometric}}/m_f}$$

$$= \frac{10.24 - 9.30}{9.30}$$

$$= 10.1 \text{ percent}$$

(*c*) To find the dew point of the gaseous products, we must determine the partial pressure of the water vapor, and this depends on the mole fraction of vapor in the gases. In part *a*, we determined m_{dg}/m_f and m_v/m_f. From the Orsat analysis we also have the molar mass of the dry gases, or 3045 kg dg/100 kmol = 30.45 kg dg/kmol. Thus,

$$y_v = \frac{N_v}{N_{dg} + N_v} = \frac{m_v/M_v}{m_{dg}/M_{dg} + m_v/M_v}$$

$$= \frac{\dfrac{m_v}{m_H}\left(\dfrac{m_H}{m_f}\right)\dfrac{1}{M_v}}{\dfrac{m_{dg}}{m_f}\left(\dfrac{1}{M_{dg}}\right) + \dfrac{m_v}{m_H}\left(\dfrac{m_H}{m_f}\right)\left(\dfrac{1}{M_v}\right)}$$

$$= \frac{9(0.05)/18}{10.64/30.45 + 9(0.05)/18}$$

$$= 0.067 \frac{\text{kmol water vapor}}{\text{kmol gas}}$$

$$p_v = y_v p_m$$

$$= 0.067(100 \text{ kPa}) = 6.7 \text{ kPa}$$

$$T_{dp} = T_{sat,6.7 \text{ kPa}} = 38.2°\text{C}$$

ALTERNATIVE SOLUTION

(*a*) The amount of dry air supplied can be determined more directly by writing the chemical equation for the actual combustion process with unknown coefficients where necessary. Then, these coefficients can be solved for using mass balances on the individual elements.

Let there be *a* moles of O_2 supplied per kilogram of fuel and *x* moles of dry gas formed per kilogram of fuel.

As a first step in the initial solution of this problem, we found that only 0.65 kg of the carbon burned (for that is all that appeared in the dry gases). The remaining carbon and the ash underwent no reaction, so they can be omitted from the chemical equation. Thus, we have

$$\frac{0.65}{12}\text{C} + \frac{0.05}{2}\text{H}_2 + \frac{0.01}{32}\text{S} + \frac{0.12}{32}\text{O}_2 + \frac{0.02}{28}\text{N}_2 + a\text{O}_2 + 3.76a\text{N}_2$$
$$\rightarrow 0.143x\text{CO}_2 + 0.04x\text{O}_2 + 0.012x\text{CO} + 0.805x\text{N}_2 + 0.025\text{H}_2\text{O}$$

where the coefficient of H_2O is established from a balance on hydrogen. x and a can be determined as follows:

$$\text{C balance:} \quad \frac{0.65}{12} = 0.143x + 0.012x$$
$$x = 0.349 \text{ kg dg/kg fuel}$$
$$\text{O balance:} \quad \frac{0.12}{32}(2) + a(2) = x(0.143(2) + 0.04(2) + 0.012) + 0.025$$
$$a = 0.0748 \text{ kmol O}_2\text{/kg fuel}$$

As a check, we use a nitrogen balance:

$$\text{N balance:} \quad \frac{0.02}{28}(2) + 3.76(a)(2) = 0.805(x)(2)$$
$$a = 0.0745 \text{ kmol O}_2\text{/kg fuel}$$

This is as close as can be expected. The mass of air supplied is then

$$\frac{m_a}{m_f} = 0.0748\frac{\text{kmol O}_2}{\text{kg fuel}}\left(\frac{32 \text{ kg O}_2}{\text{kmol}}\right)\left(\frac{4.31 \text{ kg air}}{\text{kg O}_2}\right)$$
$$= 10.3 \text{ kg air/kg fuel}$$

(*b*) Same as first solution.

(*c*) From the values of part *a* we can determine the mole fraction of water vapor in the products as

$$y_v = \frac{N_v}{N_{dg} + N_v} = \frac{0.025}{x + 0.025} = \frac{0.025}{0.349 + 0.025}$$
$$= 0.067$$

and the dew point is found as in the first solution.

Example 11·4

PROBLEM STATEMENT

A fuel gas has the following volumetric analysis:

Fuel Analysis

Constituent	*Percent*
H_2	35
CH_4	35
C_2H_6	15
N_2	15
Total	100

It is burned with air supplied at 100 kPa, 30°C, 85 percent relative humidity, and the gas chromatograph (molar) analysis of the products is as follows:

GC Analysis

Constituent	*Percent*
CO_2	9.0
O_2	2.7
CO	0.4
N_2	87.9

Determine (*a*) the air–fuel (mass) ratio and (*b*) the dew point of the products.

SOLUTION

Analysis: To determine the air–fuel ratio we need not include the atmospheric moisture in the chemical equation because the water vapor undergoes no change during the reaction. Then, to determine the dew point of the products, we can add the atmospheric moisture, calculated separately, to the water vapor formed in combustion.

Calculations:

(*a*) Let a be the number of moles of oxygen supplied per mole of fuel and let x be the number of moles of dry gaseous products, also per mole of fuel. Then, the chemical equation written per mole of fuel is

$$0.35H_2 + 0.35CH_4 + 0.15C_2H_6 + 0.15N_2 + aO_2 + 3.76aN_2 \rightarrow x(0.09CO_2 + 0.027O_2 + 0.004CO + 0.879N_2) + (0.35 + 2(0.35) + 3(0.15))H_2O$$

The coefficient of H_2O was obtained by a hydrogen balance. x and a can be determined from carbon and oxygen balances, respectively:

$$\text{C balance:} \quad 0.35 + 2(0.15) = x(0.09 + 0.004)$$
$$x = 6.91 \text{ kmol } O_2\text{/kmol fuel}$$
$$\text{O balance:} \quad 2a = 2x(0.09) + 2x(0.027) + x(0.004) + 1.5$$
$$2a = 6.91[(2(0.09) + 2(0.027) + (0.004)] + 1.5$$
$$a = 1.57 \text{ kmol } O_2\text{/kmol fuel}$$

As a check, we use the nitrogen balance:

$$\text{N balance:} \quad 2(0.15) + 2(3.76)a = 2x(0.879) = 2(6.91)(0.879)$$
$$a = 1.57 \text{ kmol } O_2\text{/kmol fuel}$$

To convert from a molar to mass basis, we need the apparent molar mass of the fuel:

$$M_{\text{fuel}} = 0.35(2) + 0.35(16) + 0.15(30) + 0.15(28)$$
$$= 15.0 \text{ kg/kmol}$$

Then, the air–fuel (mass) ratio is

$$\frac{m_a}{m_f} = \frac{N_a M_a}{N_f M_f} = \frac{a(4.76)(28.966)}{1(15.0)} = \frac{1.57(4.76)(28.97)}{15.0}$$
$$= 14.4 \text{ kg air/kg fuel}$$

(*b*) To find the dew point of the products we must find the mole fraction of water vapor in the products. The amount of water created by the combustion process was calculated above. The amount of water introduced as atmospheric moisture can be calculated as follows:

$$p_a = p_m - p_v = p_m - \phi p_g$$
$$= 100 - 0.85(4.25)$$
$$= 96.4 \text{ kPa}$$
$$v_a = \frac{RT}{p_a} = \frac{(0.287)(303)}{96.4} = 0.90 \text{ m}^3\text{/kg}$$
$$\omega = \frac{v_a}{v_g}\phi = \frac{0.90 \text{ m}^3\text{/kg}}{32.9 \text{ m}^3\text{/kg}}(0.85) = 0.023 \text{ kg vapor/kg dry air}$$

Then, the number of moles of water vapor carried in with the air per mole of fuel is

$$\frac{N_{v,\text{air}}}{N_f} = 4.76a\frac{\text{kmol air}}{\text{kmol fuel}}\left(\frac{28.97 \text{ kg air}}{\text{kmol air}}\right)0.023\frac{\text{kg vapor}}{\text{kg air}}\left(\frac{1}{18}\frac{\text{kmol vapor}}{\text{kg vapor}}\right)$$

Substituting $a = 1.57$ from above, we find

$$\frac{N_{v,\text{air}}}{N_f} = 0.28\frac{\text{kmol vapor in atmospheric air}}{\text{kmol fuel}}$$

The mole fraction of water vapor in the products is

$$y_v = \frac{N_v}{N_{dg} + N_v} = \frac{(N_{v,\text{air}} + N_{v,\text{H}_2})/N_f}{N_{dg}/N_f + (N_{v,\text{air}} + N_{v,\text{H}_2})/N_f}$$

$$= \frac{(0.28 + 1.5)}{6.91 + (0.28 + 1.5)} = 0.20$$

$$p_v = y_v p_m = 0.20(100) = 20 \text{ kPa}$$

$$T_{dp} = T_{sat,20\text{ kPa}} = 60.1°\text{C}$$

11·3 The Combustion Energy Balance

An energy balance for a reaction process is no different from any other first law energy balance, since the first law takes the same form whether a chemical reaction occurs within the system or not. For example, the first law for a stationary ($\Delta KE = 0$, $\Delta PE = 0$) closed system is

$$Q - W = U_2 - U_1 \tag{3·10}$$

where the subscripts 1 and 2 refer to the initial and final states of the system. If the closed system is composed of a pure substance, then methods we have already discussed for calculating ΔU as a function of p, v, T, and c_v apply. If, on the other hand, the chemical composition of the system changes from state 1 to state 2 (i.e., the system does not consist of a pure substance), Eq. 3·10 is not altered, but the calculation of ΔU becomes more complex; it can no longer be calculated from other system properties alone. Therefore, we next examine the determination of ΔU (and ΔH and ΔpV) for systems that vary in chemical composition. In the discussions that follow, we will assume that all substances are ideal gases. This simplifies our discussion considerably, since U and H are then functions of temperature only. This assumption is valid in many circumstances; nonetheless, later in Sections 11·3·6 and 11·7, we discuss the effect of pressure when the gases are real.

11·3·1 Internal Energy of Reaction

Consider the application of the first law to a closed system that consists initially of 1 mole of CO and $\frac{1}{2}$ mole of O_2 that undergoes the following reaction:

$$CO + \tfrac{1}{2}O_2 \rightarrow CO_2$$

so that the system consists finally of 1 mole of CO_2. The first law is

$$Q - W = U_2 - U_1$$

which can alternatively be written as

$$Q - W = U_p - U_r$$

where p refers to the products and r refers to the reactants. Suppose that this process occurs in a rigid, thermally insulated tank so that $Q = 0$ and $W = 0$; then

$$U_p = U_r \tag{11·3}$$

If the reaction occurs at low pressure, both the products and the reactants can be treated as ideal gases, in which case the internal energy is a function of temperature only. Then, Eq. 11·3 indicates that 1 mole of CO_2 at the final temperature T_p has the same internal energy as 1 mole of CO plus $\frac{1}{2}$ mole of O_2 at the initial temperature T_r. We know from experience, however, that the temperature of the products, T_p, would be appreciably greater than the temperature of the reactants, T_r. In other words, $T_p > T_r$ when $U_p = U_r$. To understand the implications of this, let us examine this result from another perspective.

Suppose that the reaction occurs in a rigid tank ($W = 0$) that is *not* insulated. Then, application of the first law shows that

$$Q = U_p - U_r$$

Since we know that without heat transfer, $T_p > T_r$, it follows that if we are to make $T_p = T_r$, we must *remove* heat generated during the process. Recall from the sign convention that heat removed from a system is negative. If $Q < 0$, then $U_p - U_r < 0$ or $U_r > U_p$. Thus, *at the same temperature* a mixture of 1 mole of CO plus $\frac{1}{2}$ mole of O_2 has a *higher* internal energy than 1 mole of CO_2. They are different by an amount equal to Q or $(U_p - U_r)$, which we call the **internal energy of reaction**, ΔU_R, where the subscript R denotes that the internal energy change is associated with a chemical reaction. You should note that ΔU_R depends on the temperature at which the reaction was conducted. Therefore, the temperature must be specified for any value of ΔU_R.

Knowledge of ΔU_R allows us to relate the internal energies of two different substances, in this case, the products and reactants of a chemical reaction.

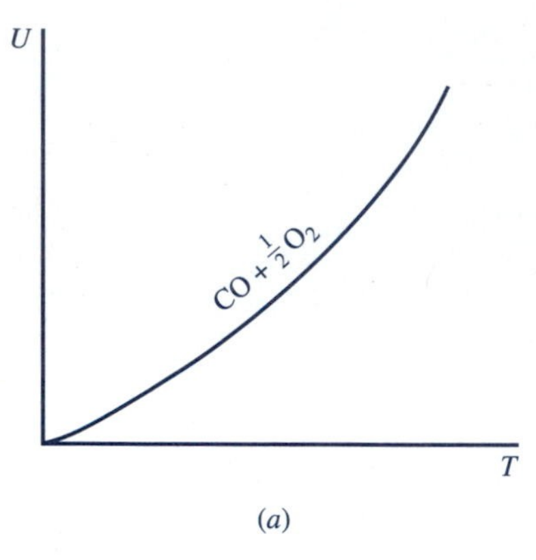

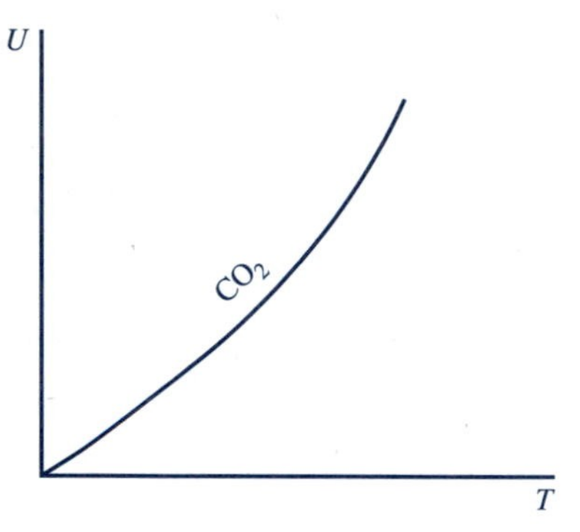

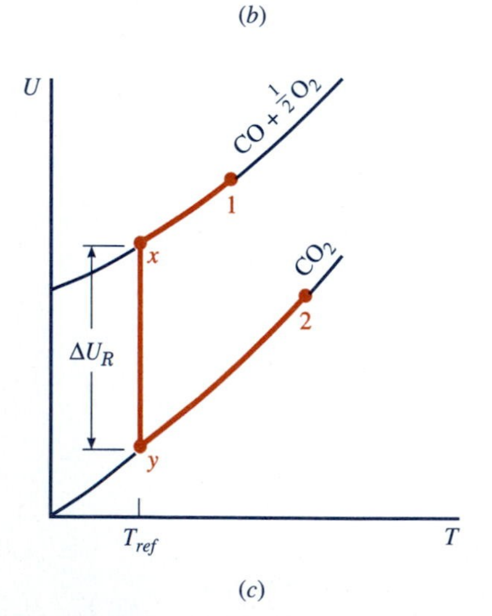

Figure 11·2 U versus T for reactants and products.

A Graphical Representation of ΔU_R. If we know the specific heats of both the products and the reactants, we can plot curves of U versus T. (Recall that we have assumed ideal-gas behavior, so U is a function of temperature only.) Figures 11·2*a* and *b* show these curves for $CO + \frac{1}{2}O_2$ and for CO_2, respectively. In each of these plots the temperature at which $U = 0$ is chosen arbitrarily. Up to now this practice has been sufficient, since we used information only about *changes* in the internal energy for pure substances. While this practice allows us to determine ΔU for $CO + \frac{1}{2}O_2$ and ΔU of CO_2 for any ΔT, it does not allow us to determine the *difference* between $U_{CO+\frac{1}{2}O_2}$ and U_{CO_2}. The internal energy of reaction, ΔU_R, solves this problem.

Recall that in the discussion above we allowed the reaction of CO and O_2 to occur in a closed rigid vessel and removed sufficient heat to bring the temperature of the products to that of the reactants. The amount of heat required to accomplish this, Q, is equal to the difference in internal energy, ΔU, between reactants and products at the same temperature, T_{ref}. We refer to this difference ΔU_R. We can use ΔU_R to replot the lines of Figs. 11·2*a* and *b* on the same axes such that the vertical distance between the two lines equals the measured value of ΔU_R at the reference temperature, T_{ref}. This fixes the curves with respect to one another, as shown in Fig. 11·2*c*. Also, Fig. 11·2*c* shows that ΔU_R changes as a function of temperature. A mental note of this point now will prevent confusion later.

Calculating ΔU for Any Process. We are now able to calculate ΔU for any process involving the reaction $CO + \frac{1}{2}O_2 \rightarrow CO_2$, regardless of the temperatures of the reactants and the products. In Fig. 11·2*c*, ΔU for a process that begins with the reactants at state 1 and ends with the products at state 2 is equivalent to

$$\begin{aligned} \Delta U &= U_2 - U_1 \\ &= (U_2 - U_y) + (U_y - U_x) + (U_x - U_1) \\ &= (U_2 - U_y) + \Delta U_R + (U_x - U_1) \end{aligned} \tag{11·4a}$$

Noting that $T_x = T_y = T_{ref}$, that is, states x and y are both at the temperature T_{ref} at which ΔU_R was measured, we may express Eq. 11·4*a* as

$$U_2 - U_1 = (U_2 - U_{ref})_p + \Delta U_R + (U_{ref} - U_1)_r \tag{11·4b}$$

If the products and reactants are mixtures of gases (see Chapter 9), then the above equation can be written as

$$U_2 - U_1 = \sum_{\text{products}} N(\bar{u}_2 - \bar{u}_{ref}) + \Delta U_R - \sum_{\text{reactants}} N(\bar{u}_1 - \bar{u}_{ref}) \tag{11·4c}$$

Internal energy change for a process of ideal gases that involves a chemical reaction

where $\bar{u}_{ref}$ is the molar internal energy of each constituent (either reactant or product) evaluated at the temperature, T_{ref}, at which ΔU_R is measured. Since we have assumed both the reactants and the products are ideal gases, we may evaluate the internal energy changes for the reactants and for the products by any of the methods listed in Section 4·2·1.

11·3·2 Enthalpy of Reaction

For a steady-flow process between states 1 and 2 in which $\Delta KE = \Delta PE = 0$, the first law states that

$$Q - W = H_2 - H_1$$

and this equation is the same whether a chemical reaction occurs or not. If a chemical reaction does occur, though, we must have data on the reaction in addition to enthalpy data on the reactants and products separately. This is directly analogous to the case for a closed system where we required ΔU_R. In the steady-flow system we require ΔH_R, the enthalpy change for the reaction at a constant temperature, T_{ref}. How then might we calculate ΔH_R?

For a system involving a chemical reaction, the above equation becomes

$$Q - W = H_p - H_r$$

where again p refers to the products and r refers to the reactants. If we fix the temperature of the reaction at T_{ref}, and measure Q and W, we can calculate the difference between the enthalpy of the products H_p and the enthalpy of the reactants H_r. This difference is called the **enthalpy of reaction**, ΔH_R. Thus

Enthalpy of reaction

$$\begin{aligned} Q - W &= H_p - H_r \\ &= (\Delta H_R)_{T_{ref}=T_p=T_r} \end{aligned}$$

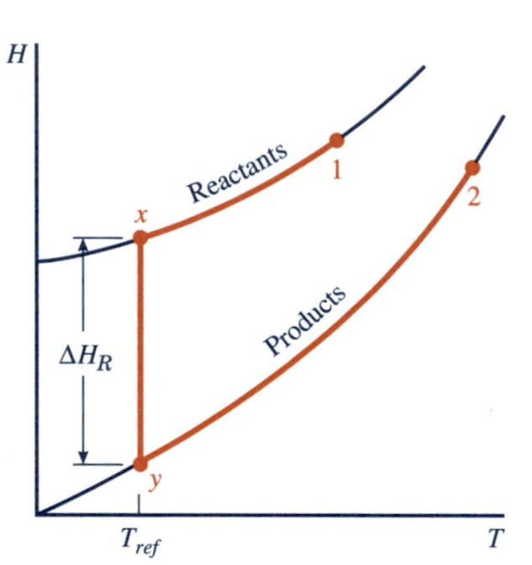

Figure 11·3 *H* versus *T* for reactants and products.

If we know the variation of H with T for the products and the reactants, then knowledge of ΔH_R allows us to plot Fig. 11·3. (Once again, note that we have assumed ideal-gas behavior, so H is a function of temperature only.) This figure shows the graphical representation of ΔH_R, and we can see clearly that ΔH_R varies with temperature. The H versus T curves are thereby fixed with respect to one another. This allows us to calculate the enthalpy change between reactants and products at arbitrary states 1 and 2, respectively, using the same method we used to find the internal energy change. Thus,

$$H_2 - H_1 = (H_2 - H_y)_p + \Delta H_R + (H_x - H_1)_r \qquad (11\cdot5a)$$

But $T_x = T_y = T_{ref}$, so

$$H_2 - H_1 = (H_2 - H_{ref})_p + \Delta H_R + (H_{ref} + H_1)_r \qquad (11\cdot5b)$$

As before, if the reactants and products are mixtures, then we can express the above equation as

Enthalpy change for a process of ideal gases that involves a chemical reaction

$$H_2 - H_1 = \sum_{\text{products}} N(\bar{h}_2 - \bar{h}_{ref}) + \Delta H_R + \sum_{\text{reactants}} N(\bar{h}_{ref} - \bar{h}_1) \qquad (11\cdot6)$$

where $\bar{h}_{ref}$ is the molar enthalpy of each constituent (either reactant or product) evaluated at the temperature, T_{ref}, at which ΔH_R is measured. The enthalpy changes for the reactants and for the products can be calculated by any of the methods listed in Section 4·2·1.

If we want to calculate the enthalpy change of a process *per mole* or *per unit mass* of one of the reactants, perhaps the fuel, we can designate the fuel by the subscript f, and write Eq. 11·6 as

$$\bar{h}_2 - \bar{h}_1 = \frac{H_2 - H_1}{N_f} = \sum_{\text{products}} \frac{N}{N_f}(\bar{h}_2 - \bar{h}_{ref}) + \Delta\bar{h}_R - \sum_{\text{reactants}} \frac{N}{N_f}(\bar{h}_1 - \bar{h}_{ref}) \quad (11\cdot6a)$$

or

$$h_2 - h_1 = \frac{H_2 - H_1}{m_f} = \sum_{\text{products}} \frac{m}{m_f}(h_2 - h_{ref}) + \Delta h_R - \sum_{\text{reactants}} \frac{m}{m_f}(h_1 - h_{ref}) \quad (11\cdot6b)$$

At this point, we reiterate: The form of the first law does not change if a chemical reaction occurs. However, *when* a reaction occurs, we need to account for the change in energy associated with the change in species. The system no longer consists of a pure substance and we account for this by using ΔU_R or ΔH_R.

11·3·3 Relating ΔU_R and ΔH_R

The relationship between ΔU_R or ΔH_R follows from the definition of enthalpy and is

$$\Delta H_R = \Delta U_R + \Delta(pV)_R$$

where

$$\Delta(pV)_R = (pV)_{\text{products}} - (pV)_{\text{reactants}}$$

at the same temperature T_{ref} at which ΔU_R and ΔH_R are known. When the reactants and the products are treated as ideal gases, the ideal-gas law allows us to write $\Delta(pV)_R$ as

$$\begin{aligned}\Delta(pV)_R &= \Delta(mRT)_R \\ &= \Delta(N\bar{R}T)_R \\ &= \bar{R}T_{ref}\,\Delta N_R\end{aligned}$$

where in the last step we have used the fact that the temperatures of the reactants and the products are the same and equal to T_{ref}.

This relationship may also be used if some of the reactants or products are solid or liquid and the rest are ideal gases if ΔN for only the gaseous constituents is used. This is possible because the molar volume of a solid or a liquid is usually negligible in comparison to that of a gas at the same pressure. Therefore, the volume change of the entire system is very nearly equal to the volume change of the gaseous constituents.

If a reaction occurs in a closed system at constant pressure and the only work done is that involved in changing the volume of the system, then

$$Q = \Delta U + W = \Delta U + p\,\Delta V$$
$$= \Delta U + \Delta\, pV = \Delta H$$

and if we use the directional notation for Q, and the reactants and products are at the reference temperature, T_{ref}, then

$$Q_{\text{out}} = -\Delta H_R$$

Consequently, $-\Delta H_R$ is also sometimes called the *heat of reaction at constant pressure*. Occasionally $-\Delta H_R$ may even be called simply the *heat of reaction*. Both of these names are somewhat misleading since $-\Delta H_R$ refers to the quantity of heat only under certain conditions. Nevertheless, this terminology is widely used.

Enthalpy of combustion

Higher heating value

Lower heating value

Other Nomenclature. For various types of reactions, *enthalpy of reaction* is frequently given other names, such as the **enthalpy of combustion** or *enthalpy of hydration*. The value of the enthalpy of combustion of any fuel containing hydrogen is called the higher enthalpy of combustion or **higher heating value** (*hhv*) if the H_2O in the products is a liquid. It is called the lower enthalpy of combustion or **lower heating value** (*lhv*) if the H_2O formed is a gas.

Example 11·5

PROBLEM STATEMENT

Verify the lower enthalpy of combustion of acetylene, C_2H_2, at 25°C as given in Table A·17 using the tabulated higher enthalpy of combustion as the starting point.

SOLUTION

The lower and higher enthalpies of combustion are given respectively by

$$-\Delta H_{lhv} = H_r - H_{p,\mathrm{H_2O}(g)} = m_{\mathrm{C_2H_2}}h_{\mathrm{C_2H_2}} + m_{\mathrm{O_2}}h_{\mathrm{O_2}} - m_{\mathrm{CO_2}}h_{\mathrm{CO_2}} - m_{\mathrm{H_2O}}h_{\mathrm{H_2O}(g)}$$
$$-\Delta H_{hhv} = H_r - H_{p,\mathrm{H_2O}(l)} = m_{\mathrm{C_2H_2}}h_{\mathrm{C_2H_2}} + m_{\mathrm{O_2}}h_{\mathrm{O_2}} - m_{\mathrm{CO_2}}h_{\mathrm{CO_2}} - m_{\mathrm{H_2O}}h_{\mathrm{H_2O}(l)}$$

where the products and reactants are described by

$$2\mathrm{C_2H_2} + 5\mathrm{O_2} \rightarrow 4\mathrm{CO_2} + 2\mathrm{H_2O}$$

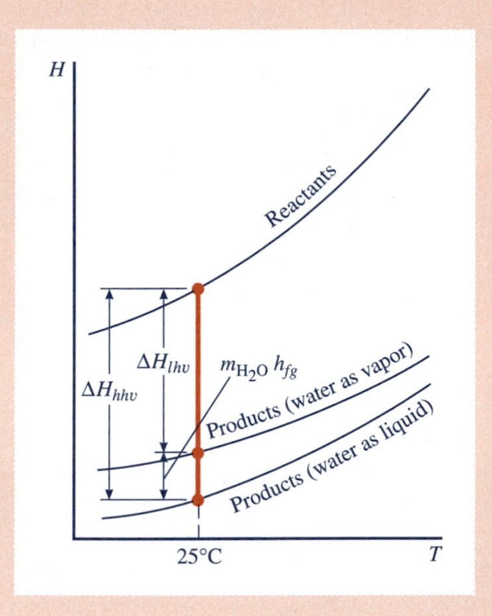

Therefore,

$$\begin{aligned}-\Delta H_{lhv} &= -\Delta H_{hhv} - m_{H_2O}[h_{H_2O}(g) - h_{H_2O}(l)]\\ &= -\Delta H_{hhv} - m_{H_2O}(h_g - h_f)\\ &= -\Delta H_{hhv} - m_{H_2O}h_{fg}\end{aligned}$$

or, per unit mass of C_2H_2,

$$-\Delta h_{lhv} = -\Delta h_{hhv} - \frac{m_{H_2O}}{m_{C_2H_2}} h_{fg}$$

Two moles of water are formed for 2 moles of acetylene; so the mass ratio is

$$\frac{m_{H_2O}}{m_{C_2H_2}} = \frac{N_{H_2O}}{N_{C_2H_2}} \frac{(M_{H_2O})}{(M_{C_2H_2})} = \frac{1}{1}\left(\frac{18\text{ kg/kmol}}{26\text{ kg/kmol}}\right) = 0.692 \frac{\text{kg } H_2O}{\text{kg } C_2H_2}$$

The value of h_{fg} at 25°C is obtained from the steam tables (or from the computer model using inputs of $T = 25°C$ and $x = 0$ and 1 to find h_f and h_g, respectively), so that

$$\begin{aligned}-\Delta h_{lhv} &= 49{,}914 - 0.692(2441.5\text{ kJ/kg})\\ &= 48{,}224\text{ kJ/kg } C_2H_2\end{aligned}$$

ALTERNATIVE SOLUTION

In the steady-flow complete combustion of acetylene with no work done and no change in kinetic or potential energy and the reactants and products both at 25°C, the heat removed is 49,914 kJ/kg C_2H_2 if all the water formed is condensed to a liquid. If all the water leaves as a vapor, the heat removed is less by the latent heat of vaporization, h_{fg}, of the water. Thus, by this direct reasoning, the lower enthalpy of combustion is

$$-\Delta h_{lhv} = -\Delta h_{hhv} - \frac{m_{H_2O}}{m_{C_2H_2}} h_{fg} = 49{,}914 - \frac{1 \text{ mol } H_2\left(18.02 \frac{\text{kg } H_2O}{\text{kmol}}\right)}{1 \text{ mol } C_2H_2\left(26.04 \frac{\text{kg } C_2H_2}{\text{kmol}}\right)} 2441.5$$

$$= 48{,}224 \text{ kJ/kg } C_2H_2$$

Example 11·6

PROBLEM STATEMENT

Starting from the higher enthalpy of combustion of gaseous *n*-octane as given in Table A·17, determine the higher enthalpy of combustion of liquid *n*-octane at the same temperature, 77 F.

SOLUTION

For liquid *n*-octane:

$$-\Delta H_{hhv,C_8H_{18}(l)} = m_{C_8H_{18}} h_{C_8H_{18}(l)} + m_{O_2} h_{O_2} - m_{CO_2} h_{CO_2} - m_{H_2O} h_{H_2O}$$

And for gaseous *n*-octane:

$$-\Delta H_{hhv,C_8H_{18}(g)} = m_{C_8H_{18}} h_{C_8H_{18}(g)} + m_{O_2} h_{O_2} - m_{CO_2} h_{CO_2} - m_{H_2O} h_{H_2O}$$

Combining these two equations, we find

$$-\Delta H_{hhv,C_8H_{18}(l)} = -\Delta H_{hhv,C_8H_{18}(g)} - m_{C_8H_{18}}(h_{C_8H_{18}(g)} - h_{C_8H_{18}(l)})$$

or, per unit mass of *n*-octane,

$$-\Delta h_{hhv,C_8H_{18}(l)} = -\Delta h_{hhv,C_8H_{18}(g)} - h_{fg,C_8H_{18}} = 20{,}745 - 156$$

$$= 20{,}589 \text{ B/lbm } C_8H_{18}$$

11·3·4 Calculating the Enthalpy of Reaction from the Enthalpy of Formation

We have demonstrated that knowledge of ΔU_R or ΔH_R is necessary for first law calculations whenever the process involves a chemical reaction. This means that to apply the first law in combustion applications we would require a ΔU_R or ΔH_R value for *every conceivable reaction* that might be of engineering use. This is an extravagant request; fortunately, there is an alternative.

In the discussion that follows, we use enthalpies only, since in most combustion applications the system is open. An important application of closed system combustion analyses, however, is the use of bomb calorimeters to measure the properties of fuels. For such analyses, we would require internal energy changes.

Enthalpy of formation

Enthalpy of formation Δh_f is the enthalpy of reaction for the formation of a substance from its elements in their most stable forms. By stable forms of elements we mean forms such as H_2 and O_2 for hydrogen and oxygen instead of H and O, the monatomic forms. The stable form of carbon, C, for example, is graphite instead of diamond. Therefore, the enthalpy of formation of any element in its most stable form is zero.

Expressing enthalpy of reaction in terms of enthalpies of formation

Suppose we are examining the energy balance for the complete combustion of methane in oxygen:

$$CH_4 + 2O_2 \rightarrow CO_2 + 2H_2O$$

The enthalpy of reaction is

$$\begin{aligned} \Delta H_R &= (H_p - H_r)_{T_p=T_r=T_{ref}} \\ &= (H_{CO_2} + H_{H_2O}) - (H_{CH_4} + H_{O_2}) \\ &= (N_{CO_2}\bar{h}_{CO_2} + N_{H_2O}\bar{h}_{H_2O}) - (N_{CH_4}\bar{h}_{CH_4} + N_{O_2}\bar{h}_{O_2}) \\ &= (1 \text{ mol } CO_2 \cdot \bar{h}_{CO_2} + 2 \text{ mol } H_2O \cdot \bar{h}_{H_2O}) \\ &\quad - (1 \text{ mol } CH_4 \cdot \bar{h}_{CH_4} + 2 \text{ mol } O_2 \cdot \bar{h}_{O_2}) \end{aligned} \qquad (b)$$

The subscript $T_p = T_r = T_{ref}$ is a reminder that all properties must be evaluated at the temperature T_{ref} and all must be on the same scale (i.e., we must have already fixed the hT scales with respect to one another.) If we know the value of $\bar{h}$ for each constituent, we can calculate ΔH_R. We use the *enthalpy of formation* to place the $\bar{h}$'s on a common scale. For each of the constituents, a reaction describing its formation from its elements in their most stable forms can be written as follows:

$$\begin{aligned} C + O_2 &\rightarrow CO_2 \\ H_2 + \tfrac{1}{2}O_2 &\rightarrow H_2O \\ C + 2H_2 &\rightarrow CH_4 \end{aligned}$$

(No equation describing the formation of the reactant oxygen is needed. By definition, the enthalpy of formation of O_2 would be based on the reaction $O_2 \rightarrow O_2$. Thus, $\Delta h_{f,O_2} = 0$ since O_2 is already an element in its most stable form.) The enthalpy of reaction for each of the equations is

$$\begin{aligned} \Delta H_{R,C+O_2\rightarrow CO_2} &= H_{CO_2} - (H_C + H_{O_2}) \\ \Delta H_{R,H_2+\frac{1}{2}O_2\rightarrow H_2O} &= H_{H_2O} - (H_{H_2} + H_{O_2}) \\ \Delta H_{R,C+2H_2\rightarrow CH_4} &= H_{CH_4} - (H_C + H_{H_2}) \end{aligned}$$

But, recall that we *defined* the *enthalpy of formation* as the enthalpy change during a reaction wherein a substance is formed from its elements in their stable forms. Therefore, the above three enthalpy of reaction equations represent identically the enthalpy of formation for CO_2, H_2O, and CH_4, respectively. Thus,

$$\Delta H_{f,CO_2} = H_{CO_2} - (H_C + H_{O_2})$$
$$\Delta H_{f,H_2O} = H_{H_2O} - (H_{H_2} + H_{O_2})$$
$$\Delta H_{f,CH_4} = H_{CH_4} - (H_C + H_{H_2})$$

and if we express the above equations using molar quantities, we have

$$\begin{aligned} \Delta H_{f,CO_2} &= N_{CO_2}\,\Delta\bar{h}_{f,CO_2} \\ &= N_{CO_2}\bar{h}_{CO_2} - (N_C\bar{h}_C + N_{O_2}\bar{h}_{O_2}) \end{aligned} \tag{c}$$

$$\begin{aligned} \Delta H_{f,H_2O} &= N_{H_2O}\,\Delta\bar{h}_{f,H_2O} \\ &= N_{H_2O}\bar{h}_{H_2O} - (N_{H_2}\bar{h}_{H_2} + N_{O_2}\bar{h}_{O_2}) \end{aligned} \tag{d}$$

$$\begin{aligned} \Delta H_{f,CH_4} &= N_{CH_4}\,\Delta\bar{h}_{f,CH_4} \\ &= N_{CH_4}\bar{h}_{CH_4} - (N_C\bar{h}_C + N_{H_2}\bar{h}_{H_2}) \end{aligned} \tag{e}$$

We can then solve equations *c*, *d*, and *e* for the quantities needed in equation *b*:

$$\begin{aligned} \bar{h}_{CO_2} &= \left(\frac{N_{CO_2}\,\Delta\bar{h}_{f,CO_2} + N_C\bar{h}_C + N_{O_2}\bar{h}_{O_2}}{N_{CO_2}}\right) \\ &= \left[\frac{(1\text{ mol }CO_2)\Delta\bar{h}_{f,CO_2} + (1\text{ mol C})\bar{h}_C + (1\text{ mol }O_2)\bar{h}_{O_2}}{(1\text{ mol }CO_2)}\right] \\ &= \Delta\bar{h}_{f,CO_2} + \bar{h}_C + \bar{h}_{O_2} \end{aligned} \tag{f}$$

$$\begin{aligned} \bar{h}_{H_2O} &= \left(\frac{N_{H_2O}\,\Delta\bar{h}_{f,H_2O} + N_{H_2}\bar{h}_{H_2} + N_{O_2}\bar{h}_{O_2}}{N_{H_2O}}\right) \\ &= \left[\frac{(1\text{ mol }H_2O)\Delta\bar{h}_{f,H_2O} + (1\text{ mol }H_2)\bar{h}_{H_2} + (\frac{1}{2}\text{ mol }O_2)\bar{h}_{O_2}}{(1\text{ mol }H_2O)}\right] \\ &= \Delta\bar{h}_{f,H_2O} + \bar{h}_{H_2} + \tfrac{1}{2}\bar{h}_{O_2} \end{aligned} \tag{g}$$

$$\begin{aligned} \bar{h}_{CH_4} &= \left(\frac{N_{CH_4}\,\Delta\bar{h}_{f,CH_4} + N_C\bar{h}_C + N_{H_2}\bar{h}_{H_2}}{N_{CH_4}}\right) \\ &= \left[\frac{(1\text{ mol }CH_4)\Delta\bar{h}_{f,CH_4} + (1\text{ mol C})\bar{h}_C + (2\text{ mol }H_2)\bar{h}_{H_2}}{(1\text{ mol }CH_4)}\right] \\ &= \Delta\bar{h}_{f,CH_4} + \bar{h}_C + 2\bar{h}_{H_2} \end{aligned} \tag{h}$$

where both sides of equation *f* are *per mole of* CO_2. Likewise, equations *g* and *h* are *per mole of* H_2O *and* CH_4, respectively. It is important to note that the numerical value substituted for the various *N*'s *must* refer to the proper reaction. Substitution into equation *b* then yields

$$\Delta H_R = (1 \text{ mol } CO_2 \cdot \bar{h}_{CO_2} + 2 \text{ mol } H_2O \cdot \bar{h}_{H_2O})$$
$$- (1 \text{ mol } CH_4 \cdot \bar{h}_{CH_4} + 2 \text{ mol } O_2 \cdot \bar{h}_{O_2})$$
$$= (1 \text{ mol } CO_2\{\Delta\bar{h}_{f,CO_2} + \bar{h}_C + \bar{h}_{O_2}\}$$
$$+ 2 \text{ mol } H_2O\{\Delta\bar{h}_{f,H_2O} + \bar{h}_{H_2} + \tfrac{1}{2}\bar{h}_{O_2}\})$$
$$- (1 \text{ mol } CH_4\{\Delta\bar{h}_{f,CH_4} + \bar{h}_C + 2\bar{h}_{H_2}\} + 2 \text{ mol } O_2\{\bar{h}_{O_2}\})$$

Cancellation of terms gives the following result:

$$\Delta H_R = (1 \text{ mol } CO_2 \cdot \Delta\bar{h}_{f,CO_2} + 2 \text{ mol } H_2O \cdot \Delta\bar{h}_{f,H_2O})$$
$$- (1 \text{ mol } CH_4 \cdot \Delta\bar{h}_{f,CH_4})$$

Thus we have expressed ΔH_R solely in terms of the $\Delta\bar{h}_f$ values of the products and the reactants. This procedure can be generalized so that for any reaction

$$\Delta H_R = \sum_{\text{products}} N\,\Delta\bar{h}_f - \sum_{\text{reactants}} N\,\Delta\bar{h}_f \qquad (11\cdot7)$$

Enthalpy of reaction in terms of enthalpy of formation

In using this relationship remember two points:

- The enthalpy of formation of an element in its most stable form is zero.
- All $\Delta\bar{h}_f$ values must be evaluated at the same temperature.

This result, although complex in its derivation, seems intuitive. It is an important relationship because it lets us evaluate the enthalpy of reaction for *any* reaction from the $\Delta\bar{h}_f$ values of the constituents involved.

11·3·5 The Enthalpy Change in a Combustion Process: General Form

We can use the result above for calculating the enthalpy of reaction from the enthalpy of formation as follows:

$$H_2 - H_1 = \sum_{\text{products}} N(\bar{h}_2 - \bar{h}_{ref}) + \Delta H_R - \sum_{\text{reactants}} N(\bar{h}_1 - \bar{h}_{ref})$$

$$= \sum_{\text{products}} N(\bar{h}_2 - \bar{h}_{ref}) + \left(\sum_{\text{products}} N\,\Delta\bar{h}_f - \sum_{\text{reactants}} N\,\Delta\bar{h}_f\right)$$

$$- \sum_{\text{reactants}} N(\bar{h}_1 - \bar{h}_{ref}) \qquad (11\cdot6)$$

Enthalpy change for a chemical reaction of ideal gases in terms of the enthalpies of formation

$$H_2 - H_1 = \sum_{\text{products}} N(\bar{h}_2 - \bar{h}_{ref} + \Delta\bar{h}_{f,ref}) - \sum_{\text{reactants}} N(\bar{h}_1 - \bar{h}_{ref} + \Delta\bar{h}_{f,ref}) \quad (11\cdot8)$$

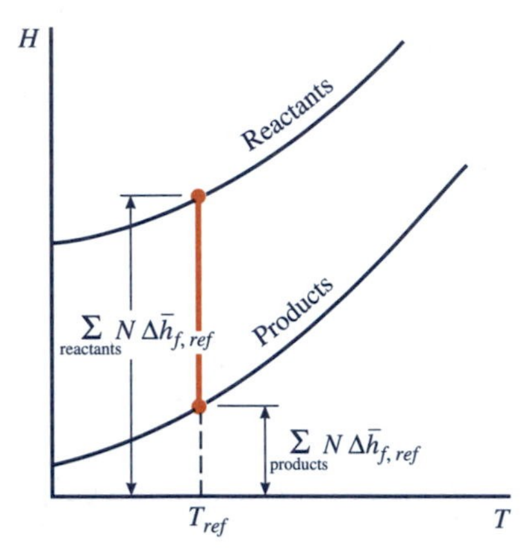

Figure 11·4 ΔH_R in terms of enthalpies of formation for an ideal-gas reaction.

which is the general form for calculating the enthalpy change for a combustion process where the products and the reactants are mixtures of ideal gases.

As we describe in the next section, the notation used in references for combustion data is frequently inconsistent and incomplete. This can be a source of frustration for students and practicing engineers as well. The published data for the enthalpy of formation appear as $\Delta\bar{h}_f^\circ$, where the superscript $^\circ$ refers to the standard pressure (usually 100 kPa), and $\Delta\bar{h}_f^\circ$ is listed as a function of temperature. To use these data, you must realize that for ideal gases, *enthalpy is a function of temperature only*. Because Eq. 11·8 is written for ideal gases, no indication of the pressure is made. Therefore, in the case of an ideal gas, the superscript $^\circ$ is superfluous.

For real gases, on the other hand, enthalpy is not a function of temperature only; therefore, Eq. 11·7 would become

General equation for enthalpy change for a chemical reaction of real gases

$$H_2 - H_1 = \sum_{\text{products}} N(\bar{h}_2 - \bar{h}^\circ_{ref} + \Delta\bar{h}^\circ_{f,ref}) - \sum_{\text{reactants}} N(\bar{h}_1 - \bar{h}^\circ_{ref} + \Delta\bar{h}^\circ_{f,ref}) \quad (11\cdot9)$$

Remember, if you find yourself confused, return to the basic definitions to analyze the problem. In such circumstances, it is often helpful to draw a picture like that shown in Fig. 11·3 or 11·4.

11·4 Combustion Data

The most common reference pressure for combustion data is 100 kPa, and we indicate this pressure with a superscript $^\circ$. Combustion notation is cumbersome even without this additional symbol, so let us review the meaning of each component.

- An overbar indicates a molar quantity.
- The subscript ***R*** refers to a ***reaction***.
- The subscript ***f*** denotes a ***formation*** reaction.
- The subscript ***ref*** indicates that the quantity is evaluated at the ***reference*** temperature, usually 25°C.
- The superscript ° indicates that the quantity is evaluated at the standard state pressure, usually 100 kPa.

Some examples are summarized below:

$\Delta \bar{h}^{\circ}_{f,ref}$	The molar enthalpy of formation at the standard pressure and reference temperature	Superscript ° is relevant only when the gases are real; it is superfluous for ideal gases.
$\bar{h}^{\circ}_{ref}$	The molar enthalpy evaluated at the standard pressure and reference temperature	Superscript ° is relevant only when the gases are real; it is superfluous for ideal gases.
$\Delta \bar{h}_{f,ref}$	The molar enthalpy of formation at the reference temperature	Used for ideal gases, where the enthalpy is a function of temperature only.
$\bar{h}_{ref}$	The molar enthalpy evaluated at the reference temperature	Used for ideal gases, where the enthalpy is a function of temperature only.
ΔH_R	The enthalpy of reaction	Used for ideal gases, where it is a function of temperature.

For any chemical reaction, ΔH_R might be determined by calorimetric measurements, either by having the reactants and products at the reference temperature or by having them at other temperatures and applying corrections to the measurements. This is called a first-law method. In practice, better methods are used. One, called the second-law method, is based on a relationship involving ΔH_R and the equilibrium constant that is introduced in Chapter 12. A third-law method, based on the third law of thermodynamics that is mentioned in Section 11·7, has further advantages. Usually, published values of ΔH_R and related quantities are based on two or more methods to ensure accuracy and consistency.

It is unlikely that for every problem all the data will be available at the same temperature and pressure. In this case, you should note that the difference between enthalpies of reaction (including enthalpies of formation) measured at temperatures within a few degrees of each other can usually be neglected and pressure increases of a few atmospheres have a negligible effect on enthalpies of reaction.

Further, in using the data from standard sources, you may find the following guidelines useful:

1. *Notation*: Enthalpy of reaction is usually expressed per mole or per unit mass of one of the reactants, but, contrary to convention, the symbol ΔH_R instead of Δh_R (or ΔH_f instead of Δh_f) is widely used. Sometimes the superscript ° that we have used to indicate the standard pressure is used to indicate a standard pressure and temperature.
2. *Heat of reaction*: The heat of reaction is usually defined as $-\Delta H_R$ because many years ago it was decided that the heat of reaction for an exothermic reaction should be a positive number. In some publications, however, $+\Delta H_R$ is called the heat of reaction. You should look at the value presented for the heat of reaction for a reaction you know is exothermic to find out which convention is used.
3. *Phases*: The enthalpy of reaction for any chemical reaction depends on the phases in which the reactants and products appear. Therefore, lists of Δh_R° values indicate the phase of each constituent by an *s*, *l*, or *g* (for solid, liquid, or gas) and such notation should always be used where there might be doubt as to the phase.
4. *Small discrepancies*: Small discrepancies among tables may be due to differences in the molar mass of constituents or the temperature scale. (Does 298 K mean exactly 298 K or 298.15 K?)
5. *Units*: Unit conversions are occasional sources of discrepancies. For example, the difference between the thermochemical calorie (= 4.186 kJ) and the international steam table (or IT) calorie (= 4.1868 kJ) must be observed. Sometimes an entire table has been converted from one set of units to another by using the original equations to recalculate values in the new units. In other cases, unit conversions were made on individual table values, and this may have involved interpolation either before or after the conversion. The interpolation may have been linear or it may have involved several neighboring values.

Ideally, documentation covering these points should accompany every published compilation of property values. However, the documentation is often overwhelmingly copious, sparse, or missing altogether.

Example 11·7

PROBLEM STATEMENT

Determine the enthalpy of reaction for $CO + \frac{1}{2}O_2 \rightarrow CO_2$ at 1000 K. How does the enthalpy of reaction vary from 500 K to 3000 K?

SOLUTION

An *HT* diagram is first made. Then, the enthalpy of reaction can be calculated from the enthalpies of formation as follows:

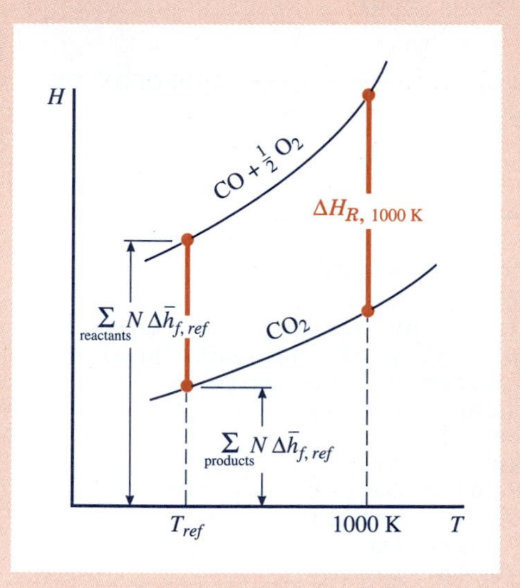

$$\Delta H_{R,1000\text{ K}} = \sum_{\text{products}} N(\bar{h}_{1000\text{ K}} - \bar{h}_{ref} + \Delta\bar{h}_{f,ref}) - \sum_{\text{reactants}} N(\bar{h}_{1000\text{ K}} - \bar{h}_{ref} + \Delta\bar{h}_{f,ref})$$

Substituting values from Tables A·15 and A·17, we find

$$\begin{aligned} &= 1(42{,}763.1 - 9364.0 - 393{,}522) - 1(30{,}359.8 - 8671.0 - 110{,}530) \\ &\quad - 0.5(31{,}384.4 - 8683.0 - 0) \\ &= -282{,}600 \text{ kJ/kmol CO} \end{aligned}$$

Alternatively, we could use a computer to solve this problem. In fact, in this case, where the solution is desired for multiple inputs, a computer solution is preferable. The Rule and Variable sheets appear below as well as a final plot of the solution.

The Rule sheet shows that we have first entered the needed equations and then loaded the data model for properties of ideal-gas mixtures to provide values of h for the three constituents. We enter values of h and $\Delta\bar{h}_{f,ref}$ from Tables A·15 and A·17. We could alternatively obtain the values from the computer model for enthalpy changes of combustion processes.

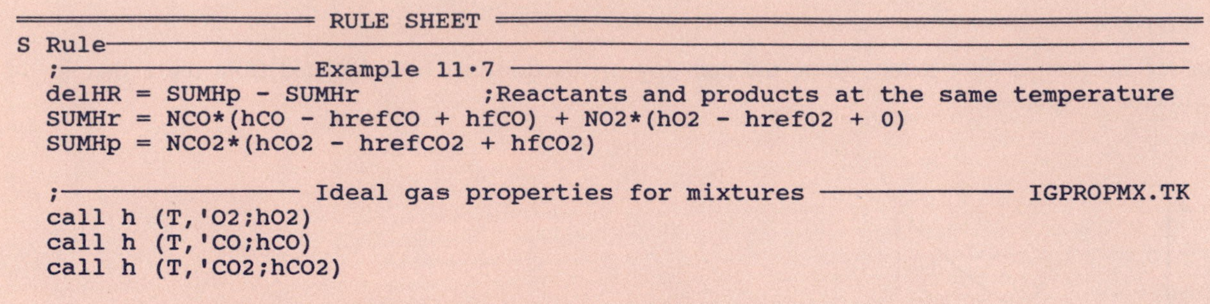

```
═══════════════════ RULE SHEET ═══════════════════════════════════════
S Rule──────────────────────────────────────────────────────────────────
  ;─────────────── Example 11·7 ─────────────────────────────────────
  delHR = SUMHp - SUMHr        ;Reactants and products at the same temperature
  SUMHr = NCO*(hCO - hrefCO + hfCO) + NO2*(hO2 - hrefO2 + 0)
  SUMHp = NCO2*(hCO2 - hrefCO2 + hfCO2)

  ;─────────────── Ideal gas properties for mixtures ──────────── IGPROPMX.TK
  call h (T,'O2;hO2)
  call h (T,'CO;hCO)
  call h (T,'CO2;hCO2)
```

```
═══════════════════ VARIABLE SHEET ═══════════════════════════════════
St Input──── Name──── Output──── Unit───── Comment─────────────────────────
                                 ───────── * E11-7.TK ────── Example 11·7 ──────
L            delHR    -282600    kJ/kmol   Enthalpy of reaction/kmol CO
L            SUMHp    -360100    kJ/kmol   Absolute enthalpy of products/kmol CO
L            SUMHr    -77490     kJ/kmol   Absolute enthalpy of reactants/kmol CO
   1         NCO2                kmol      Number of moles CO2 per mole CO
L            hCO2     42760      kJ/kmol   Enthalpy (molar basis)
   9364      hrefCO2             kJ/kmol   hCO2 at 298.15 K
   -393522   hfCO2               kJ/kmol   Enthalpy of formation of CO2
   0.5       NO2                 kmol      Number of moles O2 per mole CO
L            hO2      31380      kJ/kmol   Enthalpy (molar basis)
   1         NCO                 kmol      Number of moles CO
L            hCO      30360      kJ/kmol   Enthalpy (molar basis)
   8671      hrefCO              kJ/kmol   hCO at 298.15 K
   -110530   hfCO                kJ/kmol   Enthalpy of formation of CO
   8683      hrefO2              kJ/kmol   hO2 at 298.15 K
                                 ── Model: * IGPROPMX.TK ──── Ideal gas mixture
                                                              properties ─────
L  1000      T                   K         Temperature
```

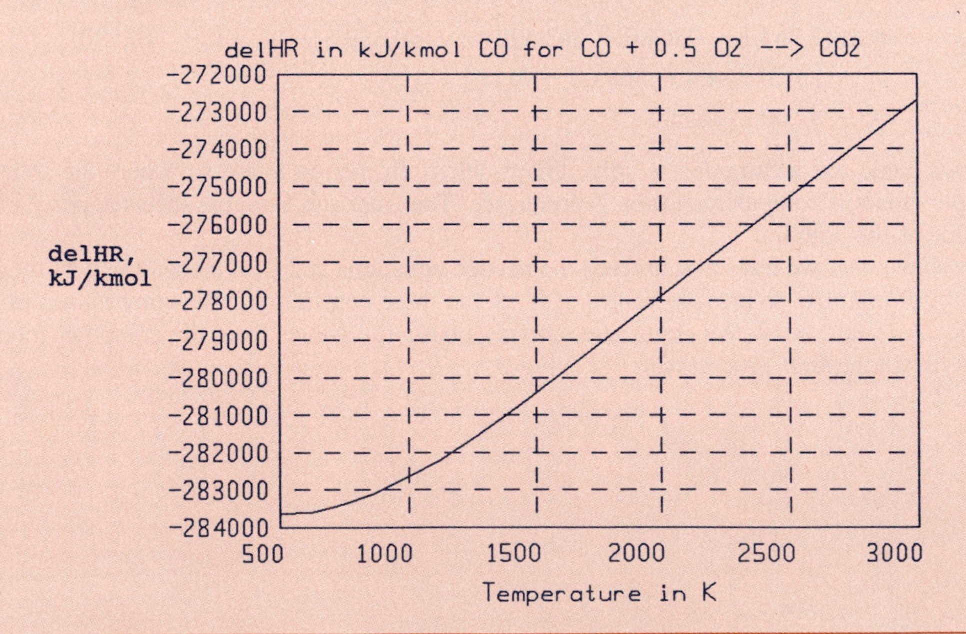

Example 11·8

PROBLEM STATEMENT

Determine the amount of heat transfer per kilogram of fuel during the complete combustion of methane, CH_4, in an open steady-flow burner without excess air if the methane enters at 100°C, the air enters at 10°C, and the products leave at 600°C.

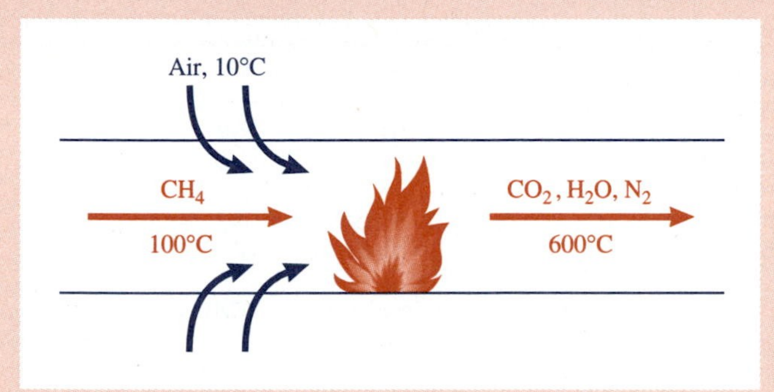

SOLUTION

The reaction equation is

$$CH_4 + 2O_2 + 2(3.76)N_2 \rightarrow CO_2 + 2H_2O + 2(3.76)N_2$$

There is no mention of pressure. Therefore, we assume ideal-gas behavior. If we neglect changes in kinetic energy, the energy balance is

$$\begin{aligned} Q &= H_2 - H_1 \\ &= \sum_{\text{products}} N(\bar{h}_2 - \bar{h}_{ref} + \Delta\bar{h}_{f,ref}) - \sum_{\text{reactants}} N(\bar{h}_1 - \bar{h}_{ref} + \Delta\bar{h}_{f,ref}) \\ &= N_{CO_2}(\bar{h}_2 - \bar{h}_{ref} + \Delta\bar{h}_{f,ref})_{CO_2} + N_{H_2O}(\bar{h}_2 - \bar{h}_{ref} + \Delta\bar{h}_{f,ref})_{H_2O} \\ &\quad + N_{N_2,p}(\bar{h}_2 - \bar{h}_{ref} + \Delta\bar{h}_{f,ref})_{N_2} - N_{CH_4}(\bar{h}_1 - \bar{h}_{ref} + \Delta\bar{h}_{f,ref})_{CH_4} \\ &\quad - N_{O_2}(\bar{h}_1 - \bar{h}_{ref} + \Delta\bar{h}_{f,ref})_{O_2} - N_{N_2,r}(\bar{h}_1 - \bar{h}_{ref} + \Delta\bar{h}_{f,ref})_{N_2} \end{aligned}$$

Since nitrogen is present in both the reactants and the products, we use symbols $N_{N_2,r}$ and $N_{N_2,p}$ to differentiate them, even though in this particular example the amount of nitrogen is the same in reactants and products. Since the products leave at 600°C, the water will be a vapor and we should use the Δh_f for the vapor phase. Obtaining enthalpy values from Tables A·15 and A·17 and canceling the appropriate nitrogen terms, we find

$$\begin{aligned} Q &= 1(35{,}978 - 9364 - 393{,}522) + 2(30{,}774 - 9904 - 241{,}826) \\ &\quad + 2(3.76)(26{,}032 - 8670.0 + 0) - 1(12{,}820 - 10{,}024 \\ &\quad - 74{,}873) - 2(8243 - 8683 + 0) - 2(3.76)(8233 - 8670 + 0) \\ &= -602{,}000 \text{ kJ} \end{aligned}$$

This calculation was made for 1 kmol CH_4, so the result is

$$q = \frac{Q}{m_{CH_4}} = \frac{Q}{N_{CH_4}M_{CH_4}} = \frac{-602{,}000}{1(16.04)} = -37{,}530 \text{ kJ/kg}$$

Alternatively, we could use the computer model for enthalpy changes of combustion processes to solve this problem. The Variable and Rule sheets are shown here.

```
==================== RULE SHEET ==========================================
S Rule----------------------------------------------------------------------
  ;---------------- Example 11·8 -------------------------------------------
  Q = delH                                  ;First law for steady flow combustion
                                            ;with W = 0, delKE = delPE = 0
  delH = Habp - Habr
  Habp = NpCO2*hbarab('CO2,Tp)+ NpH2O*hbarab('H2O,Tp)+ NpN2*hbarab('N2,Tp)
  Habr=NrCH4*hbarab('CH4,TCH4)+ NrO2*hbarab('O2,TO2)+ NrN2*hbarab('N2,TN2)
  q = Q/(NrCH4*M('CH4))
  ;------------- Absolute enthalpies of ideal gases ----------- CMBDELH1.TK

==================== VARIABLE SHEET =======================================
St Input---- Name---- Output---- Unit------ Comment---------------------------
                                 ---------- * E11-8.TK ----- Example 11·8 -----
                                 -- Model:  * CMBDELH1.TK -- Absolute enthalpies -
             q        -37530     kJ/kg      Heat transfer per unit mass
             Q        -602000    kJ         Heat transfer
             delH     -602000    kJ         Enthalpy change for the reaction
             Habp     -678286    kJ         Absolute enthalpy of products
             Habr     -76244     kJ         Absolute enthalpy of reactants
   600       Tp                  C          Temperature of products
   1         NrCH4               kmol       Amount of CH4 in reactants
   100       TCH4                C          Temperature of CH4 in reactants
   2         NrO2                kmol       Amount of O2 in reactants
   10        TO2                 C          Temperature of O2 in reactants
   7.52      NrN2                kmol       Amount of N2 in reactants
   10        TN2                 C          Temperature of N2 in reactants
   1         NpCO2               kmol       Amount of CO2 in products
   2         NpH2O               kmol       Amount of H2O in products
   7.52      NpN2                kmol       Amount of N2 in products
```

11·5 Maximum Adiabatic Combustion Temperature

Maximum adiabatic combustion temperature

Often for design purposes we need to know the temperature of the products of a combustion reaction. The ideal maximum achievable temperature is the **maximum adiabatic combustion temperature**, which is defined as the temperature that results from combustion that proceeds to the maximum extent possible combustion with no heat transfer and no work done. We can use this temperature as an upper limit for the actual combustion temperature. In some cases, when the system is adiabatic and the reaction is nearly complete, we can use the maximum adiabatic combustion temperature as an approximation of the actual combustion temperature. For example, the combustion in a rocket motor or in a gas turbine combustion chamber occurs nearly adiabatically.

In Chapter 12 we will see that the maximum adiabatic combustion temperature is not actually achieved.

We can calculate the maximum adiabatic combustion temperature as follows. For a steady-flow adiabatic combustion reaction in which $W = 0$, $\Delta KE = 0$, and $\Delta PE = 0$, application of the first law shows that

$$H_2 = H_1$$

or

$$H_2 - H_1 = 0$$

so that the states 1 and 2 of the reactants and products are as shown in Fig. 11·5. As we have shown previously, this can be expressed as

$$0 = \sum_{\text{products}} N(\bar{h}_2 - \bar{h}^\circ_{ref} + \Delta\bar{h}^\circ_{f,ref}) - \sum_{\text{reactants}} N(\bar{h}_1 - \bar{h}^\circ_{ref} + \Delta\bar{h}^\circ_{f,ref}) \quad (11\cdot9)$$

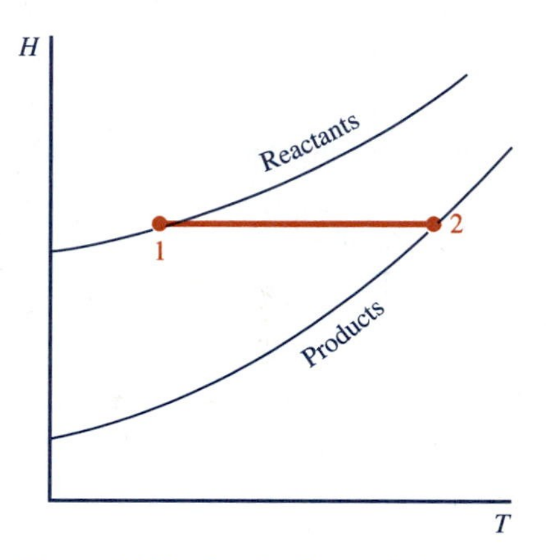

Figure 11·5 Steady-flow adiabatic combustion.

Consider a complete combustion reaction for which the initial properties of the reactants and the enthalpies of formation are known. Using the above equation, we can then solve for $\Sigma_{\text{products}} N(h_2 - \bar{h}^\circ_{ref})$. The number of moles of the product constituents can be obtained from stoichiometric calculations, and for each constituent we can write an expression for $h_2 - \bar{h}^\circ_{ref}$. If there is only one product, we can solve directly for $h_2 - \bar{h}^\circ_{ref}$ and, using the enthalpy at the end state, solve for the temperature. If there are two or more products and enthalpy tables are being used, a trial-and-error or graphical solution is necessary. If the hT equations of the products are known or equations are fitted to the tabular data in the range of interest, a direct solution is possible. The solution technique must be chosen on the basis of the computational tools available and the number of similar problems to be solved.

Actually, the maximum adiabatic combustion temperature as calculated in the following example will not be reached, even though the process is adiabatic and the reactants are well mixed, for reasons that are explained qualitatively in the following section and are treated in more detail in Chapter 12.

Example 11·9

PROBLEM STATEMENT

Calculate the maximum adiabatic combustion temperature for the steady-flow burning of methane, CH_4, at 24°C with 100 percent excess air at 150°C.

SOLUTION

Since the maximum temperature is desired, complete combustion will be considered. Water in the products will be a vapor. The reaction equation is

$$CH_4 + 2(2)O_2 + 2(2)3.76N_2 \rightarrow CO_2 + 2H_2O(g) + 1(2)O_2 + 2(2)3.76N_2$$

We know immediately that the solution to this problem involves iteration. We *could* do this computation by hand, but an iterative equation solver should be faster. Therefore, we enter the energy balance

$$\begin{aligned} Q = 0 = {} & N_{CO_2}(\bar{h}_2 - \bar{h}_{ref} + \Delta\bar{h}_{f,ref})_{CO_2} + N_{H_2O}(\bar{h}_2 - \bar{h}_{ref} + \Delta\bar{h}_{f,ref})_{H_2O} \\ & + N_{O_2,p}(\bar{h}_2 - \bar{h}_{ref} + \Delta\bar{h}_{f,ref})_{O_2} + N_{N_2,p}(\bar{h}_2 - \bar{h}_{ref} + \Delta\bar{h}_{f,ref})_{N_2} \\ & - N_{CH_4}(\bar{h}_1 - \bar{h}_{ref} + \Delta\bar{h}_{f,ref})_{CH_4} - N_{O_2,r}(\bar{h}_1 - \bar{h}_{ref} + \Delta\bar{h}_{f,ref})_{O_2} \\ & - N_{N_2,r}(\bar{h}_2 - \bar{h}_{ref} + \Delta\bar{h}_{f,ref})_{N_2} \end{aligned}$$

directly into the equation solver. The Rule and Variable sheets are shown here. The result is 1300°C.

```
==================== RULE SHEET ====================
S Rule------------------------------------------------------------------------
  ;----------------- Example 11•9 -------------------------------------------
  Q = delH          ;First law for steady flow with W = delKE = 0
  Q = 0             ;Adiabatic combustion
  delH = Habp - Habr
  Habp=NpCO2*hbarab('CO2,Tp)+NpH2O*hbarab('H2O,Tp)+NpO2*hbarab('O2,Tp)+NpN2*hba
                    ;Continuation of equation:  +NpN2*hbarab('N2,Tp)
  Habr = NrCH4*hbarab('CH4,TCH4)+NrO2*hbarab('O2,TO2)+NrN2*hbarab('N2,TN2)
                    ;Note: hbarab = absolute molar enthalpy

  ;----------------- Absolute enthalpies of ideal gases ----------- CMBDELH1.TK
```

```
==================== VARIABLE SHEET ====================
St Input---- Name--- Output--- Unit----- Comment--------------------------------
                                ---------- * E11-9.TK -------- Example 11•9 ------
   0         Q                 kJ        Heat transfer = 0, adiabatic process
                               -- Model: * CMBDELH1.TK -- Absolute enthalpies -
             delH    0         kJ        Enthalpy change for the reaction
             Habp    -5106     kJ        Absolute enthalpy of products
             Habr    -5106     kJ        Absolute enthalpy of reactants
             Tp      1300      C         Temperature of products
   4         NrO2              kmol      Amount of O2 in reactants
   150       TO2               C         Temperature of O2 in reactants
   15.04     NrN2              kmol      Amount of N2 in reactants
   150       TN2               C         Temperature of N2 in reactants
   1         NrCH4             kmol      Amount of CH4 in reactants
   25        TCH4              C         Temperature of CH4 in reactants
   1         NpCO2             kmol      Amount of CO2 in products
   2         NpH2O             kmol      Amount of H2O in products
   2         NpO2              kmol      Amount of O2 in products
   15.04     NpN2              kmol      Moles of N2 in products
```

11·6 Chemical Equilibrium

In actual combustion, the maximum adiabatic combustion temperature as calculated in Example 11·9 above is not attained for several reasons. For example, the reaction may not reach completion because the system reaches an equilibrium state in which both reactants and products are present instead of products alone. To illustrate this point, let us consider a reaction that occurs in a closed system instead of an open one. The qualitative conclusion is the same in either case, and the closed system affords a simpler explanation.

Consider a rigid thermally insulated tank that contains carbon and oxygen. Assume that all the carbon is combined with some oxygen so that the composition of the tank contents may vary from pure CO_2 to a mixture of CO and O_2. The reaction is said to be complete when all the carbon is in the form of CO_2 and none is in the form of CO. In practice we find that this does not happen, that is, some of the carbon is always in the form of CO. We can examine a plot of the entropy of the system versus chemical composition to explain why.

Since the tank is rigid and thermally insulated, there is no energy transfer across the system boundary as either work or heat, and the system is *isolated.* For an *isolated system* the internal energy must be constant. In Fig. 11·6 we have chosen an arbitrary constant-U curve. (The value of the internal energy depends on the temperature and composition of the gas when the tank was filled or when the insulation was put on the tank, so several constant-U curves are possible.) For constant internal energy, the entropy of the system changes as a function of the "reaction completion" or chemical composition. In other words, as the fraction of carbon in CO_2 changes, the entropy of the system changes. Figure 11·6 shows that the entropy is highest when the system contains carbon in both CO_2 and CO. It is lower both when the reaction has produced little CO_2 and when the reaction has produced most of the CO_2 possible.

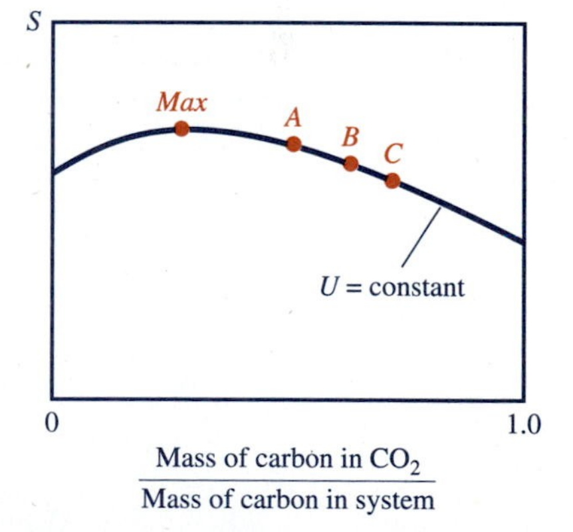

Figure 11·6 Entropy of an isolated system versus degree of reaction completion.

The second law shows that the state of the system will tend to the point of maximum entropy. For any process of an *isolated* system,

$$\Delta S_{\text{isolated system}} \geq 0$$

Therefore, if the composition of the gas mixture is represented by a point B on the figure, there can be no possible process of the isolated system that will carry the system to state C because $S_C < S_B$. Such a process would violate the increase of entropy principle. On the other hand, a process that carries the system from point B to point A *is* possible (and irreversible). Further reasoning along this line tells us that, once the system exists in the state marked *Max* (the state of maximum entropy for a given internal energy), it must remain at that state as long as the system is isolated. Furthermore, in any isolated system existing at states other than *Max*, there is always a tendency for the system to change toward the maximum entropy state for its particular

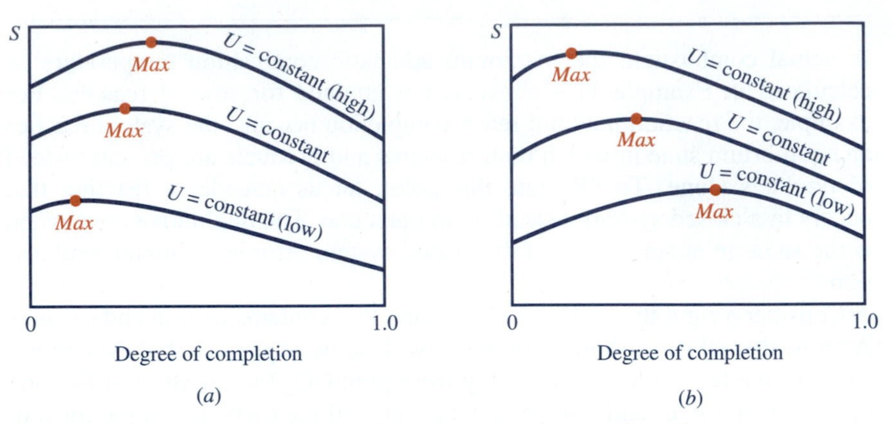

Figure 11·7 Entropy of an isolated system versus degree of reaction completion for different values of internal energy.

value of internal energy. The equilibrium chemical composition depends on the value of the internal energy. In some cases, a higher internal energy in the system will allow the reaction to proceed further toward completion, as shown in Fig. 11·7*a*. The opposite is also possible, as shown in Fig. 11·7*b*.

Now it is seen why the maximum adiabatic combustion temperature as calculated in Example 11·9 will not be attained: A state of equilibrium will be reached before the reaction is completed, and thus not all the reactants will react to form products, as was assumed in the calculation of the maximum adiabatic combustion temperature. The determination of the equilibrium composition and the temperature actually attained is treated in Chapter 12.

11·7 Second-Law Analysis of Chemical Reactions

In the preceding sections we discussed the application of the first law to chemically reacting systems. We found that it was necessary to use new methods for evaluating ΔU and ΔH because all our previous study of property relationships pertained only to pure substances. In applying the second law to chemically reacting systems, we must calculate entropy changes. However, up to this point we have calculated entropy changes for pure substances only. Therefore, we discuss next the calculation of ΔS for processes involving chemical reactions.

First, let us review the calculation of ΔS for pure substances. We can of course make use of the definition

$$\Delta S = \int \left(\frac{\delta Q}{T}\right)_{\substack{int\\rev}} \tag{6·3}$$

which is valid for any fixed quantity of matter whether a chemical reaction is involved or not. However, various relationships we have considered (such as the Maxwell equations) that relate entropy to other properties stem from the equation

$$T\,dS = dU + p\,dV \qquad (6{\cdot}7)$$

which holds for pure substances only. To illustrate that Eq. 6·7 holds only for pure substances, consider a system consisting initially of a mixture of several substances in a closed, rigid, thermally insulated vessel. Let this mixture be in mechanical and thermal equilibrium but not in chemical equilibrium; that is, the pressure and temperature are uniform throughout the system, and there is no relative motion among parts of the system, but the constituents are chemically reactive. Allow the reaction to occur so that the system finally reaches chemical equilibrium. This process, in which the system goes from a nonequilibrium to an equilibrium state, is irreversible, and since it occurs in an isolated system,

Sometimes the range of application of Eq. 6·7 is defined by saying that the relationship holds only between states in which a system is in complete (chemical, mechanical, and thermal) equilibrium. Another manner of stating the restriction on Eq. 6·7 is to limit it to states in which $S = f(U, V)$. In the process described above, $S \neq f(U, V)$ because the entropy of the system changed during the process while U and V remained constant. The relative merits of these various statements are of little concern in this book.

$$dS > 0$$

and consequently,

$$T\,dS > 0$$

In addition, U and V are constant in an isolated system so that

$$dU = 0 \quad \text{and} \quad p\,dV = 0$$

Therefore,

$$T\,dS > dU + p\,dV$$

Clearly, Eq. 6·7 does not hold for a process involving a chemical reaction.

Without the $T\,dS$ equations, how do we calculate the change in entropy for chemical reactions? Recall that we used ΔH_R to relate enthalpies on different scales to each other. We could have chosen to use a common scale for all enthalpies, but we did not exercise this option. Now we need a basis for determining entropies on a common scale. A reasonable suggestion is to make measurements of $\int (\delta Q/T)_{\substack{int\\rev}}$ during a reaction to determine ΔS_R; but finding and using a reversible path between end states involving different substances presents formidable difficulties. Therefore, we use instead the third law of thermodynamics.

The third law of thermodynamics is an independent principle that cannot be deduced from the first and second laws or from any other principles of nature. Like the second law, it can be stated in several forms that at first appear quite unrelated. For our purpose a suitable statement of the **third law of thermodynamics** is the following: *As the temperature of a pure substance approaches zero on the Kelvin scale, the entropy of the substance approaches zero.* There are important restrictions on the third law as stated

Third law of thermodynamics

here. However, the result that is important to us is that the third law makes possible the determination of *absolute* entropies based on $S = 0$ at $T = 0$. Table A·17 gives the absolute entropies of several substances, each at a standard pressure of 100 kPa and a reference temperature of 25°C (77 F). This allows them to be directly added or compared. The symbol for absolute entropy at the standard pressure is S° (or s° on a per unit mass, or $\bar{s}^\circ$ on a per unit mole basis). Consistent with our previous notation, we use the symbol s°_{ref} to indicate the entropy at the standard pressure and reference temperature. The entropy of a substance in any other state x is given by

$$s_x = s^\circ_{ref} + \int_{\text{standard state } (p^\circ, T_{ref})}^{x} ds$$

where the integral can be evaluated by the methods used for pure substances. For ideal gases,

$$s_x = s^\circ_{ref} + \int_{p^\circ, T_{ref}}^{p_x, T_x} \frac{dh - v\,dp}{T} = s^\circ_{ref} + \int_{T_{ref}}^{T_x} \frac{c_p\,dT}{T} - R\int_{p^\circ}^{p_x} \frac{dp}{p}$$

$$= s^\circ_{ref} + (\phi_x - \phi_{ref}) - R\ln\frac{p_x}{p^\circ} \tag{11·10}$$

In many references, you will find that the symbol ϕ is replaced with the symbol s°. This is possible only under certain conditions. Recall that ϕ is a function of temperature only with an arbitrarily assignable value at any one temperature. Also, *at the standard pressure*, s° is a function of temperature. If we set ϕ_{ref} equal to s°_{ref}, we will have aligned the relative scales. This is essentially equivalent to aligning the ideal-gas temperature-only function, ϕ, with the real-gas temperature-and-pressure function, s, at the standard pressure. In this case only, we can write the above equation as

$$s_x = \phi_x - R\ln\frac{p_x}{p^\circ}$$

or, equivalently, as

$$s_x = s^\circ_x - R\ln\frac{p_x}{p^\circ}$$

We are now able to apply the increase of entropy principle to determine whether a particular reaction is possible. Furthermore, we can calculate the irreversibility of a chemical reaction.

Example 11·10 illustrates a method for determining if a process is possible. Recall that in earlier chapters we presented some examples of processes where we used intuition to determine whether the process was possible. Support of these conclusions by the second law seemed unnecessary. But Example 11·10 clearly demonstrates a case in which the second law leads us to a conclusion that cannot be reached intuitively or by deductions from other principles.

The purpose of Example 11·10 is to demonstrate the usefulness of second-law calculations for determining the extent to which a reaction will proceed. However, these calculations can be done more directly, as we show in Chapter 12.

On the Standard State and Data. As mentioned earlier, the standard pressure used in this book and many modern data presentations (see, for example, Reference 11·4) is 100 kPa. This is the standard pressure recommended by the International Union of Pure and Applied Chemistry (IUPAC). Some data presentations, especially older ones, are based on a standard pressure of one atmosphere (101.325 kPa), so you must be alert when using different data sources. Symbols such as $s°$ and $p°$ indicate values at the standard pressure, and $s°$ can be tabulated against temperature. In some tables you may see $s°$ referred to as the standard *state* entropy. Entropy at a standard state, however, could not be tabulated against temperature. This is a common error and causes some confusion.

When a particular temperature, T_{ref}, is selected as a reference value, the properties at this temperature are designated h_{ref}, or s_{ref}. The combination of the standard pressure and the reference temperature may define a **standard state**. The *standard state* is often (and always in this book) set at 100 kPa, 25°C (298.15 K). In some places the standard state is defined at 1 atm (101.325 kPa), 25°C. Obviously, caution is required in using data from various sources.

Standard state

For gases, the standard state is sometimes a hypothetical state. For example, s°_{ref} for $H_2O(g)$ is listed in Table A·17. At 25°C, however, the highest pressure at which water can exist as a gas is 3.169 kPa. The s°_{ref} value in Table A·17 is therefore equal to the absolute entropy of saturated water vapor, s_g, at 25°C plus the entropy change that occurs during an isothermal compression from 3.169 kPa, 25°C, to 100 kPa. During the compression process the water is treated as an *ideal gas*.

Example 11·10

PROBLEM STATEMENT

Consider the adiabatic steady-flow burning of hydrogen with the stoichiometric amount of air. The air and the hydrogen are initially at 25°C, and the total pressure remains constant at 100 kPa. *Assuming* that it is possible for 95 percent of the hydrogen to be burned, is it possible for all of it to be burned under the conditions specified? In other words, with reference to the figure, is the process from state 2 to state 3 possible?

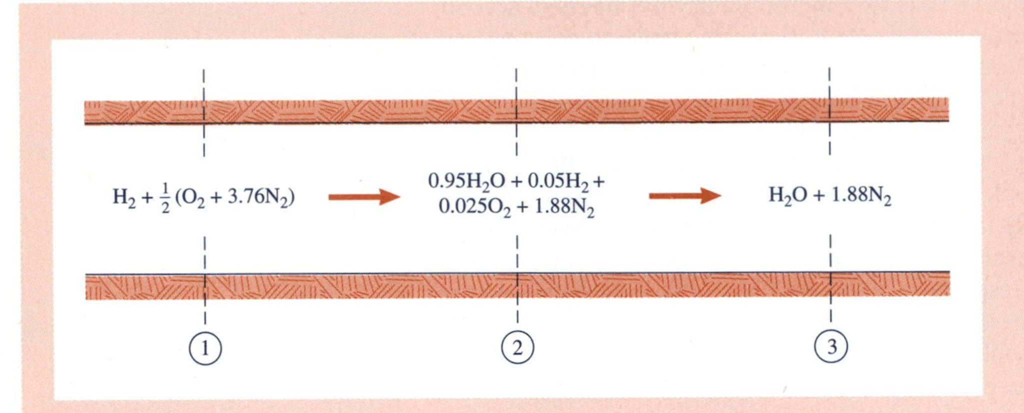

SOLUTION

An adiabatic steady-flow process is possible only if the entropy of the flowing material increases or remains constant. Therefore, we can answer the question by calculating the entropy change of the gas mixture between states 2 and 3. To do this, we must first determine T_2 and T_3. We can accomplish this using the first law for the adiabatic process 1–2–3 with no work done and $\Delta KE = 0$. We find from the first law that $H_1 = H_2 = H_3$, as shown in the HT diagram. For process 1–2, the first law is

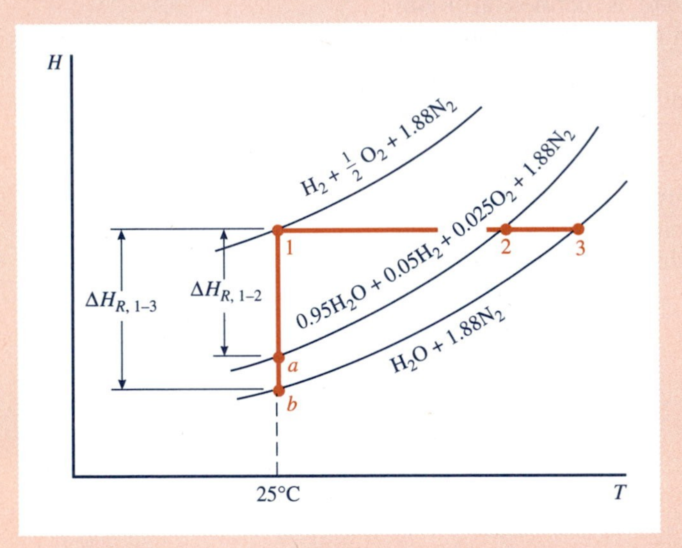

$$Q = H_2 - H_1 = \sum_2 N(\bar{h}_2 - \bar{h}_{ref} + \Delta\bar{h}_{f,ref}) - \sum_1 N(\bar{h}_1 - \bar{h}_{ref} + \Delta\bar{h}_{f,ref}) = 0$$

$$= 0.95(\bar{h}_2 - \bar{h}_{ref} + \Delta\bar{h}_{f,ref})_{H_2O} + 0.05(\bar{h}_2 - \bar{h}_{ref} + 0)_{H_2}$$
$$+ 0.025(\bar{h}_2 - \bar{h}_{ref} + 0)_{O_2} + 1.88(\bar{h}_2 - \bar{h}_{ref} + 0)_{N_2} - \Big[1(\bar{h}_1 - \bar{h}_{ref} + 0)_{H_2}$$
$$+ \tfrac{1}{2}(\bar{h}_1 - \bar{h}_{ref} + 0)_{O_2} + 1.88(\bar{h}_1 - \bar{h}_{ref} + 0)_{N_2}\Big] = 0$$

where for each element we have replaced $\Delta\bar{h}_{f,ref}$ by 0. Noting that $T_1 = T_{ref} = 25°C$ so that $\bar{h}_1 = \bar{h}_{ref}$ for each constituent and filling in numerical values from Tables A·15 and A·17, we simplify the equation to

$$= 0.95(\bar{h}_{2,H_2O} - 9904.0 - 241{,}826) + 0.05(\bar{h}_{2,H_2} - 8467.0)$$
$$+ 0.025(\bar{h}_{2,O_2} - 8683.0) + 1.88(\bar{h}_{2,N_2} - 8670.0) = 0$$

The temperature at which this equation is satisfied, using Table A·15 or the computer model for enthalpy changes of combustion processes, is $T_2 = 2149°C$.

A similar calculation is made for process 1–3, assuming such a process is possible, to determine T_3:

$$H_3 - H_1 = \sum_3 N(\bar{h}_3 - \bar{h}_{ref} + \Delta\bar{h}_{f,ref}) - \sum_1 N(\bar{h}_1 - \bar{h}_{ref} + \Delta\bar{h}_{f,ref}) = 0$$
$$= 1.0(\bar{h}_{3,H_2O} - 9904.0 - 241{,}826) + 1.88(\bar{h}_{3,N_2} - 8670.0) = 0$$

with the result, $T_3 = 2252°C$.

Now, to see if the process 2–3 is possible, we calculate ΔS:

$$S_3 - S_2 = \sum_3 N\left[\bar{s}^\circ_{ref} + (\bar{\phi} - \bar{\phi}_{ref}) - \bar{R}\ln\frac{p}{p^\circ}\right] - \sum_2 N\left[\bar{s}^\circ_{ref} + (\bar{\phi} - \bar{\phi}_{ref}) - \bar{R}\ln\frac{p}{p^\circ}\right]$$

where $\bar{s}^\circ_{ref}$ is the molar absolute entropy at the standard pressure, $p^\circ = 100$ kPa, and the reference temperature (25°C), and p is the partial pressure of each constituent. Since the mixture pressure is 100 kPa, $p/p^\circ = p/p_m = y$, where y is the mole fraction of each constituent. The equation thus becomes

$$S_3 - S_2 = \sum_3 N\left[\bar{s}^\circ_{ref} + (\bar{\phi} - \bar{\phi}_{ref}) - \bar{R}\ln y\right] - \sum_2 N\left[\bar{s}^\circ_{ref} + (\bar{\phi} - \bar{\phi}_{ref}) - \bar{R}\ln y\right]$$

$$S_3 - S_2 = \left\{1.0\left[\bar{s}^\circ_{ref} + (\bar{\phi}_3 - \bar{\phi}_{ref}) - \bar{R}\ln\left(\frac{1}{2.88}\right)\right]_{H_2O} + 1.88\left[\bar{s}^\circ_{ref} + (\bar{\phi}_3 - \bar{\phi}_{ref}) - \bar{R}\ln\left(\frac{1.88}{2.88}\right)\right]_{N_2}\right\}$$
$$- \left\{0.95\left[\bar{s}^\circ_{ref} + (\bar{\phi}_2 - \bar{\phi}_{ref}) - \bar{R}\ln\left(\frac{0.95}{2.905}\right)\right]_{H_2O} + 0.05\left[\bar{s}^\circ_{ref} + (\bar{\phi}_2 - \bar{\phi}_{ref}) - \bar{R}\ln\left(\frac{0.05}{2.905}\right)\right]_{H_2}\right.$$
$$\left. + 0.025\left[\bar{s}^\circ_{ref} + (\bar{\phi}_2 - \bar{\phi}_{ref}) - \bar{R}\ln\left(\frac{0.025}{2.905}\right)\right]_{O_2} + 1.88\left[\bar{s}^\circ_{ref} + (\bar{\phi}_2 - \bar{\phi}_{ref}) - \bar{R}\ln\left(\frac{1.88}{2.905}\right)\right]_{N_2}\right\}$$

Recall that in this book (and many others) the $\bar{\phi}$ and $\bar{s}^\circ$ scales are aligned to give equal values at the reference temperature. Therefore, the above equation simplifies to

$$S_3 - S_2 = \left\{1.0\left[\bar{\phi}_{3,H_2O} - \bar{R}\ln\left(\frac{1}{2.88}\right)\right] + 1.88\left[\bar{\phi}_{3,N_2} - \bar{R}\ln\left(\frac{1.88}{2.88}\right)\right]\right\}$$
$$- \left\{0.95\left[\bar{\phi}_{2,H_2O} - \bar{R}\ln\left(\frac{0.95}{2.905}\right)\right] + 0.05\left[\bar{\phi}_{2,H_2} - \bar{R}\ln\left(\frac{0.05}{2.905}\right)\right]\right.$$
$$\left. + 0.025\left[\bar{\phi}_{2,O_2} - \bar{R}\ln\left(\frac{0.025}{2.905}\right)\right] + 1.88\left[\bar{\phi}_{2,N_2} - \bar{R}\ln\left(\frac{1.88}{2.905}\right)\right]\right\}$$

Substituting values from Table A·15, or using the computer model, we find

$$S_3 - S_2 = -1.1 \text{ kJ/K per kmol } H_2 \text{ at section 1}$$

Process 2–3 would result in a decrease in entropy of the material flowing adiabatically through the steady-flow system. Therefore, in accordance with the increase of entropy principle, *process 2–3 is impossible*. The reaction

$$H_2 + \tfrac{1}{2}(O_2 + 3.76N_2) \rightarrow H_2O + 1.88N_2$$

will not go to completion in a steady-flow adiabatic process at 100 kPa with the reactants entering at 25°C.

Example 11·11

PROBLEM STATEMENT

Assume that the reaction

$$H_2 + \tfrac{1}{2}(O_2 + 3.76N_2) \rightarrow 0.95H_2O + 0.05H_2 + 0.025O_2 + 1.88N_2$$

occurs adiabatically at 100 kPa in a steady-flow system. The reactants enter at 25°C. (This is process 1–2 of Example 11·10.) The lowest temperature in the surroundings is 10°C. (This means that the combustion air has been preheated if the atmosphere is at 10°C, or it could mean that a large body of water is available as an energy reservoir at 10°C.) Determine the irreversibility of the process per mole of H_2 entering.

SOLUTION

The lowest temperature in the surroundings is $T_0 = 10°C$, and $I = T_0\,\Delta S_{\text{isolated system}}$. For an adiabatic steady-flow process, the total change in entropy of the universe or of an isolated system (defined as the steady-flow system plus all parts of the surroundings that interact with it) is $S_2 - S_1$. S_1 and S_2 are the entropies entering and leaving, respectively:

$$S_2 - S_1 = \sum_2 N\left[\bar{s}^\circ_{ref} + (\bar{\phi} - \bar{\phi}_{ref}) - \bar{R}\ln\frac{p}{p^\circ}\right] - \sum_1 N\left[\bar{s}^\circ_{ref} + (\bar{\phi} - \bar{\phi}_{ref}) - \bar{R}\ln\frac{p}{p^\circ}\right]$$

As before, in Example 11·10, this equation can be simplified when the $\bar{\phi}$ and $\bar{s}^\circ$ scales have been aligned at the reference temperature T_{ref}:

$$S_2 - S_1 = \sum_2 N(\bar{\phi} - \bar{R}\ln y) - \sum_1 N(\bar{\phi} - \bar{R}\ln y)$$

Using $T_2 = 2149°C$ as determined in Example 11·10, we have

$$S_2 - S_1 = \left\{0.95\left[274.804 - 8.314 \ln\left(\frac{0.95}{2.905}\right)\right] + 0.05\left[195.110 - 8.314 \ln\left(\frac{0.05}{2.905}\right)\right]\right.$$
$$\left. + 0.025\left[276.059 - 8.314 \ln\left(\frac{0.025}{2.905}\right)\right] + 1.88\left[259.016 - 8.314 \ln\left(\frac{1.88}{2.905}\right)\right]\right\}$$
$$- \left\{1.0\left[130.680 - 8.314 \ln\left(\frac{1}{3.38}\right)\right] + 0.5\left[205.147 - 8.314 \ln\left(\frac{0.5}{3.38}\right)\right]\right.$$
$$\left. + 1.88\left[191.609 - 8.314 \ln\left(\frac{1.88}{3.38}\right)\right]\right\}$$
$$= 162.3 \text{ kJ/K for each kmol } H_2 \text{ entering}$$

$$I = T_0 \, \Delta S = 283(162.3) = 45{,}900 \text{ kJ for each kmol } H_2 \text{ entering}$$

We can also use the computer model to complete the solution of this problem. The Rule and Variable sheets appear below:

```
=================== RULE SHEET ===================
S Rule-------------------------------------------------------------
  ;-------------- Example 11•11 -------------------------------------
  ;-------------- Entropy change for a combustion process ------- CMBDELS1.TK
  Np = NpH2O + NpH2 + NpO2 + NpN2        ;Number of moles of products
  Nr = NrH2 + NrO2 + NrN2                ;Number of moles of reactants

                       ;Calculation of absolute entropies:
  SrH2 = NrH2*(phibar('H2,TrH2)-Rbar*ln((pm/p0)*(NrH2/Nr)))
  SrO2 = NrO2*(phibar('O2,TrO2)-Rbar*ln((pm/p0)*(NrO2/Nr)))
  SrN2 = NrN2*(phibar('N2,TrN2)-Rbar*ln((pm/p0)*(NrN2/Nr)))
  Sabr = SrH2 + SrO2 + SrN2
  SpH2O = NpH2O*(phibar('H2O,Tp)-Rbar*ln((pm/p0)*(NpH2O/Np)))
  SpH2 = NpH2*(phibar('H2,Tp)-Rbar*ln((pm/p0)*(NpH2/Np)))
  SpO2 = NpO2*(phibar('O2,Tp)-Rbar*ln((pm/p0)*(NpO2/Np)))
  SpN2 = NpN2*(phibar('N2,Tp)-Rbar*ln((pm/p0)*(NpN2/Np)))
  Sabp = SpH2O + SpH2 + SpO2 + SpN2
  delS = Sabp - Sabr

                       ;Calculation of irreversibility:
  I = T0 * delS        ;For this adiabatic, steady-flow process, the entropy
                       ;change of the universe is the delS for the reaction.
  i = I/NrH2
```

```
═══════════════════════ VARIABLE SHEET ════════════════════════════════════════
St Input──── Name─── Output─── Unit───── Comment──────────────────────────────
                               ──────── * E11-11.TK ──── Example 11·11 ────
                               ─ Model: * CMBDELS1.TK ─── Entropy change for
                                              a combustion process ───
   1         NrH2              kmol      Number of moles of H2 in reactants
   25        TrH2              C         Temperature of H2 in reactants
   0.5       NrO2              kmol      Number of moles of O2 in reactants
   25        TrO2              C         Temperature of O2 in reactants
   1.88      NrN2              kmol      Number of moles of N2 in reactants
   25        TrN2              C         Temperature of N2 in reactants
   0.95      NpH2O             kmol      Number of moles of H2O in products
   0.05      NpH2              kmol      Number of moles of H2 in products
   0.025     NpO2              kmol      Number of moles of O2 in products
   1.88      NpN2              kmol      Number of moles of N2 in products
   100       pm                kPa       Pressure of mixture
   2149      Tp                C         Temperature of products
             Np       2.905    kmol      Number of moles of products
             Nr       3.38     kmol      Number of moles of reactants
   8.314     Rbar              kJ/(kmol* Molar gas constant
   100       p0                kPa       Standard state pressure
   10        T0                C         Sink temperature
             Sabp     783      kJ/K      Absolute entropy of products
             Sabr     620.7    kJ/K      Absolute entropy of reactants
             delS     162.3    kJ/K      Entropy change for the reaction
             I        45900    kJ        Irreversibility
             i        45900    kJ/kmol   Irreversibility per kmol of H2
```

Comment: In keeping with the meaning of I, the combustion process reduces by 45,900 kJ/kmol H_2 the amount of energy that can be converted into work. Although the heating value of hydrogen is approximately 242,000 kJ/kmol, this amount of work can never be obtained by burning hydrogen in air because the irreversible combustion process makes a significant fraction of the energy unavailable. In any engine or power plant that involves combustion, a major source of irreversibility is the combustion process itself.

11·8 Summary

In this chapter we presented some preliminary aspects of combustion analyses to refresh your memory of stoichiometry. Then we discussed the combustion mass balances and energy balances for processes involving chemical reactions. Finally, we showed how a second-law analysis of such processes can be used to determine when a reaction is possible.

The oxygen for most combustion reactions comes from air. For nearly all combustion calculations the approximate composition of air can be taken as

$$1 \text{ mole } O_2 + 3.76 \text{ moles } N_2 = 4.76 \text{ moles air}$$

$$1 \text{ kg } O_2 + 3.31 \text{ kg } N_2 = 4.31 \text{ kg air}$$

Stoichiometric air is the quantity of air required to burn a unit quantity of fuel completely with no oxygen appearing in the products of combustion. *Excess air* is the amount of air supplied to the reaction in excess of the stoichiometric amount.

Two combustion mass balance problems are encountered in engineering: *ideal* or *complete combustion* and *actual* or *incomplete combustion*.

The first law of thermodynamics has the same form whether a chemical reaction occurs within the system or not. However, when a chemical reaction occurs, the evaluation of ΔU or ΔH is more complex. To assist in the evaluation of ΔU and ΔH, we assumed that the substances were ideal gases and defined the *internal energy of reaction* ΔU_R as the difference between the internal energy of the products and of the reactants when held at the reference temperature T_{ref}. Similarly, we defined the *enthalpy of reaction* ΔH_R as the difference between the enthalpy of the products and the enthalpy of the reactants at the reference temperature, T_{ref}. $-\Delta H_R$ is often called the *heat of reaction at constant pressure*, and $-\Delta U_R$ is often called the *heat of reaction at constant volume*.

The *enthalpy of formation* Δh_f is defined as the enthalpy of reaction for the formation of a compound from its elements. It is a valuable property that allows us to express ΔH_R for many reactions as follows:

$$\Delta H_R = \sum_{\text{products}} N \Delta \bar{h}_f - \sum_{\text{reactants}} N \Delta \bar{h}_f \tag{11·7}$$

An expression for the change in enthalpy of ideal gases that undergo a process involving a chemical reaction is

$$H_2 - H_1 = \sum_{\text{products}} N(\bar{h}_2 - \bar{h}_{ref} + \Delta \bar{h}_{f,ref}) - \sum_{\text{reactants}} N(\bar{h}_1 - \bar{h}_{ref} + \Delta \bar{h}_{f,ref}) \tag{11·8}$$

where the subscript *ref* signifies evaluation at the reference temperature, usually 25°C. If the gases are real, the pressure dependence must be considered. In this case, the most general case, the above expression becomes

$$H_2 - H_1 = \sum_{\text{products}} N(\bar{h}_2 - \bar{h}^\circ_{ref} + \Delta \bar{h}^\circ_{f,ref}) - \sum_{\text{reactants}} N(\bar{h}_1 - \bar{h}^\circ_{ref} + \Delta \bar{h}^\circ_{f,ref}) \tag{11·9}$$

where the superscript ° refers to a *standard state pressure* equal to 100 kPa.

The *maximum adiabatic combustion temperature* is defined as the temperature that results from combustion that proceeds to the maximum extent possible with no heat transfer and no work done. For any fuel–oxidizer combination, it can be calculated using the first law, data on ΔH_R, and the hT relationships of the constituents involved in the reaction.

A reaction that appears possible from stoichiometric and first-law considerations may actually be impossible. This can be shown by means of the second law. To apply the second law, we must be able to calculate ΔS for the process. For a process involving a chemical reaction, the entropy change can be determined using absolute entropy values. Absolute entropy can be found on the basis of the *third law of thermodynamics*, which states that in most cases as the temperature of a pure substance approaches zero on the Kelvin scale, the entropy of the substance approaches zero. The entropy of a substance in any state x is then

$$s_x = s^\circ_{ref} + \int_{\text{standard state }(p^\circ, T_{ref})}^{x} ds$$

where the integral is evaluated by any of several methods that apply to entropy changes of pure substances.

References

11·1 Cengel, Yunus A., and Michael A. Boles, *Thermodynamics*, 2nd ed., McGraw-Hill, New York, 1994, Chapter 14. (In this book, the superscript ° designates a standard state, not just a standard pressure.)

11·2 Moran, Michael J., and Howard N. Shapiro, *Fundamentals of Engineering Thermodynamics*, 2nd ed., Wiley, New York, 1992, Chapter 13.

11·3 Van Wylen, Gordon J., Richard E. Sonntag, and Claus Borgnakke, *Fundamentals of Classical Thermodynamics*, 4th ed., Wiley, New York, 1994, Chapter 12.

11·4 *JANAF Thermochemical Tables*, 3rd ed., published by the American Chemical Society and the American Institute of Physics for the National Bureau of Standards, 1986. (Also, *Journal of Physical and Chemical Reference Data*, Vol. 14, 1985, Supplement No. 1.)

11·5 *Selected Values of Properties of Hydrocarbons and Related Compounds*, American Petroleum Institute Project 44, Thermodynamics Research Center, Texas A&M University, College Station, TX 77843.

11·6 *Selected Values of Chemical Thermodynamic Properties*, National Bureau of Standards Technical Note 270-3, 1968.

11·7 Gallant, R. W., ''Physical Properties of Hydrocarbons,'' *Hydrocarbon Processing*, Vol. 45, No. 10, October 1966.

Problems

11·1 The molar mass of air is 28.968. Explain why it is not possible to obtain this value by calculating the molar mass from the composition:

$$4.764 \text{ moles air} = 1 \text{ mole } O_2 + 3.764 \text{ moles } N_2$$

11·2 One thousand cubic meters of ethane at 95 kPa, 10°C, is burned with 20 percent excess air. Calculate (*a*) the mass of water formed, and (*b*) the dew point of the products that are at 95 kPa, 210°C.

11·3 Refer to Problem 11·2. Determine the sensitivity of the dew point to excess air. That is, determine the change in dew point temperature for one percent change in excess air or, for more convenient numbers, the change in dew point temperature for ten percent change in excess air.

11·4 Refer to Problem 11·2. Determine the dew point if the air supplied is at 30°C and has a relative humidity of 90 percent.

11·5 A gasoline with an ultimate analysis of 85 percent carbon and 15 percent hydrogen has a specific gravity of 0.72. For the complete combustion of this gasoline with 10 percent excess air, determine (*a*) the air–fuel ratio and (*b*) the number of liters of water formed per liter of gasoline burned.

11·6 A gaseous fuel has the following volumetric analysis: 60% CH_4, 20% CO, 10% O_2, 10% N_2. The fuel is to be burned with 30 percent excess air. Compute (*a*) the mass analysis of this fuel by chemical compounds, (*b*) the ultimate analysis, (*c*) the fuel specific volume at 1 atm, 40°C, (*d*) the volume of dry air at 1 atm, 40°C, required per unit volume of this fuel, (*e*) the volumetric analysis of the dry products of complete combustion, (*f*) the mass of the total products per unit mass of fuel, (*g*) the humidity ratio of the products per unit mass of fuel, (*h*) the air–fuel ratio, (*i*) the dew point of the products if dry air is supplied for combustion, and (*j*) the dew point of the products if the air supplied for combustion is at 1 atm, 40°C, 80 percent relative humidity.

11·7 A producer gas has the following volumetric analysis: 26% CO_2, 15% CO, 3% CH_4, 26% H_2, 30% N_2. Compute the same items as in Problem 11·6.

11·8 For the complete burning of 1000 cu ft (measured at 14.0 psia, 40 F) of ethane with 20 percent excess air, calculate (*a*) the amount of water formed and (*b*) the dew point of the products that are at 14.0 psia, 515 F.

11·9 A gaseous mixture at 1 atm, 130°C, has a mass analysis of 80% CH_4, 14% N_2, and 6% H_2O. Determine (*a*) the dew point of this mixture and (*b*) the air–fuel ratio for complete, stoichiometric combustion.

11·10 A coal with an ultimate analysis of 68% C, 5% H, 16% O, 2% N, 2% S, and 7% ash is to be completely burned with 30 percent excess air. Calculate (*a*) the amount of air required, (*b*) the volumetric analysis of the dry products of combustion, (*c*) the mass of the total products per unit mass of fuel, (*d*) the mass of water vapor in the products per unit mass of fuel, (*e*) the dew point of the products if dry air is supplied for combustion, and (*f*) the dew point of the products if the combustion air is at 14.5 psia, 100 F, 75 percent relative humidity.

11·11 A hydrocarbon fuel is completely burned in air. How does the dew point of the products vary with (*a*) the fraction of hydrogen in the fuel, (*b*) the humidity ratio of the combustion air, (*c*) the barometric pressure?

11·12 A powdered substance has a mass analysis of 63% carbon, 28% sulfur, 9% ash. It is burned in air, and a molar analysis of the products includes 12.3% CO_2, 4.5% O_2, 3.5% CO, 2.6% SO_2. Find the air–fuel ratio, assuming that all of the carbon and sulfur burn.

11·13 Ethyl alcohol, C_2H_5OH, is burned in air, and a volumetric analysis of the dry products includes 9.8% CO_2, 5% O_2, 0.6% CO, 0.6% HCHO, 82.5% N_2. Air for combustion is supplied at 0.92 atm, 25°C. Determine per unit mass of ethyl alcohol (*a*) the mass of air supplied and (*b*) the mass of water formed.

11·14 Gasoline is burned in an automobile engine, and a molar analysis of the products of combustion shows 12.1% CO_2, 2.6% O_2, 1.1% CO. The gasoline has a specific gravity of 0.76 and an ultimate analysis of 85% carbon and 15% hydrogen. Barometric pressure was 0.95 atm, and the combustion air had a humidity ratio of 0.030. Determine per unit mass of fuel (*a*) the amount of air supplied and (*b*) the amount of water in the products of combustion.

11·15 A coal has the following ultimate analysis: 76% C, 5% H, 6% O, 1% S, 1% N, 5% ash, 6% free moisture. This coal is burned with air at 1 atm, 40°C, and the volumetric analysis of the dry products of combustion shows 12.7% CO_2, 0.4% CO, 6.1% O_2. Assume that the solid refuse contains 25 percent carbon and that all of this material is discharged to the ash pit. Calculate (*a*) the air–fuel ratio, (*b*) the percent excess air, (*c*) the mass of carbon burned to CO per unit mass of fuel, (*d*) the dew point of the products if dry air is supplied, and (*e*) the dew point of the products if the air supplied for combustion is at 70 percent relative humidity.

11·16 Coal is burned with air at 29.2 in. Hg, 90 F, and a humidity ratio of 0.030. Determine from the following data the mass of water in the gaseous products per pound of fuel. Coal ultimate analysis: 70% C, 5% H, 10% O, 4% N, 3% S, 8% ash. The analysis of the dry gaseous products shows 13.6% CO_2, 4.5% O_2, 1.1% CO. Amount of coal fired during test: 5000 lbm. Solid refuse collected: 500 lbm with 20 percent carbon. Products temperature: 510 F.

11·17 An exothermic reaction occurs in a closed, rigid, thermally insulated vessel. The number of moles of products equals the number of moles of reactants. The specific heats are constant for both reactants and products, and they are higher for the products than for the reactants. Sketch a complete *UT* diagram for the reactants, the products, and the process.

11·18 Determine ΔU_R at 500 K for the complete combustion of acetylene.

11·19 Determine the lower heat of combustion of hydrogen at 560°C.

11·20 Methane at 25°C is burned with excess air supplied at 25°C in a steady-flow process at 1 atm. The products leave at 410°C. For excess air amounts of 20, 40, and 60 percent, calculate (*a*) the amount of heat transfer per unit mass of methane, and (*b*) the dew point of the products.

11·21 Gaseous propane at 77 F is to be mixed with air at the same temperature and burned adiabatically in a steady-flow system. Determine the air–fuel ratios required for products temperatures of 900 F, 1000 F, and 1100 F.

11·22 The heat of combustion of carbon is determined by means of a bomb calorimeter submerged in a water bath. In addition to a small amount of carbon, a considerable amount of excess air is placed in the bomb. Assuming that in any case there is sufficient air for complete combustion, how does the amount of air present affect the value obtained for the heat of combustion?

11·23 The heat of combustion of a liquid hydrocarbon is to be measured by means of a bomb calorimeter. It is proposed that a drop of liquid water be placed in the bomb so that the amount of heat removed to return the bomb and its contents to their initial temperature will be the higher heat of combustion at constant volume. The reasoning behind this proposal is that part of the water drop will evaporate and make the initial air–vapor atmosphere in the bomb saturated so that all of the water formed by combustion must eventually condense. Comment on this proposal.

11·24 Methane and four times the stoichiometric amount of oxygen are placed in a rigid, insulated container at 1 atm, 25°C, and ignited. Determine the final pressure, temperature, and mixture composition, assuming that the methane is completely burned.

11·25 Solve Problem 11·24 for an initial mixture that contains air instead of oxygen.

11·26 Nitrogen is generally considered to be inert in combustion reactions. Still, oxides of nitrogen do exist. Therefore, why do we not consider the combustion of the nitrogen in the air in combustion analyses and thereby obtain some heating value from the air as well as from the fuel being burned?

11·27 For each of the following fuels, state whether the constant-pressure heat of combustion is greater than, less than, or equal to the constant-volume heat of combustion: C, CO, C_2H_2, C_2H_4, C_3H_8.

11·28 The enthalpy of formation of butane, C_4H_{10}, is given in Table A·17. Verify this number from other data given in the same table.

11·29 Verify the higher heat of combustion given in Table A·17 for ethylene gas at 25°C from the enthalpy of formation data given in that same table.

11·30 On the basis of data in Table A·15 construct a skeleton absolute enthalpy table, h versus T, for CO_2, H_2, H_2O, and O_2, showing values at 0, 298.15 and 500 K. Assign $h = 0$ at $T = 0$ for elements.

11·31 Solve Problem 11·30 but assign $h = 0$ at $T = 0$ for CO_2 and H_2O.

11·32 How much methane (in thousands of cubic feet per hour, measured at 1 atm, 70 F) is required to heat a home that has a heat loss of 120,000 B/h, assuming that 70 percent of the heat released by combustion goes into heating the home?

11·33 Natural gas, which can be treated as methane, is to be supplied to six homes, each of which requires a maximum of 140,000 B/h. With only six homes on the supply line, the maximum flow rate in the line is the sum of the maximum for the homes. The pressure in the buried pipeline feeding the homes is 55 psig, and the gas velocity in the line is not to exceed 60 fps. Estimate the required inside diameter of the line. (State your assumptions clearly.)

11·34 Solve Example 11·8 for a range of excess air from 0 to 200 percent, assuming that the products temperature is held constant, and present the results in a plot.

11·35 Carbon monoxide and 50 percent excess air are completely burned in a steady-flow process at 100 kPa. The carbon monoxide and the air are supplied at the reference temperature. Calculate the adiabatic combustion temperature for this reaction, assuming that the reaction goes to completion.

11·36 Solve Problem 11·35 for a range of excess air from 0 to 100 percent and present the results in a plot.

11·37 Determine the adiabatic combustion temperature for the steady-flow burning of propane, C_3H_8, with 100 percent excess air with propane and air supplied at 1 bar, 25°C. Assume that the reaction goes to completion.

11·38 Plot a curve of the adiabatic combustion temperature against the percent excess air for the steady-flow burning of propane, C_3H_8. Assume that the reaction goes to completion, and that both fuel and air are supplied at 14.50 psia, 77 F.

11·39 Plot a curve of the adiabatic combustion temperature of hydrogen against percent excess air. Both fuel and air are supplied at 25°C. Let the percent excess air range from 0 to 100 percent. Combustion occurs in a steady-flow system under a constant pressure of 100 kPa. Assume that the reaction goes to completion.

11·40 Determine the maximum adiabatic combustion temperature for the combustion of hydrogen supplied at 77 F with 100 percent excess air also supplied at 77 F. Combustion occurs in a steady-flow system under a constant pressure of 14.50 psia. Assume that the reaction goes to completion.

11·41 Refer to Example 11·9 and determine the irreversibility of the process.

11·42 Refer to Example 11·8 and calculate the irreversibility if all the heat from the combustion process is used to boil water at 300°C.

11·43 Refer to Problem 11·20. All the heat released by the combustion process is absorbed by water that flows into the heat exchanger at 25 atm, 110°C, and leaves at 24 atm, 310°C. The lowest temperature in the surroundings is 25°C. Determine, per kilogram of methane, the irreversibility.

11·44 For the process described in Problem 11·21, determine the irreversibility per lbm of propane.

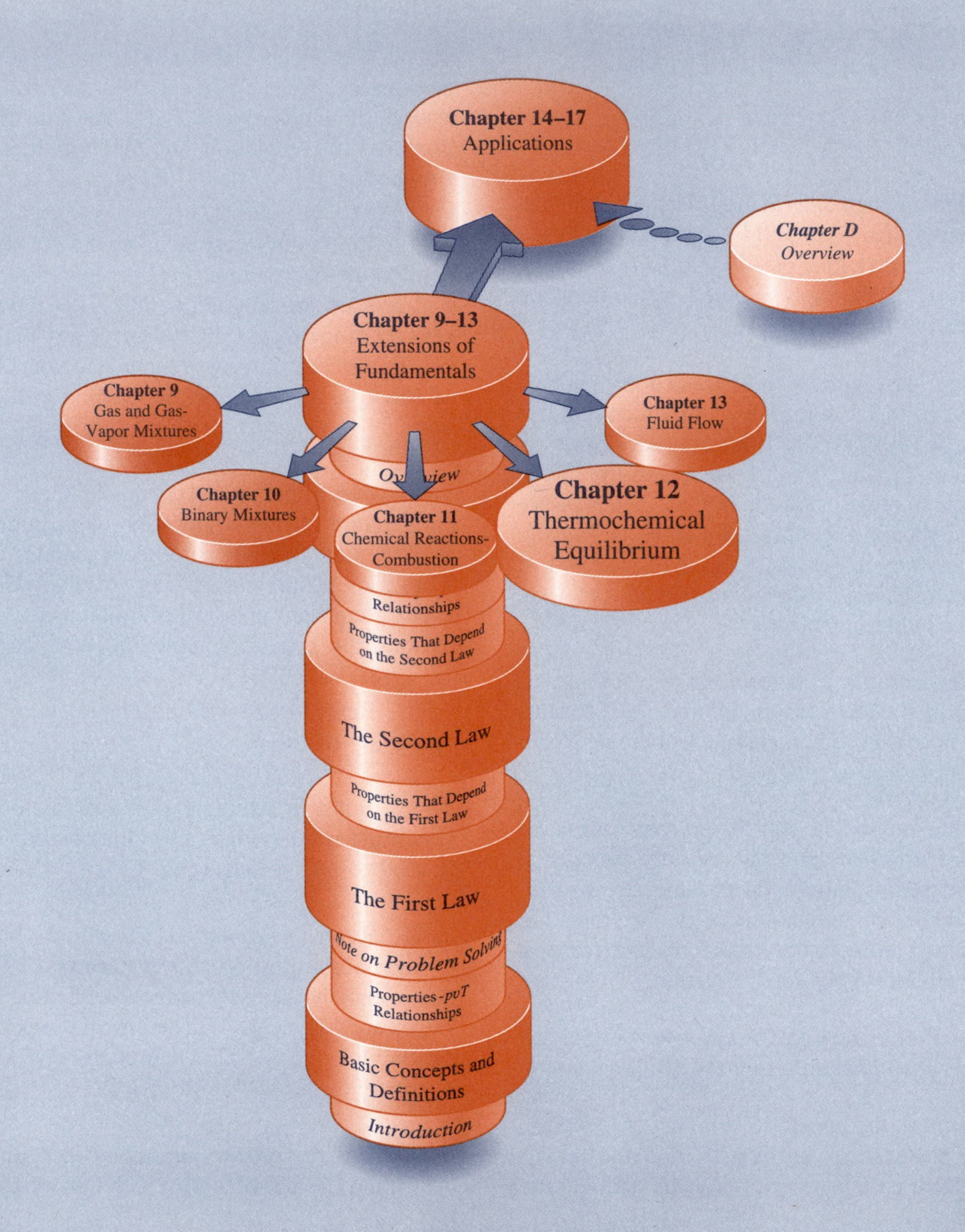

Chapter 14–17
Applications
Chapter D
Overview
Chapter 9–13
Extensions of
Fundamentals
Chapter 9
Gas and Gas-
Vapor Mixtures
Chapter 13
Fluid Flow
Chapter 10
Binary Mixtures
Chapter 11
Chemical Reactions-
Combustion
Chapter 12
Thermochemical
Equilibrium
Relationships
Properties That Depend
on the Second Law
The Second Law
Properties That Depend
on the First Law
The First Law
Note on Problem Solving
Properties-pvT
Relationships
Basic Concepts and
Definitions
Introduction

Chapter 12

Chemical Equilibrium in Ideal-Gas Reactions

In Chapter 11, we found that under specified conditions some chemical reactions are not possible and we developed techniques, based on the second law of thermodynamics, for determining whether a given reaction could occur. The technique introduced in Chapter 11 required that we first postulate a reaction and then calculate whether or not the reaction was possible. In this chapter, we introduce a parameter known as the *equilibrium constant*. This parameter allows us to calculate directly the maximum extent to which a reaction can proceed. The equilibrium constant allows us to perform calculations of great utility in engineering analyses. While the equilibrium constant allows us to determine the equilibrium composition of the reaction products, it does not allow us to determine how fast the equilibrium condition may be reached. For this purpose, we need to understand reaction rates. Thus, we also provide a brief discussion of reaction rates in this chapter. For simplicity, we confine our attention to chemical reactions in which all the constituents may be treated as ideal gases.

12·1 Criteria of Equilibrium

Complete equilibrium requires mechanical, thermal, and chemical equilibrium.

Before we can discuss chemical equilibrium, we must establish criteria of equilibrium. A system is said to be in equilibrium (or in an equilibrium state) if no changes can occur in the state of the system without the aid of an external stimulus. To test if a system is in equilibrium, we might isolate the system and observe whether any changes in its state occur. To ensure thermal and mechanical equilibrium, the pressure and temperature must be uniform throughout a system and the system must be free of velocity and concentration gradients. If these conditions are not met, spontaneous changes will occur. These conditions alone are sufficient to ensure that a system is in thermal and mechanical equilibrium, but they are not sufficient to ensure that the system is in chemical equilibrium. That is, the system may still undergo a chemical reaction. If we further restrict the state of the system to one in which no chemical reaction can occur without an external stimulus, then the system is in complete (mechanical, thermal, and chemical) equilibrium. Under such conditions, the system is chemically homogeneous and invariant; so it is by definition a pure substance and the relationship

$$T\,dS = dU + p\,dV \qquad \text{(pure substance only)} \tag{6·7}$$

applies.

By contrast, if we consider an isolated system in which a spontaneous chemical reaction occurs, the process is irreversible so that

$$dS_{\text{isolated system}} > 0$$

and because the system is isolated,

$$dU = 0 \quad \text{and} \quad p\,dV = 0$$

Therefore,

$$T\,dS > dU + p\,dV \qquad \text{(irreversible chemical reactions)}$$

Assuming that this relationship holds also for reactions in nonisolated systems (and this can be proved rigorously), we have the general relationship for all possible processes:

$$T\,dS \geq dU + p\,dV \tag{12·1}$$

where the equality holds if the system is chemically invariant and the inequality holds for processes involving chemical reactions. In the limiting case of a *reversible* chemical reaction, the equality holds.

For a process of constant U and constant V it is apparent that $T\,dS \geq 0$, or, using a brief notation,

$$(dS)_{U,V} \geq 0 \tag{12·2a}$$

Equation 12·2*a* states that for all processes at constant U and V, the entropy must increase or remain constant. Therefore, when a system is in complete equilibrium (i.e., no process is possible at constant U and V), its entropy must be a maximum. Otherwise there would be some process for which $(dS)_{U,V} > 0$, indicating that the system had not in fact been in a state of complete equilibrium (see Fig. 12·1). This is one criterion of equilibrium.

One criterion for equilibrium: *A system is in complete equilibrium when, for a given energy and volume, its entropy is at a maximum.*

Other criteria of equilibrium for different conditions can be obtained from Eq. 12·1 by rewriting it in terms of H, G, and A, as we did in deriving the Maxwell equations (Section 7·1).

As an example, we can rewrite Eq. 12·1 as

$$T\,dS \geq dU + p\,dV = dH - V\,dp \tag{12·1}$$

and combine this with the definition of the Gibbs function

$$dG = d(H - TS) = dH - T\,dS - S\,dT$$

to obtain

$$dG \leq V\,dp - S\,dT$$

From this we conclude that for a constant-pressure, constant-temperature process

$$(dG)_{p,T} \leq 0 \tag{12·2b}$$

Another criterion for equilibrium: *A system is in complete equilibrium when the Gibbs function is at a minimum for all states at the same pressure and temperature.*

Thus the Gibbs function of any system in complete equilibrium must be a minimum with regard to all states at the same pressure and temperature. Otherwise there would be some process for which $(dG)_{p,T} < 0$, indicating that the system had not in fact been in a state of complete equilibrium (see Fig. 12·2).

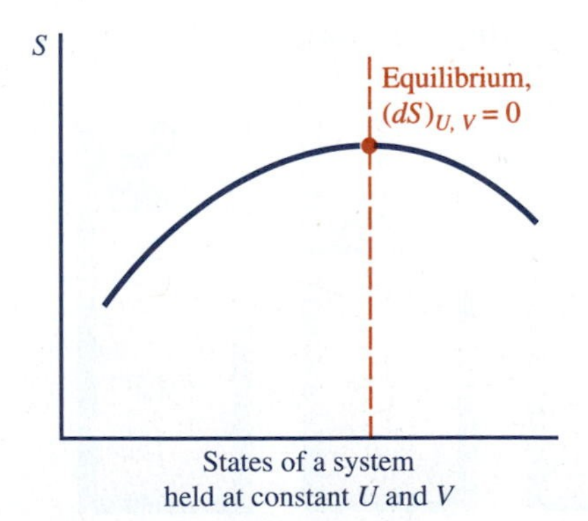

Figure 12·1 The equilibrium criterion for a system with constant energy held at constant volume.

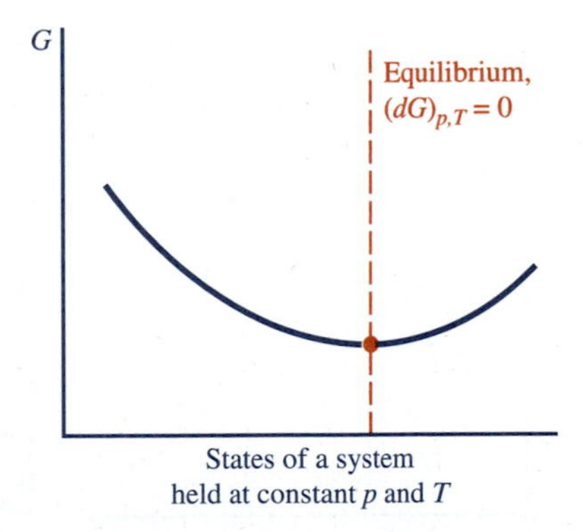

Figure 12·2 The equilibrium criterion for a system held at constant pressure and temperature.

This is a valuable criterion of equilibrium for two reasons:

If no spontaneous reaction can occur at constant pressure and temperature, then no spontaneous reaction at all can occur.

1. Many chemical reactions take place at constant pressure and temperature.
2. We can show that if no spontaneous process can occur between states at the same pressure and temperature, then no spontaneous process at all can occur.

To illustrate this latter point, suppose that a system initially at the pressure and temperature of its surroundings is in a state such that some spontaneous process changes the system pressure or temperature. This process can always be followed by a spontaneous process that restores the system to the original pressure and temperature of the surroundings.* These two processes together constitute a spontaneous process between two states at the same pressure and temperature. Using this reasoning, any spontaneous process that changes the pressure or temperature of the system can be viewed as part of a spontaneous process between two states at the same pressure and temperature. Therefore, if no spontaneous change to another state at the same pressure and temperature is possible, then no spontaneous change to a different pressure or temperature is possible, and no spontaneous process at all is possible.

***The system can be allowed to expand or contract to restore the original pressure of the system, and heat can be transferred to restore the original temperature of the system.**

Let us note in passing that the $(dG)_{p,T} \leq 0$ criterion can be used to show that when two or more phases of a pure substance coexist in equilibrium, the specific Gibbs function of each phase must be the same. Consider a mixture of liquid and vapor of a pure substance. The masses of the liquid and vapor are m_f and m_g, respectively, and g_f and g_g are the corresponding specific Gibbs functions. The Gibbs function of the system is

$$G = m_f g_f + m_g g_g$$

Now suppose that some of the liquid evaporates at constant pressure (and therefore also at constant temperature) so that the mass of the vapor changes by dm_g and the mass of the liquid changes by $dm_f = -dm_g$. (This is part of the process 1–2 shown in Fig. 12·3.) The change in the Gibbs function of the system is

$$dG = m_f\, dg_f + g_f\, dm_f + m_g\, dg_g + g_g\, dm_g$$

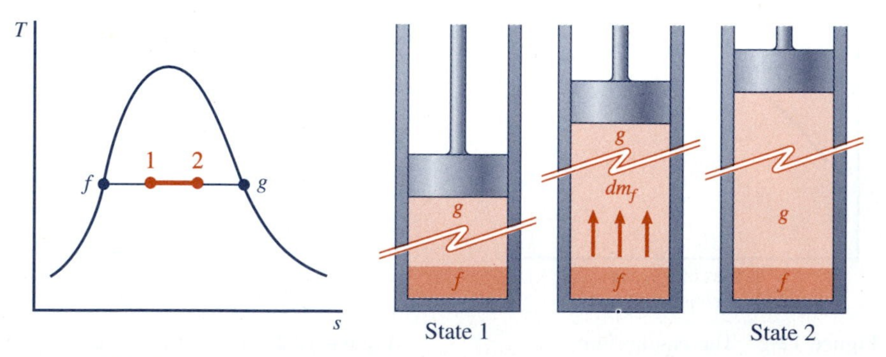

Figure 12·3 Change in state of a system of two phases of a pure substance in equilibrium.

Both g_f and g_g are constant because the pressure and temperature of the system are constant; hence

$$dG = 0 + g_f\, dm_f + 0 + g_g\, dm_g$$
$$= -g_f\, dm_g + g_g\, dm_g = (g_g - g_f)\, dm_g$$

But for this process of a pure substance at constant pressure and temperature, the system is in equilibrium; therefore, $dG = 0$ and

$$g_g = g_f$$

For any two phases (not just liquid and vapor) in equilibrium, the specific Gibbs functions are the same. At a triple state all three phases have the same g values.

For any two phases in equilibrium, the specific Gibbs functions are the same.

The Clapeyron equation that was derived in Section 7·2 from the Maxwell equations can be derived alternatively from the above conclusion that two phases of a pure substance existing together in equilibrium must have equal g values. Consider for the purpose of illustration a mixture of liquid and vapor. If the mixture is in equilibrium, then, as shown above, $g_f = g_g$. Let the temperature (and consequently the pressure) of the mixture change by an infinitesimal amount. The g values will change by dg_f and dg_g as the system goes to the new equilibrium state infinitesimally close to the original state, but, since the g values of the two phases must remain equal to each other, $dg_f = dg_g$. For each phase we can write $dg = v\, dp - s\, dT$; so, noting that dp and dT are the same for the two phases, we have

The Clapeyron equation can be derived using the Gibbs function expression describing complete equilibrium.

$$v_f\, dp - s_f\, dT = v_g\, dp - s_g\, dT$$

Rearranging gives

$$\frac{dp}{dT} = \frac{s_g - s_f}{v_g - v_f} = \frac{s_{fg}}{v_{fg}}$$

From $T\, ds = dh - v\, dp$ we find that $Ts_{fg} = h_{fg}$, which we then substitute into the equation above to give the Clapeyron equation:

$$\frac{dp}{dT} = \frac{h_{fg}}{Tv_{fg}} \qquad (7\cdot13)$$

12·2 The Equilibrium Constant, K_p

In a system that is initially in thermal and mechanical equilibrium, but not in chemical equilibrium, chemically reactive constituents will react. The reactions will proceed until the system reaches chemical equilibrium. This may or may not mean that all the reactants are converted to products. In fact, the extent to which the chemical reaction proceeds is precisely what we are interested in determining. Therefore, to determine the equilibrium state for any group of such constituents we now apply the equilibrium criterion $(dG)_{p,T} \leq 0$ to ideal-gas chemical reactions. The first step is to derive an expression for the Gibbs function of an ideal-gas mixture. Then we apply the equilibrium criterion, and finally we rearrange the resulting equation to show that for any reaction the relationship among the partial pressures of the constituents is a function of temperature only. We call this temperature function the equilibrium constant. Finally, we show how to determine the equilibrium constant for any reaction from the equilibrium constants describing the formation reactions of the constituents.

Derivation of the equilibrium constant

First Step: The Gibbs Function of an Ideal-Gas Mixture. If we assume that the expressions for the enthalpy and the entropy of a mixture of inert ideal gases apply also to a reactive mixture, we have

$$H_m = \sum N_i \overline{h}_i \tag{a}$$

and

$$S_m = \sum N_i \overline{s}_i(T, V_m) = \sum N_i \overline{s}_i(p_i, T) \tag{b}$$

where, in accordance with the Dalton model described in Chapter 3, the entropy of each constituent, i, is that of the constituent existing alone at the temperature and volume of the mixture or at the constituent partial pressure, p_i, and the mixture temperature. Recall that in Chapter 9 we dropped the subscript m for the mixture temperature since all constituents are at the same temperature and we are concerned with no other temperature. Eventually, we wish to apply the equilibrium condition to the ideal gas mixture. Therefore, it is useful to express the Gibbs function of the mixture in terms of the mixture pressure and temperature. We begin by expressing the entropy of the mixture at mixture pressure and temperature.

The entropy of each constituent at the mixture pressure and temperature can be obtained by first writing

$$\overline{s}_i(p_i, T) = \overline{s}^\circ_{ref} + \overline{\phi} - \overline{\phi}_{ref} - \overline{R} \ln \frac{p_i}{p^\circ}$$

where $\overline{s}^\circ_{ref}$ is at the standard state (see Section 11·7) and $\overline{\phi} - \overline{\phi}_{ref} = \int_{T_{ref}}^{T} \overline{c}_p \, dT/T$, as introduced in Section 6·2·1. If we then add and subtract $\overline{R} \ln y_i$ on the right side, where y_i is the mole fraction of constituent i, we have

$$\overline{s}_i(p_i, T) = \overline{s}^\circ_{ref} + \overline{\phi} - \overline{\phi}_{ref} - \overline{R} \ln \frac{p_i/y_i}{p^\circ} - \overline{R} \ln y_i$$

But $p_i/y_i = p_m$, where p_m is the mixture pressure, so

The relationship between the entropy of a constituent at its partial pressure to its entropy at the mixture pressure

$$\begin{aligned}\overline{s}_i(p_i, T) &= \overline{s}^\circ_{ref} + \overline{\phi} - \overline{\phi}_{ref} - \overline{R} \ln \frac{p_m}{p^\circ} - \overline{R} \ln y_i \\ &= \overline{s}_i(p_m, T) - \overline{R} \ln y_i\end{aligned} \tag{c}$$

Then we combine Eqs. *a*, *b*, and *c* with the defining equation for the Gibbs function to get

$$G_m = H_m - TS_m = \sum N_i\bar{h}_i - T\sum N_i\bar{s}_i(p_i, T)$$
$$= \sum N_i\bar{h}_i - T\sum N_i\bar{s}_i(p_m, T) + T\sum N_i\bar{R}\ln y_i$$
$$= \sum N_i[\bar{g}_i(p_m, T) + \bar{R}T\ln y_i] \qquad (12\cdot3)$$

The Gibbs function of the mixture in terms of the constituents at the pressure and temperature of the mixture

Equation 12·3 gives the Gibbs function of the mixture in terms of the Gibbs functions of the constituents at the pressure and temperature *of the mixture* instead of in terms of the constituent partial pressures.

Second Step: Application of the Equilibrium Criterion to an Ideal-Gas Mixture. Applying the equilibrium criterion $(dG)_{p,T} = 0$ to a mixture of ideal gases, we differentiate Eq. 12·3 to obtain

$$(dG_m)_{p_m,T} = \sum N_i d[\bar{g}_i(p_m, T) + \bar{R}T\ln y_i]$$
$$+ \sum [\bar{g}_i(p_m, T) + \bar{R}T\ln y_i]\, dN_i = 0 \qquad (d)$$

Under the conditions of constant-mixture pressure and temperature, however, the first summation term of Eq. *d* is zero because (1) the mixture pressure and temperature are constant, therefore $\bar{g}_i(p_m, T)$ is unchanged so $d\bar{g}_i(p_m, T) = 0$, and (2)

$$\sum N_i d(\bar{R}T\ln y_i) = \bar{R}T\sum N_i\frac{dy_i}{y_i} = \bar{R}T\sum N_i\frac{dy_i}{N_i/N_m}$$
$$= \bar{R}T\sum N_m\, dy_i = \bar{R}TN_m\sum dy_i$$
$$= \bar{R}TN_m(dy_1 + dy_2 + \cdots) = \bar{R}TN_m d\left(\sum y_i\right)$$
$$= 0$$

because the sum of $y_1, y_2, \ldots$ is unity by definition. Thus Eq. *d* becomes

$$(dG_m)_{p_m,T} = \sum [\bar{g}_i(p_m, T) + \bar{R}T\ln y_i]\, dN_i = 0 \qquad (e)$$

Let us now apply the equilibrium criterion in this form to an ideal-gas chemical reaction of the form

$$\nu_1 A_1 + \nu_2 A_2 \rightarrow \nu_3 A_3 + \nu_4 A_4 \qquad (f)$$

where ν_1 moles of ideal gas A_1 react with ν_2 moles of ideal gas A_2 to form ν_3 moles of ideal gas A_3 and ν_4 moles of ideal gas A_4. The ν's are the stoichiometric coefficients that satisfy the reaction equation. These coefficients are independent of the amounts of constituents actually present at any time in the system, which we represent by $N_1, N_2, \ldots$. Although the ν's are independent of the N's, in a mixture of reacting gases, the *changes* in the number of moles of the various constituents present (dN_1, dN_2, . . .) are proportional to the corresponding stoichiometric coefficients as follows:

$$-\frac{dN_1}{\nu_1} = -\frac{dN_2}{\nu_2} = \frac{dN_3}{\nu_3} = \frac{dN_4}{\nu_4} \qquad (g)$$

where the two terms in Eq. g that pertain to the reactants carry minus signs because all of the ν's are positive, and, as the reaction proceeds in the direction shown, N_1 and N_2 decrease while N_3 and N_4 increase. In general form, Eq. e can then be written as

$$\sum_{\text{products}} [\overline{g}_j(p_m, T) + \overline{R} \ln y_j]\nu_j - \sum_{\text{reactants}} [\overline{g}_i(p_m, T) + \overline{R}T \ln y_i]\nu_i = 0 \qquad (h)$$

where we use a subscript i to refer to the reactants and j to refer to the products.

Third Step: Definition of the Equilibrium Constant. The quantities within brackets in Eq. h can be reexpanded and rearranged to form the sum of a temperature function and a partial pressure function. For each constituent,

The criterion of equilibrium separated into the sum of a temperature function and a pressure function

$$\begin{aligned}
[\overline{g}_i(p_m, T) + \overline{R}T \ln y_i] &= \overline{h}_i(T) - T\overline{s}_i(p_m, T) + \overline{R}T \ln y_i \\
&= \overline{h}_i(T) - T\left(\overline{s}^{\circ}_{ref} + \overline{\phi} - \overline{\phi}_{ref} - \overline{R} \ln \frac{p_m}{p^{\circ}}\right) + \overline{R}T \ln y_i \\
&= \overline{h}_i(T) - T\overline{s}^{\circ}_{ref} - T\overline{\phi} + T\overline{\phi}_{ref} + \overline{R}T \ln \frac{y_i p_m}{p^{\circ}} \\
&= f_i(T) + \overline{R}T \ln \frac{p_i}{p^{\circ}} \qquad (i)
\end{aligned}$$

Equation h becomes

$$\sum_{\text{products}} \left[f_j(T) + \overline{R}T \ln \frac{p_j}{p^{\circ}}\right]\nu_j - \sum_{\text{reactants}} \left[f_i(T) + \overline{R}T \ln \frac{p_i}{p^{\circ}}\right]\nu_i = 0 \qquad (j)$$

Rearranging gives

$$\sum_{\text{products}} \nu_j \ln \frac{p_j}{p^{\circ}} - \sum_{\text{reactants}} \nu_i \ln \frac{p_i}{p^{\circ}} = \frac{1}{\overline{R}T}\left[\sum_{\text{reactants}} \nu_i f_i(T) - \sum_{\text{products}} \nu_j f_j(T)\right]$$

$$\sum_{\text{products}} \ln\left(\frac{p_j}{p^\circ}\right)^{\nu_j} - \sum_{\text{reactants}} \ln\left(\frac{p_i}{p^\circ}\right)^{\nu_i} = \frac{1}{\bar{R}T}\left[\sum_{\text{reactants}} \nu_i f_i(T) - \sum_{\text{products}} \nu_j f_j(T)\right]$$

$$\ln\left\{\frac{\Pi_{\text{products}}(p_j/p^\circ)^{\nu_j}}{\Pi_{\text{reactants}}(p_i/p^\circ)^{\nu_i}}\right\} = F(\nu_i, \nu_j, T)$$

Therefore,

$$\frac{\Pi_{\text{products}}(p_j/p^\circ)^{\nu_j}}{\Pi_{\text{reactants}}(p_i/p^\circ)^{\nu_i}} = F'(\nu_i, \nu_j, T)$$

For a specific reaction of ideal gases, the values of the ν's are known and fixed. Consequently,

$$\frac{\Pi_{\text{products}}(p_j/p^\circ)^{\nu_j}}{\Pi_{\text{reactants}}(p_i/p^\circ)^{\nu_i}} = \text{function of } T$$

This means that for a given group of ideal gases, the relationship among their partial pressures at equilibrium depends only on the temperature. For convenience in handling this temperature function, we define it as the equilibrium constant K_p. Thus, the equilibrium constant, valid only for ideal gases, is defined as

$$K_p(T) \equiv \frac{\Pi_{\text{products}}(p_j/p^\circ)^{\nu_j}}{\Pi_{\text{reactants}}(p_i/p^\circ)^{\nu_i}} \quad (12\cdot4)$$

The equilibrium constant for a reaction of ideal gases

For the reaction shown in Eq. *f*, this becomes

$$K_p(T) = \frac{(p_3/p^\circ)^{\nu_3}(p_4/p^\circ)^{\nu_4}}{(p_1/p^\circ)^{\nu_1}(p_2/p^\circ)^{\nu_2}}$$

Test yourself: Le Chatelier's principle states that if the number of moles of reactants exceeds that of the products, then increasing the pressure at which the reaction occurs forces the reaction toward completion. Verify this principle from the material presented here.

Let us repeat for emphasis the conclusion reached in the preceding paragraph:

If a mixture of ideal gases A_1, A_2, A_3, and A_4 undergoes the reaction

$$\nu_1 A_1 + \nu_2 A_2 \rightarrow \nu_3 A_3 + \nu_4 A_4$$

and is in equilibrium, then the equilibrium constant, K_p, defined as

$$K_p = \frac{(p_3/p^\circ)^{\nu_3}(p_4/p^\circ)^{\nu_4}}{(p_1/p^\circ)^{\nu_1}(p_2/p^\circ)^{\nu_2}}$$

is a function of temperature only.

We will use this conclusion to determine the equilibrium composition of reacting ideal-gas mixtures. Another way to determine such compositions is simply to find the composition that results in the minimum Gibbs function of the mixture at a given pressure and temperature in accordance with the equilibrium criterion

$$(dG)_{p,T} \leq 0 \qquad (12{\cdot}2b)$$

The computer program of Reference 12·4 is based on this approach.

12·2·1 The Nature of K_p

We can make the following statements regarding K_p:

1. K_p does not depend on the amount of the various constituents initially present in the mixture.
2. The value of K_p does not depend on any past state of the mixture (although some earlier state may be referred to in establishing the amount of each element present; see Examples 12·2, 12·4, and 12·5).
3. The value of K_p is not affected by the presence of constituents other than those that take part in the reaction used to define K_p (see Example 12·3).

Several different methods of determining equilibrium constants experimentally and from other thermochemical data are used. For example, K_p for a given reaction can be measured at various temperatures by analyzing the equilibrium gas mixture. For an ideal-gas reaction in which the number of moles of products is different from the number of moles of reactants, the extent to which the reaction has proceeded when equilibrium is reached can be determined by measuring the total volume of the constituents at equilibrium. This provides an indication of the extent of a given reaction.

K_p data for several ideal-gas reactions are given in Table A·18. Figure 12·4 presents K_p data graphically for some of these reactions to show typical variations of K_p with temperature. Often in equilibrium calculations, several relations, including one or more K_pT relations, must be solved simultaneously. Thus, it is convenient to have the relationships in the form of equations instead of charts or tables, and Fig. 12·4 indicates that a curve-fitting technique can be readily used to obtain K_pT equations over limited temperature ranges. (Section 12·4 gives a suggestion to help in the curve-fitting process.)

12·2·2 K_p in Terms of the Equilibrium Constant of Formation, K_f

In Chapter 11 we noted the difficulty of tabulating Δh_R for every conceivable reaction. To ease this burden, we derived an expression for Δh_R in terms of the Δh_f's of the reaction constituents. Similarly, to avoid the mammoth task

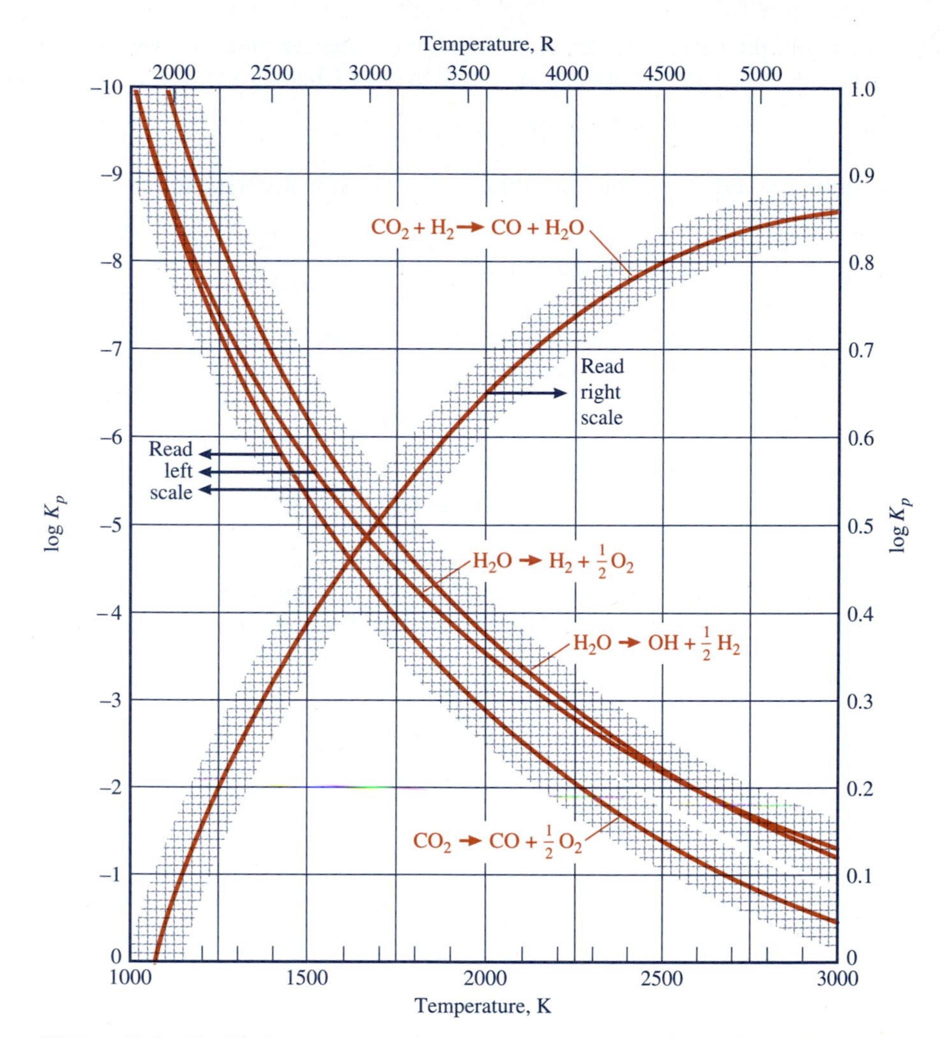

Figure 12·4 Equilibrium constants for several reactions. Data from Table A·18.

of tabulating K_p for every reaction, we now derive an expression relating the equilibrium constant to the *equilibrium constant of formation* of reacting constituents.

Standard data sources for K_p, including some listed in the references at the end of this chapter, often give K_p values for substances rather than for reactions. In such a case, K_p is interpreted as the equilibrium constant for the reaction forming that substance *from its elements* in their most stable form. It is called the **equilibrium constant of formation** and often the symbol K_f is used.

Equilibrium constant of formation

Derivation of the equilibrium constant in terms of the equilibrium constants of formation for the reaction constituents

To begin the derivation, let us examine the same reaction that we used to derive the relationship between Δh_R and Δh_f in Chapter 11:

$$CH_4 + 2O_2 \rightarrow CO_2 + 2H_2O$$

According to Eq. 12·4, the equilibrium constant for this reaction is

$$K_{p,CH_4+2O_2\rightarrow CO_2+2H_2O} = \frac{(p_{CO_2}/p^\circ)(p_{H_2O}/p^\circ)^2}{(p_{CH_4}/p^\circ)(p_{O_2}/p^\circ)^2} \tag{k}$$

For each of the constituents, a reaction describing its formation from elements in their most stable form can be written as follows:

$$C + O_2 \rightarrow CO_2$$
$$H_2 + \tfrac{1}{2}O_2 \rightarrow H_2O$$
$$C + 2H_2 \rightarrow CH_4$$

(*Note*: We need not write an equation for the formation of O_2 since it is in its most stable form. In this case, by definition, the equilibrium constant of formation for O_2 is 1.) The equilibrium constant for each of these reactions is

$$K_{p,CO_2} = \frac{(p_{CO_2}/p^\circ)}{(p_{O_2}/p^\circ)(p_C/p^\circ)}$$

$$K_{p,H_2O} = \frac{(p_{H_2O}/p^\circ)}{(p_{H_2}/p^\circ)(p_{O_2}/p^\circ)^{1/2}}$$

$$K_{p,CH_4} = \frac{(p_{CH_4}/p^\circ)}{(p_C/p^\circ)(p_{H_2}/p^\circ)^2}$$

But, recall that according to the definition of K_f, the above equations represent identically the K_f's for CO_2, H_2O, and CH_4, respectively. We then solve these equations for the relevant pressure ratios as follows:

$$\left(\frac{p_{CO_2}}{p^\circ}\right) = \left(\frac{p_{O_2}}{p^\circ}\right)\left(\frac{p_C}{p^\circ}\right)K_{f,CO_2}$$

$$\left(\frac{p_{H_2O}}{p^\circ}\right) = \left(\frac{p_{H_2}}{p^\circ}\right)\left(\frac{p_{O_2}}{p^\circ}\right)^{1/2}K_{f,H_2O}$$

$$\left(\frac{p_{CH_4}}{p^\circ}\right) = \left(\frac{p_C}{p^\circ}\right)\left(\frac{p_{H_2}}{p^\circ}\right)^2 K_{f,CH_4}$$

Substituting into Eq. *k*, we find that

$$K_{p,CH_4+2O_2\rightarrow CO_2+2H_2O} = \frac{(K_{f,CO_2})(K_{f,H_2O})^2}{(K_{f,CH_4})}$$

This equation can be generalized to

$$K_p = \frac{\Pi_{\text{products}}(K_{f,j})^{\nu_j}}{\Pi_{\text{reactants}}(K_{f,i})^{\nu_i}} \tag{12·5a}$$

The equilibrium constant in terms of the equilibrium constants of formation

or, in logarithmic form,

$$\ln K_p = \sum_{\text{products}} \nu_j \ln K_{f,j} - \sum_{\text{reactants}} \nu_i \ln K_{f,i}$$
$$\log K_p = \sum_{\text{products}} \nu_j \log K_{f,j} - \sum_{\text{reactants}} \nu_i \log K_{f,i} \tag{12·5b}$$

***Caution*: Data are usually given in terms of log K_p instead of ln K_p.**

Both equations 12·5*a* and 12·5*b* show that any K_f equal to unity does not affect the resulting value of K_p. This is a further reminder that when calculating K_p we need not consider reactants or products that are elements in their most stable form.

Figures 12·5 through 12·7 present K_f data for seven substances found in hydrocarbon combustion reactions.

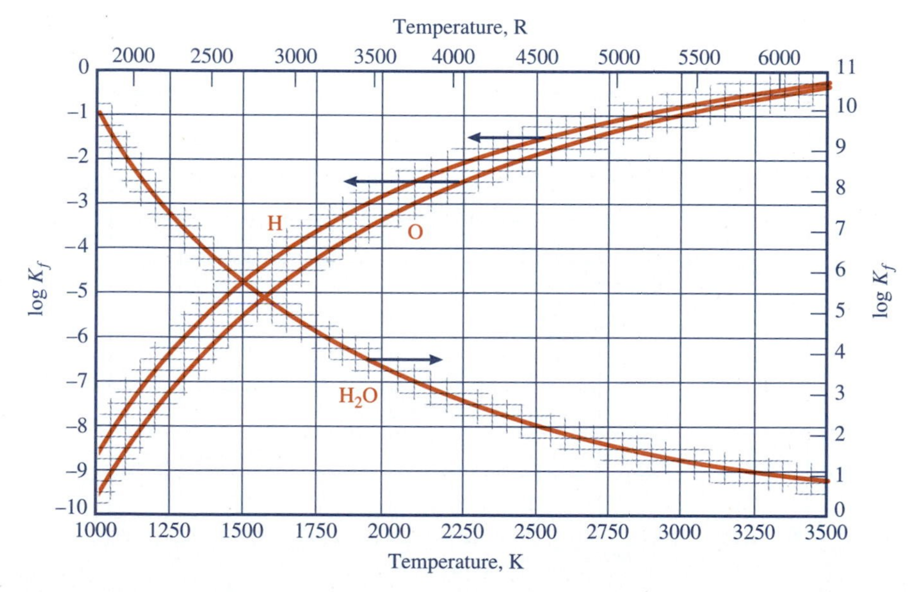

Figure 12·5 Equilibrium constants of formation for H, O, and H_2O. Data from Table A·19.

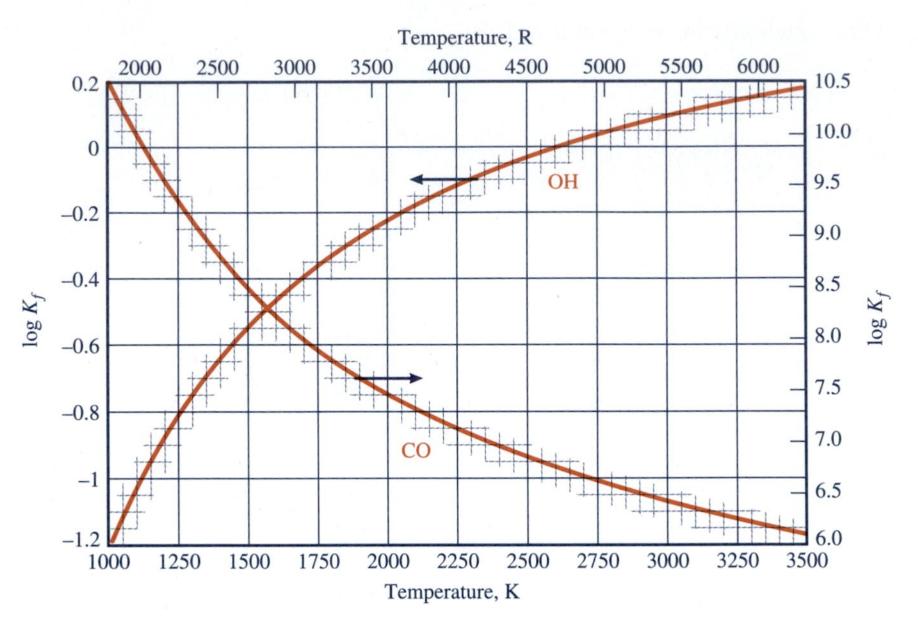

Figure 12·6 Equilibrium constants of formation for OH and CO. Data from Table A·19.

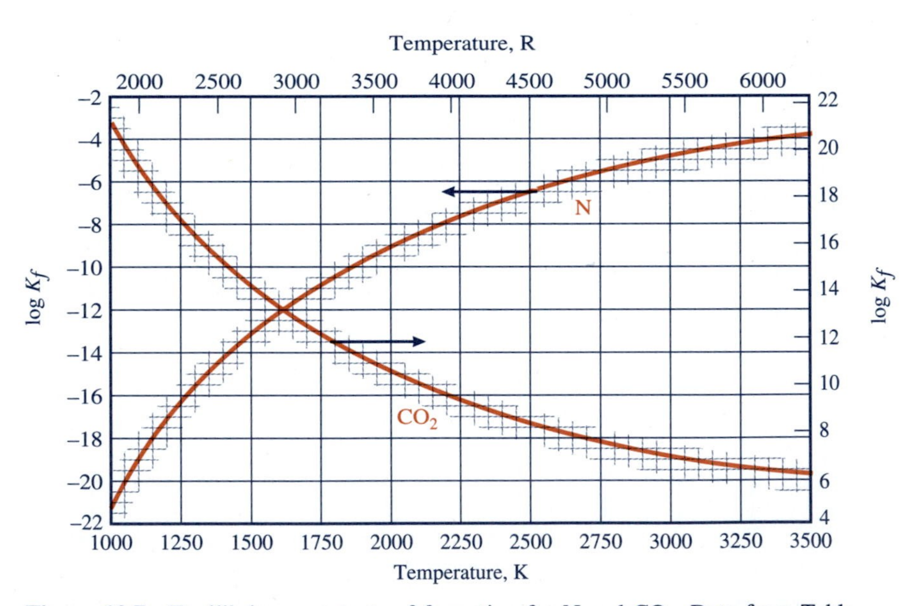

Figure 12·7 Equilibrium constants of formation for N and CO_2. Data from Table A·19.

We now take up several examples illustrating the use of the equilibrium constant. Remember that we are treating only the special case of ideal-gas reactions.

Example 12·1 Equilibrium Constant

PROBLEM STATEMENT

For the reaction $\frac{1}{2}O_2 + \frac{1}{2}N_2 \rightarrow NO$, K_p at 3300 K (or 5940 R) is 0.1694. Determine K_p at this temperature for (*a*) the reaction $O_2 + N_2 \rightarrow 2NO$ and (*b*) the reaction $NO \rightarrow \frac{1}{2}O_2 + \frac{1}{2}N_2$.

SOLUTION

The equilibrium constant given is

$$K_{p1} = \frac{p_{NO}/p^\circ}{(p_{O_2}/p^\circ)^{1/2}(p_{N_2}/p^\circ)^{1/2}} = 0.1694$$

(*a*) For the reaction $O_2 + N_2 \rightarrow 2NO$, call the equilibrium constant K_{p2}:

$$K_{p2} = \frac{(p_{NO}/p^\circ)^2}{(p_{O_2}/p^\circ)(p_{N_2}/p^\circ)} = K_{p1}^2 = (0.1694)^2 = 0.0287$$

(*b*) For the reaction $NO \rightarrow \frac{1}{2}O_2 + \frac{1}{2}N_2$ the equilibrium constant K_{p3} is

$$K_{p3} = \frac{(p_{O_2}/p^\circ)^{1/2}(p_{N_2}/p^\circ)^{1/2}}{p_{NO}/p^\circ} = \frac{1}{K_{p1}} = \frac{1}{0.1694} = 5.90$$

Comment: The equilibrium constant is a measure of the degree of completion for a reaction. The higher the value of K_p, the more complete the reaction. Thus, at 3300 K reaction *b* goes much further toward completion than reaction *a*.

Example 12·2 Equilibrium Mixture Composition

PROBLEM STATEMENT

At 2500 K (or 4500 R), $K_p = 27.5$ for the reaction $CO + \frac{1}{2}O_2 \rightarrow CO_2$. For an initial mixture of $CO + \frac{1}{2}O_2$, determine the composition of the equilibrium mixture at 2500 K (or 4500 R) and (*a*) 100 kPa, (*b*) 500 kPa.

SOLUTION

Let x be the fraction of the carbon that is in the form of CO under equilibrium conditions. Then the equilibrium mixture is

$$xCO + (1 - x)CO_2 + \tfrac{1}{2}xO_2$$

This can also be expressed by saying that the actual reaction is

$$CO + \tfrac{1}{2}O_2 \rightarrow xCO + (1 - x)CO_2 + \tfrac{1}{2}xO_2$$

but remember that the initial state has no bearing on the calculation except to specify the amounts of the various elements present. The total number of moles of mixture is $x + 1 - x + \frac{1}{2}x = 1 + \frac{1}{2}x$. The partial pressures of the three constituents are therefore

$$p_{CO} = y_{CO}p_m = \frac{x}{1 + \frac{1}{2}x}p_m = \frac{2x}{2 + x}p_m$$

$$p_{CO_2} = y_{CO_2}p_m = \frac{1 - x}{1 + \frac{1}{2}x}p_m = \frac{2 - 2x}{2 + x}p_m$$

$$p_{O_2} = x_{O_2}p_m = \frac{\frac{1}{2}x}{1 + \frac{1}{2}x}p_m = \frac{x}{2 + x}p_m$$

Substituting these values into the defining equation for K_p gives

$$K_p = \frac{p_{CO_2}/p^\circ}{(p_{CO}/p^\circ)(p_{O_2}/p^\circ)^{1/2}} = \frac{2 - 2x}{2x[x/(2 + x)]^{1/2}(p_m/p^\circ)^{1/2}}$$

$$K_p^2 = \frac{(1 - x)^2}{x^2[x/(2 + x)](p_m/p^\circ)} = \frac{(1 - x)^2(2 + x)}{x^3(p_m/p^\circ)} = (27.5)^2$$

For $p_m = p^\circ = 100$ kPa the solution of this equation is $x = 0.129$; and, for $p_m = 500$ kPa, $x = 0.0776$.

(*a*) For $p_m = 100$ kPa, $x = 0.129$ and the mole fractions are

$$y_{CO} = \frac{2x}{2 + x} = 0.121$$

$$y_{CO_2} = \frac{2 - 2x}{2 + x} = 0.818$$

$$y_{O_2} = \frac{x}{2 + x} = 0.061$$

(*b*) For $p = 500$ kPa, $x = 0.0776$ and the mole fractions are

$$y_{CO} = \frac{2x}{2 + x} = 0.075$$

$$y_{CO_2} = \frac{2 - 2x}{2 + x} = 0.888$$

$$y_{O_2} = \frac{x}{2 + x} = 0.037$$

A comparison of parts *a* and *b* shows that increasing the pressure causes the reaction $CO + \frac{1}{2}O_2 \rightarrow CO_2$ at this temperature to proceed further to the right, the direction of decreasing volume.

Example 12·3 Equilibrium Mixture Composition

PROBLEM STATEMENT

Using the K_p data of Example 12·2, determine the equilibrium mixture at 100 kPa and 2500 K (or 4500 R) for CO and the stoichiometric amount of air.

SOLUTION

The complete reaction equation is

$$CO + \tfrac{1}{2}O_2 + \tfrac{1}{2}(3.76)N_2 \rightarrow CO_2 + \tfrac{1}{2}(3.76)N_2$$

If x stands for the fraction of the carbon that is in CO under equilibrium conditions, then the equilibrium mixture is

$$xCO + (1 - x)CO_2 + \tfrac{1}{2}xO_2 + \frac{3.76}{2}N_2$$

The total number of moles of mixture is $2.88 + \frac{1}{2}x$. The partial pressures of the four constituents are

$$p_{CO} = y_{CO}p_m = \frac{x}{2.88 + \frac{1}{2}x}p_m = \frac{2x}{5.76 + x}p_m$$

$$p_{CO_2} = y_{CO_2}p_m = \frac{1 - x}{2.88 + \frac{1}{2}x}p_m = \frac{2 - 2x}{5.76 + x}p_m$$

$$p_{O_2} = y_{O_2}p_m = \frac{\frac{1}{2}x}{2.88 + \frac{1}{2}x}p_m = \frac{x}{5.76 + x}p_m$$

$$p_{N_2} = y_{N_2}p_m = \frac{1.88}{2.88 + \frac{1}{2}x}p_m = \frac{3.76}{5.76 + x}p_m$$

Substituting these values into the defining equation for K_p gives

$$K_p = \frac{p_{CO_2}/p^\circ}{(p_{CO}/p^\circ)(p_{O_2}/p^\circ)^{1/2}} = \frac{2 - 2x}{2x[x/(5.76 + x)]^{1/2}(p_m/p^\circ)^{1/2}}$$

$$K_p^2 = \frac{(1 - x)^2(5.76 + x)}{x^3(p_m/p^\circ)} = (27.5)^2$$

For $p_m = p^\circ = 100$ kPa, the solution of this equation is $x = 0.175$, so the mole fractions are

$$y_{CO} = \frac{2x}{5.76 + x} = 0.059$$

$$y_{CO_2} = \frac{2 - 2x}{5.76 + x} = 0.278$$

$$y_{O_2} = \frac{x}{5.76 + x} = 0.029$$

$$y_{N_2} = \frac{3.76}{5.76 + x} = 0.634$$

(Notice that the presence of nitrogen affects the extent to which the reaction proceeds at this temperature, not because the equilibrium constant is changed, but rather because the partial pressures change. The presence of excess oxygen would cause the reaction to go further toward completion at this temperature, but again this result follows from the same equilibrium constant value. Remember that the value of K_p is independent of the amounts of the various constituents actually present.)

12·3 Using the Equilibrium Constant in Applications

The three examples above deal with only the conditions of equilibrium at given states. The following examples involve equilibrium calculations as a necessary part of applying the first and second laws to combustion processes. In one kind of application, the temperatures and pressures of the end states are specified. Then equilibrium calculations determine the mixture composition that is subsequently used to evaluate mixture properties. Finally, these properties are used in first and second law calculations (see Example 12·4). In this type of application, to determine the mixture composition at a fixed temperature we use the following:

Equations for determining the mixture composition at a fixed temperature

1. Mass balances for individual elements in the reacting constituents.
2. The defining relationships between K_f values at a given temperature and the mole fractions or partial pressures of the constituents.

Depending on the mixture composition, more than one reaction may be involved. Solving for the mixture composition requires a number of equations equal to the number of reacting constituents. One equation is provided by the mass balance for each element involved in a reaction. One K_p equation is provided for each reaction considered. Therefore, in general, the number of reactions that must be considered is equal to the number of reacting constituents in the mixture minus the number of elements in those constituents.

In another kind of application the end states must be determined from information regarding energy transfers. An example is the determination of an adiabatic combustion temperature, where the specified energy transfers are $Q = 0$ and $W = 0$. For an exothermic reaction carried out adiabatically, the temperature of the mixture increases as the reaction proceeds, where the energy for the increased enthalpy of the constituents is supplied by the reaction. However, equilibrium calculations show that for exothermic reactions, the higher the temperature of the products, the less complete the reaction can be. The less complete the reaction is, the less energy there is available from the enthalpy of reaction for raising the products temperature, so the *equilibrium* adiabatic combustion temperature is lower than the *nonequilibrium* or maximum adiabatic combustion temperature that is calculated assuming the

The equilibrium adiabatic combustion temperature is lower than the nonequilibrium or maximum adiabatic combustion temperature.

reaction goes to completion. This is why in Chapter 11 we referred to the calculated adiabatic combustion temperature as the *maximum* adiabatic combustion temperature.

The calculation of the equilibrium adiabatic combustion temperature involves several simultaneous relationships:

Equations for determining the equilibrium adiabatic combustion temperature

1. Mass balances for the individual elements in the reacting constituents.
2. Relationships between the mixture composition and K_p.
3. K_pT relationships. These are often given in tabular or graphical form but usually can be readily fitted by equations, at least for limited temperature ranges.
4. hT relationships. These may be in the form of tables, charts, or equations.
5. The first law, which for the special case of ideal gases and steady adiabatic flow with no work through an open system is

$$Q = 0 = \sum_{\text{products}} N(\bar{h} - \bar{h}_{ref} + \Delta\bar{h}_{f,ref}) - \sum_{\text{reactants}} N(\bar{h} - \bar{h}_{ref} + \Delta\bar{h}_{f,ref}) \qquad (11\cdot9)$$

For this application, the number of reactions that must be considered is still equal to the number of reacting constituents in the mixture minus the number of elements in those constituents. However, relationships 1 and 2 cannot be solved without involving relationships 3, because the K_p values are functions of the unknown temperature. The temperature is involved also in relationships 4, which are coupled to relationship 5 through the enthalpy values. Therefore, all five of the relationships listed must be solved simultaneously (see Examples 12·5 and 12·6).

The best calculation procedure depends on (1) the data at hand, including the accuracy and form of the data; (2) the computational tools at hand, including equipment, software, and one's skill in using them; (3) the number of similar problems to be solved; and (4) the trade-off between the accuracy desired and the computational and data-gathering effort.

Example 12·4 Equilibrium Mixture, First Law

PROBLEM STATEMENT

Methane is burned with 80 percent of stoichiometric air in a steady-flow process at 100 kPa. Methane and air are both supplied at 25°C (77 F), and the products leave at 1700 K (3060 R). Assuming that no CH_4, OH, NO, or free oxygen appears in the products, determine the amount of heat transferred per kilogram of methane.

SOLUTION

The reaction is

$$CH_4 + 0.8(2)O_2 + 0.8(2)3.76N_2 \rightarrow xCO_2 + (1 - x)CO + zH_2O + (2 - z)H_2 + 0.8(2)3.76N_2$$

x and z must be determined before an energy balance can be made. One relation between x and z is given by a mass balance for oxygen

$$1.6 = x + \frac{1}{2}(1 - x) + \frac{1}{2}z \tag{A}$$

$$2.2 = x + z$$

A second relationship between x and z can be found by means of the equilibrium constant for the reaction $CO_2 + H_2 \rightarrow H_2O + CO$, which is known as the *water–gas reaction*:

$$K_p = \frac{(p_{CO}/p^\circ)(p_{H_2O}/p^\circ)}{(p_{CO_2}/p^\circ)(p_{H_2}/p^\circ)}$$

The partial pressures for substitution into this expression are obtained from the equilibrium mixture

$$xCO_2 + (1 - x)CO + zH_2O + (2 - z)H_2 + 6.02N_2$$

as

$$p_{CO} = \frac{(1 - x)p_m}{x + (1 - x) + z + (2 - z) + 6.02} = \frac{1 - x}{9.02}p_m$$

$$p_{H_2O} = \frac{z}{9.02}p_m$$

$$p_{CO_2} = \frac{x}{9.02}p_m$$

$$p_{H_2} = \frac{2 - z}{9.02}p_m$$

Thus, for $K_p = 3.22$ at 1700 K (or 3060 R) (from Table A·18 as $\log K_p = 0.508$), we have

$$K_p = \frac{(1 - x)z}{x(2 - z)} = 3.22 \tag{B}$$

Solving Eqs. *A* and *B* gives

$$x = 0.574 \qquad z = 1.626$$

Thus the actual reaction equation is

$$CH_4 + 1.6O_2 + 6.02N_2 \rightarrow 0.574CO_2 + 0.426CO + 1.626H_2O + 0.374H_2 + 6.02N_2$$

The enthalpy of reaction at 25°C for this particular reaction is

$$\Delta H_R = N_{CH_4}\,\Delta\bar{h}_R = \sum_{\text{products}} N\,\Delta\bar{h}_{f,ref} - \sum_{\text{reactants}} N\,\Delta\bar{h}_{f,ref}$$

$$\begin{aligned}\Delta\bar{h}_R &= 0.574\,\Delta\bar{h}_{f,ref,CO_2} + 0.426\,\Delta\bar{h}_{f,ref,CO} + 1.626\,\Delta\bar{h}_{f,ref,H_2O} \\ &\quad + 0 + 0 - \Delta\bar{h}_{f,ref,CH_4} - 0 - 0 \\ &= 0.574(-393{,}522) + 0.426(-110{,}530) \\ &\quad + 1.626(-241{,}826) - (-74{,}873) \\ &= -591{,}300 \text{ kJ/kmol } CH_4\end{aligned}$$

where the $\Delta\bar{h}_{f,ref}$ values are taken from Table A·17.

The first law for the steady-flow system is

$$Q = \Delta H_R + \sum_{\text{products}} N(\bar{h}_2 - \bar{h}_{ref}) - \sum_{\text{reactants}} N(\bar{h}_1 - \bar{h}_{ref})$$

Since the reactants enter at 25°C, the last term on the right-hand side is zero. Then, dividing by the number of kilomoles of CH_4, we have

$$\begin{aligned}q &= \Delta\bar{h}_R + 0.574(\bar{h}_2 - \bar{h}_{ref})_{CO_2} + 0.426(\bar{h}_2 - \bar{h}_{ref})_{CO} + 1.626(\bar{h}_2 - \bar{h}_{ref})_{H_2O} \\ &\quad + 0.374(\bar{h}_2 - \bar{h}_{ref})_{H_2} + 6.02(\bar{h}_2 - \bar{h}_{ref})_{N_2} \\ &= -591{,}300 + 0.574(82{,}855 - 9364) + 0.426(54{,}615 - 8671) \\ &\quad + 1.626(67{,}666 - 9904) + 0.374(51{,}298 - 8467) \\ &\quad + 6.02(54{,}103 - 8670) \\ &= -146{,}390 \text{ kJ/kmol } CH_4 = -9127 \text{ kJ/kg } CH_4\end{aligned}$$

Example 12·5 Equilibrium Adiabatic Combustion Temperature, Irreversibility

PROBLEM STATEMENT

Determine the equilibrium adiabatic combustion temperature for the steady-flow constant-pressure burning of a stoichiometric mixture of carbon monoxide and dry air with inlet conditions of 100 kPa, 25°C. Determine also the irreversibility of the process.

Analysis: We will assume ideal-gas behavior. Since we are looking for the temperature of the products mixture, we do not know the mixture composition because it in turn depends on the temperature. In addition to not knowing mole fractions to use in the first law, we do not have enthalpy values because they depend on the unknown temperature. Therefore, we will write the expression for K_p in terms of a composition variable.

For several values of product temperature, we will solve this equation for the composition. Then using for each temperature the corresponding composition and the enthalpy values obtained from hT tables, we will calculate the heat transfer by means of the first law. In this manner we will determine the temperature for which $Q = 0$. After the adiabatic combustion temperature and composition are determined, we will calculate the irreversibility by $i = T_0\,\Delta s_{\text{universe}}$.

SOLUTION

The reaction equation is

$$CO + \tfrac{1}{2}O_2 + \frac{3.76}{2}N_2 \rightarrow (1-x)CO_2 + xCO + \frac{x}{2}O_2 + 1.88N_2$$

The total number of moles of products is $1 - x + x + x/2 + 1.88 = 2.88 + x/2$. The partial pressures are

$$p_{CO_2} = \frac{1-x}{2.88 + x/2}p_m \qquad p_{CO} = \frac{x}{2.88 + x/2}p_m$$

$$p_{O_2} = \frac{x/2}{2.88 + x/2}p_m \qquad p_{N_2} = \frac{1.88}{2.88 + x/2}p_m$$

For the reaction $CO_2 \rightarrow CO + \tfrac{1}{2}O_2$, the equilibrium constant is

$$K_p = \frac{(p_{CO}/p^\circ)(p_{O_2}/p^\circ)^{1/2}}{p_{CO_2}/p^\circ} = \frac{y_{CO}y_{O_2}^{1/2}}{y_{CO_2}}\left(\frac{p_m}{p^\circ}\right)^{1/2} = \frac{x}{1-x}\left(\frac{x}{5.76 + x}\right)^{1/2}(1)^{1/2} \qquad (A)$$

The first law for the steady-flow system of ideal gases is

$$Q = H_{\text{products}} - H_{\text{reactants}} = \sum_{\text{products}} N(\bar{h} - \bar{h}_{ref} + \Delta\bar{h}_{f,ref}) - \sum_{\text{reactants}} N(\bar{h} - \bar{h} + \Delta\bar{h}_{f,ref})$$

where each $\Delta\bar{h}_f$ is on a per mole of constituent basis. Applying this equation to the case at hand, and dividing by the number of moles of CO entering gives

$$\begin{aligned}\bar{q} = {} & (1-x)(\bar{h} - \bar{h}_{ref} + \Delta\bar{h}_{f,ref})_{CO_2,p} + x(\bar{h} - \bar{h}_{ref} + \Delta\bar{h}_{f,ref})_{CO,p} + \frac{x}{2}(\bar{h} - \bar{h}_{ref} + \Delta\bar{h}_{f,ref})_{O_2,p} \\ & + 1.88(\bar{h} - \bar{h}_{ref} + \Delta\bar{h}_{f,ref})_{N_2,p} - (\bar{h} - \bar{h}_{ref} + \Delta\bar{h}_{f,ref})_{CO,r} - \frac{1}{2}(\bar{h} - \bar{h}_{ref} + \Delta\bar{h}_{f,ref})_{O_2,r} \\ & - 1.88(\bar{h} - \bar{h}_{ref} + \Delta\bar{h}_{f,ref})_{N_2,r}\end{aligned}$$

where the subscript r refers to the reactants and the subscript p refers to the products. Since $T_{ref} = T = 25°C$ and $\Delta\bar{h}_f$ of elements is zero, the first law simplifies to

$$\begin{aligned}\bar{q} = {} & (1-x)(\bar{h} - \bar{h}_{ref} + \Delta\bar{h}_{f,ref})_{CO_2,p} + x(\bar{h} - \bar{h}_{ref} + \Delta\bar{h}_{f,ref})_{CO,p} \\ & + \frac{x}{2}(\bar{h} - \bar{h}_{ref})_{O_2,p} + 1.88(\bar{h} - \bar{h}_{ref})_{N_2,p} - (\Delta\bar{h}_{f,ref})_{CO,r}\end{aligned}$$

Substituting from Tables A·15 and A·17 the numerical values that are independent of the products temperature gives

$$\bar{q} = (1 - x)(\bar{h} - 9364 - 393{,}522)_{CO_2,p} + x(\bar{h} - 8671 - 110{,}530)_{CO,p}$$
$$+ \frac{x}{2}(\bar{h} - 8683)_{O_2,p} + 1.88(\bar{h} - 8670)_{N_2,p} - (-110{,}530) \qquad (B)$$

Equations *A* and *B* are now entered into a computer equation solver and solved for several values of *T* and the corresponding log K_p values obtained from Table A·18 (or calculated from K_f values from Table A·19), as shown in the following table. The resulting values of *x* and *q* are also shown in the table. Fortuitously, the $\bar{q} = 0$ condition is seen to be met by a temperature close to 2400 K, one of the table values, so no closer search is needed; $x = 0.130$. (If *hT* equations were entered instead of discrete *T* and *h* values, or if an interpolation procedure is available to use with the *hT* data, the calculation can be made directly under the constraint of $\bar{q} = 0$.)

T, K	log $K_{f,CO}$	log K_{f,CO_2}	log K_p	K_p	x	$\bar{q}$, kJ/kmol CO
2100	7.319	9.846	−2.527	0.002 97	0.03623	−63,170
2200	7.190	9.398	−2.208	0.006 17	0.05828	−44,210
2300	7.072	8.988	−1.916	0.012 13	0.08938	−22,720
2400	6.961	8.613	−1.652	0.022 29	0.13030	1,483
2500	6.856	8.268	−1.412	0.038 73	0.18140	28,490
2600	6.756	7.949	−1.193	0.064 10	0.24200	58,100
2700	6.660	7.654	−0.994	0.101 39	0.30980	89,640
2800	6.576	7.380	−0.804	0.157 04	0.38540	123,300

For the steady-flow system per mole of CO entering

$$\bar{i} = T_0\,\Delta\bar{s}_{univ} = T_0(\Delta\bar{s}_{sys} + \Delta\bar{s}_{surr}) = T_0(0 + \bar{s}_2 - \bar{s}_1) = T_0(\bar{s}_p - \bar{s}_r)$$
$$= T_0\left[\sum_p N\left(\bar{s}^\circ_{ref} + \bar{\phi} - \bar{\phi}_{ref} - \bar{R}\ln\frac{p}{p^\circ}\right) - \sum_r N\left(\bar{s}^\circ_{ref} + \bar{\phi} - \bar{\phi}_{ref} - \bar{R}\ln\frac{p}{p^\circ}\right)\right]$$

where N is the number of moles of each constituent per mole of CO entering, $\bar{s}^\circ_{ref}$ is the absolute entropy of each constituent at 1 atm, 25°C; the p's are partial pressures, and p° is the standard-state pressure of 100 kPa. Because the mixture pressure is also 100 kPa, $p/p^\circ = p/p_m = y$, where y is the mole fraction. In Table A·16 and in the Keenan, Chao, and Kaye *Gas Tables*, $\bar{\phi} = \bar{s}^\circ$ for each constituent. Therefore, the last equation simplifies to

$$\bar{i} = T_0\Big[(1 - x)(\bar{\phi} - \bar{R}\ln y)_{CO_2,p} + x(\bar{\phi} - \bar{R}\ln y)_{CO,p} + \frac{x}{2}(\bar{\phi} - \bar{R}\ln y)_{O_2,p}$$
$$+ 1.88(\bar{\phi} - \bar{R}\ln y)_{N_2,p} - (\bar{\phi} - \bar{R}\ln y)_{CO,r} - \frac{1}{2}(\bar{\phi} - \bar{R}\ln y)_{O_2,r}$$
$$- 1.88(\bar{\phi} - \bar{R}\ln y)_{N_2,r}\Big]$$

For substituting mole fractions, note that per mole of CO entering there are 3.38 moles of reactants and $(2.88 + x/2) = 2.94$ moles of products:

$$\begin{aligned}\bar{i} = 298\bigg[&(1 - 0.130)\left(320.387 - 8.314 \ln \frac{1 - 0.130}{2.94}\right) + 0.130\bigg(265.366 \\ &- 8.314 \ln \frac{0.130}{2.94}\bigg) + \frac{0.130}{2}\left(275.706 - 8.314 \ln \frac{0.130}{2(2.94)}\right) \\ &+ 1.88\left(258.683 - 8.314 \ln \frac{1.88}{2.94}\right) - \left(197.653 - 8.314 \ln \frac{1}{3.38}\right) \\ &- \frac{1}{2}\left(205.147 - 8.314 \ln \frac{0.5}{3.38}\right) - 1.88\left(191.609 - 8.314 \ln \frac{1.88}{3.38}\right)\bigg] \\ = {}& 45{,}026 \text{ kJ/kmol CO entering}\end{aligned}$$

This much energy has been made unavailable by the combustion process. Thus the maximum work obtainable from this adiabatic steady-flow burning of CO is less than the enthalpy of combustion of CO (−283,022 kJ/kmol) by at least 45,026 kJ/kmol.

Example 12·6 Equilibrium Adiabatic Combustion Temperature

PROBLEM STATEMENT

Methane supplied at 100 kPa, 25°C (77 F), is burned adiabatically in a steady-flow burner with the stoichiometric amount of air supplied at the same conditions. Assuming that the products contain no CH_4 or NO, determine the temperature of the products.

SOLUTION

The reaction equation is

$$\begin{aligned}CH_4 + 2O_2 + 7.52N_2 \rightarrow {}& yCO_2 + (1 - y)CO + zH_2O \\ &+ \tfrac{1}{2}(1 - z + y + 2x)H_2 + (3 - y - 2x - z)OH + xO_2 + 7.52N_2\end{aligned}$$

The molar coefficients of the products in this equation were determined by C, O, and H mass balances. The products temperature and the values of x, y, and z must be such that they satisfy an energy balance and also satisfy the equilibrium conditions readily expressed by means of equilibrium constants. In the products are six reacting constituents involving three elements, so we must consider three independent reactions for determining the equilibrium composition. We then have an equal number of relationships and variables.

Before carrying out numerical calculations, we set up in convenient forms the relations to be used. The total number of moles of products is $y + (1 - y) + z + \frac{1}{2}(1 - z - y + 2x) + (3 - y - 2x - z) + x + 7.52$ or $(12.02 - \frac{1}{2}y - \frac{1}{2}z)$. The partial pressures of the products constituents are

$$p_{CO_2} = \frac{y}{12.02 - \frac{1}{2}y - \frac{1}{2}z} p_m$$

$$p_{CO} = \frac{1 - y}{12.02 - \frac{1}{2}y - \frac{1}{2}z} p_m$$

$$p_{H_2O} = \frac{z}{12.02 - \frac{1}{2}y - \frac{1}{2}z} p_m$$

$$p_{H_2} = \frac{\frac{1}{2}(1 - z + y + 2x)}{12.02 - \frac{1}{2}y - \frac{1}{2}z} p_m$$

$$p_{OH} = \frac{3 - y - 2x - z}{12.02 - \frac{1}{2}y - \frac{1}{2}z} p_m$$

$$p_{O_2} = \frac{x}{12.02 - \frac{1}{2}y - \frac{1}{2}z} p_m$$

$$p_{N_2} = \frac{7.52}{12.02 - \frac{1}{2}y - \frac{1}{2}z} p_m$$

We can use equilibrium constant data for any three reactions that involve constituents of the products mixture. Three such reactions for which equilibrium constant data are available from Table A·18, Table A·19, and Fig. 12·1 are

$$CO + \tfrac{1}{2}O_2 \rightarrow CO_2 \tag{1}$$

$$OH + \tfrac{1}{2}H_2 \rightarrow H_2O \tag{2}$$

$$H_2 + \tfrac{1}{2}O_2 \rightarrow H_2O \tag{3}$$

The equilibrium constants for these reactions are

$$K_{p1} = \frac{p_{CO_2}/p^\circ}{(p_{CO}/p^\circ)(p_{O_2}/p^\circ)^{1/2}} \qquad K_{p2} = \frac{p_{H_2O}/p^\circ}{(p_{OH}/p^\circ)(p_{H_2}/p^\circ)^{1/2}} \qquad K_{p3} = \frac{p_{H_2O}/p^\circ}{(p_{H_2}/p^\circ)(p_{O_2}/p^\circ)^{1/2}}$$

Substituting the partial pressure values from above into these expressions, we have

$$K_{p1} = \frac{y(12.02 - \frac{1}{2}y - \frac{1}{2}z)^{1/2}}{(1 - y)x^{1/2}(p_m/p^\circ)^{1/2}} \tag{A}$$

$$K_{p2} = \frac{z(12.02 - \frac{1}{2}y - \frac{1}{2}z)^{1/2}}{(3 - y - 2x - z)(1/2 - z/2 + y/2 + x)^{1/2}(p_m/p^\circ)^{1/2}} \tag{B}$$

$$K_{p3} = \frac{z(12.02 - \frac{1}{2}y - \frac{1}{2}z)^{1/2}}{(1/2 - z/2 + y/2 + x)x^{1/2}(p_m/p^\circ)} \tag{C}$$

The first law for this adiabatic steady-flow process is

$$Q = 0 = H_{products} - H_{reactants}$$

$$= \sum_{products} N(\bar{h} - \bar{h}_{ref} + \Delta\bar{h}_{f,ref}) - \sum_{reactants} N(\bar{h} - \bar{h}_{ref} + \Delta\bar{h}_{f,ref}) \tag{12·3}$$

For the reaction under consideration, recalling that $\Delta\bar{h}_{f,ref} = 0$ for elements and that the reactants enter at the reference temperature, we expand the last equation on a per mole of CH_4 basis to

$$\begin{aligned}\bar{q} = 0 = {} & y(\bar{h} - \bar{h}_{ref} + \Delta\bar{h}_{f,ref})_{CO_2} + (1 - y)(\bar{h} - \bar{h}_{ref} + \Delta\bar{h}_{f,ref})_{CO} \\ & + z(\bar{h} - \bar{h}_{ref} + \Delta\bar{h}_{f,ref})_{H_2O} + \tfrac{1}{2}(1 - z + y + 2x)(\bar{h} - \bar{h}_{ref})_{H_2} \\ & + (3 - y - 2x - z)(\bar{h} - \bar{h}_{ref} + \Delta\bar{h}_{f,ref})_{OH} + x(\bar{h} - \bar{h}_{ref})_{O_2} \\ & + 7.52(\bar{h} - \bar{h}_{ref})_{N_2} - \Delta\bar{h}_{f,ref,CH_4}\end{aligned}$$

Substitution of $\bar{h}_{ref}$ and $\Delta\bar{h}_{f,ref}$ values from Tables A·15 and A·17 gives

$$\begin{aligned}0 = {} & y(\bar{h}_{CO_2} - 402{,}886) + (1 - y)(\bar{h}_{CO} - 119{,}201) + z(\bar{h}_{H_2O} - 251{,}730) \\ & + \tfrac{1}{2}(1 - z + y + 2x)(\bar{h}_{H_2} - 8467) + (3 - y - 2x - z)(\bar{h}_{OH} + 29{,}815) \\ & + x(\bar{h}_{O_2} - 8683) + 7.52(\bar{h}_{N_2} - 8670) + 74{,}873 \qquad (D)\end{aligned}$$

Now we are ready to determine the final temperature as that value which satisfies Eqs. *A*, *B*, *C*, and *D*. The unknowns in these equations and supporting relationships are K_{p1}, K_{p2}, K_{p3}, x, y, z, T, and the $\bar{h}$'s of the products constituents. The supporting relationships available are the K_pT relations in Table A·18 and the various hT relations in Table A·15.

Several approaches are now possible. One that is suitable to use even with limited computer capability is to select several T values in a likely range of product temperatures and then obtain from the appropriate tables the corresponding values of K_{p1}, K_{p2}, K_{p3}, and enthalpies. Then by a suitable computer program for solving simultaneous algebraic equations, x, y, z, and q can then be calculated for each T value. This will show the temperature for which $q = 0$. The result of this procedure is an adiabatic combustion temperature of approximately 2240 K.

(So that you might check this calculation, the values for $T = 2200$ K are given here: $x = 0.050$, $y = 0.9177$, $z = 1.953$; and Eq. *D* becomes

$$\begin{aligned}\bar{q} &= 0.9177(112{,}927 - 402{,}886) + 0.082(72{,}690 - 119{,}201) + 1.953(93{,}058 \\ &\quad - 251{,}730) + 0.0328(68{,}344 - 8467) + 0.0287(69{,}925 + 29{,}815) \\ &\quad + 0.0505(75{,}451 - 8683) + 7.52(72{,}032 - 8670) + 74{,}873 \\ \bar{q} &= -20{,}240 \text{ kJ/kmol } CH_4 \text{ entering}\end{aligned}$$

This means that in order for the products to be at 2200 K, heat must be removed in the amount of 20,240 kJ/kmol. The adiabatic combustion temperature must therefore be higher than 2200 K.)

12·4 Relating K_p to Other Thermodynamic Properties

In Section 12·2 we mentioned that K_p values can be determined from other thermochemical data. In this section we derive two important relationships for this purpose, as well as for others.

12·4·1 The Relationship Between K_p and $\Delta \bar{h}_R$

If we combine Eq. *j* from Section 12·2 and the definition of K_p, we have

$$\ln K_p = \frac{1}{\bar{R}T}\left[\sum_{\text{reactants}} \nu_i f_i(T) - \sum_{\text{products}} \nu_j f_j(T)\right] \tag{l}$$

If we differentiate Eq. *j* or Eq. *l* with respect to temperature, we will have some terms $d[f(T)/T]/dT$; so let us examine these terms. In Eq. *i* of Section 12·2 we defined $f(T)$ as

$$f(T) \equiv \bar{h} - T\bar{s}^\circ_{ref} - T(\bar{\phi} - \bar{\phi}_{ref})$$

and

$$\frac{f(T)}{T} = \frac{\bar{h}}{T} - \bar{s}^\circ_{ref} - \bar{\phi} + \bar{\phi}_{ref}$$
$$= \frac{\bar{h}}{T} - \bar{s}^\circ_{ref} - \int \frac{\bar{c}_p\, dT}{T}$$

Then

$$\frac{d}{dT}\left[\frac{f(T)}{T}\right] = \left(\frac{d\bar{h}/dT}{T} - \frac{\bar{h}}{T^2}\right) - 0 - \frac{d}{dT}\left(\int \frac{\bar{c}_p\, dT}{T}\right)$$
$$= \frac{1}{T}\frac{d\bar{h}}{dT} - \frac{\bar{h}}{T^2} - \frac{\bar{c}_p}{T} = -\frac{\bar{h}}{T^2}$$

since $\bar{s}^\circ_{ref} \neq f(T)$ and the last simplification is possible because for an ideal gas, $\bar{h}$ is a function of T only so that $\bar{c}_p = d\bar{h}/dT$. Differentiation of Eq. *l* thus gives

$$\frac{d \ln K_p}{dT} = \frac{1}{\bar{R}T^2}\left(\sum_{\text{products}} \nu_j \bar{h}_j - \sum_{\text{reactants}} \nu_i \bar{h}_i\right)$$

Recall that in Chapter 11, we found that Δh_R, the enthalpy of reaction, is equal to the enthalpy of the products minus the enthalpy of the reactants when both are evaluated on a common scale. If we consider the stoichiometric reaction, then the term in parentheses above is equal to $\Delta \bar{h}_R$. Therefore,

$$\frac{d \ln K_p}{dT} = \frac{\Delta \bar{h}_R}{\bar{R}T^2} \quad \text{or} \quad \frac{d \ln K_p}{dT} = \frac{\Delta h_R}{RT^2} \tag{12·6}$$

Relationship between the equilibrium constant and the enthalpy of reaction

This equation shows that if the K_pT variation is known, $\Delta h_R(T)$ can be calculated. Notice that for exothermic reactions ($\Delta h_R < 0$), K_p decreases as the temperature increases, and for endothermic reactions ($\Delta h_R > 0$), K_p increases with increasing temperature. Equation 12·6 can also be written as

$$\frac{d \ln K_p}{d(1/T)} = -\frac{\Delta \bar{h}_R}{\bar{R}} \quad \text{or} \quad \frac{d \ln K_p}{d(1/T)} = -\frac{\Delta h_R}{R}$$

This form shows that if Δh_R varies only slightly with temperature, a plot of $\ln K_p$, or $\log K_p$ versus $1/T$ is nearly a straight line. We can use Eq. 12·5*b* to determine K_p from the constituent K_f's or Eq. 11·7 to determine Δh_R from the constituent Δh_f's. However, substituting these relationships into 12·6 is not usually of great value.

12·4·2 The Relationship Between K_p and Δg_R

Even more useful than Eq. 12·6 is an equation that relates $\ln K_p$ (not its derivative) to the Gibbs function of reaction, Δg_R. The definition of Δg_R is analogous to that of Δh_R and Δu_R. The derivation of this useful equation starts with Eq. *l* of the preceding section:

$$\ln K_p = \frac{1}{\overline{R}T}\left[\sum_{\text{reactants}} \nu_i f_i(T) - \sum_{\text{products}} \nu_j f_j(T)\right] \tag{l}$$

We refer again to Eq. *i* from Section 12·2, where

$$f(T) \equiv \overline{h} - T\overline{s}^{\circ}_{ref} - T\overline{\phi} + T\overline{\phi}_{ref}$$

Recall that $\overline{s}$ at any pressure p and temperature T can be calculated for an ideal gas by

$$\overline{s} = \overline{s}^{\circ}_{ref} + (\overline{\phi} - \overline{\phi}_{ref}) - \overline{R}\ln\frac{p}{p^{\circ}} \tag{11·10}$$

where $\overline{s}^{\circ}_{ref}$ is the molar absolute entropy evaluated at the standard state, p° = 100 kPa, T_{ref} = 25°C. If we choose to evaluate $\overline{s}$ at the standard pressure, then the last term is zero. Thus,

$$\overline{s}_{p^{\circ}} = \overline{s}^{\circ}_{ref} + (\overline{\phi} - \overline{\phi}_{ref})$$

and we can rewrite $f(T)$ as

$$f(T) = (\overline{h} - T\overline{s})_{p^{\circ}}$$
$$= \overline{g}_{p^{\circ}}$$

Thus Eq. *l* becomes

$$\ln K_p = \frac{1}{\overline{R}T}\left(\sum_{\text{reactants}} \nu_i \overline{g}_i - \sum_{\text{products}} \nu_j \overline{g}_j\right)_{p^{\circ}}$$

The relationship between the equilibrium constant and the change in the Gibbs function for the reaction is often called the van't Hoff equation.

The quantity within the brackets is $-\Delta g_R$ at p° and temperature T; so we have

$$\ln K_p = -\frac{\Delta\overline{g}_{R,p^{\circ}}}{\overline{R}T} \quad \text{or} \quad \ln K_p = \frac{-\Delta g_{R,p^{\circ}}}{RT} \tag{12·7}$$

$\Delta g_{R,p^{\circ}}$ is often represented by the symbol Δg°_R and is sometimes called the *standard free-energy* change, although the use of the name free energy for the Gibbs function is inadvisable because the same name has been used for the Helmholtz function. Δg_R° can be calculated from Δh_R° and the absolute entropy values; it can also be found in tables of thermochemical data. Some data sources provide values of Δg_R° for substances instead of for reactions, in the same manner that K_p values are sometimes presented for substances as mentioned in Section 12·2·2. These values can be designated by Δg_f°. (You should exercise caution in using various sources of thermochemical data. Usually, Δg_R° or Δg° refers only to a specified standard pressure; so that for any reaction, tables of Δg_R° or Δg° versus temperature are possible. Other times it refers to a specified pressure *and* a specified temperature so that there is but a single value for each reaction.)

This chapter is concerned with only ideal-gas reactions, but Eq. 12·7 actually applies to other reactions. In fact, it is widely used as a general definition of the equilibrium constant.

12·5 A Note on Reaction Rates

Equilibrium calculations based on classical thermodynamics provide no information on the rate at which a reaction occurs. Calculations may show that a system is not in a state of equilibrium, but the reaction may be occurring so slowly that measurements made on the system even over a period of years reveal no change in state. Such a condition is sometimes referred to as **frozen equilibrium**. A familiar example is that hydrogen and oxygen can exist together at room pressure and temperature for an indefinitely long period of time with no measurable reaction occurring, even though the equilibrium constant for the reaction $H_2 + \frac{1}{2}O_2 \rightarrow H_2O$ at room conditions shows that the equilibrium mixture consists almost entirely of H_2O. The reaction rate for the combining of hydrogen and oxygen at room conditions is so low that it can be approximated as zero for virtually all purposes. The presence of a catalyst or a stimulus such as an electric spark speeds the reaction, however, and the reaction can be observed to proceed always in the direction indicated by equilibrium calculations.

Frozen equilibrium

As we have discussed already in this book, if changes in a system occur on a time scale that is very large in comparison to the time scale we are interested in, the system may be modeled as though it is in equilibrium. For example, a system containing iron and oxygen is not in chemical equilibrium since over time the iron will spontaneously oxidize to form rust. Or, as we mentioned in Section 9·4·3, if a closed room contains a cup of water, equilibrium conditions are not reached until all the water in the cup has evaporated. Over small time periods, however, both of these systems can be treated as if they are in equilibrium with accurate results. The same can be said for the system containing H_2 and O_2 described above. Therefore, when we are analyzing chemically reactive systems, the ability to assume equilibrium conditions is determined by the *reaction rate* or, in other words, by *reaction kinet-*

If a system undergoing a chemical reaction changes very little in the time period of interest, we may model the system as if it is in chemical equilibrium.

ics. Our purpose in this section is to introduce reaction kinetics so that you will be familiar with this important concept. If you wish to delve further into this subject, refer to books such as Reference 12·3.

Consider first a reaction with only one reactant, such as

$$\nu_1 A_1 \rightarrow \nu_2 A_2 + \nu_3 A_3 \tag{m}$$

The *molar density* of constituent i is given the symbol $\bar{\rho}_i$ and the units of $\bar{\rho}$ are moles per unit volume. Then, at constant temperature, the rate at which the molar density of constituent A_1 changes, $d\bar{\rho}/dt$, can be described as follows:

$$-\frac{d\bar{\rho}_1}{dt} = \frac{\nu_1}{\nu_2}\frac{d\bar{\rho}_2}{dt} = \frac{\nu_1}{\nu_3}\frac{d\bar{\rho}_3}{dt} = f(\bar{\rho}_1, \bar{\rho}_2, \bar{\rho}_3) \tag{n}$$

The presence of the stoichiometric coefficients in this equation is in keeping with the result shown in Eq. g, Section 12·2. Namely, if in Eq. m, $\nu_1 = 1$ and $\nu_3 = 2$, then Eq. n shows that the rate of decrease in the molar density of A_1 is half the rate of increase in the molar density of A_3.

The function $f(\bar{\rho}_1, \bar{\rho}_2, \bar{\rho}_3)$ on the right-hand side of Eq. n is often a simple one. As an example, the rate of change of the reactant can often be written as

$$-\frac{d\bar{\rho}_1}{dt} = k\bar{\rho}_1 \tag{o}$$

where in this case k represents the first-order reaction rate constant and has units of inverse time. It may alternatively be true that

$$-\frac{d\bar{\rho}_1}{dt} = k\bar{\rho}_1^2 \tag{p}$$

is a better representation of the reaction rate, in which case the reaction is said to be second order, k is called the second-order reaction rate constant, and there is a concomitant change in the units of k. For a reaction involving two reactants,

$$\nu_1 A_1 + \nu_2 A_2 \rightarrow \nu_3 A_3 + \nu_4 A_4$$

a function such as

$$-\frac{d\bar{\rho}_1}{dt} = k\bar{\rho}_1\bar{\rho}_2 \tag{q}$$

may be required. This is also called a second-order reaction. Third- and higher-order reactions are possible as well. Even though the function f described in Eq. n may be quite complex, under restricted conditions it is often convenient and reasonable to use a first-order approximation.

For an ideal gas, the rate equation can be written in terms of the partial pressure of the constituents. From the ideal-gas equation of state, we find that

$$p = \frac{\rho \bar{R} T}{M} = \bar{\rho} \bar{R} T$$

Therefore, Eq. q (or any similar equation) can be written in terms of the partial pressures as

$$-\frac{d\bar{\rho}_1}{dt} = k' p_1 p_2 \qquad (r)$$

Note, however, that the relationship between Eq. q and Eq. r is not independent of temperature and k is not equal to k'.

Reaction rates depend on both pressure and temperature. The equilibrium constant for ideal gases, on the other hand, depends only on temperature.

In each of these cases we observe that the rate at which a reaction proceeds depends on the molar densities, or partial pressures of the reactants. This is in sharp contrast to the result obtained for chemical equilibrium calculations wherein we showed that for a reaction of ideal gases, the ratio of the reactant and product partial pressures (and hence the equilibrium composition) depends only on the temperature at which the reaction occurs.

In general, we find that the reaction rate increases with increasing molar density of the reactants and with increasing temperature. The effect of increasing molar density is shown clearly in Eqs. o, p, q, and r. The effect of temperature is realized through its effect on the reaction rate constant, k. The relationship between temperature and the reaction rate constant k can be described by the Arrhenius equation:

The Arrhenius equation

$$k = Ae^{-E_a/RT} \qquad (12\cdot8)$$

where E_a is called the Arrhenius activation energy, and over small temperature ranges A can be approximated as a constant, that is, independent of temperature. E_a can then be determined from the slope of ln k versus $1/T$, and then it is straightforward to calculate k at any given temperature.

Over large temperature ranges, *A* cannot be assumed to be constant. In such cases, the form of its temperature dependence can be determined from other principles or fits of the data.

Finally, it is worth noting that if the reaction rate for the forward reaction is known, then the reaction rate for the reverse reaction can be determined using the equilibrium constant as follows:

The reaction rates for the forward and reverse reactions are related by the equilibrium constant.

$$K_p = \frac{k_{\text{forward}}}{k_{\text{reverse}}} \qquad (12\cdot9)$$

Intuitively, Eq. 12·9 makes sense; it suggests that the equilibrium composition of a given reaction is the result of a balance between the forward and the reverse reactions. In practice, this reasoning introduces errors in part because reaction rates are usually measured far from equilibrium conditions. Even so, the error in this approximation seldom exceeds a factor of two or three.

12·6 Summary

A more general form of the relation

$$T\,dS = dU + p\,dV \qquad \text{(pure substances)} \tag{8·3}$$

which holds only for pure substances is

$$T\,dS \geq dU + p\,dV \tag{12·1}$$

which holds even for systems that undergo chemical reactions. The equality holds for systems in complete (mechanical, thermal, and chemical) equilibrium and reversible chemical reactions; the inequality holds for all other cases including those of irreversible chemical reactions.

From Eq. 12·1 we obtain equilibrium criteria such as

$$(dS)_{U,V} \geq 0 \quad \text{and} \quad (dG)_{p,T} \leq 0 \tag{12·2}$$

Referring to the second of these criteria, one interpretation is that when a system is in complete equilibrium (i.e., no spontaneous process is possible), its Gibbs function must be a minimum with regard to all states at the same pressure and temperature. This is a valuable criterion of equilibrium, not only because many chemical reactions occur at constant pressure and temperature, but also because it can be shown that if a system is in a state such that no spontaneous process can occur at constant pressure and temperature, then no spontaneous process at all can occur.

Application of one of the equilibrium criteria shows that for a mixture of ideal gases A_1, A_2, A_3, and A_4 that undergo a reaction

$$\nu_1 A_1 + \nu_2 A_2 \rightarrow \nu_3 A_3 + \nu_4 A_4$$

the equilibrium mixture is such that

$$\frac{(p_3/p^\circ)^{\nu_3}(p_4/p^\circ)^{\nu_4}}{(p_1/p^\circ)^{\nu_1}(p_2/p^\circ)^{\nu_2}} = K_p$$

where the p's are partial pressures and K_p, called the equilibrium constant, is a function of temperature only. K_p does not depend on the amount of the various constituents initially present. Therefore, a knowledge of equilibrium constants for an ideal-gas reaction provides us with one relationship among the partial pressures of the constituents. Stoichiometric considerations give us other relationships; so we are able to determine the extent to which a reaction can proceed at a given temperature.

Two valuable relationships between K_p and other thermochemical quantities are

$$\frac{d \ln K_p}{dT} = \frac{\Delta \bar{h}_R}{\bar{R}T^2} \tag{12·4}$$

and

$$\ln K_p = -\frac{\Delta \bar{g}_R^\circ}{\bar{R}T} \tag{12·5}$$

Classical thermodynamics gives no information on reaction rates, so we cannot predict on this basis whether a reaction will proceed to the indicated extent within a given time interval. Knowledge of reaction rates is important for this purpose.

References

12·1 Moran, Michael J., and Howard N. Shapiro, *Fundamentals of Engineering Thermodynamics*, 2nd ed., Wiley, New York, 1992, Chapter 4.

12·2 Van Wylen, Gordon J., Richard E. Sonntag, and Claus Borgnakke, *Fundamentals of Classical Thermodynamics*, 4th ed., Wiley, New York, 1994, Chapter 3.

12·3 Mulcahy, M. F. R., *Gas Kinetics*, Wiley (Halstead Press), New York, 1973. (A compact, advanced treatment of the kinetics of chemical reactions of gases. The notation is somewhat different from that in this book.)

12·4 Gordon, Sanford, and B. J. McBride, "Computer Program for Calculation of Complex Chemical Equilibrium Compositions, Rocket Performance, Incident and Reflected Shocks, and Chapman-Jouget Detonations," NASA SP-273, 1971 (and interim revision, March 1976). (This paper describes a widely used algorithm for equilibrium calculations.)

12·5 *JANAF Thermochemical Tables*, 3rd ed., published by the American Chemical Society and the American Institute of Physics for the National Bureau of Standards, 1986. (Also, *Journal of Physical and Chemical Reference Data*, Vol. 14, 1985, Supplement No. 1.)

12·6 *Selected Values of Properties of Hydrocarbons and Related Compounds,* American Petroleum Institute Project 44, Thermodynamics Research Center, Texas A&M University, College Station, TX, 77843.

12·7 *Selected Values of Chemical Thermodynamic Properties*, National Bureau of Standards Technical Notes 270-3, 1968.

12·8 Gallant, R. W., "Physical Properties of Hydrocarbons," *Hydrocarbon Processing*, Vol. 45, No. 10, October 1966.

Problems

12·1 Demonstrate that each of the following is a valid criterion of equilibrium:

$$(dU)_{S,V} \leq 0 \qquad (dS)_{H,p} \leq 0 \qquad (dA)_{T,V} \leq 0$$

12·2 Refer to Eq. *h* in Section 12·2. For a mixture of reactants and products *not* in equilibrium, the equality sign would be replaced by an inequality sign. State a rule relating the inequality to the direction (toward more products or toward more reactants) in which the reaction would proceed.

12·3 The equilibrium constant

- for the reaction $CO + \frac{1}{2}O_2 \rightarrow CO_2$ is K_{p1}
- for the reaction $2CO + O_2 \rightarrow 2CO_2$ is K_{p2}
- for the reaction $4CO_2 \rightarrow 4CO + 2O_2$ is K_{p3}
- for the reaction $2H_2 + O_2 \rightarrow 2H_2O$ is K_{p4}

Express (*a*) K_{p2} in terms of K_{p1}, (*b*) K_{p2} in terms of K_{p3}, and (*c*) K_p for the reaction $CO + H_2 + O_2 \rightarrow H_2O + CO_2$ in terms of K_{p1} and K_{p4}.

12·4 For the reaction $\frac{1}{2}Br_2 \rightarrow Br$, $K_p = 0.4099$ at 1200 K. Determine K_p at 1200 K for (*a*) the reaction $Br_2 \rightarrow 2Br$ at 500 kPa and (*b*) the reaction $2Br \rightarrow Br_2$ at 500 kPa.

12·5 Reference 12·5 lists K_f values for substances. For hydrazine, N_2H_4, $\log K_f = -15.957$ at 1200 K, according to this source. Use this information to determine at this temperature K_p for the reaction $N_2H_4 + O_2 \rightarrow 2H_2O + N_2$.

12·6 Reference 12·5 lists K_f values for substances. For ammonia, $\log K_f = -3.496$ at 1100 K. Determine the extent to which ammonia dissociates into diatomic hydrogen and nitrogen at 1100 K and (*a*) 100 kPa, (*b*) 500 kPa.

12·7 Determine the equilibrium constant for the reaction $H_2 + O_2 \rightarrow 2OH$ at 100 kPa, 2500 K.

12·8 Determine the equilibrium constant for the reaction $H_2O + \frac{1}{2}O_2 \rightarrow 2OH$ at 100 kPa, 2500 K.

12·9 Determine the equilibrium constant for the reaction $H_2O \rightarrow 0.5H_2 + OH$ at 100 kPa, 2500 K.

12·10 For an initial mixture of 1 mole each of CO and O_2, determine the composition of the equilibrium mixture at 3000 K and (*a*) 100 kPa, (*b*) 500 kPa.

12·11 Refer to Example 12·2. Using an appropriate K_p value from Table A·18, solve the problem for a temperature of 3000 K and a pressure of (*a*) 100 kPa, (*b*) 500 kPa.

12·12 CO is burned in a steady-flow process with 1.5 times the stoichiometric amount of oxygen at 100 kPa. Determine the equilibrium mixture at 2500 K.

12·13 Refer to Example 12·3. Using K_p data from Table A·18, determine the equilibrium mixture resulting from the combustion of CO and the stoichiometric amount of air. The products are at 2500 K and a pressure of (*a*) 500 kPa, (*b*) 1 MPa.

12·14 Determine the equilibrium mixture at 2500 K for the steady-flow burning of CO and 1.5 times the stoichiometric amount of air at 1 bar. Compare the results with those of Example 12·3.

12·15 In a steady-flow process, CO is burned with air to produce an equilibrium mixture at 2500 K. Plot a curve of mole fraction of CO in the mixture versus excess air in the range of 0 to 100 percent.

12·16 A system composed initially of H_2O is heated to 5000 R. Determine the composition of the equilibrium mixture at (*a*) 100 kPa, (*b*) 1 MPa.

12·17 Oxygen exists alone in a system. Calculate the fraction of oxygen that is dissociated at 100 kPa, 2500 K. Does increasing the pressure at the same temperature increase, decrease, or have no effect on the fraction dissociated?

12·18 Oxygen exists alone in a system. For equilibrium conditions, make a plot of the fraction of dissociated O_2 versus temperature in the range of 1500 K to 3000 K for pressures of 100 kPa, 500 kPa, and 2000 kPa.

12·19 Carbon dioxide enters a steady-flow system at 500 kPa, 25°C, and is heated at constant pressure until 15 percent of it is dissociated to form CO and O_2. Determine the final temperature. State the assumptions you make.

12·20 Solve Problem 12·19 for a range of pressures from 50 kPa to 1000 kPa and plot the final temperature against pressure.

12·21 Determine the temperature at which 10 percent of H_2O is dissociated into H_2 and O_2 at (*a*) 100 kPa, (*b*) 500 kPa.

12·22 At a pressure of 500 kPa, 5 percent of oxygen (O_2) has dissociated to monatomic oxygen (O). Determine (*a*) K_p for the reaction $O_2 \rightarrow 2O$ and (*b*) the amount of dissociation at the same temperature and 100 kPa.

12·23 Determine the equilibrium composition at 100 kPa, 4000 R, for an initial mixture of one mole of CO_2 and one mole of H_2O, considering CO, OH, and H_2 as possible dissociation products.

12·24 Air is heated to 4500 K at 100 kPa. Assuming that the only species present are N_2, O_2, N, O, and NO, determine the equilibrium mixture.

12·25 Solve Problem 12·23 for an initial mixture of one mole of CO_2, one mole of H_2, and one mole of N_2.

12·26 Solve Problem 12·23 for an initial mixture of one mole of CO_2 and two moles of N_2.

12·27 At what temperature is 10 percent of CO_2 dissociated at 100 kPa?

12·28 Solve Problem 12·27 if the CO_2 is initially mixed with an equal number of moles of air.

12·29 How much heat is required to heat CO_2 from 25°C to 2800 K in a steady-flow constant-pressure process at 100 kPa, 25°C?

12·30 Acetylene is burned with 20 percent excess air in a steady-flow process at 100 kPa with the reactants entering at 25°C and the products leaving at 2400 K. What is the heat transfer per unit mass of acetylene if CO, OH, and H_2 are considered as dissociation products? What does this result indicate regarding the temperature that would result from adiabatic burning of this mixture?

12·31 Ethylene is burned with 20 percent excess air in a steady-flow process at 100 kPa with the reactants entering at 25°C and the products leaving at 2400 K. Considering as dissociation products CO, OH, and H_2, determine the heat transfer per unit mass of ethane. What does this result indicate regarding the temperature that would result from adiabatic burning of this mixture?

12·32 Solve Problem 12·30 for pressures of 500 kPa and 1.5 MPa.

12·33 Solve Problem 12·31 for pressures of 500 kPa and 1500 kPa.

12·34 Solve Problem 12·30 with combustion air at 100 percent relative humidity rather than for dry air.

12·35 Solve Problem 12·30 for a reactant inlet temperature of 35°C and with the combustion air at 100 percent relative humidity.

12·36 Determine the equilibrium adiabatic combustion temperature of a stoichiometric mixture of carbon monoxide and air reacting in a steady-flow system under a pressure of 2 bar if the initial mixture temperature is 25°C.

12·37 Solve Problem 12·36 assuming a system pressure of 5 bar.

12·38 Solve Problem 12·36 for the mixture CO + O_2 instead of CO and the stoichiometric amount of air.

12·39 Solve Problem 12·36 for a mixture of carbon monoxide and 20 percent excess air.

12·40 Solve Problem 12·36 for a mixture of carbon monoxide and 80 percent of the stoichiometric amount of air.

12·41 Determine the equilibrium adiabatic combustion temperature of a stoichiometric mixture of hydrogen and air, initially at the reference temperature, reacting in a steady-flow system at 100 kPa.

12·42 Solve Problem 12·41 for a mixture of hydrogen and 10 percent excess air.

12·43 Solve Problem 12·41 for a mixture of hydrogen and 30 percent excess air.

12·44 Solve Problem 12·41 for a mixture of hydrogen and 70 percent of the stoichiometric amount of air.

12·45 Without making numerical calculations, sketch a curve of adiabatic combustion temperature versus percent stoichiometric air for the burning of hydrogen with air. Explain the shape of the curve, noting especially any extrema or limiting values.

12·46 A stoichiometric mixture of methane and air, initially at 25°C, is burned in a steady-flow process at a constant pressure of 100 kPa. Determine the equilibrium adiabatic combustion temperature, assuming that the products contain no constituents other than CO, CO_2, O_2, N_2, and H_2O.

12·47 A certain gaseous fuel is composed of 60 percent CH_4 and 40 percent C_2H_6, on a molar basis. This fuel is burned with the stoichiometric amount of air in a steady-flow system that is held at a constant pressure of 100 kPa. Find the equilibrium adiabatic combustion temperature, assuming that the reactants enter at 25°C and the products contain no constituents other than CO, CO_2, O_2, N_2, and H_2O.

12·48 Solve Problem 12·46 with consideration of additional products OH and H_2.

12·49 Solve Problem 12·47 with consideration of additional products OH and H_2.

12·50 What heat transfer is required for a steady-flow constant-pressure process in which CO_2 enters at 14.5 psia, 77 F, and is heated to 5400 R?

12·51 In a steady-flow constant-pressure process, H_2O that enters at 2 atm, 700 R, is heated to 5400 R. Determine the heat transfer.

12·52 Ethane is burned with 20 percent excess air in a steady-flow process at 14.5 psia with the reactants entering at 77 F and the products leaving at 4400 R. Considering as dissociation products CO, OH, and H_2, determine the heat transfer per unit mass of ethane. What does this result indicate regarding the temperature that would result from adiabatic burning of this mixture?

12·53 Solve Problem 12·52 for pressures of 72.5 and 217.5 psia.

12·54 A certain fuel is composed of 60 percent methane and 40 percent ethane by volume. This fuel is burned with the stoichiometric amount of air in a steady-flow system held at a constant pressure of 14.5 psia. If the reactants enter the combustion chamber at 77 F, what is the equilibrium adiabatic combustion temperature? Assume that the products contain no constituents other than CO, CO_2, O_2, N_2, and H_2O.

12·55 Solve Problem 12·53 with consideration of additional products OH and H_2.

12·56 A natural gas is composed of 83 percent methane and 17 percent ethane by volume. Determine the equilibrium adiabatic combustion temperature for constant pressure burning of this gas with 30 percent excess air at 14.5 psia with the reactants entering at 77 F.

12·57 Determine the equilibrium adiabatic combustion temperature of a stoichiometric mixture of propane and air, initially at 77 F, burning in a steady-flow system at 14.50 psia.

12·58 A stoichiometric mixture of methane and air, initially at 77 F, is to be burned adiabatically in an open, steady-flow system at 14.50 psia. Liquid water is to be injected into the combustion chamber in order to limit the temperature to a maximum of 3600 R. What amount of water per unit mass of methane is required to achieve this goal, assuming that the products contain no constituents other than CO, CO_2, O_2, N_2, and H_2O? Assume that the inlet water temperature is the same as that of the other reactants.

12·59 Calculate Δh_R at 2000 K (3600 R) for the reaction $CO + \frac{1}{2}O_2 \rightarrow CO_2$.

12·60 Calculate Δh_R at 2000 K (3600 R) for the reaction $CO_2 + H_2 \rightarrow CO + H_2O$.

12·61 Calculate Δh_R at 2000 K (3600 R) for the reaction $H_2 + \frac{1}{2}O_2 \rightarrow H_2O$.

12·62 For the reaction $\frac{1}{2}H_2 + \frac{1}{2}I_2 \rightarrow HI$, the K_pT relation in the vicinity of 1000 K is given by $\log K_p = 0.385 + 346.6/T$. From this relationship determine for this reaction at 1000 K (*a*) the enthalpy of reaction and (*b*) the Gibbs function of reaction.

12·63 Calculate $\Delta \bar{h}_R$ and $\Delta \bar{g}_R$ at 2000 K (3600 R) for the reaction $H_2O \rightarrow 2H + O$.

12·64 Determine which reaction has a greater value of the Gibbs function of reaction at 2100 K and the standard state pressure: $N_2 + O \rightarrow N_2O$ or $2N + O \rightarrow N_2O$? For N_2O at 2100 K, $\log K_f = -5.836$.

12·65 For the following two reactions: $2H + O \rightarrow H_2O$ and $OH + H \rightarrow H_2O$, at a temperature of 3000 K, determine which is more exothermic.

12·66 Calculate Δh_R at 2500 K for the reaction $N_2H_4 + O_2 \rightarrow 2H_2O + N_2$. For N_2H_4 at 2500 K, $\log K_f = -13.831$.

12·67 Calculate Δh_R and Δg_R at 1500 K for the reaction $H_2O + \frac{1}{2}O \rightarrow H_2O_2$. For H_2O_2 at 1500 K, $\log K_f = -0.959$.

12·68 For ideal gases that contain carbon, the values of K_p in Reference 12·5 are based on the constituent element C as a solid in the standard state. Therefore,

$$K_{p,CH_4} = (p_{CH_4}/p^\circ)/(p_{H_2}/p^\circ)^2$$

and

$$K_{p,CO} = (p_{CO}/p^\circ)/(p_{O_2}/p^\circ)^{1/2}$$

That is, the sum of the partial pressures of only the gaseous constituents equals the mixture pressure. Determine K_p at 1000 K for the reaction $CH_4 + H_2O \rightarrow 3H_2 + CO$ (*a*) given that the base 10 log of the equilibrium constant of formation at 1000 K is -1.011 for CH_4 and is 10.459 for CO and (*b*) using data from Tables A·1 and A·17.

12·69 For the reaction $O_2 \rightarrow 2O$ at 4000 K, $\Delta \bar{g}_R^\circ = -26{,}100$ kJ/kmol O_2. At what pressure at 4000 K would one-third of the O_2 be dissociated?

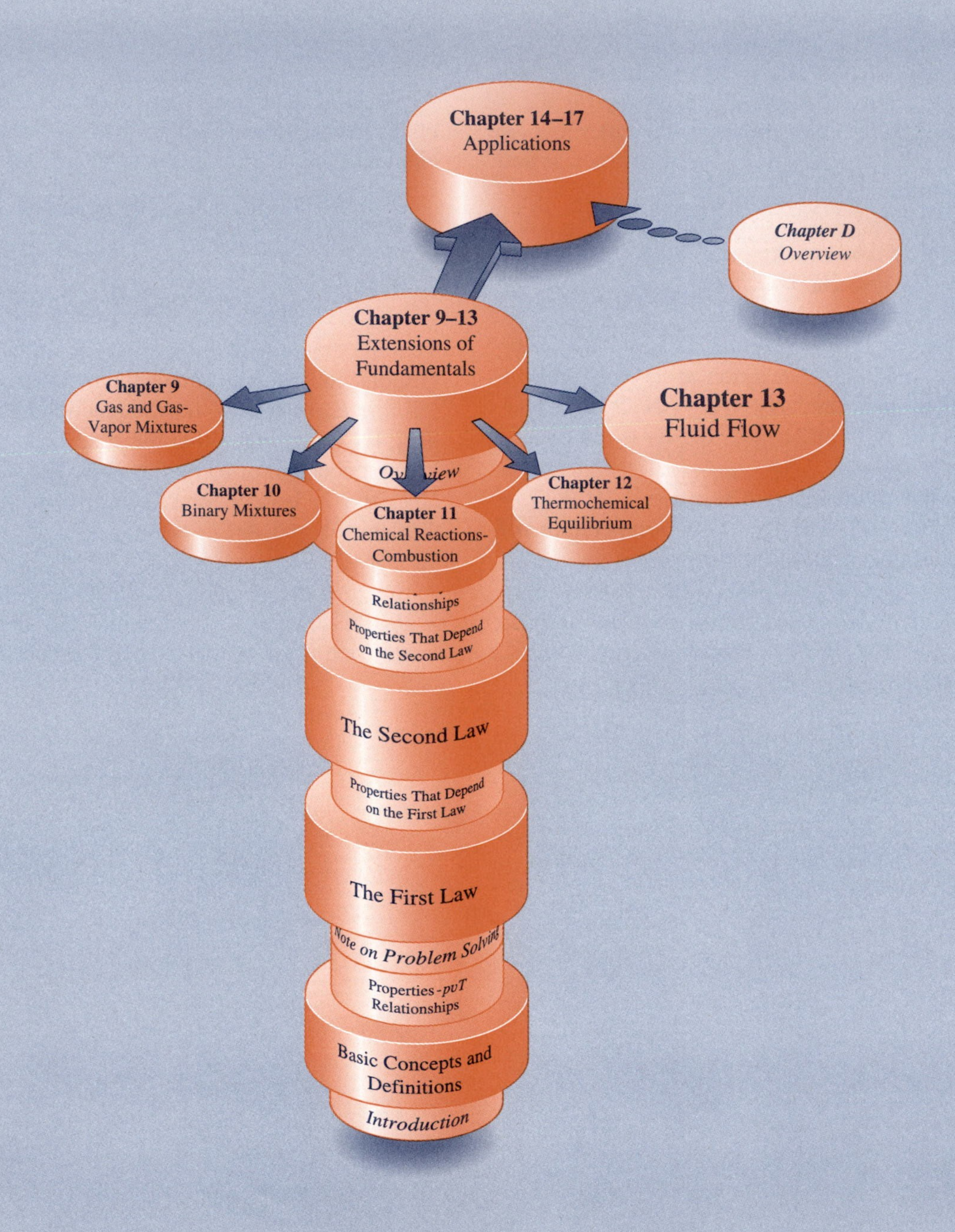
Chapter 14–17
Applications
Chapter D
Overview
Chapter 9–13
Extensions of
Fundamentals
Chapter 9
Gas and Gas-
Vapor Mixtures
Chapter 13
Fluid Flow
Chapter 10
Binary Mixtures
Chapter 11
Chemical Reactions-
Combustion
Chapter 12
Thermochemical
Equilibrium
Relationships
Properties That Depend
on the Second Law
The Second Law
Properties That Depend
on the First Law
The First Law
Note on Problem Solving
Properties-pvT
Relationships
Basic Concepts and
Definitions
Introduction

Chapter 13

Thermodynamic Aspects of Fluid Flow

In the study of fluid flow, the principles of both fluid mechanics and thermodynamics must be applied, and there is no sharp demarcation between the areas covered by these two sciences. This chapter is devoted to some of the thermodynamic aspects of fluid flows. To treat these we must touch briefly upon some points that are usually considered to be in the realm of fluid mechanics. This chapter shows how the first law, the second law, the principles of mechanics, and physical property relationships lead to many interesting and useful results when applied to flowing fluids.

13·1 Definitions

When examining fluid flows, it is important to understand the distinction between one-, two-, and three-dimensional flows. We discussed the characteristic features of these flows for the first time in Section 1·7 (see Fig. 1·15). If we restrict our attention now to steady flows, then all fluid properties are independent of time. Thus, one-dimensional steady flow is flow in which the fluid properties (pressure, temperature, velocity, etc.) depend on only one spatial coordinate. As an example, we could represent the velocity as $V(x)$, since V does not depend on y, z, or t.

One-dimensional flow exists if the velocity, temperature, and other properties are uniform at each cross section. From one cross section to another the properties may change; however, for each value of the coordinate (defined as distance in the direction of flow) there is a single value of velocity, a single value of density, and so on. The direction of flow may change, but there must be a corresponding change in the coordinate defining the single spatial dimension.

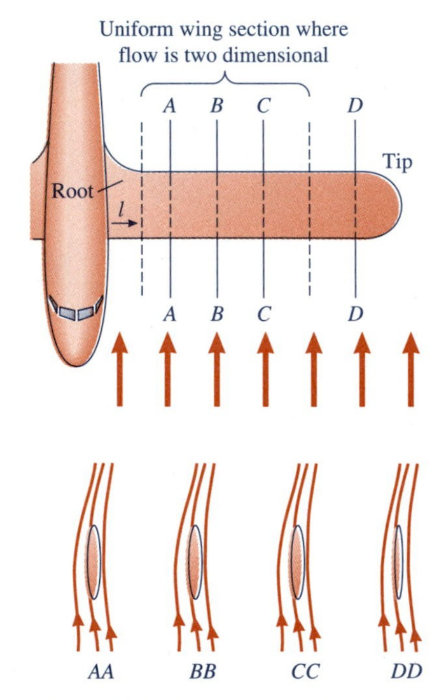

Figure 13·1 Flow across a long airplane wing illustrating the difference between two- and three-dimensional flows.

In the actual flow of a fluid through a passage, shear forces from the passage wall on the fluid cause a variation in velocity across any cross section, as shown in Fig. 1·15. If the passage is circular and the flow is axially symmetric, the flow is two dimensional because all properties can be completely described in terms of two spatial coordinates: distance along the passage and radius from the passage centerline. If in addition the flow is not symmetric, it is three dimensional. Thus, in two-dimensional steady flow the properties depend on two spatial coordinates, and in three-dimensional steady flow they depend on three.

As an illustration of the differences between two- and three-dimensional flow, consider the flow across a long airplane wing, as shown in Fig. 13·1. Over wing sections *AA*, *BB*, and *CC* in the uniform part of the wing, far from the root and the tip, the flow patterns are the same. They are independent of the distance l along the wing span; so this flow is two dimensional. Across a tapered section of the wing and near the root or tip, the flow is three dimensional.

In the following sections, we consider only one-dimensional flow. This is not always a valid assumption. For example, in the flow through an orifice as shown in Fig. 13·2, the fluid stream continues to contract after passing through the orifice until a minimum area section, called the *vena contracta*, is reached, at which point the pressure is also a minimum. Simple analysis shows that neither the pressure nor the velocity is uniform across any section between the orifice and the vena contracta. The size and location of the vena contracta can be predicted by means discussed in books on fluid mechanics and fluid metering. We mention this only as a reminder that some common flows cannot be modeled successfully on a one-dimensional basis.

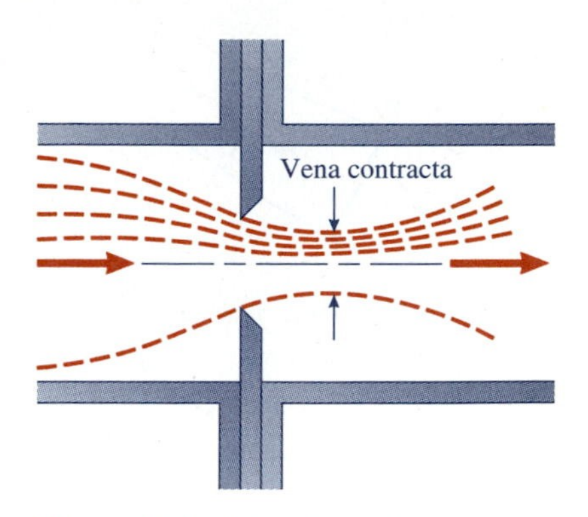

Figure 13·2 Flow through an orifice.

Nevertheless, to simplify analyses we can frequently approximate two-dimensional flow as one-dimensional flow in which the uniform velocity across a cross section is given by

$$V = \frac{\dot{m}}{\rho A}$$

Average velocity at a section

The velocity defined above is often called the **average velocity** at the section. Notice that the kinetic energy of a one-dimensional stream is not the same as that of a two-dimensional stream with the same average velocity. In many cases the difference can be neglected, but you should recognize this as an approximation, since in cases where other energy changes are small compared to this difference, serious errors can result.

13·2 The Basic Dynamic Equation for Steady, One-Dimensional Fluid Flow

Derivation of the basic dynamic equation for steady, one-dimensional flow. This derivation is for steady flow. A more general derivation can be based on the motion of a closed system passing through an open system like the derivation of the first law expression for an open system in Section 3·5·1.

Effective study of the thermodynamics of fluid flow requires attention to the mechanics or dynamics of flow. We now derive the basic dynamic equation for steady, one-dimensional flow. We begin with Newton's second law of motion,

$$F = \frac{d}{dt}(mV) \tag{1·1b}$$

where F is the resultant of all forces acting on a body that at any time t has mass m and velocity V. Equation 1·1b is a vector equation: F must be in the same direction as the change in momentum, $(d/dt)(mV)$. If we apply the above equation to any body of fixed mass m, we find

$$F = V\frac{dm}{dt} + m\frac{dV}{dt} = 0 + m\frac{dV}{dt}$$

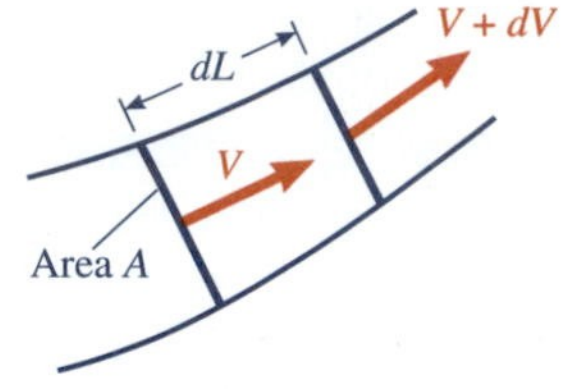

(*a*) Infinitesimal element

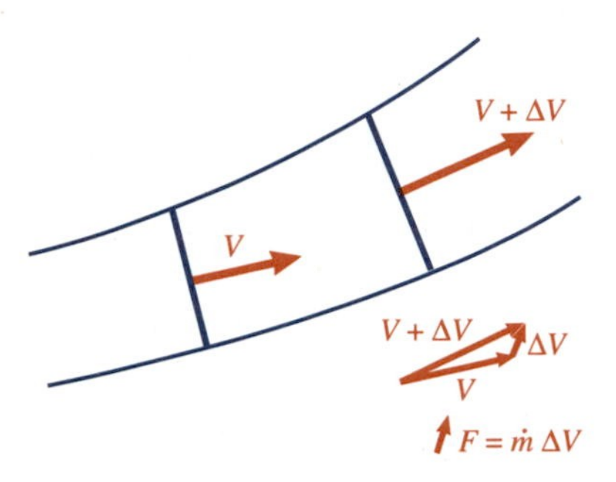

(*b*) Finite element

Figure 13·3 Element of fluid used to derive the dynamic equation for steady, one-dimensional fluid flow.

If dL is the distance traveled by the body in time dt, then $dV/dt = (dL/dt)(dV/dL) = V(dV/dL)$, and we have

$$F = mV\left(\frac{dV}{dL}\right)$$

Thus the resultant force on a mass m at any instant is proportional to the rate of change of velocity with distance. Let us consider now a fluid stream, directing our attention to an element of mass δm that has a length dL in the direction of flow (see Fig. 13·3). The mass and the length of the element are related by $\delta m = \rho A\, dL$, where A is the cross-sectional area of the element normal to the flow direction. An infinitesimal force dF causes this infinitesimal element to change velocity by dV as it moves a distance dL:

$$dF = (\delta m)V\left(\frac{dV}{dL}\right) = (\rho A\, dL)V\frac{dV}{dL} = \rho AV\, dV = \dot{m}\, dV$$

For steady flow, $\dot{m} = \rho AV =$ constant; so we can integrate to get the resultant force on a stream that changes velocity by ΔV:

The basic dynamic equation for steady, one-dimensional flow

$$F = \dot{m}\, \Delta V \qquad (13\cdot 1)$$

This is the basic dynamic equation for steady, one-dimensional flow. Remember that Eq. 13·1 is a vector equation: The resultant force is in the same direction as ΔV, as shown in Fig. 13·3. The equation holds also for corresponding components of F and ΔV.

In analyses using the basic dynamic equation for steady, one-dimensional flow, it is good practice to sketch the system and show all forces on the fluid. This is analogous to drawing a free-body diagram.

A convenient way of writing Eq. 13·1 for the resultant force on a fluid flowing steadily from a section 1 to a section 2 is

$$F = \dot{m}V_2 - \dot{m}V_1$$

Momentum flux

The vector quantity $\dot{m}V$ is called the **momentum flux** at a section. For a steady-flow system with more than one stream entering or leaving, the resultant force is the vector difference between the sum of the momentum fluxes of streams leaving and the sum of the momentum fluxes of streams entering.

In applying the dynamic equation (Eq. 13·1) it is advisable to sketch the steady-flow system under consideration, showing all of the forces *on the fluid*. The sum of these forces is then equal to $\dot{m}\, \Delta V$.

Example 13·1 Force of Air Flow on a Duct

PROBLEM STATEMENT

Air at 170 kPa, 50°C, enters a duct with a velocity of 150 m/s. The duct entrance area is 0.200 m^2. The air leaves the duct at 125 kPa with a velocity of 250 m/s through a cross-sectional area of 0.157 m^2 and in a direction 30 degrees different from the entrance direction. Determine the resultant force of the air on the duct.

SOLUTION

Analysis: First a sketch of the open system and a diagram showing the forces on the fluid within the duct are made. The only external forces on the fluid are the force of the adjacent fluid at inlet p_1A_1 and outlet p_2A_2 and the force of the duct on the fluid F_d. We assign the positive directions of velocity and force as upward and to the right in the sketch. We do not know at first the direction of F_d, so we show it on the diagram in the positive direction. Then, if we observe the sign convention we have established, the signs of the components F_{d_x} and F_{d_y} that we solve for will indicate the actual direction of the force *on the fluid.*

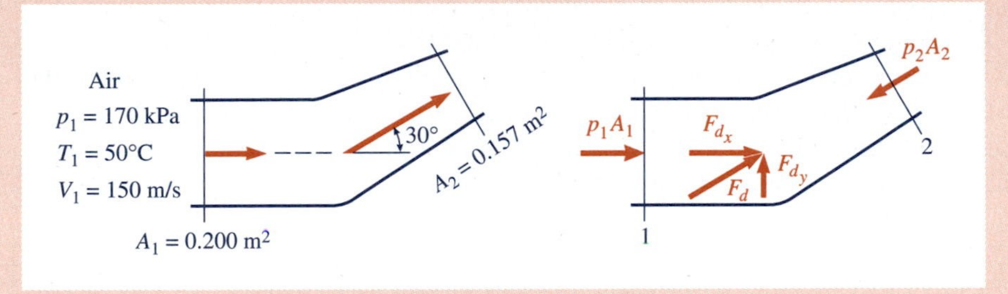

Application of the dynamic equation (Eq. 13·1) requires a knowledge of the mass rate of flow; so we use the ideal-gas equation of state to determine the inlet air density.

Calculations:

$$\rho_1 = \frac{p_1}{RT_1} = \frac{170}{0.287(323)} = 1.83 \text{ kg/m}^3$$

and then

$$\dot{m} = \rho_1 A_1 V_1 = 1.83(0.200)150 = 55.0 \text{ kg/s}$$

Equating the sum of the x components of the forces to the x component of $\dot{m}\,\Delta V$, we have

$$\begin{aligned} p_1A_1 + F_{d_x} - p_2A_2 \cos 30° &= \dot{m}(V_{2x} - V_{1x}) \\ F_{d_x} &= \dot{m}(V_2 \cos 30° - V_1) - p_1A_1 + p_2A_2 \cos 30° \\ &= 55.0(250 \cos 30° - 150) - 170(1000)0.200 \\ &\quad + 125(1000)0.157 \cos 30° \\ &= -13{,}350 \text{ N} \end{aligned}$$

Thus, the x component of the force of the duct *on the fluid* is 13,350 N to the left.

Applying the dynamic equation now in the y direction gives

$$F_{d_y} - p_2A_2 \sin 30° = \dot{m}(V_{2y} - V_{1y})$$
$$F_{d_y} = \dot{m}(V_2 \sin 30° - 0) + p_2A_2 \sin 30°$$
$$= 55.0(250 \sin 30°) + 125(1000)0.157 \sin 30°$$
$$= 16{,}690 \text{ N}$$

Adding F_{d_x} and F_{d_y} vectorially gives the resultant force of the duct *on the fluid* as 21,370 N in a direction upward to the left, 62 degrees from the vertical. The resultant force of the fluid *on the duct* is of course opposite in direction.

Example 13·2 Jet Thrust

PROBLEM STATEMENT

Air with a density of 0.0022 slug/cu ft enters a turbojet engine with a velocity of 450 fps relative to the engine. Fuel is burned at one-thirtieth of the mass rate of air flow. Products of combustion leaving the engine have a density of 0.00080 slug/cu ft. Inlet and outlet have equal cross-sectional areas of 2.2 sq ft, and the pressure is atmospheric at both inlet and outlet. Determine the thrust developed by the engine.

SOLUTION

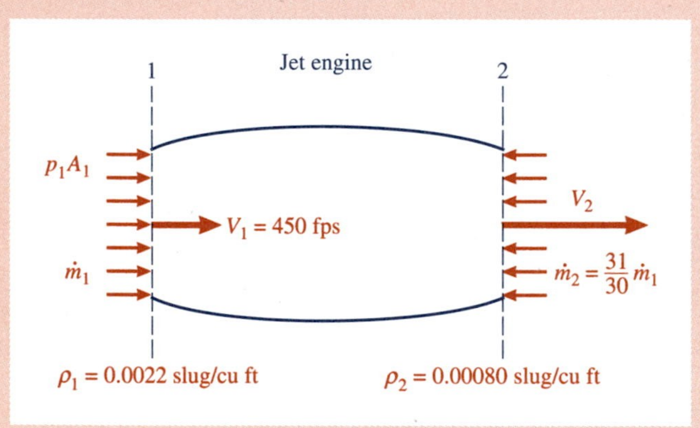

A diagram of the engine is made first. It is unnecessary to show any internal features of the engine in the sketch because application of the dynamic equation involves only the *resultant* force on the fluid and conditions at the boundary of the open system. Thrust is the net force of the fluid on the engine; so we find the equal-in-magnitude but opposite-in-direction force of the engine on the fluid by applying the dynamic equation in the axial direction, with the positive direction as shown above the sketch of the fluid:

$$F = p_1A_1 - p_2A_2 + F_e = (\dot{m}_2V_2 - \dot{m}_1V_1 - \dot{m}_fV_f)$$

where F is the resultant force on the fluid in the engine, and F_e is the net force of the engine on the fluid. Since $p_1A_1 = p_2A_2$, the resultant force is equal to F_e. Since the fuel is carried with the craft, its initial velocity has a zero component in the axial direction; so the last term on the right-hand side of the equation above is zero. Also, $\dot{m}_2 = (\dot{m}_1 + \dot{m}_f) = (\dot{m}_1 + \dot{m}_1/30) = 31\dot{m}_1/30$. Thus the dynamic equation becomes

$$F = F_e = \left(\frac{31}{30}\dot{m}_1V_2 - \dot{m}_1V_1\right) = \dot{m}_1\left(\frac{31}{30}V_2 - V_1\right)$$

We calculate $\dot{m}_1$ and V_2 for use in this equation by

$$\dot{m}_1 = \rho_1A_1V_1 = 0.0022(2.2)450 = 2.18 \text{ slug/s}$$

$$V_2 = \frac{\dot{m}_2}{\rho_2A_2} = \frac{\frac{31}{30}\dot{m}_1}{\rho_2A_2} = \left(\frac{31}{30}\right)\frac{2.18}{0.00080(2.2)} = 1279 \text{ fps}$$

Substitution of these values into the dynamic equation gives

$$F_e = \dot{m}_1\left(\frac{31}{30}V_2 - V_1\right) = 2.18\left[\frac{31}{30}(1279) - 450\right] = 1900 \text{ lbf}$$

This is the force of the engine on the fluid. It is positive, so, in accordance with the sign convention established at the beginning of this solution, this force is to the right for the diagrams as drawn. The force of the fluid on the engine, the thrust, is therefore to the left and has a magnitude of 1900 lbf.

13·3 Convenient Properties in Fluid Flows

Total Enthalpy. In the application of the first law of thermodynamics to flowing fluids, the group $h + V^2/2$ appears frequently. This group has been given the name **total enthalpy**, h_t, so that

$$h_t \equiv h + \frac{V^2}{2}$$

Total enthalpy

It is often said that total enthalpy is composed of a **static** part, h, and a **dynamic** part, $V^2/2$. Separation into a *static* and a *dynamic* part is common practice for other total properties as well.

Static and dynamic parts of a total property

The above definition of total enthalpy involves no assumptions or restrictions regarding the type of fluid or the type of process and is useful because it simplifies energy balances. The following table shows how some frequently used energy balances are simplified:

Type of Flow	*Energy Balance*	*Simplified Energy Balance*
Steady, adiabatic, one-dimensional flow through a turbine	$w = h_1 - h_2 + \frac{V_1^2 - V_2^2}{2}$	$w = h_{t1} - h_{t2}$
Steady, one-dimensional flow through a heat exchanger	$q = h_2 - h_1 + \frac{V_2^2 - V_1^2}{2}$	$q = h_{t2} - h_{t1}$
Steady, adiabatic one-dimensional flow through a nozzle	$h_1 + \frac{V_1^2}{2} = h_2 + \frac{V_2^2}{2}$	$h_{t1} = h_{t2}$

As the above table suggests, property diagrams of h_t versus s are often more valuable than hs diagrams. In the case of an adiabatic turbine, ordinate distances on an hs diagram represent work only if there is no change in kinetic energy; but on an h_ts diagram ordinate distances represent work even if the kinetic energy change is appreciable.

Stagnation enthalpy

Total enthalpy is also called **stagnation enthalpy** because the total enthalpy of a flowing fluid is equal to the enthalpy of the fluid if it were brought to rest adiabatically. (A *stagnation point* is defined as a point in a fluid-flow field where the velocity is zero.) The process of bringing the fluid to rest need not be reversible. Writing an energy balance between some point 1 in a fluid stream and the stagnation point for that fluid stream gives

$$h_1 + \frac{V_1^2}{2} = h_{\text{stagnation point}} = h_t$$

which is the same for any adiabatic no-work process, reversible or irreversible.

Total temperature

Total Temperature. The **total temperature** or **stagnation temperature** of a flowing fluid is defined as the temperature that would result if the fluid were brought to rest reversibly and adiabatically or, in other words, isentropically. (For an ideal gas, total temperature could be defined as the temperature corresponding to the total enthalpy, but such a definition cannot be used for a substance for which temperature and enthalpy are not uniquely related.)

Total pressure

Total Pressure. The pressure that results when a flowing fluid is brought to rest reversibly and adiabatically (isentropically) is called the **total pressure** or the **stagnation pressure**. In some publications there is a slight difference in definition between *total pressure* and *stagnation pressure* (and hence between *total temperature* and *stagnation temperature*), but in this book, as in many others, the two terms are synonymous. Also, in some publications what we call *stagnation pressure* is called *isentropic stagnation pressure.*

Note: Stagnation pressure and stagnation temperature are defined in terms of an isentropic deceleration. Stagnation enthalpy, on the other hand, requires only an adiabatic deceleration.

Sonic Velocity. An important property in the study of gas flow is the velocity of sound through the gas, the **sonic velocity** or **acoustic velocity**. We therefore derive now an expression for the sonic velocity in a gas.

Derivation of the sonic velocity

The plan of the derivation is to consider a pressure wave traveling through a gas and, from the viewpoint of an observer moving with the wave, to apply the continuity equation and the dynamic equation to determine the relative velocity between the wave and the fluid ahead of it. We do this first for a pressure wave of any size (i.e., the pressure behind the wave may differ appreciably from that ahead of it) and then apply the restriction that the pressure difference across the wave is very small, as it is for a sound wave.

For a physical picture of the generation and propagation of a pressure wave in a gas, refer to Fig. 13·4*a*. A gas fills a long tube of constant cross-sectional area. A piston in one end of the tube is moved as shown with a velocity u. The gas in a region near the piston face also moves with the velocity u. The fluid boundary defining the region of gas with velocity u from the quiescent gas is called the **wave**. Behind the wave (i.e., to the left of the wave in Fig. 13·4*a*) the pressure is greater than it is ahead of it. Ahead of the wave the gas is unaffected by the motion of the piston or the approaching wave. The wave moves with a velocity a that is equal to or greater than u.

The boundary that defines regions with different physical properties is called the wave.

Our problem is to determine the wave velocity a in terms of properties of the fluid ahead of the wave. To simplify the derivation, however, let us use a frame of reference in which the wave is fixed, or, in other words, let us adopt the viewpoint of an observer moving with the wave. The resulting picture is shown in Fig. 13·4*b*, which is the same as Fig. 13·4*a* but with a uniform

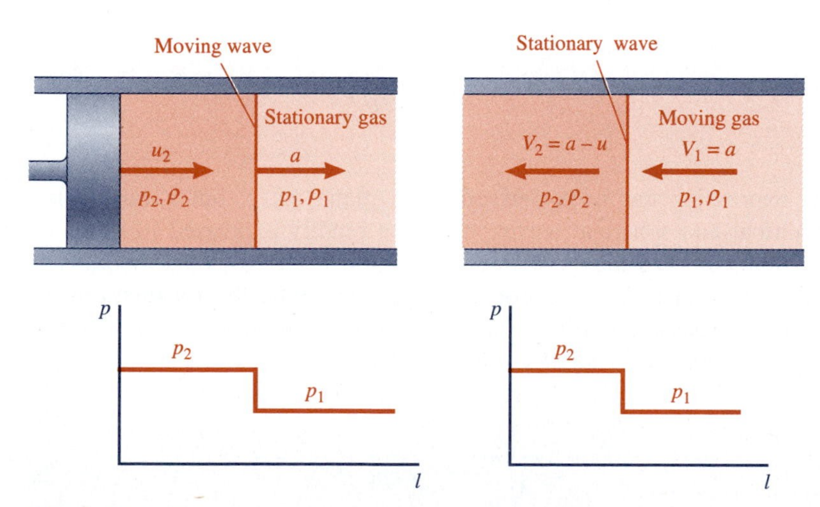

Figure 13·4 Propagation of a pressure wave in a gas.

velocity a to the left imposed on the entire system. Thus the stationary wave is approached from the right by fluid with a velocity $V_1 = a$, and fluid leaves the wave to the left with a velocity $V_2 = a - u$. The flow through the wave is steady. Since the cross-sectional area is constant, the continuity equation $\rho_1 A_1 V_1 = \rho_2 A_2 V_2$ becomes

$$\rho_1 V_1 = \rho_2 V_2$$

If we assume the thickness of the wave is so small that shearing forces at the wall are negligible, the dynamic equation becomes

$$p_1 A - p_2 A = \dot{m}\,\Delta V = \rho_1 A V_1 (V_2 - V_1)$$

Substituting V_2 from the continuity equation into this equation and solving for V_1 (which equals a) gives

$$a = V_1 = \sqrt{\frac{\rho_2}{\rho_1}\left(\frac{p_1 - p_2}{\rho_1 - \rho_2}\right)}$$

Shock wave

This is the velocity of the gas approaching the wave in Fig. 13·4*b* or the velocity of the wave in Fig. 13·4*a*. Such a wave is called a **shock wave**, and its velocity is higher for higher values of p_2/p_1.

The *sonic velocity* or velocity of sound is the velocity of a pressure wave of very small amplitude. As p_2 approaches p_1 and ρ_2 approaches ρ_1, the quantity beneath the radical sign in the equation above becomes $dp/d\rho$. Of course, the value of $dp/d\rho$ depends on the process because p and ρ can be varied independently. However, if we assume that the propagation of a sound wave is reversible and adiabatic, and hence isentropic, we restrict $dp/d\rho$ to $(\partial p/\partial \rho)_s$, which has a single value at any state. Thus the sonic velocity, which is represented by the symbol c, is

Sonic velocity

$$c = \sqrt{\left(\frac{\partial p}{\partial \rho}\right)_s} \qquad (13\cdot2)$$

This expression has been derived for a plane wave, but it holds also for cylindrical and spherical waves. c can be readily expressed in terms of the isentropic compressibility (defined in Section 7·4) or the bulk modulus of the fluid. For ideal gases with constant specific heats, the $p\rho$ relationship for isentropic processes is $p/\rho^k =$ constant, so that $(\partial p/\partial \rho)_s = kp/\rho$, and the *velocity of sound in an ideal gas* is therefore,

Sonic velocity for an ideal gas. Although this equation is simple, the consistency of units demands special attention.

$$c = \sqrt{\frac{kp}{\rho}} = \sqrt{kpv} = \sqrt{kRT} \qquad (13\cdot2a)$$

In an ideal gas the sonic velocity, which is a property of the gas, is a function of temperature only. In the derivation above we have assumed that the propagation of a sound wave is isentropic, but we have placed no restriction on the type of process, if any, that the bulk of the gas may be undergoing.

We assume that the *propagation* of the sound wave is isentropic, but we place no restriction on the process of the bulk fluid.

Mach Number. In the study of gas flow, a useful parameter is the ratio of the velocity of the gas at any point to the sonic velocity at the same point. This ratio is called the **Mach number** and is designated by the symbol M:

Mach number

$$M \equiv \frac{V}{c}$$

Fluid flows are often described qualitatively in terms of their Mach number as follows:

$M < 1$	Subsonic flow
$M \approx 1$	Transonic flow
$M > 1$	Supersonic flow
$M \gg 1$	Hypersonic flow

Just how close to 1 the Mach number must be for a flow to be called transonic is not specified, and also the boundary between supersonic and hypersonic is not sharply defined; but notice that the value of $M = 1$ denotes clearly the boundary between subsonic and supersonic flow.

Example 13·3 Calculating Total Properties

PROBLEM STATEMENT

Determine the total pressure and the total temperature of air at 100 kPa, 350°C, flowing with a velocity of 400 m/s.

SOLUTION

The total pressure and total temperature are the pressure and temperature that correspond to the total enthalpy and the same entropy as the flowing fluid. From the definition of total enthalpy,

$$h_t - h = \frac{V^2}{2}$$

For an ideal gas enthalpy is a function of temperature only; so we determine h at 350°C from Table A·12 or the computer model for air properties:

$$h_t = h + \frac{V^2}{2} = 633.4 + \frac{(400)^2}{2(1000)} = 713.4 \text{ kJ/kg}$$

Now we can use the computer model or Table A·12 to find T_t at h_t:

$$T_t = 425°\text{C}$$

Alternatively, if we assume that c_p is constant in the temperature range involved, we have

$$c_p(T_t - T) = \frac{V^2}{2}$$

and

$$T_{t,c_p=\text{constant}} = T + \frac{V^2}{2c_p} = 623.15 + \frac{(400)^2}{2(1.06)1000} = 698 \text{ K} = 425°\text{C}$$

where the value of c_p is taken from Table A·3.

The total pressure can be found as follows:

$$p_t = p\frac{p_{rt}}{p_r} = 100\frac{29.03}{19.03} = 152.5 \text{ kPa}$$

If we use constant specific heats, we can make use of the relationship derived in Section 4·2,

$$p_t = p\left(\frac{T_t}{T}\right)^{k/(k-1)}$$

and so we have a total pressure of

$$p_t = 100\left(\frac{698}{623}\right)^{1.37/0.37} = 152.3 \text{ kPa}$$

For the pressure and temperature ranges of this problem, assuming constant specific heats introduces little error.

13·3·1 Measuring Properties in Fluid Flows

In principle, the total pressure of a flowing fluid can be measured by means of a pressure probe that opens directly upstream so that the flowing fluid is brought to rest isentropically at the opening. Static pressure can be measured by means of a pressure probe that moves with the fluid, or where the fluid flows along a straight wall the static pressure can be measured by means of a small opening in the wall. It is usually assumed that the static pressure is constant along a line normal to such a wall.

Total temperature can be measured by a probe placed so that the temperature-sensitive element is in contact with only the fluid that has been brought to rest isentropically. Static temperature is measured by a thermometer that moves with the fluid. (In both cases the usual precautions must be taken with regard to radiation and other sources of error.)

In contrast to the method of measuring static pressure, a thermometer element placed in a wall along which a fluid flows does not measure the static temperature. Rather, the temperature of the fluid at the wall and its relationship to the fluid temperature far from the wall require some explanation. When a fluid flows along a wall, shear forces reduce the fluid velocity to zero right at the wall, establishing a velocity gradient between the wall and the region where the flow is virtually unaffected by the shear forces. The deceleration of the fluid layers near the wall causes an increase in enthalpy of those layers, resulting in a temperature gradient that causes heat transfer from one layer to the next. For a given velocity distribution, the temperature distribution can be determined only if the relationships between the momentum transfer and heat transfer of the flow are known. If the flow is adiabatic, there can be no heat transfer between the fluid and the wall, a condition that requires the temperature gradient at the wall to be zero. In this case, the fluid temperature at the wall is the **adiabatic wall temperature**. For many gases, the balance between the momentum transfer and heat transfer (which is determined by the relationships among properties, including c_p, dynamic viscosity, and thermal conductivity) causes the total enthalpy to be nearly constant from layer to layer. Under these conditions the temperature at the wall is the stagnation temperature corresponding to the total enthalpy and the static pressure of the flow. This is called the **approximate adiabatic wall temperature**.

Adiabatic wall temperature

Approximate adiabatic wall temperature

Figure 13·5 shows the difference between the approximate adiabatic wall temperature and the total or stagnation temperature. Part *a* of Fig. 13·5 shows the state of the flowing fluid, represented by point *f*, and the corresponding stagnation state, represented by point *s*. Process *f–s* is an isentropic deceleration to zero velocity. All adiabatic decelerations from state *f* to zero velocity must result in enthalpy equal to h_s, as shown in Fig.13·5*b*. Since the static pressure is assumed to be constant along a line normal to the wall, the approximate adiabatic wall state *w* is located on the line representing the locus of adiabatic deceleration states where $p = p_f$ (see Fig. 13·5*c*). For a real gas, lines of constant temperature show that in general, $T_w < T_t$, as illustrated in Fig. 13·5*d*. For the case of an ideal gas, however, states of equal enthalpy are states of equal temperature (regardless of the pressure); so the approximate adiabatic wall temperature equals the stagnation temperature as shown in Fig. 13·5*e*.

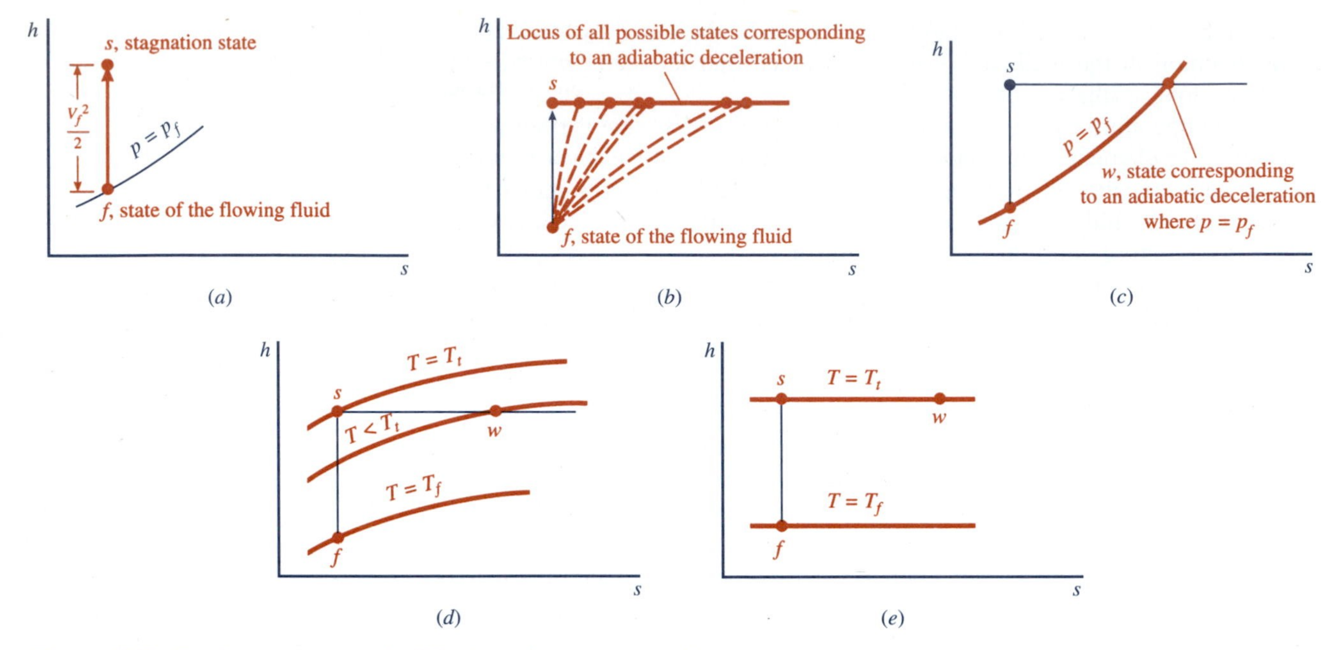

Figure 13·5 Total temperature and adiabatic wall temperature: (*a*) state of the flowing fluid and stagnation state; (*b*) adiabatic decelerations; (*c*) approximate adiabatic wall state; (*d*) comparison of T_w and T_t for a real gas; (*e*) comparison of T_w and T_t for an ideal gas.

Example 13·4

PROBLEM STATEMENT

Determine the total pressure, the total temperature, and the approximate adiabatic wall temperature of steam at 60 psia, 312 F, flowing at 1070 fps.

SOLUTION

The total pressure and total temperature are the pressure and temperature of the state defined by

$$h_t = h + \frac{V^2}{2} = 1188.2 + \frac{(1070)^2}{2(32.2)778} = 1211.1 \text{ B/lbm}$$

$$s_t = s = 1.658 \text{ B/lbm·R}$$

Accurate double interpolation in the steam tables or on a Mollier chart would be tedious, so we think of using the computer model included with this book, STEAM. However, STEAM usually requires that one of the input variables be either p or T, and here the input variables are h_t and s_t. We can still find the values sought by plotting h_t versus T_t for about four states at $s_t = 1.658$ B/lbm·R and different values of temperature within a guessed range above 312 F. The temperature corresponding to h_t = 1211.1 B/lbm on this curve is the value we are seeking. Then with either T_t and h_t or T_t and s_t as input variables, we can find the pressure.

Alternatively, in this range of properties we find that by making a reasonable guess of the temperature T_t, STEAM does solve iteratively with inputs of h_t and s_t to give

$$p_t = 78.7 \text{ psia} \quad \text{and} \quad T_t = 362.7 \text{ F}$$

The approximate adiabatic wall temperature T_w is the temperature of the state defined by

$$h_w = h_t = 1211.1 \text{ B/lbm}$$
$$p_w = p = 60 \text{ psia}$$

From STEAM we obtain

$$T_w = 356 \text{ F}$$

13·4 The Basic Relations for Steady, One-Dimensional Fluid Flow

For the remainder of this chapter we restrict our attention to steady, one-dimensional flow of fluids in which no work is done. In the analysis of such flow we have available five powerful tools:

1. The first law, which under the conditions stated can be expressed as

$$q = \Delta h + \Delta ke + \Delta pe$$

   In connection with gases and vapors, Δpe is nearly always negligible.
2. The continuity equation

$$\dot{m} = \rho_1 A_1 V_1 = \rho_2 A_2 V_2$$

3. The dynamic equation

$$F = \dot{m}\, \Delta V$$

4. The second law and its corollaries.
5. Physical property relationships, either in the form of tabular, graphical, or computer data or in the form of equations of state and other property equations resulting from the first and second laws.

In analyzing fluid flow in which no work is done, these are the basic tools that you should use. Often you will find it convenient or even necessary to combine and restrict some of the basic relations to form equations that apply only to special cases. Always remember the restrictions on these special-case equations and work as much as possible from the more general basic relations. In engineering practice, to be sure, it is the special cases that demand solution; but new special cases can be solved only by the application of the basic principles, and the great value of the basic principles is precisely that they can be applied not to just some cases, but to all cases, including ones that are as yet unthought of.

To give you an idea of the number of special cases possible, consider that within this already restricted category of steady, one-dimensional flows in which no work is done, we may have

- Adiabatic or diabatic flows
- Reversible or irreversible flows
- Constant-area or variable-area flows
- Ideal-gas or non-ideal-gas flows
- Constant-specific-heat or variable-specific-heat flows

This list is far from complete; there are turbulent and laminar flows, Newtonian and non-Newtonian fluids, and so on. It simply would not be practical to address each case individually. So instead, in the sections that follow, we show several examples of how the basic relations can be used for specific applications.

Remember, all of the equations that follow can be quickly derived. They contain no information beyond that in the basic equations, and each carries certain restrictions. Any problem that can be solved using special equations can be solved using the basic equations. Clearly, though, computational labor can be reduced by using the appropriate special-case equations. We recommend that you first develop facility in applying the basic relations, because this will allow you to use the special equations with confidence.

13·5 Applications of the Basic Relations

Figure 13·6 shows how the remainder of this chapter is structured. Our approach is to address general cases first and progress to more specific cases. As we proceed down through Fig. 13·6, all assumptions are shown in red. For example, we will first deal with reversible, adiabatic flows for a general compressible fluid. Then we will add the assumption of ideal-gas behavior and finally add the restriction of constant specific heats.

Admittedly, the equations that result when ideal-gas, constant-specific-heat behavior is assumed are much simpler. For this reason, when it was necessary to perform computations predominantly longhand, the assumption of ideal-gas, constant-specific-heat behavior was often made even when the

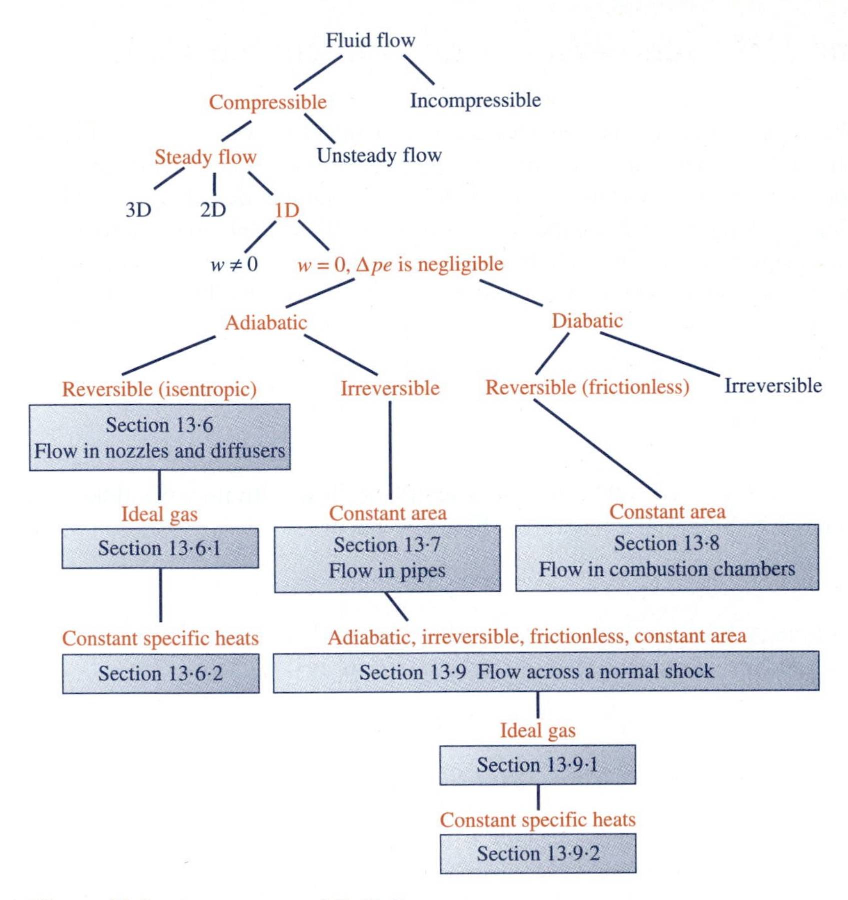

Figure 13·6 A taxonomy of fluid flow.

conditions of the flow did not warrant it. The ubiquity of computers, however, makes this approach unnecessary. Thus, in some of the examples that appear throughout the following sections, we will illustrate the potential error associated with the assumption of ideal-gas, constant-specific-heat behavior. But, you should use caution in extrapolating these errors to other situations, since they are by no means indicative of the error in general. In fact, the assumption of ideal-gas, constant-specific-heat behavior is still useful either when a quick, first-order result is desired or when the fluid property changes are sufficiently small to justify this approach (see Example 13·3). Experience will help you to determine which type of analysis is appropriate.

The "tree" shown in Fig. 13·6 is certainly not unique or complete; many other taxonomies are possible. Further, we will explore only a few of the branches. We have chosen these particular applications because the resulting formulations are commonly used or because they illustrate a particular point of interest.

13·6 Flow in Nozzles and Diffusers—Adiabatic, Reversible Flow

For many applications, the acceleration or deceleration of a fluid in a flow channel may be treated as adiabatic and reversible, hence isentropic. In this section, we show that the required shape of channels designed to change the fluid velocity depends on the conditions of the flow itself. Beginning with the basic equations, we are able to relate the variation of the cross-sectional flow area to velocity change and pressure change for the steady, one-dimensional flow of any fluid. The analysis of non-ideal-gas flows is complicated by the lack of a simple equation of state. Therefore, more complex equations of state, tabulated data, graphical data, or a computer program for properties must be used.

Area Variation Equations. For adiabatic flow with no work done and no change in potential energy, the first law becomes

$$h_t = \text{constant} \tag{13·3a}$$

Note that this equation applies to all adiabatic flows, reversible or irreversible. It can be expressed in differential form as

$$dh_t = 0$$
$$d\left(h + \frac{V^2}{2}\right) = 0$$
$$dh + V\,dV = 0 \tag{a}$$

When we further restrict the flow to reversible processes, the flow is isentropic, and the $T\,ds$ equation, Eq. 6·7, becomes

$$T\,ds = 0 = dh - \frac{dp}{\rho} \tag{b}$$

Combining these two equations gives

$$dp = -\rho V\,dV \quad \text{or} \quad \frac{dV}{V} = -\frac{dp}{\rho V^2} \tag{c}$$

Differentiating the continuity equation, $\rho AV = \text{constant}$, gives

$$\frac{d\rho}{\rho} + \frac{dA}{A} + \frac{dV}{V} = 0 \tag{d}$$

Substituting from Eq. *c* into Eq. *d* yields for the isentropic process

$$\frac{dA}{A} = \frac{dp}{\rho}\left[\frac{1}{V^2} - \frac{d\rho}{dp}\right] \tag{e}$$

We observed in Section 13·3 that $dp/d\rho = (\partial p/\partial \rho)_s = c^2$. Thus, the second term within the brackets in Eq. *e* is $1/c^2$ and

$$\frac{dA}{A} = \frac{dp}{\rho V^2}[1 - M^2] \quad \text{or} \quad \frac{dA}{dp} = \frac{A}{\rho V^2}[1 - M^2] \qquad (f)$$

Substituting from Eq. *c* into Eq. *f* gives

$$\frac{dA}{A} = -\frac{dV}{V}[1 - M^2] \quad \text{or} \quad \frac{dA}{dV} = -\frac{A}{V}[1 - M^2] \qquad (g)$$

Inspection of Eqs. *f* and *g* leads to the following conclusions:

1. When $M < 1$, $dA/dp > 0$ and $dA/dV < 0$.
2. When $M > 1$, $dA/dp < 0$ and $dA/dV > 0$.
3. When $M = 1$, $dA/dp = 0$ and $dA/dV = 0$.

These conclusions are illustrated in Fig. 13·7. A flow channel that accelerates a fluid is called a **nozzle** and a channel that decelerates a fluid is called a **diffuser**. Notice that the shape (i.e., converging or diverging) of a nozzle or of a diffuser depends on whether the flow is subsonic or supersonic.

Nozzles and diffusers

Figure 13·7 also demonstrates that once the speed of sound is reached by the fluid, further acceleration can occur only in a diverging section. Conversely, a diverging section is used to decelerate a fluid that is flowing sub-

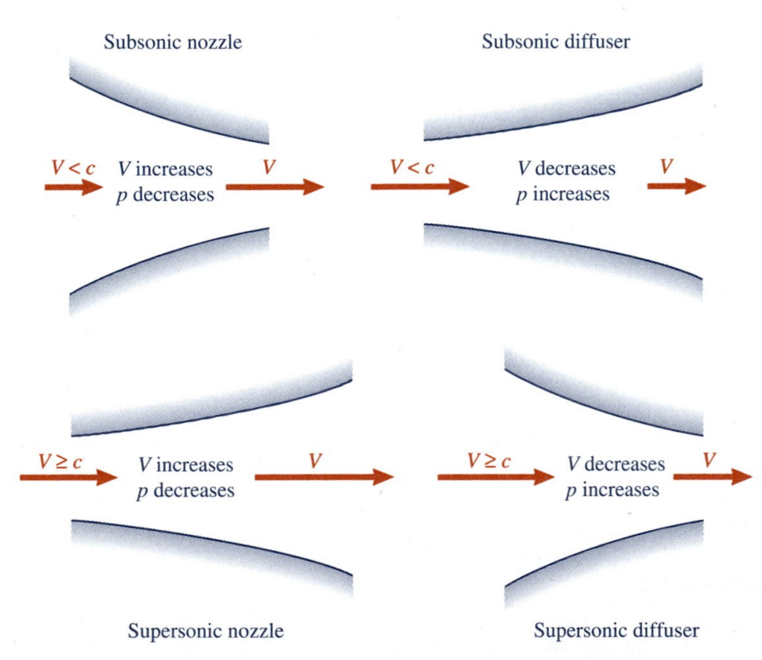

Figure 13·7 Area–velocity–pressure variation for isentropic flow.

sonically, but if the fluid were initially flowing supersonically, the diverging diffuser must be preceded by a converging diffuser to decelerate the fluid to the sonic velocity. This behavior of a supersonically flowing fluid seems strange at first because all our everyday experience is with subsonic flow.

Example 13·5 Isentropic Flow of an Ideal Gas

PROBLEM STATEMENT

Nitrogen flows isentropically through a channel of varying cross-sectional area. At one section where the area is 0.015 m^2, the nitrogen is at 5 kPa, 30°C, and has a velocity of 1500 m/s. Determine the velocity, Mach number, and cross-sectional area at a section where the pressure is 700 kPa.

SOLUTION

Analysis: Call the section where the pressure is 5 kPa section 1, and call the section where the pressure is 700 kPa section 2. Since the flow is isentropic, let us estimate T_2 from $T_2 = T_1(p_2/p_1)^{(k-1)/k}$, that is, assuming that the specific heats are constant. We get

$$T_2 = T_1\left(\frac{p_2}{p_1}\right)^{(k-1)/k} = 303\left(\frac{700}{5}\right)^{(1.4-1)/1.4} = 1244 \text{ K}$$

This temperature estimate shows that assuming constant specific heats is unjustified. Therefore, the final temperature will be found by using gas tables or the computer program for ideal-gas properties and the equation for an isentropic process,

$$\frac{p_{r2}}{p_{r1}} = \frac{p_2}{p_1}$$

Other pertinent equations for this steady-flow isentropic process of an ideal gas with no work done are the first law,

$$h_1 + \frac{V_1^2}{2} = h_2 + \frac{V_2^2}{2}$$

the continuity equation

$$\frac{A_1 V_1}{v_1} = \frac{A_2 V_2}{v_2}$$

the ideal-gas equation of state,

$$p_1 v_1 = RT_1 \quad \text{and} \quad p_2 v_2 = RT_2$$

and the defining equation for Mach number,

$$M_2 = \frac{V_2}{c_2}$$

Also, as a matter of interest, we can find the Mach number at section 1 by including

$$M_1 = \frac{V_1}{c_1}$$

Relationships among T, h, and p_r are included in the model for ideal-gas properties, and a relationship between T and c is given by the model relating the sonic velocity to the temperature of a gas. Thus, we are now ready to begin the calculations.

```
================ RULE SHEET ================
S Rule-------------------------------------------------------------
  ;------------ Example 13•5 -------------------------------------
  pr2 = pr1 * p2/p1                    ;For isentropic processes of ideal gas
  h2 + V2^2/2 = h1 + V1^2/2            ;First law, steady-flow, q=w=0, delpe=0
  M1 = V1/c1                           ;Definition of Mach number
  M2 = V2/c2                           ;   "       "    "     "
  p2 * v2 = R(gas) * T2                ;Ideal-gas equation of state
  p1 * v1 = R(gas) * T1                ;  "      "      "     "    "
  V1 * A1/v1 = V2 * A2/v2              ;Continuity equation

  ;------------ Ideal gas properties ------------------ IGPROP.TK
  call IGPR (gas, T1, h1, phi1, pr1)
  call IGPR (gas, T2, h2, phi2, pr2)

  ;------------ Ideal gas sonic velocity -------------- IGSONIC.TK
* call SONIC (gas, T1; c1)
* call SONIC (gas, T2; c2)

================ VARIABLE SHEET ================
St Input---- Name---- Output---- Unit------ Comment---------------------
                                 ---------- * E13-5.TK ---- Example 13•5 ----
                                 -- Model:  * IGPROP.TK -- Ideal gas properties --
   'N2       gas                            * Dive (>) for list of input choices
                                 -- Model:  * IGSONIC.TK—Ideal gas sonic velocity
   30        T1                  C          Temperature, state 1
   5         p1                  kPa        Pressure, state 1
             v1       18         m^3/kg     Specific volume, state 1
   1500      V1                  m/s        Velocity, state 1
   0.015     A1                  m^2        Area, section 1
             h1       314.7      kJ/kg      Enthalpy, state 1
             phi1     6.858      kJ/(kg*K)  phi = int(cp·dT/T) + C, state 1
             pr1      0.1081                Relative pressure, state 1
             c1       354.9      m/s        Sonic velocity, state 1
             M1       4.227                 Mach number at state 1

             T2       890.2      C          Temperature, state 2
   700       p2                  kPa        Pressure, state 2
             v2       0.4933     m^3/kg     Specific volume, state 2
             V2       584.4      m/s        Velocity, state 2
             A2       0.001055   m^2        Area, section 2
             h2       1269       kJ/kg      Enthalpy, state 2
             phi2     8.325      kJ/(kg*K)  phi = int(cp·dT/T) + C, state 2
             pr2      15.13                 Relative pressure, state 2
             c2       677.5      m/s        Sonic velocity, state 2
             M2       0.8625                Mach number at state 2
```

Calculations: The seven equations and the two computer models listed in the analysis are entered into our equation solver as shown in the printout of the Rule sheet. Then the known values are entered and the model is solved as shown on the Variable sheet. The results are

$$V_2 = 504 \text{ m/s} \qquad M_2 = 0.863 \qquad A_2 = 0.00106 \text{ m}^2$$

Comment: Since $M_1 > 1$ and $M_2 < 1$, there must be a throat or minimum area section between sections 1 and 2.

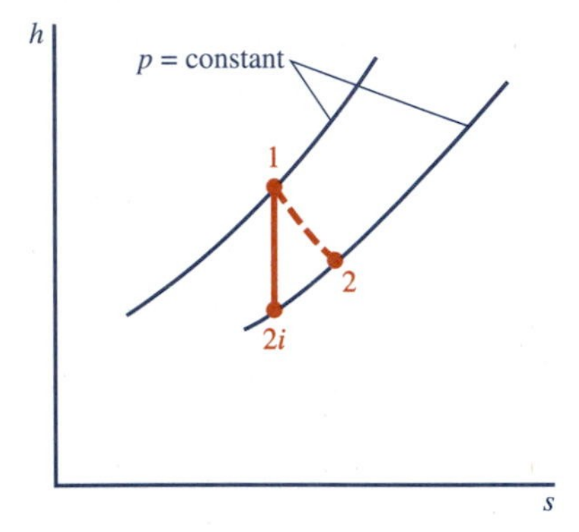

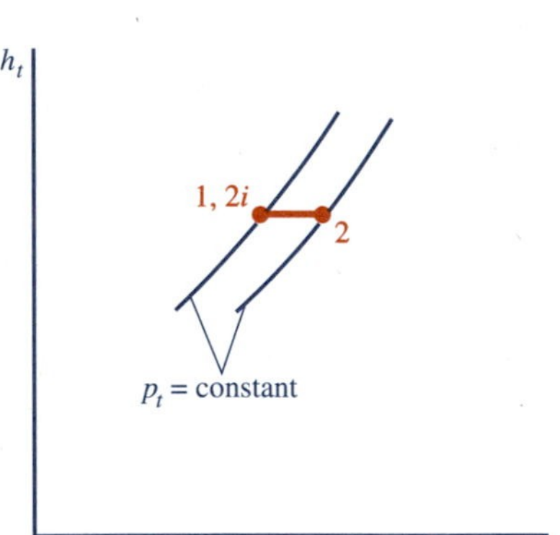

Figure 13·8 Reversible (1–2i) and irreversible (1–2) adiabatic expansions.

Nozzle Efficiency. If the adiabatic flow through a nozzle is irreversible, as it always is to some extent in an actual nozzle because of friction, the flow is not isentropic. We account for irreversibilities in a nozzle using the *nozzle efficiency*. Figure 13·8 shows reversible adiabatic (isentropic) and irreversible adiabatic expansions from the same initial state to the same final pressure. Point 2i is the isentropic (or ideal) exhaust state and point 2 is the actual one. (Notice that on an hs diagram only lines of static pressure are uniquely determined because there exists a unique hps relationship but no unique hp_ts relationship. In a like manner, a single point on an h_ts diagram can represent several static pressures but only one total pressure; so total pressure lines instead of static pressure lines are shown on h_ts diagrams.) The total or stagnation enthalpy is constant for any adiabatic expansion in a nozzle; the total or stagnation pressure is constant only for an isentropic expansion and decreases in an irreversible adiabatic expansion. **Nozzle efficiency**, η_N, is defined as the ratio of the kinetic energy at the nozzle outlet to the kinetic energy that would result at that section if the flow through the nozzle were isentropic between the same initial conditions and the same final pressure. Using this definition and then extending it by means of the first law, we have

Nozzle efficiency

$$\eta_N = \frac{V_2^2/2}{V_{2i}^2/2} = \frac{h_{t1} - h_2}{h_{t1} - h_{2i}}$$

In general, the frictional effect in a converging–diverging nozzle occurs principally between the throat and exit of the nozzle; hence, the flow from the entrance to the throat may be modeled as isentropic even in actual nozzles.

Nozzle efficiencies of 0.95 and higher are easily obtained with converging nozzles. Similar efficiencies with converging–diverging nozzles can be obtained only by careful design. As nozzle sizes increase, fluid friction has relatively less effect on the flow, so nozzle efficiencies increase.

Example 13·6

PROBLEM STATEMENT

A nozzle is to be designed to expand steam at a rate of 0.15 kg/s from 500 kPa, 350°C, to 100 kPa. Inlet velocity is to be very low. For a nozzle efficiency of 0.75, determine the exit areas for the isentropic and the actual flows.

SOLUTION

Analysis: State 1 is known, therefore we can calculate h_1 and s_1. The expansion from state 1 to state $2i$ is isentropic, thus $s_{2i} = s_1$, and since we know p_{2i}, state $2i$ is defined and we are able to determine h_{2i}. The efficiency of the nozzle is related to the ideal and actual parameters, which allows us to calculate h_2. We can use the first law to relate the change in enthalpy to the exit velocity and finally determine the area from the continuity equation.

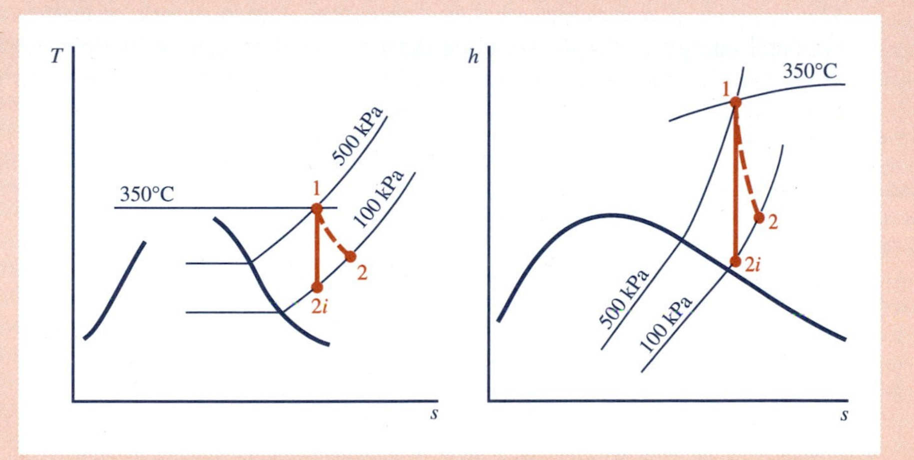

Calculations: We proceed by entering the following equations into a simultaneous equation solver:

$$s_{2i} = s_1$$

$$\eta_N = \frac{h_1 - h_2}{h_1 - h_{2i}}$$

$$h_1 = h_{2i} + \frac{V_{2i}^2}{2}$$

$$h_1 = h_2 + \frac{V_2^2}{2}$$

$$\dot{m} = \frac{A_{2i}V_{2i}}{v_{2i}}$$

$$\dot{m} = \frac{A_2V_2}{v_2}$$

In addition, we load the computer model for steam properties. The results are as follows:

	T, °C	p, kPa	h, kJ/kg	s, kJ/kg·K	v, m³/kg	V, m/s	A, cm²
State 1	350	500	3167.4	7.6324	0.57012	≈ 0	—
State 2i	154.3	100	2784.6	7.6324	1.9559	875	3.35
State 2	202.8	100	2880.3	7.8447	2.1846	758	4.32

13·6·1 Special Case: Ideal Gas

The analysis of reversible, adiabatic processes with no work or change in potential energy is simplified when the flowing fluid can be modeled as an ideal gas. For an ideal gas, the enthalpy depends only on temperature as stated by Joule's law. Therefore, the first law,

$$h_t = \text{constant} \tag{13·3a}$$

The total temperature is constant in an adiabatic process of an ideal gas.

means that T_t = constant. Equation 13·3a can be written as

$$h_2 - h_1 = \frac{V_1^2 - V_2^2}{2}$$

or

$$\int_{T_1}^{T_2} c_p\, dT = \frac{V_1^2 - V_2^2}{2} \tag{13·3b}$$

Alternatively, we can write the above equations in terms of the Mach number by substituting $(Mc)^2$ for V^2:

$$\begin{aligned} h_t &= h + \frac{(Mc)^2}{2} = \text{constant} \\ &= h + \frac{M^2kRT}{2} \end{aligned} \tag{13·3c}$$

$$h_1 + \frac{M_1^2 k_1 R T_1}{2} = h_2 + \frac{M_2^2 k_2 R T_2}{2}$$

Note that up to this point, we have not made the assumption of constant specific heats. Thus, we cannot further simplify the equation, since the specific heat ratio must be evaluated at the appropriate temperature. We indicate this by the k_1 and k_2 notation above.

For this special case of the adiabatic, reversible (isentropic) flow of an ideal gas, we can relate the pressure at any two points in the flow using the relative pressure as shown in Section 6·2:

$$\frac{p_2}{p_1} = \frac{p_{r2}}{p_{r1}}$$

13·6·2 Special Case: Ideal Gas with Constant Specific Heats

When constant specific heats can be assumed, the first law can be further simplified and Eq. 13·3*b* becomes

$$c_p(T_2 - T_1) = \frac{V_1^2 - V_2^2}{2} \qquad (13\cdot3d)$$

Notice that this equation is equivalent to

$$T_1 + \frac{V_1^2}{2c_p} = T_2 + \frac{V_2^2}{2c_p}$$

Substituting $V = Mc$ and $c_p = kR/(k-1)$, as derived in Section 4·2·1 for an ideal gas, we find that

$$T_t = T\left(1 + \frac{k-1}{2}M^2\right) = \text{constant} \qquad (13\cdot3e)$$

If the adiabatic flow of an ideal gas with constant specific heats is also reversible, we can use the pT relation for isentropic flow derived in Section 4·2·3 to obtain

$$p_t = p\left(1 + \frac{k-1}{2}M^2\right)^{k/(k-1)} = \text{constant} \qquad (13\cdot4)$$

Note that this equation shows that for isentropic flow of an ideal gas with constant specific heats, the total pressure is also constant. The last two equations and several related ones are tabulated in many references for one or more values of k; Table A·25 is an example. However, these equations can be easily entered into a computer program which allows them to be solved for any k value.

For limited ranges of pressure and temperature, the c_p and k values of vapors may sometimes be considered as constant so that relationships derived for ideal gases with constant specific heats can be used for vapors. For example, the mean value of k for slightly superheated steam at pressures less than about 1400 kPa (about 200 psia) is around 1.3 and this value is frequently used in predicting whether a converging or a converging–diverging nozzle should be used for a given expansion of steam. The same value is used for slightly wet steam because it has been observed experimentally that steam expanding isentropically (or nearly so) in a nozzle can expand for some distance into the wet region on an hs diagram before condensation actually begins.

Delay of condensation is known as *supersaturation.* Steam that is represented by a point in the wet region as far as its enthalpy and entropy are concerned and yet contains no liquid is called *supersaturated* steam. It is said to be in a metastable state, and we do not treat such states in this book.

Choked Flow. For nozzle and diffuser flows that can be modeled as ideal gases with constant specific heats, we can derive a simple and useful expression relating the temperature at any section where $M = 1$ to the total temperature of the gas. Let us denote the properties at a section where $M = 1$ by symbols such as h^* and T^*. Then using Eq. 13·3*e* with $M = 1$ we have

$$T_t = T\left(1 + \frac{k-1}{2}M^2\right) \tag{13·3e}$$

$$T_t = T^*\left(1 + \frac{k-1}{2}\right)$$

Temperature ratio for choked flow of an ideal gas with constant specific heats

$$\frac{T^*}{T_t} = \frac{2}{k+1} \tag{13·5}$$

This equation holds for the steady adiabatic (reversible or irreversible) flow of an ideal gas with constant specific heats with no work done and no change in potential energy.

One important conclusion from Eq. 13·5 is reached by first noting that at any section of the flow path where the velocity is negligibly small (i.e., $M \approx 0$), the temperature is the total temperature. Therefore, accelerating an ideal gas with constant specific heats from rest to a sonic velocity by adiabatic expansion requires that the expansion proceed until the temperature has decreased in the ratio of $2/(k + 1)$. This holds whether the expansion is reversible or irreversible. Since sonic velocity is proportional to the square root of the temperature and since T^* is less than T_t, the sonic velocity at T^* is less than the sonic velocity in the gas at rest.

If the flow is reversible as well as adiabatic, and hence isentropic, then

Pressure ratio for choked flow of an ideal gas with constant specific heats

$$\frac{p^*}{p_t} = \left(\frac{T^*}{T_t}\right)^{k/(k-1)} = \left(\frac{2}{k+1}\right)^{k/(k-1)} \tag{13·6}$$

We have shown that if a fluid enters a converging passage subsonically, it cannot reach a Mach number greater than 1 in the converging section. Therefore, for isentropic flow the pressure in a converging passage cannot be lower than a certain fraction of the total pressure, as indicated by Eq. 13·6. For gases with $k = 1.4$, this fraction is 0.53. Let us investigate this point further.

Converging Nozzles. Consider a converging nozzle as shown in Fig. 13·9. The upstream pressure p_1 and the temperature T_1, at a section where the velocity is negligibly small, are held constant. The pressure p_b (backpressure)

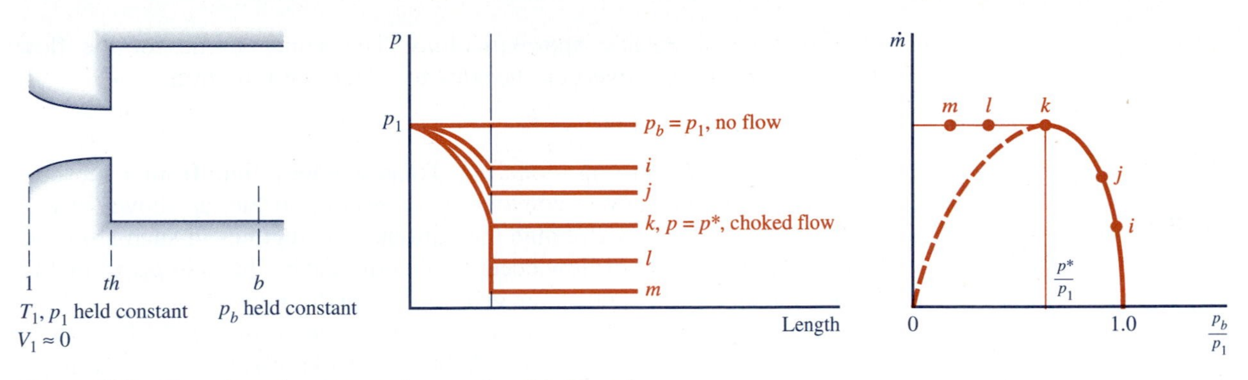

Figure 13·9 Flow through a converging nozzle with $M_1 < 1$.

in the downstream region can be varied. Designate the pressure in the minimum area section or throat as p_{th}. If $p_b = p_1$, there is no flow through the nozzle. If p_b is lowered slightly, flow occurs with $p_{th} = p_b$ and $V_{th} < c_{th}$. If p_b is lowered further, the pressure at the throat decreases to remain equal to p_b, and the mass rate of flow through the nozzle increases as shown in Fig. 13·9 until p_b and p_{th} reach the value of p^*. The velocity at the throat is then sonic, $V_{th} = c_{th} = V^*$. Under these conditions the nozzle is said to be *choked* or to have reached *limiting flow* or *critical flow*. Further reduction of p_b then has no effect on p_{th}, the velocity at the throat, or the mass rate of flow through the nozzle. As we stated above, the Mach number cannot exceed 1 in a converging passage with isentropic flow that is initially subsonic. Also, the pressure cannot be less than p^*, as given by Eq. 13·6.

Whenever $p_b < p^*$, the pressure at the throat remains equal to p^*, and the drop in pressure from p^* to p_b occurs outside the nozzle just beyond the throat. This sudden expansion outside the nozzle is not one dimensional, and it is irreversible; it is only the flow within the nozzle that can be considered isentropic. Thus we can write $p_1 v_1^k = p_{th} v_{th}^k$, but $p_1 v_1^k \neq p_b v_b^k$.

Converging nozzles with choked flow are sometimes used as flow-measuring devices, because as long as the backpressure applied is known to be sufficiently low that $p_b/p_1 < p^*/p_1$, the flow rate through a nozzle of known throat area can be determined from measurements of only p_1 and T_1. Notice that the minimum pressure at the throat is determined by only the inlet pressure and the k value of the gas, not by the shape of the nozzle. The actual flow through a well-designed smooth converging nozzle can be quite close to isentropic, so that the relations derived for isentropic flow apply accurately. Converging nozzles are used in turbines and other pieces of equipment to produce jets of subsonic or sonic velocity.

This discussion of flow in converging nozzles has been for ideal gases with constant specific heats because the quantitative relationships for this special case are quite simple. However, the physical phenomena are the same for other gases and vapors, and for limited ranges of pressure and temperature the quantitative relationships for ideal gases with constant specific heats

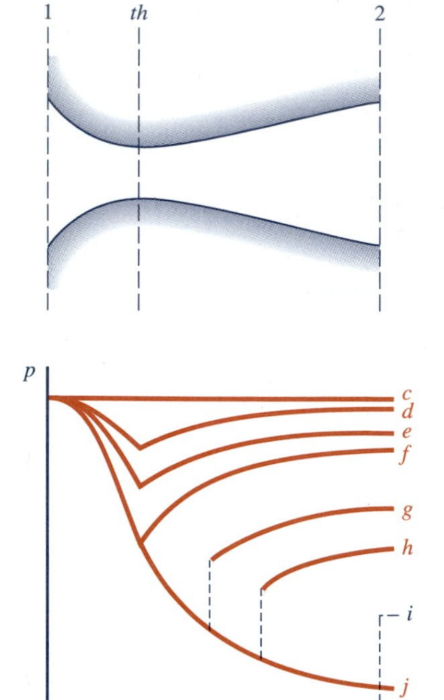

Figure 13·10 Gas flow in a converging–diverging nozzle. Solid lines represent isentropic processes; broken lines represent irreversible processes.

may provide acceptable approximations. The same is true for the flow through converging–diverging nozzles to which we now turn.

Converging–Diverging Nozzles. To accelerate a fluid from a subsonic to a supersonic velocity, a converging–diverging nozzle, as shown in Fig. 13·10, is used. We now examine the various flow regimes in such a nozzle. Which regime or kind of flow occurs is determined by the ratio p_b/p_1, that is, by the backpressure applied for given inlet conditions.

Consider a converging–diverging nozzle as shown in Fig. 13·10 with p_1 and T_1 fixed and with V_1 very low. When the backpressure p_b is equal to p_1, as indicated by point c on the pressure–length plot, there is no flow through the nozzle. As the backpressure is lowered, several different regimes of flow occur.

$p_c > p_b > p_f$: When the backpressure is just slightly below p_1 (point d), there is some flow through the nozzle. The maximum velocity and minimum pressure occur at the throat. As the backpressure is lowered from c to f, the flow rate increases. For backpressures between c and f, the flow throughout the nozzle is subsonic. The diverging section of the nozzle acts as a diffuser, with the pressure rising and the velocity decreasing in the direction of flow. In this regime the converging–diverging nozzle can serve as a venturi meter, a device used for flow-rate measurement.

$p_b = p_f$: As the backpressure is reduced, the throat pressure decreases and the throat velocity increases. The backpressure for which the throat velocity becomes sonic and the throat pressure becomes p^* is p_f. At this condition, the maximum velocity still occurs at the throat, so the diverging section is still a subsonic diffuser. However, the maximum flow rate through the nozzle has been reached, because the throat velocity is sonic. As concluded earlier, flow through a converging passage with a subsonic entrance velocity can never result in a velocity higher than sonic or a pressure lower than p^*.

$p_b < p_f$: As p_b is reduced below p_f, the flow in the *converging* portion of the nozzle is completely unaffected; the pressure at the throat remains p^*, the velocity at the throat remains sonic, and the flow rate does not change. As p_b is decreased, the character of the flow in the *diverging* portion of the nozzle does change, however, and four distinct flow regimes occur.

$p_f > p_b > p_i$: As p_b is reduced to g and h, the fluid passing through the throat continues to expand and accelerate in the diverging section of the nozzle. Its velocity just downstream of the throat is supersonic. However, at a section downstream of the throat there is an abrupt irreversible increase in pressure accompanied by a deceleration from supersonic to subsonic velocity. This discontinuity in the flow is called a **normal shock**. (The modifier *normal* indicates that the plane of the discontinuity is normal to the flow direction.) Flow through the shock is steady and adiabatic but irreversible, so it is not isentropic. Downstream of the shock the gas undergoes further isentropic deceleration as the diverging passage acts as a subsonic diffuser. As p_b

We will discuss the equations pertaining to normal shocks in Section 13·9.

is decreased, the shock moves downstream, approaching the nozzle exit plane as p_b approaches the value represented by point i.

$p_b = p_i$: When the backpressure is at the value represented by point i, the normal shock stands in the exit plane of the nozzle. The flow within the nozzle is isentropic: subsonic in the converging portion, sonic at the throat, and supersonic in the diverging portion. The jet leaving the nozzle is subsonic, however, because the fluid passes through the shock as it leaves the nozzle.

$p_i > p_b > p_j$: When the backpressure is below p_i but above the value p_j that is discussed below, the fluid expands to p_j at the nozzle exit plane and no normal shock forms within or outside the nozzle. Downstream of the exit plane the pressure increases irreversibly from p_j to the backpressure through some discontinuities called oblique shocks that destroy the one dimensionality of the flow. Flow through oblique shocks is not discussed in this book.

$p_b = p_j$: When the backpressure is maintained equal to the exit-plane pressure that causes isentropic expansion throughout the nozzle, no shocks occur within or outside the nozzle and a one-dimensional supersonic jet leaves the nozzle.

$p_b < p_j$: For any backpressure lower than p_j, the flow within and at the exit plane of the nozzle is the same as for $p_b = p_j$. Just downstream of the nozzle exit, the flow loses its one-dimensional character, and irreversible expansion and mixing occur. No matter how far the backpressure is reduced, the pressure within the nozzle never drops below p_j, and the flow rate and exit velocity do not change.

A nozzle operating with a backpressure below p_j is said to be *underexpanded*; a nozzle operating with a backpressure greater than p_j but less than p_g is said to be *overexpanded.*

A review of the flow regimes shows that isentropic subsonic flows throughout the converging–diverging nozzle are possible for flow rates from zero to the maximum. The maximum flow rate corresponds to the attainment of sonic velocity at the throat. For the maximum flow rate, only two backpressures, p_f and p_j, provide isentropic one-dimensional flow throughout the entire nozzle and the region downstream of the nozzle exit. One of these, p_f, causes subsonic diffuser flow and the other, p_j, causes supersonic nozzle flow in the diverging portion.

We can verify that two different pressures at each cross section of the diverging section are possible for the isentropic flow of an ideal gas with constant specific heats by combining the first law, the ideal-gas equation of state, the definition of the Mach number, and pv^k = constant to obtain

$$\frac{A}{A^*} = \frac{2}{M}\left[\frac{2}{k+1}\left(1 + \frac{k-1}{2}M^2\right)\right]^{(k+1)/2(k-1)} \tag{13·7}$$

where A^* is the throat area. This equation shows that there are two possible values of M, one greater than unity and the other less than unity, that satisfy the equation for any value of A greater than A^*. For the subsonic Mach number, $p > p^*$, and for the supersonic Mach number, $p < p^*$.

The precise shape of a converging nozzle or the converging portion of a

The general behavior of flow through converging and converging–diverging nozzles is independent of whether the fluid is an ideal gas with constant specific heats. However, the equations that apply to this special case are simple, and this facilitates the discussion of the physical phenomena.

converging–diverging nozzle is not crucial in nozzle design as long as the convergence is sufficiently gradual, but the geometry of a supersonic nozzle must be determined carefully if it is to produce essentially one-dimensional isentropic flow. Design methods to give the variation of area with length of supersonic nozzles are outside the scope of this textbook.

Example 13·7

PROBLEM STATEMENT

Helium at 300 kPa, 35°C, enters a converging nozzle with negligible velocity. The nozzle discharges into a receiver where the pressure is 100 kPa. The nozzle exit (or throat) cross-sectional area is 88.0 mm². Assuming isentropic flow, calculate the velocity at the nozzle exit.

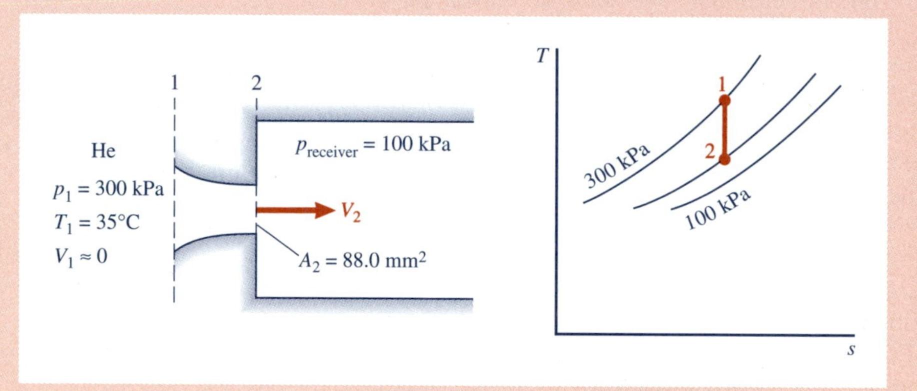

SOLUTION

The receiver pressure (or backpressure) is low enough compared with the inlet pressure that the flow may be choked; that is, the exit velocity may be sonic. We have seen that a fluid that enters subsonically cannot reach a Mach number greater than 1 in a converging section if the flow is isentropic. Therefore, the lowest temperature that can be reached by the air flowing through this converging nozzle is that corresponding to $M = 1$. This is of course the value we designate T^* and is given, for ideal gases with constant specific heats, by

$$T^* = T_t\left(\frac{2}{k+1}\right) = 308\left(\frac{2}{2.667}\right) = 231 \text{ K}$$

in which we have made use of the fact that $T_t = T_1$ since the inlet velocity is negligibly small. The lowest pressure that can exist in the converging nozzle is then

$$p^* = p_t\left(\frac{T^*}{T_t}\right)^{k/(k-1)} = 300\left(\frac{231}{308}\right)^{2.5} = 146 \text{ kPa}$$

Thus the receiver pressure is lower than p^*, so the pressure at the exit or throat of the converging nozzle is p^*, not the receiver pressure, and the velocity there is sonic:

$$V_2 = c_2 = \sqrt{kRT_2} = \sqrt{kRT^*} = \sqrt{1.667\left(2.08\frac{\text{kJ}}{\text{kg}\cdot\text{K}}\right)231\text{ K}\left(1\frac{\text{kg}\cdot\text{m}}{\text{N}\cdot\text{s}^2}\right)1000\frac{\text{N}}{\text{kN}}}$$
$$= 895 \text{ m/s}$$

(The same value of V_2 can be obtained by writing an energy balance between the inlet and the exit or throat.)

Comment: Notice that the receiver pressure does not enter the calculation once it has been determined that it is lower than the throat pressure associated with a sonic velocity.

Example 13·8

PROBLEM STATEMENT

Air at 550 kPa, 1000°C, enters a converging nozzle with negligible velocity. The nozzle discharges into a receiver where the pressure is 100 kPa. Assuming isentropic flow, calculate the velocity at the nozzle exit or throat.

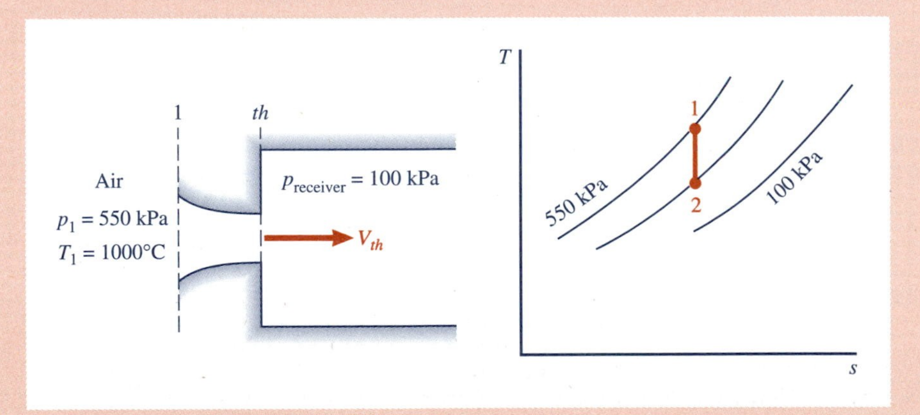

SOLUTION

The receiver pressure is so much lower than the upstream pressure that the velocity at the exit must be sonic. (The minimum value of p_{th}/p_t can be approximated by a constant-specific-heat calculation and is seen to be greater than 100/550.) Therefore, the lowest temperature that can be reached by the air flowing through this converging nozzle is that corresponding to $M = 1$. This is of course the value we designate T^*. We will assume ideal-gas behavior, but in view of the high inlet air temperature, we will not use constant specific heats. The first law is

$$h_t = h_1 = h_{th} + \frac{V_{th}^2}{2} = h_{th} + \frac{c_{th}^2}{2}$$

in which we have made use of the fact that (1) the inlet velocity is negligibly small and (2) $V_{th} = c_{th}$. We can evaluate $c_{th} = k_{th}RT_{th}$, but to do this we must find $k(T_{th})$. We could do this iteratively using air tables, but it is easier to use the computer model for air properties to relate h and T and the computer model relating c and T. The only equations we need to enter on the Rule sheet are

$$h_1 = h_{th} + \frac{V_{th}^2}{2}$$

and

$$V_{th} = c_{th}$$

The printout of the Variable sheet shows the throat or exit velocity to be 646 m/s.

```
==================== RULE SHEET ====================
S Rule---------------------------------------------------------------
  ;------------------ Example 13•8 ----------------------------------
  h1 = hth + Vth^2/2                    ;First law between inlet and throat
  Vth = cth                             ;Velocity at the throat is sonic

  ;------------------ Ideal gas properties --------------- IGPROP.TK
  call IGPR (gas, T1, h1, phi1, pr1)
  call IGPR (gas, Tth, hth, phith, prth)

  ;------------------ Ideal gas sonic velocity --------------IGSONIC.TK
* call SONIC (gas, Tth; cth)

==================== VARIABLE SHEET ====================
St Input---- Name--- Output--- Unit------ Comment---------------------------
                                ---------- * E13-8.TK ------ Example 13•8 -----
             h1      1366      kJ/kg      Enthalpy at inlet
             hth     1157      kJ/kg      Enthalpy at throat
             Vth     646       m/s        Velocity at throat
             cth     646       m/s        Sonic velocity at throat
                                -- Model: * IGPROP.TK -- Ideal gas properties --
   'Air      gas                          * Dive (>) for list of input choices
   1000      T1                C          Temperature at inlet
             phi1    8.253     kJ/(kg*K)  phi = int(cp·dT/T) + C at inlet
             pr1     308.4                Relative pressure at inlet

             Tth     821.6     C          Temperature at throat
             phith   8.077     kJ/(kg*K)  phi = int(cp·dT/T) + C at throat
             prth    166.6                Relative pressure at throat
                                -- Model: * IGSONIC.TK--Ideal gas sonic velocity
```

Comment: We can calculate $p_{th} = p_1(p_{r,th}/p_{r1}) = 297$ kPa to confirm that the receiver pressure is well below the lowest pressure that can exist in the throat for isentropic flow.

Example 13·9 Flow in a Nozzle

PROBLEM STATEMENT

A nozzle expands steam at a rate of 1.5 lbm/s from 200 psia, 800 F, to 50 psia. The initial velocity is negligible. Assuming that the flow is isentropic, compute the velocity, temperature, and area at sections along the nozzle. Plot the nozzle area as a function of pressure along the nozzle.

SOLUTION

Analysis: Since the initial kinetic energy is negligible, application of the first law shows that the kinetic energy at any section equals the decrease in enthalpy between the inlet and that section. The relationship between the velocity and the decrease in enthalpy at section j along the nozzle is given by

$$\frac{V_j^2}{2} = h_1 - h_j$$

The continuity equation, $A = \dot{m}v/V$ relates the area to other flow parameters. The solution proceeds as follows: At T_1, p_1, we calculate s_1. Then, $s_j = s_1$ combined with knowledge of p_j defines state j along the nozzle. We can then find T_j, ρ_j, and h_j. h_j is used in the first law equation above to calculate V_j, and finally the continuity equation permits calculation of A_j.

Calculations: The above equations are entered into a simultaneous equation solver along with the computer model for steam properties, which provides properties at each state. Then, for a range of pressures along the nozzle between 50 psia and 150 psia, we obtain the following results:

p_j, psia	T_j, F	V_j, ft/s	A_j, in^2
50	454.52	2864.1	0.80936
60	495.91	2690.6	0.75067
70	530.65	2535.4	0.70779
80	561.76	2387.2	0.67811
90	589.99	2243.6	0.65888
100	614.12	2113.3	0.64402
110	638.95	1968.6	0.64283
120	661.61	1826.3	0.64855
130	681.67	1690.3	0.65794
140	701.63	1542.3	0.68147
150	720.23	1389.7	0.71698

The relationship between the nozzle area and the pressure along the nozzle is shown in the plot.

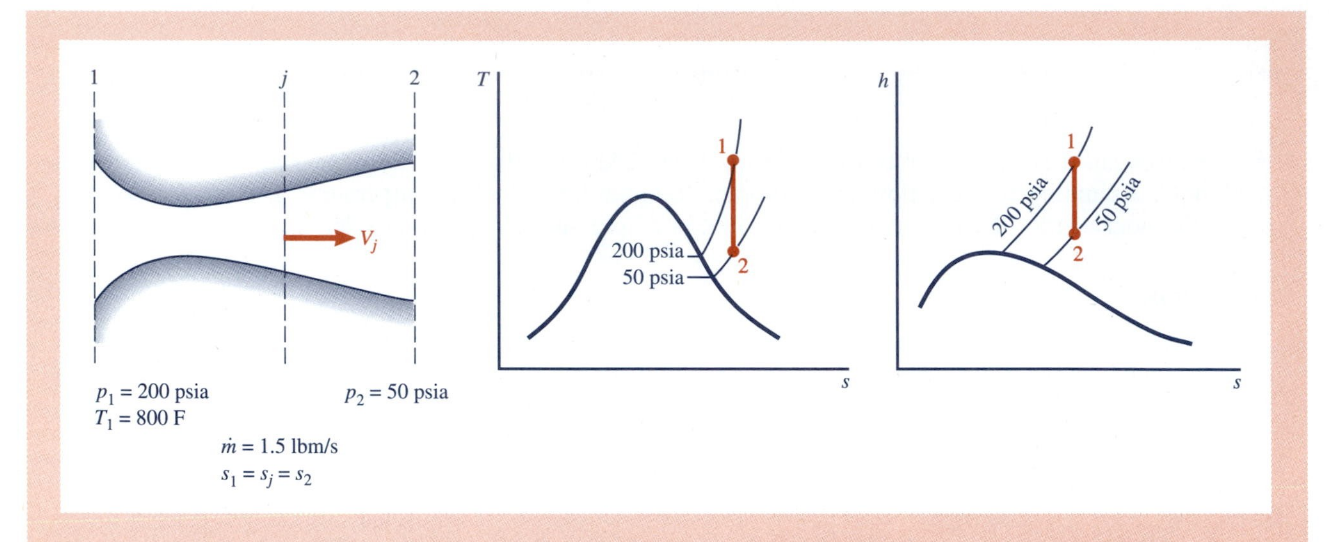

This plot shows that the nozzle area is a minimum somewhere between 100 and 110 psia.

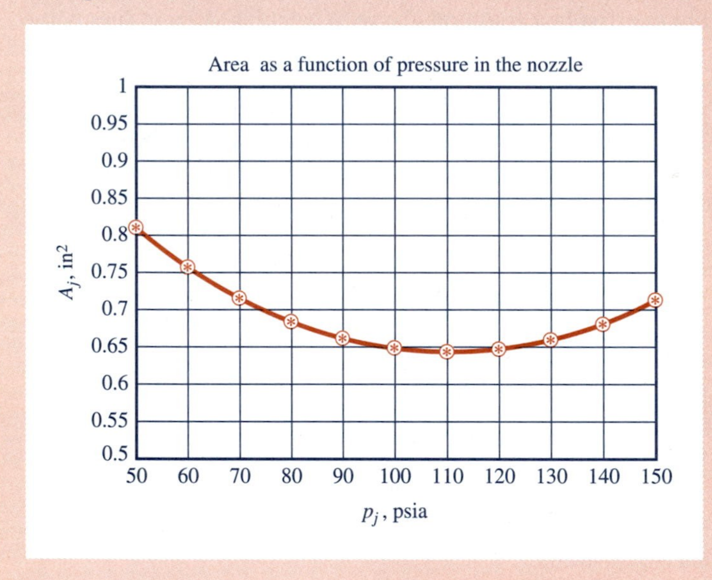

Comment: Consider how time-consuming this problem solution would be if done either by hand using a Mollier chart or even using a program with only lookup data that would require you to transcribe many numbers.

13·7 Flow in Pipes—Adiabatic, Irreversible Flow in a Constant-Area Passage

We have just completed our discussion of adiabatic, reversible flows and now proceed to another branch of the tree—adiabatic, irreversible flows (refer to Fig. 13.6). We restrict our attention in this section to flows in passages with a constant cross-sectional area; however, in this section we have dropped the restriction of ideal-gas, constant-specific-heat behavior.

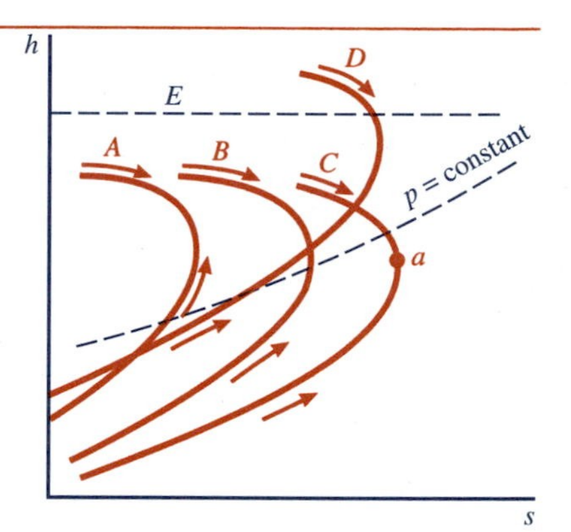

Figure 13·11 Fanno lines. Each solid line represents states of constant total enthalpy for adiabatic flow along a constant-area passage such as a pipe.

The Fanno Line. Consider the adiabatic steady flow of a fluid in a constant-area passage with no work done. The flow in an insulated pipe meets such conditions. Application of the first law and the continuity equation gives

$$h_t = h + \frac{V^2}{2} = h + \left(\frac{\dot{m}}{A}\right)^2 \frac{v^2}{2} = \text{constant} \qquad (13\cdot3f)$$

Thus, the relationship between h and v (and consequently between any two properties if we are dealing with a pure substance) is fixed; so the states that satisfy Eq. 13.3f can be plotted on an hs diagram. Figure 13·11 shows three such plots for the same total enthalpy but different values of $\dot{m}/A$ (lines A, B, and C) and one plot for a higher total enthalpy (line D). These lines are called **Fanno lines** and each one is the locus of states through which a fluid passes in adiabatic pipe flow for given entrance conditions. A limiting case Fanno line is the constant-enthalpy line E that inspection of Eq. 13·3f shows to be the limiting case as $\dot{m}/A$ approaches zero or the case for incompressible flow (v = constant).

Fanno line. Gino Fanno (1882–1962), Italian engineer, described these lines in his diploma thesis submitted to the Swiss Federal Institute of Technology, Zurich, in 1904.

We know from the increase of entropy principle that for adiabatic flow the entropy of the fluid cannot decrease; therefore, the process represented by the Fanno lines in Fig. 13·11 can proceed only in the directions shown by the arrows on the lines. Let us investigate the conditions at a maximum entropy point on a Fanno line such as point a in Fig. 13·11. The energy balance in differential form is

$$dh + V\,dV = 0$$

From the continuity equation we have ρV = constant so that $dV = -V\,d\rho/\rho$, and making this substitution in the energy balance gives

$$dh - \frac{V^2\,d\rho}{\rho} = 0$$

Then substituting for dh from the $T\,ds$ equation,

$$T\,ds = dh - v\,dp \qquad (6\cdot7)$$

gives

$$T\,ds + v\,dp - \frac{V^2\,d\rho}{\rho} = 0$$

$$T\,ds + \frac{dp}{\rho} - \frac{V^2\,d\rho}{\rho} = 0$$

At a maximum entropy point, $ds = 0$, so this last equation can be simplified and solved for V as

$$V = \sqrt{\left(\frac{\partial p}{\partial \rho}\right)_s} = c \tag{13·2}$$

Thus, the limiting state such as a in Fig. 13·11 is one for which the velocity equals the sonic velocity. Further consideration of Fig. 13·11 shows that the upper part of each Fanno line (i.e., the part above the maximum entropy point) represents states of subsonic flow, and the lower part corresponds to supersonic flow. Therefore, for subsonic adiabatic flow in a pipe, the pressure and enthalpy decrease and the velocity increases in the direction of flow, but the velocity cannot exceed the velocity of sound. For supersonic adiabatic flow in a pipe, the pressure and enthalpy increase and the velocity decreases in the direction of flow. In this case the limiting velocity again is the velocity of sound but it is approached from above. In both cases, of course, the entropy of the fluid must increase in the direction of flow. If a pipe carrying a fluid at subsonic velocities discharges into a region where the pressure is lower than that corresponding to the point of maximum entropy on the Fanno line p_a, then the expansion from p_a to the receiver pressure must occur outside the end of the pipe, because inside the pipe only states on the Fanno line that can be reached without a decrease in entropy can be attained. If the flow in the pipe is initially supersonic and the receiver pressure is higher than p_a, then a normal shock must occur inside the pipe, causing the flow to change from supersonic to subsonic. The flow is represented first by states on the lower part of the Fanno line and then, downstream of the shock, by states on the upper part (see Section 13·9). (*Caution*: The limiting pressure in a pipe, where the flow is irreversible, is not the same as the limiting pressure in a converging nozzle under isentropic flow conditions. In both cases, however, the total enthalpy is constant and the limiting velocity is sonic.)

The variation of pressure, enthalpy, velocity, and so on, of the fluid as a function of pipe length can be predicted only if information is available on friction factors or fluid shearing stresses. These matters are treated in books on fluid mechanics.

13·8 Flow with Combustion or Heat Transfer—Diabatic, Frictionless Flow

We now leave adiabatic flows briefly to consider steady frictionless flow (i.e., flows with no fluid shearing forces) in a constant-area passage with no work done.

The Rayleigh Line. The dynamic equation applied to an element of fluid between two sections 1 and 2 is

Refer to Fig. 13·6 to see how this section relates to other sections.

$$p_1 A - p_2 A = \dot{m}(V_2 - V_1) \qquad \text{(13·1, expanded)}$$

Dividing both sides of this equation by the area and substituting from the continuity equation $V = \dot{m}v/A$, we have

$$p_1 - p_2 = \left(\frac{\dot{m}}{A}\right)^2 (v_2 - v_1)$$

or

$$p + \left(\frac{\dot{m}}{A}\right)^2 v = \text{constant} \qquad \text{(13·8)}$$

This is the equation of a **Rayleigh line** that holds for steady frictionless flow in a constant-area passage with no work done. It is based on only the dynamic and continuity equations. If property relationships for the fluid are known, the equation can be plotted on various coordinates. Figure 13·12 shows a Rayleigh line on *hs* coordinates and, for comparison, a Fanno line for the same value of $\dot{m}/A$. The Fanno line is a line of constant total enthalpy, so obviously the total enthalpy varies along a Rayleigh line. This means that a fluid can follow a Rayleigh line *only* if there is heat transfer. For purposes of dynamic analysis, reacting flows can be modeled as frictionless flows with heat transfer, so a Rayleigh line can represent the flow in a combustion chamber. As in the case of the Fanno line, it can be shown that the maximum entropy point on a Rayleigh line corresponds to $M = 1$. The lower branch of the line is for supersonic flow, and the upper branch is for subsonic flow. The process is not adiabatic, so of course there is no restriction on the sign of entropy change of the fluid alone. The direction in which a process proceeds along a Rayleigh line depends on the direction of heat transfer, as indicated in Fig. 13·12. If heat is added to the fluid in this frictionless process, its entropy must increase; if heat is removed, the entropy of the fluid must decrease.

Rayleigh line. Lord Rayleigh, John William Strutt (1842–1919), English physicist, is perhaps best known for his work in acoustics but made significant contributions in optics, hydrodynamics, electromagnetism, and gas properties.

A number of interesting conclusions can be drawn from inspection of the Rayleigh line. For subsonic flow, adding heat causes the pressure to drop; for supersonic flow, adding heat causes the pressure to rise. Along the Rayleigh line between the maximum enthalpy point and the maximum entropy point, adding heat causes the enthalpy to decrease.

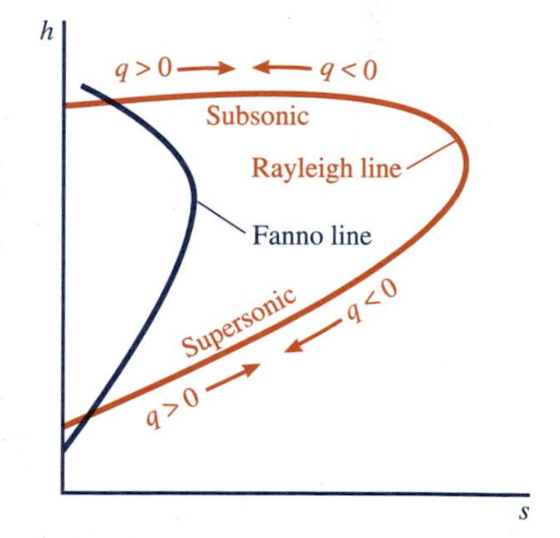

Figure 13·12 A Rayleigh line and a Fanno line for the same value of $\dot{m}/A$.

13·9 Normal Shocks—Adiabatic, Irreversible Flow in a Constant-Area Passage

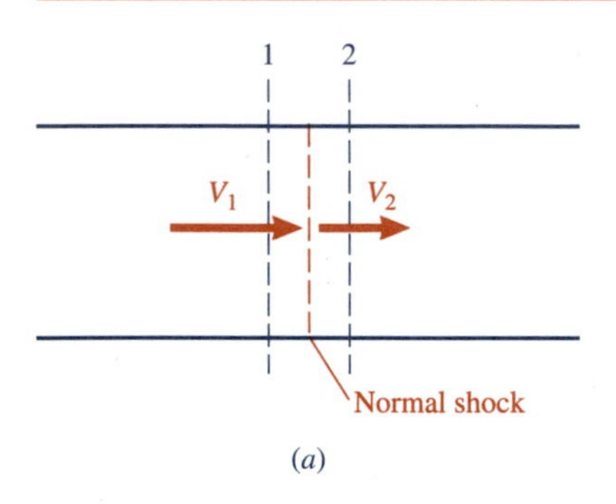

(*a*)

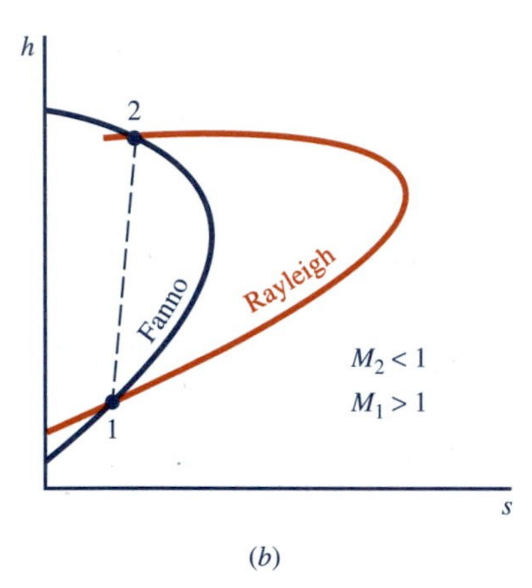

(*b*)

Figure 13·13 Use of Rayleigh and Fanno lines to determine the state change across a normal shock.

A combination of Fanno and Rayleigh lines is useful in determining the change in state that occurs across a normal shock, which was mentioned in Section 13·6 as a flow discontinuity in which the pressure rises suddenly and the velocity changes from supersonic to subsonic. The application of Fanno and Rayleigh lines to the same flow requires some explanation, since one is for frictional adiabatic flow and the other is for frictionless diabatic flow.

Referring to Fig. 13·13, fluid in state 1 approaches the shock at a supersonic velocity. The thickness of the shock is very small so that between section 1 just upstream of the shock and section 2 just downstream of it the area is virtually the same, even though the shock may occur in a diverging passage. If the process is also adiabatic, then states 1 and 2 must lie on the same Fanno line. Again, because the shock is so thin, wall friction forces can be neglected (i.e., they are very small compared to the difference in pressure forces acting on a fluid element that contains the shock). Therefore, states 1 and 2 are connected by a frictionless (but not necessarily reversible) constant-area flow and hence lie on the same Rayleigh line. The only point on the *hs* diagram of Fig. 13·13 that lies on both the same Fanno line and the same Rayleigh line as point 1 is point 2 at the other intersection of the two lines. Thus, the conditions just downstream of a shock can be determined by finding the intersection of the Fanno and Rayleigh lines that pass through the upstream state of the fluid. For an ideal gas the equation of state is simple enough so that the Fanno and Rayleigh line equations can be solved directly for the Mach number, pressure, and temperature ratios across a shock. For vapors, an empirical *hpv* relation can often be formulated that makes possible the direct solution of the two equations.

Figure 13·13 shows $s_2 > s_1$, as expected for an irreversible adiabatic process. It is possible to show that for all flows that satisfy both the Fanno and Rayleigh line conditions, the entropy of the subsonic state is always higher than that of the supersonic state. Therefore, since the shock itself is an adiabatic process, it can occur in only one direction: from a supersonic state to a subsonic state. From inspection of the Fanno and Rayleigh lines we conclude that two states satisfy the requirements of the first law, the dynamic equation, and the continuity equation, but the second law is needed to tell us in which direction a process between these two states can occur.

Shocks, as we have just shown, can occur only in supersonic flow and affect the flow significantly. Their occurrence in supersonic nozzles has already been mentioned, and is the subject of Example 13·11. They occur also in the flow ahead of a craft moving at a supersonic speed, as illustrated in Fig. 13·14. Shocks occur upstream of probes inserted into a supersonic stream, as shown in Fig. 13.14*b*; thus, the flow impinging on the probe is subsonic, not supersonic. The flow entering the engine of a supersonic craft

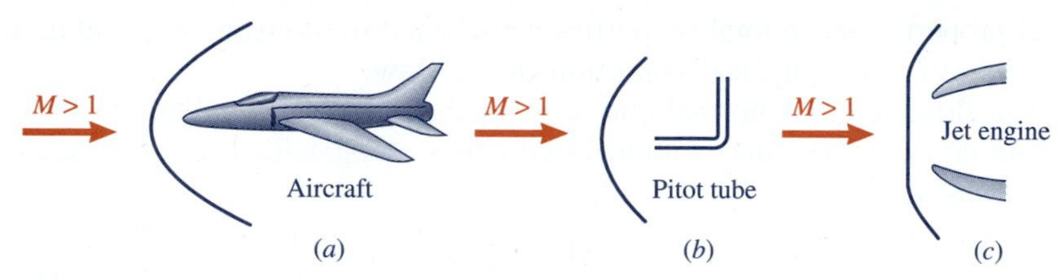

Figure 13·14 Normal shocks.

may pass through a normal shock, as shown in Fig. 13·14*c*, and the engine design may utilize the pressure rise across the shock as part of the total compression process of the engine.

We have discussed only normal shocks, which exist directly ahead of bodies in supersonic flow where the velocity is normal to the wave. The other regions of the wave structure are described by oblique shocks and are not discussed in this book.

13·9·1 Special Case: Ideal Gas

For an ideal gas, the equations describing flow across a normal shock can be collected from many of the equations already derived in earlier sections. The flow across a normal shock is restricted as mentioned above: It is adiabatic, irreversible, frictionless, and a constant-area flow. In this section we add the restriction of ideal-gas behavior to obtain the following equations:

$$\int_1^2 c_p\,dT = \frac{V_1^2 - V_2^2}{2} \qquad \text{First law}$$

$$\dot{m} = \rho_1 V_1 A = \rho_2 V_2 A \qquad \text{Continuity equation}$$

$$A(p_2 - p_1) = \dot{m}(V_1 - V_2) \qquad \text{Dynamic equation}$$

In addition to these equations, ideal-gas property relationships are available.

13·9·2 Special Case: Ideal Gas with Constant Specific Heats

If we assume constant specific heats, the above equations are further simplified. In fact, these equations can be combined explicitly and hence are convenient computationally. As before, the first law and the definition of Mach number lead to

$$T\left(1 + \frac{k-1}{2}M^2\right) = \text{constant} \qquad (13\cdot3e)$$

Equations are often written to involve either total properties or properties at a section where the velocity is sonic, because the respective Mach number values of 0 and 1 simplify them. The use of properties at the section where $M = 1$ can involve the cross-sectional area; this is not the case when using total properties since total properties are defined for convenience and do not always refer to a physical condition in the flow.

The flow across a normal shock is adiabatic but irreversible, so the dynamic equation and the continuity equation along with Eq. 13·3*e* lead to equations such as

$$M_y^2 = \frac{M_x^2 + 2/(k-1)}{[2k/(k-1)]M_x^2 - 1} \tag{13·9}$$

and

$$\frac{p_y}{p_x} = \frac{2k}{k+1} M_x^2 - \frac{k-1}{k+1} \tag{13·10}$$

where sections x and y are just upstream and just downstream of the shock, respectively.

These equations are so widely used that tabulations of them (and many others) are published in the Keenan, Chao, and Kaye *Gas Tables* and elsewhere. Most often they appear for $k = 1.4$, that is, for cold air, and the results for gases with $k \neq 1.4$ may be quite different. A greatly abridged version of one such tabulation is given in Table A·21. These equations can also be readily entered into a simultaneous equation solver program for computer solution to allow more problems for different gases to be solved.

Example 13·10 Normal Shock

PROBLEM STATEMENT

Air at 1 atm, 15°C, at a Mach number of 8 enters a normal shock. Determine the pressure and temperature following the shock, using both variable specific heats and constant specific heats. Calculate also the percent error in the constant-specific-heat results.

SOLUTION

Let us call the section just upstream of the shock section x and the section just downstream of the shock section y. Since plots that require calculations for a range of initial Mach numbers are needed, we will use a computer for the calculations.

The following equations apply to the steady flow across a shock:

1. The dynamic equation, which reduces for the case of no friction and constant area to

$$p_x - p_y = \rho_y V_y^2 - \rho_x V_x^2$$

2. The first law, which reduces for the case of steady, adiabatic flow with no work and no change in potential energy to

$$h_x + \frac{V_x^2}{2} = h_y + \frac{V_y^2}{2}$$

3. The continuity equation, which reduces for the case of constant-area flow to

$$\rho_x V_x = \rho_y V_y$$

4. The ideal-gas equation of state:

$$p_x = \rho_x R T_x$$
$$p_y = \rho_y R T_y$$

5. The definition of Mach number:

$$M_x = \frac{V_x}{c_x}$$

6. The expression for the sonic velocity, $c_x = \sqrt{kRT_x}$ for the constant-specific-heat case and the computer model that relates c and T for the variable-specific-heat case.
7. Relationships between enthalpy and temperature for an ideal gas: $\Delta h = c_p \, \Delta T$ for constant specific heats and the air properties model for the variable-specific-heat case.

The calculations are carried out within the equation solver as outlined in the analysis above. Note that for the constant-specific-heat calculation, we designate downstream properties by symbols such as T_{yc} and p_{yc}.

```
═══════════════════ RULE SHEET ═══════════════════════════════════
S Rule─────────────────────────────────────────────────────────────
  ;─────────────── Example 13•10 ──────────────────────────────────
  ;The following equations are for the variable-specific-heat case
  hx + Vx^2/2 = hy + Vy^2/2                ;First law
  px - py = rhoy*Vy^2 - rhox*Vx^2          ;Dynamic equation
  rhox*Vx = rhoy*Vy                        ;Continuity equation
  px = rhox * R * Tx                       ;Ideal-gas equation of state
  py = rhoy * R * Ty                       ;Ideal-gas equation of state
  Vx = Mx*cx
  ;─────────────── Properties of air at low pressure ──────── AIRPROP.TK
  call AIRPR (Tx, hx, phix, prx)
  call AIRPR (Ty, hy, phiy, pry)
  ;─────────────── Ideal gas sonic velocity ───────────────── IGSONIC.TK
  call SONIC (gas, Tx; cx)

  ;The equations below are for the constant-specific-heat case.  For this
  ;case, downstream property symbols carry a letter "c", e.g., pyc, rhoyc.
  delhc = Vx^2/2 - Vyc^2/2                 ;First law  (delhc = hyc - hx)
  delhc = cp * (Tyc - Tx)                  ;Enthalpy change, constant cp
  px - pyc = rhoyc*Vyc^2 - rhox*Vx^2       ;Dynamic equation
  rhox*Vx = rhoyc*Vyc                      ;Continuity equation
  px = rhox * R * Tx                       ;Ideal-gas equation of state
  pyc = rhoyc * R * Tyc                    ;Ideal-gas equation of state
  Vx = Mx*cx                               ;Also, cx = SQRT(k * R * Tx)
  perr = 100 * (pyc - py)/py               ;Error in pressure downstream of shock
  Terr = 100 * (Tyc - Ty)/Ty               ;Error in temperature downstream
```

```
==================== VARIABLE SHEET ==================================
St Input---- Name--- Output--- Unit----- Comment-----------------------
                                         * E13-10.TK ---- Example 13·10 ----
             hx      290.2     kJ/kg     Upstream enthalpy
             Vx      2722      m/s       Upstream velocity
             hy      3914      kJ/kg     Downstream enthalpy
             Vy      403       m/s       Downstream velocity
   101.3     px                kPa       Upstream pressure
             py      7839      kPa       Downstream pressure
             rhoy    8.281     kg/m^3    Downstream density
             rhox    1.226     kg/m^3    Upstream density
   0.287     R                 kJ/(kg*K) Gas constant
             Ty      3299      K         Downstream temperature
   288       Tx                K         Upstream temperature
             cx      340.3     m/s       Upstream sonic velocity
   8         Mx                          Upstream Mach number
                               -- Model: * AIRPROP.TK ---- Properties of air ---
                               -- Model: * IGSONIC.TK--Ideal gas sonic velocity
   'Air      gas
             Vyc     488.8     m/s       Downstream velocity, constant cp
             pyc     7553      kPa       Downstream pressure,     "      "
             rhoyc   6.826     kg/m^3    Downstream density,      "      "
             Tyc     3855      K         Downstream temperature,  "      "
             delhc   3585      kJ/kg     Enthalpy change,         "      "
   1.005     cp                kJ/(kg*K) Constant specific heat
             perr    -3.65               Percent error in pyc
             Terr    16.9                Percent error in Tyc
                                         (Ty and Tyc require guesses to start
                                            the iterative solution)
```

Notice that $T_y = T_x$, $p_y = p_x$, $V_y = V_x$, and so on would be one solution of the sets of equations, so guesses of values for starting the iterative solution must be selected to avoid this spurious solution. The results are shown on the Variable sheet. Notice that the constant-specific-heat solution gives an absolute temperature that is 16.9 percent too high.

Example 13·11 Isentropic Flow Through a Converging–Diverging Nozzle

PROBLEM STATEMENT

Air is to be expanded steadily and isentropically through a nozzle from an initial state of 200 kPa, 80°C, with negligible velocity, to a Mach number of 1.60 where the cross-sectional area is 0.0092 m^2. Determine (*a*) the mass rate of flow, (*b*) the discharge pressure for the design conditions, (*c*) the discharge pressure for the maximum flow rate and isentropic compression in the diverging portion of the nozzle, and (*d*) the backpressure to cause a normal shock to stand in the exit plane.

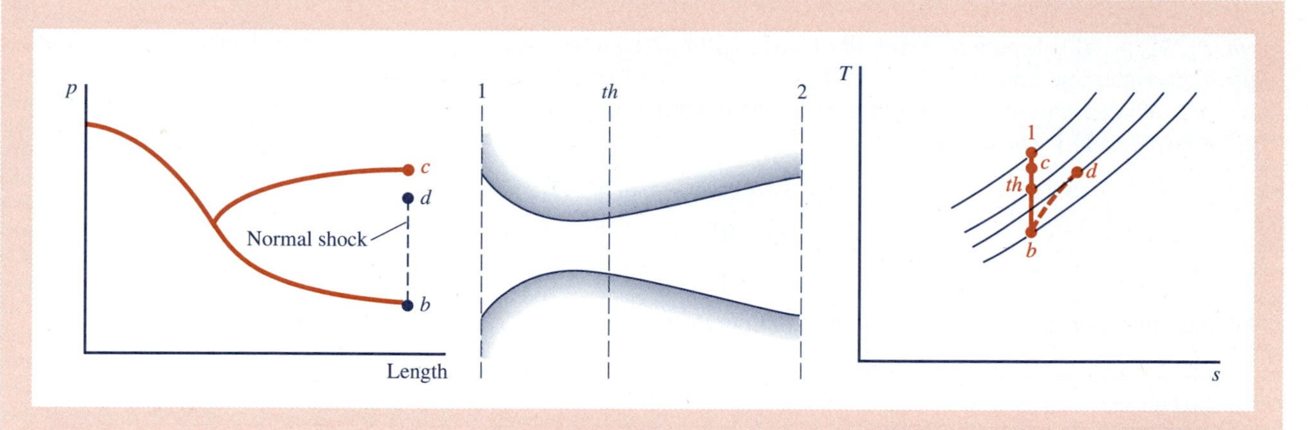

SOLUTION

Analysis: The maximum temperature in this flow is 80°C , so it is likely that constant specific heats can be used to obtain accurate results. We could use variable specific heats and employ the computer models for air properties and sonic velocity, but it is usually advisable to use the simplest approach that will give accurate results. Several solution methods are possible for various parts of the problem. To illustrate two of them, both based on constant specific heats, we will solve the problem first from basic equations and then again using published tables. As a first step for either method, we sketch pressure–length and *Ts* diagrams, labeling the different exit states to correspond to lettered sections of the problem statement.

Calculations Using Basic Equations (a, b): We can obtain the mass rate of flow from the continuity equation

$$\dot{m} = \rho_b A_2 V_b$$

We must also use the ideal-gas equation of state

$$p_b = \rho_b R T_b$$

and the first law for steady adiabatic flow of an ideal gas with constant specific heats, zero work, and zero change in potential energy

$$\frac{V_b^2}{2} = h_1 - h_b = c_p(T_1 - T_b) \qquad (T_1 = T_{t1})$$

We will also use the relation for an isentropic process between states 1 (or t) and b

$$p_b = p_1\left(\frac{T_b}{T_1}\right)^{k/(k-1)}$$

and the defining equation for Mach number combined with the expression for the sonic velocity of an ideal gas

$$V_b = M_b c_b = M_b\sqrt{kRT_b}$$

These five equations (excluding the intermediate parts involving h_1, h_b, and c_b) involve five unknowns

($\dot{m}$, ρ_b, V_b, p_b, T_b). Inputs to a simultaneous algebraic equation solver program are the data included in the problem statement and $R = 0.287$ kJ/kg·K, $c_p = 1.003$ kJ/kg·K, and $k = 1.4$. [The c_p and k values are taken from Table A·2 on the assumption that the mean temperature will be around 20°C. k, c_p, and R are related by $c_p = kR/(k-1)$.] The results are

$$\dot{m} = 3.16 \text{ kg/s} \qquad p_b = 47.0 \text{ kPa}$$
$$V_b = 490 \text{ m/s} \qquad T_b = -39.6\text{°C}$$
$$\rho_b = 0.702 \text{ kg/m}^3$$

We see that our assumption of a mean temperature around 20°C is justified.

```
=================== RULE SHEET ===================
S Rule-------------------------------------------------
  ;-------------- Example 13·11 ----------------------
  mdot = rhob * A2 * Vb               ;Continuity equation
  pb = rhob * R * Tb                  ;Ideal-gas equation of state
  Vb^2/2 = cp * (T1 - Tb)             ;First law for specified conditions
  pb = p1 * (Tb/T1)^(k/(k-1))         ;For isentropic process, ig, const sp heat
  Vb = Mb * SQRT(k * R * Tb)          ;Vb = Mb * cb, and cb = SQRT(k * R * T)

  mdot = rhoc * A2 * Vc               ;Continuity equation
  pc = rhoc * R * Tc                  ;Ideal-gas equation of state
  Vc^2/2 = cp * (T1 - Tc)             ;First law for specified conditions
  pc = p1 * (Tc/T1)^(k/(k-1))         ;For isentropic process, ig, const sp heat

  (pd - pb) * A2 = mdot * (Vb - Vd)   ;Dynamic equation applied to shock
  rhob * Vb = rhod * Vd               ;Continuity equation across shock
  pd = rhod * R * Td                  ;Ideal-gas equation of state
  (Vb^2 - Vd^2)/2 = cp * (Td - Tb)    ;First law applied across shock

================= VARIABLE SHEET =================
St Input---- Name---- Output---- Unit------- Comment--------------------
                                  ---------- * E13-11.TK ---- Example 13·11 ----
             mdot     3.16       kg/s       Mass rate of flow
             rhob     0.702      kg/m^3     Supersonic discharge density
   0.0092    A2                  m^2        Discharge cross-sectional area
             Vb       490        m/s        Supersonic discharge velocity
             pb       47         kPa        Supersonic discharge pressure
   0.287     R                   kJ/(kg*K)  Gas constant for air
             Tb       -39.6      C          Supersonic discharge temperature
   1         cp                  kJ/(kg*K)  Specific heat at constant pressure
   80        T1                  C          Inlet temperature where V1 = 0
   200       p1                  kPa        Inlet pressure where V1 = 0
   1.4       k                              Specific heat ratio
   1.6       Mb                             Supersonic Mach number at discharge
             rhoc     1.7        kg/m^3     Subsonic discharge density
             Vc       202        m/s        Subsonic discharge velocity
             pc       162        kPa        Subsonic discharge pressure
             Tc       59.6       C          Subsonic discharge temperature

             pd       133        kPa        Pressure downstream of shock
             Vd       241        m/s        Velocity downstream of shock
             rhod     1.43       kg/m^3     Density downstream of shock
             Td       51         C          Temperature downstream of shock
```

(*c*) To determine the outlet pressure, p_c, for isentropic flow at the same rate (and, therefore, with $M_{th} = 1$, still) but with subsonic flow in the diverging portion of the nozzle, we have the same equations we used for parts *a* and *b*, except that we do not need the one involving Mach number. Thus

$$\dot{m} = \rho_c A_2 V_c$$

$$p_c = \rho_c R T_c$$

$$\frac{V_c^2}{2} = c_p(T_1 - T_c)$$

$$p_c = p_1\left(\frac{T_c}{T_1}\right)^{k/(k-1)}$$

The unknowns in these four equations are ρ_c, V_c, p_c, and T_c. The results are

$$V_c = 202 \text{ m/s} \qquad T_c = 59.6°\text{C}$$

$$p_c = 162 \text{ kPa} \qquad \rho_c = 1.70 \text{ kg/m}^3$$

(*d*) For determining the pressure rise across a normal shock standing in the exit plane, we use for the properties of the fluid entering the shock those listed above for isentropic expansion to the exit plane (state *b*). The dynamic equation across a shock (inlet *b*, outlet *d*) is

$$(p_d - p_b)A = \dot{m}(V_b - V_d)$$

The continuity equation for $A_b = A_d$ is

$$\rho_b V_b = \rho_d V_d$$

The ideal-gas equation of state is

$$p_d = \rho_d R T_d$$

and the first law for the adiabatic flow of an ideal gas with constant specific heats is

$$\frac{V_b^2 - V_d^2}{2} = c_p(T_d - T_b)$$

Introducing these four equations (with unknowns p_d, V_d, ρ_d, and T_d) into our simultaneous equation solver computer program gives $p_d = 133$ kPa.

Calculations Using Tables: Using Tables A·25 and A·26, which are for $k = 1.4$, provides a quick solution to this problem. (*a*, *b*) For $M_b = 1.6$, Table A·25 shows $\rho_b/\rho_1 = 0.35573$, $p_b/p_t = 0.23527$, and $T_b/T_t = 0.66138$. Since $p_t = 200$ kPa, we immediately have the desired result for part *b*, $p_b = 47.05$ kPa. For use in the continuity equation, we calculate

$$\dot{m} = \rho_b A_2 V_b = \rho_b A_2 M_b c_b = \frac{\rho_b}{\rho_t}\rho_t A_2 M_b \sqrt{kRT_b}$$

$$= \frac{\rho_b}{\rho_t}\frac{p_t}{RT_t} A_2 M_b \sqrt{kR\frac{T_b}{T_t}T_t}$$

$$= 0.35573 \frac{200}{0.287(353)} 0.0092(1.6)\sqrt{1.4(1000)0.287(0.66138)353}$$

$$= 3.17 \text{ kg/s}$$

(*c*) Table A·25 shows that for producing $M_b = 1.6$, the nozzle has $A_2/A^* = 1.2502$. It also shows that this same value of the ratio of outlet area to throat area can result in $M_c = 0.553$ and $p_c = (p_c/p_t)p_t = 0.8124(200) = 162$ kPa.

(*d*) For $M_b = 1.6$ entering a normal shock, Table A·26 shows $p_d = (p_d/p_b)p_b = 2.820(47.05) = 133$ kPa.

Comment: In these two solutions we have illustrated both approaches: (1) using basic relations and (2) using published tables of compressible flow functions. Each approach has advantages.

13·10 Summary

This chapter deals with the thermodynamic aspects of the steady one-dimensional flow of fluids. A useful property in such studies is the *total enthalpy* that is defined as

$$h_t \equiv h + \frac{V^2}{2}$$

The *total temperature* and *total pressure* of a flowing fluid are defined as the temperature and pressure that would result if the fluid were brought to rest isentropically. They are also called the stagnation temperature and stagnation pressure.

The *basic dynamic equation* for steady one-dimensional flow is

$$F = \dot{m}\,\Delta V \tag{13·1}$$

where F is the resultant force on a stream that undergoes a velocity change ΔV and that has a flow rate $\dot{m}$. This is a vector equation: F and ΔV are in the same direction.

The *velocity of sound* in a fluid is given by

$$c = \sqrt{\left(\frac{\partial p}{\partial \rho}\right)_s} \tag{13·2}$$

which for an ideal gas becomes

$$c = \sqrt{\frac{kp}{\rho}} = \sqrt{kpv} = \sqrt{kRT}$$

A useful parameter in the study of gas flow is the *Mach number, M*, defined by

$$M \equiv \frac{V}{c}$$

The following are the basic tools used for analyzing steady, one-dimensional flow of fluids in which no work is done:

1. The first law, which under the conditions stated can be expressed as

$$q = \Delta h + \Delta ke + \Delta pe$$

In connection with gases and vapors, Δpe is nearly always negligible.

2. The continuity equation:

$$\dot{m} = \rho_1 A_1 V_1 = \rho_2 A_2 V_2$$

3. The dynamic equation:

$$F = \dot{m}\,\Delta V$$

4. The second law and its corollaries.

5. Physical property relationships, either in the form of tabular, graphical, or computer data or in the form of equations of state and other property equations resulting from the first and second laws.

Application of the first and second laws, the continuity equation, and the dynamic equation shows that for the isentropic flow of any fluid the pressure, velocity, and cross-sectional area variations depend on the Mach number as follows:

1. When $M < 1$, $dA/dp > 0$ and $dA/dV < 0$
2. When $M > 1$, $dA/dp < 0$ and $dA/dV > 0$
3. When $M = 1$, $dA/dp = 0$ and $dA/dV = 0$

Further application of the basic relations reveals that if a fluid enters a converging nozzle subsonically, the Mach number at the outlet cannot exceed unity, regardless of how far the receiver pressure is lowered. A converging–diverging nozzle must be used to accelerate a fluid from a subsonic to a supersonic velocity and a converging–diverging diffuser must be used to decelerate a fluid isentropically from a supersonic to a subsonic velocity. When $M = 1$ at the throat of a converging–diverging passage, there are, for isentropic flow, only two possible pressures for the fluid at each other section of the passage, one for subsonic and the other for supersonic flow.

The Fanno line is the locus of states of a fluid that flows adiabatically through a constant-area channel with no work done. The Rayleigh line is the locus of states of a fluid that flows frictionlessly through a constant-area passage with no work done but with heat transfer allowed. The Fanno and Rayleigh lines are useful in several types of analyses, one of which is the determination of the change of state across a normal shock. A normal shock is a flow discontinuity that occurs only if the velocity is initially supersonic and that results in a deceleration to a subsonic velocity and an abrupt pressure rise.

References and Suggested Reading

13·1 Fox, Robert W., and Alan T. McDonald, *Introduction to Fluid Mechanics*, 4th ed., Wiley, New York, 1992, Chapters 12 and 13.

13·2 Saad, Michel A., *Compressible Fluid Flow*, 2nd ed., Prentice Hall, Englewood Cliffs, N.J., 1993. (This work presents some interesting graphical aids for calculations involving ideal gases with constant specific heats.)

13·3 Shapiro, A. H., *The Dynamics and Thermodynamics of Compressible Flow*, Ronald Press, New York, 1953, Vol. 1, Chapters 3, 4, and 5. (This two-volume set is older than most references listed in this book, but it is a classic well worth studying. In keeping with the limited availability, at the time the book was written, of computational tools similar to those now widely available, ideal-gas calculations are largely limited to constant-specific-heat cases.)

13·4 Keenan, Joseph H., Jing Chao, and Joseph Kaye, *Gas Tables*, 2nd ed., Wiley, New York, 1980 (English units), 1983 (SI units).

13·5 Potter, J. H., *Steam Charts*, American Society of Mechanical Engineers, 1976. (Novel large-scale charts that can be read more accurately than most Mollier charts.)

Problems

13·1 A liquid flows steadily through a circular pipe with a velocity distribution given by $u = u_{max}[1 - (r/r_0)^2]$, where u is the velocity at any radius r, u_{max} is the velocity at the pipe axis, and r_0 is the pipe radius. Determine the ratio of the kinetic energy per unit mass of this stream to that of a one-dimensional flow in the same pipe with the same flow rate.

13·2 Consider a liquid in a circular pipe of radius r_0. At one instant the velocity is a maximum u_{max} at the pipe axis and zero at the pipe wall. Determine the ratio V/u_{max}, where V is the average velocity at one section, if the velocity u at any radius r is given by $u = u_{max}(1 - r/r_0)^2$. How might such a velocity distribution be obtained?

13·3 Consider a fluid flowing steadily in a circular pipe of radius r_0. Determine the ratio V/u_{max}, where V is the average velocity across the pipe and u_{max} is the maximum velocity, if the velocity u at any radius is given by $u = u_{max}[1 - (r/r_0)^2]$.

13·4 Air at 120 kPa, 40°C, flows in a stream with a cross-sectional area of 0.40 m^2 with a velocity of 150 m/s into a system of vanes that turns the stream through an angle of 30 degrees without changing the magnitude of the velocity or the pressure. Determine the force of the stream on the vane system.

13·5 Ethane (C_2H_6) flows at a rate of 20,000 kg/h through a long straight section of pipe that has an inside diameter of 10 cm. At the inlet the ethane is at 800 kPa, 30°C; at the outlet it is at 200 kPa, 10°C. Determine the axial force of the ethane on the pipe.

13·6 An aircraft turbojet engine is to provide a thrust of 60 kN when operating at a speed of 800 km/h through air at 40 kPa, −35°C. The inlet velocity relative to the engine will be the same as the craft speed. It is estimated that the ratio of air mass to fuel mass will be 40, the exhaust gas leaving at 40 kPa will have a density of 0.15 kg/m^3, and the exhaust velocity relative to the craft will be 420 m/s. Determine the required fuel flow rate and the diameter of the air inlet, which has a circular cross section.

13·7 Air at 20 psia, 100 F, flows in a stream of 4 sq ft cross-sectional area with a velocity of 500 fps into a system of vanes that turns the stream through an angle of 30 degrees without changing the magnitude of the velocity or the pressure. Determine the force of the stream on the vane system.

13·8 A jet of water having a cross-sectional area of 0.1 sq ft strikes a flat plate that is normal to the jet. The force on the plate is 100 lbf. If the jet velocity is doubled, what will be the magnitude of the force on the plate? What is the magnitude of the higher velocity?

13·9 Determine the stagnation pressure and temperature on a craft moving at 1000 m/s through air at 12 kPa, −20°C.

13·10 Determine the stagnation temperature and the approximate adiabatic wall temperature of steam at 3.60 MPa, 250°C, flowing with a velocity of 300 m/s.

13·11 Determine the stagnation pressure and the stagnation temperature on a craft moving at 6000 fps through air at 2 psia, −40 F. Determine also the percent errors that result from basing the calculations on specific heats that are constant at their −40 F values.

13·12 A craft flies through air at 0.5 atm, −20°C (−4 F). Plot on one set of axes the stagnation temperature on the craft versus the craft Mach number for a range of Mach numbers from 1.0 to 4.0 for (*a*) specific heats constant at their ambient values and (*b*) variable specific heats.

13·13 Determine the stagnation temperature and the approximate adiabatic wall temperature of steam at 600 psia, 500 F, flowing with a velocity of 1000 fps.

13·14 Determine the ratio of the sonic velocity in air at 2000 K (3140 F) to that in air at 300 K (80 F).

13·15 On a certain day when sea-level atmospheric temperature is 14°C, the temperature decreases with altitude at a rate of 9°C per 1000 m. How long does it take for a sound wave to travel from sea level to an altitude of 5000 m?

13·16 Determine the Mach number of the steam flow of Problem 13·10.

13·17 Determine the Mach number of the steam flow of Problem 13·13.

13·18 Argon at 600 kPa, 280°C, has a velocity of 300 m/s through a cross-sectional area of 100 cm^2. Calculate the stagnation pressure, stagnation temperature, Mach number, and mass rate of flow.

13·19 For air at 1 atm, 15°C, plot stagnation pressure and stagnation temperature versus Mach number for a range of $M = 1$ to $M = 8$ using (*a*) constant-specific-heat relations (e.g., as embodied in Table A·25) and (*b*) variable specific heats (as embodied in Table A·12). (*c*) Plot percentage error in the constant-specific-heat results versus Mach number for p_t and T_t in the same Mach number range.

13·20 If the sonic velocity of a breathing mixture used for divers is greatly different from that of air, voice intelligibility is degraded. One breathing mixture contains 20 percent oxygen by volume, 15 percent nitrogen, and the balance helium. Another contains the same fractions of oxygen and nitrogen, but the balance is a mixture of helium and neon in the same volumetric ratio they have in air (3.47 mol Ne/mol He). Compare the sonic velocities of these breathing mixtures with that of air.

13·21 A vertical pipe contains water. A pressure gage on the pipe reads 10 psi (68.9 kPa) higher than another gage located 30 ft (9.1 m) above the first. Is there upward flow, downward flow, no flow, or flow of indeterminate direction in the pipe?

13·22 Steam flowing through a pipe enters a pressure-reducing valve at 2.00 MPa, 250°C, with a velocity of 100 m/s. The pressure on the downstream side of the valve is 400 kPa. The pipe inside diameter is 10.0 cm on both sides of the valve. The flow is steady and adiabatic. Determine the exit velocity and the exit temperature.

13·23 Is the flow of Problem 13·5 adiabatic?

13·24 Helium flows steadily through an insulated passage. At one section, where the cross-sectional area is 0.080 m^2, the helium is at 70 kPa, 25°C, with a velocity of 1800 m/s. At another section the helium is at 350 kPa, 330°C. Calculate the Mach number and the cross-sectional area at this other section. Prove which one is the upstream section and sketch the variation in cross-sectional area between the two sections.

13·25 Nitrogen flows steadily and isentropically through a passage. At section 1, where the cross-sectional area is 0.186 m^2, the nitrogen is at 100 kPa, 5°C, and has a velocity of 600 m/s. At section 2, the velocity is 360 m/s. Calculate (*a*) the cross-sectional area of section 2 and (*b*) the Mach number at section 2.

13·26 At section 1 of a duct where the cross-sectional area is 0.0450 m^2, air at 50 kPa, 5°C, flows with a Mach number of 2.0. Farther downstream at section 2, where the cross-sectional area is 0.0360 m^2, the temperature is 59°C. The flow is steady and adiabatic. Determine (*a*) the pressure at section 2 and (*b*) whether the process between 1 and 2 is reversible or irreversible, showing the calculations on which you base your conclusion.

13·27 Calculate the flight Mach number that causes the stagnation density on an aircraft to differ by 5 percent from the density of the surrounding air.

13·28 Air flows through a long tapered tube that is well insulated. At section 1, where the cross-sectional area is 2.0 sq ft, the air is at 20 psia, 140 F, and has a velocity of 500 fps. Determine the cross-sectional area at section 2 where the air is at 10 psia, 60 F.

13·29 Calculate the irreversibility of the process of Problem 13·28.

13·30 Air flows steadily through an insulated passage. At one section, where the cross-sectional area is 0.500 sq ft, the air is at 10.0 psia, 40 F, with a velocity of 2000 fps. At another section the air is at 50 psia, 360 F. Calculate the Mach number and the cross-sectional area at this other section. Prove which one is the upstream section.

13·31 An airplane flies through air at 12.8 psia, 59 F, at a velocity of 600 fps. Determine the Mach number at a point on the plane where the air velocity relative to the plane is 300 fps.

13·32 Derive the following equation for the stagnation pressure of an ideal gas with constant specific heats:

$$p_t = p\left(1 + \frac{k-1}{2}M^2\right)^{k/(k-1)}$$

13·33 An insulated passage has a cross-sectional area of 0.400 m^2 at section 1 and 0.220 m^2 at section 2. Nitrogen enters at 40 kPa, 25°C, and flows reversibly through the passage. Plot the pressure at section 2 as a function of flow rate if the mean velocity at section 1 is varied from 10 m/s to 100 m/s. (Be sure to predict the shape of the curve before starting calculations.)

13·34 Air passes through a duct with a constant cross-sectional area of 6.20 sq ft. A bank of heating tubes is installed in the duct. The air enters at 2.0 inches of water above atmospheric pressure, 20 F, and is heated to 95 F. The entering flow rate is 10,500 cfm. Neglecting pressure drop due to friction, what is the exit pressure?

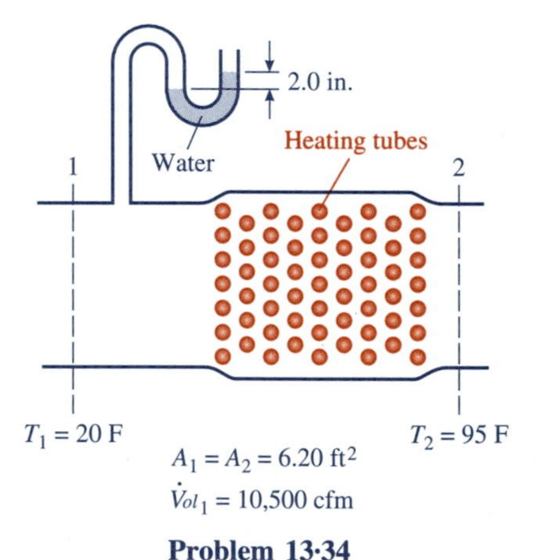

Problem 13·34

13·35 Air in a large tank is maintained by means of vacuum pumps and water cooling at 10 kPa, 15°C. The surrounding atmosphere is at 100 kPa, 25°C, and atmospheric air leaks into the tank through a converging nozzle that has a throat or discharge cross-sectional area of 2.0×10^{-6} m^2. Calculate the maximum flow rate of air through the nozzle and into the tank.

13·36 Ethylene (C_2H_4) escapes from a huge tank where the pressure and temperature are maintained at 400 kPa and 30°C into a furnace through a converging nozzle. The exit area of the nozzle is 10^{-3} m^2. The atmosphere is at 100 kPa, 25°C. List the assumptions you make and calculate the maximum flow rate through the nozzle.

13·37 Air at 850 kPa, 80°C, with negligible velocity enters a converging nozzle that discharges to a region where the pressure can be varied from 850 kPa down to 100 kPa. The flow is adiabatic. The nozzle throat has a cross-sectional area of 4 cm^2. Tabulate mass rate of flow versus backpressure, with 50-kPa increments of backpressure.

13·38 Air at 200 kPa, 93°C, enters a horizontal converging tube with a velocity of 60 m/s. The tube tapers from a cross-sectional area of 0.50 to 0.30 m^2 and is insulated. The pressure at the outlet is 70 kPa. Atmospheric temperature is 15°C. Calculate the temperature and velocity of the air at the outlet and the irreversibility of the process.

13·39 Acetylene (C_2H_2) at 100 psia and 140 F flows through a converging tube into a receiver in which the pressure is 20 psia. Assuming reversible adiabatic flow and negligible upstream velocity, calculate the velocity at the throat.

13·40 A tank with a volume of 1.35 m^3 contains air initially at 1400 kPa, 20°C. Air is allowed to escape through a converging nozzle to the atmosphere. The converging nozzle cross section is circular, and the discharge diameter is 2.0 cm. Plot a curve of pressure in the tank, from 1400 kPa down to 300 kPa, versus time, assuming that the gas remaining in the tank expands adiabatically. State all assumptions carefully.

13·41 Ethylene (C_2H_4) flows through a converging nozzle that has a throat or outlet area of 0.050 sq ft. The ethene upstream is at 14.0 psia, 80 F, with negligible velocity, and the Mach number at the throat is 0.50. Calculate the pressure at the throat.

13·42 Air is to be held in a storage tank at a maximum pressure of 1400 psig. If the 4-inch-inside-diameter discharge line should rupture, the pressure in the tank will drop rapidly, and the reaction force on the tank may damage its supporting structure. It is proposed that a converging nozzle be placed at the tank outlet as shown in the figure to limit the flow rate from the tank and the reactive force. Comment on the effectiveness of this proposal and determine the diameter of nozzle required if the reactive force is not to exceed 8000 lbf.

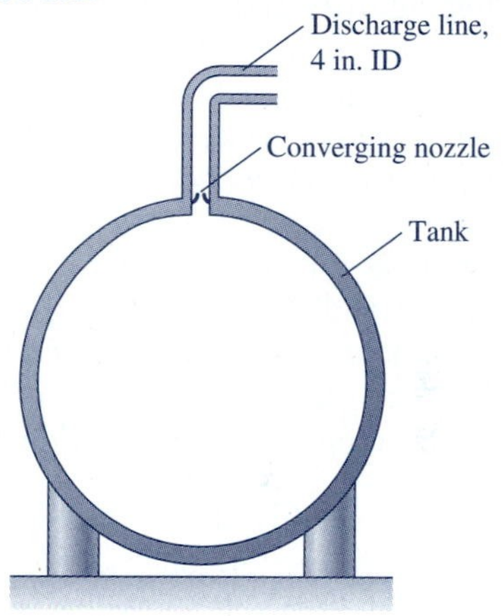

Problem 13·42

13·43 Refer to Problem 13·42. Plot a curve of pressure in the tank versus time after rupture of the discharge line, with the flow-limiting nozzle in place, assuming that the expansion is (*a*) adiabatic and (*b*) isothermal. Explain why you would choose one assumption over the other.

13·44 Air is to be expanded isentropically through a nozzle and discharged at 101.3 kPa, 35°C, with a Mach number of 2.0 through a cross-sectional area of 35.0 cm^2. Determine the initial pressure and temperature if the initial velocity is very low.

13·45 Carbon dioxide at 275 kPa, 115°C, enters a nozzle with a velocity of 180 m/s and expands reversibly and adiabatically to 135 kPa. The nozzle discharge area is 9.29 cm^2. Determine the flow rate.

13·46 Determine the inlet and throat areas of a nozzle that is to discharge hydrogen at 35 kPa, 5°C, at a

Mach number of 2.0 through an exit area of 0.05 m^2. The inlet velocity is to be 150 m/s. Assume that the flow is isentropic.

13·47 Helium at 2 MPa, 200°C, expands isentropically in a nozzle to 600 kPa. If the discharge rate is 0.15 kg/s, compute the areas at sections where the pressures are 1.9, 1.7, 1.5, 1.3, 1.1, 0.9, and 0.7 MPa. Disregard the entrance velocity.

13·48 Nitrogen at 1000 kPa, 150°C, enters a nozzle with negligible velocity and expands isentropically to 140 kPa. For a flow rate of 0.20 kg/s, determine the cross-sectional area and the Mach number at sections in the nozzle where the pressures are 900, 800, 700, 600, 500, 400, 300, and 200 kPa.

13·49 Helium is to be expanded isentropically through a nozzle and discharged at 100 kPa, 40°C, with a Mach number of 2.0 through a cross-sectional area of 0.00400 m^2. Sketch the shape of the nozzle and calculate the minimum cross-sectional area.

13·50 Air flows steadily, reversibly, and adiabatically through a passage. At one section, the cross-sectional area is 0.620 m^2, the pressure is 20 kPa, the temperature is 20°C, and the Mach number is 3.00. At a section downstream, the pressure is 30 kPa. (*a*) Determine the velocity at the second section. (*b*) Is the cross-sectional area at the second section greater than, equal to, or smaller than that of the first section? (This can be determined without calculating the area or other quantities at the second section.)

13·51 Solve Problem 13·44 for adiabatic flow with a nozzle efficiency of 90 percent. Also calculate the irreversibility of the process if the sink temperature is 15°C.

13·52 Air at 150 psia, 300 F, enters a nozzle with negligible velocity and expands isentropically to 20 psia. For a flow rate of 0.50 lbm/s, plot the cross-sectional area and the Mach number through the nozzle as a function of pressure.

13·53 Air flows through the test section of a wind tunnel at 10 psia, 40 F, with a Mach number of 1.15. At the outlet of the diffuser that follows the test section, the air is at 19 psia, 150 F. The flow is adiabatic. Determine the velocity leaving the diffuser. Is the flow reversible?

13·54 Determine the inlet and throat areas of a nozzle which is to discharge air at 5.0 psia, 40 F, at a Mach number of 2.0 through an exit area of 0.5 sq ft. The inlet velocity is to be 500 fps. Assume that the flow is isentropic.

13·55 Ethene (C_2H_4) at 125 kPa, 25°C, flows through a tube at 650 m/s. If the flow is reversible and adiabatic, does the velocity increase, decrease, remain constant, or vary indeterminately in a diverging section following the tube?

13·56 Air at 40 psia, 240 F, enters a nozzle with a velocity of 600 fps and expands reversibly and adiabatically to 20 psia. The nozzle discharge area is 1.44 sq in. Determine the flow rate.

13·57 Air flows steadily and reversibly through an insulated passage. At one section, where the cross-sectional area is 0.50 sq ft, the air is at 10 psia, 40 F, with a velocity of 2000 fps. At another section, farther downstream, the air is at 50 psia. Calculate the Mach number and the cross-sectional area at the second section.

13·58 Air enters a supersonic diffuser at 5.0 psia, 40 F, with a Mach number of 3.0. The diffuser inlet cross-sectional area is 2.0 sq ft. Sketch the diffuser and determine the temperature at the section where the Mach number is 1.0, assuming isentropic flow.

13·59 Determine the flow rate through a properly designed nozzle that steam enters at 2.0 MPa, 300°C, with negligible velocity and leaves at 0.2 MPa if the throat area is 190 mm^2.

13·60 A converging–diverging nozzle has a very large inlet area, a throat area of 0.020 sq ft, and a discharge area of 0.048 sq ft. For inlet conditions of air at 200 psia, 400 F, determine the proper backpressures for isentropic flow of air at the maximum flow rate.

13·61 The nozzle described in Problem 13·60 has the same inlet conditions applied but has a backpressure of 196 psia applied. Calculate the Mach number at the throat.

13·62 Propane (C_3H_8) with a stagnation pressure of 150 psia and stagnation temperature of 340 F enters a converging nozzle having an exit area of 1.8 sq in. and an efficiency of 94 percent. The exit pressure is 100 psia. Atmospheric temperature is 70 F. (*a*) Calculate the mass rate of flow through the nozzle. (*b*) Calculate the irreversibility per pound of propane.

13·63 Air at a stagnation pressure of 80 psia and a stagnation temperature of 160 F enters a diffuser at a Mach number of 1.0. The diffuser inlet cross-sectional area is 2.40 sq ft. The flow is adiabatic. The temperature of the surrounding atmosphere is 60 F. At the diffuser outlet, the stagnation pressure is 74 psia. Determine (*a*) the irreversibility of the diffuser flow per pound of air and (*b*) the mass rate of flow.

13·64 Air flows steadily through an insulated nozzle, entering with low velocity at 150 psia, 500 F, and leaving at 15 psia, 70 F. The mass rate of flow is 10.0 lbm/s. Atmospheric conditions are 14.7 psia, 40 F. Calculate (*a*) the exit cross-sectional area, (*b*) the exit Mach number, (*c*) the irreversibility (in B/lbm).

13·65 Derive the following equation for the mass rate of flow of an ideal gas with constant specific heats flowing isentropically from an initial condition 1 of negligible velocity:

$$\dot{m} = A\sqrt{\frac{2k}{k-1}\frac{p_1}{v_1}\left[\left(\frac{p}{p_1}\right)^{2/k} - \left(\frac{p}{p_1}\right)^{(k+1)/k}\right]}$$

13·66 Derive the following equation for p^* of an ideal gas with constant specific heats flowing isentropically:

$$\frac{p^*}{p} = \left[\frac{2}{k+1} + M^2\left(\frac{k-1}{k+1}\right)\right]^{k/(k-1)}$$

13·67 Produce a table showing, for the flow of an ideal gas with constant specific heats, A/A^*, p/p_t, ρ/ρ_t, and T/T_t, as a function of M with argument values from 0.25 to 3.00, in steps of 0.25, for (*a*) $k = 1.24$ and (*b*) $k = 1.25$.

13·68 A wind tunnel is to provide a steady supersonic stream of air of circular cross section 22.0 cm in diameter at 100 kPa, 15°C. Plot for isentropic flow the required upstream (where velocity is negligible) pressure and temperature versus stream Mach number for a range of Mach numbers from 1.0 to 2.4, using constant specific heats. For comparison, show on the plot at $M = 2.4$ the stagnation pressure and temperature if the variation of specific heats is considered.

13·69 Plot the curve called for in Problem 13·68 if the gas is a mixture having a volumetric analysis of 90 percent helium and 10 percent nitrogen.

13·70 Consider a large tank that contains air initially at 30 psig. A small converging nozzle allows air to escape to the atmosphere when it is unplugged. A certain amount of time is required for the pressure to drop to 5 psig after the plug is removed. It has been suggested that this time can be reduced by adding a diverging section to the nozzle. Write a clear statement (not longer than 600 words) explaining the effect of adding the diverging section. No equations need be used and a sketch is likely to help.

13·71 A fable tells of a traveler who was turned out of a peasant's house because he blew on his hands to warm them and blew on his soup to cool it. The peasant would not allow any person enchanted enough to be able to blow hot and cold from the same mouth to stay in his house. Explain how one can ''blow either hot or cold.''

13·72 A nozzle is to expand steam adiabatically from 700 to 550 kPa. The steam enters the nozzle with negligible velocity in a dry saturated condition. Should this be a converging nozzle, a converging–diverging nozzle, either type, or neither type?

13·73 Steam flows isentropically through a nozzle from 2000 to 170 kPa. If the initial temperature of the steam is 270°C and the flow rate is 0.20 kg/s, compute the areas at the throat and exit. Disregard the entrance velocity.

13·74 Determine the areas at the throat and outlet of a nozzle that discharges 4200 lbm of steam per hour. The steam is initially at 85 psia, 500 F, and is discharged to a vacuum of 14 in. of mercury. The barometric pressure is 30 in. of mercury. The entrance velocity is 375 fps. Assume reversible adiabatic flow conditions.

13·75 Determine the throat area of a nozzle that is to discharge dry saturated steam at 100 psia with a velocity of 2580 fps at a rate of 16.2 lbm/s. Assume that the flow is isentropic.

13·76 A nozzle with an efficiency of 90 percent is used to expand steam adiabatically from 300 psia, 550 F, to 60 psia at a rate of 2000 lbm/h. The initial velocity is negligible. The lowest temperature in the surroundings is 70 F. Compute the nozzle discharge area and the irreversibility of the process per pound of steam.

13·77 Shown in the figure is a sketch of a steam accumulator. Accumulators are used to accommodate varying steam flow demands of systems that periodically require more steam flow than can be provided by the steam generators supplying them. In the figure, steam comes from the steam generators at *A*, and the system requiring steam is connected to *B*. When the demand of the system is less than the capacity of the steam generators, the pressure in the pipeline rises and the surplus of the steam flow at *A* over that at *B* goes into the accumulator and partially condenses. When the steam flow demand at *B* exceeds the capacity of the steam generators, the pressure in the pipeline drops slightly, and some of the saturated water in the accumulator flashes into steam to provide the additional steam flow required at *B*. It has been found that a pipe rupture between *A* and *B* causes the accumulator to be driven downward. It has been proposed to solve this problem by inserting a converging nozzle within the pipe near the accumulator, as shown at *C* in the figure. Some people have been discussing this proposal and have not fully agreed on (1) why the accumulators are driven downward when the steam line ruptures and (2) whether the converging section will eliminate the problem without causing any undesirable effects. Write a clear statement (not more than 600 words long) that clarifies these points for this group. Equations should be used only sparingly if at all, but sketches may be helpful.

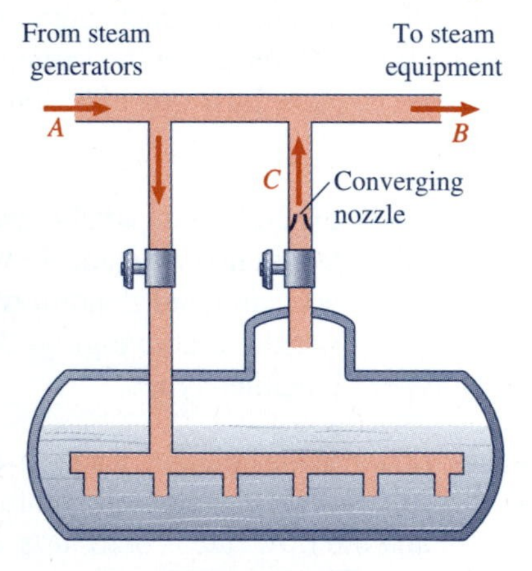

Problem 13·77

13·78 Methane and oxygen are to be burned steadily at a fixed methane flow rate of 0.22 kg/s with 20 percent excess oxygen. The two reactants are to be supplied at atmospheric temperature. It has been suggested that a simple means for maintaining the proper ratio of the two reactants is to have them supplied to the combustion chamber, which is to operate at atmospheric pressure, through converging nozzles with limiting flow, so only the supply pressures of the methane and oxygen need to be controlled. Having the reactants supplied as jets is claimed to promote turbulence and thorough mixing. Determine the required diameters of the circular nozzle exits and the supply line pressures needed, and list the benefits and drawbacks of the scheme.

13·79 As part of a preliminary design study of a self-contained thruster that is to be used to provide a force of 15 N for short periods of time, consideration is being given to using a tank of gas with a converging nozzle discharging to the atmosphere. The maximum gas pressure within the tank must not exceed 800 kPa. When the tank is fully charged, what discharge area is required if the gas used is (*a*) air, (*b*) argon?

13·80 The thrusters described in Problem 13·79 are to be used only intermittently, so that after each use there will be time for the gas remaining in the tank to come to thermal equilibrium with the surrounding atmosphere before the next use. The volume of the tank is 0.0260 m^3. Each use reduces the pressure in the tank, so the force provided decreases with each use. Plot a curve of thruster force versus pressure of the gas in the tank for such intermittent use for one of the gases suggested.

13·81 Refer to Problem 13·80. If the time between successive uses of the thruster is reduced so that there is insufficient time for the gas in the tank to come to atmospheric temperature, how would the curve of force versus pressure change?

13·82 Refer to Problem 13·80. Assume that the duration of each use of the thruster is exactly 2 seconds and the transient effects of starting and stopping are negligible. Plot a curve of thruster force versus number of use.

13·83 Under some conditions it is desirable to bleed air from a jet engine compressor, that is, exhaust some air overboard instead of having all the air that enters the compressor go to the combustion chamber(s). At a location in the compressor where the air is at approximately 440 kPa, 280°C, circular orifices are to be placed in the compressor casing. Each one, when opened, should bleed 0.70 kg/s. It is estimated that the ratio of the actual flow rate through such an orifice to the ideal flow rate through a converging nozzle is 0.6. Determine the diameter of the orifices.

13·84 Sketch h,s diagrams for an ideal gas flowing adiabatically through (*a*) a long pipe line, (*b*) a nozzle, (*c*) a diffuser.

13·85 For the steady flow of an ideal gas with constant specific heats in a long insulated pipe, sketch the variation of the following quantities versus distance along the pipe: p, T, p_t, T_t, T^*, V, ρ, s.

13·86 Refer to Figure 13·11. Which of the Fanno lines, *A*, *B*, or *C*, has the highest value of $\dot{m}/A^*$? For a fluid that is an ideal gas, compare the velocities at the maximum entropy states of the three Fanno lines.

13·87 Air at 380 kPa, 128°C, enters a 20-cm-diameter pipe at 30 m/s and leaves at 118°C. The flow is adiabatic. Calculate the Mach number at the outlet.

13·88 For an atmospheric condition of 101.3 kPa, 20°C, determine the change in stream availability between the two specified sections in Problem 13·87.

13·89 Ethane flows through a well-insulated 10-cm-diameter pipe at a rate of 275 kg/min. At one section the ethane is at 700 kPa, 25°C. Determine the minimum pressure and the maximum velocity that can occur in the pipe.

13·90 Plot the Fanno line for the flow of Problem 13·89.

13·91 At one section in an insulated pipe, steam at 2.7 MPa, 260°C, has a velocity of 120 m/s. Plot the Fanno line for this flow, and determine the minimum pressure and the maximum velocity that can occur in the pipe.

13·92 Consider the steady flow of air in a well-insulated pipe. The pipe diameter is 10 cm. At one section, called section *a*, p_a = 120 kPa, T_a = 5°C, and V_a = 500 m/s. At another section, called section *b*, V_b = 400 m/s. Determine (*a*) p_b and (*b*) the direction of flow, from *a* to *b* or from *b* to *a*.

13·93 Plot the Fanno line for the flow of Problem 13·92.

13·94 Methane is heated as it passes through a constant-area duct at a rate of 0.56 kg/s. The cross-sectional area is 0.0080 m^2. The gas enters at 110 kPa, 35°C; and a control system maintains the outlet temperature at 300°C. Assume that the flow is frictionless and one dimensional. Determine the rate of heat transfer.

13·95 Under some conditions, as a fluid flows in a pipe, friction causes the fluid velocity to increase and the fluid temperature to decrease. Also, under some conditions adding heat to a flowing fluid causes its temperature to decrease. These phenomena are not what one is likely to expect, so write a convincing physical explanation that would resolve any reasonable doubts on these matters.

13·96 State whether each of the following increases, decreases, remains constant, or varies indeterminately across a normal shock: pressure, temperature, density, velocity, Mach number, entropy, stream availability.

13·97 A gas flows steadily through a converging–diverging nozzle, entering at section *i* where velocity is low, passing through the throat *th*, and discharging at section *e*. A normal shock occurs in the nozzle between sections *x* and *y*. The flow is adiabatic and is reversible except at the shock. (*a*) Sketch the pressure–length and *Ts* diagrams. (*b*) Circle the correct symbol ($>$, $=$, $<$, or I) in each of the following statements (where I means "is indeterminate with respect to"):

$$T_e \quad > \quad = \quad < \quad \text{I} \quad T_i$$
$$s_e \quad > \quad = \quad < \quad \text{I} \quad s_i$$
$$s_e \quad > \quad = \quad < \quad \text{I} \quad s_{th}$$
$$V_e \quad > \quad = \quad < \quad \text{I} \quad V_{th}$$
$$\rho_e \quad > \quad = \quad < \quad \text{I} \quad \rho_{th}$$
$$M_e \quad > \quad = \quad < \quad \text{I} \quad M_y$$
$$c_y \quad > \quad = \quad < \quad \text{I} \quad c_x$$
$$s_x \quad > \quad = \quad < \quad \text{I} \quad s_{th}$$
$$V_y \quad > \quad = \quad < \quad \text{I} \quad V_{th}$$
$$M_e \quad > \quad = \quad < \quad \text{I} \quad M_{th}$$
$$T_y \quad > \quad = \quad < \quad \text{I} \quad T_x$$
$$p_{ty} \quad > \quad = \quad < \quad \text{I} \quad p_{tx}$$

13·98 Air flows through a long insulated pipe. At inlet, the velocity is supersonic. Inside the pipe a normal shock occurs. Sketch for this flow *hs*, *hl*, *pl*, and *sl* diagrams, where *l* is distance measured along the pipe.

13·99 Air at 95 kPa, 15°C, with a Mach number of 2.0, enters a normal shock. Determine the downstream pressure, temperature, and velocity and the irreversibility of the process if the sink temperature is taken as 15°C.

13·100 Air flows steadily through an insulated pipe that has a diameter of 0.100 m. At section 1, $M_1 = 2.00$, $V_1 = 680.3$ m/s, $p_1 = 100$ kPa, $T_1 = 15°\text{C}$. A normal shock occurs just downstream from section 1, and at section 2, which is just downstream of the shock, $V_2 = 255.1$ m/s. Determine (*a*) the pressure at section 2, p_2, and (*b*) the irreversibility of the flow between sections 1 and 2. The sink temperature is 15°C.

13·101 Consider the nozzle of Example 13·11. Air flows through the nozzle with the same inlet conditions but with a backpressure of 150 kPa applied. Determine the outlet velocity and the maximum velocity reached in the nozzle.

13·102 Refer to Example 13·11. For the same nozzle and the same inlet conditions, determine the pressure at which a normal shock occurs if a backpressure of 150 kPa is applied.

13·103 A nozzle is properly designed for isentropic flow to discharge nitrogen at 40 kPa, 20°C, with a Mach number of 1.80 through a cross-sectional area of 0.0160 m^2. Assuming that design inlet conditions are maintained, determine (*a*) the discharge pressure that causes a normal shock to stand in the exit plane of the nozzle and (*b*) the pressure at which the shock occurs if the backpressure is 175 kPa.

13·104 For the purpose of making some measurements in a normal shock, it is desired to have a flow of air such that a normal shock stands at the exit plane of a nozzle that is to have a discharge cross-sectional area of 0.0120 m^2 (18.6 in^2). The Mach number entering the shock is to be 2.00, and the conditions of the air as it leaves the shock are to be 100 kPa (14.5 psi), 25°C (77 F). Determine the throat area of the nozzle and the pressure and temperature at which air at low velocity must be supplied to the nozzle.

13·105 Air at 14.0 psia, 60 F, with a Mach number of 2.0 enters a normal shock. Determine the downstream pressure, temperature, and velocity and the irreversibility of the process, using a sink temperature of 60 F.

13·106 Air at 1 atm, 15°C, enters a normal shock. Plot the pressure and temperature following a normal shock versus Mach number for a range of $M = 2$ to

$M = 8$ using (*a*) constant-specific-heat relations (e.g., as embodied in Table A·26) and (*b*) variable specific heats (as embodied in Table A·12). (*c*) Plot percentage error in the constant-specific-heat results for downstream p and T versus Mach number in the same Mach number range.

13·107 Solve Problem 13·106 for the flow of carbon dioxide.

13·108 Helium at 1 atm, 15°C, enters a normal shock. Plot the pressure and temperature following the shock versus Mach number for a range of $M = 2$ to $M = 8$.

13·109 There is an interesting analogy between normal shocks in compressible gas flow and hydraulic jumps in open-channel liquid flow. Normal shocks occur when the velocity is higher than the velocity of sound waves and always result in a decrease in velocity to subsonic with a corresponding increase in density and pressure. Hydraulic jumps occur when the velocity is higher than the velocity of small surface waves and always result in a decrease in velocity to values lower than surface wave velocities with a corresponding increase in depth of the liquid. Describe the analogy more fully and devise some demonstrations of compressible flow phenomena that can be carried out in a flat-bottom sink.

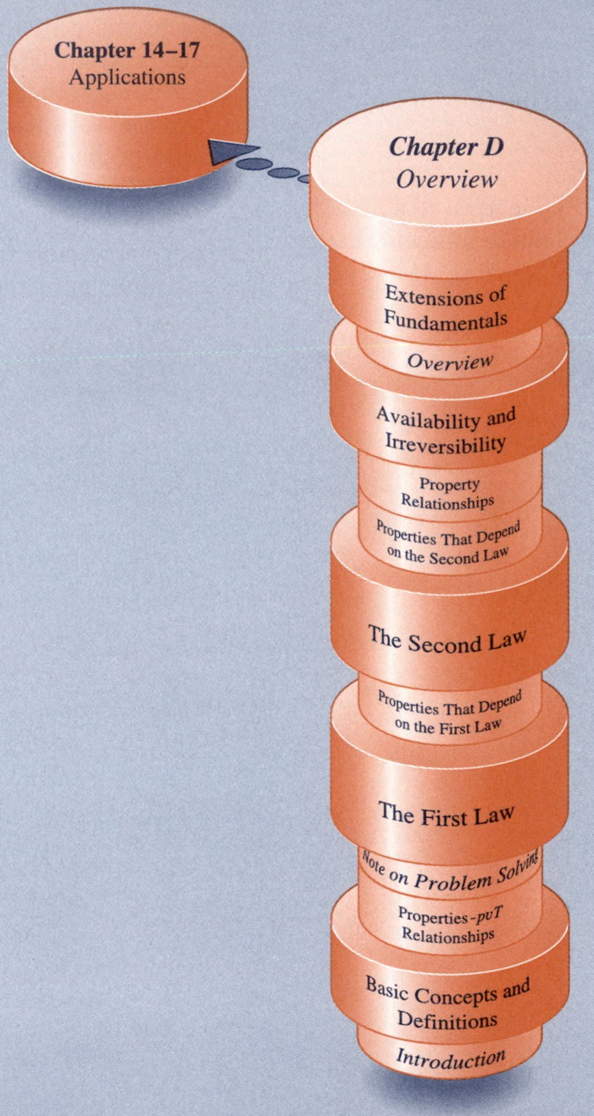

Chapter 14–17
Applications

Chapter D
Overview

Extensions of Fundamentals

Overview

Availability and Irreversibility

Property Relationships

Properties That Depend on the Second Law

The Second Law

Properties That Depend on the First Law

The First Law

Note on Problem Solving

Properties -pvT Relationships

Basic Concepts and Definitions

Introduction

Chapter D

Looking Ahead to More Extensive Modeling

The first thirteen chapters of this book cover the fundamentals of engineering thermodynamics and some extensions of them. The extensions are included so that we do not have to derive from fundamentals the equations used for applications that occur repeatedly. We have illustrated the modeling of relatively simple systems and processes, usually by considering one system or one process at a time and simplifying or restricting it to make our analysis easier. For example, we model the flow through a heat exchanger, a nozzle, or a compressor as one dimensional and frictionless, and such modeling often provides satisfactory predictions of the behavior of actual systems where we know the flow is never truly one dimensional or frictionless.

Now we go to more extensive modeling where we deal not with individual systems or processes but with several systems that interact with each other and several processes that occur simultaneously or sequentially. The result of one process provides the input conditions or otherwise influences other processes. To illustrate this more extensive modeling, we need to describe some complex engineering systems, and we do so in the ensuing chapters. Remember, though, that our goal is to illustrate modeling and the application of thermodynamic principles, not just to describe some of the complex systems that engineers have developed. Indeed, we still describe engineering systems in a simplified manner. For example, in Chapter 15 we may include in a model of a large steam power plant and show on a single diagram a steam generator, a reheater, two or more turbines, a condenser, four or more feedwater heaters, an evaporator, a steam jet ejector, a boiler feed pump, a condensate pump, several drip or transfer pumps, and other equipment, and we would still have a simplified model. We do not include in our model many auxiliary devices such as fuel pumps, lubricant pumps, lubricant coolers, other cooling systems, sealing systems, starting systems, and control systems that are essential to the operation of the plants. For the design of these auxiliary devices, of course, different models must be used.

Complex systems often must be modeled over a wide range of operating conditions, and doing so may require several models. For example, the performance of a compressor at its design flow rate is quite different from its performance at one-tenth that flow rate. Consequently, different models of the same machine may be chosen for different operating conditions.

In more extensive modeling we frequently use parametric studies and we ask more "what if" questions. We ask, for example, "What if the state of the fluid leaving one piece of equipment changes? How does this affect the performance of the next piece of equipment or of the entire plant?" Such a question is often called a *sensitivity* question. How sensitive is *this* to *that*? After all, the very purpose of modeling is either to predict and explain the behavior of systems that have not yet been built or to predict the behavior of existing systems under different conditions. An example of the latter case would be to answer the question, "How does the performance of a given jet engine change when the aircraft flies into a torrential rainstorm?"

Another facet of more extensive modeling is the need to evaluate the modeling. Often several models of a single system can be devised, and the engineer must be able to predict which models are the best from various aspects. When data on similar systems are available, the models can be tested by applying them to those systems. It is worth noting that some modeling is inherently more difficult than others. As one illustration, the transient flow involved in reciprocating machines makes their modeling more difficult than the modeling of steady-flow machines. Usually, processes that are highly irreversible are more difficult to model accurately than processes that are more closely like reversible processes. It is especially important in these cases to weigh the improved predictions of higher-fidelity models against costs and the time available.

From the myriad of engineering systems that we might use to illustrate more extensive modeling, we have selected four kinds: systems for compressing and expanding fluids (Chapter 14), power plants (Chapter 15), refrigeration plants (Chapter 16), and systems for direct conversion of fuel energy into mechanical or electrical work (Chapter 17). Throughout your study of the remaining chapters, reading more specialized books, journals, and magazines on the subjects will greatly enhance your knowledge of the systems and equipment and your understanding of the modeling process.

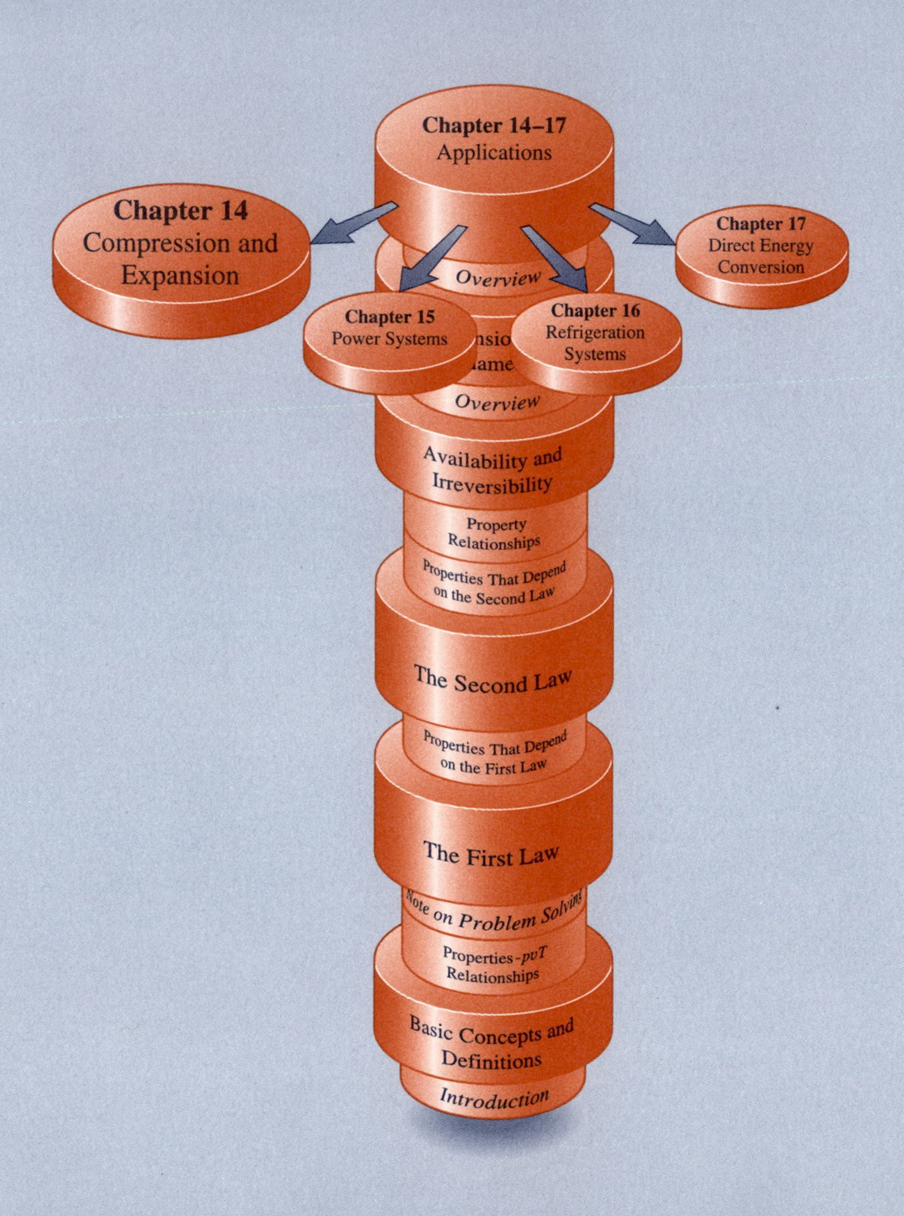

Chapter 14–17
Applications
Chapter 14
Compression and Expansion
Overview
Chapter 15
Power Systems
Chapter 16
Refrigeration Systems
Chapter 17
Direct Energy Conversion
Overview
Availability and Irreversibility
Property Relationships
Properties That Depend on the Second Law
The Second Law
Properties That Depend on the First Law
The First Law
Note on Problem Solving
Properties-pvT Relationships
Basic Concepts and Definitions
Introduction

Chapter 14

Compression and Expansion Processes and Systems

Power generation, refrigeration, and other applications involve systems that are complex: They are composed of several devices. Some of these devices have the purpose of moving fluids, increasing the pressure on fluids, or obtaining work from fluids flowing through them. To engineers designing complex systems, the pumps, fans, blowers, compressors, engines, expanders, and turbines are crucial parts of the systems. Therefore, before studying the systems for the production of power, propulsion, refrigeration, and environmental control, in this chapter we look at devices that move or compress fluids or produce work from fluids flowing through them.

We can identify three levels at which compression and expansion processes and equipment are studied:

1. *As elements of larger systems*. At this level, compression and expansion processes are analyzed on the basis of one-dimensional flow. Typical questions to be answered are: What end states are possible? How much power or heat transfer is involved? Construction details or even the type of machine is usually irrelevant.
2. *As processes and systems in themselves on an input–output basis*. Typical questions deal with parallel operation of two or more machines and startup and off-design-point operation, including no-flow or no-load operation. Thus, the type of machine is quite important.
3. *For the design of the equipment itself to meet customers' needs*. This means designing the machine for maximum work output (for engines and turbines) or minimum work input (for compressors, pumps, etc.) under nominal operating conditions as well as under other operating conditions likely to be encountered. It means many other things, too, because many other factors—size, weight, noise level, reliability, cost, delivery time, serviceability, to name a few—are important along with thermodynamic considerations. Three-dimensional flow effects and localized flow regions are often important.

In this chapter we treat compression and expansion processes chiefly on the first two levels. We first review the application of the first and second laws to all types of compression and expansion processes and define some performance parameters. Then, using principles of mechanics, we consider the energy transfer between a fluid and a rotor for various types of turbomachinery. In conclusion, we look briefly at some performance features of positive-displacement machines.

14·1 Steady-Flow Compression Processes

To any steady-flow system we can apply

- The first law for determining energy quantities
- The relationship $w_{\text{in}} = \int v\, dp + \Delta ke + \Delta pe$ for determining work if the process is reversible
- The second law and its corollaries to determine the possibility of a process and its irreversibility

We now use these tools in analyzing steady-flow compression and expansion processes.

For a compression between given end states, it is generally desirable to minimize the work input. This means that the actual process should approximate as closely as possible a reversible process between the two specified states. Often, however, the purpose of a compressor is to bring a fluid not to a specified final state but only to a specified final pressure. This is usually the case when the compressed fluid is to be stored before it is used further. Therefore, we should first investigate the various types of reversible processes between an initial state and some given final pressure to see which involves the least work. Figure 14·1 shows on pv and Ts diagrams for a gas three different reversible compression paths between a state 1 and the same final pressure p_2. Process $1-a$ is a reversible adiabatic compression, process $1-b$ is a reversible isothermal compression, and process $1-c$ is a reversible compression with some heat removed from the gas but not enough to hold the temperature constant. Process $1-c$ might be a polytropic process: one for which pv^n = constant. If it is, then $k > n > 1$ because the cases of $n = k$ and $n = 1$ are the reversible adiabatic (process $1-a$) and reversible isothermal (process $1-b$) cases, respectively.

For a reversible steady-flow process

$$w_{\text{in}} = \int v\,dp + \Delta ke + \Delta pe \tag{1·20}$$

and the integral is represented by an area on a pv diagram. Inspection of the pv diagram of Fig. 14·1 shows that, for equal changes in kinetic and potential energy, the work required to compress a gas reversibly from a state 1 to a final pressure p_2 is reduced as the compression path approaches the isothermal.

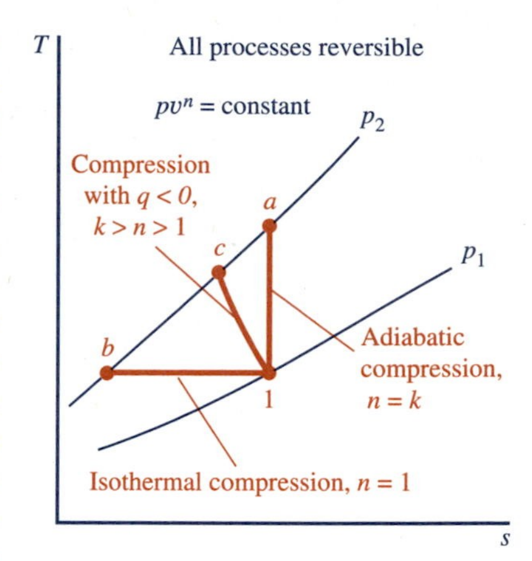

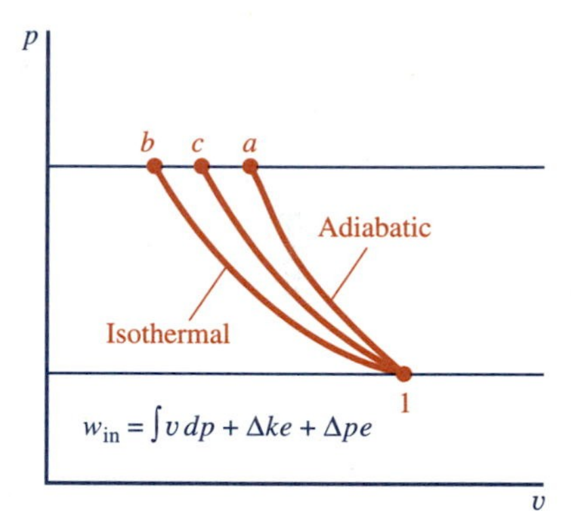

Figure 14·1 Reversible compression paths for a specified initial state and a specified final pressure.

In practice, isothermal compression is nearly impossible to achieve. Adiabatic compression can be approximated quite closely.

In practice, the amount of heat that can be transferred during the compression process is limited by both the small surface area available for heat transfer and the short time required for the gas to pass through the machine. Consequently, if we wish to model the actual process of a cooled compressor, a polytropic process with n closer to k than to 1 may be a suitable model.

Sometimes no cooling is desired during a compression process, even though the lack of cooling increases the work input for given pressure limits. This is the case in a simple gas-turbine power plant where immediately after compression the gas is to be heated by means of fuel. For this application a reversible adiabatic compression is the most desirable from the standpoint of overall plant efficiency (see Chapter 15).

Comparisons between actual and ideal compressor performance are expressed by compressor efficiencies. The ideal compression process may be either reversible isothermal, polytropic, or isentropic, so care must be taken to use the proper ideal process as a basis of comparison in any particular case. We will introduce here only one compressor efficiency, the **adiabatic compressor efficiency**, which is defined by

Adiabatic compressor efficiency

$$\eta_C \equiv \frac{\text{work of reversible adiabatic compression from state 1 to } p_2}{\text{work of actual adiabatic compression from state 1 to } p_2}$$

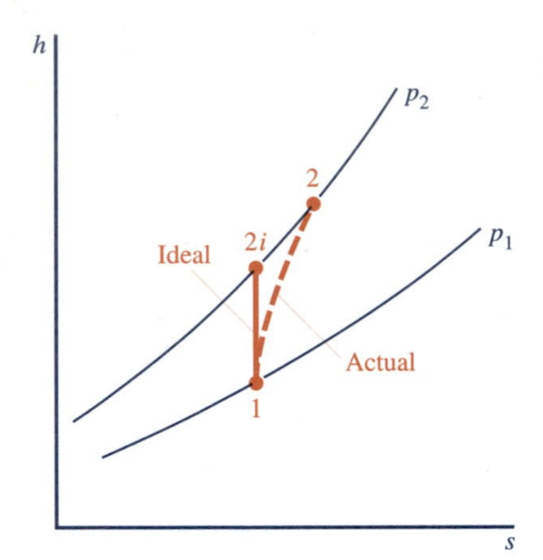

Figure 14·2 Reversible and irreversible adiabatic compressions.

Slightly different definitions are also in use. For the ideal compression the kinetic energy change is often considered to be negligible. Referring to Fig. 14·2 and applying the first law, we then have

$$\eta_C = \frac{\text{ideal input work}}{\text{input work}} = \frac{h_{2i} - h_1}{h_2 - h_1 + (V_2^2 - V_1^2)/2} = \frac{h_{2i} - h_1}{h_{t2} - h_{t1}}$$

If there is no change in kinetic energy, then

$$\eta_C = \frac{h_{2i} - h_1}{h_2 - h_1}$$

Following a practice that is quite general, especially in the gas-turbine literature, in this book we use the term *compressor efficiency*, without modifiers, to mean *adiabatic compressor efficiency* as defined above. Remember that this parameter is useful only in connection with adiabatic compressors.

Multistage compression

Multistage compressors consist of two or more compressors, each of which is called a stage, in series. If each stage operates adiabatically, and there is no heat transfer from the gas as it passes from one stage to another, the entire compressor operates adiabatically. Then we can apply the definition of (adiabatic) compressor efficiency to individual stages and to the entire compressor. However, if the efficiency of each stage is the same (and less than 100 percent), the efficiency of the entire compressor is less than that of the stages. This can be seen by reference to Fig. 14·3, which shows the actual adiabatic compression path through a four-stage compressor. Let $\Delta ke = 0$ for each stage, and therefore for the compressor. If each stage has the same efficiency, the ratio $\Delta h_i/\Delta h$ is the same for each stage. For the entire compressor the efficiency is $(h_{2i} - h_1)/(h_2 - h_1)$. Notice that $h_2 - h_1 = \Sigma\, \Delta h$, but, because of the divergence of constant-pressure lines on the diagram in the direction of increasing entropy, $h_{2i} - h_1 < \Sigma\, \Delta h_i$; therefore, the efficiency of the entire compressor is less than that of the individual stages.

$\eta_{C,\text{ entire multistage compressor}} < \eta_{C,\text{ individual stages}}$

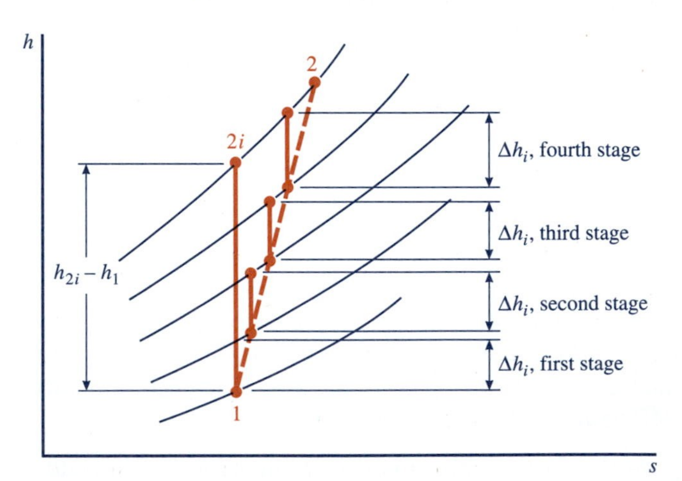

Figure 14·3 Multistage compression.

Intercooling

Multistaging of compressors is sometimes used to allow for cooling between the stages to reduce the total work input. Gas can be cooled more effectively in heat exchangers called intercoolers between stages than during the compression process. Figure 14·4 shows a polytropic compression process 1–a. If $\Delta ke = 0$, the work done on the gas is represented by the area 1–a–j–k–1 on the pv diagram. A constant-temperature line is shown as 1–x. If the polytropic compression from state 1 to p_2 is divided into two parts, 1–c and d–e, with constant pressure cooling at p_i to $T_d = T_1$ between them, the work done is represented by area 1–c–d–e–j–k–1. The area c–a–e–d–c represents the work saved by means of the two-stage compression with intercooling to the initial temperature.

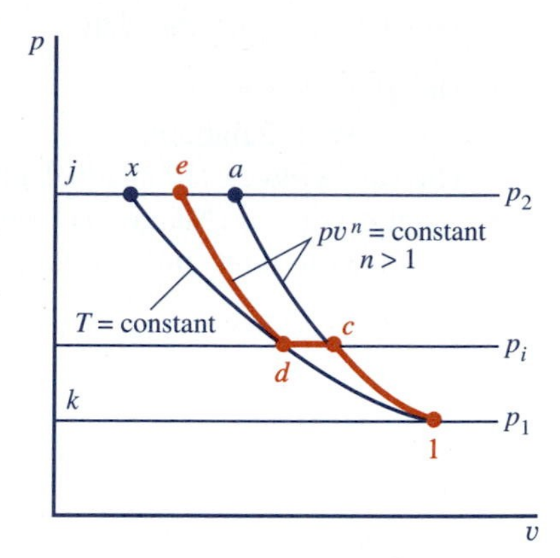

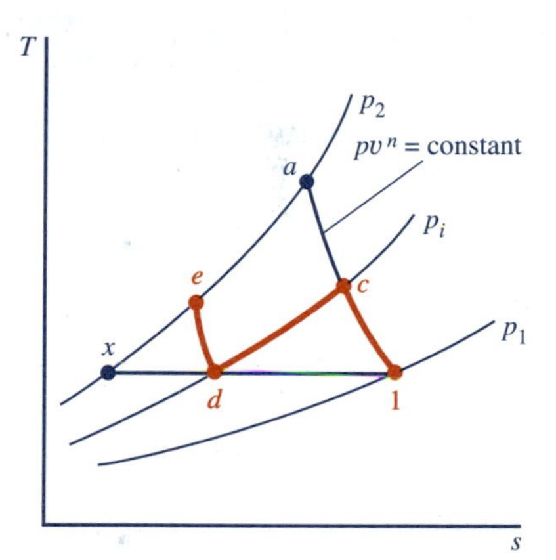

Figure 14·4 Multistage compression with intercooling.

Figure 14·4 suggests that for specified values of p_1 and p_2 there is an optimum pressure for intercooling. (This conclusion stems from reflection on extreme cases. If the process proceeds for a very short distance along the polytropic path before intercooling, or if it proceeds almost to a before intercooling, the work saved in either case is small compared to that saved with an interstage pressure such as shown in Fig. 14·4.) Call the optimum interstage pressure p_i, so that, for two stages with $\Delta ke = 0$,

$$w_{\text{in}} = \int_{p_1}^{p_i} v\,dp + \int_{p_i}^{p_2} v\,dp$$

If we expand this equation for the case of pv^n = constant and then set $dw_{\text{in}}/dp_i = 0$ to find the minimum work input, we can calculate the optimum value for p_i. If the intercooling brings the gas to its initial temperature, the minimum work input is required when $p_2/p_i = p_i/p_1$, or when the work is the same for the two stages. Further analysis shows that no matter how many stages there are, minimum work input is required when the total work is divided equally among the stages. The pressure ratio is then the same for each stage.

Another means of reducing the work required to compress a gas is to inject a liquid into the gas being compressed. As it evaporates, the liquid takes energy from the gas and thereby reduces the gas temperature rise and its specific volume.

Example 14·1 Engine Compressor

PROBLEM STATEMENT

In an aircraft jet engine, air is compressed from 1 atm, 20°C, to a pressure ranging from 4 atm to 16 atm. Plot discharge temperature against pressure ratio for adiabatic compressor efficiencies of 60 percent, 80 percent, and 100 percent.

SOLUTION

Analysis: We assume that

- The flow is steady.
- The flow is adiabatic.
- The air behaves as an ideal gas; that is, $pv = RT$ and $h = h(T)$.
- Kinetic energy changes are negligible (In a jet engine, kinetic energy changes are appreciable, but the *compressor* inlet and discharge velocities may be approximately the same.)

We first sketch a flow diagram and an *hs* diagram, showing the processes for three efficiency values.

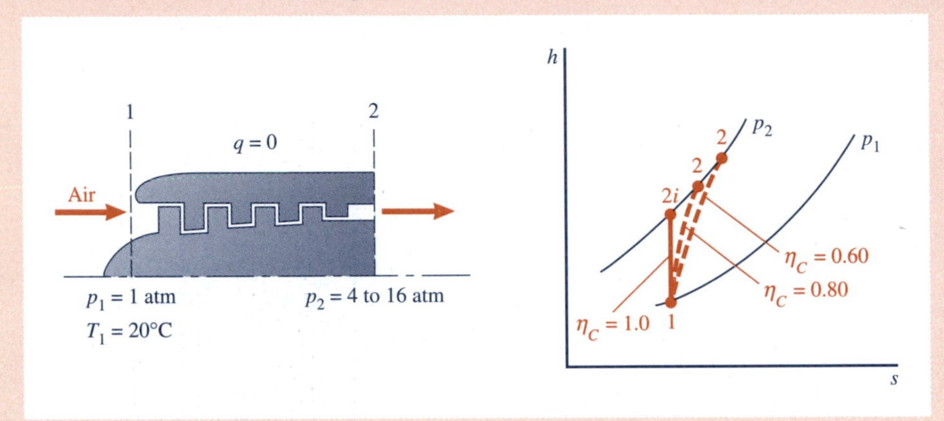

Let us see if it would be reasonable to assume constant specific heats of the air. If the specific heats were constant at inlet-condition values, for a reversible adiabatic compression (i.e., for a compressor efficiency of 100 percent), the discharge temperature for a pressure ratio of 16 would be given by

$$T_2 = T_1\left(\frac{p_2}{p_1}\right)^{(k-1)/k} = 293(16)^{(1.4-1)/1.4} = 647\text{ K} = 374°\text{C}$$

The temperature rise for this case would be 354 Celsius degrees; for a compressor efficiency of 60 percent, the temperature rise would be approximately 590 Celsius degrees, so the discharge temperature would be around 610°C. Over this temperature range, the specific heat of air varies by about 11 percent, so we will not assume constant specific heats.

The equations we will need are

$$\eta_C = \frac{h_{2i} - h_1}{h_2 - h_1}$$

$$h = f(T)$$

$$p_r = f(T)$$

$$p_2 = p_{2i}$$

$$\frac{p_{r2i}}{p_{r1}} = \frac{p_{2i}}{p_1}$$

Calculations: We enter the equations above into an equation-solving computer package, enter the specified values, and solve for a list of compressor efficiencies: 0.60, 0.80, and 1. The results are shown in the table and the plot.

```
;——————————— Example 14·1 ———————————————————————————
p2 = p2i
p2%p1 = p2/p1
pr2i/pr1 = p2i/p1                          ;For reversible adiabatic compression

etaC80 = (h2i - h1)/(h2_80 - h1)           ;For compressor efficiency = 80 percent
etaC60 = (h2i - h1)/(h2_60 - h1)           ;For compressor efficiency = 60 percent

;——————————— Ideal gas properties ——————————————— IGPROP.TK

call IGPR (gas, T1, h1, phi1, pr1)
```

p2%p1	T2i	T2_80	T2_60
4	162	197	255
6	214	262	340
8	255	312	406
10	288	354	460
12	317	389	506
14	343	420	547
16	365	448	583

Example 14·2 Ammonia Compressor

PROBLEM STATEMENT

Ammonia is compressed adiabatically at a rate of 8.50 lbm/min from 40 psia, 15 F, to 140 psia. Inlet velocity is very low, and the discharge velocity is 500 fpm. Determine the power input in horsepower if the compressor efficiency is 65 percent. Also determine the diameter of the discharge line.

SOLUTION

Analysis: We first sketch a flow diagram and a *Ts* diagram. On the *Ts* diagram we show both the actual discharge state 2 and the discharge state 2*i* for an isentropic compression. We can refer to ammonia tables or the computer model for ammonia properties to see that the specified pressure and temperature for state 1 show that the ammonia is slightly superheated.

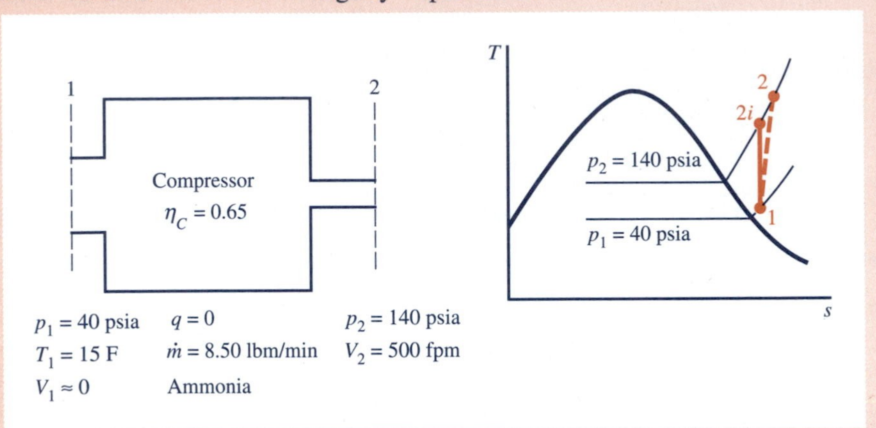

The power input is the product of the mass rate of flow and the work input. The mass rate of flow is specified, and we can find the work input by first finding the isentropic work input and using

$$w_{\text{in}} = \frac{w_{\text{in},i}}{\eta_C}$$

The isentropic work input to use in this equation can be obtained from the first law,

$$w_{\text{in},i} = h_{2i} - h_1 + \frac{V_{2i}^2 - V_1^2}{2}$$

The discharge velocity is specified, so we will use $V_{2i} = V_2$. (If the outlet diameter were specified instead of the velocity, we would use the continuity equation with v_{2i} and the same mass rate of flow to find V_{2i}.) We can find h_{2i} to use in the equation for $w_{\text{in},i}$ if we know any two properties at state $2i$. We do know $p_{2i} = p_2$ and $s_{2i} = s_1$, so we find $h_{2i} = f(p_{2i}, s_{2i})$ from ammonia tables or the computer model AMMONIA. Thus, our analysis for finding w_{in} is complete.

To find the required discharge line diameter, we find the cross-sectional area from the continuity equation, $\dot{m} = AV_2/v_2$, for which m and V_2 are specified and v_2 can be obtained as $v_2 = f(p_2, h_2)$ from ammonia tables or the computer model after we find h_2 from the first law applied to the actual compression process,

$$w_{\text{in}} = h_2 - h_1 + \frac{V_2^2 - V_1^2}{2}$$

Thus the analysis is complete, we know we can solve the problem, and the analysis has shown us what equations are needed.

Calculations: To save the time of looking up properties and interpolating in ammonia tables, we simply enter the six equations needed into our equation solver and call the computer model for online ammonia property data. The Rule sheet with the equations and the Variable sheet with variable names and values are shown in the figure.

```
═══════════════════ RULE SHEET ═══════════════════════
S Rule──────────────────────────────────────────────────
  ;──────────── Example 14•2 ────────────────────────────
  etaC = wini/win                 ;Definition of compressor efficiency
  wini = h2i - h1 + V2i^2/2       ;First law, isentropic compression
  win = h2 - h1 + V2^2/2          ;First law, irreversible adiabatic compression
  Wdotin = mdot * win             ;Relation among work, power, mass flow rate
  A2 = mdot * v2 / V2             ;Continuity equation
  A2 = pi() * d2^2 / 4            ;Area of circular-cross-section outlet

  ;──────────── Ammonia properties ──────────────────── AMMONIA.TK

═══════════════════ VARIABLE SHEET ═══════════════════
St Input─── Name─── Output─── Unit──── Comment──────────────────────────
                              ──────── * E14-2.TK ──── Example 14•2 ────
                              ─ Model: * AMMONIA.TK ── Ammonia properties ─
   15       T1                F        Temperature
   40       p1                psi      Pressure
            h1      688.11    B/lbm    Enthalpy
            s1      1.5024    B/(lbm*R) Entropy
            v1      7.1182    ft^3/lbm Specific Volume
            x1      'mngless           Quality
            phase1  'SH                Phase

            T2i     178.4     F        Temperature
   140      p2i               psi      Pressure
            h2i     768.06    B/lbm    Enthalpy
   1.5024   s2i               B/(lbm*R) Entropy
            v2i     2.7179    ft^3/lbm Specific Volume
            x2i     'mngless           Quality
            phase2i 'SH                Phase

            T2      251.9     F        Temperature
   140      p2                psi      Pressure
   811.11   h2                B/lbm    Enthalpy.  h2 is solved for on the
                                       first pass and then transferred to
                                       input status to solve for A2.
            s2      1.5733    B/(lbm*R) Entropy
            v2      3.0901    ft^3/lbm Specific Volume
            x2      'mngless           Quality
            phase2  'SH                Phase
   0.65     etaC                       Compressor efficiency
            wini    79.9      B/lbm    Isentropic work input
            win     123       B/lbm    Work input
            Wdotin  24.7      hp       Power input
   8.5      mdot              lbm/min  Mass rate of flow
   500      V2i               fpm      Outlet velocity, const s compression
   500      V2                fpm      Outlet velocity, actual compression
            A2      0.0525    ft^2     Outlet area
            d2      3.1       in       Outlet diameter
```

Comment: We see that the kinetic energy change is a very small fraction of the work input. This is often, but not always, the case when velocities are limited to values that are reasonable for velocities in pipes of refrigeration, power, and chemical processing plants.

14·2 Steady-Flow Expansion Processes

One goal in the design of a turbine or expansion engine is to maximize the work output. Heat loss to the surroundings would reduce the work output, so the machines are insulated. Furthermore, the heat addition process in a steady-flow power plant is usually separate from the expansion process. For these reasons, we treat the process in a turbine or expansion engine as adiabatic. The ideal process used as a basis of comparison is therefore the reversible adiabatic or isentropic process from the initial state to the final total pressure.

Adiabatic turbine efficiency, which we will call simply *turbine efficiency* or the efficiency of a turbine, is defined as

$$\eta_T \equiv \frac{\text{work of } \textit{actual adiabatic} \text{ expansion}}{\text{work of } \textit{reversible adiabatic} \text{ expansion between same initial state and same final total pressure}}$$

This term is also used in connection with positive-displacement expansion machines such as steam engines and compressed-air motors. It has also been called engine efficiency, and slightly different definitions are occasionally found. Applying the first law, and referring to Fig. 14·5, we have

Adiabatic turbine efficiency, usually called turbine efficiency

$$\eta_T = \frac{\text{work}}{\text{work}_i} = \frac{h_1 - h_2 + (V_1^2 - V_2^2)/2}{h_1 - h_{2i} + (V_1^2 - V_{2i}^2)/2} = \frac{h_{t1} - h_{t2}}{h_{t1} - h_{t2i}}$$

where state $2i$ is one for which the entropy is the same as it is at state 1 and the total pressure is the same as it is at the actual exhaust condition. If there is no change in kinetic energy, then

$$\eta_T = \frac{h_1 - h_2}{h_1 - h_{2i}}$$

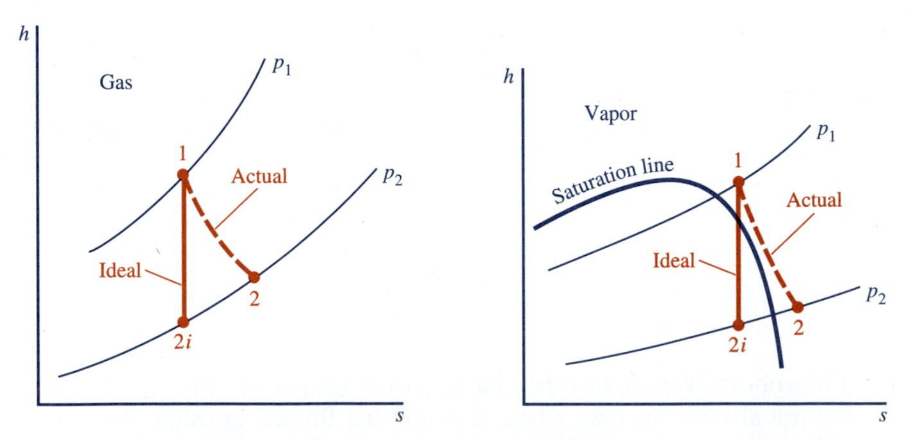

Figure 14·5 Reversible and irreversible adiabatic expansions of gases and vapors on three property diagrams.

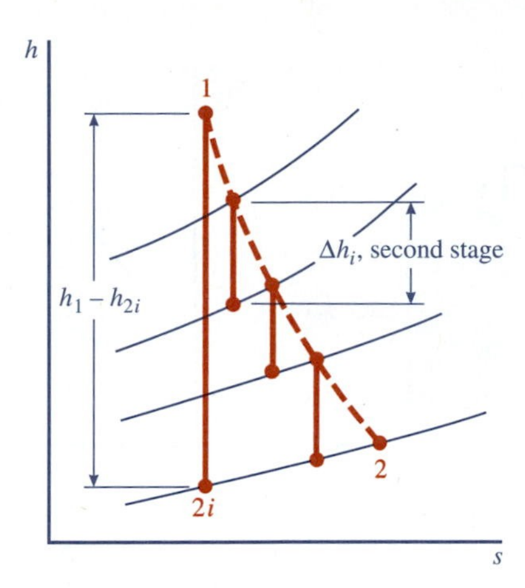

Figure 14·6 Multistage expansion.

The efficiency of a multistage turbine can be higher than the efficiency of any of its stages. For an illustration of this point, refer to Fig. 14·6, which shows the adiabatic expansion path through a four-stage turbine. Let $\Delta ke = 0$ for each stage and therefore for the entire turbine. If each stage has the same efficiency, the ratio $\Delta h/\Delta h_i$ is the same for each stage. For the entire turbine, $h_1 - h_2 = \Sigma\,|\Delta h|$, but, because of the divergence of constant-pressure lines on the diagram in the direction of increasing entropy, $|h_1 - h_{2i}| < \Sigma\,|\Delta h_i|$; consequently, the efficiency of the entire turbine is higher than that of the individual stages.

Multistage turbines

$\eta_{T,\text{ entire multistage turbine}} > \eta_{T,\text{ individual stages}}$

Adiabatic turbine effectiveness, which is often called just **turbine effectiveness**, is defined as

Turbine effectiveness

$$\text{Turbine effectiveness} \equiv \frac{\text{work}}{-\Delta b}$$

where b is the Darrieus function, $h - T_0 s$. Turbine effectiveness is sometimes called a second-law turbine efficiency. Like turbine efficiency, it can be applied to individual stages or groups of stages in a turbine as well as to the entire machine. Effectiveness gives a clearer picture of the overall effects of losses. For example, you can show that if two stages have the same efficiency, the stage at the higher pressure has a higher effectiveness and consequently a lower irreversibility. This is another case where a second-law analysis shows better than a first-law analysis where efforts to improve performance are likely to be most fruitful.

Reheat in Gas Turbines. For a given pressure ratio, the work per unit mass of fluid flowing through a gas turbine increases with increasing inlet temperature. The maximum temperature of the fluid is limited by the loss of strength of structural materials at high temperatures. One method of obtaining additional work from the same flow rate is called reheating: The gas is

Reheating

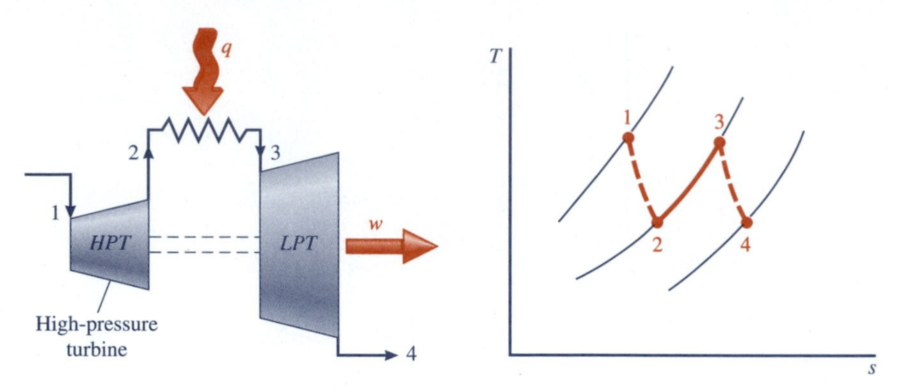

Figure 14·7 Reheating in a gas turbine.

taken from the turbine at a pressure between the inlet and exhaust pressures, is heated at constant pressure, as shown in Fig. 14·7, and then reenters the turbine or enters another turbine to expand to the exhaust pressure. From Fig. 14·7 it is clear that the work is increased by reheating. Reheating in a gas turbine plant may increase or decrease the thermal efficiency of the plant, as will be discussed in Chapter 15.

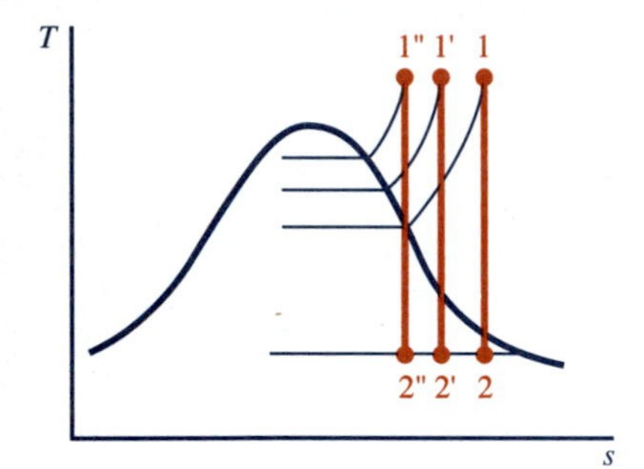

Figure 14·8 Influence on exhaust quality of increasing steam turbine throttle pressure, for a fixed throttle temperature.

Reheat in Steam Turbines. Figure 14·8 shows that for fixed throttle temperature and exhaust pressure in a steam turbine, the quality of the exhaust steam decreases with increasing throttle pressure. Low quality or high moisture in the exhaust steam causes erosion of the low-pressure turbine blades, damaging the blades and turbine performance. A means of reducing these effects is to reheat the steam after it has expanded part of the way from the throttle pressure to the exhaust pressure.

Figure 14·9 shows a reheat turbine. Steam enters the turbine at state 1, expands to state 2, leaves the turbine to pass through a constant-pressure reheater and then at state 3 reenters the same turbine or enters another turbine to expand to the exhaust pressure, state 4. There is less moisture in the

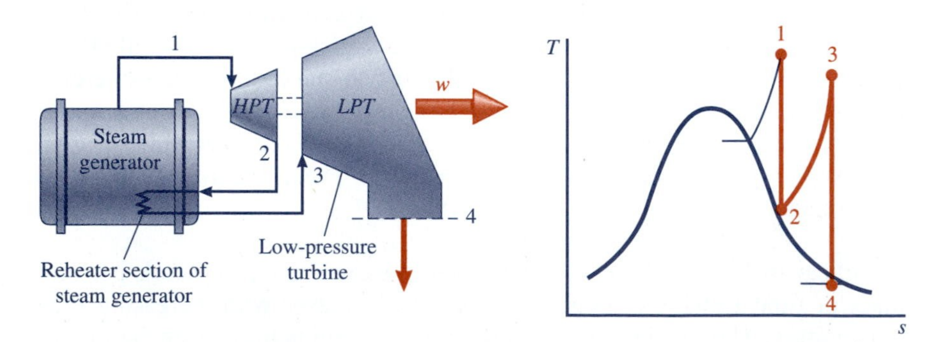

Figure 14·9 A reheat steam turbine.

exhaust than in the exhaust state n of a nonreheat turbine with the same throttle conditions. (Figure 14·9 makes the comparison for isentropic expansions, but the same conclusion is reached by comparing actual irreversible expansions.) Reheating may increase the thermal efficiency of the plant, depending on the nature of the entire power plant cycle. For reasons stated in Chapter 15, actual reheat steam turbines always have some steam bled or extracted; therefore, the flow rates entering and leaving the turbines at states 1, 2, 3, and 4 are different.

Example 14·3 Turbine Efficiency

PROBLEM STATEMENT

A steam turbine has throttle conditions of 200 psia, 600 F. The flow rate is 32,000 lbm/h. Cooling water is available at 65 F. For an exhaust pressure of 1.5 in. Hg abs., plot irreversibility and exhaust steam quality versus turbine efficiency in the range of 50 percent to 100 percent.

SOLUTION

Analysis: The specification of a flow rate implies steady flow. We will also assume that the flow is adiabatic because no indication to the contrary is given and steam turbines are usually well insulated. Lacking any information on velocities, we will assume the kinetic energy change to be negligible. We make a sketch of the system and a *Ts* diagram, showing both the ideal expansion and the actual expansion.

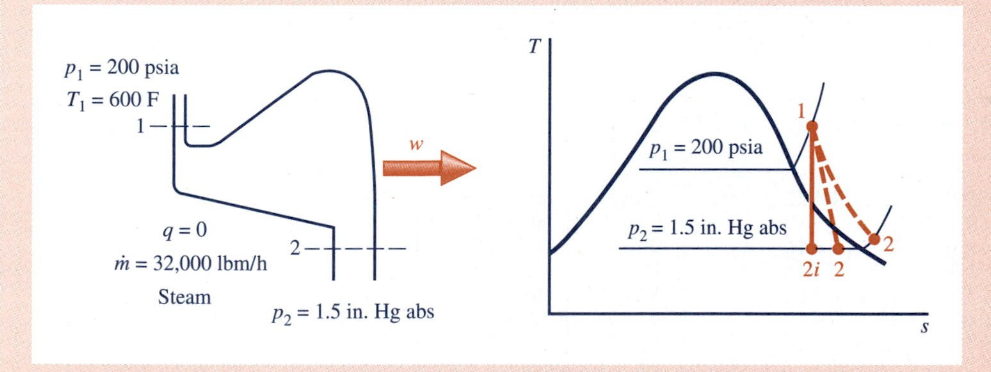

To calculate the irreversibility of the steady-flow adiabatic process, we need the entropies of the inlet and exhaust states. The inlet state is specified. The exhaust pressure is known, so we need one other independent property to define the exhaust state. We know that for a reversible adiabatic process $s_{2i} = s_1$, and since the initial state is specified, we can determine s_1. Thus, we can find the ideal exhaust state. Then from the definition of turbine efficiency we can calculate the actual exhaust enthalpy for each value of turbine efficiency.

Calculations: Our analysis shows that we need the following equations:

$$s_{2i} = s_1$$

$$i = T_0\,\Delta s_{univ} = T_0(s_2 - s_1)$$

$$\eta_T = \frac{h_1 - h_2}{h_1 - h_{2i}}$$

and, of course, we need steam properties in some form. We will use the computer model for steam, choosing the version for eight states with English units. Since we need only three states (1, 2i, and 2), we delete five states from the steam model and change the name of state 3 to 2i.

The values that will change are the turbine efficiency and all the properties at state 2 except pressure. As long as the steam at state 2 is saturated, T_2 is constant, but if it is superheated, then, of course, T_2 varies with the turbine efficiency. Therefore, for a list of turbine efficiencies ranging from 0.5 to 1, we solve the set of equations to obtain the results shown in the table and the plot.

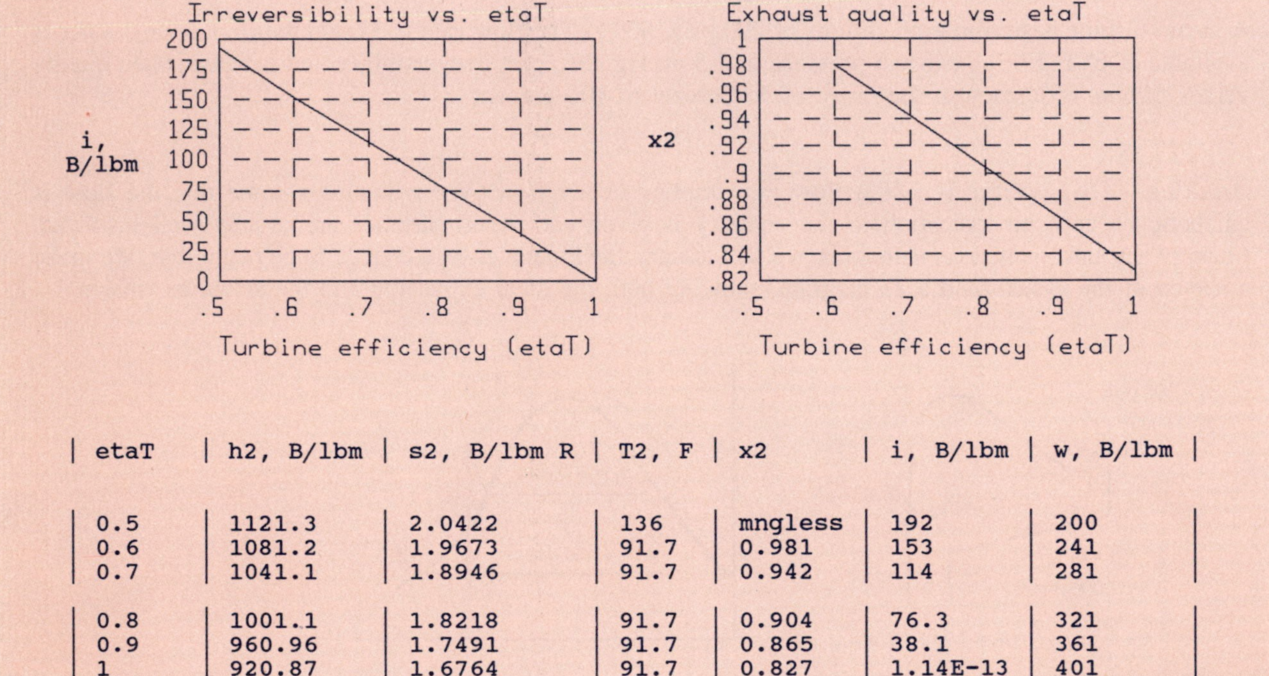

etaT	h2, B/lbm	s2, B/lbm R	T2, F	x2	i, B/lbm	w, B/lbm
0.5	1121.3	2.0422	136	mngless	192	200
0.6	1081.2	1.9673	91.7	0.981	153	241
0.7	1041.1	1.8946	91.7	0.942	114	281
0.8	1001.1	1.8218	91.7	0.904	76.3	321
0.9	960.96	1.7491	91.7	0.865	38.1	361
1	920.87	1.6764	91.7	0.827	1.14E-13	401

Comment: We have added the work of the turbine to the table to permit a comparison with irreversibility values, showing that the latter are reasonable. This comparison shows the usefulness of irreversibility in comparison with a quantity such as entropy production or entropy generation.

14·3 Incompressible Flow Through Machines

If the density of a flowing fluid remains constant, we speak of both an incompressible fluid and incompressible flow. Incompressible flow analyses are not restricted to liquids. A gas or vapor can be treated as incompressible if its density changes very little during a process. For example, atmospheric air flowing through a fan and undergoing an overall pressure change of a few inches of water can be treated as incompressible because its density change is so small. The greater the density change is, the greater are the errors that result from treating the flow as incompressible.

Even in gas flows, if the density change of the fluid is small, the fluid may be treated as incompressible.

It must also be remembered that liquids are not entirely incompressible: Their densities are affected by pressure, and for large pressure differences even isentropic pumping may cause appreciable changes in temperature. This can be seen from compressed liquid tables such as Table A·6·3.

For the steady flow of an incompressible (ρ = constant) fluid through a machine, the first law becomes

$$w = q - (u_2 - u_1) - (p_2 - p_1)v - \Delta ke - \Delta pe \qquad (a)$$

If the process is also reversible, then

$$w = -\int v\,dp - \Delta ke - \Delta pe \qquad (1\cdot 20)$$

$$= -(p_2 - p_1)v - \Delta ke - \Delta pe \qquad (b)$$

Comparison of Eqs. *a* and *b* shows that for reversible incompressible steady flow $q - \Delta u = 0$. If the process is irreversible, then $q - \Delta u \neq 0$, but the heat transfer and internal energy change are usually small and difficult to measure in comparison with the other energy quantities in Eqs. *a* and *b*; therefore in incompressible flow the quantity $(q - \Delta u)$, which is occasionally given a name such as "friction loss" or "head loss due to friction," is measured indirectly by measuring all the other quantities in an energy balance such as Eq. *a*.

Pump and fan efficiencies for incompressible flow are defined similarly to compressor efficiency. It should be noted that for incompressible flow the kinetic energy change is always the same for the actual and the ideal processes. Thus

Pump efficiency

$$\eta_P \equiv \frac{\text{ideal input work}}{\text{input work}} = \frac{v\,\Delta p + \Delta ke + \Delta pe}{\text{input work}}$$

For an actual pump the numerator is evaluated by measurements of pressure, velocity, elevation, and so on, and the denominator is obtained by power and flow rate measurements from which the actual work input can be calculated.

Notice a difference in application between the definition of compressor efficiency and the definition of pump efficiency. For a pump, the work input is the work done on the pump shaft where it is connected to its driver. That is, the system is the pump itself as well as the fluid passages of the pump. The input work to a pump exceeds the work done on the fluid at least by the work required to overcome bearing and seal friction. The definition of compressor efficiency is the ratio of work done on the fluid in an ideal compression process to the work done on the fluid in the actual case. That is, the open system is defined as the fluid passages through the compressor, not including the compressor hardware itself. The work done to drive the compressor is always greater than the work done on the fluid being compressed, the difference being the work required to overcome bearing and seal friction, to overcome fluid friction on shafts and the backs of impellers, and to drive auxiliaries such as oil pumps.

Head, or energy per unit weight

The term **head** that is frequently used in connection with pumps, fans, and hydraulic turbines refers to an energy quantity per unit weight. When energy is expressed on a unit weight basis, units such as ft·lbf/lbf can be reduced to a unit of length alone, so expressions like "a velocity head of 40 ft" (meaning kinetic energy of 40 ft·lbf/lbf weight) are common. For many engineering analyses, treating the weight of a given mass of substance as constant is satisfactory, but in a varying gravitational field (as in a high-altitude vehicle) it is much better to use a mass basis instead of a weight basis.

Cavitation

An important phenomenon in the flow of liquids is cavitation. Changes in the velocity of a liquid flowing through a passage are accompanied by pressure changes. If the pressure is lowered to the saturation pressure corresponding to the liquid temperature, vapor bubbles form. If these bubbles are then carried by the stream to a region where the pressure is higher than the vapor pressure, they collapse. This alternate formation and collapse of vapor bubbles in a liquid is called **cavitation**. When the bubbles collapse on a solid wall of a flow passage such as a pump rotor, very high impact pressures are exerted on minute areas of the wall and serious damage to the wall may result. Also, the vapor bubbles may disrupt the liquid flow to the detriment of the overall performance of the machine.

14·4 Dynamic Machines: Turbomachines

In the preceding sections we have discussed compression and expansion processes strictly from a process point of view. That is, we have made no mention of the physical form of the devices that cause these processes. We can divide them into two broad categories, dynamic devices and positive-displacement devices, and we now consider each of these categories. Dynamic devices include turbomachines, where energy is transferred between a fluid and a rotating impeller or rotor, and many other devices such as ejectors, injectors, and ion pumps that are not discussed here.

14·4·1 The Basic Relations of Turbomachines

Turbomachine performance depends on the mechanism of energy transfer between a fluid and a rotor. The flow through an actual rotor is always three dimensional and, as pointed out in the introduction to this chapter, three-dimensional effects must be considered for the optimal design of turbomachinery. For simplicity, we restrict the discussion in this section to one-dimensional flow because our goal is simply to establish some general features of energy transfer between a fluid and a rotor.

Figure 14·10*a* is a general diagram showing the velocity vectors of fluid entering and leaving a rotating rotor. Some particular types of rotors are shown in Fig. 14·11 to illustrate that the general diagram applies to various rotor types that differ markedly from each other. (In none of these rotors is the flow one dimensional except as an approximation.) The linear velocity of any point on the rotor is u. The radius to any point is r, and the angular velocity of the rotor is ω, so that $u = r\omega$ (see Fig. 14·10*b*). The fluid velocity V at any point has three mutually perpendicular components:

V_{rad}, the radial component, shown in Fig. 14·10*c* and *d*
V_{ax}, the axial component (parallel to the rotor axis of rotation), shown in Fig. 14·10*d*
V_u, the tangential component, shown in Fig. 14·10*c*

The torque of the fluid on the rotor is equal and opposite in sign to the torque of the rotor on the fluid. The net torque on the fluid is equal to the time rate of change of angular momentum of the fluid. The angular momentum with respect to the axis of rotation of a mass m of fluid entering is mr_1V_{u1}, and that of a mass m of fluid leaving is mr_2V_{u2}, and for steady flow the rate of change

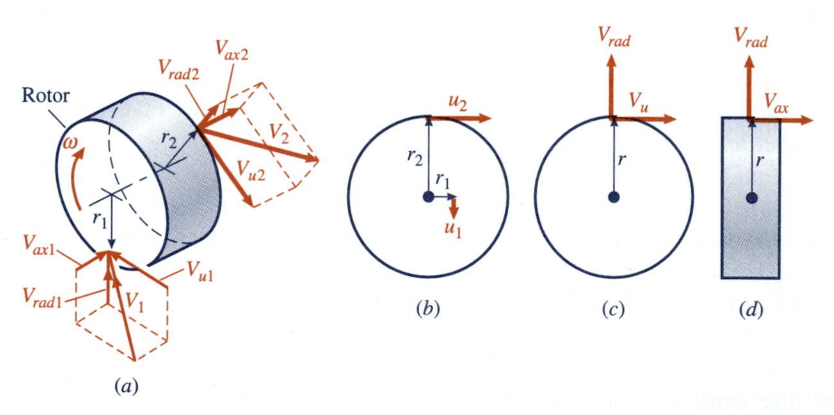

Figure 14·10 General diagram of fluid entering and leaving a turbomachine rotor.

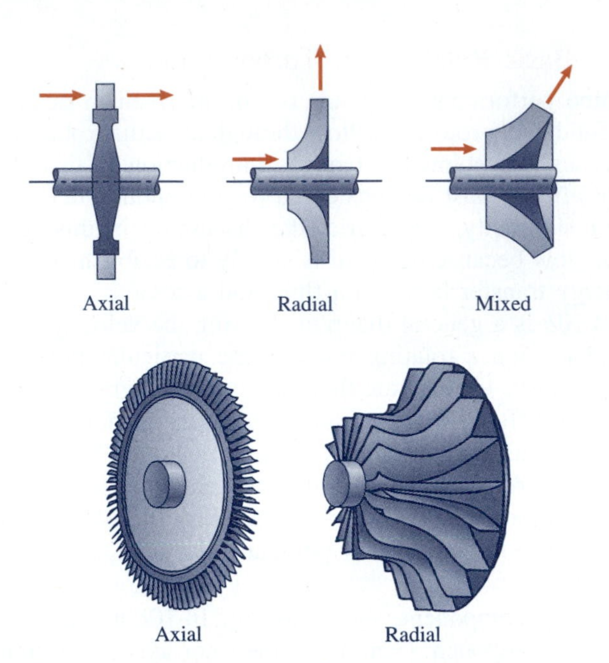

Figure 14·11 Three types of turbomachine rotors.

of angular momentum is $\dot{m}(r_2V_{u2} - r_1V_{u1})$. Thus, for the steady flow of a fluid the basic relation of mechanics,

$$\text{Torque} = \frac{d}{d\tau}(mrV_u)$$

becomes

Basic relation for steady, one-dimensional flow through a rotor

$$\text{Torque on fluid} = \dot{m}(r_2V_{u2} - r_1V_{u1})$$

This is the basic dynamic equation for steady one dimensional flow through a rotor. (Remember that one dimensional means that properties are uniform in any cross section normal to the direction of flow; it does not restrict the flow to a constant direction.) The fluid pressure does not appear in the torque equation because the pressure does not vary circumferentially. No matter what the path inside the rotor is, the net torque depends only on the flow rate, the inlet and outlet radii, and the tangential components of the fluid velocity at inlet and outlet. The inlet and outlet velocities depend on the flow rate, the rotor speed, and the rotor geometry, but this dependence is not simple because the velocities at inlet and outlet are not necessarily tangent to the rotor vanes or blades.

The power delivered to the fluid by a rotor is

$$\dot{W}_{\text{in}} = (\text{torque on fluid})\omega = \dot{m}\omega(r_2V_{u2} - r_1V_{u1})$$

and the work done on the fluid is

Work done on a fluid by a rotor

$$w_{\text{in}} = \frac{\dot{W}_{\text{in}}}{\dot{m}} = \omega(r_2V_{u2} - r_1V_{u1}) = (u_2V_{u2} - u_1V_{u1}) \tag{14·1}$$

The product uV_u is sometimes called the **whirl**, so the work equation can be written as w_{in} = change in whirl. No matter what the flow path is through the rotor, work is done only if there is a change in whirl.

Whirl

Equation 14·1 applies only to the flow through a rotor. For example, sections 1 and 2 in Eq. 14·1 cannot be taken as the inlet of one rotor and the outlet of a subsequent rotor in a multistage machine, because the torque exerted on the fluid by a stationary part of the machine does no work on the fluid. The change in the product uV_u between the inlet and outlet of an entire machine, in fact, is often zero.

In addition to the above equation for work that was derived from dynamic considerations, the first law may be used to determine the work done on a fluid by a rotor. If the process is reversible, we can also use the expression

$$w_{in} = \int v\,dp + \Delta ke + \Delta pe \tag{1·20}$$

The continuity equation can also be applied to the flow through a rotor if relative velocities are used in connection with flow areas in the rotor. Thus the basic relations (in addition to physical property relationships) that are useful in analyzing the one-dimensional steady flow through turbomachine rotors are

- The torque and work equations derived in this section
- The first law
- Equation 1·20, $w_{in} = \int v\,dp + \Delta ke + \Delta pe$
- The continuity equation

From these basic relations many special forms useful for particular phases of analysis and design can be derived.

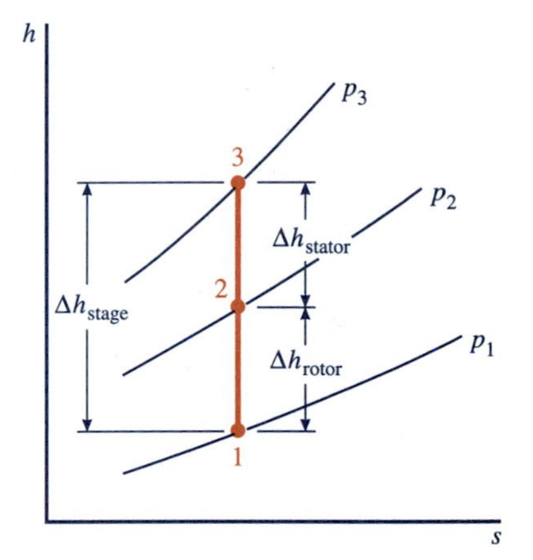

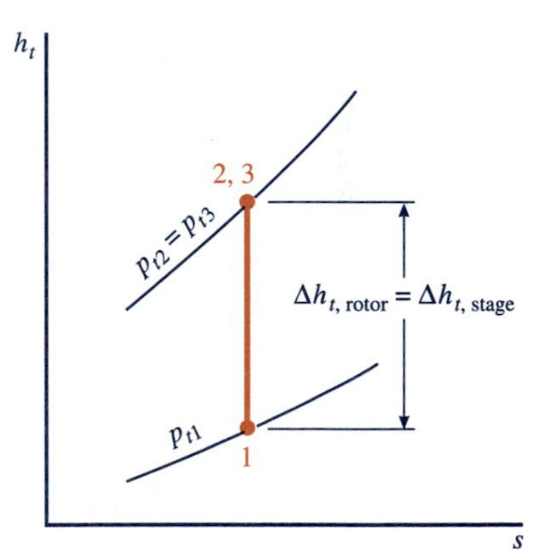

Figure 14·12 Isentropic compression in a single stage.

14·4·2 Turbomachines: Compressors and Turbines

Compressors. Dynamic compressors have two parts: a rotor that does work on the fluid, increasing its kinetic energy and its pressure, and a stator that reduces the velocity of the fluid to further increase the pressure. The stator may also serve to redirect the fluid into following rotors or into a collector that leads to the machine outlet. In a centrifugal compressor the stationary diffuser may have vanes that form a diverging passage or it may be a vaneless annular passage through which the fluid spirals outward with decreasing velocity and increasing pressure. In an axial-flow compressor the stator of each stage consists of a row of vanes that form diverging flow passages. Details can be found in the specialized literature.

Work is done on a fluid passing through a compressor only by the rotor, but in the stator or diffuser there is a decrease in kinetic energy, so the enthalpy increases in both the rotor and the stator. The fraction of the overall change in enthalpy that occurs in the rotor is called the **degree of reaction**. Figure 14·12 shows hs and h_ts diagrams for one stage, including rotor and

Degree of reaction

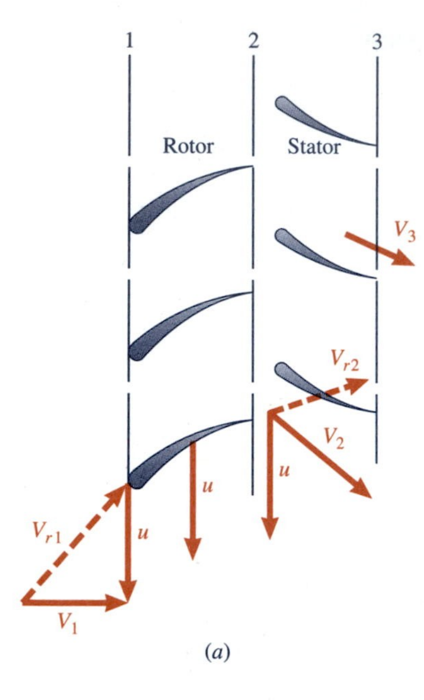

(*a*)

(*b*)

Figure 14·13 Axial-flow compressor stage.

stator, of a dynamic compressor operating isentropically. Process 1–2 occurs in the rotor, and process 2–3 occurs in the stator or stationary diffuser.

To illustrate the application of the basic dynamic equation to an axial-flow machine, a developed (or flattened) view of an axial-flow compressor stage is shown in Fig. 14·13*a*. To make this figure, we flatten the cylindrical surfaces to which the rotor and stator blades are attached (Fig. 14·13*b*) into the plane of the figure. Velocity vectors relative to the moving blades are shown as broken lines. Since the inlet and outlet of the rotor blade passages are at the same distance from the rotor axis of rotation, $u_2 = u_1$. The work done by the rotor on the fluid is then, from Eq. 14·1,

$$w_{\text{in}} = (u_2 V_{u2} - u_1 V_{u1}) = u(V_{u2} - V_{u1})$$

The work per stage is proportional to the change in the tangential component of velocity across the rotor. The more the fluid stream is turned by the rotor blades, the greater the work and the pressure rise per stage. This brings us to a fundamental difference between the flow through turbines and the flow through pumps and compressors.

In all flows of a fluid through a passage, shearing forces at the wall retard the fluid near the wall in a region called the **boundary layer**. In turbines, the pressure generally decreases in the direction of flow and thus the pressure gradient, acting oppositely to the wall shearing forces, tends to prevent the retardation of fluid in the boundary layer. In pumps and compressors, however, there is an adverse pressure gradient (i.e., pressure increases in the direction of flow) that acts with the wall shearing forces to reduce the fluid velocity near the wall. These two actions together may reduce the velocity to zero near the wall and then the adverse pressure gradient may cause the flow to reverse direction near the wall. This phenomenon is called **separation** and is highly detrimental to the performance of fluid machines because it prevents the fluid from flowing in the patterns defined by the machine flow passages. The design of pumps and compressors is therefore greatly influenced by the necessity to limit the magnitude of adverse pressure gradients, which tend to cause separation. For example, the amount of turning that can be accomplished in a blade row as shown in Fig. 14·13*a* is limited by the occurrence of separation, so axial-flow compressors for high-pressure ratios need many stages.

Turbines. A turbine consists of at least one stationary row of blades or nozzles and at least one row of moving blades that are a part of the rotor. The **degree of reaction** is defined as the fraction of the total decrease in enthalpy that occurs across the rotor. A turbine with a zero degree of reaction is called an impulse turbine or sometimes a pure impulse turbine. In such a turbine the fluid is expanded to the stage exhaust pressure in stationary nozzles; then the pressure remains constant and kinetic energy decreases as the fluid does work on the rotor blades. The blades on the rotor of an impulse turbine form flow passages of approximately constant cross-sectional area, since changes in area would involve changes in pressure.

Degree of reaction

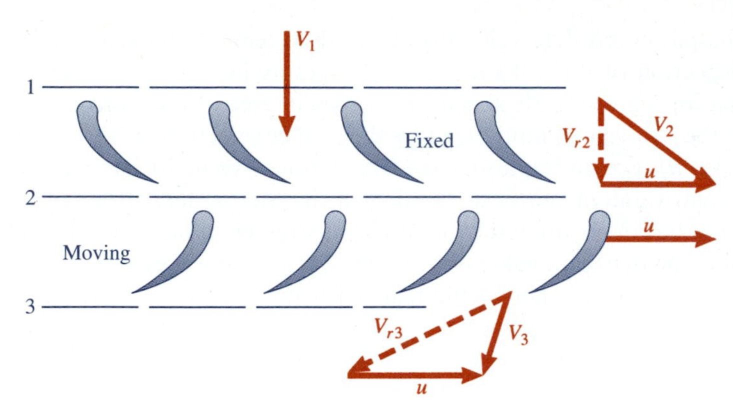

Figure 14·14 Axial-flow turbine stage.

Both axial-flow and radial-flow turbines are in use, the latter usually in the smaller sizes. Many different types of flow paths through turbines are used for various reasons, but the basic dynamic equation for turbomachines (Eq. 14·1) and the other basic relations listed in Section 14·4 apply in all cases.

To illustrate the application of the basic dynamic equation to an axial-flow turbine stage, Fig. 14·14 shows a developed view of such a stage. Fluid enters the stationary nozzles with a low velocity V_1, is expanded, and leaves with a higher velocity V_2. Relative to the moving blades, this nozzle exit velocity is V_{r2}. Within the moving blade passages, the fluid is further expanded so that its velocity relative to the rotor blades increases from V_{r2} to V_{r3}. The absolute velocity decreases across the rotor; that is, $V_3 < V_2$. The work done *by* the fluid is

$$w = (uV_{u2} - uV_{u3}) = u(V_{u2} - V_{u3}) = u(V_{ru2} - V_{ru3})$$

Figure 14·15 is a diagram of an impulse stage. If there are no frictional effects, the pressure and the relative velocity do not change across the rotor.

For given inlet conditions and exhaust pressure, the maximum amount of work is obtained from the fluid passing through a turbine stage if the kinetic

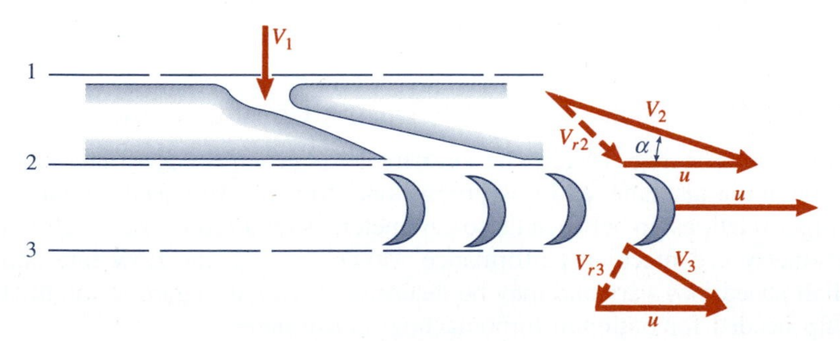

Figure 14·15 Axial-flow impulse turbine stage.

energy (based on absolute velocity) of the fluid leaving the stage is a minimum. Inspection of the velocity vector diagrams in Figs. 14·14 and 14·15 shows that for a given blade shape (or a given degree of reaction) the kinetic energy at the exit can be minimized by the proper selection of the ratio u/V_{u2} or $u/(V_2 \cos \alpha)$. For an impulse stage the optimum value of this ratio is 0.5. The optimum value increases as the degree of reaction increases. The blade speed u of a turbine is limited by centrifugal stresses in the rotor. Therefore, maintaining the optimum value of $u(V_2 \cos \alpha)$ limits the pressure or enthalpy decrease across a stage. This is the reason for multistaging of turbines. Several ingenious arrangements of fixed nozzles and moving and stationary blades are in use for increasing the allowable enthalpy drop across a single stage. Some of these, called *reentry stages* or *Curtis stages*, are described in most books on turbines.

Methods of controlling turbines. All have different thermodynamic implications.

Many turbines, such as those driving electric generators, operate at constant speed and constant inlet and exhaust pressures, so when the required power output changes, the flow rate must change. One way to do this is to partly close a throttle valve upstream of the turbine. This reduces the flow rate and also reduces the pressure at the turbine nozzle inlets so that the work per unit mass is reduced (see Fig. 14·16). Consequently, the velocity leaving the nozzles is reduced and no longer has the optimum relationship with the blade speed. This lowers the turbine efficiency. A better way to meet part-load conditions is to have a separate valve upstream of each nozzle or each group of nozzles of the first stage. The flow rate is reduced by closing some of the valves completely. Those remaining open have the same ratio of nozzle discharge velocity to blade velocity, so the efficiency of the first stage is unchanged, except that one of the valves is usually between the wide open and closed positions and hence causes throttling.

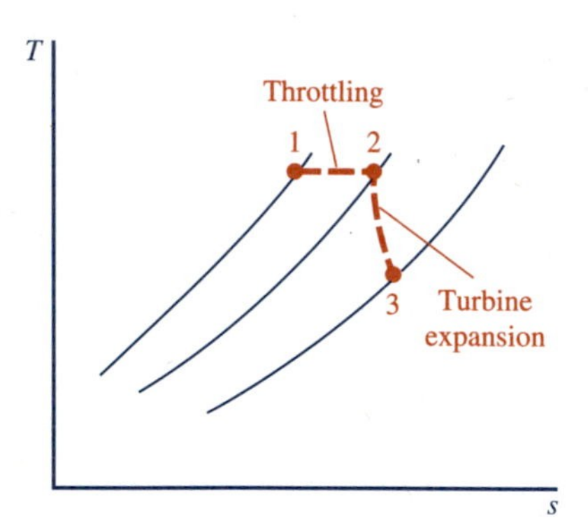

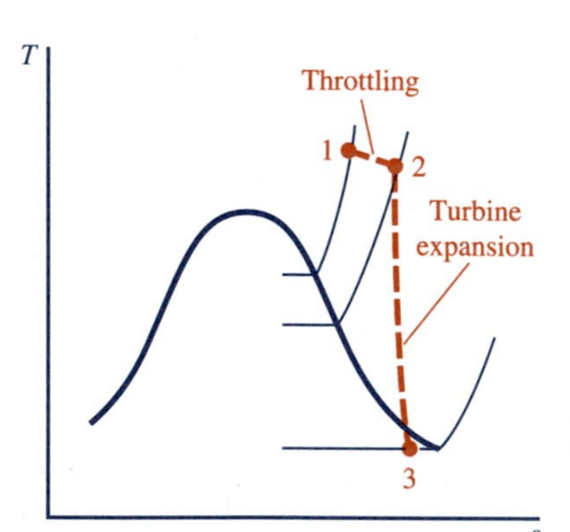

Figure 14·16 Throttling upstream of a turbine to reduce work per unit mass: (*a*) gas turbine; (*b*) steam turbine.

Variable nozzle angles or nozzle blade angles can also be used to reduce the degradation of turbine performance with varying power requirements, but an associated design challenge is to devise the mechanisms that must operate at the highest-temperature part of the machine with acceptable cost and reliability.

Still another method of meeting variable power requirements with constant-speed turbines is to vary the pressure at which the hot fluid delivered to the turbines is generated. This involves quite different considerations for gas turbines and for steam turbines and is mentioned in Chapter 15.

14·4·3 Turbomachines: Further Design Considerations

Refer again to Fig. 14·13. The velocity V_1 is proportional to the flow rate through the machine, and u is proportional to the shaft speed. Obviously, a proper relationship between these parameters is required for satisfactory flow geometry and machine performance. However, both the flow rate and the shaft speed may vary and may be determined without regard to the relationship needed for optimum turbomachine performance.

Some turbomachines operate at essentially constant speed. Examples are

- Turbines that drive electrical generators
- Pumps and compressors that are driven by synchronous or induction electric motors

The operating requirements of other turbomachines, such as aircraft jet engines as one example, however, call for variable speed. Whether operation is to be at constant or variable speed is a major factor in the design of turbomachines.

The flow rate through a machine is also determined largely by factors external to the machine. Consequently, a machine usually cannot be designed to operate at a single speed and flow rate.

Characteristic curves and performance maps

For dynamic machines the shaft speed, flow rate, pressure ratio (for compressible flow) or pressure difference (for incompressible flow), power, and efficiency are related. The relationships are frequently expressed in terms of plots called characteristic curves or performance maps. Figure 14·17 shows the characteristic curves for a centrifugal pump at a constant shaft speed. Figure 14·18 shows those of an axial-flow compressor at several shaft speeds with lines of constant efficiency (the broken lines); such a diagram is usually referred to as a performance map.

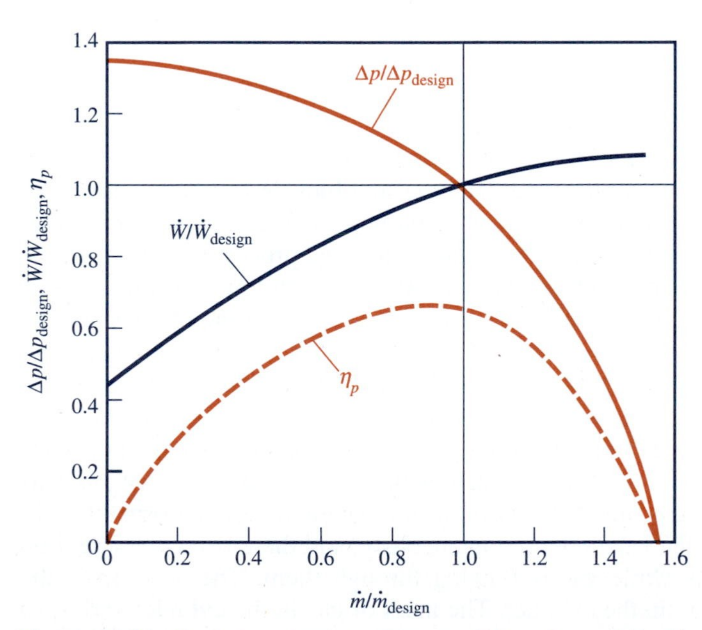

Figure 14·17 Typical performance characteristics of a centrifugal pump.

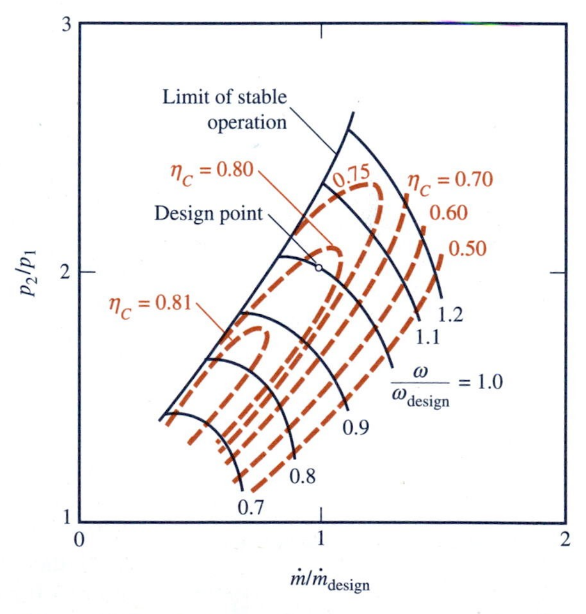

Figure 14·18 Typical performance characteristics of an axial-flow compressor.

Notice in Figs. 14·17 and 14·18 that efficiency varies markedly across the operating range. In many applications, turbomachines must operate for long periods of time under conditions removed from those that provide maximum efficiency. Therefore, having a moderately high efficiency over a wide range of operating conditions may be more important than having a very high efficiency at one operating point, the *design point*. Turbomachinery design involves many such trade-offs.

We usually think of the flow through turbomachines as steady, but for detailed flow passage design, notice that a finite number of nozzles, blades, or vanes means that the flow must be unsteady within the machine.

Again, in engineering, all factors must be considered.

Although we have emphasized the performance of machines chiefly in terms of energy produced or required and efficiencies, it must be remembered that the design and operation of equipment involves many other factors, including initial cost, maintenance, reliability, environmental impact, and safety. Also, although we have concentrated on the compression or expansion equipment itself, often a major piece of equipment involves various auxiliaries such as lubrication systems, control systems, and instrumentation that are essential to successful operation. The auxiliaries take space, require maintenance, may require energy, and are crucial to the operation of the major equipment.

14·5 Positive-Displacement Machines

A positive-displacement machine, whether for moving fluids or for producing work, is one in which a cavity expands in volume as the working fluid enters and then becomes smaller as the fluid leaves. Reciprocating machines such as gasoline and diesel piston engines and bicycle tire pumps are positive-displacement machines, but so are gear pumps, sliding-vane pumps, diaphragm pumps, and other devices. In analyzing the processes that occur in a positive-displacement machine, we deal with an open transient-flow system as fluid enters or leaves and with a closed system during the rest of the operation cycle. Yet we observed in Chapter 1 that for some purposes we can successfully treat a reciprocating machine as a steady-flow system.

Figure 14·19 pV diagram for an ideal reciprocating compressor.

To illustrate the analysis of positive-displacement machines, we will consider the operation of a reciprocating gas compressor. A convenient diagram in such analyses is a pV diagram such as the one shown in Fig. 14·19. This compressor is ideal in that the pressure and temperature are uniform throughout the cylinder, the valves act instantaneously, and there is no pressure drop across the valves while gas is flowing through them. The abscissa is the volume of gas within the cylinder. The mass of gas in the cylinder varies, so p and V do not define the state of the gas. Therefore, gas at a given pressure and temperature may be represented by many different points on the pV diagram. For example, points 1 and d represent gas at the same state.

For the compressor operation illustrated in Fig. 14·19, all valves are closed as the piston moves from point 1 to point 2, so path 1–2 is a closed-system process. When the piston reaches point 2, the pressure inside the cylinder is sufficient to open the discharge valve against the pressure of the gas in the discharge line. As the piston moves from point 2 to point c, gas is pushed out of the cylinder at constant pressure and constant temperature. Point c is the end of the piston stroke. The remaining volume in the cylinder V_c is called the **clearance volume**. The volume swept by the piston during its entire stroke $(V_1 - V_c)$ is called the **displacement**, the volume V_1 is called the **cylinder volume**, and the percent clearance is defined as the ratio of the clearance volume to the displacement. As the piston moves back to the right from point c, the gas that was trapped in the clearance volume at the discharge pressure expands. The intake valve does not open until the pressure inside the cylinder reaches the intake pressure. This occurs at point d. From point d to point 1, gas is being drawn into the cylinder.

Clearance volume, displacement, and cylinder volume

The volume of gas, measured at p_1, that is drawn in is less than the displacement. Volumetric efficiency η_{vol} is defined as

$$\eta_{vol} \equiv \frac{\text{mass of gas drawn in per stroke}}{\text{mass of gas, at intake line pressure and temperature, required to fill the displacement}}$$

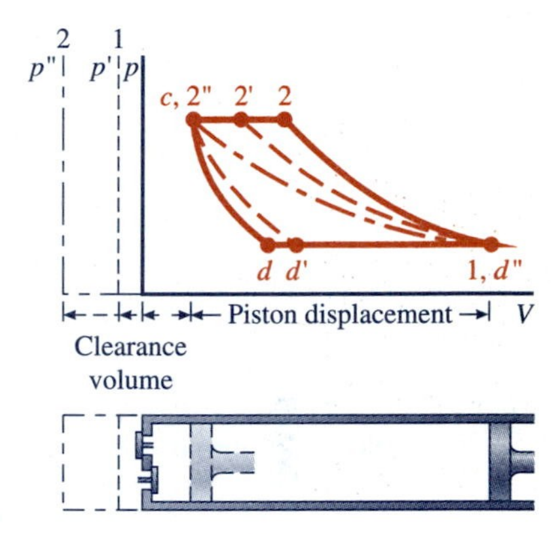

Figure 14·20 Effect of clearance volume on volumetric efficiency.

Notice that volumetric efficiency is a ratio of masses. If p_1 equals the intake line pressure, $\eta_{vol} = (V_1 - V_d)/(V_1 - V_c)$. In an actual compressor, p_1 is lower than the pressure in the intake line because of the throttling of the gas through the intake valve. (For an air compressor, the intake line pressure may be lower than the ambient pressure because of pressure drop through an intake filter or silencer.)

If the compression 1–2 and the expansion c–d are both polytropic with the same value of n in pv^n = constant, the amount of clearance has no effect on the amount of work done per unit mass of gas delivered.

Recall that a polytropic process is a reversible one for which pv^n = constant.

Clearance affects volumetric efficiency.

Increasing the clearance decreases the volumetric efficiency of a reciprocating compressor. The compressor of Fig. 14·20 has a fixed stroke and piston displacement, but the cylinder can be extended to increase the clearance volume. As the clearance volume increases, the pressure axis of the pV diagram must be moved to the left. Three different values of clearance volume are illustrated. For the minimum clearance shown, the pV diagram is 1–2–c–d–1. If the clearance is increased, the pressure axis must be moved to the left to a position such as that shown for the axis marked p'. The pV diagram is then 1–2′–c–d'–1. The volume of gas inducted per stroke has decreased from $(V_1 - V_d)$ to $(V_1 - V_{d'})$, so the volumetric efficiency has decreased. If the clearance is increased further, a value will be reached for which the pressure of the gas in the cylinder reaches the discharge pressure just at the end of the stroke. Thus no gas is delivered. If the processes are reversible, the gas expands along the same path during the return stroke of the piston and no gas is drawn in. The pressure axis for this case is marked p''. Points 1″ and c on the pV diagram coincide, and so do points d'' and 1. The

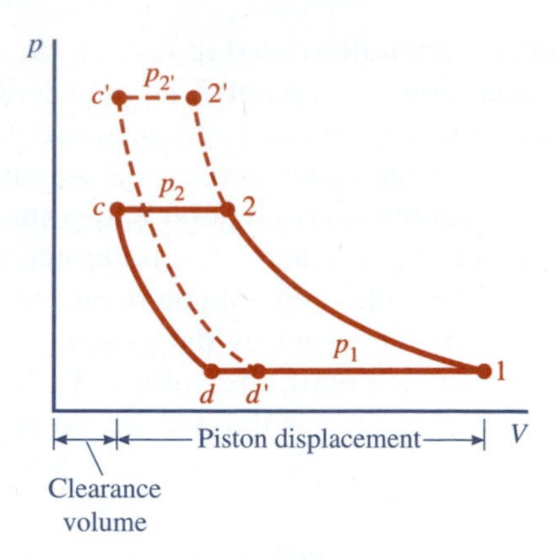

Figure 14·21 Effect of discharge pressure on volumetric efficiency.

volumetric efficiency is zero. The delivery of constant-speed compressors is sometimes regulated by changing the clearance. This is done by valves that open and close auxiliary chambers connected to the clearance volume of the cylinder.

Discharge pressure affects volumetric efficiency.

Inspection of Fig. 14·21 shows that for constant suction pressure, increasing the discharge pressure decreases the volumetric efficiency. With discharge pressure p_2, gas is inducted along the line d–1; however, with the higher discharge pressure $p_{2'}$, the induction process is shortened to d'–1.

14·6 Summary

The work required to compress a gas from a given initial state to a given final pressure is reduced by removing heat from the gas during compression. In practice, the amount of heat that can be transferred during the compression process is limited; so the ideal process for simulating an actual compression may be a polytropic process with the polytropic exponent n closer to k than to 1. In simple gas-turbine plants and in other applications where immediately after compression the gas is to be heated by means of a fuel, adiabatic compression may be desirable from the standpoint of overall plant efficiency, even though the compression work is greater than for compression with cooling.

For an adiabatic compressor, the *compressor efficiency* is defined as

$$\eta_C \equiv \frac{\text{work of } \textit{reversible adiabatic} \text{ compression from state 1 to } p_2}{\text{work of } \textit{actual adiabatic} \text{ compression from state 1 to } p_2}$$

In a multistage adiabatic compressor, the efficiency of the entire machine is lower than that of the individual stages if they have equal efficiencies of less than 100 percent.

If isothermal compression is impossible or impractical, as it usually is, a reduction in the work required for given pressure limits can be achieved by cooling the gas at constant pressure between stages. For polytropic compression with the same value of n in each stage and intercooling to the initial temperature, equal pressure ratios for all stages result in the minimum work input.

Adiabatic turbine efficiency, which we call simply *turbine efficiency* or the efficiency of a turbine, is defined as

$$\eta_T \equiv \frac{\text{work of } \textit{actual adiabatic} \text{ expansion}}{\text{work of } \textit{reversible adiabatic} \text{ expansion between same initial state and final total pressure}}$$

The efficiency of a multistage turbine can be higher than the efficiency of any of its stages.

Adiabatic turbine effectiveness, often called simply *turbine effectiveness*, is defined as

$$\text{Turbine effectiveness} \equiv \frac{\text{work}}{-\Delta b} = \frac{\text{work}}{-\Delta(h - T_0 s)}$$

To increase the work per unit mass of fluid flowing through a turbine, the fluid can be taken from the turbine at an intermediate pressure, reheated to a higher temperature, and then further expanded in the turbine to the exhaust pressure.

The basic dynamic equation for the steady, one-dimensional flow of a fluid through a turbomachine rotor is

$$\text{Torque on fluid} = \dot{m}(r_2 V_{u2} - r_1 V_{u1})$$

where r is the radial distance from the axis of rotation, V_u is the tangential component of the fluid velocity, and subscripts 1 and 2 refer to inlet and outlet of the rotor. The work done on the fluid is

$$w_{\text{in}} = (u_2 V_{u2} - u_1 V_{u1}) \tag{14·1}$$

where u is the linear velocity of a point on the rotor.

The *degree of reaction* of a turbomachine stage is defined as the fraction of the enthalpy change that occurs in the rotor.

In analyzing the performance of a reciprocating machine, it must be remembered that the piston and the cylinder walls form the boundaries of a closed system part of the time and form most of the boundary of an open transient-flow system at other times during a cycle.

References and Suggested Reading

14·1 Fox, Robert W., and Alan T. McDonald, *Introduction to Fluid Mechanics*, 4th ed., Wiley, New York, 1992, Chapter 11. (The cited chapter is an effective supplement to the material presented in this book.)

14·2 Logan, Earl, *Turbomachinery: Basic Theory and Applications*, 2nd ed., Dekker, New York, 1993. (This book gives an overview of the major kinds of turbomachinery, clear diagrams of impellers and blading, and clear thermodynamic diagrams for flow through turbomachines.)

14·3 Moss, S. A., C. W. Smith, and W. R. Foote, "Energy Transfer Between a Fluid and a Rotor for Pump and Turbine Machinery," *Transactions ASME*, Vol. 64, August 1942, pp. 567–597. (A classic paper.)

14·4 Wood, Bernard D., *Applications of Thermodynamics*, 2nd ed., Addison-Wesley, Reading, MA, 1982, Chapter 3.

Problems

14·1 Plot the air discharge temperature versus compressor efficiency, in a range of 50 percent to 100 percent, for adiabatic compression of air through a pressure ratio of 3. The air enters at 30°C (86 F).

14·2 Solve Problem 14·1 for air entering with a relative humidity of 100 percent.

14·3 Air at 1 atm, 30°C (86 F), $\phi = 0.50$, is compressed to 3 atm. Calculate the final relative humidity for (*a*) reversible adiabatic compression and (*b*) reversible isothermal compression.

14·4 Isothermal compression of a gas–vapor mixture can cause condensation; adiabatic compression does not. For processes that follow pv^n = constant, determine the limiting value of n to avoid condensation if the entering gas is atmospheric air at 1 atm, 20°C (86 F), and $\phi = 0.60$ and the pressure ratio is 3.

14·5 Solve Problem 14·1 for the compression of natural gas having a volumetric analysis of 84 percent methane (CH_4) and 16 percent ethylene (C_2H_4).

14·6 Carbon dioxide is compressed polytropically from 1 atm, 15°C (59 F), to 3 atm at a rate of 0.0960 kg/s. For a range of polytropic exponent $1 \leq n \leq k$, plot power input and discharge temperature versus n.

14·7 It is desired to produce a steady jet of air at atmospheric pressure, 10°C, 6 cm in diameter, with a velocity of 60 m/s. This is to be accomplished by a compressor inducting atmospheric air and discharging through a heat exchanger into a nozzle. The cooling water passing through the heat exchanger enters at 10°C and experiences a temperature rise of no more than 10 degrees C. For adiabatic compression, plot compressor power input and the minimum mass flow rate of cooling water versus compressor efficiency in a range of 60 percent to 100 percent.

14·8 An ideal gas with constant specific heats is to be compressed from p_1 and T_1 to a state at p_2 and the same temperature. Two methods for doing this are (1) isentropic compression from state 1 to p_2 followed by constant-pressure cooling to the initial temperature, and (2) constant-pressure cooling from state 1 to a temperature T_c, isentropic compression to p_2, and then, if needed, constant-pressure cooling to the initial temperature. Show the processes on a Ts diagram and calculate (in terms of p_1, p_2, T_1, T_2, and gas properties) the amount of work that can be saved by the proper

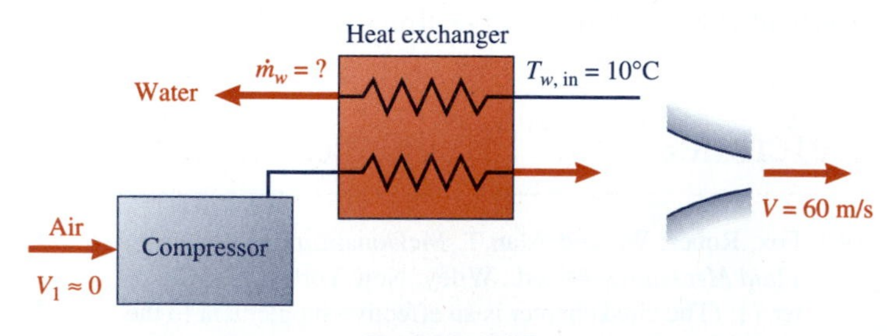

Problem 14·7

choice of method. (Do not consider the power required to refrigerate the coolant.)

14·9 Air at 200 kPa, 150°C, is to be produced from air at 100 kPa, 20°C, by isentropic compression and constant-pressure heating. Calculate the work required per kilogram of air if (*a*) the isentropic compression precedes the heating, and (*b*) the isentropic compression follows the heating.

14·10 Solve Problem 14·9 with the compression for each case being adiabatic with an efficiency of 70 percent.

14·11 Nitrogen at 96 kPa, 15°C, enters a compressor with negligible velocity and is discharged at 300 kPa, 110°C, through a cross-sectional area of 0.042 m^2. The flow rate through the compressor is 18.5 kg/s, and the power input is 2710 kW. Heat exchange is with only the surrounding atmosphere. Calculate the heat transfer and the irreversibility, both per kilogram of nitrogen.

14·12 A compressor draws in ambient air at 95 kPa, 15°C, at a rate of 0.80 lbm/min and compresses it to 440 kPa, 205°C. The compressor is uncooled except for possible heat loss to the surrounding atmosphere. Calculate the minimum possible power input to any compressor system working between these end states and exchanging heat with only the atmosphere.

14·13 Refer to Problem 13·68. Plot a curve of power required for isentropic compression versus test section Mach number if the wind tunnel is operated in a once-through manner.

14·14 Ethylene is compressed from 101.3 kPa, 20°C, to 350 kPa, 120°C, by a centrifugal compressor at a rate of 20 kg/min. The rise in temperature of the cooling water flowing through the jacket is 5 Celsius degrees for a water flow rate of 16 kg/min. Neglecting kinetic energy changes, calculate the power required to operate the compressor.

14·15 An air compressor uses a lubricating oil having a flash point of 175°C. If air enters at 105 kPa, 15°C, and is compressed polytropically with an exponent of 1.35, what is the maximum allowable discharge pressure if the maximum allowable air temperature is 25 Celsius degrees below the flash point of the oil?

14·16 An ideal centrifugal compressor takes in 180 m^3/min of air at 100 kPa, 5°C, with negligible velocity. It compresses the air polytropically with n = 1.35 and discharges it at 200 kPa with a velocity of 180 m/s. Calculate the power input to the compressor and the heat transfer rate in kJ/s.

14·17 Determine the total amount of work required by a compressor that draws in methane at 95 kPa, 25°C, and compresses it isentropically to raise the pressure to 300 kPa in a 0.5-m^3 insulated tank that initially contains methane at 95 kPa, 25°C.

14·18 Carbon monoxide is compressed adiabatically from 100 kPa, 20°C, to 225 kPa, 130°C. Inlet velocity is 8 m/s and discharge velocity is 35 m/s. The atmosphere is at 101 kPa, 20°C. Determine the compressor efficiency and the irreversibility.

14·19 Determine the irreversibility of the steady-flow adiabatic compression of air initially at 101.3 kPa, 25°C, through a pressure ratio of 5 if the compressor efficiency is 74 percent.

14·20 Oxygen is to be compressed from 100 kPa, 30°C, to 620 kPa at a rate of 8.60 kg/min. A supply of coolant is available so that the oxygen can be cooled to 20°C before being compressed reversibly and adiabatically or the coolant can be used to make the compression polytropic with a discharge temperature 20 Celsius degrees lower than the discharge temperature with isentropic compression starting at 30°C. Compare the power requirements.

14·21 A two-stage compressor draws in 8.5 m^3/min of air at 98.5 kPa and compresses it polytropically with n = 1.3 to 985 kPa. The intercooler cools the air to its initial temperature before it enters the second stage. How much work does two-stage compression save in comparison with compression in a single stage with the same value of n?

14·22 Air is to be compressed from 1 atm, 25°C (77 F), to 5 atm. For two-stage ideal compression with intercooling to the initial temperature and reversible adiabatic compression in each stage, plot work per unit mass of air against first-stage pressure ratio. (Use a constant mean specific heat.)

14·23 Solve Problem 14·22 for an adiabatic efficiency of 0.75 for each stage.

14·24 Solve Problem 14·22 for the case in which some cooling is effected during compression so that for each stage $pv^{1.38}$ = constant.

14·25 If two stages in a multistage compressor have the same work input and the same efficiency, which (the higher-pressure stage or the lower-pressure stage) has the greater irreversibility? Where would efforts to improve compressor stage efficiency likely be more fruitful?

14·26 For use in design studies, a chart that shows temperature at the end of adiabatic compression of air is needed. The chart should plot discharge temperature versus gage discharge pressure (in a range extending as high as 15 atmospheres) for three values of inlet temperature typical of construction sites in North America and for compressor efficiencies ranging from 50 to 90 percent. For clarity, more than one chart may be needed.

14·27 At what pressure should intercooling be done in air compression processes? To what temperature should the air be cooled? Does it make a difference whether the compressed air will be used in pneumatic tools immediately, stored in a receiver, or go through a heater or combustion chamber directly to a gas turbine?

14·28 Ambient air enters an axial-flow compressor at 30°C (86 F) and is compressed adiabatically through a pressure ratio of 3. Plot the work required per unit mass and the irreversibility of the process per unit mass versus compressor efficiency for an efficiency range of 50 percent to 100 percent.

14·29 A natural gas compressor station receives natural gas at 50 psia (345 kPa), 59 F (15°C), and is to discharge it at 760 psia (5240 kPa). The gas has a volumetric analysis of 84 percent methane and 16 percent ethane. The flow rate is 88,000 lbm/h (39,900 kg/h). Two-stage and three-stage compression are to be compared. In each case, the stages are to have equal pressure ratios, and each stage but the last is to be followed by an intercooler in which the gas will be cooled at essentially constant pressure to within 20 degrees F (8.3 degrees C) of the temperature of available cooling water, 59 F. Adiabatic compressor efficiency is estimated to be 65 percent for each stage. Calculate for each of the cases (*a*) the required power input and (*b*) the irreversibility per unit mass of gas.

14·30 Air is compressed from 14.7 psia, 70 F, to 50 psia, 250 F, by a centrifugal compressor at a rate of 50 lbm/min. The rise in temperature of the cooling water flowing through the jacket is 7 degrees F for a water flow rate of 50 lbm/min. Neglecting kinetic energy changes, calculate the power required to operate the compressor.

14·31 An ideal centrifugal air compressor takes in 10,000 cfm of air at 14 psia, 50 F, with negligible velocity. It compresses the air polytropically with n = 1.35 and discharges it at 30 psia with a velocity of 600 fps. Calculate the power input to the compressor.

14·32 A compressor discharges into a 20-cu-ft insulated tank that initially contains air at 14.0 psia, 80 F. Determine the total work required for the compressor that draws in air at 14.0 psia, 80 F, to compress it isentropically to raise the pressure to 42 psia in the tank.

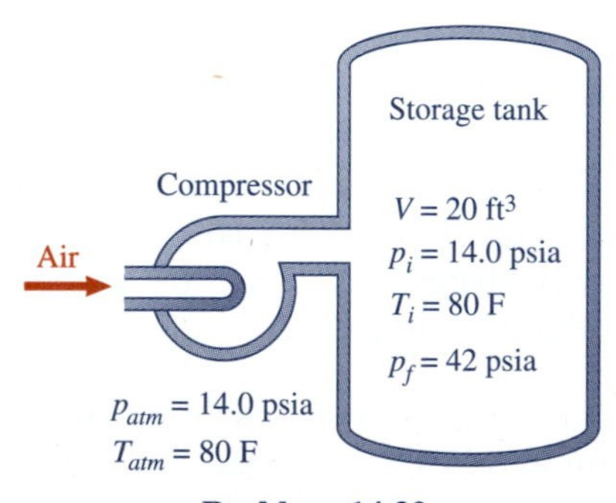

Problem 14·32

14·33 Carbon monoxide is compressed adiabatically from 14.5 psia, 80 F, to 30 psia, 255 F. Inlet velocity is 15 fps; discharge velocity is 110 fps. Determine the compressor efficiency.

14·34 Determine the irreversibility of the steady-flow adiabatic compression of air initially at 14.7 psia, 80 F, through a pressure ratio of 5 if the compressor efficiency is 74 percent.

14·35 A two-stage compressor draws in 300 cfm of air at 14.3 psia and compresses it polytropically with $n = 1.3$ to 143 psia. The intercooler cools the air to its initial temperature before it enters the second stage. How much work does two-stage compression save in comparison with compression in a single stage with the same value of n?

14·36 Atmospheric air is to be compressed reversibly and adiabatically to 6 times atmospheric pressure. For typical values of inlet conditions, make a chart showing the effect of pressure drop through an inlet filter on the required work input. Use a pressure drop range of 1 percent to 10 percent of atmospheric pressure.

14·37 Solve Problem 14·36 for a compressor efficiency of 65 percent.

14·38 What would be the effects of spraying a fine mist of water droplets into the inlet of an air compressor?

14·39 An air blower is driven by a synchronous electric motor. Describe the effect on power required of (*a*) ambient temperature, (*b*) ambient pressure, (*c*) ambient relative humidity.

14·40 Refer to Example 2·9. The data given are insufficient for an accurate calculation of the power required. Estimate the range of reasonable values of power required.

14·41 Fans, blowers, and compressors are all devices for moving gases from a region at one pressure to a region at a higher pressure. The lines of demarcation are not sharp. Investigate the use of these terms and list guidelines for their use.

14·42 A plant has a utility compressed air system operating at 1000 kPa gage. Some hand tools used in the plant require air at only 200 kPa gage. Compare the power requirements for throttling from the main system with the power requirement of a separate low-pressure system. What are the disadvantages of a separate system? Moisture separation is used in the main system. Would it be needed in the low-pressure system?

14·43 For various kinds of service, compare the use of one large compressor with the use of two or three smaller ones. (Some of the factors to consider are the power required, the total energy required during a period of time, the space required, reliability, and costs.)

14·44 A series of reciprocating compressors is to be designed for supplying utility compressed air for manufacturing plants. The compressed air will be used principally for driving tools. A typical pressure to be maintained in the compressed air header is 1000 kPa gage, with a minimum pressure of 900 kPa gage and a maximum pressure of 1100 kPa gage. Occasionally, the demand for compressed air may exceed the nominal flow rate through the compressor by as much as 15 percent. Therefore, a receiver (a storage tank connected to the compressor discharge line) is delivered with each compressor. List the information you would need so you could make a recommendation on the size of receiver for each compressor in the series. Then list the steps you would take to determine the required size of receiver.

14·45 At the end of Section 14·5 the effect of discharge pressure on the performance of a reciprocating gas compressor is analyzed. Perform a similar analysis for the effect of discharge pressure on the performance of a gear pump used to compress gas.

14·46 The compressor in a household refrigeration unit normally takes in saturated refrigerant at a high quality. Sometimes the cylinder heads are mounted on springs so that if a volume of liquid greater than the clearance volume is in the cylinder, excess liquid can be drained away without damaging the compressor. Explain the trade-offs involved in such a design.

14·47 Is it true that for a turbine operating on an ideal gas with constant specific heats the work output for a given pressure ratio is proportional to the inlet gas absolute temperature? Prove your answer.

14·48 Helium enters a turbine at 572 kPa, 830°C, and expands adiabatically to 100 kPa, 504°C, with negligible change in kinetic energy. Determine the turbine efficiency.

14·49 Determine the mass rate of flow through a turbine that takes in air at 500 kPa, 870°C, and exhausts at 100 kPa if the flow is adiabatic with a turbine efficiency of 75 percent and the power output is 1050 kW.

14·50 A three-stage gas turbine operating on air is to produce equal work per stage. Determine the intermediate stage pressures if the inlet conditions are 620 kPa, 980°C, the exhaust pressure is 100 kPa, and the adiabatic efficiency for each stage is 75 percent. Atmospheric temperature is 20°C. Calculate also the effectiveness of each stage.

14·51 Determine the irreversibility of an adiabatic expansion of air from 507 kPa, 870°C, through a pressure ratio of 5 if the turbine efficiency is 78.5 percent and the lowest temperature in the surroundings is 25°C.

14·52 Air enters a well-insulated turbine at 300 kPa, 200°C, and exhausts into the atmosphere which is at 100 kPa, 20°C. The flow rate is 2.08 kg/s. Plot the irreversibility of the turbine process in kJ/kg versus turbine efficiency for an efficiency range from 50 percent to 100 percent.

14·53 Air enters a turbine at 83.0 psia, 1540 F, and expands adiabatically to 14.6 psia, 940 F, with negligible change in kinetic energy. Determine the turbine efficiency and the turbine effectiveness.

14·54 Determine the mass rate of flow required for a turbine that takes in air at 75 psia, 1600 F, and exhausts at 15 psia if the flow is adiabatic with a turbine efficiency of 75 percent and the power output is 1500 hp.

14·55 Determine the irreversibility of an adiabatic expansion of air from 73.5 psia, 1600 F, through a pressure ratio of 5 if the turbine efficiency is 78.5 percent and the lowest temperature in the surroundings is 80 F.

14·56 Solve Problem 14·55, using constant specific heats, and compare your results with those of Problem 14·55.

14·57 Air enters a turbine at 6 atm, 1175 F (635°C). If the final pressure is 1 atm, plot the exhaust temperature against turbine efficiency for an efficiency range of 60 percent to 100 percent.

14·58 Two air streams are available for power production by expansion to 1 atm. One is at 2 atm, 500 F (260°C), and the other is at 2 atm, 1130 F (610°C). The flow rates are the same. For maximum power output, should the two streams be mixed and then expanded through a single turbine, or should each be expanded through a separate turbine without mixing? Prove your answer.

14·59 In a design study for an emergency power supply it is proposed to use a turbine driven by compressed air from a storage tank. The turbine will operate adiabatically between the decreasing pressure in the tank and atmospheric pressure. Disregard heat transfer to the air in the tank. Apply the first law to the tank to obtain a differential equation relating m (the mass of air in the tank at any instant), T (the temperature of the air in the tank at that instant), and such constant properties of the air as R, c_p, c_v, and k. Then solve the differential equation for T in terms of m, T_i (the initial temperature of the air in the tank), m_i (the initial mass of air in the tank), and constant properties of the air.

14·60 In a design study for an emergency power supply that can start quickly and produce 2 kW for 20 minutes, it is proposed to use a turbine driven by compressed air from a storage tank in which the air is initially at 15 atm, 25°C (77 F). The turbine will operate adiabatically between the decreasing pressure in the tank and atmospheric pressure of 1 atm with an expected efficiency of 55 percent until the tank pressure drops to 1.5 atm. Disregard heat transfer to the air in the tank. What size tank is needed in order for the turbine to deliver 2 kW for 20 min? Would doubling the initial pressure halve the required volume? (See figure.)

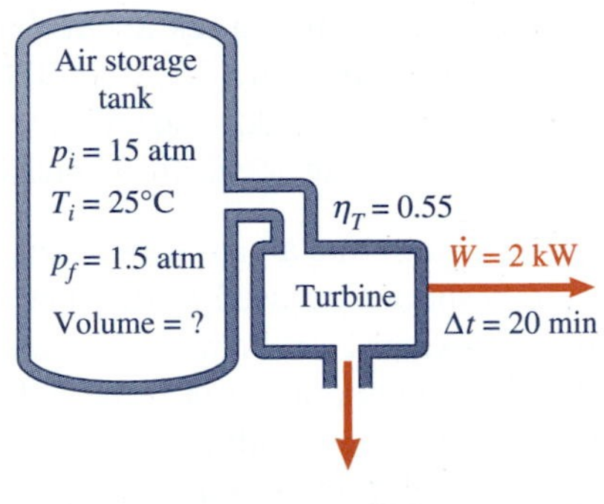

Problem 14·60

14·61 For the design study of Problem 14·60, determine the minimum tank volume for three other candidate gases: nitrogen, helium, and argon.

14·62 Solve Problem 14·60 for the case of a turbine efficiency of 100 percent.

14·63 As a means of reducing the tank volume required for the power supply described in Problem 14·60, it is suggested that a large number of short, hollow metal cylinders be placed in the tank to reduce the decrease in temperature of the air in the tank during the operation of the system. Readily available steel tubing has dimensions such that the mass of steel that can be packed into the tank is 38 times the mass of air initially in the tank. The steel has a specific heat of 0.50 kJ/kg·K (0.119 B/lbm·R). Assuming that the steel temperature follows very closely the air temperature, determine the size tank needed for the service specified in Problem 14·60.

14·64 The efficiency of the turbine for the power supply described in Problem 14·60 will actually depend on the pressure ratio and consequently will not be constant during the operation. To achieve a smaller tank volume, would it be better to design the turbine for higher efficiency at higher or lower pressure ratios?

14·65 Suppose the air is heated between the tank and the turbine of Problem 14·60 to maintain a constant turbine inlet temperature of 485°C (905 F). What size tank would be needed? How much heat would be required during the 20 min?

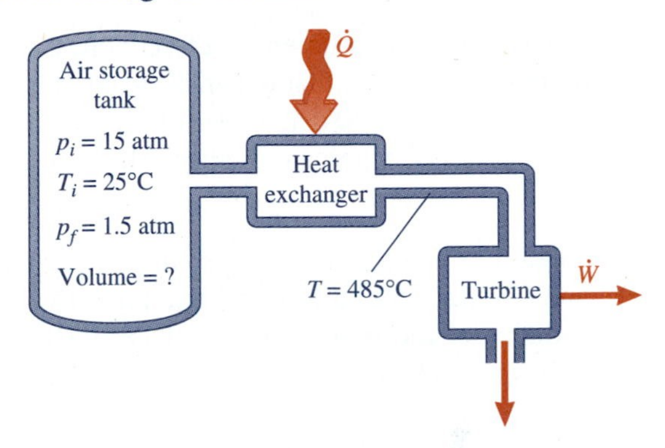

Problem 14·65

14·66 In a design study for an emergency power supply that will be able to start quickly and provide 2 kW of electrical power for 15 minutes, one proposal is for a tank of compressed gas at ambient temperature that will supply gas to a speed-controlled turbine that drives a generator. For any gas, it is estimated that the turbine requires a minimum pressure ratio of 1.9 to maintain adequate speed at full load. The electrical generator efficiency is of the order of 0.9. For a maximum pressure of 15 atmospheres, determine the minimum tank volume required for four candidate gases: air, nitrogen, helium, and argon.

14·67 Determine the efficiency and the effectiveness of a turbine that expands steam at a rate of 9000 kg/h from 1.4 MPa, 225°C, to 35 kPa, 90.5 percent quality, if the power output is 1150 kW.

14·68 Steam is supplied to a turbine at 1.4 MPa, 315°C, and is exhausted at 101.3 kPa. The flow is adiabatic, the power output of the turbine is 10,000 kW, and the turbine efficiency is 65 percent. The lowest temperature in the surroundings is 25°C. Determine (*a*) the steam rate in kg/kW·h, (*b*) the amount of energy made unavailable during the turbine expansion in kJ/kg, and (*c*) the turbine effectiveness.

14·69 An ideal steam turbine operates between pressures of 2.00 MPa and 10 kPa. The power output is 250 kW. Determine the inlet temperature needed to ensure that the exhaust steam contains not more than y percent moisture, where $y = 12$, 8, 6, 4, 2, and 0.

14·70 Determine the efficiency, effectiveness, and power output of a turbine that expands steam at a rate of 20,000 lbm/h from 200 psig, 440 F, to 27 inches of mercury vacuum, 92.5 percent quality. The sink temperature is 80 F.

14·71 Steam is supplied to a turbine at 200 psig, 600 F, and is exhausted at atmospheric pressure. The flow is adiabatic, the power output of the turbine is 10,000 kW, and the turbine efficiency is 62 percent. The lowest temperature in the surroundings is 80 F. Determine (*a*) the steam rate in lbm/kW·h, (*b*) the amount of energy made unavailable during the turbine expansion in B/lbm, and (*c*) the turbine effectiveness.

14·72 Devise a single-stage turbine that can be made to rotate in either direction, with the change in direction of rotation to be brought about by the movement of a single lever or valve. Explain its operation clearly by means of sketches or diagrams as well as words.

14·73 Prove that, if a turbine discharges wet steam at the sink temperature, the product of $(1 - \eta_T)$ and the isentropic enthalpy drop between the initial condition and the exhaust pressure is equal to the irreversibility.

14·74 What are the effects of spraying a fine mist of liquid water droplets into a gas turbine just upstream of the first-stage nozzles?

14·75 In a design study for a gas turbine with an inlet pressure of 6 atmospheres that exhausts to the atmosphere, determine the sensitivity of power output to pressure drop through exhaust mufflers. The frictional pressure drops may be as high as 10 percent of atmospheric pressure. How do the effects of exhaust muffler pressure drops vary with inlet temperature? With turbine efficiency?

14·76 In Section 14·2, it is stated that you can show that if two turbine stages have the same efficiency, the stage at the higher pressure has a higher effectiveness and consequently a lower irreversibility. Do so.

14·77 A design study for a gas turbine has indicated that four stages should be used with equal work per stage. Inlet and exhaust pressures are to be 800 kPa and 100 kPa. Inlet temperature is to be 600°C. For a continuation of the design study, specify the optimum interstage pressures for an ideal air-standard turbine.

14·78 A design study for a gas turbine has indicated that four stages should be used with equal work per stage. Inlet and exhaust pressures are to be 800 kPa and 100 kPa. Inlet temperature is to be 600°C. For a continuation of the design study, specify the optimum interstage pressures for an air-standard turbine if each stage has an efficiency of 76 percent.

14·79 Determine the sensitivity of the interstage pressures in Problem 14·78 to changes in the turbine efficiency, assuming still that the efficiencies of the four stages are equal.

14·80 Water flows at a rate of 0.06 m^3/s through a centrifugal pump running at 1200 rpm. The inlet and discharge lines are 15 cm in diameter and are at the same elevation. The inlet pressure is 200 mm Hg vacuum, and the discharge pressure is 140 kPa. Power input to the pump is 15 kW. Barometric pressure is 740 mm Hg. Determine the pump efficiency.

14·81 A fan draws air at 95 kPa, 25°C, from a room at a rate of 17 m^3/min and discharges it through a duct of 0.20 m^2 cross-sectional area at a static gage pressure of 8.0 cm of water. The fan efficiency is 60 percent. Determine the power input.

14·82 Consider an ideal single-stage impulse turbine. Steam enters the nozzles at 200°C, 700 kPa, with a velocity of 150 m/s, and the exhaust pressure is 200 kPa. Flow is frictionless and adiabatic. The blades are symmetrical, and the blade speed is 300 m/s. The flow rate is 45 kg/min. Sketch a complete accurate vector diagram for the stage, and calculate (*a*) the power output and (*b*) the kinetic energy of the steam leaving the blades. What change, if any, should be made in the blade speed to increase the stage efficiency?

14·83 Consider a steam-turbine impulse stage. Steam enters the nozzle at 60 psia, 460 F, with a velocity of 500 fps and expands adiabatically through the nozzle to 40 psia, 400 F. The nozzle angle is 20 degrees. The blade velocity is 500 fps, and the symmetrical blades are so shaped that the steam enters the blades tangentially to the blade surface. The steam flow rate is 10,000 lbm/h. Sketch a complete vector diagram, and, disregarding friction in the blade passage, calculate (*a*) the kinetic energy of the steam leaving the turbine and (*b*) the power output of the turbine.

14·84 A single-acting reciprocating compressor compresses air reversibly and adiabatically. The piston displacement is 280 cm^3, and there is 4.0 percent clearance. The speed is 60 rpm. Suction pressure is 95 kPa; discharge pressure is 480 kPa. Ambient temperature is 15°C. Disregard pressure drop across the valves. Sketch a pV diagram and determine the volume of ambient air compressed per minute.

14·85 Compute the power and the cylinder volume required to compress 15 m^3/min of ambient air to 825 kPa. Atmospheric pressure is 101.3 kPa, and the suction pressure is 95 kPa. The clearance is 2.5 percent, and the value of n during compression and expansion is 1.3. The compressor is double-acting and operates at 50 rpm.

14·86 A single-stage air compressor having a clearance of 3 percent takes in air at 14.2 psia, 60 F, and compresses it to 200 psia. The mass of air compressed per minute is 50 lbm, and n for both compression and expansion processes is 1.35. Compute (*a*) the power required, (*b*) the piston displacement if the compressor operates at 100 rpm, and (*c*) the heat transferred to the water jacket.

14·87 Consider an ideal reciprocating gas compressor that operates with polytropic compression and with polytropic expansion of the clearance gas, but the polytropic exponents are not the same for the two processes. If the cylinder walls are generally cooler than the gas, for which process will the n be greater? In such a case, how does increasing the clearance affect the work done per pound of gas delivered?

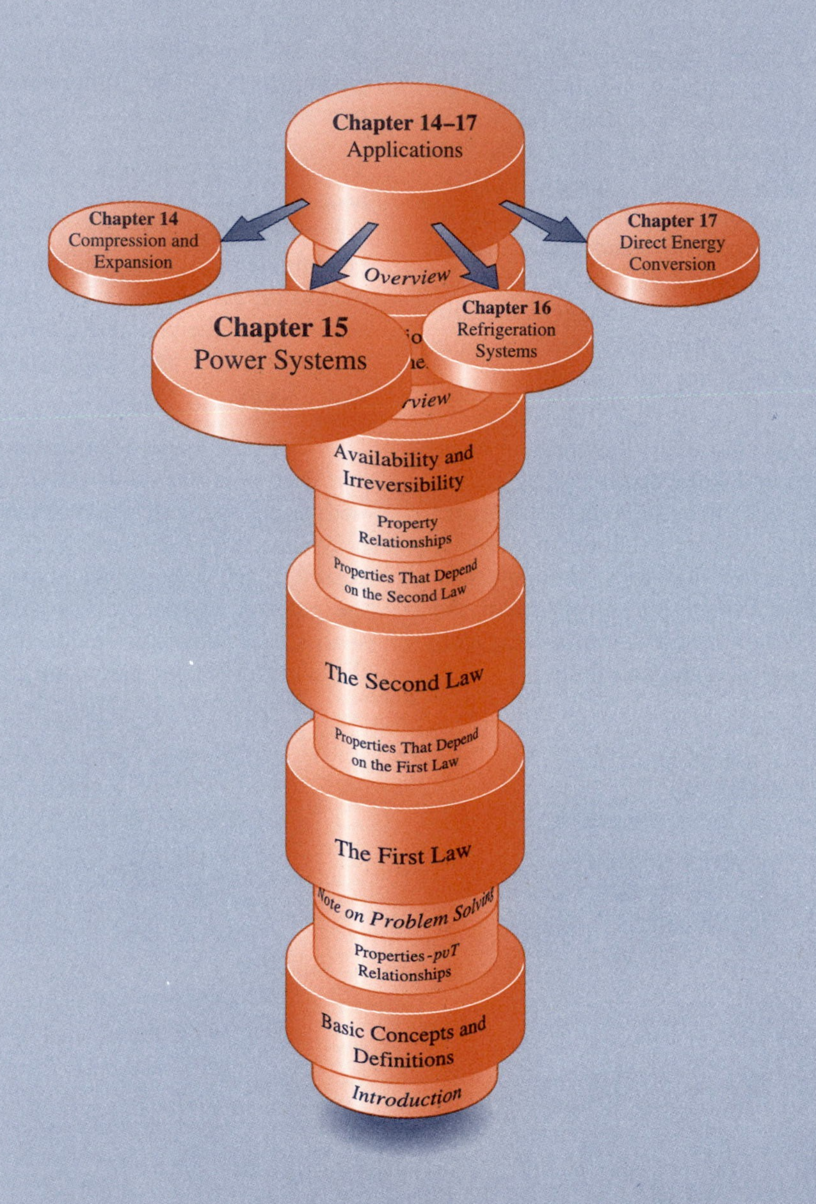
Chapter 14–17
Applications
Chapter 14
Compression and Expansion
Chapter 17
Direct Energy Conversion
Overview
Chapter 15
Power Systems
Chapter 16
Refrigeration Systems
Availability and Irreversibility
Property Relationships
Properties That Depend on the Second Law
The Second Law
Properties That Depend on the First Law
The First Law
Note on Problem Solving
Properties-pvT Relationships
Basic Concepts and Definitions
Introduction

Chapter 15

Power Systems

Power generation involves innumerable applications of thermodynamics. The power may be generated for propulsion of automobiles, aircraft, or ships, for electric power supply, or for direct application in driving pumps, mowers, materials processing equipment, and many other devices. We will categorize power-generating equipment broadly into gas power systems and vapor power systems. Gas power systems include both gas turbines and internal-combustion engines, both of which use a working substance that never changes phase; it remains a gas. In a vapor power system the phase of the working substance changes. The change in phase allows more energy to be stored in the working substance than can be stored by only sensible heating. Also, because the working substance expands as a vapor but is compressed as a liquid with a small specific volume, very little of the expansion work must be used for the compression processes.

In this chapter, we look first at gas-turbine power plants, including aircraft jet engines, where typically air is compressed, combustion takes place in a burner with no moving parts, and the combustion gases then flow through a turbine to produce work. Then we study reciprocating internal combustion engines where combustion occurs within a cylinder or other combustion chamber on an intermittent basis.

15·1 Modeling Gas Turbines

Notice that sometimes *gas turbine* refers to the entire plant that consists of at least a compressor, a combustion chamber, and a turbine; but sometimes *gas turbine* refers to only a single component of the plant, the turbine.

In a simple gas turbine, as shown in Fig. 15·1, atmospheric air is drawn into a compressor where it is compressed to a higher pressure and temperature. Fuel is then injected into the hot pressurized air stream in a combustion chamber, and the resulting hot products of combustion expand through a turbine to produce work. Most of the turbine work is used to drive the compressor, a small amount is used to drive auxiliary devices, and the remainder is the net plant work output.

Air-standard models

For gas turbines, the simplest models that provide valuable quantitative results are **air-standard models**. In an air-standard model for gas turbines

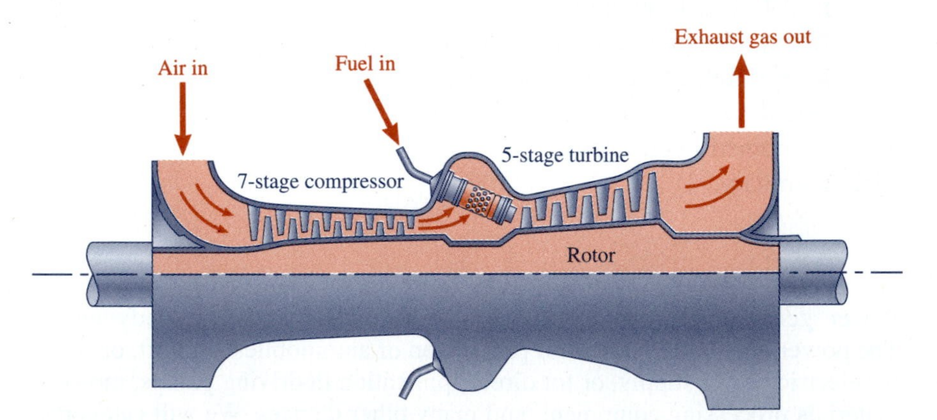

Figure 15·1 A simple gas-turbine power plant.

- The working substance is air, which is modeled as an ideal gas, throughout the cycle.
- The combustion process is replaced by a constant-pressure heat addition process.
- A constant-pressure heat rejection process replaces the actual exhaust of combustion products from the turbine and the drawing in of atmospheric air by the compressor.

Although the individual processes occur as steady-flow processes through open systems, the entire plant can be treated as a closed system.

Sometimes other features are specified for an air-standard model. As one example, all or some of the processes may be modeled as reversible. As another example, for approximate results, specific heats can be assumed to be constant at their room-temperature values, and the model is then referred to as a **cold-air-standard model**. As noted earlier, different models are used for different phases of analysis and design. An engineer does not use only models from a standard set, but rather creates models that best suit the situation at hand. Consequently, be careful: The term *air-standard model* does not always specify the same restrictions or assumptions.

Cold-air-standard model

15·1·1 The Brayton Cycle or Basic Gas-Turbine Cycle

The **Brayton cycle** is an air-standard model of a simple gas-turbine cycle. Figure 15·2 shows a flow diagram, a *pv* diagram, and a *Ts* diagram for a Brayton cycle.

George B. Brayton (1830–1892), American engineer, invented a breech-loading gun, a riveting machine, and a sectional steam generator in addition to the internal-combustion engine for which he is best remembered. The Brayton engine, developed around 1870, was a reciprocating oil-burning engine with fuel injection directly into the cylinder, and a compressor that was separate from the power cylinder. The Brayton cycle, which is now used only for gas turbines, was thus first used with reciprocating machines.

The compression and expansion processes are isentropic, and the heat addition and heat rejection occur in reversible constant-pressure processes. (In sketching *Ts* diagrams of gas-turbine cycles, show clearly the divergence of constant-pressure lines with increasing entropy to ensure that the diagram indicates the Δh of the turbine expansion to be greater in magnitude than the Δh of compression. Otherwise, the turbine cannot drive the compressor and deliver work to the surroundings.)

Application of the first law to an air-standard gas-turbine cycle numbered as in Fig. 15·2 shows the turbine work, compressor work, net work, and cycle efficiency to be given by

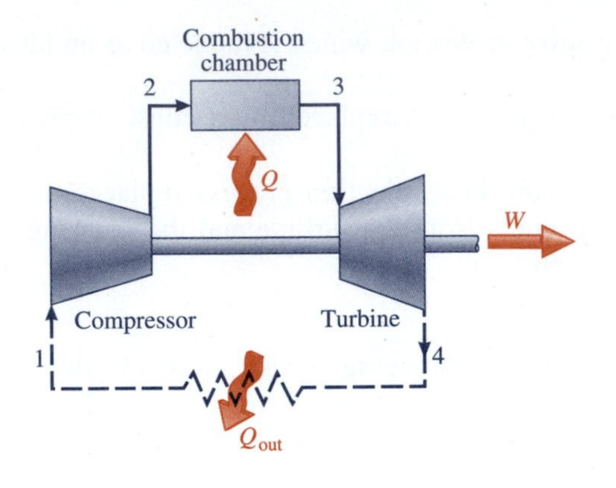

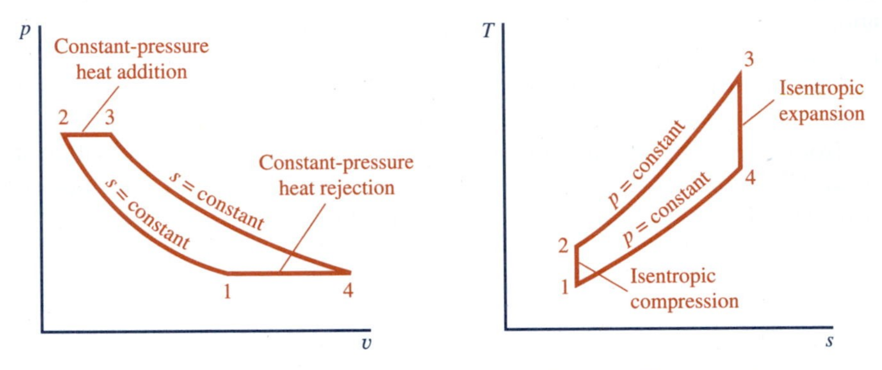

Figure 15·2 The Brayton cycle that simulates simple gas-turbine operation.

$$w_T = h_3 - h_4 + \frac{V_3^2 - V_4^2}{2} = h_{t3} - h_{t4}$$

$$w_{\text{in},C} = h_2 - h_1 + \frac{V_2^2 - V_1^2}{2} = h_{t2} - h_{t1}$$

$$w = w_T - w_{\text{in},C} = h_{t3} - h_{t4} - (h_{t2} - h_{t1})$$

$$\eta = \frac{w}{q_{2-3}} = \frac{h_{t3} - h_{t4} - (h_{t2} - h_{t1})}{h_{t3} - h_{t2}} = 1 - \frac{h_{t4} - h_{t1}}{h_{t3} - h_{t2}}$$

The thermal efficiency of a cold-air-standard Brayton cycle depends only on the pressure ratio.

For the *special case of constant specific heats* (i.e., the cold-air-standard analysis), the last equation can be reduced to show that the thermal efficiency of the cycle is a function of the total pressure ratio p_{t2}/p_{t1} only. Surprisingly, it is independent of the peak temperature. This is not true for a model that uses variable specific heats of the working fluid.

A *Ts* diagram shows why the efficiency of a Brayton cycle is significantly lower than that of a Carnot cycle operating between the same temperature limits. Figure 15·3*a* shows a Brayton cycle and a Carnot cycle superimposed. The two cycles have the same temperature limits and the same entropy change of the working fluid during the heat addition process. Figure 15·3*b* shows that a Brayton cycle can be thought of as being composed of many

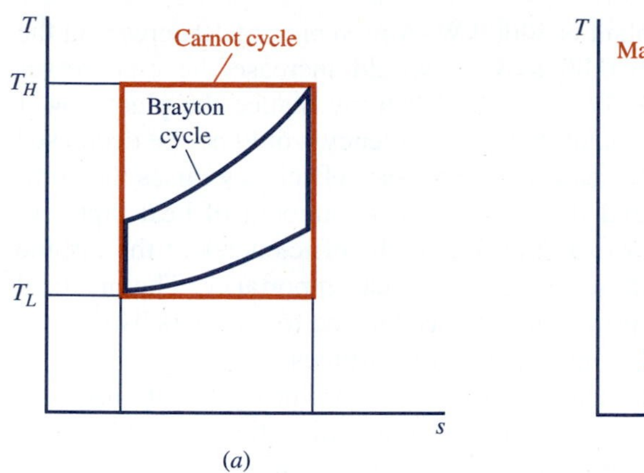

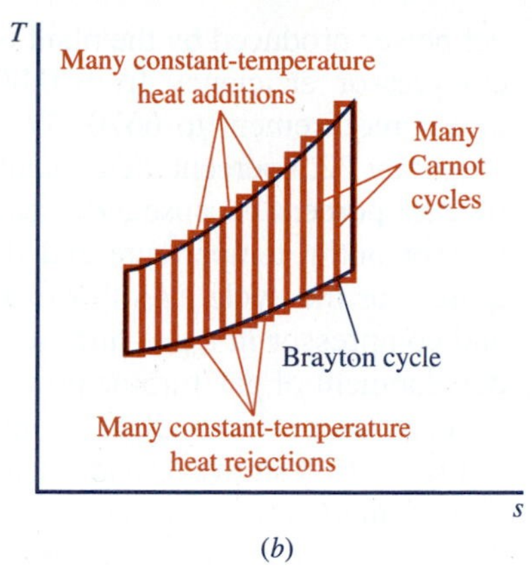

Figure 15·3 Superposition of a Brayton cycle and a Carnot cycle for a comparison of efficiencies.

Carnot cycles with so little heat added to each one that the heat is added at essentially constant temperature. Clearly, all of these Carnot cycles have efficiencies lower than the efficiency of the Carnot cycle that operates between the maximum and minimum temperature. Thus the efficiency of the Brayton cycle is less than that of the Carnot cycle with the same temperature limits. We will use this kind of reasoning in evaluating changes in various cycles, including the Brayton cycle.

Recall that any reversible cycle can be simulated by a number of Carnot cycles (see Section 6·1).

Usually the working fluid of a gas turbine is atmospheric air and the products of combustion with a high air–fuel ratio. For some applications, other gases can be used as the working fluid in a closed circuit so that the same fluid is used repeatedly. Then, as in the air-standard Brayton cycle, the combustion process is actually replaced by a constant-pressure heat addition and the exhaust and intake processes are replaced by a constant-pressure heat rejection. It is easy to show that the efficiency of a cold-air-standard Brayton cycle depends on the compressor ratio and the specific-heat ratio of the working fluid, and the cold-air-standard model indicates a trend that holds also for an actual system. Consequently, selecting a gas with a specific-heat ratio different from that of air can be advantageous.

The **backwork ratio**, defined as

Backwork ratio

$$\text{Backwork ratio} \equiv \frac{\text{compression work}_{\text{in}}}{\text{gross work of prime mover}} = \frac{w_{\text{in},C}}{w_T} = \frac{w_T - w_{\text{cycle}}}{w_T}$$

$$= 1 - \frac{w_{\text{cycle}}}{w_T}$$

is high for a gas-turbine cycle. Consequently, the efficiency of the cycle is reduced appreciably by relatively small reductions in compressor and turbine efficiencies. For example, suppose that the gross power output of a turbine is 9000 kW and the compressor requires an input power of 6000 kW so that the net power produced by the plant is 3000 kW. A reduction of 10 percent in the compressor efficiency from 0.80 to 0.72 would increase the compressor power requirement to 6670 kW and would thereby reduce the plant power output by 22.3 percent. (The plant thermal efficiency would not be decreased by 22.3 percent because a decrease in compressor efficiency raises the compressor outlet temperature and thus decreases the amount of heat input required for the cycle.) As this example shows, the efficiencies of the turbine and compressor in a gas-turbine plant are of great importance. The practical development of gas-turbine power plants was delayed many years by the low efficiencies of the available compressors and turbines.

Low efficiencies of turbines and, especially, compressors long delayed the development of gas turbine plants.

Figure 15·4 shows *hs* and *Ts* diagrams of a gas-turbine cycle with irreversible adiabatic compression and expansion. For an adiabatic steady-flow process, the entropy change of the universe equals the entropy change of the matter flowing through the steady-flow system. Therefore, for the compression process 1–2,

$$i = T_0 \, \Delta s_{\text{universe}} = T_0(s_2 - s_1)$$

and this can be represented by an area on the *Ts* diagram. The irreversibility of the adiabatic turbine expansion can likewise be shown. Remember that the irreversibility of a process is the decrease, caused by the process, in the amount of energy of the universe that can be converted into work. In an actual gas-turbine plant, frictional effects in the passages between components and in the combustion chamber also add to the irreversibility of the cycle.

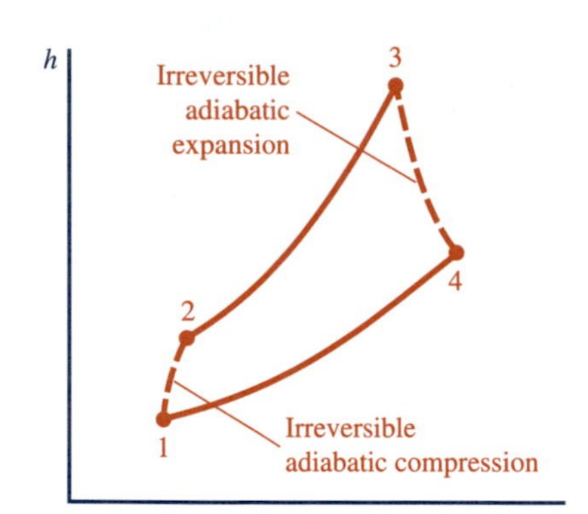

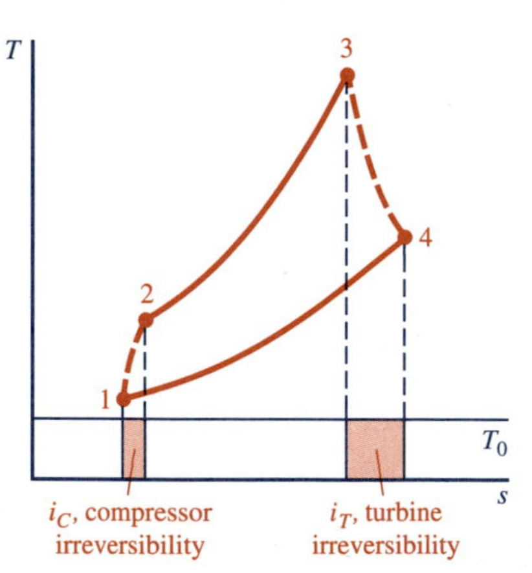

Figure 15·4 Air-standard gas-turbine cycle with irreversible compression and expansion.

Increasing the turbine inlet temperature helps to offset the effects of various irreversibilities in the cycle, but this temperature is limited by the loss of strength of turbine construction materials with increasing temperature. This temperature limitation is more stringent in a steady-flow machine like a turbine, where the gas temperature at each point is constant, than it is in a reciprocating engine, where the metal parts are in contact with gases at the maximum temperature for only an instant, and in contact with relatively cool gases for the remainder of the cycle. For this reason, much higher maximum gas temperatures can be used in reciprocating engines than in gas turbines.

To limit the maximum temperature in a gas-turbine cycle, high air–fuel ratios are used. Therefore, the products of combustion can be treated as air with little loss of accuracy. Because the backwork ratio is so high, however, the small increase of mass flowing through the turbine resulting from the addition of fuel may increase the net work of the cycle by an appreciable amount. A model that considers the mass of the fuel added but still treats the combustion products as air is called an *air-standard model with mass of fuel considered.* The air–fuel ratio required to produce a given temperature with a given fuel can be found by starting with an energy balance for adiabatic combustion in a steady-flow burner,

$$-Q = 0 = -\Delta H = \sum_{\text{reactants}} m(h - h_{ref}) - \sum_{\text{products}} m(h - h_{ref}) - \Delta H_R \qquad (11\text{·}6)$$

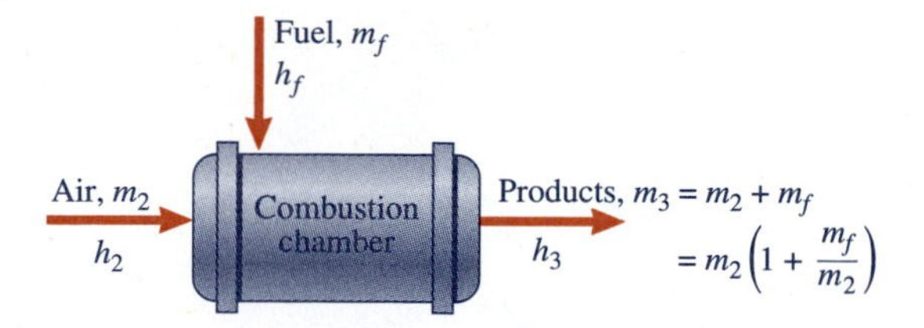

Figure 15·5 Combustion chamber mass balance.

This energy balance is based on the assumption that the change in kinetic energy is negligible. Designating the air entering by 2, the fuel entering by f, and the products leaving by 3, as shown in Fig. 15·5, we have

$$0 = m_2(h_2 - h_{ref})_{\text{air}} + m_f(h_f - h_{ref})_{\text{fuel}} - m_3(h_3 - h_{ref})_{\text{prod}} - \Delta H_R$$

Using the steady-flow mass balance, $m_3 = m_2 + m_f$, we can rearrange this to

$$\frac{m_2}{m_f} = \frac{\Delta H_R/m_f + (h_3 - h_{ref})_{\text{prod}} - (h_f - h_{ref})_{\text{fuel}}}{(h_2 - h_{ref})_{\text{air}} - (h_3 - h_{ref})_{\text{prod}}}$$

where $\Delta H_R/m_f$ is the enthalpy of combustion per unit mass of fuel. If the products are treated as air, the air–fuel ratio expression can be simplified to

$$\frac{m_2}{m_f} = \frac{\Delta H_R/m_f + (h_3 - h_{ref})_{\text{air}} - (h_f - h_{ref})_{\text{fuel}}}{h_2 - h_3}$$

where all of the h's except those marked for fuel can be obtained from air tables or from c_pT relations as in Table A·1.

Example 15·1 Air-Standard Gas Turbine

PROBLEM STATEMENT

A Brayton cycle operates with air entering the compressor at 100 kPa, 25°C, a pressure ratio of 6, and a turbine inlet temperature of 800°C. Determine the compressor work input, the turbine work, and the cycle thermal efficiency for equal compressor and turbine efficiencies of 100 percent, 90 percent, 80 percent, 70 percent, and 60 percent.

SOLUTION

Analysis: We first sketch a flow diagram and a *Ts* diagram, and on the *Ts* diagram we show two cases: (1) For machine efficiencies of 100 percent, states $2i$ and $4i$ are the states at the ends of the compression and expansion processes. (2) For machine efficiencies lower than 100 percent, the states are labeled 2 and 4.

Referring to the states on the *Ts* diagram, if we neglect changes in kinetic energy, the equations from the first law, from the definitions of compressor efficiency, turbine efficiency, and cycle efficiency, and from the

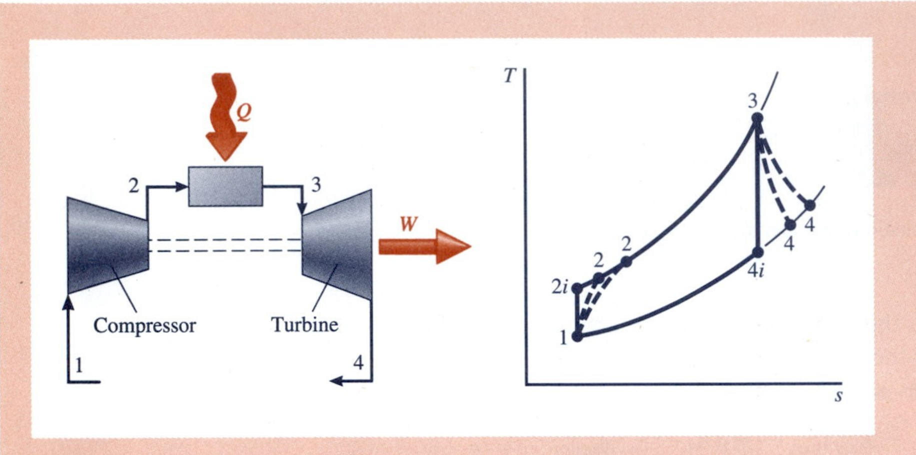

isentropic relationships for ideal gases that we need are

$$w_{\text{in},C} = h_2 - h_1$$

$$w_T = h_3 - h_4$$

$$w = w_T - w_{\text{in},C}$$

$$q_{2-3} = h_3 - h_2$$

$$\eta_C = \frac{h_{2i} - h_1}{h_2 - h_1}$$

$$\eta_T = \frac{h_3 - h_4}{h_3 - h_{4i}}$$

$$\eta = \frac{w}{q_{2-3}}$$

$$\frac{p_{r2i}}{p_{r1}} = \frac{p_2}{p_1}$$

$$\frac{p_{r3}}{p_{r4i}} = \frac{p_3}{p_4}$$

Calculations: We enter the equations into our equation solver along with the computer model for air properties and solve for the five values of compressor and turbine efficiencies to obtain the following result:

η_C and η_T	$w_{in,C}$	w_T	η
1	200	443	0.385
0.9	222.2	398.7	0.290
0.8	250	354.4	0.180
0.7	285.7	310.1	0.045
0.6	333.3	265.8	Impossible

Comment: Notice that as the compressor and turbine efficiencies drop, the plant or cycle thermal efficiency drops precipitously. This is characteristic of cycles with high backwork ratios. For equal compressor and turbine efficiencies of between 70 percent and 60 percent, the compressor input work equals the turbine work, so there is a zero net plant output. For any lower machine efficiencies, the plant will not operate because the work output of the turbine is insufficient even to drive the compressor.

Since the numerator of the expression for thermal efficiency of a Brayton cycle is the difference between two numbers of the same magnitude, inaccuracies of calculations are multiplied. For example, using constant specific heats can result in large errors in thermal efficiency calculations.

15·1·2 Regeneration in Gas Turbines

In a simple gas-turbine cycle, the turbine exhaust temperature is appreciably higher than the temperature of the air leaving the compressor and entering the combustion chamber. The amount of fuel needed can be reduced by the use of a heat exchanger that uses the hot turbine exhaust gas to preheat the air between the compressor and the combustion chamber. This heat exchanger is called a **regenerator**. In the ideal case the flows through the regenerator are at constant pressure.

Regenerator

An air-standard regenerative cycle is shown in Fig. 15·6. Application of the first law to the regenerator, assuming that kinetic energy changes are negligible and that no heat is lost to the surroundings, shows that $h_3 - h_2 =$

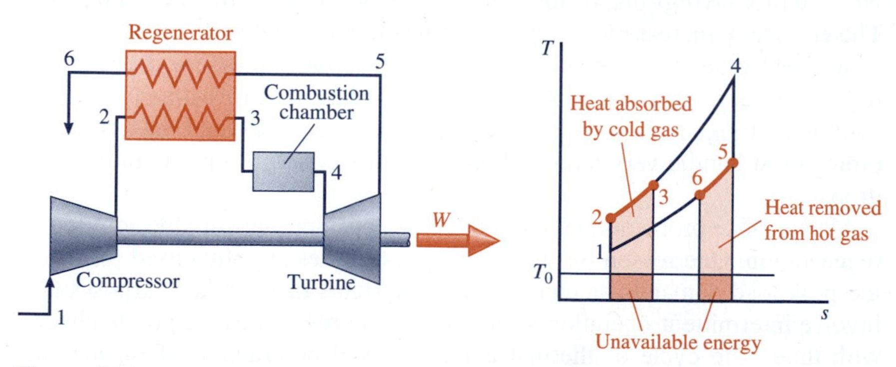

Figure 15·6 Air-standard regenerative gas-turbine cycle.

$h_5 - h_6$. Since $q_{2-3} = -q_{5-6}$, the two crosshatched areas on the Ts diagram of Fig. 15·6 are equal in magnitude. If, as shown in the Ts diagram, heat is transferred across a finite temperature difference in the regenerator, then the increase of entropy principle shows that, for this irreversible steady-flow process, $s_3 - s_2 + s_6 - s_5 > 0$, or $|\Delta s_{2-3}| > |\Delta s_{5-6}|$. The irreversibility of the regenerator process is $i = T_0(s_3 - s_2 + s_6 - s_5)$.

Another way to look at the irreversibility of the regenerator is this: The magnitude of the unavailable portion of the heat removed from the hot gas is $T_0(s_5 - s_6)$. This part of the heat removed from the hot gas could not be converted into work even by an externally reversible engine. As the heat is added to the cold gas in the regenerator, however, the unavailable part of it is $T_0(s_3 - s_2)$. The difference between $T_0(s_3 - s_2)$ and $T_0(s_5 - s_6)$ is therefore the amount of *energy made unavailable* by the irreversible transfer of heat in the regenerator. This is the same as the irreversibility of the process.

For maximum thermal efficiency of a regenerative cycle, T_3 in Fig. 15·6 should be as high as possible, its limiting value being T_5. The extent to which this limit is approached in any cycle is expressed by the **regenerator effectiveness**, which, with reference to Fig. 15·6, is defined as

Regenerator effectiveness

$$\text{Regenerator effectiveness} \equiv \frac{T_3 - T_2}{T_5 - T_2}$$

Increasing the effectiveness of a regenerator calls for more heat-transfer surface area, which increases both cost and space requirements. Using regenerators of very high effectiveness cannot be justified economically because the fixed costs (i.e., costs that are largely independent of plant output, such as depreciation, interest, taxes, insurance, and maintenance) exceed the savings in fuel cost that result from the higher plant thermal efficiency. Consequently, effectiveness values rarely exceed about 0.7 in actual plants.

Regeneration may be attractive in stationary gas turbine plants if the investment in additional equipment, including the heat exchanger and associated piping, and the increase in maintenance costs are small in comparison with the fuel savings that result. This is often not the case for several reasons. The efficiency increase with regeneration is not as great in an actual plant as in an ideal plant because of the effects of frictional pressure drops in the regenerator and additional piping of an actual plant. The investment in equipment to add regeneration is large because the regenerator and the associated piping must handle very large volumes of hot gases with minimum pressure drop.

Maintenance problems and costs increase considerably with the addition of regeneration. One reason for this is that gas turbines are often used for meeting peak load demands on electric power systems or for other purposes that involve intermittent operation so that temperatures at various points change with time. The cycle of thermal expansion and contraction of regenerator parts increases maintenance problems. Another reason is that the enormous volumes of gas passing through the plant cause fouling of heat-exchanger and other surfaces. This calls for more frequent and longer shutdowns for maintenance to prevent deterioration of performance.

A gas-turbine application where regeneration is often employed is for driving gas-line compressors, especially in locations where there is no demand for power other than that to drive the compressors, no need for steam, and limited availability of cooling water.

A more attractive means of increasing the efficiency of power generation is the combination of a gas-turbine plant with a steam plant. Energy in the gas-turbine exhaust stream, which may be at a temperature above 500°C, is used to generate steam for power generation, space heating, or process purposes.

The decision of whether to use regeneration in a gas-turbine plant is a case where factors other than increasing the thermal efficiency come strongly into play. In aircraft gas turbines, regeneration is not used because the regenerator would add to the projected frontal area of the craft and increase the drag prohibitively.

In gas turbines, shaft speeds and plant pressure ratios often vary during operation. Sometimes two shafts or spindles are used, with one shaft carrying a turbine and the compressor so that they can operate at the optimum speed to produce hot, high-pressure gas that drives the power turbine. The first turbine drives nothing but the compressor. The power turbine is on a second shaft that drives the load, and the speed of the second shaft is dictated by the load requirements. One such arrangement is shown in Fig. 15·7.

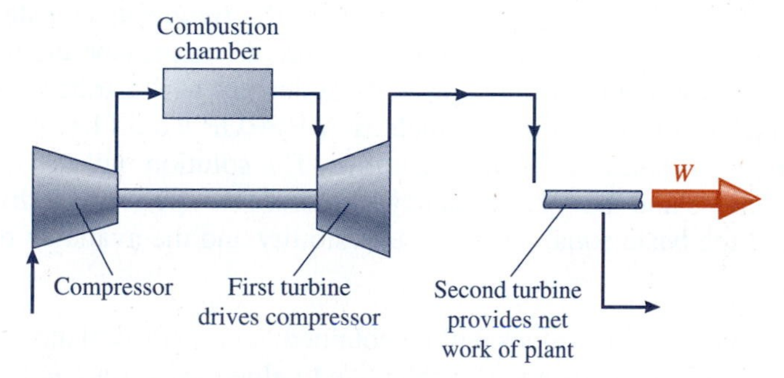

Figure 15·7 A two-shaft simple gas-turbine plant that allows the compressor and its driving turbine to rotate at optimum speed while the power turbine speed is determined independently.

Example 15·2 Regenerative Gas-Turbine Cycle

PROBLEM STATEMENT

Air from the atmosphere enters the compressor of an air-standard regenerative gas turbine plant at 95 kPa, 20°C, and is discharged into the regenerator at 475 kPa, 260°C. Air enters the combustion chamber at 425°C, enters the turbine at 875°C, and leaves the turbine at 575°C. Neglecting the mass of fuel, frictional pressure drops in the regenerator and combustion chamber, and kinetic energy changes, determine (*a*) the compressor and turbine efficiencies, (*b*) the cycle thermal efficiency, and (*c*) the irreversibilities of the compressor, turbine, and regenerator processes. (*d*) Determine also the available fractions of both the heat added to the cycle and the heat rejected and show these on a diagram with the irreversibilities determined in part *c*.

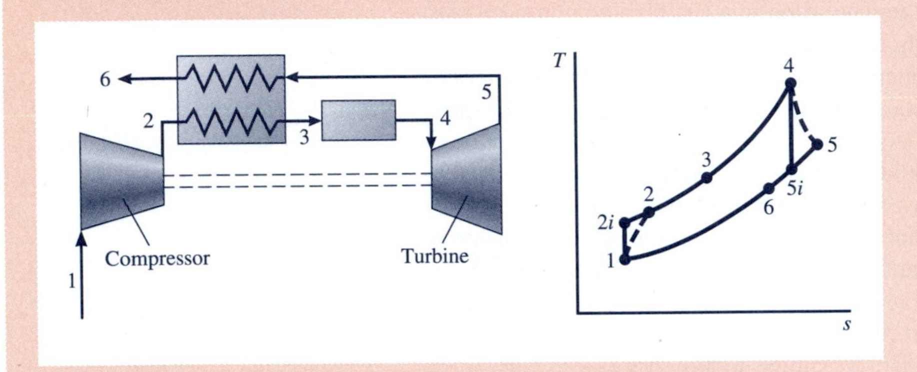

SOLUTION

Analysis: First we make a flow diagram and a *Ts* diagram. States $2i$ and $5i$ are the *isentropic* end states of the compression and expansion processes. These states are needed for calculation of compressor and turbine efficiencies. For this example problem solution, we will use printed property tables just to illustrate their use, even though using a computer and an online gas-properties model such as AIRPROP would be faster and would permit rapid additional solutions for other values of various quantities. The solution will involve the definitions of compressor and turbine efficiencies and thermal efficiency, the first law applied to individual pieces of equipment and the entire plant, and the basic equations for irreversibility and the available part of heat transfer.

Calculations: At various stages of the solution, property values will be obtained; so for convenience a table of these values is made, showing both the specified values and the calculated values, which are entered as soon as they are obtained. Air tables (A·12) are used.

State	p, kPa	T, °C	h, kJ/kg	p_r	ϕ, kJ/kg·K
1	95	20	295.4	1.301	6.684
2*i*	475		467.2	6.507	7.146
2	475	260	539.3	10.78	7.291
3	475	425	713.3		7.575
4	475	875	1218.9	202.1	8.132
5*i*	95		782.7	40.4	7.670
5	95	575	877.1		7.787
6	95		703.1		7.560

The solution of the first three parts involves the application of (1) the definitions of the three efficiencies, (2) the pressure–temperature (or p_r) relationship for isentropic processes, and (3) the first law:

(*a*)

$$p_{r2i} = p_{r1}\frac{p_2}{p_1} = 1.301\frac{475}{95} = 6.507 \qquad h_{2i} = 467.2 \text{ kJ/kg}$$

$$\eta_C = \frac{w_{\text{in},i}}{w_{\text{in}}} = \frac{h_{2i} - h_1}{h_2 - h_1} = \frac{467.2 - 295.4}{539.3 - 295.4} = 70.5 \text{ percent}$$

$$p_{r5i} = p_{r4}\frac{p_5}{p_4} = 202.1\frac{95}{475} = 40.4 \qquad h_{5i} = 782.7 \text{ kJ/kg}$$

$$\eta_T = \frac{w}{w_i} = \frac{h_4 - h_5}{h_4 - h_{5i}} = \frac{1218.9 - 877.1}{1218.9 - 782.7} = 78.4 \text{ percent}$$

(*b*)

$$\eta = \frac{w}{q_{3-4}} = \frac{w_T - w_{\text{in},C}}{h_4 - h_3} = \frac{1218.9 - 877.1 - (539.3 - 295.4)}{1218.9 - 713.3} = 19.4 \text{ percent}$$

The compressor, the turbine, and the entire regenerator (including both fluid streams) operate steadily and adiabatically, so for each one the irreversibility per kilogram of fluid is given by $i = T_0\,\Delta s$, where Δs is the entropy change of the fluid flowing through the device. (See Eq. *b* or 8·8*b* of Section 8·4.)

(*c*) Compressor:

$$i = T_0(s_2 - s_1) = T_0\left(\phi_2 - \phi_1 - R\ln\frac{p_2}{p_1}\right)$$

$$= 293\left(7.291 - 6.684 - 0.287\ln\frac{475}{95}\right)$$

$$= 42.5 \text{ kJ/kg}$$

Turbine:

$$i = T_0(s_5 - s_4) = T_0\left(\phi_5 - \phi_4 - R \ln \frac{p_5}{p_4}\right)$$
$$= 293\left(7.787 - 8.132 - 0.287 \ln \frac{95}{475}\right)$$
$$= 34.4 \text{ kJ/kg}$$

Regenerator:

$$h_6 = h_5 - (h_3 - h_2) = 877.1 - (713.3 - 539.3) = 703.1 \text{ kJ/kg}$$
$$i = T_0(\Delta s_{2-3} + \Delta s_{5-6}) = T_0\left(\phi_3 - \phi_2 - R \ln \frac{p_3}{p_2} + \phi_6 - \phi_5 - R \ln \frac{p_6}{p_5}\right)$$
$$= 293(7.575 - 7.291 - 0 + 7.560 - 7.787 - 0)$$
$$= 16.7 \text{ kJ/kg}$$

(*d*) Of the heat added:

$$q_{unav} = T_0(s_4 - s_3) = T_0\left(\phi_4 - \phi_3 - R \ln \frac{p_4}{p_3}\right) = 293(8.132 - 7.575 - 0)$$
$$= 163.3 \text{ kJ/kg}$$
$$q_{av} = q_{3-4} - q_{unav} = 505.6 - 163.3 = 342.3 \text{ kJ/kg}$$

For the heat rejected:

$$q_{\text{out}} = h_6 - h_1 = 703.1 - 295.4 = 407.7 \text{ kJ/kg}$$
$$q_{unav} = T_0(s_6 - s_1) = T_0(\phi_6 - \phi_1) = 293(7.560 - 6.684) = 256.8 \text{ kJ/kg}$$
$$q_{av,\text{out}} = q_{\text{out}} - q_{unav,\text{out}} = 407.7 - 256.8 = 150.9 \text{ kJ/kg}$$

The energy flow diagram shows graphically that the unavailable energy rejected from the cycle exceeds the unavailable energy added to the cycle by the sum of the irreversibilities in the cycle.

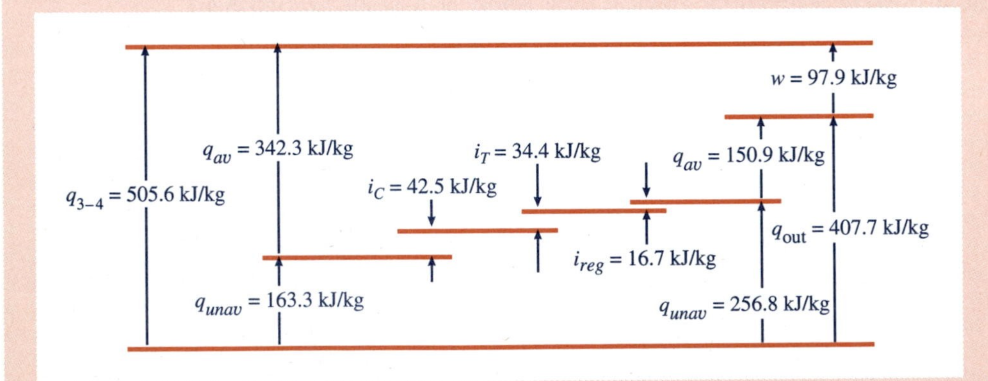

15·1·3 Intercooling in Gas Turbines

The net work of a gas-turbine cycle is given by

$$w = w_T - w_{\text{in},C}$$

and can be increased either by decreasing the compressor work or by increasing the turbine work. These are the purposes of *intercooling* and *reheating*, respectively.

It was shown in Section 14·1 that intercooling between stages reduces the work input required to compress a gas from a given initial state to a specified final pressure. Therefore, if a simple gas-turbine cycle is modified by having the compression accomplished in two or more adiabatic processes with intercooling between them, the backwork ratio is reduced and consequently the net work of the cycle is increased with no change in the work of the turbine.

Several different lines of reasoning can show that the thermal efficiency of an ideal simple gas-turbine cycle is lowered by the addition of an intercooler. One line of reasoning is based on a Ts diagram like Fig. 15·8 in which the ideal simple gas-turbine cycle is 1–2–3–4–1, and the cycle with the intercooler is 1–a–b–c–2–3–4–1. If the simple gas-turbine cycle 1–2–3–4–1 is divided into a number of cycles like i–j–k–l–i and m–n–o–p–m, these little cycles approach Carnot cycles as their number increases. Notice that *if the specific heats are constant*, then

Recall that any reversible cycle can be simulated by a number of Carnot cycles (see Section 6·1).

$$\frac{T_3}{T_4} = \frac{T_m}{T_p} = \frac{T_i}{T_l} = \frac{T_2}{T_1} = \left(\frac{p_2}{p_1}\right)^{(k-1)/k}$$

Thus all the Carnot cycles making up the simple gas-turbine cycle have the same efficiency. Likewise, all the Carnot cycles into which the cycle a–b–c–

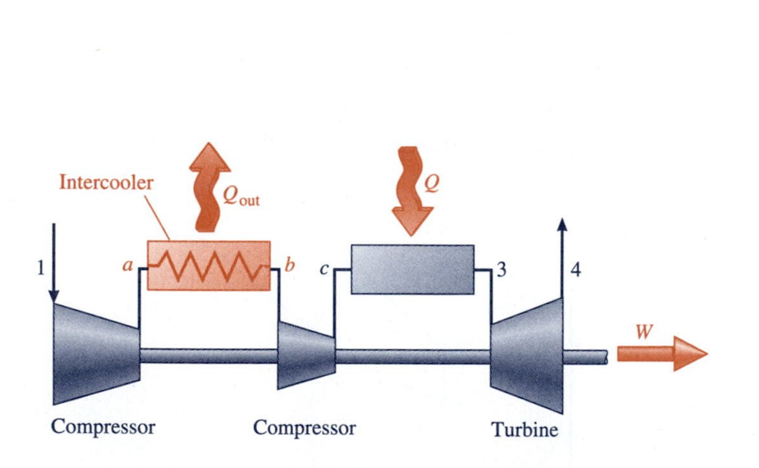

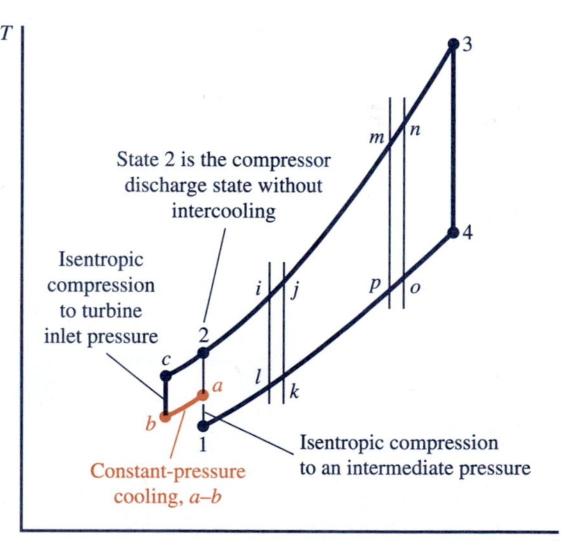

Figure 15·8 Air-standard gas-turbine cycle with intercooling.

2–*a* might similarly be divided have a common value of efficiency that is *lower* than the Carnot cycles that comprise cycle 1–2–3–4–1. Thus the addition of an intercooler, which adds *a*–*b*–*c*–2–*a* to the simple cycle, lowers the efficiency of the ideal gas-turbine cycle. We have based our reasoning here on constant specific heats (the cold-air-standard analysis), but the same conclusions can be reached for variable specific heats.

The addition of an intercooler to a regenerative gas-turbine cycle can increase the cycle thermal efficiency because the heat required for the process *c*–2 in Fig. 15·8 can be obtained from the hot turbine exhaust gas passing through the regenerator instead of from burning additional fuel (or, in the case of the air-standard cycle, from adding more heat from outside the cycle.) The use of an intercooler consequently calls for a larger regenerator.

15·1·4 Reheat in Gas Turbines

The turbine work, and consequently the net work of a gas-turbine cycle, can be increased without changing the compressor work or the maximum temperature in the cycle by dividing the turbine expansion into two or more parts with constant-pressure heating (i.e., combustion in an actual cycle, heat transfer in an air-standard cycle) before each expansion. This cycle modification is known as **reheating**. Flow and *Ts* diagrams of an ideal simple gas-turbine cycle modified by a single reheat are shown in Fig. 15·9. Reasoning similar to that used with intercooling shows that adding reheat to a simple gas-turbine cycle lowers the thermal efficiency, but that a combination of regeneration and reheating can increase the thermal efficiency.

Reheating

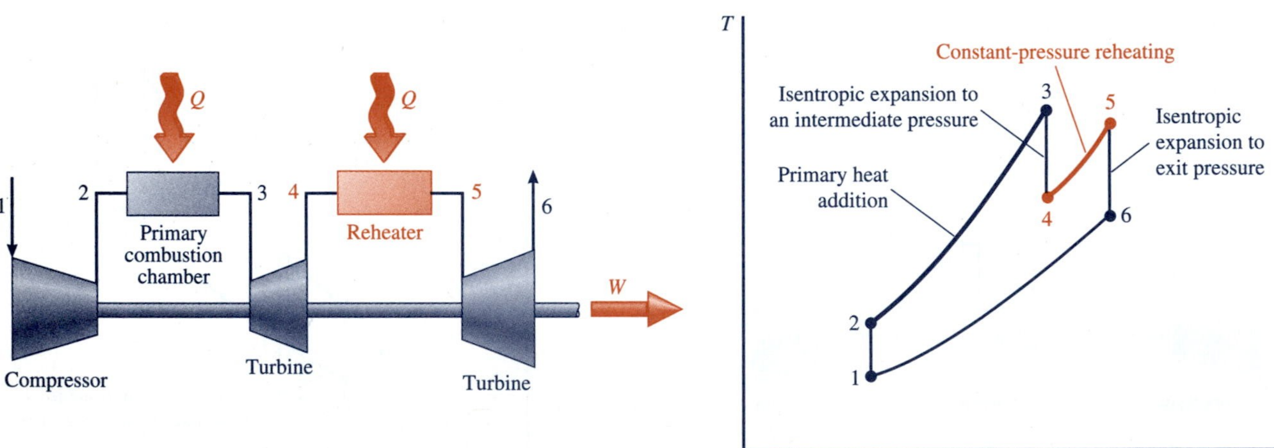

Figure 15·9 Air-standard gas-turbine cycle with reheat.

Notice that reheating increases the turbine exhaust temperature so that regeneration can bring the temperature of the air entering the combustion chamber closer to the turbine inlet temperature than is possible without reheat. As the number of reheats is increased, the exhaust temperature increases, making regeneration even more valuable.

Diagrams of an intercooled-regenerative-reheat cycle are shown in Fig. 15·10. If the number of intercools, the number of reheats, and the regenerator effectiveness are increased, this cycle approaches the Ericsson cycle discussed in Section 5·5. With adiabatic compression and expansion, the costs of equipment for multiple intercools and reheats are likely to exceed the fuel-cost savings, but the advantages of approaching isothermal compression and isothermal turbine expansion (as by burning fuel in the turbine in an actual cycle) are apparent.

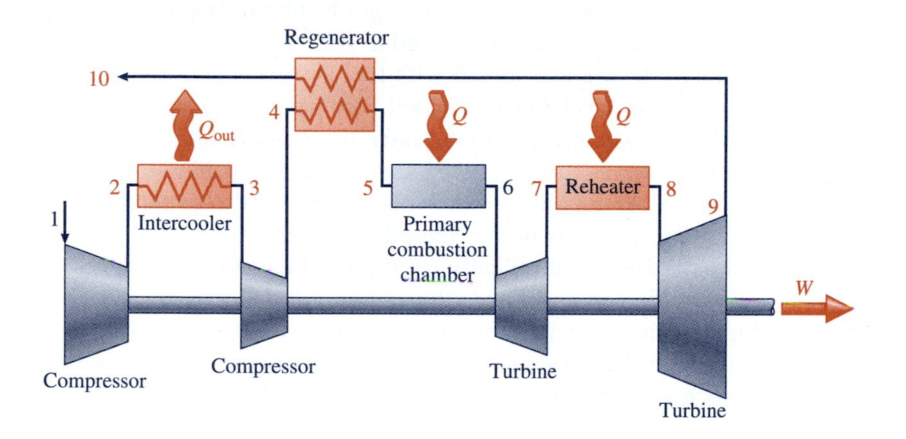

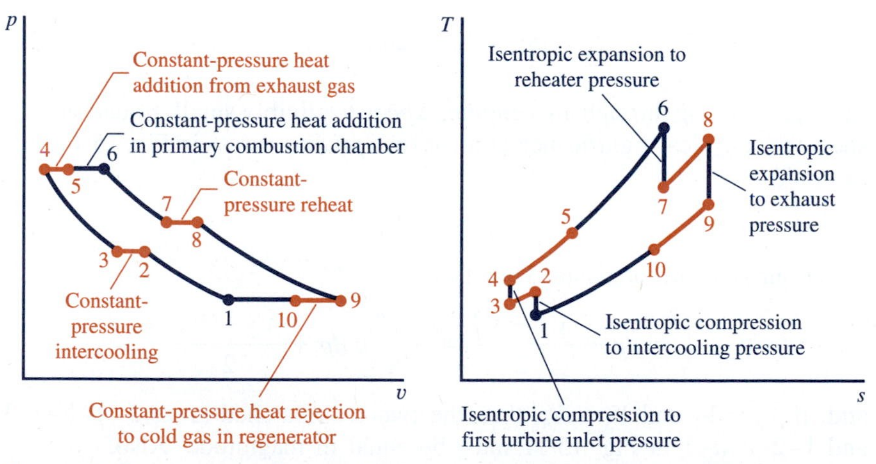

Figure 15·10 Air-standard gas-turbine cycle with intercooling, regeneration, and reheating.

15·1·5 Gas-Turbine Jet Propulsion

Powered aircraft are propelled by accelerating air rearward by exerting a force on it. The equal-magnitude and opposite-direction force of the fluid on the craft drives the craft. A propeller accelerates slightly a large mass of fluid, while an aircraft jet engine causes a relatively small mass of air to undergo a large change in velocity. In both cases the action is described by the basic dynamic equation for one-dimensional steady flow:

$$F = \dot{m}\,\Delta V \qquad (13\cdot 1)$$

where F is the resultant force on the fluid that flows at a rate $\dot{m}$. (See Example 13·2, but notice that a special case of $p_1 = p_2 = p_{atm}$ is treated there.)

From among the various jet-propulsion engines that are in use or have been proposed, we will select the gas-turbine jet engine or turbojet engine to illustrate the application of the principles that have already been introduced.

Figure 15·11 shows the flow, pv, and Ts diagrams of a turbojet engine. Air enters the engine at state 1 and is compressed to state 2. It then flows into the combustion chamber, where it is mixed with fuel and combustion occurs, or where, in the air-standard cycle, heat is added. The gas then expands adiabatically through the turbine, which drives only the compressor, so that *the turbine work equals the compressor work input*. For the turbine to produce enough work to drive the compressor, it is necessary for it to expand the gas only to a pressure p_4, which is higher than the ambient pressure p_1 $(= p_5)$. The gas is then expanded isentropically (in the ideal case) through a nozzle from state 4 to state 5. The gas leaves the engine at a velocity V_5, which is appreciably higher than V_1, and consequently a thrust is developed.

If the flow through the turbojet engine is reversible and steady, we can apply to any process the relationship developed from the principles of mechanics:

$$\int v\,dp = w_{\text{in}} - \Delta ke - \Delta pe \qquad (1\cdot 20)$$

For a gas flowing through a jet engine, Δpe is negligibly small. Equation 1·20 shows the physical significance of areas on a pv diagram as in Fig. 15·11. For example,

$$w_{\text{in},1-2} = w_{3-4}$$

for an air-standard analysis. Therefore,

$$\int_1^2 v\,dp + \frac{V_2^2 - V_1^2}{2} = -\int_3^4 v\,dp + \frac{V_3^2 - V_4^2}{2}$$

and, if $V_2 = V_1$ and $V_3 = V_4$, then the two crosshatched areas 3–4–b–c–3 and 1–2–c–a–1 in Fig. 15·11 must be equal in magnitude. Also,

$$\frac{V_5^2 - V_4^2}{2} = -\int_4^5 v\,dp$$

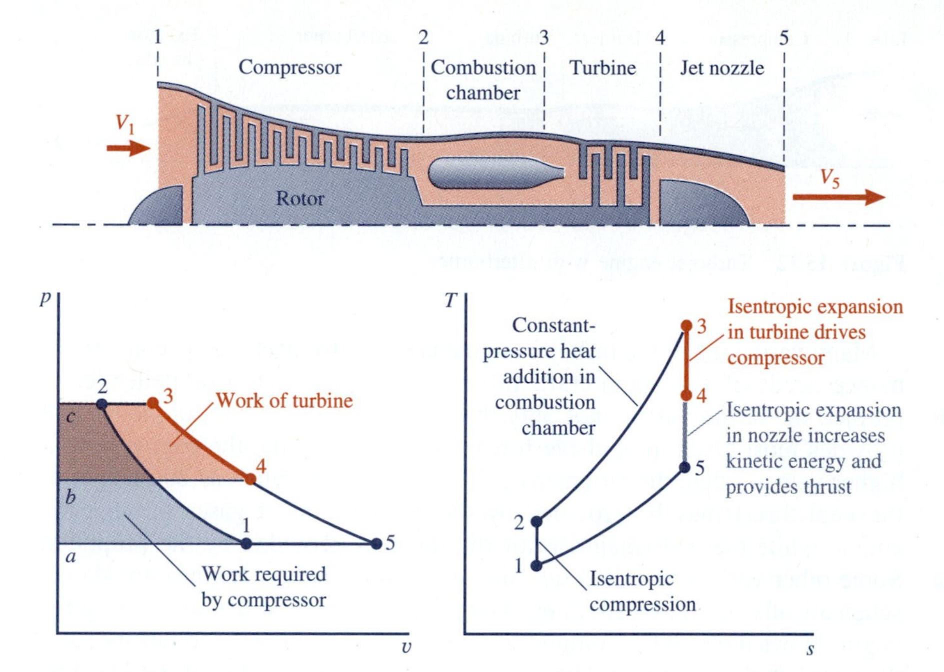

Figure 15·11 Turbojet engine.

so that the area 4–5–*a*–*b*–4 in Fig. 15·11 represents the increase in kinetic energy of the gas flowing through the turbojet nozzle.

Although for irreversible processes areas on *pv* and *Ts* diagrams have no significance, study of Fig. 15·11 indicates that low compressor or turbine efficiency will markedly decrease the thrust of a jet engine. Just as in a stationary gas-turbine plant, the net work is the difference between the turbine work and the compressor work input, two numbers close to each other. In the jet engine the difference between the turbine work and compressor work input is the energy available to produce the high-velocity exhaust jet and the thrust of the engine.

Afterburning

Space and weight limitations prohibit the use of regenerators and intercoolers on aircraft engines. The counterpart of reheating is **afterburning**. The air–fuel ratio in a jet engine is so high that the turbine exhaust gases are sufficiently rich in oxygen to support the combustion of more fuel in an afterburner. Such burning of fuel (or, in the air-standard cycle, addition of heat) raises the temperature of the gas before it expands in the turbojet nozzle, increasing the kinetic energy change in the nozzle and consequently increasing the thrust. Figure 15·12 shows a flow diagram for a turbojet engine with an afterburner. In the air-standard case, the combustion is replaced by constant-pressure heat addition. In an actual engine, the combustion process and the nozzle expansion process are not fully separate; expansion of the gas in the nozzle begins while combustion occurs, so combustion is not at constant pressure. Operation of actual engines in afterburning mode causes a huge increase in fuel consumption.

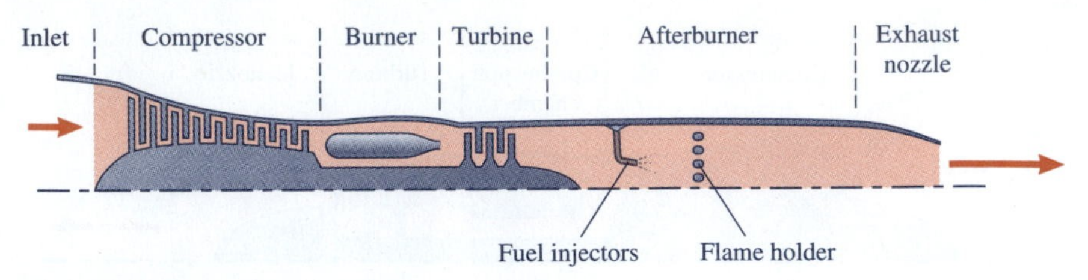

Figure 15·12 Turbojet engine with afterburner.

Many variations of the turbojet engine are used to meet the special performance needs of various aircraft. In one variation, called a propeller jet or propjet, the turbine drives not only the compressor but also a propeller that provides most of the propulsive force. The gases leaving the engine have a higher velocity than the air entering the engine, so as much as 20 percent of the total thrust may be provided by the acceleration of gases through the engine while the rest (majority) of the thrust is provided by the propeller. Some other variations called turbofan engines or bypass engines are shown schematically in Fig. 15·13. In each engine, only some of the air entering the engine flows through the compressor, combustion chamber, and turbine as in the simple turbojet engine. The rest passes through only a fan or the first few compressor stages and is then either mixed with the exhaust jet or is discharged separately at a velocity lower than that of the exhaust jet but higher than the inlet velocity, thereby producing part of the propulsive thrust. The purpose of such arrangements is to improve performance over a broad oper-

Propjet

Turbofan and bypass engines

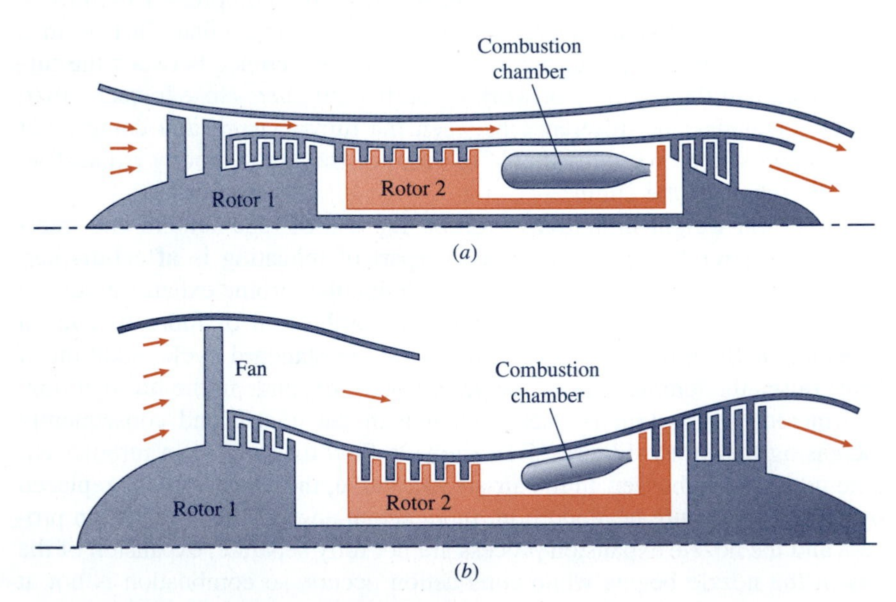

Figure 15·13 Turbofan or bypass jet engines.

ating range. The ratio of the mass flow rates of the two streams is called the **bypass ratio,**

Bypass ratio

$$\text{Bypass ratio} \equiv \frac{\dot{m}_{\text{total}} - \dot{m}_{\text{turbine}}}{\dot{m}_{\text{turbine}}}$$

The bypass ratio can be varied in flight by various means. It is apparent that the bypass ratios are much different for the engines shown in Fig. 15·13.

Many other arrangements involving various combinations of propellers, fans, and various numbers of compressors, turbines, and shafts are also used.

This brief discussion of turbojet engines has not touched on the performance characteristics of the individual components—compressors, combustion chambers, and turbines—and the important problems of matching these characteristics over a wide operating range. Another major problem is the prediction of the system behavior under transient conditions. For example, a change in fuel flow changes the turbine inlet temperature, which changes the turbine power output and hence the shaft speed. A change in shaft speed changes the flow rate and the pressure ratio, which in turn cause further changes in the turbine inlet temperature. Such problems must be thoroughly analyzed in designing an engine. We mention them as a reminder that the design of jet engines involves much more than a thermodynamic analysis of steady-flow cycles.

15·2 Modeling Reciprocating Internal Combustion Engines

With hundreds of millions of internal combustion engines in operation around the world and many millions more produced each year, there are many incentives for their continual improvement in terms of efficiency, other operating characteristics, weight, size, cost, reliability, and environmental impact. The analysis, design, and experimentation underlying improvements requires modeling that at first appears daunting and nearly impossible because of the complexity of the systems and processes involved. Despite this complexity, we can often adequately simulate the operation of such an engine by comparatively simple models.

Spark-ignition

For **spark-ignition** engines, the kind used in most automobiles, a combustible mixture of air and fuel is drawn into a cylinder and compressed. A spark ignites the mixture, raising its temperature to a high value, and the gas expands, doing work on a piston that turns the engine driveshaft. At the end of this power stroke of the piston, the exhaust gases must be expelled to complete the cycle of operations. This is indeed a complex sequence of processes involving sometimes an open system, sometimes a closed system, mixing, combustion, and complex heat transfer and fluid flow. Typically, the entire cycle is completed in a time interval of the order of a fiftieth of a second, so chemical reaction rates are important. Conditions within the cylinder are far from uniform at any instant.

Compression-ignition

Compression-ignition engine processes are just as complex and differ from spark-ignition engine processes in that only air, not an air–fuel mixture, is drawn into a cylinder and compressed. The compression raises the air temperature above the ignition temperature of the fuel. Near the end of the compression stroke, fuel is injected and quickly ignites. The gas expands as combustion proceeds, and it continues to expand until the exhaust process begins.

The simplest models for both spark-ignition and compression-ignition engines are air-standard models. In an air-standard model for internal combustion engines,

- The system is closed, so the same fluid is used repeatedly.
- Air is the working fluid throughout the cycle, and it is modeled as an ideal gas.
- Compression and expansion processes are reversible and adiabatic.
- The combustion process is replaced by a reversible heat transfer process.
- Since the system is closed, the exhaust process of the actual engine is replaced by a reversible constant-volume heat rejection to return the fluid to its initial state before compression.

Recall that reciprocating machines can even be modeled for some purposes as steady-flow devices (see Section 1·8).

For a reciprocating engine, an air-standard model is vastly different from an actual engine. Still, for some purposes an air-standard model provides adequate quantitative results, although they are not as accurate as air-standard model results for gas turbines where the model processes more closely simulate the actual processes.

15·2·1 The Air-Standard Otto Cycle

The air-standard Otto cycle is composed of four reversible processes of air in a closed system: adiabatic compression, constant-volume heat addition, adiabatic expansion, and constant-volume heat rejection. Figure 15·14 shows *pv* and *Ts* diagrams of an Otto cycle. The **compression ratio** r of the cycle is defined by

Compression ratio

$$r \equiv \frac{V_1}{V_2} = \frac{v_1}{v_2}$$

Compression ratio is thus a volume ratio that is fixed by the geometry of the engine.

Nikolaus A. Otto (1832–1891) and his partner Eugen Langen built a gas engine in 1867 in Deutz, Germany, and began commercial manufacture of it. In 1876 Otto produced a successful four-stroke cycle engine that was far superior to any internal-combustion engine previously built. The four-stroke cycle had been worked out in principle in 1862 by Alphonse Beau de Rochas in Paris.

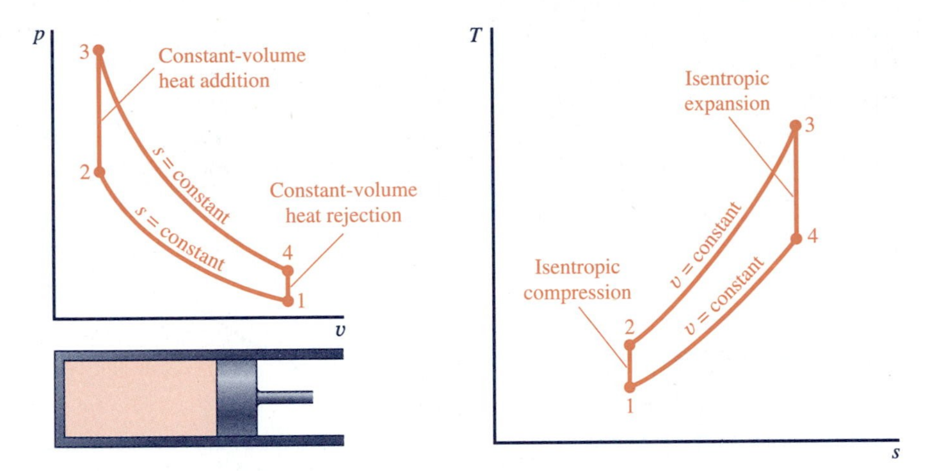

Figure 15·14 Air-standard Otto cycle.

A first-law analysis of this cycle is very simple because no work is done during the two heat-transfer processes (2–3 and 4–1) and both processes that involve work (1–2 and 3–4) are adiabatic; so for each process one term is zero in the first-law formulation

$$Q = \Delta U + W \tag{3·10}$$

A cold-air-standard analysis is of course even simpler and yields results such as $T_2/T_1 = T_3/T_4$ and $\eta = 1 - r^{1-k}$. The latter result shows that the cold-air-standard Otto cycle thermal efficiency is a function of compression ratio only. The thermal efficiency of an air-standard (i.e., variable specific heats) Otto cycle depends on the temperature limits as well as on the compression ratio.

The qualitative effects on efficiency and on work per cycle of varying the compression ratio, temperature limits, and pressure limits are similar for an air-standard Otto cycle and for an actual spark-ignition engine, and this is why the air-standard analysis is of value. The differences between the air-standard and the actual cycles are so great, however, that close quantitative agreement is not to be expected.

15·2·2 The Air-Standard Diesel Cycle

Figure 15·15 shows *pv* and *Ts* diagrams of an air-standard Diesel cycle. All four processes are reversible: 1–2, adiabatic compression; 2–3, constant-pressure heat addition; 3–4, adiabatic expansion to the initial volume; and 4–1, constant-volume heating rejection.

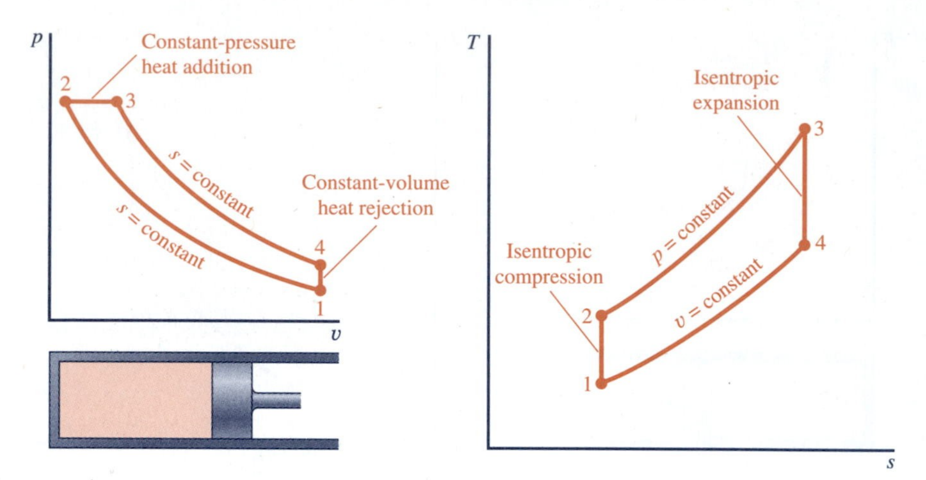

Figure 15·15 Air-standard Diesel cycle.

Rudolph Diesel (1858–1913) was born in Paris of German parents and educated in Munich. In 1893 he published a book, *The Theory and Construction of a Rational Heat Motor*, and obtained a patent on a compression-ignition engine. By 1899 he had developed this new type of engine to the point where he was able to begin commercial production in his factory at Augsburg.

The **cutoff ratio** r_c is defined by

Cutoff ratio

$$r_c \equiv \frac{V_3}{V_2} = \frac{v_3}{v_2}$$

and the **percent cutoff** is defined by

Percent cutoff

$$\text{Percent cutoff} \equiv \frac{V_3 - V_2}{V_1 - V_2} = \frac{v_3 - v_2}{v_1 - v_2}$$

that is, it is the fraction of the stroke during which heat is added.

The thermal efficiency of a cold-air-standard Diesel cycle is a function of the compression ratio and cutoff ratio only, but other factors are pertinent with variable specific heats.

Actual Diesel engines are *compression-ignition* (as distinguished from *spark-ignition*) engines. Only air is compressed in the cylinder, and, since there is no fuel in the cylinder, the compression ratio can be high enough to raise the air temperature above the fuel ignition temperature without causing ignition. Then, as the piston moves away from the cylinder head, fuel is injected and burned. The pressure variation during the first part of the expansion stroke depends on the rate at which fuel is injected and burned. In low-speed engines at least part of the fuel may be burned at approximately constant pressure as it is injected. In high-speed engines, ignition lag delays the start of combustion until most or all of the fuel is in the cylinder, so the ideal picture of the pressure being held constant by a regulated injection of fuel as the gas expands is unrealistic. Once again the differences between an actual cycle and a corresponding ideal cycle must be remembered so that conclusions based on one will not be blindly applied to the other.

Performance comparisons among various gas power cycles are readily made using *Ts* diagrams to aid in reasoning that is based on the first and second laws and ideal-gas properties. As one example, let us determine which has the higher thermal efficiency, an air-standard Otto cycle or an air-standard Diesel cycle with the same compression ratio and the same heat input. Superimposed *Ts* diagrams of the two cycles are shown in Fig. 15·16. Since the two cycles have the same compression ratio, process 1–2 is common to them. After one of the cycle *Ts* diagrams is sketched, the relative location of points 3*O* (Otto) and 3*D* are fixed, because equal heat input means equal area under the constant-volume line 2–3*O* and the constant-pressure line 2–3*D*. Thus points 3*O* and 4*O* are at a lower entropy than points 3*D* and 4*D*. Since both cycles reject heat in a constant-volume process ending at state 1, the relative location of points 4*O* and 4*D* on the *Ts* diagram shows that less heat is rejected by the Otto cycle. The heat inputs are the same; so the work of the Otto cycle exceeds the work of the Diesel cycle. Consequently, the Otto cycle has a higher thermal efficiency. *Caution*: This conclusion applies to cycles that have the same compression ratio and the same heat input. Comparisons can be made in a similar manner for various other conditions.

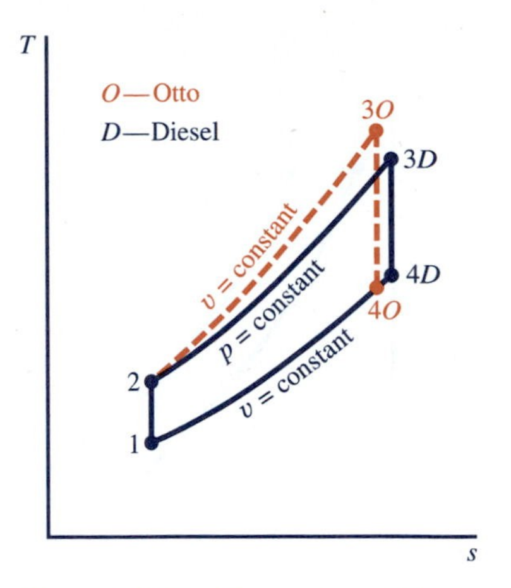

Figure 15·16 Comparison of Otto and Diesel cycles.

Example 15·3 Air-standard Diesel cycle

PROBLEM STATEMENT

The mass of air in an air-standard Diesel cycle is 0.110 lbm. Compression starts at 14.5 psia, 120 F, and ends at 550 psia. The heat added per cycle is 31.0 B/lbm. Determine (*a*) the compression ratio, (*b*) the work, and (*c*) the amount of heat rejected that is available energy.

SOLUTION

Analysis: We first sketch *pv* and *Ts* diagrams of the air-standard Diesel cycle. The temperature range is likely to be so high that assuming constant specific heats would be unsound. We could use the computer model for air properties or the air tables (Table A·12) for the problem. We will show the solution with values as taken from the table (although the computer model provides identical values).

As we sketch the *pv* and *Ts* diagrams, we can see that the specified values completely determine all four states and four processes of the cycle. Since part *c* requires determination of the heat rejected (process 4–1) and part *b* can also be solved by using the heat rejected, it is well to determine state 4 at which the heat rejection process starts. Since this involves sequential calculations "around the cycle," we make a table in which we write specified values immediately and calculated values as soon as they are obtained:

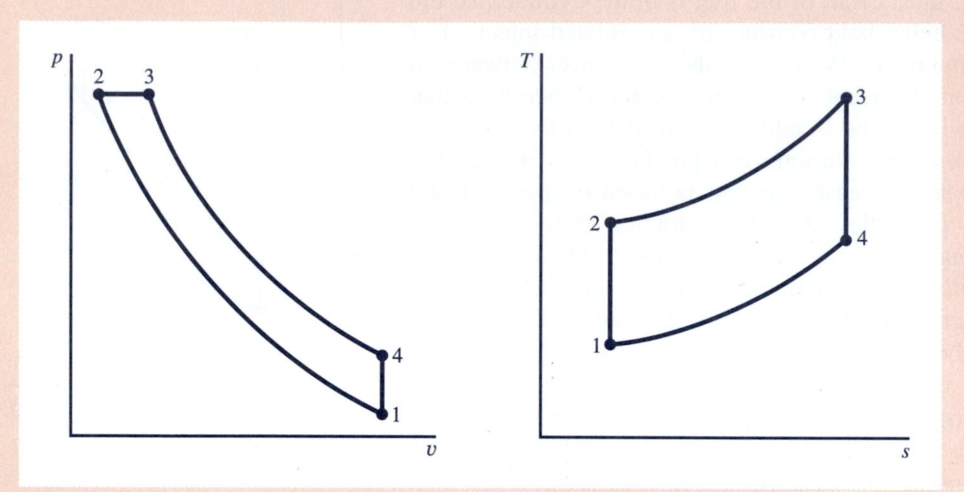

Calculations:

State	p, psia	T, F	p_r	v_r	u, B/lbm	h, B/lbm	ϕ, B/lbm·R
1	14.5	120	1.808	178.1	99.7		
2	550	1115	68.6	12.76		389.9	1.868
3	550	2128		2.811		671.7	2.006
4		801		23.88	221.2		

The calculated values are obtained as follows:

$$p_{r2} = p_{r1}\frac{p_2}{p_1} = 1.808\left(\frac{550}{14.5}\right) = 68.6$$

From p_{r2}, values of v_r, T, u, and h at state 2 can be determined from the air tables.

(*a*)

$$\text{Compression ratio} \equiv \frac{v_1}{v_2} = \frac{v_{r1}}{v_{r2}} = \frac{178.1}{12.76} = 13.96$$

(*b*) Applying the first law to process 2–3 and noting that for a reversible constant-pressure process of a closed system $q = \Delta u + w = \Delta u + \int p\,dv = \Delta u + \Delta pv = \Delta h$, we find

$$h_3 = h_2 + \frac{Q}{m} = 389.9 + \frac{31.0}{0.110} = 389.9 + 281.8 = 671.7 \text{ B/lbm}$$

Knowing h_3, we take from the air tables values of T and v at state 3:

$$v_{r4} = v_{r3}\frac{v_4}{v_3} = v_{r3}\frac{v_1}{v_3} = v_{r3}\frac{T_1 p_3}{T_3 p_1} = 2.811\left(\frac{580}{2588}\right)\frac{550}{14.5} = 23.88$$

Knowing v_{r4}, we take from the air tables the value of u for state 4:

$$W = \oint \delta Q = Q_{2-3} - Q_{\text{out},4-1} = Q_{2-3} - m(u_4 - u_1) = 31.0 - 0.11(221.2 - 99.7)$$
$$= 31.0 - 13.4 = 17.6 \text{ B}$$

(c)
$$Q_{av,4-1} = Q_{4-1} - Q_{unav,4-1} = Q_{4-1} - T_0 m(s_1 - s_4) = Q_{4-1} - T_0 m(s_2 - s_3)$$
$$= Q_{4-1} - T_0 m\left(\phi_2 - \phi_3 - R \ln \frac{p_2}{p_3}\right)$$
$$= -13.4 - 580(0.11)(1.868 - 2.006 - 0) = -4.6 \text{ B}$$

Comments: In the absence of any other information, we have used the lowest temperature in the cycle for T_0. The negative sign on $Q_{av,4-1}$ indicates that 4.6 B of available energy is *removed* from the system. This means that if the 13.4 B of heat removed from the Diesel cycle were transferred into a series of Carnot engines rejecting heat at 120 F, 4.6 B out of the 13.4 B could be converted into work by the Carnot engines. A shortcoming of a Diesel cycle in comparison with a Carnot cycle is that some of the heat rejected by the Diesel cycle *could* be converted into work, while the Carnot cycle rejects only heat that *cannot* be converted into work.

15·2·3 Other Gas Power Cycles

The Dual Cycle. Neither the air-standard Otto nor the air-standard Diesel cycle closely approximates the pressure–volume variation of the working substance in an actual engine. Some analyses require an air-standard model in which the pressure–volume variation does simulate that of an actual engine. One such model, called the **dual cycle**, involves two heat-addition processes: one at constant volume and one at constant pressure, as shown in Fig. 15·17. The relative amounts of heat added in the two processes can be adjusted to make the air-standard cycle more like an actual one.

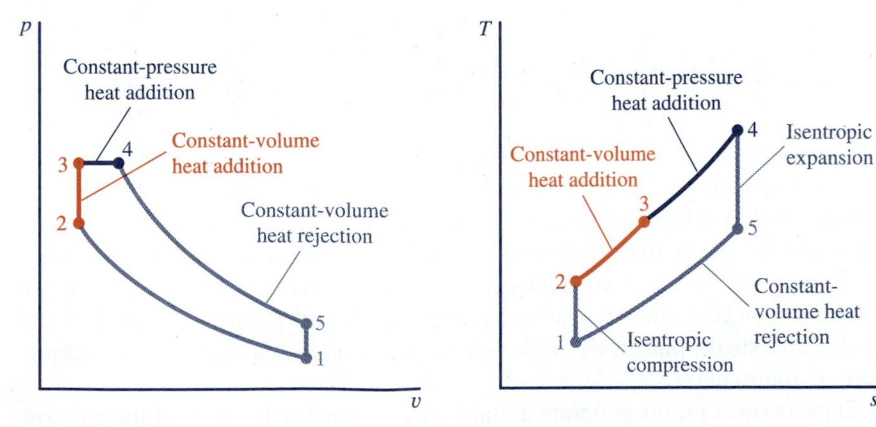

Figure 15·17 Air-standard dual cycle.

A further refinement is to add heat first at constant volume, then at constant pressure, and then at constant temperature in an air-standard cycle. Such cycles can be analyzed and evaluated by the same fundamental methods used for the simpler ones. Remember, however, that the most highly refined air-standard cycle is still much different from what occurs in an actual internal-combustion engine. An ideal system can be more confusing than helpful if the differences between it and the corresponding actual system are not kept in mind.

Compound Gas Power Cycles. As mentioned in Section 15·2, one advantage of reciprocating engines is that they can sustain much higher maximum gas temperatures than turbines. Since adiabatic expansions are used in most gas power cycles, the higher temperatures are accompanied by higher pressures. However, reciprocating engines are not as good as turbines for expanding gas to low pressures. Expanding the gas to a low pressure results in large volumes, because the engine size increases more rapidly than the power output and in a reciprocating engine this is a disadvantage. Increasing the mass flow rate, and hence the power output, through a reciprocating engine—either by increasing engine size or by increasing engine speed—exacerbates the design difficulties that stem from large reciprocating parts. In contrast, turbines are readily adaptable to handling large gas volumes. To capitalize on the advantages and minimize the drawbacks of each type of engine, *compound power plants* or *compound engines* are used in which the high-pressure and high-temperature processes occur in a reciprocating engine and the low-pressure expansion occurs in a turbine.

Three different arrangements of compound engines are shown in Fig. 15·18. In addition to the thermodynamic advantage of a compound engine, there are some mechanical advantages of a turbine, instead of a reciprocating engine, driving the plant load. In some analyses of compound power plants, the flow through reciprocating engines can often be treated as steady, using mean values of properties.

15·3 Modeling Vapor Power Plants

We turn now to power cycles where the working fluid is alternately vaporized and condensed. We recognize that if the fluid is expanded as a vapor and compressed as a liquid with a much smaller specific volume, the backwork ratio will be lower than that of a gas cycle. This is why vapor cycles were used long before the first successful gas cycles were developed. The most common working fluid is water, and the cycles and plants are called steam cycles and steam plants, even though the water is in the liquid phase during part of the cycle.

Steam power plants generate a major and increasing fraction of the electric power produced in the world. Also, steam power generation is often

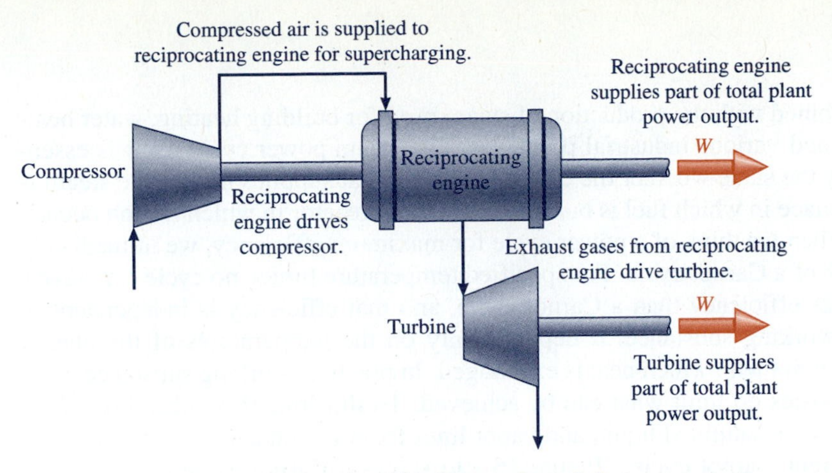

(*a*) Reciprocating engine and turbine supply total plant power output.

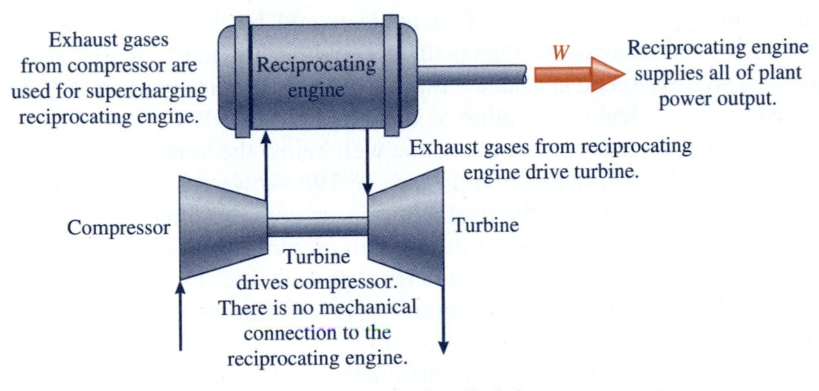

(*b*) Reciprocating engine supplies all of plant power output.

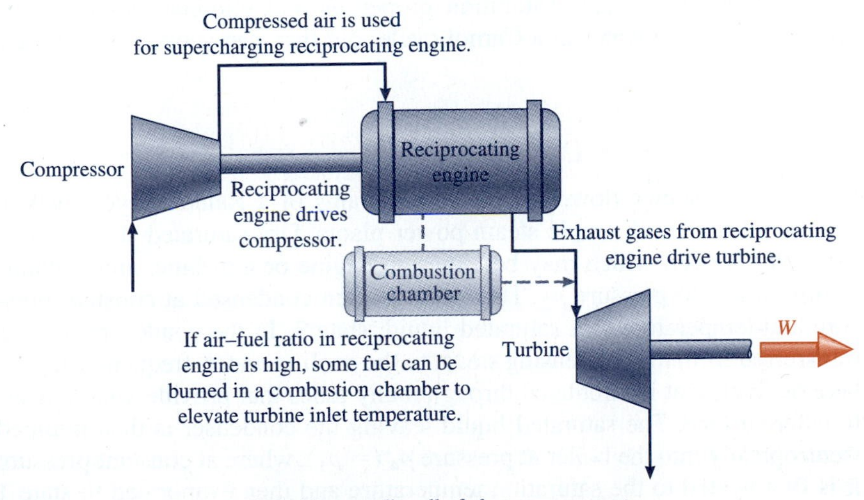

(*c*) Turbine supplies all of plant power output.

Figure 15·18 Three variations of compound gas power plants.

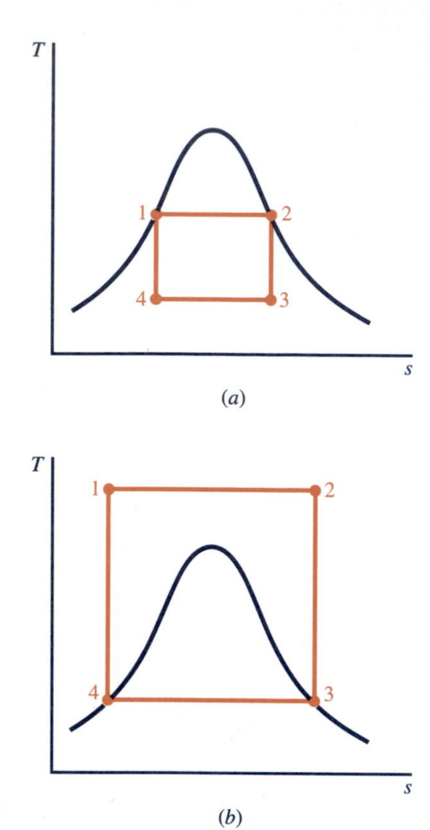

Figure 15·19 Carnot cycles using water as the working substance.

combined with the production of steam used for building heating, water heating, and various industrial processes. The steam power cycle itself is essentially the same whether the energy reservoir that supplies heat to the steam is a furnace in which fuel is burned or a nuclear reactor in which fission occurs.

When we think of a power cycle for maximum efficiency, we immediately think of a Carnot cycle. For specified temperature limits, no cycle can have a higher efficiency than a Carnot cycle, and that efficiency is independent of the working substance. It depends only on the temperatures of the energy reservoirs with which heat is exchanged. In practice, working substance characteristics do limit what can be achieved. To illustrate this point, Fig. 15·19 shows the saturated liquid and vapor lines for water on a *Ts* diagram and two different Carnot cycles. Figure 15·19*a* shows a Carnot cycle operating entirely on wet steam—a liquid–vapor mixture. Notice that the temperature of the steam during processes 1–2 and 3–4 can be maintained constant by holding the pressure constant. This cycle would be difficult to operate in practice for several reasons: One is that it would be difficult to cause the heat rejection process to end at a state 3 that is at the same entropy as state 1. Also, notice that the critical temperature of steam (374°C or 705 F) limits the maximum temperature in the cycle to a value well below the temperature at which structural materials lose strength. Figure 15·19*b* shows a Carnot cycle going to higher temperatures for the heat addition process, but note that enormous pressures are needed to achieve the maximum temperature by compressing the liquid from state 4, and the isothermal process 1–2 involves a continual pressure variation that poses a control problem not encountered with the other Carnot cycle shown in Fig. 15·19*a*.

In short, the Carnot cycle is an unsuitable model for the design of steam power plants because of difficulties in carrying out the required processes while utilizing the maximum possible temperature for heat addition. Other working substances have saturation properties that eliminate some of the problems of using water in a Carnot cycle, but they cause other difficulties.

15·3·1 The Rankine Cycle

Figure 15·20 shows flow, *pv*, and *Ts* diagrams of a Rankine cycle, which is a useful model of simple steam power plants. Dry saturated steam enters the prime mover, which may be either an engine or a turbine, and expands isentropically to pressure p_2. The steam is then condensed at constant pressure and temperature to a saturated liquid, state 3. In the condenser, heat is transferred from the condensing steam to the cooling water, frequently from a lake or river, that is circulated through many tubes that provide a large heat-transfer surface. The saturated liquid leaving the condenser is then pumped isentropically into the boiler at pressure p_4 $(= p_1)$, where at constant pressure it is first heated to the saturation temperature and then evaporated to state 1 to complete the cycle. The temperature rise that results from isentropic com-

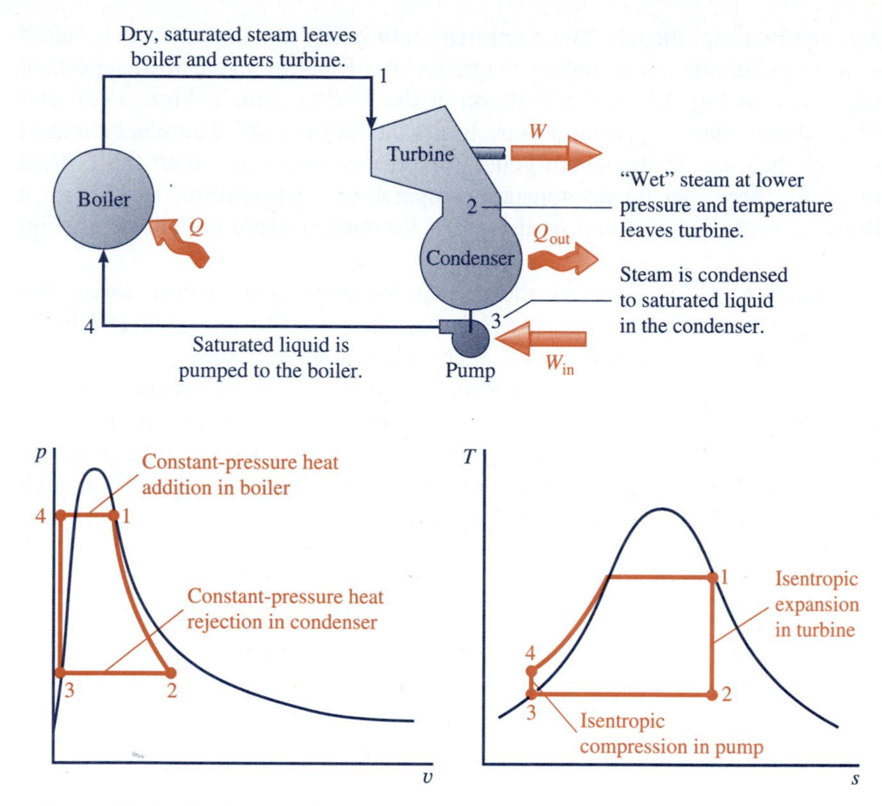

Figure 15·20 Rankine cycle.

pression of liquid water is very small, so the length of line 3–4 in the *Ts* diagram of Fig. 15·20 is greatly exaggerated.

Property diagrams for steam cycles are usually distorted in the liquid region.

To use higher temperatures without increasing the maximum pressure of the cycle, the steam after leaving the boiler is heated further at constant

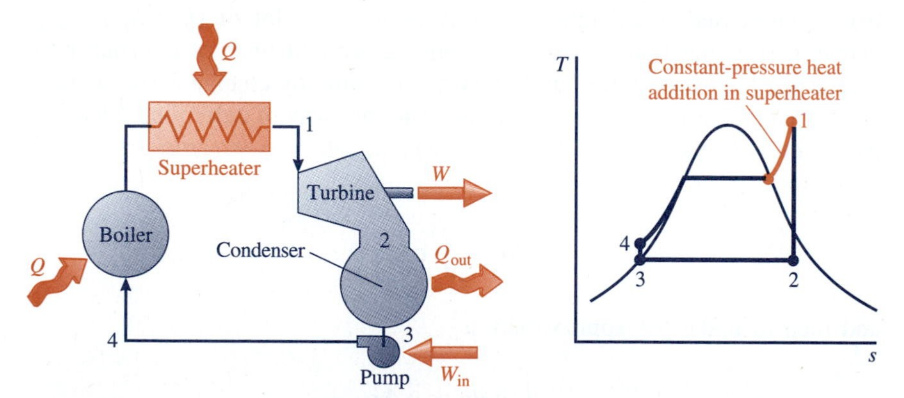

Figure 15·21 Rankine cycle with superheat.

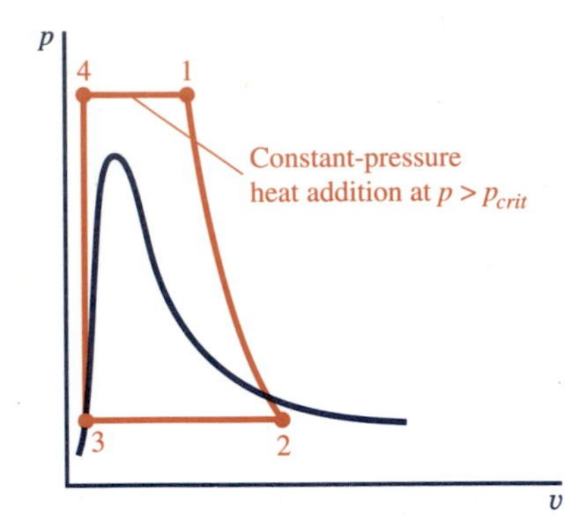

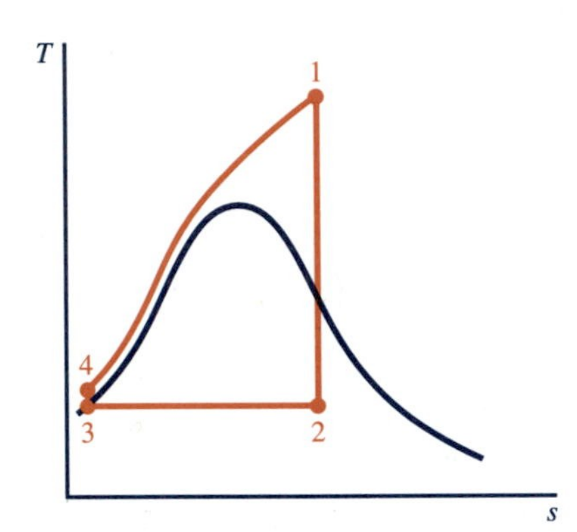

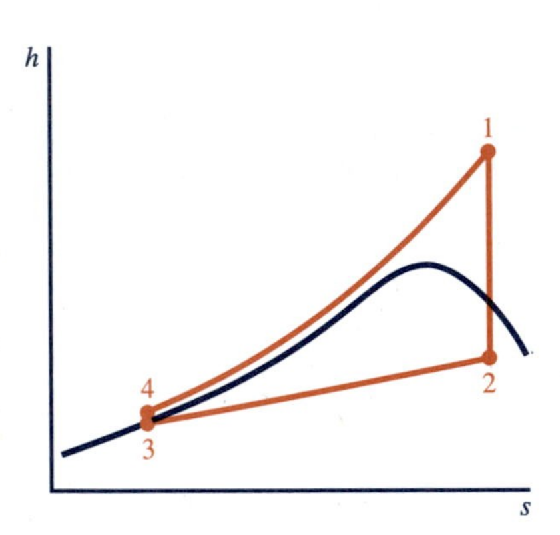

Figure 15·22 Supercritical-pressure Rankine cycle.

pressure in a superheater. The combination of boiler and superheater is called a steam generator. Flow and *Ts* diagrams of a Rankine cycle with superheat are shown in Fig. 15·21. Comparison of the *Ts* diagrams of Figs. 15·20 and 15·21 shows that for given pressure limits the thermal efficiency is increased by superheating. If the steam-generator pressure is higher than the critical pressure, there can be no constant-temperature heat-addition process in a Rankine cycle. *pv*, *Ts*, and *hs* diagrams for such a cycle are shown in Fig. 15·22.

Inspection of *Ts* diagrams shows that *for given temperature limits*, the Rankine cycle efficiency is always lower than the Carnot cycle efficiency because not all the heat is added at the highest temperature.

Performance calculations for a Rankine cycle are simple because no work is done in the two heat-transfer processes and the two processes involving work are adiabatic. Also, the changes in kinetic and potential energy across each cycle component are usually negligible. The first law applied to each piece of steady-flow equipment thus gives

Steam generator:	$q_{4-1} = h_1 - h_4$	$(p_1 = p_4)$
Turbine:	$w_T = h_1 - h_2$	$(s_2 = s_1)$
Condenser:	$q_{\text{out},2-3} = h_3 - h_2$	$(p_3 = p_2;\ h_3 = h_f)$
Pump:	$w_{\text{in},P} = h_4 - h_3$	$(s_4 = s_3)$

The cycle thermal efficiency is

$$\eta = \frac{w}{q_{4-1}} = \frac{w_T - w_{\text{in},P}}{q_{4-1}} = \frac{(h_1 - h_2) - (h_4 - h_3)}{(h_1 - h_4)}$$

At state 4 of the ideal Rankine cycle, the values of p_4 and s_4 $(= s_3)$ are usually known, but most tables, charts, and computer programs are unsuitable for determining h of a compressed liquid from values of p and s. Inspection of the compressed liquid region of a *ph* diagram shows why this is the case. (Anyone who frequently needs to make such a calculation can make, from compressed liquid tables, a convenient crossplot of $(h - h_f)_{s=\text{constant}}$ versus s, showing lines of constant temperature and lines of constant pressure. Then the ideal pump input work or the enthalpy change across a pump can be read directly from the crossplot.) Another way to determine Δh across the pump is to use $T\,ds = dh - v\,dp$, which, *for an isentropic process*, reduces to

$$\Delta h = \int v\,dp$$

and then to make the approximation

$$\int v\,dp \approx v\,\Delta p$$

This approximation is quite accurate because the specific volume of liquid water is nearly independent of pressure. (The same expression for reversible pump work can be obtained from Eq. 1·20, $w_{in} = \int v\,dp + \Delta ke + \Delta pe$.)

An advantage of the Rankine cycle over all other power cycles introduced so far in this chapter is its low backwork ratio, which is given by

Vapor power cycles have low backwork ratios.

$$\text{Backwork ratio} = \frac{w_{in,P}}{w_T}$$

where w_T and $w_{in,P}$ are respectively the turbine work and the pump work input. The analysis of a steam power cycle can easily cover deviations from the Rankine cycle caused by pressure drops in piping, stray heat loss, and turbine and pump efficiencies of less than unity.

Example 15·4 Rankine Cycle

PROBLEM STATEMENT

A Rankine cycle has steam entering the turbine at 3000 kPa, 400°C, and a condenser pressure of 5 kPa. Cooling water is available at 20°C. Determine (*a*) the cycle thermal efficiency, (*b*) the backwork ratio, and (*c*) the fraction of the heat added that is available energy.

SOLUTION

Analysis: A *Ts* diagram and a flow diagram are sketched first, and we place the specified data on the flow diagram. Calculation of thermal efficiency requires determination of the net work of the cycle (which is the turbine work less the pump work input) and the heat input of process 4–1. Each of these quantities can be obtained from the first law if we assume that kinetic energy changes across each piece of equipment are negligible.

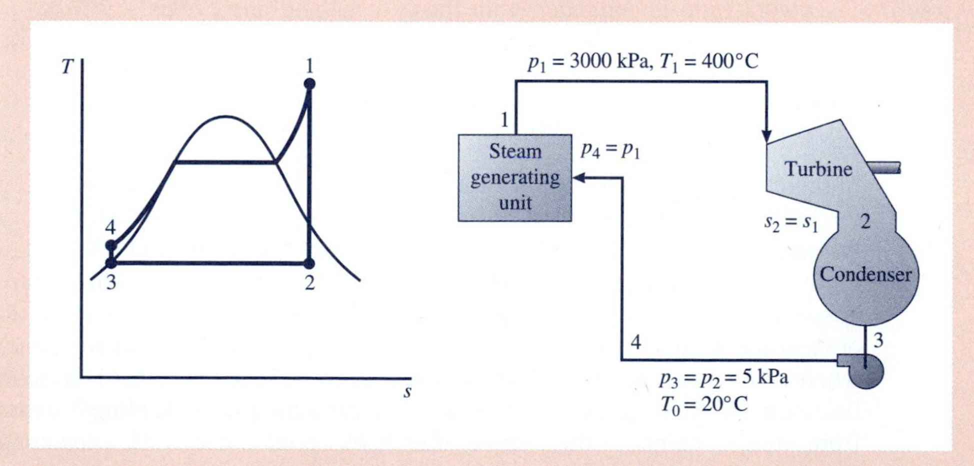

Calculations:

$$\Delta h_P = h_4 - h_3 = \int_3^4 v\,dp \approx v(p_4 - p_3) = 0.001005(3000 - 5) = 3.0 \text{ kJ/kg}$$

$$h_4 = h_3 + \Delta h_P = 137.8 + 3.0 = 140.8 \text{ kJ/kg}$$

h_2 can be determined from $p_2 = 5$ kPa and $s_2 = s_1 = 6.921$ kJ/kg·K as follows:

$$x_2 = \frac{s_2 - s_f}{s_{fg}} = \frac{6.921 - 0.4763}{7.9129} = 0.814$$

$$h_2 = h_f + x_2 h_{fg} = 137.8 + 0.814(2422.5) = 2110.1 \text{ kJ/kg}$$

or this value can be determined from an *hs* (Mollier) chart or *hv* chart for steam. Having obtained all the h and s values needed, we can now complete the solution.

(*a*)

$$\eta = \frac{w}{q_{4-1}} = \frac{w_T - w_{\text{in},P}}{q_{4-1}} = \frac{h_1 - h_2 - \Delta h_P}{h_1 - h_4}$$

$$= \frac{3230.7 - 2110.1 - 3.0}{3230.7 - 140.8} = \frac{1121 - 3}{3090} = 0.362$$

(*b*)

$$\text{Backwork ratio} = \frac{w_{\text{in},P}}{w_T} = \frac{\Delta h_P}{h_1 - h_2} = \frac{3}{1121} = 0.0027$$

(*c*)

$$q_{av} = q - q_{unav} = q - T_0(s_1 - s_4) = 3090 - 293(6.921 - 0.4763)$$
$$= 1200 \text{ kJ/kg}$$

$$\frac{q_{av}}{q} = \frac{1200}{3090} = 0.388$$

Comment: Notice how low the backwork ratio is compared with those of gas-turbine cycles.

15·3·2 Regeneration in Steam Power Plants

In searching for ways to improve the Rankine cycle, we notice that following the boiler feed pump heat is transferred to water at a low temperature from furnace gases (or from a nuclear reactor), although the second law shows that for maximum efficiency heat should be added at the highest temperature possible. One way to reduce the irreversibility of this heat transfer is to preheat the water flowing toward the steam generator with heat from somewhere else within the cycle. This is similar to the regeneration that is used in Ericsson and Sterling cycles. It is done by extracting or "bleeding" steam from various points in the turbine after it has done some work expanding from the turbine inlet pressure to the pressure at the bleed point and using this bled steam to heat water flowing from the pump to the steam generator. This energy transfer occurs in *feedwater heaters.*

Flow and *Ts* diagrams of a steam power cycle using one *open feedwater heater* (also called a *direct-contact heater*) are shown in Fig. 15·23. In an open feedwater heater, the bled steam and the condensate pumped from the condenser are mixed, and, under optimum operating conditions, saturated liquid leaves the heater. The heater operates at the bleed-point pressure (or slightly below it in an actual plant because of frictional pressure drop in the bleed line), so another pump is needed to force the water into the steam generator.

For energy and mass balances on regenerative cycles it is convenient to define the mass ratio m' for any point a in the cycle as

$$m'_a \equiv \frac{\dot{m}_a}{\dot{m}_1}$$

where $\dot{m}_a$ is the mass rate of flow at any point a, and $\dot{m}_1$ is the mass rate of flow leaving the steam generator. By definition, $m'_1 = 1$. Thus an energy balance on the open feedwater heater of Fig. 15·23 for which $w = 0$, $q = 0$, $\Delta ke = 0$, and $\Delta pe = 0$ is

$$\dot{m}_2 h_2 + \dot{m}_5 h_5 = \dot{m}_6 h_6$$
$$\dot{m}_2 h_2 = (\dot{m}_1 - \dot{m}_2) h_5 = \dot{m}_1 h_6$$

and this can also be written as

$$m'_2 h_2 + (1 - m'_2) h_5 = h_6$$

If the states of the fluid at points 2, 5, and 6 are known, this equation can be solved for m'_2, the fraction of the throttle flow that is bled to the open feedwater heater. Notice that on a *Ts* diagram for a regenerative cycle, as shown in Fig. 15·23, the state of the fluid at each point in the cycle is correctly shown but the mass of fluid is not the same at all points. Therefore, the area beneath the path of a reversible process represents heat transfer *per unit mass*

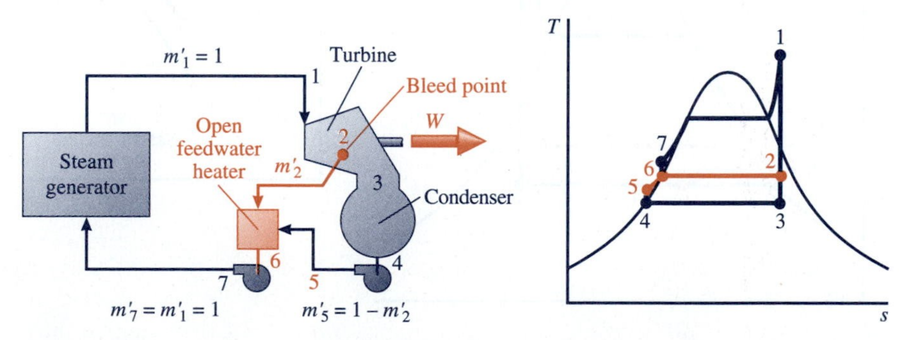

Figure 15·23 A regenerative cycle with one open feedwater heater.

of fluid undergoing that process, and care must be taken to account for variations in mass when comparing areas.

Water in a steam power plant must be deaerated to limit corrosion.

In addition to raising the temperature of feedwater to reduce the amount of heat that is added to the cycle at low temperature, an open feedwater heater in an actual cycle also serves as a *deaerator* to remove air and other noncondensable gases that would cause corrosion. This deaeration occurs because the solubility of gases in water decreases with increasing temperature as noted in Chapter 10. If the water is brought to its saturation temperature and provision is made for letting the noncondensable gases escape, they can be removed from the system.

Flow and *Ts* diagrams of a steam power cycle using one *closed feedwater heater* are shown in Fig. 15·24. The feedwater (i.e., the condensate from the condenser) is pumped through many tubes in the feedwater heater, and the bled steam condenses (after first being desuperheated, if necessary) on the outside of the tubes. The two fluid streams do not mix in the closed feedwater

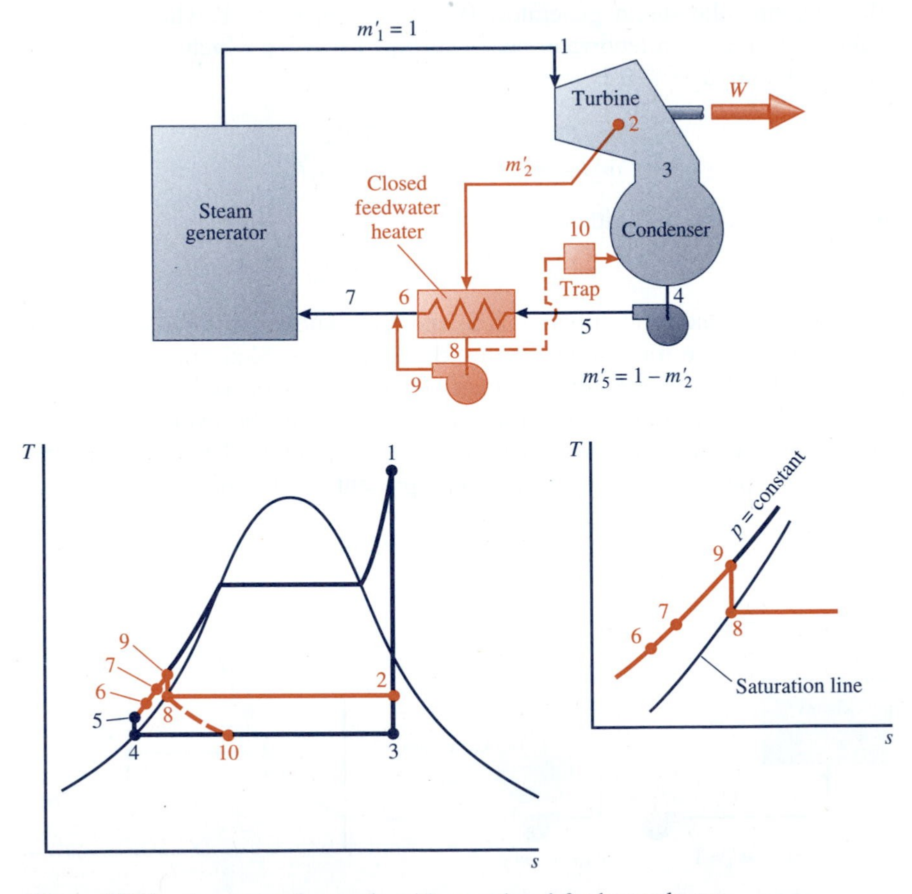

Figure 15·24 A regenerative cycle with one closed feedwater heater.

heater. In the ideal case, the condensate leaving at state 8 is saturated liquid, and the feedwater leaving at state 6 is at a temperature only infinitesimally lower than T_8. In practice, to limit the amount of heat-transfer surface area needed, the difference $(T_8 - T_6)$, called the *terminal temperature difference*, is usually of the order of 5 Celsius degrees.

Feedwater heater drain configuration

The solid lines in Fig. 15·24 show the heater drain being pumped into the feedwater line between the heater and the steam generator. The drain after being pumped to state 9 is mixed with the feedwater at state 6 to form feedwater at state 7. A magnified portion of the *Ts* diagram in Fig. 15·24 shows these processes. Another method of handling the heater drain is shown by the broken lines in Fig. 15·24. It is throttled through a trap or valve into the condenser or into a lower-pressure feedwater heater. (A trap operates so that liquid that enters is throttled to the lower pressure on the discharge side, but vapor that enters the trap is not allowed to pass. The same function is performed by a valve in the drain line that is controlled by a float to maintain a constant liquid level in the lower part of a heater.) Application of the first law to the trap results in $h_8 = h_{10}$.

A mass and energy balance on the closed feedwater heaters gives

$$\dot{m}_2 h_2 + \dot{m}_5 h_5 = \dot{m}_2 h_8 + \dot{m}_5 h_6$$
$$\dot{m}_2 h_2 + (\dot{m}_1 - \dot{m}_2) h_5 = \dot{m}_2 h_8 + (\dot{m}_1 - \dot{m}_2) h_6$$
$$m_2' h_2 + (1 - m_2') h_5 = m_2' h_8 + (1 - m_2') h_6$$

A significant advantage of closed heaters is that they do not require a separate pump to handle the feedwater flow for each heater, but a drawback of closed heaters is that they do not bring the feedwater very close to the heater saturation temperature as open heaters do.

Example 15·5 Regenerative Steam Power Cycle

PROBLEM STATEMENT

It is desired to add an open feedwater heater to the Rankine cycle described in Example 15·4 (turbine inlet or throttle conditions of 3000 kPa, 400°C, and condenser pressure of 5 kPa). Determine to within 100 kPa the optimum pressure (in terms of cycle thermal efficiency) for the heater.

SOLUTION

Analysis: We first sketch flow and *Ts* diagrams for the regenerative cycle. We have no way to express in analytical form the efficiency of the cycle in terms of heater pressure to find the maximum efficiency through setting the derivative of $d\eta/dp_2$ equal to zero. Therefore, we will calculate the efficiency for several values of p_2 to find the optimum value. For each value of p_2, the fraction of the throttle flow rate bled to the feedwater heater, m_2', will be different. Clearly, such a calculation is best made using a computer model that includes online property data as the model for steam properties does. The equations we will use are the following:

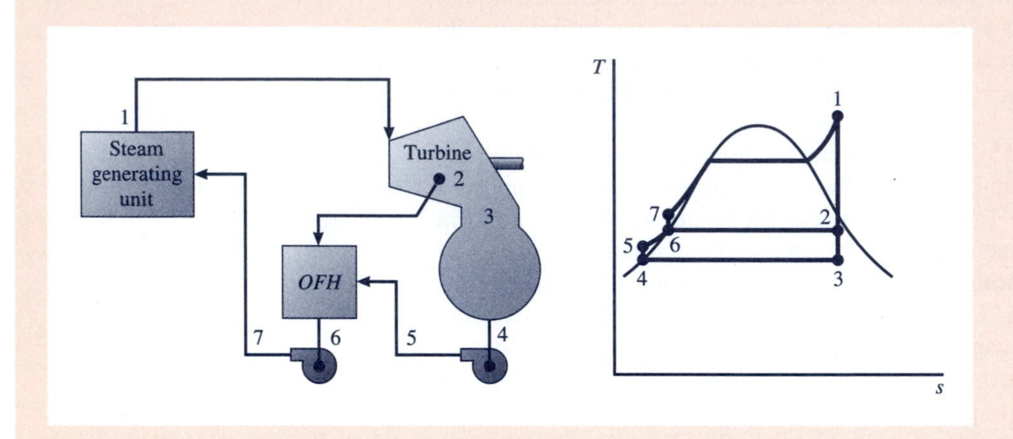

- The simplification of work $= -\int v\, dp$ for steady isentropic incompressible flow through the pumps:

$$w_{in,P,4-5} \approx v_4(p_5 - p_4)$$
$$w_{in,P,6-7} \approx v_6(p_7 - p_6)$$

- The first law for steady flow with negligible changes in kinetic and potential energy:

Pumps:	$h_5 = h_4 + w_{in,P,4-5}$
	$h_7 = h_6 + w_{in,P,6-7}$
Steam generator:	$q_{7-1} = h_1 - h_7$
Turbine:	$w_T = h_1 - h_2 + (1 - m_2')(h_2 - h_3)$
Feedwater heater:	$m_2'h_2 + (1 - m_2')h_5 = h_6$

- An expression for the net work of the plant:

$$w = w_T - (1 - m_2')w_{in,P,4-5} - w_{in,P,6-7}$$

- The definition of thermal efficiency:

$$\eta = \frac{w}{q_{7-1}}$$

- A property relationship from the second law for steady-flow reversible adiabatic processes:

$$s_1 = s_2 = s_3$$

Calculations: We enter the equations above into our equation solver along with the model for steam properties. The Rule sheet is shown here. (The equation $s_1 = s_2 = s_3$ and similar equations relating equal pressures are not entered, because the first state is solved to find s_1 and then s_2 and s_3 are entered as input values to define states 2 and 3. All the pressures are entered as input values. A resulting table and plot show that the optimum pressure is approximately 300 kPa.

```
================== RULE SHEET ====================================
S Rule-----------------------------------------------------------------
  ;------------------ Example 15•5 ---------------------------------
  winP45 = v4 * (p5 - p4)                ;Work input to condensate pump
  winP67 = v6 * (p7 - p6)                ;Work input to boiler feed pump
  h5 = h4 + winP45                       ;First law, to determine h5
  h7 = h6 + winP67                       ;First law to determine h7
  q71 = h1 - h7                          ;First law applied to steam generator
  w = wT - (1 - m2`)*winP45 - winP67     ;Net work of plant per kg at throttle
  wT = h1 - h2 + (1 - m2`) * (h2 - h3)   ;First law applied to turbine
  m2` * h2 + (1-m2`) * h5 = h6           ;First law applied to feedwater heater
  eta = w/q71                            ;Definition of thermal efficiency
  ;------------------ Steam properties --------------------------- STEAM.TK
```

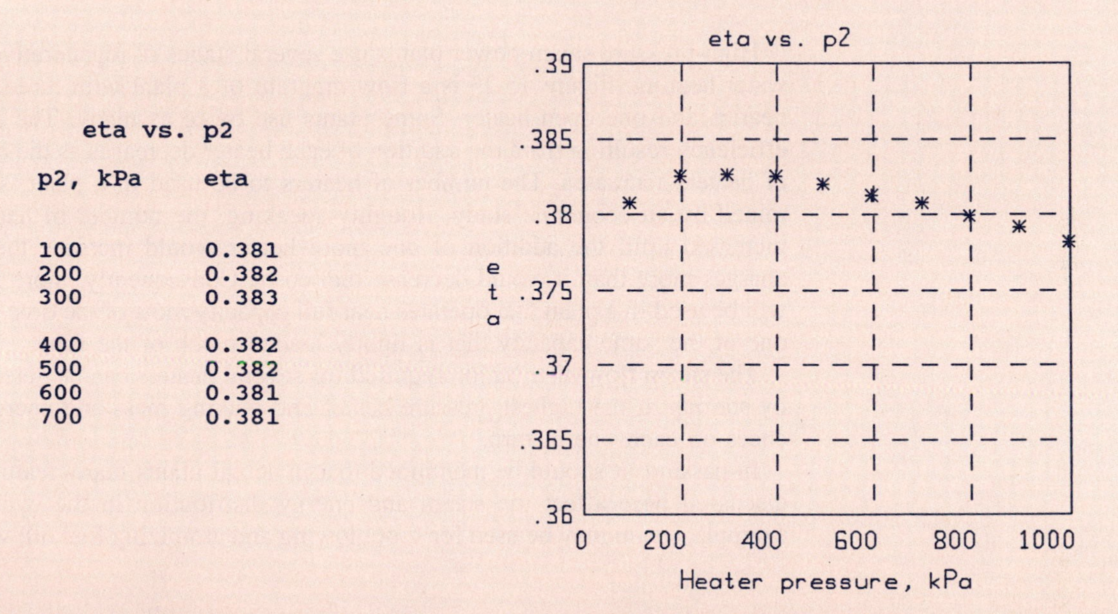

eta vs. p2

p2, kPa	eta
100	0.381
200	0.382
300	0.383
400	0.382
500	0.382
600	0.381
700	0.381

Although it is unnecessary for this solution, we present the following table of properties to help anyone checking this solution.

State	*p*, kPa	*T*, °C	*h*, kJ/kg	*s*, kJ/kg·K	*v*, m³/kg
1	3000	400	3231	6.921	
2	300	134	2696	6.921	
3	5	33	2110	6.921	
4	5	33	137.8		0.001005
5	300		138.1		
6	300	134	561.6		0.001115
7	3000		564.7		

Comments: At first it may appear that the number of variables is well above the number of equations we enter in the equation solver, but remember that the steam properties model contains several relationships among properties.

The results show that for the conditions specified here, the thermal efficiency is not highly sensitive to the heater pressure. This is not always the case.

Comparing the result of this problem with that of Example 15·4 shows that adding the feedwater heater at 300 kPa increases the cycle thermal efficiency from 0.362 to 0.383. This would reduce the amount of fuel needed by about 5.8 percent, so the saving in fuel cost must be weighed against the increased investment and operating expenses incurred by adding the heater. In most cases, the added expense is justified by the saving in fuel costs.

High-pressure steam power plants use several stages of regenerative feedwater heating. Figure 15·25 is a flow diagram of a plant with three closed heaters and one open heater. Some plants use twice as many. The gain in efficiency resulting from the addition of each heater decreases as the number of heaters increases. The number of heaters to be used in a plant is determined by an economic study. Roughly speaking, the number of heaters is increased until the addition of one more heater would increase the fixed charges more than it would decrease fuel costs. Consequently, more heaters will be used in a plant that operates near full capacity most of the time than in one of the same capacity that is lightly loaded much of the time.

The steam flow distribution required for several heaters can be determined by starting at the highest-pressure heater and making mass and energy balances on each one in turn.

In passing, it should be mentioned that in actual plants many features not discussed here affect the steam and energy distribution in the cycle. For example, steam may be used for soot blowing and atomizing fuel oil; water is

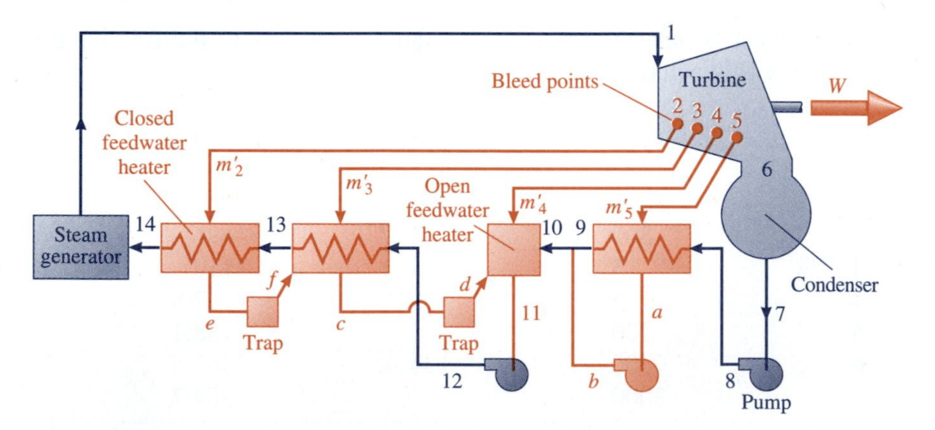

Figure 15·25 A cycle with four stages of feedwater heating.

lost through blowdown, and makeup water must be added; steam may be used to drive auxiliaries; and there are numerous components such as shaft seals, oil coolers, and fuel handling and water testing equipment that have small but significant effects on the plant energy balance. For analyzing these various effects there are no more powerful tools than the first and second laws of thermodynamics, which on account of their generality, can be studied and learned in connection with some systems and then applied to quite different ones.

Example 15·6 Regenerative Steam Cycle: Irreversibilities

PROBLEM STATEMENT

A regenerative steam power cycle is shown in the figure with property values for all states. Flow through the turbine is adiabatic and irreversible. For these calculations, consider the flow through the pumps to be

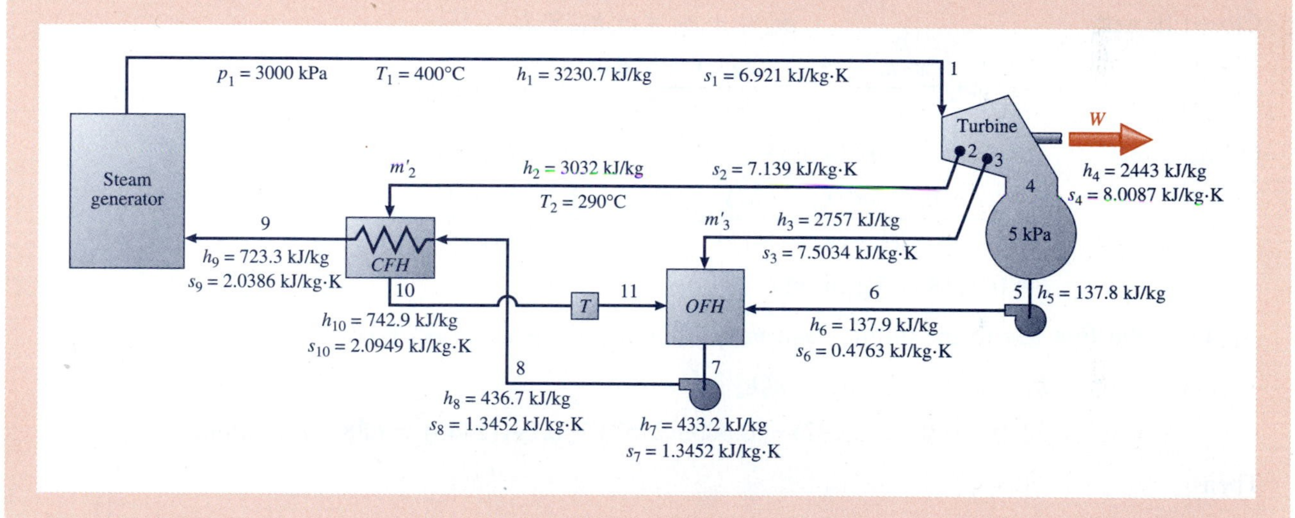

reversible and adiabatic. The lowest temperature in the surroundings is 20°C. Determine the efficiency of the cycle and the irreversibilities, per kilogram of throttle steam, of the processes in the turbine, open feedwater heater, and closed feedwater heater (including the trap). Show the results on an energy flow diagram.

SOLUTION

Analysis: We first sketch a *Ts* diagram to correlate with the specified flow diagram for the plant. To apply the first law to the various pieces of equipment we must first determine the flow distribution, that is, the values of m'_2 and m'_3. We will do this by making energy and mass balances on the two feedwater heaters, starting with the higher-pressure one, the closed heater. We will then apply the first law to the turbine, the

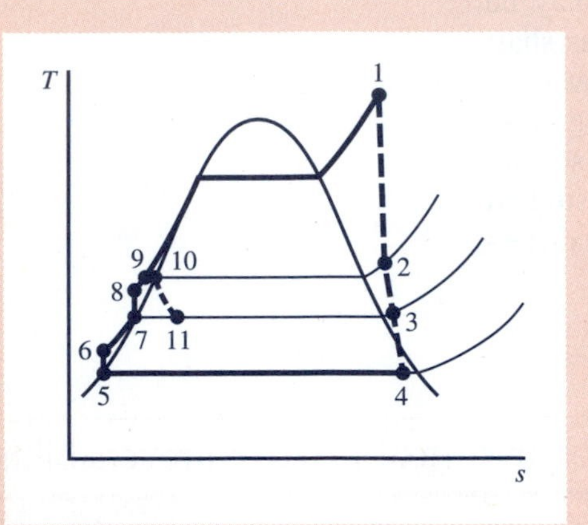

pumps, and the steam generator to find the plant work and the heat added. The ratio of these is the plant efficiency. Irreversibilities of processes will be calculated by $i = T_0\, \Delta s_{univ}$.

Calculations:

Closed heater: $$m_2' h_2 + h_8 = m_2' h_{10} + h_9$$

$$m_2' = \frac{h_9 - h_8}{h_2 - h_{10}} = \frac{723.3 - 436.4}{3032 - 742.9} = 0.125 \text{ kg/kg throttle}$$

Open heater: $$m_3' h_3 + (1 - m_2' - m_3')h_6 + m_2' h_{11} = h_7$$

$$m_3' = \frac{h_7 - h_6 - m_2'(h_{11} - h_6)}{h_3 - h_6} = \frac{433.2 - 137.9 - 0.125(742.9 - 137.9)}{2757 - 137.9}$$

$$= 0.0838 \text{ kg/kg throttle}$$

Applying the first law to the turbine, and noting that $m_4' = 1 - m_2' - m_3'$ gives

$$w_T = h_1 - m_2' h_2 - m_3' h_3 - m_4' h_4$$
$$= 3230.7 - 0.125(3032) - 0.8038(2757) - 0.791(2443) = 688 \text{ kJ/kg throttle}$$

Then

$$\eta = \frac{w_T - m_4' w_{\text{in},5\text{-}6} - w_{\text{in},7\text{-}8}}{q_{9\text{-}1}} = \frac{w_T - m_4'\, \Delta h_{5\text{-}6} - \Delta h_{7\text{-}8}}{h_1 - h_9}$$
$$= \frac{688 - 0.791(0.116) - 3.2}{3398.3 - 747.0} = 0.273$$

For each piece of adiabatic steady-flow equipment, $i = T_0\, \Delta s$, where $\Delta s = \Sigma_{\text{out}}\, m's - \Sigma_{\text{in}}\, m's$. For the turbine,

$$i = T_0(m_4' s_4 + m_3' s_3 + m_2' s_2 - s_1)$$
$$= 293[0.791(8.009) + 0.0838(7.504) + 0.125(7.139) - 6.921] = 274.5 \text{ kg/kg throttle}$$

To find s_{11}, we notice that $p_{11} = 125$ kPa and $h_{11} = h_{10} = 742.9$ kJ/kg because the flow through the trap is a throttling process. Then

$$x_{11} = \frac{h_{11} - h_f}{h_{fg}} = \frac{742.9 - 439.4}{2243.9} = 0.135$$

$$s_{11} = s_f + x_{11}s_{fg} = 1.361 + 0.135(5.9368) = 2.164 \text{ kJ/kg·K}$$

Then for the closed feedwater heater and its trap,

$$i = T_0(s_9 + m_2' s_{11} - s_8 - m_2' s_2) = T_0[s_9 - s_8 + m_2'(s_{11} - s_2)]$$
$$= 293[2.039 - 1.345 + 0.125(2.164 - 7.139)] = 20.4 \text{ kJ/kg throttle}$$

For the open feedwater heater,

$$i = T_0(s_7 - m_3' s_3 - m_4' s_6 - m_2' s_{11})$$
$$= 293[1.345 - 0.0838(7.503) - 0.791(0.4763) - 0.125(2.164)] = 20.1 \text{ kJ/kg throttle}$$

Other data needed for making the energy flow diagram are the unavailable part of the heat added,

$$q_{unav,9-1} = T_0(s_1 - s_9) = 293(6.921 - 2.039) = 1430 \text{ kJ/kg throttle}$$

the amount of heat rejected per kilogram of throttle steam,

$$q_{\text{out},4-5} = m_4'(h_4 - h_5) = 0.791(2443 - 137.8) = 1823 \text{ kJ/kg throttle}$$

and the unavailable part of the heat rejected,

$$q_{\text{out},unav,4-5} = m_4' T_0(s_4 - s_5) = 0.791(293)(8.009 - 0.4763) = 1746 \text{ kJ/kg throttle}$$

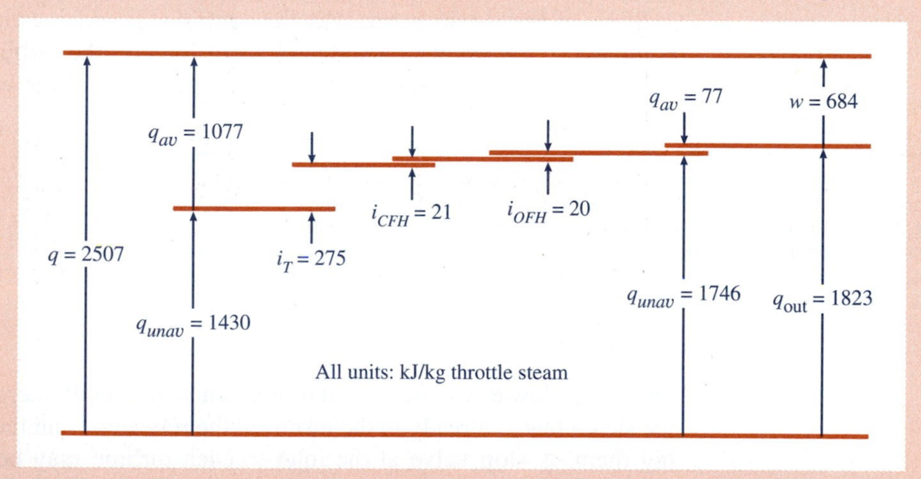

Comment: The energy flow diagram shows that the difference between the unavailable energy rejected by the cycle and the unavailable fraction of the energy added to the cycle is accounted for completely by the sum of the irreversibilities of the cycle components.

15·3·3 Reheat in Steam Power Plants

The thermal efficiency of a Rankine cycle is increased by increasing the steam-generator pressure or the maximum temperature. For a given maximum temperature, which is usually determined by the strength characteristics of construction materials, increasing the steam-generator pressure decreases the quality of the steam leaving the turbine. As mentioned in Section 14·2, this is undesirable because moisture content of more than 10 percent to 12 percent causes serious erosion of turbine blades. (Notice that the exhaust quality in Examples 15·4 and 15·5 is only 0.814, far below the erosion limit of 0.88 to 0.9. Of course, an actual turbine with an efficiency less than 100 percent has a higher exhaust quality than the corresponding ideal turbine with isentropic expansion.) One way to prevent the exhaust moisture content from exceeding the limiting value is to reheat the steam as described in Section 14·2. For very high throttle pressures, steam may be drawn from the turbine at two different pressures for reheating. This is known as a double reheat.

The effect on thermal efficiency of adding reheat to a Rankine cycle depends on cycle conditions. The higher the reheat pressures and the higher the temperatures after reheating, the higher the thermal efficiency is. In calculating the efficiency of a cycle involving reheat, remember to include the work of all turbines less the work input to all pumps in the net work and to include the heat added in the reheaters as well as that added in the initial generation of steam.

The expense of reheaters and the associated piping and control system is great. Consequently, reheat is economically justifiable only in high-capacity plants, so reheat is used only in conjunction with several stages of regeneration. Figure 15·26 is a flow diagram of a power plant with double reheat.

Study of Fig. 15·26 shows why the control system for a plant with reheat is complex. In the event of a sudden loss of load on the turbines, it is obvious that simply closing a valve at the inlet of the first turbine will not cause the

> The flow diagrams shown in this chapter are greatly simplified. Actual plants duplicate some of the equipment. For example, instead of a single boiler feed pump, there may be three operating in parallel or to provide backup. Also, plants have additional equipment for purposes of control, lubrication, treatment of both circulating water and cooling water, fuel and waste handling, environmental protection, measurements, and other purposes. For example, Fig. 15·26 shows a small number of pumps, but an actual plant with this basic flow diagram would require more than 100 pumps for successful operation.

turbine to slow down, because a tremendous amount of energy is contained in the steam that is already in the turbines, the reheaters, and the piping connecting them. A stop valve at the inlet to each turbine may be thought of as a solution, but steam flow must be maintained through the tubes of the reheater sections of the steam generator, because otherwise the tubes that are exposed to high-temperature furnace gases while no steam flows through them will overheat.

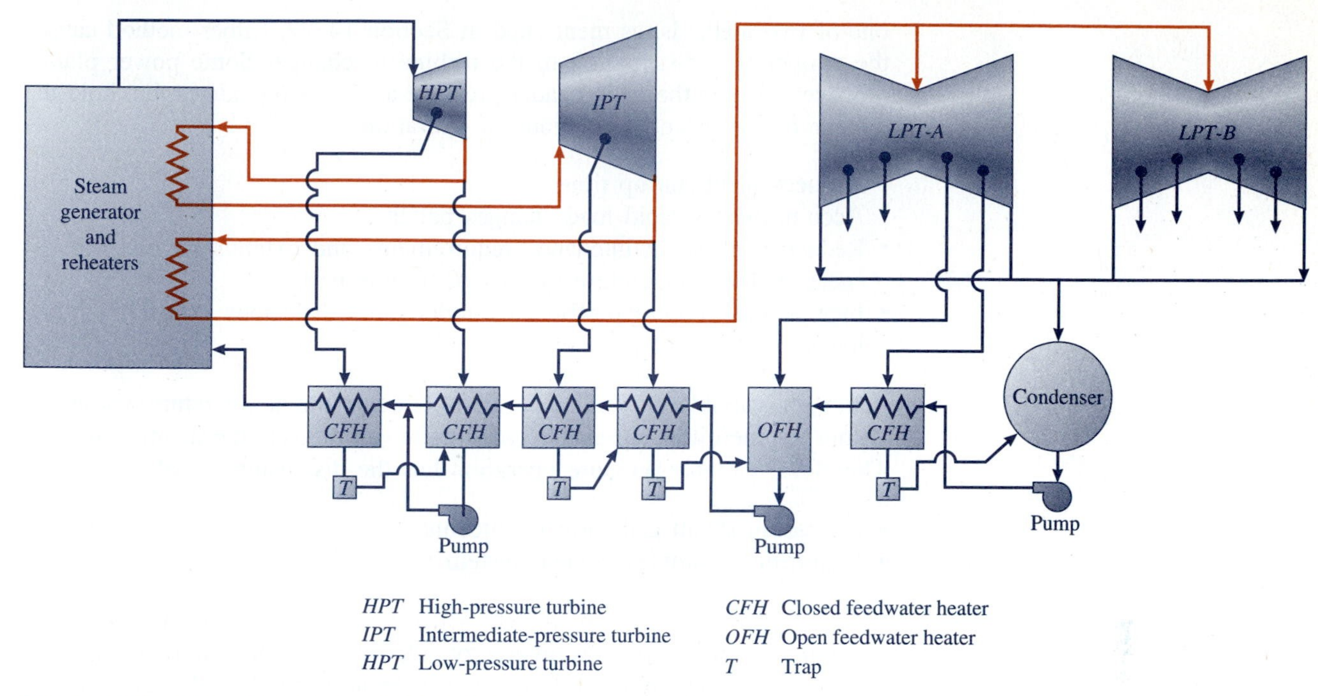

Figure 15·26 Flow diagram of a double-reheat steam power plant.

For any steam power plant, the specific volume during expansion processes is enormously larger than the specific volume during pumping or compression, and this explains the low backwork ratio of steam power cycles. Also, the increase in specific volume of steam passing through a turbine is great. Notice in Fig. 15·26 that the steam passing through a single high-pressure turbine, even when reduced by steam bled for regenerative feedwater heating, passes through four low-pressure turbines in parallel.

For a numerical example of what 0.1 percent improvement in efficiency can mean to fuel cost savings, consider a 1000-MW plant operating on base load. Base load means that the plant continually produces its maximum power output while other plants in the system carry varying loads to meet the fluctuating demand that is characteristic of electric power systems. Base load plants operate near maximum output throughout the year. Fuel costs vary widely with location, but for a cost of $1.50 per million B, a plant efficiency increase of 0.1 percent decreases annual fuel costs by around $125,000. This explains why major equipment guarantees may state efficiencies to within 0.1 percent, and it gives some idea of the size of investment that can be justified for a specified increase in efficiency. Primary factors in such calculations are the fraction of the time the plant is operating, the load carried, and the price of fuel.

15·3·4 Other Steam Power Cycles

Variable Pressure Operation. Steam power plants usually operate with constant pressure in the header between the steam generator and the turbine inlet. When the load on the plant decreases, the steam flow rate is reduced by one of two methods, as mentioned in Section 14·4·2. Either method causes the temperature distribution in the turbine to change. Some power plants, however, change the main header pressure as the load changes, but hold the temperature constant. This mode of operation

- Reduces plant startup times
- Accommodates rapid load changes better
- Reduces turbine maintenance requirements and extends turbine life because of less temperature cycling of turbine parts
- Increases plant thermal efficiency under some conditions, especially at low loads

Although variable pressure operation reduces the temperature variations within turbines, it increases the temperature variations in the steam generator. Therefore, variable pressure operation has the disadvantages of

- Increasing steam generator maintenance
- Requiring a complex control system

Cogeneration: Producing Power Plus Heat. In all the cycles discussed so far in this chapter, the useful energy output is work or power, and the heat rejected is considered to be of no value, especially since it is largely unavailable energy. There are many instances, however, where energy in the form of heat is needed, and, since there is no intention of producing power from it, its value is not lessened by the fact that its available fraction is small. Where this heat can be supplied by steam, as the heating of buildings and the heating required by many industrial processes, the functions of heating and power production can often be combined effectively. This combination is often called **cogeneration**.

As an example, consider a plant that needs 10^7 B of heat from steam per hour at 30 psia with a minimum temperature in the heating system of 150 F. Figure 15·27*a* shows how this heating load can be met by 9550 lbm of dry saturated steam per hour at 30 psia with condensate returned to the boiler at 150 F. The pump is needed to overcome frictional pressure drop in the sys-

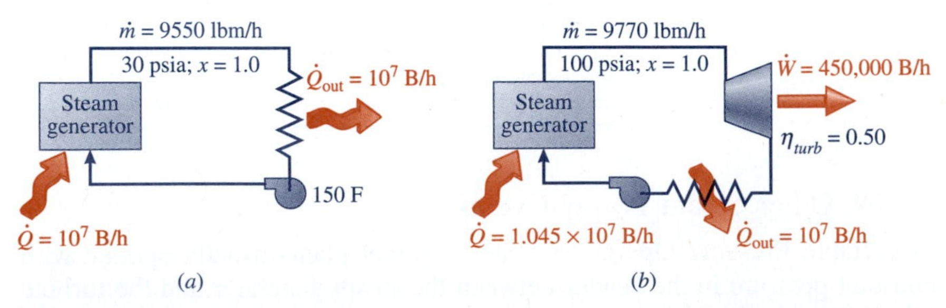

Figure 15·27 (*a*) Steam heating cycle; (*b*) steam cycle for heating and power: cogeneration.

tem. Its work input is negligibly small in comparison with the total heat output of the plant. An alternative method of meeting this heating load is shown in Fig. 15·27*b*, where dry saturated steam is generated at 100 psia, expanded through a turbine that has an efficiency of 50 percent, and gives up to 10^7 B/h as it condenses and is subcooled to 150 F at 30 psia. Analysis of this cycle shows that the same heating load is carried and also 450,000 B/h (177 hp) of power can be produced with an additional heat input of only 450,000 B/h. Thus the amount of power produced is equivalent to the additional rate of heat input; so this *by-product power* or *cogenerated power* is much cheaper than that from any cycle devised for power only.

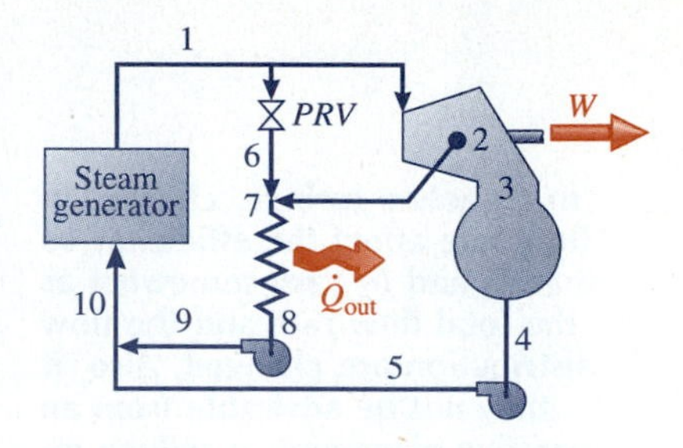

Figure 15·28 Steam cycle for heating and power: cogeneration.

The power output and heat output of the cycle in Fig. 15·27*b* are tied together because the same steam flow passes through the turbine and the heating system. Such a cycle is satisfactory where the steam flow can be controlled only by the heating needs, and whatever power produced can always be used, perhaps to supplement power that is obtained from other sources. Where varying heat and power loads must be carried by a single plant, a cycle such as shown in Fig. 15·28 is used. When the power load is zero, all the steam passes through the pressure-reducing valve (*PRV*), and none through the turbine. When the heating load is zero, all of the steam expands through the turbine and into the condenser. For meeting both heating and power demands, the steam flow distribution is determined as follows.

Assuming that all steam to the heating system comes from the turbine extraction point (state 2), the flow rate in the extraction line is determined from an energy balance on the heating system as

As used here, a turbine extraction point differs from a turbine bleed point in that the pressure at an extraction point is held constant by an automatic valve arrangement in the turbine, while a bleed point is simply an opening that takes steam from a section of the turbine where the pressure may vary slightly as the flow rate changes.

$$\dot{m}_2 = \frac{\dot{Q}_{\text{heating}}}{h_2 - h_8}$$

Then the power produced by this steam flowing through the turbine from the throttle to the extraction point is determined from an energy balance on that part of the turbine:

$$\dot{W} = \dot{m}_2(h_1 - h_2)$$

If this power is less than the total required, the remainder is produced by additional steam that expands all the way through the turbine to the condenser. If this power obtained from the flow $\dot{m}_2$ is greater than the total required, then the flow $\dot{m}_2$ into the turbine and through the extraction line must be reduced to the value that produces the required amount of power, even though it is inadequate for the heating load. The deficit in energy to the heating system is then made up by steam passed through the pressure-reducing valve. Remember that $h_6 = h_1$ and $h_2 = h_1 - w$, so that $h_6 > h_2$.

In short, the steam distribution problem is to determine the values of $\dot{m}_2$, $\dot{m}_3$, and $\dot{m}_6$ such that they satisfy the first-law equations for the power required,

$$\dot{W} = \dot{m}_2(h_1 - h_2) + \dot{m}_3(h_1 - h_3)$$

and for the heating load,

$$\dot{Q} = \dot{m}_2(h_2 - h_8) + \dot{m}_6(h_6 - h_8)$$

In an actual turbine, changes in flow rate affect the efficiency so that h_2 and h_3 vary somewhat as the total flow rate and the flow distribution are changed. Also, it may not be advisable from an operating viewpoint to reduce $\dot{m}_3$ to zero.

and their sum is minimized. For the minimum total steam flow, $\dot{m}_3$ or $\dot{m}_6$ must be zero.

When $\dot{m}_3$ is zero, all the heat added in the steam generator is utilized; none is rejected to condenser cooling water. For a given heating load, when $\dot{m}_6$ is zero, the maximum amount of by-product power is obtained. When $\dot{m}_3 = \dot{m}_6 = 0$, the heating and power loads are said to be balanced, and this is the most economical condition. Loads are usually variable, so a plant cannot be operated with balanced loads all the time; however, the plant designer must select operating conditions that most nearly achieve balanced conditions in the long run.

Binary Vapor Cycles. The only working fluid we have considered and the one that is used almost exclusively in practice for vapor power cycles is water. No better fluid has been found, although water is quite undesirable in some respects. Desirable characteristics of a vapor-cycle working fluid include the following:

- A critical temperature well above the highest temperature that can be tolerated by construction materials. This makes it possible to vaporize the fluid, and thus add a considerable amount of heat to it, at the maximum temperature.
- A saturation pressure at the maximum cycle temperature that introduces no strength problems and a saturation pressure at the minimum cycle temperature that introduces no problems of sealing against infiltration of the atmosphere.
- A high ratio of h_{fg} to c_p of the liquid so that most of the heat added in a Rankine cycle is added at the maximum temperature. This reduces the need for regeneration.
- Chemical inertness and stability throughout the cycle temperature range.
- A triple-state temperature below the expected minimum ambient temperature. This ensures that the fluid will not solidify at any point in the cycle or while being handled outside the cycle.
- A saturated vapor line (on a property diagram) that is close to a turbine expansion path. This prevents excessive moisture in the turbine exhaust, thus eliminating the need for reheating, and still permits all or nearly all the heat rejection to occur at the minimum temperature.
- Low cost and ready availability.
- Nontoxicity.

No fluid has all these desirable characteristics. Water is better than any other in an overall evaluation, but it is poor in regard to the first two characteristics listed. The critical temperature of water is 374°C (705 F), approximately 300 Celsius degrees below the temperature limit set by material

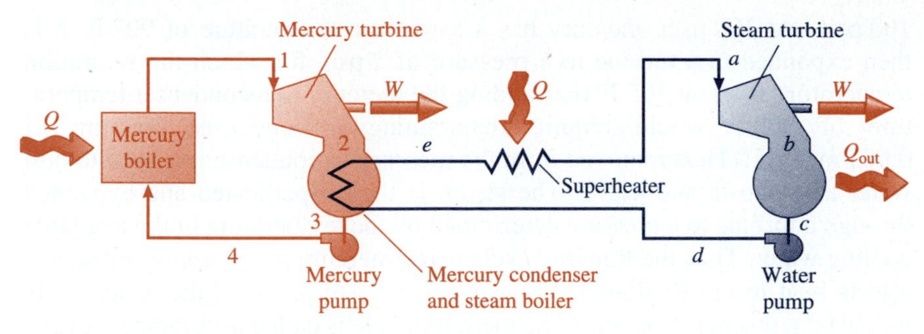

Figure 15·29 Mercury–water binary vapor cycle.

strength properties. Also, the saturation pressure of water is quite high even at moderate vaporization temperatures (8.58 MPa at 300°C and 16.5 MPa at 350°C, for example.) Thus water is especially poor at the high-temperature end of the operating range.

Since no *single* working fluid better than water has been found, searches have been made for a *combination* of fluids such that one is well suited to the high-temperature part and the other to the low-temperature part of the cycle. Mercury and water are one successful combination.

Flow and *Ts* diagrams for a mercury–water *binary vapor cycle* are shown in Figs. 15·29 and 15·30. The mercury is vaporized in a boiler, at, say,

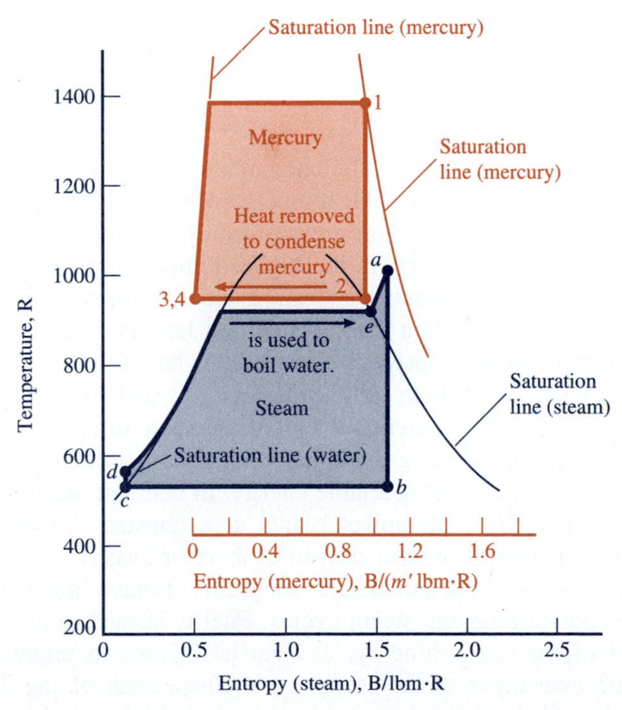

Figure 15·30 *Ts* diagram of mercury–water binary vapor cycle.

100 psia. At 100 psia, mercury has a saturation temperature of 907 F. It is then expanded in a turbine to a pressure of 2 psia for which the saturation temperature is about 505 F. (Expanding the mercury to a condenser temperature of 100 F would require maintaining a condenser pressure of 0.005 mm Hg!) Heat removed from the mercury to condense it is used to boil water at 540 psia and 475 F. The steam is then superheated and expanded through a turbine to a pressure determined by the temperature of the available cooling water. Thus the Rankine cycle using mercury as a working substance rejects heat to the Rankine cycle using water. (In practice, the water cycle would be a regenerative one.) The mercury cycle is called a ''topping'' cycle. Since h_{fg} for water at the steam-generation temperature is several times as large as h_{fg} for mercury, several pounds (m' in Fig. 15·30) of mercury must circulate per pound of water in the binary vapor cycle.

Inspection of the *Ts* diagram of Fig. 15·30 shows that the binary vapor cycle approaches a Carnot cycle more closely than a steam cycle can for the same temperature limits. Consequently, higher thermal efficiency can be reached by the binary vapor cycle. Although binary mercury–water plants have been operated successfully in the past, they are not currently cost-effective. The overriding reason is that the increased fixed costs of the more complicated equipment cannot be offset by the savings in fuel costs. Higher fuel prices would alter this economic balance.

15·4 Combined Cycle Plants

We have seen that for maximum efficiency of a power cycle, it is usually well to absorb heat at the highest possible temperature and reject heat at the lowest possible temperature. For this reason gas turbines and steam cycles are combined. The gas turbine operates in the higher temperature range and rejects heat that serves as energy input to a steam cycle operating mostly in a lower temperature range. Notice that gas-turbine inlet temperatures can be higher than steam-turbine inlet temperatures for reasons having to do with the high pressures associated with high steam temperatures and the methods of turbine-blade cooling. A consequence of high gas-turbine inlet temperatures and the behavior of gases is that gas-turbine exhaust temperatures are high, resulting in a large rejection of available energy. In contrast, we have seen that heat is rejected from steam power plants at a constant temperature only slightly above the temperature of the atmosphere or available cooling water.

Figure 15·31 shows flow and ideal *Ts* diagrams of one of the many variants of combined gas-turbine and steam cycles. Fuel is burned to heat the gas in process 2–3 of the gas-turbine cycle. Heat is rejected in process 4–5 and serves as the heat input to the steam cycle. Inspection of the *Ts* diagram shows that the efficiency of the combined cycle is higher than that of either the gas cycle or the steam cycle alone. (Of course, for areas on the *Ts* dia-

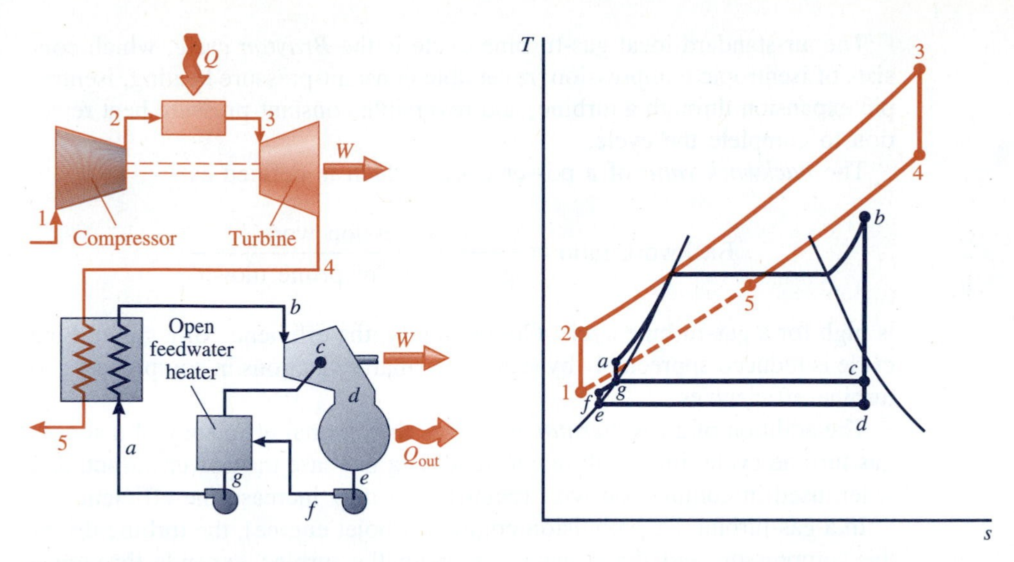

Figure 15·31 Combined-cycle power plant.

gram to be comparable between the gas and steam cycles, the entropy scale for one of the fluids must be multiplied by the mass flow ratio, as is done in the case of the binary cycle *Ts* diagram of Fig. 15·30.) Some fuel may be burned in the steam generator as well as in the gas-turbine combustion chamber, and various combinations of enhancements to both the gas and steam cycles may be incorporated.

Increasing thermal efficiency is only one motivation for combined cycles. Gas turbines can be started and brought to full-load conditions much more quickly than steam plants, so the combination of the two can meet power requirements that vary with time while minimizing the sum of fuel and fixed costs. When process steam is needed in conjunction with varying power demands, combined cycles may be especially attractive.

The first four references listed at the end of this chapter all discuss variants of combined cycles, and each one covers some aspects that are treated by none of the others. The important point is that all of the variants, including the most complex ones, can be analyzed by applying the principles we have used repeatedly in this and earlier chapters.

15·5 Summary

In an *air-standard* cycle analysis, the working substance is air that behaves like an ideal gas, and the combustion process of an actual power cycle is replaced by a heat-addition process. A *cold-air-standard* analysis involves the further simplification that the specific heats of air are constant at their room-temperature values.

The air-standard ideal gas-turbine cycle is the *Brayton cycle*, which consists of isentropic compression, reversible constant-pressure heating, isentropic expansion through a turbine, and reversible constant-pressure heat rejection to complete the cycle.

The *backwork ratio* of a power cycle, which is defined as

$$\text{Backwork ratio} \equiv \frac{\text{compression work}_{\text{in}}}{\text{gross work of prime mover}}$$

is high for a gas-turbine cycle. Consequently, the efficiency of a gas-turbine cycle is reduced appreciably by relatively small reductions in compressor and turbine efficiencies.

The addition of a *regenerator* increases the thermal efficiency of a simple gas-turbine cycle. Intercooling and reheating increase the power output, and when used in conjunction with regeneration may increase the efficiency.

In a gas-turbine jet-propulsion engine (turbojet engine), the turbine drives the compressor, and the exhaust gas from the turbine expands through a nozzle to leave at a high velocity in a rearward direction. The change in momentum of the gas passing through the engine results in a forward thrust on the engine, and thus on the aircraft. In a jet engine, the counterpart of reheating is *afterburning*, which increases the thrust by increasing the enthalpy drop and hence the velocity change across the jet nozzle.

The air-standard *Otto cycle* consists of four reversible processes of air in a closed system: adiabatic compression, constant-volume heat addition, adiabatic expansion, and constant-volume heat rejection. The *compression ratio* of a reciprocating engine is defined as the ratio of the maximum to the minimum volume of the working substance in the cylinder.

The air-standard *Diesel cycle* also consists of four reversible processes of air in a closed system: adiabatic compression, constant-pressure heat addition, adiabatic expansion to the initial volume, and constant-volume heat rejection. For a Diesel cycle the *cutoff ratio* is defined as the ratio of the cylinder volume after heat addition to that before heat addition. The *percent cutoff* is the fraction of the stroke during which heat is added.

Other air-standard cycles, such as the dual cycle in which the heat addition is divided between a constant-volume and a constant-pressure process, are used in analyses where it is desired to simulate the pressure–volume variation of actual engine operation more closely than can be done with an air-standard Otto or Diesel cycle.

Compound power plants or compound engines are combinations of reciprocating engines and turbines designed to capitalize on the advantages and minimize the drawbacks of each type of prime mover.

The Carnot cycle is an unsuitable model for the design of vapor power plants because of the difficulty of carrying out the required processes in actual machines. Actual cycles can, however, simulate the *Rankine cycle*, which consists of four reversible processes: (1) constant-pressure heat addition that takes the fluid from a compressed liquid state to a saturated or superheated vapor state, (2) isentropic expansion in a prime mover, (3) constant-pressure heat rejection to condense the vapor to a saturated liquid, and (4) isentropic pumping to a compressed liquid state at the steam-generation pressure. The backwork ratio of a Rankine cycle is low, so cycle thermal efficiency suffers relatively little from low prime-mover or pump efficiency.

The efficiency of a steam power cycle can be increased by *regeneration*, which involves *bleeding* some steam from the turbine after it has expanded part of the way to the condenser pressure and using it in *feedwater heaters* to preheat the liquid going to the steam generator. In open feedwater heaters, the bled steam is mixed with the feedwater in the heater; in closed feedwater heaters, the bled steam condenses on the outside of tubes through which the feedwater flows.

To reduce the moisture content that causes erosion in the low-pressure stages of turbines, *reheating* is used. This means that steam is removed from the turbine after part of its expansion is completed, heated at constant pressure, and returned to the turbine for the remainder of the expansion to the condenser pressure. Reheating is used only in conjunction with regeneration because economic studies always show that regeneration should be added to a Rankine cycle before reheating.

Whenever power and heat are both to be produced, a combination steam cycle, often using an *extraction turbine*, is much more economical than separate power and heating cycles.

Water has some serious shortcomings as a vapor-power-cycle working fluid, but no better fluid has been found. A *binary vapor cycle* uses two fluids: One with good high-temperature characteristics is used in a Rankine cycle that rejects heat into another Rankine cycle (or regenerative cycle) that uses a second fluid with good low-temperature characteristics.

Combined cycles combine gas turbines and steam power plants to capitalize on the desirable features of each. Many variants of combined cycles are attractive for various reasons.

It must be remembered that modifications to improve the efficiency of a power plant can be economically justified only if the saving in fuel costs exceeds the additional other costs associated with the modifications.

References

15·1 Sorensen, Harry A., *Energy Conversion Systems*, Wiley, New York, 1983. (This book provides much valuable information on equipment and performance to enhance your study of thermodynamics.)

15·2 Wood, Bernard D., *Applications of Thermodynamics*, 2nd ed., Addison-Wesley, Reading, MA, 1982. (This book provides valuable information on many variations of processes and systems.)

15·3 El-Wakil, M. M., *Power Plant Technology*, McGraw-Hill, New York, 1984.

15·4 Li, Kam W., and A. Paul Priddy, *Power Plant System Design*, Wiley, New York, 1985.

15·5 *Encyclopædia Britannica*, Encyclopædia Britannica, Inc., Chicago, periodic editions. (An excellent source of historical information on the development of internal-combustion engines and steam power plants.)

Problems

15·1 A Brayton cycle operates with air entering the compressor at 100 kPa, 20°C, at a rate of 32.0 m^3/s, and air entering the turbine at 800 kPa, 1000°C. Determine the power output and the cycle thermal efficiency.

15·2 Solve Problem 15·1 for turbine inlet temperatures of 500°C, 600°C, 700°C, 800°C, 900°C, and 1000°C and plot the results versus turbine inlet temperature.

15·3 Solve Problem 15·1 for upper pressure limits of 300 kPa, 400 kPa, 500 kPa, 600 kPa, 700 kPa, and 800 kPa and plot the results versus compressor pressure ratio.

15·4 Refer to Example 15·1 and solve the problem using a cold-air-standard analysis. Compare your results with the example solution.

15·5 Refer to Example 15·1. Solve the problem using an air-standard analysis with the mass of fuel considered. The fuel is added at 25°C and has a lower heating value of 44,000 kJ/kg at 25°C.

15·6 For a Brayton cycle with compressor inlet conditions of 100 kPa, 20°C, plot against pressure ratio in the range of 2 to 10 the throttle temperature that causes the turbine exhaust temperature to equal the compressor discharge temperature.

15·7 A simple gas-turbine cycle operates at a pressure ratio of 5. Air enters the compressor at 20°C. If the turbine and compressor efficiencies are 0.85 and 0.84, respectively, compute the cycle efficiency for a turbine inlet temperature of 700°C and sketch an energy flow diagram showing irreversibilities quantitatively. Assume that the compression and expansion are adiabatic.

15·8 For the pressure ratios and limiting temperatures as given in Problem 15·6, plot a curve of cycle thermal efficiency versus compressor efficiency with turbine efficiency as a parameter. (Use machine efficiencies from 40 percent to 100 percent, if possible.)

15·9 Solve Problem 15·7 for a pressure ratio of 8 and an inlet turbine temperature of 900°C.

15·10 A simple air-standard gas-turbine cycle has compressor inlet conditions of 90 kPa, 10°C. The pressure ratio is 5, and the maximum temperature is 800°C. What reduction in compressor efficiency from 100 percent would have the same effect on the cycle efficiency as a reduction of the turbine efficiency from 100 percent to 75 percent?

15·11 For equal pressure and temperature limits, rank in order of decreasing efficiency cold-air-standard simple gas-turbine cycles using as working fluids helium, air, and carbon dioxide. (*Note:* When the working fluid is not air, perhaps the model should be referred to as a *cold-gas-standard* model, but this terminology has not been widely adopted.)

15·12 The thermal efficiency of a cold-air-standard Brayton cycle depends only on the pressure ratio and k of the gas. Make a plot of efficiency versus pressure ratio for various k values of common gases to show the influence of these parameters. (See the note in Problem 15·11 regarding terminology.)

15·13 For an ideal simple gas-turbine cycle with compressor inlet conditions of 1 atm, 25°C, and a maximum temperature of 900°C, make a comparative plot of efficiency versus pressure ratio for a cold-air-standard analysis and an air-standard analysis.

15·14 Derive an expression for the thermal efficiency of a cold-air-standard Brayton cycle in terms of the compressor pressure ratio only.

15·15 An air-standard gas turbine operates on the Brayton cycle between pressure limits of 14.7 psia and 70 psia. The inlet air temperature to the compressor is 60 F, and the air entering the turbine is at 1500 F. Air enters the compressor at a rate of 50,000 cfm. Compute the power output and the cycle thermal efficiency.

15·16 Solve Problem 15·15 for turbine inlet temperatures ranging from 900 F to 1500 F and plot the results versus turbine inlet temperature.

15·17 Solve Problem 15·15 for upper pressure limits ranging from 40 psia to 120 psia and plot the results versus compressor pressure ratio.

15·18 Air enters the compressor of a simple gas-turbine cycle at 14.7 psia and 15 F, and leaves at 103 psia, 445 F. The temperature of the gases entering and leaving the turbine are 1600 F and 716 F, respectively. If such states are possible, compute the net work per pound of air and the efficiency of the cycle.

15·19 For a simple gas-turbine cycle with a pressure ratio of 6, peak temperature of 900°C, and turbine and compressor efficiencies of 78 percent, the sensitivity of plant thermal efficiency to compressor efficiency is to be studied. Plot a curve of the change in plant thermal efficiency versus differences in compressor efficiency that range from −8 percent to 8 percent.

15·20 For a simple gas-turbine cycle with a pressure ratio of 6, peak temperature of 900°C, and turbine and compressor efficiencies of 78 percent, the sensitivity of plant thermal efficiency to turbine efficiency is to be studied. Plot a curve of change in plant thermal efficiency versus differences in turbine efficiency that range from −8 percent to 8 percent.

15·21 Air enters the compressor of a simple gas-turbine plant at 100 kPa, 20°C, at a rate of 2.20 kg/s. The compressor efficiency is 60 percent. Discharge pressure is 450 kPa. Calculate the amount of heat (kJ/kg) that must be added to provide a turbine inlet temperature of 650°C.

15·22 A gas turbine (Brayton cycle) is used to produce hot gas to help in snow and ice removal from an airport runway. Air enters the compressor at 95 kPa, −10°C. The turbine drives only the compressor. The compressor efficiency is 65 percent and the pressure ratio is 4. Gas at 165°C leaves the turbine at a rate of 20 kg/s. Assume adiabatic compression and expansion. Using an air-standard analysis, calculate (*a*) the temperature of the gas entering the turbine and (*b*) the turbine efficiency.

15·23 A cold-air-standard Brayton cycle has a minimum temperature of T_{min} and a maximum temperature of T_{max}. Show that the compressor pressure ratio for the maximum cycle thermal efficiency equals $(T_{max}/T_{min})^{k/2(k-1)}$.

15·24 An air-standard simple gas turbine cycle operates with an inlet pressure of 1 atm, a pressure ratio of 5, and a turbine inlet temperature of 660°C (1220 F). The working fluid is air. Plot cycle thermal efficiency and backwork ratio against compressor inlet temperature for a range of −40°C (−40 F) to 40°C (104 F).

15·25 Solve Problem 15·24 with turbine and compressor efficiencies of 75 percent. Compare the results with those of Problem 15·24.

15·26 An air-standard simple gas-turbine cycle operates with inlet conditions of 1 atm and 25°C, a pressure ratio of 5, and a turbine inlet temperature of 660°C (1220 F). The working fluid is air. A filter is placed ahead of the compressor inlet, causing a reduction in compressor inlet pressure, but the pressure ratio across the compressor remains unchanged. Also, the turbine exhaust pressure remains unchanged. Plot the decrease in cycle thermal efficiency against pressure drop across the filter for a pressure drop range of 0 to 15 kPa (2.18 psi).

15·27 Solve Problem 15·26 with turbine and compressor efficiencies of 75 percent. Compare the results.

15·28 How would the performance of an ideal simple gas-turbine cycle be affected by injecting liquid water into the compressor inlet in an amount that would result in no liquid water leaving the compressor? (Assume that the mass flow rate of dry air is held fixed.) What would be the effect of injecting the same amount of liquid water at the turbine inlet? (Both of these procedures have been used in practice for certain purposes.)

15·29 For a power output of 1000 kW, determine the minimum flow rate for a simple air-standard gas-turbine cycle with minimum and maximum temperatures of 15°C (59 F) and 660°C (1220 F) if compressor and turbine efficiencies are each 75 percent.

15·30 An air-standard gas turbine operates on a regenerative cycle with a regenerator effectiveness of 100 percent. Compression and expansion are isentropic. The inlet temperature to the compressor is 15°C (59 F). Compute the air-standard cycle efficiency for the following conditions: (*a*) inlet temperature to the turbine is 635°C (1175 F), pressure ratio is 5; (*b*) inlet temperature to the turbine is 885°C (1625 F), pressure ratio is 5; and (*c*) inlet temperature to the turbine is 635°C (1175 F), pressure ratio is 10.

15·31 Solve Problem 15·30*a* for regenerator effectiveness of (*a*) 75 percent, (*b*) 50 percent.

15·32 Solve Problem 15·30*a* if there is a leak in the high-temperature end of the regenerator that allows 5 percent of the entering air to flow into the exhaust line.

15·33 Solve Problem 15·30*a* if there is a frictional pressure drop of 22 kPa (3.2 psi) in each gas stream passing through the regenerator.

15·34 Solve Problem 15·30*a* for $\eta_C = \eta_T = 0.80$.

15·35 Add to the cycle described in Example 15·1 a regenerator of 70 percent effectiveness, and for the cases of compressor and turbine efficiencies of 100 percent and 80 percent, determine the cycle efficiency.

15·36 For an ideal regenerative gas-turbine cycle, plot a curve of thermal efficiency versus pressure ratio (in a range of 2 to 20) for ratios of turbine inlet temperature to compressor inlet temperature of 3, 4, and 5. Also plot the thermal efficiency of a Brayton cycle versus pressure ratio on the same chart.

15·37 For a regenerative gas-turbine cycle as in Problem 15·35, does a frictional pressure drop that ranges from 1 kPa to 15 kPa have a more deleterious effect on cycle efficiency if it is on the hot side or the cold side of the regenerator? Use compressor and turbine efficiencies of 80 percent.

15·38 For an ideal regenerative gas-turbine cycle, plot thermal efficiency versus pressure ratio, in a range of 2 to 20, for a compressor inlet temperature of 15°C, a turbine inlet temperature of 1000°C, and regenerator effectiveness values of 0.9, 0.75, and 0.6.

15·39 Someone says that high compressor efficiency becomes less important when regeneration is added to a simple gas-turbine cycle. Is this true? If it is not generally true, is there a germ of truth in it? Does the statement relate to the fact that with inefficient compressors less regeneration can be done?

15·40 For fixed inlet temperatures, sketch a curve of regenerator irreversibility versus regenerator effectiveness. Does the maximum irreversibility correspond to the least desirable value of effectiveness from the standpoint of cycle thermal efficiency?

15·41 Plot curves of thermal efficiency versus pressure ratio (in a range of 2 to 20) for a regenerative gas-turbine cycle with turbine efficiency of 0.80, compressor efficiency of 0.77, and regenerator effectiveness values of 1.0, 0.8, 0.6, and 0 (no regeneration).

15·42 Air enters a gas-turbine plant at 95 kPa, 5°C. Compression is adiabatic with an efficiency of 0.70 and pressure ratio of 5. The regenerator effectiveness is 0.60. Turbine inlet conditions are 475 kPa, 850°C.

The turbine expansion is adiabatic with an efficiency of 0.70. The plant power output is 1500 kW. Calculate (*a*) the mass rate of flow of air through the plant and (*b*) the irreversibility (kJ/kg) of the turbine expansion.

15·43 A gas-turbine plant is used to supply compressed air by having the turbine drive only the compressor, and the air stream from the compressor is divided into two parts: that which goes to the combustion chamber and turbine, and that which is the compressed air delivered by the plant. Inlet conditions are 14.0 psia (96.5 kPa), 70 F (21°C), and the compressor discharge pressure is 70 psia (48.3 kPa). The fuel lower heating value is 17,300 B/lbm (40,240 kJ/kg). The turbine inlet temperature is 1500 F (816°C), and the exhaust products can be treated as air. All velocities are negligible except for the turbine exhaust velocity of 500 fps (152 m/s). Determine the mass ratio of compressed air delivered to air drawn in.

15·44 In the gas-turbine plant shown in the figure, the compressor is driven by one turbine, and the other turbine delivers 2000 kW as the power output of the plant. The compressor and the turbines operate adiabatically. The compressor efficiency is 80 percent; the efficiency of each turbine is 80 percent. Air drawn into the compressor is at 98.0 kPa, 15°C. Both turbines exhaust at 98.0 kPa. The inlet condition for each turbine is 410 kPa, 700°C. High-pressure air leaves the regenerator at 315°C. Disregard pressure drop in the heat exchangers and changes in kinetic energy. Treat the working fluid as air. Calculate the heat added in each combustion chamber in kJ/h.

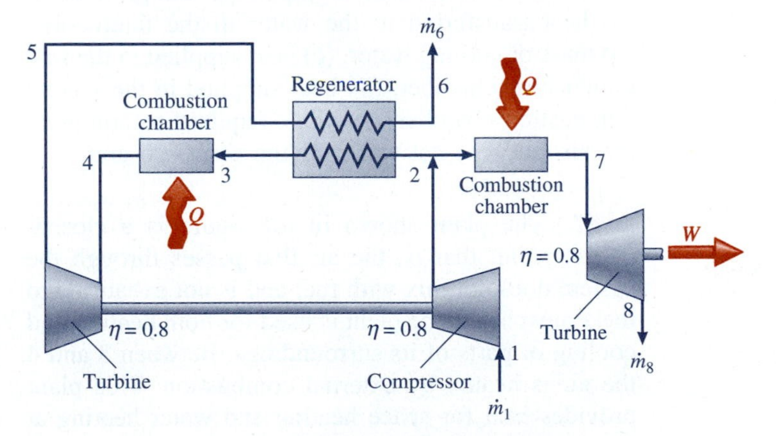

Problems 15·44 and 15·45

15·45 In the gas-turbine plant shown in the figure, the compressor is driven by one turbine, and the other turbine delivers 1000 hp as the power output of the plant. The compressor and the turbines operate adiabatically. The compressor efficiency is 80 percent; the efficiency of each turbine is 80 percent. Air drawn into the compressor is at 14.7 psia, 60 F. Both turbines exhaust at 14.7 psia. The inlet condition for each turbine is 61 psia, 1280 F. High-pressure air leaves the regenerator at 600 F. Disregard pressure drop in the heat exchangers and changes in kinetic energy. Treat the working fluid as air. Calculate the heat added in each combustion chamber in B/h.

15·46 The following data apply to an air-standard intercooled regenerative cycle; air temperatures entering and leaving the low-pressure stage of the compressor are 25°C and 90°C; air temperatures entering and leaving the high-pressure stage of the compressor are 60°C and 138°C; air temperature leaving the regenerator is 340°C; gas temperatures entering and leaving the turbine are 870°C and 410°C. The initial pressure is 101.3 kPa, and the pressure ratio is 6. Cooling water enters the intercooler at 20°C and leaves at 25°C. Compute (*a*) the net work of the cycle per kilogram of air, (*b*) the thermal efficiency, (*c*) the amount of water required per kilogram of air in the intercooler, and (*d*) the irreversibility of the regenerator per kilogram of air.

15·47 Consider an air-standard reheat cycle with the high-pressure turbine driving the compressor and the low-pressure turbine supplying the net plant power output of 9000 kW. Air enters the plant at 1 atm, 15°C, and the compressor pressure ratio is 6. The inlet temperature to each turbine is 815°C. The compressor efficiency is 0.85, and each turbine efficiency is 0.82. Determine the mass rate of flow and the cycle thermal efficiency. Make a diagram showing available and unavailable parts of the heat added and rejected and the irreversibility of various processes.

15·48 Consider an air-standard reheat cycle with the high-pressure turbine driving the compressor and the low-pressure turbine supplying the net plant power output of 5000 hp. Air enters the plant at 14.7 psia, 60 F, and the compressor pressure ratio is 6. The inlet temperature to each turbine is 1500 F. The compressor efficiency is 0.85, and each turbine efficiency is 0.82. Determine the mass rate of flow and the cycle thermal efficiency. Make a diagram showing available and unavailable parts of the heat added and rejected and the irreversibility of various processes.

15·49 The flow diagram of a gas-turbine power plant is shown in the figure. The high-pressure turbine (*HPT*) drives only the compressor, and the low-pressure turbine (*LPT*) supplies the net power output of the plant, $p_1 = p_8 = 1$ atm; $p_2 = p_3 = p_4 = 6$ atm; $T_1 = 5°C$ (41 F); $T_4 = 760°C$ (1400 F); regenerator effectiveness = 0.60; net power output = 570 kW.

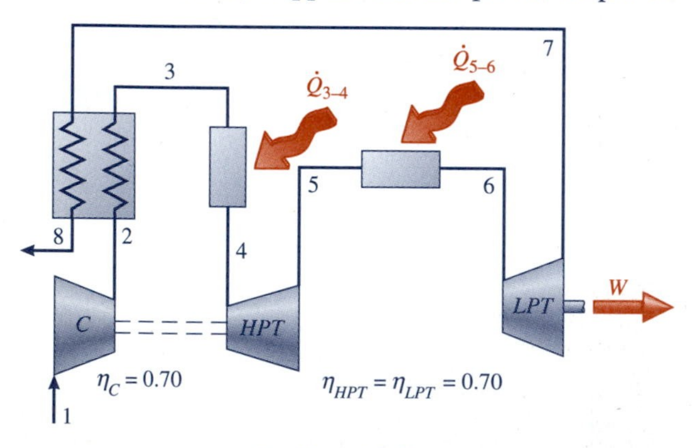

Problem 15·49

15·50 The system shown in the figure is used to provide 10 lbm/s of air at 28.0 psia, 1000 F. Pertinent data are given on the flow diagram. Notice that the turbine drives only the compressor. Assuming isentropic compression and expansion, no frictional effects, and negligible changes in kinetic energy, calculate on the basis of an air-standard analysis the rate of heat transfer in processes 2–3 and 4–5.

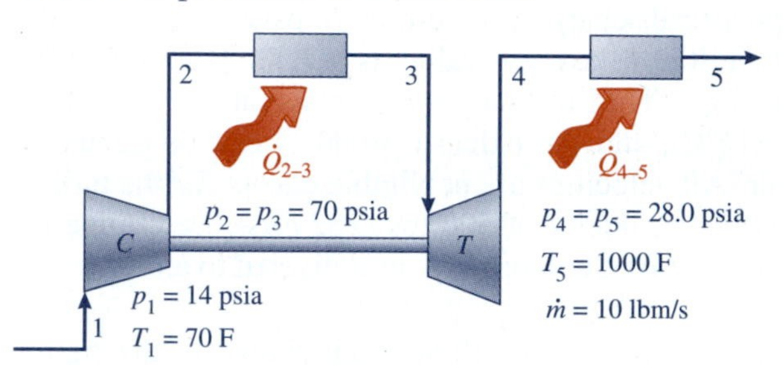

Problem 15·50

15·51 The following temperatures are for the intercooled regenerative air-standard cycle with reheat as shown in Fig. 15·10: $T_1 = 30°C$, $T_2 = 120°C$, $T_3 = 75°C$, $T_4 = 180°C$, $T_5 = 375°C$, $T_6 = 875°C$, $T_7 = 585°C$ (at 320 kPa), $T_8 = 875°C$, and $T_9 = 585°C$. The initial pressure is 1 atm and the overall pressure ratio is 10. Cooling water enters the intercooler at 20°C and leaves at 25°C. Compute per kilogram of air (*a*) heat transferred to the water in the intercooler, (*b*) mass of cooling water, (*c*) heat supplied in the first combustion chamber, (*d*) heat supplied in the second combustion chamber, (*e*) work required to compress the air, and (*f*) net work for the complete unit.

15·52 The plant shown in the figure is a closed-circuit plant; that is, the air that passes through the system does not mix with fuel and is not exhausted to the atmosphere. The plant is used for both heating and cooling of parts of its surroundings. Between 3 and 4 the air is heated by external combustion. This plant provides heat for space heating and water heating at the heat exchangers 2–8 and 6–7. In the heat exchanger 9–10, the air in the system absorbs heat from one part of the surroundings at a rate of 500 B/s. The two turbines together drive the compressor and nothing else (i.e., there is no net power output from the plant). Disregard pressure drops in heat exchangers and connecting lines. Calculate the mass rates of flow $\dot{m}_1$ and $\dot{m}_4$.

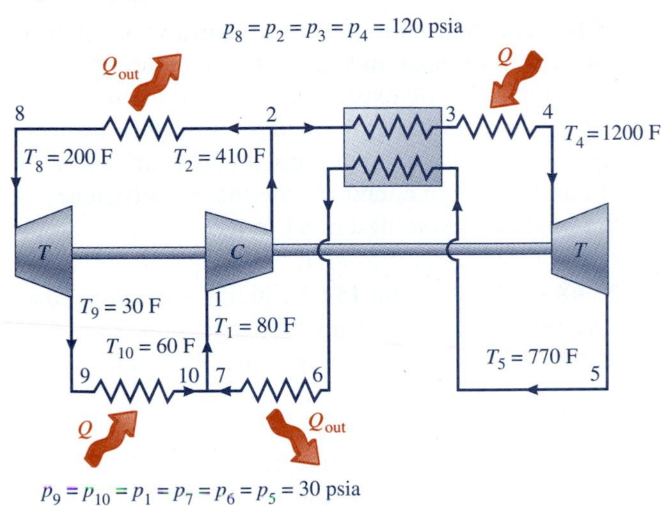

Problem 15·52

15·53 An ideal turbojet engine draws in air at 95 kPa, 25°C, with a velocity of 150 m/s at a rate of 15 kg/s. After passing through the compressor and the combustion chamber, the air enters the turbine at 400 kPa, 870°C. The velocity is 150 m/s at the compressor outlet, turbine inlet, and turbine exhaust. Calculate the velocity of the exhaust jet, assuming isentropic flow through the exhaust nozzle to a pressure of 95 kPa.

15·54 Determine the maximum thrust obtainable from an air-standard turbojet engine that takes in air at 70 kPa, 5°C, with a velocity of 250 m/s (the craft speed) at a rate of 32.0 kg/s. The compressor total pressure ratio is 3, compressor outlet and turbine inlet and outlet velocities are 150 m/s, and the maximum temperature is 980°C. Assume that the total pressure is constant from the compressor outlet to the turbine inlet. The exhaust pressure is 70 kPa.

15·55 Solve Problem 15·54 with an afterburner raising the temperature to 980°C at the entrance to the jet nozzle.

15·56 In a propjet aircraft engine, the gas turbine drives a propeller and also the compressor, and the exhaust gases from the turbine are expanded through a nozzle to form an exhaust jet that provides additional thrust. Determine the required compressor pressure ratio for an air-standard propjet engine that provides 4.5 kN of thrust in addition to 750 kW to a propeller if air at 55 kPa, −20°C, is drawn in with a velocity of 180 m/s. The velocity is also 180 m/s at the compressor outlet, turbine inlet, and turbine exhaust. The maximum temperature is 870°C, and the flow rate is not to exceed 9.0 kg/s. Assume that all processes are reversible.

15·57 In a ramjet engine, air at 70 kPa, 5°C, enters the diffuser at 460 m/s (the craft speed) through a cross-sectional area of 0.075 m^2 and is decelerated to 95 m/s relative to the engine. Fuel at 0.02 times the air flow rate is burned to bring the temperature to 870°C, and the combustion products, which can be treated as air, are then expanded through a nozzle to the ambient pressure and leave at 760 m/s. Calculate the thrust.

15·58 An ideal turbojet engine draws in air at 14.0 psia, 80 F, with a velocity of 500 fps at a rate of 32 lbm/s. After passing through the compressor and the combustion chamber, the air enters the turbine at 60 psia, 1600 F. The velocity is 500 fps at the compressor outlet, turbine inlet, and turbine exhaust. Calculate the velocity of the exhaust jet, assuming isentropic flow through the exhaust nozzle to a pressure of 14.0 psia.

15·59 The following pressure and temperature measurements were made on an aircraft jet engine during flight: at compressor inlet, 14.2 psia, 59 F (98.0 kPa, 15°C); at compressor outlet, 161.0 psia, 660 F (1110 kPa, 349°C); at turbine inlet, 154 psia, 1570 F (1062 kPa, 855°C); and at turbine outlet, 34.8 psia, 1000 F (240 kPa, 538°C). The fuel flow rate was

165 lbm/s, and a gas analysis of the exhaust showed an air–fuel ratio of 70. The thrust was 10,000 lbf. (*a*) Assuming that the jet exit-plane pressure equals the inlet-plane pressure, estimate the difference between the velocity of the exhaust jet and that of the incoming flow. (*b*) Estimate the compressor and turbine efficiencies. State any assumptions you make.

15·60 Estimate the static takeoff thrust of a turbofan engine as shown in Fig. 15·13*b* if ambient conditions are 1 atm, 59 F (15°C), the total mass rate of flow is 1500 lbm/s (680 kg/s), one-sixth of the flow passes through the turbines, the turbine exhaust jet at 850 F (455°C) has a velocity of 1190 fps (363 m/s), the fan exhaust flow at 130 F (55°C) has a velocity of 885 fps (270 m/s), and the maximum temperature in the engine is 1970 F (1077°C).

15·61 A turbofan engine is configured as shown in Fig. 15·13*a*. At takeoff, 42 percent of the total mass flow passes through the turbines. Estimate the static takeoff thrust under ambient conditions of 1 atm, 59 F, if the compressor discharge conditions are 200 psia, 715 F; turbine inlet conditions are 190 psia, 1600 F; low-pressure turbine exhaust pressure is 28 psia; the turbine exhaust jet has a velocity of 1560 fps, and the fan exhaust has a velocity of 990 fps.

15·62 What effects do you believe a heavy rain would have on the performance of a turbojet aircraft engine in flight?

15·63 List some of the "other operating characteristics" mentioned in the first paragraph of Section 15·2.

15·64 Derive the following expression for the cold-air-standard Otto-cycle thermal efficiency; $\eta = 1 - r^{1-k}$, where r is the compression ratio.

15·65 At the beginning of compression in a cold-air-standard Otto cycle, the working substance is at 95 kPa, 30°C, and has a volume of 0.030 m^3. At the end of compression, the pressure is 950 kPa, and 10 kJ is added during the constant-volume process. Calculate (*a*) the thermal efficiency, (*b*) the amount of heat added that is available energy, and (*c*) the amount of heat rejected that is available energy. (In the absence of other information, the sink temperature should be taken as equal to the lowest temperature in the cycle.)

15·66 Solve Problem 15·65 on an air-standard basis.

15·67 For the conditions of Problem 15·65, calculate (*a*) the percentage increase of stroke necessary to allow the adiabatic expansion to proceed until the initial pressure (95 kPa) is reached, causing the heat-rejection process to be at constant pressure instead of at constant volume, and (*b*) the thermal efficiency of the modified cycle described in *a*.

15·68 Solve Problem 15·67 with the adiabatic expansion proceeding to the initial temperature (30°C) so that the heat-rejection process is isothermal.

15·69 An air-standard Otto cycle has a compression ratio of 8 and a maximum temperature of 1097°C. At the beginning of the compression stroke, the air is at 100 kPa, 25°C. Determine the maximum pressure in the cycle, the amount of heat added in kJ/kg, and the available fraction of the heat added.

15·70 Consider a cold-air-standard Otto cycle with a compression ratio of 9. At the beginning of compression, the air is at 95 kPa, 5°C (13.8 psia, 41 F); during the heat-addition process, the pressure of the air is doubled. Calculate the efficiency and the backwork ratio (How do you think it should be defined?) of this cycle and the efficiency of a Carnot cycle operating between the same overall temperature limits. What would be the minimum overall volume ratio for a Carnot cycle with the same temperature limits and the same maximum specific volume?

15·71 In a cold-air-standard Otto cycle with a compression ratio of 6, the working fluid is 0.045 kg (0.099 lbm) of air that is at 95 kPa, 5°C (13.8 psia, 41 F), at the beginning of the compression stroke. The maximum temperature in the cycle is 950°C (1740 F).

Heat is received from a constant-temperature reservoir at 950°C (1740 F) and is rejected to the atmosphere at 5°C (40 F). How much energy is made unavailable by the operation of one cycle?

15·72 At the beginning of compression in a cold-air-standard Otto cycle, the working substance is at 14 psia, 90 F, and has a volume of 1 cu ft. At the end of compression, the pressure is 140 psia, and 10 B is added during the constant-volume process. Calculate (*a*) the thermal efficiency, (*b*) the amount of heat added that is available energy, and (*c*) the amount of heat rejected that is available energy. (In the absence of other information, the sink temperature should be taken as equal to the lowest temperature in the cycle.)

15·73 Solve Problem 15·72 on an air-standard basis.

15·74 For the conditions of Problem 15·72, calculate (*a*) the percentage increase of stroke necessary to allow the adiabatic expansion to proceed until the initial pressure (14 psia) is reached, causing the heat-rejection process to be at constant pressure instead of at constant volume, and (*b*) the thermal efficiency of the modified cycle described in *a*.

15·75 Solve Problem 15·72 with the adiabatic expansion proceeding to the initial temperature (90 F) so that the heat-rejection process is isothermal.

15·76 In a brake test, an automobile having a gross mass (vehicle, fuel, and payload) of 1270 kg is accelerated from rest to 100 km/h and then braked to a stop. This cycle is carried out at a rate of 70 times per hour. During an hour of operation, how much energy must be dissipated by the brakes? Estimate the amount of fuel (gasoline) required per hour.

15·77 Refer to Fig. 15·15. For the cold-air-standard Diesel cycle, determine the compression ratio that results in $T_3 - T_4 = T_2 - T_1$.

15·78 Use a *Ts* diagram to show that for fixed conditions at the beginning of compression, the thermal efficiency of a cold-air-standard Diesel cycle decreases as the maximum temperature is raised (or as the load on the engine is increased).

15·79 A cold-air-standard Diesel cycle has a compression ratio of 16 and a cutoff ratio of 2. The cylinder volume is 0.0140 m^3 (0.494 ft^3), and at the beginning of the compression stroke the air is at 95 kPa, 5°C (13.8 psia, 41 F). The lowest temperature in the surroundings is 5°C (41 F). How much of the heat added is available energy?

15·80 Solve Problem 15·79 on an air-standard basis.

15·81 Solve Example 15·3 on a cold-air-standard basis, and compare the results of the two analyses.

15·82 Describe the differences between the exhaust stroke of an internal combustion engine and the corresponding process of an air-standard model for (*a*) a gasoline engine and (*b*) a diesel engine.

15·83 Gas turbine, gasoline, and diesel engines are all open systems. Air-standard models for these engines are usually closed systems. Explain how the first law, $Q = U_2 - U_1 + W$, applies to air-standard models that simulate open systems.

15·84 An engine indicator is an instrument that measures the pressure in a reciprocating internal combustion engine cylinder and produces a plot of pressure versus the cylinder volume. State clearly the differences between a *pv* diagram as used with air-standard cycles and an indicator diagram for an actual engine.

15·85 In the closed-system *pv* diagram shown, paths 1–2 and *a*–*b*–*c*–*d* are reversible adiabatic paths. All processes shown are reversible. (*a*) Sketch the corresponding *Ts* diagram. (*b*) Four cycles are possible: 1–2–*a*–*d*–1; 1–2–*a*–*c*–1; 1–2–*b*–*c*–1; and 1–2–*b*–*d*–1. Rank these cycles in order of thermal efficiency.

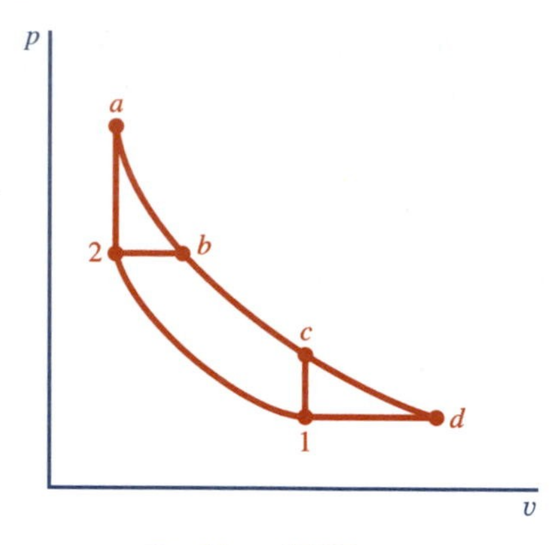

Problem 15·85

15·86 For the same temperature limits and equal amounts of heat added, which air-standard cycle has the higher thermal efficiency, Otto or Diesel? Which has the higher compression ratio?

15·87 The compression ratio for an engine operating on the cold-air-standard dual cycle is 7, the cylinder diameter is 25 cm, and the stroke is 30 cm. The air at the start of compression is at 101 kPa, 20°C. At the end of the constant-volume process, the pressure is 5600 kPa. If heat is added at constant pressure during 3 percent of the stroke, compute (*a*) the net work of the cycle, (*b*) the thermal efficiency, (*c*) the amount of heat added that is available energy, and (*d*) the amount of heat rejected that is available energy.

15·88 Compare the net work and the heat input of cold-air-standard Ericsson and Stirling cycles that have identical isothermal heat rejection processes and identical temperature limits. Superpose the two cycles on *pv* and *Ts* diagrams.

15·89 For Stirling cycles with the same temperature limits and the same heat addition, how does the work output vary with *k* of the working substance?

15·90 A compound power plant consists of a high-speed multicylinder Diesel engine that drives only a centrifugal compressor and a turbine that operates on the gases that leave the Diesel engine. Air enters the compressor at 14.7 psia, 70 F. The compressor operates adiabatically with an efficiency of 80 percent and a pressure ratio of 3. Air from the compressor is supplied to the Diesel engine. The air–fuel ratio is high because a high amount of excess air is used for scavenging and cooling. The pressure in the engine exhaust line is 43.5 psia. The brake specific fuel consumption of the Diesel engine alone when operating at 3 atm is 0.50 lbm/hp·h. The lower heating value of the fuel is 18,000 B/lbm. Twenty percent of the lower heating value of the fuel is transferred to cooling water, which is used to cool the hottest parts of the engine cylinders. Gases leaving the engine expand adiabatically to 14.7 psia through a turbine that has an efficiency of 80 percent. The turbine output is 1000 hp. Calculate the specific fuel consumption of the plant. (Note: In view of the high air–fuel ratio, treat the engine exhaust products as air. Assume that the flow through the entire plant is steady.)

15·91 Compare the thermal efficiencies of two ideal steam power plants, each operating with pressure limits of 3000 kPa and 20 kPa. One is a Carnot cycle and the other is a Rankine cycle without superheat. In each case, at the end of isothermal heat addition the steam is dry and saturated.

15·92 Determine the efficiency and the backwork ratio of a Rankine cycle operating with throttle conditions of 5.0 MPa, 400°C, and a condenser pressure of (*a*) 100 kPa, (*b*) 50 kPa, and (*c*) 10 kPa.

15·93 Determine the efficiency and the backwork ratio of a Rankine cycle operating between 5.0 MPa and 10 kPa with a throttle temperature of (*a*) 300°C, (*b*) 400°C, and (*c*) 500°C.

15·94 Solve Example 15·4 with a turbine efficiency of 0.70 and a pump efficiency of 0.65.

15·95 Solve Example 15·4 with the same throttle pressure but a 200-kPa frictional pressure drop between the steam generator and the turbine, turbine efficiency of 0.70, pump efficiency of 0.60, and a pump discharge pressure of 8300 kPa to allow for frictional pressure drops in the piping, feedwater regulator valve, and steam generator. In addition, (*d*) make an availability accounting (as in Example 8·12) and (*e*) make an energy flow diagram (as in Example 15·2).

15·96 For a Rankine cycle with throttle conditions of 10.0 MPa, 500°C, plot cycle thermal efficiency as a function of exhaust pressure in the range of 1 atm down to 0.01 atm.

15·97 For a Rankine cycle throttle temperature of 600°C, an exhaust pressure of 3.0 kPa, and a turbine efficiency of 0.74, determine the maximum throttle temperature if the moisture content of the exhaust steam is not to exceed 10 percent.

15·98 For a Rankine cycle throttle pressure of 4.0 MPa and an exhaust pressure of 4 kPa, plot thermal efficiency against throttle temperature. Vary the throttle temperature from the saturation temperature to 600°C.

15·99 For a Rankine cycle throttle temperature of 500°C and an exhaust pressure of 4 kPa, plot thermal efficiency against throttle pressure in the range of 100 kPa to 5.0 MPa. Indicate the part of the curve that involves superheated exhaust steam.

15·100 A design study is being made for a steam turbine to operate with a throttle temperature of 500°C. Cooling water is available to maintain an exhaust pressure of 3 kPa. Moisture fraction in the turbine exhaust is not to exceed 10 percent, but maximum power output for a given flow rate is desired. Determine the throttle pressure to use for turbine efficiencies of 70 percent, 80 percent, 90 percent, and 100 percent.

15·101 Steam enters a turbine at 1.50 MPa, 300°C, and leaves at 100 kPa. Assume reversible adiabatic flow. To what pressure must the incoming steam be throttled to reduce the work per unit mass to two-thirds of that obtained without throttling? Assume that the flow through the turbine remains reversible and adiabatic and that the exhaust pressure is unchanged.

15·102 Compare the thermal efficiencies of two ideal steam power plants, each operating with pressure limits of 400 psia and 1.5 inches Hg. One is a Carnot cycle, the other a Rankine cycle without superheat. In each case, the steam at the end of heat addition is dry and saturated.

15·103 Determine the efficiency and the backwork ratio of a Rankine cycle with steam entering the turbine at 400 psia, 700 F, and a condenser pressure of (*a*) 14.7 psia, (*b*) 5.0 psia, and (*c*) 1 in. Hg absolute.

15·104 A Rankine cycle operates with throttle conditions of 1000 psia, 800 F, and a condenser pressure of 1 psia. Available cooling water is at 70 F. Determine (*a*) the thermal efficiency, (*b*) the backwork ratio, and (*c*) the fraction of the heat added that is available energy.

15·105 Solve Problem 15·104 with a turbine efficiency of 0.75 and a pump efficiency of 0.60. Also, (*d*) make an availability accounting (as in Example 8·12) and (*e*) make an energy flow diagram (as in Example 15·2).

15·106 Solve Problem 15·104, allowing for a 30-psi drop in pressure between the steam-generator outlet and the turbine (i.e., the steam-generator outlet pressure is 1030 psia), a turbine efficiency of 0.70, a pump efficiency of 0.65, and a pump discharge pressure of 1300 psia to allow for pressure drops in the piping and in the steam generator.

15·107 A Rankine cycle operates with throttle conditions of 1000 psia, 800 F, and a condenser pressure of 1 psia. Available cooling water is at 70 F. (*a*) Determine the thermal efficiency. (*b*) Determine the thermal efficiency if there is a continuous blowdown of 1 percent. This means that for the purpose of water treatment, saturated water is allowed to escape from a boiler drum at a rate of 1 percent of the rate of steam generation. Makeup water in equal amount is added at the inlet to the boiler feed pump.

15·108 For a Rankine cycle with throttle conditions of 1000 psia, 900 F, plot cycle thermal efficiency as a function of exhaust pressure in the range of 1 atm down to 0.01 atm.

15·109 The thermal efficiency of a steam power plant is highly sensitive to exhaust pressure. If the cooling water temperature for a plant cannot be changed, what other steps can designers take to achieve the lowest possible pressure in the condenser?

15·110 The designers of a condenser for a steam turbine need to know what ranges of conditions are to be encountered, because the heat transfer coefficients are quite different for superheated vapor, liquid–vapor mixtures, and liquid. Sketch a surface condenser that is located directly below a steam turbine and indicate the regions where various phases of water may be present. What information can you give the designers regarding possible causes of thermal stress and extreme structural loads?

15·111 A river has limited capacity for condenser cooling because preventing environmental damage requires that the temperature rise of river water flowing through the condensers of a plant be limited. Would a combination of a cooling tower or a large spray pond with cooling water from a river be feasible for increasing the size of steam power plant that can be operated at a given site?

15·112 For a throttle temperature of 1150 F, an exhaust pressure of 0.5 psia, and a turbine efficiency of 0.76, what is the maximum throttle pressure that can be used if the moisture in the exhaust is not to exceed 10 percent?

15·113 For a throttle pressure of 600 psia and an exhaust pressure of 1 psia, plot Rankine cycle efficiency against throttle temperature in the range from the saturation temperature to 1000 F.

15·114 For a throttle temperature of 900 F and an exhaust pressure of 1 psia, plot Rankine cycle efficiency against throttle pressure in the range of 15 to 700 psia. Indicate the part of the curve that involves superheated exhaust steam.

15·115 A steam turbine is to operate with a throttle temperature of 900 F. Cooling water is available to maintain an exhaust pressure of 0.50 psia. Moisture fraction in the turbine exhaust is not to exceed 10 percent, but maximum power output for a given flow rate is desired. Determine the throttle pressure to use for turbine efficiencies of 70 percent, 80 percent, 90 percent, and 100 percent.

15·116 Steam enters a turbine at 5000 psia, 1100 F, and expands to 1 psia. At the section of the turbine where the quality reaches 90 percent, 90 percent of the liquid present is separated mechanically from the steam flow and removed from the turbine. Calculate the amount by which the ideal turbine work is changed.

15·117 Steam enters a turbine at 200 psia, 600 F, and exhausts at 14.7 psia. Assume reversible adiabatic flow. To what pressure must the incoming steam be throttled to reduce the work per pound to two-thirds of that obtained without throttling? Assume that the flow through the turbine remains reversible and adiabatic and that the exhaust pressure is unchanged.

15·118 Consider any type of steam power cycle operating with fixed conditions at the steam-generator inlet and outlet and a fixed condenser pressure. The turbine exhaust is always wet (i.e., not superheated). Demonstrate that an increase in irreversibility i anywhere in the cycle always results in an increased heat rejection in the condenser.

15·119 Mechanical design limitations are similar for both gas and steam turbines. What reasons can you give for the fact that steam power plants are built in much larger capacities than gas-turbine power plants?

15·120 Prove that the constant-pressure lines in the wet region of an *hs* diagram for steam are straight. Also prove that they are or are not parallel.

15·121 Dry saturated steam at a pressure p_1 enters a turbine at a rate of $\dot{m}_1$ lbm/h. Part way through the turbine, $\dot{m}_2$ lbm/h of dry saturated steam at a pressure p_2 is added. All of the steam is exhausted at a pressure p_3. The expansion in the turbine is adiabatic. The sink temperature is T_3. The quality of the exhaust steam is measured by means of an instrument called a throttling calorimeter. Write equations in terms of flow rates and properties at 1, 2, and 3 for (*a*) the turbine power output, (*b*) the irreversibility of the turbine process, and (*c*) the minimum possible exhaust quality if no measurement of it had been made. (Of course, without the measurement of x_3, parts *a* and *b* could not be solved.)

15·122 Liquid sodium leaving a nuclear reactor is at 20 psia, 700 F. It goes to a heat exchanger where it is cooled to 600 F before returning to the reactor. The sodium flow rate in this circuit (the *primary circuit*) is 100,000 lbm/h. In the heat exchanger, heat is transferred to liquid sodium in an *intermediate circuit,* which in turn transfers heat to a boiler that produces dry saturated steam at 1000 psia. The steam expands adiabatically through a turbine to 1 psia. The turbine efficiency is 0.60. Sketch a flow diagram showing all three (primary, intermediate, and steam) circuits. Assuming (1) negligible pressure drops due to friction, (2) negligible work required by sodium pumps, (3) negligible stray heat losses, and (4) the specific heat of sodium to be constant at 0.30 B/lbm·F, determine (*a*) the net amount of power available from the plant, (*b*) the irreversibility, in B/h, of the process of transferring energy from the primary sodium circuit through the intermediate circuit to the water.

15·123 An ideal regenerative cycle operates with throttle conditions of 3.0 MPa, 400°C; a single open feedwater heater at 250 kPa; and a condenser pressure of 30 mm Hg absolute. Compare its efficiency with that of a simple Rankine cycle having the same throttle conditions and the same exhaust pressure.

15·124 An ideal regenerative cycle operates with throttle conditions of 3.0 MPa, 400°C; one open feedwater heater at 140 kPa; one closed feedwater heater (with its drain trapped to the open heater) at 700 kPa; and a condenser pressure of 30 mm Hg absolute. The terminal temperature difference of the closed feedwater heater is 5 Celsius degrees. Determine the thermal efficiency of the cycle.

15·125 Solve Problem 15·123, assuming that the turbine efficiency between the throttle and any point in the turbine is 0.65 and any pump efficiency is 0.70. Also make an energy flow diagram similar to that of Example 15·6. Is the efficiency improvement greater with the isentropic expansions and compressions or with the irreversible ones?

15·126 An ideal regenerative steam cycle has throttle conditions of 6000 kPa, 400°C; an open feedwater heater at 170 kPa; a closed feedwater heater at 1100 kPa with a terminal temperature difference of 5 Celsius degrees and its drain pumped into the feed line on the steam-generator side of the heater; and a condenser pressure of 4 kPa. Determine the cycle efficiency and make an energy flow diagram showing the irreversibilities of various components and the fractions of the heat added and of the heat rejected that are unavailable.

15·127 An ideal regenerative steam cycle has throttle conditions of 6000 kPa, 400°C; an open feedwater heater at 105 kPa; two closed feedwater heaters, one at 400 kPa and one at 1100 kPa; and a condenser pressure of 4 kPa. Each closed heater has a terminal temperature difference of 5 Celsius degrees and its drain is pumped into the feed line on the steam generator side of the heater. Determine the cycle efficiency.

15·128 Solve Problem 15·126, assuming that the turbine efficiency between the throttle and any point in the turbine is 0.75 and that the efficiency of each pump is 0.70.

15·129 An ideal steam power plant has steam entering the turbine at 3000 kPa, 400°C, and a condenser pressure of 5 kPa. (These are the conditions of Examples 15·4 and 15·5.) A feedwater heater is to be added at 300 kPa. Determine the cycle efficiency if the feedwater heater is (*a*) an open feedwater heater, (*b*) a closed feedwater heater with its drain of saturated liquid trapped into the condenser. The terminal temperature difference of the closed heater is 5 Celsius degrees.

15·130 Solve Problem 15·129*b* if a closed heater is used and the condensate from the heater is trapped into the condenser. (*a*) Assume that the feedwater is heated to the saturation temperature of the heater. (*b*) Assume that the feedwater leaves the heater at a temperature 10 Celsius degrees lower than the saturation temperature of the heater.

15·131 Calculate the irreversibility per kilogram of throttle steam of the operation of each feedwater heater of Problem 15·129.

15·132 A regenerative steam power-plant cycle uses one open and one closed feedwater heater. The closed heater operates at a higher pressure than the open one. There are three ways that the condensate drain from the closed heater can be handled: (*a*) It can be pumped into the feedwater line between the closed heater and the boiler. (*b*) It can be trapped into the open heater. (*c*) It can be trapped into the condenser. From the standpoint of cycle efficiency, which method is the best? Which is the worst? Explain your reasoning.

15·133 Most steam power plant equipment and piping are insulated to minimize heat losses to the surroundings. For a given plant operating steadily, does a heat loss from a high-pressure steam line (for example, between a turbine bleed point and the highest-pressure heater) have the same effect on cycle efficiency as an equal magnitude of heat loss from a low-pressure line (for example, the bleed line to the lowest-pressure heater)? Explain your reasoning.

15·134 An ideal regenerative cycle operates with throttle conditions of 400 psia, 700 F; a single open feedwater heater at 35 psia; and a condenser pressure of 1 in. Hg absolute. Compare its efficiency with that of a simple Rankine cycle having the same throttle conditions and exhaust pressure.

15·135 An ideal regenerative cycle operates with throttle conditions of 400 psia, 700 F; one open feedwater heater at 20 psia; one closed feedwater heater (with its drain trapped to the open heater) at 100 psia; and a condenser pressure of 1 in. Hg absolute. The terminal temperature difference of the closed feedwater heater is 8 Fahrenheit degrees. Determine the thermal efficiency of the cycle.

15·136 Solve Problem 15·134, assuming that the turbine efficiency between the throttle and any point in the turbine is 0.65 and the boiler feed pump efficiency is 0.70. Also make an energy flow diagram similar to that of Example 15·6.

15·137 An ideal regenerative steam cycle has throttle conditions of 1000 psia, 800 F; an open feedwater heater at 25 psia; a closed feedwater heater at 150 psia with a terminal temperature difference of 10 degrees and its drain pumped into the feed line on the steam-generator side of the heater; and a condenser pressure of 1 psia. Determine the cycle efficiency and make an energy flow diagram showing the irreversibilities of various components and the fractions of the heat added and of the heat rejected that are unavailable.

15·138 Solve Problem 15·137, assuming that the turbine efficiency between the throttle and any point in the turbine is 0.75 and that the efficiency of each pump is 0.70.

15·139 From the standpoint of cycle efficiency, determine to within 20 psi the optimum pressure for a single open feedwater heater in an ideal steam power plant cycle operating between throttle conditions of 1000 psia, 800 F, and an exhaust pressure of 1 psia.

15·140 Consider a regenerative steam cycle with one open heater. If the turbine expansion is adiabatic but irreversible, the values of several flow and energy quantities in the cycle may be different from their values with isentropic turbine expansion. Explain clearly how and why each of the following values is affected: (*a*) the mass of steam bled, (*b*) the heat rejected in the condenser per unit mass of throttle steam, and (*c*) the heat rejected in the condenser per unit mass of steam entering the condenser.

15·141 An ideal reheat cycle with no regeneration has throttle conditions of 14.0 MPa, 450°C, and reheats at 3.8 MPa to 480°C. (As mentioned in Section 15·3·3, reheat is never used in practice unless accompanied by regeneration, so this problem and the following two are academic in that they are posed only to provide exercises on reheat calculations in simplified form.) The exhaust pressure is 5 kPa. Compute the thermal efficiency of the cycle. Per kilogram of throttle steam, how much unavailable energy is rejected by the cycle if the sink temperature is 25°C?

15·142 Solve Problem 15·141 for a high-pressure turbine efficiency of 0.80 and a low-pressure turbine efficiency of 0.85. Calculate also the effectiveness of each turbine.

15·143 Determine the efficiency and the required flow rate of an ideal reheat cycle that is to produce 150,000 kW at the turbine coupling if the throttle conditions are 15.0 MPa, 600°C; reheat is at 1.4 MPa to 600°C; and the condenser pressure is 5 kPa.

15·144 Solve Problem 15·143 if a closed feedwater heater with a terminal temperature difference of 3 Celsius degrees is supplied with steam from the high-pressure turbine exhaust. The heater drain is pumped into the feed line.

15·145 Determine the efficiency and the required steam generator flow rate for an ideal reheat–regenerative cycle that is to produce 150,000 kW at the turbine coupling if the throttle conditions are 15.0 MPa, 600°C; reheat is at 1.4 MPa to 600°C; there are closed feedwater heaters at 1400 kPa, 300 kPa, and 50 kPa; there is an open feedwater heater at 150 kPa; and the condenser pressure is 7 kPa. The drain from each closed heater is trapped into the next lower heater except that the drain from the 50-kPa heater is pumped into the open heater.

15·146 Steam is to be supplied to a turbine at a maximum pressure of 7000 kPa and a maximum temperature of 540°C, and exhausted to a condenser at 5 kPa. It is desired to have no moisture entering a reheater, no more than 10 percent moisture entering the condenser, a reheat temperature not exceeding 480°C, and equal enthalpy drops across the turbines. For a single reheat turbine, determine a suitable throttle pressure, throttle temperature, reheat pressure, and reheat temperature, assuming isentropic turbine expansions.

15·147 Solve Problem 15·146 using an efficiency of 80 percent for each turbine.

15·148 Steam is to be supplied to a turbine at 35.0 MPa, 600°C, and exhausted to a condenser at 5 kPa. For isentropic expansion it is desired to have no moisture entering a reheater and not more than 10 percent moisture entering the condenser. These conditions call for a double-reheat arrangement. If reheat temperatures are not to exceed 550°C and it is desired to have the same enthalpy drop across each turbine, determine the reheat pressures to be used.

15·149 Solve Problem 15·148 using an efficiency of 80 percent for each turbine.

15·150 A nuclear reactor provides saturated steam at 7000 kPa for a steam power cycle as shown in the figure. The separator mechanically removes moisture from the steam passing through it so that the steam leaving (state 6) contains only 1 percent moisture. The steam leaving the first reheater is at T_7 = 250°C, and that leaving the second reheater is at T_8 = 280°C. Saturated liquid leaves the separator and each of the reheaters and feedwater heaters. The terminal temperature difference of each of the closed feedwater heaters is 5 Celsius degrees. Immediately downstream from

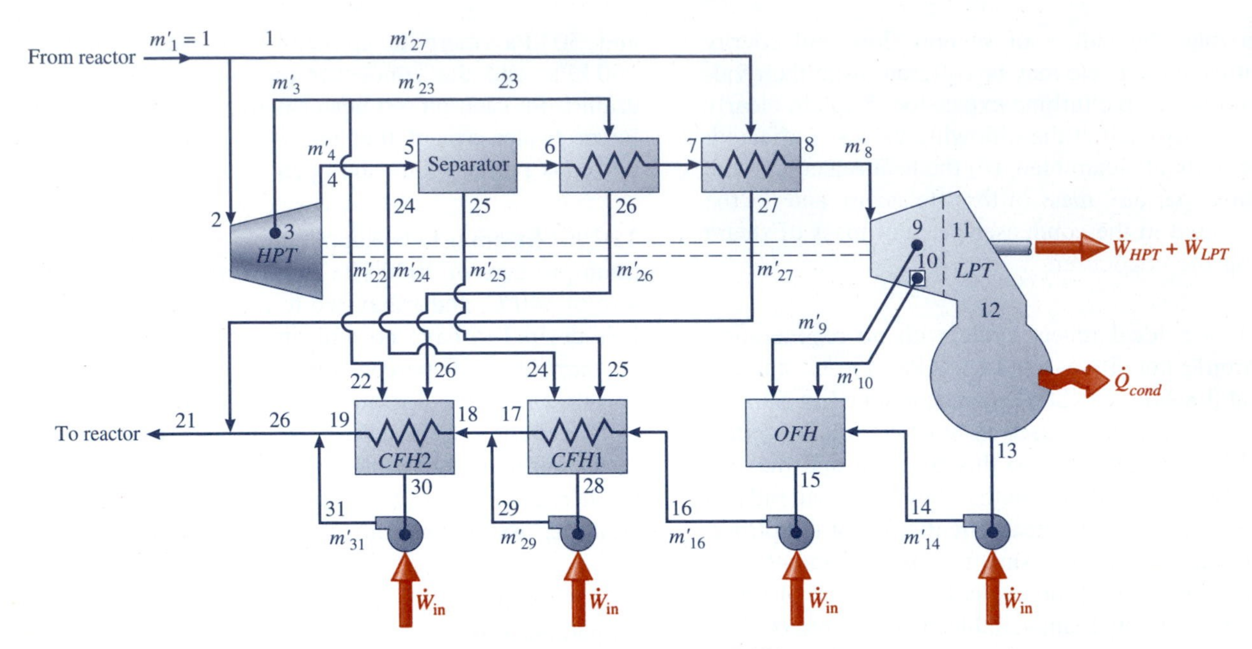

Problems 15·150 and 15·151

the bleed point (state 9) in the low-pressure turbine, a separator removes liquid water so that the quality of the steam starting the remainder of the expansion through the turbine is $x_{11} = 0.96$. Disregard frictional pressure drops in piping and heat exchangers and assume that turbine expansions are isentropic. Pressures are $p_1 = p_{27} = p_{16} = p_{21} = 7000$ kPa, $p_3 = p_{26} = p_{30} = 4200$ kPa, $p_4 = p_{24} = p_{25} = p_{28} = p_8 = 2500$ kPa, $p_9 = p_{10} = p_{11} = p_{15} = 85$ kPa, $p_{12} = p_{13} = 5.0$ kPa. For a net plant power output of 100,000 kW, determine the steam flow rate through the reactor and the cycle thermal efficiency.

15·151 A nuclear reactor provides saturated steam at 1000 psia for a steam power cycle as shown in the figure. The separator mechanically removes moisture from the steam passing through it so that the steam leaving (state 6) contains only 1 percent moisture. The steam leaving the first reheater is at $T_7 = 480$ F, and that leaving the second reheater is at $T_8 = 540$ F. Saturated liquid leaves the separator and each of the reheaters and feedwater heaters. The terminal temperature difference of each of the closed feedwater heaters is 8 Fahrenheit degrees. Immediately downstream from the bleed point (state 9) in the low-pressure turbine, a separator removes liquid water so that the quality of the steam starting the remainder of the expansion through the turbine is $x_{11} = 0.96$. Disregard frictional pressure drops in piping and heat exchangers and assume that turbine expansions are isentropic. Pressures are $p_1 = p_{27} = p_{16} = p_{21} = 1000$ psia, $p_3 = p_{26} = p_{30} = 600$ psia, $p_4 = p_{24} = p_{25} = p_{28} = p_8 = 350$ psia, $p_9 = p_{10} = p_{11} = p_{15} = 12$ psia, $p_{12} = p_{13} = 0.70$ psia. For a net plant power output of 100,000 kW, determine the steam flow rate through the reactor and the cycle thermal efficiency.

15·152 Determine the efficiency and the required steam generator flow rate for an ideal reheat–regenerative cycle that is to produce 150,000 kW at the turbine coupling if the throttle conditions are 2400 psia, 1100 F; reheat is at 200 psia to 1050 F; there are closed feedwater heaters at 200 psia, 40 psia, and 8 psia; there is an open feedwater heater at 20 psia; and the condenser pressure is 1 psia. The drain from each closed heater is trapped into the next lower heater except that the drain from the 8-psia heater is pumped into the open heater.

15·153 Steam is to be supplied to a turbine at a maximum pressure of 1000 psia and a maximum temperature of 1000 F, and exhausted to a condenser at 1 psia. It is desired to have no moisture entering a reheater, no more than 10 percent moisture entering the condenser, a reheat temperature not exceeding 900 F, and equal enthalpy drops across the turbines. For a single reheat turbine, determine a suitable throttle pressure, throttle temperature, reheat pressure, and reheat temperature, assuming isentropic turbine expansions.

15·154 Solve Problem 15·153 using an efficiency of 80 percent for each turbine.

15·155 Steam is to be supplied to a turbine at 5000 psia, 1100 F, and exhausted to a condenser at 1 psia. For isentropic expansion it is desired to have no moisture entering a reheater and not more than 10 percent moisture entering the condenser. These conditions call for a double-reheat arrangement. If reheat temperatures are not to exceed 1000 F and it is desired to have the same enthalpy drop across each turbine, determine the reheat pressures to be used.

15·156 Solve Problem 15·155 using an efficiency of 80 percent for each turbine.

15·157 Consider a reheat cycle as shown in Fig. 15·26 with four turbines on the same shaft driving a single generator. The load on the generator is suddenly reduced from full load to zero. Steam flow through the turbines must be stopped immediately or they will overspeed, but some flow of steam must be maintained through the superheater and reheater to keep them from overheating before the furnace or reactor cools. Make a sketch showing what valves are needed in the system to take care of this emergency, each one's normal position, and its action when the turbine load is suddenly dropped.

15·158 Comment on the advisability of extracting steam for regenerative feedwater heating from the line between a reheater and the following turbine.

15·159 For a minimum river flow rate of 1600 cfs and an allowable temperature rise not greater than 3 Fahrenheit degrees, what is the maximum power output from a steam power plant at a site that uses only river water (no cooling tower or cooling pond) for condenser cooling? (This kind of calculation based on assumptions of typical or general values is sometimes called a *scoping* calculation. Scoping calculations are a powerful tool in engineering.)

15·160 Does the statement of Problem 15·159 suggest that the temperature rise of the cooling water passing through the condenser is 3 Fahrenheit degrees? Explain.

15·161 Estimate the air flow rate (in m^3/min or cfm) into a modern 200,000-kW steam power plant that burns a high-grade coal as fuel. State the probable accuracy of your estimate.

15·162 Estimate the ratio of the air flow rate into a modern 60,000-kW gas-turbine power plant to that into a modern steam power plant of the same power output burning the same fuel. Indicate the probable accuracy of your result.

15·163 Estimate the air flow rate through a modern 40,000-kW power plant operating at full load if it is a (*a*) gas-turbine plant, (*b*) steam plant.

15·164 Estimate the amount of coal (in tons per day) required by a modern steam power plant that operates with an average power output of 250 MW.

15·165 An available solar collector can generate dry saturated steam at 420 F in a site where atmospheric air temperature varies from 40 F to 100 F. No cooling water is available. Estimate the steam flow rate needed to provide a power output of 5000 kW.

15·166 A steam accumulator is an energy-storage device that consists of an insulated tank that normally contains both liquid water and steam. Steam is admitted below the liquid surface, and steam is withdrawn from the highest point in the tank. A certain accumulator contains 4.0 m^3 of vapor and 20 m^3 of liquid at 1200 kPa. For how long a period, with no steam entering, can steam at 300 kPa be withdrawn through a pressure-reducing valve at a steady rate of 2000 kg/h? (One possible approximation is to use a constant mean enthalpy value for the steam leaving.)

15·167 An accumulator (see Problem 15·166) is filled with 5.0 m^3 of steam and 10.0 m^3 of liquid water at 300 kPa. The accumulator is charged from a line carrying steam at 1200 kPa, 250°C. If steam is supplied to the accumulator at a steady rate of 2000 kg/h, how long will it take to charge it to 1200 kPa?

15·168 If a steam accumulator is alternately charged with dry saturated steam at 1200 kPa and discharged to pressures as low as 300 kPa with only dry saturated vapor leaving, is it necessary to add or remove water periodically to keep the accumulator operating? Explain.

15·169 An accumulator is filled with 200 cu ft of steam and 400 cu ft of liquid water at 50 psia. The accumulator is charged from a line carrying steam at 200 psia, 500 F. If steam is supplied to the accumulator at a steady rate of 4000 lbm/h, how long will it take to charge it to 200 psia?

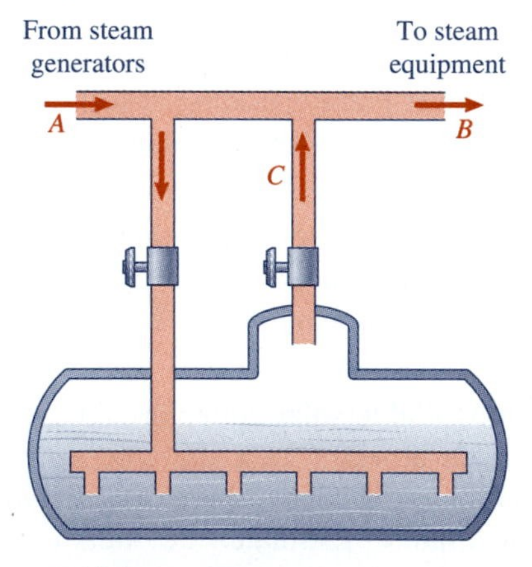

Problems 15·166 through 15·169

15·170 Steam is supplied to a turbine at 1400 kPa, 300°C. The turbine, which has an efficiency of 65 percent, exhausts at 100 kPa into a heating system that supplies 50,000 kJ/h to heat buildings. Condensate leaves the heating system at 50°C. Determine the power output of the turbine.

15·171 A large building is heated by having steam that is dry and saturated at 1.5 MPa throttled through a pressure-reducing valve into the heating system at 100 kPa, where it is condensed and cooled to 40°C. The heating system delivers 20,000 kJ/s to the air in the building at 20°C. It is proposed to replace the pressure-reducing valve with a turbine that has an efficiency of 60 percent in order to obtain some power. For the same heating system energy requirement and the same temperature of condensate leaving the heating system, determine the required mass rate of flow through the turbine in kg/s and the power output.

15·172 A steam turbine operates with throttle conditions of 1500 kPa, 250°C. At 250 kPa steam is extracted to provide 25×10^6 kJ/h for heating. Condensate from the heating system is at 200 kPa, 85°C, when it enters a trap that discharges into an open tank. The turbine exhausts into a condenser where the pressure is 5 kPa. Condensate, which is cooled to 30°C in the condenser, is pumped into the same open tank that collects the heating system condensate. From this tank, water is pumped into the steam generator. Turbine efficiency between the throttle and any point in the turbine is estimated to be 65 percent. Atmospheric conditions are 92 kPa, 20°C. The power output is 4000 kW. Determine the required throttle flow rate.

15·173 A plant is designed to supply a power load of 5000 kW and a heating load of 50 million kJ/h. Steam is generated at 4.5 MPa, 450°C, and is expanded in a turbine exhausting to a condenser at a pressure of 50 mm Hg absolute. Assume that expansion is isentropic and that the condensate leaves the condenser as saturated water. The heating load is supplied by steam at 150 kPa, which is condensed and subcooled to 50°C and returned to the boiler at that temperature. Two types of plant can be used: (1) power produced by a condensing turbine and heating steam supplied by throttling high-pressure steam, and (2) power produced by an automatic extraction turbine and heating steam supplied by bleeding 150 kPa steam from the turbine. Compute for each type (*a*) the amount of steam required per hour, (*b*) the heat input to the boiler, kJ/h, and (*c*) the heat rejected in the condenser, kJ/h.

15·174 Steam enters an extraction turbine at 200 psia, 500 F; extraction occurs at 20 psia, $h =$ 1140 B/lbm; and steam entering the condenser is at 1 psia, $h = 1005$ B/lbm. Part of the extracted steam goes to an open feedwater heater, which operates at 20 psia, and the rest goes to a heating system that provides 30 million B/h and from which condensate at 152 F is pumped into the open heater. Saturated liquid leaves the open heater. The turbine power output is 3500 kW. Determine (*a*) the efficiency of the turbine between the throttle and the extraction point, (*b*) the irreversibility of the turbine expansion upstream of the extraction point, and (*c*) the flow rate of steam into the turbine.

15·175 A plant is designed to supply a power load of 5000 kW and a heating load of 45 million B/h. Steam is generated at 650 psia, 850 F, and is expanded in a turbine exhausting to a condenser at a pressure of 2 in. Hg absolute. Assume that expansion is isentropic and that the condensate leaves the condenser as saturated water. The heating load is supplied by steam at 22 psia that is condensed and subcooled to 120 F and returned to the boiler at that temperature. Two types of plant can be used: (1) power produced by a condensing turbine and heating steam supplied by throttling high-pressure steam, and (2) power produced by an automatic extraction turbine and heating steam supplied by bleeding 22-psia steam from the turbine. Compute for each type (*a*) the amount of steam required per hour, (*b*) the heat input to boiler, B/h, and (*c*) the heat rejected in the condenser, B/h.

15·176 In a manufacturing plant, steam for heating purposes is obtained at 20 psia by drawing it through a pressure-reducing valve from a line where it is dry and saturated at 100 psia. The heating load is 20 million B/h. Condensate leaves the heating system at 20 psia, 180 F. It is proposed to use a turbine in place of the pressure-reducing valve to obtain cogenerated or ''by-product'' power while still carrying the same heating load. Assuming a turbine efficiency of 60 percent, determine (*a*) the amount of power that can be obtained with the proposed arrangement, and (*b*) the change in steam flow rate, in lbm/h, that will be required if the proposal is adopted.

15·177 A steam plant is to produce 7500 hp at the turbine coupling and supply a heating load of 70 million B/h by means of steam extracted at 30 psia, that is condensed and subcooled to 200 F in the heating system. Condenser pressure is 1 psia and the condenser flow rate must not be less than 5000 lbm/h. Determine the optimum steam-generator pressure for balanced operation under these conditions, and calculate the amount of heat that must be added in B/h. A throttle pressure below 300 psia should be accompanied by a throttle temperature of 600 F; 300 to 600 psia, by 700 F; and over 600 psia, by 800 F.

15·178 An industrial heating and power plant has a base load of approximately 15,000 kW, but for approximately five to seven hours per day the power load may be as high as 24,000 kW. In addition, a manufacturing line must be supplied with between 4000 and 7000 lbm/h of steam at 15 psig. Maintainability constraints are that steam pressures over 750 psia not be used and steam temperatures over 900 F not be used. Gas-turbine throttle temperatures over 1600 F are not to be used. It is suggested that a combined gas-turbine-steam plant be used. Investigate this possibility, using reasonable machine efficiencies, and specify gas and steam turbine throttle and exhaust conditions as well as flow rates.

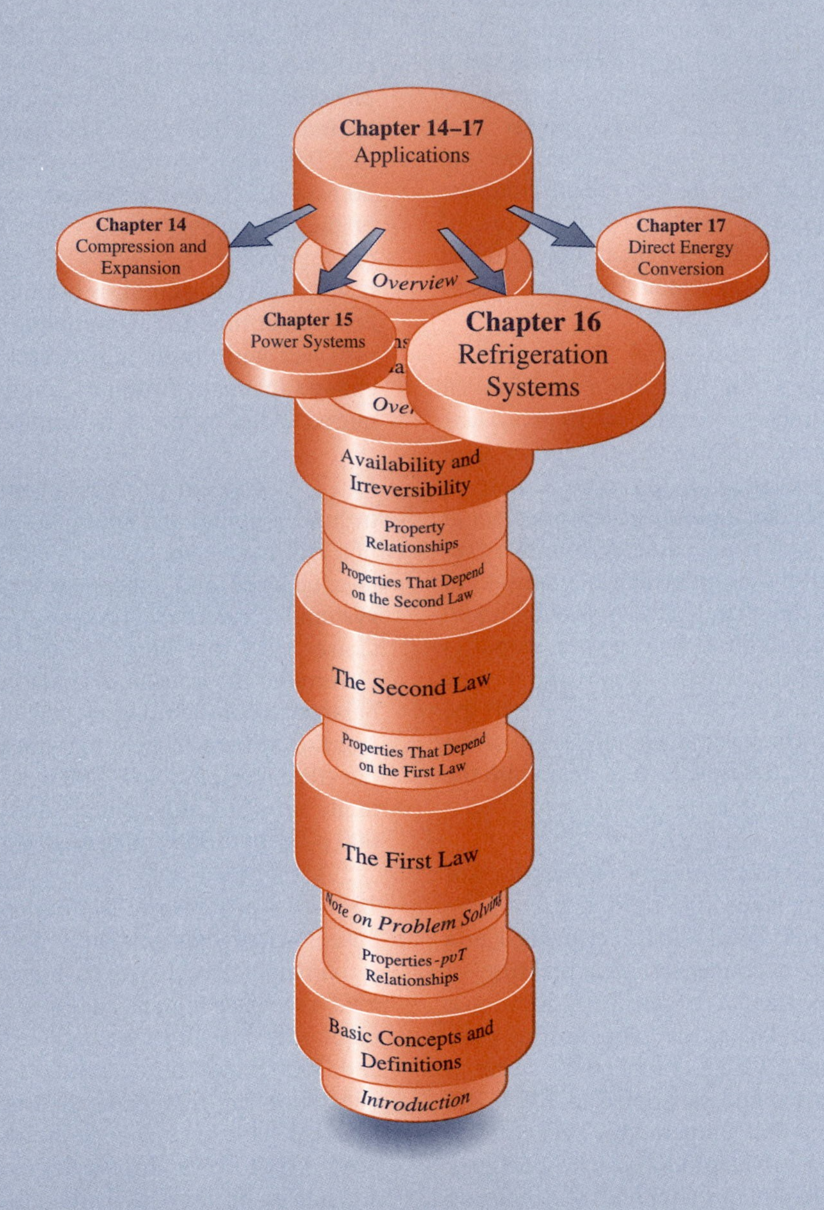

Chapter 14–17
Applications
Chapter 14
Compression and Expansion
Chapter 17
Direct Energy Conversion
Overview
Chapter 15
Power Systems
Chapter 16
Refrigeration Systems
Availability and Irreversibility
Property Relationships
Properties That Depend on the Second Law
The Second Law
Properties That Depend on the First Law
The First Law
Note on Problem Solving
Properties-pvT Relationships
Basic Concepts and Definitions
Introduction

Chapter 16

Refrigeration Systems

A refrigerating system removes heat from one part of its surroundings and discharges heat to a warmer part. One example is a household refrigerator that removes heat from a food compartment and discharges heat to the air in the room. Another example is a heat pump that removes heat from outdoor air and delivers heat to a building to keep the building warmer than the outdoor air. The only difference between a refrigerator and a heat pump is in their purpose: a refrigerator's primary purpose is to remove heat from a low-temperature region; a heat pump's primary purpose is to deliver heat to a high-temperature region. Thermodynamically, they are the same, and both come under the general classification of refrigerators. If any refrigerator operates cyclically (so that there is no net change in its stored energy), the energy transfers are as shown in Fig. 16·1*a*. The first law tells us that $W_{in} = Q_H - Q_L$ and the second law tells us that $W_{in} > 0$.

This chapter discusses a few refrigeration systems as further illustrations of the use of the first law, the second law, and physical property relationships, including the behavior of binary mixtures discussed in Chapter 10. (Your study of this chapter will be more rewarding if you refer to a handbook

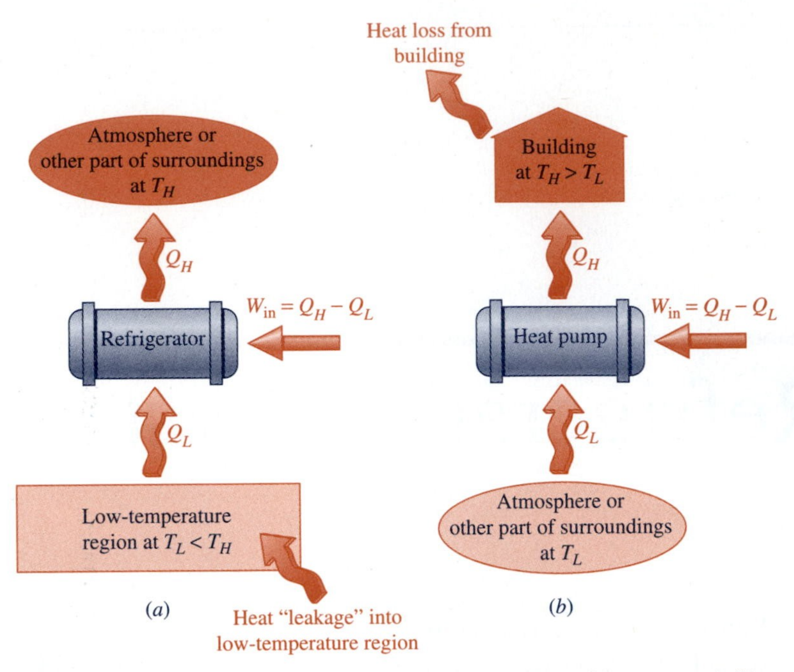

Figure 16·1 Refrigeration cycle applications as (*a*) a refrigerator and (*b*) a heat pump.

or other sources for physical descriptions, including pictures, of the various cycle components mentioned. Reference 16·1 is a good place to start.) Refrigeration systems can be classified as gas or vapor systems just as power systems were in Chapter 15.

16·1 Refrigeration Performance Parameters

The coefficient of performance β_R of any refrigerator is defined in Section 3·6 as

Refrigerator coefficient of performance

$$\beta_R \equiv \frac{Q_L}{W_{in}}$$

where Q_L is the heat absorbed from the low-temperature body. The coefficient of performance β_{HP} of a heat pump is defined as

Heat pump coefficient of performance

$$\beta_{HP} = \frac{Q_{out,H}}{W_{in}}$$

where $Q_{out,H}$ is the heat delivered to the high-temperature part of the surroundings.

The Carnot cycle is used as a standard of comparison for heat-engine cycles because its efficiency is the maximum for given temperature limits. In a similar manner, the reversed Carnot cycle introduced in Section 5·5·2 is used as a standard of comparison for refrigerators and heat pumps because for given temperature limits its coefficient of performance is a maximum.

The reversed Carnot cycle is used as a standard of comparison.

Application of the first law and the important relationship

$$\frac{Q_H}{Q_L} = \frac{T_H}{T_L} \tag{7·3}$$

(which applies only to externally reversible cycles) shows that the coefficient of performance of the Carnot refrigerator is

$$\beta_{R,\text{Carnot}} = \frac{T_L}{T_H - T_L}$$

For a Carnot heat pump, the coefficient of performance is

$$\beta_{HP,\text{Carnot}} = \frac{T_H}{T_H - T_L}$$

Figure 16·2 shows *Ts* diagrams for reversed Carnot cycles using different working substances. The cycle always involves two isothermal processes and two isentropic processes for any working substance, but Fig. 16·2 also shows the saturation line for two cycles that use a vapor. For each diagram, the heat absorbed at temperature T_L during process 1–2 is represented by area 1–2–*b*–*a*–1; the heat rejected at T_H during process 3–4 is represented by area 3–4–*a*–*b*–3; and the net work input is therefore represented by the difference between these two areas, which is area 1–2–3–4–1.

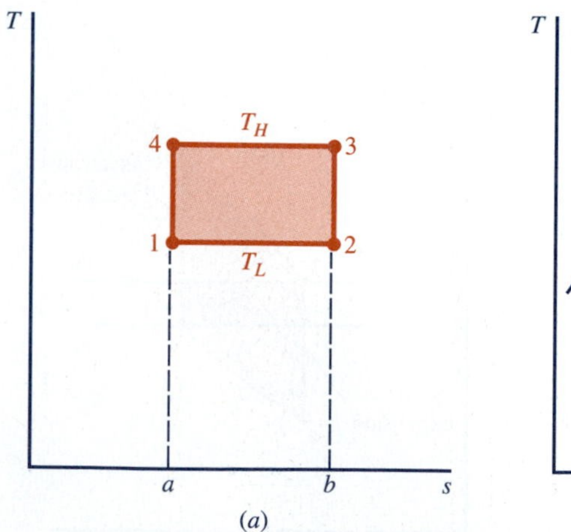

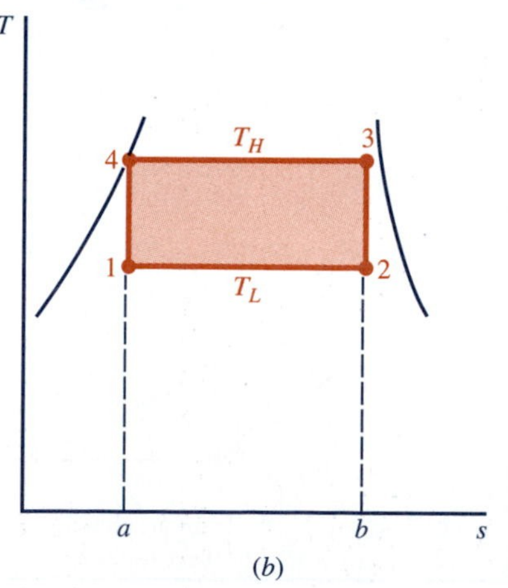

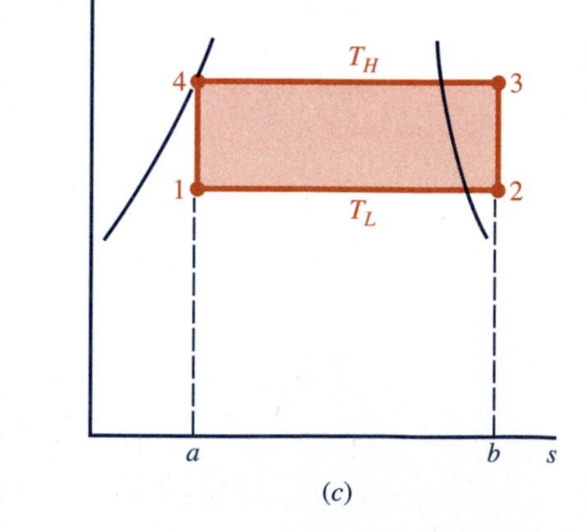

Figure 16·2 Reversed Carnot cycle *Ts* diagrams for (*a*) any working substance, (*b*) a saturated vapor, and (*c*) a vapor that is sometimes saturated, sometimes superheated.

16·2 Modeling Gas Refrigeration Systems: The Reversed Brayton Cycle

Actual refrigeration cycles using gaseous working substances are not based on the reversed Carnot cycle because of the practical difficulties involved in carrying out the processes. An ideal cycle that can be more easily simulated in practice is the reversed Brayton cycle. Figure 16·3 shows flow, *pv*, and *Ts* diagrams of a reversed Brayton cycle. T_H and T_L are the temperatures of those parts of the surroundings to which heat is to be rejected and from which heat

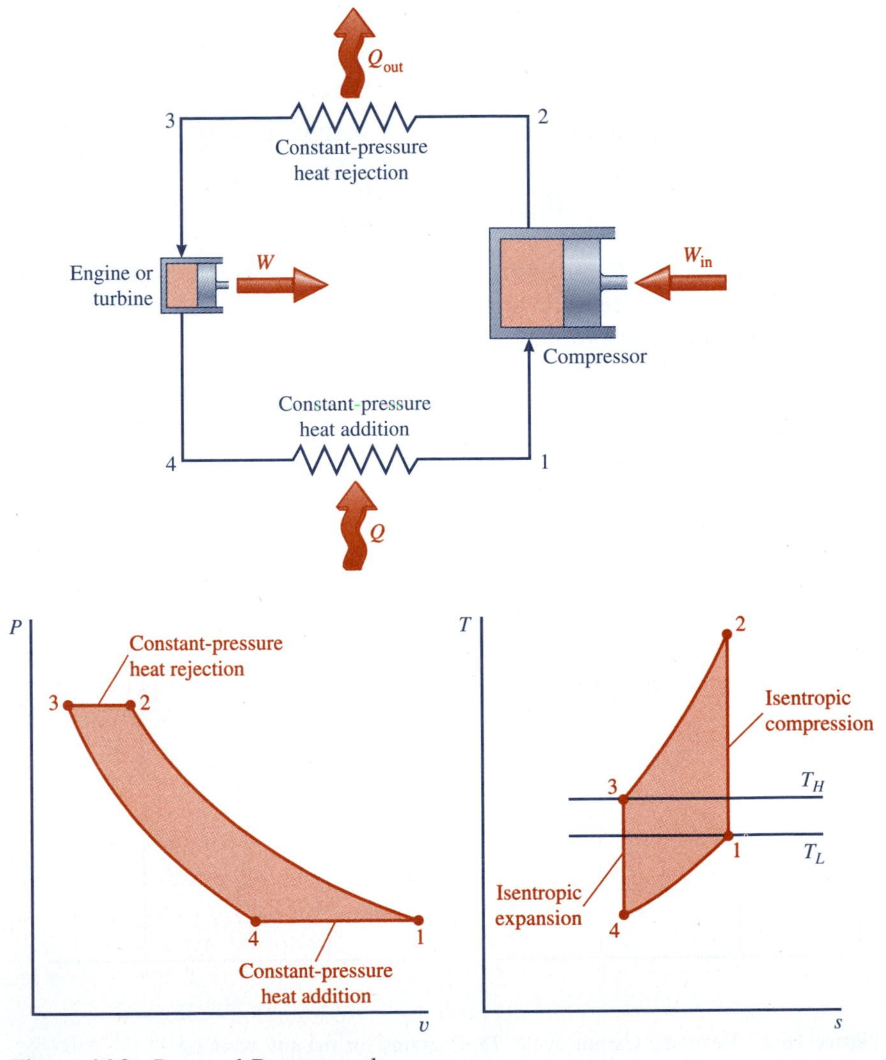

Figure 16·3 Reversed Brayton cycle.

is to be removed, respectively. The working fluid is usually air. The air is compressed isentropically to a temperature above T_H, and then it is cooled reversibly at constant pressure until its temperature is T_H or slightly higher than T_H. The air is then expanded isentropically through an engine or turbine that supplies some of the power requirement of the compressor. The temperature of the air passing through the engine drops to a value lower than T_L. Therefore, heat can be absorbed from the surroundings at T_L as the air flows at constant pressure through a heat exchanger to state 1 to complete the cycle. Application of the first law to each of the four pieces of equipment in the steady-flow cycle, assuming that $\Delta ke + \Delta pe = 0$, gives

The constant-pressure cooling process is *externally* reversible because the heat is transferred across a finite temperature difference between the gas being cooled (process 2–3) and the surroundings at T_H.

$$w_{\text{in},C} = h_2 - h_1$$
$$q_{\text{out},2-3} = h_2 - h_3$$
$$w_{eng} = h_3 - h_4$$
$$q_{4-1} = h_1 - h_4$$

There is a net work input to the cycle because the work input to the compressor is greater than the engine work. This can be seen from either the *pv* or the *Ts* diagram of Fig. 16·3 and is, of course, in accordance with the second law.

The engine of the reversed Brayton cycle cannot be successfully replaced by a throttling valve if the air in the cycle behaves as an ideal gas because for an ideal gas the Joule–Thomson coefficient, $(\partial T/\partial p)_h$, is zero. That is, there is no temperature change across a throttling valve, since there is no change in enthalpy. For an indication of how closely air at low temperatures approximates an ideal gas, refer to the *Ts* diagram for air in the appendix, Chart A·13, and recall that for an ideal gas, constant-enthalpy lines coincide with constant-temperature lines. In addition, in the temperature ranges normally encountered in refrigeration cycles, the specific heats of air can be treated as constant with very little loss in accuracy.

In a reversed Brayton cycle, the expander cannot be replaced by a throttling valve if the refrigerant is an ideal gas.

Recall that for any process the irreversibility is given by

$$I = T_0\,\Delta S_{univ} \tag{8·8}$$

where T_0 is the temperature of the atmosphere. If, instead of the atmosphere, some other energy reservoir such as the water of a river or lake is used as a source or absorber of heat, then its temperature is used as T_0. In the case of a refrigeration cycle, it is not correct to refer to T_0 as "the lowest temperature in the surroundings" unless it is understood that the temperature of the body being cooled is excluded from consideration because that body is certainly not an energy reservoir that can absorb or reject large amounts of heat without experiencing a temperature change. Referring to Fig. 16·1, notice that for the refrigerator in part *a*, $T_H = T_0$, while for the heat pump in part *b*, $T_L = T_0$.

A drawback of an air refrigeration system is that the low density of air at moderate pressures calls for either very high pressures or very high volume flow rates in order to obtain moderate refrigerating capacity. The early solution to this problem was to use very high pressures in order to limit the

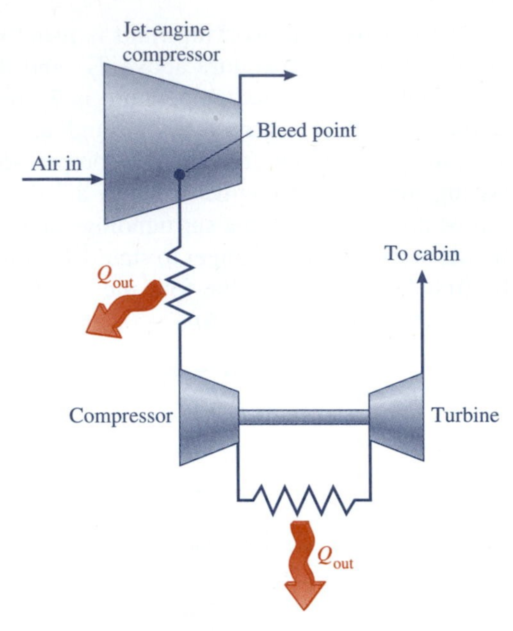

Figure 16·4 Aircraft cooling system.

physical size of the equipment. Consequently, the early systems were called "dense air refrigerating machines." Such machines have been completely displaced by vapor-compression systems, which are cheaper to build because of the lower pressures and smaller volumes involved, and cheaper to operate because their coefficients of performance are higher for any specified temperature range. The major application for air refrigerating systems is now aircraft cabin or cockpit cooling where cool air must be supplied and low system weight is desirable. Small high-speed, low-weight compressors and turbines that handle large volumes of air are used. Figure 16·4 shows an example of an aircraft cooling system. Some air is bled from the jet-engine compressor, cooled in a heat exchanger, and then compressed further by a small compressor that is driven by the turbine. The air is cooled in a heat exchanger between the compressor discharge and the turbine inlet. The cool air leaving the turbine is then ducted to the cabin. Many variations of this cycle are possible. A significant problem with such systems in high-speed aircraft is cooling the air in the two heat exchangers, because the ambient air used as the cooling fluid experiences a temperature rise when it is brought into the craft and decelerated relative to the craft.

Reversed Brayton cycle refrigerators were essentially replaced by vapor-compression machines, but then the special needs of aircraft cooling systems brought them back into practical use.

Test yourself: What is the relationship between the temperature change and the deceleration?

A *Ts* diagram shows the chief disadvantage of the reversed Brayton cycle from a thermodynamic standpoint. Figure 16·5 shows a reversed Brayton cycle 1–2–3–4–1 that absorbs the same amount of heat as the Carnot cycle a–2–b–c–a and exchanges heat with the same energy reservoirs at T_H and T_L. (The area beneath 1–2 equals the area beneath a–2.) The area within each diagram represents the work input, and it is apparent that the coefficient of

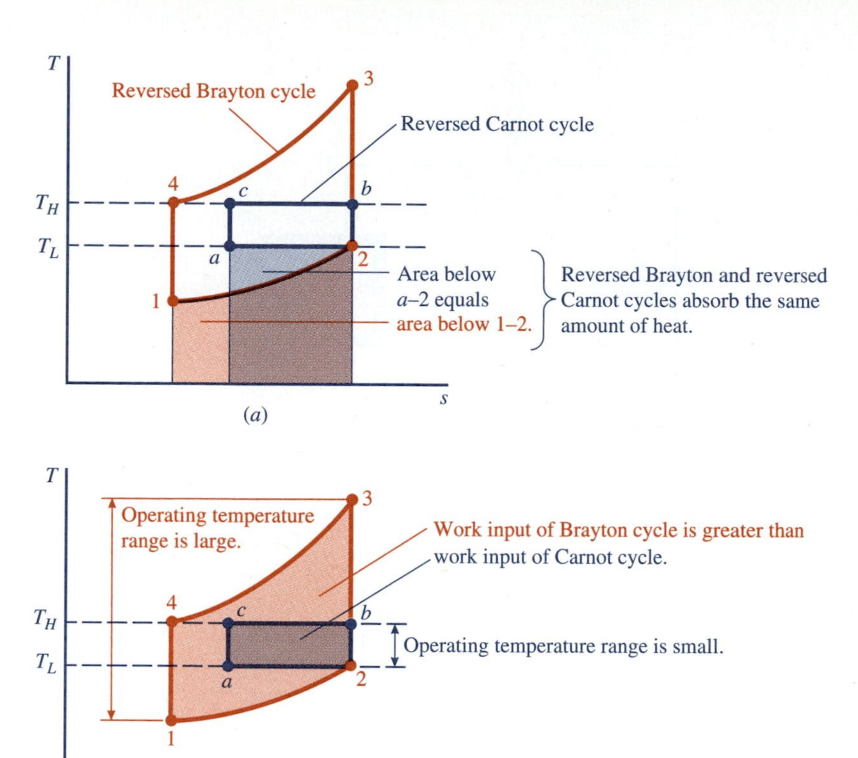

Figure 16·5 Comparison of reversed Brayton cycle and reversed Carnot cycle for the same surroundings temperatures and the same heat removal from the low-temperature region.

performance of the reversed Brayton cycle is much lower than that of the reversed Carnot cycle. The constant-pressure heat-transfer processes result in an operating temperature range much greater than the minimum range established by the temperatures of those parts of the surroundings with which heat is exchanged. In a reversed Brayton cycle, the compression is isentropic; in an actual system, the compression is certainly not reversible, and heat may be removed during compression.

16·3 Modeling Vapor Refrigeration Systems: Vapor-Compression Cycles

Figure 16·6 shows flow, *Ts*, and *ph* diagrams of a conventional vapor-compression refrigerating system. Wet vapor enters the compressor at state 1 and is compressed reversibly and adiabatically to state 2. The vapor then

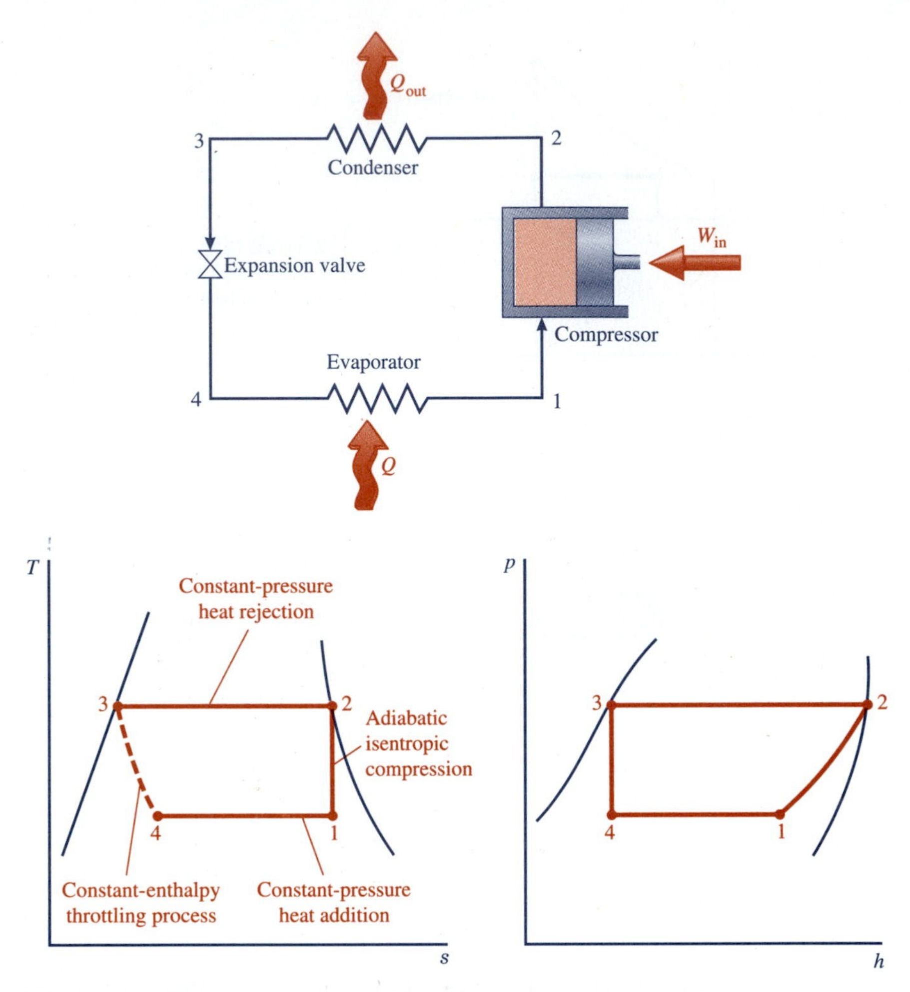

Figure 16·6 Vapor-compression refrigeration cycle: wet compression.

flows into a condenser where heat is removed to condense the vapor to a saturated liquid at the same pressure, state 3. The refrigerant then expands through a valve or capillary tube to state 4. For the throttling process, $h_4 = h_3$. Some of the liquid is vaporized as it passes through the expansion valve, so a low-quality mixture enters the evaporator, and this mixture is at a temperature lower than that of the body being refrigerated. In the evaporator, heat is absorbed from the body being refrigerated, and in this process, most of the liquid evaporates. Thus the cycle is completed when the refrigerant leaves the evaporator and enters the compressor in state 1. (More pictorial diagrams of vapor-compression refrigeration systems are given in the introductory chapter, Chapter A.)

Notice that three of the four processes in the ideal vapor-compression refrigeration cycle are reversible, but that process 3–4 is irreversible. An engine is not used in place of the expansion valve in actual cycles because the work obtained by expanding a saturated liquid in an engine is too small to justify the addition of the engine and its related equipment. Since engines are typically not used in practice, the *ideal* cycle or model includes no engine. This makes the model a better representation of the actual cycle.

Several thermodynamic property diagrams are called Mollier diagrams.

The *ph* diagram, which also is called a Mollier diagram, is convenient in analyzing vapor-compression refrigeration cycles because (1) three of the four processes appear on it as straight lines, and (2) for the evaporator and condenser processes, the heat transfer is proportional to the lengths of the process paths. Such diagrams for water, ammonia, and R134a are in the appendix (Charts A·7·3, A·9, and A·11). The pressure scale on a *ph* diagram is usually logarithmic, but the *ph* diagram of Fig. 16·6 has a linear pressure scale.

Wet compression

The cycle shown in Fig. 16·6 is called a **wet compression** cycle because it involves the compression of a liquid–vapor mixture. Notice on the *Ts* diagram how closely it approximates a reversed Carnot cycle. In fact, replacing the expansion valve throttling process by an isentropic expansion would make the two cycles identical.

Dry compression

If the vapor entering the compressor is dry and saturated or superheated, as in Fig. 16·7, the cycle is said to involve **dry compression**. Notice that dry compression, especially with superheated vapor entering the compressor, makes part of the cycle resemble a reversed Brayton cycle and usually reduces the coefficient of performance for given evaporation and condensation temperatures. Nevertheless, dry compression is often favored over wet compression because it results in higher compressor efficiency, higher compressor volumetric efficiency, and less danger of damage to the compressor caused by slugs of liquid.

In a dry-compression cycle, the temperature of the vapor during part of the compression exceeds the condensation temperature, so it is possible to cool the compressor somewhat by the same coolant used in the condenser. Doing so decreases the work required by the compressor.

For a given condensation temperature, the refrigerating capacity of a vapor-compression system is increased by subcooling the condensate before it reaches the expansion valve. In actual systems the incoming (coldest) coolant is sometimes used for this purpose before it is used for condensing the refrigerant.

Selection of refrigerants

Many factors must be considered in the selection of a refrigerant for a vapor-compression system. One of the most important properties is the saturation pressure–temperature relationship that establishes the operating pressure range for any particular application. Saturation curves for several refrigerants are shown in Fig. 16·8. Also, for given temperature limits the coefficient of performance that can be obtained varies significantly from one refrigerant to another. Tables showing the relative performance obtainable

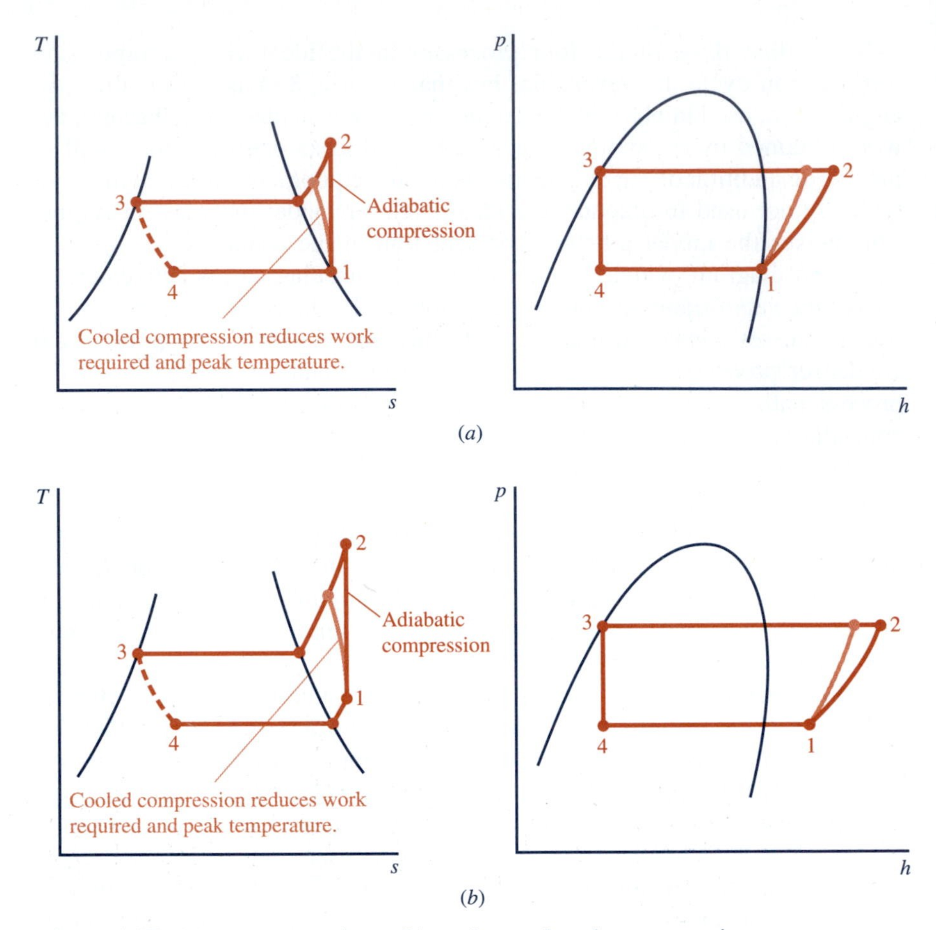

Figure 16·7 Vapor-compression refrigeration cycles: dry compression.

with various refrigerants are published in virtually all engineering handbooks that include sections on refrigeration. Other important factors in the selection of a refrigerant are its toxicity and environmental effects.

Many modifications of the basic vapor-compression cycle are possible for the purpose of increasing the coefficient of performance. In an actual plant, various refinements are used only if the saving in power costs is greater than the additional costs incurred by adding the refinements. Details of these cycle variations can be found in the literature on refrigeration. Here, simply as an example of such a variation, we describe the use of multistage compression with intercooling by means of a flash chamber.

For a fixed condensation temperature, lowering the evaporator temperature increases the compressor pressure ratio. For a reciprocating compressor, a high pressure ratio across a single stage means low volumetric efficiency. Also, with dry compression the high pressure ratio results in a high com-

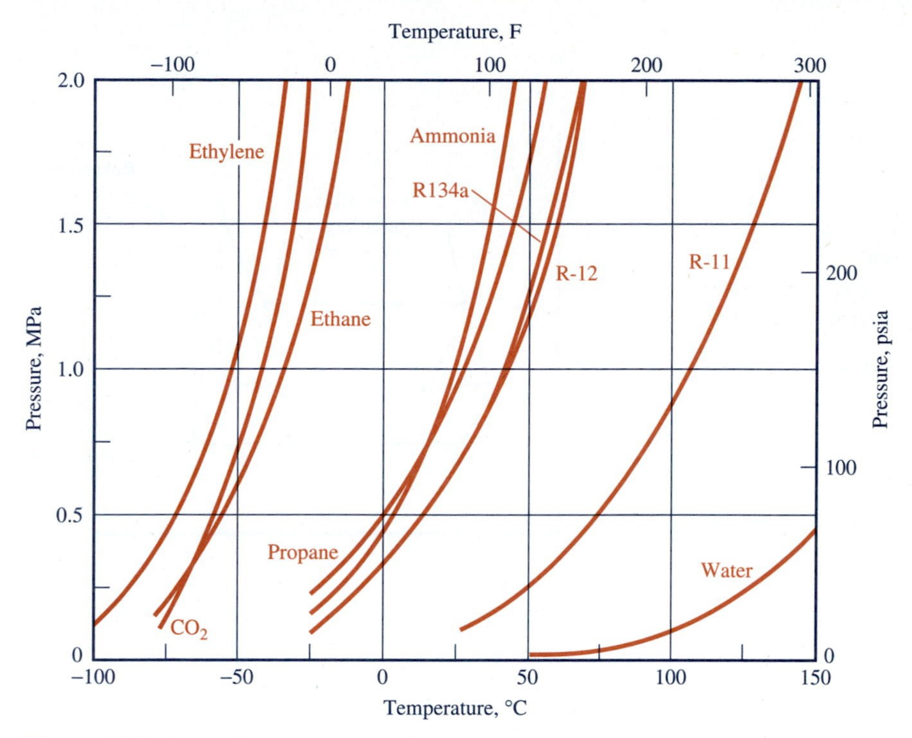

Figure 16·8 Saturation (pT) curves for some refrigerants.

pressor-discharge temperature that may damage the refrigerant. Multistage compression with intercooling is obviously called for, but effective intercooler temperatures may be well below the temperature of available cooling water that is used for the condenser. Therefore, several methods of using the refrigerant as an intercooling medium have been devised. One of these methods employs a flash chamber, as shown in Fig. 16·9.

Multistage compression with flash chamber intercooling

Liquid from the condenser in Fig. 16·9 expands through the first expansion valve into the flash chamber where the pressure is the same as the compressor interstage pressure. The refrigerant entering the flash chamber is a liquid–vapor mixture in state 6. The liquid fraction, which alone is in state 7 as shown on the *Ts* diagram, flows through the second expansion valve into the evaporator. The vapor from the flash chamber, which is in state 9, is mixed with the vapor (state 2) from the low-pressure compressor to form vapor in state 3, which enters the high-pressure compressor. For the same pressure ratio, a smaller work input to the compressor per unit mass of refrigerant is required with refrigerant entering in state 3 instead of in state 2. The ratio $\dot{m}_9/\dot{m}_7$ can be determined from an energy balance on the flash chamber. State 3 can then be determined by an energy balance on the mixing point between the compressors.

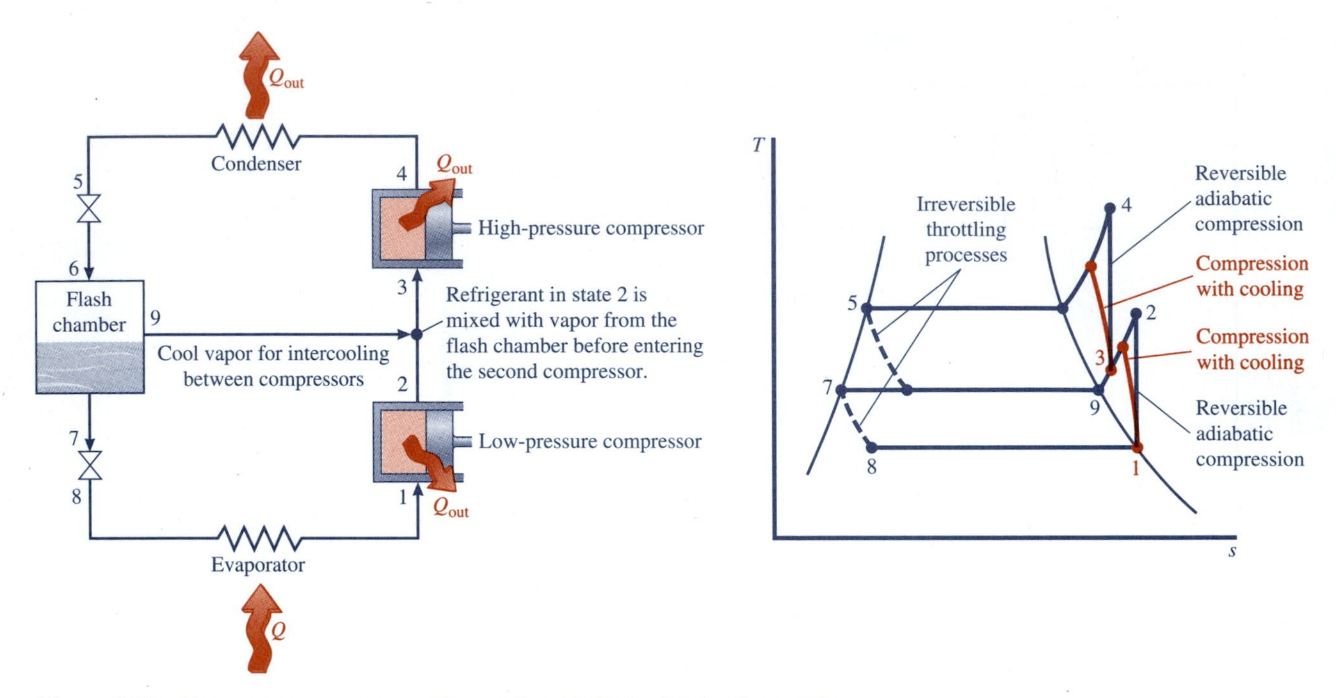

Figure 16·9 Two-stage vapor-compression cycle with flash chamber intercooler.

For very wide temperature ranges, binary vapor cycles are used in which the condenser for the lower-temperature refrigerator is the evaporator for the higher-temperature refrigerator. Figure 16·10 shows a binary vapor refrigerator. Three- and four-fluid cycles have also been used. These are all called **cascade refrigeration systems**.

Cascade refrigeration systems

Control systems for refrigerating units range from very complex ones used in large industrial installations to the simple on–off control of household refrigerators. One aspect of the control problem should be noted: The heat absorbed is (1) the product of the mass rate of flow through the evaporator and the specific enthalpy increase in the evaporator and also (2) proportional to the temperature difference between the evaporating refrigerant and the surroundings of the evaporator. Therefore, no one of these three variables ($\dot{m}$, Δh in the evaporator, and ΔT for heat transfer) can be changed without affecting the others.

Steam-Jet or Vacuum Refrigeration. One other refrigeration system that involves the compression of a vapor will be described, even though it does not operate on the usual vapor-compression refrigeration cycle. This is the *steam-jet system* or *vacuum system*. For refrigeration at temperatures above 0°C, water is a satisfactory refrigerant. Its chief drawback is the high specific volume of water vapor at low temperatures that necessitates large compressors. Centrifugal compressors have been used for this service, but steam-jet ejectors for the same flow rate and pressure ratio are cheaper and involve less maintenance expense, even though they use large quantities of steam. The application of a steam-jet ejector to a system for chilling water is shown in Fig. 16·11. Water to be chilled is sprayed into the flash chamber where part of it evaporates. The pressure is kept low by the steam ejector that removes the vapor formed. The flash chamber is thermally insulated, so the latent heat required to evaporate part of the spray water is taken from the spray water that remains a liquid. Thus the liquid is chilled. Makeup water must be supplied continually because a portion of the spray water is carried away as vapor. The ejector discharges into a condenser where the pressure is determined by the temperature of available cooling water. This pressure is usually below atmospheric, so a pump or ejector must be used to remove air and other noncondensable gases from the condenser. The steam-jet or vacuum system is well suited to air-conditioning applications where ample supplies of steam and condenser cooling water are available.

In connection with English units, a widely used unit of refrigeration capacity is the **ton of refrigeration**, which is defined as a heat absorption rate of 200 B/min. The name of this unit stems from the fact that in 24 hours this rate of heat removal will freeze approximately one ton of water at 32 F into ice at 32 F. Thus a 10-ton refrigerator can absorb 2000 B/min *when operating under design conditions*. This does not mean that the rate of heat absorption by this system at any instant actually is 2000 B/min, because many factors other than the system design affect the actual performance. (This is analogous to the fact that the power delivered by a 10-hp electric motor is not 10 hp just because the motor is running.)

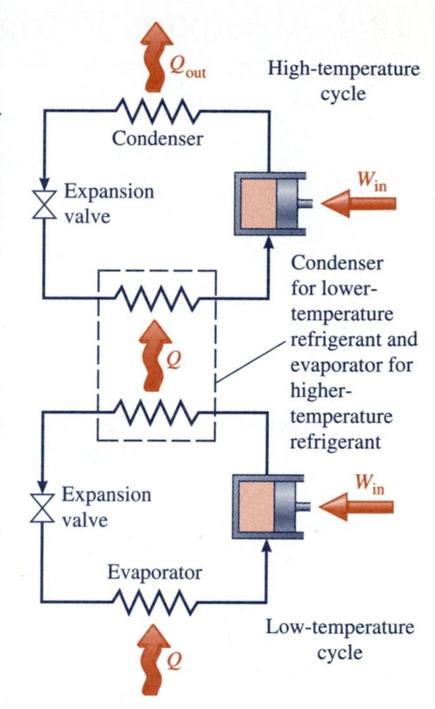

Figure 16·10 A binary vapor or cascade refrigeration cycle.

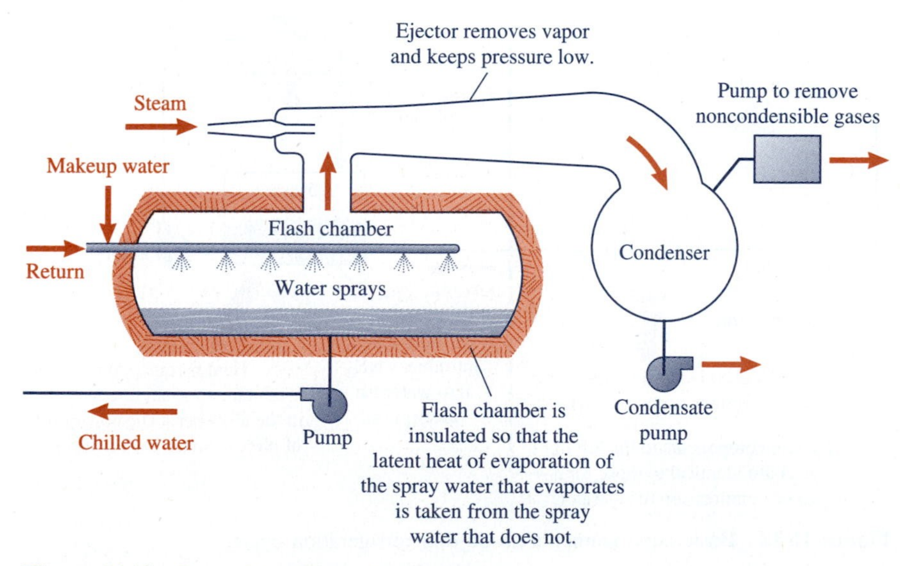

Figure 16·11 Steam-jet refrigeration system.

16·4 Absorption Refrigeration

Any vapor-compression refrigeration system requires a power input to compress the refrigerant vapor from the evaporator pressure to the condenser pressure. Pumping a liquid through the same pressure difference at the same mass flow rate would require less power, so is it possible to have the refrigerant in a liquid phase as its pressure is raised? To condense the pure refrigerant at the evaporator pressure would defeat the purpose of the refrigeration system; however, a workable arrangement is the absorption refrigeration cycle in which the refrigerant is dissolved in a liquid at the evaporator pressure, the liquid is pumped to the condenser pressure, and the refrigerant is then separated at the condenser pressure from the liquid.

Operation of an absorption refrigerator

Figure 16·12 is a flow diagram of an elementary absorption cycle that uses ammonia as the refrigerant and water (or a water–ammonia mixture) as the liquid in which the refrigerant is carried from the low pressure to the high pressure. Ammonia enters the condenser as a vapor, is condensed, and enters the expansion valve as a liquid. It partially flashes to vapor when it expands through the valve, and it is further vaporized as it absorbs heat in the evaporator. Thus the part of the flow diagram to the left of line *A–A* in Fig. 16·12 is

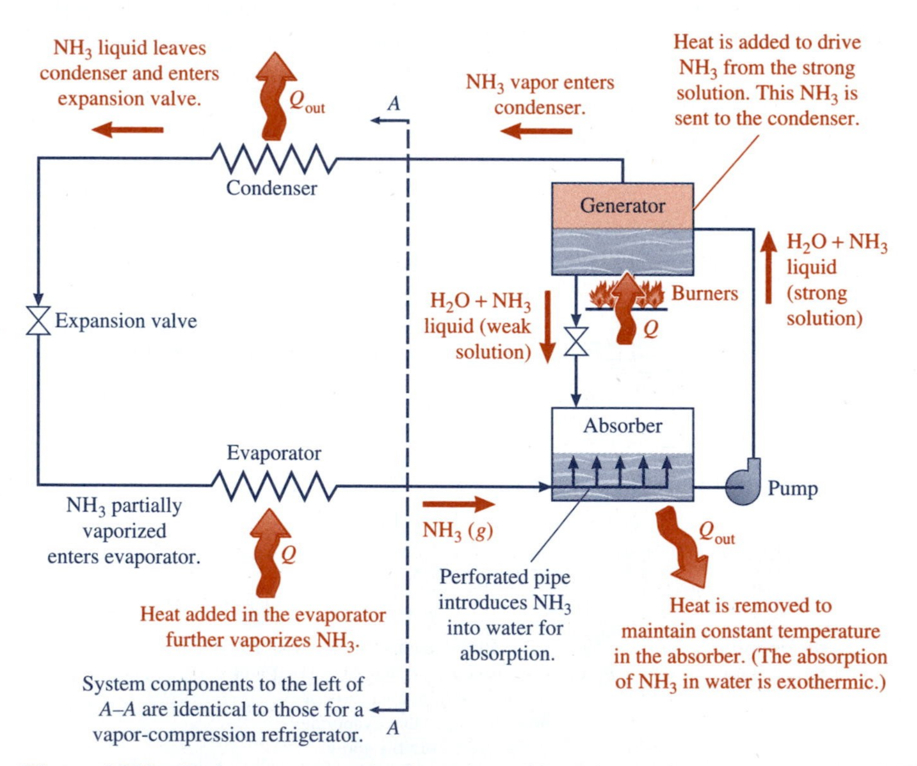

Figure 16·12 Basic aqua ammonia absorption refrigeration cycle.

identical with that of a vapor-compression cycle as shown in Fig. 16·6. However, to the right of line *A–A* we see that in place of a compressor are an absorber, a pump, a generator, and a return line for liquid. Ammonia vapor from the evaporator is dissolved in water in the absorber. This process is exothermic, just as a condensation process of a pure substance is, so heat must be removed from the absorber to keep its temperature constant. Also, at a given pressure the amount of ammonia that can be dissolved in water increases as the temperature is decreased (see Chapter 10); so the absorber temperature must be kept as low as possible. The strong ammonia–water solution is then pumped to the generator, which is at the condenser pressure. Heat is added to the solution in the generator to drive much of the ammonia out of solution (an endothermic process, just as an evaporation process of a pure substance is). Ammonia vapor, or a vapor mixture that is very rich in ammonia, then goes to the condenser, and the weak ammonia–water solution left in the generator passes through a valve back to the absorber. The liquid that travels through the absorber–pump–generator–valve cycle is simply a transport medium for the refrigerant, carrying it from the evaporator pressure to the condenser pressure in the liquid phase. Notice that the strong solution flows from the absorber that needs to be kept cold to the generator that needs to be kept hot, while weak solution flows in the opposite direction. This situation suggests that a heat exchanger that preheats the fluid flowing toward the generator by taking heat from the fluid flowing toward the absorber would certainly be desirable. In fact, such heat exchangers are usually used, although one is not shown in Fig. 16·12.

Phase equilibrium in an absorption refrigerator

An understanding of the operation of an absorption cycle requires some knowledge of the characteristics of binary mixtures as discussed in Chapter 10. A skeleton *equilibrium diagram* for aqua ammonia, as the ammonia–water mixture is called, is shown in Fig. 16·13. At any pressure, the boiling temperature of aqua ammonia liquid depends on the concentration or mass fraction x that is the mass of ammonia per unit mass of mixture. When $x = 0$, the boiling temperature is the saturation temperature of water; when $x = 1.0$, it is the saturation temperature of ammonia at the specified pressure. As you recall from Chapter 10, when liquid and vapor exist together in equilibrium, the concentration is not the same in the two phases. In Fig. 16·13, the solid curves represent saturated liquid states at two pressures, p_1 and p_2; and the broken curves represent saturated vapor states at the same pressures. p_2 is greater than p_1. Points a and b represent states of saturated liquid and saturated vapor, respectively, that can exist together in equilibrium. Notice that at any temperature the concentration of ammonia is higher in the vapor than in the liquid.

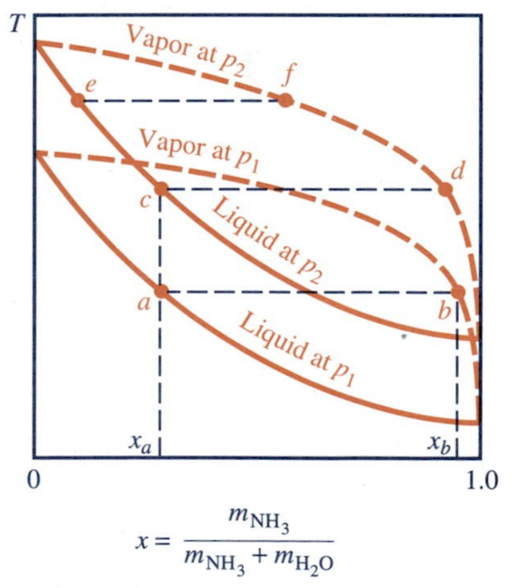

$$x = \frac{m_{NH_3}}{m_{NH_3} + m_{H_2O}}$$

Figure 16·13 Aqua ammonia equilibrium diagram.

In an absorption cycle, saturated liquid in state a might leave the absorber and enter the pump. Leaving the pump, the concentration is the same, and the temperature is very nearly the same, but the pressure is p_2. To evaporate any of the liquid, its temperature must be raised to bring the liquid to state c. This is done in the generator. Then, as more heat is added, some of the liquid is evaporated to form vapor in state *d*. Since x_d is greater than x_c, the formation of the ammonia-rich vapor reduces the concentration of ammonia in the liquid. Evaporation then ceases if the pressure and temperature are held constant, because the liquid at a lower concentration is not at its boiling point. If more heat is added so that evaporation continues, the temperature of the two phases increases, and the states of the liquid and vapor change toward *e* and *f*, respectively. Liquid *e* flows from the generator back to the absorber, and vapor *f* flows to the condenser.

Equilibrium diagrams such as Fig. 16·13 are adequate for mass balance analyses of absorption systems, but for energy balance or first law analyses, enthalpy data are also needed. These cannot be conveniently presented on a diagram such as Fig. 16·13, so other forms of equilibrium diagrams have been devised. One that relates p, T, h, x' (concentration in the liquid), and x'' (concentration in the vapor) for aqua ammonia is included in the appendix as Chart A·20. Figure 16·14 shows on such a diagram some states of the cycle shown in Fig. 16·12.

Modifications to the basic absorption cycle

Actual ammonia absorption systems always involve some modifications not shown in Fig. 16·12. The pump delivers cold liquid from the absorber to the generator where the temperature must be high, and hot liquid flows from the generator back to the absorber, which must be kept cold; so, as mentioned above, a heat exchanger is always used to transfer heat from the weak solution to the strong solution. Also, carrying water vapor into the condenser must be avoided because it will freeze in the expansion valve and evaporator; therefore, a *rectifier* (see Section 10·5) is placed between the generator and the condenser. Its function is to remove traces of water from the refrigerant

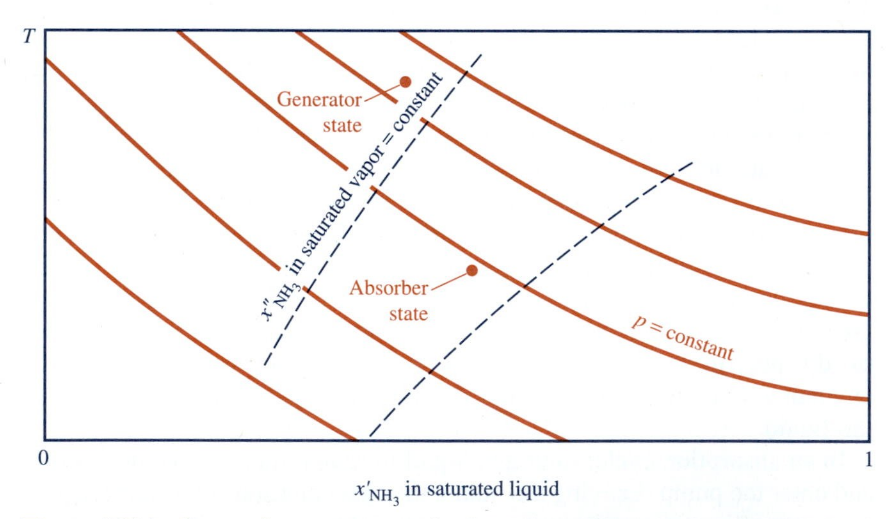

Figure 16·14 States of an aqua ammonia absorption refrigerator (see Fig. 16·12) shown on an equilibrium diagram.

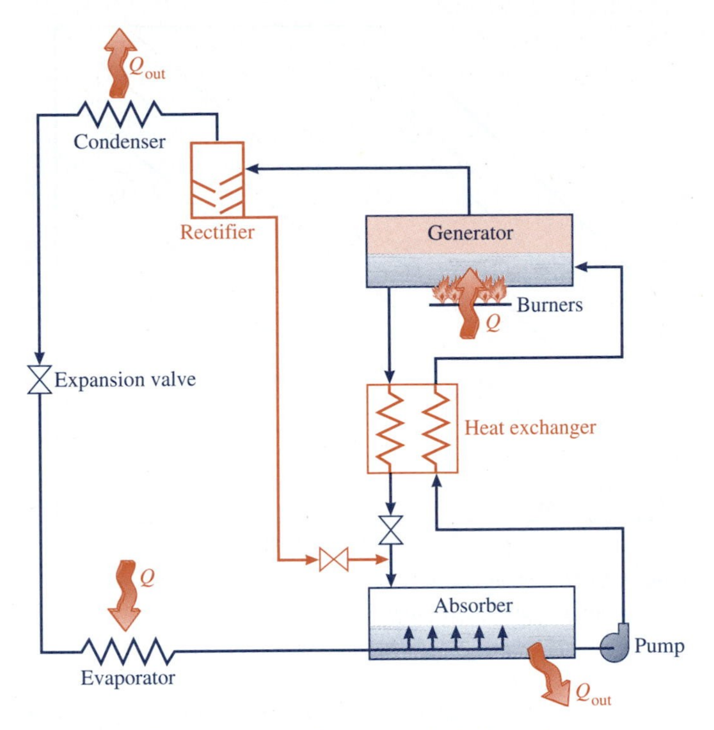

Figure 16·15 Aqua ammonia system, including a heat exchanger for liquids between the absorber and the generator and a rectifier to remove water from the ammonia entering the condenser.

before it reaches the condenser. Figure 16·15 shows an aqua ammonia refrigerating system with a heat exchanger and rectifier.

Another pair of substances commonly used for absorption refrigeration systems is lithium bromide and water. An aqueous solution of lithium bromide is the carrier and water is the refrigerant; therefore, lithium bromide absorption systems are used only for applications where the minimum temperature is above 0°C.

Differences between water–ammonia and lithium bromide–water absorption systems

The LiBr–water absorption system is strikingly different from the aqua ammonia system because LiBr has a very low vapor pressure; it is essentially nonvolatile. The water that is driven from the liquid solution of LiBr and water in the generator is therefore nearly pure water, so its properties can be determined from the steam tables. Thus the property diagrams used for LiBr absorption systems (Charts A·21 and A·22) are simpler than for aqua ammonia (Chart A·20). They give properties of only the liquid solution. The equilibrium chart (Chart A·21) does show the vapor pressure for pure water as a function of the refrigerant (water) temperature, and these values are directly

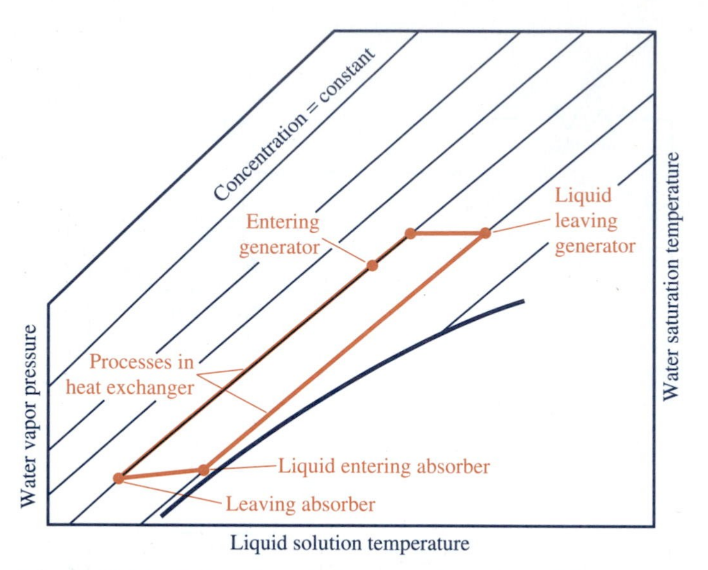

Figure 16·16 States of an LiBr–water absorption refrigerator shown on an equilibrium diagram.

from the steam tables for saturation states. The enthalpy of the liquid depends chiefly on temperature and concentration, so Chart A·22 can be used over a wide range of pressures. Figure 16·16 shows states on an LiBr–water equilibrium chart for an absorption system numbered to correspond with states indicated in Fig. 16·12, which is for an aqua ammonia system.

The crystallization line marks the maximum concentration that can be used at each temperature without causing crystallization of LiBr from the liquid solution. Crystallization refers to the formation of solid LiBr crystals when a solution rich in LiBr is suddenly cooled. It causes the solution to become a thick suspension of crystals that cannot be pumped, so the refrigerating system stops operating.

Difference between elementary flow diagrams and actual equipment arrangements

The basic pieces of equipment—absorber, generator, condenser, and evaporator—and the expansion valve and liquid pump as shown in Fig. 16·12 are parts of any absorption cycle. Flow diagrams such as Fig. 16·12 are shown in elemental form, however, and the actual equipment may be arranged much differently. For example, most lithium bromide absorption refrigerators have all the major pieces of equipment contained within only one or two cylindrical shells. Such an arrangement has many advantages over connecting the equipment by external piping. Figure 16·17 shows in simplified form the physical arrangement, within a single shell, of the components of an LiBr–water refrigeration system: absorber, generator, condenser, and evaporator. Notice that some pumps not shown on the simplified diagram of Fig. 16·12 are shown here. The extra pumps introduce no new processes, but they ensure that fluids continue to circulate through various components until the

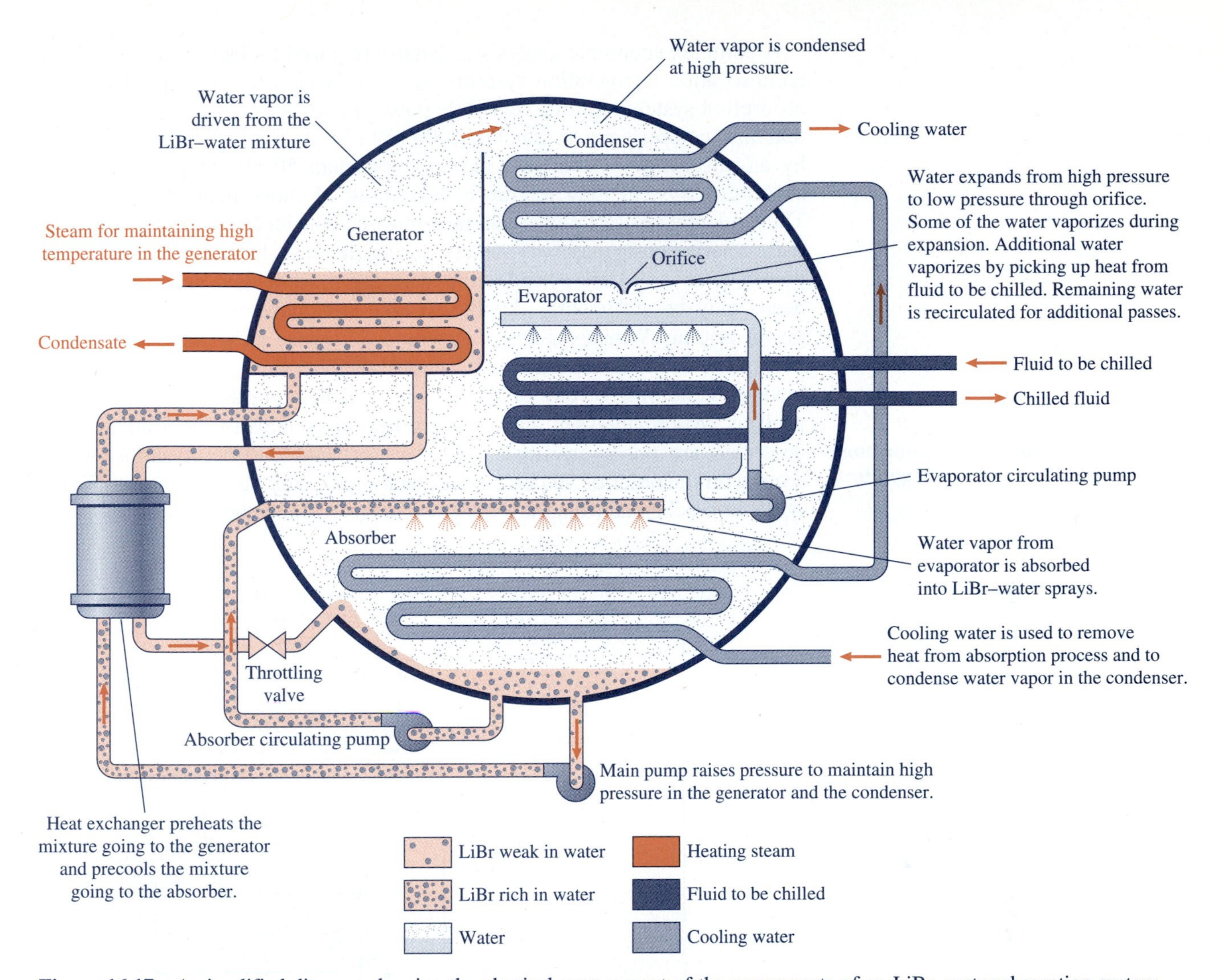
Water vapor is condensed at high pressure.
Water vapor is driven from the LiBr–water mixture
Condenser
Cooling water
Water expands from high pressure to low pressure through orifice. Some of the water vaporizes during expansion. Additional water vaporizes by picking up heat from fluid to be chilled. Remaining water is recirculated for additional passes.
Steam for maintaining high temperature in the generator
Generator
Orifice
Evaporator
Condensate
Fluid to be chilled
Chilled fluid
Evaporator circulating pump
Absorber
Water vapor from evaporator is absorbed into LiBr–water sprays.
Cooling water is used to remove heat from absorption process and to condense water vapor in the condenser.
Throttling valve
Absorber circulating pump
Main pump raises pressure to maintain high pressure in the generator and the condenser.
Heat exchanger preheats the mixture going to the generator and precools the mixture going to the absorber.
LiBr weak in water
LiBr rich in water
Water
Heating steam
Fluid to be chilled
Cooling water

Figure 16·17 A simplified diagram showing the physical arrangement of the components of an LiBr–water absorption system—absorber, generator, condenser, and evaporator—within a single cylindrical shell.

processes that should occur in them are complete. Actual installations involve various other pieces of auxiliary equipment that are essential to plant operation, although they have little effect on the thermodynamic analysis of the cycle. Notice also that the piping and throttle valve between the condenser and evaporator of Fig. 16·12 are replaced by simply an orifice in the partition separating the evaporator space from the condenser space. You should refer to books on refrigeration for information on actual systems and their components.

A complex economic analysis is usually required to decide whether to use an absorption refrigeration system and, if so, what kind to use. Clearly, an absorption system requires very little power input. A vapor-compression system that requires a power input of 1000 kW (3.6×10^6 kJ/h) can be replaced by an absorption system that requires less than 50 kW of power, but the absorption system may require a heat input of more than 20×10^6 kJ/h. Thus, relative costs of power and heat must be considered. At many locations these relative costs may vary widely during the year. In fact, in some places, power costs may vary considerably with the time of day, and refrigeration needs may be sufficiently flexible to take advantage of night or off-peak power pricing. Whenever heat that would otherwise be wasted is available, absorption systems are especially attractive. The variation of refrigeration load with the time of day or the season is also important in the economic analysis. Where analyses show that absorption systems of the kind we have described are not competitive, two-stage or double-effect systems, which require on the order of 40 percent less heat input, may be competitive. Figure 16·18 is a schematic diagram of a two-stage LiBr–water system.

Double-effect absorption refrigerators

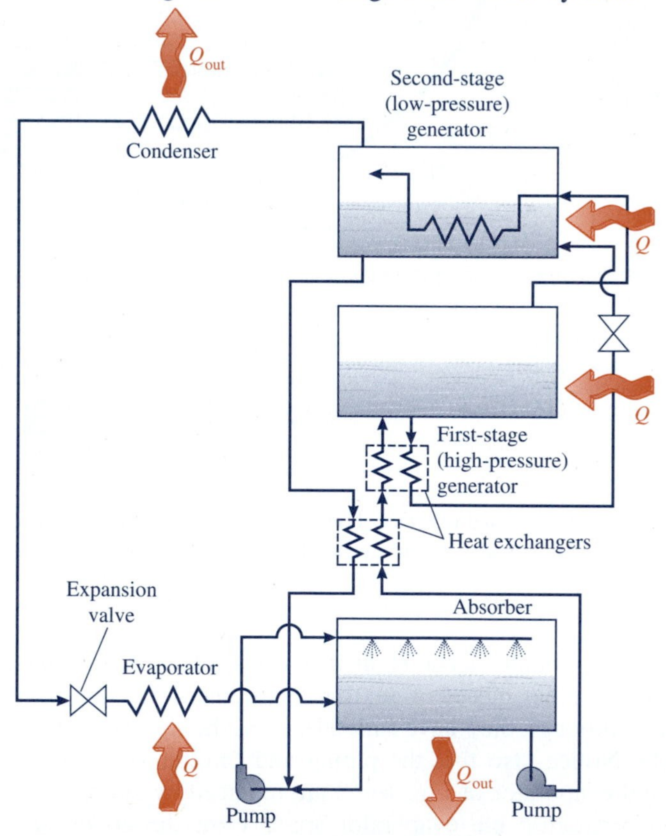

Figure 16·18 Schematic diagram of two-stage or double-effect LiBr–water absorption system.

The advantage of an absorption cycle over a compression cycle is the large reduction in power input. An ingenious method of reducing to zero the required power input of an absorption cycle was devised by two undergraduates at the Royal Institute of Technology in Stockholm, Carl G. Munters and Baltzar von Platen. It is usually referred to as the Servel system. The basic feature is that the total or mixture pressure is constant throughout the system. Condensation of the refrigerant ammonia occurs where the ammonia exists alone under the system pressure; but evaporation occurs where the ammonia is mixed with hydrogen so that the ammonia behaves as though it existed alone at a pressure approximately (exactly, if it were an ideal gas) equal to its partial pressure in the mixture. Thus the ammonia is condensed at one temperature and evaporated at a lower one, even though the total pressure is the same in the condenser and the evaporator. Refrigeration is achieved without the use of a mechanical pump or compressor.

Servel system

A highly simplified flow diagram of a Servel refrigerator is shown in Fig. 16·19. A water–ammonia solution flows down into the generator where heat is added to vaporize part of the mixture, the vapor formed being very rich in ammonia. The formation of vapor in the dome of the generator forces liquid up the tube toward the separator until the liquid level in the generator drops to the end of the tube. Then some vapor from the dome passes into the tube, the liquid level rises as more mixture enters the generator, and the process is repeated. The result is that alternate ammonia-rich bubbles and slugs of weak liquid solution flow up to the separator. From the separator, the ammonia passes to the condenser, and the water (or weak solution) flows by gravity to the absorber. In the condenser, heat is removed from the ammonia, which exists alone at the total pressure of the system, to condense it. The liquid drains from the condenser through a U-tube liquid seal or trap, which allows only liquid to pass, and into the evaporator. The total pressure is the same in both the condenser and the evaporator, so the liquid seal does not blow out. Hydrogen is present in the evaporator, however, so the ammonia passing through the liquid seal from the condenser evaporates at a low partial pressure and the corresponding low saturation temperature, thus absorbing heat from the region that is to be refrigerated. The cold ammonia–hydrogen mixture flows down from the evaporator and back up through the absorber, where the ammonia is absorbed by water that is flowing from the separator through a liquid seal that prevents the entry of hydrogen into the separator and condenser. The ammonia is thereby carried back to the generator to complete its cycle and the hydrogen is left in the evaporator and absorber circuit. Heat must be removed from the absorber to keep its temperature constant, just as in the ammonia-absorption cycle discussed earlier.

In analyzing a Servel system, notice that there are three fluids, each of which flows in a different circuit. The refrigerant ammonia passes through the generator, separator, condenser, evaporator, and absorber, and back to the generator. The water is a transport medium that carries ammonia from the

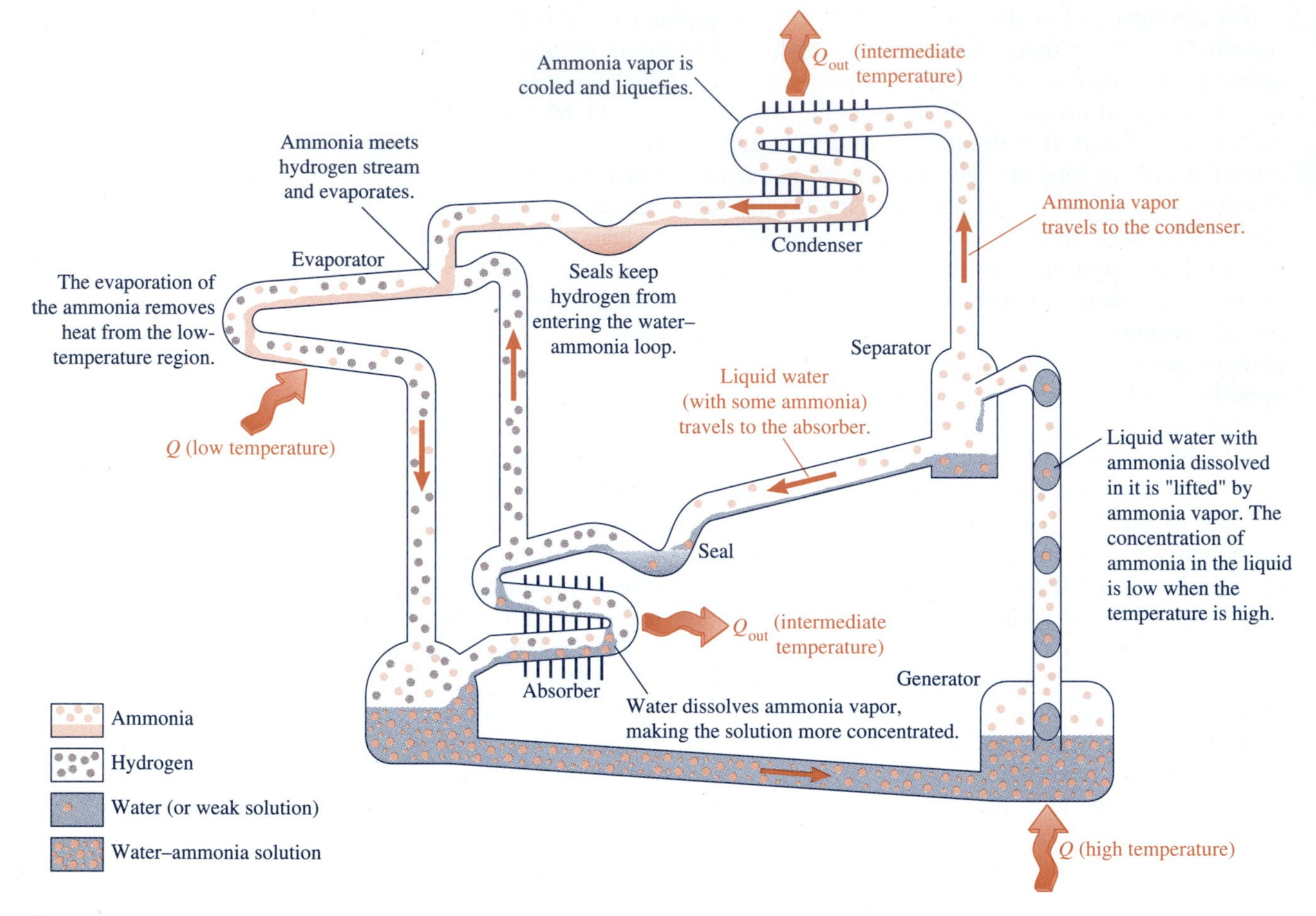

Q_{out} (intermediate temperature)
Ammonia vapor is cooled and liquefies.
Ammonia meets hydrogen stream and evaporates.
Condenser
Ammonia vapor travels to the condenser.
Evaporator
The evaporation of the ammonia removes heat from the low-temperature region.
Seals keep hydrogen from entering the water–ammonia loop.
Separator
Liquid water (with some ammonia) travels to the absorber.
Q (low temperature)
Liquid water with ammonia dissolved in it is "lifted" by ammonia vapor. The concentration of ammonia in the liquid is low when the temperature is high.
Seal
Q_{out} (intermediate temperature)
Generator
Absorber
Water dissolves ammonia vapor, making the solution more concentrated.
Ammonia
Hydrogen
Water (or weak solution)
Water–ammonia solution
Q (high temperature)

Figure 16·19 Schematic diagram of a Servel absorption refrigerator.

absorber, where the ammonia partial pressure is low, to the separator, where it is high. The function of the hydrogen is to circulate through the evaporator and absorber to hold down the ammonia partial pressure.

16·5 The Liquefaction of Gases

The liquefaction of a gas is an important step in most very-low-temperature refrigeration systems, and also in the preparation of pure oxygen and nitrogen from air.

As pointed out in Section 7·5, a gas can be cooled by throttling if its Joule–Thomson coefficient, $(\partial T/\partial p)_h$, is positive, and this means that its initial temperature must be less than its maximum inversion temperature. Further information on the conditions necessary for gas liquefaction by throttling can be obtained from a *Ts* diagram such as Fig. 16·20. (A *Ts* diagram for air is included in the appendix as Chart A·13.) A throttling process between pressures p_1 and p_2 will not result in liquefaction if it begins at state *a*, *b*, or *c*, but it will if it starts from a state such as *d*.

Hampson–Linde gas-liquefaction system

Flow and *Ts* diagrams for an ideal **Hampson–Linde gas liquefaction system** are shown in Fig. 16·21. After multistage compression, the gas is cooled from state 2 to state 3 at constant pressure in an aftercooler that is cooled either by water or by a refrigerating system. The gas is further cooled to state 4 in a regenerative heat exchanger, which is supplied with very cold gas from elsewhere in the cycle. After expansion through a throttle valve, the fluid is in the liquid–vapor mixture state 5 and is mechanically separated into liquid (state *a*) and vapor (state 6) parts. The liquid is drawn off as the desired product, and the vapor flows through the regenerative heat exchanger to cool high-pressure gas flowing toward the throttle valve. The area under line 6–7 on the *Ts* diagram equals that under 3–4 after adjustment for the smaller mass flow rate for process 6–7. The gas at state 7 is mixed with an amount of gas from outside equal to the amount of liquid removed, and this mixture in state 1 enters the compressor.

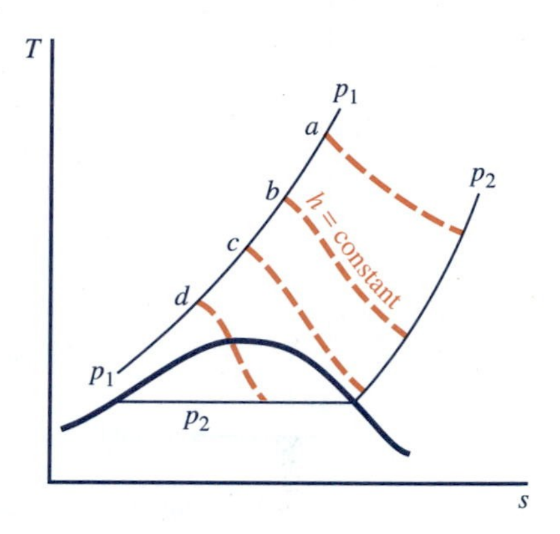

Figure 16·20 Throttling processes from gaseous states.

Claude gas-liquefaction system

From a thermodynamic viewpoint, better performance of a gas liquefaction plant could be obtained by replacing the highly irreversible throttling process by expansion in an engine. The operation of an engine at very low temperatures presents difficulties, however, especially if liquefaction occurs in the engine. A compromise solution is the **Claude system** for liquefying

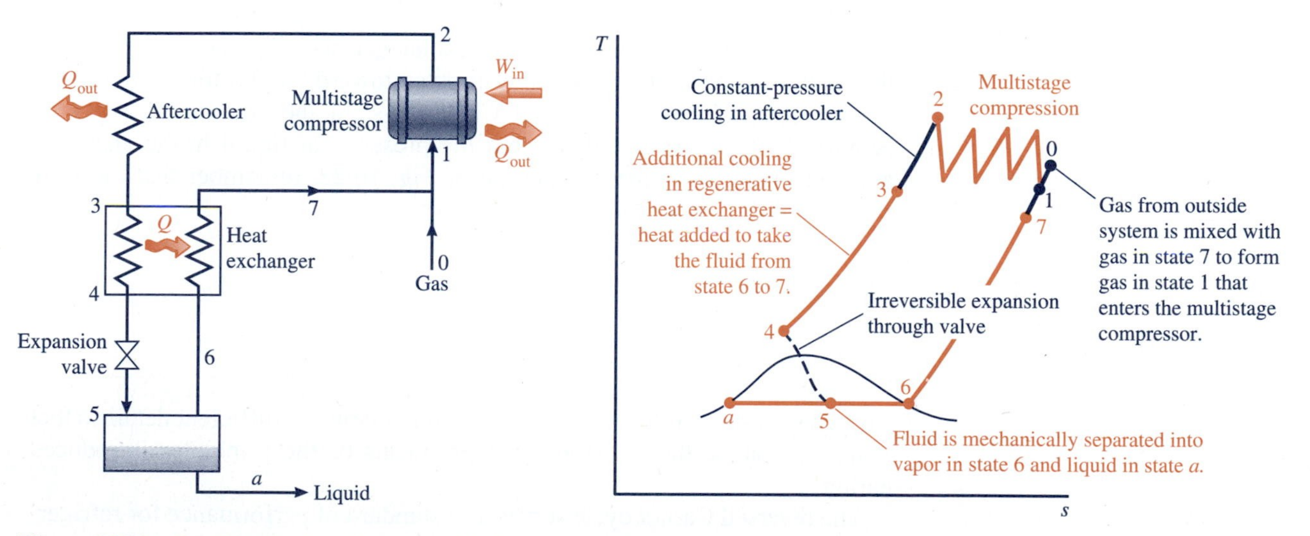

Figure 16·21 Gas liquefaction: Hampson–Linde system.

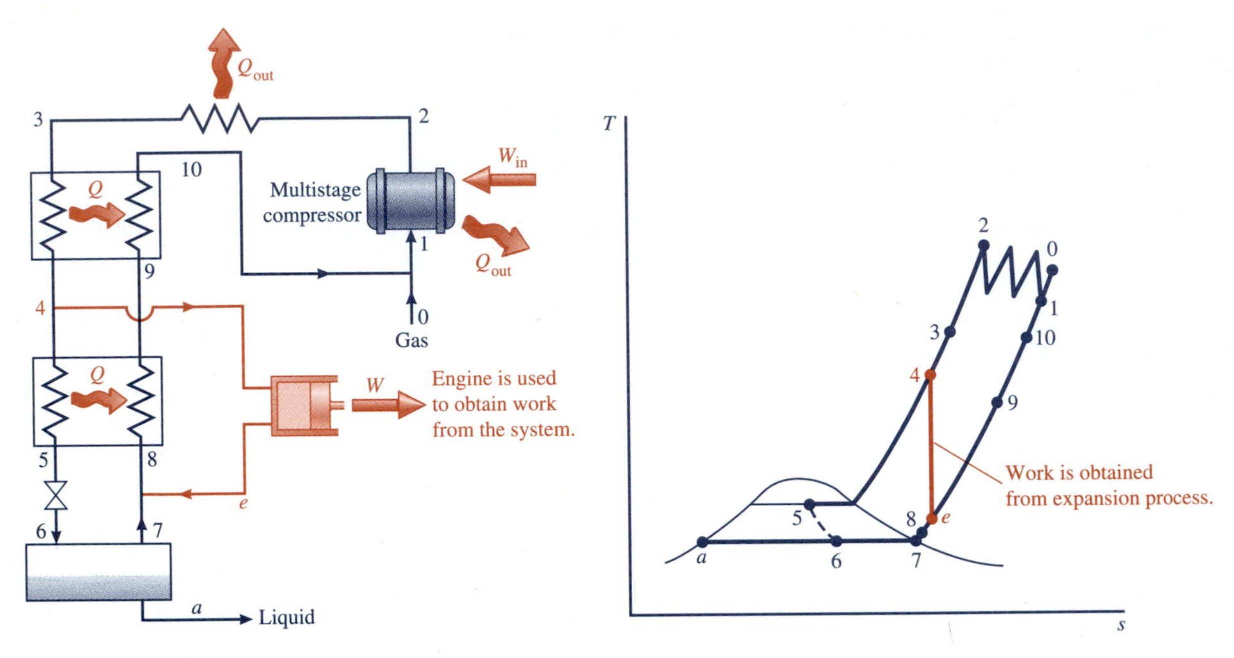

Figure 16·22 Gas liquefaction: Claude system.

gases. Simplified flow and *Ts* diagrams are shown in Fig. 16·22. From state 1 to state 4 the Claude system processes are the same as those of the Hampson–Linde system. After the gas is cooled to state 4 by the compressor aftercooler followed by a regenerative heat exchanger, most of it is expanded through an engine and then is mixed with vapor from the separator and flows back toward the compressor through a heat exchanger, which precools appreciably the small fraction of the flow that is directed toward the throttle valve instead of the engine. Thus in the ideal Claude system the cooling of a gas by isentropic expansion is utilized, but the presence of liquid in the engine is avoided. In studying the *Ts* diagram of Fig. 16·22, remember that the flow rates are different at points 3, 5, 7, and 9.

16·6 Summary

This chapter describes several refrigeration systems in sufficient detail so that you can analyze their performance by means of the principles introduced earlier.

The reversed Carnot cycle serves as a standard of performance for refrigerating cycles, but it is unsuitable for actual systems.

$Q_H - Q_L$ and the second law tells us that $W_{in} > 0$.

The reversed Brayton cycle employs a gas as working fluid and is widely applied in aircraft cooling. Shortcomings of a gas as a refrigerant are its low specific heat, its low density at moderate pressures, and the difficulty of executing isothermal heat absorption and rejection processes that are necessary for performance anywhere near that of a reversed Carnot cycle.

Vapor-compression refrigeration systems employing many different refrigerants are widely used. In large commercial or industrial applications, the basic cycle is modified in various ways to improve performance. For air conditioning, water is sometimes used as the refrigerant in a variation of the basic vapor-compression cycle known as the steam-jet or vacuum system.

Absorption refrigeration requires little power input in comparison with vapor-compression refrigeration, but a large amount of heat is required. Power is saved by dissolving the refrigerant in a liquid before pumping it from the evaporator pressure to the condenser pressure. Pumping a liquid requires less power than pumping a vapor through the same pressure range because its specific volume is much lower. The refrigerant is then driven out of the liquid by means of heat addition. Absorption refrigeration is especially attractive where large quantities of heat that otherwise would be of little value are available, and it is widely used for large air-conditioning systems.

Several methods of producing very low temperatures or liquefying gases are based on the adiabatic expansion of gases through throttles or engines. Two of these are the Hampson–Linde and Claude systems.

Suggested Reading

16·1 *ASHRAE Handbook*, American Society of Heating, Refrigerating, and Air-Conditioning Engineers, Inc., Atlanta. (This book is published in four volumes: *Fundamentals, Systems, Equipment,* and *Applications*, with one volume published per year. It is a valuable source of information on refrigeration.)

16·2 McQuiston, Faye C., and Jerald D. Parker, *Heating, Ventilating, and Air Conditioning*, 4th ed., Wiley, New York, 1994.

16·3 Pita, Edward G., *Refrigeration Principles and Systems*, Wiley, New York, 1984. (This book provides many clear diagrams of refrigerating systems and components.)

16·4 Sorensen, Harry A., *Energy Conversion Systems*, Wiley, New York, 1983, Chapter 11.

16·5 Wood, Bernard, D., *Applications of Thermodynamics*, 2nd ed., Addison-Wesley, Reading, MA, 1982, Chapter 4.

Problems

16·1 Determine the power input to a Carnot refrigerator that operates between a cold storage room at −30°C and cooling water at 15°C.

16·2 Determine the power input if the Carnot refrigerator of Problem 16·1 has a temperature difference of 4 Celsius degrees across each heat exchanger. Calculate also the amount of energy made unavailable per hour by the operation of the refrigerator.

16·3 If a Carnot refrigerator is used as a heat pump for a building maintained at 20°C with a heat loss rate of 90,000 kJ/h, and the heat source is atmospheric air at −15°C, how much power is required?

16·4 Heat leakage from the surroundings into a household refrigerator is predicted to be 500 kJ/h. The food compartment is to be maintained at 3°C; there is no freezer compartment. Estimate the power requirement if a Carnot refrigerator is used with heat transfer across only infinitesimal temperature differences.

16·5 Solve Problem 16·4 if a temperature differential of 4 Celsius degrees is maintained across each heat exchanger. Calculate also the amount of energy made unavailable per hour by the operation of the refrigerator.

16·6 A skating rink has an area of 550 m^2. Three centimeters of water at 0°C stands on a layer of ice at 0°C. If the water is all to be frozen within four hours, determine the minimum possible power input to the refrigerating unit.

16·7 Determine the coefficient of performance of a reversed Brayton cycle refrigerator that removes 9.22 kJ/s from a cold storage room while operating between pressure limits of 700 and 1400 kPa. Air leaves the cold heat exchanger at 5°C and enters the air engine at 35°C.

16·8 Consider a reversed Brayton cycle as in Problem 16·7. It has been suggested that the temperature of the region to be refrigerated can be lowered by using a heat exchanger to precool the air entering the turbine by means of the air leaving the low-temperature heat exchanger before it enters the compressor. Is this a feasible idea? What would be the effects of the additional heat exchanger on power input and coefficient of performance for a range of heat exchanger effectiveness?

16·9 A dense air refrigerating machine removes 2000 kJ/min from a cold storage room while operating between pressures of 350 kPa and 1400 kPa. Air temperatures entering the compressor and expansion cylinders are 0 and 35°C, respectively. The expansion is reversible and adiabatic; the compression is polytropic with $n = 1.35$. Determine the power input and the coefficient of performance.

16·10 An aircraft cockpit is to be supplied with 0.10 kg/s of cooling air at 70 kPa, 5°C, by bleeding air at 350 kPa, 200°C, from the jet-engine compressor, cooling it in a heat exchanger, and expanding it isentropically through a turbine that exhausts into the cockpit. Disregarding friction and kinetic energy changes, determine the heat-transfer rate in the heat exchanger and the turbine power output.

16·11 For cooling an aircraft cabin, 0.80 kg/s of air at 200 kPa, 115°C, is bled from the jet-engine compressor, cooled in a heat exchanger to 60°C, and then expanded adiabatically through a turbine to 65 kPa. The turbine efficiency is 0.70. Determine the turbine power output and explain how effective the cooling system would be if the turbine were replaced by an expansion valve.

16·12 A heat pump takes heat from the atmosphere at 20 F and supplies 50,000 B/h to a house that is to be maintained at 70 F. Determine the minimum power input to the heat pump and explain why this is the *minimum* power input.

16·13 A reversed Brayton cycle operates between 100 and 200 psia. Air leaves the cold heat exchanger at 40 F and enters the air engine at 100 F. Refrigerating capacity is one ton. Determine the coefficient of performance.

16·14 For an aircraft cabin cooling system, a turbine is to exhaust air at 10 psia, 40 F, at a rate of 4.0 lbm/s. The turbine drives a compressor that takes in air at 10 psia, 120 F. The air discharged from this compressor is mixed with air at the same state bled from the main jet-engine compressor. It is then cooled to 140 F before entering the cooling-system turbine. Assume that compression and expansion are reversible and adiabatic. Disregard pressure drops in the heat exchanger and ducting and kinetic energy changes. Determine the pressure and the flow rate of air to be bled from the main engine compressor.

16·15 Explain how a reversed Brayton cycle can be modified to approach a reversed Carnot cycle by means of multistage compression with intercooling and also multistage expansion. Sketch flow, pv, and Ts diagrams for the modified cycle.

16·16 Comment on the advisability of using a reversed regenerative gas-turbine cycle for refrigeration.

16·17 An ideal vapor-compression refrigerating cycle operates between temperature limits of 5°C and 35°C. Dry saturated vapor leaves the compressor and saturated liquid enters the expansion valve. For a refrigerating capacity of 20 kJ/s, calculate the required power input if the refrigerant is (*a*) ammonia, (*b*) water, (*c*) R134a, (*d*) propane.

16·18 Solve Problem 16·17 with an isentropic expansion in an engine replacing the throttling through the expansion valve.

16·19 In a vapor-compression refrigeration system the condenser temperature is 25°C and the evaporator temperature is −10°C. Saturated liquid enters the expansion valve, and dry saturated vapor enters the compressor, which operates reversibly and adiabatically. For a refrigeration effect of 3.5 kJ/s, determine the flow rate and the power input if the refrigerant is (*a*) ammonia, (*b*) R134a.

16·20 Refer to Example 9·12. The air conditioner described draws in 30 m^3/min of atmospheric air. Determine the minimum power to operate the refrigerating unit in the air conditioner.

16·21 A refrigerating plant uses 6450 kg/h of cooling water in the condenser. The average inlet and outlet water temperatures are 13°C and 27°C, respectively. The power required to drive the compressor is 15 kW. Compression is adiabatic. Calculate the coefficient of performance and the capacity of the plant in kilojoules per second of refrigeration.

16·22 For a condensing temperature of 20°C and adiabatic dry compression, plot the minimum compressor discharge temperature against evaporator temperature for a range of −15°C to −50°C of the latter for (*a*) ammonia, (*b*) R134a. What conclusions do you draw from these plots?

16·23 An ideal R134a compression refrigerating cycle operates between 80.7 kPa and 700 kPa. Dry saturated vapor enters the compressor, and R134a at 700 kPa, 15°C, enters the expansion valve. Compression is adiabatic. The refrigerating capacity is 35 kJ/s. Determine the power input.

16·24 An R134a compression refrigeration machine operates at rated capacity of 28 kJ/s with pressures in the evaporator and condenser of 243 kPa and 666 kPa, respectively. Dry saturated vapor enters the compressor and vapor leaves the compressor at 35°C. Heat removed from the refrigerant being compressed amounts to 2.4 kJ/kg. Liquid enters the expansion valve at 20°C. Determine the power input and the irreversibility rate.

16·25 An ammonia compression refrigeration cycle has evaporator and condenser pressures of 140 kPa and 1240 kPa, respectively. Dry saturated vapor enters the compressor, and vapor at 135°C leaves. Liquid at 25°C leaves the condenser. The heat absorption rate of the evaporator is 390,000 kJ/h. Cooling water passing through the compressor cylinder water jacket picks up heat at a rate of 420 kJ/min. (*a*) Calculate the power input. (*b*) Is the compression process reversible? Prove your answer.

16·26 In a household refrigerator, one large compartment is to be kept at a temperature of approximately 3°C while a smaller compartment is to be held at approximately −15°C. In each case, the temperature difference between the compartment and the refrigerant evaporating in the tubes in the walls of the compartment should be about 10 Celsius degrees. Devise an arrangement for providing the needed refrigeration while using only a single compressor. Show the control system you suggest.

16·27 For an ideal vapor-compression refrigerating system using R134a and dry compression, the refrigerated region is to be kept at 3°C, and the ambient air to which heat is rejected is at 30°C. The rate of heat removal from the refrigerated region is 200 kJ/min. A minimum temperature difference of 10 Celsius degrees is to be maintained for heat transfer at the condenser and at the evaporator. For a fixed condensation temperature of 40°C, plot coefficient of performance and refrigerant flow rate versus temperature difference between the evaporating refrigerant and the refrigerated space in the range of 10 Celsius degrees to 30 Celsius degrees.

16·28 For an ideal vapor-compression refrigerating system using R134a and dry compression, the refrigerated region is to be kept at 3°C, and the ambient air to which heat is rejected is at 30°C. The rate of heat removal from the refrigerated region is 200 kJ/min. A minimum temperature difference of 10 Celsius degrees is to be maintained for heat transfer at the condenser and at the evaporator. For a fixed evaporation temperature of −13°C, plot coefficient of performance and refrigerant flow rate versus temperature difference between the condensing refrigerant and the ambient air in the range of 10 Celsius degrees to 30 Celsius degrees.

16·29 Instructions on room air conditioners and other refrigeration equipment often say, "Wait two minutes before restarting." What is the basis for this instruction?

16·30 Sometimes it is said that a heat pump is a refrigerator "running backwards." Discuss the aptness of this statement. When is it true? In what respect is it true?

16·31 A portable emergency refrigerating unit includes a bottle containing 10.8 kg of liquid ammonia and a coil of tubing that is fed from the bottle through a small valve that prevents the pressure in the tubing from exceeding 140 kPa. The tubing volume is negligible compared to the bottle volume. The tubing discharges to the atmosphere through a thermostatic valve that closes only when the temperature of ammonia at the discharge end of the tubing is less than 0°C. Determine the maximum amount of heat that can be absorbed by completely discharging this unit if the ammonia in the bottle is initially at 10°C. (Some assumptions must be made. State them clearly.)

16·32 Determine the power input to an ideal flash-chamber-intercooled vapor-compression refrigerating system that has condenser, flash chamber, and evaporator pressures of 750 kPa, 243 kPa, and 66 kPa, respectively. Saturated liquid enters each expansion valve, and the refrigerating effect is 50 kJ/s.

16·33 Solve Problem 16·32 for the same capacity and the same condenser, flash chamber, and evaporator temperatures if the refrigerant is ammonia.

16·34 An ideal vapor-compression refrigerating cycle operates between temperature limits of 40 F and 100 F. Dry saturated vapor leaves the compressor and saturated liquid enters the expansion valve. For a refrigerating capacity of 5 tons, calculate the required power input if the refrigerant is (*a*) ammonia, (*b*) water, (*c*) R134a, (*d*) propane.

16·35 Solve Problem 16·34 with an isentropic expansion in an engine replacing the throttling through the expansion valve.

16·36 The condenser and evaporator temperatures in a vapor-compression refrigeration system are 80 F and 10 F, respectively. Saturated liquid enters the expansion valve and dry saturated vapor enters the compressor, which operates reversibly and adiabatically. For a refrigeration effect of five tons, determine the refrigerant flow rate and the power input if the refrigerant is (*a*) ammonia, (*b*) R134a.

16·37 Cooling water enters the condenser of a vapor-compression refrigerating plant at 55 F and leaves at 81 F. The water flow rate is 14,200 lbm/h. The power required to drive the adiabatic compressor is 20 hp. Determine the coefficient of performance and the capacity of the plant in tons of refrigeration.

16·38 For a condensing temperature of 70 F and adiabatic dry compression, plot the minimum compressor discharge temperature against evaporator temperature for a range of 10 F to −60 F for the latter for (*a*) ammonia, (*b*) R134a. What conclusions do you draw from these plots?

16·39 The evaporator and condenser pressures of an ammonia compression refrigeration system are

20 psia and 180 psia, respectively. Dry saturated vapor enters the compressor and vapor at 280 F leaves. Liquid at 75 F leaves the condenser. When the refrigerating effect is 20 tons, cooling water passing through the compressor cylinder water jacket picks up heat at a rate of 250 B/min. (*a*) Calculate the power input. (*b*) Is the compression process reversible? Prove your answer.

16·40 In an ammonia compression refrigeration cycle, saturated liquid leaves the condenser at 140 psia, and dry saturated vapor leaves the evaporator at 20 psia. These fluids then pass through a heat exchanger in which the vapor is superheated to 20 F while the liquid is subcooled before entering the expansion valve. Compression is adiabatic, and the compressor discharge temperature is 320 F. The refrigerating effect is 50 tons. Calculate the power input.

16·41 An ideal ammonia vapor-compression refrigeration cycle with two-stage compression and interstage cooling by means of a flash chamber has an evaporator pressure of 10 psia, interstage pressure of 30 psia, and condenser pressure of 150 psia. Saturated liquid leaves the condenser and saturated vapor leaves the evaporator. Both compressions are isentropic. For a refrigerating capacity of 50 tons, determine the total power input to the cycle.

16·42 An ideal cascade refrigeration system uses ammonia as the lower-temperature refrigerant and R134a as the higher-temperature refrigerant. Wet compression is used for both refrigerants. The ammonia evaporator temperature is −45°C, and the ammonia condenses at 0°C. Although the system is ideal, use a temperature differential of 10 Celsius degrees between the evaporating R134a and the condensing ammonia. The R134a condenses at 40°C. Neither condenser removes any sensible heat. The ammonia absorbs 150,000 kJ/h at −45°C. Determine the power input and the coefficient of performance of the refrigeration system.

16·43 Derive the relationship between coefficient of performance and horsepower per ton of refrigeration. For refrigerating machines used in ice making and food preservation, a rule of thumb for the power required is 1 hp/ton. What is the corresponding coefficient of performance?

16·44 Calculate the slope of an isentropic line on a pressure–enthalpy diagram of a pure substance. Check your result qualitatively by the ammonia *ph* chart in the appendix, Chart A·9.

16·45 A reciprocating ammonia compressor used in a refrigerating system operates normally between pressures of 10 psia and 180 psia (69 kPa and 1240 kPa). It is observed that when the suction pressure rises to 20 psia (138 kPa) while the discharge pressure remains 180 psia (1240 kPa), the overload protection device on the electric motor driving the compressor disconnects the motor from the power line. How do you account for this overloading of the driving motor when the compressor pressure ratio has been reduced?

16·46 A household electric refrigerator is operated in a kitchen that is closed and thermally insulated. If it is operated continuously for 2 h, will the average temperature of the air in the kitchen increase, decrease, or remain constant if the refrigerator door is kept (*a*) open, (*b*) closed?

16·47 Determine the makeup water flow rate for a system such as the one shown in Fig. 16·11 if 23,000 kg of water per hour at 7°C is to be delivered, the return water is at 20°C, and the makeup water is at 13°C. Assume that the ejector removes only vapor.

16·48 If the ejector in Problem 16·47 is replaced by a centrifugal compressor and cooling water is available to maintain a condenser temperature of 18°C, determine the minimum power input to the compressor if it operates adiabatically.

16·49 In Section 16·4 it is stated that pumping a liquid requires less power input than pumping a vapor through the same pressure ratio and at the same mass flow rate. Explain why this is true.

16·50 The Servel refrigeration system causes heat to be transferred from one body to another one at a higher temperature with zero work input. Explain any possible conflict with the Clausius statement of the second law.

16·51 Refer to Figs. 16·14 and 16·16. Expand these to show the additional state points of interest if a heat exchanger and return line with a throttle valve are placed between the absorber and the generator.

16·52 Aqua ammonia liquid that is 50 percent ammonia by mass enters a generator of an absorption refrigeration system at 2.0 MPa, 20°C. Saturated liquid and saturated vapor, both at 110°C, leave. For steady flow, determine the mass of vapor leaving per kilogram of liquid entering and the amount of heat added per kilogram of liquid entering.

16·53 Refer to Fig. 16·12. For pressure limits of 1400 kPa and 400 kPa, a generator temperature of 100°C, and an absorber temperature of 20°C, determine (*a*) the amount of liquid that must be pumped per kilogram of ammonia entering the condenser and (*b*) the amount of heat added to the generator per kilogram of ammonia entering the condenser.

16·54 A simple ideal ammonia absorption refrigeration system has condenser and evaporator pressures of 1.40 MPa and 0.30 MPa. Saturated liquid leaves the condenser and saturated vapor leaves the evaporator. Generator and absorber temperatures are maintained at 80°C and 20°C. For a refrigerating effect of 12.0 MJ/h, determine (*a*) the rate of heat supply to the generator, (*b*) the rate of heat removal from the condenser, and (*c*) the rate of heat removal from the absorber.

16·55 Aqua ammonia liquid that is 50 percent ammonia by mass enters a generator of an absorption refrigeration system at 300 psia, 80 F. Saturated liquid and saturated vapor, both at 210 F, leave. For steady flow, determine the mass of vapor leaving per pound of liquid entering and the amount of heat added per pound of liquid entering.

16·56 Refer to Fig. 16·12. For pressure limits of 215 psia and 55 psia, a generator temperature of 200 F, and an absorber temperature of 80 F, determine (*a*) the amount of liquid that must be pumped per pound of ammonia entering the condenser, and (*b*) the amount of heat added to the generator per pound of ammonia entering the condenser.

16·57 A simple ideal ammonia absorption refrigeration system has condenser and evaporator pressures of 200 psia and 40 psia. Saturated liquid leaves the condenser and saturated vapor leaves the evaporator. Generator and absorber temperatures are maintained at 180 F and 70 F. For a refrigerating effect of 80 tons, determine (*a*) the rate of heat supply to the generator, (*b*) the rate of heat removal from the condenser, and (*c*) the rate of heat removal from the absorber.

16·58 A lithium bromide–water absorption refrigeration system has evaporator and condenser temperatures of 5°C and 45°C, respectively. Saturated liquid leaves the condenser, and dry saturated vapor leaves the evaporator. The solution leaves the absorber at 45°C and enters the generator at 80°C. The solution leaving the generator is at 100°C. The refrigerating effect is 400 kJ/s. Dry saturated steam at 120 kPa enters the generator heating coils, and subcooled liquid water at 95°C leaves. Cooling water is supplied at 12°C and should undergo a temperature rise not greater than 8 Celsius degrees. Determine the flow rate of cooling water to the condenser and to the absorber.

16·59 For the conditions of Problem 16·58 determine the required flow rate of heating steam to the generator.

16·60 In a lithium bromide–water absorption refrigeration system, the evaporator and condenser temperatures are 5°C and 50°C, respectively. Condensate leaves the condenser at 45°C, and dry vapor leaves the evaporator at 8°C. The solution leaves the absorber at 30°C and enters the generator at 75°C. Solution leaving the generator is at 95°C. The refrigeration system absorbs heat from the low-temperature region at a rate of 900,000 kJ/h. Dry saturated steam at 125 kPa enters the generator heating coils, and subcooled liquid water at 95°C leaves. Cooling water is supplied at 15°C and should undergo a temperature rise not greater than 10 Celsius degrees. Determine the required flow rate of (*a*) heating steam to the generator, (*b*) cooling water to the condenser, and (*c*) cooling water to the absorber.

16·61 Evaporator and condenser temperatures of a lithium bromide–water absorption system are 40 F and 115 F, respectively. Saturated liquid leaves the condenser, and dry saturated vapor leaves the evaporator. The solution leaves the absorber at 115 F and enters the generator at 175 F. The solution leaving the generator is at 210 F. The refrigerating effect is 100 tons. Dry saturated steam at 17 psia enters the generator heating coils, and subcooled liquid water at 200 F leaves. Cooling water is supplied at 45 F and should undergo a temperature rise not greater than 20 Fahrenheit degrees. (*a*) Sketch the cycle on a property diagram. (*b*) Determine the flow rate through the pump.

16·62 For the conditions of Problem 16·61, determine the required flow rate of (*a*) heating steam to the generator, (*b*) cooling water to the condenser, and (*c*) cooling water to the absorber.

16·63 A lithium bromide–water absorption refrigeration system has evaporator and condenser temperatures of 40 F and 125 F, respectively. Condensate leaves the condenser at 110 F, and dry vapor leaves the evaporator at 45 F. The solution leaves the absorber at 85 F and enters the generator at 165 F. Solution leaving the generator is at 200 F. The refrigerating effect is 60 tons. Dry saturated steam at 20 psia enters the generator heating coils, and subcooled liquid water at 210 F leaves. Cooling water is supplied at 55 F and should undergo a temperature rise not greater than 15 Fahrenheit degrees. Determine the required flow rate of (*a*) heating steam to the generator, (*b*) cooling water to the condenser, and (*c*) cooling water to the absorber.

16·64 What are the principal advantages and disadvantages of a lithium bromide absorption system in comparison with an ammonia absorption system? Why is no rectifier used in a lithium bromide system?

16·65 Air is to be liquefied by the Hampson–Linde regenerative process operating with pressure limits of 200 atm and 1 atm. It is desired to obtain 1 mass unit of liquid from each 6 mass units of air flowing through the expansion valve. Assume that the regenerative heat exchanger is perfectly insulated and that the minimum temperature differential between the two streams of air passing through this heat exchanger is to be 10 Celsius degrees (18 Fahrenheit degrees). Neglect pressure drops in heat exchangers. To what temperature must the air be cooled by the compressor aftercooler before it enters the regenerative heat exchanger to maintain steady conditions?

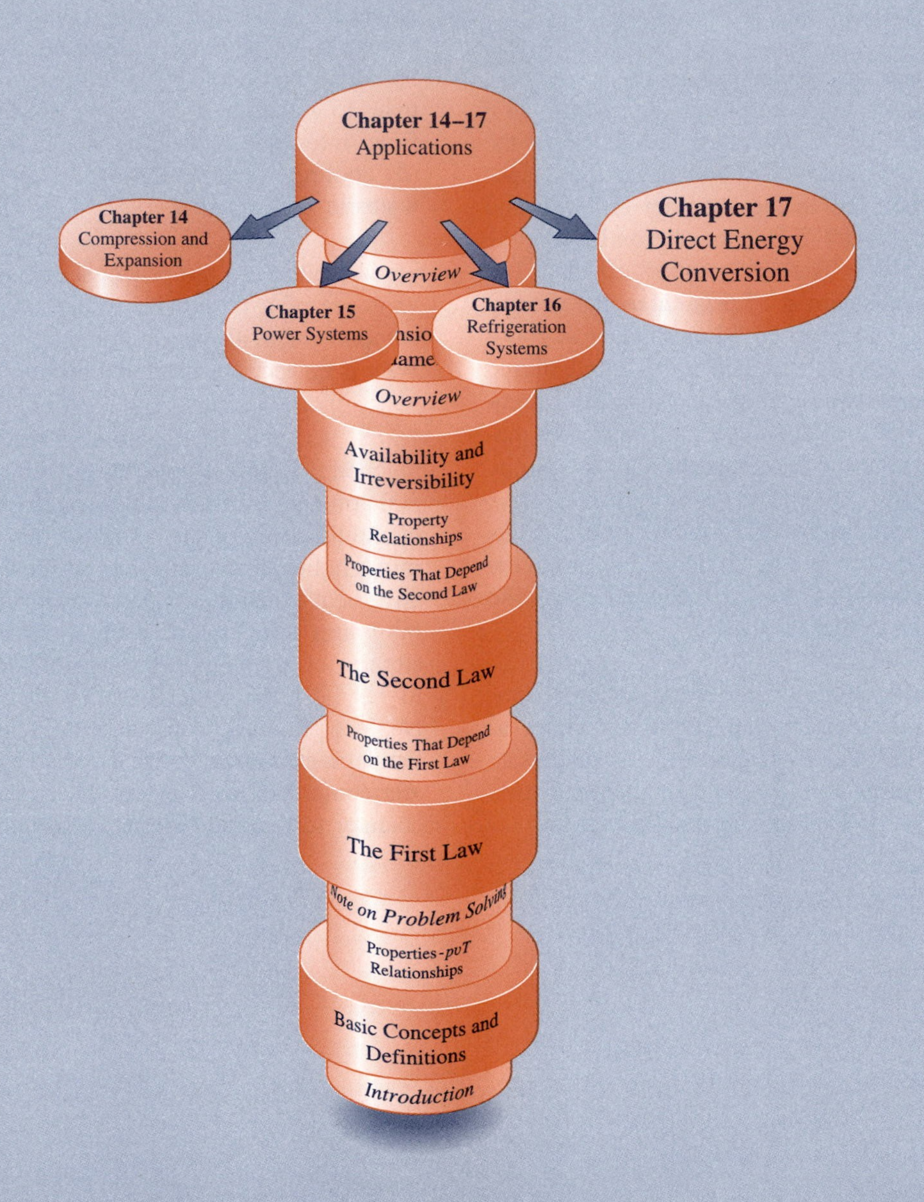
Chapter 14–17
Applications
Chapter 14
Compression and Expansion
Chapter 17
Direct Energy Conversion
Overview
Chapter 15
Power Systems
Chapter 16
Refrigeration Systems
Overview
Availability and Irreversibility
Property Relationships
Properties That Depend on the Second Law
The Second Law
Properties That Depend on the First Law
The First Law
Note on Problem Solving
Properties-pvT Relationships
Basic Concepts and Definitions
Introduction

Chapter 17

Direct Energy Conversion

In preceding chapters of this book, several different engines have been analyzed. These engines all convert heat into work by using a working substance—usually a fluid—that alternately absorbs and rejects heat as it undergoes a cycle. A device that would continually and directly convert some part of the heat supplied into work without involving a cycle of a working substance would be quite desirable. The efficiency of this type of direct energy conversion system, however, would still be limited in accordance with the

Carnot principle discussed in Chapter 5. If these engines are operated in reverse, they serve as direct energy conversion refrigerators. Although the Carnot principle limits the coefficient of performance of such systems, the simplicity of such devices makes them valuable in some applications.

A yet more desirable device would convert chemical energy directly into work, without first converting the chemical energy into heat. In this case, the Carnot principle limitation on efficiency would not apply. On this basis, we can divide direct energy devices into two general classes:

1. Direct energy conversion devices that are limited by the second law
2. Direct energy conversion devices that operate isothermally and are not limited by the second law

17·1 Direct Energy Conversion Systems

The significant advantage of direct energy conversion systems is that they avoid intermediate steps involving working substances.

Electrical energy and work are equivalent, because either can be fully converted into the other, at least in the ideal case. That is, there are no inherent limitations on such conversion.

The significant advantage of direct energy conversion systems is that they are capable of converting various forms of thermal, chemical, or nuclear energy directly into work or electrical energy. These systems operate without the use of working substances that alternately absorb, store, and reject energy. Ideally, the conversion of thermal, chemical, or nuclear energy in a fuel cell, battery, photovoltaic cell, thermoelectric system, or other direct energy conversion system can achieve a very high efficiency. However, various limitations on the system operation make it difficult to realize these efficiencies in practice.

In the selection of an energy conversion system for a given application, many factors in addition to efficiency must be considered. Some of these are size, mass, initial cost, reliability, cost of maintenance, environmental pollution, and adaptability to various fuels. Consequently, each of the direct energy conversion devices mentioned here is well suited to certain applications.

The first fuel-cell-powered vehicle was demonstrated by Harry Karl Ihrig in 1959: a 20-horsepower, fuel-cell-powered tractor (Ref. 17·1).

The pollution caused by emissions from various types of engines is of great concern today. Automobile engines, partly because they are so numerous, but also for other reasons, are responsible for a significant fraction of the air pollution, especially in large cities. Figure 17·1 compares the emissions, energy requirements, and volume of various systems for providing energy for common automotive applications. (The energy required and pollutants created in the production of the fuel for each of the systems are also included in the totals.) As Fig. 17·1 shows, in terms of environmental pollution, fuel cells are an attractive option. Other practical considerations limit their usefulness as a mobile energy source, and thus, fuel cells are more often considered for other applications, such as decentralized or centralized power generation. Nevertheless, fuel-cell technology is advancing rapidly. We now turn our attention to fuel cells.

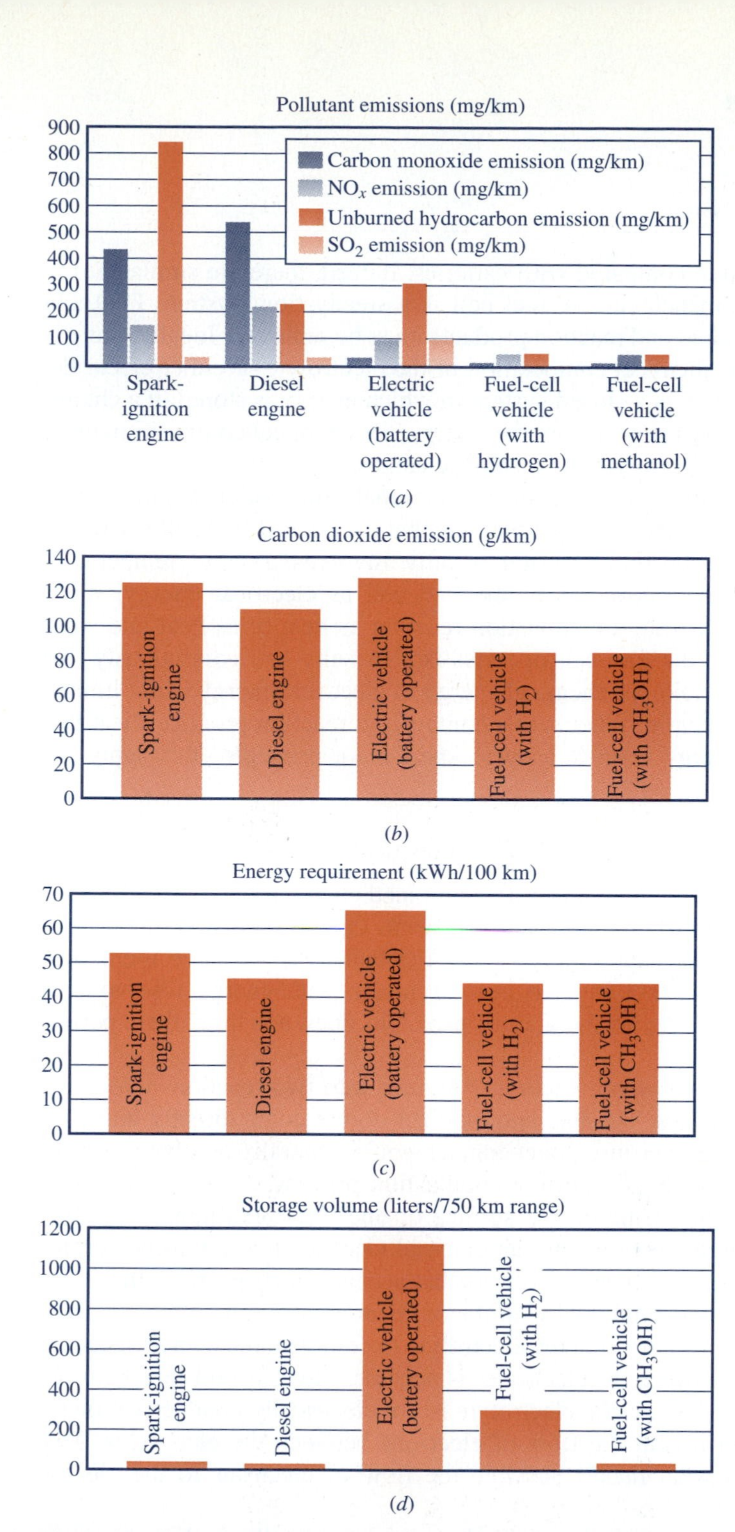

Figure 17·1 A comparison of various systems for providing energy to automobiles: (*a*) pollutant emissions, (*b*) carbon dioxide emissions, (*c*) energy requirements, and (*d*) storage volume.

17·2 Fuel Cells

The most important design parameter in power supply for spacecraft applications is power per unit mass, or power density, which is measured in kW/kg. Using 1960s technology, a hydrogen–oxygen fuel cell provided 1.6 kWh/kg of power, whereas the most advanced batteries could provide only 0.2 kWh/kg (see Ref. 17·1). In addition, a hydrogen–oxygen fuel cell provides water for the crew. For these reasons, the higher capital cost of fuel cells was easily justified for spacecraft use even if they were not cost-effective for everyday use. The U.S. space program has effectively utilized fuel cells since the 1960s.

Fuel cells are often compared with batteries. Indeed, there are similarities in structure and nomenclature. A fuel cell is a steady-flow system. Fuel and oxidizer are supplied and reaction products must be removed. Ideally, there is no change in the chemical composition of the electrolyte or either electrode. In contrast, a battery is a closed system in which energy is stored in a chemical form. As energy is taken from the battery, the chemical composition of an electrode, of the electrolyte, or of both changes.

In comparing the fuel cell with conventional combustion engine power generation, note that the chemical reaction that occurs in a fuel cell is isothermal, while a combustion reaction usually involves a large temperature change. Energy is removed from the fuel cell as electrical energy, while energy is removed from a combustion reaction as heat or as heat and work together. Because the fuel cell operates isothermally and continuously, the extent of its conversion of chemical energy to electrical energy is not limited by the Carnot principle. Whenever a combustion reaction produces heat that is used as the energy input to a cycle, the Carnot principle does apply.

17·2·1 Description of Fuel-Cell Operation

In a fuel cell, fuel and an oxidizer are combined chemically in a steady-flow, isothermal reaction to produce electrical work. Operation is continuous provided there is a constant supply of the reactants. As we have mentioned above, fuel cells do not involve heat exchange with energy reservoirs at different temperatures, so the Carnot principle does not limit their performance.

In a fuel cell, the overall reaction is divided into two reactions that occur on separate electrodes. The fuel and oxidizer do not come directly into contact with each other, because direct contact would generally involve a nonisothermal reaction such as a normal combustion process.

One reaction, occurring on the surface of one electrode, ionizes the fuel and sends released electrons into an external electric circuit. On the surface of the other electrode, a reaction occurs that accepts electrons from the external circuit and when combined with the oxidizer creates ions. The ions from each reaction are combined in the electrolyte to complete the overall reaction. The electrolyte between the electrodes is necessary to transport ions between the electrodes. However, the electrolyte is not electrically conductive; therefore, it does not permit the flow of electrons between the electrodes. This characteristic is what makes possible the flow of electrons to the electric circuit.

A hydrogen–oxygen fuel cell is shown schematically in Fig. 17·2*a* to illustrate the function of the various parts of the cell. External connections provide for flows of hydrogen and oxygen into the fuel cell and for the flow

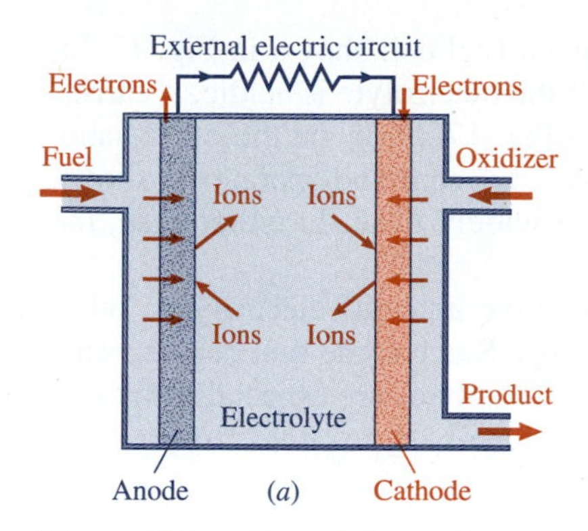

Figure 17·2*a* General fuel-cell configuration.

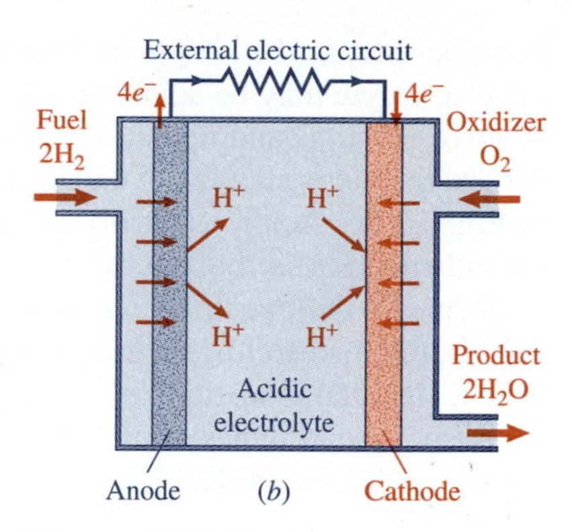

Figure 17·2*b* Hydrogen–oxygen fuel cell using an acidic electrolyte.

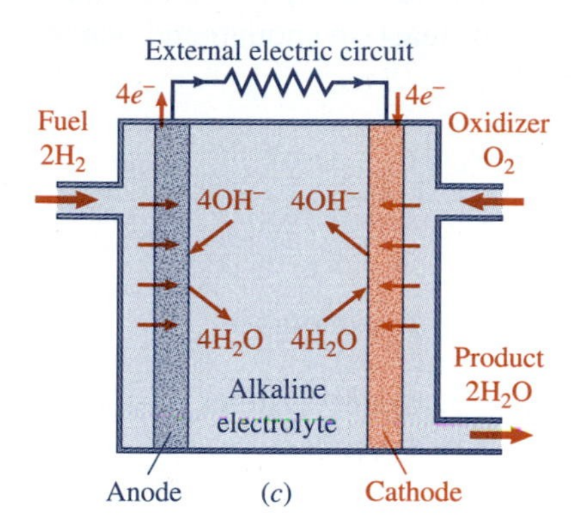

Figure 17·2*c* Hydrogen–oxygen fuel cell using an alkaline electrolyte.

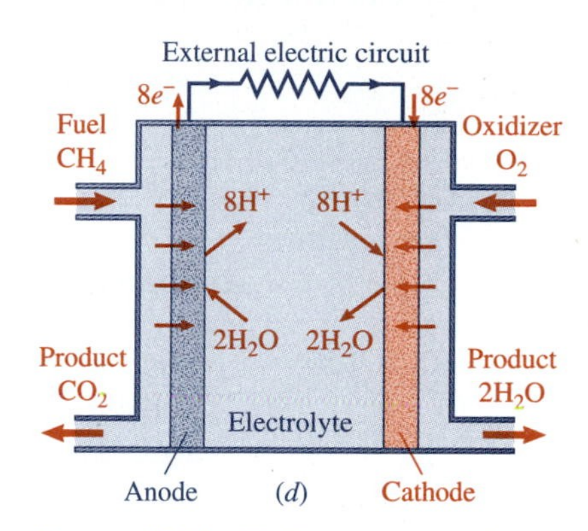

Figure 17·2*d* Methane–oxygen fuel cell.

of the reaction product, water, from the cell. The water may be in either the liquid or gaseous phase depending on the operating temperature. Other external connections permit the flow of electrons to complete the electric circuit.

The Electrodes. The **anode** is defined as the electrode that supplies electrons to the external circuit. The **cathode** is defined as the electrode that receives electrons from the external circuit. In the fuel cell shown, the electrodes also perform the important function of separating the fuel and oxidizer from the electrolyte. The electrodes are typically made of porous materials; this substantially increases the area over which reactions can take place.

Anode
Cathode

The Electrolyte. Fuel cells are often categorized by the type of electrolyte used. The electrolyte determines the flow of ions in the cell, the temperature limitations of the cell, as well as many other system operating features.

In solid oxide fuel cells, the electrolyte is alkaline but carries O^- ions instead of OH^- ions.

Solid polymer electrolytes are most common in fuel cells used for transportation; however, phosphoric acid fuel cells are the most common overall.

For example, in the simple hydrogen–oxygen fuel cell shown in Fig. 17·2*a*, the electrolyte may be acidic or alkaline. If the electrolyte is acidic, it carries positive H^+ ions and operates as shown in Fig. 17·2*b*. If, on the other hand, the electrolyte is alkaline, it carries negative OH^- ions and operates as shown in Fig. 17·2*c*. Acidic electrolytes are more widely used. In either case, the overall reaction is the same.

Although the examples we give here involve aqueous electrolytes, other liquids and also solids are used as electrolytes. Since some fuel cells operate best at elevated temperatures, nonaqueous electrolytes are especially suitable for them. Several types of fuel cells, classified by the type of electrolyte used, are described in Table 17·1.

Fuel. A wide variety of fuels may be used in fuel cells. It was recognized in the 1970s, however, that gaseous hydrogen was the only effective fuel that could be widely used. Other fuels were either not reactive enough and hence

Table 17·1 Types of Fuel Cells

Type	*Electrolyte*	*Operating Temperature* (°C)	*Special Features*	*Applications*
Alkaline fuel cell (AFC)	Diluted potassium hydroxide solution	60 to 120	High efficiency, only suitable for pure hydrogen and oxygen	Space systems, defense systems
Proton exchange membrane fuel cell (PEM)	Proton-conducting polymer membrane	20 to 120	Very flexible operating behavior, high power density	Vehicles, decentralized electricity generation (small plants)
Phosphoric acid fuel cell (PAFC)	Phosphoric acid	160 to 220	Limited efficiency, corrosion problems	Decentralized electricity generation, combined heat and power
Molten carbonate fuel cell (MCFC)	Molten carbonates	600 to 650	Complex process control, corrosion problems	Centralized and decentralized electricity generation, combined heat and power
Solid oxide fuel cell (SOFC)	Solid zirconium dioxide	850 to 1000	Electric power direct from natural gas, ceramic technology (high temperatures)	Centralized and decentralized electricity generation, combined heat and power

Source: Reference 17·4.

required expensive noble metal catalysts, or were toxic and thus difficult to handle safely. It was determined that if fuel cells were to gain acceptance, they would have to use fossil fuels such as natural gas or other light hydrocarbons which are easy to handle, inexpensive, and readily available to provide the necessary hydrogen supply (Ref. 17·1). Of course, this requires an intermediate step, called **reforming**, which generates hydrogen gas from the primary fuel. For example, the reforming of methane is achieved by the reaction

Reforming

$$CH_4 + H_2O \rightarrow CO + 3H_2$$

Since CO poisons the catalysts used in fuel cells, the reforming reaction is followed by an oxidation reaction called the **shift reaction**:

Shift reaction

$$CO + H_2O \rightarrow CO_2 + H_2$$

The hydrogen gas that is the product of the reforming and shift reactions is impure; thus, electrolytes that are not fouled by carbon dioxide are required.

If a fuel cell is to be used as a mobile energy source, then using methane or methanol instead of pure hydrogen substantially reduces the storage volume required for a given range of operation (see Fig. 17·1).

Catalysts. One of the early limitations in fuel-cell technology was caused by the catalysts. For low-temperature fuel cells using aqueous electrolytes, expensive noble metals, such as platinum, are required to increase the electrode reaction rates. In the 1960s, for example, typical fuel cells required 25 mg/cm^2 of platinum on the electrodes. This translates, in 1991 dollars, to a cost of $10,000/kW. This made the fuel cell cost-prohibitive for commercial use. In addition, the typical operating lifetime of the catalyst was on the order of hundreds, rather than thousands of hours. However, technical advances have brought the platinum requirement to below 0.1 mg/cm^2, thus reducing the cost of the catalyst several hundredfold (Ref. 17·1). This puts the capital cost per kilowatt of installed capacity within reach for public utility power plants, industrial power generation, and large public transportation systems. As Fig. 17·1 indicates, fuel cells also operate with low emissions, and this helps to offset the higher initial capital costs.

Oxidizer. Although Fig. 17·2 shows oxygen as the oxidizer, it is unnecessary to use pure oxygen. Cryogenic oxygen is used for space applications, but air is the most common oxidizer for terrestrial applications.

17·2·2 Analysis of Fuel-Cell Operation

Section 6·5 shows that for a steady-flow system with a substance both entering and leaving at the temperature of the surrounding atmosphere and exchanging heat with only the atmosphere, the maximum work that can be produced is equal to the decrease in G of the substance

$$W_{max} = -\Delta G \tag{17·1}$$

The value of ΔG for any reaction can be calculated in a manner analogous to that for calculating ΔH as given by Eq. 11·9. That is,

General equation for the Gibbs function change for a process involving a chemical reaction

$$\Delta G = G_2 - G_1 = \sum_{\text{products}} N(\bar{g}_2 - \bar{g}^\circ_{ref} + \Delta \bar{g}^\circ_{f,ref}) - \sum_{\text{reactants}} N(\bar{g}_1 - \bar{g}^\circ_{ref} + \Delta \bar{g}^\circ_{f,ref}) \tag{17·2}$$

where the subscript *ref* and the superscript ° indicate the temperature and pressure, respectively, of the reference state. Values of $\Delta \bar{g}^\circ_{f,ref}$ are often tabulated with other thermodynamic property data (see Table A·17). Values can also be calculated from a relationship based on the definition $G \equiv H - TS$. For any element in its most stable form, the Gibbs function of formation at the reference state is zero. For any substance,

$$\Delta \bar{G}^\circ_{f,ref} = \Delta \bar{H}^\circ_{f,ref} - T_{ref}\, \Delta \bar{S}^\circ_{f,ref}$$

and

$$\Delta \bar{S}^\circ_{f,ref} = \sum_{\text{products}} N\bar{s}^\circ_{ref} - \sum_{\text{reactants}} N\bar{s}^\circ_{ref}$$

where all entropies are absolute entropies. Property values for some substances used in fuel cells are given in Table 17·2.

For a reversible, isothermal, steady-flow process at T_0 (the temperature of the surrounding atmosphere), during which heat exchange occurs with only the atmosphere, the work is the maximum work and the heat transfer is given by

$$Q = H_2 - H_1 + W_{max} = H_2 - H_1 + G_1 - G_2 = T_0(S_2 - S_1) \tag{17·3}$$

The heat transfer may be greater than, equal to, or less than zero, depending on the relative magnitudes of ΔG and ΔH.

Unlike heat engines, fuel cells do not operate cyclically and exchange only heat and work with the surroundings. They diminish the amount of fuel and oxidizer and increase the amount of products in the surroundings. Consequently, no performance parameter such as a thermal efficiency that involves only heat and work transfer can be defined for fuel cells.

Table 17·2 $\Delta\bar{h}^\circ_{f,ref}$ and $\Delta\bar{g}^\circ_{f,ref}$ at 100 kPa, 25°C

Fuel-cell efficiency

	$\Delta\bar{h}^\circ_{f,ref}$		$\Delta\bar{g}^\circ_{f,ref}$	
Compound	kJ/kmol	B/lbmol	kJ/kmol	B/lbmol
$CO(g)$	−110,530	−47,519	−137,163	−58,970
$CO_2(g)$	−393,522	−169,197	−394,389	−169,557
$CH_4(g)$	−74,873	−32,190	−50,768	−21,826
$C_3H_8(g)$	−104,673	−45,001	−24,398	−10,489
$CH_3OH(l)$	−238,910	−102,713	−166,640	−71,642
$H_2O(l)$	−285,830	−122,885	−237,141	−101,952
$H_2O(g)$	−241,826	−103,966	−228,582	−98,273

Note: From Table A·17.

One performance parameter in use is

$$\epsilon \equiv \frac{\Delta G}{\Delta H} \tag{17·4}$$

Fuel-cell efficiency

The parameter ϵ has been called the **fuel-cell efficiency**. Notice that ϵ cannot be compared directly with thermal efficiency as defined in connection with other cycles. Indeed, ϵ is less than 1 for an ideal hydrogen–oxygen fuel cell, equal to 1 for an ideal methane–oxygen cell, and greater than 1 for a carbon–oxygen cell. A value of ϵ greater than 1 indicates that W_{max} exceeds the decrease in enthalpy of the stream passing through the cell; therefore, there must be heat transfer from the surroundings to the fuel cell.

The work done by a fuel cell is given by the product of the charge $\mathscr{C}$ carried by the external circuit and the cell terminal emf or voltage $\mathscr{V}$, which is the work done per unit charge. The charge delivered is the product of the number of electrons delivered to the external circuit and the charge per electron (1.6022×10^{-19} C/electron). The number of electrons is equal to the product of the number of molecules giving up electrons and the valence j. In turn, the number of molecules giving up electrons is given by Avogadro's number, 6.022169×10^{26} molecules/kmol, times the number of moles of reactants giving up electrons. Combining these parameters gives

$$W_{max} = \text{moles of reactant} \underbrace{\left(\frac{\text{molecules}}{\text{kmol}}\right)}_{\text{Avogadro's number}} \underbrace{\left(\frac{\text{electrons}}{\text{molecule}}\right)}_{\text{valence}} \left(\frac{\text{charge}}{\text{electron}}\right) \text{voltage}$$

$$= N(6.022169 \times 10^{26})j(1.6022 \times 10^{-19})\mathscr{V} = \mathfrak{F}Nj\mathscr{V} \tag{17·5}$$

where $\mathfrak{F}$ is the Faraday constant (96.487×10^6 C/kmol of electrons or 96,487 kJ/V·kmol of electrons), N is the number of moles of reactant having a valence j, and $\mathcal{V}$ is the cell terminal voltage. Thus, the ideal cell terminal voltage is given by

Ideal fuel-cell terminal voltage

$$\mathcal{V}_i = \frac{\Delta G}{\mathfrak{F} N j} \tag{17·6}$$

Remember that this is the *ideal* cell voltage. Internal cell resistance and other effects prevent this value from being realized in an actual fuel cell.

17·2·3 Actual Fuel-Cell Performance

As pointed out in Section 1·3, we often analyze an ideal system and then adjust our results in predicting an actual system's behavior. In the preceding section we have considered only ideal fuel cells. They operate reversibly, and their performance, including the effect of temperature on operating performance, can be calculated from principles introduced earlier.

Actual fuel cells do not perform as well as ideal ones. Several factors cause the differences, and we now describe briefly the major ones:

- **Interface surface area in the electrodes**. In the ideal fuel cell, we assume that fuel or oxidizer and the electrolyte are always present at the reaction surface of an electrode in the proper proportion and that there are no impediments to the reaction. In an actual fuel cell, the transport of fuel, oxidizer, electrolyte, and products to and from the electrode surfaces is complex and introduces several problems. To provide a large electrode surface for the three-phase interface of a gaseous fuel or oxidizer, a liquid electrolyte, and a solid electrode, porous electrodes are used. However, if either too much liquid or too much gas enters the electrode and displaces the other fluid phase, the interfacial area between the phases is reduced. Reducing the interfacial area impedes the reaction. Some of these problems are mitigated by the use of solid electrolytes.
- **Buildup of product**. Even if the phases are present in the proper proportions, there is a tendency for the reaction to increase the concentrations of products and therefore reduce the concentration of reactants in the reaction zone. This also retards the reaction. Agitating the fluids is one method of addressing this problem.
- **Diffusion of chemical species at the electrode surface**. The diffusion process at the electrode surface is quite complex. Diffusion occurs only when there is a concentration gradient of chemical species. The concentration gradient is sustained by chemisorption at the electrode interface. Some energy is required to cause the chemisorption process. This is called the activation energy, and however it is supplied, it reduces the work of the cell.

- **Electrical impedances in the fuel cell**. Associated with the diffusion process at the electrode surface is an impedance to electron flow. There is also ohmic resistance elsewhere in the cell. These are irreversible effects that reduce the terminal emf and work of the cell.
- **Additional reactions**. Spurious chemical and electrochemical reactions may occur at various places within the cell, and they always impair cell performance.
- **Miscellaneous**. Other losses in actual cells result from leakage of both electrical current and fluids. Also, various auxiliaries such as pumps, fans, and control systems require some energy in an actual fuel-cell power plant.

Most of the losses associated with these various effects increase with increasing current. Therefore, actual fuel-cell performance approaches ideal cell performance most closely under conditions of very low current rather than under design conditions. For the same fuel flow rate, an actual cell may provide, under low-current conditions, 80 percent of the maximum work of an ideal cell, whereas under design or rated conditions this ratio may be on the order of 50 percent or lower.

The desire to increase the work done must be balanced by considerations of other factors such as reliability, safety, initial cost, and environmental effects. For example, an electrolyte that is corrosive may be unacceptable for actual fuel cells even though its ideal thermodynamic performance would make it attractive.

Actual fuel cells may be appreciably more complex than the simple ideal ones we have discussed. For example, regenerative fuel cells, in which the reactants are at least partially regenerated by means of an energy input either to the cell or to an external regenerator, are attractive for some applications. The design and performance of actual fuel cells are discussed in some of the references at the end of this chapter as well as in the current periodical literature.

17·3 Batteries

The first batteries were operated in the laboratory early in the nineteenth century. However, batteries were not widely used until after the invention of the electric telegraph and the electric incandescent lamp (see Ref. 17·2). Today batteries are a vital source of power for many applications. Advances in the state of the art have extended their operating life, decreased their cost, and reduced their environmental impact. Modern development of batteries is driven partly by growing interest in the electric car as a means of reducing pollution. However, the battery-operated car is not new to the 1990s—more electric cars were in operation in 1915 than in the 1980s (Ref. 17·2). Of course, today's consumer expectations for automobile performance have significantly increased the design requirements for the electric car. This has placed high demands on battery performance.

A battery converts chemical energy stored in the system into electrical energy. The chemical energy may either be inherent in the chemical substances, as it is in nonrechargeable batteries, or it may be produced by an externally applied electric current, as in the case of rechargeable batteries. Nonrechargeable batteries are by far the most common of all batteries; they are inexpensive and typically have a long shelf life and high energy per unit volume. Rechargeable batteries, on the other hand, require a higher initial capital cost but offer savings in overall operating cost. Rechargeable batteries are widely used in internal combustion engines for ignition and other electrical systems.

17·3·1 Description of Battery Operation

The basic operation of a battery is shown in Fig. 17·3. The system consists of two electrodes connected externally by an electrical circuit. The negative electrode is called the *anode* and the positive electrode is called the *cathode*. As for the fuel cell, the two electrodes are placed in an electrolyte that permits the movement of ions but not the movement of electrons. Thus, there are two ''flows'' of importance: the flow of electrons and the flow of ions (see Fig. 17·3). Unlike fuel cells, however, batteries cannot be operated continuously. To be sure, some batteries can be operated over long periods of time, but eventually, the chemical energy in the system is depleted and must be restored. Eventually, additional energy must be added to the system to extend its life either by replacing a spent chemical or by recharging.

The leakage rate in a battery refers to the rate at which the electrolyte leaks from the battery. Acidic electrolytes, as are commonly found in zinc–carbon batteries, are often corrosive to metal parts and will irritate human skin.

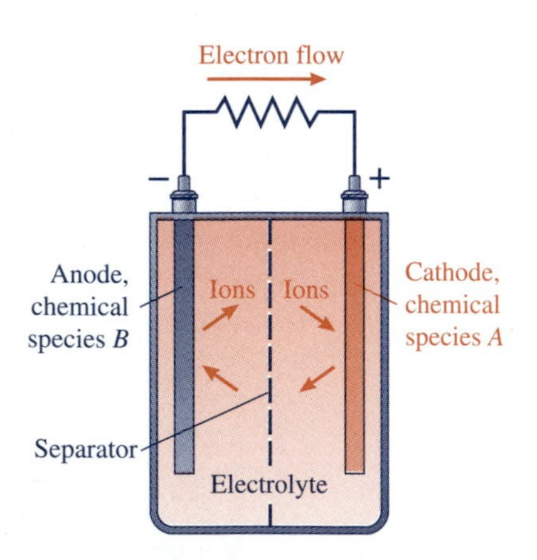

Figure 17·3 General battery configuration.

Alkaline Batteries. Alkaline batteries are commonly used in radios, toys, cameras, flashlights, calculators, pagers, alarm clocks, or any other device requiring a battery with good performance at relatively low cost. Alkaline batteries are well known for their long shelf life, ability to deliver high currents, and low leakage rates. An alkaline battery consists of a zinc anode and any of several possible materials for the cathode, although manganese oxide or carbon is the most common. In this respect, alkaline batteries are much like zinc–carbon batteries, or ''dry cells,'' which also use zinc for the anode and manganese dioxide for the cathode. Alkaline batteries, however, derive their name from the electrolyte, which is an alkaline solution of potassium hydroxide, rather than the acidic electrolytes used in zinc–carbon batteries. We discuss the alkaline–manganese battery as shown in Fig. 17·4 next.

The reactions occurring at the zinc anode and the manganese-dioxide cathode in the alkaline–manganese battery are described by the following reactions:

$$\begin{array}{ll} \text{anode:} & Zn + 2OH^- \rightarrow ZnO + H_2O + 2e^- \\ \text{cathode:} & 2e^- + 2MnO_2 + 2H_2O \rightarrow 2MnOOH + 2OH^- \\ \hline \text{overall reaction:} & Zn + 2MnO_2 + H_2O \rightarrow ZnO + 2MnOOH \end{array}$$

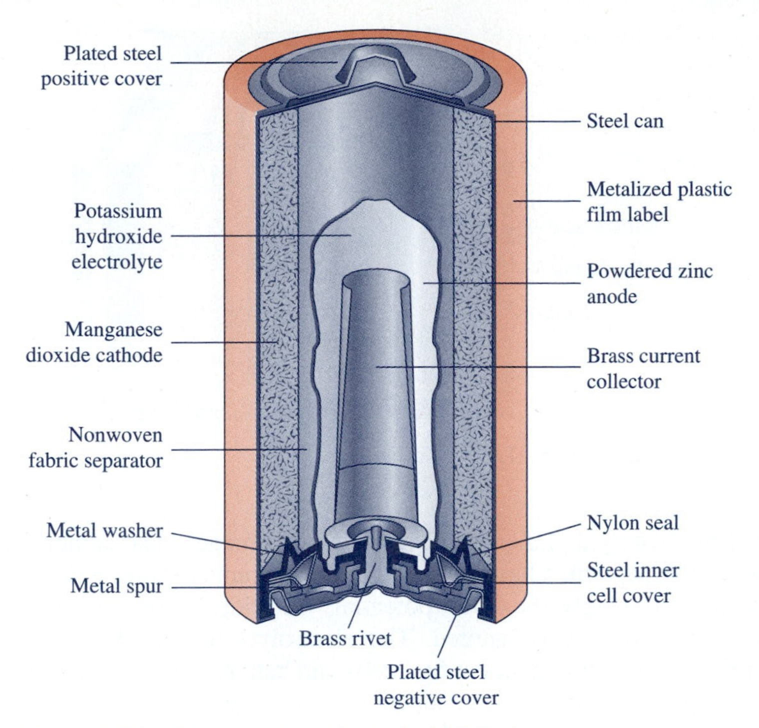

Figure 17·4 Cutaway view of a typical ''D''-size alkaline–manganese battery. (Courtesy of Eveready Battery Co.)

The voltage of an alkaline–manganese battery is 1.5 V–1.6 V at the beginning of its life; but, as the battery approaches discharge capacity, the voltage drops. The drop in voltage over the life of the battery is caused by the fouling of the cathode by product. Zinc–carbon batteries exhibit this same characteristic drop in voltage.

Lead–Acid Battery. The lead–acid battery is one of the most successfully marketed batteries in history. This success is largely the result of its use (almost to the exclusion of any other type of battery) in internal-combustion-engine automobiles. Many improvements in performance have allowed the lead–acid battery to retain its market share despite the laboratory development of other more powerful, lighter batteries. The ever-increasing amount of lead produced annually despite declining use of the metal overall, reflects the ubiquity of the automobile and the lead–acid battery. Consider that, in the United States, lead–acid battery production is the largest consumer of lead (Ref. 17·2). However, the environmental impact and hazard to human health posed by the disposal of lead–acid batteries has inspired the development of reclamation techniques.

Figure 17·5 is a schematic diagram of a lead–acid battery. The anode consists of ''sponge lead,'' which is a porous mixture of lead and lead oxide with other additives. This prevents the anode from losing porosity over the

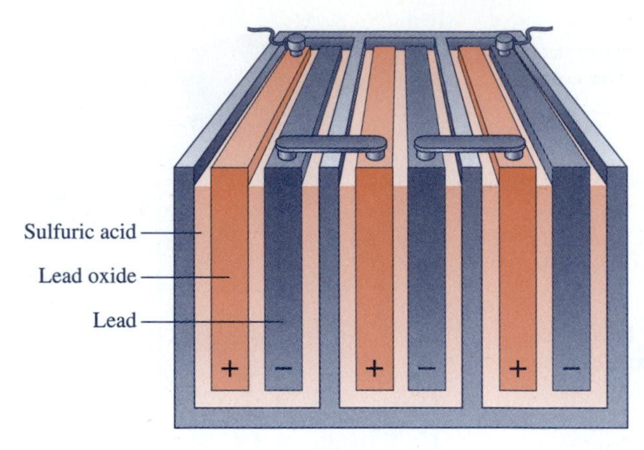

Figure 17·5 Lead–acid battery.

life of the battery and thereby degrading performance. The cathode is made of lead dioxide, which has two crystalline structures existing in equilibrium. They provide nearly the same potential, differing by only 0.01 V, and so usually this subtlety is ignored. The electrolyte in a lead–acid battery is sulfuric acid and, unlike in the fuel cells and batteries discussed to this point, it participates actively in the reaction. The concentration of sulfuric acid decreases during discharge and increases during recharge. The salient reactions for a lead–acid battery are shown below:

Up to this point the electrolyte has been merely an ion-conducting medium that does not participate in the overall reaction.

anode:	$Pb \rightarrow Pb^{2+} + 2e^-$
reaction with electrolyte:	$Pb^{2+} + SO_4^{2-} \rightarrow PbSO_4$
cathode:	$PbO_2 + 4H^+ + 2e^- \rightarrow Pb^{2+} + 2H_2O$
reaction with electrolyte:	$Pb^{2+} + SO_4^{2-} \rightarrow PbSO_4$
overall reaction:	$Pb + PbO_2 + 2H_2SO_4 \rightarrow 2PbSO_4 + 2H_2O$

You may have experienced the inability to recharge a lead–acid battery if you have tried to jumpstart a car after a prolonged period when the battery was fully discharged.

Lead–acid batteries can reliably undergo many charge–recharge cycles without significant loss of performance, although the ability to recharge is lost after prolonged periods in the fully discharged state. The lead–acid battery is self-discharging. Both the anode and cathode materials will spontaneously react with the electrolyte and over time the battery will lose capacity. Fortunately, the reaction rates are slow, so this does not significantly affect the performance of the battery.

In addition, unlike the alkaline battery, which has a discharge potential that steadily decreases over its life, the discharge potential for a lead–acid battery is nearly constant.

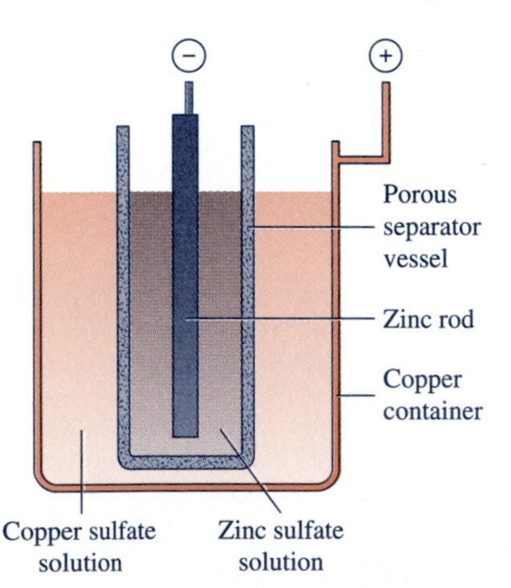

Figure 17·6 Daniell cell battery.

17·3·2 Analysis of Battery Operation

For simplicity, let us consider one of the earliest batteries, the Daniell cell, which was invented in 1836 by J. F. Daniell. Figure 17·6 shows a schematic of this battery that utilizes a zinc anode and a copper cathode. There are two electrolytes in the Daniell cell, copper sulfate, which contains copper ions, and zinc sulfate, which contains zinc ions. The reactions at the anode and cathode are, respectively,

$$Zn \leftrightarrow Zn^{2+} + 2e^-$$
$$Cu^{2+} + 2e^- \leftrightarrow Cu$$

The overall reaction for the cell at equilibrium is then

$$Zn + Cu^{2+} \leftrightarrow Zn^{2+} + Cu$$

In a conventional chemical reaction, the heat released from the overall reaction shown above would be equal to the enthalpy of reaction (see Chapter 11). The enthalpy of reaction represents the heat that would be released if zinc and copper were brought directly into contact with one another. However, in a battery, this condition does not occur; rather, the reaction is electrochemical. Under these conditions, the enthalpy of reaction is released as the sum of the heat released from the battery ($-Q_b$), and the work done on the external electric circuit. The work done on the external circuit is equal to the product of the charge carried by the external circuit $\mathscr{C}$ and the cell voltage is $\mathscr{V}$, as it was in the case of the fuel cell. As long as current is flowing, there are irreversibilities in the battery. As the current approaches zero, the irreversibilities become smaller and in the limit of zero current, the process is reversible. As we showed for a fuel cell, in this limit of a reversible reaction, the heat released from the battery is $T_0\,\Delta S$, and the maximum work that the battery can do on the external circuit is therefore equal to the change in the Gibbs function. Under these circumstances, we can find an expression describing the ideal voltage of the battery, in other words, the maximum voltage achievable in the limit of no current:

Recall that the maximum work is defined as that which is achieved for a reversible process.

The reversible case may alternatively be thought of as the case when the load on the external circuit goes to zero.

$$-\Delta H = -Q_b + W_{max}$$
$$= -T_0\,\Delta S + \mathscr{V}_i\mathscr{C}$$

Since $\Delta G = \Delta H - T\,\Delta S$, we find

$$-\Delta G = \mathscr{V}_i \mathfrak{F} N j$$

$$\mathscr{V}_i = \frac{-\Delta G}{\mathfrak{F} N j}$$

Note that the ideal voltage for a battery in the limit of no current is the same as the ideal cell voltage for a fuel cell.

Also, we have made use of Eq. 17·5 to describe the work done on the external circuit, which is equal to the product of the voltage and the charge passing.

For batteries, the ideal voltage or potential of a cell can be derived from the difference in the ideal potentials of the reactions at the individual electrodes:

$$\mathscr{V}_i = \mathscr{V}_{\text{cathode}} - \mathscr{V}_{\text{anode}} \qquad (17{\cdot}7)$$

It is common practice to provide information on battery half-reactions. This reduces the amount of data required, since in many cases, one electrode remains the same while the other changes, as would be the case for many lithium batteries. Several common battery half-reactions are shown in Table 17·3.

It would be advantageous to utilize materials that would create the highest potential. However, realistic concerns about safety, corrosion, and so forth, often limit the possible choices in battery systems.

Battery performance is usually measured in terms of energy density either by mass or by volume depending on the application. In the ideal case, the battery performance can be measured on a mass basis as

$$\epsilon_{m,i} = \frac{\mathcal{V}_i \mathfrak{F} N j}{\text{total mass of reactive components}} \tag{17·8a}$$

Ideal battery performance parameter

or on a volume basis as

$$\epsilon_{v,i} = \frac{\mathcal{V}_i \mathfrak{F} N j}{\text{total volume of reactive components}} \tag{17·8b}$$

A common unit of measure for the performance of a battery is the total energy output measured in Wh per kilogram or per liter. Of course, in an

Table 17·3 Common Battery Half-Reactions

Reaction	$\mathcal{V}_i$ (Volts)
$AgCl + e^- \rightarrow Ag + Cl^-$	0.222
$Al^{3+} + 3e^- \rightarrow Al$	−1.66
$Br_2(l) + 2e^- \rightarrow 2Br^-$	1.065
$Ca^{2+} + 2e^- \rightarrow Ca$	−2.84
$Cl_2 + 2e^- \rightarrow 2Cl^-$	1.360
$Cu^{2+} + 2e^- \rightarrow Cu$	0.337
$Fe^{2+} + 2e^- \rightarrow Fe$	−0.444
$2H_2O + 2e^- \rightarrow H_2 + 2OH^-$	−0.828
$H_2O_2 + 2H^+ + 2e^- \rightarrow 2H_2O$	1.77
$I_2 + 2e^- \rightarrow 2I^-$	0.536
$Li^+ + e^- \rightarrow Li$	−3.045
$MnO_2 + 4H^+ + 2e^- \rightarrow Mn^+ + 2H_2O$	1.23
$\gamma\text{-}MnO_2 + H_2O + e^- \rightarrow \alpha\text{-}MnOOH + OH^-$	0.30
$\gamma\text{-}MnO_2 + H_2O + e^- \rightarrow \gamma\text{-}MnOOH + OH^-$	0.36
$O_2 + 4H^+ + 4e^- \rightarrow 2H_2O$	1.229
$O_2 + H_2O + 2e^- \rightarrow HO_2^- + OH^-$	−0.076
$O_2 + 2H_2O + 4e^- \rightarrow 4OH^-$	0.401
$PbO_2 + SO_4^{2-} + 2e^- \rightarrow Pb^{2+} + 2H_2O$	1.455
$PbO_2 + SO_4^{2-} + 4H^+ + 2e^- \rightarrow PbSO_4 + 2H_2O$	1.685
$PbSO_4 + 2e^- \rightarrow Pb + SO_4^{2-}$	−0.356
$Zn^{2+} + 2e^- \rightarrow Zn$	−0.763
$Zn(OH)_2 + 2e^- \rightarrow Zn + 2OH^-$	−1.245

Source: Adapted from Reference 17·3.

actual system, the mass or volume of the reactive components is not all that contributes to the total mass of the system. Therefore, the actual performance can be measured by the following parameters:

$$\epsilon_m = \frac{\text{total discharge (Wh)}}{\text{total mass of battery}} \qquad \epsilon_v = \frac{\text{total discharge (Wh)}}{\text{total volume of battery}} \tag{17·9}$$

Actual battery performance parameter

Empirical data suggest that there is an approximate relationship between the actual and the ideal performance (Ref. 17·2):

$$\epsilon_m = \epsilon_{m,i}{}^{0.7} \tag{17·10}$$

Empirical relationship between ideal and actual battery performance parameter

Care must be taken to ensure that the units are the same when using this empirical relationship.

If we desire a battery that operates for a period of time t with a current I, we can calculate the mass of the anode required. This is accomplished using the following equation:

$$m = \frac{ItM}{j\mathfrak{F}} = \frac{\mathscr{C}M}{j\mathfrak{F}} \tag{17·11}$$

Faraday's law describing the relationship between the mass of the anode required for a given battery voltage, current, and time of operation

Table 17·4 Charge per Unit Mass or Volume for Various Battery Electrode Materials

Material	*Valence*	Ah/g	Ah/cm^3
Al	3	2.98	8.05
Cd	2	0.47	4.12
Ca	2	1.33	2.06
C (graphite)	4	8.92	20.08
Cu	2	0.84	7.49
	1	0.42	3.75
H	1	26.59	2.04(*l*)
Fe	2	0.96	7.54
Pb	2	0.26	2.93
PbO_2	2	0.22	2.10
Li	1	3.86	2.06
MnO_2	1	0.31	1.55
Ni	2	0.91	8.10
AgCl	1	0.19	1.04
Na	1	1.17	1.13
Zn	2	0.82	5.85

Source: Adapted from Reference 17·3.

where M is the molar mass of the anode material, j is the valence, and $\mathfrak{F}$ is Faraday's constant. This is a statement of one of Faraday's laws of electrolysis (see Ref. 17·2). The unit of current is typically an ampere and the unit of time is usually an hour, so that the charge passing, $\mathscr{C} = It$, would be measured in units of amp-hours (Ah). Thus, we can calculate the number of amp-hours per gram or per unit volume of electrode material as Ah/g or Ah/cm^3. Table 17·4 shows some typical values.

Finally, it is interesting to note that many of the same types of phenomena that limit the performance of fuel cells also limit the performance of batteries.

17·4 Other Direct Energy Conversion Systems

In this section, we discuss, qualitatively, other direct energy conversion systems: photovoltaic, thermoelectric, thermionic, and magnetohydrodynamic devices. Photovoltaic devices may operate isothermally, but the input radiant energy comes from a source at a different temperature. Thermoelectric, thermionic, and magnetohydrodynamic devices all involve the addition of heat from a higher-temperature reservoir and the rejection of heat to a lower-temperature reservoir. Therefore, the Carnot principle applies. Although these devices may not have high efficiencies, they may be attractive because of the simplicity of the system or for other reasons. For example, the lack of moving parts may make it possible for these devices to operate at a much higher temperature than conventional power plants. If these devices reject heat at very high temperatures, they can be combined with conventional power plants in "topping" arrangements, as mentioned in Section 15·4 for binary vapor power plants.

17·4·1 Photovoltaic Devices

Photovoltaic energy conversion devices convert the energy of light or other electromagnetic radiation directly into work or electrical energy. There is no intermediate conversion to other forms of energy such as heat. Therefore, thermal efficiency is not defined in connection with photovoltaic devices, and comparisons cannot be made with Carnot cycle performance.

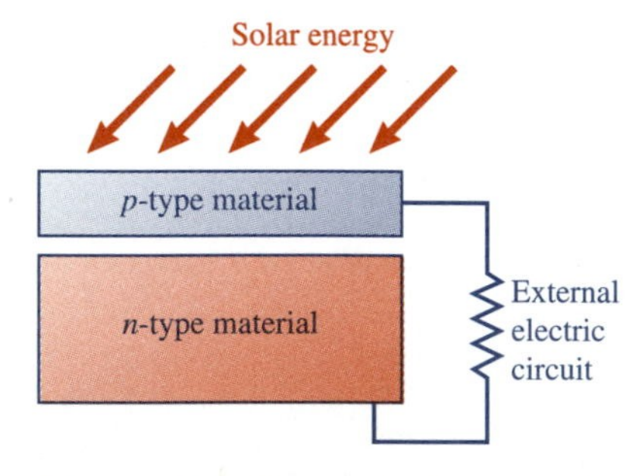

Figure 17·7 Schematic diagram of a photovoltaic cell.

Figure 17·7 is a schematic diagram of a photovoltaic cell. The *n*-type material is a semiconductor such as silicon that contains a trace of an element with one more valence electron than silicon. This addition makes the silicon conductive. The *p*-type material is formed by adding a trace of an element that contains one fewer valence electron than silicon. This also causes the silicon to become conductive. The *p*-type layer is usually very thin. When light strikes the upper surface of the *p*-type layer, some photons are absorbed near the junction of the two layers. This generates an emf, and if the two layers are connected through an external electric circuit, a current flows.

Losses are caused by the internal ohmic resistance of the cell. This applies both to the external load current and to the shunt current within the cell between the two layers. The ohmic losses tend to increase the cell temperature. Also, some of the incident radiation may be absorbed by the cell material and cause a temperature rise. Consequently, the photovoltaic cell may not operate isothermally. This is not the same as a device that exchanges heat with energy reservoirs at different temperatures.

Although the operation of photovoltaic cells involves many losses so that only a small fraction of the radiant energy incident upon the cell is converted into electrical work, these cells are simple and reliable. Also, their cost has been reduced considerably in recent years. Consequently, they are the best power source for some applications.

17·4·2 Thermoelectric Devices

Engineers and physical scientists are familiar with thermoelectric effects and the underlying Seebeck, Peltier, and Thomson effects. Figure 17·8 is a schematic diagram of a simple thermoelectric device that can be used in several ways. If the current is nulled, this device could be a thermocouple or temperature-measuring instrument when a suitable meter is used to measure the potential difference across the external circuit connections. By connecting an external electrical load and maintaining a temperature difference between the hot junctions and cold junctions, electrical work can be done on the external circuit.

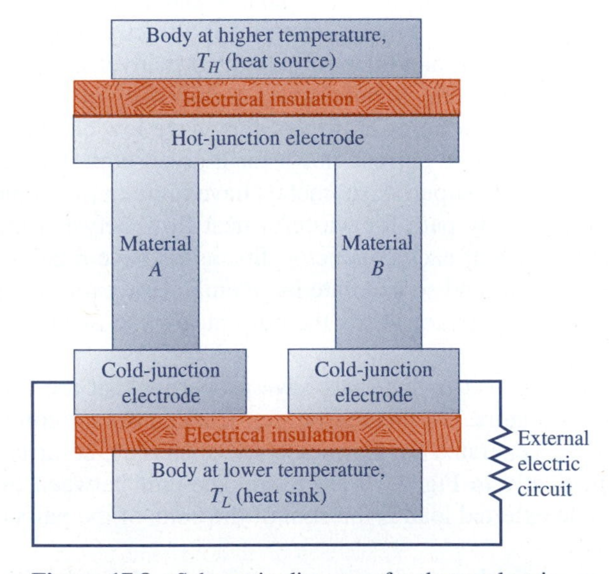

Figure 17·8 Schematic diagram of a thermoelectric generator.

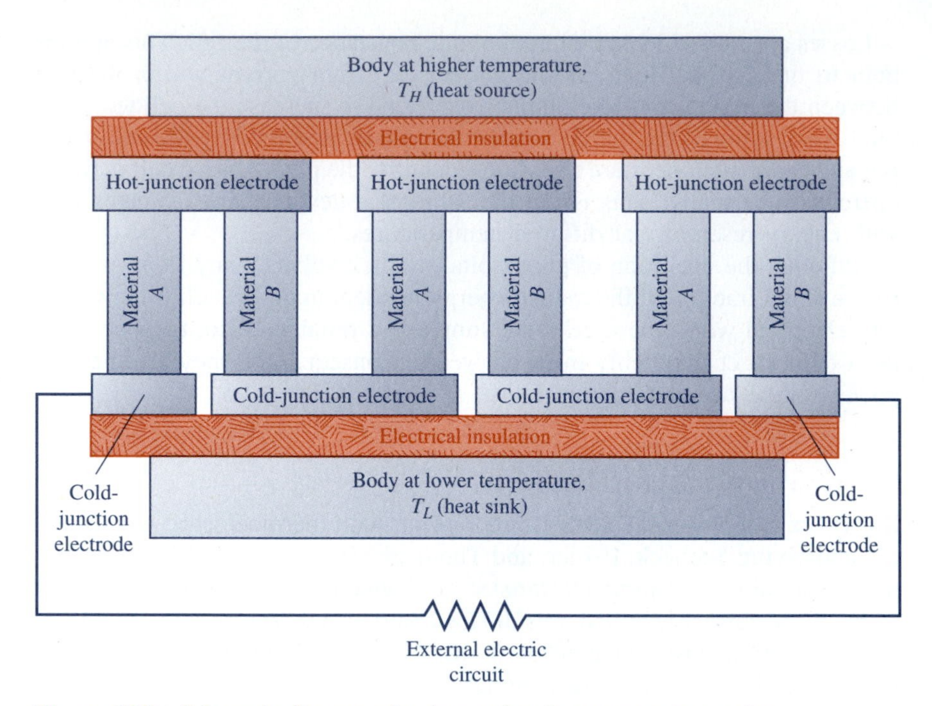

Figure 17·9 Schematic diagram of a thermoelectric generator in a series arrangement.

In selecting the thermoelectric materials, *A* and *B*, one is guided by two phenomena. The conductivity of a material is directly proportional to the number of free electrons, and the emf generated is inversely proportional to the number of free electrons. Metals have a high density of free electrons and so produce relatively low emfs. However, they have low electrical resistance and therefore permit large current flows for a given emf. This combination results in a low power output. Also, metals have high thermal conductivities, so they provide a ready path for wasteful heat flow between the high- and low-temperature ends of each conductor. Insulators have relatively low densities of free electrons and so generate large emfs. However, the high electrical resistance of an insulator keeps the current low, so again little power is generated.

Desirable thermoelectric materials have properties that are between those of metals and insulators. For this reason, semiconductors are commonly used.

Low emfs are generated by thermoelectric materials, so arrangements as shown schematically in Fig. 17·9 are used. The emf between the terminals connected to the external load is the sum of the emfs of the pairs of materials shown.

The efficiency of a thermoelectric generator depends on the operating temperatures, the material properties, and the current density. (If the device operated steadily and reversibly, the efficiency would depend only on the temperatures of the bodies the generator exchanges heat with, in accordance with the Carnot principle.) Irreversible effects include heat conduction through the thermoelectric materials from the hot ends to the cold ends, ohmic resistance throughout the device, contact resistance between materials, and current and heat losses. Efficiencies are low, but thermoelectric generators are simple and reliable and can use heat input from various sources: fuels, solar radiation, temperature differences occurring in nature, and nuclear heat sources.

Thermoelectric generators can also be operated in reverse as thermoelectric refrigerators. Coefficients of performance are low, but here again other advantages make thermoelectric refrigerators suitable for some applications.

17·4·3 Thermionic Devices

Figure 17·10 is a schematic diagram of a thermionic generator. The two electrodes, here shown as flat plates, are closely spaced and parallel to each other in a sealed enclosure that contains either a vacuum or a plasma. By means of an external heat source, the cathode is raised to such a high temperature that electrons are driven from it. The electrons flow to the anode, which is maintained at a lower temperature by heat transfer to some part of the surroundings. Electrons leave the anode to flow through an external electric circuit and back to the cathode. Cathode temperatures typically exceed 1200 K and may exceed 2000 K. Anode temperatures may be as high as 1000 K. Since heat is rejected at such a high temperature, thermionic generators may best be applied as topping units, with the heat rejected by the thermionic device being the heat input to another power system such as a gas turbine or steam power plant.

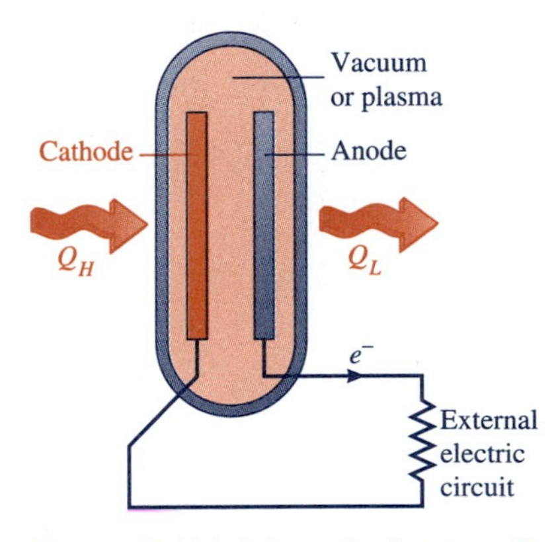

Figure 17·10 Schematic diagram of a thermionic generator.

Several factors impair the performance of thermionic generators. As electrons pass from the cathode to the anode within the thermionic generator, the mutual repulsion of the electrons in the interelectrode gap inhibits the emission of electrons from the cathode. This is called the space charge effect. One way of reducing this effect is to make the spacing between the electrodes very small. Another is to fill the device with an ionized gas or plasma to replace the vacuum. A plasma permits greater spacing between the electrodes because of the positive charges in the plasma. Other effects that impair thermionic generator performance are the radiation heat transfer between the electrodes and stray heat losses to the surroundings, including the heat transfer along lead wires. There are also ohmic losses in the electrodes and internal circuitry.

17·4·4 Magnetohydrodynamic Devices

In a conventional electric generator, electric conductors are moved through a magnetic field to induce electric currents in the conductors. The conductors must be moved against a retarding force, and this requires a work input. Magnetohydrodynamic (mhd) generators are analogous to conventional electric generators, but a conducting fluid, instead of a solid conductor, is moved through a magnetic field. A current is then induced in the conducting fluid. This current passes through electrodes in the channel walls and through an external electric circuit, as shown in Fig. 17·11. Other electromagnetic effects not mentioned here are important in the operation of an mhd generator.

Gases are made electrically conductive either by the addition of a small quantity of an easily ionized substance or by having the gas at a high temperature. The former procedure is called *seeding*, and the substance added is called *seed*.

Work must be done on the fluid in order to move it against the resistant force of the magnetic field. If the fluid is a gas, it must be compressed and heated. The temperature can be raised by burning fuel in the gas. Figure 17·12 is a schematic diagram of an mhd power plant. Clearly, the mhd generator is only a small part of the plant, and much auxiliary equipment is needed. Nevertheless, the mhd generator itself is simple and has no moving parts, so it can withstand higher gas temperatures than turbines. Because the temperature of the gas leaving is so high, mhd generators are considered for use only in topping arrangements, because the energy and availability loss involved in discarding this high-temperature gas would be unacceptable.

Factors that damage the performance of actual mhd generators include space charge effects (similar to those in thermionic converters), ohmic losses in the ionized gas, heat transfer from the gas, fluid friction losses, and some losses associated with the interactions of the electric and magnetic fields in the gas.

Remember that an mhd generator involves heat transfer with energy reservoirs at different temperatures, so the efficiency limitation of the Carnot principle applies.

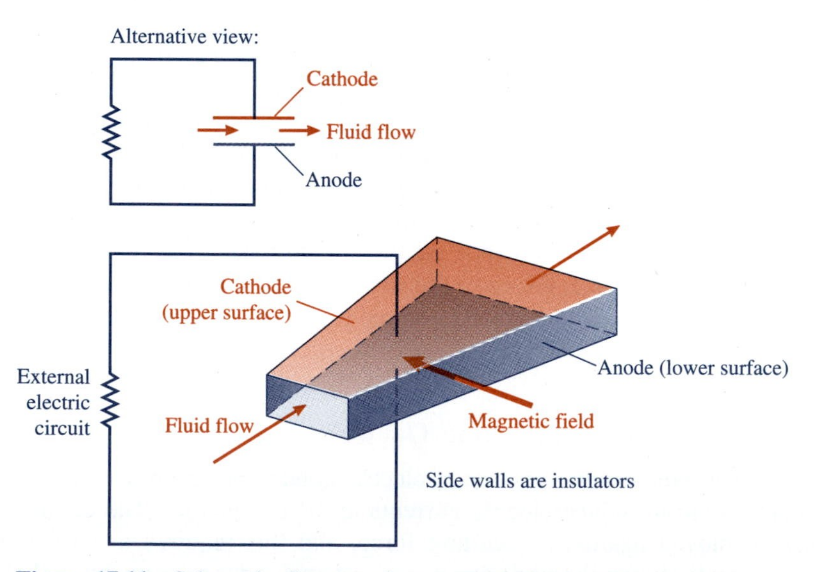

Figure 17·11 Schematic diagram of a magnetohydrodynamic generator.

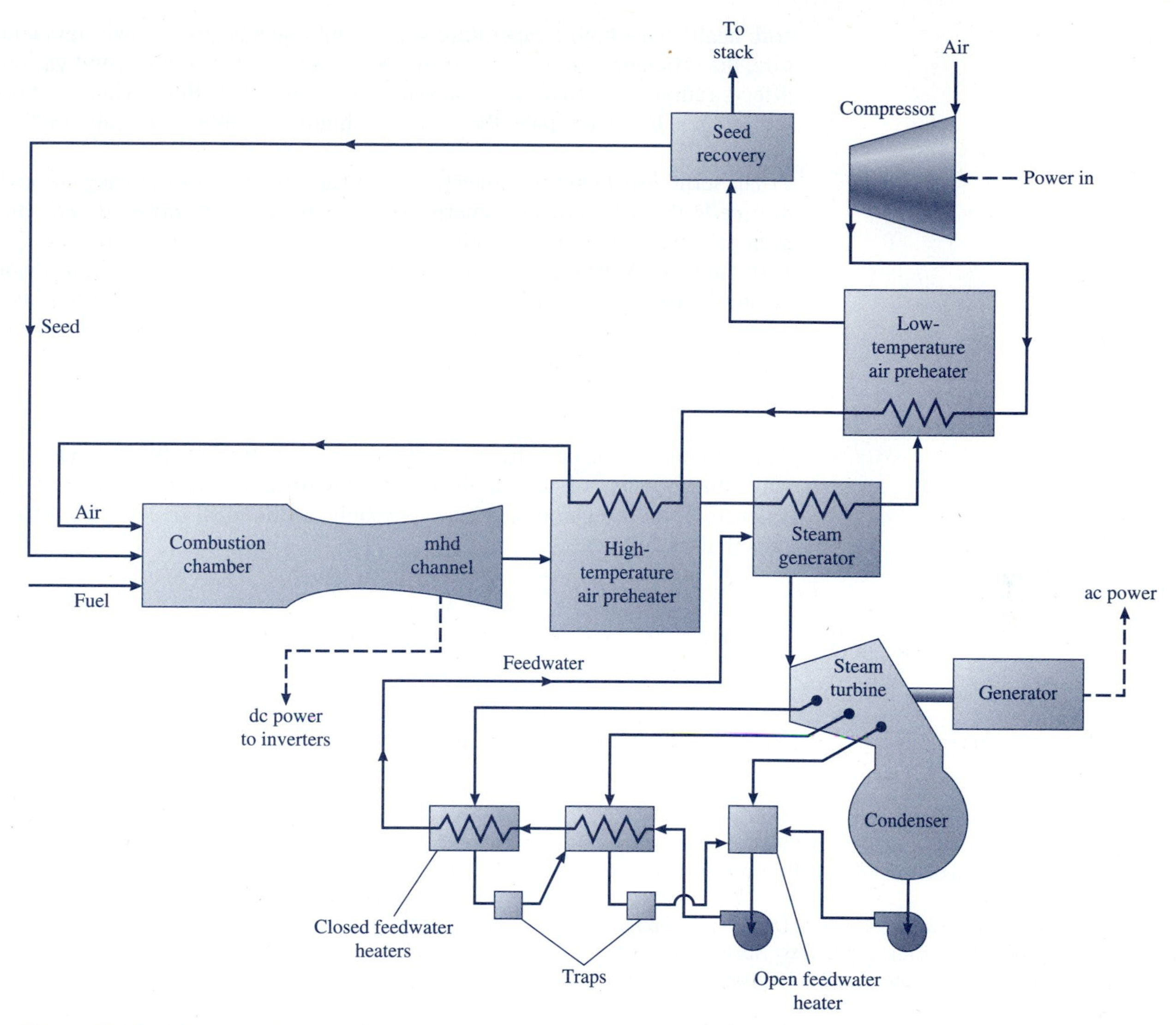

Figure 17·12 Schematic diagram of a combined mhd generator and steam cycle.

17·5 Summary

One use of the term *direct energy conversion* refers to the conversion of energy from heat to electric or mechanical work without the interposition of a working substance that undergoes a cycle. Examples are *thermoelectric*, *thermionic*, and *magnetohydrodynamic* generators. Since these devices absorb heat from a high-temperature source and reject heat at a lower temperature, the efficiency limitation imposed by the second law applies, and various effects reduce the efficiencies of actual devices to still lower values. Other advantages, however, make these devices highly suitable for certain applications.

The term *direct energy conversion* refers also to systems such as *photovoltaic cells* that convert light energy into electrical work and *fuel cells* and *batteries* that convert the chemical energy of a fuel into electrical work. The maximum work that can be obtained from either a battery or a fuel cell occurs in the limit of zero charge passing through the electric circuit. In this case the maximum work is equal to the change in the Gibbs function describing the chemical reaction:

$$W_{max} = -\Delta G$$

The maximum work can be related to the charge passing, the Faraday constant, the number of moles of reactant involved in the reaction, and the valence. Thus, we can derive an expression for the ideal fuel-cell or battery potential as follows:

$$\mathcal{V}_i = \frac{-\Delta G}{\mathfrak{F}Nj}$$

References and Suggested Readings

17·1 Appleby, A. J., and Foulkes, F. R., *Fuel Cell Handbook*, Krieger, Malabar, FL, 1993. (An interesting historical sketch of the development of fuel cells is provided in this book.)

17·2 *Modern Battery Technology*, Clive D. S. Tuck, editor, Ellis Horwood, Ltd., Chichester, West Sussex, England, 1991. (This book contains detailed explanations of the operation of many battery systems in use today and some systems sure to be in use in the near future.)

17·3 *Handbook of Chemistry and Physics*, CRC Press, Cleveland, annual editions. (This reference provides data for many battery electrode reactions.)

17·4 Daimler Benz, *High Tech Report*, March 1994.

Problems

17·1 An ideal fuel cell as shown in Fig. 17·2*c* operates at 25°C (77 F). Calculate (*a*) the ideal cell voltage, (*b*) the fuel-cell efficiency, and (*c*) the heat transfer per mole of hydrogen entering when the water leaves as a liquid. Describe how these calculations would differ if the electrolyte were acidic rather than alkaline.

17·2 Determine the ideal cell voltage at 100 kPa, 25°C (77 F) for the following reactions:

(*a*) $CO + \frac{1}{2}O_2 \rightarrow CO_2$

(*b*) $CH_4 + 2O_2 \rightarrow CO_2 + 2H_2O$

(*c*) $CH_3OH(l) + \frac{3}{2}O_2 \rightarrow CO_2 + 2H_2O$

(*d*) $2NH_3 + \frac{3}{2}O_2 \rightarrow N_2 + 3H_2O$

(*e*) $C_3H_8 + 5O_2 \rightarrow 3CO_2 + 4H_2O$

17·3 Figure 17·1 shows electric vehicles to be less desirable than fuel-cell vehicles for several reasons. Describe the potential advantages of an electric vehicle over a fuel-cell vehicle. What are some of the possible disadvantages of fuel-cell vehicles? Consider in particular the attributes you might consider when purchasing a vehicle.

17·4 An ideal fuel cell as shown in Fig. 17·2*c* operates at 700 K (1260 R). Calculate (*a*) the ideal cell voltage, (*b*) the fuel-cell efficiency, and (*c*) the heat transfer per mole of hydrogen entering.

17·5 An ideal fuel cell as shown in Fig. 17·2*d* operates at 25°C (77 F). Calculate (*a*) the ideal cell voltage, (*b*) the fuel-cell efficiency, and (*c*) the heat transfer per mole of methane entering.

17·6 An ideal fuel cell as shown in Fig. 17·2*d* operates at 700 K (1260 R). Calculate (*a*) the ideal cell voltage, (*b*) the fuel-cell efficiency, and (*c*) the heat transfer per mole of methane entering.

17·7 Determine the relationship between fuel-cell efficiency, heat transfer, and enthalpy change across the fuel cell.

17·8 An ideal fuel cell that uses carbon monoxide and oxygen is to be designed to operate somewhere between 400 K and 600 K (720 R and 1080 R). Plot as a function of operating temperature (*a*) the ideal cell voltage, (*b*) the fuel-cell efficiency, and (*c*) the heat transfer per mole of carbon monoxide entering.

17·9 For a carbon monoxide and oxygen fuel cell, determine the temperature range that would result in a fuel-cell efficiency greater than 0.75.

17·10 Calculate the ideal cell voltage of a carbon–oxygen fuel cell operating at 25°C (77 F).

17·11 Plot the ideal cell voltage and the fuel-cell efficiency against temperature for a hydrogen and oxygen cell in the temperature range of 298 to 1200 K.

17·12 Determine the temperature at which the fuel-cell efficiency of a hydrogen–oxygen fuel cell equals that of a carbon monoxide–oxygen fuel cell. What is the voltage of each cell at this temperature?

17·13 Write an equation that relates the equilibrium constant to the ideal cell voltage.

17·14 An ideal fuel cell that uses CO and O_2 is to be operated in a room where the temperature is maintained by a thermal control unit capable of removing 44,000 kJ of heat in an 8-hour day. For temperatures ranging from 300 K to 1000 K, plot the operating temperature of the fuel cell versus the number of moles of CO that can be used in the fuel cell during the 8-hour day without overheating the room.

17·15 In a bank of fuel cells, each cell has an actual voltage that is 60 percent of its ideal voltage when the current is 1.20 A. Twenty percent of the fuel passes through the cell without reacting and is wasted. Determine the fuel flow rate, the power output, and the heat transfer rate of each cell when they are operating at 25°C (77 F) and the reactants are gaseous hydrogen and oxygen.

17·16 For a specific fuel-cell design, the actual cell voltage is 80 percent of its ideal voltage when the current is 1.20 A. Twenty percent of the fuel passing through the cell does not react and is wasted. The fuel cell operates at 25°C (77 F) and the reactants are liquid methanol (CH_3OH) and gaseous oxygen. The fuel rate of flow is 2×10^{-5} mol/s. Determine the number of fuel cells required to power an ordinary household light bulb.

17·17 In a bank of fuel cells, each cell has an actual voltage of 70 percent of its ideal voltage when the current is 1.20 A. Twenty-five percent of the fuel passes through the cell without reacting and is wasted. Determine the fuel flow rate, the power output, and the heat transfer rate of each cell when they are operating at 600 K (1080 R) and the reactants are (*a*) gaseous methane and oxygen, and (*b*) gaseous propane (C_3H_8) and oxygen.

17·18 In a Daniell cell with 0.5 kg of Zn, determine the minimum amount of copper sulfate solution required if the Zn and the Cu are to be depleted simultaneously. The copper sulfate solution is 0.100 molar (moles of solute per liter of solution).

17·19 A Daniell cell has 1 kg of zinc and 0.5 L of 1.0 molar copper sulfate (molar = moles of solute per liter of solution). Determine (*a*) which will be depleted first, the zinc or the copper, (*b*) the number of moles of copper sulfate utilized per mole of electrons, and (*c*) the length of time the cell can deliver a current of 1 amp.

17·20 For the following reactions, determine whether it is possible to produce electrical energy:
(*a*) $Cl_2 + 2Br^- \rightarrow 2Cl^- + Br_2(l)$
(*b*) $Zn + Cu^{2+} \rightarrow Cu + Zn^{2+}$
(*c*) $Fe + Zn^{2+} \rightarrow Fe^{2+} + Zn$
(*d*) $I_2 + 2Br^- \rightarrow 2I^- + Br_2(l)$

17·21 Calculate ΔG, ΔH, and K_p at 25°C for the reactions of Problem 17·20.

17·22 For the following reaction at 25°C,

$$Ag^+ + \tfrac{1}{2}Cu \rightarrow Ag + \tfrac{1}{2}Cu^{2+}$$

calculate (*a*) the equilibrium constant ($\mathcal{V}_i = 0.799$ V for $Ag^+ + e^- \rightarrow Ag$), (*b*) the enthalpy of reaction, and (*c*) the change in the Gibbs energy.

17·23 For the following reaction at 25°C,

$$CH_4 + 2O_2 \rightarrow CO_2 + 2H_2O$$

calculate (*a*) the equilibrium constant, (*b*) the enthalpy of reaction, and (*c*) the change in the Gibbs function.

17·24 The overall reaction for an alkaline–manganese battery is

$$Zn + 2MnO_2 + H_2O \rightarrow ZnO + 2MnOOH$$

Determine (*a*) the amount of zinc consumed by 1 Ah of usage, and (*b*) the amount of MnO_2 consumed.

17·25 For Problem 17·24, determine the ideal and the actual battery performance as measured on a mass basis as $\varepsilon_{m,i}$.

17·26 In a zinc–silver battery (also called a voltaic pile), the silver is used only to effect the reduction of water, and the two half-cell reactions are
(*i*) Anode reaction: $Zn \rightarrow Zn^{2+} + 2e^-$
(*ii*) Cathode reaction: $2H_2O + 2e^- \rightarrow H_2 + 2OH^-$
Determine the ideal voltage of the battery.

17·27 For Problem 17·26 determine the total energy output possible if the battery contains 2 kg of zinc.

17·28 For the zinc–silver battery described in Problem 17·26, calculate the mass of zinc and of water required to operate the battery at 2 A for 2 days.

17·29 Determine the ideal battery performance for the battery described in Problem 17·26 as measured on a mass basis as $\varepsilon_{m,i}$ and on a volume basis as $\varepsilon_{v,i}$.

17·30 What mass of silver can be ''plated out'' in 1 hour at 0.5 A? (Use the following reaction to solve the problem: $AgCl + e^- \rightarrow Ag + Cl^-$.)

17·31 In designing a lead–acid battery for a typical automobile, a trade is to be made between the amount of time the radio in the car can be played when the engine is not running versus the weight of the battery. Assume that the depletion of lead limits the operating life of the battery and that a maximum of 70 percent of the total battery energy can be depleted without affecting the ability to recharge the battery. Plot a curve of the time the radio can be played versus the weight of the battery. (A number of assumptions and observations are required to solve this problem.)

Appendix I

Dimensions and Units

I·1 Dimensions, Units, and Conversion Factors

As we discussed in Section 1·4, we refer to physical quantities such as length, time, mass, and temperature as *dimensions.* We refer to a small group of dimensions from which all others can be formed as *primary dimensions.* These can be selected arbitrarily. If we denote some possible primary dimensions by the symbols L for length, M for mass, F for force, t for time, T for temperature, and Q for electric charge, we can write the dimensions of other physical quantities in terms of these:

Linear velocity	L/t	Density	M/L^3
Linear acceleration	L/t^2	Specific volume	L^3/M
Angular velocity*	1/t	Specific gravity*	—
Angular acceleration*	$1/t^2$	Energy	FL
Pressure	F/L^2	Specific heat	FL/MT
Work	FL	Electric potential	FL/Q
Power	FL/t	Electrical current	Q/t

* Angular measure is dimensionless and can be thought of as a ratio of two lengths: an arc length and a radius. Specific gravity is dimensionless because it is a ratio of two densities.

Any equation involving physical quantities, whether it is algebraic, differential, or integral, must be dimensionally homogeneous; that is, both sides of the equation must have the same dimensions and only terms with the same dimensions can be added to each other. Dimensions are not changed by the performance of mathematical operations such as differentiation or integration.

The dimensional homogeneity of an equation such as

$$pv = RT$$

is shown by substituting the dimensions of each quantity:

$$\left[\frac{F}{L^2}\right]\left[\frac{L^3}{M}\right] = \left[\frac{FL}{MT}\right]T$$

$$\left[\frac{FL}{M}\right] = \left[\frac{FL}{M}\right]$$

Also, the dimensions of

$$\left(\frac{\partial T}{\partial v}\right)_s = \left(\frac{\partial p}{\partial s}\right)_v$$

are readily checked for homogeneity by

$$\left[\frac{\mathrm{TM}}{\mathrm{L}^3}\right] = \left[\frac{\mathrm{F}}{\mathrm{L}^2}\right]\left[\frac{\mathrm{MT}}{\mathrm{FL}}\right]$$
$$\left[\frac{\mathrm{TM}}{\mathrm{L}^3}\right] = \left[\frac{\mathrm{TM}}{\mathrm{L}^3}\right]$$

Notice that dimensional homogeneity has nothing to do with numbers or numerical values. If an equation is valid, it must be dimensionally homogeneous. If it is not dimensionally homogeneous, it cannot be made valid simply by a judicious selection of numerical values or the use of conversion factors. The requirement of dimensional homogeneity thus provides a means of checking equations, and it is also useful in determining the general form of physical equations through a process known as dimensional analysis.

Units are arbitrary magnitudes of dimensions that are used for measurement. Some units of the primary dimensions are given in the following table:

Primary Dimension	*Sample Units*
Length	inch, centimeter, foot, mile, light-year
Time	hour, minute, second, day
Force	pound force, poundal, dyne, newton
Mass	pound mass, slug, gram, kilogram
Temperature	rankine, kelvin, degree Celsius, Fahrenheit
Electric current	ampere, microampere

As pointed out above, any equation involving physical quantities must be dimensionally homogeneous. For numerical computations, the equation must have consistency of units in addition to consistency of dimensions. For example, although the equation

$$v = \frac{RT}{p} \tag{a}$$

is dimensionally homogeneous, substitution of numerical values such as

$$v\,\frac{\mathrm{ft}^3}{\mathrm{lbm}} = \frac{0.0686\,\dfrac{\mathrm{B}}{\mathrm{lbm\cdot R}}\,520\ \mathrm{R}}{14.7\,\dfrac{\mathrm{lbf}}{\mathrm{in}^2}} \tag{b}$$

does not satisfy the requirement of consistent units and leads to confusion. Consistent units can be provided by means of *conversion factors* or *unitary constants* such as

$$\frac{778\ \mathrm{ft\cdot lbf}}{\mathrm{B}} = 1$$
$$\frac{144\ \mathrm{in}^2}{\mathrm{ft}^2} = 1$$

that follow from the relations

$$778\ \mathrm{ft\cdot lbf} = 1\ \mathrm{B}$$
$$144\ \mathrm{in}^2 = 1\ \mathrm{ft}^2$$

(Notice that a conversion factor is always dimensionless and has a dimensionless value of unity.) Thus instead of Eq. *b* we can write

$$v\,\frac{\mathrm{ft}^3}{\mathrm{lbm}} = \frac{0.0686\,\dfrac{\mathrm{B}}{\mathrm{lbm\cdot R}}\,778\,\dfrac{\mathrm{ft\cdot lbf}}{\mathrm{B}}\,520\ \mathrm{R}}{14.7\,\dfrac{\mathrm{lbf}}{\mathrm{in}^2}\,144\,\dfrac{\mathrm{in}^2}{\mathrm{ft}^2}} = 13.1\,\frac{\mathrm{ft}^3}{\mathrm{lbm}}$$

in which the units are consistent or homogeneous. Notice that in any equation that is dimensionally homogeneous, the units can be made consistent by judicious selection of units and the use of conversion factors. Conversion factors are necessary in numerical computations but are not written in equations, because equations should be valid independently of any set of units or numerical values.

I·2 Systems of Units; Number of Primary Dimensions

Several systems of units are in common use. An important difference among them is the number of primary dimensions on which they are based. All involve primary dimensions such as length, time, electric current, temperature, and luminous intensity. Some systems, however, use both force and mass as primary dimensions, while others take only one of these as primary and the other as a derived dimension. (Newton's second law of motion is usually used to determine the derived dimension.) If both mass and force are taken as primary, any corresponding system of units is called an FMLt system. If only one of these two is taken as primary, the system of units is called an FLt or MLt system.

This book uses two systems of units: SI (for le Système International d'Unités) and a common variant of the English system. SI is an MLt system, and it is clearly so defined. The English system is an FMLt system, but several variants are in use, and some of these are essentially MLt systems.

I·3 International System (SI)

The primary dimensions (and their base units) in SI are mass (kilogram), length (meter), time (second), electric current (ampere), temperature (kelvin), luminous intensity (candela), and quantity of matter (mole). The basic equation for defining force and its unit is $F = ma$, and the dimensions of the equation are

$$[F] = \left[\frac{ML}{t^2}\right]$$

The unit of force defined as 1 kg·m/s^2 is called a newton, so the defining equation is

$$1 \text{ newton} = 1 \text{ kilogram} \times 1 \text{ m/s}^2$$

Consequently, *conversion factors* are

$$1\frac{N\cdot s^2}{kg\cdot m} = 1 \quad \text{and} \quad 1\frac{kg\cdot m}{N\cdot s^2} = 1$$

As is the case with any conversion factor, these must be dimensionless and have a dimensionless value of unity.

The kinetic energy of an object with a mass of 10 kilograms moving in translation at a velocity of 100 meters per second is given by

$$KE = \frac{mV^2}{2} = \frac{10\text{ kg}(100)^2\text{m}^2/\text{s}^2}{2} = 50{,}000\frac{kg\cdot m^2}{s^2}$$

Note that in a system that takes as primary dimensions only three of the four dimensions of force, mass, length, and time, the basic form of Newton's equation is $F = ma$ and this equation is valid in any consistent set of units. In numerical calculations, conversion factors often must be used to obtain consistency of units, but we do not include conversion factors in symbolic equations.

Some other sets of consistent units for use with the primary dimensions of an FLt or MLt system are shown in the following table:

F	M	L	t	*Conversion Factor*
lbf	slug	foot	second	$1\frac{slug\cdot ft}{lbf\cdot s^2} = 1$
poundal	lbm	foot	second	$1\frac{lbm\cdot ft}{pdl\cdot s^2} = 1$
dyne	gram	centimeter	second	$1\frac{g\cdot cm}{dyne\cdot s^2} = 1$

I·4 English System

Force, mass, length, and time are all taken as independent primary dimensions in the English system. It is an FMLt system, which, as we discussed in Section 1·4·2, is overdetermined. In an FMLt system,

$$[\mathrm{F}] \neq \left[\frac{\mathrm{ML}}{\mathrm{t}^2}\right]$$

Therefore, Newton's second law cannot be written as $F = ma$ because this equation is not dimensionally homogeneous. We must write instead

$$F = \frac{ma}{g_c}$$

where g_c is a *dimensional constant.* To make the equation above dimensionally homogeneous, the dimensions of g_c must be ML/Ft^2.

The numerical value of g_c depends only on the units selected. The relationship among the traditional units of the English system—pound force (lbf), pound mass (lbm), foot, and second—is that a force of 1 lbf accelerates a mass of 1 lbm at a rate of 32.174 ft/s^2. Substituting values into

$$F = \frac{ma}{g_c} \qquad (c)$$

gives

$$1\ \mathrm{lbf} = \frac{1}{g_c} \times 1\ \mathrm{lbm} \times 32.174\ \mathrm{ft/s^2}$$

In this set of units, therefore, g_c has the value

$$g_c = 32.174 \frac{\mathrm{lbm{\cdot}ft}}{\mathrm{lbf{\cdot}s^2}}$$

g_c is dimensional; it is not dimensionless. Therefore, we say that g_c is a dimensional constant and *not* a conversion factor. Since g_c is a dimensional constant, it is usually written in equations to provide dimensional homogeneity *when using an FMLt system.* It is unlike conversion factors, such as 100 cm/m or 144 sq in/sq ft, that are necessary in numerical computations but are never written in equations.

All equations in this book are written for FLt or MLt systems and are dimensionally homogeneous. Each equation is valid for any consistent set of units. Therefore, *g_c does not appear in the equations in this book.* If you wish to use strictly an FMLt system, you must remember to include dimensional constants such as g_c as needed for *dimensional* homogeneity in equations.

Difficulty arises using the English system when an MLt or FLt system is used with inconsistent units. Often these units are those common in the FMLt system and this is the source of the confusion. For example, suppose an engineer wishes to use the units of lbf, lbm, foot, and second in an MLt system. These units are not a consistent set, so they cannot be used in the equation

$$F = ma$$

unless one or more conversion factors are used to convert some units. One possible conversion factor is 32.174 lbm/slug = 1. This conversion factor allows the use of pound mass as a mass unit along with pound force as the force unit:

$$\begin{aligned} F(\mathrm{lbf}) &= m(\mathrm{slug})a(\mathrm{ft/s^2}) \\ &= m(\mathrm{lbm})\left(\frac{\mathrm{slug}}{32.174\ \mathrm{lbm}}\right)a(\mathrm{ft/s^2}) \end{aligned}$$

In an MLt system, $F = ma$, and using a consistent set of units (lbf, slug, foot, second) shows a conversion factor to be

$$1\ \mathrm{slug{\cdot}ft/lbf{\cdot}s^2} = 1$$

Combining this conversion factor with that shown above yields another conversion factor,

$$32.174\ \mathrm{lbm{\cdot}ft/lbf{\cdot}s^2} = 1$$

that is useful in an MLt or FLt system of dimensions. This is *not* g_c; g_c is a dimensional constant that has meaning only in an FMLt system of dimensions where it is used to relate four independently defined dimensions. The following table reiterates this point:

For an MLt *System of Dimensions*	*For an* FMLt *System of Dimensions*
$32.174 \frac{\text{ft·lbm}}{\text{lbf·s}^2} = 1$	$32.174 \frac{\text{lbm·ft}}{\text{lbf·s}^2} = g_c$
Conversion factor	Dimensional constant

Consider calculating the kinetic energy of an object with a mass of 20 lbm translating at a velocity of 100 ft/s. For an FMLt system of dimensions, we have

$$KE = \frac{mV^2}{2g_c}$$
$$= \frac{20 \text{ lbm}(100)^2(\text{ft/s})^2}{2}\left(\frac{1}{32.174}\ \frac{\text{lbf·s}^2}{\text{lbm·ft}}\right)$$
$$= 3108 \text{ ft·lbf}$$

Alternatively, for an MLt system of dimensions, we have

$$KE = \frac{mV^2}{2}$$
$$= \frac{20 \text{ lbm}(100)^2(\text{ft/s})^2}{2}\left(\frac{\text{slug}}{32.174 \text{ lbm}}\right)$$
$$= 3108 \frac{\text{slug·ft}^2}{\text{s}^2}$$

Recall that a pound force is defined as the force required to accelerate 1 slug at a rate of 1 ft/s^2 to obtain the same final result

$$KE = 3108 \frac{\text{slug·ft}^2}{\text{s}^2} = 3108 \text{ ft·lbf}$$

Clearly, various procedures can be successful. It is essential, however, that you understand fully the procedure you are using.

Difficulty with units and dimensions in the English system also stems from the use of the term *pound* as a unit of force and as a unit of mass. (The unfortunate use of the term *kilogram* as a unit of force promotes the same difficulty in a variant of SI.) The difficulty is compounded by the fact that in some cases, and for some purposes, mass and a particular force—weight, a gravitational force—can apparently be used interchangeably. Let us look into the relationship that involves weight and mass by applying the basic equation for a body of fixed mass

$$F = \frac{ma}{g_c} \tag{d}$$

to a body that is acted upon by only a gravitational force. This force we call weight. The acceleration of a body caused by this force alone is called the acceleration of gravity g. Substituting this particular force w and the corresponding particular acceleration g into Eq. d gives

$$w = \frac{mg}{g_c}$$

Obviously, if g is constant, then since g_c is also constant, weight and mass are in a fixed proportion to each other. For accounting purposes, such as mass balances, mass and weight can actually be used interchangeably. This cannot be done if the acceleration of gravity g varies. A serious danger in this practice is that weight, which is a force, comes to be thought of as mass, and confusion then results between force and mass, which are as different from each other as length and time are.

Another factor in the confusion between weight and mass is that the operation called *weighing* is usually a determination of mass. This is certainly the case when a balance-type scale is used. It is also the case with a spring scale if the scale has been calibrated by means of standard masses (which are commonly and unfortunately called standard weights).

Remember that g_c has nothing to do with gravity or the acceleration of a freely falling body, even though historically some systems of units were founded on the basis of gravitational force measurements on bodies in certain locations.

The exercises below provide a means of checking your understanding of units and dimensions.

Exercises

1. A body weighs 360 N in a location where $g = 8.80\ m/s^2$. Determine the force required to accelerate this body at a rate of $10.0\ m/s^2$.

2. Determine the force required to accelerate a body with a mass of 85 kg at a rate of $4.2\ m/s^2$ in a location where $g = 9.10\ m/s^2$ if the acceleration is (*a*) horizontal, (*b*) vertically upward.

3. Determine the force in newtons required to accelerate a body with a mass of 68 lbm at a rate of $14\ m/s^2$ in a location where $g = 930\ cm/s^2$.

4. A body weighs 100 N in a location where $g = 890\ cm/s^2$. (*a*) What is its mass in kg? In lbm? In slugs? (*b*) What is its weight in newtons and its mass in kilograms in a location where $g = 500\ m/s^2$?

5. A liquid has a density of $880\ kg/m^3$. Determine its specific volume and its specific weight in a location where (*a*) $g = 9.80\ m/s^2$, (*b*) $g = 920\ m/s^2$.

6. A body weighs 20 lbf in a location where $g = 31.6\ ft/s^2$. Determine the force required to accelerate this body at a rate of $30.0\ ft/s^2$.

7. Determine the force required to accelerate a body with a mass of 500 lbm at a rate of $5.0\ ft/s^2$ in a location where $g = 31.6\ ft/s^2$ if the acceleration is (*a*) horizontal, (*b*) vertically upward.

8. Determine the force in pounds required to accelerate a body with a mass of 15 kg at a rate of $10\ ft/s^2$ in a location where $g = 30.0\ ft/s^2$.

9. A body weighs 30 lbf in a location where $g = 32.0\ ft/s^2$. (*a*) What is its mass in lbm? In slugs? (*b*) What are its weight in pounds, mass in pounds, and mass in slugs in a location where $g = 16.0\ ft/s^2$?

10. A liquid has a density of $55\ lbm/ft^3$. Determine its specific volume, its specific weight, and its density in $slug/ft^3$ in a location where (*a*) $g = 32.2\ ft/s^2$, (*b*) $g = 31.6\ ft/s^2$.

11. What is the value of g_c in a location where a body with a mass of 270 lbm weighs 195 lbf?

12. A certain liquid has a dynamic viscosity of 6.20×10^{-5} slug/ft·s at room temperature. Determine the value of dynamic viscosity in each of the following sets of units: $lbf{\cdot}s/ft^2$, $N{\cdot}s/m^2$, kg/m·s, and lbm/ft·s.

13. A useful dimensionless ratio in fluid mechanics, Reynolds number, is defined as $Re = \rho DV/\mu$, where ρ is the density, D is the diameter of a pipe in which the fluid flows, V is the mean velocity of the fluid, and μ is the dynamic viscosity. Determine the value of the Reynolds number for a fluid with a density of $52.0\ lbm/ft^3$ and a dynamic viscosity of $4.00 \times 10^{-2}\ lbf{\cdot}s/ft^2$ flowing at a mean velocity of 8.00 ft/s through a pipe of 6.25 in. inside diameter.

Appendix II

Note on Partial Derivatives

Partial derivatives are important in thermodynamics because several properties are defined in terms of partial derivatives (e.g., specific heats and the Joule–Thomson coefficient). Furthermore, an understanding of relationships among partial derivatives is helpful in correlating properties.

II·1 A Graphical Interpretation of Partial Derivatives

If three quantities x, y, and z are functionally related, this fact can be stated in any of the following ways:

$$f(x, y, z) = 0 \qquad x = f(y, z) \qquad y = f(x, z) \qquad z = f(x, y)$$

The differential dz is given by

$$dz = \left(\frac{\partial z}{\partial x}\right)_y dx + \left(\frac{\partial z}{\partial y}\right)_z dy$$

and similar expressions can be written for dx and dy.

Figure II·1 gives a physical picture of this differential dz where a–b–c–d is any *very small* element of a surface that satisfies the equation $z = f(x, y)$. The change in z between points a and c can be evaluated as

$$\Delta z_{a-c} = \Delta z_{a-b} + \Delta z_{b-c}$$

Since b and c both lie in a plane of y = constant,

$$\Delta z_{b-c} = \Delta x_{b-c} \times \begin{pmatrix}\text{slope of } zx \text{ curve in} \\ \text{plane of } y = \text{constant}\end{pmatrix} = \Delta x_{b-c}\left(\frac{\Delta z}{\Delta x}\right)_{y=\text{constant}}$$

and similarly, since a and b lie in a plane of x = constant,

$$\Delta z_{a-b} = \Delta y_{a-b}\left(\frac{\Delta z}{\Delta y}\right)_{x=\text{constant}}$$

We can substitute these into the first expression for Δz_{a-c} to obtain

$$\Delta z_{a-c} = \left(\frac{\Delta z}{\Delta y}\right)_x \Delta y_{a-b} + \left(\frac{\Delta z}{\Delta x}\right)_y \Delta x_{b-c}$$

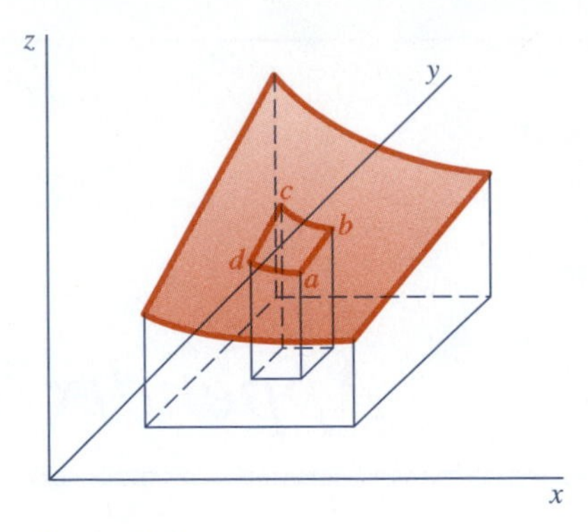

Figure II·1

If the surface element a–b–c–d is allowed to become smaller and smaller, Δz approaches dz as a limit, and

$$dz = \left(\frac{\partial z}{\partial x}\right)_y dx + \left(\frac{\partial z}{\partial y}\right)_x dy$$

Notice that, at each point on the surface, $(\partial z/\partial x)_y$ has a single fixed value, because it is the slope of a zx curve along a particular path. In this case the path is described by the intersection of the surface and a $y =$ constant plane). dz/dx, however, has many values at each point on the surface, since it can be evaluated along many different paths that pass through any point. In general,

$$\frac{dz}{dx} = \left(\frac{\partial z}{\partial x}\right)_y + \left(\frac{\partial z}{\partial y}\right)_x \frac{dy}{dx}$$

where $(\partial z/\partial x)_y$ and $(\partial z/\partial y)_x$ are point functions that have fixed values at each point. dz/dx and dy/dx, on the other hand, have many possible values at each point depending on the path along which they are evaluated.

Exercises

Figure II·2 is a contour map with lines of constant elevation H. x is distance measured east of the origin, and y is distance measured north of the origin. At each xy location there is a fixed value of H; so $H = f(x, y)$ or $f(H, z, y) = 0$. The map shows a hill with its summit at 3, 5 (i.e., $x = 3$, $y = 5$). A stream flows southwest along a path given approximately by $y = x - 11.5$.

1. Give the approximate coordinates of a point at which $(\partial H/\partial y)_x$ is (*a*) a maximum, (*b*) a minimum, and (*c*) zero.

2. Repeat Exercise 1 for $(\partial H/\partial x)_y$.

3. Give the coordinates of a point at which $(\partial y/\partial x)_H$ equals approximately (*a*) zero, (*b*) one, and (*c*) minus one.

4. A path on the ground is given by the relation $y = 1.2x$. Give the coordinates of points (if any) along this path where dH/dx is (*a*) a maximum, (*b*) a minimum, and (*c*) zero.

5. Repeat Exercise 4 for $(\partial H/\partial x)_y$.

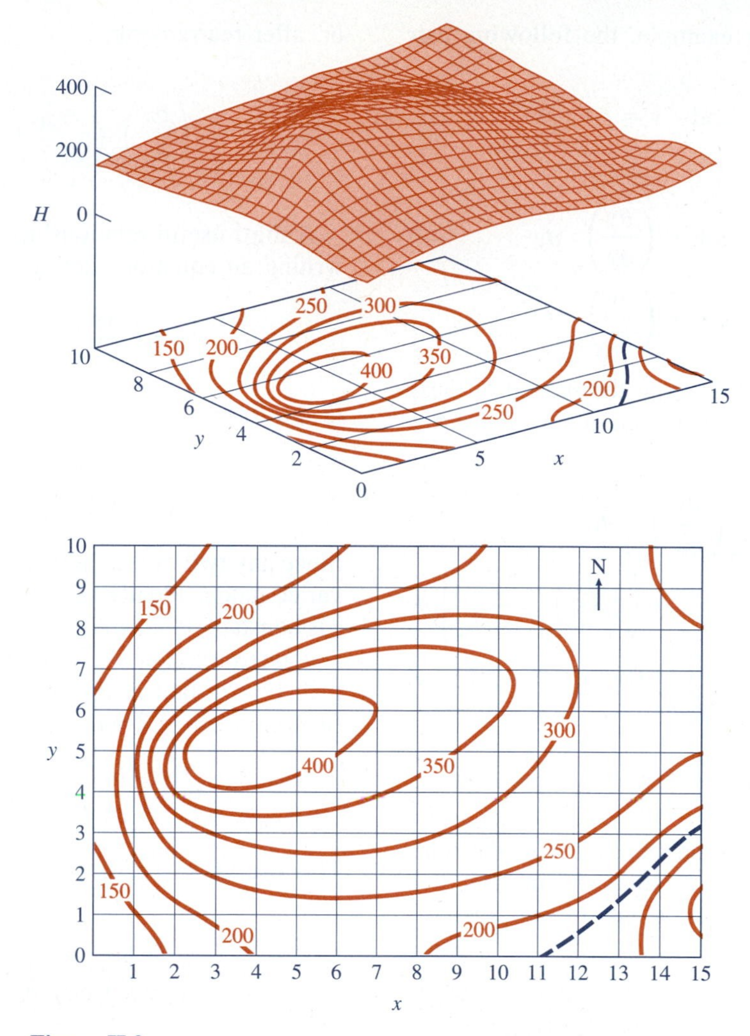

Figure II·2

II·2 Three Relations Among Partial Derivatives

If four quantities w, x, y, and z are related in such a manner that any two, but not more than two, are independent (and this is a common case in thermodynamics), then any one can be considered as a function of any other two:

$$\begin{array}{ll} w = f(x, y) & x = f(y, z) \\ w = f(x, z) & x = f(w, z) \\ w = f(y, z) & x = f(w, y) \end{array}$$

$$\begin{array}{ll} y = f(x, z) & z = f(x, y) \\ y = f(w, x) & z = f(w, x) \\ y = f(w, z) & z = f(w, y) \end{array}$$

Of course, $f(x, y)$ signifies only some function of x and y and does not represent any particular function.

Writing the differentials of two of the quantities as expressed above is the first step toward establishing three relationships that are quite useful in thermodynamics. If we take, for example, the following pair from the list above,

$$x = f(w, y) \quad \text{and} \quad y = f(w, z)$$

the differentials are

$$dx = \left(\frac{\partial x}{\partial w}\right)_y dw + \left(\frac{\partial x}{\partial y}\right)_w dy$$

$$dy = \left(\frac{\partial y}{\partial w}\right)_z dw + \left(\frac{\partial y}{\partial z}\right)_w dz$$

Substituting the value of dy from the second of these two equations into the first one and collecting terms gives

$$dx = \left[\left(\frac{\partial x}{\partial w}\right)_y + \left(\frac{\partial x}{\partial y}\right)_w \left(\frac{\partial y}{\partial w}\right)_z\right] dw + \left[\left(\frac{\partial x}{\partial y}\right)_w \left(\frac{\partial y}{\partial z}\right)_w\right] dz$$

Alternatively, we can write $x = f(w, z)$. Thus, we can write another expression for dx:

$$dx = \left(\frac{\partial x}{\partial w}\right)_z dw + \left(\frac{\partial x}{\partial z}\right)_w dz$$

But w and z can be varied independently, so the coefficients of dw and dz must be the same in these two equations. Equating the coefficients of dw gives the first useful relationship:

$$\left(\frac{\partial x}{\partial w}\right)_z = \left(\frac{\partial x}{\partial w}\right)_y + \left(\frac{\partial x}{\partial y}\right)_w \left(\frac{\partial y}{\partial w}\right)_z \qquad (a)$$

Equating the coefficients of dz gives

$$\left(\frac{\partial x}{\partial z}\right)_w = \left(\frac{\partial x}{\partial y}\right)_w \left(\frac{\partial y}{\partial z}\right)_w$$

or, after rearranging,

$$\left(\frac{\partial x}{\partial y}\right)_w \left(\frac{\partial y}{\partial z}\right)_w \left(\frac{\partial z}{\partial x}\right)_w = 1 \qquad (b)$$

A third useful relationship can be obtained by first writing an equation such as

$$dz = \left(\frac{\partial z}{\partial y}\right)_x dy + \left(\frac{\partial z}{\partial x}\right)_y dx$$

in the form

$$\frac{dz}{dy} = \left(\frac{\partial z}{\partial y}\right)_x + \left(\frac{\partial z}{\partial x}\right)_y \frac{dx}{dy}$$

Since any two of the three quantities x, y, and z can be varied independently, we can assign the condition that z is constant. [This means, referring back to the graphical interpretation in Section II·1, that we follow a path of constant z. It does not mean that $(\partial z/\partial y)$ or $(\partial z/\partial x)$ is zero.] Then

$$0 = \left(\frac{\partial z}{\partial y}\right)_x + \left(\frac{\partial z}{\partial x}\right)_y \left(\frac{\partial x}{\partial y}\right)_z$$

This useful relationship is often written as

$$\left(\frac{\partial x}{\partial y}\right)_z \left(\frac{\partial y}{\partial z}\right)_x \left(\frac{\partial z}{\partial x}\right)_y = -1 \qquad (c)$$

and is sometimes called the "rotating partials." Equation c shows the absurdity of treating ∂x, ∂y, and ∂x as quantities that can be handled alone and canceled as parts of fractions.

Equations a, b, and c are frequently used in the correlation of thermodynamic property data.

Tables and Charts

SI Units

English Units

Dimensionless

Table A·1 Properties of Gases

Gas	Molar Mass M kg/kmol	Gas Constant R kJ/kg·K	Specific Heats at 25°C c_p kJ/kg·K	c_v kJ/kg·K	k	Temperature Range	Equation Coefficients for $c_p/R = a + bT + cT^2 + dT^3 + eT^4$ a	$b \times 10^3$ K^{-1}	$c \times 10^6$ K^{-2}	$d \times 10^{10}$ K^{-3}	$e \times 10^{13}$ K^{-4}
Acetylene, C_2H_2	26.04	0.319	1.69	1.37	1.232	300–1000K	0.8021	23.51	−35.95	286.1	−87.64
						1000–3000K	3.825	6.767	−3.014	6.931	−0.6469
Air	28.97	0.287	1.01	0.718	1.400	300–1000K	3.721	−1.874	4.719	−34.45	8.531
						1000–3000K	2.786	1.925	−0.9465	2.321	−0.2229
Argon, Ar	39.95	0.208	0.520	0.312	1.667		2.50	0	0	0	0
Butane, C_4H_{10}	58.12	0.143	1.67	1.53	1.094	300–1500K	0.4756	44.65	−22.04	42.07	0
Carbon Dioxide CO_2	44.01	0.189	0.844	0.655	1.289	300–1000K	2.227	9.992	−9.802	53.97	−12.81
						1000–3000K	3.247	5.847	−3.412	9.469	−1.009
Carbon Monoxide, CO	28.01	0.297	1.04	0.744	1.399	300–1000K	3.776	−2.093	4.880	−32.71	6.984
						1000–3000K	2.654	2.226	−1.146	2.851	−0.2762
Ethane, C_2H_6	30.07	0.276	1.75	1.48	1.187	300–1500K	0.8293	20.75	-7.704	8.756	0
Ethylene, C_2H_4	28.05	0.296	1.53	1.23	1.240	300–1000K	1.575	10.19	11.25	−199.1	81.98
						1000–3000K	0.2530	18.67	−9.978	26.03	−2.668
Helium, He	4.003	2.08	5.19	3.12	1.667		2.50	0	0	0	0
Hydrogen, H_2	2.016	4.12	14.3	10.2	1.405	300–1000K	2.892	3.884	−8.850	86.94	−29.88
						1000–3000K	3.717	−0.9220	1.221	−4.328	0.5202
Hydrogen, H	1.008	8.25	20.6	12.4	1.667	300–1000K	2.496	0.02977	−0.07655	0.8238	−0.3158
						1000–3000K	2.567	−0.1509	0.1219	−0.4184	0.05182
Hydroxyl, OH	17.01	0.489	1.76	1.27	1.384	300–1000K	3.874	−1.349	1.670	−5.670	0.6189
						1000–3000K	3.229	0.2014	0.4357	−2.043	0.2696
Methane, CH_4	16.04	0.518	2.22	1.70	1.304	300–1000K	4.503	−8.965	37.38	−364.9	122.2
						1000–3000K	−0.6992	15.31	−7.695	18.96	−1.849
Neon, Ne	20.18	0.412	1.03	0.618	1.667		2.50	0	0	0	0
Nitric Oxide, NO	30.01	0.277	0.995	0.718	1.386	300–1000K	4.120	−4.225	10.77	−97.64	31.85
						1000–3000K	2.730	2.372	−1.338	3.604	−0.3743
Nitrogen, N_2	28.01	0.297	1.04	0.743	1.400	300–1000K	3.725	−1.562	3.208	−15.54	1.154
						1000–3000K	2.469	2.467	−1.312	3.401	−0.3454
Nitrogen, N	14.01	0.594	1.48	0.890	1.667	300–1000K	2.496	0.02977	−0.07655	0.8238	−0.3158
						1000–3000K	2.483	0.03033	−0.01517	0.001879	0.009657
Oxygen, O_2	32.00	0.260	0.919	0.659	1.395	300–1000K	3.837	−3.420	10.99	−109.6	37.47
						1000–3000K	3.156	1.809	−1.052	3.190	−0.3629
Oxygen, O	16.00	0.520	1.37	0.850	1.612	300–1000K	3.020	−2.176	3.793	−30.62	9.402
						1000–3000K	2.662	−0.3051	0.2250	−0.7447	0.09383
Propane, C_3H_8	44.10	0.189	1.67	1.48	1.127	300-1500K	-0.4861	36.63	-18.91	38.14	0
Water, H_2O	18.02	0.462	1.86	1.40	1.329	300–1000K	4.132	−1.559	5.315	−42.09	12.84
						1000–3000K	2.798	2.693	−0.5392	−0.01783	0.09027

Table A·1 Properties of Gases (*Continued*)

Gas	Critical State Properties p_c MPa	Critical State Properties T_c K	Triple State Properties p_t kPa	Triple State Properties T_t K	van der Waals Constants a $\frac{kPa \cdot m^6}{kmol^2}$	van der Waals Constants b $m^3/kmol$	Redlich–Kwong Constants a $\frac{kPa \cdot m^6 \cdot K^{0.5}}{kmol^2}$	Redlich–Kwong Constants b $m^3/kmol$	Acentric Factor ω	Gas
C_2H_2	6.14	308	126	192	452	0.0522	8030	0.0362	0.1873	Acetylene, C_2H_2
Air	3.77	132			136	0.0365	1580	0.0253	0.007400	Air
Ar	4.90	151	68.8	83.8	135	0.0320	1680	0.0222	0.0000	Argon, Ar
C_4H_{10}	3.80	425	0.000568	135	1390	0.116	29000	0.0806	0.1931	Butane, C_4H_{10}
CO_2	7.38	304	518	217	366	0.0428	6450	0.0297	0.2276	Carbon Dioxide CO_2
CO	3.50	133	15.4	68.1	147	0.0395	1720	0.0274	0.06630	Carbon Monoxide, CO
C_2H_6	4.88	306			557	0.0650	9860	0.0450	0.099	Ethane, C_2H_6
C_2H_4	5.03	282	0.140	104	462	0.0583	7860	0.0404	0.08520	Ethylene, C_2H_4
He	0.228	5.20			3.47	0.0238	8.00	0.0165	−0.3900	Helium, He
H_2	1.31	33.2	7.04	13.8	24.5	0.0263	143	0.0182	−0.2150	Hydrogen, H_2
H										Hydrogen, H
OH										Hydroxyl, OH
CH_4	4.60	191	11.7	90.7	894	0.0430	3210	0.0298	0.01080	Methane, CH_4
Ne	2.65	44.4	50.0	24.6	21.7	0.0174	146	0.0120	−0.04140	Neon, Ne
NO	6.48	180	21.9	110	146	0.0289	1980	0.0200	9.5846	Nitric Oxide, NO
N_2	3.39	126	12.5	63.2	137	0.0386	1550	0.0267	0.04030	Nitrogen, N_2
N										Nitrogen, N
O_2	5.04	155	0.146	54.4	138	0.0319	1740	0.0221	0.02180	Oxygen, O_2
O										Oxygen, O
C_3H_8	4.26	370			938	0.0904	18300	0.0626	0.153	Propane, C_3H_8
H_2O	22.1	647	0.612	273	554	0.0305	14300	0.0211	0.3449	Water, H_2O

Source: Chiefly *JANAF Thermochemical Tables,* 3rd ed., published by the American Chemical Society and the American Institute of Physics for the National Bureau of Standards, 1986. Data for butane, ethane, and propane from Kobe, K. A., and E. G. Long, "Thermochemistry for the Petrochemical Industry, Part II—Paraffinic Hydrocarbons, C_1-C_6," *Petroleum Refiner,* Vol. 28, No. 2, 1949, pp. 113–116. Constants for equations of state are calculated from critical state properties.

Table A·2 Benedict–Webb–Rubin Constants

Gas	Formula	R kJ/kg·K	A_0 N·m^4/kg^2	B_0 m^3/kg	C_0 N·m^4·K^2/kg^2	a N·m^7/kg^3
Methane	CH_4	0.518	731.195	2.65735×10^{-3}	0.88964×10^{7}	1.21466
Ethylene	C_2H_4	0.296	430.550	1.98649×10^{-3}	1.69071×10^{7}	1.19119
Ethane	C_2H_6	0.276	466.269	2.08914×10^{-3}	2.01509×10^{7}	1.28892
Propylene	C_3H_6	0.198	350.217	2.02308×10^{-3}	2.51642×10^{7}	1.05482
Propane	C_3H_8	0.189	358.575	2.20855×10^{-3}	2.65194×10^{7}	1.12224
i-Butylene	C_4H_8	0.148	288.571	2.06958×10^{-3}	2.98871×10^{7}	0.97316
i-Butane	C_4H_{10}	0.143	307.308	2.36826×10^{-3}	2.55256×10^{7}	1.00195
n-Butane	C_4H_{10}	0.143	302.865	2.14127×10^{-3}	2.98168×10^{7}	0.97334
i-Pentane	C_5H_{12}	0.115	249.391	2.22006×10^{-3}	3.40357×10^{7}	1.01546
n-Pentane	C_5H_{12}	0.115	237.376	2.17426×10^{-3}	4.13424×10^{7}	1.10159
n-Hexane	C_6H_{14}	0.0965	197.242	2.06498×10^{-3}	4.53487×10^{7}	1.12913
n-Heptane	C_7H_{16}	0.0830	177.041	1.98756×10^{-3}	4.79543×10^{7}	1.04602

Gas	Formula	b m^6/kg^2	c N·m^7·K^2/kg^3	α m^9/kg^3	γ m^6/kg^2
Methane	CH_4	1.31523×10^{-5}	0.62577×10^{5}	30.1853×10^{-9}	23.3469×10^{-6}
Ethylene	C_2H_4	1.09451×10^{-5}	0.97139×10^{5}	8.08173×10^{-9}	11.7469×10^{-6}
Ethane	C_2H_6	1.23191×10^{-5}	1.22361×10^{5}	8.97220×10^{-9}	13.0701×10^{-6}
Propylene	C_3H_6	1.05806×10^{-5}	1.39829×10^{5}	6.13014×10^{-9}	10.3453×10^{-6}
Propane	C_3H_8	1.15892×10^{-5}	1.52759×10^{5}	7.09776×10^{-9}	11.3317×10^{-6}
i-Butylene	C_4H_8	1.10774×10^{-5}	1.58056×10^{5}	5.16963×10^{-9}	9.41616×10^{-6}
i-Butane	C_4H_{10}	1.25806×10^{-5}	1.47891×10^{5}	5.48279×10^{-9}	10.0799×10^{-6}
n-Butane	C_4H_{10}	1.18582×10^{-5}	1.63610×10^{5}	5.62184×10^{-9}	10.0799×10^{-6}
i-Pentane	C_5H_{12}	1.28545×10^{-5}	1.87887×10^{5}	4.53682×10^{-9}	8.90805×10^{-6}
n-Pentane	C_5H_{12}	1.28545×10^{-5}	2.22807×10^{5}	4.83038×10^{-9}	9.13893×10^{-6}
n-Hexane	C_6H_{14}	1.47181×10^{-5}	2.40013×10^{5}	4.40244×10^{-9}	8.99353×10^{-6}
n-Heptane	C_7H_{16}	1.51575×10^{-5}	2.49275×10^{5}	4.33982×10^{-9}	8.97754×10^{-6}

Source: Howell, John R., and Richard O. Buckius, *Fundamentals of Engineering Thermodynamics,* McGraw-Hill, 2nd ed., 1992, as adapted from Cravalho, Ernest, and Joseph L. Smith, *Engineering Thermodynamics,* Pitman, 1981 (used with permission of both copyright owners).

Table A·3 Zero-pressure Specific Heat vs. *T* for Several Gases

T, °C	Air c_p kJ/kg·K	Air c_v kJ/kg·K	Air k	CO_2 c_p kJ/kg·K	CO_2 c_v kJ/kg·K	CO_2 k	CO c_p kJ/kg·K	CO c_v kJ/kg·K	CO k	*T*, °C
0	1.003	0.716	1.401	0.818	0.629	1.300	1.041	0.744	1.399	**0**
50	1.005	0.718	1.400	0.869	0.680	1.278	1.041	0.744	1.399	**50**
100	1.009	0.722	1.397	0.916	0.727	1.260	1.044	0.748	1.397	**100**
150	1.016	0.729	1.394	0.957	0.769	1.246	1.051	0.754	1.394	**150**
200	1.024	0.737	1.389	0.995	0.806	1.234	1.059	0.762	1.390	**200**
300	1.045	0.758	1.379	1.060	0.871	1.217	1.080	0.783	1.379	**300**
400	1.068	0.781	1.367	1.114	0.925	1.204	1.106	0.809	1.367	**400**
500	1.092	0.805	1.356	1.158	0.969	1.195	1.132	0.835	1.355	**500**
600	1.115	0.828	1.347	1.195	1.006	1.188	1.157	0.860	1.345	**600**
700	1.135	0.848	1.338	1.227	1.038	1.182	1.179	0.882	1.336	**700**
800	1.154	0.867	1.331	1.252	1.063	1.178	1.199	0.902	1.329	**800**
900	1.169	0.882	1.325	1.275	1.086	1.174	1.216	0.919	1.323	**900**
1000	1.184	0.897	1.320	1.294	1.105	1.171	1.231	0.934	1.318	**1000**
1200	1.207	0.920	1.312	1.324	1.135	1.166	1.255	0.958	1.310	**1200**
1400	1.226	0.939	1.306	1.346	1.157	1.163	1.273	0.976	1.304	**1400**
1600	1.241	0.954	1.301	1.362	1.173	1.161	1.287	0.990	1.300	**1600**
1800	1.253	0.966	1.297	1.375	1.186	1.159	1.298	1.001	1.297	**1800**
2000	1.264	0.977	1.294	1.386	1.197	1.158	1.307	1.010	1.294	**2000**
2200	1.273	0.986	1.291	1.396	1.207	1.157	1.314	1.017	1.292	**2200**
2400	1.282	0.995	1.289	1.404	1.215	1.155	1.321	1.024	1.290	**2400**
2600	1.289	1.002	1.286	1.411	1.222	1.155	1.326	1.029	1.288	**2600**

T, °C	H_2 c_p kJ/kg·K	H_2 c_v kJ/kg·K	H_2 k	N_2 c_p kJ/kg·K	N_2 c_v kJ/kg·K	N_2 k	O_2 c_p kJ/kg·K	O_2 c_v kJ/kg·K	O_2 k	*T*, °C
0	14.240	10.116	1.408	1.041	0.744	1.399	0.915	0.655	1.397	**0**
50	14.367	10.242	1.403	1.040	0.743	1.399	0.922	0.663	1.392	**50**
100	14.445	10.321	1.400	1.042	0.745	1.398	0.934	0.674	1.386	**100**
150	14.491	10.366	1.398	1.046	0.749	1.396	0.948	0.688	1.378	**150**
200	14.515	10.390	1.397	1.052	0.756	1.393	0.963	0.703	1.370	**200**
300	14.538	10.414	1.396	1.070	0.773	1.384	0.994	0.735	1.354	**300**
400	14.577	10.453	1.395	1.091	0.795	1.374	1.024	0.764	1.340	**400**
500	14.661	10.537	1.391	1.115	0.819	1.363	1.049	0.789	1.329	**500**
600	14.792	10.668	1.387	1.140	0.843	1.352	1.068	0.808	1.321	**600**
700	14.942	10.818	1.381	1.162	0.865	1.343	1.085	0.825	1.315	**700**
800	15.123	10.999	1.375	1.182	0.885	1.335	1.100	0.840	1.309	**800**
900	15.319	11.195	1.368	1.200	0.903	1.329	1.111	0.851	1.305	**900**
1000	15.526	11.402	1.362	1.216	0.919	1.323	1.122	0.862	1.301	**1000**
1200	15.954	11.830	1.349	1.241	0.944	1.314	1.140	0.880	1.295	**1200**
1400	16.379	12.254	1.337	1.261	0.964	1.308	1.156	0.896	1.290	**1400**
1600	16.778	12.654	1.326	1.276	0.979	1.303	1.170	0.911	1.285	**1600**
1800	17.139	13.015	1.317	1.288	0.991	1.299	1.184	0.925	1.281	**1800**
2000	17.459	13.335	1.309	1.298	1.001	1.297	1.198	0.939	1.277	**2000**
2200	17.740	13.616	1.303	1.306	1.010	1.294	1.212	0.953	1.273	**2200**
2400	17.995	13.871	1.297	1.313	1.017	1.292	1.226	0.966	1.269	**2400**
2600	18.245	14.121	1.292	1.319	1.022	1.290	1.238	0.979	1.266	**2600**

Table A·3 Zero-pressure Specific Heat vs. *T* for Several Gases (*Continued*)

T, °C	CH_4, Methane c_p kJ/kg·K	c_v kJ/kg·K	k	C_2H_2, Acetylene c_p kJ/kg·K	c_v kJ/kg·K	k	C_2H_4, Ethylene c_p kJ/kg·K	c_v kJ/kg·K	k	*T*, °C
0	2.160	1.642	1.316	1.621	1.302	1.245	1.434	1.137	1.261	**0**
50	2.286	1.768	1.293	1.761	1.442	1.221	1.618	1.322	1.224	**50**
100	2.438	1.919	1.270	1.880	1.560	1.205	1.798	1.502	1.197	**100**
150	2.607	2.088	1.248	1.980	1.661	1.192	1.972	1.676	1.177	**150**
200	2.786	2.268	1.229	2.066	1.747	1.183	2.139	1.842	1.161	**200**
300	3.157	2.639	1.196	2.206	1.887	1.169	2.444	2.147	1.138	**300**
400	3.516	2.997	1.173	2.320	2.001	1.160	2.709	2.413	1.123	**400**
500	3.843	3.325	1.156	2.421	2.102	1.152	2.935	2.639	1.112	**500**
600	4.136	3.618	1.143	2.515	2.195	1.145	3.129	2.832	1.105	**600**
700	4.406	3.888	1.133	2.601	2.281	1.140	3.303	3.007	1.099	**700**
800	4.648	4.129	1.126	2.678	2.358	1.135	3.455	3.159	1.094	**800**
900	4.863	4.345	1.119	2.750	2.430	1.131	3.591	3.295	1.090	**900**
1000	5.052	4.534	1.114	2.815	2.495	1.128	3.710	3.413	1.087	**1000**
1200	5.363	4.844	1.107	2.926	2.607	1.122	3.901	3.605	1.082	**1200**
1400	5.601	5.083	1.102	3.017	2.698	1.118	4.047	3.750	1.079	**1400**
1600	5.786	5.268	1.098	3.092	2.773	1.115	4.159	3.863	1.077	**1600**
1800	5.933	5.414	1.096	3.155	2.835	1.113	4.248	3.951	1.075	**1800**
2000	6.050	5.532	1.094	3.208	2.889	1.111	4.321	4.024	1.074	**2000**
2200	6.147	5.629	1.092	3.254	2.934	1.109	4.381	4.085	1.073	**2200**
2400	6.227	5.708	1.091	3.293	2.973	1.107	4.431	4.135	1.072	**2400**
2600	6.288	5.770	1.090	3.326	3.007	1.106	4.468	4.171	1.071	**2600**

Source: JANAF Thermochemical Tables, 3rd ed., published by the American Chemical Society and the American Institute of Physics for the National Bureau of Standards, 1986. Also, *Journal of Physical and Chemical Reference Data,* Vol. 14, Supplement No. 1, 1985.

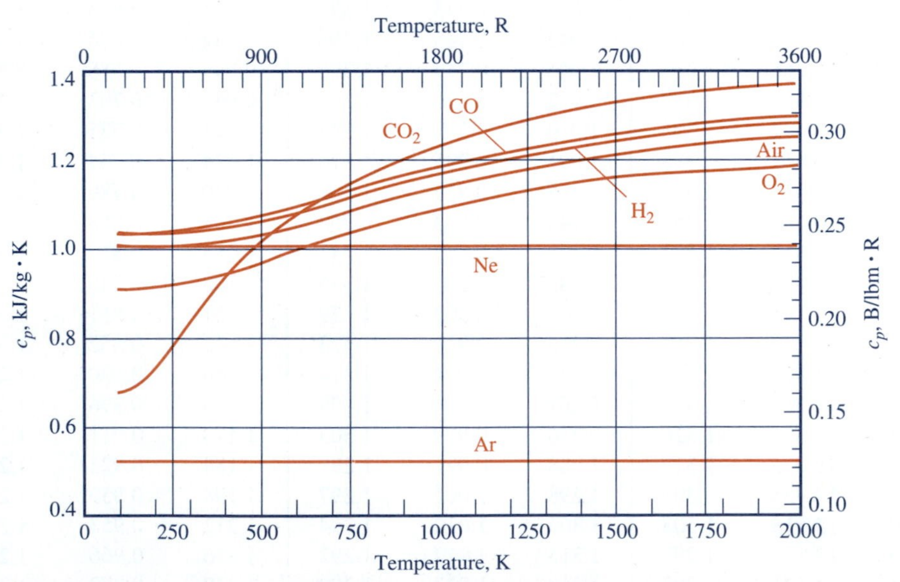

Chart A·4 Zero-pressure specific heat vs. *T* for several gases.

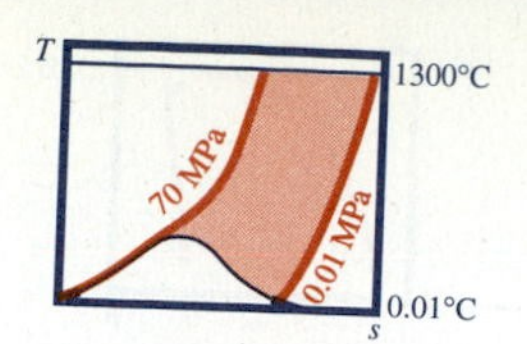

Table A·6·3 Water: Compressed Liquid and Superheated Vapor

T °C	p = 0.01 MPa v, m³/kg	h, kJ/kg	s, kJ/kg·K
0.01	0.001 000	0.0	-0.0002
25	0.001 003	104.8	0.3670
T_{sat}= 45.83 f	0.001 010	191.9	0.6494
g	14.6245	2583.8	8.1480
50	14.8703	2591.8	8.1732
75	16.0364	2639.4	8.3151
100	17.1982	2687.0	8.4471
125	18.3576	2734.6	8.5708
150	19.5156	2782.5	8.6874
175	20.6726	2830.7	8.7979
200	21.8290	2879.1	8.9030
225	22.9851	2927.8	9.0034
250	24.1409	2976.9	9.0996
275	25.2964	3026.4	9.1920
300	26.4519	3076.3	9.2809
325	27.6073	3126.5	9.3666
350	28.7627	3177.1	9.4495
375	29.9180	3228.1	9.5298
400	31.0733	3279.5	9.6076
450	33.3841	3383.6	9.7567
500	35.6950	3489.3	9.8981
550	38.0062	3596.8	10.0327
600	40.3178	3706.0	10.1615
650	42.6297	3816.9	10.2850
700	44.9420	3929.7	10.4039
800	49.5679	4160.5	10.6296
900	54.1959	4398.3	10.8414
1000	58.8262	4643.0	11.0415
1100	63.4590	4894.1	11.2313
1200	68.0944	5151.3	11.4120
1300	72.7328	5414.2	11.5846

T °C	p = 0.05 MPa v, m³/kg	h, kJ/kg	s, kJ/kg·K
0.01	0.001 000	0.0	-0.0002
25	0.001 003	104.8	0.3669
50	0.001 012	209.0	0.7039
75	0.001 026	313.6	1.0158
T_{sat}= 81.34 f	0.001 030	340.5	1.0912
g	3.2393	2645.3	7.5928
100	3.4188	2682.1	7.6941
125	3.6549	2731.0	7.8208
150	3.8895	2779.7	7.9394
175	4.1231	2828.3	8.0511
200	4.3560	2877.2	8.1572
225	4.5885	2926.2	8.2582
250	4.8206	2975.6	8.3548
275	5.0524	3025.2	8.4475
300	5.2841	3075.2	8.5367
325	5.5156	3125.5	8.6226
350	5.7470	3176.2	8.7057
375	5.9783	3227.3	8.7861
400	6.2095	3278.8	8.8640
450	6.6718	3382.9	9.0132
500	7.1339	3488.7	9.1547
550	7.5960	3596.2	9.2894
600	8.0579	3705.4	9.4182
650	8.5197	3816.4	9.5417
700	8.9815	3929.1	9.6607
800	9.9051	4159.9	9.8863
900	10.8285	4397.7	10.0981
1000	11.7519	4642.2	10.2981
1100	12.6752	4893.2	10.4878
1200	13.5986	5150.2	10.6685
1300	14.5219	5412.9	10.8410

T °C	p = 0.1 MPa v, m³/kg	h, kJ/kg	s, kJ/kg·K
0.01	0.001 000	0.0	-0.0002
25	0.001 003	104.8	0.3669
50	0.001 012	209.4	0.7037
75	0.001 026	314.0	1.0155
T_{sat}= 99.63 f	0.001 043	417.5	1.3027
g	1.6940	2675.1	7.3589
100	1.6961	2675.9	7.3609
125	1.8171	2726.3	7.4918
150	1.9364	2776.1	7.6129
175	2.0547	2825.5	7.7263
200	2.1723	2874.8	7.8335
225	2.2893	2924.3	7.9353
250	2.4061	2973.9	8.0325
275	2.5225	3023.8	8.1257
300	2.6388	3073.9	8.2152
325	2.7549	3124.4	8.3014
350	2.8709	3175.3	8.3846
375	2.9869	3226.4	8.4652
400	3.1027	3278.0	8.5432
450	3.3342	3382.3	8.6927
500	3.5656	3488.2	8.8342
550	3.7968	3595.8	8.9690
600	4.0279	3705.0	9.0979
650	4.2590	3816.1	9.2215
700	4.4900	3928.8	9.3405
800	4.9519	4159.7	9.5663
900	5.4138	4397.5	9.7781
1000	5.8755	4642.0	9.9781
1100	6.3373	4893.0	10.1678
1200	6.7990	5150.0	10.3485
1300	7.2606	5412.8	10.5210

T °C	p = 0.2 MPa v, m³/kg	h, kJ/kg	s, kJ/kg·K
0.01	0.001 000	0.2	-0.0001
25	0.001 003	104.9	0.3669
50	0.001 012	209.5	0.7037
75	0.001 026	314.1	1.0154
100	0.001 044	418.4	1.3076
T_{sat}= 120.24 f	0.001 061	504.8	1.5304
g	0.8858	2706.6	7.1272
125	0.8978	2716.6	7.1527
150	0.9597	2768.6	7.2793
175	1.0204	2819.6	7.3963
200	1.0803	2870.0	7.5059
225	1.1397	2920.3	7.6094
250	1.1988	2970.5	7.7078
275	1.2576	3020.9	7.8018
300	1.3162	3071.4	7.8920
325	1.3746	3122.2	7.9787
350	1.4329	3173.3	8.0624
375	1.4912	3224.7	8.1433
400	1.5493	3276.4	8.2216
450	1.6655	3381.0	8.3714
500	1.7814	3487.1	8.5133
550	1.8973	3594.9	8.6483
600	2.0130	3704.3	8.7773
650	2.1287	3815.4	8.9011
700	2.2443	3928.3	9.0201
800	2.4755	4159.2	9.2460
900	2.7066	4397.1	9.4579
1000	2.9376	4641.7	9.6580
1100	3.1685	4892.8	9.8478
1200	3.3994	5149.8	10.0284
1300	3.6303	5412.6	10.2010

T °C	p = 0.3 MPa v, m³/kg	h, kJ/kg	s, kJ/kg·K
0.01	0.001 000	0.3	-0.0001
25	0.001 003	105.0	0.3669
50	0.001 012	209.6	0.7036
75	0.001 026	314.2	1.0154
100	0.001 044	418.6	1.3074
125	0.001 067	525.1	1.5751
T_{sat}= 133.56 f	0.001 073	561.6	1.6721
g	0.6056	2725.3	6.9921
150	0.6339	2760.9	7.0779
175	0.6755	2813.5	7.1987
200	0.7163	2865.1	7.3108
225	0.7565	2916.2	7.4160
250	0.7963	2967.1	7.5157
275	0.8359	3017.9	7.6107
300	0.8753	3068.9	7.7015
325	0.9145	3120.0	7.7888
350	0.9536	3171.3	7.8729
375	0.9926	3223.0	7.9541
400	1.0315	3274.9	8.0327
450	1.1092	3379.7	8.1830
500	1.1867	3486.1	8.3252
550	1.2641	3594.0	8.4604
600	1.3414	3703.5	8.5895
650	1.4186	3814.7	8.7134
700	1.4958	3927.7	8.8325
800	1.6500	4158.8	9.0585
900	1.8042	4396.7	9.2705
1000	1.9582	4641.4	9.4706
1100	2.1123	4892.5	9.6605
1200	2.2662	5149.6	9.8412
1300	2.4202	5412.4	10.0137

T °C	p = 0.4 MPa v, m³/kg	h, kJ/kg	s, kJ/kg·K
0.01	0.001 000	0.4	-0.0001
25	0.001 003	105.1	0.3669
50	0.001 012	209.7	0.7036
75	0.001 026	314.3	1.0153
100	0.001 044	418.7	1.3073
125	0.001 067	525.3	1.5749
T_{sat}= 143.64 f	0.001 084	604.9	1.7770
g	0.4623	2738.5	6.8961
150	0.4708	2752.8	6.9300
175	0.5029	2807.2	7.0551
200	0.5342	2860.1	7.1699
225	0.5648	2912.1	7.2770
250	0.5951	2963.6	7.3779
275	0.6251	3015.0	7.4738
300	0.6548	3066.3	7.5654
325	0.6844	3117.8	7.6533
350	0.7139	3169.4	7.7378
375	0.7433	3221.2	7.8194
400	0.7726	3273.3	7.8982
450	0.8311	3378.5	8.0489
500	0.8894	3485.0	8.1913
550	0.9475	3593.1	8.3268
600	1.0056	3702.7	8.4561
650	1.0636	3814.1	8.5801
700	1.1215	3927.1	8.6993
800	1.2373	4158.3	8.9254
900	1.3530	4396.4	9.1375
1000	1.4686	4641.1	9.3377
1100	1.5841	4892.3	9.5275
1200	1.6997	5149.4	9.7083
1300	1.8152	5412.2	9.8809

Table A·5 Specific Heats and Densities of Liquids and Solids at 1 atm, 25°C

Substance	c_p kJ/kg·K	ρ kg/m³
Aluminum	0.900	2690
Carbon, diamond	0.519	3270
Carbon, graphite	0.712	2510
Castor oil	1.968	956
Copper	0.385	8930
Ethylene glycol	2.366	1100
Glycerine	2.626	1260
Gold	0.128	19300
Granite	1.018	2700
Iron	0.451	7850
Kerosene	2.094	820
Lead	0.199	11300
Limestone	0.909	2710
Linseed oil	1.843	929
Lithium bromide	0.750	3450
A lubricating oil	1.801	910
Mercury	0.139	13600
Mica	0.503	2920
Oak	2.094	753
Sandstone	0.921	2260
Silicon	0.708	2320
Silver	0.236	10500
Water	4.188	997
White pine	2.513	424
Zinc	0.384	6980

Table A·6·1 Water: Liquid–vapor Saturation, Temperature Table

T °C	p kPa	v_f m³/kg	v_g m³/kg	h_f kJ/kg	h_{fg} kJ/kg	h_g kJ/kg	s_f kJ/kg·K	s_{fg} kJ/kg·K	s_g kJ/kg·K	T °C
0.01	0.61173	0.001 000	205.99	0.00	2500.5	2500.5	0.0000	9.1541	9.1541	**0.01**
1	0.6508	0.001 000	193.83	4.17	2498.2	2502.3	0.0152	9.1126	9.1278	**1**
3	0.7444	0.001 000	170.79	12.59	2493.4	2506.0	0.0458	9.0295	9.0753	**3**
5	0.8579	0.001 000	149.80	21.00	2488.7	2509.7	0.0761	8.9476	9.0237	**5**
10	1.2281	0.001 000	106.32	41.99	2476.9	2518.9	0.1510	8.7476	8.8986	**10**
15	1.7029	0.001 001	77.37	62.93	2465.1	2528.1	0.2243	8.5550	8.7792	**15**
20	2.3388	0.001 002	57.78	83.84	2453.4	2537.2	0.2962	8.3689	8.6651	**20**
25	3.1656	0.001 003	43.14	104.76	2441.5	2546.3	0.3670	8.1889	8.5559	**25**
30	4.2455	0.001 004	32.90	125.67	2429.6	2555.3	0.4365	8.0148	8.4513	**30**
35	5.6225	0.001 006	25.13	146.59	2417.8	2564.4	0.5050	7.8461	8.3511	**35**
40	7.3814	0.001 008	19.528	167.50	2405.9	2573.4	0.5723	7.6827	8.2550	**40**
45	9.5844	0.001 010	15.220	188.42	2393.9	2582.3	0.6385	7.5244	8.1629	**45**
50	12.344	0.001 012	12.037	209.33	2381.9	2591.2	0.7037	7.3708	8.0745	**50**
55	15.745	0.001 015	9.552	230.24	2369.8	2600.0	0.7679	7.2216	7.9896	**55**
60	19.932	0.001 017	7.674	251.15	2357.7	2608.8	0.8312	7.0768	7.9080	**60**
65	25.016	0.001 020	6.189	272.07	2345.4	2617.5	0.8935	6.9360	7.8295	**65**
70	31.176	0.001 023	5.045	293.01	2333.1	2626.1	0.9549	6.7991	7.7540	**70**
75	38.556	0.001 026	4.1278	313.96	2320.7	2634.7	1.0155	6.6657	7.6813	**75**
80	47.373	0.001 029	3.4088	334.93	2308.2	2643.1	1.0753	6.5359	7.6112	**80**
85	57.807	0.001 032	2.8260	355.92	2295.5	2651.4	1.1343	6.4093	7.5436	**85**
90	70.117	0.001 036	2.3617	376.93	2282.7	2659.6	1.1925	6.2859	7.4784	**90**
95	84.520	0.001 040	1.9811	397.98	2269.7	2667.7	1.2501	6.1653	7.4154	**95**
100	101.32	0.001 043	1.6736	419.06	2256.6	2675.7	1.3069	6.0476	7.3545	**100**
105	120.78	0.001 047	1.4190	440.18	2243.4	2683.6	1.3631	5.9326	7.2956	**105**
110	143.24	0.001 052	1.2106	461.34	2230.0	2691.3	1.4186	5.8200	7.2386	**110**
115	169.01	0.001 056	1.0364	482.54	2216.3	2698.8	1.4735	5.7098	7.1833	**115**
120	198.48	0.001 060	0.8922	503.78	2202.4	2706.2	1.5278	5.6019	7.1297	**120**
130	270.02	0.001 070	0.6687	546.41	2174.0	2720.4	1.6346	5.3926	7.0272	**130**
140	361.19	0.001 080	0.5090	589.24	2144.6	2733.8	1.7394	5.1908	6.9302	**140**
150	475.72	0.001 090	0.3929	632.32	2114.1	2746.4	1.8421	4.9960	6.8381	**150**
160	617.66	0.001 102	0.3071	675.65	2082.4	2758.0	1.9429	4.8074	6.7503	**160**
170	791.47	0.001 114	0.2428	719.28	2049.2	2768.5	2.0421	4.6241	6.6662	**170**
180	1001.9	0.001 127	0.1940	763.25	2014.6	2777.8	2.1397	4.4456	6.5853	**180**
190	1254.2	0.001 141	0.1565	807.60	1978.2	2785.8	2.2358	4.2713	6.5071	**190**
200	1553.7	0.001 156	0.1273	852.38	1940.1	2792.5	2.3308	4.1004	6.4312	**200**
210	1906.2	0.001 173	0.1044	897.66	1900.0	2797.7	2.4246	3.9326	6.3572	**210**
220	2317.8	0.001 190	0.0862	943.51	1857.8	2801.3	2.5175	3.7672	6.2847	**220**
230	2795.1	0.001 209	0.0716	990.00	1813.1	2803.1	2.6097	3.6034	6.2131	**230**
240	3344.7	0.001 229	0.0597	1037.2	1765.8	2803.0	2.7014	3.4409	6.1423	**240**
250	3973.7	0.001 252	0.0501	1085.3	1715.4	2800.7	2.7926	3.2791	6.0717	**250**
260	4689.5	0.001 276	0.0422	1134.4	1661.8	2796.2	2.8838	3.1171	6.0009	**260**
270	5499.9	0.001 303	0.035 64	1184.6	1604.5	2789.1	2.9751	2.9542	5.9293	**270**
280	6413.2	0.001 332	0.030 16	1236.1	1543.1	2779.2	3.0669	2.7896	5.8565	**280**
290	7438.0	0.001 366	0.025 56	1289.1	1476.8	2765.9	3.1595	2.6223	5.7818	**290**
300	8583.8	0.001 404	0.021 67	1344.1	1404.6	2748.7	3.2534	2.4508	5.7042	**300**
310	9860.5	0.001 447	0.018 34	1401.2	1325.8	2727.0	3.3491	2.2735	5.6226	**310**
320	11279	0.001 498	0.015 48	1461.3	1238.4	2699.7	3.4476	2.0880	5.5356	**320**
330	12852	0.001 560	0.012 99	1525.0	1140.3	2665.3	3.5501	1.8906	5.4407	**330**
340	14594	0.001 637	0.010 79	1593.8	1027.5	2621.3	3.6587	1.6758	5.3345	**340**
350	16521	0.001 740	0.008 81	1670.4	893.1	2563.5	3.7774	1.4331	5.2105	**350**
360	18655	0.001 894	0.006 96	1761.0	721.0	2482.0	3.9153	1.1389	5.0542	**360**
365	19809	0.002 012	0.006 03	1816.7	607.9	2424.6	3.9994	0.9526	4.9520	**365**
370	21030	0.002 207	0.004 99	1889.7	450.5	2340.2	4.1094	0.7004	4.8098	**370**
373	21799	0.002 485	0.004 12	1966.6	276.4	2243.0	4.2258	0.4278	4.6536	**373**
374.14	22055	0.003 106	0.003 11	2086.0	0.0	2086.0	4.4090	0.0000	4.4090	**374.14**

Source: Based on equations and data of Haar, Lester, John S. Gallagher, and George S. Kell, *NBS/NRC Steam Tables,* Hemisphere, 1984.

H₂O, Sat

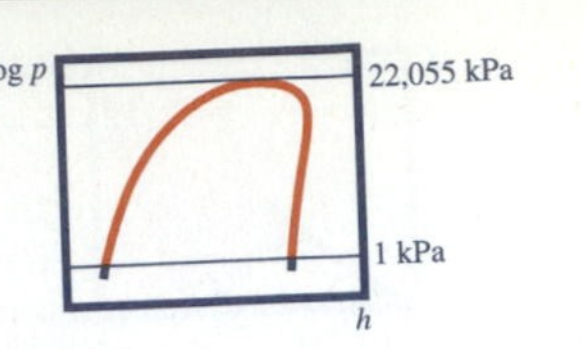

Table A·6·2 Water: Liquid–vapor Saturation, Pressure Table

p kPa	T °C	v_f m³/kg	v_g m³/kg	h_f kJ/kg	h_{fg} kJ/kg	h_g kJ/kg	s_f kJ/kg·K	s_{fg} kJ/kg·K	s_g kJ/kg·K	p kPa
1.0	7.12	0.001 000	129.77	29.92	2483.7	2513.6	0.1081	8.8618	8.9699	**1.0**
1.5	13.04	0.001 001	87.39	54.74	2469.7	2524.5	0.1958	8.6295	8.8253	**1.5**
2.0	17.51	0.001 001	66.57	73.44	2459.2	2532.7	0.2606	8.4606	8.7212	**2.0**
2.5	21.09	0.001 002	54.14	88.39	2450.8	2539.2	0.3117	8.3292	8.6409	**2.5**
3.0	24.10	0.001 003	45.40	101.00	2443.6	2544.6	0.3543	8.2208	8.5751	**3.0**
4.0	28.97	0.001 004	34.72	121.37	2432.1	2553.4	0.4223	8.0501	8.4724	**4.0**
5.0	32.89	0.001 005	28.09	137.77	2422.8	2560.5	0.4763	7.9165	8.3928	**5.0**
7.5	40.30	0.001 008	19.23	168.76	2405.2	2573.9	0.5763	7.6731	8.2494	**7.5**
10	45.83	0.001 010	14.62	191.88	2391.9	2583.8	0.6494	7.4986	8.1480	**10**
15	53.99	0.001 014	9.999	226.02	2372.2	2598.3	0.7551	7.2514	8.0065	**15**
20	60.07	0.001 017	7.649	251.46	2357.5	2608.9	0.8321	7.0747	7.9068	**20**
25	64.99	0.001 020	6.193	272.01	2345.4	2617.5	0.8933	6.9364	7.8297	**25**
30	69.11	0.001 022	5.226	289.30	2335.3	2624.6	0.9441	6.8230	7.7672	**30**
40	75.88	0.001 026	3.988	317.66	2318.5	2636.1	1.0261	6.6426	7.6687	**40**
50	81.34	0.001 030	3.239	340.55	2304.8	2645.3	1.0912	6.5016	7.5928	**50**
75	91.78	0.001 037	2.216	384.44	2278.1	2662.5	1.2131	6.2425	7.4557	**75**
100	99.63	0.001 043	1.6940	417.51	2257.6	2675.1	1.3027	6.0562	7.3589	**100**
101.325	100.00	0.001043	1.6735	419.07	2256.6	2675.7	1.3069	6.0476	7.3545	**101.325**
150	111.38	0.001 053	1.1592	467.18	2226.2	2693.4	1.4338	5.7894	7.2232	**150**
200	120.24	0.001 060	0.8858	504.81	2201.7	2706.6	1.5304	5.5968	7.1272	**200**
250	127.44	0.001 067	0.7185	535.50	2181.3	2716.8	1.6075	5.4453	7.0528	**250**
300	133.56	0.001 073	0.6056	561.62	2163.6	2725.3	1.6721	5.3200	6.9921	**300**
350	138.89	0.001 079	0.5242	584.48	2147.9	2732.4	1.7279	5.2128	6.9407	**350**
400	143.64	0.001 084	0.4623	604.91	2133.6	2738.5	1.7770	5.1191	6.8961	**400**
500	151.87	0.001 093	0.3748	640.39	2108.3	2748.6	1.8610	4.9604	6.8214	**500**
600	158.86	0.001 101	0.3156	670.71	2086.0	2756.7	1.9316	4.8285	6.7601	**600**
700	164.98	0.001 108	0.2728	697.35	2066.0	2763.4	1.9925	4.7154	6.7080	**700**
800	170.44	0.001 115	0.2404	721.23	2047.7	2768.9	2.0464	4.6161	6.6625	**800**
900	175.39	0.001 121	0.2149	742.93	2030.7	2773.7	2.0949	4.5274	6.6222	**900**
1000	179.92	0.001 127	0.194 38	762.89	2014.8	2777.7	2.1389	4.4471	6.5860	**1000**
1100	184.10	0.001 133	0.177 44	781.39	1999.8	2781.2	2.1793	4.3737	6.5529	**1100**
1200	187.99	0.001 138	0.163 26	798.67	1985.6	2784.3	2.2167	4.3059	6.5226	**1200**
1300	191.64	0.001 144	0.151 19	814.92	1972.1	2787.0	2.2515	4.2430	6.4945	**1300**
1400	195.08	0.001 149	0.140 77	830.28	1959.1	2789.4	2.2842	4.1841	6.4683	**1400**
1500	198.33	0.001 154	0.131 71	844.85	1946.6	2791.5	2.3149	4.1288	6.4438	**1500**
2000	212.42	0.001 177	0.099 59	908.69	1890.0	2798.7	2.4471	3.8924	6.3396	**2000**
2500	223.99	0.001 197	0.079 94	961.98	1840.3	2802.2	2.5544	3.7016	6.2560	**2500**
3000	233.89	0.001 217	0.066 66	1008.29	1795.0	2803.3	2.6454	3.5401	6.1855	**3000**
3500	242.60	0.001 235	0.057 05	1049.63	1753.0	2802.6	2.7251	3.3989	6.1240	**3500**
4000	250.39	0.001 252	0.049 77	1087.22	1713.3	2800.6	2.7962	3.2727	6.0689	**4000**
5000	263.98	0.001 286	0.039 44	1154.19	1639.5	2793.7	2.9201	3.0524	5.9725	**5000**
6000	275.62	0.001 319	0.032 44	1213.34	1570.6	2783.9	3.0266	2.8619	5.8886	**6000**
7000	285.86	0.001 352	0.027 37	1267.0	1504.9	2771.8	3.1211	2.6919	5.8130	**7000**
8000	295.04	0.001 384	0.023 52	1316.6	1441.2	2757.8	3.2066	2.5365	5.7431	**8000**
9000	303.38	0.001 418	0.020 48	1363.1	1378.8	2741.9	3.2855	2.3917	5.6771	**9000**
10000	311.03	0.001 452	0.018 025	1407.3	1317.2	2724.5	3.3591	2.2548	5.6139	**10000**
11000	318.11	0.001 488	0.015 985	1449.7	1255.7	2705.3	3.4288	2.1238	5.5525	**11000**
12000	324.71	0.001 526	0.014 261	1490.7	1193.8	2684.5	3.4953	1.9968	5.4921	**12000**
13000	330.89	0.001 566	0.012 779	1530.9	1131.0	2661.8	3.5595	1.8723	5.4318	**13000**
14000	336.70	0.001 610	0.011 484	1570.4	1066.7	2637.1	3.6220	1.7492	5.3712	**14000**
15000	342.19	0.001 657	0.010 339	1609.7	1000.4	2610.1	3.6836	1.6258	5.3093	**15000**
16000	347.40	0.001 709	0.009 311	1649.3	931.1	2580.4	3.7450	1.5003	5.2453	**16000**
17000	352.31	0.001 772	0.008 379	1690.2	856.2	2546.4	3.8077	1.3691	5.1769	**17000**
18000	357.00	0.001 843	0.007 511	1732.5	776.1	2508.6	3.8721	1.2320	5.1041	**18000**
19000	361.55	0.001 923	0.006 678	1776.4	690.7	2467.1	3.9386	1.0882	5.0268	**19000**
20000	365.80	0.002 036	0.005 873	1826.8	586.5	2413.3	4.0147	0.9180	4.9326	**20000**
21000	369.88	0.002 200	0.005 021	1887.6	455.2	2342.8	4.1062	0.7078	4.8140	**21000**
22000	373.77	0.002 775	0.003 595	2025.3	141.0	2166.3	4.3157	0.2178	4.5335	**22000**
22055	374.14	0.003 106	0.003 106	2086.0	0.0	2086.0	4.4090	0.0000	4.4090	**22055**

H₂O, Sat

Table A·6·3 Water: Compressed Liquid and Superheated Vapor (*Continued*)

T	*p* = 0.5 MPa			*T*	*p* = 0.6 MPa			*T*	*p* = 0.8 MPa		
°C	*v*, m³/kg	*h*, kJ/kg	*s*, kJ/kg·K	°C	*v*, m³/kg	*h*, kJ/kg	*s*, kJ/kg·K	°C	*v*, m³/kg	*h*, kJ/kg	*s*, kJ/kg·K
0.01	0.001 000	0.5	-0.0001	**0.01**	0.001 000	0.6	-0.0001	**0.01**	0.001 000	0.8	-0.0000
25	0.001 003	105.2	0.3668	**25**	0.001 003	105.3	0.3668	**25**	0.001 003	105.5	0.3668
50	0.001 012	209.7	0.7035	**50**	0.001 012	209.8	0.7035	**50**	0.001 012	210.0	0.7034
75	0.001 026	314.3	1.0152	**75**	0.001 026	314.4	1.0152	**75**	0.001 025	314.6	1.0150
100	0.001 044	418.9	1.3071	**100**	0.001 043	419.1	1.3069	**100**	0.001 043	419.4	1.3065
125	0.001 067	525.4	1.5747	**125**	0.001 067	525.6	1.5745	**125**	0.001 067	525.9	1.5742
150	0.001 091	632.0	1.8423	**150**	0.001 090	632.1	1.8422	**150**	0.001 090	632.4	1.8418
T_{sat}= *f*	0.001 093	640.4	1.1.8610	T_{sat}= *f*	0.001 101	670.7	1.9316	T_{sat}= *f*	0.001 115	721.2	2.0464
151.87 *g*	0.3748	2748.6	6.8214	**158.86** *g*	0.3156	2756.7	6.7601	**170.44** *g*	0.2404	2768.9	6.6625
175	0.3993	2800.7	6.9408	**175**	0.3302	2794.1	6.8450	**175**	0.2436	2780.0	6.6875
200	0.4249	2854.9	7.0585	**200**	0.3520	2849.7	6.9658	**200**	0.2607	2838.8	6.8151
225	0.4498	2907.9	7.1676	**225**	0.3731	2903.6	7.0769	**225**	0.2772	2894.8	6.9306
250	0.4743	2960.1	7.2699	**250**	0.3938	2956.6	7.1806	**250**	0.2931	2949.3	7.0373
275	0.4985	3012.0	7.3668	**275**	0.4142	3009.0	7.2785	**275**	0.3087	3002.9	7.1373
300	0.5226	3063.8	7.4591	**300**	0.4344	3061.2	7.3716	**300**	0.3241	3055.9	7.2319
325	0.5464	3115.5	7.5475	**325**	0.4544	3113.3	7.4605	**325**	0.3393	3108.7	7.3221
350	0.5701	3167.4	7.6324	**350**	0.4742	3165.4	7.5459	**350**	0.3544	3161.4	7.4084
375	0.5937	3219.4	7.7144	**375**	0.4940	3217.7	7.6282	**375**	0.3694	3214.1	7.4913
400	0.6173	3271.7	7.7935	**400**	0.5137	3270.2	7.7076	**400**	0.3843	3267.0	7.5713
450	0.6642	3377.2	7.9446	**450**	0.5529	3375.9	7.8591	**450**	0.4139	3373.3	7.7237
500	0.7109	3483.9	8.0873	**500**	0.5920	3482.9	8.0021	**500**	0.4433	3480.7	7.8673
550	0.7576	3592.2	8.2230	**550**	0.6309	3591.2	8.1380	**550**	0.4726	3589.4	8.0036
600	0.8041	3701.9	8.3524	**600**	0.6697	3701.2	8.2676	**600**	0.5018	3699.6	8.1335
650	0.8505	3813.4	8.4765	**650**	0.7085	3812.7	8.3918	**650**	0.5310	3811.4	8.2579
700	0.8969	3926.5	8.5959	**700**	0.7472	3925.9	8.5112	**700**	0.5601	3924.7	8.3775
800	0.9896	4157.8	8.8221	**800**	0.8246	4157.4	8.7376	**800**	0.6182	4156.5	8.6041
900	1.0822	4396.0	9.0342	**900**	0.9018	4395.6	8.9498	**900**	0.6762	4394.9	8.8165
1000	1.1748	4640.8	9.2345	**1000**	0.9789	4640.5	9.1501	**1000**	0.7341	4639.9	9.0169
1100	1.2673	4892.0	9.4244	**1100**	1.0560	4891.8	9.3401	**1100**	0.7920	4891.3	9.2070
1200	1.3597	5149.2	9.6051	**1200**	1.1331	5149.0	9.5209	**1200**	0.8498	5148.6	9.3878
1300	1.4521	5412.1	9.7778	**1300**	1.2101	5411.9	9.6935	**1300**	0.9076	5411.6	9.5605

T	*p* = 1.0 MPa			*T*	*p* = 1.2 MPa			*T*	*p* = 1.4 MPa		
°C	*v*, m³/kg	*h*, kJ/kg	*s*, kJ/kg·K	°C	*v*, m³/kg	*h*, kJ/kg	*s*, kJ/kg·K	°C	*v*, m³/kg	*h*, kJ/kg	*s*, kJ/kg·K
0.01	0.001 000	1.0	-0.0000	**0.01**	0.001 000	1.2	-0.0000	**0.01**	0.001 000	1.4	-0.0000
25	0.001 003	105.7	0.3667	**25**	0.001 002	105.9	0.3666	**25**	0.001 002	106.0	0.3666
50	0.001 012	210.2	0.7033	**50**	0.001 012	210.4	0.7032	**50**	0.001 012	210.5	0.7031
75	0.001 025	314.7	1.0149	**75**	0.001 025	314.9	1.0148	**75**	0.001 025	315.1	1.0147
100	0.001 043	419.7	1.3062	**100**	0.001 043	419.9	1.3060	**100**	0.001 043	420.0	1.3059
125	0.001 067	526.2	1.5738	**125**	0.001 066	526.3	1.5736	**125**	0.001 066	526.5	1.5735
150	0.001 090	632.6	1.8415	**150**	0.001 090	632.8	1.8413	**150**	0.001 090	632.9	1.8410
175	0.001 123	742.4	2.0861	**175**	0.001 123	742.5	2.0859	**175**	0.001 123	742.6	2.0857
T_{sat}= *f*	0.001 127	762.9	2.1389	T_{sat}= *f*	0.001 139	798.7	2.2167	T_{sat}= *f*	0.001 149	830.3	2.2842
179.92 *g*	0.1944	2777.7	6.5860	**188.00** *g*	0.1633	2784.3	6.5226	**195.08** *g*	0.1408	2789.4	6.4683
200	0.2059	2827.4	6.6932	**200**	0.1692	2815.4	6.5890	**200**	0.1430	2802.7	6.4966
225	0.2195	2885.8	6.8135	**225**	0.1811	2876.4	6.7147	**225**	0.1535	2866.6	6.6282
250	0.2326	2941.9	6.9234	**250**	0.1923	2934.3	6.8281	**250**	0.1635	2926.4	6.7454
275	0.2454	2996.6	7.0257	**275**	0.2032	2990.3	6.9327	**275**	0.1730	2983.8	6.8526
300	0.2579	3050.6	7.1219	**300**	0.2138	3045.2	7.0307	**300**	0.1823	3039.7	6.9523
325	0.2703	3104.1	7.2133	**325**	0.2242	3099.4	7.1233	**325**	0.1913	3094.7	7.0462
350	0.2825	3157.3	7.3005	**350**	0.2345	3153.3	7.2115	**350**	0.2003	3149.1	7.1354
375	0.2946	3210.5	7.3842	**375**	0.2447	3206.9	7.2959	**375**	0.2091	3203.3	7.2206
400	0.3066	3263.8	7.4648	**400**	0.2548	3260.6	7.3771	**400**	0.2178	3257.3	7.3024
450	0.3304	3370.7	7.6180	**450**	0.2748	3368.1	7.5312	**450**	0.2351	3365.5	7.4573
500	0.3541	3478.6	7.7622	**500**	0.2946	3476.4	7.6760	**500**	0.2521	3474.2	7.6028
550	0.3776	3587.6	7.8989	**550**	0.3143	3585.8	7.8131	**550**	0.2691	3584.0	7.7403
600	0.4011	3698.1	8.0292	**600**	0.3339	3696.5	7.9437	**600**	0.2860	3694.9	7.8712
650	0.4245	3810.0	8.1538	**650**	0.3535	3808.7	8.0686	**650**	0.3027	3807.3	7.9963
700	0.4478	3923.6	8.2736	**700**	0.3730	3922.4	8.1885	**700**	0.3195	3921.2	8.1164
800	0.4944	4155.5	8.5005	**800**	0.4118	4154.6	8.4156	**800**	0.3529	4153.7	8.3438
900	0.5408	4394.2	8.7130	**900**	0.4506	4393.4	8.6284	**900**	0.3861	4392.7	8.5567
1000	0.5872	4639.3	8.9136	**1000**	0.4893	4638.8	8.8290	**1000**	0.4193	4638.2	8.7575
1100	0.6335	4890.8	9.1037	**1100**	0.5279	4890.3	9.0192	**1100**	0.4525	4889.8	8.9477
1200	0.6798	5148.2	9.2846	**1200**	0.5665	5147.8	9.2002	**1200**	0.4856	5147.4	9.1288
1300	0.7261	5411.2	9.4573	**1300**	0.6051	5410.9	9.3729	**1300**	0.5187	5410.6	9.3016

Table A·6·3 Water: Compressed Liquid and Superheated Vapor (*Continued*)

T	p = 1.6 MPa		
°C	v, m^3/kg	h, kJ/kg	s, kJ/kg·K
0.01	0.000 999	1.6	-0.0000
25	0.001 002	106.2	0.3665
50	0.001 011	210.7	0.7030
75	0.001 025	315.2	1.0145
100	0.001 043	420.2	1.3057
125	0.001 066	526.6	1.5733
150	0.001 090	633.0	1.8408
175	0.001 123	742.7	2.0854
200	0.001 156	852.4	2.3300
T_{sat}= f	0.001 159	858.7	2.3440
201.41 g	0.1237	2793.3	6.4207
225	0.1328	2856.5	6.5506
250	0.1418	2918.4	6.6718
275	0.1503	2977.2	6.7817
300	0.1586	3034.1	6.8833
325	0.1667	3089.9	6.9786
350	0.1746	3145.0	7.0688
375	0.1823	3199.6	7.1547
400	0.1900	3254.1	7.2371
450	0.2053	3362.8	7.3930
500	0.2203	3472.0	7.5390
550	0.2352	3582.1	7.6770
600	0.2500	3693.4	7.8082
650	0.2647	3806.0	7.9335
700	0.2794	3920.0	8.0539
800	0.3086	4152.8	8.2815
900	0.3378	4392.0	8.4946
1000	0.3669	4637.6	8.6954
1100	0.3959	4889.3	8.8858
1200	0.4249	5147.0	9.0669
1300	0.4538	5410.2	9.2397

T	p = 1.8 MPa		
°C	v, m^3/kg	h, kJ/kg	s, kJ/kg·K
0.01	0.000 999	1.8	-0.0000
25	0.001 002	106.4	0.3665
50	0.001 011	210.9	0.7029
75	0.001 025	315.4	1.0144
100	0.001 043	420.3	1.3056
125	0.001 066	526.7	1.5731
150	0.001 090	633.1	1.8406
175	0.001 123	742.8	2.0852
200	0.001 156	852.5	2.3298
T_{sat}= f	0.001 168	884.70	2.3980
207.15 g	0.1104	2796.4	6.3781
225	0.1167	2846.0	6.4795
250	0.1249	2910.1	6.6052
275	0.1327	2970.4	6.7179
300	0.1402	3028.5	6.8214
325	0.1475	3085.1	6.9181
350	0.1546	3140.8	7.0093
375	0.1616	3196.0	7.0961
400	0.1685	3250.8	7.1792
450	0.1821	3360.2	7.3359
500	0.1955	3469.9	7.4825
550	0.2088	3580.3	7.6209
600	0.2220	3691.8	7.7524
650	0.2351	3804.6	7.8781
700	0.2482	3918.8	7.9986
800	0.2742	4151.9	8.2264
900	0.3002	4391.2	8.4397
1000	0.3261	4637.0	8.6406
1100	0.3519	4888.8	8.8311
1200	0.3777	5146.6	9.0122
1300	0.4034	5409.9	9.1851

T	p = 2.0 MPa		
°C	v, m^3/kg	h, kJ/kg	s, kJ/kg·K
0.01	0.000 999	2.0	-0.0000
25	0.001 002	106.6	0.3664
50	0.001 011	211.0	0.7028
75	0.001 025	315.5	1.0143
100	0.001 042	420.5	1.3054
125	0.001 066	526.9	1.5729
150	0.001 089	633.3	1.8404
175	0.001 123	742.9	2.0850
200	0.001 156	852.6	2.3296
T_{sat}= f	0.001 177	908.7	2.4471
212.42 g	0.0996	2798.7	6.3396
225	0.1037	2835.0	6.4133
250	0.1114	2901.6	6.5438
275	0.1186	2963.5	6.6595
300	0.1254	3022.7	6.7651
325	0.1321	3080.2	6.8633
350	0.1386	3136.6	6.9556
375	0.1449	3192.2	7.0432
400	0.1512	3247.5	7.1269
450	0.1635	3357.5	7.2845
500	0.1757	3467.7	7.4318
550	0.1877	3578.4	7.5706
600	0.1996	3690.2	7.7024
650	0.2114	3803.2	7.8283
700	0.2232	3917.6	7.9490
800	0.2467	4150.9	8.1771
900	0.2701	4390.5	8.3905
1000	0.2934	4636.4	8.5916
1100	0.3167	4888.4	8.7821
1200	0.3399	5146.2	8.9634
1300	0.3631	5409.6	9.1363

T	p = 2.2 MPa		
°C	v, m^3/kg	h, kJ/kg	s, kJ/kg·K
0.01	0.000 999	2.2	0.0000
25	0.001 002	106.8	0.3664
50	0.001 011	211.2	0.7027
75	0.001 025	315.7	1.0142
100	0.001 042	420.6	1.3053
125	0.001 066	527.0	1.5727
150	0.001 089	633.4	1.8402
175	0.001 123	743.0	2.0848
200	0.001 156	852.6	2.3294
T_{sat}= f	0.001 185	931.0	2.4924
217.29 g	0.0907	2800.5	6.3042
225	0.0931	2823.6	6.3509
250	0.1003	2892.8	6.4866
275	0.1070	2956.5	6.6056
300	0.1134	3016.9	6.7133
325	0.1195	3075.2	6.8130
350	0.1255	3132.3	6.9064
375	0.1313	3188.5	6.9949
400	0.1371	3244.2	7.0792
450	0.1483	3354.9	7.2378
500	0.1594	3465.5	7.3857
550	0.1704	3576.6	7.5249
600	0.1813	3688.6	7.6571
650	0.1921	3801.9	7.7832
700	0.2028	3916.5	7.9040
800	0.2242	4150.0	8.1325
900	0.2455	4389.8	8.3460
1000	0.2667	4635.8	8.5472
1100	0.2879	4887.9	8.7378
1200	0.3090	5145.8	8.9191
1300	0.3301	5409.2	9.0921

T	p = 2.5 MPa		
°C	v, m^3/kg	h, kJ/kg	s, kJ/kg·K
0.01	0.000 999	2.5	0.0000
25	0.001 002	107.1	0.3663
50	0.001 011	211.5	0.7026
75	0.001 025	315.9	1.0140
100	0.001 042	420.9	1.3050
125	0.001 066	527.2	1.5724
150	0.001 089	633.6	1.8398
175	0.001 122	743.2	2.0844
200	0.001 155	852.8	2.3290
T_{sat}= f	0.001 197	962.0	2.5544
223.99 g	0.0799	2802.2	6.2560
225	0.0802	2805.4	6.2625
250	0.0870	2879.1	6.4069
275	0.0931	2945.6	6.5312
300	0.0989	3008.0	6.6424
325	0.1044	3067.7	6.7444
350	0.1098	3125.8	6.8395
375	0.1150	3182.8	6.9293
400	0.1201	3239.2	7.0146
450	0.1301	3350.9	7.1746
500	0.1340	3462.2	7.3235
550	0.1497	3573.8	7.4634
600	0.1593	3686.3	7.5960
650	0.1688	3799.8	7.7225
700	0.1783	3914.7	7.8436
800	0.1972	4148.6	8.0724
900	0.2160	4388.7	8.2862
1000	0.2347	4634.9	8.4876
1100	0.2533	4887.1	8.6783
1200	0.2719	5145.2	8.8597
1300	0.2905	5408.7	9.0328

T	p = 3.0 MPa		
°C	v, m^3/kg	h, kJ/kg	s, kJ/kg·K
0.01	0.000 999	3.0	0.0000
25	0.001 002	107.5	0.3662
50	0.001 011	211.9	0.7024
75	0.001 024	316.4	1.0137
100	0.001 042	421.2	1.3046
125	0.001 065	527.6	1.5720
150	0.001 089	633.9	1.8393
175	0.001 122	743.4	2.0839
200	0.001 155	853.0	2.3285
225	0.001 205	969.1	2.5628
T_{sat}= f	0.001 217	1008.3	2.6454
233.89 g	0.0666	2803.3	6.1855
250	0.0706	2854.8	6.2857
275	0.0761	2926.7	6.4199
300	0.0811	2992.6	6.5375
325	0.0859	3054.8	6.6438
350	0.0905	3114.8	6.7420
375	0.0950	3173.2	6.8340
400	0.0994	3230.7	6.9210
450	0.1079	3344.1	7.0835
500	0.1162	3456.6	7.2339
550	0.1244	3569.1	7.3750
600	0.1324	3682.3	7.5084
650	0.1404	3796.4	7.6355
700	0.1484	3911.7	7.7571
800	0.1642	4146.3	7.9865
900	0.1799	4386.8	8.2008
1000	0.1955	4633.4	8.4024
1100	0.2110	4885.9	8.5934
1200	0.2266	5144.2	8.7749
1300	0.2421	5407.9	8.9481

Table A·6·3 Water: Compressed Liquid and Superheated Vapor (*Continued*)

T °C	p = 4.0 MPa v, m³/kg	h, kJ/kg	s, kJ/kg·K
0.01	0.000 998	4.0	0.0001
25	0.001 001	108.4	0.3659
50	0.001 010	212.8	0.7019
75	0.001 024	317.2	1.0130
100	0.001 041	422.0	1.3039
125	0.001 065	528.2	1.5710
150	0.001 088	634.5	1.8382
175	0.001 121	743.9	2.0825
200	0.001 154	853.4	2.3269
225	0.001 203	969.3	2.5603
250	0.001 252	1085.3	2.7937
T_{sat} = 250.39 *f*	0.001 252	1087.2	2.7962
g	0.0498	2800.6	6.0689
275	0.0546	2885.2	6.2269
300	0.0588	2959.7	6.3598
325	0.0627	3027.7	6.4760
350	0.0664	3091.8	6.5811
375	0.0700	3153.4	6.6780
400	0.0734	3213.4	6.7688
450	0.0800	3330.4	6.9364
500	0.0864	3445.4	7.0902
550	0.0927	3559.7	7.2335
600	0.0988	3674.3	7.3687
650	0.1049	3789.5	7.4970
700	0.1110	3905.7	7.6195
800	0.1229	4141.7	7.8503
900	0.1348	4383.1	8.0654
1000	0.1465	4630.4	8.2676
1100	0.1582	4883.5	8.4590
1200	0.1699	5142.2	8.6408
1300	0.1816	5406.2	8.8142

T °C	p = 5.0 MPa v, m³/kg	h, kJ/kg	s, kJ/kg·K
0.01	0.000 998	5.1	0.0002
25	0.001 001	109.4	0.3656
50	0.001 010	213.6	0.7014
75	0.001 024	318.0	1.0123
100	0.001 041	422.8	1.3031
125	0.001 064	528.9	1.5701
150	0.001 087	635.1	1.8371
175	0.001 120	744.5	2.0812
200	0.001 153	853.8	2.3253
225	0.001 201	969.5	2.5577
250	0.001 250	1085.3	2.7901
T_{sat} = 263.98 *f*	0.001 286	1154.2	2.9201
g	0.0394	2793.7	5.9725
275	0.0414	2837.5	6.0532
300	0.0453	2923.5	6.2067
325	0.0488	2998.7	6.3352
350	0.0519	3067.7	6.4482
375	0.0549	3132.9	6.5507
400	0.0578	3195.5	6.6456
450	0.0633	3316.3	6.8187
500	0.0686	3433.9	6.9760
550	0.0737	3550.2	7.1218
600	0.0787	3666.2	7.2586
650	0.0836	3782.6	7.3882
700	0.0885	3899.7	7.5117
800	0.0981	4137.0	7.7438
900	0.1077	4379.4	7.9598
1000	0.1171	4627.4	8.1626
1100	0.1266	4881.1	8.3543
1200	0.1359	5140.2	8.5365
1300	0.1453	5404.6	8.7101

T °C	p = 6.0 MPa v, m³/kg	h, kJ/kg	s, kJ/kg·K
0.01	0.000 997	6.1	0.0003
25	0.001 000	110.3	0.3654
50	0.001 010	214.5	0.7010
75	0.001 023	318.8	1.0118
100	0.001 040	423.5	1.3023
125	0.001 064	529.6	1.5691
150	0.001 087	635.7	1.8360
175	0.001 119	745.0	2.0799
200	0.001 152	854.2	2.3238
225	0.001 200	969.8	2.5558
250	0.001 248	1085.3	2.7878
275	0.001 339	1218.8	3.0351
T_{sat} = 275.62 *f*	0.001 319	1213.3	3.0266
g	0.0324	2783.9	5.8886
300	0.0362	2883.2	6.0659
325	0.0394	2967.4	6.2097
350	0.0422	3042.2	6.3322
375	0.0449	3111.4	6.4411
400	0.0474	3177.0	6.5404
450	0.0521	3301.9	6.7195
500	0.0566	3422.3	6.8805
550	0.0610	3540.6	7.0287
600	0.0652	3658.1	7.1673
650	0.0694	3775.6	7.2982
700	0.0735	3893.6	7.4227
800	0.0816	4132.3	7.6561
900	0.0896	4375.7	7.8730
1000	0.0976	4624.5	8.0764
1100	0.1054	4878.6	8.2686
1200	0.1133	5138.2	8.4510
1300	0.1211	5402.9	8.6249

T °C	p = 7.0 MPa v, m³/kg	h, kJ/kg	s, kJ/kg·K
0.01	0.000 997	7.1	0.0003
25	0.001 000	111.2	0.3651
50	0.001 009	215.4	0.7005
75	0.001 023	319.6	1.0112
100	0.001 040	424.3	1.3015
125	0.001 063	530.3	1.5682
150	0.001 086	636.4	1.8349
175	0.001 119	745.5	2.0786
200	0.001 151	854.6	2.3222
225	0.001 199	970.0	2.5538
250	0.001 246	1085.3	2.7854
275	0.001 335	1217.5	3.0307
T_{sat} = 285.86 *f*	0.001 352	1267.0	3.1211
g	0.027 37	2771.9	5.8130
300	0.029 46	2837.6	5.9293
325	0.032 56	2933.5	6.0931
350	0.035 23	3015.1	6.2269
375	0.037 66	3088.9	6.3430
400	0.039 93	3157.9	6.4474
450	0.044 16	3287.3	6.6329
500	0.048 13	3410.5	6.7978
550	0.051 94	3530.8	6.9486
600	0.055 65	3649.8	7.0889
650	0.059 27	3768.5	7.2211
700	0.062 84	3887.5	7.3466
800	0.069 85	4127.6	7.5815
900	0.076 75	4372.0	7.7992
1000	0.083 58	4621.5	8.0032
1100	0.090 35	4876.2	8.1958
1200	0.097 09	5136.2	8.3785
1300	0.103 81	5401.3	8.5526

T °C	p = 8.0 MPa v, m³/kg	h, kJ/kg	s, kJ/kg·K
0.01	0.000 996	8.1	0.0004
25	0.000 999	112.1	0.3649
50	0.001 009	216.2	0.7001
75	0.001 022	320.4	1.0106
100	0.001 039	425.0	1.3008
125	0.001 062	531.0	1.5673
150	0.001 085	637.0	1.8338
175	0.001 118	746.0	2.0772
200	0.001 150	855.1	2.3207
225	0.001 197	970.2	2.5519
250	0.001 244	1085.3	2.7830
275	0.001 331	1216.3	3.0264
T_{sat} = 295.04 *f*	0.001 384	1316.6	3.2066
g	0.023 52	2757.8	5.7431
300	0.024 26	2784.6	5.7901
325	0.027 38	2896.2	5.9809
350	0.029 95	2986.3	6.1286
375	0.032 22	3065.4	6.2530
400	0.034 31	3138.0	6.3630
450	0.038 17	3272.2	6.5554
500	0.041 74	3398.5	6.7243
550	0.045 15	3521.0	6.8778
600	0.048 45	3641.5	7.0200
650	0.051 66	3761.4	7.1535
700	0.054 82	3881.4	7.2800
800	0.061 01	4122.9	7.5163
900	0.067 09	4368.3	7.7349
1000	0.073 09	4618.5	7.9395
1100	0.079 04	4873.8	8.1325
1200	0.084 96	5134.2	8.3156
1300	0.090 84	5399.7	8.4899

T °C	p = 9.0 MPa v, m³/kg	h, kJ/kg	s, kJ/kg·K
0.01	0.000 996	9.1	0.0004
25	0.000 999	113.1	0.3646
50	0.001 008	217.1	0.6996
75	0.001 022	321.2	1.0100
100	0.001 039	425.8	1.3000
125	0.001 062	531.7	1.5664
150	0.001 085	637.6	1.8327
175	0.001 117	746.6	2.0759
200	0.001 149	855.5	2.3192
225	0.001 196	970.4	2.5499
250	0.001 243	1085.3	2.7807
275	0.001 325	1215.1	3.0195
300	0.001 407	1344.8	3.2583
T_{sat} = 303.34 *f*	0.001 418	1363.1	3.2855
g	0.020 48	2741.9	5.6772
325	0.023 26	2854.9	5.8696
350	0.025 79	2955.5	6.0345
375	0.027 96	3040.7	6.1687
400	0.029 93	3117.5	6.2848
450	0.033 50	3256.9	6.4847
500	0.036 77	3386.4	6.6579
550	0.039 86	3511.0	6.8141
600	0.042 84	3633.2	6.9583
650	0.045 74	3754.3	7.0932
700	0.048 58	3875.2	7.2207
800	0.054 13	4118.2	7.4584
900	0.059 57	4364.6	7.6779
1000	0.064 93	4615.5	7.8831
1100	0.070 24	4871.4	8.0766
1200	0.075 52	5132.2	8.2599
1300	0.080 76	5398.1	8.4345

Table A·6·3 Water: Compressed Liquid and Superheated Vapor (*Continued*)

T °C		p = 10.0 MPa v, m³/kg	h, kJ/kg	s, kJ/kg·K
0.01		0.000 995	10.1	0.0005
25		0.000 999	114.0	0.3643
50		0.001 008	217.9	0.6991
75		0.001 021	322.0	1.0093
100		0.001 038	426.5	1.2992
125		0.001 061	532.4	1.5654
150		0.001 084	638.3	1.8316
175		0.001 116	747.1	2.0746
200		0.001 148	855.9	2.3177
225		0.001 194	970.6	2.5480
250		0.001 241	1085.4	2.7784
275		0.001 319	1213.9	3.0127
300		0.001 397	1342.4	3.2470
T_{sat}=	f	0.001 452	1407.3	3.3591
311.03	g	0.018 02	2724.5	5.6139
325		0.019 86	2808.1	5.7555
350		0.022 42	2922.2	5.9425
375		0.024 53	3014.7	6.0882
400		0.026 41	3096.1	6.2114
450		0.029 75	3241.1	6.4193
500		0.032 78	3374.0	6.5971
550		0.035 63	3500.9	6.7561
600		0.038 36	3624.7	6.9022
650		0.041 01	3747.1	7.0385
700		0.043 59	3869.0	7.1671
800		0.048 63	4113.5	7.4062
900		0.053 56	4360.9	7.6266
1000		0.058 40	4612.6	7.8324
1100		0.063 20	4869.0	8.0263
1200		0.067 97	5130.3	8.2100
1300		0.072 70	5396.4	8.3847

T °C		p = 12.5 MPa v, m³/kg	h, kJ/kg	s, kJ/kg·K
0.01		0.000 994	12.6	0.0005
25		0.000 997	116.3	0.3636
50		0.001 007	220.1	0.6980
75		0.001 020	324.0	1.0078
100		0.001 037	428.4	1.2974
125		0.001 060	534.1	1.5632
150		0.001 083	639.8	1.8290
175		0.001 114	748.4	2.0714
200		0.001 146	857.0	2.3139
225		0.001 191	971.3	2.5434
250		0.001 237	1085.5	2.7728
275		0.001 312	1212.7	3.0047
300		0.001 388	1339.9	3.2365
325		0.001 549	1500.2	3.4842
T_{sat}=	f	0.001 546	1510.9	3.5276
327.85	g	0.013 49	2673.4	5.4619
350		0.016 12	2824.8	5.7097
375		0.018 25	2942.9	5.8957
400		0.020 01	3038.8	6.0409
450		0.022 99	3200.2	6.2724
500		0.025 60	3342.2	6.4625
550		0.028 01	3475.2	6.6291
600		0.030 29	3603.3	6.7802
650		0.032 48	3728.9	6.9202
700		0.034 61	3853.4	7.0514
800		0.038 73	4101.5	7.2942
900		0.042 73	4351.6	7.5169
1000		0.046 66	4605.1	7.7243
1100		0.050 54	4863.0	7.9192
1200		0.054 38	5125.4	8.1037
1300		0.058 19	5392.4	8.2791

T °C		p = 15.0 MPa v, m³/kg	h, kJ/kg	s, kJ/kg·K
0.01		0.000 993	15.1	0.0006
25		0.000 996	118.6	0.3630
50		0.001 006	222.2	0.6968
75		0.001 019	326.0	1.0063
100		0.001 036	430.3	1.2955
125		0.001 059	535.8	1.5609
150		0.001 081	641.4	1.8263
175		0.001 112	749.8	2.0683
200		0.001 143	858.1	2.3102
225		0.001 188	971.9	2.5387
250		0.001 233	1085.7	2.7672
275		0.001 305	1211.6	2.9967
300		0.001 378	1337.5	3.2261
325		0.001 536	1496.5	3.4765
T_{sat}=	f	0.001 657	1609.7	3.6836
342.19	g	0.010 34	2610.1	5.3093
350		0.011 47	2691.3	5.4404
375		0.013 89	2857.8	5.7029
400		0.015 65	2974.7	5.8799
450		0.018 45	3156.6	6.1410
500		0.020 80	3309.3	6.3452
550		0.022 92	3448.8	6.5201
600		0.024 90	3581.5	6.6767
650		0.026 79	3710.5	6.8204
700		0.028 62	3837.6	6.9544
800		0.032 13	4089.6	7.2009
900		0.035 52	4342.2	7.4260
1000		0.038 83	4597.7	7.6350
1100		0.042 09	4857.0	7.8310
1200		0.045 32	5120.5	8.0163
1300		0.048 51	5388.4	8.1922

T °C		p = 17.5 MPa v, m³/kg	h, kJ/kg	s, kJ/kg·K
0.01		0.000 992	17.6	0.0006
25		0.000 995	120.8	0.3623
50		0.001 005	224.4	0.6957
75		0.001 018	328.1	1.0048
100		0.001 035	432.2	1.2936
125		0.001 057	537.6	1.5586
150		0.001 079	643.0	1.8237
175		0.001 110	751.1	2.0652
200		0.001 141	859.3	2.3066
225		0.001 185	972.6	2.5342
250		0.001 229	1086.0	2.7619
275		0.001 299	1210.7	2.9884
300		0.001 369	1335.5	3.2150
325		0.001 524	1493.0	3.4683
350		0.001 680	1650.5	3.7217
T_{sat}=	f	0.001 806	1711.2	3.8397
354.67	g	0.007 94	2527.9	5.1409
375		0.010 55	2751.5	5.4919
400		0.012 45	2901.9	5.7198
450		0.015 18	3110.3	6.0191
500		0.017 36	3275.0	6.2395
550		0.019 28	3421.7	6.4234
600		0.021 06	3559.3	6.5857
650		0.022 73	3691.8	6.7334
700		0.024 34	3821.6	6.8703
800		0.027 42	4077.5	7.1206
900		0.030 37	4332.9	7.3481
1000		0.033 24	4590.3	7.5587
1100		0.036 07	4851.0	7.7558
1200		0.038 85	5115.7	7.9418
1300		0.041 60	5384.5	8.1184

T °C		p = 20.0 MPa v, m³/kg	h, kJ/kg	s, kJ/kg·K
0.01		0.000 990	20.1	0.0007
25		0.000 994	123.1	0.3616
50		0.001 003	226.5	0.6946
75		0.001 017	330.1	1.0032
100		0.001 034	434.1	1.2917
125		0.001 056	539.3	1.5564
150		0.001 078	644.6	1.8211
175		0.001 108	752.5	2.0621
200		0.001 139	860.4	2.3030
225		0.001 182	973.4	2.5298
250		0.001 225	1086.3	2.7565
275		0.001 293	1209.9	2.9802
300		0.001 361	1333.4	3.2038
325		0.001 513	1489.4	3.4602
350		0.001 665	1645.4	3.7165
T_{sat}=	f	0.002 036	1826.8	4.0147
365.80	g	0.005 87	2413.3	4.9326
375		0.007 67	2601.0	5.2246
400		0.009 95	2816.9	5.5521
450		0.012 70	3060.8	5.9026
500		0.014 77	3239.4	6.1417
550		0.016 55	3393.9	6.3355
600		0.018 17	3536.7	6.5039
650		0.019 69	3672.9	6.6556
700		0.021 13	3805.5	6.7955
800		0.023 88	4065.4	7.0498
900		0.026 51	4323.5	7.2797
1000		0.029 05	4582.8	7.4919
1100		0.031 54	4845.1	7.6901
1200		0.034 00	5110.9	7.8769
1300		0.036 42	5380.6	8.0541

T °C	p = 25.0 MPa v, m³/kg	h, kJ/kg	s, kJ/kg·K
0.01	0.000 988	25.0	0.0006
25	0.000 992	127.7	0.3602
50	0.001 001	230.8	0.6923
75	0.001 015	334.1	1.0002
100	0.001 031	437.9	1.2880
125	0.001 053	542.8	1.5520
150	0.001 075	647.8	1.8159
175	0.001 105	755.3	2.0559
200	0.001 135	862.7	2.2959
225	0.001 176	974.9	2.5211
250	0.001 218	1087.0	2.7463
275	0.001 282	1208.7	2.9683
300	0.001 345	1330.4	3.1902
325	0.001 472	1476.8	3.3297
350	0.001 598	1623.1	3.4692
375	0.001 979	1849.3	4.0340
400	0.006 001	2578.1	5.1388
450	0.009 167	2950.6	5.6755
500	0.011 123	3164.2	5.9616
550	0.012 716	3336.5	6.1778
600	0.014 126	3490.4	6.3593
650	0.015 423	3634.5	6.5198
700	0.016 644	3773.0	6.6659
800	0.018 938	4041.1	6.9282
900	0.021 104	4304.7	7.1631
1000	0.023 188	4568.0	7.3785
1100	0.025 218	4833.2	7.5790
1200	0.027 209	5101.3	7.7674
1300	0.029 173	5372.8	7.9458

Table A·6·3 Water: Compressed Liquid and Superheated Vapor (*Continued*)

T	*p* = 30.0 MPa			*T*	*p* = 35.0 MPa			*T*	*p* = 40.0 MPa		
°C	*v*, m³/kg	*h*, kJ/kg	*s*, kJ/kg·K	°C	*v*, m³/kg	*h*, kJ/kg	*s*, kJ/kg·K	°C	*v*, m³/kg	*h*, kJ/kg	*s*, kJ/kg·K
0.01	0.000 986	29.9	0.0005	**0.01**	0.000 983	34.8	0.0003	**0.01**	0.000 981	39.6	0.0003
25	0.000 990	132.2	0.3587	**25**	0.000 988	136.7	0.3573	**25**	0.000 986	141.2	0.3558
50	0.000 999	235.1	0.6900	**50**	0.000 997	239.3	0.6877	**50**	0.000 995	243.6	0.6855
75	0.001 012	338.1	0.9973	**75**	0.001 010	342.2	0.9943	**75**	0.001 008	346.2	0.9914
100	0.001 029	441.6	1.2844	**100**	0.001 027	445.4	1.2808	**100**	0.001 024	449.2	1.2772
125	0.001 050	546.3	1.5476	**125**	0.001 048	549.9	1.5433	**125**	0.001 045	553.4	1.5391
150	0.001 072	651.0	1.8109	**150**	0.001 069	654.3	1.8059	**150**	0.001 066	657.6	1.8010
175	0.001 101	758.1	2.0500	**175**	0.001 098	761.0	2.0441	**175**	0.001 094	763.9	2.0384
200	0.001 130	865.2	2.2890	**200**	0.001 126	867.6	2.2823	**200**	0.001 122	870.1	2.2758
225	0.001 171	976.6	2.5128	**225**	0.001 165	978.4	2.5047	**225**	0.001 160	980.2	2.4969
250	0.001 211	1088.0	2.7365	**250**	0.001 205	1089.1	2.7270	**250**	0.001 198	1090.3	2.7179
275	0.001 271	1208.0	2.9555	**275**	0.001 262	1207.6	2.9433	**275**	0.001 253	1207.6	2.9318
300	0.001 332	1328.0	3.1744	**300**	0.001 319	1326.1	3.1596	**300**	0.001 308	1324.8	3.1457
325	0.001 442	1468.1	3.4083	**325**	0.001 418	1461.5	3.3857	**325**	0.001 398	1456.5	3.3657
350	0.001 552	1608.1	3.6421	**350**	0.001 517	1596.8	3.6118	**350**	0.001 488	1588.1	3.5857
375	0.001 873	1791.4	3.9303	**375**	0.001 701	1761.9	3.8714	**375**	0.001 641	1742.2	3.8280
400	0.002 793	2150.7	4.4723	**400**	0.002 106	1988.3	4.2136	**400**	0.001 910	1930.8	4.1134
450	0.006 736	2822.3	5.4435	**450**	0.004 960	2672.9	5.1969	**450**	0.003 691	2513.2	4.9463
500	0.008 676	3083.5	5.7936	**500**	0.006 924	2997.3	5.6320	**500**	0.005 619	2906.7	5.4745
550	0.010 158	3276.8	6.0362	**550**	0.008 333	3215.0	5.9052	**550**	0.006 973	3151.6	5.7819
600	0.011 431	3443.1	6.2324	**600**	0.009 511	3394.7	6.1174	**600**	0.008 077	3345.8	6.0111
650	0.012 583	3595.5	6.4022	**650**	0.010 558	3556.0	6.2971	**650**	0.009 046	3516.3	6.2011
700	0.013 655	3740.1	6.5547	**700**	0.011 523	3706.9	6.4563	**700**	0.009 930	3673.8	6.3673
800	0.015 645	4016.7	6.8254	**800**	0.013 295	3992.2	6.7355	**800**	0.011 536	3967.8	6.6551
900	0.017 504	4285.9	7.0653	**900**	0.014 935	4267.2	6.9805	**900**	0.013 010	4248.5	6.9052
1000	0.019 281	4553.3	7.2840	**1000**	0.016 491	4538.6	7.2025	**1000**	0.014 401	4523.9	7.1305
1100	0.021 002	4821.4	7.4867	**1100**	0.017 993	4809.7	7.4075	**1100**	0.015 737	4798.1	7.3378
1200	0.022 686	5091.8	7.6768	**1200**	0.019 456	5082.4	7.5992	**1200**	0.017 035	5073.1	7.5311
1300	0.024 341	5365.2	7.8563	**1300**	0.020 891	5357.6	7.7799	**1300**	0.018 304	5350.1	7.7130

T	*p* = 50.0 MPa			*T*	*p* = 60.0 MPa			*T*	*p* = 70.0 MPa		
°C	*v*, m³/kg	*h*, kJ/kg	*s*, kJ/kg·K	°C	*v*, m³/kg	*h*, kJ/kg	*s*, kJ/kg·K	°C	*v*, m³/kg	*h*, kJ/kg	*s*, kJ/kg·K
0.01	0.000 977	49.2	-0.0008	**0.01**	0.000 972	58.7	-0.0018	**0.01**	0.000 968	68.0	-0.0032
25	0.000 982	150.2	0.3528	**25**	0.000 978	159.0	0.3498	**25**	0.000 974	167.9	0.3466
50	0.000 991	252.0	0.6810	**50**	0.000 988	260.5	0.6765	**50**	0.000 984	268.9	0.6720
75	0.001 004	354.2	0.9856	**75**	0.001 000	362.3	0.9799	**75**	0.000 997	370.3	0.9742
100	0.001 020	456.8	1.2702	**100**	0.001 016	464.5	1.2633	**100**	0.001 002	472.1	1.2566
125	0.001 040	560.5	1.5309	**125**	0.001 036	567.6	1.5228	**125**	0.001 026	574.8	1.5150
150	0.001 061	664.2	1.7915	**150**	0.001 056	670.8	1.7823	**150**	0.001 051	677.6	1.7733
175	0.001 088	769.8	2.0273	**175**	0.001 082	775.8	2.0167	**175**	0.001 076	781.9	2.0064
200	0.001 115	875.3	2.2631	**200**	0.001 108	880.7	2.2510	**200**	0.001 101	886.2	2.2394
225	0.001 151	984.3	2.4819	**225**	0.001 142	988.6	2.4677	**225**	0.001 134	993.2	2.4542
250	0.001 187	1093.2	2.7006	**250**	0.001 176	1096.5	2.6843	**250**	0.001 166	1100.1	2.6689
275	0.001 237	1208.2	2.9104	**275**	0.001 223	1209.6	2.8907	**275**	0.001 210	1211.6	2.8723
300	0.001 288	1323.1	3.1202	**300**	0.001 270	1322.6	3.0970	**300**	0.001 254	1323.0	3.0757
325	0.001 365	1449.2	3.3310	**325**	0.001 338	1394.7	3.3011	**325**	0.001 316	1441.9	3.2745
350	0.001 442	1575.3	3.5417	**350**	0.001 407	1466.7	3.5051	**350**	0.001 377	1560.8	3.4733
375	0.001 559	1716.1	3.7632	**375**	0.001 504	1699.7	3.7139	**375**	0.001 461	1688.2	3.6735
400	0.001 730	1874.1	4.0022	**400**	0.001 633	1843.0	3.9312	**400**	0.001 567	1822.8	3.8774
450	0.002 486	2284.7	4.5894	**450**	0.002 084	2180.0	4.4134	**450**	0.001 892	2123.6	4.3081
500	0.003 892	2724.2	5.1780	**500**	0.002 955	2571.9	4.9373	**500**	0.002 464	2466.9	4.7669
550	0.005 111	3023.3	5.5535	**550**	0.003 955	2901.0	5.3505	**550**	0.003 227	2795.0	5.1784
600	0.006 098	3247.7	5.8184	**600**	0.004 826	3152.3	5.6471	**600**	0.003 972	3063.8	5.4957
650	0.006 949	3437.2	6.0296	**650**	0.005 582	3359.9	5.8786	**650**	0.004 642	3286.8	5.7442
700	0.007 713	3607.8	6.2097	**700**	0.006 259	3543.5	6.0723	**700**	0.005 245	3481.9	5.9502
800	0.009 083	3919.5	6.5148	**800**	0.007 461	3872.3	6.3942	**800**	0.006 317	3826.7	6.2878
900	0.010 322	4211.5	6.7751	**900**	0.008 538	4175.4	6.6644	**900**	0.007 272	4140.3	6.5674
1000	0.011 479	4495.0	7.0070	**1000**	0.009 536	4466.7	6.9027	**1000**	0.008 153	4439.1	6.8118
1100	0.012 582	4775.2	7.2188	**1100**	0.010 482	4752.7	7.1189	**1100**	0.008 985	4730.7	7.0324
1200	0.013 647	5054.8	7.4154	**1200**	0.011 391	5036.7	7.3186	**1200**	0.009 781	5019.1	7.2351
1300	0.014 684	5335.3	7.5997	**1300**	0.012 273	5320.9	7.5053	**1300**	0.010 551	5306.8	7.4240

Table A·6·4 Water: Solid–vapor Saturation

T °C	p kPa	v_i m^3/kg	v_g m^3/kg	h_i kJ/kg	h_{ig} kJ/kg	h_g kJ/kg	s_i kJ/kg·K	s_{ig} kJ/kg·K	s_g kJ/kg·K	T °C
0.01	0.6117	0.001 0908	206.0	−334.30	2834.8	2500.5	−1.223	10.378	9.154	**0.01**
0	0.6112	0.001 0908	206.2	−334.33	2834.8	2500.4	−1.223	10.378	9.155	**0**
−2	0.5180	0.001 0904	241.6	−338.52	2835.3	2496.8	−1.239	10.456	9.217	**−2**
−4	0.4379	0.001 0901	283.7	−342.68	2835.7	2493.1	−1.255	10.536	9.281	**−4**
−6	0.3693	0.001 0898	334.1	−346.81	2836.2	2489.4	−1.270	10.616	9.346	**−6**
−8	0.3106	0.001 0894	394.3	−350.92	2836.6	2485.7	−1.286	10.698	9.412	**−8**
−10	0.2606	0.001 0891	466.6	−354.99	2837.0	2482.0	−1.301	10.781	9.479	**−10**
−12	0.2180	0.001 0888	553.6	−359.04	2837.3	2478.3	−1.317	10.865	9.548	**−12**
−14	0.1819	0.001 0884	658.7	−363.05	2837.6	2474.6	−1.333	10.950	9.617	**−14**
−16	0.1514	0.001 0881	785.9	−367.04	2837.9	2470.9	−1.348	11.036	9.688	**−16**
−18	0.1256	0.001 0878	940.4	−371.00	2838.2	2467.2	−1.364	11.123	9.760	**−18**
−20	0.1039	0.001 0874	1128.5	−374.93	2838.4	2463.4	−1.379	11.212	9.833	**−20**
−22	0.0857	0.001 0871	1358.3	−378.83	2838.6	2459.7	−1.395	11.302	9.907	**−22**
−24	0.0705	0.001 0868	1640.0	−382.70	2838.7	2456.0	−1.410	11.394	9.983	**−24**
−26	0.0578	0.001 0864	1986.3	−386.54	2838.9	2452.3	−1.426	11.486	10.060	**−26**
−28	0.0473	0.001 0861	2413.6	−390.35	2839.0	2448.6	−1.441	11.580	10.139	**−28**
−30	0.0385	0.001 0858	2942.9	−394.13	2839.0	2444.9	−1.457	11.676	10.219	**−30**
−32	0.0313	0.001 0854	3599.9	−397.88	2839.1	2441.2	−1.473	11.773	10.301	**−32**
−34	0.0254	0.001 0851	4418.9	−401.61	2839.1	2437.5	−1.488	11.872	10.384	**−34**
−36	0.0205	0.001 0848	5443.9	−405.30	2839.1	2433.8	−1.503	11.972	10.468	**−36**
−38	0.0165	0.001 0844	6730.9	−408.96	2839.0	2430.0	−1.519	12.073	10.554	**−38**
−40	0.0133	0.001 0841	8353.9	−412.60	2838.9	2426.3	−1.534	12.176	10.642	**−40**

Source: Based on a combination of data from Haar, Lester, John S. Gallagher, and George S Kell, *NBS/NRC Steam Tables,* Hemisphere, 1984, and Keenan, Joseph H., Frederick G. Keyes, Philip G. Hill, and Joan G. Moore, *Steam Tables,* SI Units, Wiley, New York, 1978.

Chart A·7·1 Water: *Ts Diagram*

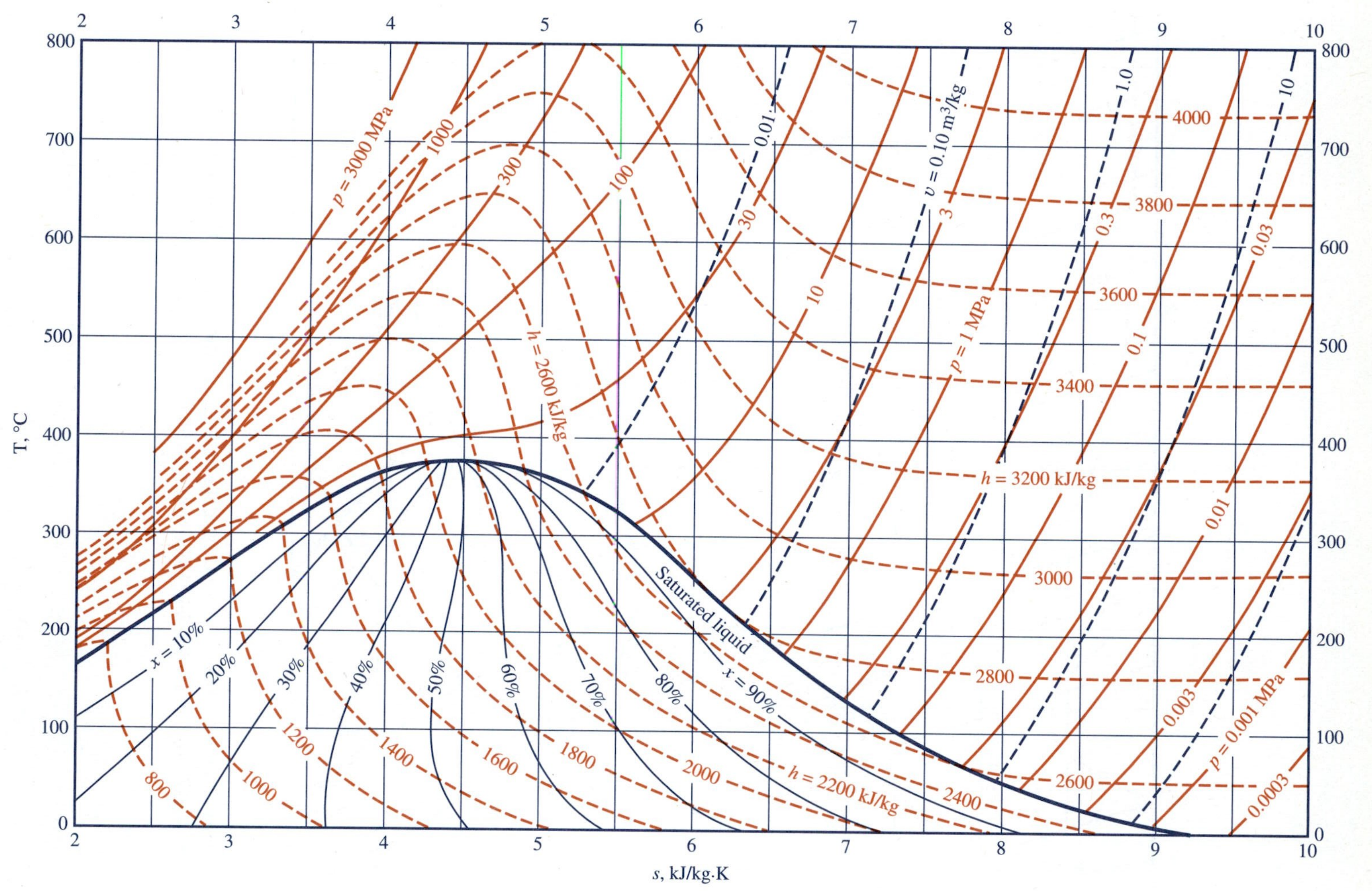

Chart A·7·1 Water *Ts* diagram. (*Source:* Based on data and formulations from Haar, Lester, John S. Gallagher, and George S. Kell, *NBS/NRC Steam Tables*, Hemisphere, 1984.)

H_2O, *Ts* diagram

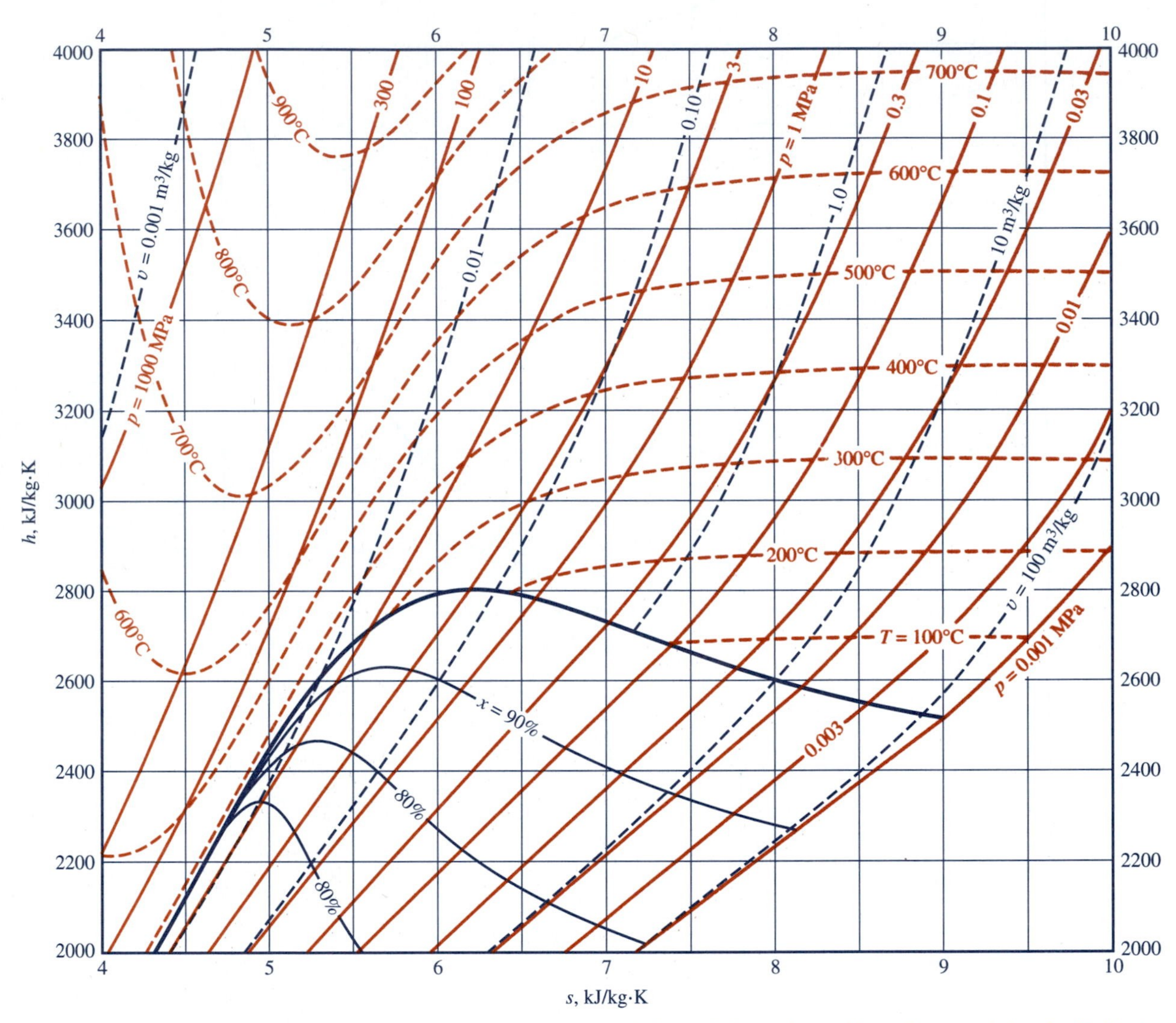

Chart A·7·2 Water *hs* diagram. (*Source:* Based on data and formulations from Haar, Lester, John S. Gallagher, and George S. Kell, *NBS/NRC Steam Tables,* Hemisphere, 1984.)

H_2O, *hs* diagram

Chart A·7·3 Water: *ph* Diagram

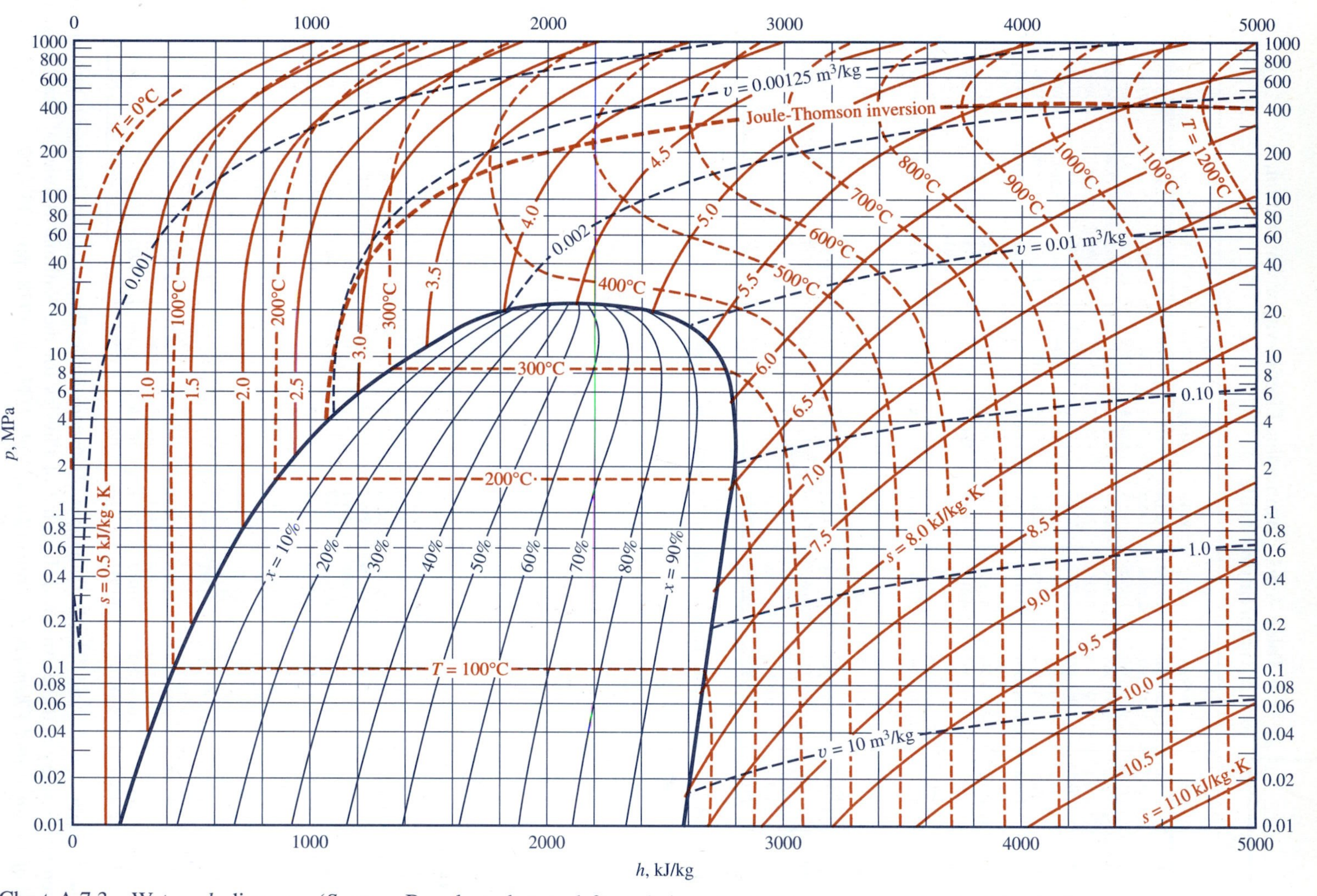

Chart A·7·3 Water *ph* diagram. (*Source:* Based on data and formulations from Haar, Lester, John S. Gallagher, and George S. Kell, *NBS/NRC Steam Tables,* Hemisphere, 1984.)

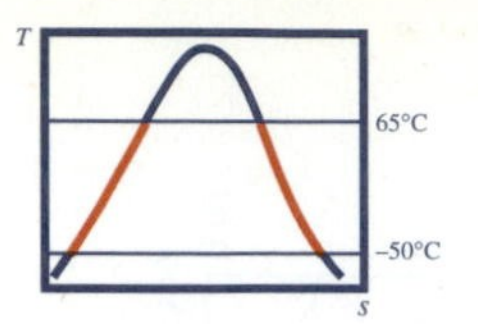

Table A·8·1 Ammonia: Liquid–vapor Saturation, Temperature Table

T	p	v_f	v_g	h_f	h_{fg}	h_g	s_f	s_{fg}	s_g
°C	kPa	m³/kg	m³/kg	kJ/kg	kJ/kg	kJ/kg	kJ/kg·K	kJ/kg·K	kJ/kg·K
-50	40.9	0.001 424	2.6270	123.4	1416.3	1539.7	0.5908	6.3468	6.9376
-45	54.5	0.001 437	2.0064	145.2	1402.7	1547.9	0.6873	6.1483	6.8356
-40	71.7	0.001 450	1.5526	167.1	1388.8	1555.9	0.7824	5.9566	6.7390
-35	93.1	0.001 463	1.2161	189.2	1374.5	1563.7	0.8758	5.7714	6.6472
-30	119.5	0.001 476	0.9634	211.4	1359.7	1571.1	0.9680	5.5920	6.5600
-25	151.5	0.001 490	0.7712	233.7	1344.6	1578.3	1.0586	5.4184	6.4770
-20	190.2	0.001 504	0.6233	256.2	1329.0	1585.1	1.1481	5.2496	6.3977
-15	236.3	0.001 519	0.5084	278.8	1312.9	1591.7	1.2362	5.0858	6.3220
-10	290.8	0.001 534	0.4181	301.6	1296.4	1597.9	1.3232	4.9263	6.2495
-5	354.9	0.001 550	0.3465	324.4	1279.4	1603.8	1.4090	4.7710	6.1800
0	429.5	0.001 566	0.2892	347.5	1261.8	1609.3	1.4938	4.6194	6.1132
5	515.9	0.001 583	0.2430	370.7	1243.7	1614.4	1.5775	4.4713	6.0488
10	615.2	0.001 600	0.2054	394.1	1225.0	1619.1	1.6603	4.3264	5.9867
15	728.6	0.001 619	0.1746	417.7	1205.8	1623.4	1.7421	4.1845	5.9266
20	857.5	0.001 638	0.1492	441.4	1185.8	1627.3	1.8232	4.0451	5.8683
25	1003.1	0.001 658	0.1281	465.4	1165.2	1630.6	1.9034	3.9081	5.8115
30	1166.9	0.001 680	0.1105	489.6	1143.9	1633.4	1.9828	3.7733	5.7561
35	1350.3	0.001 702	0.0957	513.9	1121.7	1635.7	2.0616	3.6402	5.7018
40	1554.8	0.001 725	0.0831	538.6	1098.8	1637.3	2.1397	3.5088	5.6485
45	1781.9	0.001 750	0.0725	563.4	1074.9	1638.3	2.2173	3.3785	5.5958
50	2033.0	0.001 777	0.0634	588.6	1050.0	1638.6	2.2944	3.2493	5.5437
55	2309.9	0.001 804	0.0556	614.1	1024.1	1638.1	2.3711	3.1207	5.4918
60	2614.2	0.001 834	0.0488	639.9	996.9	1636.8	2.4476	2.9924	5.4400
65	2947.6	0.001 866	0.0430	669.1	965.5	1634.6	2.5239	2.8641	5.3880

Table A·8·2 Ammonia: Liquid–vapor Saturation, Pressure Table

p	T	v_f	v_g	h_f	h_{fg}	h_g	s_f	s_{fg}	s_g
kPa	°C	m³/kg	m³/kg	kJ/kg	kJ/kg	kJ/kg	kJ/kg·K	kJ/kg·K	kJ/kg·K
50	-46.53	0.001 433	2.1746	138.5	1406.9	1545.4	0.6580	6.2081	6.8662
75	-39.15	0.001 452	1.4884	170.9	1386.4	1557.3	0.7983	5.9249	6.7232
100	-33.59	0.001 466	1.1375	195.4	1370.3	1565.8	0.9019	5.7203	6.6222
125	-29.07	0.001 479	0.9235	215.6	1356.9	1572.5	0.9850	5.5593	6.5443
150	-25.21	0.001 489	0.7784	232.8	1345.2	1578.0	1.0548	5.4257	6.4805
175	-21.86	0.001 499	0.6738	247.8	1334.8	1582.6	1.1150	5.3117	6.4267
200	-18.86	0.001 507	0.5946	261.3	1325.3	1586.7	1.1683	5.2119	6.3802
225	-16.15	0.001 515	0.5323	273.6	1316.6	1590.2	1.2161	5.1230	6.3391
250	-13.66	0.001 523	0.4821	284.9	1308.5	1593.4	1.2596	5.0427	6.3023
275	-11.37	0.001 530	0.4407	295.3	1300.9	1596.2	1.2996	4.9695	6.2690
300	-9.23	0.001 536	0.4060	305.1	1293.8	1598.8	1.3365	4.9022	6.2387
350	-5.36	0.001 548	0.3511	322.8	1280.6	1603.4	1.4029	4.7819	6.1849
400	-1.89	0.001 560	0.3094	338.8	1268.5	1607.3	1.4619	4.6763	6.1382
450	1.25	0.001 570	0.2767	353.3	1257.3	1610.6	1.5149	4.5820	6.0969
500	4.13	0.001 580	0.2503	366.7	1246.9	1613.6	1.5631	4.4968	6.0598
550	6.79	0.001 589	0.2286	379.1	1237.1	1616.2	1.6074	4.4189	6.0263
600	9.28	0.001 598	0.2104	390.7	1227.8	1618.5	1.6484	4.3472	5.9956
650	11.61	0.001 606	0.1949	401.7	1218.9	1620.6	1.6867	4.2805	5.9672
700	13.80	0.001 614	0.1815	412.0	1210.4	1622.4	1.7225	4.2183	5.9409
750	15.87	0.001 622	0.1698	421.8	1202.3	1624.1	1.7564	4.1599	5.9163
800	17.85	0.001 630	0.1596	431.2	1194.5	1625.7	1.7884	4.1048	5.8932
900	21.52	0.001 644	0.1424	448.7	1179.6	1628.3	1.8477	4.0031	5.8509
1000	24.90	0.001 658	0.1285	464.9	1165.6	1630.5	1.9018	3.9108	5.8126
1100	28.03	0.001 671	0.1171	480.0	1152.4	1632.4	1.9516	3.8263	5.7778
1200	30.94	0.001 684	0.1075	494.1	1139.7	1633.9	1.9978	3.7480	5.7458
1400	36.27	0.001 708	0.0923	520.2	1116.0	1636.2	2.0814	3.6068	5.6883
1600	41.04	0.001 730	0.0808	543.7	1093.9	1637.6	2.1559	3.4817	5.6375
1800	45.38	0.001 752	0.0717	565.3	1073.0	1638.4	2.2232	3.3687	5.5919
2000	49.37	0.001 773	0.0644	585.4	1053.2	1638.6	2.2847	3.2655	5.5503

Source: Based on Haar, Lester, and John S. Gallagher, *Thermodynamic Properties of Ammonia,* published by the American Chemical Society and the American Institute of Physics for the National Bureau of Standards, 1978. Also, *Journal of Physical and Chemical Reference Data,* Vol. 7, No. 3, 1978, pp. 635–792.

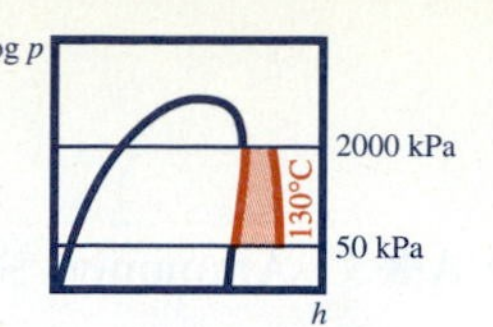

Table A·8·3 Ammonia: Superheated Vapor

T °C		p = 50 kPa v, m³/kg	h, kJ/kg	s, kJ/kg·K
T_{sat}=	f	0.001 433	138.5	0.6580
−46.53	g	2.1746	1545.4	6.8662
−20		2.4459	1601.5	7.1024
−10		2.5457	1622.5	7.1863
0		2.6454	1643.6	7.2671
10		2.7448	1664.8	7.3450
20		2.8441	1686.1	7.4202
30		2.9433	1707.5	7.4928
40		3.0422	1729.1	7.5629
50		3.1410	1750.7	7.6307
60		3.2396	1772.4	7.6964
70		3.3380	1794.3	7.7601
80		3.4363	1816.2	7.8220
90		3.5344	1838.3	7.8823
100		3.6323	1860.5	7.9410
110		3.7300	1882.7	7.9984
120		3.8276	1905.1	8.0546
130		3.9250	1927.6	8.1098

T °C		p = 75 kPa v, m³/kg	h, kJ/kg	s, kJ/kg·K
T_{sat}=	f	0.001 452	170.9	0.7983
−39.15	g	1.4884	1557.3	6.7232
−20		1.6217	1598.3	6.8939
−10		1.6896	1619.6	6.9792
0		1.7573	1641.0	7.0616
10		1.8248	1662.5	7.1412
20		1.8920	1684.1	7.2180
30		1.9591	1705.8	7.2921
40		2.0259	1727.5	7.3637
50		2.0925	1749.4	7.4328
60		2.1589	1771.3	7.4995
70		2.2251	1793.3	7.5638
80		2.2910	1815.4	7.6260
90		2.3568	1837.5	7.6860
100		2.4223	1859.8	7.7440
110		2.4876	1882.1	7.8000
120		2.5527	1904.5	7.8542
130		2.6176	1927.0	7.9067

T °C		p = 100 kPa v, m³/kg	h, kJ/kg	s, kJ/kg·K
T_{sat}=	f	0.001 466	195.4	0.9019
−33.59	g	1.1375	1565.8	6.6222
−20		1.2103	1595.2	6.7435
−10		1.2618	1616.9	6.8297
0		1.3131	1638.6	6.9131
10		1.3642	1660.3	6.9936
20		1.4151	1682.2	7.0714
30		1.4658	1704.1	7.1466
40		1.5163	1726.0	7.2191
50		1.5666	1748.1	7.2890
60		1.6167	1770.1	7.3564
70		1.6666	1792.3	7.4214
80		1.7162	1814.5	7.4840
90		1.7657	1836.7	7.5442
100		1.8150	1859.1	7.6022
110		1.8640	1881.5	7.6579
120		1.9129	1903.9	7.7115
130		1.9616	1926.4	7.7630

T °C		p = 125 kPa v, m³/kg	h, kJ/kg	s, kJ/kg·K
T_{sat}=	f	0.001 479	215.6	0.9850
−29.07	g	0.9235	1572.5	6.5443
−20		0.9635	1592.2	6.6255
−10		1.0051	1614.2	6.7123
0		1.0465	1636.2	6.7963
10		1.0877	1658.2	6.8775
20		1.1287	1680.3	6.9560
30		1.1696	1702.4	7.0318
40		1.2102	1724.6	7.1050
50		1.2507	1746.8	7.1756
60		1.2909	1769.0	7.2436
70		1.3310	1791.3	7.3091
80		1.3709	1813.6	7.3722
90		1.4106	1836.0	7.4328
100		1.4501	1858.4	7.4910
110		1.4895	1880.8	7.5469
120		1.5286	1903.3	7.6005
130		1.5676	1925.8	7.6519

T °C		p = 150 kPa v, m³/kg	h, kJ/kg	s, kJ/kg·K
T_{sat}=	f	0.001 489	232.8	1.0548
−25.21	g	0.7784	1578.0	6.4805
−20		0.7989	1589.3	6.5282
−10		0.8339	1611.6	6.6154
0		0.8687	1633.8	6.6999
10		0.9033	1656.1	6.7815
20		0.9377	1678.4	6.8605
30		0.9720	1700.7	6.9369
40		1.0060	1723.1	7.0106
50		1.0399	1745.5	7.0817
60		1.0737	1767.9	7.1502
70		1.1072	1790.3	7.2162
80		1.1406	1812.7	7.2797
90		1.1738	1835.2	7.3408
100		1.2068	1857.7	7.3995
110		1.2397	1880.2	7.4557
120		1.2724	1902.7	7.5096
130		1.3049	1925.2	7.5612

T °C		p = 200 kPa v, m³/kg	h, kJ/kg	s, kJ/kg·K
T_{sat}=	f	0.001 507	261.3	1.1683
−18.86	g	0.5946	1586.7	6.3802
−10		0.6198	1606.5	6.4608
0		0.6464	1629.3	6.5458
10		0.6727	1652.0	6.6281
20		0.6989	1674.8	6.7078
30		0.7249	1697.5	6.7849
40		0.7508	1720.2	6.8593
50		0.7765	1742.9	6.9312
60		0.8021	1765.6	7.0006
70		0.8275	1788.3	7.0675
80		0.8527	1810.9	7.1319
90		0.8778	1833.6	7.1940
100		0.9027	1856.2	7.2536
110		0.9275	1878.8	7.3109
120		0.9521	1901.4	7.3658
130		0.9766	1924.0	7.4185

T °C		p = 250 kPa v, m³/kg	h, kJ/kg	s, kJ/kg·K
T_{sat}=	f	0.001 523	284.9	1.2596
−13.66	g	0.4821	1593.4	6.3023
−10		0.4914	1601.7	6.3394
0		0.5129	1624.9	6.4248
10		0.5343	1648.1	6.5075
20		0.5556	1671.3	6.5876
30		0.5767	1694.4	6.6651
40		0.5977	1717.4	6.7401
50		0.6185	1740.4	6.8126
60		0.6392	1763.4	6.8827
70		0.6597	1786.3	6.9503
80		0.6801	1809.2	7.0156
90		0.7003	1832.0	7.0786
100		0.7204	1854.8	7.1393
110		0.7403	1877.5	7.1977
120		0.7601	1900.2	7.2540
130		0.7797	1922.8	7.3081

T °C		p = 300 kPa v, m³/kg	h, kJ/kg	s, kJ/kg·K
T_{sat}=	f	0.001 536	305.1	1.3365
−9.23	g	0.4060	1598.8	6.2387
0		0.4239	1620.7	6.3248
10		0.4421	1644.3	6.4078
20		0.4601	1667.8	6.4882
30		0.4779	1691.3	6.5661
40		0.4956	1714.7	6.6415
50		0.5132	1738.0	6.7145
60		0.5306	1761.2	6.7851
70		0.5479	1784.3	6.8534
80		0.5650	1807.4	6.9195
90		0.5820	1830.4	6.9834
100		0.5988	1853.3	7.0451
110		0.6155	1876.2	7.1048
120		0.6321	1898.9	7.1624
130		0.6485	1921.6	7.2180

T °C		p = 350 kPa v, m³/kg	h, kJ/kg	s, kJ/kg·K
T_{sat}=	f	0.001 548	322.8	1.4029
−5.36	g	0.3511	1603.4	6.1849
0		0.3604	1616.6	6.2396
10		0.3762	1640.6	6.3228
20		0.3918	1664.5	6.4034
30		0.4073	1688.3	6.4815
40		0.4227	1711.9	6.5572
50		0.4379	1735.5	6.6306
60		0.4530	1759.0	6.7018
70		0.4680	1782.4	6.7707
80		0.4828	1805.6	6.8375
90		0.4975	1828.8	6.9023
100		0.5121	1851.9	6.9650
110		0.5265	1874.8	7.0258
120		0.5407	1897.7	7.0848
130		0.5549	1920.4	7.1420

Table A·8·3 Ammonia: Superheated Vapor (*Continued*)

T °C		*p* = 400 kPa *v*, m³/kg	*h*, kJ/kg	*s*, kJ/kg·K
T_{sat} =	*f*	0.001 560	338.8	1.4619
−1.89	*g*	0.3094	1607.3	6.1382
0		0.3127	1612.5	6.1653
10		0.3267	1636.9	6.2485
20		0.3406	1661.1	6.3293
30		0.3544	1685.3	6.4076
40		0.3680	1709.2	6.4836
50		0.3815	1733.1	6.5574
60		0.3949	1756.8	6.6289
70		0.4081	1780.4	6.6984
80		0.4212	1803.9	6.7659
90		0.4342	1827.2	6.8315
100		0.4470	1850.4	6.8952
110		0.4597	1873.5	6.9572
120		0.4722	1896.4	7.0175
130		0.4847	1919.2	7.0762

T °C		*p* = 450 kPa *v*, m³/kg	*h*, kJ/kg	*s*, kJ/kg·K
T_{sat} =	*f*	0.001 570	353.3	1.5149
1.25	*g*	0.2767	1610.6	6.0969
10		0.2883	1633.2	6.1826
20		0.3008	1657.8	6.2635
30		0.3132	1682.3	6.3420
40		0.3255	1706.6	6.4182
50		0.3377	1730.7	6.4922
60		0.3497	1754.6	6.5642
70		0.3616	1778.4	6.6342
80		0.3733	1802.1	6.7023
90		0.3849	1825.6	6.7687
100		0.3964	1848.9	6.8334
110		0.4078	1872.1	6.8965
120		0.4190	1895.1	6.9581
130		0.4301	1917.9	7.0183

T °C		*p* = 500 kPa *v*, m³/kg	*h*, kJ/kg	*s*, kJ/kg·K
T_{sat} =	*f*	0.001 580	366.7	1.5631
4.13	*g*	0.2503	1613.6	6.0598
10		0.2575	1629.6	6.1233
20		0.2690	1654.5	6.2043
30		0.2803	1679.3	6.2829
40		0.2915	1703.9	6.3593
50		0.3026	1728.3	6.4336
60		0.3135	1752.5	6.5059
70		0.3243	1776.5	6.5764
80		0.3350	1800.3	6.6451
90		0.3455	1824.0	6.7122
100		0.3560	1847.4	6.7778
110		0.3662	1870.7	6.8419
120		0.3764	1893.8	6.9048
130		0.3864	1916.7	6.9665

T °C		*p* = 600 kPa *v*, m³/kg	*h*, kJ/kg	*s*, kJ/kg·K
T_{sat} =	*f*	0.001 598	390.7	1.6484
9.28	*g*	0.2104	1618.5	5.9956
10		0.2114	1622.3	6.0200
20		0.2212	1647.9	6.1010
30		0.2309	1673.3	6.1798
40		0.2405	1698.5	6.2565
50		0.2500	1723.4	6.3312
60		0.2593	1748.0	6.4042
70		0.2685	1772.5	6.4755
80		0.2775	1796.7	6.5453
90		0.2865	1820.7	6.6138
100		0.2953	1844.4	6.6811
110		0.3040	1867.9	6.7473
120		0.3125	1891.2	6.8125
130		0.3209	1914.2	6.8771

T °C		*p* = 700 kPa *v*, m³/kg	*h*, kJ/kg	*s*, kJ/kg·K
T_{sat} =	*f*	0.001 614	412.0	1.7225
13.80	*g*	0.1815	1622.4	5.9409
20		0.1871	1641.2	6.0129
30		0.1956	1667.2	6.0918
40		0.2041	1692.9	6.1687
50		0.2124	1718.4	6.2438
60		0.2205	1743.6	6.3173
70		0.2286	1768.5	6.3894
80		0.2365	1793.0	6.4602
90		0.2443	1817.4	6.5300
100		0.2520	1841.4	6.5988
110		0.2595	1865.1	6.6669
120		0.2669	1888.5	6.7344
130		0.2742	1911.7	6.8015

T °C		*p* = 800 kPa *v*, m³/kg	*h*, kJ/kg	*s*, kJ/kg·K
T_{sat} =	*f*	0.001 630	431.2	1.7884
17.85	*g*	0.1596	1625.7	5.8932
20		0.1615	1634.2	5.9359
30		0.1692	1660.9	6.0149
40		0.1768	1687.3	6.0920
50		0.1842	1713.3	6.1675
60		0.1915	1739.0	6.2415
70		0.1987	1764.3	6.3142
80		0.2057	1789.3	6.3859
90		0.2127	1814.0	6.4568
100		0.2195	1838.3	6.5271
110		0.2261	1862.3	6.5969
120		0.2327	1885.9	6.6666
130		0.2391	1909.2	6.7362

T °C		*p* = 900 kPa *v*, m³/kg	*h*, kJ/kg	*s*, kJ/kg·K
T_{sat} =	*f*	0.001 644	448.7	1.8477
21.52	*g*	0.1424	1628.3	5.8509
30		0.1486	1654.4	5.9466
40		0.1555	1681.5	6.0239
50		0.1623	1708.1	6.0996
60		0.1689	1734.3	6.1740
70		0.1754	1760.2	6.2474
80		0.1818	1785.6	6.3200
90		0.1881	1810.6	6.3920
100		0.1942	1835.2	6.4636
110		0.2002	1859.4	6.5351
120		0.2061	1883.2	6.6067
130		0.2119	1906.6	6.6786

T °C		*p* = 1000 kPa *v*, m³/kg	*h*, kJ/kg	*s*, kJ/kg·K
T_{sat} =	*f*	0.001 658	464.9	1.9018
24.90	*g*	0.1285	1630.5	5.8126
30		0.1322	1647.7	5.8852
40		0.1385	1675.5	5.9625
50		0.1447	1702.7	6.0385
60		0.1508	1729.5	6.1134
70		0.1568	1755.9	6.1873
80		0.1627	1781.7	6.2606
90		0.1684	1807.1	6.3336
100		0.1740	1832.1	6.4065
110		0.1795	1856.5	6.4796
120		0.1848	1880.5	6.5530
130		0.1901	1904.1	6.6272

T °C		*p* = 1200 kPa *v*, m³/kg	*h*, kJ/kg	*s*, kJ/kg·K
T_{sat} =	*f*	0.001 684	494.1	1.9978
30.94	*g*	0.1075	1633.9	5.7458
40		0.1130	1662.9	5.8555
50		0.1184	1691.5	5.9319
60		0.1237	1719.6	6.0075
70		0.1289	1747.0	6.0824
80		0.1340	1773.9	6.1572
90		0.1389	1800.1	6.2320
100		0.1437	1825.7	6.3072
110		0.1484	1850.7	6.3832
120		0.1529	1875.2	6.4601
130		0.1574	1899.0	6.5383

Table A·8·3 Ammonia: Superheated Vapor (*Continued*)

T °C		p = 1400 kPa v, m^3/kg	h, kJ/kg	s, kJ/kg·K
T_{sat}=	f	0.001 708	520.2	2.0814
36.27	g	0.0923	1636.2	5.6883
40		0.0948	1649.8	5.7642
50		0.0997	1679.9	5.8409
60		0.1044	1709.3	5.9171
70		0.1090	1737.9	5.9930
80		0.1135	1765.7	6.0690
90		0.1178	1792.9	6.1455
100		0.1221	1819.2	6.2228
110		0.1262	1844.8	6.3013
120		0.1301	1869.7	6.3813
130		0.1340	1893.8	6.4633

T °C		p = 1600 kPa v, m^3/kg	h, kJ/kg	s, kJ/kg·K
T_{sat}=	f	0.001 730	543.7	2.1559
41.04	g	0.0808	1637.6	5.6375
50		0.0856	1668.1	5.7615
60		0.0899	1698.7	5.8381
70		0.0940	1728.5	5.9149
80		0.0981	1757.3	5.9920
90		0.1020	1785.4	6.0700
100		0.1058	1812.5	6.1492
110		0.1095	1838.8	6.2301
120		0.1131	1864.2	6.3130
130		0.1165	1888.8	6.3984

T °C		p = 1800 kPa v, m^3/kg	h, kJ/kg	s, kJ/kg·K
T_{sat}=	f	0.001 752	565.3	2.2232
45.38	g	0.0717	1638.4	5.5919
50		0.0746	1656.7	5.6909
60		0.0786	1688.3	5.7680
70		0.0824	1719.0	5.8455
80		0.0861	1748.7	5.9237
90		0.0897	1777.6	6.0030
100		0.0932	1805.5	6.0840
110		0.0965	1832.5	6.1671
120		0.0998	1858.6	6.2527
130		0.1029	1883.7	6.3412

T °C		p = 2000 kPa v, m^3/kg	h, kJ/kg	s, kJ/kg·K
T_{sat}=	f	0.001 773	585.4	2.2847
49.37	g	0.0644	1638.6	5.5503
50		0.0659	1646.2	5.6274
60		0.0696	1678.4	5.7049
70		0.0731	1709.6	5.7831
80		0.0766	1740.0	5.8622
90		0.0799	1769.5	5.9429
100		0.0831	1798.2	6.0255
110		0.0862	1825.9	6.1106
120		0.0891	1852.8	6.1987
130		0.0920	1878.8	6.2901

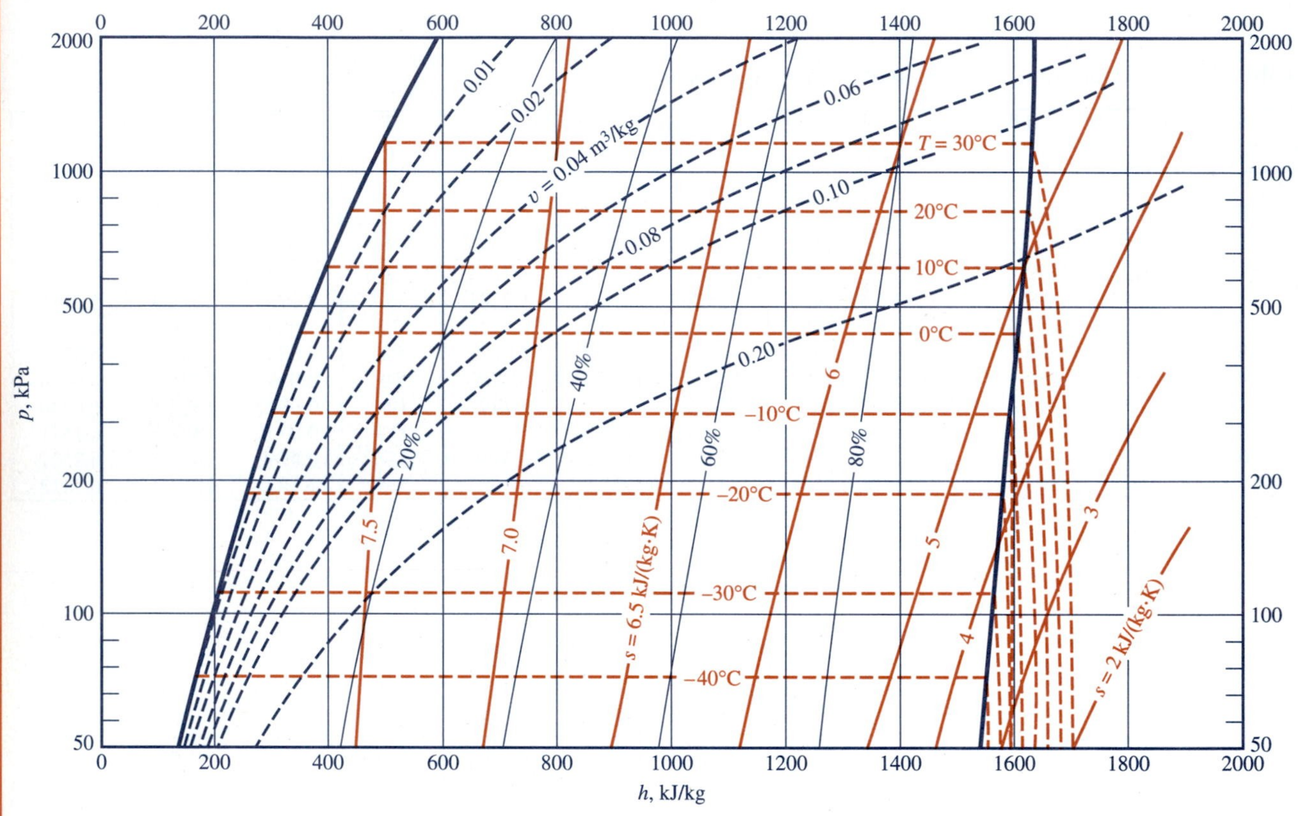

Chart A·9 Ammonia *ph* diagram. (*Source:* Based on Haar, Lester, and John S. Gallagher, *Thermodynamic Properties of Ammonia,* published by the American Chemical Society and the American Institute of Physics for the National Bureau of Standards, 1978. Also, *Journal of Physical and Chemical Reference Data,* Vol. 7, No. 3, 1978. pp. 635–792.)

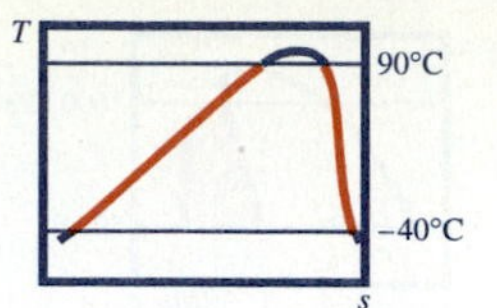

Table A·10·1 R134a: Liquid–vapor Saturation, Temperature Table

T °C	P kPa	v_f m³/kg	v_g m³/kg	h_f kJ/kg	h_{fg} kJ/kg	h_g kJ/kg	s_f kJ/kg·K	s_{fg} kJ/kg·K	s_g kJ/kg·K
-40	51.1	0.000 705	0.361 483	0.000	226.087	226.087	0.0000	0.9697	0.9697
-35	66.1	0.000 717	0.284 244	6.280	222.928	229.208	0.0264	0.9361	0.9624
-30	84.3	0.000 721	0.226 069	12.560	219.762	232.323	0.0525	0.9038	0.9563
-25	106.3	0.000 730	0.181 675	18.828	216.608	235.436	0.0786	0.8723	0.9509
-20	132.7	0.000 738	0.147 400	25.357	213.143	238.501	0.1037	0.8424	0.9461
-15	163.9	0.000 743	0.120 655	31.866	209.805	241.671	0.1290	0.8131	0.9420
-10	200.6	0.000 754	0.099 570	38.325	206.277	244.602	0.1543	0.7839	0.9382
-5	243.4	0.000 765	0.082 770	45.031	202.502	247.533	0.1789	0.7559	0.9347
0	292.9	0.000 774	0.069 273	51.730	198.733	250.464	0.2034	0.7283	0.9317
5	349.9	0.000 782	0.058 343	58.429	194.965	253.394	0.2279	0.7013	0.9292
10	414.9	0.000 793	0.049 418	65.128	191.197	256.325	0.2520	0.6749	0.9270
15	488.8	0.000 804	0.042 072	72.257	186.809	259.066	0.2762	0.6485	0.9247
20	572.3	0.000 815	0.035 980	79.117	182.710	261.828	0.3000	0.6228	0.9229
25	666.1	0.000 829	0.030 903	86.235	178.092	264.327	0.3238	0.5974	0.9212
30	771.0	0.000 844	0.026 637	93.602	173.237	266.839	0.3476	0.5716	0.9192
35	887.9	0.000 855	0.023 036	100.948	168.402	269.351	0.3710	0.5468	0.9177
40	1017.6	0.000 873	0.019 971	108.245	163.389	271.634	0.3947	0.5215	0.9163
45	1161.0	0.000 890	0.017 353	115.773	158.012	273.785	0.4185	0.4958	0.9143
50	1319.0	0.000 908	0.015 099	123.650	151.979	275.629	0.4420	0.4706	0.9127
55	1492.6	0.000 926	0.013 150	131.605	145.848	277.453	0.4657	0.4447	0.9104
60	1682.8	0.000 949	0.011 456	139.560	139.327	278.887	0.4899	0.4178	0.9077
65	1890.6	0.000 976	0.009 973	147.934	132.221	280.155	0.5140	0.3908	0.9048
70	2117.4	0.001 003	0.008 662	156.434	124.467	280.901	0.5383	0.3627	0.9010
75	2364.3	0.001 037	0.007 499	165.317	115.924	281.241	0.5633	0.3332	0.8966
80	2633.0	0.001 079	0.006 450	174.571	106.386	280.957	0.5890	0.3012	0.8902
85	2925.2	0.001 130	0.005 494	184.452	95.366	279.818	0.6159	0.2663	0.8822
90	3242.9	0.001 195	0.004 604	195.064	82.348	277.412	0.6443	0.2264	0.8707

Table A·10·2 R134a: Liquid–vapor Saturation, Pressure Table

p kPa	T °C	v_f m³/kg	v_g m³/kg	h_f kJ/kg	h_{fg} kJ/kg	h_g kJ/kg	s_f kJ/kg·K	s_{fg} kJ/kg·K	s_g kJ/kg·K
50	-40.43	0.000 705	0.369 168	-0.537	226.341	225.804	-0.0022	0.9726	0.9703
75	-32.43	0.000 718	0.252 286	9.511	221.244	230.755	0.0400	0.9193	0.9593
100	-26.34	0.000 730	0.192 485	17.153	217.401	234.555	0.0716	0.8807	0.9523
125	-21.37	0.000 735	0.155 930	23.571	214.146	237.717	0.0969	0.8505	0.9474
150	-17.12	0.000 743	0.131 239	29.017	211.228	240.245	0.1184	0.8252	0.9436
200	-10.08	0.000 754	0.099 855	38.223	206.335	244.558	0.1539	0.7843	0.9382
250	-4.29	0.000 767	0.080 672	45.982	201.967	247.949	0.1823	0.7520	0.9343
300	0.66	0.000 774	0.067 699	52.613	198.237	250.850	0.2066	0.7247	0.9313
350	5.01	0.000 782	0.058 323	58.443	194.957	253.401	0.2280	0.7012	0.9292
400	8.91	0.000 790	0.051 218	63.651	192.035	255.685	0.2468	0.6807	0.9275
500	15.71	0.000 806	0.041 142	73.242	186.194	259.436	0.2796	0.6448	0.9244
600	21.54	0.000 819	0.034 320	81.321	181.267	262.588	0.3073	0.6149	0.9222
700	26.68	0.000 837	0.029 388	88.637	176.533	265.169	0.3316	0.5890	0.9207
800	31.29	0.000 847	0.025 651	95.433	172.054	267.487	0.3536	0.5652	0.9188
900	35.49	0.000 857	0.022 715	101.647	167.912	269.559	0.3733	0.5443	0.9176
1000	39.35	0.000 871	0.020 343	107.258	164.047	271.305	0.3917	0.5248	0.9165
1200	46.28	0.000 893	0.016 746	117.725	156.569	274.293	0.4245	0.4895	0.9139
1400	52.39	0.000 914	0.014 135	127.453	149.131	276.584	0.4535	0.4581	0.9116
1600	57.88	0.000 939	0.012 145	136.176	142.191	278.367	0.4796	0.4294	0.9090
1800	62.87	0.000 962	0.010 578	144.372	135.239	279.611	0.5037	0.4027	0.9064
2000	67.47	0.000 989	0.009 308	152.050	128.578	280.628	0.5258	0.3771	0.9029
2200	71.72	0.001 015	0.008 246	159.464	121.576	281.040	0.5469	0.3528	0.8996
2400	75.69	0.001 042	0.007 347	166.588	114.652	281.240	0.5669	0.3289	0.8958
2600	79.41	0.001 073	0.006 569	173.451	107.533	280.984	0.5860	0.3050	0.8910
2800	82.91	0.001 106	0.005 886	180.228	100.337	280.565	0.6045	0.2815	0.8860
3000	86.22	0.001 143	0.005 273	186.969	92.382	279.351	0.6226	0.2571	0.8797
3200	89.35	0.001 185	0.004 718	193.628	84.193	277.821	0.6405	0.2320	0.8725

Source: Based on *Thermodynamic Properties of HFC-134a (1,1,1, 2-tetrafluoroethane),* DuPont Company, Wilmington, Delaware, 1993, with permission.

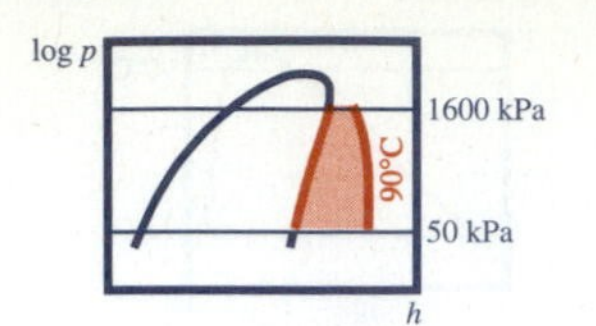

Table A·10·3 R134a: Superheated Vapor

T °C		p = 0.05 MPa v, m³/kg	h, kJ/kg	s, kJ/kg·K
T_{sat} = −40.43	f	0.000 705	-0.5373	-0.0022
	g	0.369 168	225.804	0.9703
-40		0.370 209	226.150	0.9715
-35		0.378 848	229.862	0.9867
-30		0.387 468	233.614	1.0112
-25		0.396 070	237.407	1.0180
-20		0.404 654	241.239	1.0338
-15		0.413 220	245.112	1.0495
-10		0.421 767	249.025	1.0649
-5		0.430 297	252.978	1.0801
0		0.438 808	256.971	1.0950
5		0.447 301	261.005	1.1095
10		0.455 776	265.079	1.1239
15		0.464 233	269.193	1.1379
20		0.472 672	273.347	1.1518
25		0.481 093	277.541	1.1656
30		0.489 495	281.776	1.1793
35		0.497 880	286.051	1.1930
40		0.506 246	290.366	1.2068
45		0.514 594	294.721	1.2206
50		0.522 924	299.117	1.2345
55		0.531 236	303.552	1.2484
60		0.539 529	308.028	1.2624
65		0.547 805	312.544	1.2763
70		0.556 062	317.101	1.2901
75		0.564 301	321.697	1.3036
80		0.572 523	326.334	1.3166
85		0.580 725	331.011	1.3287
90		0.588 910	335.728	1.3398

T °C		p = 0.10 MPa v, m³/kg	h, kJ/kg	s, kJ/kg·K
T_{sat} = −26.34	f	0.000 730	17.154	0.0716
	g	0.192 485	234.555	0.9523
-25		0.193 933	235.744	0.9555
-20		0.198 417	239.659	0.9716
-15		0.202 885	243.612	0.9879
-10		0.207 337	247.601	1.0040
-5		0.211 772	251.628	1.0198
0		0.216 191	255.692	1.0353
5		0.220 594	259.793	1.0503
10		0.224 981	263.931	1.0650
15		0.229 351	268.106	1.0793
20		0.233 705	272.318	1.0933
25		0.238 043	276.568	1.1072
30		0.242 365	280.854	1.1210
35		0.246 671	285.178	1.1349
40		0.250 960	289.539	1.1489
45		0.255 233	293.937	1.1631
50		0.259 489	298.372	1.1775
55		0.263 730	302.844	1.1922
60		0.267 954	307.353	1.2070
65		0.272 162	311.900	1.2220
70		0.276 354	316.483	1.2370
75		0.280 530	321.104	1.2517
80		0.284 689	325.762	1.2658
85		0.288 832	330.457	1.2789
90		0.292 959	335.189	1.2904

T °C		p = 0.15 MPa v, m³/kg	h, kJ/kg	s, kJ/kg·K
T_{sat} = −17.12	f	0.000 743	29.017	0.1184
	g	0.131 239	240.245	0.9436
-15		0.132 784	242.118	0.9502
-10		0.135 862	246.183	0.9667
-5		0.138 926	250.281	0.9829
0		0.141 974	254.414	0.9987
5		0.145 008	258.581	1.0140
10		0.148 027	262.783	1.0289
15		0.151 031	267.018	1.0433
20		0.154 021	271.288	1.0573
25		0.156 995	275.591	1.0712
30		0.159 955	279.929	1.0849
35		0.162 900	284.301	1.0988
40		0.165 831	288.707	1.1128
45		0.168 747	293.148	1.1270
50		0.171 648	297.622	1.1416
55		0.174 534	302.131	1.1566
60		0.177 405	306.674	1.1719
65		0.180 262	311.251	1.1874
70		0.183 104	315.862	1.2029
75		0.185 931	320.507	1.2182
80		0.188 744	325.186	1.2328
85		0.191 541	329.900	1.2462
90		0.194 324	334.647	1.2578

T °C		p = 0.20 MPa v, m³/kg	h, kJ/kg	s, kJ/kg·K
T_{sat} = −10.08	f	0.000 754	38.223	0.1539
	g	0.099 855	244.558	0.9382
-10		0.100 108	244.763	0.9396
-5		0.102 483	248.933	0.9560
0		0.104 844	253.134	0.9721
5		0.107 192	257.367	0.9876
10		0.109 525	261.630	1.0025
15		0.111 846	265.925	1.0169
20		0.114 152	270.252	1.0310
25		0.116 445	274.609	1.0448
30		0.118 725	278.998	1.0584
35		0.120 990	283.418	1.0722
40		0.123 242	287.870	1.0861
45		0.125 481	292.352	1.1004
50		0.127 706	296.866	1.1151
55		0.129 917	301.412	1.1302
60		0.132 115	305.988	1.1457
65		0.134 298	310.596	1.1615
70		0.136 469	315.235	1.1774
75		0.138 625	319.905	1.1930
80		0.140 768	324.607	1.2079
85		0.142 898	329.340	1.2214
90		0.145 014	334.104	1.2329

T °C		p = 0.25 MPa v, m³/kg	h, kJ/kg	s, kJ/kg·K
T_{sat} = −4.29	f	0.000 767	45.982	0.1823
	g	0.080 672	247.949	0.9343
0		0.082 541	251.847	0.9510
5		0.084 477	256.144	0.9666
10		0.086 401	260.470	0.9816
15		0.088 311	264.825	0.9961
20		0.090 209	269.208	1.0101
25		0.092 094	273.619	1.0238
30		0.093 966	278.059	1.0374
35		0.095 825	282.527	1.0510
40		0.097 671	287.024	1.0649
45		0.099 505	291.549	1.0791
50		0.101 325	296.103	1.0938
55		0.103 133	300.685	1.1090
60		0.104 928	305.296	1.1247
65		0.106 710	309.935	1.1407
70		0.108 480	314.603	1.1568
75		0.110 236	319.299	1.1727
80		0.111 980	324.024	1.1877
85		0.113 711	328.777	1.2014
90		0.115 429	333.558	1.2127

T °C		p = 0.30 MPa v, m³/kg	h, kJ/kg	s, kJ/kg·K
T_{sat} = 0.66	f	0.000 774	52.613	0.2066
	g	0.067 699	250.850	0.9313
5		0.069 310	254.910	0.9492
10		0.070 961	259.299	0.9643
15		0.072 600	263.713	0.9787
20		0.074 226	268.152	0.9927
25		0.075 840	272.618	1.0064
30		0.077 442	277.109	1.0198
35		0.079 032	281.626	1.0334
40		0.080 609	286.169	1.0472
45		0.082 174	290.737	1.0614
50		0.083 726	295.331	1.0761
55		0.085 267	299.951	1.0913
60		0.086 795	304.597	1.1071
65		0.088 311	309.268	1.1232
70		0.089 814	313.965	1.1395
75		0.091 305	318.687	1.1555
80		0.092 784	323.436	1.1707
85		0.094 251	328.210	1.1844
90		0.095 705	333.010	1.1956

Table A·10·3 R134a: Superheated Vapor (*Continued*)

T °C	*p* = 0.40 MPa *v*, m³/kg	*h*, kJ/kg	*s*, kJ/kg·K
T_{sat} = 8.91 *f*	0.000 790	63.651	0.2468
T_{sat} = 8.91 *g*	0.051 218	255.685	0.9275
10	0.051 607	256.907	0.9364
15	0.052 910	261.442	0.9509
20	0.054 202	265.998	0.9648
25	0.055 482	270.573	0.9783
30	0.056 750	275.170	0.9916
35	0.058 006	279.786	1.0049
40	0.059 250	284.423	1.0186
45	0.060 483	289.080	1.0326
50	0.061 703	293.757	1.0473
55	0.062 912	298.454	1.0625
60	0.064 109	303.172	1.0784
65	0.065 295	307.910	1.0948
70	0.066 468	312.668	1.1113
75	0.067 630	317.447	1.1275
80	0.068 779	322.245	1.1429
85	0.069 917	327.064	1.1565
90	0.071 044	331.904	1.1674

T °C	*p* = 0.50 MPa *v*, m³/kg	*h*, kJ/kg	*s*, kJ/kg·K
T_{sat} = 15.71 *f*	0.000 806	73.242	0.2796
T_{sat} = 15.71 *g*	0.041 142	259.436	0.9244
20	0.042 138	263.764	0.9426
25	0.043 223	268.455	0.9560
30	0.044 295	273.162	0.9691
35	0.045 356	277.882	0.9823
40	0.046 405	282.618	0.9958
45	0.047 441	287.368	1.0097
50	0.048 466	292.133	1.0243
55	0.049 479	296.913	1.0395
60	0.050 480	301.708	1.0555
65	0.051 469	306.517	1.0719
70	0.052 446	311.341	1.0886
75	0.053 411	316.179	1.1050
80	0.054 364	321.032	1.1204
85	0.055 305	325.900	1.1340
90	0.056 234	330.783	1.1445

T °C	*p* = 0.60 MPa *v*, m³/kg	*h*, kJ/kg	*s*, kJ/kg·K
T_{sat} = 21.54 *f*	0.000 819	81.321	0.3073
T_{sat} = 21.54 *g*	0.034 320	262.588	0.9222
25	0.035 015	266.244	0.9375
30	0.035 962	271.067	0.9505
35	0.036 895	275.900	0.9635
40	0.037 817	280.742	0.9768
45	0.038 726	285.592	0.9906
50	0.039 623	290.451	1.0051
55	0.040 507	295.319	1.0203
60	0.041 378	300.197	1.0363
65	0.042 238	305.082	1.0528
70	0.043 085	309.977	1.0696
75	0.043 919	314.881	1.0861
80	0.044 741	319.794	1.1015
85	0.045 550	324.715	1.1150
90	0.046 347	329.645	1.1253

T °C	*p* = 0.70 MPa *v*, m³/kg	*h*, kJ/kg	*s*, kJ/kg·K
T_{sat} = 26.68 *f*	0.000 837	88.637	0.3316
T_{sat} = 26.68 *g*	0.029 388	265.169	0.9207
30	0.029 983	268.869	0.9344
35	0.030 829	273.823	0.9473
40	0.031 663	278.780	0.9605
45	0.032 483	283.739	0.9741
50	0.033 290	288.701	0.9886
55	0.034 084	293.665	1.0038
60	0.034 865	298.632	1.0197
65	0.035 632	303.602	1.0363
70	0.036 387	308.574	1.0531
75	0.037 128	313.549	1.0697
80	0.037 856	318.526	1.0852
85	0.038 571	323.506	1.0985
90	0.039 273	328.488	1.1085

T °C	*p* = 0.80 MPa *v*, m³/kg	*h*, kJ/kg	*s*, kJ/kg·K
T_{sat} = 31.29 *f*	0.000 847	95.433	0.3536
T_{sat} = 31.29 *g*	0.025 651	267.487	0.9188
35	0.026 258	271.636	0.9331
40	0.027 028	276.719	0.9461
45	0.027 783	281.798	0.9597
50	0.028 525	286.872	0.9740
55	0.029 253	291.942	0.9892
60	0.029 967	297.008	1.0052
65	0.030 667	302.069	1.0218
70	0.031 353	307.126	1.0386
75	0.032 025	312.178	1.0552
80	0.032 683	317.226	1.0707
85	0.033 327	322.270	1.0840
90	0.033 957	327.309	1.0937

T °C	*p* = 0.90 MPa *v*, m³/kg	*h*, kJ/kg	*s*, kJ/kg·K
T_{sat} = 35.49 *f*	0.000 857	101.647	0.3733
T_{sat} = 35.49 *g*	0.022 715	269.559	0.9176
40	0.023 404	274.546	0.9333
45	0.024 111	279.757	0.9468
50	0.024 803	284.956	0.9610
55	0.025 481	290.142	0.9762
60	0.026 144	295.317	0.9921
65	0.026 793	300.479	1.0087
70	0.027 426	305.629	1.0257
75	0.028 045	310.766	1.0423
80	0.028 650	315.892	1.0577
85	0.029 240	321.005	1.0709
90	0.029 815	326.106	1.0804

T °C	*p* = 1.00 MPa *v*, m³/kg	*h*, kJ/kg	*s*, kJ/kg·K
T_{sat} = 39.35 *f*	0.000 871	107.258	0.3917
T_{sat} = 39.35 *g*	0.020 343	271.305	0.9165
40	0.020 484	272.248	0.9218
45	0.021 154	277.606	0.9351
50	0.021 809	282.942	0.9493
55	0.022 449	288.258	0.9644
60	0.023 073	293.552	0.9803
65	0.023 682	298.825	0.9969
70	0.024 275	304.077	1.0139
75	0.024 853	309.309	1.0305
80	0.025 415	314.519	1.0460
85	0.025 962	319.708	1.0590
90	0.026 494	324.876	1.0683

T °C	*p* = 1.20 MPa *v*, m³/kg	*h*, kJ/kg	*s*, kJ/kg·K
T_{sat} = 46.28 *f*	0.000 893	117.725	0.4245
T_{sat} = 46.28 *g*	0.016 746	274.293	0.9139
50	0.017 264	278.583	0.9287
55	0.017 852	284.200	0.9436
60	0.018 423	289.775	0.9595
65	0.018 978	295.306	0.9762
70	0.019 516	300.795	0.9931
75	0.020 037	306.242	1.0098
80	0.020 541	311.645	1.0252
85	0.021 029	317.006	1.0381
90	0.021 499	322.325	1.0468

T °C	*p* = 1.40 MPa *v*, m³/kg	*h*, kJ/kg	*s*, kJ/kg·K
T_{sat} = 52.39 *f*	0.000 914	127.453	0.4535
T_{sat} = 52.39 *g*	0.014 135	276.584	0.9116
55	0.014 494	279.706	0.9258
60	0.015 038	285.622	0.9416
65	0.015 564	291.468	0.9583
70	0.016 072	297.243	0.9753
75	0.016 562	302.948	0.9919
80	0.017 033	308.581	1.0072
85	0.017 485	314.144	1.0199
90	0.017 920	319.636	1.0283

T °C	*p* = 1.60 MPa *v*, m³/kg	*h*, kJ/kg	*s*, kJ/kg·K
T_{sat} = 57.88 *f*	0.000 939	136.176	0.4796
T_{sat} = 57.88 *g*	0.012 145	287.266	0.9090
60	0.012 423	281.041	0.9259
65	0.012 941	287.266	0.9425
70	0.013 439	293.385	0.9595
75	0.013 916	299.397	0.9762
80	0.014 372	305.302	0.9914
85	0.014 808	311.100	1.0039
90	0.015 223	316.791	1.0119

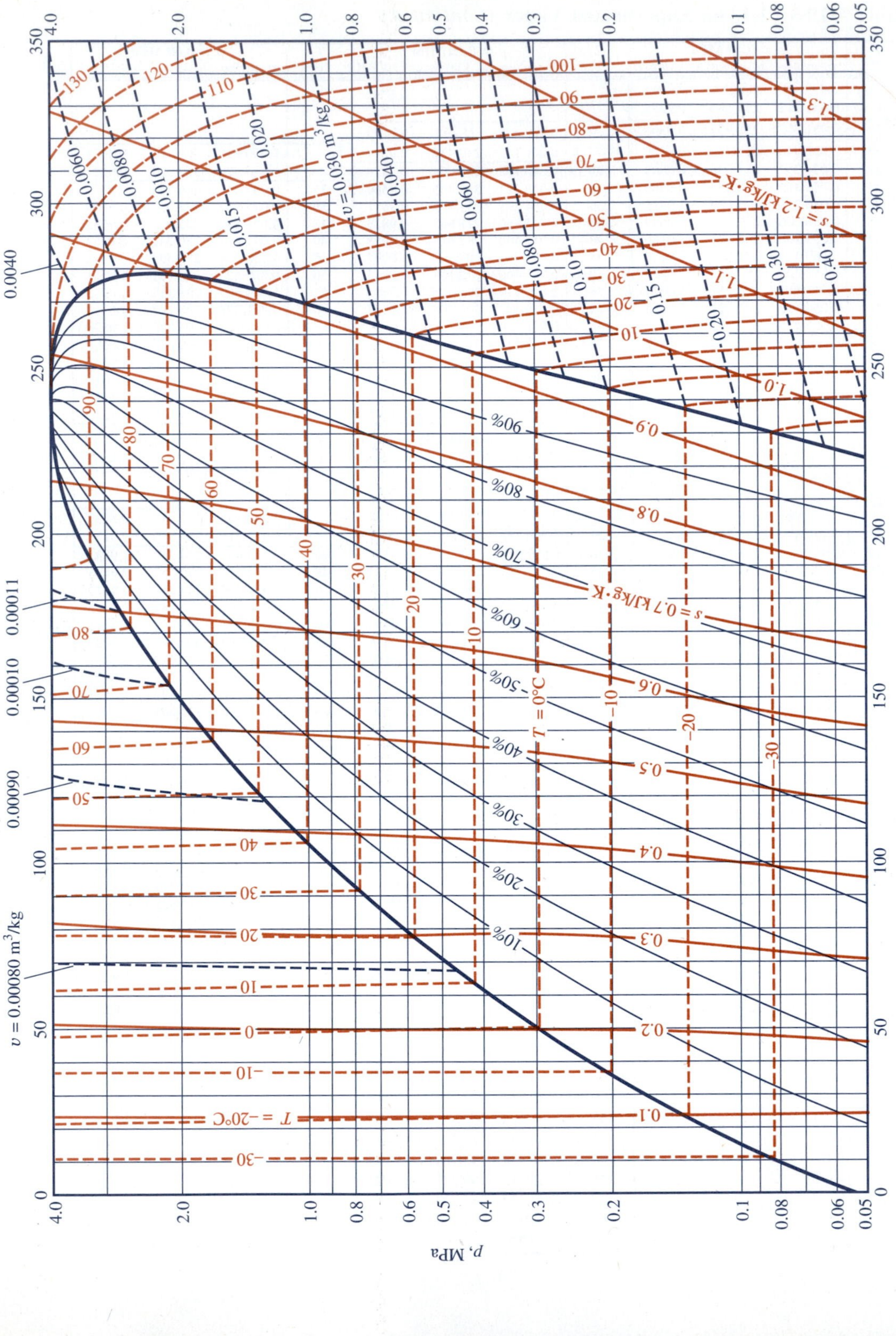

Chart A·11 R134a *ph* diagram. (*Source:* Based on *Thermodynamic Properties of HFC-134a (1,1,1, 2-tetrafluoroethane)*, DuPont Company, Wilmington, Delaware, 1993, with permission.)

Table A·12 Properties of Air

T K	T °C	h kJ/kg	p_r	v_r	ϕ kJ/kg·K
200	**−73.15**	201.87	0.3414	585.82	6.3000
210	**−63.15**	211.94	0.4051	518.39	6.3491
220	**−53.15**	221.99	0.4768	461.41	6.3959
230	**−43.15**	232.04	0.5571	412.85	6.4406
240	**−33.15**	242.08	0.6466	371.17	6.4833
250	**−23.15**	252.12	0.7458	335.21	6.5243
260	**−13.15**	262.15	0.8555	303.92	6.5636
270	**−3.15**	272.19	0.9761	276.61	6.6015
280	**6.85**	282.22	1.1084	252.62	6.6380
290	**16.85**	292.25	1.2531	231.43	6.6732
300	**26.85**	302.29	1.4108	212.65	6.7072
310	**36.85**	312.33	1.5823	195.92	6.7401
320	**46.85**	322.37	1.7682	180.98	6.7720
330	**56.85**	332.42	1.9693	167.57	6.8029
340	**66.85**	342.47	2.1865	155.50	6.8330
350	**76.85**	352.54	2.4204	144.60	6.8621
360	**86.85**	362.61	2.6720	134.73	6.8905
370	**96.85**	372.69	2.9419	125.77	6.9181
380	**106.85**	382.79	3.2312	117.60	6.9450
390	**116.85**	392.89	3.5407	110.15	6.9713
400	**126.85**	403.01	3.8712	103.33	6.9969
410	**136.85**	413.14	4.2238	97.069	7.0219
420	**146.85**	423.29	4.5993	91.318	7.0464
430	**156.85**	433.45	4.9989	86.019	7.0703
440	**166.85**	443.62	5.4234	81.130	7.0937
450	**176.85**	453.81	5.8739	76.610	7.1166
460	**186.85**	464.02	6.3516	72.423	7.1390
470	**196.85**	474.25	6.8575	68.538	7.1610
480	**206.85**	484.49	7.3927	64.929	7.1826
490	**216.85**	494.76	7.9584	61.570	7.2037
500	**226.85**	505.04	8.5558	58.440	7.2245
510	**236.85**	515.34	9.1861	55.519	7.2449
520	**246.85**	525.66	9.8506	52.789	7.2650
530	**256.85**	536.01	10.551	50.232	7.2847
540	**266.85**	546.37	11.287	47.843	7.3040
550	**276.85**	556.76	12.062	45.598	7.3231
560	**286.85**	567.16	12.877	43.488	7.3418
570	**296.85**	577.59	13.732	41.509	7.3603
580	**306.85**	588.04	14.630	39.645	7.3785
590	**316.85**	598.52	15.572	37.889	7.3964
600	**326.85**	609.02	16.559	36.234	7.4140
610	**336.85**	619.54	17.593	34.673	7.4314
620	**346.85**	630.08	18.676	33.198	7.4486
630	**356.85**	640.65	19.810	31.802	7.4655
640	**366.85**	651.24	20.995	30.483	7.4821
650	**376.85**	661.85	22.234	29.235	7.4986
660	**386.85**	672.49	23.528	28.052	7.5148
670	**396.85**	683.15	24.880	26.929	7.5309
680	**406.85**	693.84	26.291	25.864	7.5467
690	**416.85**	704.55	27.763	24.853	7.5623
700	**426.85**	715.28	29.298	23.892	7.5778
710	**436.85**	726.04	30.898	22.979	7.5931
720	**446.85**	736.82	32.565	22.110	7.6081
730	**456.85**	747.63	34.301	21.282	7.6230
740	**466.85**	758.46	36.109	20.494	7.6378
750	**476.85**	769.32	37.989	19.743	7.6523
760	**486.85**	780.19	39.945	19.026	7.6667
770	**496.85**	791.10	41.978	18.343	7.6810
780	**506.85**	802.02	44.092	17.690	7.6951
790	**516.85**	812.97	46.288	17.067	7.7090
800	**526.85**	823.94	48.568	16.472	7.7228
810	**536.85**	834.94	50.935	15.903	7.7365
820	**546.85**	845.96	53.392	15.358	7.7500
830	**556.85**	857.00	55.941	14.837	7.7634
840	**566.85**	868.06	58.584	14.338	7.7767
850	**576.85**	879.15	61.325	13.861	7.7898
860	**586.85**	890.26	64.165	13.403	7.8028
870	**596.85**	901.39	67.107	12.964	7.8156
880	**606.85**	912.54	70.155	12.544	7.8284
890	**616.85**	923.72	73.310	12.140	7.8410
900	**626.85**	934.91	76.576	11.753	7.8535
910	**636.85**	946.13	79.956	11.381	7.8659
920	**646.85**	957.37	83.452	11.024	7.8782
930	**656.85**	968.63	87.067	10.681	7.8904
940	**666.85**	979.90	90.805	10.352	7.9024
950	**676.85**	991.20	94.667	10.035	7.9144
960	**686.85**	1002.52	98.659	9.7305	7.9262
970	**696.85**	1013.86	102.78	9.4376	7.9380
980	**706.85**	1025.22	107.04	9.1555	7.9496
990	**716.85**	1036.60	111.43	8.8845	7.9612
1000	**726.85**	1047.99	115.97	8.6229	7.9727
1010	**736.85**	1059.41	120.65	8.3713	7.9840
1020	**746.85**	1070.84	125.49	8.1281	7.9953
1030	**756.85**	1082.30	130.47	7.8945	8.0065
1040	**766.85**	1093.77	135.60	7.6696	8.0175
1050	**776.85**	1105.26	140.90	7.4521	8.0285
1060	**786.85**	1116.76	146.36	7.2424	8.0394
1070	**796.85**	1128.28	151.98	7.0404	8.0503
1080	**806.85**	1139.82	157.77	6.8454	8.0610
1090	**816.85**	1151.38	163.74	6.6569	8.0716
1100	**826.85**	1162.95	169.88	6.4752	8.0822
1110	**836.85**	1174.54	176.20	6.2997	8.0927
1120	**846.85**	1186.15	182.71	6.1299	8.1031
1130	**856.85**	1197.77	189.40	5.9662	8.1134
1140	**866.85**	1209.40	196.29	5.8077	8.1237
1150	**876.85**	1221.06	203.38	5.6544	8.1339
1160	**886.85**	1232.72	210.66	5.5065	8.1440
1170	**896.85**	1244.41	218.15	5.3633	8.1540
1180	**906.85**	1256.10	225.85	5.2247	8.1639
1190	**916.85**	1267.82	233.77	5.0905	8.1738

Source: Calculated from *JANAF Thermochemical Tables,* 3rd ed., published by the American Chemical Society and the American Institute of Physics for the National Bureau of Standards, 1986.

Table A·12 Properties of Air (*Continued*)

T K	T °C	h kJ/kg	p_r	v_r	ϕ kJ/kg·K
1200	**926.85**	1279.54	241.90	4.9607	8.1836
1210	**936.85**	1291.28	250.25	4.8352	8.1934
1220	**946.85**	1303.04	258.83	4.7135	8.2031
1230	**956.85**	1314.81	267.64	4.5957	8.2127
1240	**966.85**	1326.59	276.69	4.4815	8.2222
1250	**976.85**	1338.39	285.98	4.3709	8.2317
1260	**986.85**	1350.20	295.51	4.2638	8.2411
1270	**996.85**	1362.03	305.30	4.1598	8.2504
1280	**1006.85**	1373.86	315.33	4.0592	8.2597
1290	**1016.85**	1385.71	325.63	3.9616	8.2690
1300	**1026.85**	1397.58	336.19	3.8669	8.2781
1310	**1036.85**	1409.45	347.02	3.7750	8.2872
1320	**1046.85**	1421.34	358.13	3.6858	8.2963
1330	**1056.85**	1433.24	369.51	3.5994	8.3052
1340	**1066.85**	1445.16	381.19	3.5153	8.3142
1350	**1076.85**	1457.08	393.15	3.4338	8.3230
1360	**1086.85**	1469.02	405.40	3.3547	8.3318
1370	**1096.85**	1480.97	417.96	3.2778	8.3406
1380	**1106.85**	1492.93	430.82	3.2032	8.3493
1390	**1116.85**	1504.91	444.00	3.1306	8.3579
1400	**1126.85**	1516.89	457.49	3.0602	8.3665
1410	**1136.85**	1528.89	471.30	2.9917	8.3751
1420	**1146.85**	1540.89	485.44	2.9252	8.3836
1430	**1156.85**	1552.91	499.92	2.8605	8.3920
1440	**1166.85**	1564.94	514.74	2.7975	8.4004
1450	**1176.85**	1576.98	529.90	2.7364	8.4087
1460	**1186.85**	1589.03	545.42	2.6768	8.4170
1470	**1196.85**	1601.09	561.29	2.6190	8.4252
1480	**1206.85**	1613.17	577.53	2.5626	8.4334
1490	**1216.85**	1625.25	594.13	2.5079	8.4415
1500	**1226.85**	1637.34	611.12	2.4545	8.4496
1510	**1236.85**	1649.44	628.48	2.4026	8.4577
1520	**1246.85**	1661.56	646.24	2.3521	8.4657
1530	**1256.85**	1673.68	664.38	2.3029	8.4736
1540	**1266.85**	1685.81	682.94	2.2550	8.4815
1550	**1276.85**	1697.95	701.89	2.2083	8.4894
1560	**1286.85**	1710.10	721.27	2.1629	8.4972
1570	**1296.85**	1722.26	741.06	2.1186	8.5050
1580	**1306.85**	1734.43	761.29	2.0754	8.5127
1590	**1316.85**	1746.61	781.94	2.0334	8.5204
1600	**1326.85**	1758.80	803.04	1.9924	8.5280
1610	**1336.85**	1771.00	824.59	1.9525	8.5356
1620	**1346.85**	1783.21	846.60	1.9135	8.5432
1630	**1356.85**	1795.42	869.06	1.8756	8.5507
1640	**1366.85**	1807.64	892.00	1.8386	8.5582
1650	**1376.85**	1819.88	915.41	1.8025	8.5656
1660	**1386.85**	1832.12	939.31	1.7673	8.5730
1670	**1396.85**	1844.37	963.70	1.7329	8.5804
1680	**1406.85**	1856.63	988.59	1.6994	8.5877
1690	**1416.85**	1868.89	1013.99	1.6667	8.5950
1700	**1426.85**	1881.17	1039.9	1.6348	8.6022
1725	**1451.85**	1911.89	1107.0	1.5583	8.6201
1750	**1476.85**	1942.66	1177.4	1.4863	8.6378
1775	**1501.85**	1973.48	1251.4	1.4184	8.6553
1800	**1526.85**	2004.34	1329.0	1.3544	8.6726
1825	**1551.85**	2035.25	1410.4	1.2940	8.6897
1850	**1576.85**	2066.21	1495.6	1.2370	8.7065
1875	**1601.85**	2097.20	1584.9	1.1830	8.7231
1900	**1626.85**	2128.25	1678.4	1.1320	8.7396
1925	**1651.85**	2159.33	1776.2	1.0838	8.7558
1950	**1676.85**	2190.45	1878.4	1.0381	8.7719
1975	**1701.85**	2221.61	1985.3	0.99481	8.7878
2000	**1726.85**	2252.82	2096.9	0.95379	8.8035
2025	**1751.85**	2284.05	2213.5	0.91484	8.8190
2050	**1776.85**	2315.33	2335.1	0.87791	8.8344
2075	**1801.85**	2346.64	2461.9	0.84284	8.8495
2100	**1826.85**	2377.99	2594.2	0.80950	8.8646
2125	**1851.85**	2409.38	2732.0	0.77782	8.8794
2150	**1876.85**	2440.80	2875.6	0.74767	8.8941
2175	**1901.85**	2472.25	3025.0	0.71901	8.9087
2200	**1926.85**	2503.73	3180.6	0.69169	8.9230
2225	**1951.85**	2535.25	3342.5	0.66567	8.9373
2250	**1976.85**	2566.80	3510.8	0.64088	8.9514
2275	**2001.85**	2598.38	3685.8	0.61723	8.9654
2300	**2026.85**	2630.00	3867.6	0.59468	8.9792
2325	**2051.85**	2661.64	4056.5	0.57315	8.9929
2350	**2076.85**	2693.32	4252.6	0.55260	9.0064
2375	**2101.85**	2725.02	4456.2	0.53297	9.0198
2400	**2126.85**	2756.75	4667.4	0.51420	9.0331
2425	**2151.85**	2788.51	4886.5	0.49627	9.0463
2450	**2176.85**	2820.31	5113.7	0.47911	9.0593
2475	**2201.85**	2852.12	5349.2	0.46269	9.0722
2500	**2226.85**	2883.97	5593.2	0.44697	9.0851
2525	**2251.85**	2915.84	5846.0	0.43192	9.0977
2550	**2276.85**	2947.74	6107.7	0.41751	9.1103
2575	**2301.85**	2979.67	6378.7	0.40369	9.1228
2600	**2326.85**	3011.63	6659.2	0.39044	9.1351
2625	**2351.85**	3043.61	6949.4	0.37773	9.1474
2650	**2376.85**	3075.61	7249.5	0.36554	9.1595
2675	**2401.85**	3107.64	7559.8	0.35385	9.1715
2700	**2426.85**	3139.70	7880.6	0.34261	9.1835
2725	**2451.85**	3171.78	8212.1	0.33183	9.1953
2750	**2476.85**	3203.88	8554.7	0.32146	9.2070
2775	**2501.85**	3236.01	8908.4	0.31150	9.2186
2800	**2526.85**	3268.16	9273.7	0.30193	9.2302
2825	**2551.85**	3300.33	9650.9	0.29272	9.2416
2850	**2576.85**	3332.53	10040.0	0.28386	9.2530
2875	**2601.85**	3364.75	10441.6	0.27534	9.2642
2900	**2626.85**	3396.99	10855.8	0.26714	9.2754
2925	**2651.85**	3429.25	11283.0	0.25924	9.2865

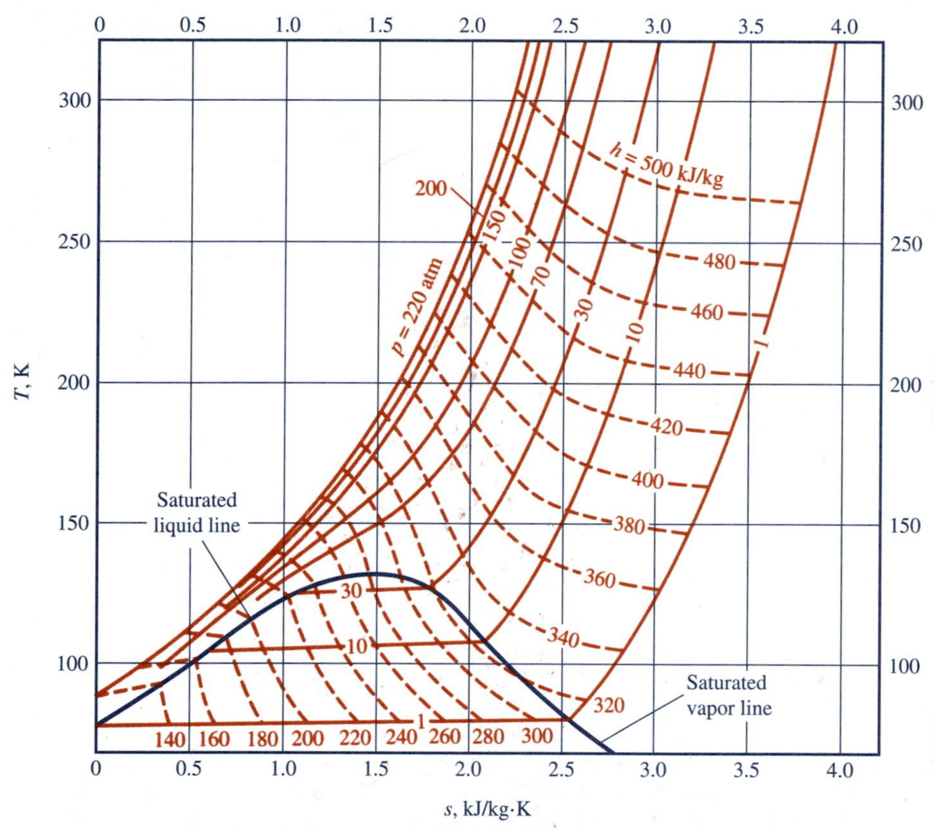

Chart A·13 Air *Ts* diagram. (*Source:* Data from "Thermodynamic Properties of Air at Low Temperatures," by V. C. Williams, *Transactions of the AIChE,* Vol. 39, no. 1, Feb. 1943.)

Chart A·14 Psychrometric Chart

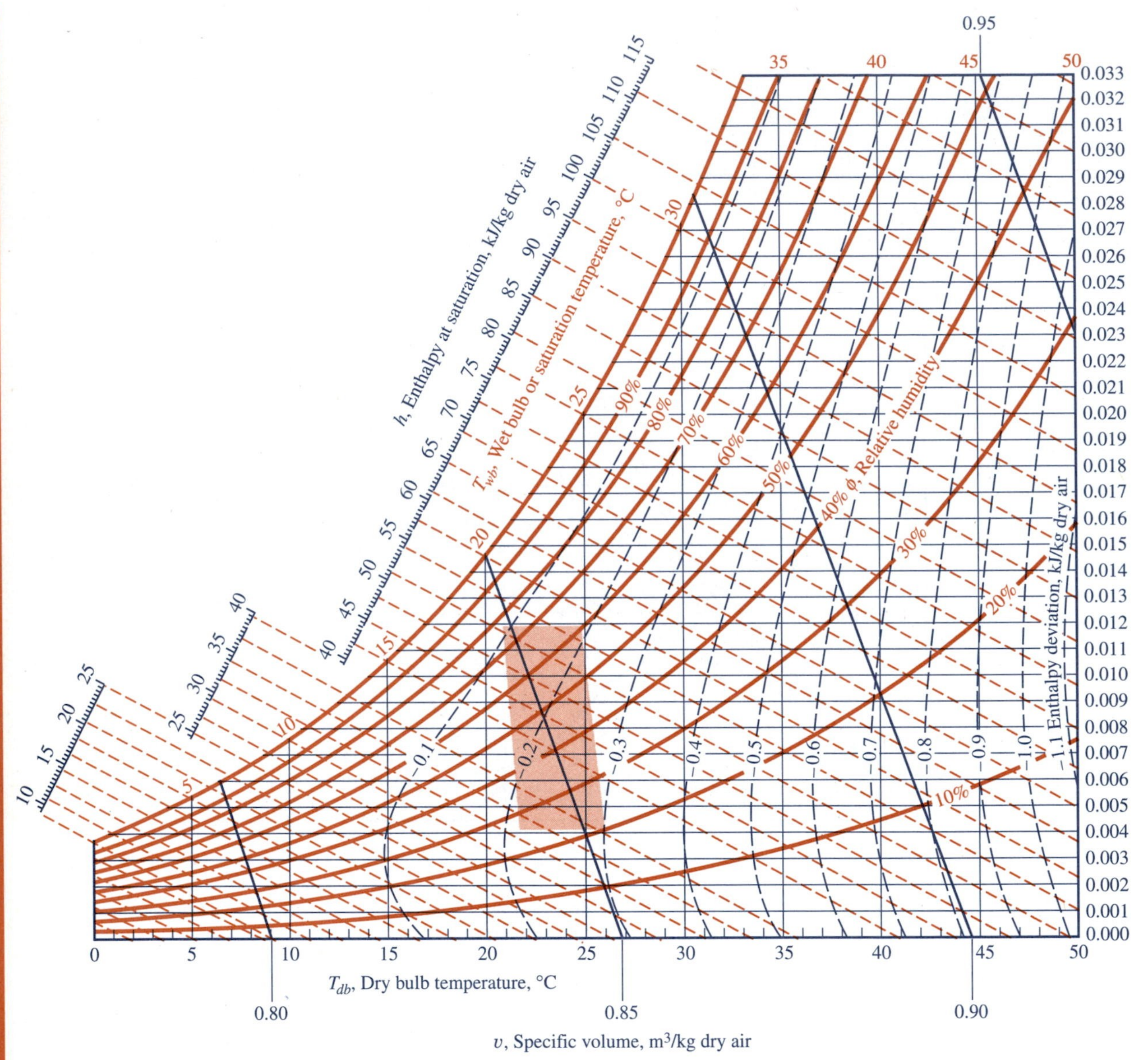

Chart A·14 Psychrometric chart. (*Source:* Based on Trane SI Metric Psychrometric Chart, The Trane Co., La Crosse, Wisconsin, with permission.)

Table A·15 Properties of Ideal Gases. $\bar{h}$ in kJ/kmol, $\bar{\phi}$ in kJ/kmol·K

	CO		CO_2		H_2		H_2O		N_2		
T, K	$\bar{h}$	$\bar{\phi}$	$\bar{h}$	$\bar{\phi}$	$\bar{h}$	$\bar{\phi}$	$\bar{h}$	$\bar{\phi}$	$\bar{h}$	$\bar{\phi}$	T, K
200	5804.9	185.991	5951.8	199.980	5667.8	119.303	6626.8	175.506	5803.1	179.944	**200**
298.15	8671.0	197.653	9364.0	213.795	8467.0	130.680	9904.0	188.834	8670.0	191.609	**298.15**
300	8724.9	197.833	9432.8	214.025	8520.4	130.858	9966.1	189.042	8723.9	191.789	**300**
400	11646.2	206.236	13366.7	225.314	11424.9	139.212	13357.0	198.792	11640.4	200.179	**400**
500	14601.4	212.828	17668.9	234.901	14348.6	145.736	16830.2	206.538	14580.2	206.737	**500**
600	17612.7	218.317	22271.3	243.284	17278.6	151.078	20405.9	213.054	17564.2	212.176	**600**
700	20692.6	223.063	27120.0	250.754	20215.1	155.604	24096.2	218.741	20606.6	216.865	**700**
800	23845.9	227.273	32172.6	257.498	23166.4	159.545	27907.2	223.828	23715.2	221.015	**800**
900	27070.6	231.070	37395.9	263.648	26141.9	163.049	31842.5	228.461	26891.8	224.756	**900**
1000	30359.8	234.535	42763.1	269.302	29147.3	166.215	35904.6	232.740	30132.2	228.169	**1000**
1100	33705.1	237.723	48248.2	274.529	32187.4	169.112	40094.1	236.732	33428.8	231.311	**1100**
1200	37099.6	240.676	53836.7	279.391	35266.4	171.791	44412.4	240.489	36778.0	234.225	**1200**
1300	40537.1	243.428	59512.8	283.934	38386.7	174.289	48851.4	244.041	40173.0	236.942	**1300**
1400	44012.0	246.003	65263.1	288.195	41549.8	176.633	53403.6	247.414	43607.8	239.487	**1400**
1500	47519.3	248.422	71076.3	292.206	44756.1	178.845	58061.6	250.628	47077.1	241.881	**1500**
1600	51054.8	250.704	76943.0	295.992	48005.7	180.942	62818.1	253.697	50576.6	244.139	**1600**
1700	54614.9	252.862	82855.4	299.576	51297.9	182.937	67666.1	256.636	54102.5	246.276	**1700**
1800	58196.5	254.909	88807.3	302.978	54631.5	184.843	72599.1	259.455	57651.4	248.305	**1800**
1900	61796.9	256.856	94793.9	306.215	58004.8	186.666	77610.7	262.165	61220.9	250.235	**1900**
2000	65413.9	258.711	100811.2	309.301	61416.1	188.416	82694.7	264.772	64808.5	252.075	**2000**
2100	69045.8	260.483	106856.4	312.250	64863.3	190.098	87845.6	267.285	68412.5	253.833	**2100**
2200	72691.0	262.179	112927.3	315.075	68344.2	191.717	93057.9	269.710	72031.5	255.517	**2200**
2300	76348.3	263.805	119022.2	317.784	71856.8	193.279	98326.7	272.052	75664.0	257.132	**2300**
2400	80016.7	265.366	125139.6	320.387	75399.0	194.786	103647.3	274.316	79309.1	258.683	**2400**
2500	83695.4	266.867	131278.2	322.893	78969.1	196.243	109015.5	276.508	82965.7	260.176	**2500**
2600	87383.4	268.314	137436.6	325.308	82565.7	197.654	114427.7	278.630	86632.9	261.614	**2600**
2700	91080.2	269.709	143613.0	327.639	86187.7	199.021	119880.3	280.688	90309.8	263.001	**2700**
2800	94785.0	271.056	149805.2	329.891	89834.8	200.347	125370.6	282.685	93995.3	264.342	**2800**
2900	98497.0	272.359	156010.3	332.069	93507.0	201.636	130896.2	284.624	97688.2	265.638	**2900**
3000	102215.2	273.620	162224.3	334.175	97205.4	202.890	136455.2	286.508	101387.1	266.892	**3000**

	O_2		H		N		O		NO		
T, K	$\bar{h}$	$\bar{\phi}$	$\bar{h}$	$\bar{\phi}$	$\bar{h}$	$\bar{\phi}$	$\bar{h}$	$\bar{\phi}$	$\bar{h}$	$\bar{\phi}$	T, K
200	5814.7	193.481	4157.1	106.417	4157.1	145.001	4545.4	152.183	6253.1	198.797	**200**
298.15	8683.0	205.147	6197.0	114.716	6197.0	153.300	6725.0	161.058	9192.0	210.758	**298.15**
300	8737.3	205.329	6235.5	114.845	6235.5	153.429	6765.5	161.193	9247.1	210.942	**300**
400	11708.9	213.872	8314.1	120.824	8314.1	159.408	8932.8	167.431	12234.2	219.534	**400**
500	14767.3	220.693	10392.7	125.463	10392.7	164.047	11068.7	172.198	15262.9	226.290	**500**
600	17926.1	226.449	12471.3	129.253	12471.3	167.837	13186.7	176.060	18358.2	231.931	**600**
700	21181.4	231.466	14549.9	132.457	14549.9	171.041	15294.6	179.310	21528.3	236.817	**700**
800	24519.3	235.922	16628.5	135.232	16628.5	173.816	17395.9	182.116	24770.9	241.146	**800**
900	27924.0	239.931	18707.1	137.680	18707.1	176.264	19492.3	184.585	28079.3	245.042	**900**
1000	31384.4	243.576	20785.8	139.871	20785.8	178.455	21585.0	186.790	31449.2	248.591	**1000**
1100	34893.5	246.921	22864.8	141.852	22864.2	180.436	23675.9	188.783	34871.9	251.853	**1100**
1200	38441.1	250.007	24943.1	143.661	24942.9	182.244	25764.2	190.600	38339.5	254.870	**1200**
1300	42022.9	252.874	27021.3	145.324	27021.7	183.908	27850.8	192.270	41845.3	257.676	**1300**
1400	45635.9	255.551	29099.3	146.864	29100.5	185.449	29936.0	193.815	45383.8	260.298	**1400**
1500	49277.4	258.064	31177.4	148.298	31179.3	186.883	32020.4	195.253	48950.2	262.759	**1500**
1600	52945.4	260.431	33255.7	149.639	33258.0	188.224	34104.2	196.598	52540.4	265.076	**1600**
1700	56638.3	262.670	35334.3	150.899	35336.6	189.485	36187.8	197.861	56151.1	267.265	**1700**
1800	60355.1	264.794	37413.0	152.087	37415.2	190.673	38271.3	199.052	59779.5	269.339	**1800**
1900	64094.9	266.816	39491.9	153.211	39493.8	191.797	40354.7	200.179	63423.4	271.309	**1900**
2000	67857.5	268.746	41570.9	154.278	41572.5	192.863	42438.2	201.247	67081.0	273.185	**2000**
2100	71642.6	270.593	43649.9	155.292	43651.5	193.877	44521.8	202.264	70750.9	274.975	**2100**
2200	75450.0	272.364	45728.8	156.259	45730.8	194.844	46605.6	203.233	74432.0	276.688	**2200**
2300	79279.7	274.066	47807.6	157.183	47810.8	195.769	48689.7	204.160	78123.4	278.329	**2300**
2400	83131.9	275.706	49886.3	158.068	49891.6	196.655	50774.1	205.047	81824.5	279.904	**2400**
2500	87006.2	277.287	51964.7	158.916	51973.6	197.505	52859.2	205.898	85534.6	281.418	**2500**
2600	90902.6	278.815	54043.0	159.731	54057.2	198.322	54945.0	206.716	89253.1	282.877	**2600**
2700	94820.6	280.294	56121.1	160.516	56142.7	199.109	57032.1	207.504	92979.3	284.283	**2700**
2800	98759.4	281.726	58199.2	161.271	58230.5	199.868	59120.8	208.263	96712.4	285.641	**2800**
2900	102717.9	283.115	60277.6	162.001	60321.3	200.602	61211.8	208.997	100451.2	286.953	**2900**
3000	106694.5	284.464	62356.4	162.706	62415.5	201.312	63305.8	209.707	104194.5	288.222	**3000**

Source: JANAF Thermochemical Tables, 3rd ed., National Bureau of Standards, NSRDS-NBS 37, 1986.

Table A·15 Properties of Ideal Gases, $\bar{h}$ in kJ/kmol, $\bar{\phi}$ in kJ/kmol·K (*Continued*)

	OH		CH_4		C_2H_2		C_2H_4		Air		
T, K	$\bar{h}$	$\bar{\phi}$	$\bar{h}$	$\bar{\phi}$	$\bar{h}$	$\bar{\phi}$	$\bar{h}$	$\bar{\phi}$	$\bar{h}$	$\bar{\phi}$	*T*, K
200	6206.5	171.638	6691.7	172.733	6076.7	185.062	6818.6	204.417	5848.2	182.511	**200**
298.15	9172.0	183.708	10018.7	186.233	10005.4	200.936	10511.6	219.308	8703.5	194.128	**298.15**
300	9227.5	183.893	10089.9	186.471	10093.7	201.231	10597.4	219.595	8757.2	194.308	**300**
400	12207.1	192.468	13888.9	197.367	14843.4	214.853	15406.8	233.362	11675.2	202.701	**400**
500	15164.3	199.067	18225.3	207.019	20118.2	226.605	21188.4	246.224	14631.0	209.294	**500**
600	18115.1	204.447	23151.4	215.984	25783.2	236.924	27850.1	258.347	17643.3	214.784	**600**
700	21073.7	209.007	28659.1	224.463	31759.0	246.130	35281.9	269.789	20721.8	219.529	**700**
800	24052.2	212.984	34704.6	232.528	38003.7	254.465	43372.8	280.584	23869.7	223.731	**800**
900	27060.5	216.527	41232.6	240.212	44496.0	262.109	52027.1	290.771	27084.6	227.517	**900**
1000	30106.8	219.736	48200.7	247.550	51217.3	269.188	61180.4	300.411	30360.5	230.968	**1000**
1100	33197.0	222.681	55567.3	254.568	58143.2	275.788	70773.3	309.551	33691.0	234.142	**1100**
1200	36333.7	225.410	63290.1	261.285	65261.1	281.980	80761.2	318.239	37068.6	237.080	**1200**
1300	39516.3	227.957	71325.4	267.716	72552.1	287.815	91092.2	326.506	40488.0	239.817	**1300**
1400	42743.3	230.349	79634.7	273.872	79999.2	293.333	101721.0	334.382	43944.6	242.379	**1400**
1500	46013.1	232.604	88183.9	279.770	87587.2	298.568	112608.4	341.893	47434.0	244.786	**1500**
1600	49323.9	234.741	96943.5	285.422	95302.5	303.547	123720.8	349.064	50952.8	247.057	**1600**
1700	52673.7	236.772	105887.7	290.844	103133.1	308.294	135029.5	355.919	54497.7	249.206	**1700**
1800	56060.0	238.707	114994.3	296.049	111068.3	312.829	146510.3	362.481	58066.0	251.245	**1800**
1900	59480.7	240.556	124244.4	301.050	119098.8	317.171	158142.9	368.770	61655.6	253.186	**1900**
2000	62933.2	242.327	133621.7	305.860	127216.2	321.334	169910.5	374.805	65264.4	255.037	**2000**
2100	66415.3	244.026	143112.5	310.490	135413.3	325.334	181799.2	380.606	68890.8	256.807	**2100**
2200	69924.5	245.659	152705.1	314.952	143683.8	329.181	193797.1	386.187	72533.5	258.501	**2200**
2300	73458.6	247.230	162389.7	319.257	152022.0	332.887	205894.6	391.564	76191.4	260.127	**2300**
2400	77015.6	248.743	172157.5	323.414	160422.9	336.463	218082.9	396.752	79863.5	261.690	**2400**
2500	80593.6	250.204	182001.0	327.432	168882.0	339.916	230354.4	401.761	83549.0	263.194	**2500**
2600	84191.0	251.615	191913.1	331.320	177395.1	343.254	242701.4	406.603	87247.2	264.645	**2600**
2700	87806.6	252.979	201886.9	335.084	185958.3	346.486	255115.9	411.289	90957.4	266.045	**2700**
2800	91439.5	254.301	211915.6	338.731	194567.7	349.617	267589.1	415.825	94679.0	267.398	**2800**
2900	95089.2	255.581	221991.7	342.267	203219.6	352.653	280110.9	420.219	98411.2	268.708	**2900**
3000	98755.9	256.824	232106.8	345.696	211909.8	355.599	292669.1	424.476	102153.2	269.977	**3000**

Table A·16 Enthalpy of Eight Gaseous Fuels at Low Pressure

T K	Methane CH_4 kJ/kmol	Ethane C_2H_6 kJ/kmol	Propane C_3H_8 kJ/kmol	*n*-Butane C_4H_{10} kJ/kmol	*n*-Octane C_8H_{18} kJ/kmol	Acetylene C_2H_2 kJ/kmol	Ethylene C_2H_4 kJ/kmol	Methanol CH_3OH kJ/kmol	*T* K
0	0	0	0	0	0	0	0	0	**0**
200	6644	7259	8414	10744	21581	6077	6819		**200**
298.15	10016	11874	14740	19276	37782	10012	10518	11427	**298.15**
300	10083	11975	14874	19460	38116	10094	10597	11510	**300**
400	13887	17874	23276	30648	59580	14843	15407	16263	**400**
500	18238	25058	33623	44350	85939	20118	21188	21811	**500**
600	23192	33430	45689	60250	116650	25783	27850	28146	**600**
700	28727	42844	59287	78115	151084	31759	35282	35192	**700**
800	34819	53220	74182	97613	188698	38004	43373	42865	**800**
900	41417	64434	90207	118533	228466	44496	52027	51087	**900**
1000	48492	76358	107194	140666	270705	51217	61180	59810	**1000**
1100	55982	88910	125102	163929		58143	70773		**1100**

Source: Data for all gases but ammonia from *Selected Values of Properties of Hydrocarbons and Related Compounds,* Thermodynamics Research Center Data Project, Thermodynamics Research Center, Texas A & M University, College Station, Texas 77843; data for ammonia from *JANAF Thermochemical Tables,* National Bureau of Standards, 3rd ed., published by the American Chemical Society and the American Institute of Physics for the National Bureau of Standards, 1986.

Table A·17 Thermochemical Properties of Substances at 25°C, 100 kPa

Name	Molar Mass M kg/kmol	Phase	$\bar{h}_{fg}$ kJ/kmol	Boiling Point (1 atm) °C	Enthalpy of Combustion at 25°C, 100 kPa: $H_2O(l)$ Δh_c kJ/kg	$H_2O(l)$ $\Delta \bar{h}_c$ kJ/kmol	$H_2O(g)$ Δh_c kJ/kg	$H_2O(g)$ $\Delta \bar{h}_c$ kJ/kmol	Enthalpy of Formation $\Delta \bar{h}^\circ_{f,ref}$ kJ/kmol	Gibbs Energy of Formation $\Delta \bar{g}^\circ_{f,ref}$ kJ/kmol	Absolute Entropy $\Delta \bar{s}^\circ_{ref}$ kJ/kmol·K	Formula
Hydrogen, H_2	2.016	gas	...	−252.77	−141779	−285812	−119959	−241826	0	0	130.680	H_2
Water, H_2O	18.015	gas	43986	100.00	...	...	...	...	−241826	−228582	188.834	H_2O
Water, H_2O	18.015	liquid	43986	100.00	...	...	...	...	−285830	−237141	69.950	H_2O
Carbon, C	12.011	solid	...	...	−32766	−393552	−32766	−393552	0	0	5.740	C
Carbon monoxide, CO	28.010	gas	...	−191.45	−10104	−283022	−10104	−283022	−110530	−137163	197.650	CO
Carbon dioxide, CO_2	44.010	gas	...	−78.48	...	...	...	...	−393522	−394389	213.795	CO_2
Nitrogen, N_2	28.013	gas	...	−195.81	...	...	...	...	0	0	191.609	N_2
Oxygen, O_2	31.999	gas	...	−182.96	...	...	...	...	0	0	205.147	O_2
Methane, CH_4	16.043	gas	...	−161.52	−55496	−890303	−50012	−802231	−74873	−50768	186.251	CH_4
Acetylene, C_2H_2	26.038	gas	...	−84.00	−49914	−1299650	−48224	−1255660	226731	248163	200.958	C_2H_2
Ethylene, C_2H_4	28.054	gas	...	−103.71	−50303	−1411200	−47167	−1323220	52467	68421	219.330	C_2H_4
Ethane, C_2H_6	30.069	gas	5021	−88.61	−51903	−1560690	−47514	−1428730	−83851	−31951	229.447	C_2H_6
Propene, C_3H_6	42.080	gas	...	−47.70	−48902	−2057800	−45766	−1925840	19710	62719	265.007	C_3H_6
Propane, C_3H_8	44.093	gas	14820	−42.08	−50331	−2219230	−46341	−2043290	−104673	−24398	270.696	C_3H_8
***n*-Butane, C_4H_{10}**	58.123	gas	21066	−0.51	−49509	−2877620	−45725	−2657690	−125652	−16560	310.464	C_4H_{10}
***n*-Pentane, C_5H_{12}**	72.150	gas	26426	36.07	−61198	−3556980	−56657	−3293060	−146712	−8770	350.120	C_5H_{12}
***n*-Hexane, C_6H_{14}**	86.177	gas	31552	68.73	−48680	−4195050	−45107	−3887150	−166942	−150	389.777	C_6H_{14}
***n*-Heptane, C_7H_{16}**	100.203	gas	36547	98.42	−48439	−4853710	−44927	−4501820	−187654	8151	428.887	C_7H_{16}
***n*-Octane, C_8H_{18}**	114.230	gas	41484	125.67	−48253	−5511900	−44787	−5116030	−208824	15921	468.243	C_8H_{18}
***n*-Octane, C_8H_{18}**	114.230	liquid	41484	125.67	−47889	−5470420	−44424	−5074540	−250308	5939	362.583	C_8H_{18}
Benzene, C_6H_6	78.113	gas	33849	80.10	−42268	−3301680	−40579	−3169730	82935	129661	269.760	C_6H_6
Benzene, C_6H_6	78.113	liquid	33849	80.10	−41835	−3267830	−40145	−3135880	49086	124532	173.433	C_6H_6
Methanol, CH_3OH	32.042	gas	39790	64.70	−23851	−764242	−21104	−676234	−200940	−162240	239.880	CH_3OH
Methanol, CH_3OH	32.042	liquid	39790	64.70	−22666	−726272	−19920	−638264	−238910	−166640	127.270	CH_3OH
Ethanol, C_2H_5OH	46.069	gas	42560	78.29	−30597	−1409584	−27732	−1277572	−234950	−167730	280.640	C_2H_5OH
Ethanol, C_2H_5OH	46.069	liquid	42560	78.29	−29673	−1367024	−26808	−1235012	−277510	−174350	160.100	C_2H_5OH
Hydrazine, N_2H_4	32.045	liquid	44760	114.15	−19418	−622250	−16673	−534278	50626	149440	121.544	N_2H_4
Hydrogen, H	1.008	gas	...	...	−358076	−360905	−336256	−338912	217999	203278	114.716	H
Nitrogen, N	14.007	gas	...	...	...	...	...	...	472680	455540	153.300	N
Oxygen, O	15.999	gas	...	...	...	...	...	...	249170	231736	161.058	O
Hydroxyl, OH	17.007	gas	...	...	...	...	...	...	38987	34277	183.708	OH

Source: JANAF Thermochemical Tables, 3rd ed., published by the American Chemical Society and the American Institute of Physics for the National Bureau of Standards, 1986. Also, *Journal of Physical and Chemical Reference Data,* Vol. 14, Supplement No. 1, 1985.

Table A·18 Log K_p for Seven Ideal-gas Reactions, $p° = 100$ kPa

T, K	$H_2 \rightarrow 2H$	$N_2 \rightarrow 2N$	$O_2 \rightarrow 2O$	$CO_2 \rightarrow CO + \frac{1}{2}O_2$	$H_2O \rightarrow OH + \frac{1}{2}H_2$	$\frac{1}{2}N_2 + \frac{1}{2}O_2 \rightarrow NO$	$CO_2 + H_2 \rightarrow CO + H_2O$	T, K
298.15	-71.30	-159.7	-81.28	-45.05	-46.10	-15.15	-4.950	**298.15**
300	-70.83	-158.7	-80.73	-44.74	-45.80	-15.07	-4.905	**300**
400	-51.73	-117.4	-58.91	-32.43	-33.48	-11.14	-3.215	**400**
500	-40.26	-92.63	-45.82	-25.03	-26.09	-8.784	-2.193	**500**
600	-32.62	-76.12	-37.10	-20.10	-21.16	-7.210	-1.506	**600**
700	-27.16	-64.33	-30.86	-16.57	-17.64	-6.086	-1.014	**700**
800	-23.06	-55.48	-26.19	-13.92	-15.00	-5.243	-0.642	**800**
900	-19.88	-48.60	-22.55	-11.86	-12.95	-4.587	-0.352	**900**
1000	-17.33	-43.10	-19.64	-10.21	-11.31	-4.063	-0.120	**1000**
1100	-15.18	-38.54	-17.21	-8.843	-9.922	-3.633	0.040	**1100**
1200	-13.40	-34.75	-15.20	-7.739	-8.784	-3.275	0.152	**1200**
1300	-11.89	-31.54	-13.49	-6.802	-7.821	-2.972	0.251	**1300**
1400	-10.60	-28.79	-12.03	-6.004	-6.996	-2.712	0.330	**1400**
1500	-9.474	-26.41	-10.76	-5.315	-6.280	-2.487	0.397	**1500**
1600	-8.492	-24.32	-9.657	-4.711	-5.654	-2.290	0.456	**1600**
1700	-7.626	-22.48	-8.680	-4.175	-5.102	-2.116	0.510	**1700**
1800	-6.856	-20.85	-7.811	-3.697	-4.611	-1.962	0.560	**1800**
1900	-6.168	-19.39	-7.033	-3.268	-4.172	-1.824	0.608	**1900**
2000	-5.548	-18.07	-6.334	-2.879	-3.777	-1.699	0.652	**2000**
2100	-4.987	-16.88	-5.701	-2.527	-3.419	-1.587	0.692	**2100**
2200	-4.477	-15.79	-5.125	-2.207	-3.094	-1.484	0.729	**2200**
2300	-4.012	-14.81	-4.600	-1.917	-2.797	-1.391	0.761	**2300**
2400	-3.585	-13.90	-4.118	-1.652	-2.525	-1.306	0.788	**2400**
2500	-3.192	-13.06	-3.675	-1.412	-2.274	-1.227	0.810	**2500**
2600	-2.830	-12.29	-3.266	-1.194	-2.043	-1.154	0.828	**2600**
2700	-2.495	-11.58	-2.887	-0.995	-1.829	-1.087	0.840	**2700**
2800	-2.183	-10.92	-2.536	-0.813	-1.631	-1.025	0.849	**2800**
2900	-1.893	-10.30	-2.208	-0.646	-1.446	-0.967	0.855	**2900**
3000	-1.622	-9.729	-1.903	-0.491	-1.273	-0.913	0.859	**3000**
3100	-1.369	-9.191	-1.617	-0.347	-1.111	-0.863	0.863	**3100**
3200	-1.131	-8.686	-1.349	-0.208	-0.960	-0.815	0.869	**3200**
3300	-0.908	-8.213	-1.097	-0.073	-0.818	-0.771	0.881	**3300**
3400	-0.698	-7.767	-0.860	0.062	-0.684	-0.729	0.900	**3400**
3500	-0.501	-7.346	-0.637	0.202	-0.558	-0.690	0.929	**3500**

Source: Calculated from Table A·19.

Table A·19 Log K_f for Substances, $p° = 100$ kPa

T, K	CO_2	CO	H	OH	H_2O	N	O	NO	T, K
298.15	69.09	24.04	−35.65	−6.001	40.10	−79.85	−40.64	−15.15	**298.15**
300	68.66	23.92	−35.42	−5.959	39.84	−79.34	−40.37	−15.07	**300**
400	51.54	19.11	−25.86	−4.266	29.21	−58.70	−29.46	−11.14	**400**
500	41.26	16.23	−20.13	−3.250	22.84	−46.32	−22.91	−8.784	**500**
600	34.41	14.31	−16.31	−2.573	18.59	−38.06	−18.55	−7.210	**600**
700	29.51	12.94	−13.58	−2.089	15.55	−32.16	−15.43	−6.086	**700**
800	25.83	11.91	−11.53	−1.726	13.28	−27.74	−13.09	−5.243	**800**
900	22.97	11.11	−9.940	−1.444	11.51	−24.30	−11.28	−4.587	**900**
1000	20.68	10.47	−8.666	−1.218	10.09	−21.55	−9.822	−4.063	**1000**
1100	18.80	9.961	−7.592	−1.039	8.883	−19.27	−8.605	−3.633	**1100**
1200	17.25	9.511	−6.700	−0.8920	7.892	−17.37	−7.598	−3.275	**1200**
1300	15.92	9.119	−5.945	−0.7678	7.053	−15.77	−6.746	−2.972	**1300**
1400	14.78	8.777	−5.298	−0.6612	6.334	−14.40	−6.015	−2.712	**1400**
1500	13.79	8.480	−4.737	−0.5689	5.711	−13.20	−5.382	−2.487	**1500**
1600	12.93	8.220	−4.246	−0.4881	5.166	−12.16	−4.829	−2.290	**1600**
1700	12.17	7.993	−3.813	−0.4168	4.685	−11.24	−4.340	−2.116	**1700**
1800	11.49	7.793	−3.428	−0.3535	4.258	−10.42	−3.905	−1.962	**1800**
1900	10.88	7.617	−3.084	−0.2968	3.875	−9.693	−3.517	−1.824	**1900**
2000	10.34	7.460	−2.774	−0.2458	3.531	−9.034	−3.167	−1.699	**2000**
2100	9.846	7.319	−2.494	−0.1996	3.219	−8.439	−2.850	−1.587	**2100**
2200	9.398	7.190	−2.239	−0.1576	2.936	−7.897	−2.563	−1.484	**2200**
2300	8.988	7.072	−2.006	−0.1193	2.677	−7.403	−2.300	−1.391	**2300**
2400	8.613	6.961	−1.793	−0.08421	2.440	−6.949	−2.059	−1.306	**2400**
2500	8.268	6.856	−1.596	−0.05189	2.222	−6.532	−1.838	−1.227	**2500**
2600	7.949	6.756	−1.415	−0.02206	2.021	−6.147	−1.633	−1.154	**2600**
2700	7.654	6.660	−1.247	0.005554	1.835	−5.791	−1.444	−1.087	**2700**
2800	7.380	6.567	−1.091	0.03120	1.662	−5.460	−1.268	−1.025	**2800**
2900	7.125	6.479	−0.9464	0.05508	1.501	−5.152	−1.104	−0.967	**2900**
3000	6.887	6.396	−0.8111	0.07736	1.350	−4.864	−0.9514	−0.913	**3000**
3100	6.665	6.318	−0.6844	0.09821	1.210	−4.595	−0.8084	−0.863	**3100**
3200	6.456	6.248	−0.5657	0.1178	1.078	−4.343	−0.6744	−0.815	**3200**
3300	6.260	6.187	−0.4542	0.1361	0.9538	−4.106	−0.5485	−0.771	**3300**
3400	6.075	6.138	−0.3492	0.1534	0.8372	−3.883	−0.4301	−0.729	**3400**
3500	5.901	6.104	−0.2503	0.1697	0.7272	−3.673	−0.3184	−0.690	**3500**

Source: JANAF Thermochemical Tables, 3rd ed., published by the American Chemical Society and the American Institute of Physics for the National Bureau of Standards, 1986.

Chart A·20 Aqua–ammonia Chart

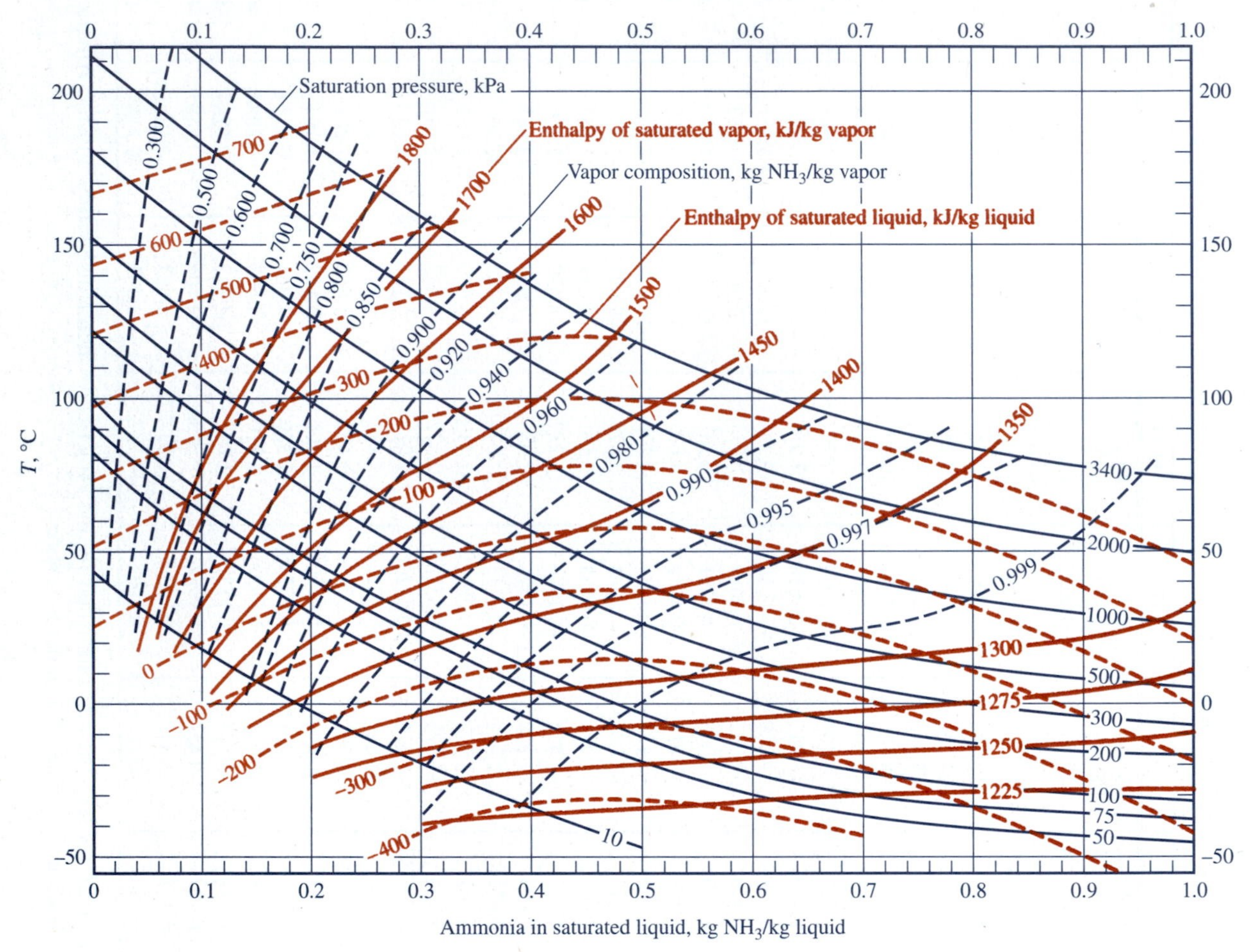

Chart A·20 Aqua–ammonia chart. (Adapted from "Equilibrium Properties of Aqua–ammonia in Chart Form," by F. H. Kohloss, Jr., and G. L. Scott, *Refrigeration Engineering,* Vol. 58, no. 10, Oct. 1950, p. 970.) Note: The state of zero enthalpy is not the same for this chart and for Tables A·8.

Chart A·21 Aqueous Lithium Bromide: Equilibrium Chart

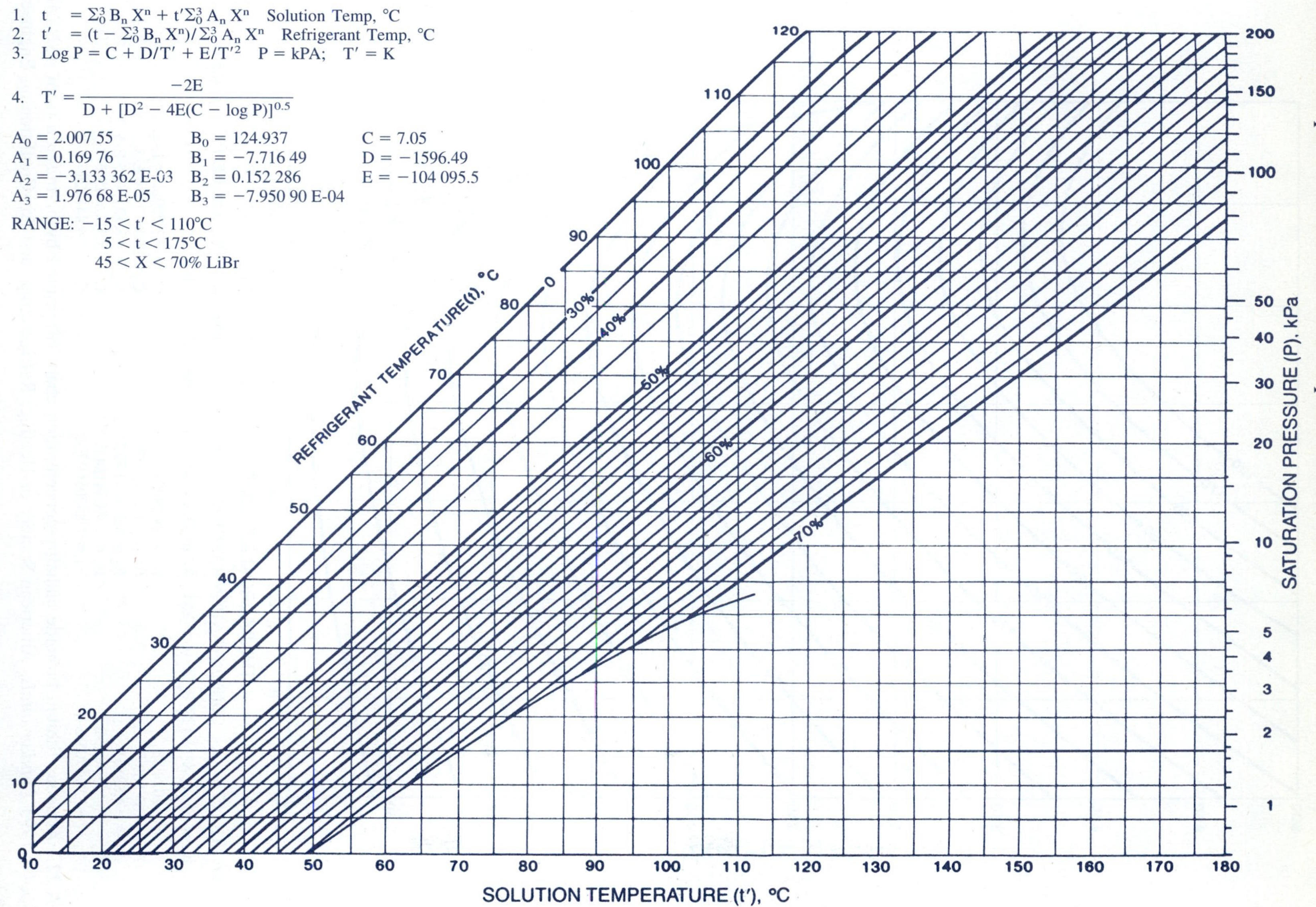

Chart A·21 Aqueous lithium bromide equilibrium chart. (Reprinted by permission from *ASHRAE Handbook—1993 Fundamentals,* American Society of Heating, Refrigerating, and Air-Conditioning Engineers, Atlanta.)

Chart A·22 Aqueous Lithium Bromide: Enthalpy-concentration Chart

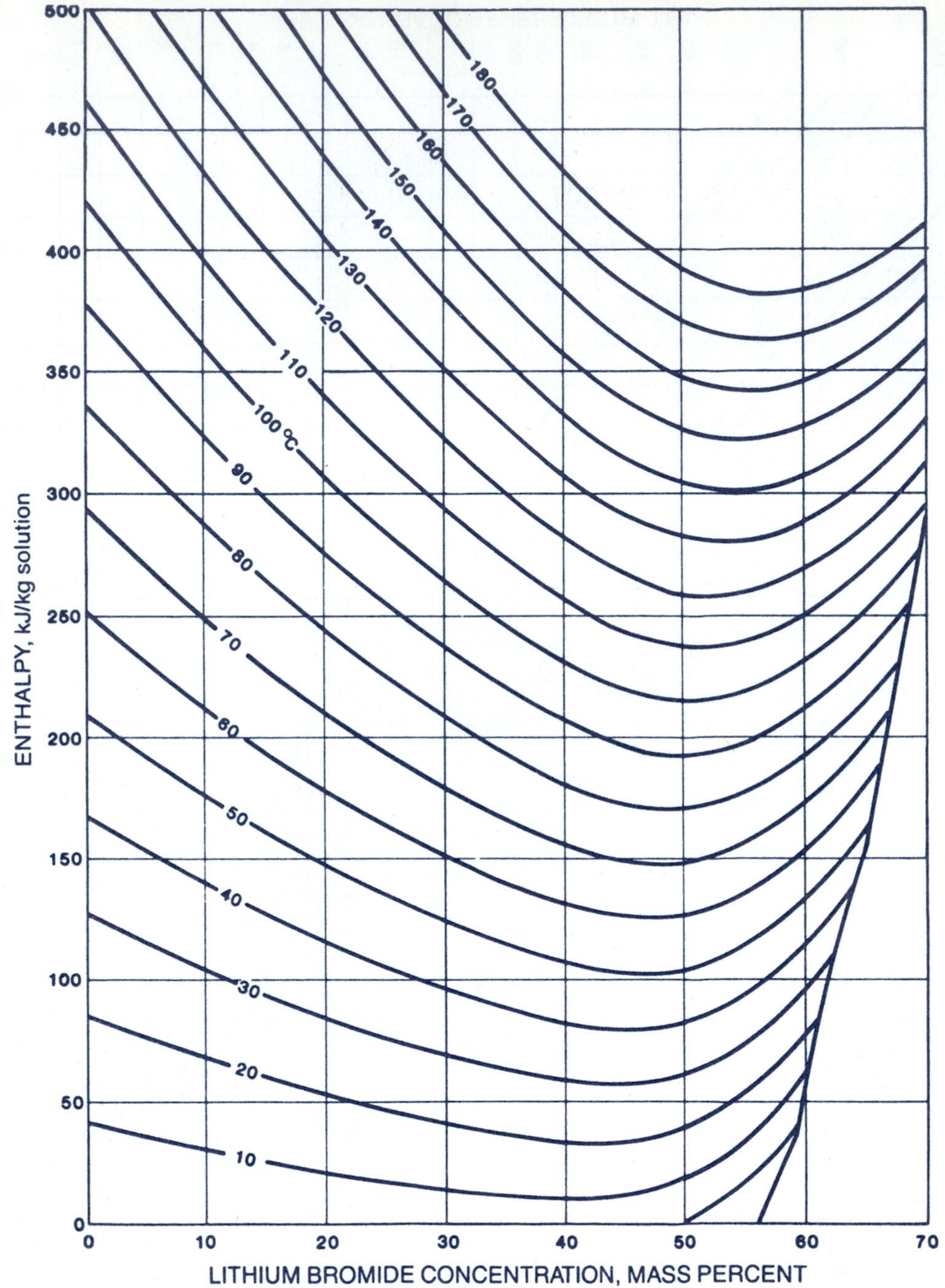

EQUATIONS CONCENTRATION RANGE 40 < X < 70% LiBr TEMPERATURE RANGE 15 < t < 165°C

$h = \Sigma_0^4 A_n X^n + t\, \Sigma_0^4 B_n X^n + t^2\, \Sigma_0^4 C_n X^n$ in kJ/kg, where t = °C and X = %LiBr

$A_0 = -2024.33$	$B_0 = 18.2829$	$C_0 = -3.7008214$ E-2
$A_1 = 163.309$	$B_1 = -1.1691757$	$C_1 = 2.8877666$ E-3
$A_2 = -4.88161$	$B_2 = 3.248041$ E-2	$C_2 = -8.1313015$ E-5
$A_3 = 6.302948$ E-2	$B_3 = -4.034184$ E-4	$C_3 = 9.9116628$ E-7
$A_4 = -2.913705$ E-4	$B_4 = 1.8520569$ E-6	$C_4 = -4.4441207$ E-9

Chart A·22 Aqueous lithium bromide enthalpy-concentration chart. (Reprinted by permission from *ASHRAE Handbook—1993 Fundamentals,* American Society of Heating, Refrigerating, and Air-Conditioning Engineers, Atlanta.)

Table A·1E Properties of Gases

Gas	Molar Mass M	Gas Constant R	Specific Heats at 77 F c_p	c_v	k	Temperature Range	Equation Coefficients for $c_p/R = a + bT + cT^2 + dT^3 + eT^4$ a	$b\times10^3$	$c\times10^6$	$d\times10^{10}$	$e\times10^{13}$
	lbm/lbmol	ft·lbf/lbm·R	B/lbm·R	B/lbm·R				R^{-1}	R^{-2}	R^{-3}	R^{-4}
Acetylene, C_2H_2	26.04	59.3	0.404	0.327	1.232	540–1800R	0.8021	13.06	−11.10	49.06	−8.349
						1800–5400R	3.825	3.759	−0.9303	0.3981	−0.02123
Air	28.97	53.3	0.241	0.171	1.400	540–1800R	3.721	−1.041	1.465	−5.906	0.8126
						1800–5400R	2.786	1.069	−0.2921	0.3981	−0.02123
Argon, Ar	39.95	38.7	0.124	0.0745	1.667		2.50	0	0	0	0
Butane, C_4H_{10}	58.12	26.6	0.399	0.365	1.094	540–2700R	0.4756	24.81	−6.802	7.214	0
Carbon Dioxide CO_2	44.01	35.1	0.202	0.156	1.289	540–1800R	2.227	5.551	−3.025	9.253	−1.220
						1800–5400R	3.247	3.248	−1.053	1.624	−0.09608
Carbon Monoxide, CO	28.01	55.2	0.248	0.178	1.399	540–1800R	3.776	−1.163	1.506	−5.690	0.6652
						1800–5400R	2.654	1.237	−0.3538	0.4888	−0.02631
Ethane, C_2H_6	30.07	51.4	0.418	0.353	1.184	540–2700R	0.8293	11.53	−2.378	1.501	0
Ethylene, C_2H_4	28.05	55.1	0.365	0.294	1.240	540–1800R	1.575	5.66	3.472	−34.14	7.809
						1800–5400R	0.2530	10.37	−3.080	4.463	−0.2541
Helium, He	4.003	386	1.24	0.745	1.667		2.50	0	0	0	0
Hydrogen, H_2	2.016	767	3.42	2.44	1.405	540–1800R	2.892	2.158	−2.731	14.91	−2.847
						1800–5400R	3.717	−0.5122	0.3767	−0.7421	0.04956
Hydrogen, H	1.008	1530	4.92	2.96	1.667	540–1800R	2.496	0.01654	−0.02363	0.1413	−0.03009
						1800–5400R	2.567	−0.08384	0.03761	−0.07175	0.004937
Hydroxyl, OH	17.01	90.9	0.420	0.303	1.384	540–1800R	3.874	−0.7494	0.5154	−0.9722	0.05896
						1800–5400R	3.229	0.1119	0.1345	−0.3504	0.02568
Methane, CH_4	16.04	96.3	0.530	0.406	1.304	540–1800R	4.503	−4.981	11.54	−62.58	11.64
						1800–5400R	−0.6992	8.505	−2.375	3.251	−0.1762
Neon, Ne	20.18	76.6	0.246	0.148	1.667		2.50	0	0	0	0
Nitric Oxide, NO	30.01	51.5	0.238	0.171	1.386	540–1800R	4.120	−2.347	3.325	−16.74	3.034
						1800–5400R	2.730	1.318	−0.4129	0.6180	−0.03566
Nitrogen, N_2	28.01	55.2	0.248	0.177	1.400	540–1800R	3.725	−0.868	0.9900	−2.664	0.1100
						1800–5400R	2.469	1.371	−0.4049	0.5832	−0.03290
Nitrogen, N	14.01	110	0.353	0.213	1.667	540–1800R	2.496	0.01654	−0.02363	0.1413	−0.03009
						1800–5400R	2.483	0.01685	−0.004683	0.0003222	0.0009200
Oxygen, O_2	32.00	48.3	0.219	0.157	1.395	540–1800R	3.837	1.900	3.392	−18.79	3.569
						1800–5400R	3.156	1.005	−0.3245	0.5470	−0.03457
Oxygen, O	16.00	96.6	0.327	0.203	1.612	540–1800R	3.020	1.209	1.171	−5.250	8.956
						1800–5400R	2.662	−0.1695	0.06944	−0.1277	0.008938
Propane, C_3H_8	44.10	35.0	0.399	0.353	1.127	540–2700R	−0.4861	20.35	−5.833	6.540	0
Water, H_2O	18.02	85.8	0.444	0.334	1.329	540–1800R	4.132	−0.8659	16.40	−7.218	1.223
						1800–5400R	2.798	1.496	−0.1664	−0.003057	0.008599

Table A·1E Properties of Gases (*Continued*)

Gas	Critical State Properties		Triple State Properties		van der Waals Constants		Redlich–Kwong Constants		Acentric Factor	Gas
	p_c	T_c	p_t	T_t	a	b	a	b	ω	
	psi	R	psi	R	$\frac{psi \cdot ft^6}{lbmol^2}$	$ft^3/lbmol$	$\frac{psi \cdot ft^6 R^{0.5}}{lbmol^2}$	$ft^3/lbmol$		
C_2H_2	891	554	18.3	346	16800	0.836	401000	0.580	0.1873	Acetylene, C_2H_2
Air	547	238			5060	0.585	78900	0.405	0.007400	Air
Ar	711	273	9.98	151	5020	0.513	83900	0.356	0.0000	Argon, Ar
C_4H_{10}	551	675	0.0000978	243	51700	1.86	1450000	1.29	0.1931	Butane, C_4H_{10}
CO_2	1070	547	75.1	391	13600	0.686	322000	0.476	0.2276	Carbon Dioxide, CO_2
CO	508	239	2.23	123	5470	0.663	85900	0.439	0.06630	Carbon Monoxide, CO
C_2H_6	708	550			20700	1.040	492000	0.721	0.099	Ethane, C_2H_6
C_2H_4	730	508	0.0203	187	17200	0.934	392000	0.647	0.08520	Ethylene, C_2H_4
He	33.1	9.36			129	0.381	399	0.264	−0.3900	Helium, He
H_2	190	59.8	1.02	24.8	912	0.421	7140	0.292	−0.2150	Hydrogen, H_2
H										Hydrogen, H
OH										Hydroxyl, OH
CH_4	667	344	1.70	16.3	33300	0.689	160000	0.477	0.01080	Methane, CH_4
Ne	384	79.9	7.25		808	0.279	7290	0.192	−0.04140	Neon, Ne
NO	940	324	3.18		5430	0.463	98900	0.320	0.5846	Nitric Oxide, NO
N_2	492	227	1.81		5100	0.618	77400	0.428	0.04030	Nitrogen, N_2
N										Nitrogen, N
O_2	731	279	0.0212		5140	0.511	86900	0.354	0.02180	Oxygen, O_2
O										Oxygen, O
C_3H_8	618	666			34900	1.45	914000	1.000	0.153	Propane, C_3H_8
H_2O	3210	1160	0.0888		20600	0.489	71400	0.338	0.3449	Water, H_2O

Source: Chiefly *JANAF Thermochemical Tables,* 3rd ed., published by the American Chemical Society and the American Institute of Physics for the National Bureau of Standards, 1986. Data for butane, ethane, and propane from Kobe, K. A., and E. G. Long, ''Thermochemistry for the Petrochemical Industry, Part II—Paraffinic Hydrocarbons, C_1-C_6,'' *Petroleum Refiner,* Vol. 28, No. 2, 1949, pp. 113–116. Constants for equations of state are calculated from critical state properties.

Table A·2E Benedict–Webb–Rubin Constants

Gas	Formula	R ft·lbf/lbm·R	A_0 lbf·ft^4/lbm^2	B_0 ft^3/lbm	C_0 lbf·ft^4·R^2/lbm^2	a lbf·ft^7/lbm^3
Methane	CH_4	96.3	3918.49	4.25667×10^{-2}	1.54469×10^8	104.270
Ethylene	C_2H_4	55.1	2307.33	3.18205×10^{-2}	2.93562×10^8	102.256
Ethane	C_2H_6	51.4	2498.75	3.34648×10^{-2}	3.49885×10^8	110.645
Propylene	C_3H_6	36.7	1876.82	3.24066×10^{-2}	4.36932×10^8	90.5491
Propane	C_3H_8	35.0	1921.61	3.53776×10^{-2}	4.60462×10^8	96.3367
i-Butylene	C_4H_8	56.1	1546.46	3.31515×10^{-2}	5.18937×10^8	83.539
i-Butane	C_4H_{10}	26.6	1646.87	3.79359×10^{-2}	4.43207×10^8	86.0106
n-Butane	C_4H_{10}	26.6	1623.06	3.43000×10^{-2}	5.17716×10^8	83.555
i-Pentane	C_5H_{12}	21.4	1336.49	3.55620×10^{-2}	5.90970×10^8	87.1704
n-Pentane	C_5H_{12}	21.4	1272.10	3.48283×10^{-2}	7.17838×10^8	94.5640
n-Hexane	C_6H_{14}	17.9	1057.02	3.30778×10^{-2}	7.87400×10^8	96.9282
n-Heptane	C_7H_{16}	15.4	948.77	3.18377×10^{-2}	8.32642×10^8	89.7937

Gas	Formula	b ft^6/lbm^2	c lbf·ft^7·R^2/lbm^3	α ft^9/lbm^3	γ ft^6/lbm^2
Methane	CH_4	3.37476×10^{-3}	1.74047×10^7	12.4068×10^{-4}	5.99061×10^{-3}
Ethylene	C_2H_4	2.80842×10^{-3}	2.70175×10^7	3.32175×10^{-4}	3.01415×10^{-3}
Ethane	C_2H_6	3.16097×10^{-3}	3.40325×10^7	3.68775×10^{-4}	3.35367×10^{-3}
Propylene	C_3H_6	2.71489×10^{-3}	3.88910×10^7	2.51961×10^{-4}	2.65451×10^{-3}
Propane	C_3H_8	2.97369×10^{-3}	4.24872×10^7	2.91732×10^{-4}	2.90761×10^{-3}
i-Butylene	C_4H_8	2.84236×10^{-3}	4.39605×10^7	2.12482×10^{-4}	2.41610×10^{-3}
i-Butane	C_4H_{10}	3.22807×10^{-3}	4.11333×10^7	2.25353×10^{-4}	2.58641×10^{-3}
n-Butane	C_4H_{10}	3.04271×10^{-3}	4.55052×10^7	2.31069×10^{-4}	2.58641×10^{-3}
i-Pentane	C_5H_{12}	3.29835×10^{-3}	5.22574×10^7	1.86472×10^{-4}	2.28573×10^{-3}
n-Pentane	C_5H_{12}	3.29835×10^{-3}	6.19698×10^7	1.98538×10^{-4}	2.34497×10^{-3}
n-Hexane	C_6H_{14}	3.77654×10^{-3}	6.67554×10^7	1.80949×10^{-4}	2.30766×10^{-3}
n-Heptane	C_7H_{16}	3.88928×10^{-3}	6.93314×10^7	1.78375×10^{-4}	2.30356×10^{-3}

Source: Howell, John R., and Richard O. Buckius, *Fundamentals of Engineering Thermodynamics,* McGraw-Hill, 2nd ed., 1992, as adapted from Cravalho, Ernest, and Joseph L. Smith, *Engineering Thermodynamics,* Pitman, 1981 (used with permission of both copyright owners).

Table A·3E Zero Pressure Specific Heats for Various Gases

T, F	Air c_p B/lbm·R	Air c_v B/lbm·R	Air k	CO_2 c_p B/lbm·R	CO_2 c_v B/lbm·R	CO_2 k	CO c_p B/lbm·R	CO c_v B/lbm·R	CO k	T, F
32	0.240	0.171	1.401	0.195	0.150	1.300	0.249	0.178	1.399	**32**
100	0.240	0.171	1.400	0.205	0.160	1.283	0.249	0.178	1.399	**100**
200	0.241	0.172	1.398	0.217	0.172	1.262	0.249	0.178	1.397	**200**
300	0.243	0.174	1.394	0.228	0.183	1.246	0.251	0.180	1.394	**300**
400	0.245	0.176	1.389	0.238	0.193	1.233	0.253	0.182	1.389	**400**
500	0.247	0.179	1.383	0.247	0.202	1.223	0.256	0.185	1.383	**500**
600	0.250	0.182	1.377	0.255	0.210	1.215	0.259	0.188	1.377	**600**
800	0.257	0.188	1.364	0.269	0.224	1.202	0.266	0.195	1.364	**800**
1000	0.263	0.194	1.353	0.280	0.235	1.192	0.273	0.202	1.351	**1000**
1200	0.269	0.200	1.342	0.289	0.244	1.185	0.279	0.208	1.341	**1200**
1400	0.274	0.205	1.334	0.297	0.252	1.179	0.285	0.214	1.332	**1400**
1600	0.278	0.210	1.327	0.303	0.258	1.175	0.289	0.218	1.325	**1600**
1800	0.282	0.214	1.321	0.308	0.263	1.171	0.293	0.222	1.319	**1800**
2000	0.285	0.217	1.316	0.313	0.268	1.169	0.297	0.226	1.314	**2000**
2400	0.291	0.222	1.308	0.319	0.274	1.165	0.302	0.231	1.306	**2400**
2800	0.295	0.227	1.302	0.324	0.279	1.162	0.306	0.236	1.301	**2800**
3200	0.299	0.230	1.298	0.328	0.283	1.160	0.309	0.239	1.297	**3200**
3600	0.302	0.233	1.294	0.331	0.286	1.158	0.312	0.241	1.294	**3600**
4000	0.304	0.236	1.291	0.333	0.288	1.156	0.314	0.243	1.292	**4000**
4400	0.306	0.238	1.288	0.336	0.291	1.155	0.316	0.245	1.290	**4400**
4800	0.308	0.240	1.286	0.337	0.292	1.155	0.317	0.246	1.288	**4800**

T, F	H_2 c_p B/lbm·R	H_2 c_v B/lbm·R	H_2 k	N_2 c_p B/lbm·R	N_2 c_v B/lbm·R	N_2 k	O_2 c_p B/lbm·R	O_2 c_v B/lbm·R	O_2 k	T, F
32	3.401	2.416	1.408	0.249	0.178	1.399	0.218	0.156	1.397	**32**
100	3.425	2.440	1.404	0.248	0.178	1.399	0.220	0.158	1.394	**100**
200	3.448	2.463	1.400	0.249	0.178	1.399	0.223	0.161	1.387	**200**
300	3.461	2.476	1.398	0.250	0.179	1.396	0.226	0.164	1.378	**300**
400	3.467	2.482	1.397	0.252	0.181	1.393	0.230	0.168	1.369	**400**
500	3.470	2.485	1.396	0.254	0.183	1.388	0.235	0.172	1.360	**500**
600	3.473	2.488	1.396	0.256	0.185	1.383	0.239	0.177	1.351	**600**
800	3.486	2.501	1.394	0.262	0.191	1.371	0.246	0.184	1.337	**800**
1000	3.512	2.527	1.390	0.269	0.198	1.359	0.252	0.190	1.326	**1000**
1200	3.551	2.566	1.384	0.275	0.204	1.348	0.257	0.195	1.318	**1200**
1400	3.594	2.609	1.378	0.280	0.209	1.339	0.261	0.199	1.311	**1400**
1600	3.645	2.660	1.370	0.285	0.214	1.331	0.265	0.203	1.306	**1600**
1800	3.699	2.714	1.363	0.290	0.219	1.324	0.267	0.205	1.302	**1800**
2000	3.756	2.771	1.356	0.293	0.223	1.319	0.270	0.208	1.298	**2000**
2400	3.870	2.885	1.341	0.299	0.228	1.310	0.275	0.212	1.292	**2400**
2800	3.978	2.993	1.329	0.304	0.233	1.305	0.278	0.216	1.287	**2800**
3200	4.077	3.092	1.319	0.307	0.236	1.300	0.282	0.220	1.282	**3200**
3600	4.164	3.179	1.310	0.310	0.239	1.297	0.286	0.224	1.277	**3600**
4000	4.239	3.253	1.303	0.312	0.241	1.294	0.290	0.228	1.273	**4000**
4400	4.306	3.321	1.297	0.314	0.243	1.292	0.293	0.231	1.268	**4400**
4800	4.373	3.388	1.291	0.315	0.244	1.290	0.296	0.234	1.265	**4800**

Table A·3E Zero Pressure Specific Heats for Various Gases (*Continued*)

T, F	CH_4, Methane c_p B/lbm·R	c_v B/lbm·R	k	C_2H_2, Acetylene c_p B/lbm·R	c_v B/lbm·R	k	C_2H_4, Ethylene c_p B/lbm·R	c_v B/lbm·R	k	T, F
32	0.516	0.392	1.316	0.387	0.311	1.245	0.342	0.272	1.261	**32**
100	0.538	0.414	1.299	0.413	0.337	1.226	0.376	0.305	1.232	**100**
200	0.577	0.453	1.273	0.445	0.369	1.207	0.424	0.353	1.201	**200**
300	0.622	0.498	1.249	0.472	0.396	1.192	0.470	0.399	1.177	**300**
400	0.669	0.546	1.227	0.495	0.419	1.182	0.514	0.443	1.160	**400**
500	0.719	0.595	1.208	0.515	0.438	1.174	0.556	0.485	1.146	**500**
600	0.768	0.644	1.192	0.532	0.455	1.168	0.594	0.523	1.135	**600**
800	0.861	0.738	1.168	0.561	0.485	1.157	0.662	0.591	1.120	**800**
1000	0.945	0.821	1.151	0.587	0.511	1.149	0.719	0.648	1.109	**1000**
1200	1.020	0.896	1.138	0.611	0.535	1.143	0.768	0.697	1.102	**1200**
1400	1.088	0.964	1.128	0.632	0.556	1.137	0.811	0.740	1.096	**1400**
1600	1.147	1.024	1.121	0.652	0.576	1.132	0.849	0.778	1.091	**1600**
1800	1.199	1.075	1.115	0.670	0.593	1.129	0.881	0.811	1.087	**1800**
2000	1.244	1.120	1.111	0.685	0.609	1.125	0.909	0.838	1.084	**2000**
2400	1.316	1.192	1.104	0.712	0.636	1.120	0.953	0.882	1.080	**2400**
2800	1.369	1.246	1.099	0.733	0.657	1.116	0.986	0.915	1.077	**2800**
3200	1.411	1.287	1.096	0.751	0.674	1.113	1.011	0.940	1.075	**3200**
3600	1.443	1.319	1.094	0.765	0.689	1.111	1.031	0.960	1.074	**3600**
4000	1.469	1.345	1.092	0.777	0.701	1.109	1.047	0.976	1.073	**4000**
4400	1.489	1.366	1.091	0.788	0.711	1.107	1.060	0.989	1.072	**4400**
4800	1.505	1.381	1.090	0.796	0.720	1.106	1.069	0.998	1.071	**4800**

Source: *JANAF Thermochemical Tables,* 3rd ed., published by the American Chemical Society and the American Institute of Physics for the National Bureau of Standards, 1986. Also, *Journal of Physical and Chemical Reference Data,* Vol. 14, Supplement No. 1, 1985.

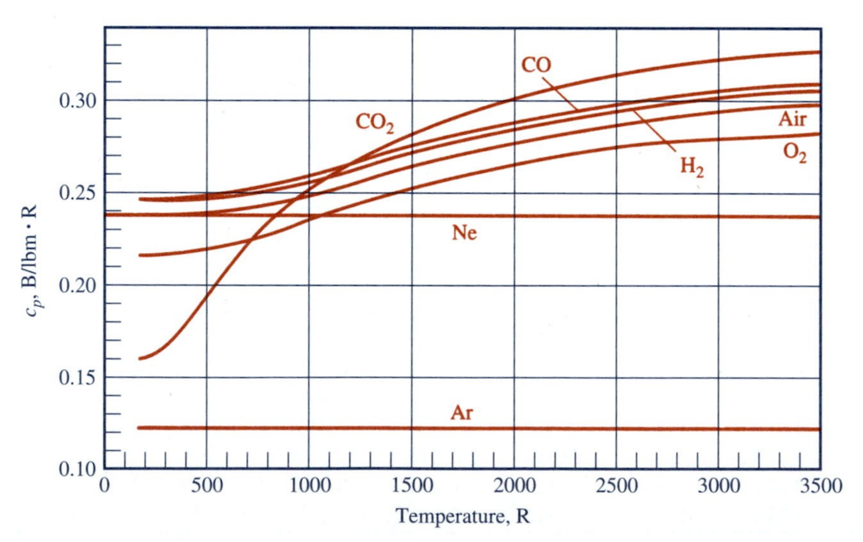

Chart A·4E Zero-pressure specific heat vs. T for several gases.

Table A·5E Specific Heats and Densities of Liquids and Solids at 1 atm, 77 F

Substance	c_p B/lbm·R	ρ lbm/ft^3
Aluminum	0.215	168
Carbon, diamond	0.124	204
Carbon, graphite	0.170	157
Castor oil	0.470	59.7
Copper	0.092	558
Ethylene glycol	0.565	68.5
Glycerine	0.627	78.6
Gold	0.030	1200
Granite	0.243	169
Iron	0.108	490
Kerosene	0.500	51.2
Lead	0.047	706
Limestone	0.217	169
Linseed oil	0.440	58
Lithium bromide	0.179	216
A lubricating oil	0.430	56.8
Mercury	0.033	846
Mica	0.120	183
Oak	0.500	47
Sandstone	0.220	141
Silicon	0.169	145
Silver	0.056	654
Water	1.000	62.2
White pine	0.600	26.5
Zinc	0.092	436

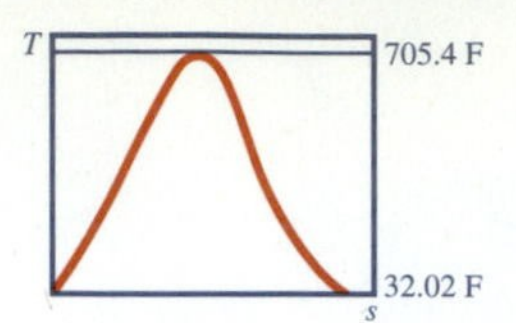

Table A·6·1E Water: Liquid–Vapor Saturation, Temperature Table

T F	p lbf/in^2	v_f ft^3/lbm	v_g ft^3/lbm	h_f B/lbm	h_{fg} B/lbm	h_g B/lbm	s_f B/lbm·R	s_{fg} B/lbm·R	s_g B/lbm·R	T F
32.02	0.08873	0.016 02	3299	0.00	1075.0	1075.0	0.0000	2.1864	2.1864	**32.02**
35	0.09859	0.016 02	2978	3.00	1073.3	1076.3	0.0061	2.1699	2.1759	**35**
40	0.11957	0.016 02	2490	8.02	1070.5	1078.5	0.0162	2.1425	2.1587	**40**
45	0.14608	0.016 02	2065	13.04	1067.7	1080.7	0.0262	2.1157	2.1418	**45**
50	0.17812	0.016 02	1703.1	18.05	1064.9	1082.9	0.0361	2.0893	2.1254	**50**
60	0.2560	0.016 03	1198.1	28.05	1059.3	1087.3	0.0555	2.0383	2.0938	**60**
70	0.3631	0.016 05	866.1	38.04	1053.6	1091.7	0.0745	1.9892	2.0637	**70**
77	0.4591	0.01607	691.0	45.04	1049.7	1094.7	0.0877	1.9559	2.0435	**77**
80	0.5069	0.016 07	629.7	48.03	1048.0	1096.0	0.0932	1.9419	2.0351	**80**
90	0.6984	0.016 10	466.6	58.02	1042.3	1100.3	0.1116	1.8962	2.0078	**90**
100	0.9497	0.016 13	349.1	68.02	1036.6	1104.6	0.1296	1.8522	1.9818	**100**
110	1.2753	0.016 17	264.5	78.01	1030.9	1108.9	0.1473	1.8096	1.9569	**110**
120	1.6937	0.016 21	202.9	88.00	1025.2	1113.2	0.1647	1.7685	1.9332	**120**
130	2.224	0.016 25	156.89	97.99	1019.4	1117.4	0.1817	1.7288	1.9105	**130**
140	2.891	0.016 29	122.92	107.97	1013.6	1121.6	0.1985	1.6903	1.8888	**140**
150	3.719	0.016 34	96.86	117.97	1007.8	1125.7	0.2151	1.6530	1.8680	**150**
160	4.743	0.016 39	77.23	127.97	1001.9	1129.8	0.2313	1.6168	1.8481	**160**
170	5.994	0.016 45	61.97	137.98	995.9	1133.9	0.2473	1.5817	1.8290	**170**
180	7.512	0.016 51	50.19	148.00	989.9	1137.9	0.2631	1.5475	1.8107	**180**
190	9.340	0.016 57	40.93	158.03	983.8	1141.9	0.2787	1.5144	1.7930	**190**
200	11.526	0.016 63	33.62	168.08	977.7	1145.7	0.2940	1.4821	1.7761	**200**
210	14.122	0.016 70	27.82	178.15	971.4	1149.6	0.3091	1.4506	1.7598	**210**
212	14.695	0.016 71	26.81	180.16	970.2	1150.3	0.3122	1.4444	1.7566	**212**
220	17.183	0.016 77	23.14	188.23	965.1	1153.4	0.3241	1.4200	1.7441	**220**
230	20.78	0.016 84	19.39	198.34	958.7	1157.0	0.3388	1.3901	1.7289	**230**
240	24.96	0.016 92	16.323	208.47	952.2	1160.6	0.3534	1.3609	1.7143	**240**
250	29.81	0.017 00	13.828	218.62	945.5	1164.1	0.3678	1.3323	1.7001	**250**
260	35.41	0.017 08	11.767	228.80	938.8	1167.6	0.3820	1.3044	1.6864	**260**
270	41.84	0.017 17	10.066	239.00	931.9	1170.9	0.3960	1.2771	1.6732	**270**
280	49.17	0.017 26	8.650	249.23	924.8	1174.1	0.4099	1.2504	1.6603	**280**
290	57.52	0.017 35	7.466	259.49	917.7	1177.2	0.4237	1.2241	1.6478	**290**
300	66.97	0.017 45	6.472	269.79	910.4	1180.2	0.4373	1.1984	1.6356	**300**
310	77.62	0.017 55	5.631	280.11	902.9	1183.0	0.4507	1.1731	1.6238	**310**
320	89.58	0.017 65	4.919	290.48	895.2	1185.7	0.4641	1.1482	1.6123	**320**
330	102.97	0.017 76	4.311	300.88	887.4	1188.3	0.4773	1.1238	1.6010	**330**
340	117.91	0.017 87	3.792	311.33	879.4	1190.7	0.4904	1.0997	1.5900	**340**
350	134.50	0.017 99	3.345	321.82	871.1	1193.0	0.5033	1.0759	1.5792	**350**
360	152.89	0.018 11	2.960	332.36	862.7	1195.1	0.5162	1.0525	1.5687	**360**
370	173.21	0.018 23	2.627	342.95	854.0	1197.0	0.5289	1.0294	1.5583	**370**
380	195.58	0.018 36	2.338	353.60	845.1	1198.7	0.5416	1.0065	1.5481	**380**
390	220.2	0.018 50	2.086	364.31	836.0	1200.3	0.5542	0.9839	1.5381	**390**
400	247.1	0.018 64	1.8652	375.08	826.6	1201.6	0.5667	0.9615	1.5282	**400**
410	276.5	0.018 78	1.6720	385.92	816.9	1202.8	0.5791	0.9393	1.5184	**410**
420	308.5	0.018 94	1.5017	396.84	806.9	1203.7	0.5915	0.9173	1.5087	**420**
430	343.4	0.019 09	1.3515	407.84	796.6	1204.5	0.6038	0.8954	1.4992	**430**
440	381.2	0.019 26	1.2185	418.93	786.0	1204.9	0.6160	0.8737	1.4897	**440**
450	422.2	0.019 43	1.1005	430.11	775.1	1205.2	0.6282	0.8520	1.4802	**450**
460	466.4	0.019 62	0.9956	441.39	763.8	1205.2	0.6404	0.8305	1.4708	**460**
470	514.2	0.019 81	0.9020	452.78	752.1	1204.8	0.6525	0.8090	1.4614	**470**
480	565.6	0.020 01	0.8183	464.29	740.0	1204.2	0.6646	0.7875	1.4521	**480**
490	620.9	0.020 22	0.7433	475.92	727.4	1203.4	0.6767	0.7660	1.4427	**490**
500	680.2	0.020 44	0.6759	487.70	714.5	1202.1	0.6888	0.7445	1.4333	**500**
520	811.7	0.020 92	0.5603	511.70	687.0	1198.7	0.7130	0.7012	1.4143	**520**
540	961.7	0.021 46	0.4657	536.43	657.3	1193.7	0.7374	0.6575	1.3949	**540**
560	1132.2	0.022 07	0.3875	562.00	624.9	1186.9	0.7621	0.6128	1.3749	**560**
580	1324.8	0.022 78	0.3224	588.63	589.2	1177.8	0.7871	0.5668	1.3539	**580**
600	1541.8	0.023 62	0.2675	616.58	549.6	1166.2	0.8129	0.5187	1.3316	**600**
620	1785.4	0.024 64	0.2207	646.28	504.9	1151.2	0.8396	0.4677	1.3073	**620**
640	2058.3	0.025 92	0.1803	678.39	453.2	1131.6	0.8679	0.4121	1.2801	**640**
660	2364	0.027 66	0.14454	714.23	391.1	1105.3	0.8989	0.3492	1.2481	**660**
680	2706	0.030 33	0.11152	757.09	310.0	1067.1	0.9352	0.2720	1.2072	**680**
700	3091	0.036 36	0.07574	821.41	173.7	995.1	0.9890	0.1497	1.1388	**700**
705.4	3199	0.049 75	0.04975	896.8	0.0	896.8	1.0531	0.0000	1.0531	**705.4**

Source: Based on equations and data of Haar, Lester, John S. Gallagher, and George S. Kell, *NBS/NRC Steam Tables,* Hemisphere, 1984.

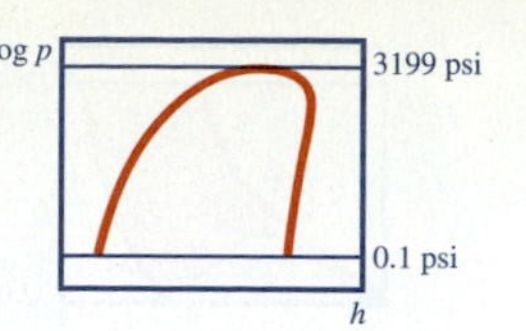

Table A·6·2E Water: Liquid–vapor Saturation, Pressure Table

p lbf/in^2	T F	v_f ft^3/lbm	v_g ft^3/lbm	h_f B/lbm	h_{fg} B/lbm	h_g B/lbm	s_f B/lbm·R	s_{fg} B/lbm·R	s_g B/lbm·R	p lbf/in^2
0.1	35.38	0.016 020	2939	3.38	1073.1	1076.5	0.0069	2.1677	2.1746	0.1
0.2	53.16	0.016 026	1520	21.22	1063.1	1084.3	0.0423	2.0730	2.1152	0.2
0.3	64.48	0.016 041	1034	32.53	1056.7	1089.3	0.0641	2.0161	2.0801	0.3
0.4	72.87	0.016 057	788.4	40.91	1052.0	1092.9	0.0799	1.9754	2.0554	0.4
0.5	79.58	0.016 072	637.8	47.62	1048.2	1095.8	0.0925	1.9438	2.0363	0.5
1.0	101.72	0.016 137	332.9	69.73	1035.6	1105.4	0.1326	1.8448	1.9774	1.0
1.5	115.68	0.016 188	227.0	83.68	1027.7	1111.3	0.1572	1.7861	1.9433	1.5
2.0	126.06	0.016 230	173.5	94.05	1021.7	1115.7	0.1750	1.7443	1.9193	2.0
3.0	141.45	0.016 299	118.7	109.42	1012.8	1122.2	0.2009	1.6848	1.8857	3.0
4.0	152.95	0.016 357	90.5	120.92	1006.0	1126.9	0.2199	1.6422	1.8620	4.0
5.0	162.22	0.016 406	73.5	130.20	1000.6	1130.7	0.2349	1.6089	1.8438	5.0
7.5	179.93	0.016 506	50.27	147.93	990.0	1137.9	0.2630	1.5478	1.8108	7.5
10	193.20	0.016 589	38.42	161.25	981.9	1143.1	0.2836	1.5039	1.7875	10
14.696	212.00	0.016 714	26.81	180.17	970.2	1150.3	0.3122	1.4444	1.7566	14.696
15	213.04	0.016 721	26.30	181.21	969.5	1150.7	0.3137	1.4412	1.7549	15
20	227.97	0.016 829	20.09	196.29	960.0	1156.3	0.3358	1.3961	1.7319	20
25	240.09	0.016 921	16.30	208.56	952.1	1160.7	0.3535	1.3606	1.7141	25
30	250.36	0.017 003	13.75	218.98	945.3	1164.3	0.3683	1.3313	1.6996	30
35	259.31	0.017 077	11.90	228.09	939.2	1167.3	0.381 0	1.3064	1.6874	35
40	267.28	0.017 145	10.50	236.22	933.8	1170.0	0.3922	1.2845	1.6767	40
45	274.47	0.017 208	9.40	243.57	928.7	1172.3	0.4023	1.2651	1.6674	45
50	281.05	0.017 268	8.52	250.30	924.1	1174.4	0.4114	1.2476	1.6590	50
60	292.75	0.017 377	7.175	262.32	915.7	1178.0	0.4274	1.2170	1.6444	60
70	302.97	0.017 477	6.208	272.85	908.2	1181.0	0.4413	1.1908	1.6321	70
80	312.08	0.017 569	5.472	282.27	901.3	1183.6	0.4535	1.1679	1.6214	80
90	320.33	0.017 655	4.898	290.82	895.0	1185.8	0.4645	1.1474	1.6119	90
100	327.87	0.017 735	4.432	298.66	889.1	1187.8	0.4745	1.1289	1.6034	100
110	334.84	0.017 812	4.050	305.93	883.6	1189.5	0.4836	1.1121	1.5957	110
120	341.32	0.017 885	3.729	312.71	878.3	1191.0	0.4921	1.0965	1.5886	120
130	347.39	0.017 956	3.455	319.07	873.3	1192.4	0.4999	1.0821	1.5820	130
140	353.10	0.018 023	3.2202	325.08	868.6	1193.6	0.5073	1.0686	1.5759	140
150	358.49	0.018 088	3.0151	330.77	864.0	1194.7	0.5143	1.0560	1.5702	150
160	363.61	0.018 151	2.8345	336.18	859.6	1195.8	0.5201	1.0441	1.5649	160
170	368.48	0.018 212	2.6747	341.34	855.3	1196.7	0.5270	1.0328	1.5599	170
180	373.14	0.018 272	2.5323	346.29	851.2	1197.5	0.5329	1.0222	1.5551	180
190	377.59	0.018 329	2.4040	351.03	847.3	1198.3	0.5386	1.0120	1.5505	190
200	381.87	0.018 386	2.2881	355.59	843.4	1199.0	0.5440	1.0022	1.5462	200
225	391.87	0.018 522	2.0425	366.32	834.2	1200.5	0.5565	0.9797	1.5362	225
250	401.04	0.018 651	1.8440	376.20	825.6	1201.8	0.5680	0.9592	1.5271	250
275	409.52	0.018 776	1.6807	385.40	817.3	1202.7	0.5785	0.9403	1.5189	275
300	417.42	0.018 895	1.5435	394.02	809.5	1203.5	0.5883	0.9229	1.5112	300
350	431.81	0.019 124	1.3262	409.84	794.7	1204.6	0.6060	0.8915	1.4974	350
400	444.69	0.019 341	1.1615	424.15	780.9	1205.1	0.6217	0.8635	1.4852	400
450	456.38	0.019 548	1.0322	437.29	767.9	1205.2	0.6359	0.8383	1.4742	450
500	467.11	0.019 749	0.9280	449.47	755.5	1205.0	0.6490	0.8152	1.4642	500
550	477.04	0.019 945	0.8421	460.87	743.6	1204.4	0.6610	0.7938	1.4548	550
600	486.31	0.020 136	0.7700	471.61	732.1	1203.7	0.6722	0.7739	1.4462	600
700	503.20	0.020 509	0.6558	491.49	710.2	1201.7	0.6927	0.7376	1.4303	700
800	518.33	0.020 874	0.5691	509.68	689.4	1199.0	0.7110	0.7049	1.4159	800
900	532.08	0.021 235	0.5009	526.55	669.3	1195.9	0.7277	0.6749	1.4026	900
1000	544.71	0.021 594	0.4459	542.37	649.9	1192.3	0.7432	0.6470	1.3902	1000
1250	572.51	0.022 503	0.3454	578.53	603.0	1181.5	0.7777	0.5842	1.3619	1250
1500	596.33	0.023 455	0.2769	611.33	557.3	1168.6	0.8081	0.5277	1.3358	1500
1750	617.24	0.024 483	0.2268	642.05	511.5	1153.5	0.8359	0.4750	1.3108	1750
2000	635.91	0.025 630	0.1881	671.58	464.5	1136.1	0.8620	0.4240	1.2860	2000
2250	652.81	0.026 957	0.1570	700.75	415.0	1115.8	0.8873	0.3731	1.2604	2250
2500	668.18	0.028 636	0.1309	730.93	360.1	1091.0	0.9131	0.3194	1.2324	2500
2750	682.47	0.030 741	0.1075	762.95	298.5	1061.4	0.9401	0.2613	1.2014	2750
3000	695.49	0.034 269	0.0849	802.48	215.5	1017.9	0.9732	0.1865	1.1597	3000
3199	705.40	0.049 750	0.0498	896.80	0.0	896.8	1.0531	0.0000	1.0531	3199

H_2O, Sat

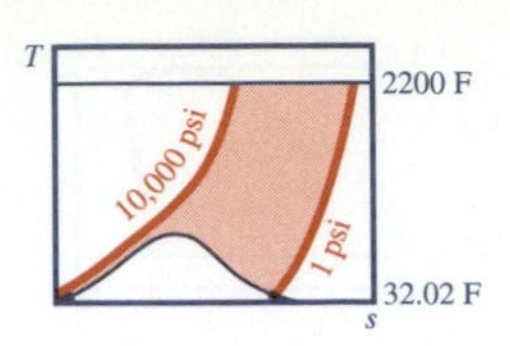

Table A·6·3E Water: Compressed Liquid and Superheated Vapor

T F	p = 1.0 psi v, ft³/lbm	h, B/lbm	s, B/lbm·R
32.02	0.016 02	-0.0	0.0000
50	0.016 04	18.0	0.0350
100	0.016 15	67.8	0.1288
T_{sat}= 101.72 *f*	0.016 14	69.7	0.1326
g	332.89	1105.4	1.9774
150	362.1	1127.3	2.0184
200	392.2	1149.9	2.0542
250	422.3	1172.6	2.0875
300	452.3	1195.5	2.1187
350	482.1	1218.4	2.1480
400	511.9	1241.6	2.1757
450	541.7	1264.9	2.2020
500	571.5	1288.4	2.2272
550	601.3	1312.1	2.2512
600	631.2	1336.1	2.2744
650	661.1	1360.2	2.2966
700	691.0	1384.5	2.3181
750	721.0	1409.0	2.3388
800	749.9	1433.9	2.3587
850	780.5	1458.8	2.3783
900	809.8	1484.1	2.3970
1000	869.8	1535.2	2.4333
1100	929.8	1587.2	2.4679
1200	989.9	1640.1	2.5007
1300	1049.4	1694.1	2.5323
1400	1107.1	1749.2	2.5621
1600	1204.9	1856.5	2.6197
1800	1267.1	1960.1	2.6736
2000	1458.1	2095.3	2.7244
2200	1585.1	2219.3	2.7724

T F	p = 5.0 psi v, ft³/lbm	h, B/lbm	s, B/lbm·R
32.02	0.016 02	-0.0	0.0000
50	0.016 04	18.0	0.0350
100	0.016 14	67.9	0.1288
150	0.016 35	117.8	0.2145
T_{sat}= 162.22 *f*	0.016 41	130.2	0.2349
g	73.5	1130.7	1.8438
200	77.93	1148.3	1.8989
250	84.06	1171.5	1.9328
300	90.14	1194.6	1.9644
350	96.16	1217.7	1.9939
400	102.15	1241.0	2.0217
450	108.13	1264.4	2.0482
500	114.11	1288.0	2.0734
550	120.09	1311.8	2.0975
600	126.07	1335.7	2.1207
650	132.06	1359.9	2.1430
700	138.04	1384.2	2.1645
750	144.03	1408.8	2.1852
800	149.81	1433.6	2.2051
850	155.91	1458.5	2.2247
900	161.77	1483.8	2.2435
1000	173.74	1534.9	2.2798
1100	185.71	1586.9	2.3143
1200	197.70	1639.8	2.3473
1300	209.55	1693.7	2.3787
1400	221.05	1748.8	2.4087
1600	245.00	1861.2	2.4662
1800	268.98	1977.2	2.5201
2000	292.99	2096.6	2.5709
2200	316.88	2219.4	2.6189

T F	p = 10.0 psi v, ft³/lbm	h, B/lbm	s, B/lbm·R
32.02	0.016 02	0.0	0.0000
50	0.016 04	18.0	0.0350
100	0.016 14	68.0	0.1288
150	0.016 35	117.9	0.2145
T_{sat}= 193.20 *f*	0.016 59	161.3	0.2836
g	38.42	1143.1	1.7875
200	38.75	1146.3	1.7986
250	41.87	1170.1	1.8334
300	44.95	1193.5	1.8655
350	47.98	1216.9	1.8953
400	50.99	1240.3	1.9233
450	53.99	1263.9	1.9499
500	56.99	1287.5	1.9752
550	59.99	1311.4	1.9994
600	62.99	1335.4	2.0226
650	65.99	1359.6	2.0450
700	68.98	1383.9	2.0665
750	71.98	1408.5	2.0873
800	74.88	1433.4	2.1071
850	77.93	1458.3	2.1268
900	80.86	1483.6	2.1455
1000	86.85	1534.7	2.1819
1100	92.84	1586.7	2.2164
1200	98.83	1639.6	2.2494
1300	104.76	1693.6	2.2808
1400	110.51	1748.7	2.3108
1600	122.48	1861.1	2.3683
1800	134.47	1977.1	2.4222
2000	146.47	2096.5	2.4730
2200	158.41	2219.2	2.5210

T F	p = 14.7 psi v, ft³/lbm	h, B/lbm	s, B/lbm·R
32.02	0.016 02	0.0	0.0000
50	0.016 04	18.0	0.0350
100	0.016 14	68.0	0.1288
150	0.016 35	118.0	0.2144
200	0.016 65	167.9	0.2937
T_{sat}= 212.00 *f*	0.016 71	180.2	0.3122
g	26.81	1150.3	1.7566
250	28.39	1168.7	1.7834
300	30.51	1192.5	1.8159
350	32.59	1216.1	1.8460
400	34.64	1239.7	1.8743
450	36.69	1263.3	1.9010
500	38.74	1287.1	1.9264
550	40.78	1311.0	1.9507
600	42.83	1335.0	1.9739
650	44.87	1359.3	1.9963
700	46.91	1383.7	2.0178
750	48.95	1408.3	2.0386
800	50.92	1433.2	2.0585
850	53.00	1458.2	2.0782
900	54.99	1483.5	2.0970
1000	59.07	1534.6	2.1333
1100	63.15	1586.6	2.1679
1200	67.23	1639.5	2.2008
1300	71.26	1693.5	2.2323
1400	75.17	1748.6	2.2623
1600	83.32	1861.0	2.3198
1800	91.47	1977.0	2.3737
2000	99.64	2096.4	2.4245
2200	107.76	2219.1	2.4725

T F	p = 15.0 psi v, ft³/lbm	h, B/lbm	s, B/lbm·R
32.02	0.016 02	0.0	0.0000
50	0.016 04	18.0	0.0350
100	0.016 14	68.1	0.1288
150	0.016 35	118.0	0.2144
200	0.016 65	167.9	0.2937
T_{sat}= 213.04 *f*	0.016 72	181.2	0.3137
g	26.30	1150.7	1.7549
250	27.81	1168.6	1.7817
300	29.89	1192.4	1.8143
350	31.93	1216.1	1.8444
400	33.94	1239.7	1.8726
450	35.95	1263.3	1.8994
500	37.96	1287.1	1.9248
550	39.96	1311.0	1.9491
600	41.97	1335.0	1.9723
650	43.97	1359.3	1.9947
700	45.97	1383.7	2.0162
750	47.97	1408.3	2.0371
800	49.90	1433.2	2.0569
850	51.94	1458.2	2.0766
900	53.89	1483.5	2.0954
1000	57.89	1534.6	2.1317
1100	61.88	1586.6	2.1663
1200	65.88	1639.5	2.1993
1300	69.83	1693.5	2.2307
1400	73.67	1748.6	2.2607
1600	81.65	1861.0	2.3182
1800	89.64	1977.0	2.3721
2000	97.64	2096.4	2.4229
2200	105.60	2219.1	2.4709

T F	p = 20.0 psi v, ft³/lbm	h, B/lbm	s, B/lbm·R
32.02	0.016 02	0.0	0.0000
50	0.016 04	18.0	0.0350
100	0.016 14	68.1	0.1287
150	0.016 35	118.0	0.2144
200	0.016 65	167.9	0.2937
T_{sat}= 227.97 *f*	0.016 83	196.3	0.3358
g	20.09	1156.3	1.7319
250	20.73	1167.1	1.7537
300	22.32	1191.3	1.7868
350	23.87	1215.2	1.8173
400	25.39	1239.0	1.8457
450	26.91	1262.7	1.8726
500	28.42	1286.6	1.8981
550	29.93	1310.6	1.9224
600	31.44	1334.7	1.9457
650	32.95	1359.0	1.9682
700	34.45	1383.4	1.9897
750	35.95	1408.0	2.0106
800	37.40	1433.0	2.0305
850	38.94	1458.0	2.0501
900	40.40	1483.3	2.0689
1000	43.40	1534.4	2.1053
1100	46.40	1586.5	2.1399
1200	49.40	1639.4	2.1729
1300	52.37	1693.4	2.2043
1400	55.24	1748.6	2.2343
1600	61.23	1861.0	2.2919
1800	67.23	1977.0	2.3458
2000	73.23	2096.4	2.3966
2200	79.20	2219.1	2.4446

Table A·6·3E Water: Compressed Liquid and Superheated Vapor (*Continued*)

T F	p = 40.0 psi: v, ft³/lbm	*h*, B/lbm	*s*, B/lbm·R
32.02	0.016 02	0.1	0.0000
50	0.016 04	18.1	0.0350
100	0.016 14	68.1	0.1287
150	0.016 35	118.1	0.2144
200	0.016 64	168.0	0.2937
250	0.017 04	218.6	0.3663
$T_{sat}=$ *f*	0.017 15	236.2	0.3922
267.28 *g*	10.50	1170.0	1.6767
300	11.028	1186.7	1.7008
350	11.829	1211.7	1.7326
400	12.611	1236.2	1.7619
450	13.384	1260.5	1.7893
500	14.152	1284.7	1.8153
550	14.915	1309.0	1.8399
600	15.676	1333.3	1.8634
650	16.434	1357.8	1.8860
700	17.191	1382.4	1.9077
750	17.947	1407.1	1.9287
800	18.675	1432.1	1.9486
850	19.444	1457.2	1.9684
900	20.180	1482.6	1.9872
1000	21.683	1533.9	2.0237
1100	23.186	1586.0	2.0583
1200	24.689	1639.0	2.0914
1300	26.173	1693.1	2.1229
1400	27.613	1748.3	2.1528
1600	30.611	1860.7	2.2104
1800	33.611	1976.8	2.2644
2000	36.613	2096.2	2.3152
2200	39.600	2219.0	2.3632

T F	p = 60.0 psi: v, ft³/lbm	*h*, B/lbm	*s*, B/lbm·R
32.02	0.016 02	0.2	0.0000
50	0.016 04	18.2	0.0350
100	0.016 14	68.2	0.1287
150	0.016 35	118.1	0.2144
200	0.016 64	168.1	0.2936
250	0.017 04	218.7	0.3662
$T_{sat}=$ *f*	0.017 38	262.3	0.4274
292.75 *g*	7.175	1178.0	1.6444
300	7.255	1181.9	1.6498
350	7.811	1208.0	1.6832
400	8.346	1233.3	1.7134
450	8.872	1258.1	1.7415
500	9.392	1282.8	1.7678
550	9.907	1307.3	1.7928
600	10.419	1331.9	1.8166
650	10.929	1356.5	1.8393
700	11.437	1381.3	1.8612
750	11.943	1406.2	1.8822
800	12.431	1431.3	1.9022
850	12.945	1456.5	1.9221
900	13.438	1481.9	1.9410
1000	14.443	1533.3	1.9775
1100	15.447	1585.6	2.0122
1200	16.451	1638.6	2.0453
1300	17.442	1692.7	2.0768
1400	18.403	1748.0	2.1068
1600	20.404	1860.5	2.1644
1800	22.405	1976.6	2.2184
2000	24.408	2096.1	2.2692
2200	26.400	2218.9	2.3172

T F	p = 80.0 psi: v, ft³/lbm	*h*, B/lbm	*s*, B/lbm·R
32.02	0.016 02	0.2	0.0000
50	0.016 04	18.2	0.0350
100	0.016 14	68.2	0.1287
150	0.016 35	118.2	0.2144
200	0.016 64	168.1	0.2936
250	0.017 03	218.8	0.3661
300	0.017 45	269.7	0.4372
$T_{sat}=$ *f*	0.017 57	282.3	0.4535
312.08 *g*	5.472	1183.6	1.6214
350	5.797	1204.2	1.6479
400	6.211	1230.3	1.6792
450	6.614	1255.7	1.7079
500	7.010	1280.8	1.7347
550	7.401	1305.7	1.7600
600	7.789	1330.5	1.7840
650	8.175	1355.3	1.8069
700	8.559	1380.2	1.8289
750	8.941	1405.2	1.8500
800	9.308	1430.5	1.8701
850	9.696	1455.7	1.8900
900	10.066	1481.2	1.9090
1000	10.823	1532.8	1.9456
1100	11.578	1585.1	1.9804
1200	12.332	1638.2	2.0135
1300	13.076	1692.4	2.0451
1400	13.798	1747.7	2.0751
1600	15.300	1860.3	2.1328
1800	16.802	1976.4	2.1867
2000	18.305	2095.9	2.2376
2200	19.800	2218.7	2.2856

T F	p = 100 psi: v, ft³/lbm	*h*, B/lbm	*s*, B/lbm·R
32.02	0.016 02	0.3	0.0000
50	0.016 03	18.3	0.0350
100	0.016 14	68.3	0.1287
150	0.016 34	118.2	0.2143
200	0.016 64	168.2	0.2935
250	0.017 03	218.9	0.3661
300	0.017 45	269.8	0.4371
$T_{sat}=$ *f*	0.017 74	298.7	0.4745
327.87 *g*	4.432	1187.8	1.6034
350	4.583	1200.2	1.6200
400	4.926	1227.2	1.6525
450	5.256	1253.3	1.6819
500	5.579	1278.8	1.7092
550	5.897	1304.0	1.7348
600	6.210	1329.1	1.7590
650	6.522	1354.1	1.7821
700	6.831	1379.2	1.8043
750	7.138	1404.3	1.8255
800	7.434	1429.6	1.8457
850	7.745	1455.0	1.8657
900	8.043	1480.6	1.8847
1000	8.650	1532.2	1.9214
1100	9.255	1584.6	1.9562
1200	9.860	1637.8	1.9894
1300	10.457	1692.0	2.0210
1400	11.034	1747.3	2.0510
1600	12.237	1860.0	2.1087
1800	13.440	1976.2	2.1627
2000	14.643	2095.8	2.2136
2200	15.840	2218.6	2.2616

T F	p = 120 psi: v, ft³/lbm	*h*, B/lbm	*s*, B/lbm·R
32.02	0.016 02	0.3	0.0000
50	0.016 03	18.3	0.0350
100	0.016 14	68.3	0.1287
150	0.016 34	118.3	0.2143
200	0.016 63	168.3	0.2935
250	0.017 03	219.0	0.3660
300	0.017 45	269.9	0.4371
$T_{sat}=$ *f*	0.017 89	312.7	0.4921
341.32 *g*	3.729	1191.0	1.5886
350	3.779	1196.0	1.5951
400	4.073	1224.1	1.6287
450	4.355	1250.8	1.6589
500	4.628	1276.8	1.6867
550	4.896	1302.3	1.7126
600	5.160	1327.6	1.7371
650	5.421	1352.9	1.7604
700	5.681	1378.1	1.7827
750	5.938	1403.3	1.8041
800	6.185	1428.8	1.8243
850	6.446	1454.2	1.8444
900	6.695	1479.9	1.8634
1000	7.202	1531.6	1.9002
1100	7.708	1584.1	1.9351
1200	8.213	1637.4	1.9683
1300	8.711	1691.7	1.9999
1400	9.193	1747.0	2.0300
1600	10.196	1859.8	2.0878
1800	11.199	1976.1	2.1418
2000	12.202	2095.6	2.1926
2200	13.199	2218.5	2.2407

T F	p = 140 psi: v, ft³/lbm	*h*, B/lbm	*s*, B/lbm·R
32.02	0.016 01	0.4	0.0000
50	0.016 03	18.4	0.0350
100	0.016 13	68.4	0.1287
150	0.016 34	118.3	0.2143
200	0.016 63	168.4	0.2934
250	0.017 03	219.1	0.3660
300	0.017 45	269.9	0.4370
350	0.018 03	322.3	0.5022
$T_{sat}=$ *f*	0.018 02	325.1	0.5073
353.10 *g*	3.220	1193.6	1.5759
400	3.461	1220.8	1.6088
450	3.708	1248.3	1.6398
500	3.947	1274.7	1.6681
550	4.180	1300.6	1.6944
600	4.408	1326.2	1.7192
650	4.634	1351.6	1.7426
700	4.858	1377.0	1.7650
750	5.080	1402.4	1.7865
800	5.293	1427.9	1.8069
850	5.518	1453.5	1.8270
900	5.732	1479.2	1.8461
1000	6.168	1531.1	1.8830
1100	6.602	1583.7	1.9179
1200	7.036	1637.0	1.9512
1300	7.463	1691.3	1.9828
1400	7.877	1746.7	2.0129
1600	8.738	1859.6	2.0707
1800	9.598	1975.9	2.1247
2000	10.458	2095.5	2.1756
2200	11.314	2218.4	2.2237

Table A·6·3E Water: Compressed Liquid and Superheated Vapor (*Continued*)

T	*p* = 160 psi			*T*	*p* = 180 psi			*T*	*p* = 200 psi		
F	*v*, ft³/lbm	*h*, B/lbm	*s*, B/lbm·R	F	*v*, ft³/lbm	*h*, B/lbm	*s*, B/lbm·R	F	*v*, ft³/lbm	*h*, B/lbm	*s*, B/lbm·R
32.02	0.016 01	0.5	0.0000	**32.02**	0.016 01	0.5	0.0000	**32.02**	0.016 01	0.6	0.0000
50	0.016 03	18.5	0.0350	**50**	0.016 03	18.5	0.0350	**50**	0.016 03	18.6	0.0350
100	0.016 13	68.4	0.1287	**100**	0.016 13	68.5	0.1287	**100**	0.016 13	68.5	0.1286
150	0.016 34	118.4	0.2143	**150**	0.016 34	118.4	0.2143	**150**	0.016 34	118.5	0.2142
200	0.016 63	168.5	0.2934	**200**	0.016 63	168.5	0.2934	**200**	0.016 63	168.5	0.2934
250	0.017 02	219.1	0.3659	**250**	0.017 02	219.2	0.3659	**250**	0.017 02	219.2	0.3659
300	0.017 44	270.0	0.4370	**300**	0.017 44	270.0	0.4369	**300**	0.017 44	270.1	0.4369
350	0.018 03	322.3	0.5021	**350**	0.018 03	322.4	0.5021	**350**	0.018 03	322.4	0.5020
T_{sat}= *f*	0.018 15	336.2	0.5208	T_{sat}= *f*	0.018 27	346.3	0.5329	T_{sat}= *f*	0.018 39	355.6	0.5440
363.61 *g*	2.835	1195.8	1.5649	**373.14** *g*	2.532	1197.5	1.5551	**381.87** *g*	2.2881	1199.0	1.5462
400	3.001	1217.5	1.5910	**400**	2.644	1214.0	1.5746	**400**	2.357	1210.4	1.5596
450	3.223	1245.7	1.6229	**450**	2.846	1243.0	1.6074	**450**	2.544	1240.3	1.5934
500	3.435	1272.6	1.6518	**500**	3.038	1270.5	1.6369	**500**	2.720	1268.3	1.6234
550	3.642	1298.9	1.6785	**550**	3.224	1297.1	1.6639	**550**	2.890	1295.4	1.6509
600	3.844	1324.7	1.7035	**600**	3.406	1323.3	1.6892	**600**	3.056	1321.8	1.6764
650	4.044	1350.4	1.7271	**650**	3.585	1349.1	1.7131	**650**	3.218	1347.8	1.7005
700	4.241	1375.9	1.7497	**700**	3.762	1374.8	1.7358	**700**	3.378	1373.7	1.7233
750	4.437	1401.4	1.7713	**750**	3.937	1400.5	1.7575	**750**	3.536	1399.5	1.7452
800	4.624	1427.0	1.7917	**800**	4.104	1426.2	1.7780	**800**	3.687	1425.3	1.7658
850	4.821	1452.7	1.8119	**850**	4.280	1451.9	1.7983	**850**	3.847	1451.2	1.7861
900	5.009	1478.5	1.8311	**900**	4.447	1477.8	1.8175	**900**	3.998	1477.1	1.8054
1000	5.392	1530.5	1.8681	**1000**	4.789	1529.9	1.8545	**1000**	4.306	1529.4	1.8425
1100	5.773	1583.2	1.9030	**1100**	5.128	1582.7	1.8896	**1100**	4.612	1582.2	1.8776
1200	6.153	1636.6	1.9363	**1200**	5.467	1636.2	1.9229	**1200**	4.917	1635.8	1.9110
1300	6.528	1691.0	1.9680	**1300**	5.800	1690.7	1.9547	**1300**	5.218	1690.3	1.9428
1400	6.890	1746.4	1.9982	**1400**	6.123	1746.1	1.9848	**1400**	5.509	1745.8	1.9730
1600	7.644	1859.3	2.0560	**1600**	6.794	1859.1	2.0427	**1600**	6.113	1858.9	2.0308
1800	8.398	1975.7	2.1100	**1800**	7.464	1975.5	2.0967	**1800**	6.717	1975.3	2.0850
2000	9.151	2095.3	2.1609	**2000**	8.134	2095.2	2.1477	**2000**	7.320	2095.1	2.1359
2200	9.899	2218.3	2.2090	**2200**	8.799	2218.1	2.1958	**2200**	7.919	2218.0	2.1840

T	*p* = 225 psi			*T*	*p* = 250 psi			*T*	*p* = 275 psi		
F	*v*, ft³/lbm	*h*, B/lbm	*s*, B/lbm·R	F	*v*, ft³/lbm	*h*, B/lbm	*s*, B/lbm·R	F	*v*, ft³/lbm	*h*, B/lbm	*s*, B/lbm·R
32.02	0.016 01	0.7	0.0000	**32.02**	0.016 01	0.7	0.0000	**32.02**	0.016 01	0.8	0.0000
50	0.016 03	18.6	0.0350	**50**	0.016 03	18.7	0.0350	**50**	0.016 03	18.8	0.0350
100	0.016 13	68.6	0.1286	**100**	0.016 13	68.7	0.1286	**100**	0.016 13	68.7	0.1286
150	0.016 34	118.5	0.2142	**150**	0.016 34	118.6	0.2142	**150**	0.016 34	118.6	0.2142
200	0.016 63	168.6	0.2933	**200**	0.016 63	168.7	0.2933	**200**	0.016 62	168.7	0.2933
250	0.017 02	219.3	0.3658	**250**	0.017 02	219.3	0.3658	**250**	0.017 02	219.4	0.3658
300	0.017 44	270.1	0.4368	**300**	0.017 44	270.1	0.4368	**300**	0.017 44	270.2	0.4368
350	0.018 03	322.4	0.5020	**350**	0.018 02	322.5	0.5020	**350**	0.018 02	322.5	0.5019
T_{sat}= *f*	0.018 52	366.3	0.5665	**400**	0.018 67	375.4	0.5665	**400**	0.018 66	375.4	0.5664
391.87 *g*	2.043	1200.5	1.5362	T_{sat}= *f*	0.018 65	376.2	0.5680	T_{sat}= *f*	0.018 78	385.4	0.5785
400	2.068	1205.7	1.5423	**401.04** *g*	1.8440	1201.8	1.5271	**409.52** *g*	1.681	1202.7	1.5189
450	2.241	1236.8	1.5774	**450**	1.998	1233.1	1.5626	**450**	1.798	1229.4	1.5488
500	2.401	1265.6	1.6083	**500**	2.146	1262.7	1.5944	**500**	1.937	1259.9	1.5815
550	2.555	1293.1	1.6362	**550**	2.287	1290.8	1.6229	**550**	2.068	1288.5	1.6106
600	2.704	1319.9	1.6622	**600**	2.423	1318.0	1.6492	**600**	2.194	1316.1	1.6373
650	2.850	1346.2	1.6865	**650**	2.556	1344.6	1.6738	**650**	2.316	1343.0	1.6621
700	2.994	1372.3	1.7095	**700**	2.687	1370.9	1.6970	**700**	2.435	1369.5	1.6856
750	3.136	1398.3	1.7315	**750**	2.815	1397.1	1.7191	**750**	2.553	1395.9	1.7078
800	3.271	1424.2	1.7522	**800**	2.937	1423.2	1.7399	**800**	2.665	1422.1	1.7287
850	3.413	1450.2	1.7726	**850**	3.067	1449.2	1.7605	**850**	2.783	1448.3	1.7494
900	3.548	1476.3	1.7920	**900**	3.189	1475.4	1.7799	**900**	2.894	1474.5	1.7689
1000	3.823	1528.7	1.8292	**1000**	3.437	1528.0	1.8173	**1000**	3.121	1527.2	1.8064
1100	4.096	1581.7	1.8644	**1100**	3.684	1581.1	1.8525	**1100**	3.346	1580.5	1.8417
1200	4.368	1635.3	1.8979	**1200**	3.929	1634.8	1.8860	**1200**	3.569	1634.3	1.8753
1300	4.636	1689.9	1.9297	**1300**	4.170	1689.4	1.9179	**1300**	3.789	1689.0	1.9072
1400	4.895	1745.5	1.9599	**1400**	4.403	1745.1	1.9481	**1400**	4.001	1744.7	1.9374
1600	5.433	1858.6	2.0178	**1600**	4.888	1858.3	2.0061	**1600**	4.443	1858.0	1.9954
1800	5.970	1975.1	2.0719	**1800**	5.372	1974.9	2.0602	**1800**	4.883	1974.6	2.0496
2000	6.506	2094.9	2.1229	**2000**	5.855	2094.7	2.1112	**2000**	5.323	2094.5	2.1006
2200	7.039	2217.9	2.1710	**2200**	6.335	2217.7	2.1594	**2200**	5.759	2217.6	2.1488

Table A·6·3E Water: Compressed Liquid and Superheated Vapor (*Continued*)

T F	*p* = 300 psi *v*, ft³/lbm	*h*, B/lbm	*s*, B/lbm·R	*T* F	*p* = 350 psi *v*, ft³/lbm	*h*, B/lbm	*s*, B/lbm·R	*T* F	*p* = 400 psi *v*, ft³/lbm	*h*, B/lbm	*s*, B/lbm·R
32.02	0.016 01	0.9	0.0000	**32.02**	0.016 00	1.0	0.0000	**32.02**	0.016 00	1.2	0.0000
50	0.016 02	18.9	0.0350	**50**	0.016 02	19.0	0.0350	**50**	0.016 02	19.2	0.0350
100	0.016 13	68.8	0.1286	**100**	0.016 12	68.9	0.1286	**100**	0.016 12	69.1	0.1285
150	0.016 33	118.7	0.2141	**150**	0.016 33	118.8	0.2141	**150**	0.016 33	119.0	0.2141
200	0.016 62	168.8	0.2932	**200**	0.016 62	168.9	0.2932	**200**	0.016 62	169.0	0.2931
250	0.017 02	219.4	0.3657	**250**	0.017 01	219.5	0.3657	**250**	0.017 01	219.6	0.3656
300	0.017 43	270.2	0.4367	**300**	0.017 43	270.3	0.4366	**300**	0.017 43	270.4	0.4365
350	0.018 02	322.6	0.5019	**350**	0.018 01	322.6	0.5018	**350**	0.018 01	322.7	0.5017
400	0.018 66	375.4	0.5664	**400**	0.018 65	375.5	0.5663	**400**	0.018 65	375.5	0.5662
T_{sat} = *f*	0.018 90	394.0	0.5883	T_{sat} = *f*	0.019 12	409.8	0.6060	T_{sat} = *f*	0.019 34	424.2	0.6217
417.42 *g*	1.5435	1203.5	1.5112	**431.81** *g*	1.3262	1204.6	1.4974	**444.69** *g*	1.1615	1205.1	1.4852
450	1.6322	1225.5	1.5358	**450**	1.3689	1217.4	1.5117	**450**	1.1681	1208.5	1.4892
500	1.7628	1256.9	1.5694	**500**	1.4878	1250.8	1.5475	**500**	1.2796	1244.3	1.5277
550	1.8851	1286.2	1.5992	**550**	1.5970	1281.3	1.5785	**550**	1.3797	1276.3	1.5602
600	2.0019	1314.1	1.6263	**600**	1.7003	1310.2	1.6065	**600**	1.4732	1306.1	1.5891
650	2.1152	1341.4	1.6514	**650**	1.7997	1338.0	1.6322	**650**	1.5625	1334.6	1.6154
700	2.2259	1368.1	1.6750	**700**	1.8965	1365.3	1.6562	**700**	1.6489	1362.4	1.6399
750	2.3348	1394.6	1.6974	**750**	1.9913	1392.1	1.6790	**750**	1.7332	1389.6	1.6630
800	2.4377	1421.0	1.7185	**800**	2.0805	1418.8	1.7002	**800**	1.8122	1416.5	1.6844
850	2.5465	1447.3	1.7392	**850**	2.1749	1445.4	1.7212	**850**	1.8959	1443.4	1.7056
900	2.6491	1473.6	1.7588	**900**	2.2636	1471.9	1.7409	**900**	1.9742	1470.1	1.7254
1000	2.8577	1526.5	1.7964	**1000**	2.4438	1525.1	1.7787	**1000**	2.1332	1523.6	1.7635
1100	3.0644	1579.9	1.8318	**1100**	2.6221	1578.7	1.8143	**1100**	2.2902	1577.5	1.7992
1200	3.2698	1633.8	1.8654	**1200**	2.7990	1632.8	1.8480	**1200**	2.4458	1631.8	1.8331
1300	3.4717	1688.6	1.8973	**1300**	2.9727	1687.7	1.8800	**1300**	2.5984	1686.8	1.8652
1400	3.6665	1744.3	1.9276	**1400**	3.1402	1743.5	1.9104	**1400**	2.7454	1742.8	1.8956
1600	4.0716	1857.7	1.9857	**1600**	3.4883	1857.1	1.9685	**1600**	3.0508	1856.5	1.9538
1800	4.4756	1974.4	2.0399	**1800**	3.8352	1973.9	2.0228	**1800**	3.3549	1973.5	2.0082
2000	4.8788	2094.3	2.0909	**2000**	4.1814	2094.0	2.0739	**2000**	3.6583	2093.6	2.0593
2200	5.2794	2217.4	2.1391	**2200**	4.5251	2217.1	2.1220	**2200**	3.9594	2216.8	2.1075

T F	*p* = 450 psi *v*, ft³/lbm	*h*, B/lbm	*s*, B/lbm·R	*T* F	*p* = 500 psi *v*, ft³/lbm	*h*, B/lbm	*s*, B/lbm·R	*T* F	*p* = 600 psi *v*, ft³/lbm	*h*, B/lbm	*s*, B/lbm·R
32.02	0.016 00	1.3	0.0000	**32.02**	0.015 99	1.5	0.0000	**32.02**	0.015 99	1.8	0.0000
50	0.016 02	19.3	0.0350	**50**	0.016 01	19.4	0.0350	**50**	0.016 01	19.7	0.0350
100	0.016 12	69.2	0.1285	**100**	0.016 12	69.3	0.1285	**100**	0.016 11	69.6	0.1284
150	0.016 33	119.1	0.2140	**150**	0.016 32	119.2	0.2140	**150**	0.016 32	119.4	0.2139
200	0.016 62	169.1	0.2931	**200**	0.016 61	169.2	0.2930	**200**	0.016 61	169.5	0.2929
250	0.017 01	219.7	0.3655	**250**	0.017 00	219.8	0.3654	**250**	0.017 00	220.0	0.3653
300	0.017 42	270.5	0.4364	**300**	0.017 42	270.6	0.4363	**300**	0.017 41	270.8	0.4362
350	0.018 00	322.8	0.5016	**350**	0.018 00	322.9	0.5015	**350**	0.017 99	323.0	0.5013
400	0.018 64	375.6	0.5661	**400**	0.018 63	375.7	0.5659	**400**	0.018 62	375.8	0.5656
450	0.019 53	431.1	0.6282	**450**	0.019 52	431.1	0.6280	**450**	0.019 50	431.1	0.6275
T_{sat} = *f*	0.019 55	437.3	0.6359	T_{sat} = *f*	0.019 75	449.5	0.6490	T_{sat} = *f*	0.020 14	471.6	0.6722
456.38 *g*	1.0322	1205.2	1.4742	**467.11** *g*	0.9280	1205.0	1.4642	**486.31** *g*	0.7700	1203.7	1.4462
500	1.1169	1237.5	1.5090	**500**	0.9835	1230.1	1.4918	**500**	0.7884	1214.6	1.4577
550	1.2105	1271.1	1.5432	**550**	1.0729	1265.6	1.5279	**550**	0.8708	1254.3	1.4982
600	1.2966	1301.9	1.5730	**600**	1.1536	1297.6	1.5589	**600**	0.9427	1288.8	1.5315
650	1.3780	1331.2	1.6000	**650**	1.2292	1327.6	1.5866	**650**	1.0089	1320.4	1.5608
700	1.4564	1359.4	1.6250	**700**	1.3014	1356.4	1.6121	**700**	1.0714	1350.4	1.5872
750	1.5326	1387.1	1.6484	**750**	1.3712	1384.5	1.6358	**750**	1.1314	1379.3	1.6117
800	1.6037	1414.3	1.6701	**800**	1.4361	1412.0	1.6578	**800**	1.1867	1407.4	1.6341
850	1.6790	1441.4	1.6915	**850**	1.5048	1439.4	1.6794	**850**	1.2453	1435.4	1.6562
900	1.7493	1468.3	1.7115	**900**	1.5687	1466.5	1.6995	**900**	1.2995	1462.9	1.6766
1000	1.8917	1522.2	1.7498	**1000**	1.6981	1520.7	1.7381	**1000**	1.4089	1517.8	1.7157
1100	2.0321	1576.3	1.7857	**1100**	1.8254	1575.0	1.7741	**1100**	1.5162	1572.6	1.7521
1200	2.1711	1630.8	1.8196	**1200**	1.9511	1629.7	1.8082	**1200**	1.6219	1627.7	1.7864
1300	2.3073	1685.9	1.8518	**1300**	2.0742	1685.1	1.8405	**1300**	1.7252	1683.3	1.8188
1400	2.4383	1742.0	1.8823	**1400**	2.1925	1741.2	1.8710	**1400**	1.8244	1739.7	1.8495
1600	2.7106	1855.9	1.9406	**1600**	2.4382	1855.4	1.9294	**1600**	2.0301	1854.2	1.9081
1800	2.9814	1973.0	1.9950	**1800**	2.6825	1972.6	1.9839	**1800**	2.2344	1971.6	1.9627
2000	3.2514	2093.2	2.0462	**2000**	2.9259	2092.9	2.0351	**2000**	2.4378	2092.1	2.0139
2200	3.5194	2216.6	2.0944	**2200**	3.1674	2216.3	2.0834	**2200**	2.6394	2215.7	2.0622

H₂O, CL/SH

Table A·6·3E Water: Compressed Liquid and Superheated Vapor (*Continued*)

p = 700 psi

T, F	v, ft^3/lbm	h, B/lbm	s, B/lbm·R
32.02	0.015 98	2.1	0.0000
50	0.016 00	20.0	0.0349
100	0.016 11	69.9	0.1283
150	0.016 31	119.7	0.2138
200	0.016 60	169.7	0.2927
250	0.016 99	220.2	0.3651
300	0.017 40	271.0	0.4360
350	0.017 98	323.2	0.5010
400	0.018 61	375.9	0.5653
450	0.019 47	431.2	0.6271
500	0.020 63	489.8	0.6904
T_{sat} = 503.20, f	0.020 51	491.5	0.6927
T_{sat} = 503.20, g	0.6558	1201.7	1.4303
550	0.7233	1241.9	1.4713
600	0.7901	1279.4	1.5076
650	0.8501	1312.9	1.5385
700	0.9060	1344.1	1.5661
750	0.9593	1374.0	1.5914
800	1.0078	1402.7	1.6143
850	1.0593	1431.3	1.6368
900	1.1066	1459.3	1.6576
1000	1.2019	1514.8	1.6972
1100	1.2950	1570.2	1.7339
1200	1.3865	1625.6	1.7685
1300	1.4758	1681.5	1.8011
1400	1.5612	1738.2	1.8318
1600	1.7385	1853.0	1.8906
1800	1.9143	1970.7	1.9454
2000	2.0891	2091.4	1.9967
2200	2.2623	2215.1	2.0451

p = 800 psi

T, F	v, ft^3/lbm	h, B/lbm	s, B/lbm·R
32.02	0.015 98	2.4	0.0000
50	0.016 00	20.3	0.0349
100	0.016 10	70.1	0.1283
150	0.016 31	119.9	0.2137
200	0.016 60	169.9	0.2926
250	0.016 98	220.5	0.3650
300	0.017 40	271.2	0.4358
350	0.017 97	323.3	0.5008
400	0.018 60	376.0	0.5651
450	0.019 45	431.2	0.6267
500	0.020 59	489.6	0.6898
T_{sat} = 518.33, f	0.020 87	509.7	0.7110
T_{sat} = 518.33, g	0.5691	1199.0	1.4159
550	0.6097	1228.1	1.4454
600	0.6741	1269.3	1.4854
650	0.7301	1305.0	1.5183
700	0.7814	1337.6	1.5472
750	0.8297	1368.5	1.5733
800	0.8732	1397.9	1.5968
850	0.9195	1427.1	1.6198
900	0.9617	1455.5	1.6409
1000	1.0465	1511.9	1.6810
1100	1.1290	1567.7	1.7181
1200	1.2099	1623.5	1.7529
1300	1.2886	1679.8	1.7856
1400	1.3639	1736.6	1.8165
1600	1.5198	1851.8	1.8755
1800	1.6742	1969.8	1.9304
2000	1.8276	2090.7	1.9818
2200	1.9795	2214.5	2.0302

p = 1000 psi

T, F	v, ft^3/lbm	h, B/lbm	s, B/lbm·R
32.02	0.015 97	3.0	0.0000
50	0.015 99	20.9	0.0349
100	0.016 09	70.7	0.1282
150	0.016 30	120.4	0.2135
200	0.016 59	170.4	0.2924
250	0.016 97	220.9	0.3647
300	0.017 38	271.5	0.4354
350	0.017 95	323.6	0.5004
400	0.018 57	376.2	0.5645
450	0.019 42	431.3	0.6260
500	0.020 53	489.4	0.6888
T_{sat} = 544.71, f	0.021 59	542.4	0.7432
T_{sat} = 544.71, g	0.4459	1192.3	1.3902
550	0.4476	1195.8	1.3936
600	0.5111	1247.3	1.4437
650	0.5622	1288.2	1.4815
700	0.6072	1324.1	1.5132
750	0.6486	1357.2	1.5412
800	0.6852	1388.1	1.5658
850	0.7242	1418.6	1.5899
900	0.7592	1448.0	1.6117
1000	0.8293	1505.8	1.6529
1100	0.8969	1562.7	1.6907
1200	0.9629	1619.4	1.7260
1300	1.0268	1676.2	1.7591
1400	1.0877	1733.5	1.7903
1600	1.2137	1849.5	1.8496
1800	1.3381	1968.0	1.9047
2000	1.4615	2089.2	1.9563
2200	1.5836	2213.3	2.0048

p = 1250 psi

T, F	v, ft^3/lbm	h, B/lbm	s, B/lbm·R
32.02	0.015 95	3.7	0.0000
50	0.015 97	21.6	0.0349
100	0.016 08	71.3	0.1280
150	0.016 29	121.0	0.2133
200	0.016 57	170.9	0.2921
250	0.016 96	221.4	0.3643
300	0.017 36	272.0	0.4350
350	0.017 93	324.0	0.4998
400	0.018 55	376.5	0.5639
450	0.019 38	431.4	0.6251
500	0.020 45	489.0	0.6873
550	0.021 95	551.2	0.7512
T_{sat} = 572.51, f	0.022 50	578.5	0.7777
T_{sat} = 572.51, g	0.3454	1181.5	1.3619
600	0.3737	1213.7	1.3926
650	0.4245	1264.6	1.4398
700	0.4658	1305.8	1.4762
750	0.5025	1342.2	1.5070
800	0.5337	1375.1	1.5333
850	0.5672	1407.6	1.5589
900	0.5966	1438.3	1.5817
1000	0.6551	1498.1	1.6243
1100	0.7109	1556.4	1.6630
1200	0.7651	1614.1	1.6990
1300	0.8172	1671.7	1.7326
1400	0.8666	1729.6	1.7640
1600	0.9688	1846.5	1.8239
1800	1.0693	1965.7	1.8793
2000	1.1687	2087.4	1.9311
2200	1.2668	2211.9	1.9798

p = 1500 psi

T, F	v, ft^3/lbm	h, B/lbm	s, B/lbm·R
32.02	0.015 94	4.5	0.0001
50	0.015 96	22.3	0.0348
100	0.016 07	72.0	0.1279
150	0.016 27	121.7	0.2130
200	0.016 56	171.5	0.2918
250	0.016 94	221.9	0.3639
300	0.017 35	272.5	0.4346
350	0.017 91	324.4	0.4993
400	0.018 52	376.8	0.5632
450	0.019 34	431.6	0.6243
500	0.020 37	488.7	0.6858
550	0.021 75	550.0	0.7479
T_{sat} = 596.33, f	0.023 46	611.3	0.8081
T_{sat} = 596.33, g	0.2769	1168.6	1.3358
600	0.2713	1168.8	1.3377
650	0.3279	1236.5	1.3989
700	0.3695	1285.5	1.4422
750	0.4038	1326.1	1.4766
800	0.4318	1361.5	1.5049
850	0.4620	1396.2	1.5322
900	0.4877	1428.2	1.5560
1000	0.5387	1490.2	1.6002
1100	0.5868	1550.0	1.6399
1200	0.6331	1608.7	1.6765
1300	0.6774	1667.1	1.7106
1400	0.7192	1725.6	1.7424
1600	0.8055	1843.5	1.8028
1800	0.8900	1963.3	1.8585
2000	0.9735	2085.6	1.9105
2200	1.0557	2210.4	1.9593

p = 1750 psi

T, F	v, ft^3/lbm	h, B/lbm	s, B/lbm·R
32.02	0.015 93	5.2	0.0001
50	0.015 95	23.1	0.0348
100	0.016 06	72.6	0.1277
150	0.016 26	122.3	0.2128
200	0.016 55	172.1	0.2915
250	0.016 93	222.4	0.3636
300	0.017 33	272.9	0.4341
350	0.017 89	324.8	0.4988
400	0.018 49	377.1	0.5626
450	0.019 30	431.7	0.6235
500	0.020 31	488.6	0.6847
550	0.021 66	549.5	0.7463
600	0.023 87	619.2	0.8102
T_{sat} = 617.24, f	0.024 48	642.1	0.8359
T_{sat} = 617.24, g	0.2268	1153.5	1.3108
650	0.2562	1202.5	1.3559
700	0.3000	1263.0	1.4095
750	0.3332	1308.9	1.4484
800	0.3589	1347.1	1.4789
850	0.3869	1384.3	1.5082
900	0.4100	1417.8	1.5332
1000	0.4556	1482.1	1.5789
1100	0.4982	1543.4	1.6197
1200	0.5389	1603.3	1.6570
1300	0.5776	1662.5	1.6916
1400	0.6140	1721.7	1.7237
1600	0.6889	1840.5	1.7846
1800	0.7620	1961.0	1.8406
2000	0.8341	2083.8	1.8928
2200	0.9050	2209.0	1.9418

Table A·6·3E Water: Compressed Liquid and Superheated Vapor (*Continued*)

T F	p = 2000 psi v, ft³/lbm	h, B/lbm	s, B/lbm·R
32.02	0.015 91	6.0	0.0001
50	0.015 94	23.8	0.0348
100	0.016 04	73.3	0.1276
150	0.016 25	122.9	0.2126
200	0.016 53	172.6	0.2912
250	0.016 91	222.9	0.3632
300	0.017 31	273.4	0.4337
350	0.017 86	325.2	0.4982
400	0.018 46	377.4	0.5619
450	0.019 26	431.9	0.6227
500	0.020 25	488.5	0.6836
550	0.021 57	548.9	0.7448
600	0.023 74	618.2	0.8088
T_{sat}= 635.91 f	0.025 63	671.6	0.8620
T_{sat}= 635.91 g	0.1881	1136.1	1.2860
650	0.1910	1163.9	1.3008
700	0.2448	1236.8	1.3760
750	0.2788	1290.2	1.4213
800	0.3033	1331.7	1.4544
850	0.3300	1371.8	1.4861
900	0.3513	1407.0	1.5123
1000	0.3931	1473.9	1.5599
1100	0.4316	1536.8	1.6018
1200	0.4682	1597.8	1.6398
1300	0.5027	1657.9	1.6749
1400	0.5350	1717.7	1.7073
1600	0.6014	1837.5	1.7687
1800	0.6661	1958.7	1.8251
2000	0.7295	2082.0	1.8775
2200	0.7919	2207.5	1.9266

T F	p = 3000 psi v, ft³/lbm	h, B/lbm	s, B/lbm·R
32.02	0.015 86	8.9	0.0002
50	0.015 88	26.6	0.0346
100	0.016 00	75.9	0.1269
150	0.016 20	125.3	0.2117
200	0.016 48	174.9	0.2900
250	0.016 85	225.0	0.3618
300	0.017 24	275.3	0.4320
350	0.017 78	326.8	0.4961
400	0.018 36	378.7	0.5594
450	0.019 12	432.6	0.6196
500	0.020 04	488.3	0.6794
550	0.021 23	547.2	0.7388
600	0.023 23	614.4	0.8029
650	0.025 87	688.2	0.8706
T_{sat}= 695.49 f	0.034 27	802.5	0.9732
T_{sat}= 695.49 g	0.0849	1017.9	1.1597
700	0.0647	1016.8	1.1581
750	0.1420	1191.5	1.3080
800	0.1680	1257.0	1.3605
850	0.1958	1316.4	1.4076
900	0.2134	1360.2	1.4402
1000	0.2471	1439.3	1.4966
1100	0.2764	1509.5	1.5433
1200	0.3033	1575.5	1.5844
1300	0.3282	1639.1	1.6216
1400	0.3509	1701.5	1.6554
1600	0.3975	1825.5	1.7190
1800	0.4422	1949.5	1.7767
2000	0.4857	2074.8	1.8300
2200	0.5282	2201.9	1.8797

T F	p = 4000 psi v, ft³/lbm	h, B/lbm	s, B/lbm·R
32.02	0.015 81	11.8	0.0001
50	0.015 83	29.5	0.0344
100	0.015 95	78.5	0.1263
150	0.016 16	127.8	0.2108
200	0.016 43	177.2	0.2889
250	0.016 79	227.1	0.3604
300	0.017 18	277.2	0.4303
350	0.017 70	328.4	0.4941
400	0.018 26	380.0	0.5570
450	0.018 99	433.4	0.6166
500	0.019 85	488.3	0.6758
550	0.020 95	546.0	0.7343
600	0.022 61	609.7	0.7880
650	0.024 72	678.1	0.8379
700	0.030 16	771.7	0.9387
750	0.057 89	1003.3	1.1366
800	0.083 06	1134.4	1.2422
850	0.1261	1249.2	1.3338
900	0.1425	1306.1	1.3761
1000	0.1735	1402.0	1.4447
1100	0.1986	1480.9	1.4972
1200	0.2208	1552.5	1.5419
1300	0.2409	1620.0	1.5813
1400	0.2589	1685.1	1.6166
1600	0.2956	1813.4	1.6824
1800	0.3304	1940.3	1.7415
2000	0.3639	2067.6	1.7956
2200	0.3964	2196.3	1.8460

T F	p = 5000 psi v, ft³/lbm	h, B/lbm	s, B/lbm·R
32.02	0.015 76	14.7	0.0000
50	0.015 78	32.3	0.0342
100	0.015 90	81.1	0.1257
150	0.016 11	130.2	0.2099
200	0.016 38	179.5	0.2877
250	0.016 74	229.3	0.3590
300	0.017 11	279.2	0.4287
350	0.017 62	330.1	0.4922
400	0.018 16	381.4	0.5547
450	0.018 86	434.3	0.6138
500	0.019 68	488.6	0.6723
550	0.020 70	545.3	0.7297
600	0.022 15	606.5	0.7888
650	0.023 93	671.4	0.8490
700	0.027 15	750.2	0.9186
750	0.034 23	857.6	1.0090
800	0.049 99	1018.8	1.1380
850	0.083 79	1168.1	1.2565
900	0.099 62	1243.9	1.3131
1000	0.129 69	1362.4	1.3980
1100	0.152 25	1451.4	1.4573
1200	0.171 63	1529.1	1.5057
1300	0.188 78	1600.6	1.5475
1400	0.203 91	1668.6	1.5844
1600	0.234 65	1801.3	1.6525
1800	0.263 38	1931.2	1.7129
2000	0.290 87	2060.6	1.7680
2200	0.317 37	2190.7	1.8189

T F	p = 8000 psi v, ft³/lbm	h, B/lbm	s, B/lbm·R
32.02	0.015 61	23.2	-0.0003
50	0.015 65	40.5	0.0333
100	0.015 78	88.9	0.1239
150	0.015 98	137.5	0.2073
200	0.016 24	186.4	0.2844
250	0.016 58	235.6	0.3550
300	0.016 94	285.0	0.4240
350	0.017 40	335.3	0.4866
400	0.017 90	385.8	0.5483
450	0.018 53	437.6	0.6060
500	0.019 24	490.3	0.6629
550	0.020 09	544.8	0.7181
600	0.021 20	595.0	0.7732
650	0.022 50	641.9	0.8282
700	0.024 32	724.5	0.8866
750	0.026 77	796.0	0.9447
800	0.031 17	884.3	1.0151
850	0.037 29	974.9	1.0863
900	0.045 78	1074.0	1.1599
1000	0.065 87	1239.6	1.2784
1100	0.083 67	1361.4	1.3596
1200	0.098 50	1458.7	1.4204
1300	0.111 11	1542.9	1.4696
1400	0.121 78	1619.5	1.5112
1600	0.143 42	1765.6	1.5862
1800	0.163 06	1904.2	1.6508
2000	0.181 48	2039.8	1.7085
2200	0.198 97	2174.6	1.7612

T F	p = 10000 psi v, ft³/lbm	h, B/lbm	s, B/lbm·R
32.02	0.015 52	28.8	-0.0007
50	0.015 56	46.0	0.0327
100	0.015 69	94.0	0.1226
150	0.015 89	142.4	0.2056
200	0.016 04	190.9	0.2823
250	0.016 39	239.9	0.3524
300	0.016 82	289.0	0.4210
350	0.017 27	338.9	0.4832
400	0.017 74	388.9	0.5443
450	0.018 33	440.1	0.6013
500	0.018 98	492.0	0.6573
550	0.019 77	545.3	0.7114
600	0.020 74	599.3	0.7648
650	0.021 85	653.7	0.8177
700	0.023 32	719.2	0.8727
750	0.025 12	782.0	0.9253
800	0.027 82	854.4	0.9830
850	0.031 26	928.9	1.0415
900	0.036 16	1012.1	1.1033
1000	0.048 98	1171.9	1.2171
1100	0.062 91	1305.9	1.3064
1200	0.075 41	1414.3	1.3741
1300	0.086 12	1506.3	1.4278
1400	0.095 12	1588.2	1.4724
1600	0.113 40	1742.8	1.5518
1800	0.129 84	1886.9	1.6189
2000	0.145 14	2026.5	1.6783
2200	0.159 56	2164.1	1.7322

Source: Based on a combination of data from Haar, Lester, John S. Gallagher, and George S. Kell, *NBS/NRC Steam Tables,* Hemisphere, 1984, and Keenan, Joseph H., Frederick G. Keyes, Philip G. Hill, and Joan G. Moore, *Steam Tables,* English Units, Wiley, New York, 1969.

Table A·6·4E Water: Solid–Vapor Saturation

T F	p lbf/in³	v_i ft³/lbm	v_g ft³/lbm	h_i B/lbm	h_{ig} B/lbm	h_g B/lbm	s_i B/lbm·R	s_{ig} B/lbm·R	s_g B/lbm·R	T F
32.02	0.08872	0.017472	3300	-143.73	1218.7	1075.0	-0.2921	2.4787	2.1866	32.02
32	0.08865	0.017472	3303	-143.74	1218.7	1075.0	-0.2921	2.4787	2.1867	32
28	0.07375	0.017465	3939	-145.74	1219.0	1073.3	-0.2963	2.4995	2.2031	28
24	0.06117	0.017460	4711	-147.72	1219.2	1071.5	-0.3005	2.5207	2.2202	24
20	0.05058	0.017454	5654	-149.69	1219.4	1069.7	-0.3046	2.5421	2.2375	20
16	0.04169	0.017448	6804	-151.65	1219.6	1068.0	-0.3087	2.5640	2.2551	16
12	0.03424	0.017443	8215	-153.59	1219.8	1066.2	-0.3128	2.5861	2.2732	12
8	0.02804	0.017436	9954	-155.51	1219.9	1064.4	-0.3171	2.6086	2.2915	8
4	0.02288	0.017430	12101	-157.42	1220.0	1062.6	-0.3211	2.6313	2.3102	4
0	0.01861	0.017425	14763	-159.31	1220.2	1060.9	-0.3253	2.6544	2.3292	0
-4	0.01507	0.017418	18076	-161.19	1220.3	1059.1	-0.3293	2.6779	2.3486	-4
-8	0.01217	0.017413	22214	-163.05	1220.4	1057.3	-0.3336	2.7019	2.3683	-8
-12	0.00979	0.017407	27403	-164.90	1220.4	1055.5	-0.3376	2.7263	2.3885	-12
-16	0.00785	0.017400	33937	-166.73	1220.5	1053.8	-0.3418	2.7508	2.4091	-16
-20	0.00627	0.017395	42203	-168.54	1220.6	1052.0	-0.3458	2.7760	2.4301	-20
-24	0.00498	0.017389	52698	-170.34	1220.6	1050.2	-0.3501	2.8016	2.4516	-24
-28	0.00396	0.017383	66077	-172.13	1220.6	1048.5	-0.3542	2.8277	2.4736	-28
-32	0.00312	0.017377	83221	-173.90	1220.6	1046.7	-0.3582	2.8541	2.4958	-32
-36	0.00246	0.017371	105281	-175.65	1220.6	1044.9	-0.3624	2.8809	2.5185	-36
-40	0.00193	0.017365	133813	-177.39	1220.5	1043.1	-0.3664	2.9082	2.5418	-40

Source: Based on a combination of data from Haar, Lester, John S. Gallagher, and George S. Kell, *NBS/NRC Steam Tables,* Hemisphere, 1984, and Keenan, Joseph H., Frederick G. Keyes, Philip G. Hill, and Joan G. Moore, *Steam Tables,* SI Units, Wiley, New York, 1978.

Chart A·7·1E Water: *Ts* diagram

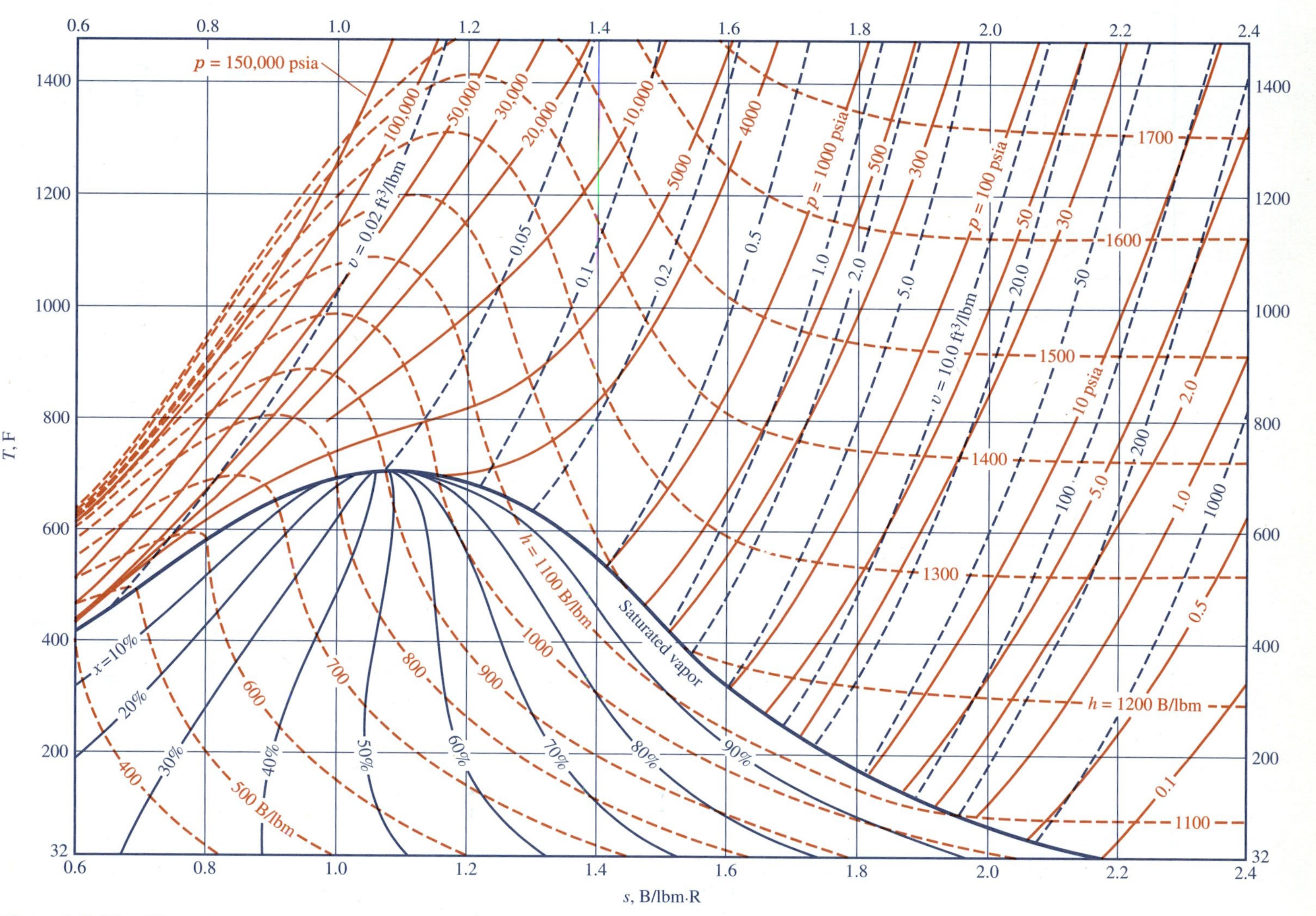

Chart A·7·1E Water *Ts* diagram. (*Source:* Based on data and formulations from Haar, Lester, John S. Gallagher, and George S. Kell, *NBS/NRC Steam Tables*, Hemisphere, 1984.)

H_2O, *Ts* diagram

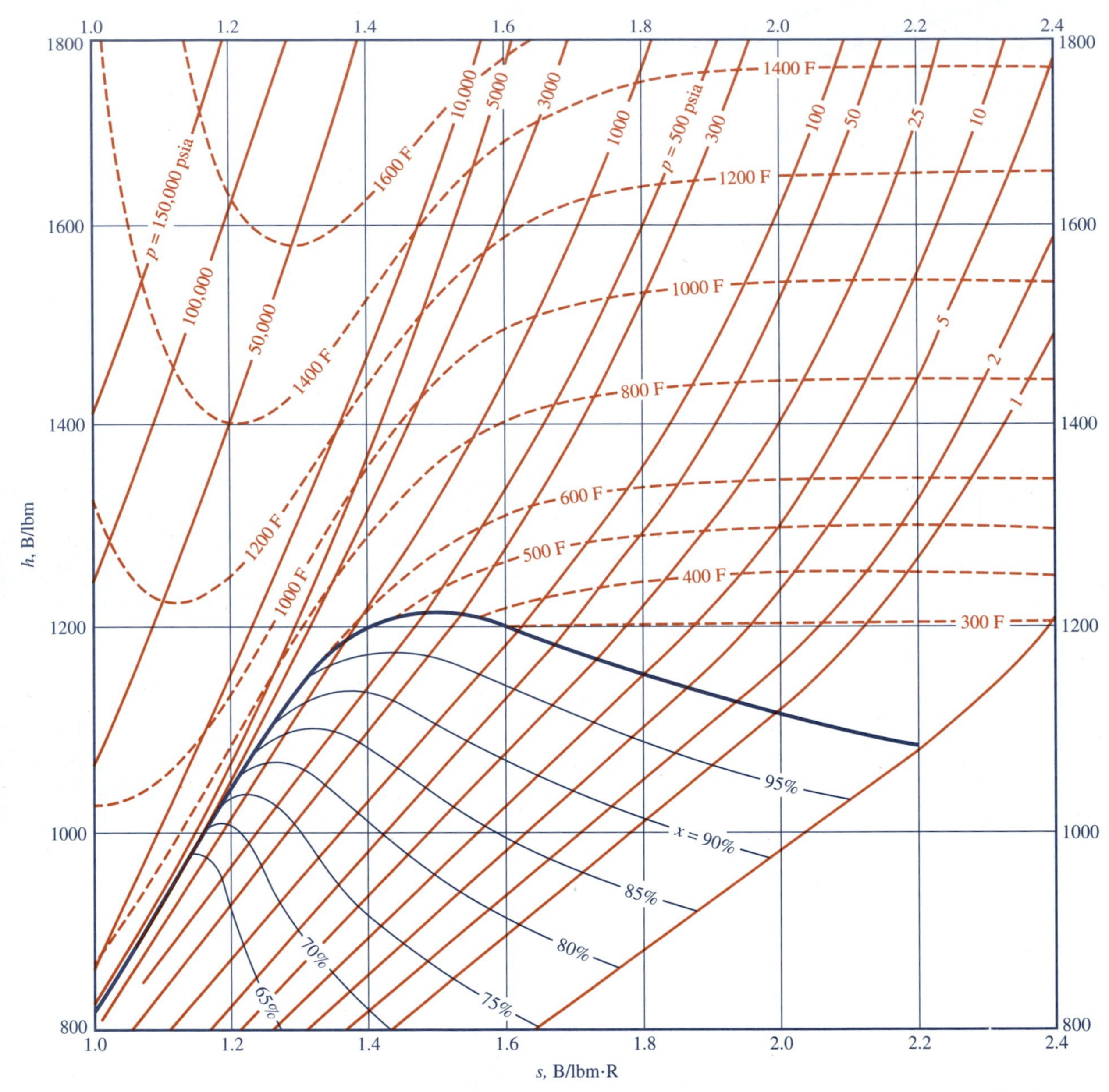

Chart A·7·2E Water *hs* diagram. (*Source:* Based on data and formulations from Haar, Lester, John S. Gallagher, and George S. Kell, *NBS/NRC Steam Tables,* Hemisphere, 1984.)

Chart A·7·3E Water: *ph* Diagram

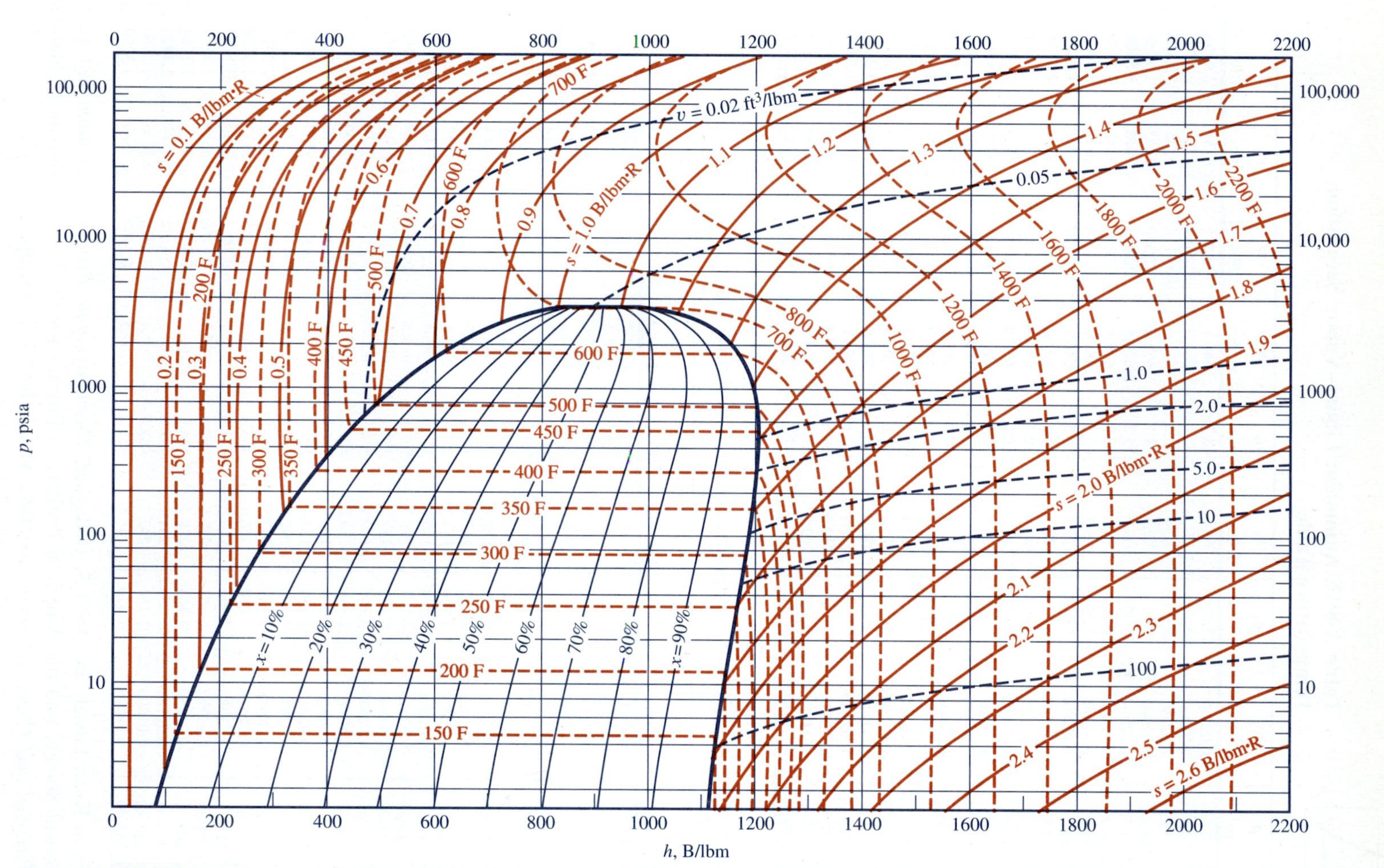

Chart A·7·3E Water *ph* diagram. (*Source:* Based on data and formulations from Haar, Lester, John S. Gallagher, and George S. Kell, *NBS/NRC Steam Tables*, Hemisphere, 1984.)

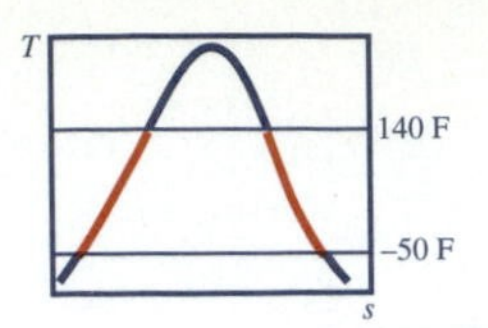

Table A·8·1E Ammonia: Liquid–Vapor Saturation, Temperature Table

T F	p psi	v_f ft³/lbm	v_g ft³/lbm	h_f B/lbm	h_{fg} B/lbm	h_g B/lbm	s_f B/lbm·R	s_{fg} B/lbm·R	s_g B/lbm·R
-50	7.66	0.02299	33.09	61.4	603.7	665.1	0.1616	1.4737	1.6353
-40	10.40	0.02322	24.87	71.9	597.1	668.9	0.1869	1.4227	1.6096
-30	13.89	0.02345	18.97	82.4	590.2	672.6	0.2116	1.3736	1.5853
-20	18.29	0.02369	14.67	93.0	583.1	676.2	0.2360	1.3263	1.5623
-10	23.73	0.02394	11.49	103.7	575.8	679.5	0.2600	1.2806	1.5406
0	30.42	0.02420	9.109	114.5	568.3	682.8	0.2836	1.2363	1.5199
10	38.50	0.02446	7.297	125.3	560.5	685.8	0.3068	1.1934	1.5003
20	48.21	0.02474	5.903	136.2	552.5	688.7	0.3297	1.1518	1.4815
30	59.75	0.02502	4.819	147.2	544.2	691.4	0.3523	1.1113	1.4636
32	62.29	0.02508	4.633	149.4	542.5	691.9	0.3568	1.1033	1.4601
40	73.34	0.02532	3.967	158.3	535.6	693.8	0.3746	1.0718	1.4464
50	89.23	0.02564	3.290	169.4	526.7	696.1	0.3966	1.0333	1.4299
60	107.6	0.02597	2.748	180.7	517.4	698.1	0.4183	0.9957	1.4140
70	128.8	0.02631	2.310	192.1	507.9	699.9	0.4397	0.9588	1.3986
77	145.5	0.02657	2.052	200.1	501.0	701.0	0.4546	0.9334	1.3881
80	153.1	0.02668	1.953	203.5	497.9	701.5	0.4610	0.9226	1.3836
90	180.7	0.02706	1.659	215.1	487.6	702.7	0.4820	0.8871	1.3690
100	211.9	0.02747	1.417	226.8	476.8	703.6	0.5028	0.8520	1.3548
110	247.1	0.02790	1.215	238.7	465.6	704.2	0.5234	0.8173	1.3407
120	286.5	0.02836	1.046	250.6	453.8	704.5	0.5439	0.7829	1.3269
130	330.4	0.02885	0.903	262.8	441.5	704.3	0.5643	0.7488	1.3131
140	379.2	0.02938	0.782	275.1	428.6	703.7	0.5846	0.7147	1.2993

Table A·8·2E Ammonia: Liquid–Vapor Saturation, Pressure Table

p psi	T F	v_f ft³/lbm	v_g ft³/lbm	h_f B/lbm	h_{fg} B/lbm	h_g B/lbm	s_f B/lbm·R	s_{fg} B/lbm·R	s_g B/lbm·R
5	-63.13	0.02271	49.37	47.7	612.2	659.9	0.1277	1.5438	1.6715
10	-41.31	0.02319	25.79	70.5	598.0	668.4	0.1836	1.4293	1.6129
15	-27.26	0.02352	17.66	85.3	588.3	673.6	0.2184	1.3605	1.5789
20	-16.62	0.02378	13.49	96.6	580.7	677.3	0.2442	1.3107	1.5549
25	-7.94	0.02399	10.95	105.9	574.3	680.2	0.2649	1.2714	1.5363
30	-0.57	0.02418	9.228	113.8	568.7	682.6	0.2823	1.2388	1.5211
35	5.89	0.02435	7.984	120.8	563.7	684.6	0.2973	1.2109	1.5082
40	11.67	0.02451	7.040	127.1	559.2	686.3	0.3107	1.1864	1.4971
45	16.89	0.02465	6.300	132.8	555.0	687.8	0.3226	1.1646	1.4873
50	21.66	0.02478	5.704	138.0	551.1	689.1	0.3335	1.1450	1.4785
60	30.20	0.02503	4.800	147.4	544.0	691.4	0.3528	1.1105	1.4633
70	37.68	0.02525	4.147	155.7	537.6	693.3	0.3694	1.0809	1.4503
80	44.38	0.02546	3.652	163.1	531.7	694.9	0.3842	1.0549	1.4391
90	50.45	0.02565	3.263	169.9	526.3	696.2	0.3975	1.0316	1.4292
100	56.03	0.02583	2.949	176.2	521.1	697.4	0.4097	1.0106	1.4202
110	61.18	0.02601	2.691	182.0	516.3	698.4	0.4208	0.9913	1.4121
120	65.99	0.02617	2.474	187.5	511.7	699.2	0.4312	0.9735	1.4047
130	70.51	0.02633	2.290	192.6	507.4	700.0	0.4408	0.9570	1.3978
140	74.76	0.02648	2.131	197.5	503.2	700.7	0.4499	0.9415	1.3914
160	82.62	0.02678	1.870	206.6	495.3	701.8	0.4665	0.9133	1.3798
180	89.76	0.02705	1.666	214.8	487.8	702.7	0.4815	0.8879	1.3694
200	96.32	0.02732	1.501	222.5	480.8	703.3	0.4952	0.8648	1.3600
220	102.40	0.02757	1.365	229.7	474.2	703.8	0.5078	0.8436	1.3514
240	108.07	0.02782	1.251	236.4	467.8	704.2	0.5195	0.8239	1.3434
260	113.40	0.02805	1.154	242.7	461.6	704.4	0.5304	0.8056	1.3360
280	118.43	0.02829	1.070	248.8	455.7	704.5	0.5407	0.7883	1.3290

Source: Adapted from Haar, Lester, and John S. Gallagher, *Thermodynamic Properties of Ammonia,* published by the American Chemical Society and the American Institute of Physics for the National Bureau of Standards, 1978. Also, *Journal of Physical and Chemical Reference Data,* Vol. 7, No. 3, 1978, pp. 635–792.

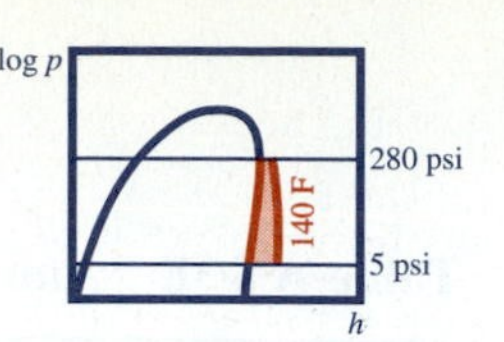

Table A·8·3E Ammonia: Superheated Vapor

T F	$p = 5$ psia v, ft^3/lbm	h, B/lbm	s, B/lbm·R
T_{sat}= −63.13 f	0.02271	47.7	0.1277
T_{sat}= −63.13 g	49.37	659.9	1.6715
−50	51.31	666.8	1.6870
−40	52.56	671.7	1.6993
−30	53.82	676.6	1.7113
−20	55.07	681.5	1.7230
−10	56.33	686.4	1.7345
0	57.58	691.4	1.7456
10	58.84	696.3	1.7566
20	60.09	701.3	1.7673
30	61.35	706.3	1.7777
40	62.60	711.3	1.7879
50	63.86	716.3	1.7980
60	65.11	721.4	1.8078
70	66.37	726.5	1.8174
80	67.62	731.5	1.8269
90	68.88	736.6	1.8361
100	70.13	741.7	1.8452
110	71.39	746.9	1.8542
120	72.65	752.0	1.8630
130	73.90	757.2	1.8717
140	75.16	762.3	1.8803

T F	$p = 10$ psia v, ft^3/lbm	h, B/lbm	s, B/lbm·R
T_{sat}= −41.31 f	0.02319	70.5	0.1836
T_{sat}= −41.31 g	25.79	668.4	1.6129
−40	25.92	669.3	1.6141
−30	26.58	674.3	1.6263
−20	27.24	679.4	1.6383
−10	27.90	684.4	1.6501
0	28.55	689.5	1.6616
10	29.21	694.6	1.6729
20	29.86	699.7	1.6839
30	30.52	704.8	1.6948
40	31.17	709.9	1.7054
50	31.82	715.0	1.7158
60	32.47	720.1	1.7261
70	33.12	725.3	1.7361
80	33.77	730.4	1.7459
90	34.41	735.6	1.7556
100	35.06	740.8	1.7650
110	35.70	746.0	1.7743
120	36.34	751.2	1.7834
130	36.99	756.4	1.7923
140	37.63	761.6	1.8011

T F	$p = 15$ psia v, ft^3/lbm	h, B/lbm	s, B/lbm·R
T_{sat}= −27.26 f	0.02352	85.3	0.2184
T_{sat}= −27.26 g	17.66	673.6	1.5789
−20	18.02	677.4	1.5875
−10	18.47	682.5	1.5994
0	18.91	687.7	1.6111
10	19.35	692.9	1.6225
20	19.80	698.1	1.6338
30	20.24	703.3	1.6448
40	20.68	708.5	1.6556
50	21.12	713.7	1.6662
60	21.56	718.9	1.6767
70	22.00	724.1	1.6869
80	22.43	729.4	1.6969
90	22.87	734.6	1.7067
100	23.30	739.9	1.7164
110	23.74	745.1	1.7258
120	24.17	750.4	1.7351
130	24.60	755.7	1.7441
140	25.03	761.0	1.7530

T F	$p = 20$ psia v, ft^3/lbm	h, B/lbm	s, B/lbm·R
T_{sat}= −16.62 f	0.02378	96.6	0.2442
T_{sat}= −16.62 g	13.49	677.3	1.5549
−10	13.75	680.7	1.5629
0	14.09	686.0	1.5747
10	14.43	691.3	1.5862
20	14.76	696.6	1.5976
30	15.10	701.8	1.6087
40	15.43	707.1	1.6196
50	15.77	712.4	1.6304
60	16.10	717.7	1.6409
70	16.43	723.0	1.6512
80	16.76	728.3	1.6614
90	17.09	733.7	1.6713
100	17.42	739.0	1.6810
110	17.75	744.3	1.6906
120	18.07	749.6	1.7000
130	18.40	754.9	1.7091
140	18.73	760.3	1.7181

T F	$p = 25$ psia v, ft^3/lbm	h, B/lbm	s, B/lbm·R
T_{sat}= −7.94 f	0.02399	105.9	0.2649
T_{sat}= −7.94 g	10.95	680.2	1.5363
0	11.20	684.3	1.5461
10	11.47	689.7	1.5577
20	11.74	695.1	1.5691
30	12.01	700.5	1.5803
40	12.28	705.8	1.5913
50	12.55	711.2	1.6021
60	12.82	716.6	1.6127
70	13.09	722.0	1.6231
80	13.36	727.3	1.6333
90	13.62	732.7	1.6434
100	13.89	738.1	1.6532
110	14.15	743.5	1.6628
120	14.42	748.8	1.6723
130	14.68	754.2	1.6815
140	14.94	759.6	1.6906

T F	$p = 30$ psia v, ft^3/lbm	h, B/lbm	s, B/lbm·R
T_{sat}= −0.57 f	0.02418	113.8	0.2823
T_{sat}= −0.57 g	9.228	682.6	1.5211
0	9.267	682.8	1.5225
10	9.497	688.2	1.5342
20	9.726	693.7	1.5456
30	9.954	699.1	1.5569
40	10.18	704.6	1.5679
50	10.41	710.0	1.5788
60	10.64	715.5	1.5894
70	10.86	720.9	1.5999
80	11.09	726.3	1.6102
90	11.31	731.8	1.6203
100	11.53	737.2	1.6301
110	11.75	742.7	1.6399
120	11.98	748.1	1.6494
130	12.20	753.5	1.6587
140	12.42	758.9	1.6678

T F	$p = 40$ psia v, ft^3/lbm	h, B/lbm	s, B/lbm·R
T_{sat}= 11.67 f	0.02451	127.1	0.3107
T_{sat}= 11.67 g	7.040	686.3	1.4971
20	7.206	690.9	1.5082
30	7.381	696.5	1.5195
40	7.556	702.1	1.5306
50	7.730	707.7	1.5415
60	7.903	713.3	1.5522
70	8.075	718.9	1.5628
80	8.247	724.4	1.5731
90	8.418	730.0	1.5833
100	8.588	735.5	1.5933
110	8.758	741.1	1.6031
120	8.927	746.6	1.6127
130	9.095	752.1	1.6221
140	9.262	757.6	1.6314

T F	$p = 50$ psia v, ft^3/lbm	h, B/lbm	s, B/lbm·R
T_{sat}= 21.66 f	0.02478	138.0	0.3335
T_{sat}= 21.66 g	5.704	689.1	1.4785
30	5.837	694.0	1.4901
40	5.980	699.8	1.5012
50	6.122	705.5	1.5122
60	6.264	711.2	1.5229
70	6.404	716.9	1.5335
80	6.544	722.6	1.5439
90	6.684	728.2	1.5541
100	6.823	733.9	1.5642
110	6.961	739.5	1.5741
120	7.098	745.1	1.5838
130	7.235	750.7	1.5933
140	7.371	756.3	1.6027

T F	$p = 60$ psia v, ft^3/lbm	h, B/lbm	s, B/lbm·R
T_{sat}= 30.20 f	0.02503	147.4	0.3528
T_{sat}= 30.20 g	4.800	691.4	1.4633
40	4.929	697.5	1.4770
50	5.050	703.3	1.4880
60	5.171	709.1	1.4987
70	5.290	714.9	1.5094
80	5.409	720.7	1.5198
90	5.528	726.5	1.5301
100	5.646	732.2	1.5401
110	5.763	738.0	1.5501
120	5.879	743.7	1.5598
130	5.995	749.4	1.5694
140	6.110	755.0	1.5789

Table A·8·3E Ammonia: Superheated Vapor (*Continued*)

Ammonia, SH

T F	p = 70 psia v, ft³/lbm	h, B/lbm	s, B/lbm·R
T_{sat} = 37.68 f	0.02525	155.7	0.3694
g	4.147	693.3	1.4503
40	4.179	695.2	1.4563
50	4.285	701.1	1.4673
60	4.390	707.1	1.4781
70	4.495	713.0	1.4887
80	4.599	718.9	1.4992
90	4.702	724.8	1.5095
100	4.805	730.6	1.5196
110	4.907	736.4	1.5296
120	5.009	742.2	1.5394
130	5.109	748.0	1.5491
140	5.210	753.7	1.5586

T F	p = 80 psia v, ft³/lbm	h, B/lbm	s, B/lbm·R
T_{sat} = 44.38 f	0.02546	163.1	0.3842
g	3.652	694.9	1.4391
50	3.710	699.0	1.4493
60	3.804	705.0	1.4601
70	3.898	711.1	1.4707
80	3.991	717.1	1.4812
90	4.083	723.0	1.4915
100	4.175	729.0	1.5017
110	4.266	734.9	1.5117
120	4.356	740.8	1.5216
130	4.446	746.6	1.5313
140	4.535	752.4	1.5409

T F	p = 90 psia v, ft³/lbm	h, B/lbm	s, B/lbm·R
T_{sat} = 50.45 f	0.02565	169.9	0.3975
g	3.263	696.2	1.4292
60	3.349	703.0	1.4441
70	3.434	709.1	1.4547
80	3.518	715.2	1.4652
90	3.601	721.3	1.4756
100	3.684	727.3	1.4857
110	3.767	733.3	1.4958
120	3.848	739.3	1.5057
130	3.929	745.2	1.5155
140	4.010	751.1	1.5251

T F	p = 100 psia v, ft³/lbm	h, B/lbm	s, B/lbm·R
T_{sat} = 56.03 f	0.02583	176.2	0.4097
g	2.949	697.4	1.4202
60	2.985	700.9	1.4297
70	3.063	707.1	1.4403
80	3.140	713.3	1.4508
90	3.216	719.5	1.4612
100	3.292	725.6	1.4714
110	3.367	731.7	1.4815
120	3.442	737.8	1.4914
130	3.516	743.8	1.5012
140	3.590	749.8	1.5109

T F	p = 120 psia v, ft³/lbm	h, B/lbm	s, B/lbm·R
T_{sat} = 65.99 f	0.02617	187.5	0.4312
g	2.474	699.2	1.4047
70	2.506	703.0	1.4152
80	2.572	709.5	1.4257
90	2.639	715.9	1.4361
100	2.704	722.2	1.4464
110	2.769	728.5	1.4565
120	2.833	734.7	1.4665
130	2.897	740.9	1.4764
140	2.960	747.1	1.4862

T F	p = 140 psia v, ft³/lbm	h, B/lbm	s, B/lbm·R
T_{sat} = 74.76 f	0.02648	197.5	0.4499
g	2.131	700.7	1.3914
80	2.167	705.4	1.4043
90	2.226	712.0	1.4147
100	2.284	718.6	1.4250
110	2.341	725.1	1.4351
120	2.398	731.6	1.4452
130	2.454	738.0	1.4551
140	2.510	744.3	1.4650

T F	p = 160 psia v, ft³/lbm	h, B/lbm	s, B/lbm·R
T_{sat} = 82.62 f	0.02678	206.6	0.4665
g	1.870	701.8	1.3798
90	1.916	708.1	1.3960
100	1.969	714.9	1.4063
110	2.021	721.6	1.4165
120	2.072	728.3	1.4266
130	2.123	734.9	1.4366
140	2.173	741.4	1.4466

T F	p = 180 psia v, ft³/lbm	h, B/lbm	s, B/lbm·R
T_{sat} = 89.76 f	0.02705	214.8	0.4815
g	1.666	702.7	1.3694
90	1.676	703.9	1.3794
100	1.724	711.0	1.3897
110	1.771	718.0	1.3999
120	1.818	724.9	1.4101
130	1.865	731.7	1.4201
140	1.910	738.4	1.4302

T F	p = 200 psia v, ft³/lbm	h, B/lbm	s, B/lbm·R
T_{sat} = 96.32 f	0.02732	222.5	0.4952
g	1.501	703.3	1.3600
100	1.528	707.0	1.3748
110	1.572	714.2	1.3850
120	1.615	721.3	1.3952
130	1.658	728.4	1.4053
140	1.701	735.3	1.4154

T F	p = 220 psia v, ft³/lbm	h, B/lbm	s, B/lbm·R
T_{sat} = 102.40 f	0.02757	229.7	0.5078
g	1.365	703.8	1.3514
110	1.409	710.4	1.3715
120	1.449	717.8	1.3817
130	1.489	725.1	1.3918
140	1.529	732.2	1.4020

T F	p = 240 psia v, ft³/lbm	h, B/lbm	s, B/lbm·R
T_{sat} = 108.07 f	0.02782	236.4	0.5195
g	1.251	704.2	1.3434
110	1.273	706.7	1.3590
120	1.311	714.3	1.3693
130	1.349	721.7	1.3795
140	1.386	729.1	1.3896

T F	p = 260 psia v, ft³/lbm	h, B/lbm	s, B/lbm·R
T_{sat} = 113.40 f	0.02805	242.7	0.5304
g	1.154	704.4	1.3360
120	1.194	710.9	1.3578
130	1.230	718.5	1.3680
140	1.265	726.0	1.3783

T F	p = 280 psia v, ft³/lbm	h, B/lbm	s, B/lbm·R
T_{sat} = 118.43 f	0.02829	248.8	0.5407
g	1.070	704.5	1.3290
120	1.094	707.7	1.3471
130	1.128	715.4	1.3574
140	1.161	723.0	1.3677

Chart A·9E Ammonia: *ph* Diagram

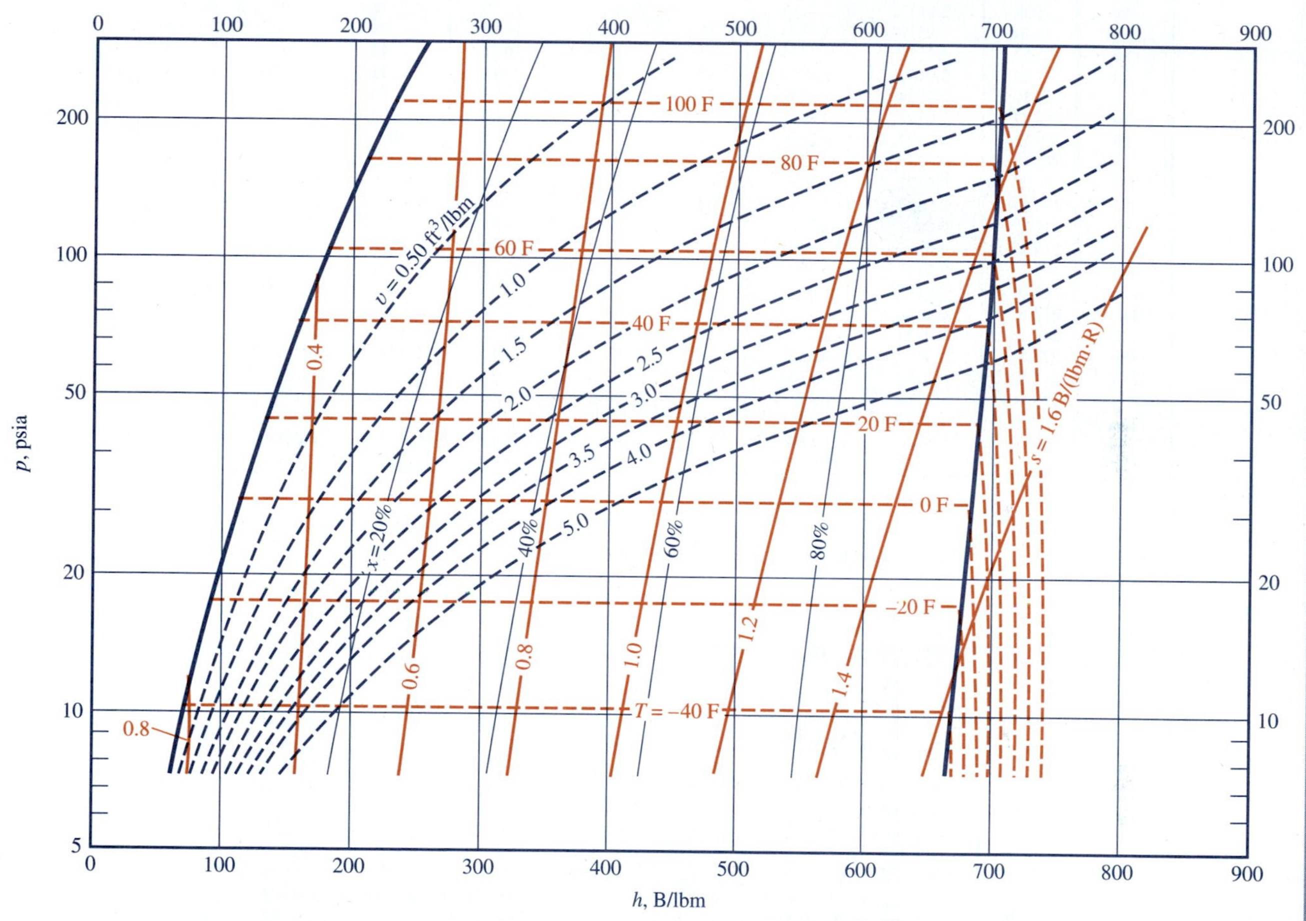

Chart A·9E Ammonia *ph* diagram. (*Source:* Based on Haar, Lester, and John S. Gallagher, *Thermodynamic Properties of Ammonia,* published by the American Chemical Society and the American Institute of Physics for the National Bureau of Standards, 1978. Also, *Journal of Physical and Chemical Reference Data,* Vol. 7, No. 3, 1978, pp. 635–792.)

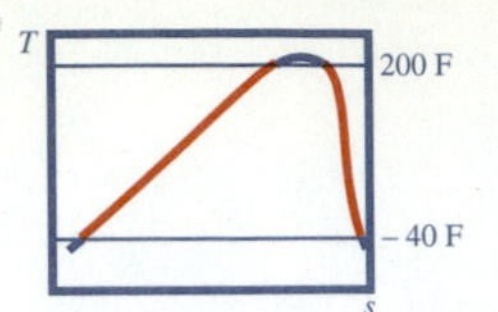

Table A·10·1E R134a: Liquid–vapor Saturation, Temperature Table

T	p	v_f	v_g	h_f	h_{fg}	h_g	s_f	s_{fg}	s_g
F	psi	ft³/lbm	ft³/lbm	B/lbm	B/lbm	B/lbm	B/lbm·R	B/lbm·R	B/lbm·R
-40	7.417	0.0113	5.7904	0.0	97.2	97.2	0.0000	0.2316	0.2316
-30	9.851	0.0115	4.4366	3.0	95.7	98.7	0.0070	0.2227	0.2297
-20	12.89	0.0116	3.4471	6.0	94.2	100.2	0.0139	0.2142	0.2281
-10	16.62	0.0117	2.7115	9.0	92.7	101.7	0.0208	0.2059	0.2267
0	21.16	0.0119	2.1580	12.1	91.0	103.1	0.0275	0.1980	0.2255
10	26.62	0.0120	1.7355	15.2	89.4	104.6	0.0342	0.1902	0.2244
20	33.13	0.0122	1.4090	18.4	87.6	106.0	0.0408	0.1827	0.2235
30	40.80	0.0124	1.1538	21.6	85.8	107.4	0.0473	0.1754	0.2227
40	49.77	0.0125	0.9523	24.8	84.0	108.8	0.0538	0.1682	0.2220
50	60.18	0.0127	0.7916	28.0	82.2	110.2	0.0602	0.1612	0.2214
60	72.17	0.0129	0.6622	31.4	80.1	111.5	0.0666	0.1542	0.2208
70	85.89	0.0131	0.5570	34.7	78.1	112.8	0.0729	0.1474	0.2203
80	101.5	0.0134	0.4709	38.1	75.9	114.0	0.0792	0.1407	0.2199
90	119.1	0.0136	0.3999	41.6	73.6	115.2	0.0855	0.1339	0.2194
100	139.0	0.0139	0.3408	45.1	71.2	116.3	0.0918	0.1272	0.2190
110	161.2	0.0142	0.2912	48.7	68.7	117.4	0.0981	0.1204	0.2185
120	186.0	0.0145	0.2494	52.4	65.9	118.3	0.1043	0.1138	0.2181
130	213.6	0.0148	0.2139	56.2	63.0	119.2	0.1106	0.1069	0.2175
140	244.1	0.0152	0.1835	60.0	59.9	119.9	0.1170	0.0998	0.2168
150	277.7	0.0157	0.1573	64.0	56.5	120.5	0.1234	0.0926	0.2160
160	314.8	0.0162	0.1344	68.1	52.7	120.8	0.1299	0.0851	0.2150
170	355.5	0.0168	0.1143	72.4	48.5	120.9	0.1366	0.0771	0.2137
180	400.3	0.0176	0.0964	76.9	43.8	120.7	0.1435	0.0684	0.2119
190	449.4	0.0186	0.0800	81.8	38.0	119.8	0.1508	0.0585	0.2093
200	503.4	0.0201	0.0645	87.3	30.7	118.0	0.1589	0.0465	0.2054

Table A·10·2E R134a: Liquid–vapor Saturation, Pressure Table

p	T	v_f	v_g	h_f	h_{fg}	h_g	s_f	s_{fg}	s_g
psi	F	ft³/lbm	ft³/lbm	B/lbm	B/lbm	B/lbm	B/lbm·R	B/lbm·R	B/lbm·R
5	-53.01	0.011 140	8.376 051	-3.855	99.085	95.230	-0.0093	0.2436	0.2343
10	-29.46	0.011 501	4.374 382	3.163	95.614	98.778	0.0074	0.2222	0.2296
15	-14.09	0.011 702	2.986 985	7.769	93.275	101.043	0.0180	0.2092	0.2273
20	-2.39	0.011 859	2.276 472	11.384	91.376	102.760	0.0259	0.1999	0.2258
25	7.21	0.011 938	1.842 468	14.362	89.853	104.215	0.0323	0.1924	0.2247
30	15.40	0.012 108	1.549 035	16.928	88.428	105.356	0.0378	0.1862	0.2240
35	22.59	0.012 252	1.336 788	19.230	87.133	106.363	0.0425	0.1808	0.2233
40	29.03	0.012 385	1.175 985	21.289	85.975	107.264	0.0467	0.1761	0.2228
50	40.24	0.012 505	0.948 071	24.876	83.957	108.833	0.0540	0.1680	0.2220
60	49.84	0.012 697	0.793 922	27.947	82.230	110.177	0.0601	0.1613	0.2214
70	58.29	0.012 866	0.682 384	30.826	80.468	111.294	0.0655	0.1554	0.2209
80	65.87	0.013 017	0.597 870	33.291	79.018	112.308	0.0703	0.1502	0.2205
90	72.76	0.013 149	0.531 568	35.638	77.493	113.131	0.0747	0.1455	0.2202
100	79.09	0.013 365	0.478 032	37.788	76.104	113.891	0.0786	0.1413	0.2199
110	84.97	0.013 500	0.433 901	39.890	74.707	114.597	0.0824	0.1372	0.2196
120	90.46	0.013 608	0.396 945	41.764	73.492	115.257	0.0858	0.1336	0.2194
130	95.61	0.013 723	0.365 415	43.609	72.253	115.862	0.0890	0.1302	0.2192
140	100.48	0.013 911	0.338 234	45.269	71.087	116.356	0.0921	0.1269	0.2190
160	109.48	0.014 180	0.293 586	48.511	68.838	117.349	0.0978	0.1208	0.2185
180	117.67	0.014 407	0.258 576	51.507	66.606	118.113	0.1028	0.1154	0.2182
200	125.20	0.014 607	0.230 284	54.377	64.441	118.817	0.1076	0.1102	0.2178
250	141.83	0.015 268	0.178 394	60.727	59.283	120.010	0.1182	0.0985	0.2167
300	156.12	0.015 962	0.142 945	66.466	54.262	120.728	0.1273	0.0881	0.2154
350	168.69	0.016 716	0.116 809	71.821	49.089	120.910	0.1357	0.0782	0.2139
400	179.94	0.017 595	0.096 502	76.872	43.830	120.702	0.1435	0.0685	0.2119
450	190.12	0.018 615	0.079 811	81.860	37.926	119.786	0.1509	0.0584	0.2093
500	199.41	0.019 993	0.065 424	86.939	31.219	118.158	0.1584	0.0473	0.2057

Source: Adapted from *Thermodynamic Properties of HFC-134a (1,1,1, 2-tetrafluoroethane),* DuPont Company, Wilmington, Delaware, 1993, with permission.

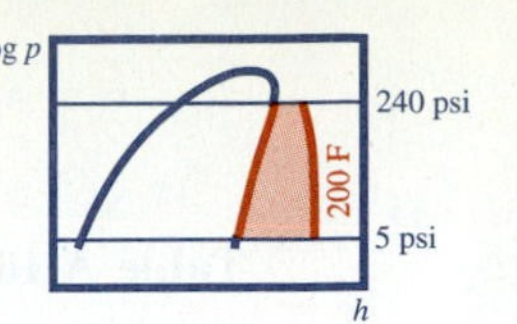

Table A·10·3E R134a: Superheated Vapor

T F	$p = 5$ psi: v, ft³/lbm	h, B/lbm	s, B/lbm·R
$T_{sat}=$ −53.01 *f*	0.0111	-3.86	-0.0093
$T_{sat}=$ −53.01 *g*	8.376	95.2	0.2343
-40	8.671	97.5	0.2397
-30	8.890	99.2	0.2438
-20	9.109	101.0	0.2480
-10	9.327	102.8	0.2521
0	9.546	104.7	0.2562
10	9.763	106.5	0.2602
20	9.981	108.4	0.2642
30	10.198	110.3	0.2681
40	10.414	112.2	0.2719
50	10.631	114.1	0.2756
60	10.847	116.1	0.2793
70	11.062	118.1	0.2830
80	11.277	120.1	0.2866
90	11.492	122.1	0.2902
100	11.707	124.1	0.2937
110	11.921	126.2	0.2973
120	12.134	128.3	0.3009
130	12.348	130.4	0.3045
140	12.561	132.5	0.3080
150	12.773	134.7	0.3115
160	12.985	136.8	0.3150
170	13.197	139.0	0.3183
180	13.409	141.3	0.3215
190	13.620	143.5	0.3244
200	13.830	145.8	0.3271

T F	$p = 10$ psi: v, ft³/lbm	h, B/lbm	s, B/lbm·R
$T_{sat}=$ −29.46 *f*	0.0115	3.16	0.0074
$T_{sat}=$ −29.46 *g*	4.3744	98.8	0.2296
-20	4.4855	100.5	0.2334
-10	4.5984	102.3	0.2376
0	4.7110	104.2	0.2418
10	4.8233	106.1	0.2460
20	4.9352	108.0	0.2502
30	5.0468	109.9	0.2542
40	5.1581	111.8	0.2582
50	5.2689	113.8	0.2620
60	5.3795	115.8	0.2658
70	5.4897	117.8	0.2695
80	5.5996	119.8	0.2732
90	5.7091	121.8	0.2768
100	5.8183	123.9	0.2805
110	5.9271	126.0	0.2842
120	6.0356	128.1	0.2879
130	6.1437	130.2	0.2917
140	6.2516	132.3	0.2956
150	6.3590	134.5	0.2994
160	6.4661	136.7	0.3032
170	6.5729	138.9	0.3069
180	6.6793	141.1	0.3104
190	6.7854	143.3	0.3136
200	6.8912	145.6	0.3164

T F	$p = 15$ psi: v, ft³/lbm	h, B/lbm	s, B/lbm·R
$T_{sat}=$ −14.09 *f*	0.0117	7.77	0.0180
$T_{sat}=$ −14.09 *g*	2.9870	101.0	0.2273
-10	3.0227	101.9	0.2287
0	3.0999	103.7	0.2331
10	3.1769	105.6	0.2374
20	3.2535	107.6	0.2416
30	3.3299	109.5	0.2457
40	3.4058	111.5	0.2498
50	3.4815	113.4	0.2537
60	3.5569	115.4	0.2575
70	3.6319	117.5	0.2612
80	3.7066	119.5	0.2649
90	3.7810	121.5	0.2685
100	3.8551	123.6	0.2722
110	3.9289	125.7	0.2760
120	4.0023	127.8	0.2798
130	4.0755	130.0	0.2837
140	4.1483	132.1	0.2876
150	4.2208	134.3	0.2916
160	4.2929	136.5	0.2956
170	4.3648	138.7	0.2995
180	4.4363	140.9	0.3031
190	4.5075	143.2	0.3064
200	4.5784	145.5	0.3092

T F	$p = 20$ psi: v, ft³/lbm	h, B/lbm	s, B/lbm·R
$T_{sat}=$ −2.39 *f*	0.0119	11.4	0.0259
$T_{sat}=$ −2.39 *g*	2.2765	102.8	0.2258
0	2.2945	103.3	0.2267
10	2.3537	105.2	0.2310
20	2.4127	107.2	0.2353
30	2.4713	109.1	0.2396
40	2.5296	111.1	0.2436
50	2.5876	113.1	0.2476
60	2.6454	115.1	0.2514
70	2.7028	117.1	0.2551
80	2.7599	119.2	0.2588
90	2.8167	121.3	0.2624
100	2.8733	123.4	0.2661
110	2.9295	125.5	0.2699
120	2.9855	127.6	0.2737
130	3.0411	129.8	0.2777
140	3.0964	131.9	0.2817
150	3.1515	134.1	0.2858
160	3.2062	136.3	0.2899
170	3.2607	138.5	0.2938
180	3.3148	140.8	0.2976
190	3.3687	143.0	0.3009
200	3.4222	145.3	0.3036

T F	$p = 30$ psi: v, ft³/lbm	h, B/lbm	s, B/lbm·R
$T_{sat}=$ 15.40 *f*	0.0121	16.9	0.0378
$T_{sat}=$ 15.40 *g*	1.5490	105.4	0.2240
20	1.5714	106.3	0.2263
30	1.6122	108.3	0.2306
40	1.6528	110.4	0.2347
50	1.6932	112.4	0.2387
60	1.7333	114.5	0.2425
70	1.7731	116.5	0.2462
80	1.8126	118.6	0.2499
90	1.8519	120.7	0.2535
100	1.8909	122.9	0.2572
110	1.9296	125.0	0.2609
120	1.9681	127.2	0.2648
130	2.0063	129.3	0.2688
140	2.0443	131.5	0.2729
150	2.0819	133.7	0.2771
160	2.1193	135.9	0.2813
170	2.1565	138.2	0.2854
180	2.1933	140.4	0.2893
190	2.2299	142.7	0.2926
200	2.2663	145.0	0.2952

T F	$p = 40$ psi: v, ft³/lbm	h, B/lbm	s, B/lbm·R
$T_{sat}=$ 29.03 *f*	0.0124	21.3	0.0467
$T_{sat}=$ 29.03 *g*	1.1760	107.3	0.2228
30	1.1820	107.6	0.2240
40	1.2138	109.6	0.2282
50	1.2453	111.7	0.2322
60	1.2766	113.8	0.2361
70	1.3076	115.9	0.2398
80	1.3384	118.1	0.2434
90	1.3689	120.2	0.2470
100	1.3992	122.3	0.2506
110	1.4292	124.5	0.2543
120	1.4590	126.7	0.2582
130	1.4885	128.9	0.2622
140	1.5178	131.1	0.2664
150	1.5469	133.3	0.2706
160	1.5757	135.6	0.2749
170	1.6042	137.8	0.2791
180	1.6325	140.1	0.2830
190	1.6606	142.4	0.2864
200	1.6884	144.7	0.2888

Table A·10·3E R134a: Superheated Vapor (*Continued*)

T F	p = 50 psi v, ft³/lbm	h, B/lbm	s, B/lbm·R
T_{sat}= f	0.0125	24.9	0.0540
40.24 g	0.9481	108.8	0.2220
50	0.9759	111.0	0.2271
60	1.0020	113.2	0.2309
70	1.0278	115.3	0.2346
80	1.0533	117.5	0.2382
90	1.0787	119.6	0.2418
100	1.1038	121.8	0.2454
110	1.1286	124.0	0.2491
120	1.1532	126.2	0.2529
130	1.1776	128.4	0.2570
140	1.2017	130.7	0.2611
150	1.2256	132.9	0.2655
160	1.2493	135.2	0.2698
170	1.2727	137.5	0.2740
180	1.2959	139.7	0.2780
190	1.3189	142.0	0.2813
200	1.3416	144.3	0.2836

T F	p = 60 psi v, ft³/lbm	h, B/lbm	s, B/lbm·R
T_{sat}= f	0.0127	27.9	0.0601
49.84 g	0.7939	110.2	0.2214
50	0.7958	110.3	0.2229
60	0.8184	112.5	0.2267
70	0.8408	114.7	0.2304
80	0.8629	116.9	0.2339
90	0.8848	119.1	0.2374
100	0.9065	121.3	0.2410
110	0.9279	123.5	0.2447
120	0.9491	125.7	0.2485
130	0.9701	128.0	0.2526
140	0.9908	130.3	0.2568
150	1.0113	132.5	0.2611
160	1.0316	134.8	0.2655
170	1.0516	137.1	0.2698
180	1.0715	139.4	0.2737
190	1.0910	141.7	0.2770
200	1.1104	144.0	0.2792

T F	p = 80 psi v, ft³/lbm	h, B/lbm	s, B/lbm·R
T_{sat}= f	0.0130	33.3	0.0703
65.87 g	0.5979	112.3	0.2206
70	0.6061	113.3	0.2235
80	0.6240	115.6	0.2270
90	0.6418	117.9	0.2305
100	0.6593	120.2	0.2340
110	0.6765	122.5	0.2376
120	0.6935	124.8	0.2414
130	0.7103	127.1	0.2454
140	0.7268	129.4	0.2496
150	0.7431	131.7	0.2540
160	0.7592	134.0	0.2585
170	0.7750	136.3	0.2628
180	0.7906	138.7	0.2667
190	0.8059	141.0	0.2699
200	0.8210	143.4	0.2720

T F	p = 100 psi v, ft³/lbm	h, B/lbm	s, B/lbm·R
T_{sat}= f	0.013 365	37.8	0.0786
79.10 g	0.4780	113.9	0.2199
80	0.4798	114.3	0.2215
90	0.4952	116.6	0.2249
100	0.5103	119.0	0.2284
110	0.5251	121.4	0.2320
120	0.5397	123.7	0.2357
130	0.5540	126.1	0.2397
140	0.5680	128.5	0.2440
150	0.5818	130.8	0.2484
160	0.5954	133.2	0.2528
170	0.6087	135.6	0.2571
180	0.6217	137.9	0.2611
190	0.6345	140.3	0.2642
200	0.6470	142.7	0.2661

T F	p = 120 psi v, ft³/lbm	h, B/lbm	s, B/lbm·R
T_{sat}= f	0.013 608	41.8	0.0858
90.46 g	0.3970	115.3	0.2194
100	0.4103	117.7	0.2237
110	0.4236	120.2	0.2272
120	0.4366	122.6	0.2310
130	0.4494	125.1	0.2350
140	0.4618	127.5	0.2392
150	0.4740	129.9	0.2436
160	0.4859	132.4	0.2481
170	0.4975	134.8	0.2524
180	0.5088	137.2	0.2563
190	0.5199	139.6	0.2594
200	0.5306	142.0	0.2612

T F	p = 140 psi v, ft³/lbm	h, B/lbm	s, B/lbm·R
T_{sat}= f	0.013 911	45.3	0.0921
100.48 g	0.3382	116.4	0.2190
110	0.3506	118.9	0.2232
120	0.3626	121.4	0.2269
130	0.3742	124.0	0.2309
140	0.3856	126.5	0.2351
150	0.3966	129.0	0.2395
160	0.4073	131.5	0.2440
170	0.4178	133.9	0.2484
180	0.4279	136.4	0.2522
190	0.4378	138.9	0.2553
200	0.4473	141.3	0.2569

T F	p = 160 psi v, ft³/lbm	h, B/lbm	s, B/lbm·R
T_{sat}= f	0.0142	48.5	0.0978
109.48 g	0.2936	117.4	0.2185
110	0.2953	117.5	0.2197
120	0.3065	120.2	0.2233
130	0.3174	122.8	0.2273
140	0.3279	125.4	0.2315
150	0.3382	128.0	0.2359
160	0.3481	130.5	0.2404
170	0.3578	133.1	0.2448
180	0.3671	135.6	0.2486
190	0.3760	138.1	0.2516
200	0.3847	140.6	0.2531

T F	p = 180 psi v, ft³/lbm	h, B/lbm	s, B/lbm·R
T_{sat}= f	0.0144	51.5	0.1029
117.67 g	0.2586	118.1	0.2182
120	0.2622	118.8	0.2201
130	0.2726	121.5	0.2241
140	0.2827	124.2	0.2283
150	0.2924	126.9	0.2327
160	0.3018	129.5	0.2372
170	0.3108	132.2	0.2415
180	0.3195	134.8	0.2454
190	0.3279	137.3	0.2483
200	0.3359	139.9	0.2497

T F	p = 200 psi v, ft³/lbm	h, B/lbm	s, B/lbm·R
T_{sat}= f	0.0146	54.4	0.1076
125.20 g	0.2303	118.8	0.2178
130	0.2362	120.2	0.2211
140	0.2459	123.0	0.2253
150	0.2553	125.8	0.2298
160	0.2643	128.5	0.2343
170	0.2730	131.2	0.2386
180	0.2813	133.9	0.2424
190	0.2892	136.5	0.2453
200	0.2968	139.1	0.2466

T F	p = 220 psi v, ft³/lbm	h, B/lbm	s, B/lbm·R
T_{sat}= f	0.0149	57.0	0.1120
132.19 g	0.2068	119.4	0.2174
140	0.2153	121.7	0.2227
150	0.2245	124.6	0.2271
160	0.2333	127.4	0.2316
170	0.2418	130.2	0.2359
180	0.2498	132.9	0.2398
190	0.2575	135.6	0.2426
200	0.2648	138.3	0.2437

T F	p = 240 psi v, ft³/lbm	h, B/lbm	s, B/lbm·R
T_{sat}= f	0.0151	59.5	0.1162
138.70 g	0.1871	119.8	0.2169
140	0.1890	120.3	0.2202
150	0.1983	123.3	0.2246
160	0.2071	126.2	0.2291
170	0.2154	129.1	0.2335
180	0.2234	132.0	0.2373
190	0.2309	134.8	0.2400
200	0.2380	137.5	0.2411

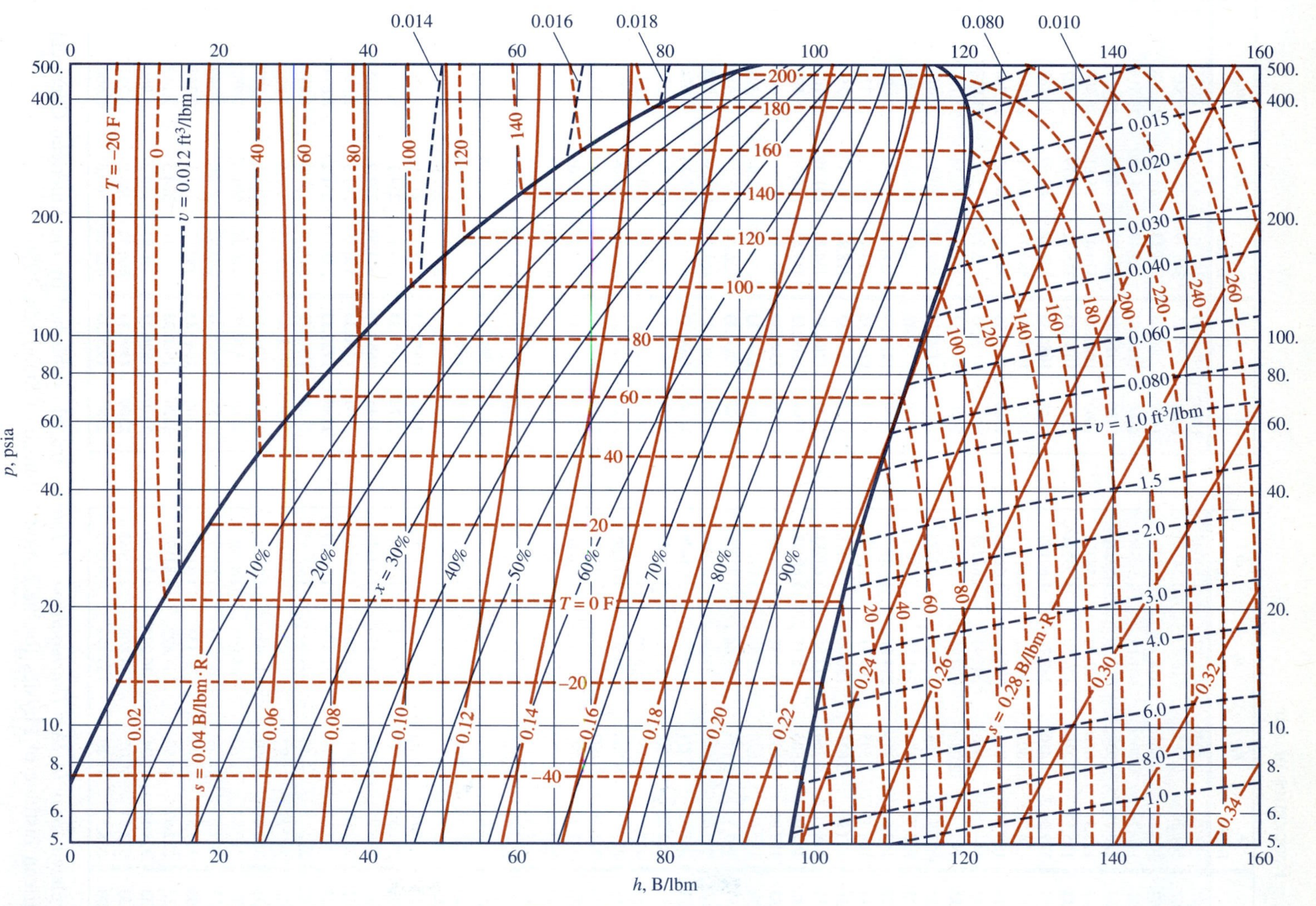

Chart A·11E R134a *ph* diagram. (*Source:* Based on *Thermodynamic Properties of HFC-134a (1,1,1,2-tetrafluoroethane)*, DuPont Company, Wilmington, Delaware, 1993, with permission.)

Table A·12E Properties of Air

T R	T F	h B/lbm	p_r	v_r	ϕ B/lbm·R
400	**–59.67**	96.40	0.4939	809.88	1.53005
420	**–39.67**	101.20	0.5859	716.85	1.54175
440	**–19.67**	105.99	0.6895	638.14	1.55291
460	**0.33**	110.79	0.8054	571.14	1.56356
480	**20.33**	115.58	0.9346	513.59	1.57376
500	**40.33**	120.37	1.0780	463.82	1.58354
520	**60.33**	125.17	1.2364	420.58	1.59294
540	**80.33**	129.96	1.4108	382.76	1.60199
560	**100.33**	134.76	1.6022	349.52	1.61071
580	**120.33**	139.55	1.8115	320.18	1.61913
600	**140.33**	144.35	2.040	294.12	1.62727
620	**160.33**	149.16	2.288	270.98	1.63514
640	**180.33**	153.97	2.558	250.20	1.64278
660	**200.33**	158.78	2.850	231.58	1.65019
680	**220.33**	163.60	3.165	214.85	1.65738
700	**240.33**	168.43	3.505	199.71	1.66438
720	**260.33**	173.26	3.871	186.00	1.67118
740	**280.33**	178.10	4.264	173.55	1.67781
760	**300.33**	182.95	4.686	162.19	1.68428
780	**320.33**	187.81	5.138	151.81	1.69058
800	**340.33**	192.67	5.620	142.35	1.69674
820	**360.33**	197.54	6.136	133.64	1.70276
840	**380.33**	202.42	6.686	125.64	1.70864
860	**400.33**	207.32	7.271	118.28	1.71439
880	**420.33**	212.22	7.894	111.48	1.72003
900	**440.33**	217.13	8.556	105.19	1.72555
920	**460.33**	222.05	9.258	99.374	1.73095
940	**480.33**	226.98	10.003	93.972	1.73626
960	**500.33**	231.93	10.792	88.955	1.74146
980	**520.33**	236.88	11.627	84.287	1.74657
1000	**540.33**	241.85	12.510	79.936	1.75159
1020	**560.33**	246.82	13.442	75.882	1.75652
1040	**580.33**	251.81	14.427	72.087	1.76136
1060	**600.33**	256.82	15.465	68.542	1.76612
1080	**620.33**	261.83	16.559	65.221	1.77081
1100	**640.33**	266.86	17.711	62.108	1.77542
1120	**660.33**	271.89	18.924	59.184	1.77996
1140	**680.33**	276.95	20.20	56.436	1.78443
1160	**700.33**	282.01	21.54	53.853	1.78883
1180	**720.33**	287.08	22.95	51.416	1.79317
1200	**740.33**	292.17	24.42	49.140	1.79745
1220	**760.33**	297.28	25.97	46.977	1.80166
1240	**780.33**	302.39	27.60	44.928	1.80582
1260	**800.33**	307.52	29.30	43.003	1.80992
1280	**820.33**	312.66	31.08	41.184	1.81397
1300	**840.33**	317.81	32.95	39.454	1.81797
1320	**860.33**	322.97	34.90	37.822	1.82191
1340	**880.33**	328.15	36.94	36.275	1.82580
1360	**900.33**	333.34	39.07	34.809	1.82965
1380	**920.33**	338.55	41.29	33.422	1.83345
1400	**940.33**	343.76	43.62	32.095	1.83720
1420	**960.33**	348.99	46.04	30.843	1.84091
1440	**980.33**	354.23	48.57	29.648	1.84457
1460	**1000.33**	359.48	51.20	28.516	1.84819
1480	**1020.33**	364.75	53.95	27.433	1.85178
1500	**1040.33**	370.03	56.81	26.404	1.85532
1520	**1060.33**	375.32	59.79	25.422	1.85882
1540	**1080.33**	380.62	62.89	24.487	1.86229
1560	**1100.33**	385.93	66.11	23.597	1.86571
1580	**1120.33**	391.26	69.47	22.744	1.86910
1600	**1140.33**	396.59	72.95	21.933	1.87246
1620	**1160.33**	401.94	76.58	21.154	1.87578
1640	**1180.33**	407.30	80.34	20.413	1.87907
1660	**1200.33**	412.67	84.24	19.706	1.88232
1680	**1220.33**	418.05	88.30	19.026	1.88555
1700	**1240.33**	423.44	92.51	18.376	1.88874
1720	**1260.33**	428.84	96.87	17.756	1.89190
1740	**1280.33**	434.26	101.39	17.161	1.89503
1760	**1300.33**	439.68	106.08	16.591	1.89812
1780	**1320.33**	445.11	110.94	16.045	1.90119
1800	**1340.33**	450.56	115.97	15.521	1.90424
1820	**1360.33**	456.01	121.18	15.019	1.90725
1840	**1380.33**	461.47	126.58	14.536	1.91023
1860	**1400.33**	466.95	132.16	14.074	1.91319
1880	**1420.33**	472.43	137.94	13.629	1.91613
1900	**1440.33**	477.92	143.91	13.203	1.91903
1920	**1460.33**	483.42	150.09	12.792	1.92191
1940	**1480.33**	488.93	156.47	12.399	1.92477
1960	**1500.33**	494.45	163.06	12.020	1.92760
1980	**1520.33**	499.98	169.88	11.655	1.93040
2000	**1540.33**	505.52	176.91	11.305	1.93318
2020	**1560.33**	511.06	184.18	10.968	1.93594
2040	**1580.33**	516.61	191.68	10.643	1.93868
2060	**1600.33**	522.18	199.42	10.330	1.94139
2080	**1620.33**	527.75	207.4	10.029	1.94408
2100	**1640.33**	533.32	215.6	9.7403	1.94675
2120	**1660.33**	538.91	224.1	9.4601	1.94940
2140	**1680.33**	544.50	232.9	9.1885	1.95202
2160	**1700.33**	550.10	241.9	8.9293	1.95463
2180	**1720.33**	555.71	251.2	8.6783	1.95722
2200	**1740.33**	561.33	260.8	8.4356	1.95978
2220	**1760.33**	566.95	270.6	8.2040	1.96232
2240	**1780.33**	572.59	280.8	7.9772	1.96485
2260	**1800.33**	578.22	291.2	7.7610	1.96736
2280	**1820.33**	583.87	302.0	7.5497	1.96984
2300	**1840.33**	589.52	313.1	7.3459	1.97231
2320	**1860.33**	595.18	324.5	7.1495	1.97476
2340	**1880.33**	600.85	336.2	6.9601	1.97719
2360	**1900.33**	606.52	348.2	6.7777	1.97961
2380	**1920.33**	612.20	360.6	6.6001	1.98201

Source: Calculated from *JANAF Thermochemical Tables,* 3rd ed., published by the American Chemical Society and the American Institute of Physics for the National Bureau of Standards, 1986.

Table A·12E Properties of Air (*Continued*)

T R	T F	h B/lbm	p_r	v_r	ϕ B/lbm·R
2400	**1940.33**	617.89	373.4	6.4274	1.98438
2420	**1960.33**	623.58	386.5	6.2613	1.98675
2440	**1980.33**	629.28	399.9	6.1015	1.98909
2460	**2000.33**	634.99	413.7	5.9463	1.99142
2480	**2020.33**	640.70	427.9	5.7957	1.99373
2500	**2040.33**	646.42	442.5	5.6497	1.99603
2520	**2060.33**	652.15	457.5	5.5082	1.99831
2540	**2080.33**	657.88	472.9	5.3711	2.00058
2560	**2100.33**	663.61	488.6	5.2395	2.00283
2580	**2120.33**	669.36	504.8	5.1109	2.00506
2600	**2140.33**	675.10	521.4	4.9866	2.00728
2620	**2160.33**	680.86	538.5	4.8654	2.00949
2640	**2180.33**	686.62	556.0	4.7482	2.01168
2660	**2200.33**	692.38	573.9	4.6350	2.01385
2680	**2220.33**	698.15	592.3	4.5247	2.01601
2700	**2240.33**	703.93	611.1	4.4183	2.01816
2720	**2260.33**	709.71	630.4	4.3147	2.02029
2740	**2280.33**	715.50	650.2	4.2141	2.02241
2760	**2300.33**	721.29	670.5	4.1163	2.02452
2780	**2320.33**	727.09	691.3	4.0214	2.02661
2800	**2340.33**	732.89	712.6	3.9293	2.02869
2820	**2360.33**	738.70	734.4	3.8399	2.03076
2840	**2380.33**	744.51	756.8	3.7526	2.03281
2860	**2400.33**	750.33	779.6	3.6685	2.03485
2880	**2420.33**	756.15	803.0	3.5866	2.03688
2900	**2440.33**	761.98	827.0	3.5067	2.03890
2920	**2460.33**	767.81	851.5	3.4292	2.04090
2940	**2480.33**	773.64	876.7	3.3535	2.04289
2960	**2500.33**	779.48	902.3	3.2805	2.04487
2980	**2520.33**	785.33	928.6	3.2091	2.04684
3000	**2540.33**	791.18	955.5	3.1397	2.04880
3020	**2560.33**	797.03	983.0	3.0722	2.05074
3040	**2580.33**	802.89	1011.1	3.0066	2.05268
3060	**2600.33**	808.76	1039.9	2.9426	2.05460
3080	**2620.33**	814.62	1069.3	2.8804	2.05651
3100	**2640.33**	820.49	1099.4	2.8197	2.05841
3120	**2660.33**	826.37	1130.1	2.7608	2.06030
3140	**2680.33**	832.25	1161.5	2.7034	2.06218
3160	**2700.33**	838.13	1193.6	2.6475	2.06405
3180	**2720.33**	844.02	1226.3	2.5932	2.06590
3200	**2740.33**	849.91	1259.8	2.5401	2.06775
3220	**2760.33**	855.81	1294.1	2.4882	2.06959
3240	**2780.33**	861.71	1329.0	2.4379	2.07141
3260	**2800.33**	867.61	1364.7	2.3888	2.07323
3280	**2820.33**	873.52	1401.1	2.3410	2.07504
3300	**2840.33**	879.43	1438.3	2.2944	2.07683
3320	**2860.33**	885.35	1476.3	2.2489	2.07862
3340	**2880.33**	891.27	1515.1	2.2045	2.08040
3360	**2900.33**	897.19	1554.7	2.1612	2.08217
3380	**2920.33**	903.12	1595.1	2.1190	2.08393
3400	**2940.33**	909.05	1636	2.0782	2.08567
3450	**2990.33**	923.89	1743	1.9793	2.09001
3500	**3040.33**	938.75	1855	1.8868	2.09428
3550	**3090.33**	953.63	1973	1.7993	2.09851
3600	**3140.33**	968.54	2097	1.7167	2.10268
3650	**3190.33**	983.46	2227	1.6390	2.10679
3700	**3240.33**	998.40	2363	1.5658	2.11086
3750	**3290.33**	1013.3	2505	1.4970	2.11488
3800	**3340.33**	1028.3	2655	1.4313	2.11884
3850	**3390.33**	1043.3	2811	1.3696	2.12277
3900	**3440.33**	1058.3	2975	1.3109	2.12664
3950	**3490.33**	1073.4	3146	1.2556	2.13047
4000	**3540.33**	1088.4	3324	1.2034	2.13426
4050	**3590.33**	1103.5	3511	1.1535	2.13800
4100	**3640.33**	1118.6	3706	1.1063	2.14171
4150	**3690.33**	1133.7	3909	1.0617	2.14537
4200	**3740.33**	1148.8	4121	1.0192	2.14899
4250	**3790.33**	1163.9	4342	0.97881	2.15257
4300	**3840.33**	1179.1	4573	0.94030	2.15612
4350	**3890.33**	1194.2	4813	0.90380	2.15962
4400	**3940.33**	1209.4	5062	0.86922	2.16309
4450	**3990.33**	1224.6	5323	0.83599	2.16653
4500	**4040.33**	1239.8	5593	0.80458	2.16993
4550	**4090.33**	1255.1	5875	0.77447	2.17329
4600	**4140.33**	1270.3	6167	0.74591	2.17662
4650	**4190.33**	1285.6	6471	0.71859	2.17992
4700	**4240.33**	1300.8	6787	0.69250	2.18319
4750	**4290.33**	1316.1	7115	0.66760	2.18642
4800	**4340.33**	1331.4	7455	0.64386	2.18963
4850	**4390.33**	1346.7	7808	0.62116	2.19280
4900	**4440.33**	1362.0	8175	0.59939	2.19594
4950	**4490.33**	1377.4	8555	0.57861	2.19906
5000	**4540.33**	1392.7	8948	0.55878	2.20214
5050	**4590.33**	1408.1	9357	0.53970	2.20520
5100	**4640.33**	1423.5	9779	0.52153	2.20823
5150	**4690.33**	1438.8	10217	0.50406	2.21123
5200	**4740.33**	1454.2	10670	0.48735	2.21420
5250	**4790.33**	1469.6	11139	0.47132	2.21715
5300	**4840.33**	1485.1	11624	0.45595	2.22008
5350	**4890.33**	1500.5	12126	0.44120	2.22297
5400	**4940.33**	1515.9	12645	0.42705	2.22585
5450	**4990.33**	1531.4	13182	0.41344	2.22869
5500	**5040.33**	1546.8	13736	0.40041	2.23152
5550	**5090.33**	1562.3	14309	0.38787	2.23432
5600	**5140.33**	1577.8	14900	0.37584	2.23709
5650	**5190.33**	1593.3	15511	0.36426	2.23985
5700	**5240.33**	1608.8	16141	0.35314	2.24258
5750	**5290.33**	1624.3	16791	0.34245	2.24528
5800	**5340.33**	1639.8	17462	0.33215	2.24797
5850	**5390.33**	1655.3	18154	0.32224	2.25063

Chart A·13E Air: *Ts* Diagram

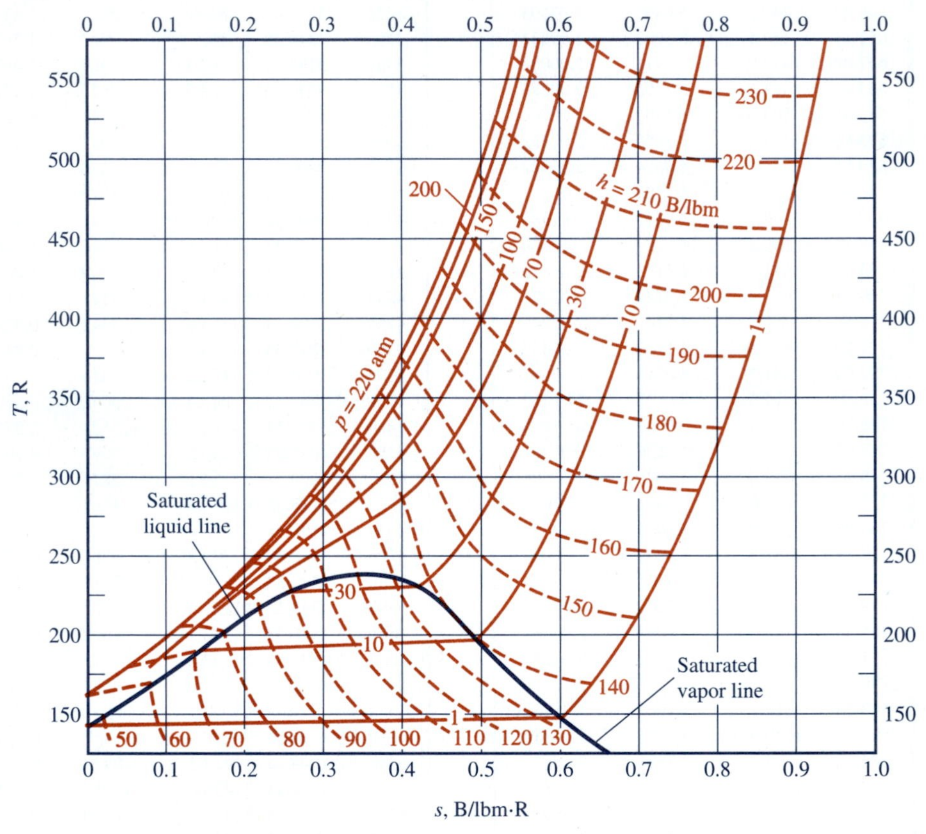

Chart A·13E Air *Ts* diagram. (*Source:* Data from "Thermodynamic Properties of Air at Low Temperatures," by V. C. Williams, *Transactions of the AIChE,* Vol. 39, no. 1, Feb. 1943.)

Chart A·14E Psychrometric Chart

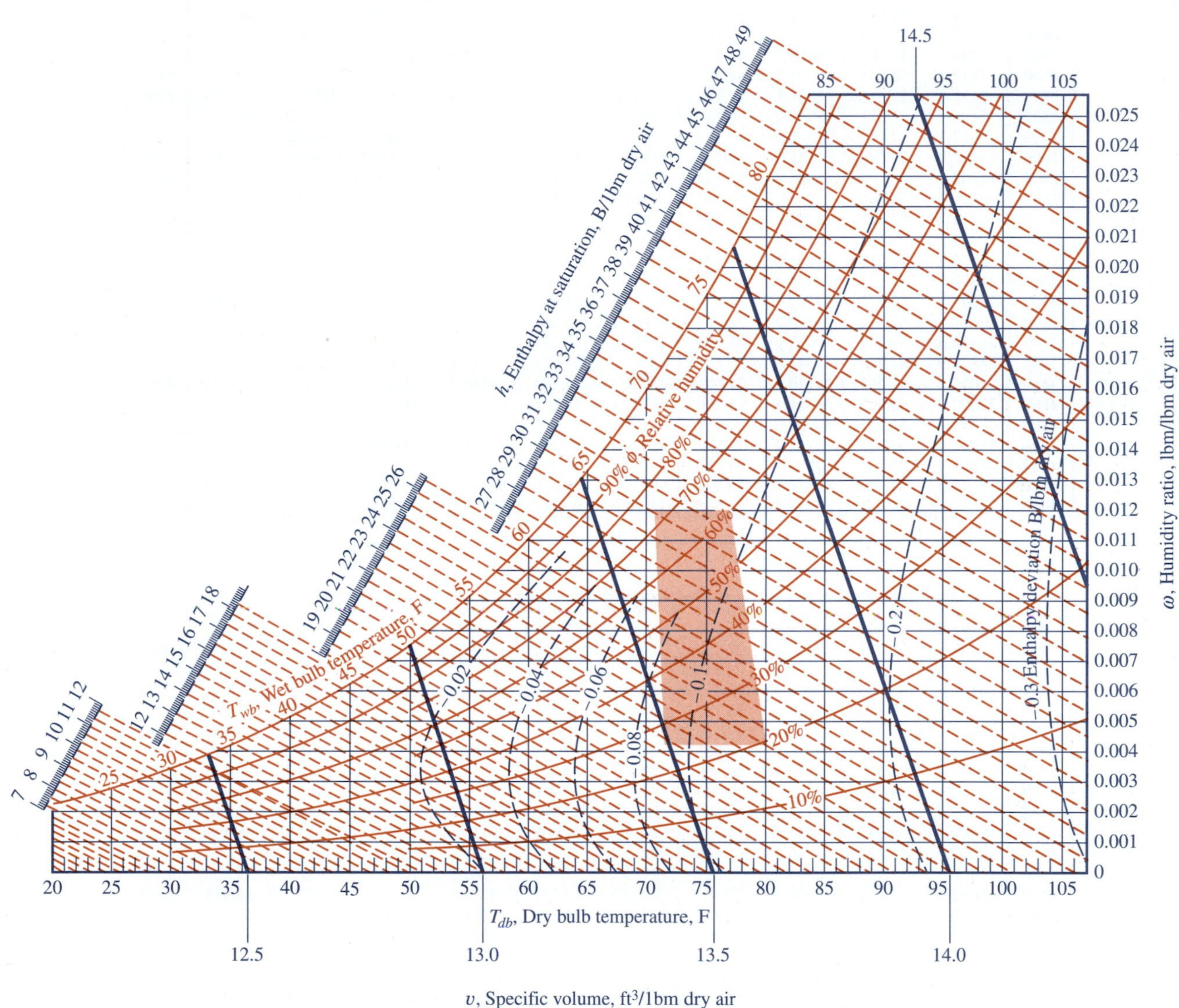

Chart A·14E Psychrometric chart. (*Source:* Based on Carrier Psychrometric Chart, 1959, Carrier Corporation, Syracuse, New York, with permission.)

Table A·15E Properties of Ideal Gases, $\bar{h}$ in B/lbmol, $\bar{\phi}$ in B/lbmol · R

T, R	CO $\bar{h}$	CO $\bar{\phi}$	CO_2 $\bar{h}$	CO_2 $\bar{\phi}$	H_2 $\bar{h}$	H_2 $\bar{\phi}$	H_2O $\bar{h}$	H_2O $\bar{\phi}$	N_2 $\bar{h}$	N_2 $\bar{\phi}$	T, R
300	2074.1	43.142	2110.7	46.404	2036.8	27.280	2372.3	40.470	2073.4	41.698	300
400	2775.5	45.160	2872.4	48.590	2706.2	29.205	3167.0	42.757	2774.8	43.716	400
500	3472.6	46.716	3704.5	50.444	3388.0	30.726	3964.3	44.535	3472.2	45.272	500
536.67	3727.9	47.209	4025.8	51.064	3640.2	31.212	4258.0	45.102	3727.4	45.765	536.67
600	4168.8	47.985	4599.5	52.074	4077.8	31.983	4767.4	45.999	4168.1	46.541	600
700	4866.9	49.061	5550.3	53.539	4772.5	33.054	5578.9	47.250	4864.8	47.615	700
800	5569.2	49.999	6551.1	54.874	5469.9	33.985	6401.1	48.348	5564.4	48.549	800
900	6277.5	50.833	7596.3	56.105	6168.8	34.808	7235.7	49.331	6268.4	49.378	900
1000	6993.3	51.587	8681.2	57.248	6868.4	35.545	8083.9	50.224	6978.3	50.126	1000
1100	7717.7	52.278	9801.7	58.315	7568.5	36.213	8946.8	51.047	7695.2	50.810	1100
1200	8451.4	52.916	10954.1	59.318	8269.6	36.823	9825.0	51.811	8420.2	51.440	1200
1300	9194.7	53.511	12135.2	60.263	8972.2	37.385	10719.0	52.526	9153.7	52.027	1300
1400	9947.9	54.069	13342.3	61.158	9677.0	37.907	11629.3	53.201	9896.2	52.578	1400
1600	11482.8	55.094	15824.6	62.814	11096.3	38.855	13499.2	54.449	11408.3	53.587	1600
1800	13052.4	56.018	18384.8	64.322	12531.1	39.700	15436.2	55.589	12954.5	54.497	1800
2000	14651.7	56.860	21007.9	65.703	13984.3	40.465	17441.0	56.645	14530.7	55.327	2000
2200	16276.9	57.635	23684.9	66.979	15458.4	41.168	19513.6	57.632	16134.4	56.092	2200
2400	17924.1	58.351	26406.7	68.163	16954.6	41.819	21649.5	58.561	17761.7	56.800	2400
2600	19590.3	59.018	29165.8	69.267	18473.6	42.426	23844.0	59.439	19409.2	57.459	2600
2800	21272.7	59.641	31955.9	70.301	20015.6	42.998	26093.0	60.273	21073.9	58.076	2800
3000	22968.9	60.226	34772.1	71.272	21580.3	43.537	28392.2	61.066	22753.4	58.655	3000
3200	24677.1	60.778	37610.3	72.188	23167.3	44.050	30737.7	61.823	24445.8	59.201	3200
3400	26395.5	61.298	40467.4	73.054	24775.6	44.537	33125.6	62.546	26149.3	59.717	3400
3600	28122.9	61.792	43341.0	73.875	26404.2	45.002	35552.3	63.240	27862.6	60.207	3600
3800	29858.2	62.261	46229.5	74.656	28051.8	45.448	38014.5	63.905	29584.7	60.673	3800
4000	31600.5	62.708	49131.4	75.400	29717.2	45.875	40509.0	64.545	31314.5	61.116	4000
4200	33349.1	63.135	52046.0	76.111	31399.1	46.285	43033.0	65.161	33051.5	61.540	4200
4400	35103.4	63.543	54972.2	76.792	33096.5	46.680	45583.6	65.754	34794.9	61.945	4400
4600	36862.9	63.934	57909.4	77.445	34808.3	47.060	48158.7	66.326	36544.2	62.334	4600
4800	38627.3	64.309	60856.5	78.072	36533.8	47.428	50756.0	66.879	38298.9	62.708	4800
5000	40396.0	64.670	63812.5	78.675	38272.6	47.783	53373.8	67.413	40058.3	63.067	5000

T, R	O_2 $\bar{h}$	O_2 $\bar{\phi}$	H $\bar{h}$	H $\bar{\phi}$	N $\bar{h}$	N $\bar{\phi}$	O $\bar{h}$	O $\bar{\phi}$	NO $\bar{h}$	NO $\bar{\phi}$	T, R
300	2080.9	44.939	1489.4	24.512	1489.4	33.728	1628.8	35.359	2253.3	46.160	300
400	2778.6	46.946	1985.8	25.940	1985.8	35.156	2168.8	36.914	2976.2	48.240	400
500	3476.2	48.503	2482.2	27.048	2482.2	36.264	2698.9	38.097	3690.7	49.835	500
536.67	3733.0	48.999	2664.2	27.399	2664.2	36.615	2891.2	38.468	3951.8	50.339	536.67
600	4178.8	49.784	2978.6	27.953	2978.6	37.169	3221.2	39.049	4402.7	51.133	600
700	4890.3	50.880	3475.1	28.719	3475.1	37.934	3737.7	39.846	5116.5	52.233	700
800	5613.1	51.845	3971.6	29.382	3971.6	38.597	4249.8	40.529	5835.4	53.193	800
900	6348.8	52.712	4468.1	29.966	4468.1	39.182	4758.7	41.129	6561.9	54.048	900
1000	7097.9	53.501	4964.5	30.489	4964.5	39.705	5265.3	41.663	7297.3	54.823	1000
1100	7860.4	54.227	5461.0	30.963	5461.0	40.178	5770.1	42.144	8042.4	55.533	1100
1200	8635.6	54.902	5957.5	31.395	5957.5	40.610	6273.8	42.582	8797.6	56.190	1200
1300	9422.4	55.532	6453.9	31.792	6453.9	41.008	6776.5	42.984	9562.7	56.803	1300
1400	10219.8	56.122	6950.4	32.160	6950.4	41.375	7278.4	43.356	10337.2	57.377	1400
1600	11841.3	57.205	7943.3	32.823	7943.3	42.038	8280.1	44.025	11912.6	58.428	1600
1800	13492.9	58.177	8936.3	33.408	8936.3	42.623	9279.9	44.614	13520.7	59.375	1800
2000	15170.2	59.061	9929.4	33.931	9929.1	43.146	10278.6	45.140	15157.0	60.237	2000
2200	16867.7	59.870	10922.2	34.404	10922.1	43.619	11276.0	45.615	16816.6	61.028	2200
2400	18583.0	60.616	11914.9	34.836	11915.1	44.051	12272.6	46.049	18495.9	61.758	2400
2600	20314.3	61.309	12907.5	35.233	12908.2	44.449	13268.5	46.448	20191.6	62.437	2600
2800	22060.2	61.956	13900.3	35.601	13901.2	44.817	14264.0	46.816	21901.1	63.070	2800
3000	23819.7	62.563	14893.1	35.943	14894.1	45.159	15259.4	47.160	23622.3	63.664	3000
3200	25592.0	63.134	15886.1	36.264	15887.1	45.480	16254.6	47.481	25353.3	64.223	3200
3400	27376.7	63.675	16879.1	36.565	16880.0	45.781	17249.9	47.783	27092.8	64.750	3400
3600	29173.5	64.189	17872.3	36.849	17873.0	46.064	18245.2	48.067	28839.6	65.249	3600
3800	30982.2	64.678	18865.4	37.117	18866.1	46.333	19240.5	48.336	30593.0	65.723	3800
4000	32802.7	65.145	19858.5	37.372	19859.4	46.588	20235.9	48.592	32352.3	66.174	4000
4200	34635.1	65.592	20851.5	37.614	20853.1	46.830	21231.5	48.834	34117.0	66.605	4200
4400	36479.4	66.021	21844.4	37.845	21847.3	47.061	22227.3	49.066	35886.7	67.016	4400
4600	38335.4	66.433	22837.2	38.066	22842.2	47.283	23223.5	49.287	37661.0	67.411	4600
4800	40203.0	66.831	23829.9	38.277	23838.0	47.494	24220.2	49.500	39439.6	67.789	4800
5000	42081.8	67.214	24822.6	38.479	24835.1	47.698	25217.7	49.703	41222.0	68.153	5000

Source: JANAF Thermochemical Tables, 3rd ed., National Bureau of Standards, NSRDS-NBS 37, 1986.

Table A·15E Properties of Ideal Gases, $\bar{h}$ in B/lbmol, $\bar{\phi}$ in B/lbmol·R (*Continued*)

T, R	OH		CH_4		C_2H_2		C_2H_4		Air		T, R
	$\bar{h}$	$\bar{\phi}$	$\bar{h}$	$\bar{\phi}$	$\bar{h}$	$\bar{\phi}$	$\bar{h}$	$\bar{\phi}$	$\bar{h}$	$\bar{\phi}$	
300	2229.9	39.663	2411.5	39.843	2127.7	42.731	2489.8	47.484	2095.3	42.319	**300**
400	2958.9	41.760	3191.3	42.085	2963.2	45.125	3253.6	49.672	2792.7	44.325	**400**
500	3680.3	43.370	4001.3	43.890	3924.7	47.265	4156.2	51.680	3487.2	45.875	**500**
536.67	3943.3	43.878	4309.5	44.485	4304.4	47.998	4521.9	52.386	3741.8	46.367	**536.67**
600	4395.7	44.675	4859.4	45.453	4990.3	49.206	5196.9	53.574	4182.0	47.142	**600**
700	5106.4	45.770	5779.0	46.870	6142.2	50.980	6372.7	55.384	4879.4	48.217	**700**
800	5813.9	46.715	6769.2	48.191	7365.6	52.612	7679.0	57.126	5581.6	49.155	**800**
900	6519.5	47.546	7835.5	49.446	8649.3	54.124	9109.4	58.810	6290.2	49.989	**900**
1000	7224.2	48.289	8980.5	50.651	9984.3	55.530	10656.7	60.439	7006.3	50.743	**1000**
1100	7929.2	48.961	10204.3	51.817	11364.1	56.845	12312.9	62.017	7730.8	51.434	**1100**
1200	8635.4	49.575	11505.1	52.949	12784.0	58.080	14069.7	63.545	8464.3	52.072	**1200**
1300	9343.7	50.142	12879.6	54.048	14240.9	59.246	15918.6	65.024	9207.0	52.667	**1300**
1400	10055.1	50.669	14324.0	55.118	15732.6	60.351	17851.7	66.456	9958.8	53.224	**1400**
1600	11489.5	51.627	17405.4	57.174	18814.7	62.408	21943.3	69.186	11489.4	54.245	**1600**
1800	12943.6	52.483	20722.6	59.126	22019.5	64.295	26302.8	71.752	13052.7	55.166	**1800**
2000	14421.0	53.261	24251.4	60.984	25333.2	66.040	30896.2	74.171	14644.9	56.004	**2000**
2200	15923.1	53.977	27966.5	62.754	28747.6	67.666	35696.1	76.457	16261.8	56.775	**2200**
2400	17449.3	54.641	31842.9	64.440	32251.9	69.191	40672.4	78.622	17900.4	57.488	**2400**
2600	18998.9	55.261	35858.4	66.047	35836.2	70.625	45799.7	80.673	19557.9	58.151	**2600**
2800	20570.7	55.843	39994.0	67.579	39492.1	71.980	51056.0	82.621	21231.9	58.771	**2800**
3000	22163.8	56.393	44233.6	69.041	43211.9	73.263	56422.9	84.472	22920.6	59.354	**3000**
3200	23776.6	56.913	48563.0	70.438	46989.0	74.481	61885.3	86.235	24622.2	59.903	**3200**
3400	25408.0	57.408	52970.8	71.774	50817.7	75.642	67430.6	87.915	26335.2	60.422	**3400**
3600	27056.4	57.879	57447.0	73.053	54693.1	76.749	73048.4	89.521	28058.6	60.915	**3600**
3800	28720.5	58.329	61983.5	74.280	58610.8	77.808	78730.5	91.057	29791.4	61.383	**3800**
4000	30398.9	58.759	66573.4	75.457	62567.1	78.823	84470.0	92.529	31532.7	61.830	**4000**
4200	32090.2	59.172	71211.0	76.588	66558.7	79.797	90261.2	93.941	33282.0	62.256	**4200**
4400	33793.4	59.568	75891.4	77.677	70582.8	80.733	96099.4	95.299	35038.7	62.665	**4400**
4600	35507.3	59.949	80610.3	78.725	74636.8	81.634	101979.8	96.606	36802.3	63.057	**4600**
4800	37231.1	60.316	85363.7	79.737	78718.2	82.502	107898.0	97.866	38572.4	63.434	**4800**
5000	38964.2	60.669	90147.3	80.713	82824.9	83.340	113849.0	99.080	40348.7	63.796	**5000**

Table A·16E Enthalpy of Eight Gaseous Fuels at Low Pressure, $\bar{h}$

T R	Methane CH_4 B/lbmol	Ethane C_2H_6 B/lbmol	Propane C_3H_8 B/lbmol	n-Butane C_4H_{10} B/lbmol	n-Octane C_8H_{18} B/lbmol	Acetylene C_2H_2 B/lbmol	Ethylene C_2H_4 B/lbmol	Methanol CH_3OH B/lbmol	T R
0	0	0	0	0	0	0	0	0	**0**
536.67	4306	5105	6337	8287	16243	4304	4522	4913	**536.67**
600	4853	5929	7499	9841	19215	4990	5197	5589	**600**
800	6782	9005	11903	15693	30446	7366	7679	8011	**800**
1000	8999	12717	17261	22769	44087	9984	10657	10853	**1000**
1200	11525	17018	23466	30927	59839	12784	14070	14085	**1200**
1400	14359	21850	30412	40028	77374	15733	17852	17671	**1400**
1600	17482	27148	37993	49932	96340	18815	21943	21564	**1600**
1800	20857	32837	46105	60501	116354	22020	26303	25712	**1800**
2000	24427	38831	54646	71594	137009	25333	30896	30050	**2000**

Source: Data for all gases but ammonia from *Selected Values of Properties of Hydrocarbons and Related Compounds,* Thermodynamics Research Center Data Project, Thermodynamics Research Center, Texas A & M University, College Station, Texas 77843; data for ammonia from *JANAF Thermochemical Tables,* National Bureau of Standards, 3rd ed., published by the American Chemical Society and the American Institute of Physics for the National Bureau of Standards, 1986.

Table A·17E Properties of Substances at 77F, 14.50 psia

Name	Molar Mass M	Phase	$\bar{h}_{fg}$	Boiling Point (1 atm)	Enthalpy of Combustion at 77F, 14.50 psia: $H_2O(l)$		$H_2O(g)$		Enthalpy of Formation	Gibbs Energy of Formation	Absolute Entropy	Formula
					Δh_c	$\Delta \bar{h}_c$	Δh_c	$\Delta \bar{h}_c$	$\Delta \bar{h}^\circ_{f,ref}$	$\Delta \bar{g}^\circ_{f,ref}$	$\Delta \bar{s}^\circ_{ref}$	
	lbm/lbmol		B/lbmol	F	B/lbm	B/lbmol	B/lbm	B/lbmol	B/lbmol	B/lbmol	B/lbmol·R	
Hydrogen, H_2	2.016	gas	···	-422.99	-60954	-122877	-51573	-103966	0	0	31.2124	H_2
Water, H_2O	18.015	gas	18911	212.00	0	0	0	0	-103966	-98273	45.1022	H_2O
Water, H_2O	18.015	liquid	18911	212.00	0	0	0	0	-122885	-101952	16.7073	H_2O
Carbon, C	12.011	solid	···	···	-14087	-169197	-14087	-169197	0	0	1.3710	C
Carbon monoxide, CO	28.010	gas	···	-312.61	-4344	-121678	-4344	-121678	-47519	-58970	47.2079	CO
Carbon dioxide, CO_2	44.010	gas	···	-109.26	0	0	0	0	-169197	-169557	51.0641	CO_2
Nitrogen, N_2	28.013	gas	···	-320.46	0	0	0	0	0	0	45.7650	N_2
Oxygen, O_2	31.999	gas	···	-297.33	0	0	0	0	0	0	48.9985	O_2
Methane, CH_4	16.043	gas	···	-258.74	-23859	-382761	-21501	-344897	-32190	-21826	44.4853	CH_4
Acetylene, C_2H_2	26.038	gas	···	-119.20	-21459	-558749	-20733	-539837	97477	106691	47.9980	C_2H_2
Ethene, C_2H_4	28.054	gas	···	-154.68	-21626	-606707	-20278	-568882	22557	29416	52.3861	C_2H_4
Ethane, C_2H_6	30.069	gas	2159	-127.50	-22314	-670976	-20427	-614243	-36049	-13737	54.8025	C_2H_6
Propene, C_3H_6	42.080	gas	···	-53.86	-21024	-884695	-19676	-827962	8474	26964	63.2958	C_3H_6
Propane, C_3H_8	44.093	gas	6371	-43.74	-21638	-954097	-19923	-878457	-45001	-10489	64.6546	C_3H_8
n-Butane, C_4H_{10}	58.123	gas	9057	31.08	-21285	-1237154	-19658	-1142601	-54021	-7120	74.1531	C_4H_{10}
n-Pentane, C_5H_{12}	72.150	gas	11361	96.93	-26310	-1529226	-24358	-1415761	-63075	-3770	83.6247	C_5H_{12}
n-Hexane, C_6H_{14}	86.177	gas	13565	155.71	-20929	-1803547	-19393	-1671174	-71772	-64	93.0966	C_6H_{14}
n-Heptane, C_7H_{16}	100.203	gas	15712	209.16	-20825	-2086720	-19315	-1935434	-80677	3504	102.4379	C_7H_{16}
n-Octane, C_8H_{18}	114.230	gas	17835	258.21	-20745	-2369690	-19255	-2199497	-89778	6845	111.8379	C_8H_{18}
n-Octane, C_8H_{18}	114.230	liquid	17835	258.21	-20589	-2351857	-19099	-2181660	-107613	2553	86.6015	C_8H_{18}
Benzene, C_6H_6	78.113	gas	14552	176.18	-18172	-1419467	-17446	-1362739	35656	55744	64.4311	C_6H_6
Benzene, C_6H_6	78.113	liquid	14552	176.18	-17986	-1404914	-17259	-1348186	21103	53539	41.4238	C_6H_6
Methanol, CH_3OH	32.042	gas	17107	148.46	-10254	-328565	-9073	-290728	-86389	-69751	57.2944	CH_3OH
Methanol, CH_3OH	32.042	liquid	17107	148.46	-9745	-312241	-8564	-274404	-102713	-71642	30.3979	CH_3OH
Ethanol, C_2H_5OH	46.069	gas	18298	172.92	-13154	-606012	-11923	-549257	-101010	-72111	67.0297	C_2H_5OH
Ethanol, C_2H_5OH	46.069	liquid	18298	172.92	-12757	-587715	-11525	-530960	-119308	-74957	38.2392	C_2H_5OH
Hydrazine, N_2H_4	32.045	liquid	19243	237.47	-8348	-267519	-7168	-229698	21765	64248	29.0303	N_2H_4
Hydrogen, H	1.008	gas	···	···	-153945	-155161	-144564	-145706	93723	87394	27.3995	H
Nitrogen, N	14.007	gas	···	···	···	···	···	···	203216	195847	36.6151	N
Oxygen, O	15.999	gas	···	···	···	···	···	···	107124	99629	38.4680	O
Hydroxyl, OH	17.007	gas	···	···	···	···	···	···	16761	14736	43.8779	OH

Source: JANAF Thermochemical Tables, 3rd ed., published by the American Chemical Society and the American Institute of Physics for the National Bureau of Standards, 1986. Also, *Journal of Physical and Chemical Reference Data,* Vol. 14, Supplement No. 1, 1985.

Table A·18E Log K_p for Seven Ideal-gas Reactions

T, R	$H_2 \to 2H$	$N_2 \to 2N$	$O_2 \to 2O$	$CO_2 \to CO + \frac{1}{2}O_2$	$H_2O \to OH + \frac{1}{2}H_2$	$\frac{1}{2}N_2 + \frac{1}{2}O_2 \to NO$	$CO_2 + H_2 \to CO + H_2O$	T, R
536.67	−71.31	−159.7	−81.27	−45.05	−46.10	−15.15	−4.947	**536.67**
600	−63.19	−142.2	−72.00	−39.82	−40.87	−13.68	−4.231	**600**
700	−53.36	−120.9	−60.78	−33.49	−34.54	−11.54	−3.361	**700**
800	−45.99	−105.0	−52.37	−28.73	−29.78	−9.947	−2.705	**800**
900	−40.26	−92.63	−45.82	−25.03	−26.09	−8.784	−2.193	**900**
1000	−35.68	−82.72	−40.59	−22.07	−23.13	−7.834	−1.781	**1000**
1200	−28.80	−67.87	−32.73	−17.63	−18.70	−6.421	−1.162	**1200**
1400	−23.88	−57.25	−27.12	−14.45	−15.53	−5.411	−0.717	**1400**
1600	−20.20	−49.29	−22.92	−12.07	−13.16	−4.652	−0.381	**1600**
1800	−17.33	−43.10	−19.64	−10.21	−11.31	−4.063	−0.120	**1800**
2000	−14.97	−38.08	−16.97	−8.726	−9.785	−3.590	0.038	**2000**
2200	−13.04	−33.99	−14.79	−7.517	−8.556	−3.203	0.177	**2200**
2400	−11.44	−30.58	−12.98	−6.522	−7.532	−2.881	0.279	**2400**
2600	−10.08	−27.69	−11.45	−5.686	−6.665	−2.608	0.361	**2600**
2800	−8.913	−25.22	−10.13	−4.970	−5.923	−2.374	0.430	**2800**
3000	−7.903	−23.07	−8.992	−4.347	−5.279	−2.172	0.492	**3000**
3200	−7.020	−21.20	−7.995	−3.799	−4.715	−1.995	0.549	**3200**
3400	−6.241	−19.54	−7.116	−3.313	−4.218	−1.839	0.602	**3400**
3600	−5.548	−18.07	−6.334	−2.879	−3.777	−1.699	0.652	**3600**
3800	−4.928	−16.75	−5.634	−2.490	−3.381	−1.575	0.697	**3800**
4000	−4.370	−15.57	−5.004	−2.140	−3.025	−1.463	0.736	**4000**
4200	−3.865	−14.49	−4.435	−1.826	−2.704	−1.362	0.770	**4200**
4400	−3.407	−13.52	−3.917	−1.543	−2.411	−1.270	0.799	**4400**
4600	−2.988	−12.63	−3.444	−1.288	−2.144	−1.186	0.820	**4600**
4800	−2.604	−11.81	−3.011	−1.059	−1.899	−1.109	0.836	**4800**
5000	−2.250	−11.06	−2.612	−0.852	−1.673	−1.038	0.847	**5000**
5200	−1.924	−10.37	−2.244	−0.664	−1.465	−0.973	0.854	**5200**
5400	−1.622	−9.729	−1.903	−0.491	−1.273	−0.913	0.859	**5400**
5600	−1.342	−9.133	−1.586	−0.331	−1.094	−0.858	0.864	**5600**
5800	−1.081	−8.579	−1.292	−0.178	−0.928	−0.805	0.871	**5800**
6000	−0.837	−8.061	−1.017	−0.028	−0.772	−0.757	0.886	**6000**
6200	−0.609	−7.577	−0.759	0.124	−0.627	−0.711	0.911	**6200**
6400	−0.395	−7.123	−0.518	0.283	−0.490	−0.669	0.952	**6400**

Source: Calculated from Table A·19.

Table A·19E Log K_f for Substances

T, R	CO_2	CO	H	OH	H_2O	N	O	NO	T, R
536.67	69.09	24.04	-35.65	-6.001	40.10	-79.85	-40.64	-15.15	**536.67**
600	61.81	22.00	-31.60	-5.281	35.59	-71.08	-36.00	-13.68	**600**
700	53.01	19.53	-26.68	-4.411	30.12	-60.47	-30.39	-11.54	**700**
800	46.40	17.67	-23.00	-3.758	26.03	-52.51	-26.18	-9.947	**800**
900	41.26	16.23	-20.13	-3.250	22.84	-46.32	-22.91	-8.784	**900**
1000	37.15	15.08	-17.84	-2.844	20.29	-41.36	-20.29	-7.834	**1000**
1200	30.98	13.35	-14.40	-2.234	16.47	-33.93	-16.37	-6.421	**1200**
1400	26.56	12.11	-11.94	-1.799	13.73	-28.63	-13.56	-5.411	**1400**
1600	23.25	11.19	-10.10	-1.472	11.68	-24.65	-11.46	-4.652	**1600**
1800	20.68	10.47	-8.666	-1.218	10.09	-21.55	-9.822	-4.063	**1800**
2000	18.63	9.908	-7.485	-1.021	8.764	-19.04	-8.484	-3.590	**2000**
2200	16.94	9.419	-6.521	-0.8627	7.694	-16.99	-7.397	-3.203	**2200**
2400	15.52	9.000	-5.718	-0.7305	6.802	-15.29	-6.490	-2.881	**2400**
2600	14.33	8.640	-5.039	-0.6186	6.047	-13.85	-5.723	-2.608	**2600**
2800	13.30	8.331	-4.456	-0.5227	5.400	-12.61	-5.066	-2.374	**2800**
3000	12.41	8.065	-3.952	-0.4397	4.839	-11.54	-4.496	-2.172	**3000**
3200	11.63	7.836	-3.510	-0.3669	4.348	-10.60	-3.998	-1.995	**3200**
3400	10.95	7.636	-3.120	-0.3028	3.916	-9.770	-3.558	-1.839	**3400**
3600	10.34	7.460	-2.774	-0.2458	3.531	-9.034	-3.167	-1.699	**3600**
3800	9.794	7.304	-2.464	-0.1948	3.186	-8.376	-2.817	-1.575	**3800**
4000	9.304	7.163	-2.185	-0.1488	2.877	-7.783	-2.502	-1.463	**4000**
4200	8.860	7.034	-1.933	-0.1073	2.596	-7.247	-2.217	-1.362	**4200**
4400	8.456	6.913	-1.703	-0.06952	2.341	-6.760	-1.958	-1.270	**4400**
4600	8.088	6.799	-1.494	-0.03503	2.109	-6.315	-1.722	-1.186	**4600**
4800	7.750	6.691	-1.302	-0.00342	1.895	-5.907	-1.505	-1.109	**4800**
5000	7.439	6.588	-1.125	0.02566	1.699	-5.532	-1.306	-1.038	**5000**
5200	7.153	6.489	-0.9621	0.05250	1.518	-5.185	-1.122	-0.973	**5200**
5400	6.887	6.396	-0.8111	0.07736	1.350	-4.864	-0.9514	-0.913	**5400**
5600	6.641	6.310	-0.6709	0.1004	1.194	-4.567	-0.7931	-0.858	**5600**
5800	6.411	6.233	-0.5403	0.1219	1.049	-4.289	-0.6458	-0.805	**5800**
6000	6.197	6.169	-0.4185	0.1420	0.9141	-4.030	-0.5083	-0.757	**6000**
6200	5.997	6.121	-0.3045	0.1608	0.7875	-3.788	-0.3796	-0.711	**6200**
6400	5.809	6.093	-0.1977	0.1783	0.6688	-3.561	-0.2590	-0.669	**6400**

Source: JANAF Thermochemical Tables, 3rd ed., published by the American Chemical Society and the American Institute of Physics for the National Bureau of Standards, 1986.

Chart A·20E Aqua–ammonia Chart

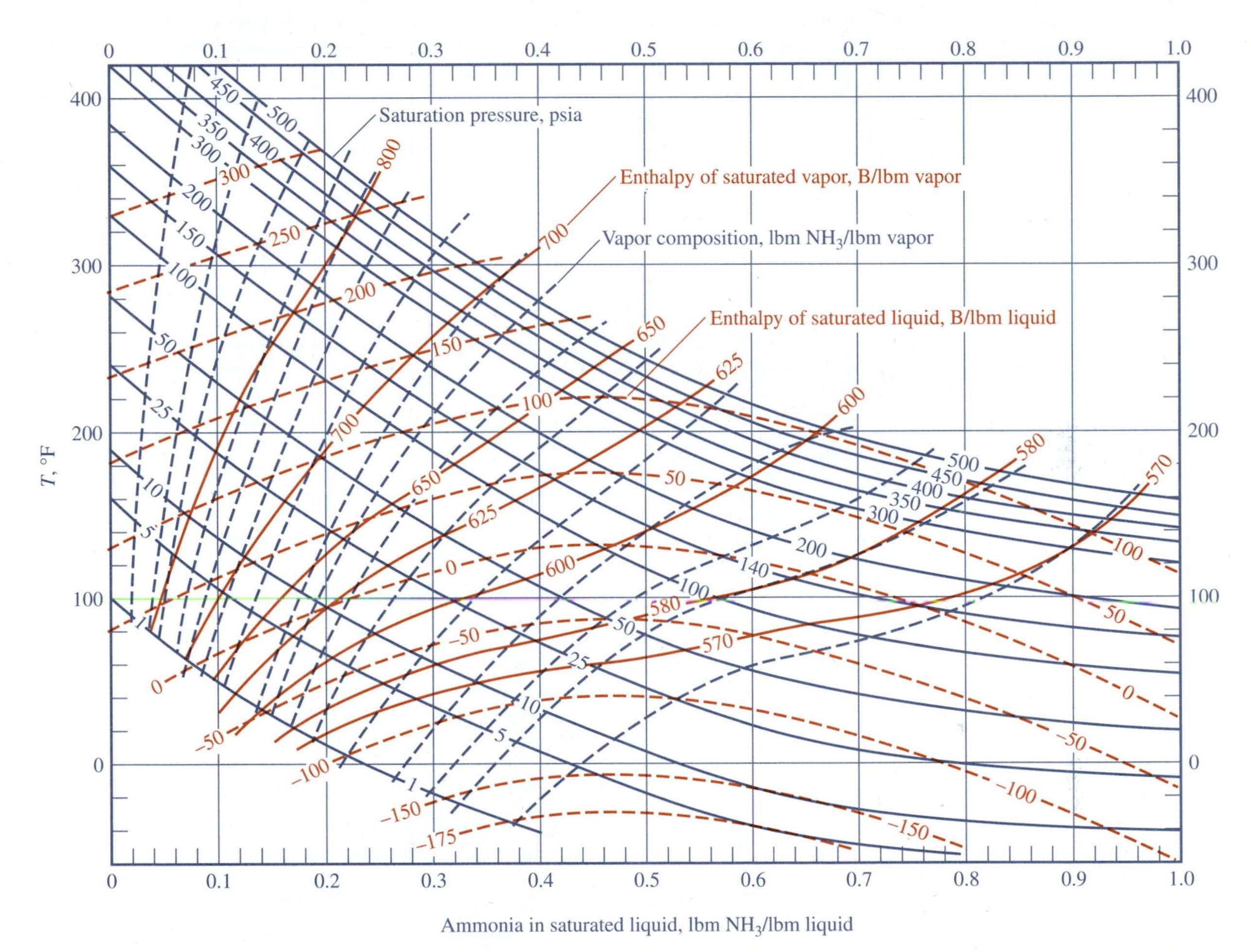

Chart A·20E Aqua–ammonia chart. (Adapted from ''Equilibrium Properties of Aqua–ammonia in Chart Form,'' by F. H. Kohloss, Jr., and G. L. Scott, *Refrigeration Engineering,* Vol. 58, no. 10, Oct. 1950, p. 970.) Note: The state of zero enthalpy is not the same for this chart and for Tables A8.

Chart A·21E Aqueous Lithium Bromide: Equilibrium Chart

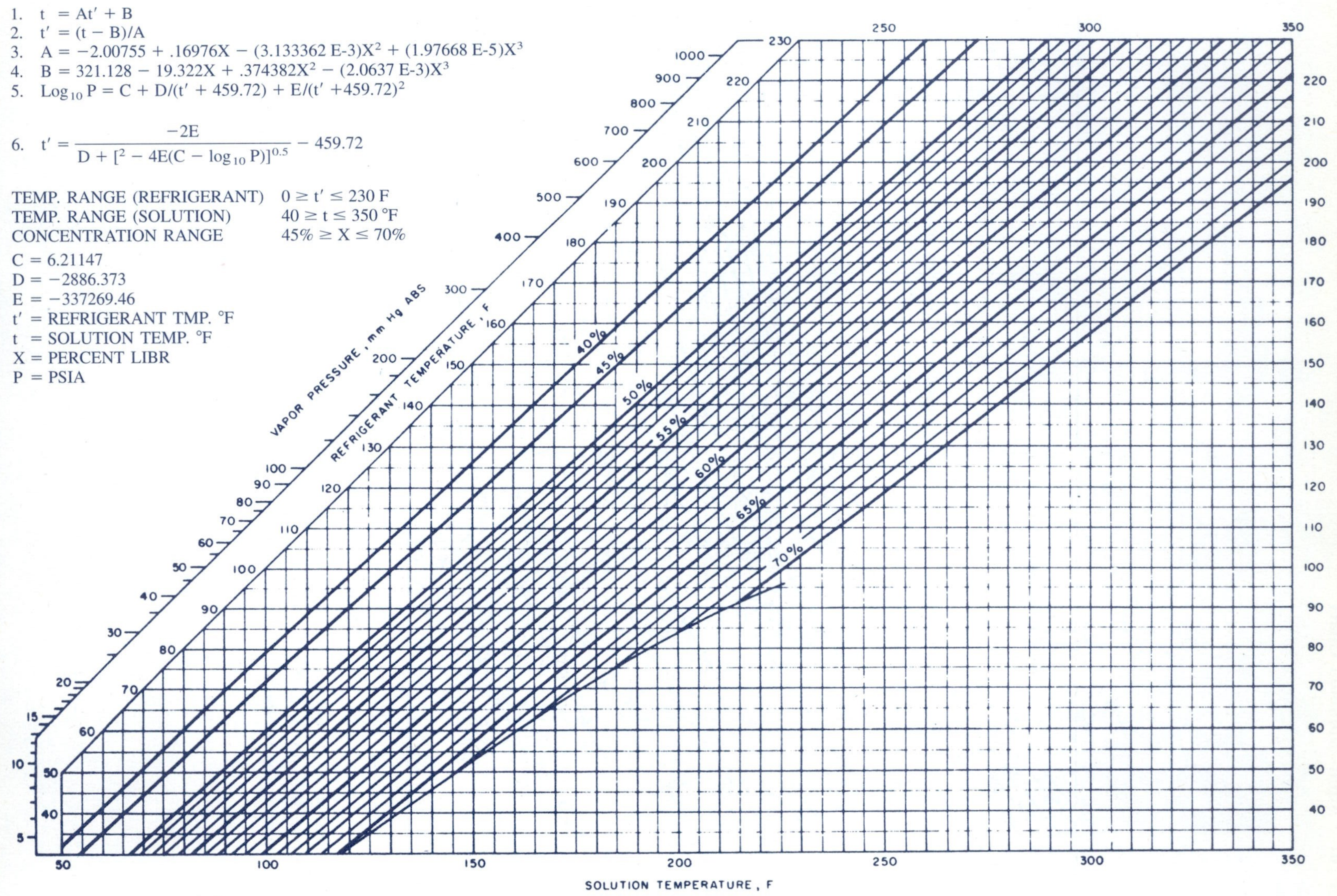

EQUATIONS
1. t = At′ + B
2. t′ = (t − B)/A
3. A = −2.00755 + .16976X − (3.133362 E-3)X² + (1.97668 E-5)X³
4. B = 321.128 − 19.322X + .374382X² − (2.0637 E-3)X³
5. Log₁₀ P = C + D/(t′ + 459.72) + E/(t′ + 459.72)²
6. t′ = −2E / (D + [² − 4E(C − log₁₀ P)]^0.5) − 459.72
TEMP. RANGE (REFRIGERANT) 0 ≥ t′ ≤ 230 F
TEMP. RANGE (SOLUTION) 40 ≥ t ≤ 350 °F
CONCENTRATION RANGE 45% ≥ X ≤ 70%
C = 6.21147
D = −2886.373
E = −337269.46
t′ = REFRIGERANT TMP. °F
t = SOLUTION TEMP. °F
X = PERCENT LIBR
P = PSIA
VAPOR PRESSURE, mm Hg ABS
REFRIGERANT TEMPERATURE, F
SOLUTION TEMPERATURE, F
40%
45%
50%
55%
60%
65%
70%
*Reprinted by permission of Carrier Corp.

Chart A·21E Aqueous lithium bromide equilibrium chart. (Reprinted by permission from *ASHRAE Handbook—1993 Fundamentals,* American Society of Heating, Refrigerating, and Air-Conditioning Engineers, Atlanta.)

Chart A·22E Aqueous Lithium Bromide: Enthalpy-concentration Chart

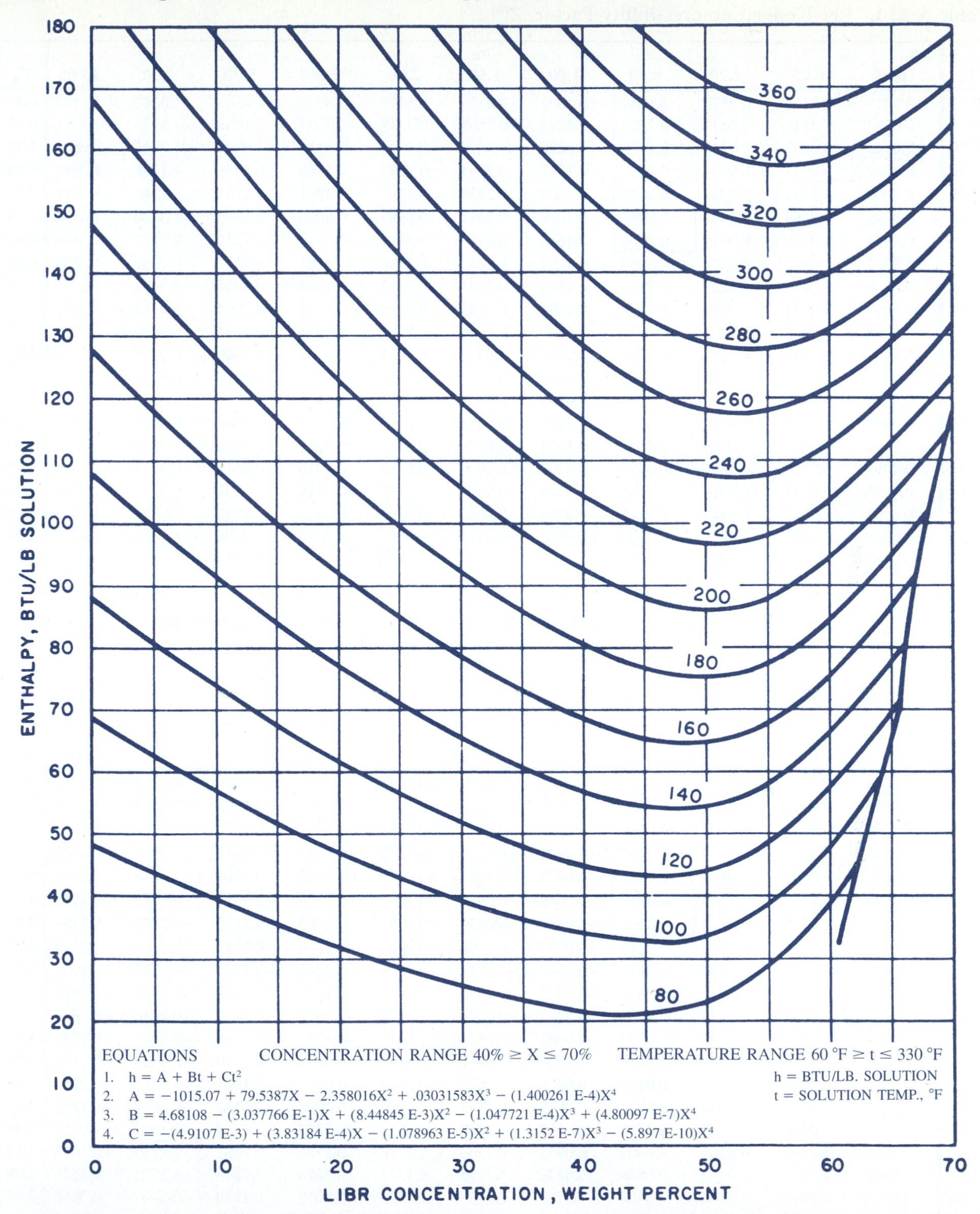

Chart A·22E Aqueous lithium bromide enthalpy-concentration chart. (Reprinted by permission from *ASHRAE Handbook—1993 Fundamentals,* American Society of Heating, Refrigerating, and Air-Conditioning Engineers, Atlanta.)

Table A·23·1 Lee-Kesler Compressibility Factor, $Z^{(0)}$

T_R	0.01	0.05	0.20	0.60	1.00	1.50	2.00	3.00	5.00	7.00	10.00	T_R
0.30	0.0029	0.0145	0.0579	0.1737	0.2892	0.4335	0.5775	0.8648	1.4366	2.0048	2.8507	0.30
0.40	0.0024	0.0119	0.0477	0.1429	0.2379	0.3563	0.4744	0.7095	1.1758	1.6373	2.3211	0.40
0.50	0.0021	0.0103	0.0413	0.1236	0.2056	0.3077	0.4092	0.6110	1.0094	1.4017	1.9801	0.50
0.60	0.9849	0.0093	0.0371	0.1109	0.1842	0.2753	0.3657	0.5446	0.8959	1.2398	1.7440	0.60
0.70	0.9904	0.9504	0.0344	0.1027	0.1703	0.2538	0.3364	0.4991	0.8161	1.1241	1.5729	0.70
0.80	0.9935	0.9669	0.8539	0.0985	0.1626	0.2411	0.3182	0.4690	0.7598	1.0400	1.4456	0.80
0.90	0.9954	0.9768	0.9015	0.1006	0.1630	0.2383	0.3114	0.4527	0.7220	0.9793	1.3496	0.90
0.95	0.9961	0.9803	0.9174	0.6967	0.1705	0.2432	0.3138	0.4501	0.7092	0.9561	1.3108	0.95
0.98	0.9965	0.9821	0.9253	0.7360	0.1844	0.2503	0.3182	0.4508	0.7035	0.9442	1.2901	0.98
1.00	0.9967	0.9832	0.9300	0.7574	0.2901	0.2583	0.3229	0.4522	0.7004	0.9372	1.2772	1.00
1.01	0.9968	0.9837	0.9322	0.7671	0.4648	0.2640	0.3260	0.4533	0.6991	0.9339	1.2710	1.01
1.02	0.9969	0.9842	0.9343	0.7761	0.5146	0.2715	0.3297	0.4547	0.6980	0.9307	1.2650	1.02
1.05	0.9971	0.9855	0.9401	0.8002	0.6026	0.3131	0.3452	0.4604	0.6956	0.9222	1.2481	1.05
1.10	0.9975	0.9874	0.9485	0.8323	0.6880	0.4580	0.3953	0.4770	0.6950	0.9110	1.2232	1.10
1.20	0.9981	0.9904	0.9611	0.8779	0.7858	0.6605	0.5605	0.5425	0.7069	0.8990	1.1844	1.20
1.40	0.9988	0.9942	0.9768	0.9298	0.8827	0.8256	0.7753	0.7202	0.7761	0.9112	1.1419	1.40
1.60	0.9993	0.9964	0.9856	0.9575	0.9308	0.9000	0.8738	0.8410	0.8617	0.9518	1.1320	1.60
1.80	0.9995	0.9977	0.9910	0.9739	0.9583	0.9413	0.9275	0.9118	0.9297	0.9961	1.1391	1.80
2.00	0.9997	0.9986	0.9944	0.9842	0.9754	0.9664	0.9599	0.9550	0.9772	1.0328	1.1516	2.00
2.50	1.0000	0.9997	0.9989	0.9974	0.9967	0.9973	0.9993	1.0070	1.0390	1.0860	1.1760	2.50
3.00	1.0000	1.0002	1.0008	1.0030	1.0057	1.0101	1.0153	1.0284	1.0635	1.1075	1.1848	3.00
4.00	1.0001	1.0005	1.0021	1.0066	1.0115	1.0179	1.0249	1.0401	1.0747	1.1136	1.1773	4.00

Column headings are values of p_R.

Source: Abstracted from Lee, Byung Ik, and Michael G. Kesler, ''A Generalized Thermodynamic Correlation Based on Three-Parameter Corresponding States,'' *AIChE Journal,*

Table A·23·2 Lee-Kesler Compressibility Factor, $Z^{(1)}$

T_R	0.01	0.05	0.20	0.60	1.00	1.50	2.00	3.00	5.00	7.00	10.00	T_R
0.30	−0.0008	−0.0040	−0.0161	−0.0484	−0.0806	−0.1207	−0.1608	−0.2407	−0.3996	−0.5572	−0.7915	0.30
0.40	−0.0010	−0.0048	−0.0191	−0.0570	−0.0946	−0.1414	−0.1879	−0.2799	−0.4603	−0.6365	−0.8936	0.40
0.50	−0.0009	−0.0045	−0.0181	−0.0539	−0.0930	−0.1330	−0.1762	−0.2611	−0.4253	−0.5831	−0.8099	0.50
0.60	−0.0205	−0.0041	−0.0164	−0.0487	−0.0803	−0.1192	−0.1572	−0.2312	−0.3718	−0.5047	−0.6928	0.60
0.70	−0.0093	−0.0507	0.0148	−0.0438	−0.0718	−0.1057	−0.1385	−0.2013	−0.3184	−0.4270	−0.5785	0.70
0.80	−0.0044	−0.0228	−0.1160	−0.0401	−0.0648	−0.0940	−0.1217	−0.1736	−0.2682	−0.3545	−0.4740	0.80
0.90	−0.0019	−0.0090	−0.0442	−0.0396	−0.0604	−0.0840	−0.1059	−0.1463	−0.2195	−0.2862	−0.3788	0.90
0.95	−0.0012	−0.0062	−0.0262	−0.1110	−0.0607	−0.0788	−0.0967	−0.1310	−0.1943	−0.2526	−0.3339	0.95
0.98	−0.0009	−0.0044	−0.0184	−0.0641	−0.0641	−0.0740	−0.0893	−0.1202	−0.1783	−0.2322	−0.3075	0.98
1.00	−0.0007	−0.0034	−0.0140	−0.0435	−0.0879	−0.0678	−0.0824	−0.1118	−0.1672	−0.2185	−0.2902	1.00
1.01	−0.0006	−0.0030	−0.0120	−0.0351	−0.0223	−0.0621	−0.0778	−0.1072	−0.1615	−0.2116	−0.2816	1.01
1.02	−0.0005	−0.0026	−0.0102	−0.0277	−0.0062	−0.0524	−0.0722	−0.1021	−0.1556	−0.2047	−0.2731	1.02
1.05	−0.0003	−0.0015	−0.0054	−0.0097	0.0220	0.0451	−0.0432	−0.0838	−0.1370	−0.1835	−0.2476	1.05
1.10	0.0000	0.0000	0.0007	0.0106	0.0476	0.1630	0.0698	−0.0373	−0.1021	−0.1469	−0.2056	1.10
1.20	0.0004	0.0019	0.0084	0.0326	0.0719	0.1477	0.1990	0.1095	−0.0141	−0.0678	−0.1231	1.20
1.40	0.0007	0.0036	0.0147	0.0477	0.0857	0.1383	0.1894	0.2397	0.1737	0.1008	0.0350	1.40
1.60	0.0008	0.0040	0.0162	0.0501	0.0855	0.1303	0.1720	0.2381	0.2631	0.2255	0.1673	1.60
1.80	0.0008	0.0040	0.0162	0.0488	0.0836	0.1216	0.1593	0.2224	0.2846	0.2871	0.2576	1.80
2.00	0.0008	0.0039	0.0155	0.0464	0.0767	0.1133	0.1476	0.2069	0.2819	0.3097	0.3096	2.00
2.50	0.0007	0.0034	0.0135	0.0399	0.0655	0.0959	0.1246	0.1759	0.2543	0.3049	0.3467	2.50
3.00	0.0006	0.0029	0.0117	0.0345	0.0565	0.0828	0.1076	0.1529	0.2268	0.2817	0.3385	3.00
4.00	0.0005	0.0023	0.0091	0.0270	0.0443	0.0651	0.0849	0.1219	0.1057	0.2378	0.2994	4.00

Column headings are values of p_R.

Table A·23·3 Lee-Kesler Enthalpy Departure, $h^{(0)}$

T_R	p_R 0.01	0.05	0.20	0.60	1.00	1.50	2.00	3.00	5.00	7.00	10.00	T_R
0.30	6.045	6.043	6.034	6.011	5.987	5.957	5.927	5.868	5.748	5.628	5.446	**0.30**
0.40	5.769	5.761	5.751	5.726	5.700	5.668	5.636	5.572	5.442	5.311	5.113	**0.40**
0.50	5.465	5.463	5.453	5.427	5.401	5.369	5.336	5.270	5.135	4.999	4.791	**0.50**
0.60	0.027	5.162	5.153	5.129	5.104	5.073	5.041	4.976	4.842	4.704	4.492	**0.60**
0.70	0.020	0.101	4.848	4.828	4.808	4.781	4.752	4.693	4.566	4.432	4.221	**0.70**
0.80	0.015	0.078	0.345	4.504	4.494	4.478	4.459	4.413	4.303	4.178	3.974	**0.80**
0.90	0.012	0.062	0.264	4.074	4.108	4.127	4.132	4.119	4.043	3.935	3.744	**0.90**
0.95	0.011	0.056	0.235	0.885	3.825	3.904	3.940	3.958	3.910	3.815	3.634	**0.95**
0.98	0.010	0.053	0.221	0.797	3.544	3.736	3.806	3.854	3.829	3.742	3.569	**0.98**
1.00	0.010	0.051	0.212	0.750	2.584	3.598	3.706	3.782	3.774	3.695	3.526	**1.00**
1.01	0.010	0.050	0.208	0.728	1.796	3.516	3.652	3.744	3.746	3.671	3.505	**1.01**
1.02	0.010	0.049	0.203	0.708	1.627	3.422	3.595	3.705	3.718	3.647	3.484	**1.02**
1.05	0.009	0.046	0.192	0.654	1.359	3.030	3.398	3.583	3.632	3.575	3.420	**1.05**
1.10	0.008	0.042	0.175	0.581	1.120	2.203	2.965	3.353	3.484	3.453	3.315	**1.10**
1.20	0.007	0.036	0.148	0.474	0.857	1.443	2.079	2.807	3.166	3.202	3.107	**1.20**
1.40	0.005	0.027	0.110	0.341	0.588	0.915	1.253	1.857	2.486	2.679	2.692	**1.40**
1.60	0.004	0.021	0.086	0.261	0.440	0.667	0.894	1.318	1.904	2.177	2.285	**1.60**
1.80	0.003	0.017	0.068	0.206	0.344	0.515	0.683	0.996	1.476	1.751	1.908	**1.80**
2.00	0.003	0.014	0.056	0.167	0.276	0.411	0.541	0.782	1.167	1.411	1.577	**2.00**
2.50	0.002	0.009	0.035	0.1045	0.171	0.252	0.328	0.468	0.691	0.843	9.958	**2.50**
3.00	0.001	0.006	0.023	0.067	0.109	0.159	0.205	0.288	0.415	0.495	0.545	**3.00**
4.00	0.000	0.002	0.009	0.026	0.041	0.058	0.072	0.095	0.116	0.110	0.061	**4.00**

Table A·23·4 Lee-Kesler Enthalpy Departure, $h^{(1)}$

T_R	p_R 0.01	0.05	0.20	0.60	1.00	1.50	2.00	3.00	5.00	7.00	10.00	T_R
0.30	11.098	11.096	11.091	11.076	11.062	11.044	11.027	10.992	10.935	10.872	10.781	**0.30**
0.40	10.121	10.121	10.120	10.121	10.121	10.121	10.122	10.123	10.128	10.135	10.150	**0.40**
0.50	8.868	8.869	8.872	8.880	8.888	8.899	8.909	8.932	8.978	9.030	9.111	**0.50**
0.60	0.059	7.568	7.573	7.585	7.596	7.614	7.632	7.669	7.745	7.824	7.950	**0.60**
0.70	0.034	0.185	6.360	6.373	6.388	6.407	6.429	6.475	6.574	6.677	6.837	**0.70**
0.80	0.021	0.110	0.542	5.271	5.285	5.306	5.330	5.385	5.506	5.632	5.824	**0.80**
0.90	0.014	0.070	0.308	4.254	4.249	4.268	4.298	4.371	4.530	4.688	4.916	**0.90**
0.95	0.011	0.056	0.241	0.994	3.712	3.730	3.773	3.873	4.068	4.248	4.497	**0.95**
0.98	0.010	0.050	0.209	0.776	3.332	3.363	3.443	3.568	3.795	3.992	4.257	**0.98**
1.00	0.009	0.046	0.191	0.675	2.471	3.065	3.186	3.358	3.615	3.825	4.100	**1.00**
1.01	0.009	0.044	0.183	0.632	1.375	2.880	3.051	3.251	3.525	3.742	4.023	**1.01**
1.02	0.008	0.042	0.175	0.594	1.180	2.650	2.906	3.142	3.435	3.661	3.947	**1.02**
1.05	0.007	0.037	0.153	0.498	0.877	1.496	2.381	2.800	3.167	3.418	3.722	**1.05**
1.10	0.006	0.030	0.123	0.381	0.617	0.617	1.261	2.167	2.720	3.023	3.362	**1.10**
1.20	0.004	0.020	0.080	0.232	0.349	0.381	0.361	0.934	1.840	2.273	2.692	**1.20**
1.40	0.002	0.008	0.032	0.083	0.111	0.108	0.070	0.044	0.504	1.012	1.547	**1.40**
1.60	0.000	0.002	0.007	0.013	0.005	−0.023	−0.065	−0.151	−0.082	0.217	0.689	**1.60**
1.80	0.000	−0.001	−0.006	−0.025	−0.051	−0.094	−0.143	−0.241	−0.317	−0.203	0.112	**1.80**
2.00	−0.001	−0.003	−0.015	−0.047	−0.085	−0.136	−0.190	−0.295	−0.428	−0.424	−0.255	**2.00**
2.50	−0.001	−0.006	−0.025	−0.075	−0.125	−0.188	−0.250	−0.366	−0.551	−0.659	−0.700	**2.50**
3.00	−0.001	−0.007	−0.029	−0.086	−0.142	−0.211	−0.278	−0.403	−0.611	−0.763	−0.899	**3.00**
4.00	−0.002	−0.008	−0.032	−0.096	−0.158	−0.233	−0.306	−0.442	−0.680	−0.874	−1.097	**4.00**

Lee-Kesler, $s^{(0)}$, $s^{(1)}$

Table A·23·5 Lee-Kesler Entropy Departure, $s^{(0)}$

T_R	p_R 0.01	0.05	0.20	0.60	1.00	1.50	2.00	3.00	5.00	7.00	10.00	T_R
0.30	11.614	10.008	8.635	7.574	7.099	6.740	6.497	6.182	5.847	5.683	5.578	**0.30**
0.40	10.802	9.196	7.821	6.755	6.275	5.909	5.660	5.330	4.967	4.772	4.619	**0.40**
0.50	10.137	8.531	7.156	6.089	5.608	5.240	4.989	4.656	4.282	4.074	3.899	**0.50**
0.60	0.029	7.983	6.610	5.544	5.066	4.700	4.451	4.120	3.747	3.537	3.353	**0.60**
0.70	0.018	0.096	6.140	5.082	4.610	4.250	4.007	3.684	3.322	3.117	2.935	**0.70**
0.80	0.013	0.064	0.294	4.649	4.191	3.846	3.605	3.310	2.970	2.777	2.605	**0.80**
0.90	0.009	0.046	0.199	4.145	3.738	3.434	3.231	2.964	2.663	2.491	2.334	**0.90**
0.95	0.008	0.039	0.168	0.671	3.433	3.193	3.023	2.790	2.520	2.362	2.215	**0.95**
0.98	0.007	0.036	0.153	0.580	3.142	3.019	2.884	2.682	2.436	2.287	2.148	**0.98**
1.00	0.007	0.034	0.144	0.532	2.178	2.879	2.784	2.609	2.380	2.239	2.105	**1.00**
1.01	0.007	0.033	0.139	0.510	1.391	2.798	2.730	2.571	2.352	2.215	2.083	**1.01**
1.02	0.006	0.032	0.135	0.491	1.225	2.706	2.673	2.533	2.325	2.191	2.062	**1.02**
1.05	0.006	0.030	0.124	0.439	0.965	2.328	2.483	2.415	2.242	2.121	2.001	**1.05**
1.10	0.005	0.026	0.108	0.371	0.742	1.557	2.081	2.202	2.104	2.007	1.903	**1.10**
1.20	0.004	0.021	0.085	0.277	0.512	0.890	1.308	1.727	1.827	1.789	1.722	**1.20**
1.40	0.003	0.014	0.056	0.174	0.303	0.478	0.663	0.990	1.303	1.386	1.402	**1.40**
1.60	0.002	0.010	0.039	0.120	0.204	0.312	0.421	0.628	0.913	1.050	1.130	**1.60**
1.80	0.001	0.007	0.029	0.088	0.147	0.222	0.296	0.438	0.661	0.799	0.908	**1.80**
2.00	0.001	0.006	0.022	0.067	0.111	0.167	0.221	0.325	0.497	0.620	0.733	**2.00**
2.50	0.001	0.004	0.013	0.039	0.064	0.095	0.125	0.183	0.283	0.364	0.456	**2.50**
3.00	0.000	0.002	0.008	0.025	0.041	0.061	0.080	0.116	0.181	0.236	0.303	**3.00**
4.00	0.000	0.001	0.004	0.013	0.021	0.031	0.041	0.059	0.093	0.123	0.162	**4.00**

Table A·23·6 Lee-Kesler Entropy Departure, $s^{(1)}$

T_R	p_R 0.01	0.05	0.20	0.60	1.00	1.50	2.00	3.00	5.00	7.00	10.00	T_R
0.30	16.782	16.774	16.744	16.665	16.586	16.488	16.390	16.195	15.837	15.468	14.925	**0.30**
0.40	13.990	13.986	13.972	13.934	13.896	13.849	13.803	13.714	13.541	13.376	13.144	**0.40**
0.50	11.202	11.200	11.192	11.172	11.153	11.129	11.107	11.063	10.985	10.920	10.836	**0.50**
0.60	0.078	8.828	8.823	8.811	8.799	8.787	8.777	8.760	8.736	8.723	8.720	**0.60**
0.70	0.040	0.216	6.951	6.941	6.933	6.926	6.922	6.919	6.929	6.952	7.002	**0.70**
0.80	0.022	0.116	0.578	5.468	5.458	5.453	5.452	5.461	5.501	5.555	5.648	**0.80**
0.90	0.013	0.068	0.301	4.269	4.238	4.230	4.236	4.267	4.351	4.442	4.578	**0.90**
0.95	0.010	0.053	0.228	0.961	3.658	3.648	3.669	3.728	3.851	3.966	4.125	**0.95**
0.98	0.009	0.046	0.196	0.734	3.264	3.268	3.318	3.412	3.569	3.701	3.875	**0.98**
1.00	0.008	0.042	0.177	0.632	2.399	2.967	3.067	3.200	3.387	3.532	3.717	**1.00**
1.01	0.008	0.040	0.169	0.590	1.306	2.784	2.933	3.094	3.297	3.450	3.640	**1.01**
1.02	0.008	0.039	0.161	0.552	1.113	2.557	2.790	2.986	3.209	3.369	3.565	**1.02**
1.05	0.007	0.034	0.140	0.460	0.820	1.443	2.283	2.655	2.949	3.134	3.348	**1.05**
1.10	0.005	0.028	0.112	0.350	0.577	0.618	1.241	2.067	2.534	2.767	3.013	**1.10**
1.20	0.004	0.019	0.075	0.220	0.343	0.412	0.447	0.991	1.767	2.115	2.430	**1.20**
1.40	0.002	0.010	0.037	0.104	0.158	0.200	0.220	0.290	0.730	1.138	1.544	**1.40**
1.60	0.001	0.005	0.021	0.057	0.086	0.112	0.129	0.159	0.334	0.604	0.969	**1.60**
1.80	0.001	0.003	0.013	0.035	0.053	0.070	0.083	0.105	0.195	0.355	0.628	**1.80**
2.00	0.000	0.002	0.008	0.023	0.035	0.048	0.058	0.077	0.136	0.238	0.434	**2.00**
2.50	0.000	0.001	0.004	0.011	0.017	0.024	0.031	0.045	0.080	0.132	0.232	**2.50**
3.00	0.000	0.001	0.002	0.006	0.010	0.015	0.020	0.031	0.058	0.092	0.150	**3.00**
4.00	0.000	0.000	0.001	0.003	0.006	0.009	0.012	0.020	0.038	0.060	0.100	**4.00**

Chart A·24·1 Lee-Kesler Compressibility Factor $Z^{(0)}$

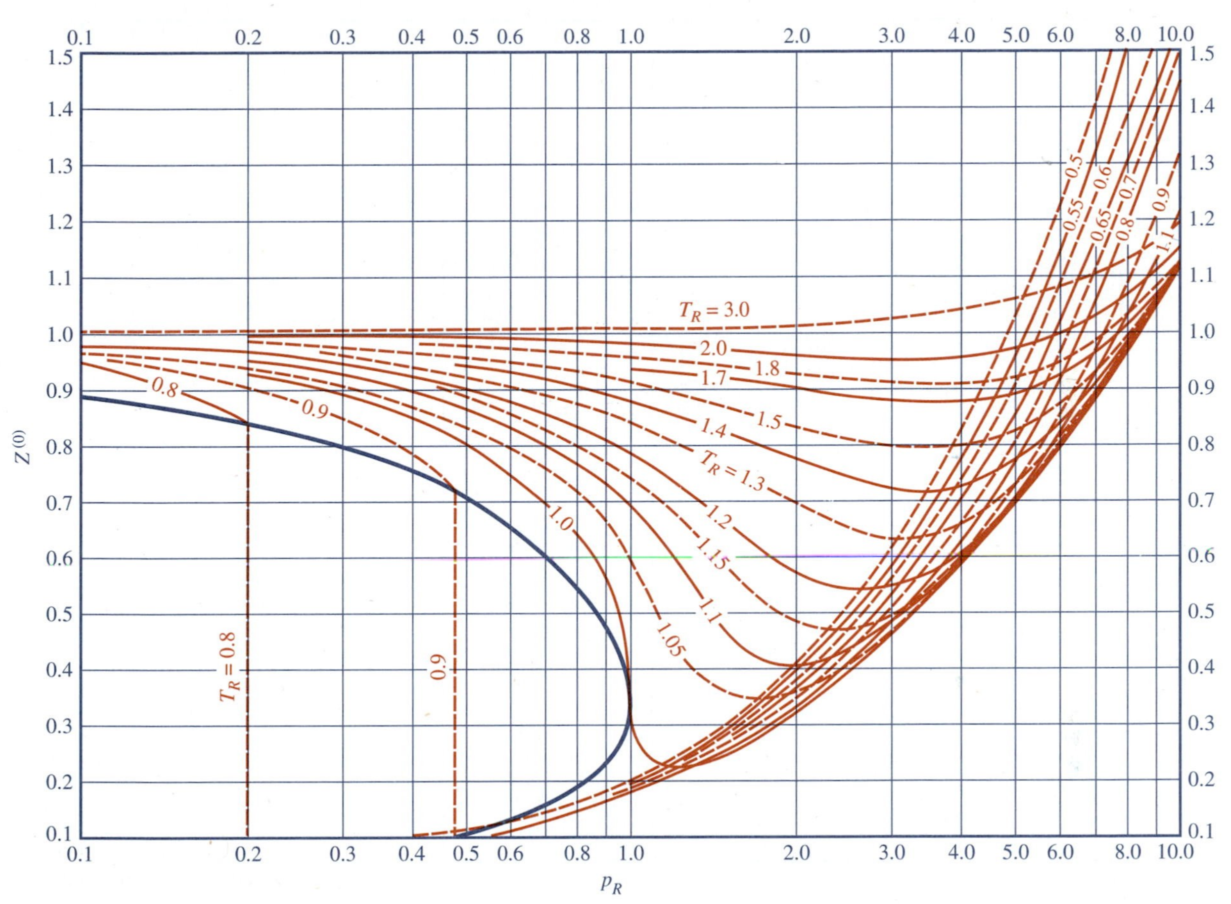

Chart A·24·1 Lee-Kesler compressibility factor, $Z^{(0)}$. (*Source:* Plotted from Tables A·23.)

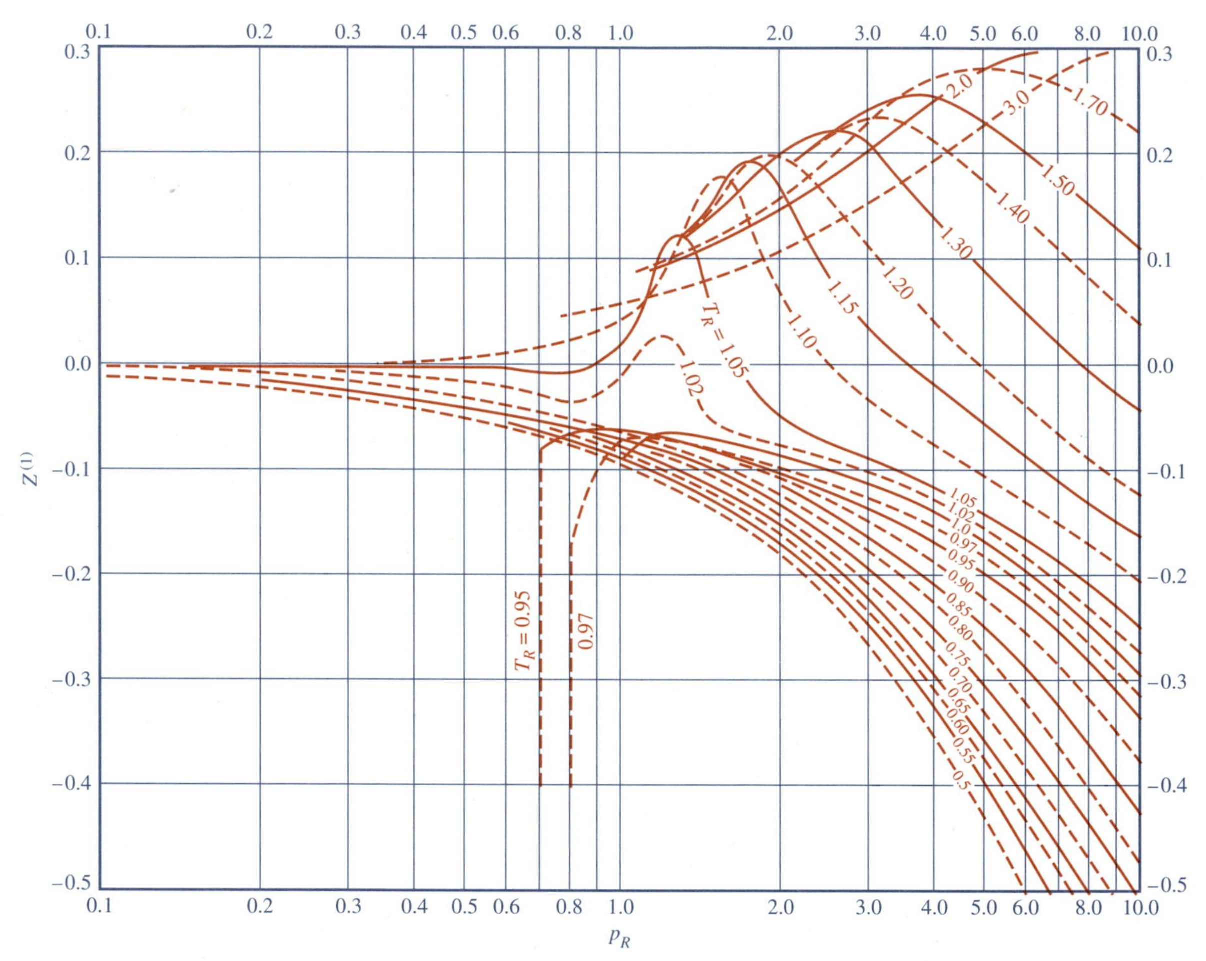

Chart A·24·2 Lee-Kesler compressibility factor, $Z^{(1)}$. (*Source:* Plotted from Tables A·23.)

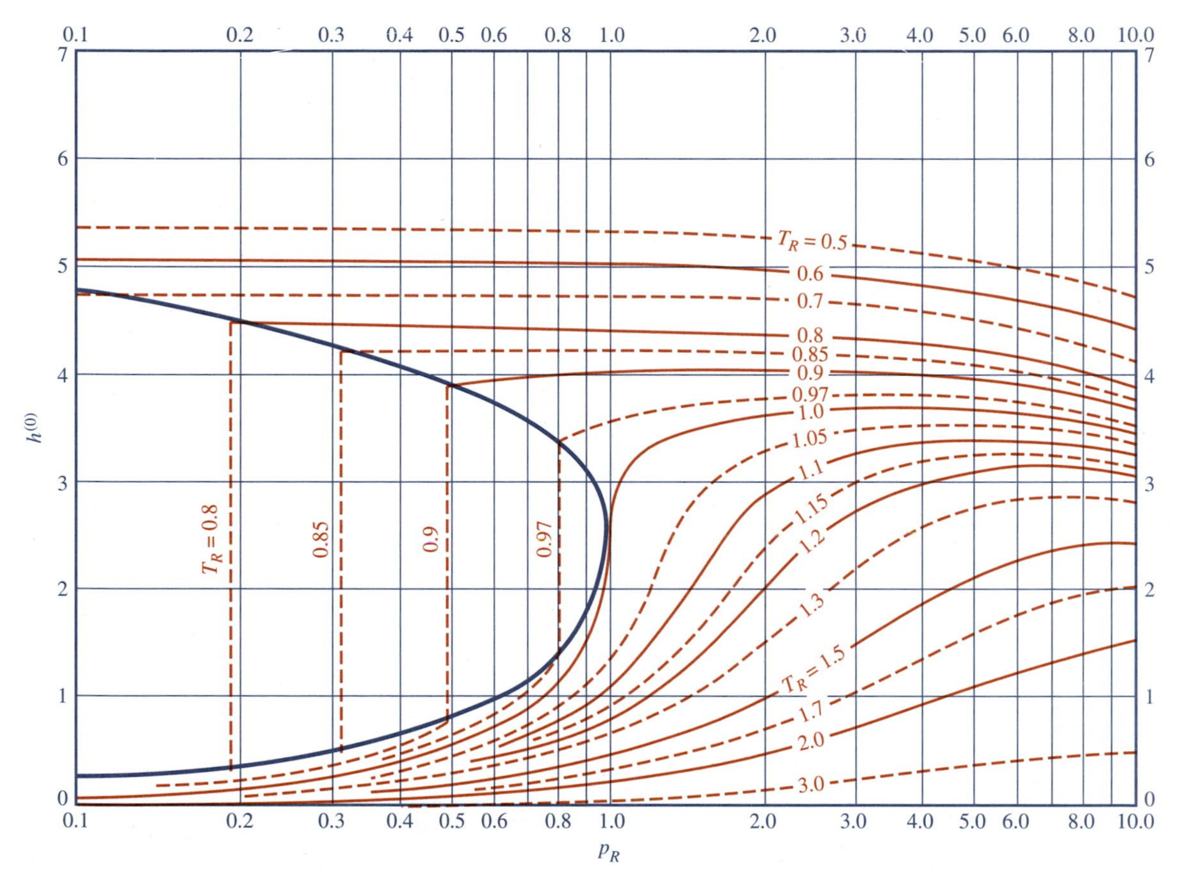

Chart A·24·3 Lee-Kesler enthalpy departure, $h^{(0)}$. (*Source:* Plotted from Tables A·23.)

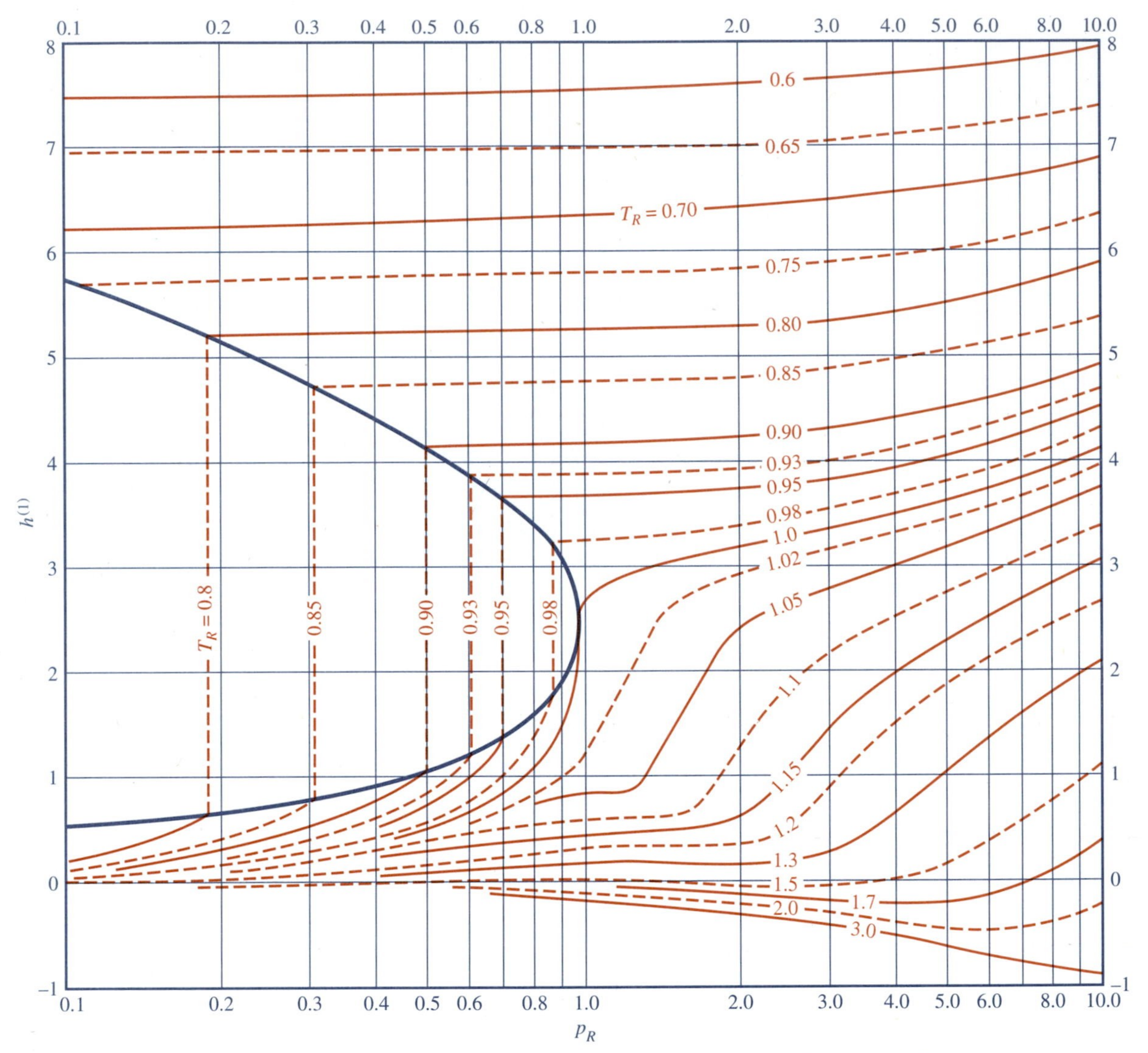

Chart A·24·4 Lee-Kesler enthalpy departure, $h^{(1)}$. (*Source:* Plotted from Tables A·23.)

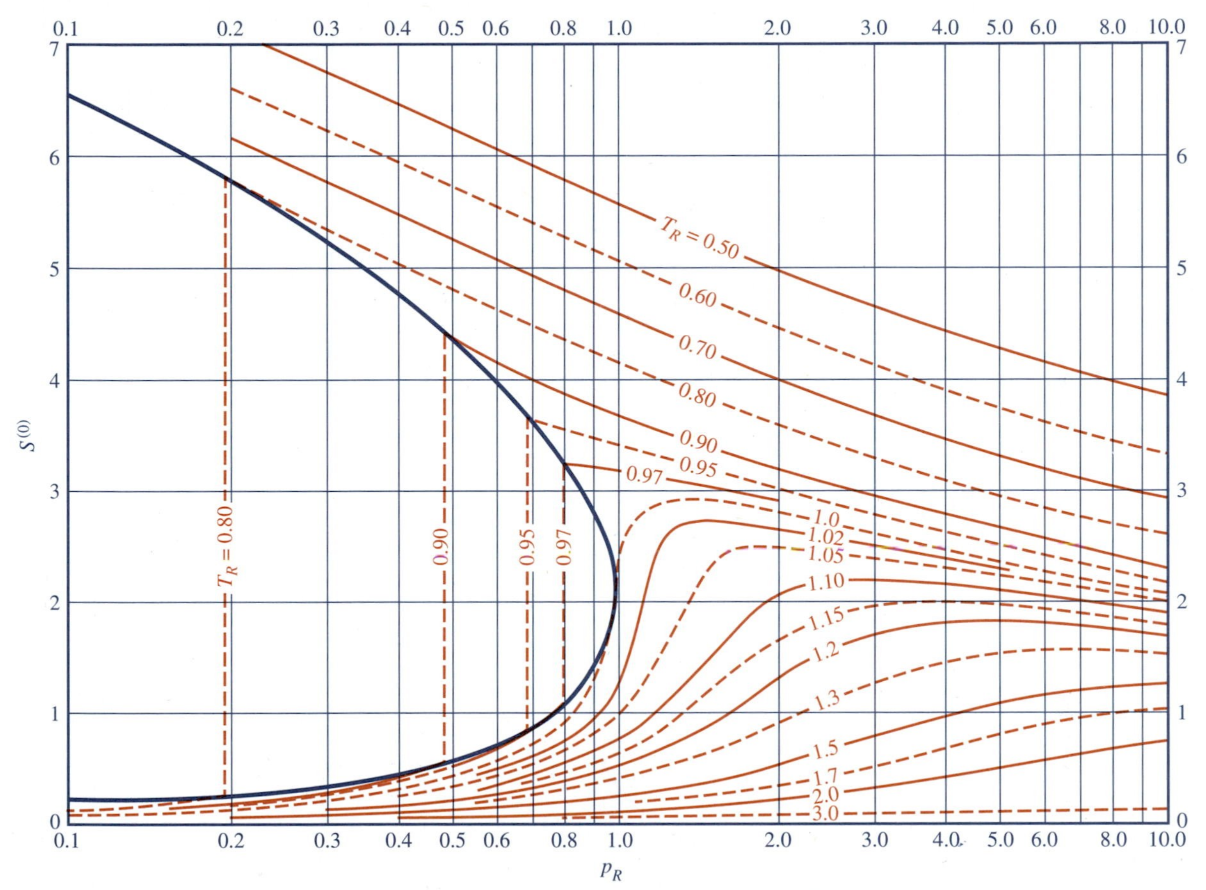

Chart A·24·5 Lee-Kesler entropy departure, $s^{(0)}$. (*Source:* Plotted from Tables A·23.)

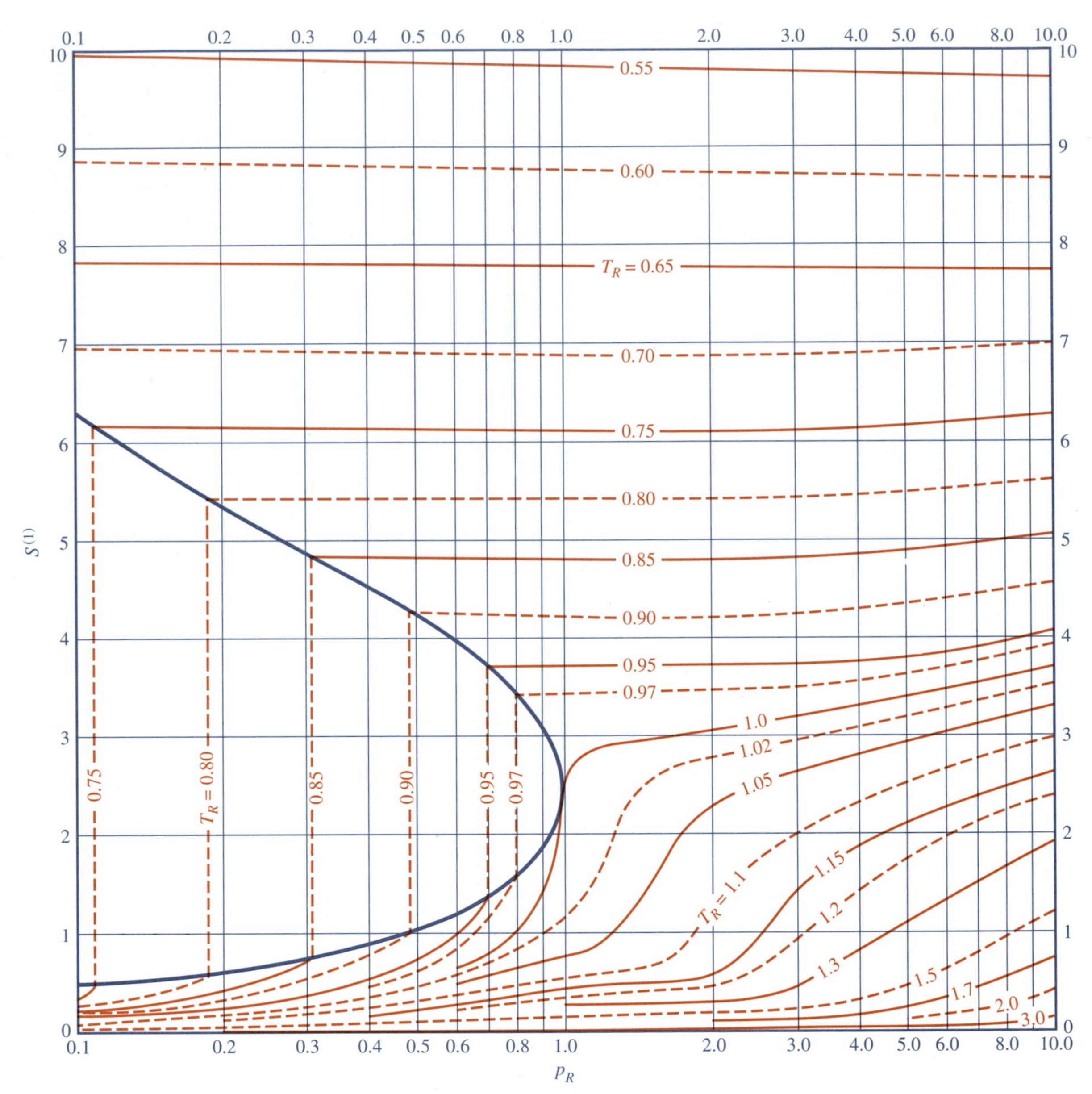

Chart A·24·6 Lee-Kesler entropy departure, $s^{(1)}$. (*Source:* Plotted from Tables A·23.)

Table A·25 Compressible Flow: 1-D Isentropic, Constant Specific Heat, $k = 1.4$

M	T/T_t	p/p_t	ρ/ρ_t	A/A^*
0.00	1.00000	1.00000	1.00000	∞
0.10	0.99800	0.99303	0.99502	5.82183
0.20	0.99206	0.97250	0.98028	2.96352
0.30	0.98232	0.93947	0.95638	2.03507
0.40	0.96899	0.89561	0.92427	1.59014
0.50	0.95238	0.84302	0.88517	1.33984
0.60	0.93284	0.78400	0.84045	1.18820
0.70	0.91075	0.72093	0.79158	1.09437
0.80	0.88652	0.65602	0.73999	1.03823
0.90	0.86059	0.59126	0.68704	1.00886
1.00	0.83333	0.52828	0.63394	1.00000
1.10	0.80515	0.46835	0.58170	1.00793
1.20	0.77640	0.41238	0.53114	1.03044
1.30	0.74738	0.36091	0.48290	1.06630
1.40	0.71839	0.31424	0.43742	1.11493
1.50	0.68966	0.27240	0.39498	1.17617
1.60	0.66138	0.23527	0.35573	1.25024
1.70	0.63371	0.20259	0.31969	1.33761
1.80	0.60680	0.17404	0.28682	1.43898
1.90	0.58072	0.14924	0.25699	1.55526

M	T/T_t	p/p_t	ρ/ρ_t	A/A^*
2.00	0.55556	0.12780	0.23005	1.6875
2.10	0.53135	0.10935	0.20580	1.8369
2.20	0.50813	0.09352	0.18405	2.0050
2.30	0.48591	0.07997	0.16458	2.1931
2.40	0.46468	0.06840	0.14720	2.4031
2.50	0.44444	0.05853	0.13169	2.6367
2.60	0.42517	0.05012	0.11787	2.8960
2.70	0.40683	0.04295	0.10557	3.1830
2.80	0.38941	0.03685	0.09463	3.5001
2.90	0.37286	0.03165	0.08489	3.8498
3.00	0.35714	0.02722	0.07623	4.2346
3.50	0.28986	0.01311	0.04523	6.7896
4.00	0.23810	6.59×10^{-3}	0.02766	10.719
4.50	0.19802	3.46×10^{-3}	0.01745	16.562
5.00	0.16667	1.89×10^{-3}	0.01134	25.000
6.00	0.12195	6.33×10^{-4}	5.19×10^{-3}	53.180
7.00	0.09259	2.42×10^{-4}	2.61×10^{-3}	104.14
8.00	0.07246	1.02×10^{-4}	1.41×10^{-3}	190.11
9.00	0.05814	4.74×10^{-5}	8.15×10^{-4}	327.19
10.00	0.04762	2.36×10^{-5}	4.95×10^{-4}	535.94

Table A·26 Compressible Flow: 1-D Normal Shock, Constant Specific Heat, k = 1.4

M_x	M_y	p_y/p_x	ρ_y/ρ_x	T_y/T_x	p_{oy}/p_{ox}	p_{oy}/p_x	M_x
1.00	1.00000	1.0000	1.0000	1.0000	1.00000	1.8929	**1.00**
1.10	0.91177	1.2450	1.1691	1.0649	0.99893	2.1328	**1.10**
1.20	0.84217	1.5133	1.3416	1.1280	0.99280	2.4075	**1.20**
1.30	0.78596	1.8050	1.5157	1.1909	0.97937	2.7136	**1.30**
1.40	0.73971	2.1200	1.6897	1.2547	0.95819	3.0492	**1.40**
1.50	0.70109	2.4583	1.8621	1.3202	0.92979	3.4133	**1.50**
1.60	0.66844	2.8200	2.0317	1.3880	0.89520	3.8050	**1.60**
1.70	0.64054	3.2050	2.1977	1.4583	0.85572	4.2238	**1.70**
1.80	0.61650	3.6133	2.3592	1.5316	0.81268	4.6695	**1.80**
1.90	0.59562	4.0450	2.5157	1.6079	0.76736	5.1418	**1.90**
2.00	0.57735	4.5000	2.6667	1.6875	0.72087	5.6404	**2.00**
2.10	0.56128	4.9783	2.8119	1.7705	0.67420	6.1654	**2.10**
2.20	0.54706	5.4800	2.9512	1.8569	0.62814	6.7165	**2.20**
2.30	0.53441	6.0050	3.0845	1.9468	0.58329	7.2937	**2.30**
2.40	0.52312	6.5533	3.2119	2.0403	0.54014	7.8969	**2.40**
2.50	0.51299	7.1250	3.3333	2.1375	0.49901	8.5261	**2.50**
2.60	0.50387	7.7200	3.4490	2.2383	0.46012	9.1813	**2.60**
2.70	0.49563	8.3383	3.5590	2.3429	0.42359	9.8624	**2.70**
2.80	0.48817	8.9800	3.6636	2.4512	0.38946	10.569	**2.80**
2.90	0.48138	9.6450	3.7629	2.5632	0.35773	11.302	**2.90**
3.00	0.47519	10.333	3.8571	2.6790	0.32834	12.061	**3.00**
3.50	0.45115	14.125	4.2609	3.3151	0.21295	16.242	**3.50**
4.00	0.43496	18.500	4.5714	4.0469	0.13876	21.068	**4.00**
4.50	0.42355	23.458	4.8119	4.8751	0.09170	26.539	**4.50**
5.00	0.41523	29.000	5.0000	5.8000	0.06172	32.654	**5.00**
6.00	0.40416	41.833	5.2683	7.9406	0.02965	46.815	**6.00**
7.00	0.39736	57.000	5.4444	10.469	0.01535	63.553	**7.00**
8.00	0.39289	74.500	5.5652	13.387	0.00849	82.863	**8.00**
9.00	0.38980	94.333	5.6512	16.693	0.00496	104.75	**9.00**
10.00	0.38758	116.500	5.7143	20.388	0.00304	129.22	**10.00**

Table A.27 Periodic Table of the Elements

Key: Atomic number, Symbol, Element, Molar mass, Valence

Light metals		Brittle metals						Ductile metals			Low melting	Nonmetallic elements					Inert gases
																1 H *Hydrogen* 1.00797 1	2 He *Helium* 4.0026 0
3 Li *Lithium* 6.939 1	4 Be *Beryllium* 9.0122 2											5 B *Boron* 10.811 3	6 C *Carbon* 12.01115 2,4	7 N *Nitrogen* 14.0067 3,5	8 O *Oxygen* 15.9994 2	9 F *Fluorine* 18.9984 1	10 Ne *Neon* 20.183 0
11 Na *Sodium* 22.9898 1	12 Mg *Magnesium* 24.312 2											13 Al *Aluminum* 26.9815 3	14 Si *Silicon* 28.086 4	15 P *Phosphorus* 30.9738 3,5	16 S *Sulphur* 32.064 2,4,6	17 Cl *Chlorine* 35.453 1,3,5,7	18 Ar *Argon* 39.948 0
19 K *Potassium* 39.102 1	20 Ca *Calcium* 40.08 2	21 Sc *Scandium* 44.956 3	22 Ti *Titanium* 47.90 3,4	23 V *Vanadium* 50.942 1,2,3,4,5	24 Cr *Chromium* 51.996 2,3,6	25 Mn *Manganese* 54.938 2,3,4,6,7	26 Fe *Iron* 55.847 2,3	27 Co *Cobalt* 58.9332 2,3	28 Ni *Nickel* 58.71 2,3,4	29 Cu *Copper* 63.546 1,2	30 Zn *Zinc* 65.37 2	31 Ga *Gallium* 69.72 2,3	32 Ge *Germanium* 72.59 2,4	33 As *Arsenic* 74.9216 3,5	34 Se *Selenium* 78.96 2,4,6	35 Br *Bromine* 79.904 1,3,5	36 Kr *Krypton* 83.80 0
37 Rd *Rubidium* 85.47 1	38 Sr *Strontium* 87.62 2	39 Y *Yttrium* 88.905 3	40 Zr *Zirconium* 91.22 4	41 Nb *Niobium* 92.906 2,3,4,5	42 Mo *Molybdenum* 95.94 3,4,5,6	43 Tc *Technetium* (99)	44 Ru *Ruthenium* 101.07 3,4,6,8	45 Rh *Rhodium* 103.905 3,4	46 Pd *Palladium* 106.4 2,4	47 Ag *Silver* 107.868 1	48 Cd *Cadmium* 112.40 2	49 In *Indium* 114.82 1,2,3	50 Sn *Tin* 118.69 2,4	51 Sb *Antimony* 121.75 3,5	52 Te *Tellurium* 127.60 2,4,6	53 I *Iodine* 126.9044 1,3,5,7	54 Xe *Xenon* 131.30 0
55 Cs *Cesium* 132.905 1	56 Ba *Barium* 137.34 2	57-71 *Rare earth elements*	72 Hf *Hafnium* 178.49 4	73 Ta *Tantalum* 180.948 4,5	74 W *Tungsten* 183.85 3,4,5,6	75 Re *Rhenium* 186.2 1,4,7	76 Os *Osmium* 190.2 2,3,4,6,8	77 Ir *Iridium* 192.2 2,3,4,6	78 Pt *Platinum* 195.09 2,4	79 Au *Gold* 196.967 1,3	80 Hg *Mercury* 200.59 1,2	81 Tl *Thallium* 204.37 1,3	82 Pb *Lead* 207.19 2,4	83 Bi *Bismuth* 208.98 3,5	84 Po *Polonium* (210) 2,4	85 At *Astatine* (210)	86 Rn *Radon* (212) 0
87 Fr *Francium* (223)	88 Ra *Radium* (227) 2																

Answers to Selected Problems

1·3 11.25 m^3, 0.67 kg/m^3
1·6 15 ft^3/lbm, 0.067 lbm/ft^3, 2.7 ft^3/lbm, 0.38 lbm/ft^3
1·9 479 kPa
1·15 16.2 kPa
1·18 624 lbf
1·21 0.72%
1·27 (a) 32 F, (b) −40 F, (c) 77 F, (d) 266 F, (e) 1112 F
1·33 (a) I = 76,000, J = −100, K = 1600; (b) I = 76,300, J = 2200, K = 1533; (c) I = 74,133, J = 2200, K = 1867
1·36 0.552 kg/s
1·42 (a) 0.21 cm^2, (b) 44.7 m/s
1·45 21.1 m/s
1·48 150 MW
1·51 8.2 MW
1·54 1000 hp
1·57 1.1×10^5 ft · lbf
1·63 250 kN
1·66 (a) 19.6 kJ, (b) 19.6 kJ
1·69 93.5 kJ
1·72 0
1·75 9.5 kJ
1·78 −64 kJ
1·84 70 kJ
1·87 27 kJ/kg
1·93 1.85 kg/s
1·96 7.85 B/lbm
1·99 3.3 kPa, 186 kPa, 19.4 m above gage
2·6 (a) 6.7 MPa, (b) 1 MPa, (c) 670 kPa
2·9 1 MPa
2·12 ≈ 25 F
2·30 120°C, 0.709 m^3/kg; I, I; M, 0.125 m^3/kg; 476 kPa, 91.9%; M, 0.00104 m^3/kg
2·33 108 psia, 7.86%; M, 6.26 ft^3/lbm; 23.7 psia, 3.92 ft^3/lbm; −0.57 F, 9.23 ft^3/lbm; 89.8 psia, 4.76%
2·36 1.394 m^3/kg
2·39 0.962 ft^3/lbm
2·42 (a) 102°C, (b) 70.1 kPa, (c) 33.7%, (d) 0
2·45 55.4 kJ/kg
2·48 0.0266 kg/s, 19.5 m/s
2·51 19.0 m^2
2·54 1.05 m^3, 1350 kPa
2·57 25.9 kg
2·60 0.646 N
2·63 569 lbm
2·66 300 psi
2·69 0.0326 kmol
2·72 0.586 m^3
2·75 3.4% escapes
2·78 (1) (a) −3.41 B, (2) (a) −1.71 B
2·81 (a) 15.0 psi, (b) 0.162 lbm
2·84 41.6 ft/s
2·87 1.85% error
3·6 −20kJ
3·9 −4000kJ
3·12 2900 kJ
3·15 −10.0 kJ/kg
3·18 −10 B
3·21 −1850 B
3·24 40 B
3·27 464 B
3·30 −0.23 kJ
3·33 (a) −16.0 B, (b) 23.6 B/lbm
3·39 −0.010 kJ
3·51 291 W
3·54 −709 kJ/kg
3·57 3980 hp
3·60 −240 kJ/kg
3·63 4090 ft/s
3·66 255 kW
3·69 7480 kW
3·72 82.7 kW, 17.8 in^2, 5.51 in^2
3·75 71 kPa
3·78 225 B/lbm
3·81 −29.9 kJ/kg
3·84 10.7 kW
3·87 41,400 kJ/kg
3·90 299 kJ/kg
3·99 −12.4 kJ/kg
3·102 305 kJ/kg
3·108 320 lbm
3·120 (a) 1.88 kW, (b) 4.37

3·123 4.72
4·3 (a) 4.25 kPa, (b) 67.7 kJ
4·6 −8.86 kJ/kg
4·9 31.1 kJ
4·12 −20.3 B
4·15 1.07×10^8 B/h
4·18 371°C
4·21 708 kJ/kg
4·24 −29.6 B/lbm
4·27 351 B/lbm
4·30 0.074 kg/s
4·33 31.3°C
4·36 174 kJ/kg
4·39 (a) 41.7 kPa, 0.035 (b) 8.81 kPa, 0.105
4·42 1349 kJ
4·45 174°C, 609 kPa
4·48 25.8 s
4·54 7.17 lbm/lbm of ice
4·57 4940 kJ/m^3
4·60 0.115 kg/s
4·63 3
4·72 199 B
4·81 224 F
4·84 (a) 0.0199 kg/s, 0.0165 m^3/s; (b) 4.10×10^{-3} kg/s, 4.11 10^{-6} m^3/s
4·87 (a) −0.627 kJ, (b) Not possible, (c) Not possible
4·90 −108 kJ/kg
4·93 −911 kJ/kg
4·96 −563 kJ/kg
4·99 1.53 kg/s
4·108 57.0 psia
4·111 2.77×10^{-3}m^2 (ΔT = 40°C)
4·120 1.62 kW, 269 K
4·123 −12.5 kJ
4·129 0.286, 0.158, 0.400
4·132 131 kJ
4·138 403 m/s
4·141 17.4 kW
4·144 −63.0 kJ, 16.2 kJ, −46.8 kJ
4·147 172 B, 240 B, −172 B
4·150 26.2 hp
4·162 6.22 min
4·165 6.75×10^{-3} kg/s
5·12 (a) −66.5 J/g, (b) −73.2 J/g
5·15 (a) 75.1 B, (b) 60.1 B
5·18 1.63, 14.8 kJ/kg
5·21 (a) 69.3 kJ, (b) 51.4 kJ
5·24 0.575, 56°C
5·27 47.8%
5·30 1.79 kW
5·33 8.5 kW
5·36 (a) 2.16 kW, (b) 1.12 kW
5·39 0.22 kW
5·42 (a) 89.0 kJ, (b) 23.9 kJ
5·48 366 F
5·51 343 B/min, 311 F
5·54 4660 B, 2460 B
5·57 8.3
5·60 150 B/h
5·69 36,700 B/h
5·78 9.7%
6·6 350 kW
6·9 0.0854 kJ/K, 0, −0.0692 kJ/K, 0.0162 kJ/K
6·12 Δh =189 kJ/kg, Δu = 145 kJ/kg, Δs = 0.81 kJ/kg·K
6·15 497 hp
6·18 Δh = 198 B/lbm, Δu = 119 B/lbm, Δs = 0.551 B/lbm·R
6·21 13%
6·24 (a) 2.34 kJ, (b) 0.0118 kJ/K
6·42 (a) 770 K, (b) 398 kJ (c) 0
6·45 190 kJ/kg
6·48 760 R, 23 psia
6·51 120 B/lbm
6·54 13 kJ/kg
6·57 92 kJ/kg
6·60 7.2 kJ
6·63 680 kJ, 409 kJ, 271 kJ, 0.399; 680 kJ, 394 kJ, 286 kJ, 0.421
6·66 3.74 kJ/K
6·69 83.9 B/lbm
6·72 39.2 B/lbm
6·75 0.189 B/R
6·93 (a) −13 kJ/kg, (b) 0.17 kJ/kg·K
6·96 (b) 38 kW
6·102 219 F
7·3 0.51 kN/m^2·K
7·6 (a) −0.042 psi/R, (b) 0.048 psi/R
7·12 2055 kJ/kg (Clapeyron Equation)
7·15 10.6 Pa
7·18 94.6 B/lbm
7·33 (a) 0.0024 K^{-1}, (b) 0.0077 K^{-1}, (c) 9.6×10^{-4} K^{-1}
7·36 (a) 6.4 F/atm, (b) 1.05 F/atm
7·51 (a) 0.00277 m^3/kg, (b) 0.00356 m^3/kg
7·54 0.0088 m^3/kg
7·57 2120 psi
7·60 1540 psi
7·63 92.0 kJ/kg
7·66 15,600 kJ
7·69 112 kJ
7·72 43.5 B/lbm
7·75 104 B
8·9 367 kJ
8·12 16.3 kJ
8·15 41.4 kJ
8·21 867 kJ/kg, 99 kJ/kg, 768 kJ/kg
8·24 80.2 kJ/kg
8·27 10.1 B/lbm
8·30 43.9 B
8·42 (a) 819 kJ, (b) 421 kJ
8·45 −48.3 kJ, −0.888 kJ/K, 0.157 kJ/K, 0
8·48 (a) 0.334, (b) 0.178
8·51 (a) 11 B, (b) 114 B
8·54 63,700 kJ/h
8·57 (a) 14.7 kJ, (b) −8.6 kJ, (c) 4.8 kJ
8·60 (a) 15.1 kJ/kg, (b) 850 kJ/K, (c) 4.8 kJ
8·63 25.8 kJ/kg
8·69 126 kJ/kg, 138 kJ/kg
8·75 60,000 B/h
8·78 (a) 1980 lbm/h, (b) −15.0 kJ/kg desuperheated steam
9·3 5.31 kg/kmol, 1.57 kJ/kg·K

9·9 0.545 kJ/kg·K
9·15 0.216 kg
9·21 39.6 B
9·24 18.5 kJ/kg
9·36 0.0192 kg/kg da
9·39 Approximately 12
9·42 17°C, $\phi = 1.0$
9·51 0.00845 lbm/lbm da
9·63 0.0263 lbm/lbm da, 78 F
9·66 101 F
9·81 (a) 2.0 kPa, (b) 0.64, (c) 0.0128 kg/kg da, (d) 0.87 m^3/kg, (e) 57.5 kJ/kg da
9·87 0.0126 lbm/lbm da, 0.29 psi
9·93 (a) 25°C, (b) 3.166 kPa, (c) 0.0201 kg/kg da, (d) 0.0201 kg/kg da
9·96 0.68 kW
9·102 0.00463 lbm/lbm da
9·105 (a) 17.0 lbm/h, (b) 28.1 B/lbm
9·108 29.1 B/lbm
10·3 2
10·6 2
10·12 2930 lbm/h
10·21 10,700 kg/h for $x = 1$
11·3 −0.4 Celsius degrees per 10% change in excess air (20 to 22%)
11·6 (a) 0.453 CH_4, 0.264 CO, 0.151 O_2, 0.132 N_2; (b) 0.453 C, 0.114 H, 0.302 O, 0.132 N; (c) 1.21 m^3/kg; (d) 7.43 m^3/m^3; (e) 0.112 CO_2, 0.051 O_2, 0.837 N_2; (f) 11.1 kg/kg; (g) 1.0 kg H_2O/kg fuel; (h) 10.1 kg/kg; (i) 53°C; (j) 58°C
11·9 (a) 35.6°C, 13.8 kg air/kg fuel
11·12 9.42 kg air/kg fuel
11·15 (a) 13.9 kg air/kg fuel, (b) 0.381, (c) 0.023, (d) 35.3°C, (e) 46.6°C
11·18 -1.26×10^6 kJ/kmol C_2H_2
11·21 94.7 kg air/kg fuel (900 F), 83.6 kg air/kg fuel (1000 F), 74.7 kg air/ kg fuel (1100 F)
11·24 850 kPa, 2260°C, 0.111 CO_2, 0.222 H_2O, 0.667 O_2
11·42 25,000 kJ/kg CH_4
12·6 (a) 99.98%, (b) 99.90%
12·9 0.00532
12·12 0.718 CO_2, 0.058 CO, 0.224 O_2
12·21 (a) 2560°C, (b) 2790°C
12·24 0.0077 O_2, 0.6469 N_2, 0.3097 O, 0.0072 N, 0.0284 NO
12·27 2140°C
12·36 2150°C
12·39 2040°C
12·57 3640 F
12·63 (a) 956,000 kJ/kmol, (b) 469,000 kJ/kmol
12·69 440 kPa
13·6 288 kg/s, 1.68 m
13·9 549 kPa, 464°C
13·15 15.4 s
13·18 862 kPa, 366°C, 0.685, 15.6 kg/s
13·24 0.186, 0.216 m^2
13·27 0.314
13·30 0.263, 0.892 ft^2
13·36 0.876 kg/s
13·39 1123 ft/s
13·42 2.63 in.
13·45 0.717 kg/s
13·51 1102 kPa, 278°C, 26.9 kJ/kg
13·54 0.535 ft^2, 0.296 ft^2
13·57 0.513, 0.448 ft^2
13·60 13.8 psia, 192 psia
13·63 (a) 2.78 B/lbm, (b) 588 lbm/s
13·75 3.20 in^2
13·78 CH_4: 2.98 cm, 186 kPa; O_2: 5.35 cm, 191 kPa
13·87 0.368
13·99 428 kPa, 486 K, 255 m/s, 27.1 kJ/kg
13·105 63.0 psia, 878 R, 838 ft/s, 11.7 B/lbm
14·9 (a) 65 kJ/kg, (b) 77 kJ/kg
14·12 1.03 kW
14·15 462 kPa
14·18 (a) 69%, (b) 26.8 kJ/kg CO
14·21 13.2%
14·30 54.7 hp
14·33 71%
14·48 58.8%
14·69 12%, 438°C; 8%, 542°C; 6%, 598°C; 4%, 658°C; 2%, 722°C; 0%, 787°C
14·81 0.37 kW
14·84 0.015 m^3/min
15·15 9120 hp, 0.345
15·21 402 kJ/kg
15·30 (a) 0.507, (b) 0.616
15·33 0.436
15·42 (a) 22.9 kg/s, (b) 44.4 kJ/kg
15·45 8.03×10^6 B/h, 5.46×10^6 B/h
15·48 45.1 lbm/s, 0.212
15·51 (a) 45.4 kJ/kg, (b) 2.17 kg water/kg air, (c) 559 kJ/kg, (d) 331 kJ/kg, (e) 197 kJ/kg, (f) 464 kJ/kg
15·54 18.1 kN
15·57 9.54 kN
15·60 194 kN
15·66 (a) 0.464, (b) 6.00 kJ, (c) 1.36 kJ
15·69 3680 kPa, 598 kJ/kg, 0.698
15·72 (a) 0.481, (b) 6.19 B, (c) 1.38 B
15·90 0.472 lbm/hp·h
15·93 (a) 0.350, 0.00583; (b) 0.362, 0.00513; (c) 0.376, 0.00457
15·102 0.390, 0.339
15·105 (a) 0.289, (b) 0.0148
15·117 71.5 psia
15·123 0.389, 0.368
15·126 0.431
15·129 (a) 0.383, (b) 0.375

15·135 0.393
15·138 0.316
15·141 0.436, 1960 kJ/kg
15·144 0.491; 324,000 kg/h
16·3 3.0 kW
16·6 28 kW for 20°C ambient
16·9 14.8 kW, 2.25
16·12 1.85 hp
16·18 (a) 2.16 hp
16·21 6, 90 kJ/s
16·36 (a) 2.10 lbm/min, 4.2 hp; (b) 15.0 lbm/min, 4.2 hp
16·39 (a) 35 hp
16·48 14.3 kW
16·54 (a) 55,000 kJ/h, (b) 14,000 kJ/h, (c) 53,000 kJ/h
16·57 (a) 81,900 B/min, (b) 19,300 B/min, (c) 78,500 B/min
16·63 (a) 980 lbm/min, (b) 49,600 lbm/h, (c) 58,500 lbm/h
17·6 (a) 1.04 V, (b) 1.0
17·9 < 740 K
17·18 76.5 L
17·24 (a) 1.22 g Zn, (b) 3.33 g MnO_2
17·30 2 g Ag

Index